Plant Proteomic Research 2.0

Plant Proteomic Research 2.0

Special Issue Editor

Setsuko Komatsu

MDPI • Basel • Beijing • Wuhan • Barcelona • Belgrade

MDPI

Special Issue Editor
Setsuko Komatsu
Fukui University of Technology
Japan

Editorial Office
MDPI
St. Alban-Anlage 66
4052 Basel, Switzerland

This is a reprint of articles from the Special Issue published online in the open access journal *International Journal of Molecular Sciences* (ISSN 1422-0067) from 2018 to 2019 (available at: https: //www.mdpi.com/journal/ijms/special_issues/plant-proteomic_2)

For citation purposes, cite each article independently as indicated on the article page online and as indicated below:

LastName, A.A.; LastName, B.B.; LastName, C.C. Article Title. *Journal Name* **Year**, *Article Number*, Page Range.

ISBN 978-3-03921-062-6 (Pbk)
ISBN 978-3-03921-063-3 (PDF)

Contents

About the Special Issue Editor

Setsuko Komatsu is a Professor at Fukui University of Technology, Japan. She has been working at Meiji Pharmaceutical University since 1980, and then also at Keio University School of Medicine. During this period, her research has been focused on the role of protein kinase-dependent phosphorylation during fertilization in mammals. She began her work on plant proteomics using protein sequencers and mass spectrometry as a unit leader at the National Agriculture and Food Research Organization, and a Professor at the University of Tsukuba. Her main research interests are within the field of proteomics, biochemistry, and molecular biology, with a special focus on signal transduction in crops under environmental stress. Prof. Komatsu is section Editor-in-Chief for International Journal of Molecular Sciences, and contributes towards improving the quality of the research related to agricultural proteomics. Furthermore, as the president of Asia Oceania Agricultural Proteomics Organization and as council member of Japanese Proteomics Society, she is involved in promoting agriculture proteomics.

International Journal of
Molecular Sciences

MDPI

Editorial

Plant Proteomic Research 2.0: Trends and Perspectives

Setsuko Komatsu

Faculty of Environmental and Information Sciences, Fukui University of Technology, Fukui 910-8505, Japan;
skomatsu@fukui-ut.ac.jp

Received: 20 May 2019; Accepted: 20 May 2019; Published: 21 May 2019

Plants being sessile in nature are constantly exposed to environmental challenges resulting in substantial yield loss. To cope with the harsh environment, plants have developed a wide range of adaptation strategies involving morpho-anatomical, physiological, and biochemical traits [1]. In recent years, there has been phenomenal progress in the understanding of plant responses to environmental cues at the protein level. Advancements in the high-throughput "Omics" technique have revolutionized plant molecular biology research. Proteomics offers one of the best options for the functional analysis of translated regions of the genome and generates much detailed information about the intrinsic mechanisms of plant stress response. This special issue has 29 articles, which includes one review and 28 original articles on proteomic and transcriptomic studies. Various proteomic approaches are being exploited extensively for elucidating master regulatory proteins, which play key roles in stress perception and signaling. They largely involve gel-based and gel-free techniques, including both label-based and label-free protein quantification.

In this special issue, out of the 27 original proteomic publications, 21 articles use the gel-free technique, in which nine are label-free and 12 are label-based. Progress has been fueled by the advancement in mass spectrometry techniques, complemented with genome-sequence data and modern bioinformatic analysis; however, until now the two-dimensional electrophoresis based proteomic technique was used [2] as shown in six articles of this special issue. The review by Ray et al. [3] summarized the potential and limitations of the proteomic approaches and focused on *Quercus ilex* as a model species for other forest tree species. Regarding the progress of techniques in proteomics with other plant species, the research in *Q. ilex* moved from a gel-based strategy to a gel-free shotgun workflow. New directions in *Q. ilex* research leads to the identification of allergens in pollen grains/acorns and the characterization of wood materials, which are objectives clearly approached by proteomics [3]

The impact of diseases on crop production negatively reflects on sustainable food production and the overall economic health of the world. Five publications focus on biotic stress using various proteomic techniques. Khoza et al. [4] used a proteomic technique to identify *Arabidopsis* plasma-membrane associated candidate proteins in response to fungal treatment as well as those possibly interacting with the microbe-associated molecular pattern as ligands. They identified defense-related proteins and elucidated unknown signaling responses to this microbe-associated molecular pattern, including endocytosis. Furthermore, proteomic techniques were used to identify the mechanism in crops such as tomato [5], sugarcane [6], potato [7], and wheat [8] under biotic stress. Plants and pathogens are entangled in a continual arms race. Because plants have evolved dynamic defense and immune mechanisms to resist infection and enhance immunity for second wave attacks from the same or different types of pathogenic species, proteomics is a very useful technique for comprehensive analysis.

Wang et al. [9] and Gao et al. [10] performed proteomic analysis using the isobaric tag for relative and absolute quantification of castor and jojoba, respectively, under cold stress. Wang et al. [9] summarized that certain processes they identified cooperatively work together to establish the beneficial equilibrium of physiological and cellular homeostasis under cold stress. Gao et al. [10] indicated that photosynthesis suppression, cytoskeleton and cell wall adjustment, lipid metabolism/transport, reactive oxygen species scavenging, and carbohydrate metabolism were closely associated with the cold stress response. On the other hand, Inomata et al. [11] and Hao et al. [12] performed proteomics to identify the mechanisms in rice and lettuce, respectively, under high temperature. Inomata et al. [11] suggested that their results provide additional insights into carbohydrate metabolism regulation under ambient and adverse conditions. Hao et al. [12] indicated that a high temperature enhances the function of photosynthesis and auxin biosynthesis to promote the process of bolting, which is in line with the physiology and transcription levels of auxin metabolism. Furthermore, drought stress [13] and ultraviolet-B stress [14] were also used for mechanism analyses in maize and *Clematis terniflora* DC, respectively.

To facilitate the biotechnological improvement of crop productivity, genes, and proteins that control crop adaptation to a wide range of environments will need to be identified. This special issue includes many functional mechanisms of plants with nitrogen utilization [15], ammonium nutrition [16], cadmium exposure [17], nanoparticle treatment [18], and plant-derived smoke treatment [19]. Furthermore, various plants were used such as rice mutants [20], barley [21], *Morus alba* [22], pea cultivars [23], maize [24], tea [25], *Brunfelsia acuminate* [26], potato [27], and *Phalaenopsis* [28]. Due to the challenges faced in text/data mining, there is a large gap between the data available to researchers and the hundreds of published plant stress proteomic articles. PlantPReS is a valuable database for most researchers working in proteomics and plant stress areas [29].

Despite recent advancements, more emphasis needs to be given to the protein-extraction protocols, especially for proteins that are not abundant. Matsuta et al. [30] and Nishiyama et al. [31] used the mass spectrometry technique to identify heterotrimeric G γ4 and γ3 subunit proteins that are not abundant. As *RGG4/DEP1/DN1/qPE9-1/OsGGC3* mutants exhibited dwarfism, the tissues that accumulated Gγ4 corresponded to the abnormal tissues observed in *RGG4/DEP1/DN1/qPE9-1/OsGGC3* mutants [30]. On the other hand, as *RGG3/GS3/Mi/OsGGC1* mutants show the characteristic phenotype in flowers and consequently in seeds, the tissues that accumulated Gγ3 corresponded to the abnormal tissues observed in *RGG3/GS3/Mi/OsGGC1* mutants [31]. An amalgamation of diverse mass spectrometry technique, complemented with genome-sequence data and modern bioinformatics analysis, offers a powerful tool to identify and characterize novel proteins. This allows for researchers to follow temporal changes in relative protein abundances in developing/growing plant stage or under adverse environmental conditions.

Furthermore, organelle function, post-translational modifications, and protein-protein interactions, which are progress of proteomic research, provide deeper insight into protein molecular function. The major subcellular organelles and compartments in plant cells are nucleus, mitochondria, chloroplasts, endoplasmic reticulum, Golgi apparatus, vacuoles, and plasma membrane. The intracellular organelles and their interactions during stressful conditions represent the primary defense response. Subcellular proteomics has the potential to elucidate localized cellular responses and investigate communications among subcellular compartments during plant development and in response to biotic and abiotic stresses. This special issue includes the proteomic results in plasma membrane [4,30,31], chloroplast [11], and cell wall [17]. Additionally, the progress of proteomic research is understanding the post-translational modification such as phosphorylation [11,21,27].

Furthermore, proteomic data will be improved with convention regarding metabolomics and transcriptomics [32]. Although there have been significant advances over the years, a big gap still exists between the number of protein-coding genes and proteins detected with sufficient experimental evidence [33]. The guest editor hopes that proteomic data can detect the proteins with less experimental evidence and identify the missing proteins, which mainly use mass spectrometry-based experimental approaches. Although proteomic articles are independently published, the systematic collaborative network will be useful for further functional analyses in the near future. The articles in this special issue will be of general interest to proteomic researchers, plant biologists, and environmental scientists.

The guest editor hopes that this special issue will provide readers with a framework for understanding plant proteomics and insights into new research directions within this field. The guest editor thanks all of the authors for their contributions and thanks the reviewers for their critical assessments of these articles. Moreover, the guest editor renders heartiest thanks to the Assistant Editor, Ms. Chaya Zeng for giving me the opportunity to serve "Plant Proteomic Research 2.0" as guest editor.

Author Contributions: S.K. has made substantial, direct and intellectual contributions to the work, and approved it for publication.

Acknowledgments: In this section you can acknowledge any support given which is not covered by the author contribution or funding sections. This may include administrative and technical support, or donations in kind (e.g., materials used for experiments).

Conflicts of Interest: The authors declare no conflict of interest.

References

1. Komatsu, S.; Hossain, Z. Preface—Plant Proteomic Research. *Int. J. Mol. Sci.* **2017**, *18*, 88. [CrossRef]
2. Jorrin-Novo, J.V.; Komatsu, S.; Sanchez-Lucas, R.; Rodríguez de Francisco, L.E. Gel electrophoresis-based plant proteomics: Past, present, and future. Happy 10th anniversary Journal of Proteomics! *J. Proteomics* **2019**, *198*, 1–10. [CrossRef]
3. Rey, M.; Castillejo, M.; Sánchez-Lucas, R.; Guerrero-Sanchez, V.; López-Hidalgo, C.; Romero-Rodríguez, C.; Valero-Galván, J.; Sghaier-Hammami, B.; Simova-Stoilova, L.; Echevarría-Zomeño, S.; et al. Proteomics, Holm oak (*Quercus ilex* L.) and other recalcitrant and orphan forest tree species: How do they see each other? *Int. J. Mol. Sci.* **2019**, *20*, 692. [CrossRef]
4. Khoza, T.; Dubery, I.; Piater, L. Identification of candidate ergosterol-responsive proteins associated with the plasma membrane of *Arabidopsis thaliana*. *Int. J. Mol. Sci.* **2019**, *20*, 1302. [CrossRef]
5. Fan, K.; Wang, K.; Chang, W.; Yang, J.; Yeh, C.; Cheng, K.; Hung, S.; Chen, Y. Application of data-independent acquisition approach to study the proteome changes from early to later phases of tomato pathogenesis responses. *Int. J. Mol. Sci.* **2019**, *20*, 863. [CrossRef] [PubMed]
6. Singh, P.; Song, Q.; Singh, R.; Li, H.; Solanki, M.; Malviya, M.; Verma, K.; Yang, L.; Li, Y. Proteomic analysis of the resistance mechanisms in sugarcane during *Sporisorium scitamineum* infection. *Int. J. Mol. Sci.* **2019**, *20*, 569.
7. Xiao, C.; Gao, J.; Zhang, Y.; Wang, Z.; Zhang, D.; Chen, Q.; Ye, X.; Xu, Y.; Yang, G.; Yan, L.; et al. Quantitative proteomics of potato leaves infected with phytophthora infestans provides insights into coordinated and altered protein expression during early and late disease stages. *Int. J. Mol. Sci.* **2019**, *20*, 136. [CrossRef] [PubMed]
8. Balakireva, A.; Deviatkin, A.; Zgoda, V.; Kartashov, M.; Zhemchuzhina, N.; Dzhavakhiya, V.; Golovin, A.; Zamyatnin, A. Proteomics analysis reveals that caspase-like and metacaspase-like activities are dispensable for activation of proteases involved in early response to biotic stress in *Triticum aestivum* L. *Int. J. Mol. Sci.* **2018**, *19*, 3991. [CrossRef] [PubMed]
9. Wang, X.; Li, M.; Liu, X.; Zhang, L.; Duan, Q.; Zhang, J. Quantitative proteomic analysis of castor (*Ricinus communis* L.) seeds during early imbibition provided novel insights into cold stress response. *Int. J. Mol. Sci.* **2019**, *20*, 355. [CrossRef]
10. Gao, F.; Ma, P.; Wu, Y.; Zhou, Y.; Zhang, G. Quantitative proteomic analysis of the response to cold stress in jojoba, a tropical woody crop. *Int. J. Mol. Sci.* **2019**, *20*, 243. [CrossRef]
11. Inomata, T.; Baslam, M.; Masui, T.; Koshu, T.; Takamatsu, T.; Kaneko, K.; Pozueta-Romero, J.; Mitsui, T. Proteomics analysis reveals non-controlled activation of photosynthesis and protein synthesis in a rice npp1 mutant under high temperature and elevated CO_2 conditions. *Int. J. Mol. Sci.* **2018**, *19*, 2655. [CrossRef]
12. Hao, J.; Zhang, L.; Li, P.; Sun, Y.; Li, J.; Qin, X.; Wang, L.; Qi, Z.; Xiao, S.; Han, Y.; et al. Quantitative proteomics analysis of lettuce (*Lactuca sativa* L.) reveals molecular basis-associated auxin and photosynthesis with bolting induced by high temperature. *Int. J. Mol. Sci.* **2018**, *19*, 2967. [CrossRef]
13. Zenda, T.; Liu, S.; Wang, X.; Jin, H.; Liu, G.; Duan, H. Comparative proteomic and physiological analyses of two divergent maize inbred lines provide more insights into drought-stress tolerance mechanisms. *Int. J. Mol. Sci.* **2018**, *19*, 3225. [CrossRef]
14. Chen, X.; Yang, B.; Huang, W.; Wang, T.; Li, Y.; Zhong, Z.; Yang, L.; Li, S.; Tian, J. Comparative proteomic analysis reveals elevated capacity for photosynthesis in polyphenol oxidase expression-silenced *Clematis terniflora* DC. leaves. *Int. J. Mol. Sci.* **2018**, *19*, 3897. [CrossRef]

15. Waqas, M.; Feng, S.; Amjad, H.; Letuma, P.; Zhan, W.; Li, Z.; Fang, C.; Arafat, Y.; Khan, M.; Tayyab, M.; et al. Protein phosphatase (PP2C9) induces protein expression differentially to mediate nitrogen utilization efficiency in rice under nitrogen-deficient condition. *Int. J. Mol. Sci.* **2018**, *19*, 2827. [CrossRef]

16. Coleto, I.; Vega-Mas, I.; Glauser, G.; González-Moro, M.; Marino, D.; Ariz, I. New Insights on *Arabidopsis thaliana* root adaption to ammonium nutrition by the use of a quantitative proteomic approach. *Int. J. Mol. Sci.* **2019**, *20*, 814. [CrossRef]

17. Gutsch, A.; Zouaghi, S.; Renaut, J.; Cuypers, A.; Hausman, J.; Sergeant, K. Changes in the proteome of medicago sativa leaves in response to long-term cadmium exposure using a cell-wall targeted approach. *Int. J. Mol. Sci.* **2018**, *19*, 2498. [CrossRef]

18. Jhanzab, H.; Razzaq, A.; Bibi, Y.; Yasmeen, F.; Yamaguchi, H.; Hitachi, K.; Tsuchida, K.; Komatsu, S. Proteomic analysis of the effect of inorganic and organic chemicals on silver nanoparticles in wheat. *Int. J. Mol. Sci.* **2019**, *20*, 825. [CrossRef]

19. Aslam, M.; Rehman, S.; Khatoon, A.; Jamil, M.; Yamaguchi, H.; Hitachi, K.; Tsuchida, K.; Li, X.; Sunohara, Y.; Matsumoto, H.; et al. Molecular responses of maize shoot to a plant derived smoke solution. *Int. J. Mol. Sci.* **2019**, *20*, 1319. [CrossRef]

20. Yang, X.; Meng, W.; Zhao, M.; Zhang, A.; Liu, W.; Xu, Z.; Wang, Y.; Ma, J. Proteomics analysis to identify proteins and pathways associated with the novel lesion mimic mutant E40 in rice using iTRAQ-based strategy. *Int. J. Mol. Sci.* **2019**, *20*, 1294. [CrossRef]

21. Ishikawa, S.; Barrero, J.; Takahashi, F.; Peck, S.; Gubler, F.; Shinozaki, K.; Umezawa, T. Comparative phosphoproteomic analysis of barley embryos with different dormancy during imbibition. *Int. J. Mol. Sci.* **2019**, *20*, 451. [CrossRef]

22. Zhu, W.; Zhong, Z.; Liu, S.; Yang, B.; Komatsu, S.; Ge, Z.; Tian, J. Organ-Specific Analysis of *Morus alba* using a gel-free/label-free proteomic technique. *Int. J. Mol. Sci.* **2019**, *20*, 365. [CrossRef]

23. Mamontova, T.; Lukasheva, E.; Mavropolo-Stolyarenko, G.; Proksch, C.; Bilova, T.; Kim, A.; Babakov, V.; Grishina, T.; Hoehenwarter, W.; Medvedev, S.; et al. Proteome map of pea (*Pisum sativum* L.) embryos containing different amounts of residual chlorophylls. *Int. J. Mol. Sci.* **2018**, *19*, 4066. [CrossRef]

24. Liu, B.; Shan, X.; Wu, Y.; Su, S.; Li, S.; Liu, H.; Han, J.; Yuan, Y. iTRAQ-Based Quantitative proteomic analysis of embryogenic and non-embryogenic calli derived from a maize (*Zea mays* L.) inbred line Y423. *Int. J. Mol. Sci.* **2018**, *19*, 4004. [CrossRef]

25. Dong, F.; Shi, Y.; Liu, M.; Fan, K.; Zhang, Q.; Ruan, J. iTRAQ-based quantitative proteomics analysis reveals the mechanism underlying the weakening of carbon metabolism in chlorotic tea leaves. *Int. J. Mol. Sci.* **2018**, *19*, 3943. [CrossRef]

26. Li, M.; Sun, Y.; Lu, X.; Debnath, B.; Mitra, S.; Qiu, D. Proteomics reveal the profiles of color change in *Brunfelsia acuminata* flowers. *Int. J. Mol. Sci.* **2019**, *20*, 2000. [CrossRef]

27. Bernal, J.; Mouzo, D.; López-Pedrouso, M.; Franco, D.; García, L.; Zapata, C. The major storage protein in potato tuber is mobilized by a mechanism dependent on its phosphorylation status. *Int. J. Mol. Sci.* **2019**, *20*, 1889. [CrossRef]

28. Chen, C.; Zeng, L.; Ye, Q. Proteomic and biochemical changes during senescence of phalaenopsis 'Red Dragon' petals. *Int. J. Mol. Sci.* **2018**, *19*, 1317. [CrossRef]

29. Mousavi, S.A.; Pouya, F.M.; Ghaffari, M.R.; Mirzaei, M.; Ghaffari, A.; Alikhani, M.; Ghareyazie, M.; Komatsu, S.; Haynes, P.A.; Salekdeh, G.H. PlantPReS: A database for plant proteome response to stress. *J. Proteomics* **2016**, *143*, 69–72. [CrossRef]

30. Matsuta, S.; Nishiyama, A.; Chaya, G.; Itoh, T.; Miura, K.; Iwasaki, Y. Characterization of heterotrimeric G protein γ4 subunit in rice. *Int. J. Mol. Sci.* **2018**, *19*, 3596. [CrossRef]

31. Nishiyama, A.; Matsuta, S.; Chaya, G.; Itoh, T.; Miura, K.; Iwasaki, Y. Identification of heterotrimeric G protein γ3 subunit in rice plasma membrane. *Int. J. Mol. Sci.* **2018**, *19*, 3591. [CrossRef]

Int. J. Mol. Sci. **2019**, *20*, 2495

32. Shi, J.; Zhao, L.; Yan, B.; Zhu, Y.; Ma, H.; Chen, W.; Ruan, S. Comparative transcriptome analysis reveals the transcriptional alterations in growth- and development-related genes in sweet potato plants infected and non-infected by SPFMV, SPV2, and SPVG. *Int. J. Mol. Sci.* **2019**, *20*, 1012. [CrossRef]
33. Rahiminejad, M.; Ledari, M.T.; Mirzaei, M.; Ghorbanzadeh, Z.; Kavousi, K.; Ghaffari, M.R.; Haynes, P.A.; Komatsu, S.; Salekdeh, G.H. The quest for missing proteins in rice. *Mol. Plant.* **2019**, *12*, 4–6. [CrossRef]

International Journal of
Molecular Sciences

MDPI

Review

Proteomics, Holm Oak (*Quercus ilex L.*) and Other Recalcitrant and Orphan Forest Tree Species: How do They See Each Other?

María-Dolores Rey [1], María Ángeles Castillejo [1], Rosa Sánchez-Lucas [1],
Victor M. Guerrero-Sanchez [1], Cristina López-Hidalgo [1], Cristina Romero-Rodríguez [2],
José Valero-Galván [3], Besma Sghaier-Hammami [1], Lyudmila Simova-Stoilova [4],
Sira Echevarría-Zomeño [1], Inmaculada Jorge [5], Isabel Gómez-Gálvez [1], María Eugenia Papa [1],
Kamilla Carvalho [1], Luis E. Rodríguez de Francisco [6], Ana María Maldonado-Alconada [1],
Luis Valledor [7] and Jesús V. Jorrín-Novo [1,*]

[1]   Department of Biochemistry and Molecular Biology, Agrifood Campus of International Excellence,
      University of Cordoba, Carretera Nacional IV, km 396, 14014 Córdoba, Spain; b52resam@uco.es (M.-D.R.);
      bb2casam@uco.es (M.Á.C.); g82salur@uco.es (R.S.-L.); b12gusav@uco.es (V.M.G.-S.);
      n12lohic@uco.es (C.L.-H.); sghaierbesma@yahoo.fr (B.S.-H.); sira@biognosys.com (S.E.-Z.);
      isa.m.goga@gmail.com (I.G.-G.); mepv33@gmail.com (M.E.P.); carvalho.kmll@gmail.com (K.C.);
      bb2maala@uco.es (A.M.M.-A.)
[2]   Departamento de Fitoquímica, Dirección de Investigación de la Facultad de Ciencias Químicas
      de la Universidad Nacional de Asunción, Asunción 1001-1925, Paraguay; mromero@rec.una.py
[3]   Department of Chemical and Biological Science, Biomedicine Science Institute,
      Autonomous University of Ciudad Juárez, Anillo Envolvente del Pronaf y Estocolmo s/n,
      Ciudad Juarez 32310, Mexico; jose.valero@uacj.mx
[4]   Plant Molecular Biology Department, Institute of Plant Physiology and Genetics,
      Bulgarian Academy of Sciences, Acad. G. Bonchev Str. Bl 21, 1113 Sofia, Bulgaria; lsimova@mail.bg
[5]   Department of Vascular Biology and Inflammation (BVI), Spanish National Centre
      for Cardiovascular Research, Melchor Fernández Almagro 3, 28029 Madrid, Spain; inmaculada.jorge@cnic.es
[6]   Laboratorio de Biología, Instituto Tecnológico de Santo Domingo, República Dominicana; luisrod95@gmail.com
[7]   Department of Organisms and Systems Biology and University Institute of Biotechnology (IUBA),
      University of Oviedo, Santiago Gascón Building, 2nd Floor (Office 2.9), 33006 Oviedo, Spain;
      valledorluis@uniovi.es
*   Correspondence: bf1jonoj@uco.es; Tel.: +34-957-218-609

Received: 10 January 2019; Accepted: 30 January 2019; Published: 6 February 2019

Abstract: Proteomics has had a big impact on plant biology, considered as a valuable tool for several forest species, such as *Quercus*, *Pines*, *Poplars*, and *Eucalyptus*. This review assesses the potential and limitations of the proteomics approaches and is focused on *Quercus ilex* as a model species and other forest tree species. Proteomics has been used with *Q. ilex* since 2003 with the main aim of examining natural variability, developmental processes, and responses to biotic and abiotic stresses as in other species of the genus *Quercus* or *Pinus*. As with the progress in techniques in proteomics in other plant species, the research in *Q. ilex* moved from 2-DE based strategy to the latest gel-free shotgun workflows. Experimental design, protein extraction, mass spectrometric analysis, confidence levels of qualitative and quantitative proteomics data, and their interpretation are a true challenge with relation to forest tree species due to their extreme orphan and recalcitrant (non-orthodox) nature. Implementing a systems biology approach, it is time to validate proteomics data using complementary techniques and integrate it with the -omics and classical approaches. The full potential of the protein field in plant research is quite far from being entirely exploited. However, despite the methodological limitations present in proteomics, there is no doubt that this discipline has contributed to deeper knowledge of plant biology and, currently, is increasingly employed for translational purposes.

Int. J. Mol. Sci. **2019**, *20*, 692

Keywords: holm oak; *Quercus ilex*; 2-DE proteomics; shotgun proteomics; non-orthodox seed; population variability; stresses responses

---

1. Introduction

Quercus ilex is the dominant tree species in natural forest ecosystems over large areas of the Western Mediterranean Basin, as well as in the agrosilvopastoral Spanish "dehesa", with relevance from an environmental, economic, and social point of view [1–3]. These ecosystems are currently subjected to different threats including: very old individuals, overexploitation and poor regeneration, inappropriate livestock management, and the severe effect of forest decline attributed to fungal attack (such as *Hypoxylon mediterraneum* or *Phytophthora cinnamomi*), extreme temperatures and extended drought periods, among other factors [3–5]. This already worrisome situation could become even worse under the threat of the foreseen climate change scenario [6,7]. In order to preserve such an invaluable ecosystem, these problems must be faced, and biotechnology is a valid alternative that could contribute to resolving some of these problems. However, the development of biotechnological approaches for the conservation, sustainable management and regeneration of *Q. ilex*, and other forest ecosystems is hampered by the limited knowledge of their biology, especially at the molecular level. Biochemical and molecular biology research is a priority for designing biotechnological approaches for simultaneously conserving and exploiting forest ecosystems. Plausible, realistic, and impactful first steps to ameliorate this situation could include the characterization of its biodiversity and the selection of elite genotypes based on molecular markers. In this context, protein profiling through different proteomic approaches would be highly useful [8].

Since 2003, our research group has worked on the proteomics of forest tree species, with a first publication in 2005 [9]. Our investigations have been focused mostly on *Quercus ilex* subsp. *ballota* [Desf.] Samp., and, to a lesser extent, on various *Pinus* spp., including *P. radiata* [10], *P. occidentalis* [11], *P. halepensis* [12], and *P. pinea* [13]. All these forest tree species can be classified and catalogued as orphan due to the absence of molecular studies and, depending on their seed characteristics, properties, and maturation, as highly recalcitrant (non-orthodox) plant systems [8], because unlike orthodox seeds, non-orthodox seeds are damaged by loss of water and are unstorable for practical purposes.

So far, our proteomics-based research on *Q. ilex* has focused on descriptive and comparative proteomics sub-areas (Figure 1). In addition, we have begun to explore the field of posttranslational protein modifications, specifically phosphorylation [14]. Using 2-DE based strategy coupled with mass spectrometry (MS), and, to a lesser extent, shotgun approaches, the proteome of seeds, pollen, roots, and leaves, both in adult plants and plantlets, have been partially characterized, and differences in protein profiles among provenances have been identified [9,15–17]. In an attempt to study the non-orthodox character of the species, we have further investigated the proteome of mature acorns, as well as the differences between developmental stages of seed maturation and germination [18,19]. Furthermore, our research has been focused on studying plant responses to abiotic and biotic stresses related to decline syndrome, mainly drought and *P. cinnamomi* infestation, as well as differences among *Q. ilex* provenances from Andalusia combining proteomics, morphometry, and physiological analysis [17,20–23]. This manuscript does not intend to be a review of the field of proteomics, because there are already a high number of publications available in the literature [24–32], or discuss terminology or scientific standards mandated by the corresponding Minimum Information about Proteomics Experiment (MIAPE) guidelines [33]. On the other hand, the application of proteomics in forest tree research has also been the subject of some previous reviews, with a quite descriptive point of view [34,35]. So, in this review, we intend to emphasize all the lessons learnt through fifteen years working on *Q. ilex*, which includes everything from the experimental design, protein extract preparation, MS analysis, confident identification and quantification of protein species to data interpretation from a biological perspective [36,37]. For most of the mentioned issues, all the studies carried out on an orphan and extremely recalcitrant experimental system such as *Q. ilex* have been highly challenging. Proteomics is more than only a single table of possible protein identifications, i.e., database matches, or, even in the best of cases, ortholog identifications and their technical validation. Literature, including our own publications, may contain errors, speculations, and incorrect interpretations, which are waiting to be revised.

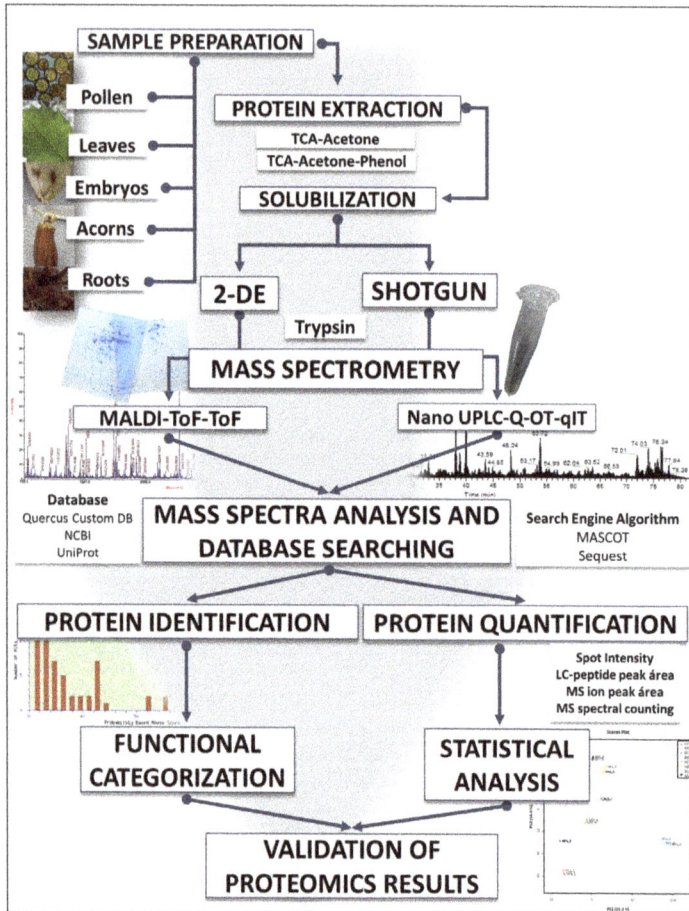

Figure 1. Workflow of a proteomics experiment, from sample preparation to data analysis and validation. It includes alternative, complementary approaches or strategies, based on MS analysis of proteins (top-down) or tryptic peptides (bottom-up), either gel-based or gel-free. LC: liquid chromatography; MS: mass spectrometry.

2. How *Quercus ilex* Is Seen by Proteomics

2.1. 'Only a Small Percentage of the Total Protein Is Extracted and Solubilized, So We Deal with the Extractome Rather Than with the Real Proteome'

There are two major approaches for making protein extracts, independently of the subcellular compartment, based on either precipitation or solubilization. Both approaches are the most common protocols to extract proteins and these should be optimized in each organism. In our hands, precipitation methods have always given the best results in terms of protein yield as determined by colorimetric methods, generally using Bradford assay (20 mg g^{-1} fresh weight from *Q. ilex* leaf, as an example) [17]. Depending on the chemical composition and protein content of the organ analyzed and the amount of tissue available, trichloroacetic acid (TCA)-acetone precipitation alone or combined with phenol partitioning, followed by ammonium acetate-methanol precipitation, have consistently yielded the best results [37]. Table 1 collects the main features of the *Q. ilex* publications cited in this review. Protein yield and even recovery across a wide range of proteins is a constant concern in protein biochemistry. Remarkably, the protein concentrations of extracts are commonly absent in many publications, although the protein quantification of the extracts has been expressly stated in the material and methods section.

Table 1. Relevant results concerning proteomics research on *Quercus ilex* carried out by our group.

Author	Year	Plant Organ	Protein Yield (mg g⁻¹ DW Tissue) a	Proteomic Strategy	Features c	Identified Proteins	Proteome Database e
Jorge [9]	2005	Leaf	Data not reported *; L	2-DE MALDI TOF/TOF	350	20 out of 100 spots	NCBI: restriction to Viridiplantae
Jorge [15]	2006	Leaf	Data not reported *; L	2-DE MALDI TOF/TOF	400	24 out of 100 spots	NCBI: restriction to Viridiplantae
Echevarría-Zomeño [20]	2009		7 *; L		390	12 out of 46 spots	SwissProt, trEMBL, and NCBI: restriction to Viridiplantae
Valero-Galván [16]	2011	Seed	6 *; B		240	16 out of 56 spots	NCBI: restriction to Viridiplantae
Valero-Galván [38]	2012	Pollen	15 §; B	2-DE MALDI-TOF/TOF	600	77 out of 100 spots	UniProtKB restricted to *Arabidopsis*; Phytozome restricted to *Populus* and *Eucaliptus*; Custom-build database from *Quercus* ESTs f
				Shotgun (nLC-MS/MS) b	Data not reported	273	
Valero-Galván [21]	2013	Leaf	10 §; B		230	18 out of 28 spots	NCBI: restriction to Viridiplantae
Sghaier-Hammami [17]	2013	Leaf	40 §; B		480	80 out of 480 spots	NCBI: restriction to Viridiplantae
Simova-Stoilova [22]	2015	Root	3 §; B	2-DE MALDI-TOF/TOF	360	79 out of 90 spots	NCBI and UniProtKB: restriction to Viridiplantae
Romero-Rodríguez [14]	2015	Embryo	150 §; B		480	20 d out of 55 spots	NCBI, UniProtKB: restriction to Viridiplantae and Custom *Quercus* database f
Sghaier-Hammami [18]	2016	Cotyledon	2 §; B		440	50 out of 153 spots	NCBI: restriction to Viridiplantae
		Embryo	80 §; B		470	50 out of 153 spots	
		Tegument	0.4 §; B		420	40 out of 153 spots	
López-Hidalgo [39]	2018	Pool of tissues: acorn, embryo, cotyledon, leaf and root	40 §; B	Shotgun (nLC-MS/MS) b	58600	2830	SwissProt: restriction to Viridiplantae / Custom-build specie database f
Romero-Rodríguez [19]	2018	Seed	25 §; B	2-DE MALDI-TOF/TOF	540	90 out of 103 spots	NCBI, UniProtKB/TrEMBL and UniProtKB/SwissProt restricted to Viridiplantae; Custom-build *Q. ilex* database f
				Shotgun (nLC-MS/MS) b	3113	1650	

a Approximated values have been adjusted to the unit. * = TCA extraction method and § = TCA-Phenol extraction method. Final pellet was resuspended in a solution containing 9 M urea, 4% CHAPS, 0.5% Triton X100, and 100 mM DTT. Proteins were quantified using the Lowry (L) or Bradford (B) protocols; b The equipment used in the shotgun strategy was nLC-MS/MS (orbitrap, Q-OT-qIT); c Spots resolved using 2-DE or peptides identified using shotgun LC-MS/MS; d This value corresponds to identified phosphoproteins; e MASCOT and SEQUEST search engines were used with MALDI-TOF/TOF and shotgun LC-MS/MS data, respectively; f The custom-build databases from the genus Quercus and Q. ilex have been published by Guerrero-Sanchez et al. [40] and Romero-Rodriguez et al. [41].

It is true that absolute quantification by current protocols (Lowry, Bradford, bicinchoninic acid (BCA), amido black) is not always reliable, as up to ten-fold difference may be observed between different protocols. Still, they may be valuable for comparative purposes and reproducibility [42].

We have extracted proteins from different organs of more than 25 different plant species, both woody and herbaceous. Protein content in those extracts was consistently lower than 10% (in the 1-20 mg g^{-1} Dry Weight (DW) range [43]) of the total as determined using the Kjeldahl method [44], with some legume species having the highest values [43]. For the acorns, pollen, and leaves of *Q. ilex*, values of 3–6, 8-14, and 10–40 mg g^{-1} DW were reported, respectively [16,17,38,45] (Table 1). Even when applying Osborn's sequential extraction protocol to *Q. ilex* seeds [46], the total protein content obtained was around 15 mg g^{-1} DW as determined using Bradford assay, which represents around 30% of the total protein as determined using near-infrared spectroscopy (NIRS) [47]. These data lead us to estimate mistakes and make speculations while interpreting our proteomics data from a biological point of view, as we are clearly not recovering and therefore not examining the huge submerged part of the proteome iceberg.

2.2. The Plant Proteome is Highly Variable and Therefore Requires Careful Experimental Design

This was one of the first major lessons that we learnt when working with *Q. ilex*. We have observed that the 2-DE protein profile of leaf samples collected from field trees is not reproducible. Only after systematic analysis of the protein pattern obtained, we could show that results strongly depended on leaf position (top, bottom), leaf orientation (north, south, east, west) and sampling time (morning, afternoon, evening) [9]. These observations were more than obvious considering the sessile and plastic nature of plants, but they were not considered when the experiments were designed. The average value of the coefficient of biological variance (CV) for protein abundance (spot intensity) was found close to 60 % for field samples and close to 45% for plantlets grown under controlled conditions, while values of 20-25 % were found for analytical variability [9,15]. The average standard error of spot intensity decreased by a factor of two when the number of biological replicates increased from two to twelve (from an average of 120 to 60 ng protein per spot) [9,15]. High variability is a common feature for plants. Plant organs are complex mixtures of tissue and cell types, each with their own protein signature. In addition, individuals of non-domesticated plants exhibit high variability. Because of these issues, a significant number of biological replicates should be considered to decrease the effect of variability in our results. The direct consequence of this is the need to characterize the variability beforehand using test measurements and then perform an exhaustive analysis to determine the number of required replicates. Alternatively, the analytical approach may have to be refined. Due to obvious limitations (space, time, equipment, and costs), it is not always possible to perform experiments based on a large number of replicates. However, the actual concern is how the data are interpreted. For comparative purposes, we only consider as variable spots those that are consistent (present in all the replicates), and with lower CVs than the average of the sample [9,48]. A higher ratio between samples makes more confident those quantitative differences observed, although sometimes only qualitative differences may be trusted. All these issues, together with tips to be considered for proper experimental designs and statistical tests (mostly multivariate and clustering), should be contemplated when a 2-DE based proteomics experiment is planned. Moreover, the correct analysis and interpretation of the data should be contemplated, thus, both are discussed in more detail in this review [36,48].

Generally, the proteome is discussed as a sum of the individual proteins identified and analyzed using a univariate approach, such as ANOVA, instead of being considered globally as a part of a biological entity and analyzed using a multivariate approach. Since univariate approaches are negatively affected by the raw structure of the data, they do not detect trends or groups increasing the false positives. On the other hand, multivariate analyses such as principal component analysis (PCA), partial least squares (PLS), principal coordinate analysis (PCoA), or partial least squares-discriminant analysis (PLS-DA), should be employed because they describe trends and reduce the complexity of the data [49]. Despite these multivariate approaches being intended to reduce data dimensionality, PCA seeks a few linear combinations of variables that can be used to summarize data while PLS considers how each predictive variable may be related to the dependent variable [49]. In any case, the combination of both univariate and multivariate approaches that provide a comprehensive overview of the data with single protein analyses and multiprotein tendency maximize the information obtained from the datasets [36].

2.3. Only a Small Fraction of the Present Protein Species Is Visualized and Identified by Any Given Approach

The number of spots resolved in different *Q. ilex* samples subjected to 2-DE analysis was in the range of 200-600 spots, depending on organ of the plant (seed, pollen, or leaf), range of isoelectric focusing (IEF) pH (5-8 as a general strategy), and staining protocol. Of the total spots subjected to mass spectrometry less than 50% of hits could be identified, depending on the database used (see above section on protein identification, Table 1) [15–18,20–22,45,47]. However, assuming the possibility of spot comigration, the maximum number of resolved proteins is below 1000. This amount of protein is notably increased into the thousands when a nLC-ESI-MS/MS shotgun approach is employed. Thus, up to 4500 peptides could be resolved in germinating seeds through LC-MS/MS shotgun analysis [19]. Assuming a theoretical calculation based on 3 peptides per protein, around 1650 protein species could be resolved. Thus, the use of a shotgun approach and a huge growth in bioinformatics has led to an explosion of data in the field of proteomics. Nevertheless, although the integration of both approaches is expanding their application in the identification of a higher number of peptides, their focus and strengths remain in the analysis of DNA sequences and genomes of plant species. The sequencing of the *Q. ilex* genome, which is indeed one of our next objectives, would be considered as a final step to integrate all the proteomics data obtained so far. However, this issue can currently be solved using the recently published genomic data available for other species of the genus *Quercus*, such as. *Q. robur* [50], *Q. lobata* [51], and *Q. suber* [52]. The genome of *Q. robur* has an estimated size of 740 Mb/C [53] and consists of 17,910 scaffolds, of length 2 kb or longer, with a total length of 1.3 Gb [50]. On the other hand, the first draft of the genome of *Q. lobata* has a genome size of approximately 730 Mb/C and 18 512 scaffolds (> 2 kb) [51]. A comparison of nuclear sequences between both *Quercus* species indicated 93% similarity [51]. Lesur et al. [54] have reported the most comprehensive transcript catalog assembled to date for the genus *Quercus*, with 91,000 annotated contigs. With the aim of sequencing the *Q. ilex* genome, our group has started to address basic aspects of the genome, such as estimation of the nuclear DNA content and the number of chromosomes of *Q. ilex*. The estimated genome size was approximately 930 Mb/C with a total length of 1.87 Gb, as assessed using flow cytometry [55] (Figure 2A). Zoldos et al. [56] and Chen et al. [57], using the same methodology as with *Q. ilex*, reported a higher *Q. robur* genome size than the data reported in Plomión et al. [50] (approximately 914 Mb/C and 890 Mb/C, respectively). Previous cytological studies established that the number of chromosomes in the genus *Quercus* has remained stable over time, being mainly $2n = 24$. Cytogenetic methods were used for chromosome count in root tip squashes of *Q. ilex* [58]. As expected, *Q. ilex* had the same chromosome number as *Quercus* spp. (Figure 2B). All chromosomes are quite similar morphologically, so that other cytogenetic methods should be used to identify all the chromosomes individually.

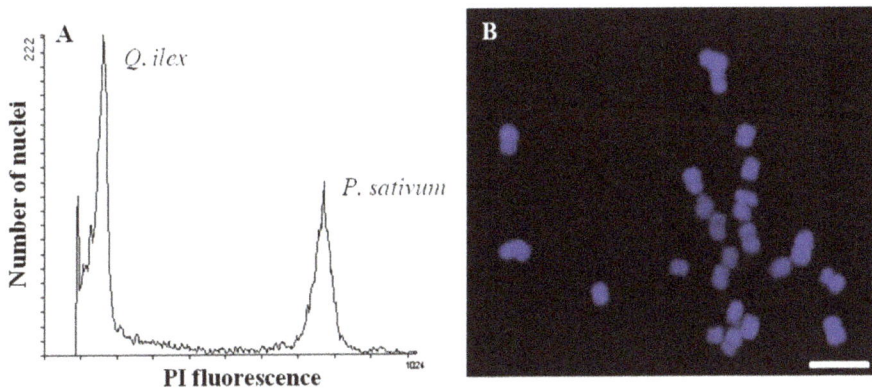

Figure 2. (**A**) Uniparametric histograms of fluorescence intensities of the nuclei of *Q. ilex* and *Pisum sativum*, used as a control, after staining with propidium iodide (PI). The 2C nuclear DNA content of *P. sativum* is 9.09 pg. (**B**) Somatic chromosomes in root tip cells of *Q. ilex*. Scale bar = 10 μm.

The proteome data can also be complemented using a transcriptomic approach. The first *de novo* assembled transcriptome of the non-conventional plant *Q. ilex* has recently been published [39,40,59]. The transcriptome of a mixture of different tissues of *Q. ilex* using two sequencing platforms, Illumina and Ion Torrent, and three different algorithms, MIRA, RAY, and TRINITY, was analyzed. Firstly, around 62,628 transcripts were identified using the Illumina platform (Illumina HiSeq 2500) [39]. Then, in a revised version of the *de novo* assembled transcriptome, the Ion Torrent sequencing platform was used, and 74,058 transcripts were identified [59]. The data reported for *Q. robur* and *Q. lobata* genomes and for the *Q. ilex* transcriptome express at least one order of magnitude higher than the number of expressed, visualized, and identified protein species in 2-DE or shotgun observed in our experiments—even without considering possible posttranslational modifications (PTMs)—although the non-consolidated nature of our data is considered. With these values in mind, we should only deal with a minimum fraction of the total proteome and any biological interpretation of the data should be made with caution, being as conservative as possible and avoiding speculations, especially if data are not validated.

The integration of omics approaches (genomics, transcriptomics, proteomics, and metabolomics) are commonly used to further our knowledge about plant biology. The data identified in each approach is quite variable, which depends on the available databases. For example, a total of 62,629 transcripts, 2380 protein species, and 62 metabolites were recently described in *Q. ilex* [39]. In spite of having a considerably lower number of proteins and metabolites than transcripts, proteomics and metabolomics could give a more connected understanding of the phenotype of the plant species. Thus, the integration of multi-omics studies with phenotypic and physiological data in the systems biology direction are necessary to obtain a better understanding of the molecular mechanisms underlying phenotypes of interest.

2.4. Gene Product Identification? Or Just Hits or Matches to Orthologs?

Proteome analysis of *Q. ilex* has been prevented for a long time due to the almost total absence of DNA or protein sequence entries in the available databases and, possibly, errors in the deposited sequences themselves. Consequently, protein identification from MS data usually had low peptide-to-spectra matching, even using *de novo* sequencing and sequence similarity searching (i.e., [9,15]). The concern that proteomics was only possible with organisms whose genome are properly sequenced and annotated, was a recurrent matter of discussion with Dr. Juan Pablo Albar (1953-2014, R.I.P.). Even considering that the possibility of orthologs identification already provided useful information on mechanisms and metabolism in many cases, some issues remained unresolved.

In parallel, plant breeding programs request increasingly accurate gene information rather than just the ortholog approximation. For this reason, we changed our strategy and decided to build a custom *Quercus* protein sequence database to improve the success rate of peptide and protein identifications and assignments [41]. This database is continuously updated and allows successful reviewing of existing data sets for the scientific community. The latest version of our custom *Q. ilex* database contained 3541 annotated proteins from the Ion Torrent platform [59]. At this moment, the number and confidence of the identifications can be carried out using the presence of whole genome sequencing of several forest tree species [60]. However, despite admitting positive identification (matches in some cases), the confidence value is not the same for all the proteins, although we assigned them the same probabilistic value when the data were interpreted from a biological point of view. Thus, the shotgun strategy in the proteome analysis of a pool of tissues (embryo, cotyledon, leaves, and root) from *Q. ilex* resulted in 7000 peptides and 1600 putative protein identifications when the species-specific database created from the *Q. ilex* transcriptome was used [40]. The confidence values obtained in this study was in the range 1-35 peptides per protein, 1-93 % sequence coverage, and 1-335 score values (using SEQUEST algorithm) [61]. However, almost 50% of identifications showed at least one parameter of low confidence (1 peptide per protein, sequence coverage <10%, or score value <2). These issues, although relevant, were rarely discussed openly, as blind acceptance of the results provided by the matching algorithm was in many cases easier and considered enough. However, publication of a list of sequence assignments is no longer enough to justify it. In the case of orphan species, ortholog identification does also not resolve the doubts about what protein species (different products of the same gene), isoforms, or allelic variants are present in a biological system nor indicate what they signify. If the aim is to obtain biological understanding of the data beyond description, proteomics data must be validated, especially in the case of orphan species; otherwise it remains largely speculative.

2.5. Methods and Protocols Must Be Validated and Optimized for Each Experimental System

The final goal of a proteomics experiment is to identify, characterize, and quantify as many protein species as possible. Different workflows, protocols, technology platforms, and algorithms are available, each one with its own signature and characteristics [27]. Small variations in a protocol used, such as different gel stains, may result in a different partial view of the protein 'firmament'. In our experience with different biological systems, including plants, bacteria, yeast, fungi, and animal cells [27,62,63], each protocol should be optimized for the experimental system under investigation, due to the presence of polysaccharides, phenolics, nucleic acids, salts, and other small metabolites in each biological sample.

Biologists are often far away from an analytical chemist's orthodox thinking, and this sometimes leads us to commit important errors in our biological interpretation of analytical results. It is of paramount importance to understand the properties of the analytical techniques employed, including selectivity, precision, accuracy, recovery, linearity range, limit of detection and quantification, robustness, and stability. Both the linearity and the limit of detection, outside of their working range, are of special relevance considering that the comparisons are not valid. This is equally applied to 2-DE and shotgun approaches [41,61,64–66]. Nevertheless, the output of analytical proteomics workflows should never be taken at face value, but they must be validated and corroborated for each experimental system. Both for 2-DE and shotgun, we usually perform a calibration curve based on different dilutions of a sample; from these serial dilution assays and depending on the protein concentration of the sample, we will see how many proteins are identified (major and minor proteins) and how many are confidently identified, proven using similar ratios in dilution and protein or peptide amount [41,61,64–66].

2.6. 2-DE and Shotgun Platforms Are Complementary

Roughly up to the year 2000, 2-DE based workflows were the predominant platforms employed in plant proteome analysis, and since then, analytical technology has been progressing to second (isotopic or isobaric labelling) and third generation (shotgun, gel-free label-free) approaches, with the latter nowadays being dominant [26]. Considered as an obsolete technique by some scientists, 2-DE based workflows are still valid for some purposes such as top-down proteomics and the identification of protein species or proteoforms of the same gene [32,67]. In our investigations on *Q. ilex*, we have followed the same tendency. The choice of one or other strategy depends on different factors, such as equipment availability, expertise, technical skills, and cost, among others. It is outside the scope of this paper to discuss the potential and limitations of the different techniques; for that, we refer the reader to previously published literature [24,27,30]. Usually, thousands of proteins are identified using a shotgun approach versus hundreds when using a 2-DE based strategy (Table 1). However, both approaches are complementary as the number of common proteins identified using each approach is not always high. Thus, we have used both approaches in parallel (2-DE/MALDI-TOF/TOF, and nLC-ESI-LTQ Orbitrap) in the analysis of seed extracts at different times after germination [19]. The *Quercus*_DB protein database [41], combined with UniProtKB/TrEMBL, UniProtKB/SwissPrto and NCBInr databases, the taxonomy restriction to *Viridiplantae*, and the SEQUEST algorithm were used. A total of 540 consistent spots were resolved using 2-DE in the 5-8 pH range. Out of the 103 variable spots subjected to MALDI-TOF/TOF analysis, 90 were identified [19]. On the other hand, up to 1650 protein species were identified using nLC/MSMS, with 25% of them not annotated. Both proteomics approaches (gel-based and shotgun) were complementary, with shotgun increasing the coverage of the proteome analyzed by over two-fold, and both providing similar results and supporting the same conclusions on the metabolic switch experienced by the seed upon germination [19]. The highest number of matches was obtained when 1-D SDS-PAGE was combined with nLC/Orbitrap/MS (Q- Exactive), with up to 9000 peptides and 1800 proteins identified at an estimated 1 % FDR from a *Q. ilex* extract obtained from a mixture of organs (seeds, leaves, roots, and pollen) [65]. The number of identified proteins depended on the algorithm (Mascot, ProteinPilot, and Maxquant) and database (NCBInr with restrictions to Viridiplantae, Fabids, Rosids, or *Quercus*) [65].

2.7. How Proteomics Sees Quercus ilex

Proteomics has been a helpful approach for our current research projects with *Q. ilex*, both from a basic research and from a translational point of view. Below, we will briefly summarize what contributions have been made with references to original articles for deeper discussion.

2.8. Characterizing Biodiversity

One of our first objectives was to characterize and catalog Andalusian *Q. ilex* populations and provenances based on the leaf 2-DE profile, using field and greenhouse samples [9,15]. Due to the high variability existing in this species, we failed with the leaf proteome, so we decided to analyze different plant tissues with a more stable proteome, such as seed and pollen. Protein extracts from these tissues were subjected to 1-DE (SDS-PAGE) or 2-DE (IEF/SDS-PAGE) protein separation, and variable bands or spots among the provenances were analyzed using MALDI-TOF/TOF MS after tryptic digestion [16,68,69]. In seed extracts, 1-DE data allowed the grouping of populations defined by their geographical location (North, South, East, West) and climate conditions (mesic and xeric). Thus, acorn flour extracts from the most distant populations were analyzed using 2-DE, and 56 differential spots were proposed as markers of variability (Table 1) [16]. A comparison of 1-DE and 2-DE protein profiles of pollen extracts from four provenances in Andalusia revealed significant differences, both qualitative and quantitative (18 bands and 16 spots, respectively), with most of them related to metabolism, defense/stress processes, and cytoskeleton [69]. Similar results have been found when triploid and tetraploid *Populus deltoids* pollen were compared [70].

A multivariate statistical analysis carried out on bands and spots clearly showed distinct associations between provenances, which highlighted their geographical origins. Other complementary approaches, including morphometric, NIRS, and microsatellite analysis, have been used for cataloguing *Q. ilex* populations, with good agreements between the different techniques [16,38,45,69,71].

2.9. Adaptation to Biotic and Abiotic Stresses

Responses to biotic and abiotic stresses are considered as the most covered topic in plant research, in general, and forest tree research, in particular. For instance, nutritional deficiency studies have been approached using proteomics in *Fagus sylvatica* and *P. massoniana* [72,73], oxidative stress in *Populus simonii x P. nigra* [74], salt in *Robinia pseudoacacia* and *Paulownia fortune* [75,76], drought in *Platycladus orientalis* [77], *P. halepensis* and *Larix olgensis* [78,79], UV light in *P. cathayana*, and *P. radiata* [80–82], heavy metals in *P. yunnanensis* [83], and pathogens in *P. tomentosa* [84]. *Quercus ilex* responses to abiotic (drought) and biotic (*P. cinnamomi)* stresses and the variability in such response among populations are a key objective of our research, ultimately aimed at characterizing and selecting elite genotypes with high levels of tolerance and resistance to both stresses, conferring fitness advantages in a climate change scenario.

For that purpose, changes in the leaf protein profile occurring in drought stressed or fungal inoculated plants were analyzed using 1-DE and 2-DE coupled twith MALDI-TOF/TOF MS [15,17,20,21,69]. The resulting proteomics data were correlated with drought tolerance, plantlet growth, presence of toxicity symptoms, and physiological (water regime and photosynthesis) parameters.

Plantlets from seven *Q. ilex* provenances distributed all over the Andalusian geography showed different levels of tolerance to drought as well as differential changes in their 1-DE and 2-DE protein profiles upon water withholding [21]. Variable spots in leaf extracts from the most contrasting populations in terms of drought tolerance were subjected to 2-DE MALDI-TOF/TOF MS analysis, resulting in 28 consistent spots varying in abundance, with 18 unique protein species identified (Table 1) [21]. A general tendency of reduction in protein abundance, especially in proteins related to ATP synthesis and photosynthesis, was observed upon water withholding. The most dramatic decrease was observed in the less tolerant seedling population [21]. The same trend was observed in sunflower plants subjected to drought stress [85]. Upon water availability reduction, changes in the protein profile were observed in two sunflower genotypes, a susceptible and a tolerant one. Two genotype-dependent, and 23 (susceptible genotype) and 5 (tolerant genotype) stress-responsive variable proteins were identified. A general decrease in enzymes of the photosynthesis and carbohydrate metabolism was observed in the susceptible genotype, suggesting inhibition of energetic metabolism. Such changes were not observed in the tolerant genotype, indicating a normal metabolism under drought stress [85].

In a similar study, responses to the fungal pathogen *Phytophthora cinnamomi*, one of the agents that triggers the decline syndrome in *Quercus* spp., were studied by our research group using one-year old seedlings from two Andalusian provenances with different levels of susceptibility [17]. Leaf protein profiles were analyzed in non-inoculated and inoculated seedlings using a 2-DE coupled with MS proteomics strategy. Seventy-nine protein species that changed in abundance upon inoculation were identified after MALDI-TOF/TOF analyses (Table 1) [17]. Out of them, 35 were chloroplastic, with 7 being a part of the photosynthetic electron transport chain and ATP synthesis, 19 belonged to the Calvin cycle and carbohydrate metabolism (with 8 large RubisCO protein spots), and 10 involved in other carbon and nitrogen pathways [17]. A general decrease in protein abundance was observed, being less pronounced in the least susceptible provenance [17]. The same trend clearly manifested in their photosynthesis, amino acid metabolism, and stress/defense proteins. On the contrary, some proteins related to starch biosynthesis, glycolysis, and stress related peroxiredoxin showed an increase upon inoculation [17]. These changes in protein abundance correlated with the estimated physiological parameters and were frequently observed in plants subjected to drought stress [17].

2.10. Development: Seed Maturation and Germination

Last but not least, proteomics has been employed to analyze the proteome of seeds and changes associated to seed maturation and germination in an attempt to characterize and differentiate, at the molecular level, orthodox and non-orthodox species and zygotic and somatic embryos ([18,19,86–93]; this study is of great importance for propagation and seed conservation programs.

Sghaier-Hammami et al. [18] reported on the 1-DE and 2-DE protein profile of the different parts of the seed: embryonic axis, cotyledons, and tegument. One hundred and ninety variable proteins among the three parts of the seed analyzed were identified using MALDI-TOF/TOF (Table 1). Cotyledon presented the highest number of metabolic and storage proteins (89% of legumins), while the embryonic axis and tegument had the largest number of fate group and defense-/stress-related proteins, respectively. This distribution was in good agreement with the biological role of the tissues and demonstrated a compartmentalization of pathways and a division of metabolic tasks between the embryonic axis, cotyledon, and tegument.

Romero-Rodríguez et al. [19] analyzed changes in the protein profile of *Q. ilex* seeds upon germination using complementary 2-DE coupled with MALDI-TOF/TOF and shotgun nLC-ESI-MS/MS approaches. Proteins from embryos at 0 h and 24 h post imbibition, as well as from shoot seedlings at 1 and 4 cm stages were separated using 2-DE, resulting in a total of 540 spots resolved, 103 of which were changes between developmental stages. Ninety differentially accumulated proteins were identified after MALDI-TOF/TOF analysis (Table 1). Proteins related to energy metabolism and photosynthesis were accumulated during seedling establishment. Few proteins showed quantitative differences during the germination period (0 to 24 h post imbibition). When a gel-free shotgun approach was used, 153 differentially accumulated proteins between non-germinated and germinated seeds were identified. Data suggested that the mature non-orthodox seeds of *Q. ilex* have the mechanisms necessary to ensure the rapid resumption of the metabolic activities required to start the germination process and to *de novo* synthesize the biomolecules required for growth, and this makes a big difference from orthodox seeds [19].

3. Conclusions and Perspectives

With this review, we aimed to illustrate the potential and limitations of a proteomics approach applied to non-model forest tree species. These species are considered experimental system that have been quite challenging due to their biological characteristics, recalcitrant nature, and the lack of phenotypic, physiological, or molecular information. The full potential of proteomics has been far from fully exploited in investigations in most plant biology research such as *Q. ilex*. In order to obtain a deeper coverage of the *Q. ilex* proteome, subcellular fractionation techniques or protein depletion and fractionation based on physicochemical or biological properties should be implemented. Apart from proteome subfractionation (e.g., [94]), future research will go in the direction of selected reaction monitoring (SRM), multiple reaction monitoring (MRM), and MS-western or data independent searches based on proteotypic peptides [95]. Some areas of proteomics, such as PTMs and interactomics, have not been approached so far in *Q. ilex* studies, the latter being necessary for understanding the mechanisms that result in a phenotype from the genotype. The lack of an accurate and annotated sequenced genome of *Q. ilex* is an important gap in our research because this is essential for obtaining confident gene product identification and describing protein species or forms as a result of alternative splicing and posttranslational events. Moreover, a sequenced genome would open the door to the application of newly developed approaches such as targeted proteomics.

We have learnt the importance of a proper experimental design and statistical analysis of the data, as well as the relevance of optimizing and validating the techniques employed in each experimental system, plant species, organ, and tissue. We have the possibility of using a range of platforms, methods, and protocols that are complementary, helping us to acquire broader proteome knowledge. In some regards, we may have to broaden our biologist mentality and assume the mindset of an analytical chemist. Plant biologists publishing papers on proteomics should go beyond the blind acceptance of the data provided by the algorithms that come from proteomics services; we should not expect proteomic technicians to be familiar with plant biology. Proteomics by itself may be considered mostly descriptive, and the biological interpretations following, to some extent, as just speculations. Thus, it is necessary to integrate proteomics research with other techniques, including morphometry phenotyping, physiology, classical biochemistry, and other -omics in order to validate the data and procure a more realistic and non-biased view of living organisms [96–100]. It is still astounding how in some publications the whole biology of an organism is discussed and compared with others using data from a poorly designed experiment with a small number of replicates and a minimum fraction of the proteome covered.

Even so, proteomics is making important contributions to the knowledge of living organisms and can be confidently employed for translational purposes. By using proteomics, we have been able to discriminate provenances of *Q. ilex* from Andalusia, find out the differential responses to biotic and abiotic stresses among them, and establish some of the differences existing between orthodox and non-orthodox plant species. New directions in *Q. ilex* research will lead to the identification of allergens in pollen grains and acorns and the characterization of wood materials, which are objectives clearly approached by proteomics [101–103].

Author Contributions: The list of authors includes undergraduate, master's and PhD students, and post doc researchers who contribute or have contributed to the Quercus ilex proteome project, under the supervision of Jorrín-Novo, in the Agroforestry and Plant Biochemistry, Proteomics and System Biology lab, at the University of Cordoba, Spain.

Funding: This research was funded by the Ministerio de Economía y Competitividad -Programa Estatal de I+D+i Orientada a los Retos de la Sociedad (AGL2009-12243-C02-02; BIO2015-64737-R).

Acknowledgments: Jorrín-Novo wishes to express his appreciation to the most important person in the group who does not appear as a coauthor, Mari Carmen Molina-Gómez. "I thank Mari Carmen and apologize for always missing you. We have been together for more than thirty years; you facing administrative and bureaucracy issues and I trying to do the best science and enjoying it. This would be impossible without you. All the authors in this review are expendable, except you. Please, do not retire before me yet, although when you make this decision, I will make it one minute after you". Moreover, the authors also thank Diana Badder for her help with the English editing, the SCAI at the University of Cordoba for the MS analysis in most of the work published and referenced, and the University of Córdoba, UCO-CeiA3 (Programas de Ayuda a los Grupos y Fortalecimiento).

Conflicts of Interest: The authors declare no conflict of interest.

References

1. Olea, L.; San Miguel-Ayanz, A. The Spanish dehesa. A traditional Mediterranean silvopastoral system linking production and nature conservation. *Grassl. Sci. Eur.* **2006**, *11*, 3–13.

2. De Rigo, D.; Gaudullo, G. Quercus ilex in Europe: Distribution, habitat, usage and threats. In *European Atlas of Forest Tree Species*; San-Miguel-Ayanz, J., Ed.; European Union: Luxembourg, 2016; pp. 130–131.

3. Surová, D.; Ravera, F.; Guiomar, N.; Martínez Sastre, R.; Pinto-Correia, T. Contributions of Iberian Silvo-Pastoral Landscapes to the Well-Being of Contemporary Society. *Rangel. Ecol. Manag.* **2017**, *71*, 560–570. [CrossRef]

4. Plieninger, T. Constructed and degraded? Origin and Development of the Spanish Dehesa Landscape, with a a Case Study on Two Municipalities. *Die Erde* **2007**, *138*, 25–46.

5. Guzmán Álvarez, J.R. The image of a tamed landscape: Dehesa through History in Spain. *Cult. Hist. Digit. J.* **2016**, *5*, e003. [CrossRef]

6. Lloret, F.; Siscart, D.; Dalmases, C. Canopy recovery after drought dieback in holm-oak Mediterranean forests of Catalonia (NE Spain). *Glob. Chang. Biol.* **2004**, *10*, 2092–2099. [CrossRef]

7. Natalini, F.; Alejano, R.; Vázquez-Piqué, J.; Cañellas, I.; Gea-Izquierdo, G. The role of climate change in the widespread mortality of holm oak in open woodlands of Southwestern Spain. *Dendrochronologia* **2016**, *38*, 51–60. [CrossRef]

8. Jorrín-Novo, J.V.; Navarro-Cerrillo, R.M. Variabilidad y respuesta a distintos estreses en poblaciones de encina (*Quercus ilex* L.) en Andalucía mediante una aproximación proteómica. *Rev. Ecosistemas* **2014**, *23*, 99–107.

9. Jorge, I.; Navarro-Cerrillo, R.M.; Lenz, C.; Ariza, D.; Porras, C.; Jorrín-Novo, J.V. The holm oak leaf proteome: Analytical and biological variability in the protein expression level assessed by 2-DE and protein

identification tandem mass spectrometry *de novo* sequencing and sequence similarity searching. *Proteomics* **2005**, *5*, 222–234. [CrossRef]

10. Valledor, L.; Castillejo, M.Á.; Lenz, C.; Rodríguez, R.; Cañal, M.J. Proteomic analysis of *Pinus radiata* needles: 2-DE map and protein identification by LC/MS/MS and substitution-tolerant database searching. *J. Proteome Res.* **2008**, *7*, 2616–2631. [CrossRef]

11. De Francisco, L.; Romero-Rodríguez, M.C.; Navarro-Cerrillo, R.M.; Miniño, V.; Perdomo, O.; Jorrín-Novo, J.V. Characterization of the orthodox *Pinus occidentalis* seed and pollen proteomes by using complementary gel-based and gel-free approaches. *J. Proteom.* **2016**, *143*, 382–389. [CrossRef]

12. Ariza, D.; Navarro-Cerrillo, R.M.; del Campo, A. Influencia de la fecha de plantación al establecimiento de *Pinus halapensis* aplicación de la proteómica a estudios de Ecofisiología en campo. *Soc. Española Cienc.* **2008**, *28*, 111–117.

13. Loewe, V.; Navarro-Cerrillo, R.M.; Sánchez-Lucas, R.; Ruiz Gómez, F.J.; Jorrín-Novo, J.V. Variability studies of allochthonous stone pine (*Pinus pinea L.*) plantations in Chile through nut protein profiling. *J. Proteom.* **2018**, *175*, 95–104. [CrossRef] [PubMed]

14. Romero-Rodríguez, M.C.; Abril, N.; Sánchez-Lucas, R.; Jorrín-Novo, J.V. Multiplex staining of 2-DE gels for an initial phosphoproteome analysis of germinating seeds and early grown seedlings from a non-orthodox specie: *Quercus ilex L.* subsp. *ballota* [Desf.] Samp. *Front. Plant Sci.* **2015**, *6*, 620. [CrossRef] [PubMed]

15. Jorge, I.; Navarro-Cerrillo, R.M.; Lenz, C.; Ariza, D.; Jorrín-Novo, J.V. Variation in the holm oak leaf proteome at different plant developmental stages, between provenances and in response to drought stress. *Proteomics* **2006**, *6*, 207–214. [CrossRef]

16. Valero-Galván, J.; Valledor, L.; Navarro-Cerrillo, R.M.; Pelegrín, E.G.; Jorrín-Novo, J.V. Studies of variability in Holm oak (*Quercus ilex* subsp. *ballota* [Desf.] Samp.) through acorn protein profile analysis. *J. Proteom.* **2011**, *74*, 1244–1255.

17. Sghaier-Hammami, B.; Valero-Galván, J.; Romero-Rodríguez, M.C.; Navarro-Cerrillo, R.M.; Abdelly, C.; Jorrín-Novo, J.V. Physiological and proteomics analyses of Holm oak (*Quercus ilex* subsp. *ballota* [Desf.] Samp.) responses to *Phytophthora cinnamomi. Plant Physiol. Biochem.* **2013**, *71*, 191–202. [PubMed]

18. Sghaier-Hammami, B.; Redondo-López, I.; Valero-Galván, J.; Jorrín-Novo, J.V. Protein profile of cotyledon, tegument, and embryonic axis of mature acorns from a non-orthodox plant species: *Quercus ilex. Planta* **2016**, *243*, 369–396. [CrossRef]

19. Romero-Rodríguez, M.C.; Jorrín-Novo, J.V.; Castillejo, M.Á. Towards characterizing seed germination and seedling establishment in the non- orthodox forest tree species *Quercus ilex* through complementary gel and gel-free proteomic approaches. *J. Proteom.* **2018**. [CrossRef]

20. Echevarría-Zomeño, S.; Ariza, D.; Jorge, I.; Lenz, C.; Del Campo, A.; Jorrín-Novo, J.V.; Navarro-Cerrillo, R.M. Changes in the protein profile of *Quercus ilex* leaves in response to drought stress and recovery. *J. Plant Physiol.* **2009**, *166*, 233–245. [CrossRef]

21. Valero-Galván, J.; González-Fernández, R.; Navarro-Cerrillo, R.M.; Gil-Pelegrín, E.; Jorrín-Novo, J.V. Physiological and proteomic analyses of drought stress response in Holm oak provenances. *J. Proteome Res.* **2013**, *12*, 5110–5123. [CrossRef]

22. Simova-Stoilova, L.P.; Romero-Rodríguez, M.C.; Sánchez-Lucas, R.; Navarro-Cerrillo, R.M.; Medina-Aunon, J.A.; Jorrín-Novo, J.V. 2-DE proteomics analysis of drought treated seedlings of *Quercus ilex* supports a root active strategy for metabolic adaptation in response to water shortage. *Front. Plant Sci.* **2015**, *6*, 627. [CrossRef] [PubMed]

23. Simova-Stoilova, L.P.; López-Hidalgo, C.; Sánchez-Lucas, R.; Valero-Galván, J.; Romero-Rodríguez, M.C.; Jorrín-Novo, J.V. Holm oak proteomic response to water limitation at seedling establishment stage reveals specific changes in different plant parts as well as interaction between roots and cotyledons. *Plant Sci.* **2018**, *276*, 1–13. [CrossRef] [PubMed]

24. Abril, N.; Gion, J.M.; Kerner, R.; Müller-Starck, G.; Navarro-Cerrillo, R.M.; Plomion, C.; Renaut, J.; Valledor, L.; Jorrín-Novo, J.V. Proteomics research on forest trees, the most recalcitrant and orphan 2plant species. *Phytochemistry* **2011**, *72*, 1219–1242. [CrossRef] [PubMed]

25. Jorrín-Novo, J.V.; Maldonado-Alconada, A.M.; Castillejo, M.Á. Plant proteome analysis: A 2006 update. *Proteomics* **2007**, *7*, 2947–2962. [CrossRef] [PubMed]

26. Jorrín-Novo, J.V.; Maldonado-Alconada, A.M.; Echevarría-Zomeño, S.; Valledor, L.; Castillejo, M.Á.; Curto, M.; Valero-Galván, J.; Sghaier-Hammami, B.; Donoso, G.; Redondo-López, I. Plant proteomics update (2007–2008): Second-generation proteomic techniques, an appropriate experimental design, and data analysis to fulfill MIAPE standards, increase plant proteome coverage and expand biological knowledge. *J. Proteom.* **2009**, *72*, 285–314. [CrossRef]

27. Jorrín-Novo, J.V.; Komatsu, S.; Weckwerth, W.; Wienkoop, S. *Plant Proteomics—Methods and Protocols*, 2nd ed.; Humana Press: Totowa, NJ, USA, 2014; ISBN 9781627036306.

28. Jorrín-Novo, J.V.; Pascual, J.; Sánchez-Lucas, R.; Romero-Rodríguez, M.C.; Rodríguez-Ortega, M.J.; Lenz, C.; Valledor, L. Fourteen years of plant proteomics reflected in Proteomics: Moving from model species and 2DE-based approaches to orphan species and gel-free platforms. *Proteomics* **2015**, *15*, 1089–1112. [CrossRef] [PubMed]

29. Jorrín-Novo, J.V.; Valledor, L. Translational proteomics special issue. *J. Proteom.* **2013**, *93*, 1–4. [CrossRef] [PubMed]

30. Sánchez-Lucas, R.; Mehta, A.; Valledor, L.; Cabello-Hurtado, F.; Romero-Rodríguez, M.C.; Simova-Stoilova, L.P.; Demir, S.; Rodríguez de Francisco, L.; Maldonado-Alconada, A.M.; Jorrín-Prieto, A.L.; et al. A year (2014–2015) of plants in Proteomics journal. Progress in wet and dry methodologies, moving from protein catalogs, and the view of classic plant biochemists. *Proteomics* **2016**, *16*, 866–876. [CrossRef]

31. Komatsu, S.; Jorrín-Novo, J.V. Food and Crop Proteomics. *J. Proteom.* **2016**, *143*, 1–2. [CrossRef]

32. Jorrín-Novo, J.V.; Komatsu, S.; Sánchez-Lucas, R.; de Francisco, L.E.R. Gel electrophoresis-based plant proteomics: Past, present, and future. Happy 10th anniversary Journal of Proteomics! *J. Proteom.* **2018**. [CrossRef]

33. Jorrín-Novo, J.V. Scientific standards and MIAPEs in plant proteomics research and publications. *Front. Plant Sci.* **2015**, *6*, 473. [CrossRef] [PubMed]

34. Song, H.; Lei, Y.; Zhang, S. Differences in resistance to nitrogen and phosphorus deficiencies explain male-biased populations of poplar in nutrient-deficient habitats. *J. Proteom.* **2018**, *178*, 123–127. [CrossRef] [PubMed]

35. Holliday, J.A.; Aitken, S.N.; Cooke, J.E.K.; Fady, B.; Gonzalez-Martinez, S.C.; Heuertz, M.; Jaramillo-Correa, J.P.; Lexer, C.; Staton, M.; Whetten, R.W.; et al. Advances in ecological genomics in forest trees and applications to genetic resources conservation and breeding. *Mol. Eco.* **2017**, *26*, 706–717. [CrossRef] [PubMed]

36. Valledor, L.; Jorrín-Novo, J.V. Back to the basics: Maximizing the information obtained by quantitative two dimensional gel electrophoresis analyses by an appropriate experimental design and statistical analyses. *J. Proteom.* **2011**, *74*, 1–18. [CrossRef] [PubMed]

37. Maldonado-Alconada, A.M.; Echevarría-Zomeño, S.; Jean-Baptiste, S.; Hernández de la Torre, M.; Jorrín-Novo, J.V. Evaluation of three different protocols of protein extraction for *Arabidopsis thaliana* leaf proteome analysis by two-dimensional electrophoresis. *J. Proteom.* **2008**, *71*, 461–472. [CrossRef] [PubMed]

38. Valero-Galván, J.; Valledor, L.; González-Fernández, R.; Navarro-Cerrillo, R.M.; Jorrín-Novo, J.V. Proteomic analysis of Holm oak (*Quercus ilex* subsp. *ballota* [Desf.] Samp.) pollen. *J. Proteom.* **2012**, *75*, 2736–2744.

39. López-Hidalgo, C.; Guerrero-Sanchez, V.M.; Gómez-Gálvez, I.; Sánchez-Lucas, R.; Castillejo, M.Á.; Maldonado-Alconada, A.M.; Valledor, L.; Jorrín-Novo, J.V. A multi-omics analysis pipeline for the metabolic pathway reconstruction in the orphan species *Quercus ilex*. *Front. Plant Sci.* **2018**, *9*, 935. [CrossRef]

40. Guerrero-Sanchez, V.M.; Maldonado-Alconada, A.M.; Amil-Ruiz, F.; Jorrín-Novo, J.V. Holm Oak (*Quercus ilex*) transcriptome. De novo sequencing and assembly analysis. *Front. Mol. Biosci.* **2017**, *4*, 70. [CrossRef]

41. Romero-Rodríguez, M.C.; Pascual, J.; Valledor, L.; Jorrín-Novo, J.V. Improving the quality of protein identification in non-model species. Characterization of *Quercus ilex* seed and *Pinus radiata* needle proteomes by using SEQUEST and custom databases. *J. Proteom.* **2014**, *105*, 85–91. [CrossRef]

42. Kao, S.H.; Wong, H.K.; Chiang, C.Y.; Chen, H.M. Evaluating the compatibility of three colorimetric protein assays for two-dimensional electrophoresis experiments. *Proteomics* **2008**, *8*, 2178–2184. [CrossRef]

43. Jorrín-Novo, J.V.; Ardilla, H.; Castillejo, M.Á.; Curto, M.; Echevarría-Zomeño, S.; Hernández de la Torre, M.; Gómez-Gálvez, F.; González-Fernández, R. From 2003 to 2011: Proteomics investigation at the agroforestry and plant biochemistry and proteomics research group (University of Cordoba, Spain). In *3rd International Symposium on Frontiers in Agriculture Proteome Research: Contribution of Proteomics Technology in Agricultural Sciences. Frontiers*; Mock, H.P., Wang, Z.Y., Komatsu, S., Eds.; Frontiers in Agriculture Proteome Research. Contribution of Proteomics Technology in Agricultural Sciences; NARO Institute of Crop Science: Ibaraki, Japan, 2011; pp. 130–137.

44. Yeoh, H.H.; Wee, Y.C. Leaf protein contents and nitrogen-to-protein conversion factors for 90 plant species. *Food Chem.* **1994**, *49*, 245–250. [CrossRef]

45. Valero-Galván, J.; González-Fernández, R.; Valledor, L.; Navarro-Cerrillo, R.M.; Jorrín-Novo, J.V. Proteotyping of Holm oak (*Quercus ilex* subsp. *ballota*) provenances through proteomic analysis of acorn flour. In *Plant Proteomics—Methods and Protocols*; Humana Press: Totowa, NJ, USA, 2014; pp. 709–724. ISBN 9780470988879.

46. Romero-Rodríguez, M.C.; Maldonado-Alconada, A.M.; Valledor, L.; Jorrín-Novo, J.V. Back to Osborne. Sequential protein extraction and LC-MS analysis for the characterization of the Holm oak seed proteome. In *Plant Proteomics—Methods and Protocols*; Humana Press: Totowa, NJ, USA, 2007; pp. 379–390. ISBN 9780470988879.

47. Valero-Galván, J.; Jorrín-Novo, J.V.; Gómez Cabrera, A.; Ariza, D.; García-Olmo, J.; Navarro-Cerrillo, R.M. Population variability based on the morphometry and chemical composition of the acorn in Holm oak (*Quercus ilex* subsp. *ballota* [Desf.] Samp.). *Eur. J. Res.* **2012**, *131*, 893–904.

48. Valledor, L.; Romero-Rodríguez, M.C.; Jorrín-Novo, J.V. Standardization of data processing and statistical analysis in comparative plant proteomics experiment. In *Plant Proteomics—Methods and Protocols*; Humana Press: Totowa, NJ, USA, 2014; pp. 51–60. ISBN 9780470988879.

49. Maitra, S.; Yan, J. Principle component analysis and partial least squares: Two dimension reduction techniques for regression. *Appl. Multivar. Statist. Models* **2008**, *79*, 79–90.

50. Plomion, C.; Aury, J.M.; Amselem, J.; Alaeitabar, T.; Barbe, V.; Belser, C.; Bergès, H.; Bodénès, C.; Boudet, N.; Boury, C.; et al. Decoding the oak genome: Public release of sequence data, assembly, annotation and publication strategies. *Mol. Ecol. Resour.* **2016**, *16*, 254–265. [CrossRef] [PubMed]

51. Sork, V.L.; Fitz-Gibbon, S.T.; Puiu, D.; Crepeau, M.; Gugger, P.F.; Sherman, R.; Stevens, K.; Langley, C.H.; Pellegrini, M.; Salzberg, S.L. First draft assembly and annotation of the genome of a california endemic oak *Quercus lobata* Nee (Fagaceae). *G3 (Bethesda)* **2016**, *6*, 3485–3495. [CrossRef] [PubMed]

52. Ramos, A.M.; Usié, A.; Barbosa, P.; Barros, P.M.; Capote, T.; Chaves, I.; Simões, F.; Abreu, I.; Carrasquinho, I.; Faro, C.; et al. The draft genome sequence of cork oak. *Sci. Data* **2018**, *5*, 180069. [CrossRef] [PubMed]

53. Kremer, A.; Casasoli, M.; Barreneche, T.; Bódenès, C.; Sisco, P.; Kubisiak, T.; Scalfi, M.; Leonardi, S.; Bakker, E.; Buiteveld, J.; et al. *Genome Mapping and Molecular Breeding in Plants*; Chittaranjan, K., Ed.; Springer: Heidelberg/Berlin, Germany, 2007; Volume 7, pp. 165–187.

54. Lesur, I.; Le Provost, G.; Bento, P.; Da Silva, C.; Leplé, J.C.; Murat, F.; Ueno, S.; Bartholomé, J.; Lalanne, C.; Ehrenmann, F.; et al. The oak gene expression atlas: Insights into Fagaceae genome evolution and the discovery of genes regulated during bud dormancy release. *BMC Genom.* **2015**, *16*. [CrossRef]

55. Rey, M.D.; Guerrero-Sánchez, V.M.; Sánchez-Lucas, R.; López-Hidalgo, C.; Maldonado-Alconada, A.M.; Jorrín-Novo, J.V. *The Use of -Omics Technologies to Progress in the Quercus Ilex Biology*; XIV RBMP: Salamanca, Spain, 2018; p. 20.

56. Zoldos, V.; Papes, D.; Brown, S.; Panaud, O.; Siljak-Yakovlev, S. Protocol for flow cytometric assay Genome size and base composition of seven *Quercus* species: Inter- and intra-population variation. *Genome* **1998**, *41*, 162–168. [CrossRef]

57. Chen, S.C.; Cannon, C.H.; Kua, C.S.; Liu, J.J.; Galbraith, D.W. Genome size variation in the Fagaceae and its implications for trees. *Tree Genet. Genomes* **2014**, *10*, 977–988. [CrossRef]

58. Rey, M.D.; Moore, G.; Martin, A.C. Identification and comparison of individual chromosomes of three *Hordeum chilense* accessions, *Hordeum vulgare* and *Triticum aestivum* by FISH. *Genome* **2018**, *61*, 387–396. [CrossRef]

59. Guerrero-Sanchez, V.M.; Maldonado-Alconada, A.M.; Amil-Ruiz, F.; Verardi, A.; Jorrín-Novo, J.V.; Rey, M.D. Ion torrent and Illumina, two complementary RNA-Seq platforms fo constructing the holm oak (*Quercus ilex*) transcriptome. *PLoS ONE* **2019**, *14*, e0210356. [CrossRef] [PubMed]

60. Neale, D.B.; McGuire, P.E.; Wheeler, N.C.; Stevens, K.A.; Crepeau, M.W.; Cardeno, C.; Zimin, A.V.; Puiu, D.; Pertea, G.M.; Sezen, U.U.; et al. The Douglas-Fir Genome Sequence Reveals Specialization of the Photosynthetic Apparatus in Pinaceae. *G3 (Bethesda)* **2017**, 3157–3167. [CrossRef] [PubMed]

61. Gómez-Gálvez, I.M.; Castillejo, M.A.; Márquez, C.; Jorrín-Novo, J.V. *Unravelling Mechanisms of Tolerance in Holm Oak through a Physiological and Molecular Approach*; INPPO: Padova, Italy, 2018.

62. Curto, M.; Valledor, L.; Navarrete, C.; Gutiérrez, D.; Sychrova, H.; Ramos, J.; Jorrín-Novo, J.V. 2-DE based proteomic analysis of *Saccharomyces cerevisiae* wild and K$^+$ transport-affected mutant *(trk1, 2)* strains at the growth exponential and stationary phases. *J. Proteom.* **2010**, *73*, 2316–2335. [CrossRef] [PubMed]

63. Maldonado-Alconada, A.M.; Echevarría-Zomeño, S.; Lindermayr, C.; Redondo-López, I.; Durner, J.; Jorrín-Novo, J.V. Proteomic analysis of *Arabidopsis* protein S-nitrosylation in response to inoculation with *Pseudomonas syringae*. *Acta Physiol. Plant.* **2011**, *33*, 1493–1514. [CrossRef]

64. González-Fernández, R.; Aloria, K.; Arizmendi, J.M.; Jorrín-Novo, J.V. Application of label-free shotgun nUPLC-MSEand 2-DE approaches in the study of *Botrytis cinerea* mycelium. *J. Proteome Res.* **2013**, *12*, 3042–3056. [CrossRef] [PubMed]

65. Lenz, C.; Seymour, S.; Shilov, I.; Jorrín-Novo, J.V. Protein Pilot: Accommodating genetic diversity in mass spectrometry-based plant proteome research. In *II "Plant Proteomics in Europe" Meeting (Working Group 2): Will Plant Proteomics Research Help in Facing Food, Health and Environmental Concerns? Changes Induced by the Pepper Mild Mottle Tobamovirus on the Chloroplast Proteome of Nicotiana Benthamiana*; University of Córdoba (UCO): Córdoba, Spain, 2008.

66. Romero-Rodríguez, M.C.; Valledor, L.; Lenz, C.; Hurlaub, H.; Jorrín-Novo, J.V. Proteomics workflows and protocols for the study of orphan species. In *HUPO 2014. The Proteome Quest to Understand Biology and Disease. Non-Human and Food Proteomics Section*; HUPO: Madrid, Spain, 2014.

67. Tong, Z.; Wang, D.; Sun, Y.; Yang, Q.; Meng, X.R.; Wang, L.M.; Feng, W.Q.; Li, L.; Wurtele, E.S.; Wang, X.C. Comparative Proteomics of Rubber Latex Revealed Multiple Protein Species of REF/SRPP Family Respond Diversely to Ethylene Stimulation among Different Rubber Tree Clones. *Int. J. Mol. Sci.* **2017**, *18*, 958. [CrossRef] [PubMed]

68. Lenz, C.; Jorrín-Novo, J.V.; Urlaub, H. *Quercus ilex*: Protein identification strategies for an orphan tree species. In Proceedings of the 61st ASMS Conference on Mass Spectrometry and Allied Topics, Baltimore, MD, USA, 6–13 June 2013.

69. Valero-Galván, J.; Sghaier-Hammami, B.; Navarro-Cerrillo, R.M.; Jorrín-Novo, J.V. Natural variability and responses to stresses in andalusia Holm oak (*Quercus ilex* subsp. *ballota*) populations. In *Oak: Ecology, Types and Management*; Chuteira, C.A., Grão, A.B., Eds.; Nova Science Publishers, Inc.: New York, NY, USA, 2012; pp. 193–206. ISBN 978-1-61942-492-0.

70. Zhang, X.L.; Zhang, J.; Guo, Y.H.; Sun, P.; Jia, H.X.; Fan, W.; Lu, M.Z.; Hu, J.J. Comparative proteomic analysis of mature pollen in triploid and diploid *Populus deltoids*. *Int. J. Mol. Sci.* **2016**, *17*, 1475. [CrossRef]

71. Fernández i Marti, A.; Romero-Rodríguez, C.; Navarro-Cerrillo, R.; Abril, N.; Jorrín-Novo, J.V.; Dodd, R. Population genetic diversity of *Quercus ilex* subsp. ballota (Desf.) Samp. reveals divergence in recent and evolutionary migration rates in the Spanish dehesas. *Forests* **2018**, *9*, 337. [CrossRef]

72. Geilfus, C.M.; Carpentier, S.C.; Zavisic, A.; Polle, A. Changes in the fine root proteome of *Fagus sylvatica* L. trees associated with P-deficiency and amelioration of P-deficiency. *J. Proteom.* **2017**, *169*, 33–40. [CrossRef]

73. Fan, F.H.; Ding, G.J.; Wen, X.P. Proteomic analyses provide new insights into the responses of *Pinus massoniana* seedlings to phosphorus deficiency. *Proteomics* **2015**, *16*, 504–515. [CrossRef]

74. Yu, J.J.; Jin, X.; Sun, X.M.; Gao, T.X.; Chen, X.M.; She, Y.M.; Jiang, T.B.; Chen, S.X.; Dai, S.J. Hydrogen peroxide response in leaves of Poplar (*Populus simonii* × *Populus nigra*) revealed from physiological and proteomic analyses. *Int. J. Mol. Sci.* **2017**, *18*, 2085. [CrossRef] [PubMed]

75. Luo, Q.X.; Peng, M.; Zhang, X.L.; Lei, P.; Ji, X.M.; Chow, W.; Meng, F.J.; Sun, G.Y. Comparative mitochondrial proteomic, physiological, biochemical and ultrastructural profiling reveal factors underpinning salt tolerance in tetraploid black locust (*Robinia pseudoacacia* L.). *Bmc Genom.* **2017**, *18*, 648. [CrossRef] [PubMed]

76. Deng, M.J.; Dong, Y.P.; Zhao, Z.L.; Li, Y.; Fan, G.Q. Dissecting the proteome dynamics of the salt stress induced changes in the leaf of diploid and autotetraploid *Paulownia fortunei*. *PLoS ONE* **2017**, *12*, e0181937. [CrossRef] [PubMed]

77. Zhang, S.; Zhang, L.L.; Zhou, K.K.; Li, Y.M.; Zhao, Z. Changes in protein profile of *Platycladus orientalis* (L.) roots and leaves in response to drought stress. *Tree Genet. Genomes* **2017**, *4*, 76. [CrossRef]

78. Taibi, K.; del Campo, A.D.; Vilagrosa, A.; Belles, J.M.; Lopez-Gresa, M.P.; Pla, D.; Calvete, J.J.; Lopez-Nicolas, J.M.; Mulet, J.M. Drought tolerance in *Pinus halepensis* seed sources as identified by distinctive physiological and molecular markers. *Front. Plant Sci.* **2017**, *8*, 1202. [CrossRef] [PubMed]

79. Zhang, L.; Zhang, H.G.; Pang, Q.Y. Physiological evaluation of the responses of *Larix olgensis* families to drought stress and proteomic analysis of the superior family. *Genet. Mol. Res.* **2015**, *14*, 15577–15586. [CrossRef] [PubMed]

80. Zhang, Y.X.; Feng, L.H.; Jiang, H.; Zhang, Y.B.; Zhang, S. Different proteome profiles between male and female *Populus cathayana* exposed to UVB radiation. *Front. Plant Sci.* **2018**, *8*, 320.

81. Pascual, J.; Canal, M.J.; Escandon, M.; Meijon, M.; Weckwerth, W.; Valledor, L. Integrated physiological, proteomic, and metabolomic analysis of ultra violet (UV) stress responses and adaptation mechanisms in *Pinus radiata*. *Mol. Cell. Proteom.* **2018**, *16*, 485–501. [CrossRef] [PubMed]

82. Pascual, J.; Alegre, S.; Nagler, M.; Escandon, M.; Annacondia, M.L.; Weckwerth, W.; Valledor, L.; Canal, M.J. The variations in the nuclear proteome reveal new transcription factors and mechanisms involved in UK stress response in *Pinus radiata*. *J. Proteom.* **2016**, *143*, 390–400. [CrossRef] [PubMed]

83. Yang, Y.Q.; Li, X.; Yang, S.H.; Zhou, Y.L.; Dong, C.; Ren, J.; Sun, X.D.; Yang, Y.P. Comparative physiological and proteomic analysis reveals the leaf response to cadmium-induced stress in Poplar (*Populus yunnanensis*). *PLoS ONE* **2015**, *10*, e0137396. [CrossRef]

84. Cao, X.B.; Fan, G.Q.; Dong, Y.P.; Zhao, Z.L.; Deng, M.J.; Wang, Z.; Liu, W.S. Proteome profiling of Paulownia seedlings infected with phytoplasma. *Front. Plant Sci.* **2017**, *8*, 342. [CrossRef] [PubMed]

85. Castillejo, M.Á.; Maldonado-Alconada, A.M.; Ogueta, S.; Jorrín-Novo, J.V. Proteomic analysis of responses to drought stress in sunflower (*Helianthus annuus*) leaves by 2DE gel electrophoresis and mass spectrometry. *Open Proteom. J.* **2008**, *1*, 59–71. [CrossRef]

86. Sghaier-Hammami, B.; Valledor, L.; Redondo-López, I.; Weckwerth, W.; Jorrín-Novo, J.V. 2-DE-based and LC-label-free proteomics studies of seed development in holm oak (*Quercus ilex*). In Proceedings of the HUPO 2011 10th World Congress, Geneva, Switzerland, 3–7 September 2011; p. 277.

87. Moothoo-Padayachie, A.; Macdonald, A.; Varghese, B.; Pammenter, N.W.; Govender, P. Uncovering the basis of viability loss in desiccation sensitive *Trichilia dregeana* seeds using differential quantitative protein expression profiling by iTRAQ. *J. Plant Physiol.* **2018**, *221*, 119–131. [CrossRef] [PubMed]

88. Zhang, P.; Liu, D.; Shen, H.L.; Li, Y.H.; Nie, Y.Z. Proteome analysis of dormancy-released seeds of *Fraxinus mandshurica* Rupr. in response to re-dehydration under different conditions. *Int. J. Mol. Sci.* **2015**, *16*, 4713–4730. [CrossRef] [PubMed]

89. Liu, C.P.; Yang, L.; Shen, H.L. Proteomic analysis of immature *Fraxinus mandshurica* cotyledon tissues during somatic embryogenesis: Effects of explant browning on somatic embryogenesis. *Int. J. Mol. Sci.* **2015**, *16*, 13692–13713. [CrossRef] [PubMed]

90. Zhang, H.; Wang, W.Q.; Liu, S.J.; Moller, I.M.; Song, S.Q. Proteome analysis of Poplar seed vigor. *PLoS ONE* **2015**, *10*, e0132509. [CrossRef] [PubMed]

91. Ratajczak, E.; Kalemba, E.M.; Pukacka, S. Age-related changes in protein metabolism of beech (*Fagus sylvatica* L.) seeds during alleviation of dormancy and in the early stage of germination. *Plant Physiol. Biochem.* **2015**, *94*, 114–121. [CrossRef]

92. Pawlowski, T.A.; Staszak, A.M. Analysis of the embryo proteome of sycamore (*Acer pseudoplatanus* L.) seeds reveals a distinct class of proteins regulating dormancy release. *J. Plant Physiol.* **2016**, *195*, 9–22.

93. Jing, D.L.; Zhang, J.W.; Xia, Y.; Kong, L.S.; Ou Yang, F.Q.; Zhang, S.G.; Zhang, H.G.; Wang, J.H. Proteomic analysis of stress-related proteins and metabolic pathways in *Picea asperata* somatic embryos during partial desiccation. *Plant Biotechnol. J.* **2017**, *15*, 27–38. [CrossRef]

94. Loijon, F.; Melzer, M.; Zhou, Q.; Srivastava, V.; Bulone, V. Proteomic analysis of plasmodesmata from populus cell suspension cultures in relation with callose biosynthesis. *Front. Plant Sci.* **2018**, *9*, 1681. [CrossRef]

95. Zhang, X.Y.; Dominguez, P.G.; Kumar, M.; Bygdell, J.; Miroshnichenko, S.; Sundberg, B.; Wingsle, G.; Niittyla, T. Cellulose synthase stoichiometry in aspen differs from Arabidopsis and Norway spruce. *Plant Physiol.* **2018**, *177*, 1096–1107. [CrossRef] [PubMed]

96. Guzicka, M.; Pawlowski, T.A.; Staszak, A.; Rozkowski, R.; Chmura, D.J. Molecular and structural changes in vegetative buds of Norway spruce during dormancy in natural weather conditions. *Tree Physiol.* **2018**, *38*, 721–734. [CrossRef] [PubMed]

97. Li, Q.F.; Zhang, S.G.; Wang, J.H. Transcriptomic and proteomic analyses of embryogenic tissues in *Picea balfouriana* treated with 6-benzylaminopurine. *Physiol. Plant.* **2015**, *154*, 95–113. [CrossRef] [PubMed]

98. Correia, B.; Valledor, L.; Hancock, R.D.; Renaut, J.; Pascual, J.; Soares, A.M.V.M.; Pinto, G. Integrated proteomics and metabolomics to unlock global and clonal responses of *Eucalyptus globulus* recovery from water deficit. *Metabolomics* **2016**, *12*, 141. [CrossRef]

99. Zheng, W.; Komatsu, S.; Zhu, W.; Zhang, L.; Li, X.M.; Cui, L.; Tian, J.K. Response and Defense Mechanisms of *Taxus chinensis* leaves Under UVA radiation are revealed using comparative proteomics and metabolomics analyses. *Plant Cell Physiol.* **2016**, *57*, 1839–1853. [CrossRef] [PubMed]

100. Liu, X.; Yu, W.; Wang, G.; Cao, F.; Cai, J.F.; Wang, H. Comparative Proteomic and Physiological analysis reveals the variation mechanisms of leaf coloration and carbon fixation in a Xantha Mutant of *Ginkgo biloba* L. *Int. J. Mol. Sci.* **2016**, *17*, 1794. [CrossRef] [PubMed]

101. McKenna, O.E.; Posselt, G.; Briza, P.; Lackner, P.; Schmitt, A.O.; Gadermaier, G.; Wessler, S.; Ferreira, F. Multi-approach analysis for the identification of proteases within birch pollen. *Int. J. Mol. Sci.* **2017**, *7*, 1433. [CrossRef]

102. Mousavi, F.; Majd, A.; Shahali, Y.; Ghahremaninejad, F.; Shoormasti, R.S.; Pourpak, Z. Immunoproteomics of tree of heaven (*Ailanthus atltissima*) pollen allergens. *J. Proteom.* **2017**, *154*, 94–101. [CrossRef]

103. Bygdell, J.; Srivastava, V.; Obudulu, O.; Srivastava, M.K.; Nilsson, R.; Sundberg, B.; Trygg, J.; Mellerowicz, E.J.; Wingsle, G. Protein expression in tension wood formation monitored at high tissue resolution in *Populus. J. Exp. Bot.* **2017**, *13*, 3405–3417. [CrossRef]

International Journal of
Molecular Sciences

MDPI

Article

Identification of Candidate Ergosterol-Responsive Proteins Associated with the Plasma Membrane of *Arabidopsis thaliana*

Thembisile G. Khoza, Ian A. Dubery and Lizelle A. Piater *

Department of Biochemistry, University of Johannesburg, Auckland Park, Johannesburg 2006, South Africa;
tkhoza03@gmail.com (T.K.); idubery@uj.ac.za (I.D.)
* Correspondence: lpiater@uj.ac.za; Tel.: +27-11-559-2403

Received: 27 January 2019; Accepted: 3 March 2019; Published: 14 March 2019

Abstract: The impact of fungal diseases on crop production negatively reflects on sustainable food production and overall economic health. Ergosterol is the major sterol component in fungal membranes and regarded as a general elicitor or microbe-associated molecular pattern (MAMP) molecule. Although plant responses to ergosterol have been reported, the perception mechanism is still unknown. Here, *Arabidopsis thaliana* protein fractions were used to identify those differentially regulated following ergosterol treatment; additionally, they were subjected to affinity-based chromatography enrichment strategies to capture and categorize ergosterol-interacting candidate proteins using liquid chromatography coupled with tandem mass spectrometry (LC-MS/MS). Mature plants were treated with 250 nM ergosterol over a 24 h period, and plasma membrane-associated fractions were isolated. In addition, ergosterol was immobilized on two different affinity-based systems to capture interacting proteins/complexes. This resulted in the identification of defense-related proteins such as chitin elicitor receptor kinase (CERK), non-race specific disease resistance/harpin-induced (NDR1/HIN1)-like protein, Ras-related proteins, aquaporins, remorin protein, leucine-rich repeat (LRR)- receptor like kinases (RLKs), G-type lectin S-receptor-like serine/threonine-protein kinase (GsSRK), and glycosylphosphatidylinositol (GPI)-anchored protein. Furthermore, the results elucidated unknown signaling responses to this MAMP, including endocytosis, and other similarities to those previously reported for bacterial flagellin, lipopolysaccharides, and fungal chitin.

Keywords: affinity chromatography; ergosterol; fungal perception; innate immunity; pattern recognition receptors; plasma membrane; proteomics

---

1. Introduction

Plants lack an adaptive immune system and solely depend on a multi-complex innate immunity to defend themselves. The first line of defense occurs on the plant cell surface, where membrane-bound pattern recognition receptors (PRRs) recognize conserved motifs within microbes. These microbe-associated molecular patterns (MAMPs) are typically essential components for microorganism functioning and include the bacterial flagellin epitope, flg22. This MAMP is recognized by the PRR receptor, flagellin sensitive 2 (FLS2), which was proven by showing that mutated epitope residues did not lead to flagellin perception but instead, susceptibility and infection was observed [1,2]. Similarly, a lipopolysaccharide (LPS) receptor was identified in the Brassicaceae family. It was found that *Arabidopsis thaliana* detected LPS of *Xanthomonas campestris* and *Pseudomonas* species using a bulb-type (B-type) lectin S-domain (SD)-1 receptor like kinase (RLK) termed lipooligosaccharide-specific reduced elicitation (LORE) [3]. The recognition of MAMPs by PRRs leads to activation of the primary defense termed microbe-triggered immunity (MTI). Due to the co-evolution of both microbes and host, several organisms have the ability to suppress MTI components by releasing virulent molecules called effectors, which leads to effector-triggered susceptibility (ETS). This marks the second line of defense, known as effector-triggered immunity (ETI), where these effectors are recognized by intracellular nucleotide-binding leucine-rich repeat (NB-LRR) proteins [4–6]. Subsequent processes include the transcription of defense genes and expression of pathogenesis-related (PR) proteins. General cellular events associated with MTI and ETI include changes in cytoplasmic Ca^{2+} levels, activation of mitogen-activated protein kinase (MAPK) cascades, bursts of reactive oxygen species (ROS) and nitric oxide (NO), deposition of callose to reinforce the cell wall, production of anti-microbial compounds such as phytoalexins, and often, localized cell death [4,7–10].

Currently, crop yield and food security are global concerns due to often devastating fungal–plant interactions [11], which also impact economies, particularly those of third world countries. Fungal MAMP molecules such as chitin and β-glucan have been shown to possess a common elicitor activity in various hosts irrespective of the different molecular structures. Here, the MAMP specific to this investigation is ergosterol, which is the major sterol component of the phospholipid bilayer of fungal cell membranes and functions in membrane stability and signaling. Ergosterol is found in several pathogens such as *Cladosporium fulvum* and *Botrytis cinerea,* but surprisingly some biotrophic fungi, including the powdery mildew (*Erysiphe cichoracearum*) and rust (*Puccinia triticina*) fungi, lack ergosterol [12]. Ergosterol contains two additional double bonds when compared to cholesterol and β-sitosterol, the most abundant phytosterol that is also an analogue of cholesterol [11,13]. Even with the aforementioned similarities of ergosterol to sitosterol, it is still perceived as a "non-self" MAMP [14], as has previously been shown in plant studies. Intracellular defense occurs within minutes in response to sub-nanomolar concentrations of ergosterol in tobacco and tomato cells. Included here is an increase of cytosolic Ca^{2+} levels, production of ROS, ion fluxes across the plasma membrane, protein phosphorylation, and production of phytoalexins [15–22]. It has been found that inhibiting the ergosterol biosynthesis pathway in colonizing fungi not only reduces fungal growth but also alters the sterol composition [12]. According to Dohnal et al. [23], ergosterol can be used as a fungal marker to evaluate infection levels in barley and corn crops, while treatment was also found to increase the expression of genes for PR1a, PR1b, PR3Q, and PR5 [16], acidic PR proteins used as markers for systemic acquired resistance (SAR) in host plants. Additionally, ergosterol elicitation has also shown expression of proteinase inhibitors, phenylalanine-ammonia lyase and sesquiterpene cyclase [16]. Although the perception mechanism is unknown, it is hypothesized that plants may possess an ergosterol receptor/receptor complex, or ergosterol penetrates the lipid bilayer and leads to perturbations of the plant cell system due to its ability to form stable microdomains in the plasma membrane [24,25]. In this study, we describe the use of proteomic approaches to identify differentially regulated plasma membrane-associated proteins following ergosterol treatment, as well as subsequent affinity-based chromatographic strategies of the said fraction to capture and enrich ergosterol-interacting candidate proteins so as to shed light on the unknown perception mechanism(s).

2. Results

2.1. Plasma Membrane (PM)-Associated Fraction Isolation and Verification

The plasma membrane (PM) outlines the interface between the cell and extracellular environment and is also the primary unit for signal recognition and transduction. Thus, elucidating and characterizing changes in the PM-associated proteome could identify possible receptor(s) and interacting/complementary complexes that are involved in immune responses to ergosterol. A challenge faced when extracting the PM proteome is the highly hydrophobic integral proteins that have a tendency of precipitating out of solution [26]. The conventional method of isolating PM proteins is the two-phase partitioning system, which requires 100–150 g of plant material [26]. However, the small-scale procedure has been found to result in PM-associated proteins comparable to the conventional method while employing much less starting material [26] and was the method followed in this investigation. The successful isolation of the

PM-associated fraction during the ergosterol-treatment time course was routinely verified using Western Blot analysis (Figure S1) and the H^+-ATPase assay. Furthermore, any non-PM-associated proteins were eliminated in the sequencing data analysis, as well as non-specific interacting proteins by the inclusion of control samples where no ergosterol was immobilized to the capture resins. Figure S2 shows the different isolated fractions with differentially regulated band intensities for each lane, thus implying successful enrichment of the PM-associated fraction.

2.2. PM-Associated Ergosterol-Responsive Candidate Protein Identification

Data analysis was initially conducted on the ergosterol-induced PM-associated fractions subsequent to isolation and prior to enrichment. The results are shown for the 1D and 2D SDS-PAGE gels (Figures 1 and 2) where differentially (densitometrically/electrophoretically) regulated bands/spots were selected for identification.

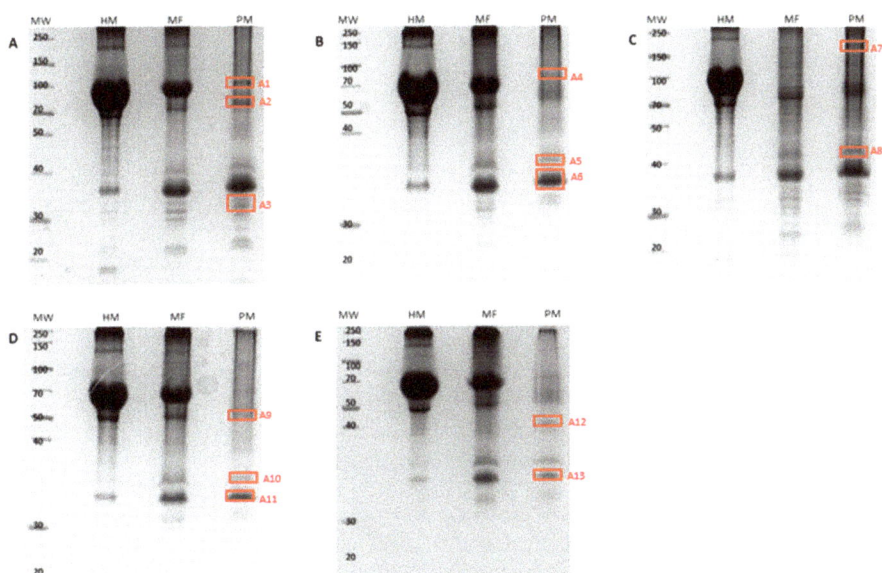

Figure 1. Representative 12% 1D-SDS PAGE gels stained with the Fairbanks method and showing the homogenate (HM), microsomal (MF), and plasma membrane (PM)-associated fractions subsequent to isolation. Gels represent all time point treatments with ergosterol, where **A** = control, **B** = 0 h treated, **C** = 6 h treated, **D** = 12 h treated, and **E** = 24 h treated. Equal volumes (20 μL) of the samples were mixed with 2X sample buffer, and electrophoresis was carried out at 90 V for 3 h. The red blocks indicate bands that were excised (A1–A13) for liquid chromatography coupled with tandem mass spectrometry (LC-MS/MS) identification.

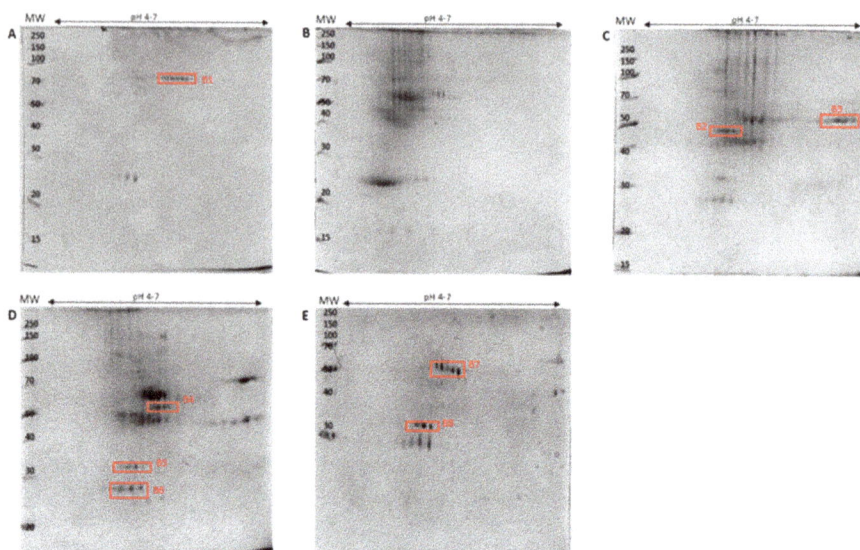

Figure 2. Comparative 2D-SDS-PAGE analysis for ergosterol-treated *Arabidopsis thaliana* PM-associated extracts. Proteins were precipitated with acetone, and 100 µg total protein was loaded onto immobilized pH gradient (IPG) strips, pH 4–7, for isoelectric focusing (IEF). The protein regulation differences are shown for **A** = control, and **B** = 0 h -, **C** = 6 h -, **D** = 12 h -, and **E** = 24 h-treated samples. The red blocks (B1–B8) indicate the protein spots excised for LC-MS/MS identification.

As previously mentioned, one band on a 1D gel may consist of multiple proteins. This emphasizes the need to identify the proteins affected/induced by ergosterol treatment and the role in perception of/response to this MAMP. Selected bands/spots from both the 1D- (Figure 1, A1–A13) and 2D SDS-PAGE (Figure 2, B1–B8) gels subsequent to ergosterol treatment were excised and prepared for liquid chromatography coupled with tandem mass spectrometry (LC-MS/MS) identification. The LC-MS/MS sequencing runs were repeated (separate experiments) for confirmation of protein lists obtained. The resulting spectra of the peptides were analyzed using the Byonic™ software (Protein Metrics, Cupertino, CA, USA). The program produces two plots, a protein score plot and mass error loadings plot (Figures S3 and S4). The protein score plot was used for the selection of proteins showing differential abundance or variable selection. This is known as the variable importance in projection (VIP) method and ranks proteins based on their contribution to the total variation of the samples. Differentially abundant proteins/peptides were selected on the VIP score where the set threshold was equal to one [27], and this value was presented as the log probability in all tables. The latter (as well as the Byonic score) determined the significance of the identified proteins. Even though these two said parameters could have been used individually, the values would have been less dependable. However, used together, they increased the significance. The dataset acquired was then normalized to the peptides of *Arabidopsis* proteins using the UniprotKB database. The identified *A. thaliana* PM-associated responsive proteins are summarized according to functional categories in Table 1 for the 1D SDS-PAGE bands and Table 2 for the 2D SDS-PAGE spots, respectively. There was better qualitative resolution for protein identification from the former to the latter. Furthermore, the differences between the theoretical and the experimental molecular weights (MW) for all proteins (low and high abundant) could be justified by the existence of structured water layers on the protein surface that affected the experimental MW determination on the SDS- PAGE [28].

Table 1. LC-MS/MS identification of *A. thaliana* PM-associated responsive proteins from selected 1D SDS-PAGE bands of control, 0-, 6-, 12-, and 24 h fractions subsequent to ergosterol treatment and organized according to functional categories (Supplementary Data Sheet 1).

Sample No.	Protein Name	Accession No.	Biological GO Term	Molecular GO Term	Calculated Mass [a] (M + H)	Mass Error [b] (ppm)	Byonic ™ Score [c]	ǀLog Probǀ [d]
			Perception and signaling (17)					
A5	Calcium-dependent lipid-binding (CaLB domain) family protein At3g61050	Q9LEX1	Response Signaling	DNA-binding	1214.699	−0.6	422.1	8.18
A7	Non-lysosomal glucosylceramidase At4g10060	F4JLJ2	Lipid Metabolism	Glycosidase	1294.627	−1.9	395.8	7.88
A10	G-type lectin S-receptor-like serine/threonine-protein kinase CES101 At3g16030	Q9LW83	Perception Response	Transferase	1113.626	−0.6	350.0	3.23
A5	Nicalin At3g44330	Q9M292	Signaling	—	1142.642	0.4	335.6	5.34
A7, A12	Cysteine-rich receptor-like protein kinase 41 At4g00970	O23081	Signaling	Transferase	973.531	0.3	328.0	1.53
A3	Axi 1 protein-like protein At2g44500	O64884	Biosynthesis Metabolism	Transferase	928.535	−2.9	289.7	2.72
A7	Cysteine-rich receptor-like protein kinase 10 At4g23180	Q8GYA4	Signaling	Transferase	1223.667	0.0	285.9	6.63
A7	PQQ_DH domain-containing protein At5g11560	F4JXW9	Biosynthesis	—	992.541	1.2	251.0	5.58
A8	Probable serine/threonine-protein kinase At4g35230	Q944A7	Defense	Transferase	1269.741	−2.3	236.2	6.31
A4	14-3-3-like protein GF14 epsilon At1g22300	P48347	Signaling	Protein binding	1229.580	−1.5	230.0	5.62
A9	Phosphoinositide phospholipase C 2 At3g08510	Q39033	Defense	Hydrolase	996.645	−0.5	228.4	4.97
A7	AMP deaminase At2g38280	O80452	Response	Hydrolase	1123.563	0.9	224.0	4.69
A10	Probable inactive leucine-rich repeat receptor-like protein kinase At3g03770	Q8LFN2	Signaling	Kinase	1041.515	0.4	217.2	1.30
A13	Mitogen-activated protein kinase 8 At1g18150	Q9LM33	Signaling	Kinase	1028.537	0.4	200.2	8.87
A7	Putative leucine-rich repeat receptor-like serine/threonine-protein kinase At2g24130	Q9ZUI0	Signaling	Transferase	1149.626	2.2	174.2	1.02
A7	Leucine-rich repeat receptor-like protein kinase At2g01210	Q9ZU46	Signaling	Transferase	870.541	0.1	164.6	0.9
A7	Receptor-like kinase TMK4 At3g23750	Q9LK43	Signaling	Kinase	1020.572	0.6	121.5	1.15
			Membrane trafficking and transport (16)					
A5	V-type proton ATPase subunit B2 At4g38510	Q9SZN1	Transport	Hydrolase	1563.801	−1.4	574.5	9.38
A7	Patellin-1 At1g72150	Q56WK6	Growth	Lipid binding	1231.689	−0.7	515.8	7.93
A3	Ras-related protein RABE1c At3g46060	P28186	Signaling Transport	GTPase	1071.641	−0.9	412.5	8.36
A7	ATPase 1, plasma membrane-type At2g18960	P20649	Transport	Translocase	1040.574	0.5	401.7	7.98
A6	Ras-related protein RABA1g At3g15060	Q9LK99	Signaling Transport	GTPase	1043.610	−0.1	384.7	8.14
A7	Clathrin heavy chain 1 At3g11130	Q0WNJ6	Transport	Clathrin binding	992.578	0.5	289.6	5.65
A3, A7	Probable aquaporin PIP1-5 At4g23400	Q8LAA6	Transport	Water transport	1049.599	−0.5	288.9	6.62

Table 1. *Cont.*

Sample No.	Protein Name	Accession No.	Biological GO Term	Molecular GO Term	Calculated Mass [a] (M + H)	Mass Error [b] (ppm)	Byonic ™ Score [c]	ǀLog Prob ǀ [d]
A5	Aquaporin PIP1-2 At2g45960	Q06611	Transport	Water transport	1033.604	−0.7	282.3	6.57
A7	CSC1-like protein ERD4 At1g30360	Q9C8G5	Transport	Ion channel	1251.612	0.5	271.5	7.51
A4	Probable ADP, ATP carrier protein At5g56450	Q9FM86	Transport	ATP:ADP transport	1021.531	−0.3	254.1	5.24
A3	Ras-related protein RABA1e At4g18430	O49513	Signaling Transport	GTPase	1274.612	−1.4	240.9	7.40
A8	Aquaporin TIP1-2 At3g26520	Q41963	Transport	Water transport	1980.030	0.0	239.9	6.69
A4, A5, A8	Aquaporin PIP2-1 At3g53420	P43286	Transport	Water transport	1069.568	0.2	215.3	5.98
A5	Probable aquaporin PIP2-6 At2g39010	Q9ZV07	Transport	Water transport	1311.669	−0.8	214.5	1.65
A7	Exocyst complex component SEC3A At1g47550	Q9SX85	Transport	GTP-Rho binding	1015.578	−1.7	183.9	1.26
A1	Aluminum-activated malate transporter 6 At2g17470	Q9SHM1	Transport	Malate transporter	1606.832	2.8	40	1.29
Defense (6)								
A5	Trans-cinnamate 4-monooxygenase At2g30490	P92994	Biosynthesis Defense	Monooygenase activity	1271.721	−0.3	377.3	8.02
A9	Protein BONZAI 2 At5g07300	Q5S1W2	Response	Phospholipid binding	1199.663	0.2	340.2	7.66
A3	Temperature-induced lipocalin-1 At5g58070	Q9FGT8	Response	Storage protein	1110.531	−0.5	329.0	7.88
A7	Disease resistance protein RPP8 At5g43470	Q8W4J9	Defense	ATP:ADP binding	1140.557	−2.1	267.7	6.39
A4	Hypersensitive-induced response protein 3 At3g01290	Q9SRH6	Response	—	949.547	−1.6	237.0	5.91
A4	Uncharacterized protein (LOW PSII ACCUMULATION-like protein) At4g28740	F4JM22	Chloroplast	—	995.600	-0.1	131.8	1.24
Structure (1)								
A2	Putative clathrin assembly protein At1g14910	P94017	Transport	Clathrin binding	1314.742	−1.3	122.2	1.22
Unknown (8)								
A11	Triacylglycerol lipase-like 1 At1g45200	Q8L7S1	Metabolism	Hydrolase	1222.622	−0.2	336.6	6.54
A11	TNF receptor associated factor (TRAF)-like family protein At1g58270/F19C14_8	Q9SLV3	Signal transduction	—	1434.722	−0.8	286.2	6.92
A7	Uncharacterized protein At4g16180	F4JLQ2	—	—	1293.669	−2.2	235.3	5.38
A1	Putative uncharacterized protein At3g19340	Q8RWC3	—	Aminopeptidase	1219.632	−0.9	229.5	5.94
A12	Putative uncharacterized protein F14P22.240 At3g58650	Q9M2F2	Growth		472.288	−1.1	160.4	1.04
A12	Putative uncharacterized protein F3A4.21 At3g50130	Q9SN05	—	—	472.288	−1.1	160.4	0.92
A2	Uncharacterized protein At4g38260	F4JTM0	—	—	1245.520	−5.2	142.3	0.98
A2	EMBǀCAB72473.1 At5g22560	Q9FK83	—	—	1467.731	0.1	133.6	1.21

a = the computed M + H precursor mass for the peptide spectrum matches (PSMs); b = a calculated mass error (parts per million) after correcting the observed M + H (single charged) precursor mass and the computed M + H precursor mass; c = Byonic score, and primary indicator of PSM correctness. A score of 300 is considered to be a significant hit [29]; d = the log p-value of the PSM, of which the value should be ≥ 1 for a hit to be significant. Proteins highlighted in red are known plasma membrane (PM) markers.

Table 2. LC-MS/MS identification of *A. thaliana* PM-associated responsive proteins from selected 2D SDS-PAGE spots of control, 0- , 6- , 12-, and 24 h fractions subsequent to ergosterol treatment and arranged according to functional categories (Supplementary Data Sheet 2).

Sample No.	Protein Name	Accession No.	Biological GO Term	Molecular GO Term	Calculated Mass [a] (M + H)	Mass Error [b] (ppm)	Byonic ™ Score [c]	I Log Prob I [d]
				Perception and signaling (10)				
B7	Probable serine/threonine-protein kinase At4g35230	Q944A7	Signaling	Transferase	1269.741	0.1	480.6	7.29
B4	At2g34560 protein (P-loop containing nucleoside triphosphate hydrolase) At2g34560	B9DGC0	Transport	ATPactivity	1156.672	−0.8	401.3	8.95
B3	Aspartyl aminopeptidase At5g60160/f15 I 12_20	Q9LST0	Biosynthesis	Metalloaminopeptidase	1148.679	0.3	382.0	8.65
B8	Probable protein phosphatase 2C 20 At2g20630	Q9SIU8	Signaling	Hydrolase	1288.711	−0.1	363.1	7.51
B6	Abscisic acid receptor PYL1 At5g46790	Q8VZS8	Signaling	Receptor	1442.760	−0.5	357.2	9.01
B7	Phosphotidylinositol 4-kinase alpha 1 At1g49340	Q9SXA1	Signaling	Kinase	1964.041	0.1	345.5	6.85
B2	Protein SGT1 homolog B At4g11260	Q9SUT5	Signaling	—	1435.709	−1.4	351.9	8.86
B1	Fasciclin-like arabinogalactan protein 7 At2g04780	Q9SJ81	Biosynthesis	—	981.500	0.9	322.0	6.49
B2	1-Phosphotidylinositol-3-phosphate 5-kinase FAB1A At4g33240	Q0WUR5	Signaling	Kinase	1470.816	−1.8	303.8	7.75
B8	Plasma membrane-associated cation-binding protein 1 At4g20260	Q96262	Response	Ion binding	1146.641	0.0	281.2	7.00
				Membrane trafficking and transport (16)				
B2	V-type proton ATPase subunit B3 At1g20260	Q8W4E2	Transport	Hydrolase	1563.801	−1.9	442.2	9.76
B8	Alpha-soluble NSF attachment protein 2 At3g56190	Q9SPE6	Transport	—	1259.684	−0.8	426.3	8.35
B6	Ras-related protein RABA1d At4g18800	Q9SN35	Signaling	GTPase	1043.610	0.0	414.7	7.96
B4, B7	Patellin-2 At1g22530	Q56ZI2	Transport	Lipid-binding	1520.784	−0.7	391.9	9.24
B4, B7	Patellin-1 At1g72150	Q56WK6	Transport	Lipid-binding	1231.689	−2.0	372.2	5.53
B2	Clathrin light chain 3 At3g51890	F4J5M9	Transport	Clathrin binding	855.530	0.0	363.4	4.96
B6	Ras-related protein RABA5b At3g07410	Q9SRS5	Signaling	GTPase	1071.641	−0.9	357.3	8.16
B3	SNAP25 homologous protein SNAP33 At5g61210	Q9S7P9	Transport	SNAP receptor	1302.715	−1.3	352.6	6.82
B1, B4, B8	V-type ATPase catalytic subunit A At1g78900	O23654	Transport	Hydrolase	1019.552	−1.6	338.6	5.50
B7	Sugar transport protein 7 At4g02050	O04249	Transport	Transmembrane transporter	1006.469	0.9	338.1	6.27
B3	Auxin transport protein BIG At3g02260	Q9SRU2	Signaling	Zinc binding	589.356	−1.4	331.2	5.39
B3	Protein NETWORKED 1C At4g02710	Q9ZQX8	—	Actin binding	478.251	0.1	311.4	6.18
B3	ABC transporter C family member 8 At3g21250	Q8LGU1	Transport	Translocase	530.330	−0.3	303.2	5.82
B8	Syntaxin-71 At3g09740	Q9SF29	Transport	SNAP receptor	1081.636	0.7	299.7	7.30
B4, B7	Flotillin-like protein 1 At5g25250	Q501E6	Transport	—	1526.909	−1.8	273.6	6.93
				Defense (9)				
B2, B4, B7	Jacalin-related lectin 35 At3g16470	O04309	Perception Response	Carbohydrate binding	1469.763	−2.5	518.7	9.21
B6	Aluminium induced protein with YGL and LRDR motifs At5g19140	Q94BR2	—	—	1439.738	−1.6	420.4	8.77

Table 2. *Cont.*

Sample No.	Protein Name	Accession No.	Biological GO Term	Molecular GO Term	Calculated Mass [a] (M + H)	Mass Error [b] (ppm)	Byonic ™ Score [c]	ǀLog Probǀ [d]
B6, B8	At3g11930 protein (Adenine nucleotide alpha hydrolases-like) At3g11930	B9DG73	—	Hydrolase	1189.631	−2.2	380.6	8.02
B4	Callose synthase 9 At3g07160	Q9SFU6	Biosynthesis Defense	Transferase	557.402	−0.9	334.8	2.44
B8	Hypersensitive-induced response protein 4 At5g51570	Q9FHM7	Defense Signaling	—	1466.764	1.7	345.6	8.36
B8	Binding partner of ACD (accelerated cell death)11 1 At5g16840	Q9LFD5	Signaling	RNA-binding	1132.621	−1.0	332.1	7.69
B5, B8	Hypersensitive-induced response protein 2 At1g69840	Q9CAR7	Defense Signaling	Kinase binding	871.500	−1.3	327.2	4.57
B6	Dessication responsive protein At2g21620	Q94II5	—	Hydrolase	980.614	−0.8	293.5	6.84
B8	Hypersensitive-induced response protein 1 At5g62740	Q9FM19	Defense Signaling	Kinase-binding	949.547	−0.5	281.6	7.29

a = the computed M + H precursor mass for the peptide spectrum matches (PSMs); b = a calculated mass error (parts per million) after correcting the observed M + H (single charged) precursor mass and the computed M + H precursor mass; c = Byonic score, and primary indicator of PSM correctness. A score of 300 is considered to be a significant hit [29]; d = the log p-value of the PSM, of which the value should be ≥ 1 for a hit to be significant. Proteins highlighted in red are known PM markers.

2.3. Identification of PM-Associated Ergosterol-Interacting Candidate Proteins

2.3.1. Epoxide Magnetic Microspheres-Based Ergosterol Immobilization

In order to capture and enrich ergosterol-interacting candidate proteins from the PM-associated leaf tissue fraction, MagResyn™ magnetic microspheres were used. The binding and elution events that showed the resulting protein elution to changing in eluents is represented in Figure 3 for the PM-associated proteins following a 6 h treatment. The elution profiles for the other time points are presented in the Supplementary Data as Figures S5–S9. The NaCl and SDS fractions for each time study were analyzed by SDS-PAGE and are illustrated as Figure 4. Proteins eluted with 0.5 M NaCl were not detectable in contrast to those eluted with 1% SDS, which disrupted non-covalent interactions between native proteins and the ligand. Table 3 lists the ergosterol-interacting candidate proteins that were identified following LC-MS/MS according to functional categories, while proteins with low scores are presented in Table S1. The negative control (no ergosterol immobilized) protein list is given in Table S2.

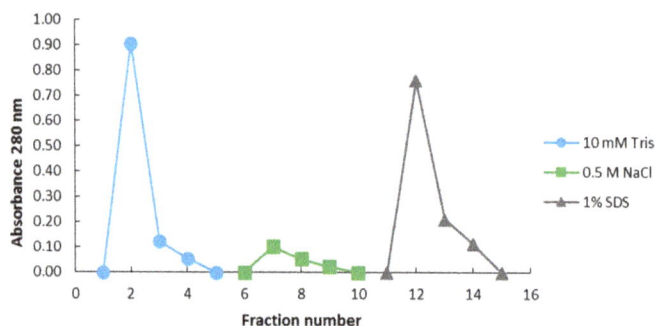

Figure 3. Representative elution profile of binding events between ergosterol-immobilized MagResyn™ magnetic microspheres and *A. thaliana* PM-associated proteins at 6 h following treatment. The blue curve represents the absorbance of the flow-through (unbound) fractions eluted with 10 mM Tris-HCl, pH 7.5. The green curve is the absorbance of the weakly bound proteins removed with 0.5 M NaCl, and the grey curve represents absorbance of proteins desorbed from the column with 1% SDS solution.

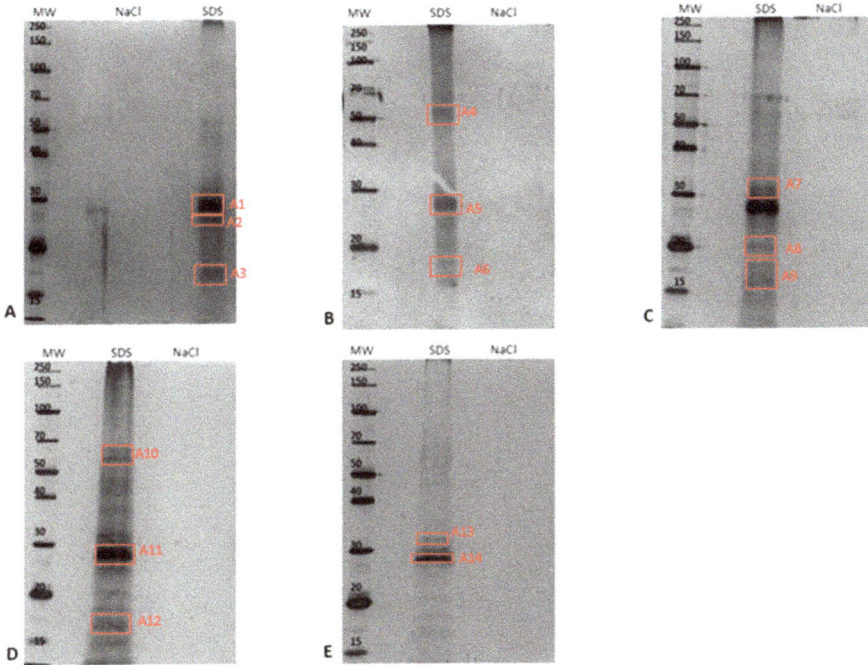

Figure 4. Comparative 12% 1D-SDS-PAGE analysis of ergosterol-interacting candidate proteins eluted with 0.5 NaCl and 1% SDS during the affinity-capture procedure using epoxide magnetic microspheres, where **A** = control, and **B** = 0 h-, **C** = 6 h-, **D** = 12 h-, and **E** = 24 h-treated samples. For each fraction, 20 μg total protein was loaded and electrophoresed at constant 90 V at room temperature. The red blocks (A1–A14) were excised subsequent to silver staining and analyzed using LC-MS/MS.

Table 3. LC-MS/MS identification of *A. thaliana* PM-associated candidate proteins interacting with ergosterol immobilized on epoxide magnetic microspheres for control, 0-, 6-, 12-, and 24 h subsequent to treatment and listed according to functional categories (Supplementary Data Sheet 3).

Sample No.	Protein Name	Accession No.	Biological GO Term	Molecular GO Term	Calculated Mass [a] (M + H)	Mass Error [b] (ppm)	Byonic™ Score [c]	∣Log Prob∣ [d]
				Signaling				
A12	Uncharacterized glycosylphophatidylinositol (GPI)-anchored protein At5g19250	P59833	—	—	1910.898	−3.1	464.4	8.69
A13	Binding partner of ACD (accelerated cell death)11 1 At5g16840	Q9LFD5	Signaling Response	RNA-binding	1132.621	−1.0	428.4	8.21
A4	Probable inactive receptor kinase At3g02880	Q9M8T0	Response	Receptor	1426.706	−3.0	392.2	6.99
A13	Uncharacterized GPI-anchored protein At5g19250	P59833	—	—	1910.898	−3.5	388.2	7.32
A11	Leucine-rich repeat-containing protein At5g07910	Q8RWI2	Response	—	1269.727	−0.5	336.8	7.18
A4	Probable inactive receptor kinase At5g16590	Q9FMD7	Response	Receptor	2127.170	−1.5	324.5	8.27
A11	Leucine-rich repeat protein kinase-like protein At1g10850	Q940B9	Signaling Response	Kinase	984.584	1.0	322.0	6.56

Table 3. *Cont.*

Sample No.	Protein Name	Accession No.	Biological GO Term	Molecular GO Term	Calculated Mass [a] (M + H)	Mass Error [b] (ppm)	Byonic™ Score [c]	lLog Prob l [d]
A11, A14	Chitin elicitor receptor kinase 1 At3g21630	A8R7E6	Perception Signaling	Kinase	1132.596	−0.6	313.8	6.87
			Membrane trafficking and transport					
A1, A5	Aquaporin PIP2-7 At4g35100	P93004	Transport	Water channel	1312.653	−1.4	541.6	7.44
A1	Aquaporin PIP1-2 At2g45960	Q06611	Transport	Water channel	1017.548	−0.3	509.0	6.59
A1	Aquaporin PIP2-1 At3g53420	P43286	Transport	Water channel	2000.996	−1.2	473.2	10.44
A1	Probable aquaporin PIP1-5 At4g23400	Q8LAA6	Transport	Water channel	1230.632	−1.1	464.3	9.05
A8	Plasma membrane-associated cation-binding protein 1 At4g20260	Q96262	Response	Ion-binding	1425.711	−2.6	418.8	8.56
A4	ATPase 2, plasma membrane-type At4g30190	P19456	Transport	Translocase	1040.574	−0.2	412.0	7.60
A6	At2g34250 protein At2g34250	O80774	Transport	Protein transport	1164.601	−2.0	408.2	6.50
A1	Ras-related protein RABE1c At3g46060	P28186	Signaling	GTPase	1071.641	−0.7	394.8	8.40
A9, A12	CASP-like protein 1D1 At4g15610	Q9FE29	—	—	1127.657	−0.3	389.4	9.41
A11	Plasma membrane ATPase At4g30190	F4JPJ7	Transport	Translocase	1040.574	−0.2	381.3	8.04
A11	ATPase 5, plasma membrane-type At2g24520	Q9SJB3	Transport	Translocase	1040.574	0.0	362.8	7.00
A11	Fasciclin-like arabinogalactan protein 8 At2g45470	O22126	—	—	967.484	−0.8	348.8	7.71
A11	Patellin-1 At1g72150	Q56WK6	Transport	Lipid-binding	1078.589	−1.2	326.8	7.67
A1	F-box/LRR-repeat protein At3g60040	Q8GWI2	Response	—	784.529	−0.3	309.1	1.05
A9	ABC transporter G family member 41 At4g15215	Q7PC83	Transport	ATP-binding	543.386	−0.9	301.8	4.38
A6	CSC1-like protein ERD4 At1g30360	Q9C8G5	Transport	Ion channel	1583.850	2.7	301.4	7.40
			Defense response					
A5	Hypersensitive-induced response protein 3 At3g01290	Q9SRH6	Defense Signaling Response	—	1519.775	−1.6	489.0	8.32
A6	Syntaxin-121 At3g11820	Q9ZSD4	Defense	SNAP receptor	1329.701	−1.4	439.9	7.40
A10	NDR1/HIN1-like protein 3 At5g06320	Q9FNH6	Defense	—	1496.843	−1.8	394.3	8.73
A10	Protein BONZAI 2 At5g07300	Q5S1W2	Defense	Phospholipid-binding	1060.615	0.0	372.4	7.25
A11, A14	Remorin At2g45820	O80837	—	—	617.409	−1.8	358.2	4.82
A11, A14	Blue copper protein At5g20230	Q07488	Transport	Electron transfer	1425.664	−0.4	337.8	9.34
			Unknown					
A11	At1g55160/T7N22.11 At1g55160	Q9C542	—	—	1174.631	−0.2	384.7	8.38
A11	At3g08600/F17014_7 At3g08600	Q9C9Z6	—	—	903.453	−0.2	352.4	7.93
A14	Expressed protein At2g18690	Q9ZV49	—	—	1115.606	−1.0	319.2	6.03

a = the computed M + H precursor mass for the peptide spectrum matches (PSMs); b = a calculated mass error (parts per million) after correcting the observed M + H (single charged) precursor mass and the computed M + H precursor mass; c = Byonic score, primary indicator of PSM correctness. Score of 300 is considered to be a significant hit [29]; d = the log p-value of the PSM, which the value should be ≥ 1 for hit to be significant.

2.3.2. EAH Sepharose 4B Immobilized with Ergosterol-Hemisuccinate

Ergosterol contains a diene group within its structure that is very reactive and requires protection by treatment with 4-phenyl-1,2,4-triazoline-3,5-dione (PTAD) prior to derivatization. Following protection, ergosterol was derivatized and validated using thin-layer chromatography (TLC) (shown in Figure S10). Figure 5 along with Figures S11–S15 show the binding events of the plasma membrane (PM)-associated fraction to the column immobilized with ergosterol-hemisuccinate. The NaCl and SDS fractions were analyzed using sodium dodecyl sulfate polyacrylamide gel electrophoresis (SDS-PAGE) and are illustrated in Figure 6. Selected bands were excised and analyzed using LC-MS/MS-based proteomics. The identified proteins are listed in Table 4, and proteins with low scores are in Table S3. The negative control (no ergosterol immobilized) protein list is presented on Table S4.

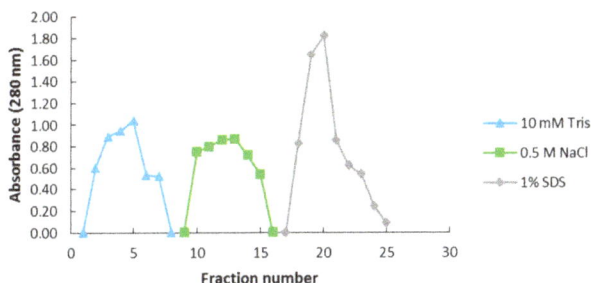

Figure 5. Representative elution profile of binding events between ergosterol-hemisuccinate immobilized on EAH Sepharose 4B resin and *A. thaliana* PM-associated proteins for the 6 h time point. The blue curve represents the flow-through fractions removed with 10 mM Tris-HCl, pH 7.5 buffer. The green curve represents the non-specifically bound fractions removed with 0.5 M NaCl in buffer, and the grey curve represents the proteins of interest eluted with 1% SDS in buffer.

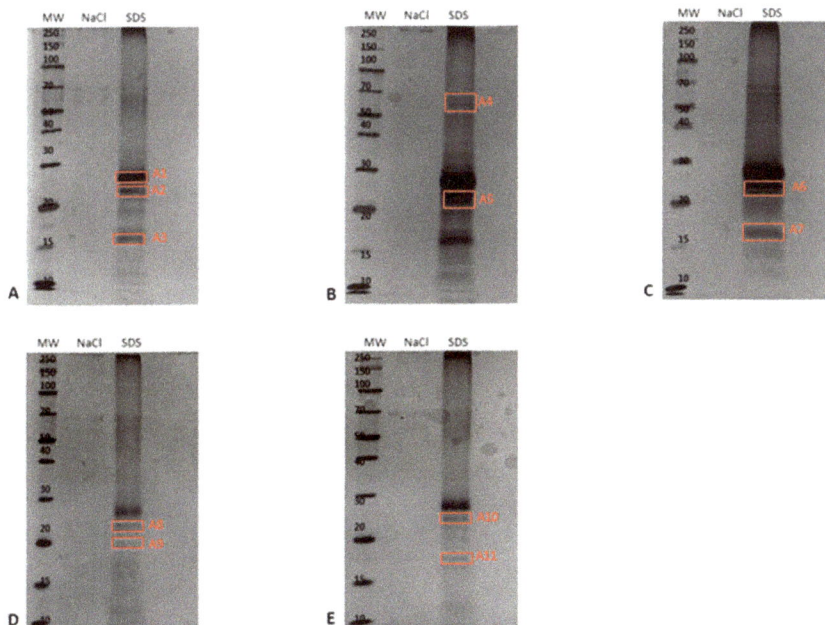

Figure 6. Comparative 12% 1D-SDS-PAGE analysis of ergosterol-interacting candidate proteins eluted with 0.5 M NaCl and 1% SDS during the affinity-capture procedure using EAH Sepharose 4B resin, where **A** = control, **B** = 0 h-, **C** = 6 h-, **D** = 12 h-, and **E** = 24 h-treated samples. For each fraction, 20 µg total protein was loaded and electrophoresed at constant 90 V at room temperature. The red blocks (A1–A11) were excised subsequent to silver staining and analyzed using LC-MS/MS for protein identification.

Table 4. LC-MS/MS identification of *A. thaliana* PM-associated candidate proteins interacting with ergosterol-hemisuccinate immobilized on EAH Sepharose 4B resin for the time study (Supplementary Data Sheet 4).

Sample No.	Protein Name	Accession No.	Biological GO Term	Molecular GO Term	Calculated Mass [a] (M + H)	Mass Error [b] (ppm)	Byonic™ Score [c]	ǀLog Probǀ[d]
A7	Ras-related protein RABG1 At5g39620	Q948K6	Signaling	GTP-binding	1071.641	0.0	515.00	6.92
A4	Aquaporin PIP1-2 At2g45960	Q06611	Transport	Water channel	1033.604	−1.2	336.3	6.55
A10	1-Phosphotidylinositol-3-phosphate-5-kinase FAB1B At3g14270	Q9LUM0	—	Kinase	956.480	−0.5	328.4	6.19
A6	Ras-related protein RABE1c At3g46060	P28186	Signaling	GTP-binding	1164.590	0.6	319.2	6.34
A1	Aquaporin PIP2-1 At3g53420	P43286	Transport	Water channel	1069.568	0.4	284.5	5.82

a = the computed M + H precursor mass for the peptide spectrum matches (PSMs); b = a calculated mass error (parts per million) after correcting the observed M + H (single charged) precursor mass and the computed M + H precursor mass; c = Byonic score, primary indicator of PSM correctness. Score of 300 is considered to be a significant hit [29]; d = the log p-value of the PSM, which the value should be ≥ 1 for hit to be significant.

3. Discussion

3.1. Functional Classification of Identified Ergosterol-Responsive – and Interacting Candidate PM-Associated Proteins from A. thaliana Leaf Tissue

The PM is known to participate in a wide spectrum of important functions, including transport of ions across the membrane, communication with the extracellular environment, cell wall biosynthesis, and defense against invading microorganisms. These functions are achieved by transport and membrane trafficking proteins and receptor kinases [30–32]. As seen with most biochemical processes, proteins are not limited to one functional group, e.g., a transport protein may also be regulated during a defense response event. Such proteomic approaches (prior to enrichment and subsequent to affinity-based strategies) aimed to provide a comprehensive understanding of both ergosterol-responsive and interacting candidate proteins at the PM-localized interface, as well as those possibly associated with the PM subsequent to MAMP treatment.

3.1.1. Membrane Trafficking and Transporters

In a plant cell, responses to a MAMP occurs within minutes [33]. As mentioned, ergosterol treatment causes ion fluxes across the PM and intracellular increase of Ca^{2+} levels. These changes are due to transport proteins and those involved in endocytosis/exocytosis. Aquaporins, identified in Table 1, Table 3, and Table 4 (i.e., both non-enriched and enriched PM-associated fractions), are water carrier proteins identified within all the time study samples and are also considered as PM markers. The PM intrinsic proteins (PIP) were differentially regulated in the samples, likely due to a defense response, and isolated during affinity chromatography. The *Arabidopsis* aquaporin AtPIP1 and AtPIP2 groups are well-known to be localized in PMs and are involved in defense responses within the plant [34].

ATP-dependent binding cassette (ABC) transporters have previously been shown to be involved in various processes such as transport of phytohormones, surface lipid deposition, and pathogen response during plant-microbe interactions [35]. In this study, an ABC transporter was identified in Table 3 (enriched PM-associated fraction) following capture affinity. The G family (AtABCG) group is the largest subfamily of ABC transporters in *A. thaliana*, and evidence was found by Ji et al. [36] that AtABCG16 is involved in basal resistance and abscisic acid (ABA) tolerance against the virulent bacterial pathogen *Pseudomonas syringae* pv. *tomato* (Pst) DC3000. Additionally, patellin-1 (Table 1, non-enriched PM-associated fraction) is a carrier protein involved in membrane trafficking by binding to hydrophobic molecules (such as the steroid-like ergosterol) and promoting their transfer between different cellular sites [37,38]. Vilakazi et al. [39] also identified this phosphoinositide-binding protein, patellin-1, in the study of capturing LPS-binding PM-associated proteins in *A. thaliana*.

Lastly, clathrin-dependent membrane trafficking is critical for determining cell polarity, and clathrin light chains are predominantly localized at the PM and early endosome compartments [40]. Both light chains (CLCs) and clathrin heavy chains (CHCs), including CHC1 and CLC3, were identified in the non-enriched PM-associated fraction (Tables 1 and 2). In this regard, Mgcina et al. [41] also speculated that the binding-site of lipopolysaccharide (LPS) as a bacterial MAMP to *A. thaliana* protoplasts is internalized into the cell by endocytosis, thus leading to the reduced level of receptors on the surface.

3.1.2. Signaling

Pathogens that successfully overcome the initial physical defense barrier are mostly recognized by PRR proteins on the cell membrane. Recognition at the PM is immediately transmitted internally to activate other defense factors [42]. Some of the proteins involved during basal resistance fall within the signaling category and are associated with the PM during a defense response event. Here, a GPI-anchored protein was identified in the 12 h- and 24 h-treated PM-associated samples, as listed in Table 3 of the enriched fractions. These proteins are known to exist independently in a soluble form and are also associated with the PM [43]. GPI anchoring acts as a PM targeting signal, either in a localized or a polarized manner, by transferring signals from activated transmembrane receptors to various constituents inside the cell [44]. Due to these targeting mechanisms, GPI-anchored proteins are associated with lipid rafts/microdomains [43,45,46] and, since Peskan et al. [44] found evidence for such rafting in plants, these proteins have been used as a model or marker for raft sorting [47].

A LRR protein kinase-like protein and LRR-containing protein were identified in non-enriched as well as enriched PM-associated proteins (Tables 1 and 3). LRR-containing RLKs are well known to confer resistance to bacterial and fungal pathogens [48]. A well-studied LRR-containing receptor is the FLS2 from *A. thaliana* that perceives the bacterial flagellin and triggers the binding of brassinosteroid intensive 1 (BRI1)-associated kinase (BAK1) to the receptor and acts as a signal enhancer [49]. Furthermore, FLS2 is said to migrate to highly organized membrane raft compartments of the PM where interaction with BAK1 takes place, forming a heterodimer [50]. Another protein kinase identified includes the G-type lectin S-receptor-like serine/threonine-protein kinase (GsSRK) listed in Table 1 (non-enriched fraction). Sanabria et al. [51] proposed a role for S-domain RLKs in M/PAMP perception, specifically for LPS. In the study, it was shown that LPS perception transiently up-regulates the expression of a G-type lectin receptor kinase in tobacco. This was also seen in the study of LPS perception in *A. thaliana* by Baloyi et al. [52], as the GsSRK protein was up-regulated during the time study. Plant lectins are proteins that are known to reversibly bind carbohydrates and are assumed to play a role in plant resistance and development. It was shown by Esch and Schaffrath [53] that the lectin domain of a jacalin-related lectin protein was responsible for relocating the protein towards the site of pathogen attack, and jacalin-related lectin 35 was identified in the 2D set of proteins analyzed subsequent to isolation (Table 2). The 14-3-3-like protein, GF14 epsilon, was also identified in the non-enriched fraction (Table 1). These proteins are known to be important components in biological pathways involved in signal transduction in response to biotic and abiotic stresses. In rice, 14-3-3 proteins regulate complex defense responses and interact with cellular components; 14-3-3 genes have also been found to be expressed in response to inoculation with rice fungal pathogens, thus suggesting functions in defense signaling [54].

CERK1 is a PM protein with three LysM motifs in the extracellular domain that was identified following affinity-capture (Table 3). LysM proteins have been shown to play a vital role in basal immunity by recognizing peptidoglycan and chitin via the *N*-acetylglucosamine (GlcNAc) moiety [52]. The *Arabidopsis* CERK1 (AtCERK1) is said to function as a ligand-binding protein and as a signaling molecule with kinase activity [55]. Lastly, the binding partner of accelerated cell death (ACD) 11 was identified for the first affinity-based approach (Table 3) and is known to mediate sphingolipid metabolism and regulate programmed cell death (PCD) upon pathogen infection in plants [56]. It has also been shown that *Arabidopsis* ACD mutant plants displayed excessive cell death upon infection with bacterial *P. syringae* [57].

3.1.3. Defense responses

During the early stages of M/PTI, upon pathogen recognition, defense-related proteins are either activated, enhanced, or transcribed. Microbes can also deliver effectors into the cytosolic space of the plant cell during ETI, thus challenging the plant's defense proteins [50]. The NDR1/HIN1-like protein 3 (NHL3), listed in the enriched fraction (Table 3), is predicted to be a membrane protein that has been shown to be triggered by avirulent *Pst* instead of the virulent strains. Hitherto, Varet et al. [58] reported that the expression of NHL3 is suppressed by virulent bacteria, and therefore the protein is hypothesized to participate in disease resistance. SNARE (soluble *N*-ethylmaleimide sensitive factor attachment protein receptors) complexes are also known to be necessary for immune responses and have been associated with targeted exocytosis of various antimicrobial compounds and proteins. Multiple SNARE complex constituents have been identified in previous studies, including the syntaxin of plants 122 (SYP122), and soluble *N*-ethylmaleimide-sensitive factor adaptor protein 23 (SNAP33) was identified in the 2DE samples (Table 2) in this study. These proteins were previously found to be highly enriched at the PM during an immune response [59].

Remorins, identified during the affinity-capture (Table 3), are proteins that play a role in cell-to-cell signaling and plant defense and have been shown to be associated with the PM in potato leaves. Furthermore, remorin 1.2 from tobacco (NtREM1.2) revealed primary accumulation in isolated DRMs and showed distinct localization in domains in the PM when expressed as a green fluorescent protein (GFP) fusion protein. These experiments showed that remorins are marker proteins for DRMs in plants that form higher order oligomers, impacting the binding affinity to these microdomains [46,60]. In *A. thaliana*, remorins are differentially phosphorylated, and this event is dependent on the presence of the NBR-LRR resistance protein RPM1. This is triggered upon perception of various M/PAMPs [61]. Within the *Arabidopsis* genome genes named, *AtONB1, AtBON2,* and *AtBON3* (bonzai, also known as copine) were shown to be regulators of plant immunity. The identified ergosterol-interacting candidate BONZAI-2 (enriched fraction in Table 3) plays a role in suppressing programmed cell death and defense in plants during pathogen attack [62]. This was supported by Zhou et al. [63], where *Arabidopsis* and rice plants were inoculated with *Pst* DC3000, and the pathogen's interaction with the plant was limited to the PM. Hypersensitive-induced response (HIR) proteins are found on the PM and interact with LRR proteins during a defense response. The *A. thaliana* AtHIR1, AtHIR2, AtHIR3, and AtHIR4, identified in both non-enriched and enriched fractions (Tables 1–3), are associated with the intracellular side of the PM and involved in the development of programmed cell death during pathogen attack [39]. Baloyi et al. [52] and Vilakazi et al. [39] identified HIR protein 1, HIR protein 2, HIR protein 3, and HIR protein 4 in their studies pertaining to LPS as a M/PAMP.

Lastly, plant disease resistance (R) proteins are quantitative and rate-limiting regulators. Disease resistance protein RPP8, listed in the non-enriched fraction (Table 1), has been seen to be up-regulated in response to multiple avirulent pathogens and by wounding. It is also suggested that RPP8 is connected to multiple pathways [64].

4. Materials and Methods

4.1. Plant Growth and Elicitor Treatment

For the study, *A. thaliana* seedlings were grown in Culterra™ Germination Mix (Culterra, Johannesburg, South Africa) soil in trays placed in a plant growth room at 20–24 °C under a 12-h light/12-h dark cycle until mature. Plants were routinely watered and fertilized with 1:300 (*v/v*) diluted Nitrosol™ Natural (Nitrosol, Manukau City, New Zealand). Mature plants with fully developed rosettes (~2 months old) were treated with 250 nM ergosterol (Sigma, Steinheim, Germany) during the day cycle using gentle pressure infiltration into the abaxial side of the leaves. An elicitor stock solution was prepared in absolute ethanol and diluted in dH_2O to a working solution containing less than 0.2% ethanol, and elicitation included a time study of 0, 6, 12, and 24 h, respectively, in accordance with related citations with untreated plants as the control. To eliminate any variation, all experiments included 3 biologicals and 3 repeats of each experiment, including sequencing. The raw data files containing the most significant proteins (Section 4.7) were merged to produce Supplementary Data Sheets 1–4 and compile Tables 1–4.

4.2. Small-Scale Isolation of the Plasma Membrane(PM)-Associated Fraction

The isolation protocol was taken from Giannini et al. [26] and modified to optimize the yield of isolated fractions. Approximately 20 g of leaf tissue was homogenized in 60 mL homogenizing buffer containing 250 mM sucrose (Merck, Darmstadt, Germany), 3 mM ethylenediaminetetraacetic acid (EDTA) (MerckDarmstadt, Germany), 10% (*v/v*) glycerol, 0.5% (*w/v*) poly(vinylpolypyrrolidine) (PVPP) (Sigma, St. Louis, MO, USA), 2 mM phenylmethane sulfonyl fluoride (PMSF) (Boehringer Mannheim, Mannheim, Germany), 15 mM β-mercaptoethanol (Sigma, St. Louis, MO, USA), 4 mM 1,4-dithiothreitol (DTT) (Fisher Chemicals, Loughborough, UK), 250 mM potassium iodide (KI) (Saarchem, Johannesburg, South Africa), and 70 mM tris(hydroxymethyl)aminomethane (Tris) (Merck, Modderfontein, South Africa) using an Ultraturax homogenizer. Homogenates (HM) were filtered through 2 layers of miracloth (Millipore/Merck, Darmstadt, Germany) and centrifuged at $6000 \times g$ for 4 min at 4 °C using a Beckman Coulter™ Avanti™ J-20 I centrifuge. Cell debris was discarded, and the supernatants were collected and centrifuged at $13,000 \times g$ for 25 min at 4 °C. After centrifugation, the supernatants were discarded, and the pellets were resuspended in 800 µL microsomal resuspension buffer containing 250 mM sucrose, 10% (*v/v*) glycerol, 1 mM DTT, and 1 mM PMSF. Five hundred µL of the microsomal fraction was layered onto a sucrose gradient containing 700 µL of 25% (*w/v*) and 38% (*w/v*) sucrose each to create a discontinuous gradient in 1 mM Tris-HCl, 1 mM EDTA, and 0.1 mM DTT, pH 7.2. The gradients with the microsomal fractions were centrifuged at $13,000 \times g$ for 1 h, after which the PM-associated fraction formed an interface within the gradient and was aspirated using a pipette and transferred into a new tube. To validate the successful isolation of the said fractions, MAPK Western blot analysis (Figure S1) and plasma membrane H^+-ATPase assays were routinely conducted.

4.3. Identification of PM-Associated Ergosterol-Responsive Candidate Proteins

Prior to affinity chromatography, SDS-PAGE was performed of the homogenates, microsomal, and PM-associated fractions in order to identify proteins that could be categorized as ergosterol-responsive candidates. The 12% 1D-SDS-PAGE gels were visualized using the Fairbanks staining protocol [65], and differentially (densitometrically) regulated protein bands were excised for identification. The PM-associated fractions also underwent 2D-SDS-PAGE with immobilized pH gradient (IPG) strips of narrow range (pH 4–7) in order to identify elecrophoretically distinct spots that could be responsive candidates to ergosterol treatment. Samples were prepared for the first dimension of separation with a concentration of 100 µg total protein. A final volume of 120 µL sample was prepared containing 2 µL 50% DTT, 1.3 µL ampholyte (Bio-Rad, Hercules, CA, USA), x µL sample, y µL urea buffer with trace amounts of bromophenol blue. The samples were loaded onto the Immobiline™ Reswelling Dry-strip tray, and non-linear IPG strips [pH 3–10 or 4–7, 7 cm ReadyStrip™ IPG, Bio-Rad, Hercules, CA, USA] were gently laid on top of the sample with the gel side down. The strips with samples were overlaid with mineral oil to prevent drying, and the tray was covered with foil. The strips were left to hydrate overnight at RT. Following hydration, the strips were placed on the Etthan IPGphorII electrophoresis unit (Amersham Bioscience, Buckinghamshire ,UK) with the gel side facing up. Electrode wicks were soaked with dH_2O and placed on opposite ends of the strips. Conditions for isoelectric focusing (IEF) included step 1 at 250 V for 15 min, step 2 at 4000 V for 1 h, and step 3 was 4000 V for 12,000 V/h. Once IEF was completed, strips were rinsed with dH$_2$O to remove excess mineral oil and then with 1X tank buffer for 5 min. The strips were incubated in DTT equilibration buffer (0.8 g DTT, 6 M urea, 30% glycerol, 2% SDS, 50 mM Tris-HCl, pH 8.8) for 20 min with constant shaking. Strips were rinsed again with 1X tank buffer and then incubated in iodoacetamide (IAA) (Sigma, St. Louis, MO, USA) equilibration buffer (0.2 g IAA, 6 M urea, 30% glycerol, 2% SDS, 50 mM Tris-HCl, pH 8.8) for 20 min with constant shaking, followed by 1X tank buffer prior to loading the strip on top of a 12% resolving gel, as previously prepared. Protein spots that showed differential regulation were excised and analyzed by LC-MS.

4.4. Affinity Chromatography

4.4.1. Magnetic Epoxide Microspheres

The MagReSyn^{TM} magnetic epoxide microspheres (ReSyn Biosciences, AEC-Amersham, Midrand, South Africa) were supplied as a 20 mg/mL suspension in 20% (*v/v*) ethanol. The microspheres contain high functional group intensity throughout the fiber surface network, which allows a ligand (in this case, ergosterol) to react and be immobilized to the lattice by covalent bonding [66]. Microspheres were resuspended in the shipping solution, and 50 µL (±1 mg) was collected with a pipette and transferred to a new tube. The tube was placed on a magnetic separator, and once microspheres were clear, the shipping solution was discarded. Microspheres were equilibrated with three washes of 200 µL milliQ H$_2$O. The activation solution supplied by the manufacturer consisting of 5.2 M 1,4-butanediol diglycidyl ether was diluted 4× to a working solution. Microspheres were resuspended in 500 µL activation solution and continuously agitated for 48 h at RT before removal thereof and washing of the microspheres two times with 200 µL 90% (*v/v*) tetrahydrofuran (THF) (Sigma, Steinheim, Germany), which was also used as the coupling buffer. Thirty mg/mL of ergosterol was prepared with 90% (*v/v*) coupling buffer. Five hundred µL of the ergosterol was added to the activated microspheres and continuously agitated for 48 h at 4 °C. On the magnetic separator, the unbound ergosterol was removed, and microspheres were washed three times with 200 µL coupling buffer. Epoxide residues that did not bind to the ergosterol were quenched with 500 µL ethanolamine, pH 8.5 blocking solution for 24 h at RT. The blocking agent was discarded, and 1 mg/mL (~1.25 mL) of PM-associated protein sample was added. The microspheres with the samples were incubated for 24 h at RT with constant agitation followed by removal of the liquid fraction and washing of the microspheres five times with 1 mL 10 mM Tris-HCl, pH 7.5 to remove unbound proteins. The microspheres were then washed with 1 mL 0.5 M NaCl in 10 mM Tris-HCl, pH 7.5 to remove non-covalently bound proteins. Ergosterol-interacting candidate proteins were subsequently eluted with five washes 1 mL 1% (*w/v*) SDS in 10 mM Tris-HCl, pH 7.5. The absorbance of the collected fractions was measured spectrophotometrically at 280 nm for monitoring absorption and desorption reactions.

4.4.2. EAH Seharose 4B

This approach required the derivatization of ergosterol to the hemisuccinate for affinity chromatographic applications, as reported by Tejada-Simon and Pestka [67]. Thirty mg/mL ergosterol was treated with 4-phenyl-1,2,3-triazoline-3,5-dione (PTAD) (Sigma, Steinheim, Germany), resulting in cyclic adducts. For the formation of the adducts to ergosterol-hemisuccinate, 2.5 mM succinic anhydride (Sigma, Steinheim, Germany) was dissolved in 800 µL pyridine (Sigma, St. Louis, MO, USA) in a reaction vessel. The ergosterol adducts were added to the mixture and refluxed for 60 min. The reaction was allowed to cool down for ±5 min before the reaction vessel was submerged in boiling water and a nitrogen steam was applied to remove pyridine. Excess succinic anhydride was removed by portioning the mixture in equal volumes of water and chloroform. The chloroform phase containing ergosterol-hemisuccinate (Erg-HS) was dried under nitrogen. To authenticate the successful derivatization of ergosterol, 30 mg/mL ergosterol and 30 mg/mL Erg-HS were separately dissolved in toluene:acetone (70:30, *v/v*). High-performance thin layer chromatography (HP-TLC) was conducted with a mobile phase of toluene:acetone (70:30, *v/v*), and plates were visualized under UV at 254 nm.

One mL of 1,6-diaminohexane (EAH) Sepharose 4B beads were swollen in 10 mL 0.5 M NaCl and washed two times with 5 mL water, pH 4.5. Erg-HS (0.05 g) was dissolved in 2 mL 50% (*v/v*) 1,4-dioxane (Sigma, Steinheim, Germany) and added to the beads. One-ethyl-3(3-dimethylaminopropyl)-carbodiimide (Sigma, Steinheim, Germany) was added to the beads to a final concentration of 0.13 M and manually inverted for 60 min at RT. The carbodiimide-promoted condensation reaction was then continuously inverted for 24 h at 4 °C to allow the coupling of Erg-HS to the beads. This was followed by washing with 10 mL 100% 1,4-dioxane, 10 mL 80% (*v/v*) ethanol, 20 mL water, and 10 mM Tris-HCl pH 7.5, respectively. Beads were transferred into a column, 1 mg/mL of the PM-associated extract was added on the resin bed, and the column was blocked for 1 h at RT to allow binding of proteins to Erg-HS. The column was then washed six times with 50 mM Tris-HCl pH 7.5 to remove all unbound proteins followed by six washes with 0.5 M NaCl in 50 mM Tris-HCl 7.5 to remove non-specifically bound proteins. Lastly, ergosterol-interacting candidate proteins were eluted with 1% (*w/v*) SDS in 50 mM Tris-HCl pH 8.0. The absorbance of the collected fractions was spectrophotometrically measured at 280 nm to determine the elution profile.

4.5. Protein Precipitation and SDS-PAGE

Proteins from the desorbed affinity fractions were precipitated with absolute acetone at a 1:2 (*v/v*) ratio at −20 °C overnight to remove any substances that may have interfered with SDS-PAGE and mass-spectrometry and to achieve maximum protein yield. Fractions with acetone were briefly vortexed prior to centrifugation at $13,000\times g$ for 10 min at 4 °C, after which the supernatant was carefully discarded. Pelleted proteins were washed twice with ice cold 80% (*v/v*) acetone, centrifuged at $13,000\times g$ for 10 min between each wash step, and solubilized in SDS sample buffer. SDS-PAGE was performed by resolving ergosterol-interacting candidate proteins on 12% gels using the Hoefer Scientific miniVE vertical electrophoresis system at constant voltage of 90 V for 3 h at RT (Section 4.2). Gels were stained using the silver staining protocol adapted from Switzer III et al. [68] and Blum et al. [69].

4.6. In-Gel Trypsin Digestion

Coomassie-stained gel slices were destained twice in a solution containing 100 mM ammonium bicarbonate (NH_4HCO_3) (Sigma, Steinheim, Germany) and 50% (*v/v*) acetonitrile (ACN) for 45 min at RT with constant agitation, while silver-stained gel pieces were covered with a solution of 30 mM potassium ferricyanide and 100 mM sodium thiosulfate and agitated until clear at RT. Solutions were discarded, and gel pieces were washed with milliQ H_2O followed by an addition of 2 mM Tris(2-carboxyethyl) phosphine (TCEP) (Sigma, Steinheim, Germany) that was made up in 25 mM NH_4HCO_3 and incubated for 15 min at RT with constant agitation to reduce proteins. Excess TCEP was removed, and gel pieces were washed three times with 500 μL NH_4HCO_3 for 15 min. Concentrated ACN was added, and samples underwent desiccation under vacuum centrifugation. Thereafter, gel pieces were covered with sequencing grade trypsin (20 ng/μL) (Promega, Madison, WI, USA) dissolved in 50 mM NH_4HCO_3 and incubated on ice for 1 h. Excess trypsin solution was removed, and sufficient 50 mM NH_4HCO_3 was added, followed by incubation for 18 h at 37 °C. Peptides were extracted from gel pieces using 0.1% (*v/v*) trifluoroacetic acid (TFA) (Sigma, St. Louis, MO, USA) and incubation for 1 h at 37 °C. Peptides were then collected by centrifugation at $40\times g$ for 5 min and dried under vacuum, followed by resuspension in 12 μL loading buffer [2% ACN, 0.1% formic acid (FA) (Sigma, St. Louis, MO, USA)] prior to analysis.

4.7. LC-MS/MS Analysis

Analysis was conducted at the Centre for Proteomics and Genomic Research (CPGR) (Cape Town, South Africa). Nano-RP LC was performed on a Dionex Ultimate 3000 nano-HPLC system coupled with a Q-Extractive Quadrupole-Orbitrap mass spectrometer (Thermo Fisher Scientific, Waltham, MA, USA) for LC-MS/MS analysis. The mobile phase solvent system employed was solvent A:

water/0.1% FA and solvent B: 100% ACN/0.1% FA. Solubilized peptides were loaded onto a C18 trap column (300 μm × 5 mm × 5 μm). Separation was performed on a C18 column (75 μm × 20 cm × 1.7 μm), and a linear gradient was generated at 300 nL/min with a change of 2–60% solvent B over 52 min. The mass spectrometer was operated in positive ion mode with a capillary temperature of 320 °C. The applied electrospray voltage was 1.95 kV. Lastly, mass spectrometry was performed using data-dependent acquisition MS/MS scans with a mass range of 350–2000 *m/z*.

4.8. Data analysis

Data analysis was performed using the Byonic software (Protein Metrics, Cupertino, CA, USA), product version PMI-Byonic-Com: v2.6.46. The Arabidopsis UniProt Knowledgebase (UniprotKB) database [70] was used to match peptide fragments resulting from the MS/MS. Peptides were fragmented using the collision-induced dissociation (CID) low energy, and the parameters were as follows: trypsin was a C-terminal cutter end, and carbamidomethyl (C) and deamination (NQ) were set for fixed and variable modification, respectively. The precursor tolerance was 7 ppm, and the fragment tolerance was 20 ppm. Maximum number of missed cleavage was 2, and the protein false discovery rate (FDR) cut-off was 1% with the best score range of 0–1000 *m/z*. A target protein with a best score of >300 was considered significant.

5. Conclusions

The purpose of the study was to identify *Arabidopsis* PM-associated candidate proteins in the response to ergosterol treatment as well as those possibly interacting with the MAMP as ligands. A small-scale isolation protocol was used to fractionate the *Arabidopsis* leaf tissue and resulted in the successful isolation of said fraction for different time points. This was confirmed by identification of PM and DRM markers subsequent to LC-MS/MS-based proteomics in excised 1D- and 2D-SDS-PAGE protein bands and spots from the non-enriched fraction (i.e., responsive candidate proteins). Thereafter, enrichment of PM-associated proteins (i.e., ergosterol-interacting candidate proteins) resulted in some that had been identified in previous studies using different elicitors, such as the bacterial flg22 and LPS, as well as fungal chitin. The perception mechanism of ergosterol in *Arabidopsis* is still unclear, but the identified candidate proteins show that there could possibly be a receptor complex (including non-PM yet associated proteins) involved in signaling the recognition of this MAMP to the intracellular components of the plant cell, and that is similar to other reported elicitors. Additionally, the second affinity-enrichment, which was meant to reduce non-specific binding, yielded very few ergosterol-interacting PM-associated candidate proteins, and this may suggest that derivatization of ergosterol resulted in critical alterations to molecular features that could have affected the association/interaction of the MAMP molecule with proteins.

Lastly, this study is a first of its kind because affinity chromatography has not yet been employed for capturing ergosterol-interacting PM-associated candidate proteins. As previously mentioned, the PM is the recognition site of many microorganisms and associated MAMPs, therefore understanding or identifying the interface proteomic changes involved during ergosterol-induced MTI may assist in further elaborating the protein network and pathways regulated during activation of immune responses that form part of plant defense.

Supplementary Materials: Supplementary materials can be found at http://www.mdpi.com/1422-0067/20/6/1302/s1.

Author Contributions: L.P. and I.D. conceptualized and designed the research; T.K. performed the research; and L.P., I.D. and T.K. analyzed data and prepared the manuscript.

Funding: L.P. would like to acknowledge financial support received from the Faculty Research Council, University of Johannesburg. The financial assistance of the South African National Research Foundation (NRF) is also acknowledged for bursary support to T.K.

References

1. Gómez-Gómez, L.; Boller, T. Flagellin perception: A paradigm for innate immunity. *Trends Plant Sci.* **2002**, *7*, 251–256. [CrossRef]

2. Jones, J.D.G.; Dangl, J.L. The plant immune system. *Nature* **2006**, *444*, 323–329. [CrossRef] [PubMed]

3. Ranf, S.; Gisch, N.; Schäffer, M.; Illig, T.; Westphal, L.; Knirel, Y.A.; Sánchez-Carballo, P.M.; Zähringer, U.; Hückelhoven, R.; Lee, J.; et al. A lectin S-domain receptor kinase mediates lipopolysaccharide sensing in *Arabidopsis thaliana*. *Nat. Immunol.* **2015**, *16*, 426–433. [CrossRef]

4. Erbs, G.; Molinaro, A.; Dow, J.M.; Newman, M.-A. Lipopolysaccharides and plant innate immunity. *Subcell. Biochem.* **2010**, *53*, 387–403. [CrossRef] [PubMed]

5. Henry, G.; Thonart, P.; Ongena, M. PAMPs, MAMPs, DAMPs and others: An update on the diversity of plant immunity elicitors. *Biotechnol. Agron. Soc. Environ.* **2012**, *16*, 257–268.

6. Fu, Z.Q.; Dong, X. Systemic acquired resistance: Turning local infection into global defense. *Annu. Rev. Plant Biol.* **2013**, *64*, 839–863. [CrossRef] [PubMed]

7. Dodds, P.N.; Rathjen, J.P. Plant immunity: Towards an integrated view of plant-pathogen interactions. *Nat. Rev. Genet.* **2010**, *11*, 539–548. [CrossRef] [PubMed]

8. Boyd, L.A.; Ridout, C.; O'Sullivan, D.M.; Leach, J.E.; Leung, H. Plant-pathogen interactions: Disease resistance in modern agriculture. *Trends Genet.* **2013**, *29*, 233–240. [CrossRef]

9. De Wit, P.J.G.M. How plants recognize pathogens and defend themselves. *Cell. Mol. Life Sci.* **2007**, *64*, 2726–2732. [CrossRef] [PubMed]

10. Pritchard, L.; Birch, P.R.J. The zig-zag model of plant-microbe interactions: Is it time to move on? *Mol. Plant Pathol.* **2014**, *15*, 865–870. [CrossRef]

11. Klemptner, R.L.; Sherwood, J.S.; Tugizimana, F.; Dubery, I.A.; Piater, L.A. Ergosterol, an orphan fungal microbe-associated molecular pattern (MAMP). *Mol. Plant Pathol.* **2014**, *15*, 747–761. [CrossRef] [PubMed]

12. Granado, J.; Felix, G.; Boller, T. Perception of fungal sterols in plants. *Plant Physiol.* **1995**, *107*, 485–490. [CrossRef] [PubMed]

13. Mannock, D.A.; Lewis, R.N.A.H.; McElhaney, R.N. A calorimetric and spectroscopic comparison of the effects of ergosterol and cholesterol on the thermotropic phase behavior and organization of dipalmitoylphosphatidylcholine bilayer membranes. *Biochim. Biophys. Acta* **2010**, *1798*, 376–388. [CrossRef] [PubMed]

14. Sanabria, N.M.; Huang, J.-C.; Dubery, I.A. Self/nonself perception in plants in innate immunity and defense. *Self/Nonself* **2010**, *1*, 40–54. [CrossRef] [PubMed]

15. Kasparovsky, T.; Milat, M.-L.; Humbert, C.; Blein, J.-P.; Havel, L.; Mikes, V. Elicitation of tobacco cells with ergosterol activates a signal pathway including mobilization of internal calcium. *Plant Physiol. Biochem.* **2003**, *41*, 495–501. [CrossRef]

16. Lochman, J.; Mikes, V. Ergosterol treatment leads to the expression of a specific set of defence-related genes in tobacco. *Plant Mol. Biol.* **2006**, *62*, 43–51. [CrossRef] [PubMed]

17. Rossard, S.; Luini, E.; Pérault, J.-M.; Bonmort, J.; Roblin, G. Early changes in membrane permeability, production of oxidative burst and modification of PAL activity induced by ergosterol in cotyledons of *Mimosa pudica*. *J. Exp. Bot.* **2006**, *57*, 1245–1252. [CrossRef] [PubMed]

18. Rossard, S.; Roblin, G.; Atanassova, R. Ergosterol triggers characteristic elicitation steps in *Beta vulgaris* leaf tissues. *J. Exp. Bot.* **2010**, *61*, 1807–1816. [CrossRef]

19. Gaulin, E.; Bottin, A.; Dumas, B. Sterol biosynthesis in oomycete pathogens. *Plant Signal. Behav.* **2010**, *5*, 258–260. [CrossRef]

20. Keinath, N.F.; Kierszniowska, S.; Lorek, J.; Bourdais, G.; Kessler, S.A.; Shimosato-Asano, H.; Grossniklaus, U.; Schulze, W.X.; Robatzek, S.; Panstruga, R. PAMP (pathogen-associated molecular pattern)-induced changes in plasma membrane compartmentalization reveal novel components of plant immunity. *J. Biol. Chem.* **2010**, *285*, 39140–39149. [CrossRef]

21. Vatsa, P.; Chiltz, A.; Luini, E.; Vandelle, E.; Pugin, A.; Roblin, G. Cytosolic calcium rises and related events in ergosterol-treated Nicotiana cells. *Plant Physiol. Biochem.* **2011**, *49*, 764–773. [CrossRef]

22. Tugizimana, F.; Steenkamp, P.A.; Piater, L.A.; Dubery, I.A. Ergosterol-induced sesquiterpenoid synthesis in tobacco cells. *Molecules* **2012**, *17*, 1698–1715. [CrossRef]

23. Dohnal, V.; Jezkova, A.; Pavlikova, L.; Musilek, K.; Jun, D.; Kuca, K. Fluctuation in the ergosterol and deoxynivalenol content in barley and malt during malting process. *Anal. Bioanal. Chem.* **2010**, *397*, 109–114. [CrossRef] [PubMed]

24. Amborabé, B.-E.; Rossard, S.; Pérault, J.-M.; Roblin, G. Specific perception of ergosterol by plant cells. *C. R. Biol.* **2003**, *326*, 363–370. [CrossRef]

25. Xu, X.; Bittman, R.; Duportail, G.; Heissler, D.; Vilcheze, C.; London, E. Effect of the structure of natural sterols and sphingolipids on the formation of ordered sphingolipid/sterol domains (rafts): Comparison of cholesterol to plant, fungal, and disease-associated sterols and comparison of sphingomyelin, cerebrosides and ceramide. *J. Biol. Chem.* **2001**, *276*, 33540–33546. [CrossRef] [PubMed]

26. Giannini, J.L.; Ruiz-Cristin, J.; Briskin, D.P. A small scale procedure for the isolation of transport competent vesicles from plant tissues. *Anal. Biochem.* **1988**, *174*, 561–567. [CrossRef]

27. Tsolis, K.C.; Economou, A. Quantitative proteomics of the *E. coli* membranome. *Methods Enzymol.* **2017**, *586*, 15–36. [CrossRef] [PubMed]

28. Fischer, H.; Polikarpov, I.; Craievich, A.F. Average protein density is a molecular-weight-dependent function. *Protein Sci.* **2004**, *13*, 2825–2828. [CrossRef]

29. Bern, M.; Kil, Y.J. Comments on "unbiased statistical analysis for multi-stage proteomic search strategies. *J. Proteome Res.* **2011**, *10*, 2123–2127. [CrossRef]

30. Santoni, V.; Doumas, P.; Rouquié, D.; Mansion, M.; Rabilloud, T.; Rossignol, M. Large scale characterization of plant plasma membrane proteins. *Biochimie* **1999**, *81*, 655–661. [CrossRef]

31. Alexandersson, E.; Saalbach, G.; Christer Larsson, C.; Kjellbom, P. *Arabidopsis* plasma membrane proteomics identifies components of transport, signal transduction and membrane trafficking. *Plant Cell Physiol.* **2004**, *45*, 1543–1556. [CrossRef] [PubMed]

32. Ephritikhine, G.; Ferro, M.; Rolland, N. Plant membrane proteomics. *Plant Physiol. Biochem.* **2004**, *42*, 943–962. [CrossRef]

33. He, P.; Shan, L.; Lin, N.-C.; Martin, G.B.; Kemmerling, B.; Nürnberger, T.; Sheen, J. Specific bacterial suppressors of MAMP signaling upstream of MAPKKK in *Arabidopsis* innate immunity. *Cell* **2006**, *125*, 563–575. [CrossRef] [PubMed]

34. Liu, C.; Fukumoto, T.; Matsumoto, T.; Gena, P.; Frascaria, D.; Kaneko, T.; Katsuhara, M.; Zhong, S.; Sun, X.; Zhu, Y.; et al. Aquaporin OsPIP$_1$ promotes rice salt resistance and seed germination. *Plant Physiol. Biochem.* **2013**, *63*, 151–158. [CrossRef]

35. Kang, J.; Park, J.; Choi, H.; Burla, B.; Kretzschmar, T.; Lee, Y.; Martinoia, E. Plant ABC transporters. *Arabidopsis Book* **2011**, *9*, e0153. [CrossRef]

36. Ji, H.; Peng, Y.; Meckes, N.; Allen, S.; Stewart, C.N., Jr.; Traw, M.B. ATP-dependent binding cassette transporter G family member 16 increases plant tolerance to abscisic acid and assists in basal resistance against *Pseudomonas syringae* DC3000. *Plant Physiol.* **2014**, *166*, 879–888. [CrossRef] [PubMed]

37. Zhang, J.; Ma, H.; Feng, J.; Zeng, L.; Wang, Z.; Chen, S. Grape berry plasma membrane proteome analysis and differential expression during ripening. *J. Exp. Bot.* **2008**, *59*, 2979–2990. [CrossRef] [PubMed]

38. Nouri, M.-Z.; Komatsu, S. Comparative analysis of soybean plasma membrane proteins under osmotic stress using gel-based and LC MS/MS-based proteomics approaches. *Proteomics* **2010**, *10*, 1930–1945. [CrossRef]

39. Vilakazi, C.S.; Dubery, I.A.; Piater, L.A. Identification of lipopolysaccharide-interacting plasma membrane-type proteins *in Arabidopsis thaliana*. *Plant Physiol. Biochem.* **2017**, *111*, 155–165. [CrossRef]

40. Wang, C.; Yan, X.; Chen, Q.; Jiang, N.; Fu, W.; Ma, B.; Liu, J.; Li, C.; Bednarek, S.Y.; Pan, J. Clathrin light chains regulate clathrin-mediated trafficking, auxin signaling, and development in *Arabidopsis*. *Plant Cell* **2013**, *25*, 499–516. [CrossRef]

41. Mgcina, L.S.; Dubery, I.A.; Piater, L.A. Comparative conventional- and quantum dot-labeling strategies for LPS binding sites detection in *Arabidopsis thaliana* mesophyll protoplasts. *Front. Plant Sci.* **2015**, *6*, 335. [CrossRef] [PubMed]

42. Pandey, D.; Rajendran, S.R.C.K.; Gaur, M.; Sajeesh, P.K.; Kumar, A. Plant defense signaling and response against necrotrophic fungal pathogens. *J. Plant Growth Regul.* **2016**, *35*, 1159–1174. [CrossRef]

43. Borner, G.H.H.; Liley, K.S.; Stevens, T.J.; Dupree, P. Identification of glycosylphosphatidylinositol-anchored proteins in *Arabidopsis*. A proteomic and genomic analysis. *Plant Physiol.* **2003**, *132*, 568–577. [CrossRef]

44. Peskan, T.; Westermann, M.; Oelmüller, R. Identification of low-density Triton X-100-insoluble plasma membrane microdomains in higher plants. *Eur. J. Biochem.* **2000**, *267*, 6989–6995. [CrossRef] [PubMed]

45. Borner, G.H.H.; Sherrier, D.J.; Stevens, T.J.; Arkin, I.T.; Dupree, P. Prediction of glycosylphosphatidylinositol-anchored proteins in *Arabidopsis*. A genomic analysis. *Plant Physiol.* **2002**, *129*, 489–499. [CrossRef]

46. Jarsch, I.K.; Ott, T. Perspectives on remorin proteins, membrane rafts, and their role during plant-microbe interactions. *Mol. Plant Microbe Interact.* **2011**, *24*, 7–12. [CrossRef] [PubMed]

47. Ikonen, E. Roles of lipid rafts in membrane transport. *Curr. Opin. Cell Biol.* **2001**, *13*, 470–477. [CrossRef]

48. Sharma, M.; Pandey, G.K. Expansion and function of repeat domain proteins during stress and development in plants. *Front. Plant Sci.* **2015**, *6*, 1218. [CrossRef] [PubMed]

49. Monaghan, J.; Zipfel, C. Plant pattern recognition receptor complexes at the plasma membrane. *Curr. Opin. Cell Biol.* **2012**, *15*, 349–357. [CrossRef]

50. Tang, D.; Wang, G.; Zhou, J.-M. Receptor kinases in plant-pathogen interactions: More than pattern recognition. *Plant Cell* **2017**, *29*, 618–637. [CrossRef] [PubMed]

51. Sanabria, N.M.; van Heerden, H.; Dubery, I.A. Molecular characterization and regulation of a *Nicotiana tabacum* S-domain receptor-like kinase gene induced during an early rapid response to lipopolysaccharides. *Gene* **2012**, *501*, 39–48. [CrossRef] [PubMed]

52. Baloyi, N.M.; Dubery, I.A.; Piater, L.A. Proteomic analysis of Arabidopsis plasma membranes reveals lipopolysaccharide-responsive changes. *Biochem. Biophys. Res. Commun.* **2017**, *486*, 1137–1142. [CrossRef] [PubMed]

53. Esch, L.; Schaffrath, U. An update on jacalin-related lectins and their role in plant defense. *Int. J. Mol. Sci.* **2017**, *18*, 1592. [CrossRef]

54. Keller, C.K.; Radwan, O. The functional role of 14-3-3 proteins in plant stress interactions. *i-ACES* **2015**, *1*, 100–110.

55. Hayafune, M.; Berisio, R.; Marchetti, R.; Silipo, A.; Kayama, M.; Desaki, Y.; Arima, S.; Squeglia, F.; Ruggiero, A.; Tokuyasu, K.; et al. Chitin-induced activation of immune signaling by the rice receptor CEBiP relies on a unique sandwich-type dimerization. *Proc. Natl. Acad. Sci. USA* **2014**, *111*, E404–E413. [CrossRef]

56. Simanshu, D.K.; Zhai, X.; Munch, D.; Hofius, D.; Markham, J.E.; Bielawski, J.; Bielawska, A.; Malinina, L.; Molotkovsky, J.G.; Mundy, J.W.; et al. *Arabidopsis* accelerated-cell-death11, ACD11, is a ceramide-1-phosphate transfer protein and intermediary regulator of phytoceramide levels. *Cell Rep.* **2014**, *6*, 388–399. [CrossRef]

57. Liang, H.; Yao, N.; Song, J.T.; Luo, S.; Lu, H.; Greenberg, J.T. Ceramides modulate programmed cell death in plants. *Genes Dev.* **2003**, *17*, 2636–2641. [CrossRef]

58. Varet, A.; Hause, B.; Hause, G.; Scheel, D.; Lee, J. The *Arabidopsis* NHL3 gene encodes a plasma membrane protein and its overexpression correlates with increased resistance to *Pseudomonas syringae* pv. tomato DC3000. *Plant Physiol.* **2003**, *132*, 2023–2033. [CrossRef]

59. Elmore, J.M.; Liu, J.; Smith, B.; Phinney, B.; Coaker, G. Quantitative proteomics reveals dynamic changes in the plasma membrane during *Arabidopsis* immune signaling. *Mol. Cell. Proteom.* **2012**, *11*, M111–014555. [CrossRef]

60. Raffaele, S.; Bayer, E.; Lafarge, D.; Cluzet, S.; Retana, S.G.; Boubekeur, T.; Leborgne-Castel, N.; Carde, J.-P.; Lherminier, J.; Noirot, E.; et al. Remorin, a *Solanaceae* protein in membrane rafts and plasmodesmata, impairs *Potato virus X* movement. *Plant Cell* **2009**, *21*, 1541–1555. [CrossRef]

61. Marin, M.; Ott, T. Phosphorylation of intrinsically disordered regions in remorin proteins. *Front. Plant Sci.* **2012**, *3*, 86. [CrossRef]

62. Zou, B.; Hong, X.; Ding, Y.; Wang, X.; Liu, H.; Hua, J. Identification and analysis of copine/BONZAI proteins among evolutionarily diverse plant series. *Genome* **2016**, *59*, 565–573. [CrossRef]

63. Zhou, L.; Cheung, M.Y.; Li, M.-W.; Fu, Y.; Sun, Z.; Sun, S.-M.; Lam, H.-M. Rice hypersensitive induced reaction protein 1 (OsHIR1) associates with plasma membrane and triggers hypersensitive cell death. *BMC Plant Biol.* **2010**, *10*, 1–10. [CrossRef]

64. Mohr, T.J.; Mammarelia, N.D.; Hoff, T.; Woffenden, B.J.; Jelesko, J.G.; McDowell, J.M. The *Arabidopsis* downy mildew resistance gene RPP8 is induced by pathogens and salicylic acid and is regulated by W box *cis* elements. *Mol. Plant Microbe Interact.* **2010**, *23*, 1303–1315. [CrossRef]

65. Fairbanks, G.; Steck, T.L.; Wallach, D.F.H. Electrophoretic analysis of the major polypeptides of the human erythrocyte membrane. *Biochemistry* **1971**, *10*, 2606–2617. [CrossRef]

66. Jordaan, J.; Simpson, C.; Brady, D.; Gardiner, N.S.; Gerber, I.B. Emulsion-Derived Particles. U.S. Patent 2013149730 A1, 13 June 2013.

67. Tejada-Simon, M.V.; Pestka, J.J. Production of polyclonal antibody against ergosterol hemisuccinate using Freund's and Titermax adjuvants. *J. Food Protect.* **1998**, *61*, 1060–1063. [CrossRef]
68. Switzer, R.C., III; Merril, C.R.; Shifrin, S. A highly sensitive silver stain for detecting proteins and peptides in polyacrylamide gels. *Anal. Biochem.* **1979**, *98*, 231–237. [CrossRef]
69. Blum, H.; Beier, H.; Gross, H.J. Improved silver staining of plant proteins, RNA and DNA in polyacrylamide gels. *Electrophoresis* **1987**, *8*, 93–99. [CrossRef]
70. UniProt Knowledgebase (UniprotKB) Database. Available online: https://www.uniprot.org/ (accessed on 2017–2018).

International Journal of
Molecular Sciences

MDPI

Article

Application of Data-Independent Acquisition Approach to Study the Proteome Change from Early to Later Phases of Tomato Pathogenesis Responses

Kai-Ting Fan [1], Kuo-Hsin Wang [1], Wei-Hung Chang [1], Jhih-Ci Yang [2,3], Ching-Fang Yeh [1], Kai-Tan Cheng [1], Sheng-Chi Hung [3,4] and Yet-Ran Chen [1,2,3,4,*]

[1]   Agricultural Biotechnology Research Center, Academia Sinica, Taipei 11529, Taiwan;
      kaitingfan@sinica.edu.tw (K.-T.F.); khwang@gate.sinica.edu.tw (K.-H.W.);
      whchang@gate.sinica.edu.tw (W.-H.C.); cfyeh@gate.sinica.edu.tw (C.-F.Y.);
      ktc77123@gate.sinica.edu.tw (K.-T.C.)
[2]   Sustainable Chemical Science and Technology, Taiwan International Graduate Program,
      Institute of Chemistry, Academia Sinica, Taipei 11529, Taiwan; jhihciy@gate.sinica.edu.tw
[3]   Sustainable Chemical Science and Technology, Taiwan International Graduate Program,
      Department of Applied Chemistry, National Chiao Tung University, Hsinchu 30010, Taiwan;
      kempis710165@gate.sinica.edu.tw
[4]   Institute of Biotechnology, National Taiwan University, Taipei 10617, Taiwan
*   Correspondence: yetran@gate.sinica.edu.tw; Tel.: +886-02-2707-2050

Received: 27 December 2018; Accepted: 12 February 2019; Published: 17 February 2019

Abstract: Plants and pathogens are entangled in a continual arms race. Plants have evolved dynamic defence and immune mechanisms to resist infection and enhance immunity for second wave attacks from the same or different types of pathogenic species. In addition to evolutionarily and physiological changes, plant-pathogen interaction is also highly dynamic at the molecular level. Recently, an emerging quantitative mass spectrometry-based proteomics approach named data-independent acquisition (DIA), has been developed for the analysis of the proteome in a high-throughput fashion. In this study, the DIA approach was applied to quantitatively trace the change in the plant proteome from the early to the later stage of pathogenesis progression. This study revealed that at the early stage of the pathogenesis response, proteins directly related to the chaperon were regulated for the defence proteins. At the later stage, not only the defence proteins but also a set of the pathogen-associated molecular pattern-triggered immunity (PTI) and effector triggered immunity (ETI)-related proteins were highly induced. Our findings show the dynamics of the plant regulation of pathogenesis at the protein level and demonstrate the potential of using the DIA approach for tracing the dynamics of the plant proteome during pathogenesis responses.

Keywords: plant pathogenesis responses; data-independent acquisition; quantitative proteomics; *Pseudomonas syringae*

---

1. Introduction

Plants are sessile organisms that are in close contact with a variety of organisms, including pathogens and have thus evolved an efficient innate immune system to defend themselves from those biotic stresses. The typical passive defence of plants actually starts from physical barriers, including the trichomes, waxy cuticle and cell wall [1]. On the other hand, successful active defence responses are initiated from the ability of plants to sense pathogens using extracellular and intracellular innate immune receptors, to induce subsequent cellular reprogramming for defence. Plants have evolved to express receptors that recognize conserved pathogen-associated molecular patterns (PAMPs) or microbial associated molecular patterns (MAMPs) such as FLS2 receptor that perceives bacterial flagellin via the minimal epitope flg22, one of conserved pathogen molecules essential for its reproduction [2]. Another type of plant sensing ability is initiated by endogenous damaged-associated molecular patterns (DAMPs) which are molecules produced by cell-damage or necrosis caused by pathogen invasion or herbivore attack, including fragments of cell wall structure, signalling peptides/polypeptides from cleaved precursor proteins like systemin [3], PLANT ELICITOR PEPTIDES (PEPs) [4] and CAP-derived peptide 1 (CAPE1) [5] and extracellular molecules like nucleotides [6]. Currently known DAMPs have been demonstrated to be able to induce similar innate immune responses in plants as microbe-derived PAMPs/MAMPs and it has been proposed that DAMPs could be used to amplify the responses triggered by PAMPs [7].

After perceiving the danger signals of pathogen attacks, plants initiate a series of defence responses. The first one, often thought as the basal pathogen resistance of plants, is triggered by the binding of PAMPs/MAMPs by the plasma membrane-localized receptor (pattern recognition receptors; PRRs) and is therefore called PAMP-triggered immunity (PTI) [8]. PTI includes the increased ion influx, a burst of reactive oxygen species, activation of the mitogen-activated protein kinase (MAPK) cascade and increased level of defence phytohormones like salicylic acid (SA), jasmonic acid (JA) or ethylene (ET). However, pathogens that have become successfully adapted though evolution can suppress or bypass the PTI responses by injecting proteins—termed effectors through a type III secretion system to the apoplast or cellular region of the host, resulting in effector-triggered susceptibility (ETS) [9,10]. To counter-attack ETS, plant species have also evolved another kind of defence response, effector-triggered immunity (ETI), which uses intracellular or transmembrane receptors (R proteins) to specifically target the effector proteins of pathogens, resulting in much stronger resistance responses, usually leading to hypersensitive response (HR) cell death. A "zig-zag" model has been proposed to interpret how/why the plant immune system is made up of this complex, multi-layered innate immune system of PTI and ETI responses through evolutionary development [10]. Although from the transcriptomics data, a significant number of genes could both be regulated by PTI and ETI, the later causes faster and greater amplitude of induction, leading scientists to speculate that PTI and ETI could result in synergistic effects [11–13]. However, when examining specific defence-related phytohormones (SA, JA and ET) in Arabidopsis, it has been shown that in PTI there are more evident synergistic relationships; while in ETI there are more compensatory relationships among the signalling sectors [14–16]. There are still gaps in understanding how these sophisticated mechanisms are regulated such as how PTI and ETI together form an effective defence network, since a large number of the transcripts involved in the regulation of immune and defence response are known to be time-dependent [17]. This is also related to the fact that regulation at the protein level should also be time dependent and highly dynamic.

The isobaric labelling approach has been used to study the dynamics of proteome regulation in plants during pathogenesis responses. This approach has been used to trace the change in the proteome of tomato in response to infection with *Pseudomonas syringae* pv. *tomato* (*Pst*) DC3000 at early and late time points. Using a combination of strong off-line cation exchange chromatography and liquid chromatography coupled with mass spectrometry (LC-MS), a total of ~2300 proteins were identified [18]. Although the isobaric approach is a promising technology to perform multiplex quantitative proteomics analysis, it is often required to fractionate the sample to minimize the effect of ratio compression [19]. The use of an additional fractionation step will require a higher quantity of sample to compensate for sample loss during the fractionation step and reduce the throughput of the proteome analysis. This limits its application to the analysis of a large quantity of the proteomics samples. Recently, a new MS analysis approach called data independent acquisition (DIA) was proposed to solve this issue. The DIA approach is considered a promising approach for performing quantitative proteome profiling in a high-throughput manner. The DIA approach is based on the acquisition of fragment-ion information for all precursor ions within a certain range of m/z values, as demonstrated by the sequential window acquisition of all theoretical mass spectra (SWATH) approach [20]. It has been demonstrated that the application of the DIA approach was able to identify and quantify thousands of proteins without performing fractionation and only a few micrograms of the protein sample is required [20]. In the DIA approach, selected reaction monitoring (SRM)-like extracted ion chromatography (XIC) on sequence specific ion transitions can be used for the identification and SRM-like quantification of the peptides after the data acquisition. To demonstrate this technology in the study of the plant pathogenesis responses, here we used the tomato proteome regulated by infection of *Pst* DC3000. *P. syringae*, a hemi-biotroph bacteria, is one of the most studied bacterial pathogens due to its ability to infect a great variety of plant species including reference plants like Arabidopsis and tobacco and crops like tomato and potato. *P. syringae* relies on effector proteins belonging to the bacterial type III secretion system to suppress the plant defence system thus achieving pathogenesis. More than 40 *P. syringae* effectors have been identified with their host targets in plants, demonstrating complex plant-pathogen interactions which makes *P. syringae* an important model system for examining the molecular mechanisms of plant pathogen defence [21]. We quantitatively profiled the total tomato proteomes regulated by the infection of *Pst* DC3000 from early to later time points using the LC-MS/MS operated in DIA mode. Using this approach, the change in the proteomes of mock and *Pst* DC3000-inoculated samples were analysed to identify the proteins with significant change in abundance in the early to the later stage of the pathogenesis responses. Without the hideous labour requirement for sample preparation that is required for peptide fractionation, the one-shot sample analysis using the DIA approach should provide researchers an efficient way to identify not only well-known but also potential protein markers for further biological studies.

2. Results and Discussion

2.1. Experimental Design and the Identification and Quantification Result

To mimic natural bacterial infection of tomato leaves without causing physical damage in cells, we optimized the dipping method without using the vacuum infiltration. *Pst* DC3000 infection symptoms and bacterial growth were recorded until 7 days post-inoculation (dpi) (Figure 1, Supplementary Figure S1). The growth of *Pst* DC3000 started 4 h post-inoculation (hpi) and dramatically increased at 3 dpi (Figure 1B) although the disease phenotype became more obvious at 3, 5 or 7 dpi.

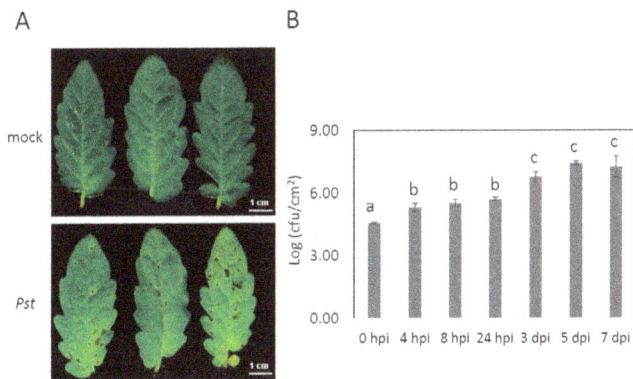

Figure 1. Disease phenotypes and population dynamics of *Pst* DC3000 on tomato leaves. (**A**) Visible disease symptoms in tomato leaves after leaves were dipped into bacterial suspension at an OD_{600} of 0.02. Photos were taken at 5 days post-inoculation. (**B**) Bacteria titres in leaves at 0, 4, 8, 24 h or 3, 5, 7 days after inoculation with *Pst* DC3000. Error bars represent the standard deviation of three biological replicates. A statistical significance difference is shown between different time points using analysis of variance (ANOVA) with post-hoc Tukey's honestly significant difference (HSD) test by labelling bars with different lowercase letter (a, b, and c).

Mysore et al. [22] have examined the transcriptomic change in tomato upon 0 to 8 hpi of *Pst* and showed that the majority of defence-related transcriptional responses were significantly activated at 8 hpi. In addition, Parker et al. have examined the proteome change of tomato after 4 and 24 h inoculation of *Pst* using isobaric tags for relative and absolute quantitation (iTRAQ) approach and the 24 hpi of *Pst* was considered as the later time point for detecting the ETI responses [18]. In this study, in order to cover early to the later phases of the pathogenesis responses in protein level, 4, 8 and 24 hpi of *Pst* DC3000 were selected for the proteomics analysis. Three different pooled tryptic peptide samples, each combining *Pst* DC3000-inoculated and mock-treated samples from the same time point, were used in the data-dependent acquisition (DDA) method and results were searched from 3 different programs, Mascot, X!tandem and Comet, which were then merged to construct the spectral library. In total, 2174 proteins were identified by DIA analysis across the 3 time points of 4, 8 and 24 hpi and the mock group, at 4, 8, 24 h-post treatment (hpt). Of 1472 proteins being quantified in 3 biological replicates, 114, 147 and 337 proteins had a change in quantity between inoculation and mock at 4, 8, 24 h time points, respectively using the Student's t-test (Supplementary Tables S1 and S2). Among those, 20, 65 and 189 proteins had significant change in abundance with fold change greater than 1.5 or less than 0.67 at 4, 8 and 24 hpi, respectively, compared to the mock ($p < 0.05$). Comparing to the previous protein quantitation results using iTRAQ approach, even though > 1.2- or < 0.83-fold change with *p*-value < 0.05 was used to identify differentially expressed proteins, there were 128 tomato proteins found to be regulated at 24 hpi of *Pst* DC3000 [18]. In this study, we identified 189 regulated proteins at 24 hpi using the protein fold change >1.5 or <0.67 with *p*-value < 0.05 as the criteria. The higher number of regulated proteins identified in this study as compared to previous iTRAQ study at 24 hpi may possibly due to different experimental design or the ratio compression issue of the use of isobaric tagging proteomics approach.

The majority of proteins (~90%) were identified in all 3 time points and only about 1 to 2% of the proteins were identified at one time point (Figure 2A). A similar situation was observed with the quantified proteins, most of which were shared between all the time points. However, there were no proteins with significant up- (>1.5-fold change) or down- (<0.67-fold change) regulation found at all the time points and only ~8% of up-regulated or ~24% of down-regulated proteins were shared between the 8 and 24 h time points. There were also more proteins with significant change, either up- or down-regulated, at 24 hpi than 8 or 4 hpi (Figure 2B), indicating that between 4, 8 and 24 hpi, a different set of proteins were regulated in tomato leaves. More proteins showed a change in quantity at both 8 and 24 hpi, than between 4 and 24 or 4 and 8 hpi.

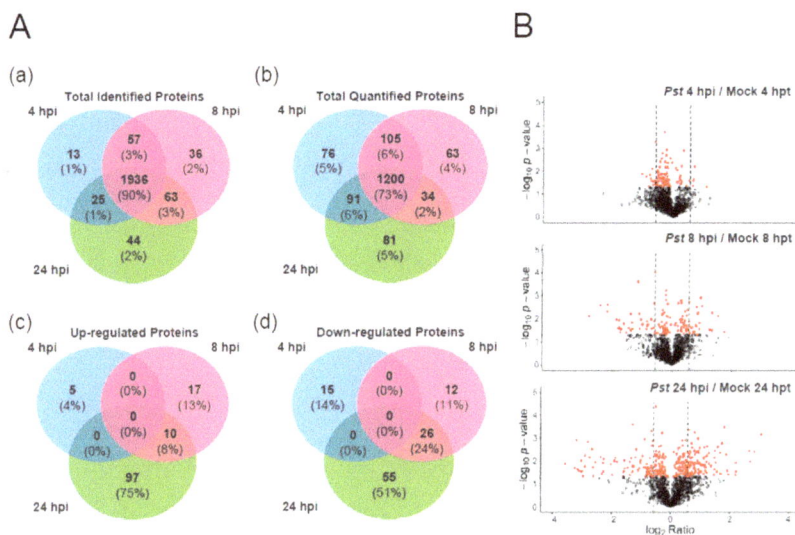

Figure 2. Proteomic change in tomato leaf between time points of *Pst* DC3000 inoculation 4, 8 or 24 hpi compared to each mock: 4, 8 or 24 hpt, respectively. (**A**) Venn diagrams showing unique and shared proteins between 3 time points which are (a) identified across replicates and (b) quantified in all 3 biological replicates or with significant increase, with fold change (c) greater than 1.5 or (d) less than 0.67 in quantity ($p < 0.05$), due to inoculation. (**B**) Volcano plots showing protein abundance ratio of *Pst* DC3000-inoculated over mock group at 4 hpi, 8 hpi and 24 hpi. Following LC-MS analysis and DIA quantification, *t* test-based significance values ($\log_{10}$ (*p*-value)) were plotted versus $\log_2$ (protein quantity ratio for all proteins between infected and mock). Differentially regulated proteins with $p < 0.05$ are plotted in red. Proteins with $p > 0.05$ are plotted in black. The level of protein abundance change with 1.5 or 0.67-fold is marked by a dashed line. The data of quantified proteins in 3 biological replicates were listed in Supplementary Table 1 while the data of each replicate were listed in Supplementary Table 2.

2.2. Functional Classification of the Proteins Regulated by Pst DC3000

2.2.1. Function Categories of Proteins That are Significantly up- or down-Regulated during Pst DC3000 Inoculation

Because there are more thorough gene annotations in the Arabidopsis Information Resource (TAIR) database, using Protein Basic Local Alignment Search Tool (BLASTP), Arabidopsis homolog proteins were identified in TAIR from the tomato proteins that had a significant change between the inoculated and mock group. Regulated proteins with >1.5 or <0.67-fold change were categorized by protein function (Figure 3). More unique function categories were identified for the up-regulated proteins than the down-regulated proteins, including "response to other organism," "response to external stimulus," "response to biotic stimulus," "organic substance metabolic process," "multi-organism reproductive process," "immune response" and "developmental process involved in reproduction." This suggests that the proteins in these categories could be positive regulators in pathogen defence and immune responses. On the other hand, only one category, "response to endogenous stimulus," was identified in the down-regulated proteins.

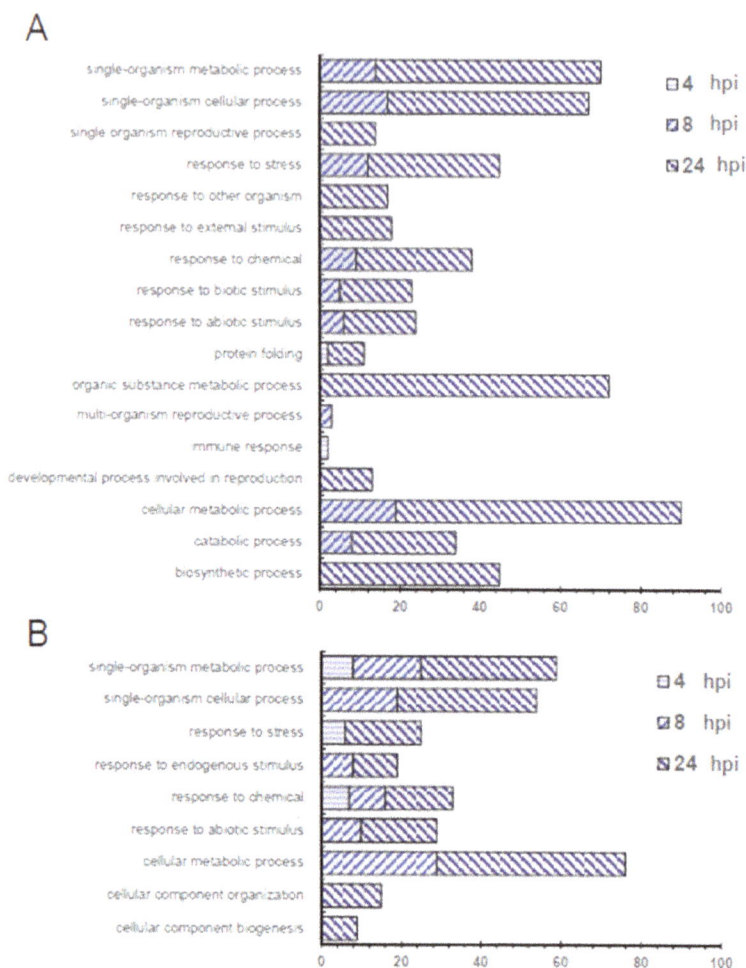

Figure 3. Biological process function analysis for tomato leaf proteins with a (**A**) significant increase or (**B**) decrease in quantity at different infection time points, 4, 8 or 24 hpi compared to the mock group. Only proteins with fold change greater than 1.5 or less than 0.67 ($p < 0.05$) were analysed and used for searching the Arabidopsis homolog proteins by Protein Basic Local Alignment Search Tool (BLASTP) against the Arabidopsis Information Resource (TAIR) database. Gene Ontology (GO) categorization of these Arabidopsis homolog proteins was performed by using the Database for Annotation, Visualization and Integrated Discovery (DAVID) v6.8.

2.2.2. Proteins Related to the Key Mechanisms of Pathogenesis

About 70 proteins with significant change in protein abundance upon Pst DC3000 inoculation across 4, 8 and 24 hpi were selected and grouped with their protein function category (Tables 1–3). Table 1 summarizes the proteins involved in defence, immune response and ROS (reactive oxygen species)/redox metabolism. Proteins involved in these categories mostly were up-regulated at 8 and/or 24 hpi, except zeaxanthin epoxidase (ZEP; Solyc02g090890) and one peroxisome (Solyc09g072700). Interestingly, none of these were up-regulated at the early time point 4 hpi. Several pathogenesis-responsive proteins with known ROS/redox function were found to be differentially expressed in our results. However, these proteins did not show significant change in quantities in either the Pst-resistant (PtoR) or susceptible (Prf3) tomato genotypes at 4 hpi and 24 hpi in the iTRAQ analysis [18], suggesting our findings could help researchers identify more protein candidates in response to pathogenesis. Table 2 summarizes proteins participating in protein translation, protein folding, degradation and transportation. Most proteins with function in translation and folding were up-regulated at 24 hpi. However, Heat shock proteins 90 (HSP90) and calnexin were the two early up-regulated proteins at 4 hpi. The protein product of Solyc04g080960, pre-pro-cysteine proteinase, also named as RD19A-like and DNA damage-inducible protein 1 (Solyc10g005890) were the ones with down-regulated protein level. DNA damage-inducible protein 1 belongs to a family of shuttle proteins targeting polyubiquitinated substrates for proteasomal degradation. Table 3 summarizes proteins involved in carbohydrate and energy metabolisms, including glycolysis/tricarboxylic acid (TCA) cycle, pentose phosphate pathway and carbon fixation (photosynthesis). Generally speaking, proteins involved in energy-generation, such as glycolysis and the TCA cycle, were up-regulated at 8 and 24 hpi; while proteins involved in photosynthesis and ATP synthesis were all down-regulated. Our data indicate that within the first 24 h of pathogenesis progression in plants, the plant required more energy for defence response and sacrificed the carbon fixation process. Table 4 summarizes proteins involved in oxidation phosphorylation, amino acid, fatty acid and secondary metabolisms like polyamine synthesis and shikimate pathway. Proteins grouped in amino acid metabolisms were up-regulated at 24 hpi and the downstream compounds of these metabolic pathways are involved in the stress response, like polyamine, flavonoids and aromatic amino acid and phytohormones.

Table 1. Proteins involved in defence, immunity and ROS/redox mechanism with significant change in abundance at 4, 8, 24, hpi of *Pst* DC3000.

Protein Description	Gene Accession	4 hpi		8 hpi		24 hpi	
		log$_2$ Ratio [a]	*p*-Value [b]	log$_2$ Ratio [a]	*p*-Value [b]	log$_2$ Ratio [a]	*p*-Value [b]
Defence							
Pathogenesis-related protein 4 (PR-4)	Solyc01g097240	−0.183	0.496	0.885	0.049	NQ	NQ
PIN-I protein (PR-6)	Solyc09g083440	−0.023	0.971	−1.163	0.052	1.484	0.004
Wound-induced proteinase inhibitor 1	Solyc09g084465	NQ	NQ	NQ	NQ	1.183	0.019
Major allergen Pru ar 1 (PR-STH 2)	Solyc09g090970	0.239	0.038	0.827	0.003	1.106	0.036
Biotin-binding protein	Solyc09g065540	0.067	0.695	0.093	0.794	0.615	0.048
Chitinase Z15140	Solyc10g055810	−0.195	0.487	0.869	0.006	0.101	0.907
Beta-1,3-glucanase (PR-2)	Solyc01g008620	0.123	0.669	0.291	0.290	1.111	0.008
Immune Regulation							
Pathogenesis-related protein 1 (PR-1)	Solyc00g174340	−0.182	0.153	0.949	0.052	3.059	0.001
1-aminocyclopropane-1-carboxylate oxidase 1	Solyc07g049530	0.132	0.705	NQ	NQ	2.831	0.004
Arginase 2 (ARG2)	Solyc01g091170	0.407	0.356	0.289	0.466	2.691	0.011
Cathepsin D Inhibitor	Solyc03g098780	NQ	NQ	NQ	NQ	1.947	0.040
Activator of 90 kDa heat shock ATPase	Solyc10g078930	0.204	0.442	0.544	0.065	0.936	0.005
FKBP-like peptidyl-prolyl cis-trans isomerase family protein	Solyc09g008650	0.084	0.827	0.851	0.006	0.925	0.035
Kunitz trypsin inhibitor	Solyc03g098730	−0.355	0.187	NQ	NQ	0.714	0.046
Zeaxanthin epoxidase (ZEP)	Solyc02g090890	0.112	0.353	−0.799	0.067	−0.711	0.047
ROS/Redox							
Glutathione S-transferase/peroxidase	Solyc01g056480	0.041	0.814	NQ	NQ	1.870	0.005
Peroxidase	Solyc09g072700	−0.345	0.089	−0.500	0.336	−0.814	0.029
Thioredoxin reductase	Solyc02g082250	0.064	0.930	0.813	0.022	NQ	NQ
Peroxidase	Solyc04g071900	NQ	NQ	0.906	0.016	NQ	NQ
Glutathione Reductase (GR)	Solyc09g091840	0.476	0.086	0.368	0.290	0.902	0.010
Glutathione S-transferase-like protein	Solyc09g011570	0.158	0.430	0.765	0.038	0.812	0.001
Glutathione S-transferase	Solyc06g009040	0.201	0.401	0.865	0.049	0.429	0.210

[a] The average log$_2$ ratio of protein quantity representing (inoculated/mock) from 3 biological replicates. [b] *p*-value calculated from Student t-test. Color-coded: red, significant quantity change greater than 0.58 of log$_2$ ratio; blue, significant quantity change less than −0.58 of log$_2$ ratio; grey, no significant change between the inoculated and mock group ($p \geq 0.05$); white, non-quantifiable (NQ).

Table 2. Proteins involved in translation, protein folding, degradation and transportation with significant change in abundance at 4, 8, 24, hpi of *Pst* DC3000.

Protein Description	Gene Accession	4 hpi		8 hpi		24 hpi	
		log$_2$ Ratio [a]	*p*-Value [b]	log$_2$ Ratio [a]	*p*-Value [b]	log$_2$ Ratio [a]	*p*-Value [b]
Protein Translation, Folding, Degradation and Transportation							
Importin subunit alpha	Solyc01g060470	−0.122	0.473	0.185	0.449	1.138	0.016
Protein transport protein sec23, putative	Solyc05g053830	−0.023	0.894	0.262	0.271	1.012	0.016
Golgin candidate 6	Solyc08g081410	−0.087	0.655	−0.095	0.612	0.680	0.040
Cathepsin B-like cysteine protease	Solyc02g069110	NQ	NQ	NQ	NQ	1.802	0.000
T-complex protein 1 subunit beta	Solyc11g069000	−0.487	0.460	0.101	0.597	1.352	0.017
26S proteasome non-ATPase regulatory subunit 3	Solyc01g111700	0.116	0.811	0.300	0.034	0.960	0.013
DNA damage-inducible protein 1	Solyc01g005890	−0.612	0.038	−0.417	0.298	−0.021	0.862
Pre-pro-cysteine proteinase	Solyc04g080960	0.313	0.193	−0.634	0.025	−0.388	0.316
Chaperonin 60 alpha subunit	Solyc06g075010	−0.128	0.579	−0.080	0.607	1.506	0.012
Heat shock 70 kDa protein, putative	Solyc07g043560	0.233	0.194	0.251	0.298	1.223	0.002
Peptidyl-prolyl cis-trans isomerase	Solyc06g070670	0.211	0.697	0.759	0.500	1.212	0.033
Heat shock protein 90	Solyc06g036290	1.140	0.050	1.475	0.109	0.841	0.050
Calnexin	Solyc03g118040	0.672	0.028	0.256	0.576	−0.067	0.881

[a] The average log$_2$ ratio of protein quantity representing (inoculated/mock) from 3 biological replicates. [b] *p*-value calculated from Student t-test. Color-coded: red, significant quantity change greater than 0.58 of log$_2$ ratio; blue, significant quantity change less than −0.58 of log$_2$ ratio; grey, no significant change between the inoculated and mock group ($p \geq 0.05$); white, non-quantifiable (NQ).

Table 3. Proteins involved in carbohydrate and energy metabolisms with significant change in abundance at 4, 8, 24, hpi of *Pst* DC3000.

Protein Description	Gene Accession	4 hpi		8 hpi		24 hpi	
		log$_2$ Ratio [a]	*p*-Value [b]	log$_2$ Ratio [a]	*p*-Value [b]	log$_2$ Ratio [a]	*p*-Value [b]
Metabolism-Primary-Carbohydrate Metabolisms-Glycolysis & TCA							
Glucose-6-phosphate 1-dehydrogenase	Solyc02g093830	0.289	0.111	0.752	0.146	0.955	0.013
Pyruvate kinase family protein	Solyc10g083720	−0.104	0.326	0.316	0.208	0.915	0.001
Glyceraldehyde 3-phosphate dehydrogenase	Solyc05g014470	0.219	0.501	0.155	0.400	0.773	0.022
Glyceraldehyde-3-phosphate dehydrogenase	Solyc04g009030	0.258	0.027	0.196	0.208	0.663	0.004
2-oxoglutarate dehydrogenase E1 component family protein	Solyc05g054640	−0.193	0.360	0.490	0.007	0.654	0.033
Fructokinase 2	Solyc06g073190	0.030	0.805	0.254	0.149	0.624	0.019
Metabolism-Primary-Carbohydrate metabolism-PPP							
6-phosphogluconate dehydrogenase, decarboxylating	Solyc04g005160	−0.372	0.392	0.919	0.016	1.221	0.006
Transaldolase	Solyc00g006800	−0.183	0.302	0.159	0.641	0.604	0.039
Metabolism-Primary-Carbon fixation							
Photosystem II oxygen-evolving complex protein 3	Solyc02g079950	−0.739	0.057	−0.694	0.028	−0.250	0.157
Chlororespiratory reduction31	Solyc08g082400	−0.529	0.136	−1.402	0.033	−1.397	0.042
ATP-dependent zinc metalloprotease FTSH protein	Solyc07g055320	0.118	0.646	−1.340	0.029	−2.270	0.008
Protein CURVATURE THYLAKOID 1A, chloroplastic	Solyc10g011770	−0.072	0.814	−1.558	0.040	−2.373	0.007
Cytochrome b6-f complex iron-sulphur subunit	Solyc12g005630	−0.087	0.801	−1.758	0.039	−2.812	0.002
Chlorophyll a-b binding protein, chloroplastic	Solyc07g063600	−0.353	0.601	−1.891	0.011	−2.993	0.047
Photosystem I reaction centre subunit III	Solyc02g069450	0.220	0.625	−1.496	0.079	−3.137	0.026
Chlorophyll a/b-binding protein	Solyc03g005760	−0.540	0.493	−1.788	0.024	−3.159	0.024
Metabolism-Primary-Carbohydrate Metabolism-Others							
Sucrose synthase	Solyc07g042550	0.250	0.297	1.258	0.031	1.542	0.031
Beta-fructofuranosidase	Solyc04g081440	0.261	0.367	NQ	NQ	0.818	0.005
xyloglucan endotransglucosylase-hydrolase 7	Solyc02g091920	0.066	0.888	0.590	0.022	0.573	0.323
Starch synthase, chloroplastic/amyloplastic	Solyc03g083095	−0.134	0.339	−0.601	0.013	−0.231	0.588

[a] The average log$_2$ ratio of protein quantity representing (inoculated/mock) from 3 biological replicates. [b] *p*-value calculated from Student t-test. Color-coded: red, significant quantity change greater than 0.58 of log$_2$ ratio; blue, significant quantity change less than −0.58 of log$_2$ ratio; grey, no significant change between the inoculated and mock group ($p \geq 0.05$); white, non-quantifiable (NQ).

Table 4. Protein involved in other primary and secondary metabolisms with significant change in abundance at 4, 8, 24, hpi of *Pst* DC3000.

Protein Description	Gene Accession	4 hpi		8 hpi		24 hpi	
		log$_2$ Ratio [a]	*p*-Value [b]	log$_2$ Ratio [a]	*p*-Value [b]	log$_2$ Ratio [a]	*p*-Value [b]
Metabolism-Primary-Oxidative phosphorylation							
ATP synthase subunit alpha, chloroplastic	Solyc06g072540	−0.244	0.124	−0.708	0.011	−0.879	0.001
ATP synthase delta-subunit protein	Solyc12g056830	−0.546	0.011	−1.014	0.018	−1.061	0.020
Metabolism-Primary-Amino Acid & Polyamine Metabolism							
S-adenosylmethionine synthase 2 (SAM2)	Solyc12g099000	0.045	0.924	−0.345	0.564	1.194	0.006
Chorismate synthase 1 precursor	Solyc04g049350	0.151	0.259	0.335	0.072	0.830	0.030
S-adenosylmethionine synthase 1 (SAM1)	Solyc01g101060	0.224	0.091	0.190	0.034	0.681	0.012
S-adenosylmethionine synthase	Solyc10g083970	NQ	NQ	NQ	NQ	0.630	0.007

Table 4. *Cont.*

Protein Description	Gene Accession	4 hpi		8 hpi		24 hpi	
		$\log_2$ Ratio [a]	*p*-Value [b]	$\log_2$ Ratio [a]	*p*-Value [b]	$\log_2$ Ratio [a]	*p*-Value [b]
Methylthioribose kinase	Solyc01g107550	0.125	0.156	0.333	0.169	0.626	0.005
Dehydroquinate dehydratase/shikimate: NADP oxidoreductase	Solyc01g067750	−0.191	0.086	0.090	0.755	0.598	0.014
Ornithine decarboxylase	Solyc04g082030	0.153	0.211	0.728	0.293	1.673	0.037
Spermidine synthase	Solyc05g005710	−0.494	0.048	−0.020	0.955	1.007	0.014
Metabolism Primary-Fatty Acid/Lipids							
Fatty acid beta-oxidation multifunctional protein	Solyc12g007170	0.340	0.573	0.663	0.204	2.662	0.004
Acetoacetyl-CoA thiolase	Solyc07g045350	0.210	0.020	0.849	0.050	2.195	0.030
12-oxophytodienoate reductase 3	Solyc07g007870	−0.045	0.860	0.347	0.198	1.468	0.023
4-coumarate-CoA ligase	Solyc03g117870	0.004	0.987	0.382	0.233	1.282	0.005
ATP-citrate synthase, putative	Solyc05g005160	0.091	0.649	0.587	0.045	0.998	0.117
Lipoxygenase	Solyc01g006560	−0.437	0.184	NQ	NQ	0.767	0.038
Metabolism-Secondary							
5-enolpyruvylshikimate-3-phosphate synthase	Solyc01g091190	0.166	0.744	0.351	0.242	1.442	0.043

[a] The average $\log_2$ ratio of protein quantity representing (inoculated/mock) from 3 biological replicates. [b] *p*-value calculated from Student t-test. Color-coded: red, significant quantity change greater than 0.58 of $\log_2$ ratio; blue, significant quantity change less than −0.58 of $\log_2$ ratio; grey, no significant change between the inoculated and mock group ($p \geq 0.05$); white, non-quantifiable (NQ).

2.3. Changes Associated with Defence and Immune Regulation

In this study, several proteins involved in the defence of pathogenesis were observed to be regulated mainly at 8 and 24 hpi. The two chitinases, pathogenesis-related protein 4 (PR-4) and chitinase Z15140 and major allergen Pru ar 1 (also known as Pathogenesis-Related Protein-STH 2-like; PR-STH 2-like) were found to be regulated at 8 hpi. Chitinases are enzymes catalysing the hydrolysis of chitin into *N*-acetyl-D-glucosamine monomers therefore this enzyme can damage fungal cell walls and the exoskeleton of arthropods. PR-4 was found to be induced not only by pathogen attack but also induced by ethylene in Arabidopsis [23,24]. Although PR-4 was classified as an endochitinase, its chitinase activity was found to be weak [25]. A more recent study indicated that PR-4 is a bifunctional enzyme with both RNase and DNase activity [26,27], which suggests that PR-4 may be involved in the regulation of HR. It was reported that the transient expression of PR-4b triggers hypersensitive cell death in Arabidopsis and the induction of PR4b is necessary to defend *P. syringae* and *Hyaloperonospora arabidopsidis* (Hpa) infection [28]. For major allergen Pru ar 1, this protein was also named as PR-10a and found to be induced after wounding, elicitor treatment or infection by *Phytophthora infestans* of the plants [29]. It has been shown that the parsley homologs of PR-l0a are highly similar to a partial sequence of a ginseng ribonuclease [30], suggesting the protein encoded by the PR-10a domain could exhibit ribonuclease activity. Moreover, many of these genes have been shown to be controlled at the transcriptional level and to be expressed during the defence response or after wounding [24,31,32]. The function of the PR-10a protein is currently unknown.

At the later time point, 24 hpi, immune-related proteins, including the well-known SA-responsive marker genes/proteins, pathogenesis-related protein 1 (PR-1; Solyc00g174340), Kunitz trypsin inhibitor (KTI; Solyc03g098730) and arginase 2 (ARG2; Solyc01g091170) appeared to be dramatically up-regulated with up to 8-fold increase compared to the mock. PR-1 was observed to be highly up-regulated specifically at 24 hpi with a fold change of ~8.33. PR-1 is triggered by SA, a hormone in the deployment of systemic acquired resistance (SAR). SAR is the defence response occurring at the non-infected site (and thus called "systemic") to establish a long-term and broad-spectrum resistance throughout the entire plant after local pathogen infection is activated by avirulent pathogens [33]. Activation of the PR-1 gene requires the recruitment of a transcriptional enhanceosome to its promoter for which the SA receptor, Non-Expressor of Pathogenesis-Related Gene 1 (NPR1), is the key regulator [34]. Our previous study demonstrated the function of the PR-1 in the regulation of the immunity, in which PR-1 acts as a precursor for the signalling peptide CAPE1 to trigger defence responses regulated by methyl-jasmonic acid (MeJA) and SA [5]. We also demonstrated that the production of CAPE1 in tomato is regulated by the wounding response and MeJA may further enhance the production of this peptide elicitor [5]. This implies that the plant defence mechanism had started to significantly activate the SAR at 24 hpi. The wound- or JA-dependent responses during the later stage of pathogenesis response may further trigger CAPE1 to induce different sets of the defence responses. On the other hand, the protease KTI, which belongs to the Kunitz-type protease inhibitor (PI) whose serine protease activity specifically inhibits trypsin proteases which cleave polypeptides at the C-terminus of lysine and arginine, was up-regulated ~1.64-fold at 24 hpi of *Pst* DC3000. The PI family are usually induced by JA-mediated response to wounding, pathogens or herbivore attack. The Arabidopsis homolog of Kunitz trypsin inhibitor, ARABIDOPSIS THALIANA KUNITZ TRYPSIN INHIBITOR 1 (AtKTI1), is a functional serine protease inhibitor that antagonizes pathogen-associated programmed cell death [35]. The gene expression of AtKTI1 is increased by 24 hpi of *Pst* DC3000 (especially with the effector avrB) in the microarray database [36] and also induced by H_2O_2, wounding, SA and some programmed cell death (PCD)-eliciting toxins from necrotrophic fungal pathogens. Since PCD, a cell-suicide act that needs to be tightly controlled for plant development and pathogen defence, it is speculated that plant induces KTI1 in the infected tissue in order to finely control the fate of plant cells under hemi-biotrophic pathogen inoculation. Previously researchers also found that AtKTI1 in plants can trigger both SA- and JA/ET-dependent defence gene expression, indicating the diverse role of this protein in defence responses including PTI and ETI [35]. In addition, another immune related protein ARG2 was observed to be ~ 6.5-fold expressed when inoculated by *Pst DC3000* only at 24 hpi. ARG2 targets the mitochondria, hydrolyses the first step of arginine degradation to ornithine and urea, also provides the upstream production of proline, histidine and the polyamine biosynthetic pathway [37]. It has been found that the level protein expression and enzyme activity can be induced by wounding, JA treatment and *Pst* DC3000 phytotoxin coronatine in tomato [38]. In our study, ARG2 was severely up-regulated; however, in previously published research, the RNA level was induced as early as 1 hpi and peaked at 8 hpi [38]. This difference could be due to the difference between the different pathogen inoculation methods, as our method could avoid causing wounding compared to the traditional vacuum infiltration in the early stage of the treatment and thus delay the overexpression at the translational level. Besides polyamine, ARG2 may be involved in regulating nitric oxide (NO) accumulation. ARG2 has been shown to be important in the defence against necrotrophic pathogen *Botrytis elliptica* [39]. NO-mediated defence and immunity could be related to SA-JA antagonism as NO induces the accumulation of SA while inhibiting the expression of JA-responsive genes [40]. The late-induction level of ARG2 suggests that the HR-directed PCD by NO signalling could be triggered at 24 hpi.

2.4. Changes Associated with the Reactive Oxygen Species (ROS) and Oxidation-Reduction Reactions

ROS, including hydrogen peroxide (H_2O_2) and superoxide (O_2^-) are involved in the early signalling for defence responses and also has direct toxicity against pathogen function, underlying PTI, ETI and SAR. The activity for mediating rapid accumulation of ROS during the oxidative burst is dependent on two classes of enzymes: NADPH oxidases and class III heme peroxidases [41,42]. NADPH oxidases synthesize the superoxide and peroxidases convert superoxide into a more stable form of ROS, H_2O_2. One peroxidase (Solyc04g071900) which belongs to the class III heme peroxidases was observed to be ~1.8-fold up-regulated at 8 hpi, suggesting a diverse ROS-related mechanism could be triggered at 8 hpi. Glutathione S-transferases (GSTs) are known for their function in anti-oxidative reactions to eliminate ROS and lipid hydroperoxides that accumulate in infected tissues thus limiting the excessive spread of HR- associated cell death [43]. Several GST family members have been shown to be PAMP-responsive genes in Arabidopsis [44]. In our study, the major antioxidant enzymes including several GSTs and GST-like proteins and glutathione reductase (GR) were up-regulated at 8 and/or 24 hpi upon *Pst* DC3000 inoculation.

In addition to glutathione and ascorbate, the most abundant antioxidant compounds in plant cells are oxidoreduction-active proteins called redoxins [45]. Thioredoxin (Trx) is a multigenic superfamily of ubiquitous redox proteins with multiple functions. Thioredoxin reductase (Solyc02g082250), one of ROS-detoxifying enzymes, was also up-regulated upon *Pst* DC3000 infection at 8 hpi, confirming that plants also trigger the redox change by thioredoxin reductase during pathogen inoculation.

Although the up-regulation of plant GST proteins as a consequence of bacterium-induced oxidative stress was recognized as above mentioned, there were also some predicted peroxidases, GST proteins and ascorbate peroxidase (Table 1 & Supplementary Table 1) were down-regulated at *Pst* DC3000 at 24 hpi. This is probably due to the effect of SA mediating ROS accumulation in cells by first promoting ROS accumulation (as these ROS are essential in the early defence response) and then inhibiting catalase and cytosolic ascorbate peroxidase, the main H_2O_2-detoxifying enzymes, to promote further accumulation of ROS which triggers the ETI response [37,46]. The effectiveness of SA-mediating defence response is also supported by our observation that between 4 and 24 hpi, the colony growth did not show significant change in our experiments (Figure 1B).

2.5. Changes Associated with Protein Folding, Transportation and Degradation

2.5.1. Protein Folding: Heat Shock Proteins and Chaperones

Heat shock proteins (HSPs) are a huge protein family participating in plant growth, development and fitness, mainly by helping mature protein folding or degrading mis-folded proteins. HSPs like HSP70, HSP90 and HSP60 belong to molecular chaperone families. Molecular chaperones bind and catalytically unfold misfolded and aggregated proteins as a primary cellular defensive and housekeeping function [47]. Plastidic chaperonin 60 (CPN60) alpha and beta are required for plastid division in Arabidopsis. CPN60 proteins are required to be maintained at a proper level for folding stromal plastid division proteins and are essential for the development of chloroplasts [48]. In our study, CPN60 alpha subunit (Solyc06g075010) was up-regulated by ~2.8-fold, suggesting its important role in bacteria defence response.

The HSP90 family have been shown to be involved in regulating drought, salt and oxidative stress and involved in ATP-dependent assembly of the 26S proteasome [49]. We found that at *Pst* DC3000 4 hpi in tomato, HSP90 (Solyc06g036290) was induced ~2.2-fold compared to the mock. It has been suggested that HSP90 protein could be an important helper to the regulation of the receptor proteins involved in plant immunity, nucleotide-binding site leucine-rich repeat (NB-LRR) type R proteins. This interaction is apparently required for the NB-LRR type R proteins to maintain the protein stability, especially their sensor signal-competent state [50]. HSP90 could physically interact with various R proteins, including RPM1 (RESISTANCE TO PSEUDOMONAS MACULICOLA 1), RPS2 (RESISTANCE TO P. SYRINGAE 2) and RPS4 (RESISTANCE TO PSEUDOMONAS SYRINGAE 4) in Arabidopsis [51,52]. In *Nicotiana benthamiana*, the complex formed by HSP90 and Suppressor of the G2 Allele of *skp1* (SGT1) or Required for Mla12 Resistance 1 (RAR1) which is required for resistance gene *Mla12* or both, is essential for plant survival against tobacco mosaic virus [53,54]. Our data showed that the quick induction of HSP90 at 4 hpi may function in improving the protein stability of NB-LRR, possibly by helping the protein folding, thus enhancing pathogen recognition and signal transduction.

2.5.2. Protein Degradation

The papain-like Cathepsin B-like cysteine protease (CathB), localized in the apoplast region, has proven to be a positive regulator of the HR defence [55]. One predicted gene product of Solyc02g069110, CathB protein, was dramatically increased in protein quantity at 24 hpi with ~3.5-fold. This protease family has endopeptidase activity at the C-terminus of the YVAD substrate and the enzyme activity was restricted predominately to acidic pH [56]. At 8 h of *Pst* DC3000 inoculation in tomato, Solyc04g080960 protein product, a cysteine protease (also known as Response to Dehydration 19A-like proteases; RD19A-like) was down-regulated with 0.64-fold change compared to mock treatment (Table 2). Liu and colleagues recently found that the expression level of Solyc04g080960 was reduced after 24 h of inoculation of the hemi-biotrophic pathogen *Fusarium oxysporum* f. sp. *lycopersici* (FOL) in tomato [57]. On the other hand, RD19 protease activity is required for RPS1-R-dependent immune activation to enhance resistance against the necrotrophic pathogen *Ralstonia solanacearum* by being the interacting protein of *R. solanacearum* type III effector, Pseudomonas outer protein P2 (PopP2), thus initiating the downstream resistance response [58]. In Arabidopsis, AtRD19A also acts as an important protein marker for dehydration stress adaptation and could be highly induced by drought and salt stresses [59]. Taking our quantitative results into consideration, RD19A-like protein may be one of the responsive ends to the danger signals from necrotrophic pathogen attack and abiotic stress like drought or salt stress. Therefore, during the infection of *Pst* DC3000, plants could down-regulate RD19A-like protein in order to be more cost-effective in defence response.

2.6. Changes Associated with Phytohormone Synthesis and Fatty Acid Metabolism

In our data, Zeaxanthin epoxidase (ZEP; Solyc02g090890), participating in the biosynthetic pathway upstream of ABA (abscisic acid), was down-regulated at 24 hpi by ~0.61-fold (Table 1). This kind of down-regulation could be the result of the antagonism between ABA and SA [60], since the phytohormone SA is induced by diverse biotrophic/hemi-biotrophic pathogens to activate SAR responses [54].

It has been long believed that there is also antagonism between SA and JA, including the control of each one's biosynthetic pathway. In our data, several proteins possibly involved upstream of the JA-biosynthetic pathway have been identified, such as fatty acid beta-oxidation multifunctional protein (Absent In Melanoma 1-like; AIM1-like; Solyc12g007170) which showed a ~6.3-fold increase at 24 hpi and lipoxygenase (Solyc01g006560) with ~1.7-fold increase at 24 hpi. The results of 24 h-time point suggest that the JA/ET-mediated defence response could be induced. Our results showed the ET biosynthetic enzyme, 1-aminocyclopropane-1-carboxylate oxidase 1 (ACO; Solyc07g049530) was dramatically up-regulated at the 24 h-time point by ~7-fold. Several S-adenosylmethionine synthase family proteins were also up-regulated by ~2.3-fold at 24 hpi and they are involved in the shikimic acid pathway, which is the upstream of ET and polyamine biosynthesis, suggesting the level of ET should be hugely increased at this late time point.

3. Materials and Methods

3.1. Plant Materials and Growth Condition

Tomato seeds (*Solanum lycopersicum* cv CL5915, originally provided by AVRDC—The World Vegetable Centre at Tainan, Taiwan) were germinated in soil and grown in a growth chamber for 4–5 weeks. The growth chamber condition was set as 25 °C/22 °C (day/night) temperature, 50%/70% (day/night) humidity and 16 h/8 h (day/night) photoperiod using the light source providing photosynthetic photon flux density (PPFD) 80 μmole/m^2·s. Leaflets with similar size from the 3rd to 5th pair of true leaves were collected for the following inoculation experiment.

3.2. Pseudomonas Preparation and Inoculation Assays

Pst DC3000 from B. N. Kunkel (Washington University, St. Louis, MO, USA) [61] was used as a pathogenic strain on tomato plants. The growth and isolation method were referred to the work published by Desclos-Theveniau et al. [62]. The dipping inoculation method for detached leaf is adapted from previous studies [18,63]. *Pst* DC3000 was grown overnight in liquid King's B medium [64] with 50 μg/mL rifampicin at 28 °C then resuspended in 10 mM MgSO$_4$ and *Pst* DC3000 suspension was adjusted to OD$_{600}$ of 0.02 (~10^7 cfu/mL). The detached leaves from tomato plants were dipped in *Pst* DC3000 suspension solution with 0.005% Silwet L-77 for 2 min and then placed on water-saturated paper in a petri dish. The dishes were covered and incubated in the growth chamber. The control inoculum as the mock group contained the same components as the bacteria inoculum but without *Pst* DC3000. Disease symptoms and bacterial population were evaluated 0, 4, 8, 24 h and 3, 5, 7 days after inoculation. To estimate the internal bacterial population, leaves were surface sterilized with 70% ethanol, washed twice with sterile distilled water and then homogenized in 10 mM MgSO$_4$. The solution was diluted and spotted onto the King's B agar medium [64] with 50 μg/mL rifampicin. Colonies were counted and reported as means and standard deviations of results for three biological replicates.

3.3. Sample Preparation: Protein Extraction and Digestion

Three biological replicates of mock-treated (4, 8, 24 hpt) and *Pst* DC3000 inoculated (4, 8, 24 hpi) leaves were prepared for the proteomics experiments. For each sample, leaves were ground into powder in the chilled mortar and pestle with liquid N_2 then 0.5 g of powder was collected. The sample was then homogenized with 2.5 mL of ice-cold homogenization buffer (50 mM HEPES-KOH, pH 7.5, 250 mM sucrose, 5% glycerol stock, 10 mM EDTA, pH 8.0, 0.5% Soluble polyvinylpyrrolidone (PVP-10), 3 mM dithiothreitol, 1 mM phenylmethylsulfonyl fluoride and 1× protease inhibitor cocktail) by vortexing for at least 3 min or until completely homogenized. The sample was further incubated in the buffer by Intelli Mixer RM-2L (ELMI Ltd., Riga, Latvia) in the cold room for 30 min. The homogenate was filtered through two layers of miracloth. The volume of filtrate should be relatively close to the initial amount of added homogenization buffer. The supernatant of the filtrate was then collected by centrifugation at 15,000× *g* for 10 min under 4 °C. The protein concentration of each sample was measured by the Bradford assay to quantify the protein amount of total protein.

For each sample, 100 μg of the total protein was precipitated by addition of acetone to 80% with incubation at −20 °C overnight and recovered by centrifugation at 16,000× *g* for 15 min under 4 °C. The reduction and alkylation steps were adapted from the literature as previously described [65]. Re-solubilized proteins were reduced in 50 mM ammonium bicarbonate (ABC) buffer and 8 M urea with 5 mM tris(2-carboxyethyl)phosphine hydrochloride (TCEP) for 1 h at 37 °C and alkylated using 20 mM iodoacetamide for 45 min in the dark at room temperature. Each sample was diluted 4-fold using 50 mM ABC to decrease the urea concentration to less than 2 M then digested with lysyl endopeptidase (LysC; Wako Chemicals, Japan) to a final ratio of 1:50 at room temperature for 3 h. Next the sample was diluted using 50 mM ABC to decrease the urea concentration to ~1 M before digestion with sequencing grade trypsin (Promega, Madison, WI, USA) to a final ratio of 1:50. The proteolysis was continued overnight (14 h) at room temperature and terminated by addition of formic acid to a final concentration of 1% (vol/vol). The digest sample was then desalted using the 50-mg tC18 SepPak cartridge (Waters Corporation, Milford, MA, USA) as described previously [66]. The tryptic peptides from individual sample was dissolved by deionized water containing 2% acetonitrile and 0.1% (v/v) formic acid to the concentration of 500 ng/μL. Three different pooled tryptic peptide samples for DDA analyses in order to construct the DIA spectral library were prepared by combining 4 μg peptides from the same time point of *Pst* DC3000-inoculated and mock-treated sample. For the purpose of retention time calibration, the iRT-standard peptides (Biognosys, Schlieren, Switzerland) were added into the pooled sample and also each individual sample at 1/10 by volume.

3.4. Liquid Chromatography-Mass Spectrometry Analysis

The nanoLC-MS/MS was equipped with a self-packed tunnel-frit [67] analytical column (ID 75 μm × 50 cm length) packed with ReproSil-Pur 120A C18-AQ 1.9 μm (Dr. Maisch GmbH, Ammerbuch-Entringen, Germany) at 40 °C on a nanoACQUITY UPLC System (Waters Corporation, Milford, MA, USA) connected to a Q Exactive HF Hybrid Quadrupole-Orbitrap mass spectrometer (Thermo Scientific, Bellefonte, PA, USA). The peptides were separated by a 135-min gradient using the mobile phases including Solvent A (0.1% (*v/v*) formic acid) and Solvent B (acetonitrile with 0.1% formic acid). With a flow rate of 250 nL/min, the gradient started with a 40 min equilibration maintained at 2% of B and set as the following segments: 2 to 8% of B in 8 min, 8 to 25% of B in 90 min, then 25% to 48% of B in 5 min, 48 to 80% of B in another 5 min followed by 80% of B wash 10 min and the last equilibrium to 2% B in the last 20 min.

The instrumentation and parameters for DDA and DIA analysis were referred to the previous studies using Q Exactive HF Hybrid Quadrupole-Orbitrap mass spectrometer [68,69]. Two micrograms of the pooled and individual tryptic peptide samples were analysed by DDA and DIA mode, respectively. For DDA analysis, the MS instrument was operated in the positive ion mode and DDA methods for detection of proteome. The instrument was configured to collect high resolution ($R = 60,000$ at m/z 200 at an automatic gain control target of 3.0×10^6) broadband mass spectra (m/z 350–1650 Da) with a maximum IT of 20 ms and MS/MS events ($R = 15,000$ at an automatic gain control target of 1.0×10^5) with a dd-MS^2 IT of 25 ms when a precursor ion charge was 2+, 3+, 4+ and 5+ and an intensity greater than 1.0×10^4, isolation window was set to 1.6 m/z, was detected. The 15 most abundant peptide molecular ions, dynamically determined from the MS1 scan, were selected for MS/MS using a relative higher energy collisional dissociation (HCD) energy of 28% with the dynamic exclusion was 35 s. For DIA analysis, MS/MS proteome profiling, was analysed by the same LC-MS/MS system. The instrument was operated in the positive ion mode and configured to collect high resolution ($R = 120,000$ at m/z 200 at an automatic gain control target of 3.0×10^6) broadband mass spectra (m/z 350–1650 Da) with a maximum IT of 60 ms and MS/MS events ($R = 30,000$ at an automatic gain control target of 3.0×10^6) with an auto MS^2 IT, isolation window was set to 52.0 m/z, fixed first mass was set to 200 m/z. The 25 segments were selected for MS/MS using a relative higher energy collisional dissociation (HCD) energy of 28%. The acquisition window covered a mass range from 350 to 1650 m/z through 25 consecutive isolation windows.

3.5. Data Analysis for LC-MS

With the DDA data files, the Mascot (ver. 2.3, http://www.matrixscience.com/), X!Tandem (ver. 2013.06.15.1) [70] and Comet (ver. 2017.01 rev.1) [71] were used to do a protein database search against a combined database of ITAG (ver. 3.1, https://solgenomics.net/organism/ Solanumlycopersicum/genome; 34881 entries reverse sequence generate as the decoy database) and the iRT standard peptides and BSA (SwissProt Accession: P02769) sequence. Search parameters were set as follows: MS tolerance, 20 ppm, allow precursor monoisotopic mass isotope error; number of trypsin missed cleavage: 2; Fragment Mass tolerance, 0.2 Da; enzyme, trypsin; static modifications, carbamidomethyl (Cys, + 57.021 Da); dynamic modifications, oxidation (+15.995 Da) of methionine. Next the software on the Trans-Proteomic Pipeline (TPP, ver. 5.1) [72] was used to combine the search result from different search engines and different repeats; there were a total 116,422 peptide-spectrum matches. In the constructed library, there were a total 67,536 transitions, 9343 peptides and 3070 proteins.

OpenMS (ver. 2.2.0) [73] was utilized for decoyed spectral library construction. We employed the OpenSWATH (ver. 0.1.2) [74] to search the DIA files against the spectral library we constructed. The retention time alignment used the information of iRT transitions. In addition to the chromatogram alignment, the spike-in iRT peptide standards were also used for the quality control of the DDA and DIA analyses. In all DDA and DIA analyses across the sequence of the instrument in the study, the coefficient of variation (CV) of iRT peptide retention time should be less than 3% and the CV of iRT peptide peak intensity should be less than 20%. Search parameters were set as follows: peptide false-discovery rate (FDR), 0.05; protein FDR, 0.01; alignment method, LocalMST; re-alignment method, lowest; retention time (RT) difference, 60; alignment score, 0.05. The ratios of protein quantitation between the *Pst* DC3000-inoculated and mock-treated sample in each replicate were normalized by the most-likely ration normalization principle as previously applied in the DIA study [75].

3.6. Quantitation Data Analysis

Only proteins detected and quantified in all runs (3 biological replicates) were included in the data set. To perform a significance test, the students' t-test was calculated. Any protein with differential abundance with a *p*-value of less than 0.05 and fold change greater than 1.5 or less than 0.67 was defined as being "significantly" regulated in protein quantity. Functional annotations of the quantified tomato proteins were obtained via PANTHERN (ver. 13.1) [76,77]. To show the functional distribution of the regulated proteins, the up- and down-regulated protein sequences were searched against the *Arabidopsis thaliana* TAIR10 database (http://arabidopsis.org) using

BLASTP with E-value $< 1.0 \times 10^{-5}$ (https://blast.ncbi.nlm.nih.gov/Blast.cgi) first and the matched Arabidopsis homolog proteins were categorized by the GO biological function level 2 using DAVID v6.8 (https://david.ncifcrf.gov/) [78,79].

4. Conclusions

This study demonstrated a successful example of using the DIA approach for a time course analysis of plant pathogenesis proteomics. A total of ~2200 proteins were identified and quantified from the tomato subjected to different treatments and 90% of the totally identified proteins were commonly observed across all the treatments. This study indicates different sets of proteins are regulated from the early to the later stage of the *Pst* DC3000 infection. We showed that no defence-related protein was observed to be up-regulated but the chaperone proteins for helping the activity of R proteins was induced at 4 hpi of *Pst* DC3000. Several major defence and immune-related proteins were found to be up-regulated at 8 and 24 hpi. One of the peroxidase proteins related to the production of H_2O_2 was up-regulated at 8 hpi. We have shown that plants do not only express proteins for accumulating H_2O_2 but also detoxification proteins to avoid the over-accumulation of ROS. The proteins involved in the later stage of the pathogenesis which are related to the HR and PCD were up-regulated at 24 hpi. We also discovered that the proteins involved in the biosynthesis of JA and ET were induced at 24 hpi, indicating ET/JA may be induced in the later pathogenesis response. More time points and treatments can be further analysed and compared with the current DIA datasets based on the library established in this study. The number of proteins identified and quantified with the use of the current DIA approach can also be increased when a more comprehensive spectra library for the tomato proteome is established.

Supplementary Materials: Supplementary materials can be found at http://www.mdpi.com/1422-0067/20/4/863/s1. Supplemental Figure S1, disease phenotypes and population dynamics of *Pseudomonas syringae* pv. *tomato* on the tomato leaves at 24 hpi, 3 dip and 7 dpi. Supplemental Table 1, list of quantified proteins with significant change in abundance at 4 hpi, 8 hpi or 24 hpi compared to mock. Supplemental Table 2, full list of quantified proteins and peptides in three biological replicates of the *Pst* DC3000-inoculated (4 hpi, 8 hpi, 24 hpi) and mock-treated (4 hpt, 8 hpt and 24 hpt) experiments. All the mass spectrometry raw data files were deposited to the ProteomeXchange Consortium via the PRIDE [80] partner repository with the dataset identifier PXD012226.

Author Contributions: Conceptualization, K.-T.F. and Y.-R.C.; Data curation, W.-H.C.; Funding acquisition, Y.-R.C.; Methodology, K.-T.F., K.-H.W., C.-F.Y., K.-T.C. and S.-C.H.; Project administration, Y.-R.C.; Resources, Y.-R.C.; Software, W.-H.C. and J.-C.Y.; Supervision, Y.-R.C.; Visualization, K.-T.F., K.-H.W. and S.-C.H.; Writing—original draft, K.-T.F., S.-C.H. and Y.-R.C.; Writing—review & editing, K.-T.F., S.-C.H. and Y.-R.C.

Funding: This work was financially supported by the Ministry of Science and Technology, Taiwan (Project number 107-2113-M-001-006-); Academia Sinica, Taiwan (Project number 105-L03); Innovative Translational Agricultural Research Program, Taiwan (Project ID AS-KPQ-108-ITAR-07).

Acknowledgments: The MS analysis was supported by the Metabolomics Facilities of the Scientific Instrument Center at Academia Sinica. We thank Chia-Wei Hsu for helping monitor the quality of LC-MS data acquisition.

Conflicts of Interest: The authors declare no conflict of interest. The funders had no role in the design of the study; in the collection, analyses or interpretation of data; in the writing of the manuscript or in the decision to publish the results.

Abbreviations

DIA	data-independent acquisition
DDA	data-dependent acquisition
PAMP	pathogen-associated molecular pattern
MAMP	microbial associated molecular pattern
DAMP	damaged-associated molecular pattern
ETS	effector-triggered susceptibility
PTI	PAMP-triggered immunity

ETI	effector-triggered immunity
HR	hypersensitive response
PCD	programmed cell death
SA	salicylic acid
JA	jasmonic acid
ABA	abscisic acid
ET	ethylene
Pst DC3000	*Pseudomonas syringae* pv. *tomato* DC3000
SAR	systemic acquired resistance

References

1. Underwood, W. The Plant Cell Wall: A Dynamic Barrier against Pathogen Invasion. *Front. Plant Sci.* **2012**, *3*, 1–6. [CrossRef]
2. Takai, R.; Isogai, A.; Takayama, S.; Che, F.-S. Analysis of Flagellin Perception Mediated by Flg22 Receptor Osfls2 in Rice. *Mol. Plant-Microbe Interact.* **2008**, *21*, 1635–1642. [CrossRef]
3. Pearce, G.; Strydom, D.; Johnson, S.; Ryan, C.A. A Polypeptide from Tomato Leaves Induces Wound-Inducible Proteinase Inhibitor Proteins. *Science* **1991**, *253*, 895–897. [CrossRef]
4. Huffaker, A.; Pearce, G.; Ryan, C.A. An Endogenous Peptide Signal in Arabidopsis Activates Components of the Innate Immune Response. *Proc. Natl. Acad. Sci. USA* **2006**, *103*, 10098–10103. [CrossRef]
5. Chen, Y.-L.; Lee, C.-Y.; Cheng, K.-T.; Chang, W.-H.; Huang, R.-N.; Nam, H.G.; Chen, Y.-R. Quantitative Peptidomics Study Reveals That a Wound-Induced Peptide from PR-1 Regulates Immune Signaling in Tomato. *Plant Cell* **2014**, *26*, 4135–4148. [CrossRef]
6. Tanaka, K.; Choi, J.; Cao, Y.; Stacey, G. Extracellular Atp Acts as a Damage-Associated Molecular Pattern (DAMP) Signal in Plants. *Front. Plant Sci.* **2014**, *5*, 1–9. [CrossRef]
7. Choi, H.W.; Klessig, D.F. Damps, Mamps and Namps in Plant Innate Immunity. *BMC Plant Biol.* **2016**, *16*, 1–10. [CrossRef]
8. Peng, Y.; Wersch, R.V.; Zhang, Y. Convergent and Divergent Signaling in PAMP-Triggered Immunity and Effector-Triggered Immunity. *Mol. Plant Microbe Interact.* **2018**, *31*, 403–409. [CrossRef]
9. Dangl, J.L. Pivoting the Plant Immune System from Dissection to Deployment. *Science* **2013**, *341*, 745–751. [CrossRef]
10. Jones, J.D.G.; Dangl, J.L. The Plant Immune System. *Nature* **2006**, *444*, 323–329. [CrossRef]
11. Tao, Y.; Xie, Z.; Chen, W.; Glazebrook, J.; Chang, H.-S.; Han, B.; Zhu, T.; Zou, G.; Katagiri, F. Quantitative Nature of Arabidopsis Responses during Compatible and Incompatible Interactions with the Bacterial Pathogen Pseudomonas syringae. *Plant Cell Online* **2003**, *15*, 317–330. [CrossRef]
12. Navarro, L.; Zipfel, C.; Rowland, O.; Keller, I.; Robatzek, S.; Boller, T.; Jones, J.D.G. The Transcriptional Innate Immune Response to Flg22. Interplay and Overlap with Avr Gene-Dependent Defense Responses and Bacterial Pathogenesis. *Plant Physiol.* **2004**, *135*, 1113–1128. [CrossRef]
13. Mukhtar, M.S.; McCormack, M.E.; Argueso, C.T.; Pajerowska-Mukhtar, K.M. Pathogen Tactics to Manipulate Plant Cell Death. *Curr. Biol.* **2016**, *26*, R608–R619. [CrossRef]
14. Kim, Y.; Tsuda, K.; Igarashi, D.; Hillmer, R.A.; Sakakibara, H.; Myers, C.L.; Katagiri, F. Mechanisms Underlying Robustness and Tunability in a Plant Immune Signaling Network. *Cell Host Microbe* **2014**, *15*, 84–94. [CrossRef]
15. Bozso, Z.; Ott, P.G.; Kaman-Toth, E.; Bognar, G.F.; Pogany, M.; Szatmari, A. Overlapping yet Response-Specific Transcriptome Alterations Characterize the Nature of Tobacco-Pseudomonas syringae Interactions. *Front. Plant Sci.* **2016**, *7*, 251. [CrossRef]
16. Tsuda, K.; Sato, M.; Stoddard, T.; Glazebrook, J.; Katagiri, F. Network Properties of Robust Immunity in Plants. *PLoS Genet.* **2009**, *5*, e1000772. [CrossRef]
17. Moore, J.; Penfold, C.A.; Jenkins, D.J.; Hill, C.; Baxter, L.; Kulasekaran, S.; Truman, W.; Littlejohn, G.; Prusinska, J.; Mead, A.; et al. Transcriptional Dynamics Driving MAMP-Triggered Immunity and Pathogen Effector-Mediated Immunosuppression in Arabidopsis Leaves Following Infection with *Pseudomonas syringae* pv. *tomato* DC3000. *Plant Cell* **2015**, *27*, 3038–3064.

18. Parker, J.; Koh, J.; Yoo, M.J.; Zhu, N.; Feole, M.; Yi, S.; Chen, S. Quantitative Proteomics of Tomato Defense against *Pseudomonas syringae* Infection. *Proteomics* **2013**, *13*, 1934–1946. [CrossRef]

19. Ow, S.Y.; Salim, M.; Noirel, J.; Evans, C.; Wright, P.C. Minimising Itraq Ratio Compression through Understanding LC-MS Elution Dependence and High-Resolution Hilic Fractionation. *Proteomics* **2011**, *11*, 2341–2346. [CrossRef]

20. Gillet, L.C.; Navarro, P.; Tate, S.; Röst, H.; Selevsek, N.; Reiter, L.; Bonner, R.; Aebersold, R. Targeted Data Extraction of the MS/MS Spectra Generated by Data-Independent Acquisition: A New Concept for Consistent and Accurate Proteome Analysis. *Mol. Cell. Proteom.* **2012**, *11*, O111.016717. [CrossRef]

21. Xin, X.F.; Kvitko, B.; He, S.Y. Pseudomonas syringae: What It Takes to Be a Pathogen. *Nat. Rev. Microbiol.* **2018**, *16*, 316–328. [CrossRef]

22. Mysore, K.S.; Crasta, O.R.; Tuori, R.P.; Folkerts, O.; Swirsky, P.B.; Martin, G.B. Comprehensive Transcript Profiling of Pto- and Prf-Mediated Host Defense Responses to Infection by *Pseudomonas syringae* pv. *tomato*. *Plant J.* **2002**, *32*, 299–315. [CrossRef]

23. Midoh, N.; Iwata, M. Cloning and Characterization of a Probenazole-Inducible Gene for an Intracellular Pathogenesis-Related Protein in Rice. *Plant Cell Physiol.* **1996**, *37*, 9–18. [CrossRef]

24. Somssich, I.E.; Schmelzer, E.; Kawalleck, P.; Hahlbrock, K. Gene Structure and in Situ Transcript Localization of Pathogenesis-Related Protein 1 in Parsley. *Mol. Gen. Genet.* **1988**, *213*, 93–98. [CrossRef]

25. Brunner, F.; Stintzi, A.; Fritig, B.; Legrand, M. Substrate Specificities of Tobacco Chitinases. *Plant J.* **1998**, *14*, 225–234. [CrossRef]

26. Guevara-Morato, M.A.; de Lacoba, M.G.; Garcia-Luque, I.; Serra, M.T. Characterization of a Pathogenesis-Related Protein 4 (PR-4) Induced in Capsicum Chinense L3 Plants with Dual Rnase and Dnase Activities. *J. Exp. Bot.* **2010**, *61*, 3259–3271. [CrossRef]

27. Caporale, C.; Di Berardino, I.; Leonardi, L.; Bertini, L.; Cascone, A.; Buonocore, V.; Caruso, C. Wheat Pathogenesis-Related Proteins of Class 4 Have Ribonuclease Activity. *FEBS Lett.* **2004**, *575*, 71–76. [CrossRef]

28. Hwang, I.S.; Choi, D.S.; Kim, N.H.; Kim, D.S.; Hwang, B.K. Pathogenesis-Related Protein 4b Interacts with Leucine-Rich Repeat Protein 1 to Suppress PR4b-Triggered Cell Death and Defense Response in Pepper. *Plant J.* **2014**, *77*, 521–533. [CrossRef]

29. Despres, C.; Subramaniam, R.; Matton, D.P.; Brisson, N. The Activation of the Potato PR-Loa Gene Requires the Phosphorylation of the Nuclear Factor PBF-1. *Plant Cell* **1995**, *7*, 589–598. [CrossRef]

30. Moiseyev, G.P.; Beintema, J.J.; Fedoreyeva, L.I.; Yakovlev, G.I. High Sequence Similarity between a Ribonuclease from Ginseng Calluses and Fungus-Elicited Proteins from Parsley Indicates That Intracellular Pathogenesis-Related Proteins Are Ribonucleases. *Planta* **1994**, *193*, 470–472. [CrossRef]

31. Chiang, C.C.; Hadwiger, L.A. Cloning and Characterization of a Disease Resistance Response Gene in Pea Inducible by Fusarium Solani. *Mol. Plant-Microbe Interact.* **1990**, *3*, 78–85. [CrossRef]

32. Warner, S.A.J.; Scott, R.; Draper, J. Characterisation of a Wound-Induced Transcript from the Monocot Asparagus That Shares Similarity with a Class of Intracellular Pathogenesis-Related (PR) Proteins. *Plant Mol. Biol.* **1992**, *19*, 555–561. [CrossRef]

33. Vlot, A.C.; Dempsey, D.M.A.; Klessig, D.F. Salicylic Acid, a Multifaceted Hormone to Combat Disease. *Ann. Rev. Phytopathol.* **2009**, *47*, 177–206. [CrossRef]

34. Rochon, A.; Boyle, P.; Wignes, T.; Fobert, P.R.; Despres, C. The Coactivator Function of Arabidopsis NPR1 Requires the Core of Its Btb/Poz Domain and the Oxidation of C-Terminal Cysteines. *Plant Cell* **2006**, *18*, 3670–3685. [CrossRef]

35. Li, J.; Brader, G.; Palva, E.T. Kunitz Trypsin Inhibitor: An Antagonist of Cell Death Triggered by Phytopathogens and Fumonisin B1 in Arabidopsis. *Mol. Plant* **2008**, *1*, 482–495. [CrossRef]

36. Zimmermann, P.; Hirsch-Hoffmann, M.; Hennig, L.; Gruissem, W. Genevestigator. Arabidopsis Microarray Database and Analysis Toolbox. *Plant Physiol.* **2004**, *136*, 2621–2632. [CrossRef]

37. Chen, Z.; Silva, H.; Klessig, D.F. Active Oxygen Species in the Induction of Plant Systemic Acquired Resistance by Salicylic Acid. *Science* **1993**, *262*, 1883–1886. [CrossRef]

38. Chen, H.; McCaig, B.C.; Melotto, M.; He, S.Y.; Howe, G.A. Regulation of Plant Arginase by Wounding, Jasmonate and the Phytotoxin Coronatine. *J. Biol. Chem.* **2004**, *279*, 45998–46007. [CrossRef]

39. Delledonne, M. No News Is Good News for Plants. *Curr. Opin. Plant Biol.* **2005**, *8*, 390–396. [CrossRef]

40. Huang, X.; Stettmaier, K.; Michel, C.; Hutzler, P.; Mueller, M.J.; Durner, J. Nitric Oxide Is Induced by Wounding and Influences Jasmonic Acid Signaling in Arabidopsis Thaliana. *Planta* **2004**, *218*, 938–946. [CrossRef]

41. O'Brien, J.A.; Daudi, A.; Butt, V.S.; Paul Bolwell, G.J.P. Reactive Oxygen Species and Their Role in Plant Defence and Cell Wall Metabolism. *Planta* **2012**, *236*, 765–779. [CrossRef]

42. Bolwell, G.P.; Butt, V.S.; Davies, D.R.; Zimmerlin, A. The Origin of the Oxidative Burst in Plants. *Free Radic. Res.* **1995**, *23*, 517–532. [CrossRef]

43. Gullner, G.; Komives, T.; Király, L.; Schröder, P. Glutathione S-Transferase Enzymes in Plant-Pathogen Interactions. *Front. Plant Sci.* **2018**, *9*, 1–19. [CrossRef]

44. Jones, A.M.E.; Thomas, V.; Bennett, M.H.; Mansfield, J.; Grant, M. Modifications to the Arabidopsis Defense Proteome Occur Prior to Significant Transcriptional Change in Response to Inoculation with Pseudomonas syringae. *Plant Physiol.* **2006**, *142*, 1603–1620. [CrossRef]

45. Serrato, A.J.; Fernández-Trijueque, J.; Barajas-López, J.-D.-D.; Chueca, A.; Sahrawy, M. Plastid Thioredoxins: A "One-for-All" Redox-Signaling System in Plants. *Front. Plant Sci.* **2013**, *4*, 463. [CrossRef]

46. Dubreuil-Maurizi, C.; Vitecek, J.; Marty, L.; Branciard, L.; Frettinger, P.; Wendehenne, D.; Meyer, A.J.; Mauch, F.; Poinssot, B. Glutathione Deficiency of the Arabidopsis Mutant Pad2-1 Affects Oxidative Stress-Related Events, Defense Gene Expression and the Hypersensitive Response. *Plant Physiol.* **2011**, *157*, 2000–2012. [CrossRef]

47. Mattoo, R.U.H.; Goloubinoff, P. Molecular Chaperones Are Nanomachines That Catalytically Unfold Misfolded and Alternatively Folded Proteins. *Cell. Mol. Life Sci.* **2014**, *71*, 3311–3325. [CrossRef]

48. Suzuki, K.; Nakanishi, H.; Bower, J.; Yoder, D.W.; Osteryoung, K.W.; Miyagishima, S.Y. Plastid Chaperonin Proteins Cpn60 Alpha and Cpn60 Beta Are Required for Plastid Division in Arabidopsis Thaliana. *BMC Plant Biol.* **2009**, *9*, 38. [CrossRef]

49. Oh, S.E.; Yeung, C.; Babaei-Rad, R.; Zhao, R. Cosuppression of the Chloroplast Localized Molecular Chaperone Hsp90.5 Impairs Plant Development and Chloroplast Biogenesis in Arabidopsis. *BMC Res. Notes* **2014**, *7*, 643. [CrossRef]

50. Schulze-Lefert, P. Plant Immunity: The Origami of Receptor Activation. *Curr. Biol.* **2004**, *14*, R22–R24. [CrossRef]

51. Hubert, D.A. Cytosolic Hsp90 Associates with and Modulates the Arabidopsis RPM1 Disease Resistance Protein. *EMBO J.* **2003**, *22*, 5679–5689. [CrossRef]

52. Liu, Y.; Burch-Smith, T.; Schiff, M.; Feng, S.; Dinesh-Kumar, S.P. Molecular Chaperone Hsp90 Associates with Resistance Protein N and Its Signaling Proteins SGT1 and Rar1 to Modulate an Innate Immune Response in Plants. *J. Biol. Chem.* **2004**, *279*, 2101–2108. [CrossRef]

53. Shirasu, K. Complex Formation, Promiscuity and Multi-Functionality: Protein Interactions in Disease-Resistance Pathways. *Trends Plant Sci.* **2003**, *8*, 252–258. [CrossRef]

54. Kadota, Y.; Shirasu, K. The Hsp90 Complex of Plants. *Biochim. Biophys. Acta* **2012**, *1823*, 689–697. [CrossRef]

55. Thomas, E.L.; van der Hoorn, R.A.L. Ten Prominent Host Proteases in Plant-Pathogen Interactions. *Int. J. Mol. Sci.* **2018**, *19*, 639. [CrossRef]

56. Shompole, S.; Jasmer, D.P. Cathepsin B-Like Cysteine Proteases Confer Intestinal Cysteine Protease Activity in Haemonchus Contortus. *J. Biol. Chem.* **2001**, *276*, 2928–2934. [CrossRef]

57. Zhao, M.; Ji, H.M.; Gao, Y.; Cao, X.X.; Mao, H.Y.; Ouyang, S.Q.; Liu, P. An Integrated Analysis of Mrna and Srna Transcriptional Profiles in Tomato Root: Insights on Tomato Wilt Disease. *PLoS ONE* **2018**, *13*, 1–18. [CrossRef]

58. Bernoux, M.; Timmers, T.; Jauneau, A.; Briere, C.; de Wit, P.J.G.M.; Marco, Y.; Deslandes, L. RD19, an Arabidopsis Cysteine Protease Required for RRS1-R-Mediated Resistance, Is Relocalized to the Nucleus by the Ralstonia Solanacearum Popp2 Effector. *Plant Cell Online* **2008**, *20*, 2252–2264. [CrossRef]

59. Koizumi, M.; Yamaguchi-Shinozaki, K.; Tsuji, H.; Shinozaki, K. Structure and Expression of Two Genes That Encode Distinct Drought-Inducible Cysteine Proteinases in Arabidopsis Thaliana. *Gene* **1993**, *129*, 175–182. [CrossRef]

60. Pieterse, C.M.; Leon-Reyes, A.; Van der Ent, S.; Van Wees, S.C. Networking by Small-Molecule Hormones in Plant Immunity. *Nat. Chem. Biol.* **2009**, *5*, 308–316. [CrossRef]

61. Brooks, D.M.; Hernandez-Guzman, G.; Kloek, A.P.; Alarcon-Chaidez, F.; Sreedharan, A.; Rangaswamy, V.; Penaloza-Vazquez, A.; Bender, C.L.; Kunkel, B.N. Identification and Characterization of a Well-Defined Series of Coronatine Biosynthetic Mutants of Pseudomonas syringae Pv. Tomato DC3000. *Mol. Plant Microbe Interact.* **2004**, *17*, 162–174. [CrossRef]

62. Desclos-Theveniau, M.; Arnaud, D.; Huang, T.Y.; Lin, G.J.; Chen, W.Y.; Lin, Y.C.; Zimmerli, L. The Arabidopsis Lectin Receptor Kinase Lecrk-V.5 Represses Stomatal Immunity Induced by Pseudomonas syringae Pv. Tomato DC3000. *PLoS Pathog.* **2012**, *8*, e1002513. [CrossRef]

63. Chakravarthy, S.; Butcher, B.G.; Liu, Y.; D'Amico, K.; Coster, M.; Filiatrault, M.J. Virulence of Pseudomonas syringae Pv. Tomato DC3000 Is Influenced by the Catabolite Repression Control Protein Crc. *Mol. Plant Microbe Interact.* **2017**, *30*, 283–294. [CrossRef]

64. King, E.O.; Ward, M.K.; Raney, D.E. Two Simple Media for the Demonstration of Pyocyanin and Fluorescin. *J. Lab. Clin. Med.* **1954**, *44*, 301–307.

65. Betancourt, L.H.; Sanchez, A.; Pla, I.; Kuras, M.; Zhou, Q.; Andersson, R.; Marko-Varga, G. Quantitative Assessment of Urea in-Solution Lys-C/Trypsin Digestions Reveals Superior Performance at Room Temperature over Traditional Proteolysis at 37 °C. *J. Proteome Res.* **2018**, *17*, 2556–2561. [CrossRef]

66. Mertins, P.; Tang, L.C.; Krug, K.; Clark, D.J.; Gritsenko, M.A.; Chen, L.; Clauser, K.R.; Clauss, T.R.; Shah, P.; Gillette, M.A.; et al. Reproducible Workflow for Multiplexed Deep-Scale Proteome and Phosphoproteome Analysis of Tumor Tissues by Liquid Chromatography–Mass Spectrometry. *Nat. Protoc.* **2018**, *13*, 1632–1661. [CrossRef]

67. Chen, C.J.; Chen, W.Y.; Tseng, M.C.; Chen, Y.R. Tunnel Frit: A Nonmetallic in-Capillary Frit for Nanoflow Ultra High-Performance Liquid Chromatography-Mass Spectrometryapplications. *Anal. Chem.* **2012**, *84*, 297–303. [CrossRef]

68. Scheltema, R.A.; Hauschild, J.P.; Lange, O.; Hornburg, D.; Denisov, E.; Damoc, E.; Kuehn, A.; Makarov, A.; Mann, M. The Q Exactive HF, a Benchtop Mass Spectrometer with a Pre-Filter, High-Performance Quadrupole and an Ultra-High-Field Orbitrap Analyzer. *Mol. Cell Proteom.* **2014**, *13*, 3698–3708. [CrossRef]

69. Bruderer, R.; Bernhardt, O.M.; Gandhi, T.; Xuan, Y.; Sondermann, J.; Schmidt, M.; Gomez-Varela, D.; Reiter, L. Optimization of Experimental Parameters in Data-Independent Mass Spectrometry Significantly Increases Depth and Reproducibility of Results. *Mol. Cell Proteom.* **2017**, *16*, 2296–2309. [CrossRef]

70. Craig, R.; Beavis, R.C. Tandem: Matching Proteins with Tandem Mass Spectra. *Bioinformatics* **2004**, *20*, 1466–1467. [CrossRef]

71. Eng, J.K.; Jahan, T.A.; Hoopmann, M.R. Comet: An Open-Source MS/MS Sequence Database Search Tool. *Proteomics* **2013**, *13*, 22–24. [CrossRef]

72. Deutsch, E.W.; Mendoza, L.; Shteynberg, D.; Slagel, J.; Sun, Z.; Moritz, R.L. Trans-Proteomic Pipeline, a Standardized Data Processing Pipeline for Large-Scale Reproducible Proteomics Informatics. *Proteom. Clin. Appl.* **2015**, *9*, 745–754. [CrossRef]

73. Röst, H.L.; Sachsenberg, T.; Aiche, S.; Bielow, C.; Weisser, H.; Aicheler, F.; Andreotti, S.; Ehrlich, H.C.; Gutenbrunner, P.; Kenar, E.; et al. OpenMS: A Flexible Open-Source Software Platform for Mass Spectrometry Data Analysis. *Nat. Methods* **2016**, *13*, 741–748. [CrossRef]

74. Röst, H.L.; Rosenberger, G.; Navarro, P.; Gillet, L.; Miladinović, S.M.; Schubert, O.T.; Wolski, W.; Collins, B.C.; Malmström, J.; Malmström, L.; et al. OpenSWATH Enables Automated, Targeted Analysis of Data-Independent Acquisition MS Data. *Nat. Biotechnol.* **2014**, *32*, 219–223. [CrossRef]

75. Lambert, J.P.; Ivosev, G.; Couzens, A.L.; Larsen, B.; Taipale, M.; Lin, Z.Y.; Zhong, Q.; Lindquist, S.; Vidal, M.; Aebersold, R.; et al. Mapping Differential Interactomes by Affinity Purification Coupled with Data-Independent Mass Spectrometry Acquisition. *Nat. Methods* **2013**, *10*, 1239–1245. [CrossRef]

76. Mi, H.; Huang, X.; Muruganujan, A.; Tang, H.; Mills, C.; Kang, D.; Thomas, P.D. Panther Version 11: Expanded Annotation Data from Gene Ontology and Reactome Pathways and Data Analysis Tool Enhancements. *Nucleic Acids Res.* **2017**, *45*, 183–189. [CrossRef]

77. Mi, H.; Muruganujan, A.; Casagrande, J.T.; Thomas, P.D. Large-Scale Gene Function Analysis with the Panther Classification System. *Nat. Protoc.* **2013**, *8*, 1551–1566. [CrossRef]

78. Huang, D.W.; Sherman, B.T.; Lempicki, R.A. Bioinformatics Enrichment Tools: Paths toward the Comprehensive Functional Analysis of Large Gene Lists. *Nucleic Acids Res.* **2009**, *37*, 1–13. [CrossRef]

Int. J. Mol. Sci. **2019**, *20*, 863

79. Huang, D.W.; Sherman, B.T.; Lempicki, R.A. Systematic and Integrative Analysis of Large Gene Lists Using David Bioinformatics Resources. *Nat. Protoc.* **2009**, *4*, 44–57. [CrossRef]

80. Vizcaino, J.A.; Cote, R.G.; Csordas, A.; Dianes, J.A.; Fabregat, A.; Foster, J.M.; Griss, J.; Alpi, E.; Birim, M.; Contell, J.; et al. The Proteomics Identifications (PRIDE) Database and Associated Tools: Status in 2013. *Nucleic Acids Res.* **2013**, *41*, D1063–D1069. [CrossRef]

International Journal of
Molecular Sciences

MDPI

Article

Proteomic Analysis of the Resistance Mechanisms in Sugarcane during *Sporisorium scitamineum* Infection

Pratiksha Singh [1,†], Qi-Qi Song [1,†], Rajesh Kumar Singh [2], Hai-Bi Li [1], Manoj Kumar Solanki [3], Mukesh Kumar Malviya [2], Krishan Kumar Verma [2], Li-Tao Yang [1,*] and Yang-Rui Li [1,2,*]

[1] Agricultural College, State Key Laboratory of Subtropical Bioresources Conservation and Utilization, Guangxi University, Nanning 530005, China; singh.pratiksha23@gmail.com (P.S.); alice771992@126.com (Q.-Q.S.); lihaibi@gxaas.net (H.-B.L.)

[2] Guangxi Key Laboratory of Sugarcane Genetic Improvement, Key Laboratory of Sugarcane Biotechnology and Genetic Improvement (Guangxi), Ministry of Agriculture, Sugarcane Research Center, Chinese Academy of Agricultural Sciences; Sugarcane Research Institute, Guangxi Academy of Agricultural Sciences, Nanning 530007, China; rajeshsingh999@gmail.com (R.K.S.); mukeshmicro@rediffmail.com (M.K.M.); drvermakishan@gmail.com (K.K.V.)

[3] Department of Food Quality and Safety, Institute for Post-harvest and Food Sciences, The Volcani Center, Agricultural Research Organization, Rishon LeZion 7528809, Israel; mkswings321@gmail.com

* Correspondence: liyr@gxaas.net (Y.-R.L.); liyr@gxu.edu.cn (L.-T.Y.); Tel./Fax: +86-771-3899033 (Y.-R.L.); +86-771-3236405 (L.-T.Y.)

† These authors contributed equally to this work.

Received: 11 December 2018; Accepted: 27 January 2019; Published: 29 January 2019

Abstract: Smut disease is caused by *Sporisorium scitamineum*, an important sugarcane fungal pathogen causing an extensive loss in yield and sugar quality. The available literature suggests that there are two types of smut resistance mechanisms: external resistance by physical or chemical barriers and intrinsic internal resistance mechanisms operating at host–pathogen interaction at cellular and molecular levels. The nature of smut resistance mechanisms, however, remains largely unknown. The present study investigated the changes in proteome occurring in two sugarcane varieties with contrasting susceptibility to smut—F134 and NCo310—at whip development stage after *S. scitamineum* infection. Total proteins from pathogen inoculated and uninoculated (control) leaves were separated by two-dimensional gel electrophoresis (2D-PAGE). Protein identification was performed using BLASTp and tBLASTn against NCBI nonredundant protein databases and EST databases, respectively. A total of thirty proteins spots representing differentially expressed proteins (DEPs), 16 from F134 and 14 from NCo310, were identified and analyzed by MALDI-TOF/TOF MS. In F134, 4 DEPs were upregulated and nine were downregulated, while, nine were upregulated and three were downregulated in NCo310. The DEPs were associated with DNA binding, metabolic processes, defense, stress response, photorespiration, protein refolding, chloroplast, nucleus and plasma membrane. Finally, the expression of CAT, SOD, and PAL with recognized roles in *S. scitamineum* infection in both sugarcane verities were analyzed by real-time quantitative PCR (RT-qPCR) technique. Identification of genes critical for smut resistance in sugarcane will increase our knowledge of *S. scitamineum*-sugarcane interaction and help to develop molecular and conventional breeding strategies for variety improvement.

Keywords: sugarcane; *Sporisorium scitamineum*; smut; proteomics; RT-qPCR; ISR

---

1. Introduction

Sugarcane (*Saccharum* spp.) is one of the most important industrial sources for crystal sugar, and it is cultivated across the world in tropical and subtropical countries. It is also the second largest biofuel crop and is an important source for many biomaterials. China is the fourth largest sugar producer in the world and the Guangxi province accounts for 92% of sugar production in China [1,2]. Sugarcane diseases are caused by bacteria, fungus, nematodes, virus, protozoa, phytoplasma, etc. with fungal diseases becoming a dominant group. Sugarcane smut disease, caused by the basidiomycete fungus *Sporisorium scitamineum* (Syn. *Ustilago scitaminea*), is a major sugarcane disease worldwide, and it can cause a 20–50% loss in cane yield [3], and up to 75% reduction in sugar production [4,5]. The fungus infects plants mainly through germinating buds in the soil or buds on standing stalks and grows in the plant in close association with the growing points or meristems, showing the presence of elongated whip, thin stalks, profuse tillering, and small narrow leaves.

Smut infection might also take place through the open stomata in leaves and the open areas in buds or wound in plant tissues and it is hard to control with chemicals in commercial crops [6]. To date, resistant varieties are the only practical, environmentally benign solution for managing sugarcane smut [7]. Breeding of smut-resistant sugarcane cultivars is a more economical and efficient approach to control the disease as compared to chemical treatments and agronomic practices [5,8,9]. Crop protection by modern genetic engineering technology is a potential tool to generate smut resistant varieties [10,11]. However, the complexity of sugarcane genetic background and the commercial viability of transgenic solutions for smut resistance make the genetic modification option unattractive [9,12]. Therefore, a better understanding of the biology, genetics, and molecular biology of smut resistance will greatly facilitate breeding sugarcane varieties for smut resistance.

Resistance mechanisms of sugarcane to smut involve external and internal disease resistance [13,14], and both mechanisms may confer resistance individually or in combination [15]. The external resistance is achieved by a physical barrier resulting from a mixture of bud structural characteristics [16], the thickness of the bud scales and chemicals such as phenyl-propanoids and glycosyl-flavonoids [17–19]. In the case of internal resistance, expressed after the pathogen attacks and penetrates through the bud scale: this is by several defense responses as well as increased lignin concentration [20], production of glycoprotein, phytoalexins, and polyamines [21–24]. At present, the exact nature of internal molecular defense mechanisms induced by smut remains less studied [15].

In previous reports, biochemical and genomics aspects of smut resistance were investigated, but not so much from a host–pathogen interaction perspective. A series of biochemical and molecular responses, such as triggering of specific defense signal transduction pathways, secretion of pathogenesis-related (PR) proteins and phytoalexins, and oxidative bursts occur in plants, during the stage of pathogen attack and the subsequent plant–pathogen interaction [25–27]. Next-generation sequencing based approaches were used to analyze the total changes in transcripts of resistant and susceptible sugarcane cultivars during sugarcane–*S. scitamineum* interaction [28,29]. Proteomic approaches offer powerful tools to study the expression of proteins and their function associated with plant-microbe interactions etc. [30]. One-dimensional gradient polyacrylamide gels (1DE), 2DE, and MS methods have been previously utilized to analyze the sugarcane proteome under various abiotic and biotic stresses [31]. Barnabas et al. [32] reported a total of 53 sugarcane differentially expressed proteins (DEPs) related to defense, stress, protein folding, and cell division. In addition, a putative effector of *S. scitamineum* pathogenesis, chorismate mutase, was found in sugarcane after *S. scitamineum* inoculation. The identification of DEGs during pathogen interaction is essential to enhance our understanding of plant resistance and offer clues as to what kind of defensive and biochemical mechanisms being regulated in a particular situation.

Therefore, the present study was mainly aimed at understanding the proteomic changes in the leaf of two sugarcane varieties with contrasting smut resistance (F134- resistant to *S. scitamineum* race 1 but susceptible to race 2 and NCo310- resistant to *S. scitamineum* race 2 but susceptible to race 1), which were planted after artificial inoculation with smut pathogen (susceptible to race 1 and race 2). The changes in contents of phyto-hormones (cytokinin (CYT) and ethylene (ETH)), as well as changes in the activity and expression of antioxidant enzymes (superoxide dismutase (SOD), catalase (CAT), and phenylalanine ammonia lyase (PAL)), were analyzed at different time intervals during the interaction. This also included a detailed study of proteome level alterations, gene expression, as well as biochemical changes sugarcane and smut pathogen interaction. To the best of our knowledge, the current study is the first comprehensive comparative proteomic analysis of sugarcane–smut pathogen interaction process in varieties with contrasting smut resistance after whip appearance stage. Expression of genes thought to be important for the pathogenesis is quantified to validate the proteomic data. The results obtained from this study clearly advance our molecular understanding of smut resistance in sugarcane, providing leads for identifying candidate genes and molecular markers for smut resistance.

2. Results

2.1. DE Analysis of Differentially Expressed Proteins in Sugarcane after S. scitamineum Inoculation

The infection of sugarcane plants inoculated with teliospores was first detected by a positive PCR reaction for molecular detection of *S. scitamineum* with pathogen species-specific primer in both varieties F134 and NCo310 (Figure S1), and it was further confirmed by anatomical changes observed between the healthy and infected plantlets under TEM. Figure 1 presents a 2D gel profile showing DEPs observed in control and smut-inoculated sugarcane plantlets, in which alteration was apparent for in the order of 80% of the spots. In the present study, a total of thirty DEPs were found, sixteen in F134 and fourteen in NCo310. Out of these proteins, the expression of four of them (spots 3, 4, 6, and 9) were upregulated and twelve (spots 1, 2, 7, 8, 16–19, and 21–24) were downregulated, as shown in Table 1. Whereas, in NCo310, the expression of eleven proteins (spots 1–4, 6, 7, 9, 10, 21, 23, and 24) were upregulated and three (spots 11, 20, and 22) were downregulated, as shown in Table 2. The identity of all the above proteins was established except for those located in spots 3, 8, and 21 in F134 and spots 3 and 21 in NCo310. All DEPs were individually collected for matrix assisted laser desorption ionization time-of-flight mass spectrometry (MALDI-TOF-TOF/MS) analysis.

Table 1. Identification of differentially expressed proteins in variety F134 during the sugarcane–*Sporisorium scitamineum* interaction.

Spot No.	Accession No.	Identified Protein	Species	Protein				Expression
				Mw (Da)	(pI)	Score	Score C.I. (%)	
1	108796050	Pyruvate orthophosphate dikinase	*Saccharum officinarum*	102,293.9	5.5	977	100	Down
2	6911551	Heat shock protein 70	*Cucumis sativus*	71,444.1	5.07	111	99.994	Down
4	25067747	Thioredoxin/transketolase fusion protein	synthetic construct	86,860.6	5.59	173	100	UP
6	147843754	Hypothetical protein	*Vitis vinifera*	42,395.7	6.29	158	100	UP
7	17225494	Translational elongation factor Tu	*Oryza sativa*	50,381.8	6.19	477	100	Down
9	162463757	Nucleic acid binding protein1	*Zea mays*	33,096.6	4.6	87	98.709	UP
16, 17, 19	56182370	Putative thioredoxin peroxidase 1	*Saccharum officinarum*	10,780.7	4.9	317	100	Down
18	125543336	Hypothetical protein OsI_010708	*Oryza sativa* (indica cultivar group)	22,687.1	5.79	283	100	Down
22	1568639	Cu/Zn superoxide dismutase	*Triticum aestivum*	20,310.4	5.35	310	100	Down
23, 24	3914607	Ribulose bisphosphate carboxylase small chain, chloroplast precursor	*Saccharum officinarum*	19,023.5	9.04	464	100	Down

Spot numbers matched up to 2-DE gel in Figure 1; Expression relation was measured comparative to protein expression in control sample.

Table 2. Identification of differentially expressed proteins in variety NCo310 during sugarcane–*Sporisorium scitamineum* interaction.

Spot No.	Accession No.	Identified Protein	Species	Protein				Expression
				Mw (Da)	(pI)	Score	Score C.I. (%)	
1	108796050	Pyruvate orthophosphate dikinase	*Saccharum officinarum*	102,293.9	5.5	977	100	UP
2	6911551	Heat shock protein 70	*Cucumis sativus*	71,444.1	5.07	111	99.994	UP
4	25067747	Thioredoxin/transketolase fusion protein	synthetic construct	86,860.6	5.59	173	100	UP
6	56554972	Heat shock protein 70	*Medicago sativa*	70,952.1	5.08	144	100	UP
7, 10	17225494	Translational elongation factor Tu	*Oryza sativa*	50,381.8	6.19	334	100	UP
9	162463757	Nucleic acid binding protein1	*Zea mays*	33,096.6	4.6	87	98.709	UP
11	115488160	Os12g0277500	*Oryza sativa* (japonica cultivar-group)	61,092.6	5.12	316	100	Down
20	13160411	Adenosine diphosphate glucose pyrophosphatase	*Hordeum vulgare* subsp. *vulgare*	21,787.1	5.68	101	99.945	Down
22	115439131	Os01g0675100	*Oryza sativa* (japonica cultivar-group)	17,280.1	5.58	100	99.929	Down
23, 24	3914607	Ribulose bisphosphate carboxylase small chain, chloroplast precursor	*Saccharum officinarum*	19,023.5	9.04	464	100	UP

Spot numbers matched up to 2-DE gel in Figure 1; Expression relation was measured comparative to protein expression in control sample.

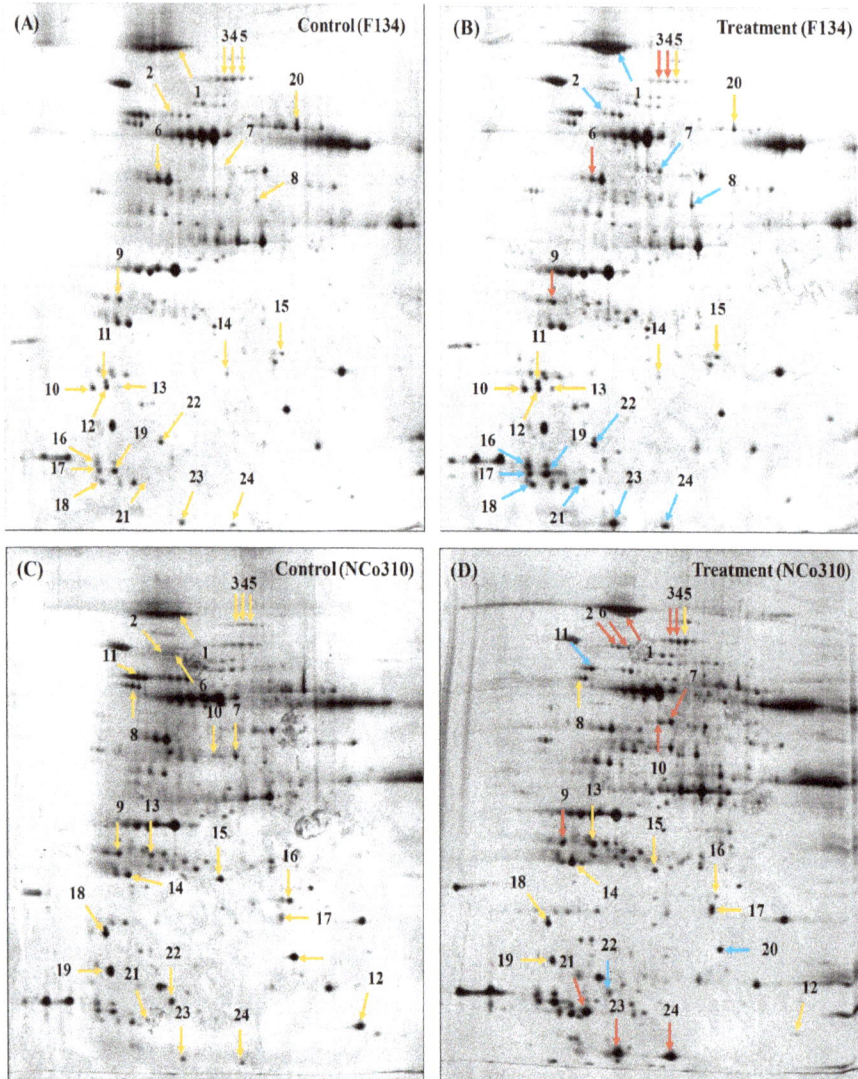

Figure 1. 2-DE SDS-PAGE gel pictures of sugarcane varieties F134 and NCo310 with their controls. (**A**) control (F134), (**B**) treatment (F134), (**C**) control (NCo310), and (**D**) treatment (NCo310). Yellow for the protein spots of interest, red for upregulated proteins, and blue for downregulated proteins.

2.2. MALDI-TOF-TOF/MS Analysis of Differentially Expressed Proteins

The results of MALDI-TOF-TOF/MS analysis of sixteen and fourteen differentially expressed proteins in F135 and NCo310 sugarcane varieties are shown in Figure S2 and S3. In both varieties, the function of some proteins was not identified. In infected F134, three spots (16, 17, and 19), and two spots (23 and 24) were identified as the same protein, whereas three spots (3, 8, and 21) were not identified (Table 1). However, in infected NCo310 variety two spots number (7 and 10), (23 and 24) were identified as the same protein, whereas two spots (3 and 21) were not identified (Table 2).

Peptide mass fingerprinting and tandem mass spectra of thirty proteins were achieved (Figure 2). Thirteen and twelve proteins had known functions and the sequence similarity was known to those proteins in both varieties. Based on bioinformatics analysis these proteins were found to be related to molecular processes, cellular components and categorized into numerous functional groups, i.e., peroxidase activity, DNA binding, metabolic processes, defense and stress responses, photorespiration, protein refolding, chloroplast thylakoid membrane, nucleus, plasma membrane, chloroplast, and proton-transporting ATP synthase complex (Table S1).

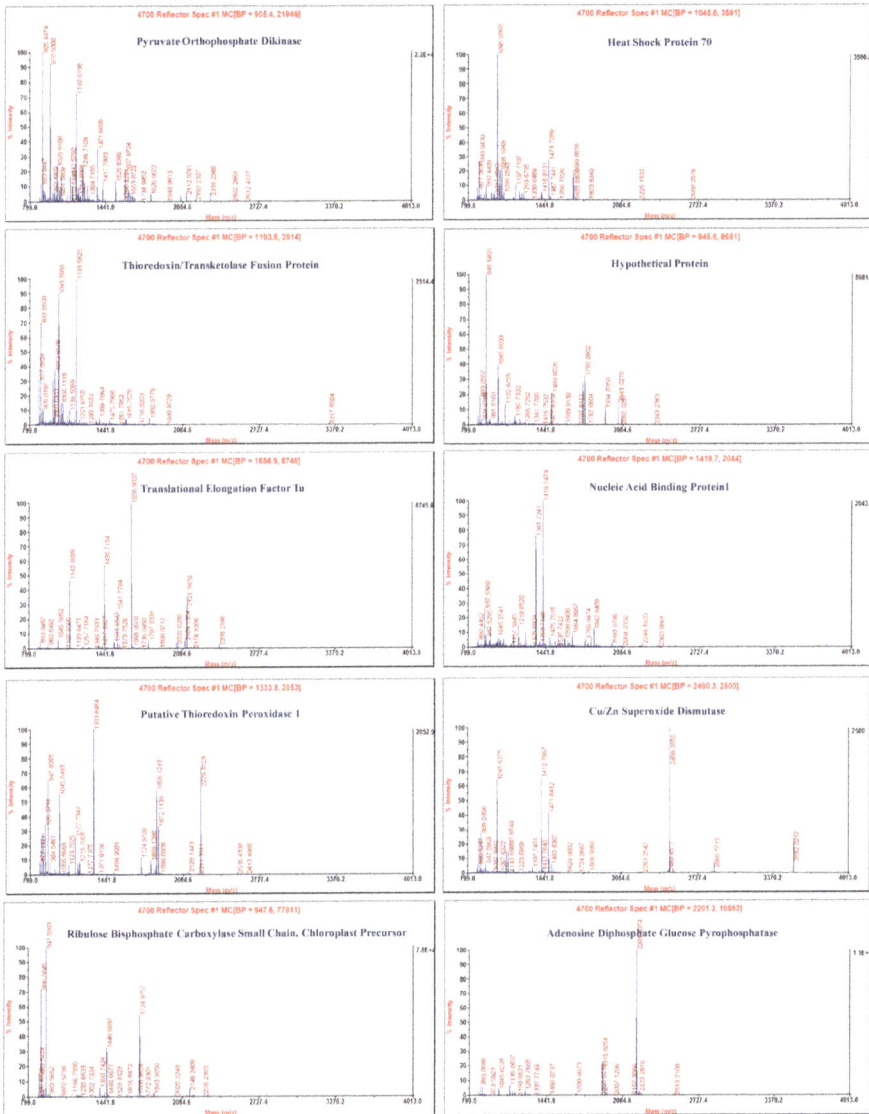

Figure 2. Peptide mass fingerprinting of differentially expressed proteins associated with sugarcane varieties F134 and NCo310 after *Sporisorium scitamineum* inoculation.

2.3. Genes Expression Analysis by Real-Time Quantitative PCR (qRT-PCR)

In the present study, gene expression of CAT, SOD, and PAL were studied by qRT-PCR in leaf tissues at different stages of crop growth (Figure 3). The data showed a significant change in the expression level of all the three selected genes in both sugarcane varieties in response to smut infection. In comparison, the expression level of CAT increased significantly during the 180 days after planting in both F134 and NCo310. The maximum difference in CAT expression was observed between the two varieties at 30 days as compared to 60 and 90 days after planting. The difference in CAT expression between 60 and 90 days after planting in both varieties was not significant (Figure 3A). In response to smut infection, the expression patterns of SOD were similar in F134 and NCo310 varieties and continuously increased from day 30 to 180 sampling period after planting, although there was a dip in the expression on day 90 in both varieties (Figure 3B). There was no consistent expression pattern for PAL in both sugarcane varieties (Figure 3C). For F134, a significant increase in PAL expression was observed at 30 and 180 days, while it was decreased on day 60 and 90. There was no significant change in the level of PAL expression in NCo310 (Figure 3C). The comparative expression level of all the three genes was calculated and the results indicate that the highest expression of CAT, SOD, and PAL was approximately 20, 3, and 28 times at 180, 60, and 30 days, respectively, while the lowest (3, 2, and 3 times) was observed on day 90, 180, and 90 days in F134 as compared to NCo310.

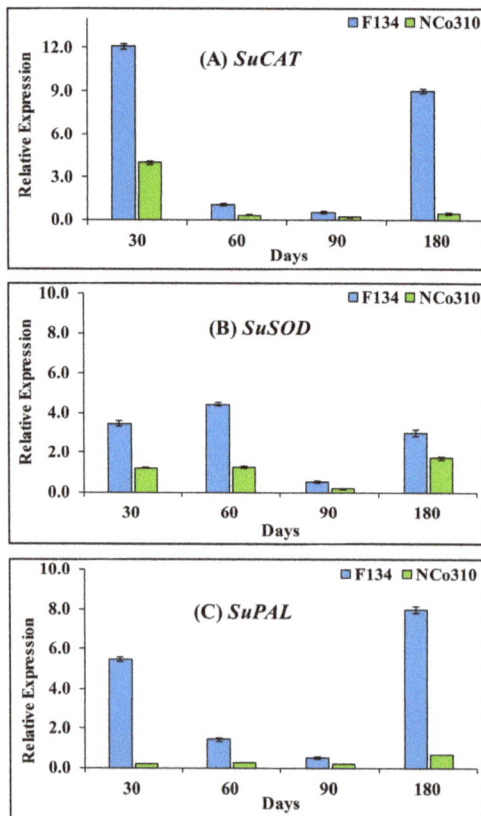

Figure 3. qRT-PCR analysis of differentially expressed genes in leaf tissue of sugarcane varieties F134 and NCo310 during sugarcane–*S. scitamineum* interaction. (**A**) Catalase (*SuCAT*), (**B**) superoxide dismutase (*SuSOD*), and (**C**) phenylalanine ammonia lyase (*SuPAL*). Data were normalized to the GAPDH expression level. All data points are the mean $\pm$ SE ($n = 3$).

2.4. Induction of Antioxidant Defense System by S. Scitamineum Infection

The activities of antioxidant enzymes SOD, PAL and CAT, and hormone contents were calculated for different time intervals (30–180 days) in both F134 and NCo310 following smut infection. Both leaf and root tissues were separately tested to measure the enzyme activities (Figure 4).

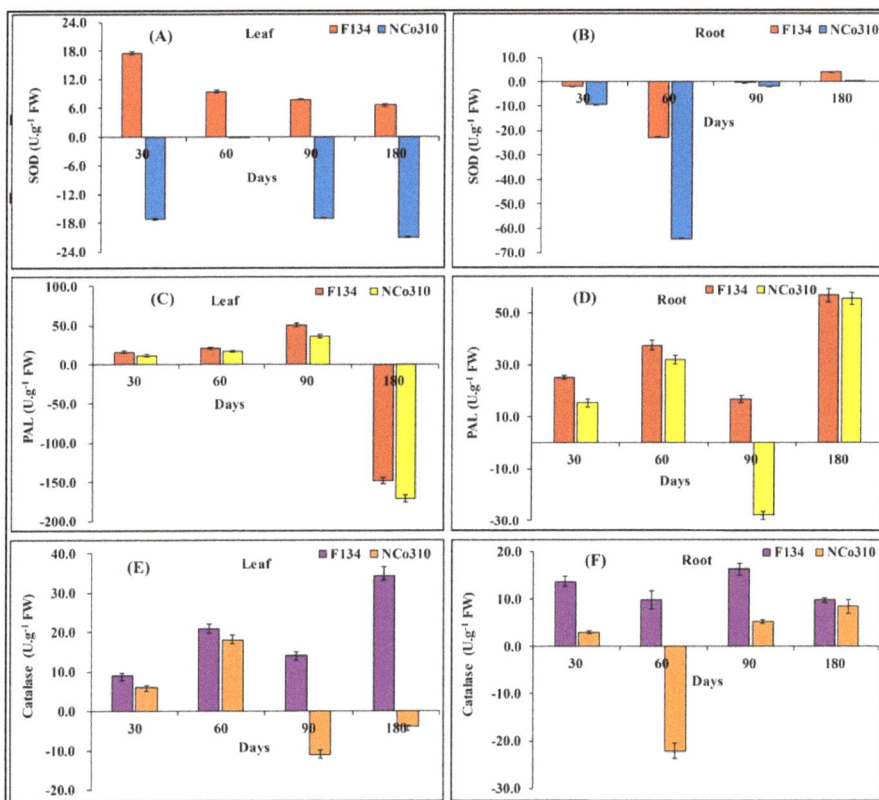

Figure 4. Analysis of enzyme activities in leaf and root tissues of sugarcane varieties F134 and NCo310 infected with *S. scitamineum* stress. (**A,B**) Superoxide dismutase, (**C,D**) phenylalanine ammonia lyase, and (**E,F**) Catalase. All data points (with the subtraction of their controls) are the mean ± SE (*n* = 3).

The data revealed that *S. scitamineum* inoculation caused an increase in the SOD activity in F134 sugarcane variety. In the leaf tissue of F134, the SOD activity was highest at 30 days (17.54 U.g^{-1}FW) after planting and then it decreased with time. In NCo310, the activity of SOD was lower as compared to the control (Figure 4A). For root, the activity of SOD in F134 variety showed a small increase at 180 days, but in general, both varieties showed a reduction in SOD activity (Figure 4B). The SOD activity was higher in leaf, after the smut pathogen treatment for F134 than the control and NCo310.

After *S. scitamineum* infection, the initial activity of PAL in the leaf showed a significant increase in both varieties and that trend was observed up to 90 days as compared to control (Figure 4C). The PAL activity was significantly higher in root for both varieties and peaked at 90 and 180 days after infection in NCo310 and F134, respectively. (Figure 4D).

The CAT activity in the leaf did not significantly differ in the first 30 days following inoculation but it increased subsequently with the maximum activity occurring on day180 in F134 (34.26 U.g^{-1}FW) and on day 60 in NCo310 (18.10 U.g^{-1}FW) compared with the control (Figure 4E). In the root system of both sugarcane varieties, a significant difference in CAT activity was found compared to the control. In F134, the maximum CAT activity was observed on day 90 following smut. But in the case of NCo310, except at 60 days, an increasing trend in the CAT activity was noticed for the other time intervals (Figure 4F).

2.5. Phytohormone Levels as Affected by S. Scitamineum Infection

The levels of different hormones (ethylene and cytokinin) in roots and leaves of infected and control sugarcane plants showed a large, significant difference between treatments. The treatment of plants with *S. scitamineum* led to elevated cytokinin levels in leaves at different time intervals. In comparison with the control, the highest cytokinin activity was recorded at 90 and 180 days in F134, and it was increased with time reaching a maximum of 9.87 ng.g^{-1}FW; whereas, in the infected NCo310, a lower level of cytokinin was produced as compared to control (Figure 5A). In the root tissue, change in cytokinin content was less and nonsignificant as compared to control except for that at 60 days in F134 and at 60 and 180 days in NCo310 (Figure 5B).

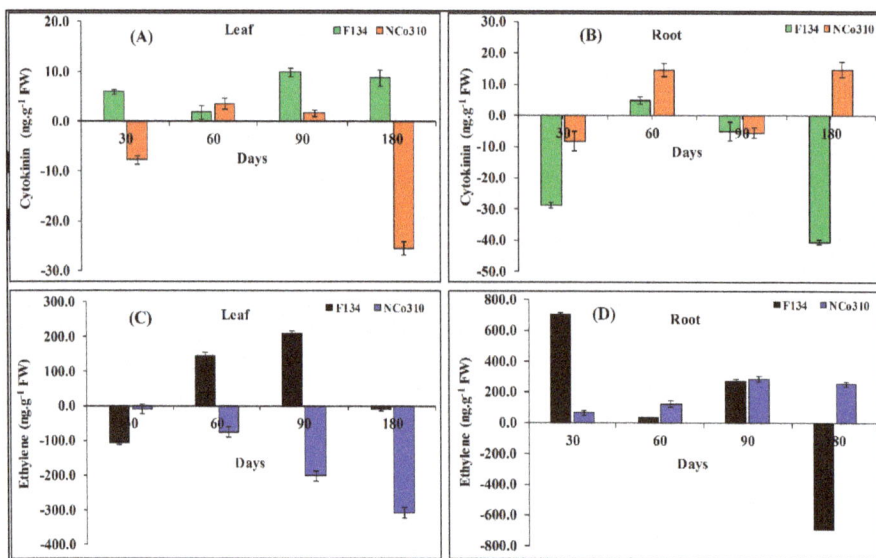

Figure 5. Analysis of phytohormone levels in leaf and root tissues of sugarcane varieties F134 and NCo310 infected with *S. scitamineum* stress. (**A,B**) Cytokinin and (**C,D**) ethylene. All data points (with the subtraction of their controls) are the mean ± SE (*n* = 3).

For ethylene, the level was lower in the leaves and higher in the roots as compared with the control (Figure 5C,D). The ethylene level was significantly higher in the leaf of F134 at 60 and 90 days than that in the control and showed a maximum (208.24 ng.g^{-1}FW) at 90 days. In NCo310, the level of ethylene was not significantly different to that of control (Figure 5C). In the root, the level of ethylene was increased in both varieties as compared to the control at each time interval expect for that at 180 days in F134, A maximum hormone level was observed at 30 days (704.26 ng.g^{-1}FW) followed by 90 days in F134 (Figure 5D). Whereas, in NCo310, the ethylene level showed a maximum value (284.30 ng.g^{-1}FW) at 90 days.

3. Discussion

Proteomics based studies of plant–pathogen interaction contributes to a better understanding of the molecular and biochemical aspects of plant diseases. Plant–pathogen interaction is very complex due to the interaction of morphological, environmental, physiological, molecular, and metabolic factors with the pathogen. [31]. In the present research, for the most part, discusses the proteomic and gene expression responses occurring between two sugarcane varieties—F134 and NCo310—after the appearance of smut symptoms. A total of 16 DEPs in F134 and 14 in NCo310 were observed compared to their controls. After MALDI-TOF/TOF analyses these proteins were classified into different categories based on their association with various molecular, biological, and cellular processes.

Defense associated proteins: Heat shock protein 70 (HSp70), thioredoxin/transketolase fusion protein, putative thioredoxin peroxidase 1, and Cu/Zn superoxide dismutase are defense-related proteins that were differentially expressed and identified in this study. Upregulation of both HSp70 (spots 2 and 6) and thioredoxin/transketolase fusion protein (spot 4) was shown in NCo310. However, downregulation of HSp70 (spot 2), putative thioredoxin peroxidase 1 (spots 16, 17, and 19), and Cu/Zn superoxide dismutase (spot 22) was observed in F134, which suggest that in smut resistant sugarcane stress and defense-related proteins were upregulated during *S. scitamineum* infection as a defense strategy. HSP is a group of specific, conserved, and ubiquitous proteins distributed in the nucleus, cytoplasm, endoplasmic reticulum, mitochondria, and chloroplasts, and play an essential function in maintaining cellular functions when plants are subject to a variety of biotic and abiotic stresses such as heat stress, high salt, and heavy metal contamination [33,34]. In this research, the upregulation of two HSP70 proteins in NCo310 may be related to sugarcane defense response to smut invasion to protect the cellular structure, participate in denatured proteins refolding, maintain cell homeostasis and repair cellular dysfunction [35]. Previous studies also found that Hsp70s were upregulated under stress conditions during sugarcane–*S. scitamineum* interaction [6,36]. Downregulation of defense responsive proteins HSP-70 and DNAK-HSP 70 was observed in smut infected meristem cells at whip emergence stage by Barnabas et al. [32].

Peroxidases played important functions in defense and stress mechanisms and could be induced by several other physiological processes such as auxin catabolism, biosynthesis of lignin, cell wall stability, and senescence [37,38]. Many previous studies confirmed that reactive oxygen species (ROS), for example, hydrogen peroxide (H_2O_2), hydroxyl free radical (OH), and superoxide anion (O_2^-), are involved in the early resistance response in plants against pathogen attack [6]. In this study, the induction of three oxidative stress associated proteins—thioredoxin/transketolase fusion protein, putative thioredoxin peroxidase 1, and Cu/Zn superoxide dismutase—in F134 were observed. Likewise, the involvement and increased abundance of defense-related proteins—NTR, GST 1, STP, MDH, BQR, and SOD—accumulated probably because of damage related by means of extreme intra- and intercellular colonization, oxidative burst reaction of the host, and perturbance of normal cellular processes of sugarcane in infected meristem cells during whip emergence [32]. Thioredoxin-dependent peroxidases scavenge the excessive ROS and defend the sugarcane from smut pathogen attack [6].

Photosynthesis-associated proteins: Photosynthesis is an essential physiological process of the plant which plays a vital role in the development of the C_4 crop. RuBisCO is a key enzyme in photosynthesis and a heteropolymer consisting of eight large subunits (RbCLs) and eight small subunits (RbCSs), which regulates photosynthesis and light respiration [39]. RuBisCO activity could be induced in several biotic and abiotic stress conditions [40,41] and increases photorespiration plus ROS production: the essential component of the hypersensitive defense response. The addition of these toxic components impairs cell death suppression and counteracts the efficiency of plant defenses to control pathogen infection [42]. In the present research, the expression of RuBisCO small subunit protein (spots 23 and 24) was upregulated in F134 and downregulated in NCo310, which suggested that upregulation of this enzyme may improve the growth of the NCo310 sugarcane variety with increased smut resistance. Similarly, two RuBisCO large subunits and one RuBisCO small subunit were upregulated after smut pathogen infection [6]. The expression of photosynthesis-related proteins was upregulated during the sugarcane and *S. scitamineum* interaction, which was favorable for the protection of the photosynthetic system in opposition to pathogen attack [36]. A *Nicotiana benthamiana* RuBisCO small subunit also played a vital role in tobacco virus movement and plant antiviral defense [43]. RuBisCO was upregulated significantly in *Rosa roxburghii* Tratt resistance to powdery mildew infection [44].

Pyruvate, orthophosphate (Pi) dikinase (PPDK) (spot 1), which is another photosynthesis-related protein, was also found to be upregulated in NCo310, whereas it was downregulated in F134 after *S. scitamineum* interaction. PPDK is a chloroplastic C_4 cycle enzyme, catalyzes the ATP- and Pi-dependent formation of phosphoenolpyruvate (PEP), the primary CO_2 acceptor molecule [45]. Chen et al. [46] observed that PPDK protein was considerably downregulated in maize responding to sugarcane mosaic virus (SCMV) infection using the first systemically infected leaves.

Other functional proteins: Translational elongation factor Tu (EF-Tu) is a protein that plays an essential function in the elongation phase of protein synthesis in plastids in plants. Spots 7 and 10 in NCo310 and spot 7 in F134 were identified as translational elongation factor Tu, and upregulation was observed in both. Fu et al. [47] indicated that EF-Tu plays a key role in the mechanisms of disease resistance and heat tolerance in plants. Nucleic acid binding protein 1 (spot 9) was upregulated in both varieties. In NCo310 variety, spot 20 identified as adenosine diphosphate glucose pyrophosphatase protein was downregulated but not expressed in F134. AGPPase catalyzes the hydrolytic breakdown of ADP glucose (ADPG) to produce equimolar amounts of glucose-1-phosphate and AMP in both mono- and dicotyledonous plants [48]. The induced expression of all the above proteins in sugarcane was useful for resisting the *S. scitamineum* infection.

Unknown and hypothetical proteins: The expression of two unknown proteins (spot 3 and 21) were also observed in both varieties, which may play a role in smut resistance in sugarcane. One hypothetical protein (spot 18) in F134 along with Os12g0277500 (spot 11) and Os01g0675100 (spot 22) were also identified in NCo310.

The RT-qPCR method was used to compare the expression of antioxidant enzymes (CAT, SOD, and PAL) at different developmental stages in both sugarcane varieties after smut pathogen interaction. The expression of all these enzymes was constantly elevated in F134 than NCo310, showing a positive response against disease resistance. In previous reports, PAL, which catalyzes an important step in the phenylpropanoid pathway, participated in sugarcane resistance to smut [32,49] and also played a role in resistance to chilling, drought, and salt stresses in sugarcane [50]. The activity of catalase, an iron porphyrin enzyme, was always higher in a smut resistant sugarcane variety than a susceptible variety, which protected sugarcane against reactive oxygen-related stimuli [51]. The expression of three different maize catalase genes was regulated differentially in response to fungal toxin [52]. Jain et al. [53] reported higher activity of SOD protects cells against ROS in water deficit conditions.

Phytohormones cytokinin and ethylene play an essential role in plants against the pathogen attacks [54]. In the present study, the different levels of both hormones were observed in leaves and roots, and their increased levels in F134 variety suggest their possible involvement in defense response. According to Rivero et al. [55], transgenic plants over producing cytokinins protected the plants from the harmful effects of abiotic stresses. Ethylene synthesis as a response to different stresses [56] is typically associated with various environmental stresses including in the resistance response of sugarcane to *S. scitamineum* [57,58].

In conclusion, the present study reports the proteomic responses of two sugarcane varieties with contrasting resistance to smut infection, F134 and NCo310, to *S. scitamineum* infection. The results showed significant DEPs expression in both varieties, and also in the plants inoculated with *S. scitamineum*. A total of 30 proteins including four upregulated and nine downregulated in F134, and nine upregulated and three downregulated in NCo310 after smut infection were identified. The protein peptide mass finger printing and tandem mass spectra of these proteins were successfully obtained in both verities. Bioinformatics investigation discovered that the functions of these 30 DEPs were related to various molecular and cellular functions associated with pathogenesis and plant defense mechanisms. The identified proteins were categorized into functional groups involved in peroxidase activity, DNA binding, metabolic processes, defense, stress responses, photorespiration, protein refolding, chloroplast thylakoid membrane, nucleus, plasma membrane, chloroplast, and proton-transporting ATP synthase complex. This is the first report of the proteomic exploration of the interactions between sugarcane interactions and *S. scitamineum*.

4. Materials and Methods

4.1. Plant Material, Source of Inoculum, and Inoculation

Two sugarcane varieties (F134 and NCo310) with contrasting *Sporisorium scitamineum* susceptibility were used in this study. F134 is resistant to *S. scitamineum* race 1 but susceptible to race 2 and NCo310 is resistant to *S. scitamineum* race 2 but susceptible to race 1. For the isolation of teliospores, mature plants of sugarcane variety ROC22 (susceptible to *S. scitamineum* race 1 and 2) were cut into 10 to 20 cm below the shoot top and placed in a sterile polythene bag. The plantlets of all sugarcane varieties were provided by Sugarcane Research Institute, Guangxi Academy of Agricultural Sciences, Nanning, China. The suspensions of teliospore were made by adding 0.1 g of *S. scitamineum* teliospores into 100 mL of sterile distilled water with a drop of Tween 20 and mixed properly with a magnetic stirrer [15]. The suspension of spore concentration was maintained to 5×10^6 teliospores mL^{-1} by counting with a hemocytometer. The teliospores were incubated on potato dextrose agar (2%) at $30 \pm 2\,°C$ for 5 to 6 h to evaluate the germination rate. The viability of teliospores before inoculation was tested to confirm the sprouting ratio of > 90% [59]. The seedcanes of sugarcane varieties used in this study were cut into single-bud setts after removal of all the leaves and grown in the trays under controlled conditions ($30 \pm 2\,°C$, >80% relative humidity) for one month. For inoculation, one group of 30 healthy sugarcane plantlets was selected and immersed in *S. scitamineum* teliospores suspension for 2 h (treatment), and the other group of 30 sugarcane plantlets was treated with water as a control [1]. Then planted in pots (35 cm in diameter, 40 cm in depth) containing soil and sand mixture (3:1 w/w) in the greenhouse at Guangxi University, Nanning, China. Each variety had five sets of biological replicates with three plantlets in each replicate. Plants were arranged in a completely randomized design in the greenhouse. All sugarcane plants were irrigated once a day. Leaf and root samples were collected after 30, 60, 90, and 180 days of infection. Plant infection was confirmed by PCR based method and microscopic examination in both F134 and NCo310 [1]. All collected samples were immediately stored at $-80\,°C$, until used for protein and RNA extraction.

4.2. Protein Extraction and Quantification

Sugarcane leaf samples were ground to fine powder under liquid nitrogen using a pestle and mortar. Extraction of total proteins followed the modified procedure described earlier [30]. Two grams of test sample powder were homogenized with 4 mL of cold extraction buffer ($4\,°C$)([containing ($g·L^{-1}$) Tris-HCl 0.25 M (pH-7.5), Sucrose 24%, EDTA-Na_2 (Ethylene diamine tetra acetic acid disodium salt) 50 mM, SDS 2%, β-mercaptoethanol 2%, and PVP (Polyvinylpyrrolidone) 2%) and 8 mL of saturated phenol with Tris-HCl (0.25 M, pH 8.0) and β-mercaptoethanol (2%) was added before completing the maceration. Again 6 mL of extraction buffer was added and continued homogenizing till the preparation became a fine slurry. The homogenates were transferred to the centrifuged tube and mix properly for 1 min, and then centrifuged at room temperature at 14,000× g for 30 min. The supernatant was collected and re-extracted twice by adding an equal volume of extraction buffer without PVP in the centrifuged tube, followed by centrifugation at 14,000× g for 15 min. The supernatants were combined and proteins were precipitated overnight at $-20\,°C$ with cold methanol solution (1:5 v/v) (containing ammonium acetate 100 mM and β-mercaptoethanol 10 mM), then centrifuged at 12,000 × g for 10 min at $4\,°C$. The protein pellets were washed with cold methanol solution and centrifuged again at $4\,°C$ to collect the protein pellets which were air dried under ice. The protein pellets were solubilized in rehydration solution (containing urea 8 M, CHAPS [3-(3-cholamidopropyl) dimethylammonio-1-propanesulfonate] 2% (w/v), thiourea 2 M, DTT (dithiothreitol) 40 mM, EDTA-Na_2 5 mM, and IPG buffer 1% (pH 4–7)) for 4 h at $28\,°C$. Finally, the proteins were centrifuged at 12,000 × g for 30 min at $4\,°C$ to remove all undissolved particles and kept them at −80 °C. Total protein concentration was estimated according to the method described by Bradford [60] using bovine serum albumin (BSA) as the standard.

4.3. 2-DE, Image Acquisition, and Analysis

The immobilized pH gradient (IPG) gel strip (17 cm, pH 4–7, Bio-Rad) was carried out with first dimensional isoelectric focusing (IEF) on a PROTEAN IEF Cell apparatus (Bio-Rad). The protein extracts were dissolved at room temperature before rehydrating with rehydration solution containing urea 8 M, CHAPS 4%, DTT 65 mM, IPG buffer (pH 4–7) 0.2%, and Bromophenol blue 0.001%. The IPG gel strips were rehydrated at 18 ± 2 °C for 10 ± 12 h with 200 mL of rehydration solution and mixed with 500 µg protein. The protein was focused at 20 °C: 50 V for 12 h, 250 V for 30 min, 1000 V for 1 h, 3000 V for 2 h, 10,000 V for 2 h, to provide an overall of 60 kVh. When IEF was complete, the IPG strips were incubated for 15 min in the equilibrated buffer solutions containing DTT for reduction (Buffer I) and then the strips were re-equilibrated for 20 min in the equilibrated buffer containing iodoacetamide for alkylation (Buffer II) respectively, following Bio-Rad protocol.

The second-dimensional separation (sodium dodecyl sulfate polyacrylamide gel electrophoresis, SDS-PAGE) was implemented with the gel concentration of 12.5% T (Bio-Rad), and after solidification of the gel, the strips were placed and 0.5% of low melting agarose containing a drop of bromophenol blue was used for gel sealing. The gels were run at 40 V for 30 min and then 100 V until the bromophenol blue dye reached at the end of the gel on a Bio-Rad PROTEAN II system, and the plate temperature was maintained at 18 ± 20 °C by water flow from thermostatic circulator. When the SDS-PAGE was completed, the gel was kept out from the tank, stained with the BioSafe Coomassie (Bio-Rad), and destained with clean dH_2O according to the manufacturer's procedure. Stained gels were imaged by a Gel Doc 2000 (Bio-Rad) image scanner. Protein spots detection, spot matching, background subtraction, normalization, quantitative intensity, and statistical analysis were accomplished by using PD-Quest advanced 2D analysis software (version 8.0 Bio-Rad). The spots that exhibited as a minimum 2-fold-change were taken for the further experiment and differences were considered significant at $p \leq 0.05$ level based on Student's *t*-tests.

4.4. Mass Spectrometry and Data Analysis

The protein spots of interest on 2-DE gels were excised using Proteome Works Spot Cutter (Bio-Rad) with a 1.5 mm cutting diameter. After three times washing with Milli-Q water, peptide samples were prepared as by Song et al. [6]. The eluted samples were suspended in 0.1 % trifluoroacetic acid and spotted on a 384 well MALDI target plate through air drying until all solvent was evaporated. The peptides analysis was completed by 4700 MALDI TOF/TOF plus analyzer (Applied Biosystems Sciex, Foster, California, USA). The initial MS data was observed via reflector mode with the 4000-laser intensity. The MS spectra were collected in 2kV positive mode with fragments produced by collision induced dissociation. The scope of Peptide Mass Fingerprinting ranged from 800 to 4000 Da. GPS Explorer (Applied Biosystems Foster, California, USA) software was used for raw data search and Mascot as a search engine by NCBI (nr) protein database. Product mass tolerance was set at ± 0.3 Da with trypsin as a search parameter; alkylation and phosphorylation modifications were accepted. The match peptides with confidence intervals more than 98 were considered to be statistically significant, while peptides with lower scores were excluded. The function of identified proteins was determined by the Gene Ontology.

4.5. RNA Extraction, cDNA Synthesis, and qRT-PCR

To investigate the expression patterns of the associated genes, i.e., the gene for CAT, SOD, and PAL enzymes, at the time points of 30, 60, 90, and 180 days following *S. scitamineum* inoculation, tissue samples were collected from the two test varieties—F134 and NCo310—for RT-qPCR analysis. The glyceraldehyde-3-phosphate dehydrogenase (GAPDH) gene served as the reference gene. Total RNA was extracted from100 mg leaf tissue collected in triplicate from control and infected sugarcane varieties after symptom appearance with trizol reagent (Tiangen, Beijing, China) following the manufacturer's guidelines. DNase I (Promega, Fitchburg, Wisconsin, USA) was used to eliminate the DNA impurity of RNA; the extracted RNA sample yield and purity were tested by a Nano photometer (Pearl, Implen-3780, Westlake, California, USA). One microgram of total RNA was used for single stranded cDNA synthesis by the Prime-ScriptTM RT Reagent Kit (TaKaRa, Dalian, China). Primers were designed as previously described with a reference gene [61] (Table 3). The specificity of primer sets was tested by melt curve examination and relative gene expression was determined by $2^{-(\Delta Ct \text{ target gene} - \Delta Ct \text{ reference gene})}$ method [62]. The relative expression of both genes was calculated by the expression level of the infected sample minus the expression level of control at individual time intervals. Quantitative Real-Time PCR analysis was carried out in a Real-Time PCR Detection System (Bio-Rad, Hercules, California, USA) in SYBR Premix Ex Tap™ II (TaKaRa, Kyoto, Japan) with five replicates. Each 20 µL reaction mixture contained 2 µL template (10 x diluted cDNA), 10 µL SYBR Premix Ex Tap™ II, 0.8 µL of each primer (10 µM), and 6.4 µL ddH$_2$O. For control, no RNA sample was used as the template. PCR conditions were 95 °C for 30 s, followed by 40 cycles of 95 °C for 5 s and 60 °C for 20 s in 96-well optical reaction plates. To confirm the specificity and amplification, a melting curve analysis was conducted. The relative quantification of *SuSOD*, *SuPAL*, and *SuCAT* genes to *GAPDH* was calculated by the $2^{-\Delta\Delta Ct}$ methods [62].

Table 3. Primers used in this study.

Gene	Primer	Sequence (5'-3')	Strategy	Reference
GAPDH	GAPDH-F	CTCTGCCCCAAGCAAAGATG	RT-qPCR	[61]
	GAPDH-R	TGTTGTGCAGCTAGCATTG		
SuPAL	PAL-F	CTCGAGGAGAACATCAAGAC	RT-qPCR	[50]
	PAL-R	GTGATGAGCTCCTTCTCG		
SuCAT	CAT-F	CTTGTCTGGAGCACATACACTTGGA	RT-qPCR	[63]
	CAT-R	TTCTCCGCATAGACCTTGAACTTTG		
SuSOD	SOD-F	TTTGTCCAAGAGGGAGATGG	RT-qPCR	[53]
	SOD-R	CTTCTCCAGCGGTGACATTT		
*b*East mating-type	*b*E4-F	CGCTCTGGTTCATCAACG	Genome PCR	[64]
	*b*E8-R	TGCTGTCGATGGAAGGTGT		

4.6. Determination of Biochemical Changes in Sugarcane

The quantitative changes in hormone (ethylene and cytokinin) content and enzyme (superoxide dismutase, phenylalanine ammonia lyase, and catalase) activity were estimated at 30, 60, 90, and 180 days after smut inoculation. Samples were randomly collected from both sugarcane varieties. Three replicates were used for all analyses. Fresh tissue samples were ground to make a fine powder under liquid nitrogen by using prechilled pestle and mortar. The measurements of hormone concentration level and enzyme activities were conducted by plant enzyme linked immune sorbent assays (ELISA) kit (Wuhan Colorful Gene Biological Technology Co., Ltd., Wuhan, China), following the manufacturer's procedure [1].

4.6.1. Antioxidant Enzyme Activities

Two grams of pulverized tissue samples from both treatment and control were homogenized in 9 mL of a 0.05 M phosphate buffer (pH 6.6) in a prechilled pestle and mortar. The homogenates were filtered through a C-18 extraction column and the filtrates were centrifuged at 15,000 for 20 min at 4 °C. The supernatant was collected and used for different enzymes activity analyses by plant ELISA kits. The whole extraction method was done at 4 °C. Briefly, ELISA was performed in 96-well microtiter plates coated with antigens against the selected enzymes. Forty microliters of test samples and 10 µL of antibodies were added in the antigen-coated wells and mixed gently. Along with, all three enzymes standard, blank and control wells was prepared separately according to manufacturer instructions, afterward, the plate was incubated for 30 min at 37 °C. The liquid in the plates was discarded after incubation, washed five times with washing buffer, and plates were air dried. Fifty microliters of

HRP-Conjugate (different for each enzyme) reagent was added to each well of all plates, except blank well. Later, another time incubated for 30 min at 37 °C and washed as described above. After adding a chromogen solution, A and B (50 µL) to each well, kept in dark condition for color development for 15 min at 37 °C. The reaction was stopped by adding 50 µL stop solution to each well and color change was observed, i.e., from blue to yellow. The standard and control wells showed appropriate color development. For the assay of all enzymes, the plates were read on an ELISA Reader (Thermo Scientific, Multiskan GO, Waltham, Massachusetts, USA) at 450 nm within 15 min after the addition of stop solution. Enzyme activities were calculated with a standard curve and represented as $U.g^{-1}FW$.

4.6.2. Hormone Extraction and Assay

One gram each of powdered test samples (leaf and root) was mixed with 1 mL of 80% chilled methanol slowly in mortar and pestle and homogenized thoroughly and then kept at 4 °C for overnight. The mixture was then centrifuged at FF [12,000 rpm for 15 min at 4 °C. The collected supernatant was filtered through a C-18 extraction column. The extracted samples from the column were vaporized to remove the methanol under vacuum condition with ice, the samples were dissolved in phosphate buffer (pH 7.5), and the level of different hormone concentrations was measured by plant ELISA kit as described above.

4.7. Statistical Analysis

All biochemical activities were measured by the concentration level of treatment with the subtraction of their controls at each time interval. Standard errors were calculated for all mean values of three replicates and differences were considered significant at the $p \le 0.05$ level by Student's t-test.

Supplementary Materials: Supplementary materials can be found at http://www.mdpi.com/1422-0067/20/3/569/s1.

Author Contributions: L.-T.Y., Y.-R.L., P.S., and R.K.S. proposed the idea and designed the experiments. P.S., Q.-Q.S., R.K.S., and H.-B.L. performed experimental works. M.K.S., M.K.M., and K.K.V. participated in data analysis. P.S., R.K.S., and Q.-Q.S. wrote the original draft preparation. LTY and YRL revised and finalized this article.

Funding: The authors thank Prakash Lakshmanan for manuscript editing. The present study was supported by the International Cooperation Program Project of China (2013DFA31600), the National High Technology Research and Development Program ("863" Program) of China (2013AA102604), Guangxi Special Funds for Bagui Scholars and Distinguished Experts (2013), Guangxi R & D Program Fund (GKG1222009-1B, GKN14121008-2-1, GK17195100), the Fund for Guangxi Innovation Teams of Modern Agriculture Technology (gjnytxgxcxtd-03-01), and the Fund of Guangxi Academy of Agricultural Sciences (2015YT02).

Conflicts of Interest: The authors declare no conflicts of interest.

References

1. Singh, P.; Song, Q.Q.; Singh, R.K.; Li, H.B.; Solanki, M.K.; Yang, L.T.; Li, Y.R. Physiological and Molecular Analysis of Sugarcane (Varieties-F134 and NCo310) During *Sporisorium scitamineum* Interaction. *Sugar Tech.* **2018**. [CrossRef]
2. Su, Y.; Wang, Z.; Xu, L. Early selection for smut resistance in sugarcane using pathogen proliferation and changes in physiological and biochemical indices. *Front. Plant Sci.* **2016**, *7*, 1133. [CrossRef] [PubMed]
3. Wang, B.H. Current status of sugarcane diseases and research progress in China. *Sugar Crops China* **2007**, *3*, 48–51.
4. Su, Y.C.; Wang, S.S.; Guo, J.L.; Xue, B.T.; Xu, L.; Que, Y. A TaqMan real-time PCR assay for detection and quantification of *Sporisorium scitamineum* in sugarcane. *Sci. World J.* **2013**, *2013*, 942682. [CrossRef] [PubMed]
5. Sundar, A.R.; Barnabas, E.L.; Malathi, P.; Viswanathan, R. A mini-review on smut disease of sugarcane caused by *Sporisorium scitamineum*. In *Botany*; Mworia, J., Ed.; InTech Press: Rijeka, Croatia, 2012; pp. 109–128.
6. Song, X.; Huang, X.; Tian, D.; Yang, L.; Li, Y. Proteomic analysis of sugarcane seedling in response to *Ustilago scitaminea* infection. *Life Sci. J.* **2013**, *10*, 3026–3035.

7. Croft, B.J.; Braithwaite, K.S. Management of an incursion of sugarcane smut in Australia. *Austral. Plant Pathol.* **2006**, *35*, 113–122. [CrossRef]

8. Xu, L.P.; Chen, R.K.; Chen, P.H. Analysis on infection index of smut caused by *Ustilago scitaminea* in sugarcane segregated population. *Chin. J. Trop. Crops* **2004**, *25*, 33–36.

9. Scortecci, K.C.; Creste, S.; Calsa, T.J.; Xavier, M.A.; Landell, M.G.A.; Figueira, A.; Benedito, V.A. Challenges, opportunities and recent advances in sugarcane breeding. In *Plant Breeding*; Abdurakhmonov, I.Y., Ed.; InTech Publisher: Rijeka, Croatia, 2012; pp. 267–296.

10. Lakshmanan, P.; Geijskes, R.J.; Aitken, K.S.; Grof, C.L.P.; Bonnett, G.D.; Smith, G.R. Sugarcane biotechnology: The challenges and opportunities. *In Vitro Cell. Dev. Biol. Plant.* **2005**, *41*, 345–363. [CrossRef]

11. Singh, A.; Isaac-Kirubakaran, S.; Sakthivel, N. Heterologous expression of new antifungal chitinase from wheat. *Protein Exp. Purif.* **2007**, *56*, 100–109. [CrossRef] [PubMed]

12. Chao, C.P.; Hoy, J.W.; Saxton, A.M.; Martin, F.A. Heritability of resistance and repeatability of clone reactions to sugarcane smut in Louisiana. *Phytopathology* **1990**, *80*, 622–626. [CrossRef]

13. Dean, J.L. The effect of wounding and high-pressure spray inoculation on the smut reactions of sugarcane clones. *Phytopathology* **1982**, *72*, 1023–1025. [CrossRef]

14. Whittle, A.M.; Walker, D.I.T. Interpretation of sugarcane smut susceptibility trials. *Trop. Pest Manag.* **1982**, *28*, 228–237. [CrossRef]

15. McNeil, M.D.; Bhuiyan, S.A.; Berkman, P.J.; Croft, B.J.; Aitken, K.S. Analysis of the resistance mechanisms in sugarcane during *Sporisorium scitamineum* infection using RNA-seq and microscopy. *PLoS ONE* **2018**, *13*, e0197840. [CrossRef] [PubMed]

16. Waller, D.I.T. Sugarcane smut (*Ustilago scitaminea*) in Kenya. II. Infection and Resistance. *Trans. Br. Mycol. Soc.* **1970**, *54*, 405–414. [CrossRef]

17. Lloyd, H.L.; Naidoo, M. Chemical assay potentially suitable for determination of smut resistance of sugarcane cultivars. *Plant Dis.* **1983**, *67*, 1103–1105. [CrossRef]

18. Fontaniella, B.; Marquez, A.; Rodriguez, C.W.; Pinon, D.; Solas, M.T.; Vicente, C.; Legaz, M.E. A role for sugarcane glycoproteins in the resistance of sugarcane to *Ustilago scitaminea*. *Plant Physiol. Biochem.* **2002**, *40*, 881–889. [CrossRef]

19. Millanes, A.M.; Fontaniella, B.; Legaz, M.E.; Vicente, C. Glycoproteins from sugarcane plants regulate cell polarity of *Ustilago scitaminea* teliospores. *J. Plant Physiol.* **2005**, *162*, 253–265. [CrossRef]

20. Santiago, R.; Alarcon, B.; de Armas, R.; Vicente, C.; Legaz, M.E. Changes in cinnamoyl alcohol dehydrogenase activities from sugarcane cultivars inoculated with *Sporisorium scitamineum* sporidia. *Physiol. Plant.* **2012**, *145*, 245–259. [CrossRef]

21. Legaz, M.E.; Pedrosa, M.M.; de Armas, R.; Rodrõguez, C.W.; de los, V.; Vicente, C. Separation of soluble glycoproteins from sugarcane juice by capillary electrophoresis. *Anal. Chim. Acta* **1998**, *372*, 201–208. [CrossRef]

22. Borras-Hidalgo, O.; Thomma, B.P.H.J.; Carmona, E.; Borroto, C.J.; Pujol, M.; Arencibia, A.; Lopez, J. Identification of sugarcane genes induced in disease-resistant somaclones upon inoculation with *Ustilago scitaminea* or *Bipolaris sacchari*. *Plant Phys. Biochem.* **2005**, *43*, 1115–1121. [CrossRef]

23. Blanch, M.; Legaz, M.E.; Milanes, A.M.; Vicente, C. Glycoproteins of sugarcane plants facilitate the infectivity of *Ustilago scitaminea* and *Xanthomonas albilineans*, two sugarcane pathogens. In *Communicating Current Research and Educational Topics and Trends in Applied Microbiology*; Mendez-Vilas, A., Ed.; Formatex Research Centre: Badajoz, Spain, 2007; pp. 163–169.

24. Bhuiyan, S.A.; Croft, B.J.; Cox, M.C.; Bade, G. Varietal resistance of sugarcane to natural infection of smut ± preliminary results. *Proc. Aust. Soc. Sug. Technol.* **2010**, *32*, 355–365.

25. Pinon, D.; de Armas, R.; Vicente, C.; Legaz, M.E. Role of polyamines in the infection of sugarcane buds by *Ustilago scitaminea* spores. *Plant Physiol. Biochem.* **1999**, *37*, 57–64. [CrossRef]

26. Torres, M.A.; Jones, J.D.G.; Dangl, J.L. Reactive oxygen species signaling in response to pathogens. *Plant Physiol.* **2006**, *141*, 373–378. [CrossRef] [PubMed]

27. De Armas, R.; Santiago, R.; Legaz, M.E.; Vicente, C. Levels of phenolic compounds and enzyme activity can be used to screen for resistance of sugarcane to smut (*Ustilago scitaminea*). *Austral. Plant Pathol.* **2007**, *36*, 32–38. [CrossRef]

28. Doehlemann, G.; Wahl, R.; Horst, R.J.; Voll, L.M.; Usadel, B.; Poree, F.; Stitt, M.; Pons-Kühnemann, J.; Sonnewald, U.; Kahmann, R.; et al. Reprogramming a maize plant: Transcriptional and metabolic changes induced by the fungal biotroph *Ustilago maydis*. *Plant J.* **2008**, *56*, 181–195. [CrossRef]

29. Que, Y.X.; Su, Y.C.; Guo, J.L.; Wu, Q.B.; Xu, L.P. A global view of transcriptome dynamics during *Sporisorium scitamineum* challenge in sugarcane by RNA-Seq. *PLoS ONE* **2014**, *9*, e106476. [CrossRef]

30. Yang, L.; Lin, H.; Takahashi, Y.; Chen, F.; Andrew, W.M.; Edwin, L.C. Proteomic analysis of grapevine stem in response to *Xylella fastidiosa* inoculation. *Physiol. Mol. Plant Pathol.* **2011**, *75*, 90–99. [CrossRef]

31. Barnabas, L.; Ramadass, A.; Amalraj, R.S.; Palaniyandi, M.; Rasappa, V. Sugarcane proteomics: An update on current status, challenges, and future prospects. *Proteomics* **2015**, *15*, 1658–1670. [CrossRef] [PubMed]

32. Barnabas, L.; Ashwin, N.M.R.; Kaverinathan, K.; Trentin, A.R.; Pivato, M.; Sundar, A.R.; Malathi, P.; Viswanathan, R.; Rosana, O.B.; Neethukrishna, K.; et al. Proteomic analysis of a compatible interaction between sugarcane and *Sporisorium scitamineum*. *Proteomics* **2016**, *16*, 1111–1122. [CrossRef]

33. Brandalise, M.; Severino, F.E.; Maluf, M.P.; Maia, I.G. The promoter of a gene encoding an isoflavone reductase-like protein in coffee (*Coffea arabica*) drives a stress-responsive expression in leaves. *Plant Cell Rep.* **2009**, *28*, 1699–1708. [CrossRef]

34. Eilers, M.; Schatz, G. Binding of a specific ligand inhibits import of a purified precursor protein into mitochondria. *Nature* **1986**, *322*, 228–232. [CrossRef] [PubMed]

35. Shoji, T.; Winz, R.; Iwase, T.; Nakajima, K.; Yamada, Y.; Hashimoto, T. Expression patterns of two tobacco isoflavone reductase-like genes and their possible roles in secondary metabolism in tobacco. *Plant Mol. Biol.* **2002**, *50*, 427–440. [CrossRef] [PubMed]

36. Que, Y.; Xu, L.; Lin, J.; Ruan, M.; Zhang, M.; Chen, R. Differential protein expression in sugarcane during sugarcane—*Sporisorium scitamineum* interaction revealed by 2-DE and MALDI-TOF-TOF/MS. *Comp. Funct. Genom.* **2011**, *2011*, 989016. [CrossRef] [PubMed]

37. Hiraga, S.; Sasaki, K.; Ito, H.; Ohashi, Y.; Matsui, H. A large family of class III plant peroxidases. *Plant Cell Physiol.* **2001**, *42*, 462–468. [CrossRef]

38. Passardi, F.; Cosio, C.; Penel, C.; Dunand, C. Peroxidases have more functions than a Swiss army knife. *Plant Cell Rep.* **2005**, *24*, 255–265. [CrossRef]

39. Chen, W.J.; Zhao, G.W.; Gu, Y.H. Advance of ribulose- 1, 5-bisphosphate carboxylase/oxygenase (RubisCO). *Prog. Biochem. Biophys.* **1999**, *26*, 433–436.

40. DeRocher, E.J.; Quigley, F.; Mache, R.; Bohnert, H.J. The six genes of the Rubisco small subunit multigene family from *Mesembryanthemum crystallinum*, a facultative CAM plant. *Mol. Gen. Genet.* **1993**, *239*, 450–462. [CrossRef]

41. Carmo-Silva, A.E.; Keys, A.J.; Andralojc, P.J.; Powers, S.J.; Arrabaca, M.C.; Parry, M.A. Rubisco activities, properties, and regulation in three different C4 grasses under drought. *J. Exp. Bot.* **2010**, *61*, 2355–2366. [CrossRef]

42. Moreno, J.I.; Martin, R.; Castresana, C. Arabidopsis SHMT1, a serine hydroxymethyltransferase that functions in the photorespiratory pathway influences resistance to biotic and abiotic stress. *Plant J.* **2005**, *41*, 451–463. [CrossRef] [PubMed]

43. Zhao, J.; Liu, Q.; Zhang, H.; Jia, Q.; Hong, Y.; Liu, Y. The rubisco small subunit is involved in tobamovirus movement and Tm-2(2)-mediated extreme resistance. *Plant Physiol.* **2013**, *161*, 374–383. [CrossRef]

44. Qiang, X. The Molecular Mechanism of Powdery Mildew Resistance in Chestnut Rose (Rosa Roxburghii Tratt). Ph.D. Dissertation, Huangzhong Agricultural University, Wuhan, China, 2007.

45. Chastain, C.J.; Fries, J.P.; Vogel, J.A.; Randklev, C.L.; Vossen, A.P.; Dittmer, S.K.; Watkins, E.E.; Fiedler, L.J.; Wacker, S.A.; Meinhover, K.C.; et al. Pyruvate, Orthophosphate Dikinase in Leaves and Chloroplasts of C$_3$ Plants Undergoes Light-/Dark-Induced Reversible Phosphorylation. *Plant Physiol.* **2002**, *128*, 1368–1378. [CrossRef]

46. Chen, H.; Cao, Y.; Li, Y.; Xia, Z.; Xie, J.; Carr John, P.; Wu, B.; Fan, Z.; Zhou, T. Identification of differentially regulated maize proteins conditioning Sugarcane mosaic virus systemic infection. *New Phytol.* **2017**, *215*, 1156–1172. [CrossRef]

47. Fu, J.; Momcilovic, I.; Vara Prasad, P.V. Roles of Protein Synthesis Elongation Factor EF-Tu in Heat Tolerance in Plants Hindawi Publishing Corporation. *J. Bot.* **2012**, *2012*, 835836.

48. Rodrı´guez-Lo´pez, M.; Baroja-Ferna´ndez, E.; Zandueta-Criado, A.; Pozueta-Romero, J. Adenosine diphosphate glucose pyrophosphatase: A plastidial phosphodiesterase that prevents starch biosynthesis Adenosine diphosphate glucose pyrophosphatase: A plastidial phosphodiesterase that prevents starch biosynthesis. *Proc. Natl. Acad. Sci. USA* **2000**, *97*, 8705–8710. [CrossRef] [PubMed]

49. Lao, M.; Arencibia, A.D.; Carmona, E.R.; Acevedo, R.; Rodríguez, E.; León, O.; Santana, I. Differential expression analysis by cDNA-AFLP of *Saccharum* spp. after inoculation with the host-pathogen *Sporisorium scitamineum*. *Plant Cell Rep.* **2008**, *27*, 1103–1111. [CrossRef]

50. Song, X.-P.; Huang, X.; Mo, F.-L.; Tian, D.-D.; Yang, L.-T.; Li, Y.-R.; Chen, B.-S. Cloning and Expression Analysis of Sugarcane Phenylalanin Ammonia-lyase (PAL) Gene. *Sci. Agric. Sin.* **2013**, *46*, 2856–2868.

51. Su, Y.; Guo, J.; Ling, H.; Chen, S.; Wang, S.; Xu, L.; Allan, A.C.; Que, Y. Isolation of a Novel Peroxisomal Catalase Gene from Sugarcane, Which Is Responsive to Biotic and Abiotic Stresses. *PLoS ONE* **2014**, *9*, e84426. [CrossRef]

52. Williamson, J.D.; Scandalios, J.G. Differential response of maize catalases and superoxide dismutases to the photoactivated fungal toxin cercosporin. *Plant J.* **1992**, *2*, 351–358. [PubMed]

53. Jain, R.; Chandra, A.; Venugopalan, V.K.; Solomon, S. Physiological Changes and Expression of SOD and P5CS Genes in Response to Water Deficit in Sugarcane. *Sugar Tech.* **2015**, *17*, 276–282. [CrossRef]

54. Bari, R.; Jones, J.D. Role of plant hormones in plant defense responses. *Plant Mol. Biol.* **2009**, *69*, 473–488. [CrossRef]

55. Rivero, R.M.; Kojima, M.; Gepstein, A.; Sakakibara, H.; Mittler, R.; Gepstein, S.; Blumwald, E. Delayed leaf senescence induces extreme drought tolerance in a Flowering plant. *Proc. Natl. Acad. Sci. USA* **2007**, *104*, 19631–19636. [CrossRef] [PubMed]

56. Abeles, F.B.; Morgan, P.W.; Saltveit, M.E., Jr. *Ethylene in Plant Biology*, 2nd ed.; Academic Press: New York, NY, USA, 1992.

57. Morgan, P.W.; Drew, M.C. Ethylene and plant responses to stress. *Physiol. Plant.* **1997**, *100*, 620–630. [CrossRef]

58. Su, Y.C.; Yang, Y.T.; Peng, Q.; Zhou, D.G.; Chen, Y.; Wang, Z.; Liping, X.; Youxiong, Q. Development and application of a rapid and visual loop-mediated isothermal amplification for the detection of *Sporisorium scitamineum* in sugarcane. *Sci. Rep.* **2016**, *6*, 23994. [CrossRef] [PubMed]

59. Bhuiyan, S.A.; Croft, B.J.; Deomano, E.C.; James, R.S.; Stringer, J.K. Mechanism of resistance in Australian sugarcane parent clones to smut and the effect of hot water treatment. *Crop Past. Sci.* **2013**, *64*, 892–900. [CrossRef]

60. Bradford, M.M. A rapid and sensitive method for the quantitation of microgram quantities of protein utilizing the principle of protein-dye binding. *Anal. Biochem.* **1976**, *72*, 255–260. [CrossRef]

61. Niu, J.Q.; Wang, A.Q.; Huang, J.L.; Yang, L.T.; Li, Y.R. Isolation, characterization and promoter analysis of cell wall invertase gene SoCIN1 from sugarcane (*Saccharum* spp.). *Sugar Tech.* **2015**, *17*, 65–76. [CrossRef]

62. Livak, K.J.; Schmittgen, T.D. Analysis of relative gene expression data using real-time quantitative PCR and the 2(-Delta Delta C(T)) method. *Methods* **2001**, *25*, 402–408. [CrossRef]

63. Chen, S.S. Cloning and Expression Analysis of ROS Metabolism Pathway Key Genes from Sugarcane. Master's Thesis, Fujian Agriculture and Forestry University, Fuzhou, China, 2012.

64. Albert, H.H.; Schenck, S. PCR amplification from a homolog of the bE mating-type gene as a sensitive assay for the presence of *Ustilago scitaminea* DNA. *Plant Dis.* **1996**, *80*, 1189–1192. [CrossRef]

International Journal of
Molecular Sciences

MDPI

Article

Quantitative Proteomics of Potato Leaves Infected with *Phytophthora infestans* Provides Insights into Coordinated and Altered Protein Expression during Early and Late Disease Stages

Chunfang Xiao [1,2], Jianhua Gao [1,2], Yuanxue Zhang [1,2], Zhen Wang [1,2], Denghong Zhang [1,2], Qiaoling Chen [1,2], Xingzhi Ye [1,2], Yi Xu [1,2], Guocai Yang [1,2], Lei Yan [1,2], Qun Cheng [1,2], Jiaji Chen [1,2] and Yanfen Shen [1,2,*]

[1]   Southern Potato Research Center of China, Academy of Agricultural Sciences, Enshi 445000, Hubei, China;
      15971827976@163.com (C.X.); wend315@126.com (J.G.); 13636289689@163.com (Y.Z.);
      wangzhen20093332@163.com (Z.W.); zhangdenghong2@126.com (D.Z.); 18627798503@163.com (Q.C.);
      yxz161718@163.com (X.Y.); enshizizi@126.com (Y.X.); ygcdxs@126.com (G.Y.); 18727670407@163.com (L.Y.);
      enshicq@126.com (Q.C.); chenjia-ji@163.com (J.C.)
[2]   Enshi Tujia and Miao Autonomous Prefecture Academy of Agricultural Sciences, Enshi 445000, Hubei, China
*   Correspondence: 13872728746@163.com

Received: 11 November 2018; Accepted: 24 December 2018; Published: 1 January 2019

Abstract: In order to get a better understanding of protein association during *Solanum tuberosum* (cv. Sarpo Mira)–*Phytophthora infestans* incompatible interaction, we investigated the proteome dynamics of cv. Sarpo Mira, after foliar application of zoospore suspension from *P. infestans* isolate, at three key time-points: zero hours post inoculation (hpi) (Control), 48 hpi (EI), and 120 hpi (LI); divided into early and late disease stages by the tandem mass tagging (TMT) method. A total of 1229 differentially-expressed proteins (DEPs) were identified in cv. Sarpo Mira in a pairwise comparison of the two disease stages, including commonly shared DEPs, specific DEPs in early and late disease stages, respectively. Over 80% of the changes in protein abundance were up-regulated in the early stages of infection, whereas more DEPs (61%) were down-regulated in the later disease stage. Expression patterns, functional category, and enrichment tests highlighted significant coordination and enrichment of cell wall-associated defense response proteins during the early stage of infection. The late stage was characterized by a cellular protein modification process, membrane protein complex formation, and cell death induction. These results, together with phenotypic observations, provide further insight into the molecular mechanism of *P. infestans* resistance in potatos.

Keywords: late blight disease; potato proteomics; *Phytophthora infestans*; Sarpo Mira; early and late disease stages

1. Introduction

Potato late blight disease caused by *Phytophthora infestans* is one of the most critical crop diseases in the world. Late blight was responsible for the European potato famine in the 19th century [1]. It poses a severe threat to potato production worldwide, with estimated annual economic losses of over six billion dollars, mainly due to yield loss and the high cost of fungicide [2]. Management of the late blight disease pathogen is challenged by global warming, environmental regulations against the use of chemical fungicides, and *P. infestans* remarkable pathogenicity [3].

P. infestans shows an initial asymptomatic biotrophic phase of infection followed by a necrotrophic phase. During the biotrophic stage, *P. infestans* forms appressoria, primary and secondary hyphae, and specialized structures called haustoria, through which effectors are delivered into the host apoplast or adjacent cells [4]. Plants have evolved an array of innate immune systems to detect and respond to a wide range of these *P. infestans* effectors. For example, PTI (Pathogen-associated molecular pattern-triggered immunity), which uses transmembrane pattern recognition receptors (PRRs) that respond to evolving microbial- or pathogen-associated molecular patterns (MAMPS or PAMPs). PRRs include a class of leucine-rich repeat (LRR)-receptor kinases (RK), for example CEBiP [5] and OsCERK1 [6,7] in rice; AtCERK1 [8–11], LYM2 [12–14], RBGP1 [15], and RLP30 [15] in Arabidopsis; and EIX2 [16], Ve1 [17,18], and Cf-9 [15]. However, knowledge about PRRs in the potato host is scarce. PTI has the potential to fend off various microbes, pathogenic or not, due to the conserved nature of PAMPs (e.g., fungal chitin) across species, genera, family, or class. Thus, PRRs can provide resistance to most non-adapted pathogens, as well as contribute to basal immunity during infection or disease process. In response to these PTI defense systems, pathogens that could breach PTI successfully deploy a huge number of effectors to render pathogen virulence. Such effectors change the normal function of PTI, resulting in effector-triggered susceptibility (ETS) [19]. A large number of extracellular and cytoplasmic effectors in the *P. infestans* genome have been identified and increasing evidence for their role in establishing ETS exists [20,21].

However, to combat pathogens with established ETS, host plants have evolved a race-specific immunity, a well-described host resistance mechanism that is governed by dominant R-genes. Many R-genes have been cloned, and most of them encode proteins with N-terminal nucleotide-binding sites (NBSs) and C-terminal leucine-rich repeats (LRRs). R-genes encode proteins that recognize pathogen effectors to establish effector-triggered immunity (ETI). This recognition triggers a cascade of defense responses, mediated by a complex-signaling network, in which plant hormones, like salicylic acid (SA) and jasmonic acid (JA), play a significant role and the resistance is manifested as a localized hypersensitive cell death response (HR) at the site of infection.

Recently, next-generation sequencing (NGS) technologies are transforming biology research [22,23]. Genome sequences of potato and *P. infestans* have been published [2,24], making sequencing-based "omics" studies more accessible to potato late blight researchers. Proteomics has become a viable alternative for molecular analysis, providing information and tools for a better understanding of the plant-pathogen relationship. Recently, proteomics has dramatically evolved in the pursuit of large-scale functional assignment of candidate proteins and, by using this approach, several proteins expressed during potato–*P. infestans* interaction have been identified [25–27]. Two-dimensional electrophoresis (2-DE) based proteomics [25], gel-based protein shotgun mass spectrometer [28], and, most recently, label-free proteomics analysis [26,27], have shed light on our understanding of compatible and incompatible interactions between *P. infestans* and potato.

The recently developed isobaric label proteomics, such as isobaric tags for relative and absolute quantification (ITRAQ) [29] and tandem mass tags (TMT) [30], are chemically conjugated to the primary amines of peptides after tryptic digestion and are compatible with samples from multiple sources [29]. Therefore, in this study, we used the TMT method to measure and compare the changes in protein abundance of potato cv. Sarpo Mira after foliar application of zoospore suspension of *P. infestans* at three key timepoints, covering potato–*P. infestans* oomycete early (EI) and late (LI) stage of interaction. A total of 1229 differentially-expressed proteins (DEP) were identified, 75 DEPs at the early stage and 723 DEPs at the late stage of the disease process. The proteins identified at the early and late stage could play an essential role in early pathogen recognition, signal transduction, disease resistance processes against *P. infestans*, and possibly disease pathogenesis. This study will contribute to a better understanding of the molecular mechanism of *P. infestans* interaction with potato.

2. Results and Discussion

2.1. Subsection Phenotypic Differences between the Three Stages of Disease Conditions

A time series assessment of Sarpo Mira leaf phenotype challenged with *P. infestans* is shown in Figure 1a,b. Three replicates were used for each treatment in these tests. After spraying the whole potato plant with *P. infestans* zoospores, there was no observable microscopic hypersensitive reaction (HR) lesions at 0 Hours post inoculation hpi (Control); however, by 48 hpi (EI), HR lesions had appeared, which resulted in a localized necrosis that resembled a *P. infestans*-induced hypersensitive response, and at 120 hpi (LI) the leaves had developed much larger H -induced necrosis, consistent with a typical R gene-mediated HR lesions expansion, as previously reported in Sarpo Mira [31]. There was a significant difference in the lesion size of LI relative to EI, as seen in (Figure 1b), and consistent with previous studies [32,33]. Based on these results, we designated the interval between time-points Control and EI as early disease stage and timepoint EI to LI as late disease stage.

(a)   (b)

Figure 1. Phenotypic observation of the three time-points. Control corresponds to time-point 0 Hours post inoculation, EI corresponds to 48 hpi, and LI corresponds to 120 hpi. (a) Sizes of lesions induced by the *P. infestans* zoospores at different exposure times. The diameter of each lesion was measured at 2 and 5 days after inoculation (b). Three replicates were used for each treatment in these tests. Bars represent the standard deviation of three replicates. Statistical significance was analyzed using Student's *t*-test. The asterisk indicates the significant difference (* $p < 0.05$).

2.2. Overview of Protein Expression in Potato Leaves Challenged with P. infestans Oomycete

Potato cv. Sarpo Mira was previously reported to have incompatible interaction with *P. infestans* [29]; however, systematic analysis of protein association during Sarpo Mira–*P. infestans* interaction, and their resultant changes in abundance leading to incompatibility, are incompletely understood. Therefore, to shed more light on the changes in protein abundance during the early and late stages of late blight disease, we performed a comparative proteome survey by TMT method [29] at three key time-points (Control, EI, and LI time-points. See method section) on leaves of potato clone Sarpo Mira inoculated with *P. infestans* oomycete. Three biological replicates were collected at the same time. With these measurements, a total of 15,813 high-quality peptides (in at least two replicates per time-point) corresponding to 4643 proteins were identified in the time series analysis (Table S1). Of the 4643 proteins, 1229 (at least one unique peptides) were found to be differentially expressed proteins (DEP) in a pairwise comparison between the time-points. Among the 1229 DEPs, 952 had functional annotations (Table S2). In all, a total of 1082 DEP could be classified into 70 different significantly-enriched protein domains and features by InterProScan analysis, 21 PFAM protein domains, and 56 significantly-enriched Kyoto Encyclopedia of Genes and Genomes pathways (Table S2) [34]. The total number of DEPs observed in each pairwise comparison is shown in Figure 2a and Tables S3–S5. For example, 1022 DEPs were observed in the group pair of H/L, covering the whole-time course. The pair of EI/Control had 75 proteins representing DEPs in the early stage of pathogen invasion, and LI/EI had 723 DEPs active during the late stage of the disease process.

Large scale comparative quantitative proteomic studies produce numerous lists of proteins containing biological identifiers, and often it is useful to highlight the overlapping sets between groups of biological data, enabling quick and easy observation of the similarities and differences between the data sets. In this study, dataset overlap between the early and late stage revealed about 2.3% (18/780) of the DEPs were commonly-shared induced proteins throughout the time course (Figure 2b), suggesting that these proteins could be necessary for the sustained HR phenotype observed in the later stages of the disease condition. More strikingly, 90.4% (705/780) of DEPs were specific to LI/EI, while 7.3% (57/780) was unique to EI/Control (Figure 2b).

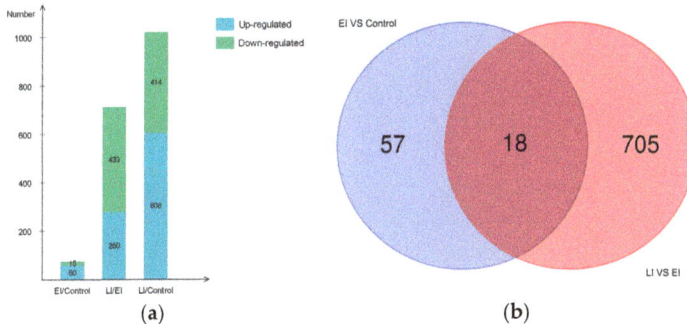

Figure 2. Differentially expressed proteins between the early and late stage (**a**). Bar chart showing the number of up- and down-regulated proteins in each of the pairwise comparisons: EI/Control, LI/EI, and LI/Control. green color indicates down-regulated proteins and blue color indicates up-regulated proteins. Overlaps among differentially expressed proteins (DEPs) between the early and late stage (**b**). Venn diagrams depict overlap of DEPs from each pairwise comparison between the timepoints, EI/Control (early) and LI/EI (late).

2.3. Gene Ontology Classification of Differentially Expressed Proteins

Gene ontology classification and KEGG (Kyoto Encyclopedia of Genes and Genomes) analysis was used to reveal the implication of all DEPs identified in this study (Figure 3a). Gene Ontology analysis showed that much more biological process categories were highly abundant in the DEPs. The protein classes "Metabolic processes" (41.3%), "Cellular processes" (32%), and "Response to stimulus" (6.1%) were the most abundant categories (Figure 3a). The label "Metabolic process" covers several sub-biological process categories; notable among them were primary and secondary metabolism, such as carbohydrate, lipid, protein, amino acid, and nucleobase-containing compound metabolism. Also, within this category are systemic-acquired resistance and defense response to fungus. The proteins involved in this group were mostly up-regulated in the early disease stage. The "Cellular processes" includes protein folding, microtubule-based process, signal transduction, cell death, protein secretion cellular homeostasis, and oxidant detoxification. "Response to stimulus" covered response to stress, response to biotic and abiotic stimulus, response to endogenous stimulus, and response to chemical. Several of these proteins encoded peroxidase, pectinesterase, and endochitinase-like compounds, and were also up-regulated in both disease stages (Figure 3a and Table S6).

Molecular function classification highlighted over-representation of "Catalytic Activity" (45%) and "Binding" (40%) as the major function of the DEPs. The "Catalytic Activity" heading included oxidoreductase activity, transferase activity, hydrolase activity, lyase activity, isomerase activity, and ligase activity. The "Binding" category encompassed protein binding, protein-containing complex binding, ion binding, drug and cofactor binding, iron-sulfur cluster binding, heterocyclic compound binding, glutathione binding, and carbohydrate derivative binding (Figure 3a and Table S6). The majority of the proteins in this category were uncharacterized. Other molecular function categories worthy of mention include antioxidant activity, enzyme regulator activity, and nuclear import signal receptor activity. About half of the DEPs were in the cellular component, of which the majority were in the "Cell" (24.10%), which included the extracellular region and apoplast. "Cell part"

(23.96%), covering intracellular plasma membrane and chloroplast envelope; "Organelle" (17.65%), covering organelle membrane, intracellular organelle, and organelle lumen; "Protein complex" (11.68%), including THO complex, U1 snRNP, nucleosome, transcription factor complex, chloroplastic endopeptidase Clp complex, and eukaryotic translation elongation factor 1 complex; "Membrane" (8.39%) and "Membrane part" (5.28%), which encompasses membrane protein complex and inner mitochondrial membrane protein complex (Figure 3a and Table S6).

A functional enrichment test was used to identify over-represented proteins that may have an association with early and late disease phenotypes, by interrogating the data for the GO enrichment of protein sets. The KEGG database was used to determine significantly enriched pathways in the early or late disease stage (Table S6).

In the early disease stage, the GO enrichment test revealed response to stimulus and detoxification as the most significantly-enriched biological processes (Figure 3b). The significantly-enriched molecular functions were antioxidant activity and catalytic activity, which included peroxidase activity (Figure 3b), while several of these enriched proteins were located within the extracellular region of the cellular component (Table S6). Considering the late disease stage, we observed that metabolic process and single-organism process were highly significantly-enriched biological processes (Figure 3c). Other enriched biological processes included detoxification, cellular component and biogenesis, and positive regulation of biological processes (Table S6). Enriched molecular functions included structural molecular activity, antioxidant activity, and catalytic activity (Figure 3c). In the cellular component, we noticed overrepresentation of proteins in the cell, cell part, organelle, and organelle part (Figure 3c).

Pathway coverage analysis using the KEGG database found that the phenylpropanoid biosynthetic pathway was the most significantly-enriched in the early disease stage (Figure 3d); most of these proteins encoded peroxidase/peroxidase-like and hydroxycinnamyl proteins. Other enriched pathways included fatty acid metabolism and biosynthesis of unsaturated fatty acid (Figure 3d). As reported by others, the phenylpropanoid pathway is essential to plants because of its role in the production of the hydroxycinnamyl alcohols, which serve as the building blocks of lignin, and confers structural support, vascular integrity, and pathogen resistance to plants [35]. Additionally, high induction of several genes mapped to the phenylpropanoid pathway has been reported following *P. infestans* invasion [36]. Meanwhile, biosynthesis of secondary metabolism, ribosome, glutathione metabolism, biosynthesis of amino acid, porphyrin and chlorophyll metabolism, ribosome biogenesis, valine, leucine and isoleucine degradation, synthesis and degradation of ketone bodies, and fatty acid metabolic pathways were identified as the most significantly-enriched pathways in the late disease stage (Figure 3d). Together these results suggested that the identified proteins represent a functionally-active subset of the entire proteome associated with the potato response to *P. infestans* oomycete infection.

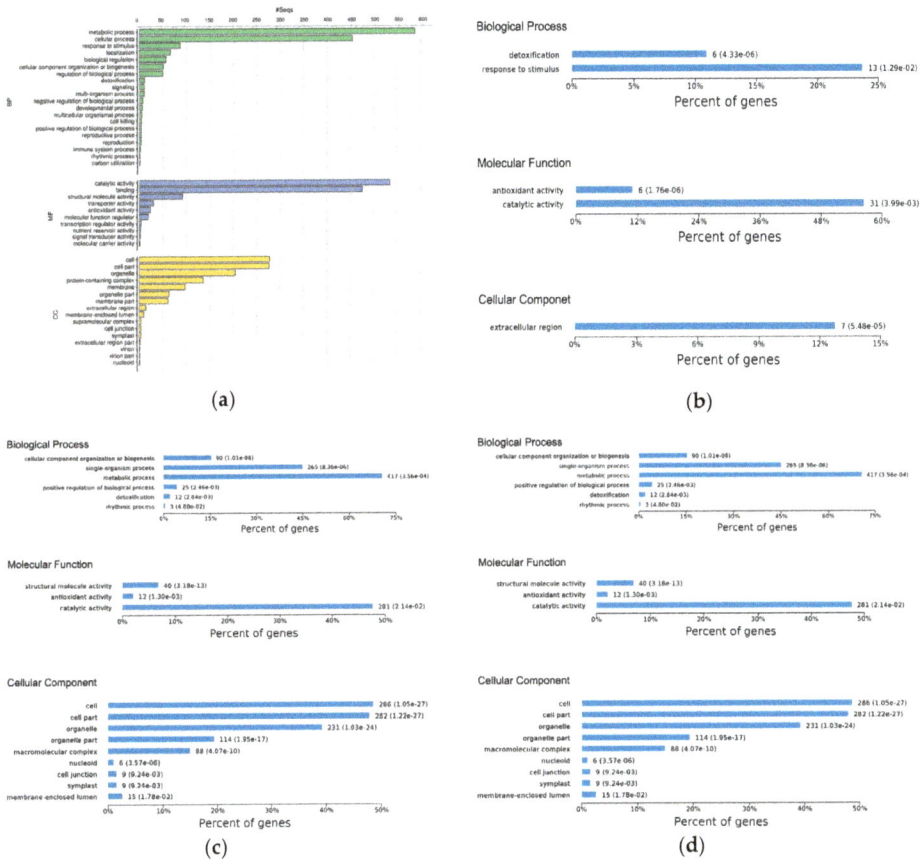

Figure 3. (**a**) Gene Ontology functional classification of all DEPs; bar chart shows the distribution of differentially-expressed proteins among the GO biological process (BP), molecular function (MF), and cellular component (CC). (**b**) GO-based functional enrichment analysis of DEPs at early disease stage. (**c**) GO-based functional enrichment analysis of DEPs at late disease stage. (**d**) Kyoto Encyclopedia of Genes and Genomes pathway enrichment of DEPs at the early and late disease stages.

2.4. Differential Expression Pattern of Proteins Involved at the Early and Late Disease Stages

To better understand potato–*P. infestans* interaction, it is important to distinguish the potato's specific response to the invading pathogen and protein signatures at various stages of the disease process, which can help shed light on the pathogenic life style, whether in biotrophic relationships (in which the pathogen feeds from living host cells) to necrotrophic associations (in which the microbe feeds on nutrients released from killed cells [37]). Therefore, we examined the expression pattern of proteins involved at the early and late disease stages.

2.4.1. Early Disease Stage Response Proteins

As stated earlier, 75 proteins showed a significant difference in protein abundance at the early stage of the disease process (EI/Control-compared proteins, Table S3), of which 60 DEPs were up-regulated, and 15 DEPs were down-regulated (Figure 2a); their expression profile is shown in (Figure 4). GO enrichment test results (False Discovery Rate < 0.05, $p < 0.01$) revealed these proteins were mostly related to detoxification and response to a stimulus, which covers stress response, defense response, oxidative burst, and cellular catabolic process (Figure 3b). Further analysis uncovered enriched proteins were most active in the apoplast, cell wall, cell periphery, and external encapsulating structure during pathogen invasion (Table S6). To elucidate cv. Sarpo Mira response to *P. infestans* at the initial stages of infection, we further examined the expression pattern and the role of DEPs in EI/Control comparison (Table S3). We found a general trend among the DEPs in EI/Control comparison, here most of the proteins significantly increased in abundance from timepoint L to M (Figure 4). Also, we noticed specific enrichment of positively regulated (from time-point Control to EI) functional categories related to defense and oxidative stress response. The most prominent early response proteins present in this group were cell wall degrading enzymes, for example, wall-associated kinase, annexin, osmotins, osmotin-like proteins, serine protease, and proteinase inhibitors, pectinesterase and putative endochitinase; the later are cell-wall degrading enzymes (CWDEs). We also observed five highly up-regulated peroxidase proteins involved in the reactive oxygen species (ROS) metabolic process. The rapid production of reactive oxygen species (ROS) upon pathogen attack has been associated to the defense mechanism for microbial killing and early initiation of host defense responses in plants [38]. In the present study, the up-regulation of these five peroxidase proteins suggests the induction of oxidative burst, which is typically associated with PTI and HR-specific induced cell death, which is consistent with the phenotypic observation (Figure 1a) and previous studies [33].

We also noticed other proteins, including the specific up-regulation of glucan endo-1,3-beta-glucosidase and serine carboxypeptidase-like-33. In this regard, it is well known that extracellular enzymes of plant pathogenic fungi (e.g., glucanases) may have a diversity of roles in host invasion and pathogenesis, either as an inducer or suppressor [39]. Indeed, there are reports that cell walls of oomycetes consist mainly of (1/3)-b-D-glucans, (1/6)-b-D-glucans, and cellulose, which might be required for normal appressorium formation and successful infection of the potato [39]. However, the specific up-regulation of glucan endo-1,3-beta-glucosidase suggests that it either played a significant role in facilitating *P. infestans* penetration of the host cell wall, or the destruction of papillae, blocking the invading pathogen by releasing glucans from the host wall polymers or by hydrolyzing biologically-active glucans, which could act as elicitor. In contrast it may digest wall components of the invading fungal pathogen [40]. Serine carboxypeptidase belonged to a large family of hydrolyzing enzymes, which are believed to play roles in processing and degradation of proteins/peptides, and studies have shown that this protein family are typically up-regulated during pathogen invasion [39]. In the current study, it may be part of the fungal mechanisms of efficient protein digestion during invasion or a past host cell proteolytic machinery against *P. infestans* [41,42]. Additionally, we identified another set of DEPs that showed significant up-regulation at the early stage, whose domains possess a binding function possibly involved in the production of antimicrobial compounds. They include the pathogenesis-related protein PR-10 family and the NtPRp27-like protein, which suggests a distinct counter-defense mechanism because most PR proteins are reported to exhibit direct antimicrobial activities and may play a role in both constitutive and induced basal defense responses [43].

Contrastingly, few DEPs were repressed at this stage; notable among them are significantly down-regulated proteins such as dehydrin, pyruvate dehydrogenase, light-inducible tissue-specific protein, and three uncharacterized proteins, M0ZKB0, M1AM40, and M1BUI4 (Table S3). These proteins have functions related to amino acid metabolism and transport, sterol biosynthesis, and abscisic acid signaling. To put these results into perspective, the increased abundance of the majority (60/75) of DEPs in EI/Control indicated an early activation of cell wall-associated defense proteins involved in signal transduction, deployment of basal resistance, and initiation of R gene-mediated resistance processes, which reflects a coordinated activation and repression of specific cell wall-associated proteins to correlate with the precise cellular defense requirement to restrict *P. infestans* invasion.

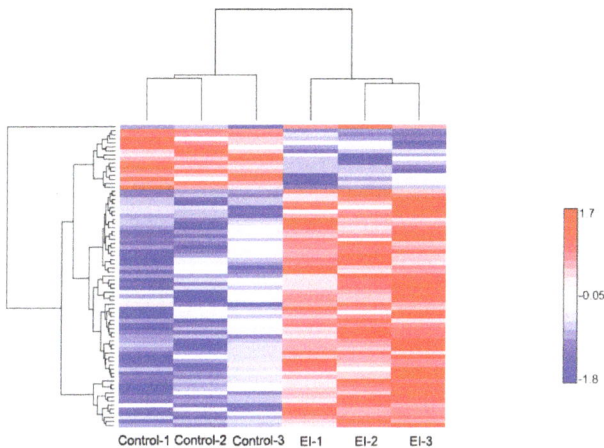

Figure 4. Hierarchical clustering of differentially-expressed proteins at the early disease stage. Heat map showing the changes in protein expression: Proteins with high expression levels (red); proteins with low expression level (blue).

2.4.2. Late Disease Stage Response Proteins

The profile of the 723 DEPs identified during the late stage is reported in Figure 5. Within this group, a total of 280 DEPs were up-regulated and 443 were significantly down-regulated, as shown in Figure 2a and Table S4. Functional enrichment showed that the majority of the proteins were involved in cellular component organization, metabolic process, and single-organism process, which encompasses cellular protein modification process and membrane protein complex formation (Figure 3c and Table S6). We noticed a consecutive up-regulation of several proteins with the binding function, including the resistance (R) gene product containing CC-NB-ARC and LRR domains, as well as chitin-binding domains. Typically, R proteins are conserved across the plant kingdom, and have been shown to mediate the resistances of race-specific diseases in plants by recognizing effectors and initiating effector triggered immunity (ETI) [20,21,38]. Several other proteins that were significantly up-regulated were identified within this group, which included proteins possessing acid phosphatase-class B domain, osmotin/thaumatin-like domain, endochitinase activity, and kunitz proteinase inhibitor. The latter is a part of potato proteolytic enzyme inhibitors, which may play an important role in the natural defense mechanisms of the potato plant against phytopathogen attack [27,14] In addition to the above were hyoscyamine-beta-hydroxylase (H6H), Puroindoline-A (PIN-A), Puroindoline-I (PIN-I), lipoxygenase, and hydroxy-methylglutaryl coenzyme A reductase (HMGR). H6H is an enzyme belonging to the family of oxidoreductases, and the last rate-limiting enzyme directly catalyzing the formation of scopolamine in the tropane alkaloids (TAs) biosynthesis pathway [45]. Earlier studies have shown that H6H was concurrently significantly up-regulated, among other R gene proteins, following *P. infestans* treatment in resistant potato cultivars relative to susceptible genotypes [46]. PIN-A and PIN-I are transmembrane proteins involved in auxin efflux [47]; their specific role in potato–*P. infestans* interaction is not clear. Similarly, HMGR is known to be strongly induced by fungal elicitors in rice [45]. In the present study, and generally, it is likely that these proteins together might play a role in the production of highly complex toxic anti-microbial or anti-fungal compounds as defense against invading pathogens [46].

To survive in a peroxidative environment, microbes produce natural antioxidants within host cells, or modulate host cells to produce protectants, including vitamin C, glutathione (GSH) carotenoids, reductases, peroxidases, and several others [48,49]. Here, we identified proteins that are possibly associated with these processes mentioned above and may be implicated in the expansion of late-stage disease processes. These included: peroxidase family proteins and threonine dehydratase, an enzyme involved in isoleucine biosynthesis by catalyzing the deamination of threonine. It was reported to have similar function to serine-threonine dehydratase, a versatile catalyst that functions as a coenzyme in a multitude of reactions, including amino–sugar breakdown [50]. Others were: sucrose synthase, formate dehydrogenase, biotin, and lipoic acid-binding proteins, which may serve as sources of energy for the pathogen. Additional up-regulated proteins were serine-threonine kinases, possibly *P. infestans* secreted kinases [51]. Calcium-dependent protein kinase, arginine N-methyltransferase (their function in the late stage disease process is unknown) and Tyrosine Phosphatase (reported in bacteria is an effector protein), when overexpressed, significantly increases bacterium *Pseudomonas syringae* virulence [52,53].

Several studies have reported that genes encoding hydrolytic enzymes, such as serine protease, glucosidases, glucanases, and lyases, constituted a major portion of Phytophthora potential pathogenicity factors [48]. Among the significantly induced pathogenesis factors were enzymes: beta-glucanase, glycosyl hydrolases, cysteine protease, and pectate lyase. We also noticed ATP synthase and proton ATPase transmembrane transporter; their specific role in pathogenesis is unknown.

Notable significantly down-regulated proteins were those involved in the structural integrity of the ribosome, such as ribosome recycling factor domain proteins, 60S ribosomal protein L18a, acidic ribosomal protein P1a-like, 50S ribosomal protein L32, and ribosomal-L12 proteins. We noticed other proteins with catalytic and binding domain functions, such as cellulose synthase, histone H2A, Nicotinamide adenine dinucleotide phosphate (NADPH)-protochlorophyllide oxidoreductase, pectinesterase, endoglucanase, FK506-binding protein, and PPM-type phosphatase domain. The latter being dephosphorylate serine and threonine residues. Among the down-regulated proteins were also: thioredoxin-like protein CITRX a chloroplastic protein, demonstrated to be involved in the negative regulation of cell death and tomato Cf-9 resistance protein function by specifically interacting with Cf-9 [54]. Silencing of CITRX accelerated the Cf-9/Avr9-triggered hypersensitive response in both tomato and *Nicotiana benthamiana*, together with the enhanced high accumulation of reactive oxygen species, and the induction of down-stream defense-related genes. In the same study, silencing of CITRX also conferred increased resistance to the fungal pathogen *Cladosporium fulvum* in susceptible Cf0 tomato [54,55]. Several uncharacterized proteins with acetyltransferase-like domain proteins, including M1BC65, a member of the chloramphenicol acetyltransferase-like domain superfamily, was also identified. In addition to enzyme inhibitor and peptide regulator proteins, such as Clone PI9149 apoptosis inhibitor 5-like protein and Proteinase inhibitor I, the high number of down-regulated DEPs (443/723) among the proteins in the LI/EI seems consistent with a shut-down of the cellular metabolic process caused by HR induced necrosis. Collectively, the cellular metabolic process, protein folding, and modification processes, including cell wall re-organization, seem to be the most common roles of the LI/EI-DEPs; therefore, we speculate that these proteins play an essential role in the biological processes possibly involved in disease conditioning and the shut-down of cell metabolic processes at the late infection stage, which are geared towards containment of the invading pathogen.

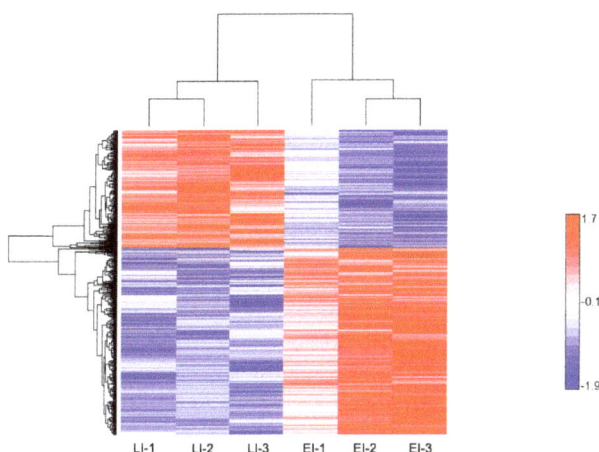

Figure 5. Hierarchical clustering of differentially-expressed proteins at the late disease stage. Heat map showing the changes in protein expression: proteins with high expression levels (red); proteins with low expression level (blue).

2.4.3. Differential Regulation of Commonly Shared Proteins in Response to *P. infestans* during Early and Late Disease Stages

Our comparative analysis identified 18 proteins shared between the EI/Control comparison and LI/EI comparison (Figure 2b, Table S7). We reasoned that these common set of proteins might respond to the same signal that controls the switch from general plant defense induction, based on PAMPs (PTI), to effector-triggered immunity (ETI), and might have a similar pattern of expression. Indeed, 10 out of the 18 common DEPs showed a similar expression profile throughout the time course. For instance, they were up-regulated explicitly from Control to EI and reached their maximum expression level by time-point LI. Based on GO analysis, these 10 proteins were mostly located within the extracellular region, cell membrane, and protein-containing complexes, and are involved in energy production, vesicle-mediated transport, and response to oxidative stress. From this group we noticed two uncharacterized proteins, M0ZTQ4 (FC = 2.40) and M1CUM0 (FC = 5.6), with at least two- and five-fold increase in abundance. The M0ZTQ4 contained the osmotin/thaumatin-like domain, and osmotins are members of pathogenesis-related proteins, secreted into the cell wall to promote basal resistance responses [56]. Whereas M1CUM0 has a domain function related to terpene biosynthesis, a part of antifungal phytoalexins shown to limit the growth hypha during pathogen invasion [57,58].

Remarkably, six out of the remaining eight DEPs (M1ABL9, M1D1L9, M1A3A0, M1A4R1, M1C8Q0, M0ZZ55) showed a dynamic reprogramming in response to *P. infestans*. It is noteworthy that after inoculation these proteins were up-regulated from timepoint Control to EI and reached their highest expression level at time-point EI, afterward they were significantly repressed (Table S7). Further analysis showed that they contained vacuolar protein sorting-associated VPS4 binding domain, ribosomal protein L24 binding domain, pentatricopeptide repeats, histone H1/H5 globular binding domain, 26S proteasome subunit RPN7 domain, and vesicle transport protein, respectively. Generally, plants activate numerous defense mechanisms that can contribute to resistance against pathogen invasion. Considering that a significant increase in lesions' size was observed on the leaves of cv. Sarpo Mira at the LI time-point, compared to the EI timepoint, it follows that the dynamic reprogramming of these proteins might be very significant. This highlights the tight regulation of defense activities within the host cell and demonstrates the urgency for cell wall modification and reinforcement, as well as an efficient transport mechanism for defense-related compounds upon pathogen invasion consistent with the incompatible pathogen–host interaction and observed phenotype in Figure 1a,b.

Contrastingly, two DEPs, K7WNX1 and M1C639, were consecutively down-regulated throughout the whole-time course; K7WNX1 is a light-inducible tissue-specific ST-LS1 protein and M1C639 contained the transmembrane helix domain. Their role in potato–*P. infestans* interaction is unknown.

2.5. Validation of Differentially-Expressed Proteins in Early and Late Disease Stages

To validate DEPs from the time series proteomics experiments, a total of seven proteins were selected, of which four were randomly selected from the early disease stage and the remaining three were selected from late stage, to verify the expression level, via western blot analysis (Figure 6). Osmotin (fold change = 1.62, p = 0.00471), pectinesterase (fold change = 2.53, p = 0.0083), endochitinase (fold change = 1.94, p = 0.0453), and annexin (fold change = 2.37, p = 0.0264) were significantly up-regulated in EI relative to the Control (Figure 6(A1)–(A4)). In the late stage, peptidyl-prolyl cis-trans isomerase (fold change = 1.74, p = 0.0051) and type I serine protease inhibitor (fold change = 2.63, p = 0.0382) were significantly increased in LI relative to EI (B1, B3). In contrast, photosystem I assembly protein Ycf4 (fold change = -1.67, p = 0.0036) was significantly down-regulated relative to EI (Figure 6(B2)). Potato actin represented loading control. The western blot results were consistent with the times series proteomics data, which strongly support the reliability of the results reported in this paper.

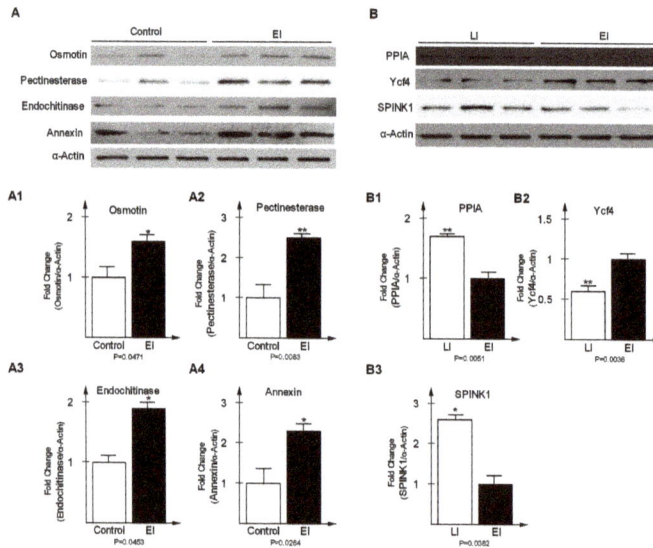

Figure 6. Western blot analysis of differentially expressed proteins at early disease stage (**A**) and late disease stage (**B**). Expression levels of Osmotin (fold change = 1.62, p = 0.00471), Pectinesterase (fold change = 2.53, p = 0.0083), Endochitinase (fold change = 1.94, p = 0.0453) and Annexin (fold change = 2.37, p = 0.0264) were significantly increased in EI relative to Control (*A1, A2, A3, A4*); PPIA (fold change = 1.74, p = 0.0051) and SPINK1 (fold change = 2.63, p = 0.0382) were significantly increased in LI relative to EI (*B1, B3*); Ycf4 (fold change = -1.67, p = 0.0036) was significantly decreased in LI relative to EI and (*B2*). Potato actin represented loading control.

2.6. Essential Proteins for Early and Late Disease Stages

In this study, we systematically identified several proteins that are important for the potato defense response against *P. infestans* at the early and late stages of infection, including those which may otherwise contribute to pathogenesis. The criteria described in the Methods allowed us to select several proteins that play a significant role in both disease stages, which are listed in Table 1 and some of which are described hereafter.

Table 1. Proposed candidate proteins that played an essential role at the early and late disease stages.

Accession	Description	Fold Change	*p* Value	GO Names
Early stage up-regulated proteins				
M0ZZ55	Uncharacterized protein containing longin N-terminal domains with C-terminal coiled-coil/SNARE motif.	1.77	0.02	C: integral component of membrane; P: vesicle-mediated transport
M0ZTM7	Osmotin	1.58	0.03	
Q5XUH0	Osmotin-like protein	1.54	0.002	
M0ZMK7	Uncharacterized protein with domain named as Bet v 1. Bet v 1 belongs to family 10 of plant pathogenesis-related proteins (PR-10)	1.51	0.004	P: defense response; P: response to biotic stimulus
P52402	Glucan endo-1,3-beta-glucosidase, basic isoform 3 (Fragment)	1.45	0.03	F: hydrolase activity, hydrolyzing O-glycosyl compounds; P: carbohydrate metabolic process
Q2HPK8	Putative endochitinase (Fragment)	1.43	0.04	F: chitinase activity; P: chitin catabolic process; P: cell wall macromolecule catabolic process
M1D1L9	Uncharacterized protein containing domain 2 of Ribosomal protein L2,	1.36	0.02	F: structural constituent of ribosome; C: ribosome; P: translation
M1CUM0	Uncharacterized protein containing a domain found in the isoprenoid synthase family	1.34	0.04	F: magnesium ion binding; P: metabolic process; F: terpene synthase activity
M1D578	Peroxidase	1.34	0.04	F: peroxidase activity; P: response to oxidative stress; F: heme binding; P: hydrogen peroxide catabolic process; P: oxidation-reduction process
M1BQC2	Pectinesterase	1.32	0.03	F: enzyme inhibitor activity; C: cell wall; F: pectinesterase activity; P: cell wall modification
M1A3A0	Uncharacterized protein containing Pentatricopeptide repeat (PPR)	1.29	0.02	F: protein binding
M1A035	Carboxypeptidase	1.28	0.03	F: serine-type carboxypeptidase activity; P: proteolysis
M1D051	Wall-associated kinase	1.28	0.03	F: protein kinase activity; F: ATP binding; P: protein phosphorylation
Q9M3H3	Annexin	1.27	0.01	F: calcium ion binding; F: calcium-dependent phospholipid binding
M0ZXE4	Uncharacterized protein contains a domain found in serine peptidases	1.27	0.02	F: serine-type endopeptidase activity; P: proteolysis
Q4FE26	Proteinase inhibitor 1 PPI3B2	1.26	0.03	F: serine-type endopeptidase inhibitor activity; P: response to wounding
M1ABL9	Uncharacterized protein contains AAA+ ATPase domain	1.25	0.003	F: ATP binding and hydrolysis
Q84XQ4	NtPRp27-like protein	1.25	0.02	
M1A4R1	Uncharacterized protein with AT hooks a DNA-binding motif	1.24	0.004	C: nucleosome; F: DNA binding; C: nucleus; P: nucleosome assembly
M1C8Q0	Uncharacterized protein contains Proteasome component Initiation (PCI) domain	1.22	0.02	F: enzyme regulator activity

Table 1. *Cont.*

Accession	Description	Fold Change	*p* Value	GO Names
		Early stage down-regulated proteins		
K7WNX1	Light-inducible tissue-specific ST-LS1	0.83	0.001	C: photosystem II oxygen-evolving complex; P: photosynthesis
M0ZJ30	Uncharacterized protein contains Major facilitator domain	0.74	0.002	C: membrane; F: transmembrane transporter activity; P: transmembrane transport
M1BNJ1	Uncharacterized protein contains Carotenoid oxygenase binding site.	0.68	0.05	F: oxidoreductase activity, acting on single donors with incorporation of molecular oxygen, incorporation of two atoms of oxygen; P: oxidation-reduction process
		Late stage up-regulated proteins		
M1BD67	Uncharacterized protein with Acid phosphatase, class B-like domain	5.68	0.03	F: acid phosphatase activity
M0ZTQ4	Uncharacterized protein predicted with Osmotin/thaumatin-like domain	5.62	0.05	
M1CJS7	Uncharacterized protein contains CC-NB-ARC and LRR Domains.	4.00	0.01	F: ADP binding
M1B864	Uncharacterized protein contains Kunitz inhibitor STI-like domain.	3.65	0.04	F: endopeptidase inhibitor activity
M1AU65	Peroxidase	2.94	0.01	F: peroxidase activity; P: response to oxidative stress; F: heme binding; P: hydrogen peroxide catabolic process; P: oxidation-reduction process
E0WCF2	PIN-A	2.92	0.0001	F: serine-type endopeptidase inhibitor activity; P: response to wounding
A0A097H185	PIN-I-Protein	2.52	0.01	F: serine-type endopeptidase inhibitor activity; P: response to wounding
M0ZZF1	Peptidyl-prolyl cis-trans isomerase	1.64	0.02	P: protein peptidyl-prolyl isomerization; F: peptidyl-prolyl cis-trans isomerase activity
M1BAS6	Uncharacterized protein (CC-NB-ARC and LRR Domains)	1.62	0.03	F: ADP binding
Q8H9B9	Hyoscyamine 6-beta-hydroxylase-like protein (Fragment)	1.62	0.03	F: oxidoreductase activity; P: oxidation-reduction process
A5A7I8	Calcium-dependent protein kinase 5	1.32	0.04	F: protein kinase activity; F: calcium ion binding; F: ATP binding; P: protein phosphorylation
M1AKD9	Uncharacterized protein contains a Histidine kinase/HSP90-like ATPase domain	1.30	0.02	F: ATP binding; P: protein folding; P: response to stress; F: unfolded protein binding
Q8VX50	Putative receptor-like serine-threonine protein kinase	1.22	0.05	F: protein kinase activity; F: ATP binding; P: protein phosphorylation
M1D3S7	Uncharacterized protein contains a CC-NB-ARC and LRR Domain.	1.22	0.05	F: ADP binding
M0ZWQ0	V-type proton ATPase proteolipid subunit	1.21	0.03	F: proton transmembrane transporter activity; P: ATP hydrolysis coupled proton transport; C: proton-transporting V-type ATPase, V0 domain

Table 1. *Cont.*

Accession	Description	Fold Change	*p* Value	GO Names
		Late stage down-regulated proteins		
K7WNV9	Thioredoxin	0.82	0.05	P: glycerol ether metabolic process; F: protein disulfide oxidoreductase activity; P: cell redox homeostasis
G1CCA4	Photosystem I assembly protein Ycf4	0.71	0.03	C: photosystem I; P: photosynthesis; C: integral component of membrane
M1D260	FK506-binding protein	0.63	0.03	P: histone peptidyl-prolyl isomerization; F: peptidyl-prolyl cis-trans isomerase activity; F:FK506 binding; C: nucleolus
M1B8N6	NADPH-protochlorophyllide oxidoreductase	0.37	0.04	F: protochlorophyllide reductase activity; P: oxidation-reduction process

2.6.1. Cv. Sarpo Mira Protein–Protein Interaction during the Early Stages of Infection

Most proteins carry out essential biological functions, like signal transduction, protein modification, cellular metabolism, cytokinesis, DNA replication, RNA transcription, and targeted degradation, by interacting with other proteins in a protein complex. To uncover functional interactions among proteins during the early stages of *P. infestans* invasion, we analyzed the 75 DEPs in the EI/Control comparison by Search Tool for the Retrieval of Interacting Genes/Proteins (https://string-db.org) in Cytoscape (Figure 7a). Of interest, was the association among nine proteins: wall-associated kinase protein (WAKP, M1D051), an integral component of the cell membrane; a ubiquitin-conjugating enzyme protein (M1BLH0); a chloroplast nucleoid DNA-binding protein (M1B6K1); a ribosomal L24 domain (M1D1L9); a 26S proteasome subunit-containing RPN7 protein domain (M1C8Q0); endochitinase (Q2HPK8); osmotin (M0ZTM9); a pathogenesis-related PR1 protein containing a RlpA-like double-psi beta-barrel domain (M1A2A4); and finally, protein disulfide isomerase inhibitor (PDI, M1C517).

We reasoned that *P. infestans* might directly or indirectly elicit the induction of these proteins, and their association was important to promote coordination of several processes, including signal transduction, cellular metabolic process, and defense against *P. infestans*. For example, WAKP (M1D051, FC = 1.28) was predicted as a major functional node having multiple protein interactions, and it was specifically significantly up-regulated from Control to EI time-points, through to LI. Previous studies have shown that WAK proteins are important oligogalacturonide receptors, required for the activation of the plant immune response, as well as for growth and development [59]. Additionally, there are reports of the arabidopsis RFO1 gene coding for a wall-associated kinase protein that conferred quantitative resistance against *Fusarium oxysporum* [60]. In the present study, it is likely that WAKP, in association and coordination with other proteins, acted as an elicitor by binding to damage-associated molecular patterns (DAMPs) or cell wall defense proteins [61,62]. We also noticed that a putative ribosomal protein in the network was highly abundant (M1D1L9, FC = 1.36), and a protein possessing ubiquitin-binding domain (M1C8Q0, FC = 1.22). Previous studies have reported the involvement of these two proteins in *P. parasitica*–tomato interaction [63]. Next are the constitutively highly-expressed Endochitinase proteins (Q2HPK8, FC = 1.43; M1AH25, FC = 1.33), which have common function in defense-related signaling to boost the non-specific defense response by releasing elicitor-active chitin oligomers [60]. It has been shown that basic chitinase and osmotin-like protein possess actin-binding capabilities and cooperate to promote cytoplasmic aggregation in the potato cells, as a defense against penetration of Phytophthora [64]. In the present study, the significant up-regulation of these several osmotins, (M0ZTM9, FC = 1.34) from Control to EI, and through to LI, suggest significant coordination among pathogenesis-related proteins within the cell wall to promote basal resistance responses. Indeed, the PR1 gene was implicated in basal resistance against *P. infestans* oomycete [65]. In this study, PR1 protein (FC = 1.30) was up-regulated, its expression level was highest at the late stage (1.40), which indicates that this protein was significantly induced throughout the time course, which is consistent with previous studies [66]. Also, the KiTH-2 protein was strongly up-regulated (FC =1.82) upon infection with *P. infestans*; its specific role in potato–*P. infestans* interaction or pathogen resistance is not clear [67]. Although it belongs to the Kiwellin family and contains the rare lipoprotein A (RlpA) domain, which has been shown to act as a prc mutant suppressor in *Escherichia coli* [68]. Here, we speculate that it may function as part of elicitor machinery during pathogen infection.

Meanwhile, studies have shown that the PDI gene (FC = 1.31, M1C517) plays a crucial role in host–pathogen interaction and that PDI protein is localized to the haustoria of Phytophthora, a major site of pathogen protein export into the host cell during infection [69]. In the present study, the significant up-regulation of PDI from Control to EI, and subsequent decrease in abundance at LI, suggests this protein might be acting as a virulence factor of *P. infestans* at the early infection stage, and potentially contributes to plant infection and the late-stage disease process [70]. Additionally, the plant's secretory system is crucial for building resistance at the cell periphery. It also enables attacked cells to transport antimicrobial compounds and cell wall material to the site of attempted penetration [38]. In this study, we found a vesicle-mediated transport protein (M0ZZ55, FC = 1.77) containing the v- SNARE domain, significantly up-regulated from Control to EI, and with stable expression afterward, which suggests a tightly coordinated transport process during early stages of pathogen attack, which has been previously reported [69]. Together, these results show that during the early stages of disease infection there was a coordinated up-regulation of defense arsenal, like the pathogen recognition proteins, signaling molecules, antimicrobial compounds, cellular trafficking, primed to initiate a broad-based resistance against *P. infestans*.

2.6.2. Cv. Sarpo Mira Protein–Protein Interaction during the Late Stages of Infection

We analyzed the protein interaction of DEPs identified in LI/EI comparison and found multifactorial interactions with several hub proteins in the network (Figure 7b). Prominent among the proteins in the network was urease (Q93WI8, FC = 1.27), with over 67 connections. In this study, urease increased slightly in abundance from Control (avg. = 0.86) to EI (avg. = 0.90), but was significantly induced at LI (avg. = 1.14). Studies have shown that urease inhibits the growth of phytopathogenic fungi [71,72]. We also found an uncharacterized protein, M1E0E5 (FC = 1.22), with 57 connections, which contained a pyridoxal-phosphate-binding site, and was predicted to participate in aminotransferase activity, oxidoreductase, and mononucleotide binding. Similarly, we observed two other proteins (57 connections) coding for Hydroxyacyl-CoA dehydrogenase (M0ZHQ8, FC = 1.26), with functions related to oxidoreductase activity. Other proteins included metalloenzyme (39 connections, O81394, FC = −1.20), involved in catalytic conversion of 2-phosphoglycerate to phosphoenolpyruvate; GA3PDH enzyme (34 connections, Q8LK04, FC = 1.20), involved in glycolysis and gluconeogenesis; citrate synthase protein (32 connections, M1AD15, FC = 1.38); histidine kinase/HSP90-like ATPase (32 connections, M1C5D1, FC = 1.21); thioredoxin (28 connections, M1CXH6, FC = 1.24); and glutathione peroxidase (20 connections, M1AWZ7, FC = 1.37).

The late stage of the *P. infestans* disease process is usually characterized by colonization and necrotrophy [48]. At this stage, it is also likely that the pathogen faced a shortage of nutrients, like glucose, fatty acid, amino acids, and energy, from the shut-down of the host's cellular metabolism by HR-induced cell death, and may attempt to induce host metabolic changes to enable nutrient supply and growth. For example, GA3PDH was up-regulated and is required for the breakdown of glucose for energy and carbon molecules. Hydroxyacyl-CoA dehydrogenase is essential for fatty acid beta-oxidation, which can also provide energy. While pyridoxal phosphate P5P is an active form of vitamin B6, which is known to be involved in various reactions, including amino acid break-down, transamination, decarboxylation, and racemization reactions. It is likely that P5P acts to counteract the toxic effects of the oxidative burst and, alternatively, to break down amino acids for energy, possibly contributing to pathogenesis. Additionally, for the pathogen colonization to succeed, it must inactivate or remove the host-induced reactive oxygen species (ROIs), and thioredoxin proteins can counteract or protection the pathogen against ROIs [48]. Other proteins that showed sparse interaction included: cytochrome P450 proteins, possibly involved in efflux and detoxification [62]; and anamorsin; interestingly, anamorsin has been reported to be involved in negative regulation of apoptosis as well as cellular iron homeostasis [73]. Likewise, we noticed glycosyl hydrolase, a cell wall-degrading enzyme, was probably involved in virulence as well as nutrition or growth at the late disease stage.

We also identified glutathione S-transferase and glutathione peroxidase, which were previously reported to be involved in pathogen antioxidant defenses and host cell detoxication [62,74]. Based on these results we hypothesize that in an incompatible interaction some of the DEPs at the late stage may have a function-related late-stage disease-susceptible process. Notwithstanding, a strong domain self-interaction was observed among the cell death regulator proteins (M1D3S7, M1CJS7, M1BAS6) at this stage.

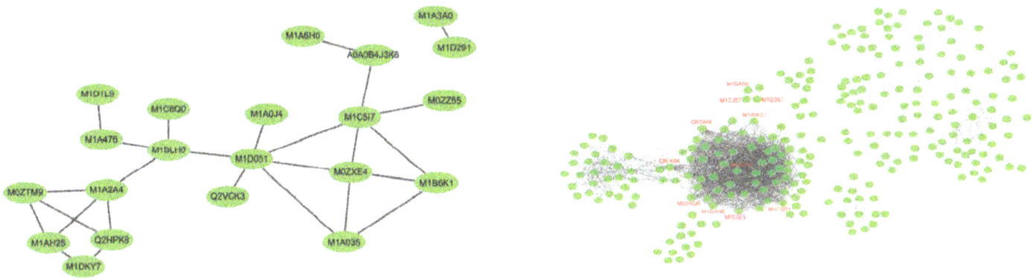

Figure 7. Network analysis results for significantly changed proteins in the (**a**) early and (**b**) late stages of infection.

2.6.3. Correlation Analysis of Protein Expression and mRNA by Real Time-qPCR

In order to evaluate the correlation between mRNA and protein levels, we analyzed the relative expression pattern of genes encoding eight representative proteins in the context of the three time-points by RT-qPCR. For each of the eight genes, we found that the transcript level increased in the early and late stage of the disease process, in agreement with their protein expression levels, as revealed by TMT data. For example, in the early stages of disease infection, three of the eight genes, coding for proteins involved in cell wall structure modification, had a higher expression level (Figure 8a). A similar pattern was observed for the remaining five genes at the later stage of the infection cycle (Figure 8b). These results validate the increased levels of the encoded proteins that were observed in the proteomic analysis and suggest that the abundance of these proteins is likely regulated at the transcriptional level.

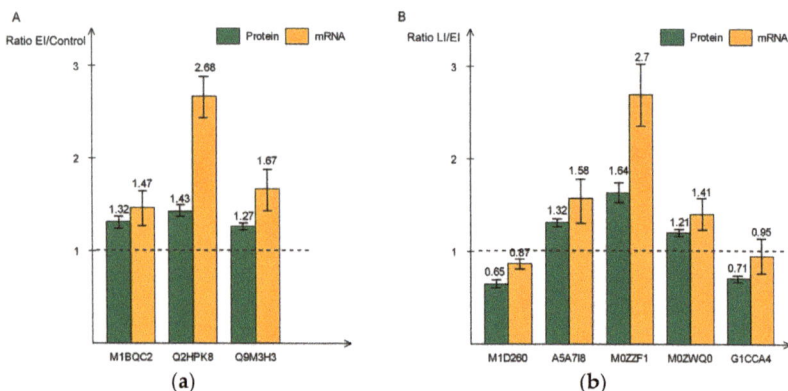

Figure 8. Real-time polymerase chain reaction (PCR) quantitative analysis of eight differentially-expressed proteins and mRNA at early (**a**) and late (**b**) disease stages. The green bar indicates the expression level determined by Tandem mass tag (TMT) and orange bar shows relate expression of mRNA. All data are presented as mean ± SD (*n* = 3 in each time-point).

In summary, this study sheds light on the changes in protein abundance and physiological roles of proteins in potato during the early and late stages of *P. infestans* oomycete infection. We have also inferred protein interaction that occurred during the two disease stages, either physical interactions verified through experiments, or predictions to further understand the disease process. Overall, our analysis suggests that differentially-expressed proteins, identified at the early stage of infection, played significant roles in signal transduction and the basal defense response, while the late stage disease process was characterized by the significant abundance of R proteins related to disease resistance processes and cell death. However, some of proteins in the late disease stage could be related to late-stage disease-susceptible processes. Therefore, the data reported here is a valuable resource for practical use to further characterize the mechanisms that are potentially involved in the potato late blight disease resistance process.

3. Materials and Methods

3.1. Plant Material and Growth Conditions

Potato plants of cultivar Sarpo Mira were grown in a greenhouse with controlled conditions set at 20 °C, 16:8 light to dark cycle, and 70% relative humidity. Five-week-old plants were transferred to an infection chamber with 100% humidity and 10:14 light:dark cycle. After 6 h, plants were sprayed with an encysted zoospore suspension from *P. infestans* isolate until the leaf surfaces were fully saturated with the zoospore suspension (15,000 sporangia/mL). Samples were collected at 0, 48, and 120 h post inoculation (hpi) according to [75], and labelled as Control, EI, and LI, respectively. The 0 dpi samples were collected immediately after inoculation with a contact time of less than one minute [75]. Afterwards, the relative humidity was maintained at 100% for two days after inoculation and then adjusted to 90% for the rest of the experiment. For each time-point, samples of fully-expanded upper leaves were collected from three independent biological experiments. All the materials were frozen in liquid nitrogen and stored at −80 °C until use [31].

3.2. Protein Extraction

For each sample, 1 g was weighed and homogenized by grinding in liquid nitrogen and transferred to a 50 mL precooled test-tube. Afterwards 25 mL of precooled acetone (−20 °C), containing 10% (*v/v*) trichloroacetic acid (TCA) and 65 mM dithiothreitol (DTT), was added. After thorough mixing, the homogenate was precipitated for 2 h at −20 °C and then centrifuged for 30 min at 16,000× *g* at 4 °C.

The supernatant was carefully removed, and the pellet was rinsed three times with 20 mL of cold acetone (-20 °C), followed by centrifugation ($20{,}000\times g$ for 30 min at 4 °C). The precipitation was collected and vacuum freeze-dried. A 250 mg sample of the freeze-dried pellets was weighed and placed in a 1.5 mL Eppendorf tube. The pellets were dissolved in SDT lysis buffer (4% SDS, 100 mM Tris-HCl, 100 mM DTT, pH 8.0) and then boiled for 5 min. After boiling and vortex mixing for 30 s, the mixture was intermittently sonicated in an ice bath, with 5 s sonication followed by 10 s break, for 5 min at 100 W. The mixture was then boiled again for 5 min, followed by 30 min centrifugation ($12{,}000\times g$, 20 °C). The supernatant was collected in a new 1.5 mL Eppendorf tube, filtered through a 0.22-µm Millipore filter and collected as lysate. Protein concentration in the lysate was determined using the bicinchoninic acid (BCA) protein assay reagent (Beyotime Institute of Biotechnology, Shanghai, China). The rest of the lysate was frozen at -80 °C until use.

3.3. Protein Digestion

TMT analysis was performed according to the method described by [76], Briefly, protein concentrates (300 µg) in an ultrafiltration filtrate tube (30 kDa cut-off, Sartorius, Gottingen, Germany) was mixed with 200 µL Urea buffer (8 M urea, 150 mM Tris-HCl, pH 8.0) and the sample was centrifuged at $14{,}000\times g$ at 20 °C for 30min. The sample was washed twice by adding 200 µL UA and centrifuged at $14{,}000\times g$ at 20 °C for 30min. The flow-through from the collection tube was discarded. Next, 100 µL Indole-3-acetic acid (IAA) solution (50 mM IAA in UA buffer) was added to the filter tube and vortexed at 600 rpm in a thermomixer comfort incubator (Eppendorf, Germany) for 1 min. Subsequently, the sample was incubated at room temperature for 30 min in the dark and spun at 14,000 g for 30 min at 20 °C. Next 100 µL UA was added to the filter unit and centrifuged at 14,000 g for 20 min, and this step was repeated twice. The protein suspension in the filtrate tube was subjected to enzyme digestion with 40 µL of trypsin (Promega, Madison, WI, USA) buffer (4 µg trypsin in 40 µL of dissolution buffer) for 16–18 h at 37 °C. Finally, the filter unit was transferred to a new tube and spun at 14,000 g for 30 min. Peptides were collected in the filtrate and concentration of the peptides was measured by optical density with a wavelength of 280 nm (OD280).

3.4. TMT Labeling and LC-MS/MS Analysis

Approximately 50ug of digested peptides from each sample, including the internal standard, were labeled with TMT reagents (Thermo Fisher Scientific, San Jose, CA, USA) following procedures recommended by the manufacturer. Briefly, peptides from the samples LI1, LI2, LI3, Control 1, Control 2, Control 3, EI 1, EI 2, and EI3 were labeled with TMT reagents 126, 127N, 127C, 128N, 128C, 129N, 130N, 130C, and 131, respectively. All labeled peptides were pooled together. Labeled and mixed peptides were subjected to high-pH reversed-phase fractionation in the 1100 Series High-performance liquid chromatography Value System (Agilent, Palo Alto, CA, USA) equipped with a Gemini-NX (Phenomemex, 00F-4453-E0) column (4.6 × 150 mm, 3 µm, 110 Å). Peptides were eluted at a flow rate of 0.8 mL/min. Buffer A consisted of 10 mM ammonium acetate (pH10.0) and buffer B consisted of 10 mM ammonium acetate and 90% *v/v* Acetonitrile (pH 10.0). Buffer A and B were both filter-sterilized. The following gradient was applied to perform separation: 100% buffer A for 40 min, 0%–5% buffer B for 3 min, 5%–35% buffer B for 30 min, and 35%–70% buffer B for 10 min. Then, 70%–75% buffer B for 10 min, 75%–100% buffer B for 7 min, 100% buffer B for 15 min, and 100% buffer A for 15 min. The elution process was monitored by measuring absorbance at 214 nm, and fractions were collected every 75 s. Finally, the collected fractions (approximately 40) were combined into 10 pools. Each fraction was concentrated via vacuum centrifugation and was reconstituted in 40 µL of 0.1% *v/v* trifluoroacetic acid. All samples were stored at -80 °C until further analysis.

The TMT-labeled samples were analyzed using easy-nLC nanoflow HPLC system connected to an Orbitrap Elite mass spectrometer (Thermo Fisher Scientific, San Jose, CA, USA). A total of 1 µg of each peptide sample was loaded onto Thermo Scientific EASY column (two columns) using an autosampler at a flow rate of 150 nL/min. The sequential separation of peptides on Thermo Scientific

EASY trap column (100 μm × 2 cm, 5 μm, 100Å, C18) and analytical column (75 μm × 25 cm, 5 μm, 100 Å, C18) was accomplished using a segmented 2 h gradient from Solvent A (0.1% formic acid in water) to 35% Solvent B (0.1% formic acid in 100% Acetonitrile) for 100 min. Followed by 35%–90% Solvent B for 12 min and then 90% Solvent B for 8 min. The mass spectrometer was operated in positive ion mode, and MS spectra were acquired over a range of 350–2000 m/z. Resolving powers of the MS scan and MS/MS at 100 m/z for the Orbitrap Elite were set as 60,000 and 15,000, respectively. The top sixteen most intense signals in the acquired MS spectra were selected for further MS/MS analysis. The isolation window was 1 m/z, and ions were fragments through higher energy collisional dissociation with normalized collision energies of 35 eV. The maximum ion injection time was set at 50 ms for the survey scan, and 150 ms for the MS/MS scans, and the automatic gain control target values for full san modes was set to 10×10^{-6}, and for MS/MS it was 5×10^4. The dynamic exclusion duration was 30 s.

3.5. Database Search, Protein Identification, and Quantification

The Proteome Discoverer 2.1 (Thermo Fisher Scientific) was used to analyze raw data. The Mascot 2.1 (Matrix Science) embedded in Proteome Discoverer was used to search raw data against the UniProt potato database (December 21, 2017; 55,715 sequences). Search parameters were as follows: monoisotopic mass; trypsin as cleavage enzyme; two max missed cleavages; TMT 10plex (N-term), TMT 10plex (K), and carbamidomethylation of cysteine as fixed modifications; and oxidation of methionine as variable modifications. Peptide mass tolerance of ±20 ppm and fragment mass tolerance of 0.1 Da were used for parent and monoisotopic fragment ions, respectively. Results were filtered based on a false discovery rate of (FDR) ≤0.01. Relative quantitative analyses of proteins were based on ratios of TMT reporter ions from all unique peptides representing each protein. For protein quantitation, each reporter ion channel was summed across all quantified proteins and normalized assuming equal protein loading of all ten samples. The protein ratios of each sample were normalized to the TMT-126 label [77]. The mass spectrometry proteomics data are available at ProteomeXchange Consortium via the PRIDE [78] partner repository with identifier PXD010045.

3.6. Bioinformatics and Statistical Analysis

Proteome Discoverer 2.1 Protein quantitation values were exported for further analysis in Excel. Proteins of *p*-values <0.05 by Student *t*-test and a fold-change of >1.20 or <0.83 in expression between any two groups were considered significant. Differentially expressed proteins (DEPs) were classified by their gene functions and also by biological pathways using the publicly available gene ontology (GO) database provided by the Gene Ontology Consortium (http://geneontology.org/) [79]. The identified protein sequence information was extracted from the UniProt knowledge base and retrieved in FASTA format. The functional information of the homologous proteins was used to annotate targeted proteins. Top 10 blast hits with E-values of less than 1e-3, for each of the query proteins, were retrieved and loaded into Blast2GO (Version 2.7.2) [80], a high-throughput online tool for gene ontology (GO) analysis, for GO mapping, and annotation. Enriched GO terms were identified with Fisher's exact test and hypergeometric distribution test cutoff of 0.05. Information on the biological pathways was obtained from the Kyoto Encyclopedia of Genes and Genomes pathways database (http://www.genome.jp/kegg/pathway.html) [81]. Visualization of these pathways and enrichment analysis was performed using the KOBAS 2.0 software [82–84]. *p* < 0.05 was set as the threshold used for enrichment analysis of KEGG pathways. Interactions among differentially-expressed proteins in early and late disease stages were analyzed by Cytoscape software, and were used to draw the protein interaction network [82].

3.7. Antibodies and Western Blot Analysis

For western blot analysis, the procedures of electrophoresis, transfer, and immunodetection were performed according to Howden et al. [85]. The primary antibodies used were as follows:

antibody for the Osmotin-like protein (Q5XUH0, Biorbyt orb27915, 1:1000); pectinesterase (M1BQC2, PLLABS PL0304687, 1:500); endochitinase (Q2HPK8, PhytoAB PHY1514S, 1:2000); annexin (Q9M3H3, PhytoAB PHY0729S, 1:2000); peptidyl-prolyl cis-trans isomerase (M0ZZF1, PhytoAB PHY0920S, 1:2000); photosystem I assembly protein Ycf4 (G1CCA4, PhytoAB PHY1363S, 1:1500); type I serine protease inhibitor (E0WCF2, PhytoAB PHY0146S, 1:2000). Horseradish peroxidase-conjugated anti-rabbit IgG (dilution 1:15,000, Bio-Rad, Hercules, CA, USA) were used as secondary antibodies. After immunodetection, the intensity of the immuno-stained bands were normalized for the total protein intensities measured by Coomassie blue from the same [86]. The images were subjected to densitometric analysis performed using Quantity One software (Bio-Rad).

3.8. Criteria for Selecting Essential Proteins

Candidate proteins were selected belonging to the following groups: (a) Common and unique DEPs belonging to the early disease stage, chosen based on their protein abundance, fold change value, and functional category enrichment. Additionally, candidates were further selected based on the inferred protein interaction as well as those with putative function that are relevant to late blight disease infection response during early infection. Furthermore, previously reported proteins were also considered in candidate gene selection in both stages. (b) Common and unique DEPs from the late disease stage, chosen based on expression profile, fold change value, functional category enrichment, and putative function relevant to the late disease stage disease process, in addition to the inferred protein interactions in the network.

3.9. Correlation between mRNA and Protein Levels by Quantitative Real-Time Polymerase Chain Reaction (qRT-PCR)

Leaf samples were collected at 4, 48, and 120 h for protein and total RNA isolation. Total RNA was extracted from each sample using TRIzol Reagent Kit (AmbionTM) according to the manufacturer's instructions. Next, each of the RNA samples was treated with RNase-free DNase (Takara, Dalian, China). Complementary DNA (cDNA) was retro-transcribed from 2 µg of total RNA using the Thermo Scientific RevertAid Kit according to the manufacturer's instructions. Quantitative real-time polymerase chain reaction (qPCR) was performed on a Bio-Rad real-time detection system (Bio-Rad). PCR conditions were 95 °C for 1 min, followed by 44 cycles at 95 °C, 12 s, 60 °C, 30 s, and 72 °C, 30 s. After cycling, melting curves of the reaction were run from 55 °C to 95 °C. Each reaction was performed in three technical replicates, and the expression profiles of 8 genes were analyzed, with potato ef1a gene used as the constitutive gene for normalization. The quantification of gene expression levels was calculated relative to ef1a with the $2^{-\Delta\Delta CT}$ method [87]. Primer sets used for qRT-PCR are reported in Table S8.

Supplementary Materials: Supplementary materials can be found at http://www.mdpi.com/1422-0067/20/1/136/s1. Table S1: List of all proteins identified in the time series analysis. Table S2: List of all differentially expressed proteins, functional annotation, significantly enriched INTERPRO protein domains and features, PFAM domains, and KEGG pathways classification. Table S3: List of differentially-expressed proteins during the early stage of *P. infestans* infection. (EI/Control). Table S4: List of differentially-expressed proteins during the late stage of *P. infestans* infection. (LI/EI). Table S5: List of differentially-expressed proteins from time-point LI/Control. Table S6: Gene Ontology classification of differentially-expressed proteins. Table S7: List of common differentially-expressed proteins throughout the whole time course of the experiment. Table S8: List of primer sets used in the qRT-PCR correlation analysis of protein expression and mRNA.

Author Contributions: Conceptualization, S.Y.; methodology, Q.C. (Qiaoling Chen), L.Y., J.G., and C.X.; validation, J.G., Y.Z., G.Y., C.X., and L.Y.; formal analysis, Z.W., D.Z., Y.X., Y.Z., and Q.C. (Qun Cheng); investigation, C.X., J.G., X.Y., G.Y., and Y.X.; resources, S.Y., C.X., and J.C.; data curation, L.Y., C.X., Z.W., and D.Z.; writing—original draft preparation, C.X. and J.C.; writing—review and editing, S.Y., J.C., C.X., Y.X., X.Y., and Q.C. (Qiaoling Chen); visualization, X.Y., G.Y., Q.C. (Qun Cheng), and Y.X.; supervision, S.Y. and J.C.; project administration, S.Y.; funding acquisition, S.Y., C.X., and J.C.

Funding: This research was funded by Hubei Provincial Specialized Technological Innovation Project (Special Project of Western Hubei Ethnic Group2016AKB052); China Agriculture Research System (CARS-09); Central Guidance for Local Science and Technology Development; Ministry of Agriculture Central China Potato

Int. J. Mol. Sci. **2019**, *20*, 136

Observation and Experiment Station; Ministry of Agriculture Key Laboratory of Agro-Products Postharvest Handing (KLAPPH2-2017-05); Innovation Team of Hubei Agricultural Science and Technology Innovation Center (2016-620-000-001-061); and Enshi Prefecture Science and Technology Plan Research and Development Project (D20170003).

Acknowledgments: The authors gratefully acknowledge Shanghai Omicsspace Biotechnology Co., Ltd. and Charles. C. Nwafor for technical assistance with peptide and protein identification and the discussion of the results.

Conflicts of Interest: The authors have declared that no competing interest exist.

References

1. Song, J.; Bradeen, J.M.; Naess, S.K.; Raasch, J.A.; Wielgus, S.M.; Haberlach, G.T.; Lui, J.; Kuang, H.; Austin-Phillips, S.; Buell, C.R.; et al. Gene RB cloned from Solanum bulbocastanum confers broad spectrum resistance to potato late blight. *Proc. Natl. Acad. Sci. USA* **2003**, *100*, 9128–9133. [CrossRef] [PubMed]
2. Haas, B.J.; Kamoun, S.; Zody, M.C.; Jiang, R.H.Y.; Handsaker, R.E.; Cano, L.M.; et al. Genome sequence and analysis of the Irish potato famine pathogen Phytophthora infestans. *Nature* **2009**, *461*, 393–398. [CrossRef] [PubMed]
3. Widmark, A.K.; Andersson, B.; Cassel-Lundhagen, A.; Sandström, M.; Yuen, J.E. Phytophthora infestans in a single field in southwest Sweden early in spring: Symptoms, spatial distribution, and genotypic variation. *Plant Pathol.* **2007**, *56*, 573–579. [CrossRef]
4. Zuluaga, A.P.; Vega-Arreguín, J.C.; Fei, Z.; Ponnala, L.; Lee, S.J.; Matas, A.J.; Patev, S.; Fry, W.E.; Rose, J.K. Transcriptional dynamics of *Phytophthora infestans* during sequential stages of hemibiotrophic infection of tomato. *Mol. Plant Pathol.* **2016**, *17*, 29–41. [CrossRef]
5. Kaku, H.; Nishizawa, Y.; Ishii-Minami, N.; Akimoto-Tomiyama, C.; Dohmae, N.; Takio, K.; Minami, E.; Shibuya, N. Plant cells recognize chitin fragments for defense signaling through a plasma membrane receptor. *Proc. Natl. Acad. Sci. USA* **2006**, *103*, 11086–11091. [CrossRef]
6. Hayafune, M.; Berisio, R.; Marchetti, R.; Silipo, A.; Kayama, M.; Desaki, Y.; Arima, S.; Squeglia, F.; Ruggiero, A.; Tokuyasu, K.; et al. Chitin-induced activation of immune signaling by the rice receptor CEBiP relies on a unique sandwich-type dimerization. *Proc. Natl. Acad. Sci. USA* **2014**, *111*, E404–E413. [CrossRef]
7. Shimizu, T.; Nakano, T.; Takamizawa, D.; Desaki, Y.; Ishii-Minami, N.; Nishizawa, Y.; Minami, E.; Okada, K.; Yamane, H.; Kaku, H.; et al. Two LysM receptor molecules, CEBiP and OsCERK1, cooperatively regulate chitin elicitor signaling in rice. *Plant J.* **2010**, *64*, 204–214. [CrossRef]
8. Miya, A.; Albert, P.; Shinya, T.; Desaki, Y.; Ichimura, K.; Shirasu, K.; Narusaka, Y.; Kawakami, N.; Kaku, H.; Shibuya, N. CERK1, a LysM receptor kinase, is essential for chitin elicitor signaling in Arabidopsis. *Proc. Natl. Acad. Sci. USA* **2007**, *104*, 19613–19618. [CrossRef]
9. Liu, T.; Liu, Z.; Song, C.; Hu, Y.; Han, Z.; She, J.; Fan, F.; Wang, J.; Jin, C.; Chang, J.; et al. Chitin-induced dimerization activates a plant immune receptor. *Science* **2012**, *336*, 1160–1164. [CrossRef]
10. Wan, J.; Zhang, X.C.; Neece, D.; Ramonell, K.M.; Clough, S.; Kim, S.Y.; Stacey, M.G.; Stacey, G. A LysM receptor-like kinase plays a critical role in chitin signaling and fungal resistance in Arabidopsis. *Plant Cell* **2008**, *20*, 471–481. [CrossRef]
11. Petutschnig, E.K.; Jones, A.M.E.; Serazetdinova, L.; Lipka, U.; Lipka, V. The lysin motif receptor-like kinase (LysM-RLK) CERK1 is a major chitin-binding protein in Arabidopsis thaliana and subject to chitin-induced phosphorylation. *J. Biol. Chem.* **2010**, *285*, 28902–28911. [CrossRef] [PubMed]
12. Narusaka, Y.; Shinya, T.; Narusaka, M.; Motoyama, N.; Shimada, H.; Murakami, K.; Shibuya, N. Presence of LYM2 dependent but CERK1 independent disease resistance in Arabidopsis. *Plant Signal. Behav.* **2013**, *8*, e25345. [CrossRef] [PubMed]
13. Shinya, T.; Motoyama, N.; Ikeda, A.; Wada, M.; Kamiya, K.; Hayafune, M.; Kaku, H.; Shibuya, N. Functional characterization of CEBiP and CERK1 homologs in Arabidopsis and rice reveals the presence of different chitin receptor systems in plants. *Plant Cell Physiol.* **2012**, *53*, 1696–1706. [CrossRef] [PubMed]
14. Faulkner, C.; Petutschnig, E.; Benitez-Alfonso, Y.; Beck, M.; Robatzek, S.; Lipka, V.; Maule, A.J. LYM2-dependent chitin perception limits molecular flux via plasmodesmata. *Proc. Natl. Acad. Sci. USA* **2013**, *110*, 9166–9170. [CrossRef] [PubMed]

15. Zhang, L.; Kars, I.; Essenstam, B.; Liebrand, T.W.H.; Wagemakers, L.; Elberse, J.; Tagkalaki, P.; Tjoitang, D.; van den Ackerveken, G.; van Kan, J.A. Fungal endopolygalacturonases are recognized as microbe-associated molecular patterns by the Arabidopsis receptor-like protein responsiveness to botrytis polygalacturonases1. *Plant Physiol.* **2014**, *164*, 352–364. [CrossRef] [PubMed]

16. Ron, M.; Avni, A. The receptor for the fungal elicitor ethylene-inducing xylanase is a member of a resistance-like gene family in tomato. *Plant Cell Am. Soc. Plant Biol.* **2004**, *16*, 1604–1615. [CrossRef] [PubMed]

17. De Jonge, R.; Van Esse, H.P.; Maruthachalam, K.; Bolton, M.D.; Santhanam, P.; Saber, M.K.; Zhang, Z.; Usami, T.; Lievens, B.; Subbarao, K.V.; et al. Tomato immune receptor Ve1 recognizes effector of multiple fungal pathogens uncovered by genome and RNA sequencing. *Proc. Natl. Acad. Sci. USA* **2012**, *109*, 5110–5115. [CrossRef]

18. Fradin, E.F.; Zhang, Z.; Ayala, J.C.J.; Castroverde, C.D.M.; Nazar, R.N.; Robb, J.; Liu, C.M.; Thomma, B.P. Genetic dissection of Verticillium wilt resistance mediated by tomato Ve1. *Plant Physiol. Am. Soc. Plant Biol.* **2009**, *150*, 320–332. [CrossRef]

19. Jones, J.D.G.; Dangl, J.L. The plant immune system. *Nature* **2006**, *444*, 323–329. [CrossRef]

20. Kamoun, S. A catalog of the effector secretome of plant pathogenic oomycetes. *Annu. Rev. Phytopathol.* **2006**, *44*, 41–60. [CrossRef]

21. Tian, M.; Benedetti, B.; Kamoun, S. A Second Kazal-like protease inhibitor from Phytophthora infestans inhibits and interacts with the apoplastic pathogenesis-related protease P69B of tomato. *Plant Physiol.* **2005**, *138*, 1785–1793. [CrossRef] [PubMed]

22. Metzker, M.L. Sequencing technologies—The next generation. *Nat. Rev. Genet.* **2010**, *11*, 31–46. [CrossRef] [PubMed]

23. Goodwin, S.; McPherson, J.D.; McCombie, W.R. Coming of age: Ten years of next-generation sequencing technologies. *Nat. Rev. Genet.* **2016**, *17*, 333–351. [CrossRef] [PubMed]

24. Potato Genome Sequencing Consortium. Genome sequence and analysis of the tuber crop potato. *Nature* **2011**, *475*, 189–195. [CrossRef] [PubMed]

25. Laurindo, B.S.; Laurindo, R.D.F.; Fontes, P.P.; Vital, C.E.; Delazari, F.T.; Baracat-Pereira, M.C.; da Silva, D.J.H. Comparative analysis of constitutive proteome between resistant and susceptible tomato genotypes regarding to late blight. *Funct. Integr. Genom.* **2018**, *18*, 11–21. [CrossRef] [PubMed]

26. Larsen, M.K.G.; Jorgensen, M.M.; Bennike, T.B.; Stensballe, A. Time-course investigation of Phytophthora infestans infection of potato leaf from three cultivars by quantitative proteomics. *Data Br.* **2016**, *6*, 238–248. [CrossRef] [PubMed]

27. Ali, A.; Alexandersson, E.; Sandin, M.; Resjö, S.; Lenman, M.; Hedley, P.; Levander, F.; Andreasson, E. Quantitative proteomics and transcriptomics of potato in response to *Phytophthora infestans* in compatible and incompatible interactions. *Bmc Genom.* **2014**, *15*, 497. [CrossRef] [PubMed]

28. Abdallah, C.; Dumas-Gaudot, E.; Renaut, J.; Sergeant, K. Gel-Based and Gel-Free Quantitative Proteomics Approaches at a Glance. *Int. J. Plant Genom.* **2012**, *2012*, 494572. [CrossRef] [PubMed]

29. Lim, S.; Borza, T.; Peters, R.D.; Coffin, R.H.; Al-mughrabi, K.I.; Pinto, D.M.; Wang-Pruski, G. Proteomics analysis suggests broad functional changes in potato leaves triggered by phosphites and a complex indirect mode of action against Phytophthora infestans. *J. Proteom.* **2013**, *93*, 207–223. [CrossRef]

30. Werner, T.; Sweetman, G.; Savitski, M.F.; Mathieson, T.; Bantscheff, M.; Savitski, M.M. Ion coalescence of neutron encoded TMT 10-plex reporter ions. *Anal. Chem.* **2014**, *86*, 3594–3601. [CrossRef]

31. Rietman, H.; Bijsterbosch, G.; Cano, L.M.; Lee, H.R.; Vossen, J.H.; Jacobsen, E.; Visser, R.G.; Kamoun, S.; Vleeshouwers, V.G. Qualitative and quantitative late blight resistance in the potato cultivar Sarpo Mira is determined by the perception of five distinct RXLR effectors. *Mol. Plant-Microbe Interact.* **2012**, *25*, 910–919. [CrossRef]

32. Guo, T.; Wang, X.W.; Shan, K.; Sun, W.; Guo, L.Y. The Loricrin-Like Protein (LLP) of Phytophthora infestans is Required for Oospore Formation and Plant Infection. *Front. Plant Sci.* **2017**, *8*, 142. [CrossRef] [PubMed]

33. Du, Y.; Mpina, M.H.; Birch, P.R.; Bouwmeester, K.; Govers, F. Phytophthora infestans RXLR Effector AVR1 Interacts with Exocyst Component Sec5 to Manipulate Plant Immunity. *Plant Physiol.* **2015**, *169*, 1975–1990. [CrossRef]

34. Quevillon, E.; Silventoinen, V.; Pillai, S.; Harte, N.; Mulder, N.; Apweiler, R.; Lopez, R. InterProScan: Protein domains identifier. *Nucleic Acids Res.* **2005**, *33*, W116–W120. [CrossRef] [PubMed]

35. Yogendra, K.N.; Kumar, A.; Sarkar, K.; Li, Y.; Pushpa, D.; Mosa, K.A.; Duggavathi, R.; Kushalappa, A.C. Transcription factor StWRKY1 regulates phenylpropanoid metabolites conferring late blight resistance in potato. *J. Exp. Bot.* **2015**, *66*, 7377–7389. [CrossRef] [PubMed]

36. Fraser, C.M.; Chapple, C. The phenylpropanoid pathway in Arabidopsis. *Arab. Book* **2011**, *9*, E0152. [CrossRef]

37. Ah-Fong, A.; Shrivastava, J.; Judelson, H.S. Lifestyle, gene gain and loss, and transcriptional remodeling cause divergence in the transcriptomes of Phytophthora infestans and Pythium ultimum during potato tuber colonization. *Bmc Genom.* **2017**, *18*, 764. [CrossRef]

38. Hückelhoven, R. Cell Wall-Associated Mechanisms of Disease Resistance and Susceptibility. *Annu. Rev. Phytopathol.* **2007**, *45*, 101–127. [CrossRef]

39. Sanchez, L.M.; Ohno, Y.; Miura, Y.; Kawakita, K.; Doke, N. Host selective suppression by water-soluble glucans from *Phytophthora* spp. of hypersensitive cell death of suspension-cultured cells from some solanaceous plants caused by hyphal wall elicitors of the fungi. *Ann. Phytopathol. Soc. Jpn.* **1992**, *58*, 664–670. [CrossRef]

40. Andreu, A.; Tonón, C.; Van Damme, M.; Huarte, M.; Daleo, G. Effect of glucans from different races of Phytophthora infestans on defense reactions in potato tuber. *Eur. J. Plant Pathol.* **1998**, *104*, 777–783. [CrossRef]

41. Liu, H.; Wang, X.; Zhang, H.; Yang, Y.; Ge, X.; Song, F. A rice serine carboxypeptidase-like gene OsBISCPL1 is involved in regulation of defense responses against biotic and oxidative stress. *Gene* **2008**, *420*, 157–165. [CrossRef] [PubMed]

42. Jashni, M.K.; Mehrabi, R.; Collemare, J.; Mesarich, C.H.; de Wit, P.J. The battle in the apoplast: Further insights into the roles of proteases and their inhibitors in plant-pathogen interactions. *Front. Plant Sci.* **2015**, *6*, 584. [CrossRef] [PubMed]

43. Avrova, A.O.; Taleb, N.; Rokka, V.M.; Heilbronn, J.; Campbell, E.; Hein, I.; Gilroy, E.M.; Cardle, L.; Bradshaw, J.E.; Stewart, H.E.; et al. Potato oxysterol binding protein and cathepsin B are rapidly up-regulated in independent defence pathways that distinguish R gene-mediated and field resistances to *Phytophthora infestans*. *Mol. Plant Pathol.* **2004**, *5*, 45–56. [CrossRef] [PubMed]

44. Kim, J.Y.; Park, S.C.; Kim, M.H.; Lim, H.T.; Park, Y.K.; Hahm, K.S. Antimicrobial activity studies on a trypsin-chymotrypsin protease inhibitor obtained from potato. *Biochem. Biophys. Res. Commun.* **2005**, *330*, 921–927. [CrossRef] [PubMed]

45. Nelson, A.J.; Doerner, P.W.; Zhu, Q.; Lamb, C.J. Isolation of a monocot 3-hydroxy-3-methylglutaryl coenzyme A reductase gene that is elicitor-inducible. *Plant Mol. Biol.* **1994**, *25*, 401–412. [CrossRef] [PubMed]

46. Fernández, M.B.; Pagano, M.R.; Daleo, G.R.; Guevara, M.G. Hydrophobic proteins secreted into the apoplast may contribute to resistance against *Phytophthora infestans* in potato. *Plant Physiol. Biochem.* **2012**, *60*, 59–66. [CrossRef]

47. Roumeliotis, E.; Kloosterman, B.; Oortwijn, M.; Visser, R.G.; Bachem, C.W. The PIN family of proteins in potato and their putative role in tuberization. *Front. Plant Sci.* **2013**, *4*, 524. [CrossRef]

48. Torto-Alalibo, T.A.; Tripathy, S.; Smith, B.M.; Arredondo, F.D.; Zhou, L.; Li, H.; Chibucos, M.C.; Qutob, D.; Gijzen, M.; Mao, C.; et al. Expressed sequence tags from *Phytophthora sojae* reveal genes specific to development and infection. *Mol. Plant-Microbe Interact.* **2007**, *20*, 781–793. [CrossRef]

49. Wang, J.; Cheng, Y.; Wu, R.; Jiang, D.; Bai, B.; Tan, D.; Yan, T.; Sun, X.; Zhang, Q.; Wu, Z. Antibacterial Activity of Juglone against Staphylococcus aureus: From Apparent to Proteomic. *Int. J. Mol. Sci.* **2016**, *17*, 965. [CrossRef]

50. Eliot, A.C.; Kirsch, J.F. Pyridoxal phosphate enzymes: Mechanistic, structural, and evolutionary considerations. *Annu. Rev. Biochem.* **2004**, *73*, 383–415. [CrossRef]

51. Van Damme, M.; Bozkurt, T.O.; Cakir, C.; Schornack, S.; Sklenar, J.; Jones, A.M.; Kamoun, S. The Irish Potato Famine Pathogen Phytophthora infestans Translocates the CRN8 Kinase into Host Plant Cells. *PLoS Pathog.* **2012**, *8*, E1002875. [CrossRef] [PubMed]

52. Bretz, J.R.; Mock, N.M.; Charity, J.C.; Zeyad, S.; Baker, C.J.; Hutcheson, S.W. A translocated protein tyrosine phosphatase of Pseudomonas syringae pv. tomato DC3000 modulates plant defence response to infection. *Mol. Microbiol.* **2003**, *49*, 389–400. [CrossRef] [PubMed]

53. Underwood, W.; Zhang, S.; He, S.Y. The Pseudomonas syringae type III effector tyrosine phosphatase HopAO1 suppresses innate immunity in Arabidopsis thaliana. *Plant J.* **2007**, *52*, 658–672. [CrossRef] [PubMed]

54. Rivas, S.; Rougon-Cardoso, A.; Smoker, M.; Schauser, L.; Yoshioka, H.; Jones, J.D.G. CITRX thioredoxin interacts with the tomato Cf-9 resistance protein and negatively regulates defense. *EMBO J.* **2004**, *23*, 2156–2165. [CrossRef] [PubMed]

55. Nekrasov, V.; Ludwig, A.A.; Jones, J.D.G. CITRX thioredoxin is a putative adaptor protein connecting Cf-9 and the ACIK1 protein kinase during the Cf-9/Avr9-induced defense response. *Febs Lett.* **2006**, *580*, 4236–4241. [CrossRef] [PubMed]

56. Burra, D.D.; Vetukuri, R.R.; Resjö, S.; Grenville-Briggs, L.J.; Andreasson, E. RNAseq and Proteomics for Analysing Complex Oomycete Plant Interactions. *Curr. Issues Mol. Biol.* **2016**, *19*, 73–88. [PubMed]

57. Akino, S.; Takemoto, D.; Hosaka, K. *Phytophthora infestans*: A review of past and current studies on potato late blight. *J. Gen. Plant Pathol.* **2014**, *80*, 24–37. [CrossRef]

58. Shibata, Y.; Kawakita, K.; Takemoto, D. Age-related resistance of *Nicotiana benthamiana* against hemibiotrophic pathogen *Phytophthora infestans* requires both ethylene- and salicylic acid-mediated signaling pathways. *Mol. Plant Microbe Interact.* **2010**, *23*, 1130–1142. [CrossRef]

59. D'Ovidio, R.; Mattei, B.; Roberti, S.; Bellincampi, D. Polygalacturonases, polygalacturonase-inhibiting proteins and pectic oligomers in plant-pathogen interactions. *Biochim. Biophys. Acta* **2004**, *1696*, 237–244. [CrossRef]

60. Diener, A.C.; Ausubel, F.M. Resistance to Fusarium Oxysporum 1, a dominant Arabidopsis disease-resistance gene, is not race specific. *Genetics* **2005**, *171*, 305–321. [CrossRef]

61. Brutus, A.; Sicilia, F.; Macone, A.; Cervone, F.; De Lorenzo, G. A domain swap approach reveals a role of the plant wall-associated kinase 1 (WAK1) as a receptor of oligogalacturonides. *Proc. Natl. Acad. Sci. USA* **2010**, *107*, 9452–9457. [CrossRef] [PubMed]

62. Boller, T.; Felix, G. A Renaissance of Elicitors: Perception of microbe-associated molecular patterns and danger signals by pattern-recognition receptors. *Annu. Rev. Plant Biol.* **2009**, *60*, 379–407. [CrossRef] [PubMed]

63. Le Berre, J.Y.; Engler, G.; Panabières, F. Exploration of the late stages of the tomato–*Phytophthora parasitica* interactions through histological analysis and generation of expressed sequence tags. *New Phytol.* **2008**, *177*, 480–492. [CrossRef] [PubMed]

64. Takemoto, D.; Furuse, K.; Doke, N.; Kawakita, K. Identification of Chitinase and Osmotin-Like Protein as Actin-Binding Proteins in Suspension-Cultured Potato Cells. *Plant Cell Physiol.* **1997**, *38*, 441–448. [CrossRef]

65. Vleeshouwers, V.G.A.A.; Dooijeweert, W.V.; Govers, F.; Kamoun, S.; TColona, L.T. Does basal PR gene expression in Solanum species contribute to non-specific resistance to *Phytophthora infestans*? *Physiol. Mol. Plant Pathol.* **2000**, *57*, 35–42. [CrossRef]

66. Draffehn, A.M.; Li, L.; Krezdorn, N.; Ding, J.; Lübeck, J.; Strahwald, J.; Muktar, M.S.; Walkemeier, B.; Rotter, B.; Gebhardt, C. Comparative transcript profiling by SuperSAGE identifies novel candidate genes for controlling potato quantitative resistance to late blight not compromised by late maturity. *Front. Plant Sci.* **2013**, *4*, 423. [CrossRef]

67. Mosquera, T.; Alvarez, M.F.; Jiménez-Gómez, J.M.; Muktar, M.S.; Paulo, M.J.; Steinemann, S.; Li, J.; Draffehn, A.; Hofmann, A.; Lübeck, J.; et al. Targeted and Untargeted Approaches Unravel Novel Candidate Genes and Diagnostic SNPs for Quantitative Resistance of the Potato (*Solanum tuberosum* L.) to *Phytophthora infestans* Causing the Late Blight Disease. *PLoS ONE* **2016**, *11*, E0156254. [CrossRef]

68. Bass, S.; Gu, Q.; Christen, A. Multicopy suppressors of prc mutant *Escherichia coli* include two HtrA (DegP) protease homologs (HhoAB), DksA, and a truncated R1pA. *J. Bacteriol.* **1996**, *178*, 1154–1161. [CrossRef]

69. Whisson, S.C.; Boevink, P.C.; Wang, S.; Birch, P.R.J. The cell biology of late blight disease. *Curr. Opin. Microbiol.* **2016**, *34*, 127–135. [CrossRef]

70. Meng, Y.; Zhang, Q.; Zhang, M.; Gu, B.; Huang, G.; Wang, Q.; Shan, W. The protein disulfide isomerase 1 of Phytophthora parasitica (PpPDI1) is associated with the haustoria-like structures and contributes to plant infection. *Front. Plant Sci.* **2015**, *6*, 632. [CrossRef]

71. Oliveira, A.E.; Gomes, V.M.; Sales, M.P.; Fernandes, K.V.; Carlini, C.R.; Xavier-Filho, J. The toxicity of jack bean [*Canavalia ensiformis* (L.) DC.] canatoxin to plant pathogenic fungi. *Rev. Bras. Biol.* **1999**, *59*, 59–62. [CrossRef]

72. Becker-Ritta, A.B.; Martinelli, H.S.; Mitidieri, S.; Feder, V.; Wassermann, G.E.; Santi, L.; Vainstein, M.H.; Oliveira, J.T.A.; Fiuza, L.M.; Pasquali, G.; et al. Antifungal activity of plant and bacterial ureases. *Toxicon* **2007**, *50*, 971–983. [CrossRef] [PubMed]

73. Resjö, S.; Brus, M.; Ali, A.; Meijer, H.J.G.; Sandin, M.; Govers, F.; Levander, F.; Grenville-Briggs, L.; Andreasson, E. Proteomic Analysis of *Phytophthora infestans* Reveals the Importance of Cell Wall Proteins in Pathogenicity. *Mol. Cell Proteom.* **2017**, *16*, 1958–1971. [CrossRef] [PubMed]

74. Tanimura, A.; Kondo, Y.; Tanaka, H.; Matsumura, I.; Ishibashi, T.; Sudo, T.; Satoh, Y.; Yokota, T.; Ezoe, S.; Oritani, K.; et al. An Anti-Apoptotic Molecule, Anamorsin, Functions in Both Iron-Sulfur Protein Assembly and Cellular Iron Homeostasis. *Blood* **2011**, *118*, 2106.

75. Gyetvai, G.; Sonderkaer, M.; Gobel, U.; Basekow, R.; Ballvora, A.; Imhoff, M.; Kersten, B.; Kare-Lehman Nielsen, K.L.; Gebhardt, C. The Transcriptome of Compatible and Incompatible Interactions of Potato (*Solanum tuberosum*) with Phytophthora infestans Revealed by DeepSAGE Analysis. *PLoS ONE* **2012**, *7*, E31526. [CrossRef] [PubMed]

76. Chen, L.; Huang, Y.; Xu, M.; Cheng, Z.; Zhang, D.; Zheng, J. iTRAQ-Based Quantitative Proteomics Analysis of Black Rice Grain Development Reveals Metabolic Pathways Associated with Anthocyanin Biosynthesis. *PLoS ONE* **2016**, *7*, e0159238. [CrossRef] [PubMed]

77. Liu, Y.; Hüttenhain, R.; Collins, B.; Aebersold, R. Mass spectrometric protein maps for biomarker discovery and clinical research. *Expert Rev. Mol. Diagn.* **2013**, *13*, 811–825. [CrossRef] [PubMed]

78. Vizcaino, J.A.; Csordas, A.; Del-Toro, N.; Dianes, J.A.; Griss, J.; Lavidas, I.; Mayer, G.; Perez-Riverol, Y.; Reisinger, F.; Ternent, T.; et al. 2016 update of the PRIDE database and its related tools. *Nucleic Acids Res.* **2015**, *44*, D447–D456. [CrossRef]

79. Consortium TGO. Gene Ontology Consortium: Going forward. *Nucleic Acids Res.* **2015**, *43*, D1049–D1056. [CrossRef]

80. Conesa, A.; Götz, S. Blast2GO: A Comprehensive Suite for Functional Analysis in Plant Genomics. *Int. J. Plant Genom.* **2008**, *2008*, 619832. [CrossRef]

81. Kanehisa, M.; Furumichi, M.; Tanabe, M.; Sato, Y.; Morishima, K. KEGG: New perspectives on genomes, pathways, diseases, and drugs. *Nucleic Acids Res.* **2017**, *45*, D353–D361. [CrossRef] [PubMed]

82. Smoot, M.E.; Ono, K.; Ruscheinski, J.; Wang, P.L.; Ideker, T. Cytoscape 2.8: New features for data integration and network visualization. *Bioinformatics* **2010**, *27*, 431–432. [CrossRef] [PubMed]

83. Pang, C.Y.; Wang, H.; Pang, Y.; Xu, C.; Jiao, Y.; Qin, Y.M.; Western, T.L.; Yu, S.X.; Zhu, Y.X. Comparative proteomics indicates that biosynthesis of pectic precursors is important for cotton fiber and Arabidopsis root hair elongation. *Mol. Cell Proteom.* **2010**, *9*, 2019–2033. [CrossRef] [PubMed]

84. Xie, C.; Mao, X.; Huang, J.; Ding, Y.; Wu, J.; Dong, S.; Kong, L.; Gao, G.; Li, C.Y.; Wei, L. KOBAS 2.0: A web server for annotation and identification of enriched pathways and diseases. *Nucleic Acids Res.* **2011**, *39*, 316–322. [CrossRef] [PubMed]

85. Howden, A.J.M.; Stam, R.; Heredia, V.M.; Motion, G.B.; Have, S.T.; Hodge, K.; Amaro, T.M.M.M.; Huitema, E. Quantitative Analysis of the Tomato Nuclear Proteome during Phytophthora Capsici Infection Unveils Regulators of Immunity. *New Phytol.* **2017**, *215*, 309–322. [CrossRef] [PubMed]

86. Aldridge, G.M.; Podrebarac, D.M.; Greenough, W.T.; Weiler, I.J. The Use of Total Protein Stains as Loading Controls: An Alternative to High-Abundance Single-Protein Controls in Semi-Quantitative Immunoblotting. *J. Neurosci. Methods* **2008**, *172*, 250–254. [CrossRef] [PubMed]

87. Livak, K.J.; Schmittgen, T.D. Analysis of relative gene expression data using real-time quantitative PCR and the $2^{-\Delta\Delta CT}$ method. *Methods* **2001**, *25*, 402–408. [CrossRef]

International Journal of
Molecular Sciences

MDPI

Article

Proteomics Analysis Reveals That Caspase-Like and Metacaspase-Like Activities Are Dispensable for Activation of Proteases Involved in Early Response to Biotic Stress in *Triticum aestivum* L.

Anastasia V. Balakireva [1,†], Andrei A. Deviatkin [1,†], Victor G. Zgoda [2], Maxim I. Kartashov [3], Natalia S. Zhemchuzhina [3], Vitaly G. Dzhavakhiya [3], Andrey V. Golovin [1,4] and Andrey A. Zamyatnin Jr. [1,5,*]

[1] Sechenov First Moscow State Medical University, Institute of Molecular Medicine, Trubetskaya str., 8, bld. 2, Moscow 119991, Russia; balakireva.anastacia@gmail.com (A.V.B.); andreideviatkin@gmail.com (A.A.D.); golovin.andrey@gmail.com (A.V.G.)
[2] Institute of Biomedical Chemistry, Pogodinskaya str., 10, bld. 8, Moscow 119121, Russia; victor.zgoda@gmail.com
[3] All Russian Research Institute of Phytopathology, VNIIF, Bolshie Vyazemi, Odintsovsky distr., Moscow region 143050, Russia; maki505@mail.ru (M.I.K.); zhemch@mail.ru (N.S.Z.); dzhavakhiya@yahoo.com (V.G.D.)
[4] Faculty of Bioengineering and Bioinformatics, Moscow State University, Moscow 119992, Russia
[5] Belozersky Institute of Physico-Chemical Biology, Lomonosov Moscow State University, Moscow 119992, Russia
*   Correspondence: zamyat@belozersky.msu.ru; Tel.: +7-495-622-98-43
†   These authors contributed equally to this work.

Received: 8 November 2018; Accepted: 8 December 2018; Published: 11 December 2018

Abstract: Plants, including *Triticum aestivum* L., are constantly attacked by various pathogens which induce immune responses. Immune processes in plants are tightly regulated by proteases from different families within their degradome. In this study, a wheat degradome was characterized. Using profile hidden Markov model (HMMer) algorithm and Pfam database, comprehensive analysis of the *T. aestivum* genome revealed a large number of proteases (1544 in total) belonging to the five major protease families: serine, cysteine, threonine, aspartic, and metallo-proteases. Mass-spectrometry analysis revealed a 30% difference between degradomes of distinct wheat cultivars (Khakasskaya and Darya), and infection by biotrophic (*Puccinia recondita* Rob. ex Desm f. sp. tritici) or necrotrophic (*Stagonospora nodorum*) pathogens induced drastic changes in the presence of proteolytic enzymes. This study shows that an early immune response to biotic stress is associated with the same core of proteases from the C1, C48, C65, M24, M41, S10, S9, S8, and A1 families. Further liquid chromatography-mass spectrometry (LC-MS) analysis of the detected protease-derived peptides revealed that infection by both pathogens enhances overall proteolytic activity in wheat cells and leads to activation of proteolytic cascades. Moreover, sites of proteolysis were identified within the proteases, which probably represent targets of autocatalytic activation, or hydrolysis by another protease within the proteolytic cascades. Although predicted substrates of metacaspase-like and caspase-like proteases were similar in biotrophic and necrotrophic infections, proteolytic activation of proteases was not found to be associated with metacaspase-like and caspase-like activities. These findings indicate that the response of *T. aestivum* to biotic stress is regulated by unique mechanisms.

Keywords: degradome; wheat; cultivar; protease; papain-like cysteine protease (PLCP); subtilase; metacaspase; caspase-like; wheat leaf rust; *Puccinia recondita*; *Stagonospora nodorum*

1. Introduction

Wheat (*Triticum aestivum* L.) is a major grain species of value to both industry and biotechnology. One of the main factors that influence the use of wheat is its resistance to pathogens. Resistance itself is defined by the fact that wheat is a hexaploid organism, which harbors three genomes with an overall haploid size of more than 15 Gbp [1]. Genome A was obtained from *Triticum urartu*, Genome B from an unknown grass related to *Aegilops speltoides* and Genome D from *Aegilops tauschii*. In addition to the complexity of wheat genomes, there are also variations between wheat cultivars. Differences in protein levels expressed amongst the cultivars have previously been analyzed and assessed as being up to 30% [2,3]. These differences determine wheat resistance to pathogens and growth conditions, and the suitability of wheat cultivars for different applications.

In fact, wheat constantly suffers from various pathogens: bacterial (*Pseudomonas* spp. [4], *Xanthomonas translucens* [5], etc.), fungal (*Puccinia recondita*, *Fusarium* spp., *Blumeria graminis*, *Zymoseptoria tritici* [6], etc.), viral (barley stripe mosaic virus [7], wheat streak mosaic virus [8], yellow leaf mosaic virus [9], etc.), herbivorous insects (*Sitobion avenae* [10]) and even nematodes (*Heterodera avenae* [11]). Pathogens have distinct strategies during plant cell infection: necrotrophic pathogens (*Botrytis cinerea* [12]) and herbivores [13] promote plant growth and lead to necrosis of the infected cell through consumption of its content, whilst biotrophic pathogens (*Pseudomonas syringae* [14]) feed on living cells, suppressing plant growth and launching programmed cell death (PCD) of infected cells.

PCD is regulated by proteolytic enzymes in living organisms and the degradome is an overall complex consisting of all the proteases in a cell. It is known that the human genome contains 588 genes for proteases, including about 150 proteases that are transcriptionally active, depending on the type of tissue [15]. The number of encoded proteases amounts to 723 in *Arabidopsis thaliana* L. [16], 997 for rice [17–19] and 901 for tomatoes [20].

The most widely studied form of PCD (apoptosis in humans) is regulated by proteolytic cascades, which include caspases—Asp-specific cysteine proteases. Caspases are divided into upstream initiator caspases (-2, -8, -9 and -10), and downstream executioner caspases (-3, -6 and -7) [21]. Caspases are synthesized as inactive procaspases. Upstream procaspases undergo autocatalytic processing and activate downstream procaspases through limited proteolysis. Downstream caspases activate proapoptotic proteins, including other proteases, through the proteolytic cascades that lead to massive degradation of proteins and cell death.

As with humans, plant PCD is also regulated through proteolytic cascades [21]. Although caspases are absent in plants, caspase-like activity can be detected in vegetation after induction of PCD with DEVDase [22,23], VEIDase [24] and YVADase [25] activities. These are attributed to proteases from different families, e.g., subtilases (S8) [26], proteasome subunits (T1) [23] and vacuolar processing enzymes (VPEs, C13) [27]. As with caspases, these proteases are synthesized as zymogens or preproenzymes and require proteolytic activation. They usually contain three regions: a signal peptide, an inhibitory prodomain and a proteolytic domain. A prodomain is often (autocatalytically) processed, resulting in the release of an active, mature enzyme. The other main characteristic of PCD in plants is metacaspase activity. Metacaspases cleave after R and K, and have been shown to participate in a variety of processes associated with the death of cells [28–32]. The nature of such processes in metacaspases has been described earlier [33]. It is worth noting that although a number of substrates are known for having caspase-like proteases and metacaspases [34], no data exist on their exact substrate specificity and the cellular proteolytic cascades with which they are associated.

In the case of wheat, its response to pathogens, e.g., to *Blumeria graminis* f. sp. *tritici* [35], or *Fusarium graminearum* [36], and to numerous abiotic stresses [37], has been characterized only in general terms without focusing on particular groups of proteins. Moreover, the degradome of wheat has not yet been characterized despite its importance for immunity and PCD-related processes. It is known that expression of the wheat metacaspase 1 gene (TaMCA1) increases when infected with *Puccinia striiformis* [38]. However, there are no data available on proteolytic cascades and proteases that regulate wheat immunity in healthy and infected plants.

Liquid chromatography-mass spectrometry (LC-MS) is a powerful technique, which has been successfully used to study the properties of different wheat cultivars [2,3,39], plant development [40,41], and wheat responses to biotic [35,36] and abiotic stresses [37]. An LC-MS approach has been adopted in this study for characterization of all wheat proteases that were identified using Pfam identificators attributed to all the proteases with the use of profile hidden Markov model (HMMER) algorithm [42] that was previously successfully used for identification of proteins [43]. We performed classification of identified proteases. Comparison of two wheat cultivars in terms of proteolytic activity was done. LC-MS enabled identification of changes in the wheat degradome upon infection by biotrophic (*Puccinia recondita* Rob. ex Desm f. sp. tritici) and necrotrophic (*Stagonospora nodorum*) pathogens, and assessment of the impact of these infections on proteolytic activity in wheat cells. LC-MS also facilitated analysis of proteases present in healthy and infected wheat plants, determining the role of caspase-like and metacaspase-like proteases in proteolytic cascades, and the overall proteolytic activity in wheat cells.

2. Results

2.1. Degradome of T. aestivum Is Represented by Diverse Protease Families

First, proteases encoded in the wheat genome were identified. Release 39 proteome of *T. aestivum* was obtained from the Ensembl genomes database (ftp://ftp.ensemblgenomes.org/) and used in identification of the wheat degradome. Protein families included cysteine, serine, aspartic, threonine, and metallo-proteases. The HMMER algorithm [42] was used to search sequence homologs for the domain IDs of each peptidase family, obtained from the Pfam database [44]. Several protease families were identified (Table 1, Figure 1) in the wheat proteome.

In total, 1544 proteases were discovered: 459 cysteine proteases from 12 families, 275 metallo-proteases from 17 families, 336 aspartic proteases from two families, 446 serine proteases from five families, and 28 threonine proteases.

The C1 family consisted of 181 members. This family was represented by the C1A subfamily of papain-like cysteine proteases (PLCPs) according to MEROPS nomenclature [45]. It is noteworthy that some proteases were simultaneously annotated with two distinct Pfam IDs: Peptidase_C1 and Peptidase_C1_2 with E value in the order of e-100 and e-10, respectively. These peptidases form an independent group with no homologs characterized across *A. thaliana* papain-like cysteine proteases (PLCPs). Peptidase_C1_2 corresponds to the C1B subfamily (according to MEROPS nomenclature). Thus, it was concluded that the *T. aestivum* genome contained no C1B peptidases. Nevertheless, C1 peptidases formed two distinct branches on the phylogenetic tree, divided by several proteases from families C2 and C78, indicating possible divergent processes within the C1A subfamily of cysteine proteases (Figure 1, "Cysteine"). There were some peculiarities on this vast tree of distantly related cysteine proteases. For example, sequences annotated to the C48 family according to its Pfam domain were widely "dispersed", whilst some "C48 sequences" were closer to other families than to each other (Figure 1, "Cysteine").

Serine proteases formed distinct family groups containing close homologs, whereas the other groups were distant from each other (Figure 1, "Serine"). Metallo-proteases were represented in a variety of families (Table 1), which were mixed up on the phylogenetic tree (Figure 1, "Metallo-proteases"). Proteases from M1 and M3 families occurred in different parts of the tree, whereas M10, M16, M24, M28, M41, and M48 formed distinct. Aspartic proteases from different families (the A1 family and A22B subfamily) did not align properly with each other, indicating an independent evolutionary relationship within the group of aspartic proteases (Figure 1, "Aspartic"). Proteases from the A1 family are unique for *T. aestivum* and are composed of *Triticum aestivum* xylanase inhibitor N (TAXi_N) and TAXi_C domains, which are distant but homologous to each other's domains. It is noteworthy that the TAXi_N domains from different proteases are closer to each other than the

TAXi_N and TAXi_C domains within one protease. Thus, the phylogenetic tree (Figure 1, "Aspartic") indicates the divergent processes between the TAXi domains.

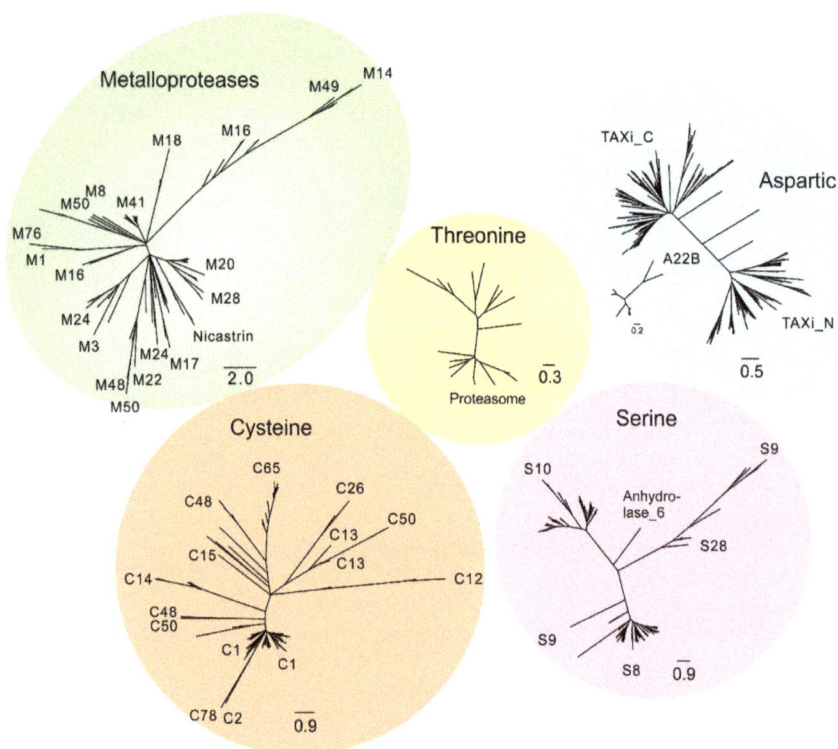

Figure 1. The *T. aestivum* degradome landscape. Metallo-proteases are indicated in green, threonine–in yellow, aspartic–in blue, cysteine–in orange, serine–in purple. Unrooted phylogenetic trees display diversity across protease families. The trees are drawn to scale, with branch lengths in the same units as those of the evolutionary distances used to infer the phylogenetic tree. Evolutionary distances were computed using the p-distance method [45] and their units correspond to the number of amino acid differences per site.

The phylogenetic trees based on the alignment of relevant domains were constructed for families currently known to be associated with plant immunity such as C1, C13, C14 (Figure 2), and S8 (Figure 4) [34]. Domain structures were identified for all proteases. Some criteria were introduced to demarcate discrete subgroups the following criteria were used based on the "from leaves to roots" principle:

1. Check the common node for two leaves. If the bootstrap value for this node is more than 70, then merge these leaves into one subgroup. Otherwise, these leaves should be considered as distinct, independent groups. Repeat this step as many times as needed.
2. Check the common node for the node with a low bootstrap value. If the bootstrap value for the upper node is more than 70, then subgroups should be considered as members of a larger group. However, the topology of tree nodes with a low bootstrap value cannot be resolved, although a highly credible, common node clearly indicates the shared origin of these subgroups.
3. If uniquely specific features are shown as representative of some subgroups, they should be considered as independent.

Table 1. Representation of different protease families in the wheat genome.

Protease Family	Protease Families and Subfamilies (According to Pfam Nomenclature)
Cysteine	C1, C2, C12, C13, C14, C15, C26, C48, C50, C54, C65, C78
Serine	S8, S9, S10, S15, S28
Aspartic	A1, A22B
Metallo-proteases	M1, M3, M8, M10, M14, M16_M, M16_C, M17, M18, M20, M22, M24, M28, M41, M48, M49, M50B, M50
Threonine	T1

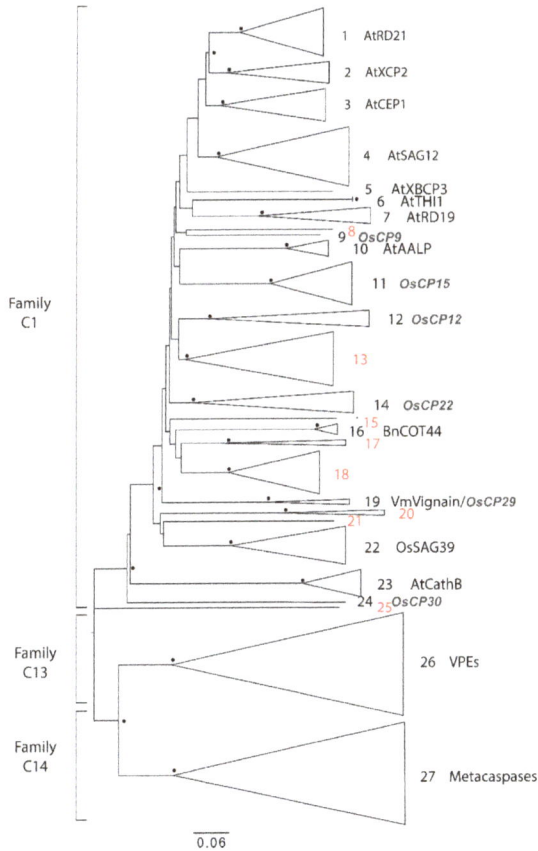

Figure 2. Phylogenetic tree of families C1, C13, and C14 of cysteine proteases. Triangles represent the contracted clades and the size of each protease group. Bootstrap values above 70% are shown with black circles at the relevant nodes. The closest homologs of the groups of proteases are indicated on the right of the phylogenetic tree: At—*A. thaliana*, Os—*Oryza sativa*, Bn—*Brassica napus* and Vm—*Vigna mungo*. Unique for *T. aestivum* groups are shown in red.

As proteases from the C1A, C13, and C14 families are closely related, a single phylogenetic tree was constructed for these families (Figure 2). In total, 25 clusters of C1A proteases were identified. Nine clusters corresponded to the nine groups identified by Richau et al. [46] for the A. thaliana PLCP subfamilies (Figure 2): Responsive to Dehydration 21-like (RD21), Xylem cysteine peptidase 2-like (XCP2), Cysteine EnsoPeptidase 1-like (CEP1), Senescence-Associated Gene 12-like (SAG12), Xylem Bark Cysteine Peptidase 3-like (XBCP3); and Responsive to Dehydration 19-like (RD19), Arabidopsis Aleurain-like Protease-like (AALP), cathepsin B-like and THIol protease 1-like (THI1). In addition, protease groups homologous to those from rice (*Oryza sativa*) were also identified (indicated as *O. sativa* cysteine proteases (OsCPs) in Figure 2). Moreover, proteases from other groups had previously undescribed homologs (shown in red in Figure 2). Several of the homologous proteases closest to some of these newly identified groups were indicated (Figure 2): 16 were COTyledon abundant protease 44-like (COT44), 19 were vignain-like and 22 were Senescence-Associated Gene 39-like (SAG39).

Typically, C1A proteases consist of a prodomain and catalytic domain, and, in the case of XBCP3-like and RD21-like proteases, a granulin domain (Figure 3). Some of the other clades possess distinct features: group 15 contains proteins with Domain of Unknown Function 4371 (DUF4371); and the SAG39-like cluster (group 22) contains the No Apical Meristem-associated (NAM-associated) domain.

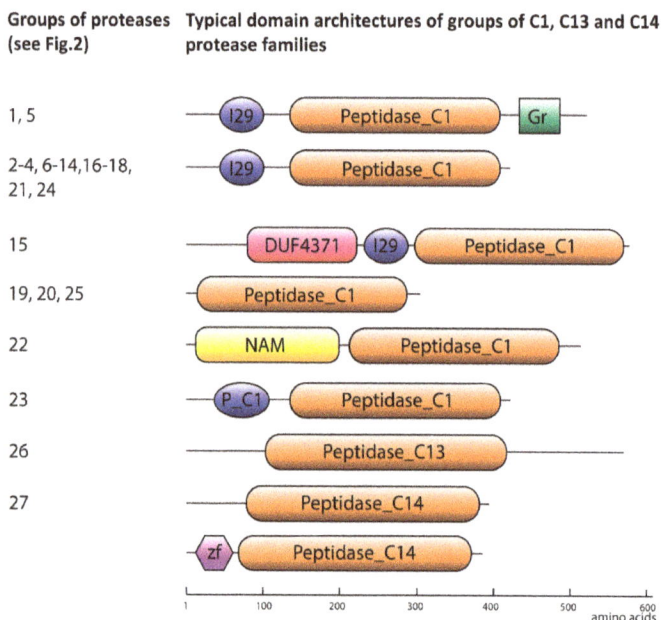

Figure 3. Typical domain organization of proteases from families C1, C13, and C14. The numbers of groups correspond to those indicated in Figure 2. The length of all the schemes are indicated on a scale according to the bar below the schemes. Pfam ID: Peptidase_C in orange—proteolytic domain of families C1, C13, and C14 of proteases; I29 in blue—Inhibitor_I29; NAM in yellow—NAM-associated; P_C1 in blue—Propeptide_C1; zf in purple—zf-LSD1 domains.

The C13 peptidases indicated on the phylogenetic tree (group 26, Figure 2) were classified into five closely related groups encompassing the homologs of VPEs α, β, γ, and δ. Analysis of the domain architecture of proteases from the C13 family revealed only the presence of the catalytic domain. Metacaspases from the C14 family formed three clades, which included Type I and Type II metacaspases. Type I metacaspases had an N-terminal zinc-finger domain (zf-Lysine-specific histone demethylase 1 (LSD1)), which coincided with the already described features of Type I metacaspases [46].

Eighty-two serine proteases from the S8 family, also known as subtilases, were classified into nine groups, for which the closest homologs from *A. thaliana* were identified: 1—subtilase 3.8-like (SBT3.8-like), 2—SBT1.7-like, 3—SBT1.4-like, 4—SBT5.3-like, 5—CO_2-response secreted protease-like, 6—SBT1.8-like, 7—SBT1.7-like, 8—SBT2.5-like, 9—tripeptidyl peptidase 2-like (Figure 4). Proteases from almost all the groups carried not only a catalytic domain (Peptidase_S8), but also a conservative prodomain (Inhibitor_I9, Figure 5). Groups three to six and eight contained proteases with a Protease-associated (PA) domain. Subtilases from groups eight and nine contained a fibronectin Type III-like (fn3_5) domain.

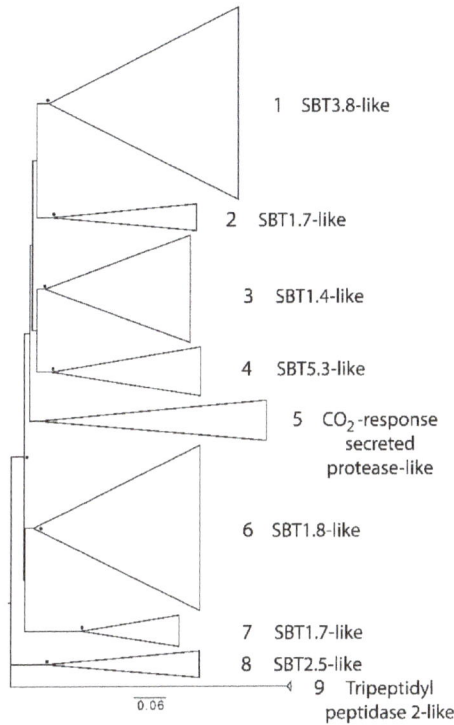

Figure 4. Phylogenetic tree of S8 family of proteases from wheat. Triangles represent the contracted clades and the size of each protease group. Bootstrap values above 70% are shown with black circles at the relevant nodes. The closest homologs of the groups of proteases are indicated on the right.

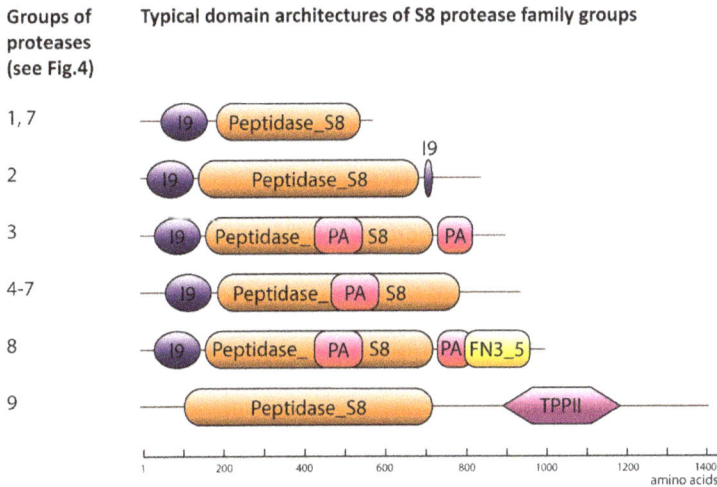

Figure 5. Domain architecture of S8 proteases from wheat. The numbers of groups correspond to those indicated in Figure 4. The length of all the schemes are indicated on a scale according to the bar below the schemes. Pfam IDs: Peptidase_S8 in orange—proteolytic domain of S8 proteases; I9 in blue—Inhibitor_I9; PA in magenta—Protease-associated; FN3_5 in yellow—fibronectin Type III-like; TPPII in purple—Tripeptidyl Peptidase 2 domains.

2.2. Differences in Degradomes of Two Wheat Cultivars Were Revealed

Using LC-MS data, proteases present in healthy plants of two cultivars, Khakasskaya and Darya, were quantified on the basis of full-specific peptides (full-tryptic and full-AspN peptides) [47]. This indicated that the degradome of wheat varies within the species (Figure 6). According to a comparative analysis, the Khakasskaya and Darya cultivars of wheat express an almost equal number of proteases (94 and 79 proteases, respectively) and share 49 proteases, representing 52,1% and 62% of the overall number of proteases in Khakasskaya (Table S1) and Darya (Table S2) cultivars, respectively. Serine proteases are relatively abundant in both cultivars: half that amount of cysteine, metallo- and aspartic proteases were detected. The most prevalent families of proteases in both cultivars are C1, S8, S9, S10, M20, M24, and A1 (Tables S1 and S2).

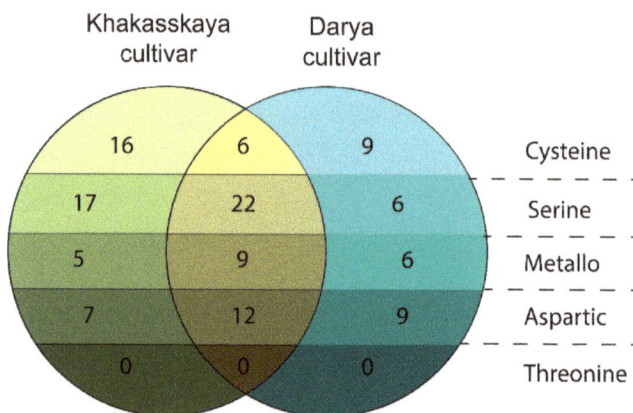

Figure 6. Venn diagrams representing the number of reproducibly quantified proteases for Khakasskaya (green shades) and Darya (cyan shades) wheat cultivars. Proteases found in both cultivars are indicated in yellow. Dotted lines divide proteases of different catalytic types.

2.3. Infection by Pathogens Leads to an Increase in the Number of Expressed Proteases

Based on the LC-MS approach, a comparison was made of proteases expressed in healthy plants and those infected by *P. recondita* (biotrophic pathogen) and *S. nodorum* (necrotrophic pathogen) plants at 24 h post inoculation (hpi). Identification of proteases was based on quantification of full-specific tryptic and full-specific AspN peptides. An increased number of detected proteases was found upon both infections (Figure 7): 117 and 77 proteases were detected in plants infected by *P. recondita* (Table S3) and *S. nodorum* (Table S4), respectively, which share 55 (58,5%) and 45 (60%) proteases with controls (healthy Khakasskaya (Table S1) and Darya (Table S2) plants, respectively).

In the case of *P. recondita* infection, induction of proteases occurred mostly from families C1, C13, C48, C65, M3, M41, M20, M24, M17, S8, S9, S10, A1, whilst in the case of *S. nodorum* infection, induction occurred mostly from families C1, C14, C26, C48, C65, M3, M41, M24, S8, S9, S10, A1 (Tables S1–S4).

2.4. The Pool of Substrates Cleaved by Proteases In Vivo Expands upon Both Types of Infection

On the basis of these LC-MS data, a search was undertaken for potential substrates cleaved by proteases in healthy plants of both cultivars. This was done through identification of semi-specific peptides released after digestion with trypsin, or AspN proteases used in the present study. One terminal of the peptide corresponds to a specific hydrolysis site (cleavage after R and K for trypsin, and before C and D for AspN), and the other corresponds to non-specific hydrolysis sites. It is worth stating that semi-specific peptides may arise from two main sources: as a result of hydrolysis by endogenous proteases within the sample, or as a result of the truncation of regular tryptic peptides through in-source fragmentation (ISF) within the electrospray ionization (ESI) source [48]. The impact of ISF in generation of semi-specific peptides was assessed earlier, and it depended to a large extent on the complexity of the biological sample, ranging from 1% (in a mouse-brain sample) to 57% (in a standard protein mixture) of the total amount of tryptic peptides (including both full-specific and semi-specific peptides) [48]. How ISF affects truncation of the peptides in plant samples has not yet been analyzed. However, despite the described limitations, identification of semi-specific peptides was originally used to analyze complex protein samples in shotgun proteomics [49]. Identifying more peptides, e.g., non-tryptic peptides, may increase the peptide coverage and improve protein identification and/or quantification. Moreover, the semi-specific peptide approach identified processing patterns of N-terminal signal peptide in human aspartyl-tRNA synthetase, by combining immuno-enrichment of the protein samples with classical shotgun LC-MS/MS analysis and semi-tryptic database searching [50]. Similarly, as semi-specific peptides are likely to be a product of in vivo hydrolysis by endogenous proteases, this method may be successfully used for studying both the maturation of proteases in vivo and their involvement in cellular proteolytic cascades.

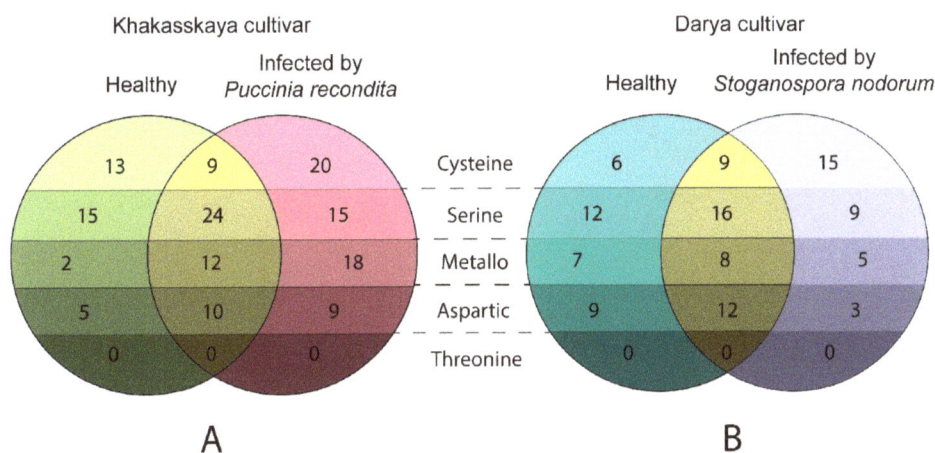

Figure 7. Venn diagrams representing the number of reproducibly quantified proteases in proteomes of (**A**) healthy Khakasskaya (green shades) plants and Khakasskaya plants infected by *P. recondita* (rose shades); (**B**) healthy Darya (cyan shades) and Darya plants infected by *S. nodorum* (purple shades). Proteases found in intersection between different samples are indicated in yellow.

The search for potential substrates of endogenous proteases in healthy plants revealed an almost equal number of cleaved substrates: 220 (Table S5) and 178 (Table S6) proteins were identified (85 shared) in healthy Khakasskaya and Darya plants (Figure 8A). The presence of substrates presumed to be cleaved by caspase-like and metacaspase-like proteases was also investigated. As stated earlier, plant immune responses and PCD are associated with caspase-like and metacaspase activities, which include hydrolysis of substrates at very specific sites such as VEID, DEVD and YVAD sites. However, it should be noted that there are limited number of known proteolytic sites for such proteases and overall substrate specificities of caspase-like proteases [51]. Most metacaspases also remain uncharacterized. A generalized rule for identifying substrates was subsequently applied: XXXD for caspase-like and XXXR, XXXK for metacaspase-like proteases, where X is any amino acid. The number of probable substrates of caspase-like proteases was found to be almost equal amongst healthy Khakasskaya and Darya plants (10 and nine, respectively, with five shared), whereas slightly more substrates of metacaspase-like proteases were found in healthy Khakasskaya and Darya plants (11 and 19, respectively, with eight shared; Figure 8A).

As the number of expressed proteases increased upon both infections (see previous section), the number of substrates cleaved by endogenous proteases was checked to determine whether they also increased upon infection. A significant increase in the number of substrates was observed. In the case of *P. recondita* infection (Figure 8B, Table S7) this number increased from 220 to 518 (121 shared), in the case of *S. nodorum* infection (Figure 8C, Table S8)—from 178 to 300 (77 shared. Substrates of caspase-like and metacaspase-like proteases were also detected. Both healthy and infected plants are characterized by increased number of proteins cleaved at XXXD sites. 10 and 28 (four shared) sites were identified in the case of *P. recondita* infection (Figure 8B); nine and 15 (three shared)—in the case of *S. nodorum* infection (Figure 8C). In addition, both types of infection are characterized by an increase in metacaspase-like activity. 11 and 22 (seven shared) sites were identified in the case of *P. recondita* infection (Figure 8B); 19 and 17 (11 shared) sites—in the case of *S. nodorum* infection (Figure 8C).

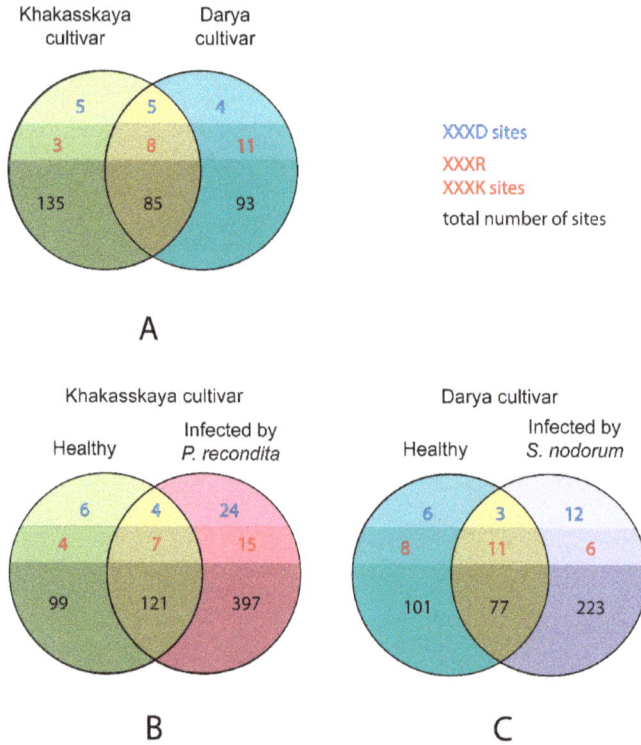

Figure 8. Venn diagrams representing the number of reproducibly quantified in vivo cleavage sites (detected with LC-MS) in all proteins across proteomes of (**A**) healthy Khakasskaya (green shades) and healthy Darya (cyan shades); (**B**) healthy Khakasskaya (green shades) and wheat plants infected with *P. recondita* (rose shades); and (**C**) healthy Darya (cyan shades) and wheat plants infected with *S. nodorum* (purple shades). Proteases found in intersection between different samples are indicated in yellow. XXXD—caspase-like site; XXXR, XXXK—metacaspase-like sites; X—any amino acid; numbers in bold and black—total number of cleavage sites across the proteomes; number of caspase-like sites are indicated by blue text, number of metacaspase sites—by red text.

Potential substrates of caspase-like and metacaspase-like proteases cleaved in response to infections were also investigated to ascertain which were caused by both *P. recondita* and *S. nodorum* 24 hpi. Among these substrates, seven caspase-like and three metacaspase-like substrates were found cleaved upon both infections. These substrates are summarized in Table 2.

Table 2. Probable substrates of caspase-like and metacaspase-like proteases detected only in samples upon both *P. recondita* and *S. nodorum* infections. Annotation of proteins were made with Blast: the best hits with E-values $< 1 \times 10^{-50}$ are shown. The sites of hydrolysis are indicated by arrows.

Protein	Site of Hydrolysis
Potential substrates of caspase-like proteases	
Uncharacterized protein	VPTD↓AQLE
Protein Translocon at the inner envelope membrane of chloroplasts 22 (TIC 22), chloroplastic-like	ITLD↓QVYM
Disease resistance protein Rho-type GTPase-activating protein 2-like (RGA2)	VSAD↓GVTR
5′-3′ exoribonuclease 2-like	ILRD↓MVPL
Homeobox-DDT (DNA-binding homeobox-containing proteins and the different transcription and chromatin remodeling factors) domain protein Ringlet 3-like (RLT3)	KPED↓LTEY
Protein Fatty Acyl-CoA Reductase 1 (FAR1)-Related Sequence 5-like	LAAD↓HPRR
Constitutive Photomorphogenic1 (COP1)-interacting protein 7	IDID↓AELG
Potential substrates of metacaspase-like proteases	
Adenine/guanine permease Azaguanine Resistant 2 (AZG2)	CLAR↓TKSD
Wall-associated receptor kinase 5-like	LSTR↓NELI
Protein FAR1-Related Sequence 5-like	LFKK↓GVGA

2.5. Infection Induces Synthesis and/or Proteolytic Activation of Proteases through Recruitment of Proteases Other Than Caspase-Like and Metacaspase-Like Proteases

It has been shown that caspase-like and metacaspase-like activities are present in wheat cells after infection by both *P. recondita* and *S. nodorum*. The impact of such activities on proteolytic activation of all proteases was examined in healthy plants and those infected by both pathogens 24 hpi. To do this, the same approach was implemented, based on identification of semi-specific peptides after digestion by trypsin or AspN for LC-MS. A search was undertaken for peptides mapped around the bordering prodomain-proteolytic domain (±40 amino acids from the border), since a large number of proteases contain autoinhibitory prodomain, which needs to be cleaved to obtain an active protease. In the case of metacaspases, semi-specific peptides at known processing sites for homologous enzymes [46] were selected. The processing status of the proteases was, therefore, identified in healthy plants and those infected by *P. recondita* 24 hpi (Table 3, Tables S9 and S10) and *S. nodorum* (Table 3, Tables S11 and S12).

Upon biotrophic infection, a significant decrease in the number of processed proteases was detected, in comparison to the control (Table 3). Healthy and infected plants share six processed proteases (Tables S9 and S10); eight proteases from families M17, M20, M24, M41, S8, S10, A1 were detected, presumably activated upon infection (Table S10); healthy plants are characterized only by 1 unique protease (Table S9). Upon necrotrophic infection, three processed proteases were detected both in healthy and infected plants (Tables S11 and S12). However, infection caused processing of more proteases from families C1, C13, S8, S49, M24, A1 (six proteases; Table S12) in comparison to the control (four proteases; Table S12).

It is worth mentioning that no recognition sites of metacaspases (XXXR, XXXK) was found among detected activation sites of proteases (Table 3). Moreover, the number of caspase-like sites (XXXD) has not changed upon *S. nodorum* infection (1 activation site, shown in Table 3, Tables S11 and S12) and slightly increased after infection with *P. recondita* (from zero to three, shown in Table 3, Tables S9 and S10).

Table 3. Proposed activation sites in proteases from different families based on LC-MS data in healthy Khakasskaya plants (exact sites and names of proteases are summarized in Table S9); Khakasskaya plants infected with *P. recondita* (Table S10); healthy Darya plants (Table S11); and Darya plants infected with *S. nodorum* (Table S12).

Catalytic Type of proteases	Family	Number of Detected by LC-MS Sites of Cleavage * (Number of XXXD Sites, If Any), [Number of XXXR or XXXK Sites, If Any]			
		Healthy Khakasskaya	Khakasskaya infected with *P. recondita*	Healthy Darya	Darya infected with *S. nodorum*
Cysteine	Peptidase_C1	2	1	1	2
	Peptidase_C13		1 (1)		1 (1)
Serine	Peptidase_S8		2	1	1
	Peptidase_S10		1		
	Peptidase_S49				1
Metallo-	Peptidase_M17		1		
	Peptidase_M20		1	1	
	Peptidase_M24		1		1
	Peptidase_M41	1	2	2	2
Aspartic	A1 (TAXi_C)	4	6 (2)	2 (1)	1
Total		7	16 (3)	7 (1)	9 (1)

* Each site of cleavage was found only in one protease, i.e., the number of detected cleavage sites coincides with the number of proteases in which they were found.

3. Discussion

In total, 1,544 proteases were found, encoded in the whole genome of *T. aestivum*. This constitutes a significant number of encoded proteases, for example, when compared with diploid dicotyledon plants, such as *Nicotiana benthamiana* L., with predicted proteases measured at around 1,245. Similarly, 796 proteases have been found in *A. thaliana* and 901 in tomatoes. Monocotyledon plants, such as rice, contain 997 [20].

The wheat degradome is represented by a large number of protease families, which have been subclassified into various groups. The most striking example is the C1A subfamily, i.e., PLCPs. In this group, 181 members were identified. With *A. thaliana*, 31 proteases were reported [52], 33 in rice [17], 43 in rubber [53] and 33 in papaya [43]. Subfamilies of PLCPs (already described for *A. thaliana* [52] and rice [17]) were also identified. Moreover, given that no known homologs from other organisms were found, eight groups of PLCPs were defined as being unique to the species subfamilies (Figure 3, in red). Classification of 82 subtilases (S8 protease family) resulted in identification of groups homologous to *A. thaliana* proteases and their features have been reported earlier [54–58].

LC-MS has shown that different sets of proteases can be detected in different wheat cultivars. Proteomes of Khakasskaya and Darya cultivars share 58,5% and 60% of all proteases, respectively, which is slightly less than already reported difference of 79.3% between proteomes of wheat cultivars [2,3].

The use of LC-MS in this study has shown for the first time that biotrophic and necrotrophic infections induce expression of proteases in plants and that these proteases are different. Responses included induction of M17 and M20 proteases in the case of biotrophic infection, and C14, C26 proteases in necrotrophic infection. It should be noted that most proteases induced by both biotrophic and necrotrophic pathogens are attributed to the same families, such as C1, C13, C48, C65, M24, M41, S10, S9, S8 and A1. The same C13 protease was activated upon both infections, but at different sites (TRIAE_AA0544900.2). C13 proteases include VPEs previously associated with responses to viruses [27] and *P. syringae* [25], which are biotrophic pathogens as well as *P. recondita*. C14 proteases include metacaspases, already widely proven to mediate plant immune responses [46]. The M17, M20, M24, M41, C26 protease families remain to be uncharacterized groups of proteases. It has been concluded that these families are likely to be associated with a response to different types of biotic stress in wheat.

A novel approach has been proposed in this study for the identification of in vivo proteolysis events, based on a search for semi-specific peptides in LC-MS data. This has facilitated assessment of overall proteolytic activity for the first time in healthy and stressed plants. These results indicate that infections by *P. recondita* and *S. nodorum* pathogens are associated with an increase of between one and a half to two times the number of substrates potentially cleaved by endogenous proteases in plant cells. However, the assessments undertaken in this study were qualitative (not quantitative), precluding assumptions about the physical amount of each substrate that was possibly cleaved. Nevertheless, it can be concluded that the number of proteins cleaved increased significantly in response to biotic stress in comparison to healthy plants, indicating the involvement of proteolytic cascades.

Identification of substrates for caspase-like and metacaspase-like proteases revealed that, despite a significant increase in the overall number of proteins cleaved after induction of biotic stress, the quantity of substrates for caspase-like proteases increased slightly. However, substrates for metacaspase-like proteases increased from one and a half to two times in infections of both pathogens. This result confirms the involvement of metacaspases in plant response to pathogens described earlier [29,32,58] and either indicates that caspase-like proteases were not active at 24 hpi and PCD-associated processes were not induced, or that caspase-like enzymes are not implemented in response to biotic stress in wheat.

It is interesting that upon infection by different pathogens in different wheat cultivars, the same substrates were cleaved by caspase-like and metacaspase-like proteases (Table 2) such as receptor-like protein kinase, permease, disease resistance proteins, and other proteins. Identified receptor-like protein kinases may be directly involved in the establishment of microbe-associated molecular patterns (MAMP)-triggered immunity (MTI) or effector-triggered immunity (ETI) [59]. The relationship between such a receptor-like protein kinase and the protease that cleaves it, is a completely unstudied field of research in plant immunity. One of the few known examples is that of ectodomain shedding, which has been identified as cleavage of the chitin receptor of *A. thaliana*, Chitin Elicitor Receptor Kinase 1 (CERK1), by an obscure protease [60]. These substrates may become potential targets for further research. The other substrate—FAR1-like protein—was found to be cleaved by both caspase-like and metacaspase-like proteases. In *Arabidopsis*, it was associated with light control and plant development [61]. The role of this protein in plant immunity needs to be studied.

The pool of substrates cleaved by metacaspase 9 from *A. thaliana* has been identified earlier [62]. It can be seen that some substrates are attributed to the same class of enzymes, for example, helicases. However, it is worth mentioning that cellular pathways are more conservative than the proteins involved in them. Hence, differences in identified potential substrates of metacaspases may be due to the fact that wheat is a monocot organism, whereas *A. thaliana* is a dicot. This study has found some potential substrates that are completely uncharacterized proteins. This result, therefore, requires confirmation and more targeted research.

The use of LC-MS analysis has made it possible to define proteolytic activation of proteases in vivo. These findings show that proteases from M17, M20, M41, S10, A1 families are likely to have been activated upon biotrophic infection, as opposed to the S49 family in the case of necrotrophic infection. The core proteases activated by both infections are from C1, C13, S8, M24, and A1 families. The C1 family includes PLCPs, C13 is a VPEs containing family, S8 family includes subtilases associated with plant immunity, which have been well-characterized in previous research [34]. However, M24 and A1 are very poorly described families of proteases, linked to plant immune responses for the first time in this study.

Our results indicate that in infected wheat, most proteolytic sites located within proteases significantly differ from the sites commonly recognized by caspases or metacaspases. Only 0/7 and 1/7 caspase-like cleavage sites (XXXD) were detected within the proteases of healthy Khakasskaya and Darya plants, respectively, whereas in infected plants, 3/16 and 1/9 cleavage sites were found for *P. recondita* and *S. nodorum* infections, respectively. No activation sites recognized by metacaspases were identified in neither healthy, nor infected by pathogens samples.

Data in this study suggest that a number of proteases are presumably activated upon both types of infection. However, this activation requires conditions other than those of caspase-like or metacaspase-like activities. This is a striking result, which gives an insight into the previously unexplored field of proteolytic cascades in plants and wheat, in particular. This study attempted to provide a comprehensive investigation of proteolytic cascades, triggered in by *P. recondita* and *S. nodorum* infections 24 hpi in wheat, but these proteases remain uncharacterized and require further consideration.

4. Materials and Methods

4.1. Phylogenetic Analysis of the Wheat Protease Families

A complete set of protein sequences for *T. aestivum* (n = 154,140) was obtained from the Ensembl genome database (ftp://ftp.ensemblgenomes.org/) [63]. The most recent release (release 39) was used. Protein sequences shorter than 200 amino acid residues were omitted. Proteins that contained more than four unspecified amino acid residues (i.e., XXXX, where X is any amino acid residue) were also omitted. Any sequences in the data set differing by less than 5% of the nucleotide sequence were omitted. This curated data set (n = 66,615) was analyzed using the HMMER 3.1b2 package (http://eddylab.org/software/hmmer/hmmer-3.1b2.tar.gz) [64] in order to obtain the distribution of Pfam domains in the wheat proteome.

Protein sequences were divided into protease families according to their Pfam domain annotation: "Peptidase_C" was considered as a cysteine peptidase; "Peptidase_M" was considered as a metallo-protease; "Peptidase_S" was considered as a serine peptidase; "Peptidase_A" and "TAXi_" were considered as aspartic peptidases; and "proteasome" was considered as a threonine peptidase. The sequences containing these domains were analyzed by the HMMER 3.1b2 package using the E-value threshold of 10^{-4}. Relevant domains were excised from complete protein sequences. Various threshold lengths were applied for different protease families: 200 amino acid residues for cysteine and serine protease families; 150 amino acid residues for metallo-protease aspartate protease families; and 100 amino acid residues for threonine protease families. Relevant domain sequences were aligned using the MAFFT server (https://mafft.cbrc.jp/alignment/software/) [65]. The evolutionary history was inferred using the neighbor-joining method implemented in MEGA7 (https://www.megasoftware.net/) [66]. The confidence in the trees [66] was estimated by 100 bootstrap replicates. Bootstrap values greater than 70 are shown in Figures 2 and 3. Phylogenetic trees were visualized with FigTree (v. 1.4.2) (http://tree.bio.ed.ac.uk/software/figtree/).

4.2. Plant Growth and Infection by Different Pathogens

Wheat seeds from the universally susceptible Khakasskaya and Darya lines were washed with water and held for 15 minutes (min) in a 5% solution of $KMnO_4$, then placed on wet filter paper in Petri dishes and incubated for two days at 22 °C. Sprouted seeds were planted in vegetation vessels with earth. The plants were grown in a climatic chamber for eight to 10 days with a 16-h light schedule, at a temperature between 20 °C and 22 °C. Plants with a fully-developed, first real leaf were used for the experiments.

A culture of the *Stagonospora nodorum* fungus (Berk; strain B-24/MS2 from the State Collection of Phytopathogenic Microorganisms) was grown on a sterile wheat grain [67]. A shaking flask was filled to two thirds of its volume with grain and water was added, corresponding to half the weight of the grain. The flask was autoclaved for one hour (0.5 atm, 114 °C). The resulting mass of grain was cooled and infected with pieces of mycelium collected from the surface, then covered with mycelium from the fungus of the agar medium in Petri dishes. The flasks of grain were incubated with a thermostat set at 26 °C for seven days. Spores were washed from the surface of the grain with distilled water. Spore suspension in a concentration of 1×10^6 spores/mL was used in the experiments. Spore suspension of *S. nodorum* fungus was used to infect wheat plants of the Darya line. The suspension was applied by means of an atomizer: consumption—10 mL/100 plants. Tween-40 was added dropwise to the suspension in 100 mL of slurry. The inoculated plants were kept in a moist chamber for two days at a temperature between 20 °C and 22 °C, then returned to the climatic chamber with the original regime.

Samples of brown rust (*P. recondita* Rob. ex Desm f. sp. Tritici; Lower Volga population 757) were suspended in water to a concentration of 1×10^6 spores/mL. Tween-40 (one drop per 100 mL) was added. Wheat plants (Khakasskaya line) at the stage when the first real leaf had unfolded, were inoculated by rubbing a suspension of spore leaves until the surface was completely moistened and inoculated. The plants were then transferred to a moist chamber at a temperature between 20 °C and 22 °C. After two days, the plants were returned to the climatic chamber with the original regime [68].

4.3. Plant Sample Preparation for Mass Spectrometry

The chloroform/methanol precipitation method was used for extraction of total protein. Enzymatic hydrolysis of the proteins was performed according to the procedure previously described [69]. Three biological and three technical replicates were produced in each experiment.

4.4. LC-MS/MS Analysis

One microgram of peptides in a volume of 1–4 µL was loaded onto the Acclaim µ-Precolumn (0.5 mm × 3 mm, 5 µm particle size, Thermo Scientific, Rockwell, IL, USA) at a flow rate of 10 µL/min for 4 min in an isocratic mode of Mobile Phase C (2% acetonitrile, 0.1% formic acid). Then the peptides were separated with high-performance liquid chromatography (HPLC, Ultimate 3000 Nano LC System, Thermo Scientific, Rockwell, IL, USA) in a 15-cm long C18 column (Acclaim® PepMap™ RSLC inner diameter of 75 µm, Thermo Fisher Scientific, Rockwell, IL, USA). The peptides were eluted with a gradient of buffer B (80% acetonitrile, 0.1% formic acid) at a flow rate of 0.3 µL/min. Total run time including initial 4 min of column equilibration to buffer A (0.1% formic acid), then gradient from 5–35% buffer B over 65 min, 6 min to reach 99% buffer B, flushing 10 min with 99% buffer B, and 5 min re-equilibration to buffer A amounted to 90 min.

Mass spectrometric analysis was performed at least in triplicate with a Q Exactive High-Field (HF) mass spectrometer (Q Exactive HF Hybrid Quadrupole-OrbitrapTM Mass spectrometer, Thermo Fisher Scientific, Rockwell, IL, USA). The temperature of capillary was 240 °C and the voltage at the emitter was 2.1 kV. Mass spectra were acquired at a resolution of 120,000 (MS) in a range of $300-1500$ mass-to-charge ratio (m/z). Tandem mass spectra of fragment were acquired at a resolution of 15,000 (MS/MS) in the range from 100 m/z to m/z value determined by a charge state of the precursor, but no more than 2000 m/z. The maximum integration time was 50 ms and 110 ms for precursor and fragment ions, respectively. Automatic Gain Control (AGC) target for precursor and fragment ions were set to 1 $\times 10^6$ and 2×10^5, respectively. An isolation intensity threshold of 50,000 counts was determined for precursor's selection, and up to top 20 precursors were chosen for fragmentation with high-energy collisional dissociation (HCD) at 29 Normalized Collision Energy (NCE). Precursors with a charged state of +1 and more than +5 were rejected and all measured precursors were dynamically excluded from triggering of a subsequent MS/MS for 20 s.

4.5. Protein Identification and Determination of Sites Hydrolyzed In Vivo by Endogenous Proteases

For the analysis of the mass spectrometry data, a database was built using *T. aestivum* proteome (release 39, Ensembl genome database (ftp://ftp.ensemblgenomes.org/) [63]) merged with pathogen proteomes. In the case of *P. recondita* infection, assembly *Puccinia triticina* 1-1 BBBD Race 1 (NCBI ID 1628) [70] was used, while in the case of *S. nodorum* infection, *Parastagonospora nodorum* SN15 (assembly ASM14691v2) [71] was used. In our study, we performed 3 biological and 3 technical replicates for each experiment, however, we used only 2-3 biological 2-3 technical replicates each. It depended on the fact that some of the replicates were not successful and, thus, were omitted.

Raw data were processed using IdentiPy search algorithm [47]. Peptide scoring was based on Hyperscore from X!Tandem (Tables S1–S12) [72]. IdentiPy algorithm suggests autotune feature that allow to reprocess spectra and to optimize initial search parameters. Initial values for parameters were set by default as 100 ppm for precursor mass error, 500 ppm for fragment mass error, and 5 for allowed miscleavages. The results of preliminary searching were filtered to 1% False Discovery Rate (FDR) using the target-decoy approach and analyzed statistically to derive the optimal parameters that were adjusted by the algorithm for all searches performed in this study, 20 parts per million (ppm) for precursor mass error, 10–11 ppm for fragment mass error, and 2 for allowed miscleavages.

IdentiPy was used for protein identification. Post-search analysis and FDR filtering relied on multiparameter (MP) score algorithm [73]. Peptide-Spectrum Matches (PSMs), peptides, and proteins were validated at a 1.0% FDR estimated using the decoy hit distribution. Only proteins having at least two unique peptides were considered as positively identified.

In order to reveal differences in number of proteases in cultivars and between healthy and infected plants, full-specific (full-tryptic and full-AspN) peptides were identified in sequences from the database containing all 1544 wheat proteases (Tables S1–S4). Proteases that were covered by at least two unique peptides that were present in all biological and all technical replicates for each type of samples were taken for further analysis.

In order to identify the sites of in vivo hydrolysis in all proteins from the samples, the search of semi-specific (semi-tryptic and semi-AspN) peptides in the sequences from the database containing whole wheat proteome was conducted (Tables S5–S8). Semi-specific peptides present in all biological and all technical replicates for each type of sample were taken for further analysis.

In order to identify the sites of in vivo processing of proteases, the search of semi-specific peptides within the areas that contained 40 amino acid residues, extracted before and after the first peptidase domain amino acid site, was conducted (Tables S9–S12). For this search, the database containing wheat proteases that were identified with the use of full-specific peptides (Tables S1–S4) were used. In the case of metacaspases, known sites [33] were used for the search. Peptidases containing semi-specific peptides (100% identical, 100% covered), released after digestion by AspN or Trypsin, as defined by IdentiPy, were identified using the BlastP algorithm [74].

5. Conclusions

This article has characterized the degradomes of wheat. Comprehensive analysis of the *T. aestivum* genome revealed a relatively large number of proteases (1544 in total) belonging to the five major protease families: serine, cysteine, threonine, aspartic, and metallo-proteases. Unique protease groups from the C1 family were identified. Analysis of the LC-MS data obtained for degradomes of distinct wheat cultivars (Khakasskaya and Darya) revealed significant differences ~40%, equally distributed between different catalytic types of proteases, which could indicate the complementation of different proteases by one another. These findings underline the importance of interactions between components of the degradomes of any organism and wheat, in particular. In turn, these should provide insights into the physiology and biochemistry of plant immunity about which little is currently known. Infection by biotrophic (*P. recondita*) or necrotrophic (*S. nodorum*) pathogens induced drastic changes in the presence of proteolytic enzymes, as observed in the LC-MS data. However, the immune response is associated with the same core, consisting of proteases from C1, C13, C65, M16, M50, S8, S10, and A1 families. This indicates that the immune response is associated with well-known proteases. However, numerous uncharacterized protease families still require careful consideration and study. Infection by both pathogens enhances overall proteolytic activity in wheat cells. The potential for proteolysis in vivo by endogenous protease substrates increases in response to both types of infection. Biotic stress seems to activate proteolytic cascades and overall degradation of proteins. Analysis of protease-derived peptides detected by LC-MS revealed proteolysis sites, which are probably the targets of autocatalytic activation or hydrolysis by other proteases within the proteolytic cascades, involving neither caspase-like proteases nor metacaspases. However, this needs to be corroborated using activity-based approaches to validate the method for determining processed proteases.

Supplementary Materials: Supplementary materials can be found at http://www.mdpi.com/1422-0067/19/12/3991/s1.

Author Contributions: A.A.Z. conceived the original idea; A.V.B., A.A.D. and A.A.Z. conceived the original screening and research plans; A.A.D. carried out the genome and phylogenetic analysis; V.G.D. provided the two cultivars of wheat; M.I.K. and N.S.Z. performed the plant infection by biotrophic and necrotrophic pathogens, respectively; A.V.B. prepared the samples for mass spectrometry; V.G.Z. performed mass spectrometry experiments; A.V.B, A.A.D. and A.V.G. analyzed the mass spectrometry data; A.V.B. conceived the project and wrote the article with contributions from the other authors; A.A.D. and A.A.Z. supervised the project and assisted in the writing. A.A.Z. has agreed to be the contact for further communication.

Funding: This research was funded by the Russian Science Foundation (grant # 16-15-10410).

Acknowledgments: We would like to extend our thanks to Roman A. Zubarev for his critical reading of the manuscript and to Yuri B. Porozov for his technical assistance. Authors are grateful to the "Human Proteome" Core Facility, Institute of Biomedical Chemistry (IBMC), which is supported by Ministry of Education and Science of the Russian Federation (agreement 14.621.21.0017, unique project ID RFMEFI62117X0017).

Conflicts of Interest: The authors declare no conflict of interest.

Abbreviations

AALP	Arabidopsis Aleurain-Like Protease
AGC	Automatic Gain Control
CEP1	Cysteine EndoPeptidase 1
COT44	COTyledon abundunt protease 44
CYP1	CYsteine Protease 1
DUF	Domain of Unknown Function
EPF2	Epidermal Patterning Factor 2
ESI	Electrospray Ionization
ET	Ethylene
FDR	False Discovery Rate
HCD	High-energy Collisional Dissociation
HMMER	Hidden Markov Models
ISF	In-Source Fragmentation
JA	Jasmonic Acid
LC-MS	Liquid Chromatography-Mass Spectrometry
MCA1	Metacaspase 1
MMP	Matrix Metallo-protease
NAM	No Apical Meristem
NCE	Normalized Collision Energy
PA	Protease Associated
PBA1	Proteasome Subunit Type
PCD	Programmed Cell Death
PLCP	Papain-Like Cysteine Protease
RD19	Responsive To Dehydration 19
RD21	Responsive To Dehydration 21
RISC	RNA-Induced Silencing Complex
SAG12	Senescence-Associated Gene 12
SAG39	Senescence-Associated Gene 39
SBT	Subtilase
THI1	Thiol Protease 1
TPP2	Tripeptidyl Peptidase 2
TSN	Tudor Staphylococcal Nuclease
VPE	Vacuolar Processing Enzyme
XBCP3	Xylem Bark Cysteine Peptidase 3
XCP2	Xylem Cysteine Peptidase 2

References

1. Zimin, A.V.; Puiu, D.; Hall, R.; Kingan, S.; Clavijo, B.J.; Salzberg, S.L. The first near-complete assembly of the hexaploid bread wheat genome, Triticum aestivum. *Gigascience* **2017**, *6*, 1–7. [CrossRef] [PubMed]

2. Vu, L.D.; Verstraeten, I.; Stes, E.; Van Bel, M.; Coppens, F.; Gevaert, K.; De Smet, I. Proteome Profiling of Wheat Shoots from Different Cultivars. *Front. Plant Sci.* **2017**, *8*, 1–11. [CrossRef] [PubMed]

3. Faghani, E.; Gharechahi, J.; Komatsu, S.; Mirzaei, M.; Khavarinejad, R.A.; Najafi, F.; Farsad, L.K.; Salekdeh, G.H. Comparative physiology and proteomic analysis of two wheat genotypes contrasting in drought tolerance. *J. Proteom.* **2015**, *114*, 1–15. [CrossRef] [PubMed]

4. Mauchline, T.H.; Chedom-Fotso, D.; Chandra, G.; Samuels, T.; Greenaway, N.; Backhaus, A.; McMillan, V.; Canning, G.; Powers, S.J.; Hammond-Kosack, K.E.; et al. An analysis of Pseudomonas genomic diversity in take-all infected wheat fields reveals the lasting impact of wheat cultivars on the soil microbiota. *Environ. Microbiol.* **2015**, *17*, 4764–4778. [CrossRef] [PubMed]

5. Gardiner, D.M.; Upadhyaya, N.M.; Stiller, J.; Ellis, J.G.; Dodds, P.N.; Kazan, K.; Manners, J.M. Genomic analysis of Xanthomonas translucens pathogenic on wheat and barley reveals cross-kingdom gene transfer events and diverse protein delivery systems. *PLoS ONE* **2014**, *9*, e84995. [CrossRef]

6. Duba, A.; Goriewa-Duba, K.; Wachowska, U. A Review of the Interactions between Wheat and Wheat Pathogens: Zymoseptoria tritici, *Fusarium* spp. and Parastagonospora nodorum. *Int. J. Mol. Sci.* **2018**, *19*, 1138. [CrossRef]

7. Zhang, K.; Zhang, Y.; Yang, M.; Liu, S.; Li, Z.; Wang, X.; Han, C.; Yu, J.; Li, D. The Barley stripe mosaic virus gammab protein promotes chloroplast-targeted replication by enhancing unwinding of RNA duplexes. *PLoS Pathog.* **2017**, *13*, e1006319. [CrossRef]

8. Lee, S.; Kim, J.H.; Choi, J.Y.; Jang, W.C. Loop-mediated Isothermal Amplification Assay to Rapidly Detect Wheat Streak Mosaic Virus in Quarantine Plants. *Plant Pathol. J.* **2015**, *31*, 438–440. [CrossRef]

9. Ogasawara, Y.; Kaya, H.; Hiraoka, G.; Yumoto, F.; Kimura, S.; Kadota, Y.; Hishinuma, H.; Senzaki, E.; Yamagoe, S.; Nagata, K.; et al. Synergistic activation of the Arabidopsis NADPH oxidase AtrbohD by Ca^{2+} and phosphorylation. *J. Biol. Chem.* **2008**, *283*, 8885–8892. [CrossRef]

10. Li, F.; Kong, L.; Liu, Y.; Wang, H.; Chen, L.; Peng, J. Response of wheat germplasm to infestation of English grain aphid (Hemiptera: Aphididae). *J. Econ. Entomol.* **2013**, *106*, 1473–1478. [CrossRef]

11. Kong, L.A.; Wu, D.Q.; Huang, W.K.; Peng, H.; Wang, G.F.; Cui, J.K.; Liu, S.M.; Li, Z.G.; Yang, J.; Peng, D.L. Large-scale identification of wheat genes resistant to cereal cyst nematode Heterodera avenae using comparative transcriptomic analysis. *BMC Genom.* **2015**, *16*, 801. [CrossRef] [PubMed]

12. Jia, H.; Zhang, C.; Pervaiz, T.; Zhao, P.; Liu, Z.; Wang, B.; Wang, C.; Zhang, L.; Fang, J.; Qian, J. Jasmonic acid involves in grape fruit ripening and resistant against Botrytis cinerea. *Funct. Integr. Genom.* **2016**, *16*, 79–94. [CrossRef] [PubMed]

13. Bosch, M.; Wright, L.P.; Gershenzon, J.; Wasternack, C.; Hause, B.; Schaller, A.; Stintzi, A. Jasmonic acid and its precursor 12-oxophytodienoic acid control different aspects of constitutive and induced herbivore defenses in tomato. *Plant Physiol.* **2014**, *166*, 396–410. [CrossRef] [PubMed]

14. Choudhary, D.K.; Prakash, A.; Johri, B.N. Induced systemic resistance (ISR) in plants: Mechanism of action. *Indian J. Microbiol.* **2007**, *47*, 289–297. [CrossRef] [PubMed]

15. Kappelhoff, R.; Puente, X.S.; Wilson, C.H.; Seth, A.; López-Otín, C.; Overall, C.M. *Overview of Transcriptomic Analysis of all Human Proteases, Non-Proteolytic Homologs and Inhibitors: Organ, Tissue and Ovarian Cancer Cell Line Expression Profiling of the Human Protease Degradome by the CLIP-CHIP*^{TM} *DNA Microarray*; Elsevier, B.V.: Amsterdam, The Netherlands, 2017; Volume 1864, ISBN 1604822295.

16. García-Lorenzo, M.; Sjödin, A.; Jansson, S.; Funk, C. Protease gene families in Populus and Arabidopsis. *BMC Plant Biol.* **2006**, *6*, 1–24. [CrossRef] [PubMed]

17. Wang, W.; Zhou, X.; Xiong, H.; Mao, W.; Zhao, P.; Sun, M. Papain-like and legumain-like proteases in rice: Genome-wide identification, comprehensive gene feature characterization and expression analysis. *BMC Plant Biol.* **2018**, *18*, 1–16. [CrossRef] [PubMed]

18. Tripathi, L.P.; Sowdhamini, R. Cross genome comparisons of serine proteases in Arabidopsis and rice. *BMC Genom.* **2006**, *7*, 1–31. [CrossRef]

19. Chen, J.; Ouyang, Y.; Wang, L.; Xie, W.; Zhang, Q. Aspartic proteases gene family in rice: Gene structure and expression, predicted protein features and phylogenetic relation. *Gene* **2009**, *442*, 108–118. [CrossRef]

20. Grosse-Holz, F.; Kelly, S.; Blaskowski, S.; Kaschani, F.; Kaiser, M.; van der Hoorn, R.A.L. The transcriptome, extracellular proteome and active secretome of agroinfiltrated N. benthamiana uncover a large, diverse protease repertoire. *Plant Biotechnol. J.* **2018**, *16*, 1068–1084. [CrossRef]

21. Zamyatnin, A.A., Jr. Plant Proteases Involved in Regulated Cell Death. *Biochemtry* **2015**, *80*, 1701–1715. [CrossRef]

22. Meyer, M.; Huttenlocher, F.; Cedzich, A.; Procopio, S.; Stroeder, J.; Pau-Roblot, C.; Lequart-Pillon, M.; Pelloux, J.; Stintzi, A.; Schaller, A. The subtilisin-like protease SBT3 contributes to insect resistance in tomato. *J. Exp. Bot.* **2016**, *67*, 4325–4338. [CrossRef] [PubMed]

23. Hatsugai, N.; Iwasaki, S.; Tamura, K.; Kondo, M.; Fuji, K.; Ogasawara, K.; Nishimura, M.; Hara-Nishimura, I. A novel membrane fusion-mediated plant immunity against bacterial pathogens. *Genes Dev.* **2009**, *23*, 2496–2506. [CrossRef] [PubMed]

24. Chichkova, N.V.; Shaw, J.; Galiullina, R.A.; Drury, G.E.; Tuzhikov, A.I.; Kim, S.H.; Kalkum, M.; Hong, T.B.; Gorshkova, E.N.; Torrance, L.; et al. Phytaspase, a relocalisable cell death promoting plant protease with caspase specificity. *EMBO J.* **2010**, *29*, 1149–1161. [CrossRef] [PubMed]

25. Zhang, H.; Zheng, X.; Zhang, Z. The role of vacuolar processing enzymes in plant immunity. *Plant Signal. Behav.* **2010**, *5*, 1565–1567. [CrossRef] [PubMed]

26. Rose, R.; Schaller, A.; Ottmann, C. Structural features of plant subtilases. *Plant Signal. Behav.* **2010**, *5*, 180–183. [CrossRef] [PubMed]

27. Hatsugai, N.; Kuroyanagi, M.; Yamada, K.; Meshi, T.; Tsuda, S.; Kondo, M.; Nishimura, M.; Hara-Nishimura, I. A plant vacuolar protease, VPE, mediates virus-induced hypersensitive cell death. *Science* **2004**, *305*, 855–858. [CrossRef] [PubMed]

28. Sundstrom, J.F.; Vaculova, A.; Smertenko, A.P.; Savenkov, E.I.; Golovko, A.; Minina, E.; Tiwari, B.S.; Rodriguez-Nieto, S.; Zamyatnin, A.A., Jr.; Valineva, T.; et al. Tudor staphylococcal nuclease is an evolutionarily conserved component of the programmed cell death degradome. *Nat. Cell Biol.* **2009**, *11*, 1347–1354. [CrossRef] [PubMed]

29. Bozhkov, P.V.; Suarez, M.F.; Filonova, L.H.; Daniel, G.; Zamyatnin, A.A., Jr.; Rodriguez-Nieto, S.; Zhivotovsky, B.; Smertenko, A. Cysteine protease mcII-Pa executes programmed cell death during plant embryogenesis. *Proc. Natl. Acad. Sci. USA* **2005**, *102*, 14463–14468. [CrossRef] [PubMed]

30. Coll, N.S.; Vercammen, D.; Smidler, A.; Clover, C.; Van Breusegem, F.; Dangl, J.L.; Epple, P. Arabidopsis type I metacaspases control cell death. *Science* **2010**, *330*, 1393–1397. [CrossRef] [PubMed]

31. Bollhoner, B.; Zhang, B.; Stael, S.; Denance, N.; Overmyer, K.; Goffner, D.; Van Breusegem, F.; Tuominen, H. Post mortem function of AtMC9 in xylem vessel elements. *New Phytol.* **2013**, *200*, 498–510. [CrossRef] [PubMed]

32. Bollhoner, B.; Jokipii-Lukkari, S.; Bygdell, J.; Stael, S.; Adriasola, M.; Muniz, L.; Van Breusegem, F.; Ezcurra, I.; Wingsle, G.; Tuominen, H. The function of two type II metacaspases in woody tissues of Populus trees. *New Phytol.* **2018**, *217*, 1551–1565. [CrossRef] [PubMed]

33. Fagundes, D.; Bohn, B.; Cabreira, C.; Leipelt, F.; Dias, N.; Bodanese-Zanettini, M.H.; Cagliari, A. Caspases in plants: Metacaspase gene family in plant stress responses. *Funct. Integr. Genom.* **2015**, *15*, 639–649. [CrossRef] [PubMed]

34. Balakireva, A.V.; Zamyatnin, A.A. Indispensable role of proteases in plant innate immunity. *Int. J. Mol. Sci.* **2018**, *19*, 629. [CrossRef] [PubMed]

35. Xin, M.; Wang, X.; Peng, H.; Yao, Y.; Xie, C.; Han, Y.; Ni, Z.; Sun, Q. Transcriptome Comparison of Susceptible and Resistant Wheat in Response to Powdery Mildew Infection. *Genom. Proteom. Bioinf.* **2012**, *10*, 94–106. [CrossRef] [PubMed]

36. Erayman, M.; Turktas, M.; Akdogan, G.; Gurkok, T.; Inal, B.; Ishakoglu, E.; Ilhan, E.; Unver, T. Transcriptome analysis of wheat inoculated with Fusarium graminearum. *Front. Plant Sci.* **2015**, *6*, 1–17. [CrossRef] [PubMed]

37. Komatsu, S.; Kamal, A.H.M.; Hossain, Z. Wheat proteomics: Proteome modulation and abiotic stress acclimation. *Front. Plant Sci.* **2014**, *5*, 1–19. [CrossRef] [PubMed]

38. Hao, Y.; Wang, X.; Wang, K.; Li, H.; Duan, X.; Tang, C.; Kang, Z. TaMCA1, a regulator of cell death, is important for the interaction between wheat and Puccinia striiformis. *Sci. Rep.* **2016**, *6*, 1–11. [CrossRef] [PubMed]

39. Yan, L.; Liu, Z.; Xu, H.; Zhang, X.; Zhao, A.; Liang, F.; Xin, M.; Peng, H.; Yao, Y.; Sun, Q.; et al. Transcriptome analysis reveals potential mechanisms for different grain size between natural and resynthesized allohexaploid wheats with near-identical AABB genomes. *BMC Plant Biol.* **2018**, *18*, 1–15. [CrossRef]

40. Yu, Y.; Zhu, D.; Ma, C.; Cao, H.; Wang, Y.; Xu, Y.; Zhang, W.; Yan, Y. Transcriptome analysis reveals key differentially expressed genes involved in wheat grain development. *Crop J.* **2016**, *4*, 92–106. [CrossRef]

41. Wan, Y.; Poole, R.L.; Huttly, A.K.; Toscano-Underwood, C.; Feeney, K.; Welham, S.; Gooding, M.J.; Mills, C.; Edwards, K.J.; Shewry, P.R.; et al. Transcriptome analysis of grain development in hexaploid wheat. *BMC Genom.* **2008**, *9*, 1–16. [CrossRef]

42. Wheeler, T.J.; Eddy, S.R. Nhmmer: DNA homology search with profile HMMs. *Bioinformatics* **2013**, *29*, 2487–2489. [CrossRef] [PubMed]

43. Liu, J.; Sharma, A.; Niewiara, M.J.; Singh, R.; Ming, R.; Yu, Q. Papain-like cysteine proteases in Carica papaya: Lineage-specific gene duplication and expansion. *BMC Genom.* **2018**, *19*, 1–12. [CrossRef] [PubMed]

44. Finn, R.D.; Coggill, P.; Eberhardt, R.Y.; Eddy, S.R.; Mistry, J.; Mitchell, A.L.; Potter, S.C.; Punta, M.; Qureshi, M.; Sangrador-Vegas, A.; et al. The Pfam protein families database: Towards a more sustainable future. *Nucleic Acids Res.* **2016**, *44*, D279–D285. [CrossRef] [PubMed]

45. Rawlings, N.D.; Waller, M.; Barrett, A.J.; Bateman, A. MEROPS: The database of proteolytic enzymes, their substrates and inhibitors. *Nucleic Acids Res.* **2014**, *42*, D503–D509. [CrossRef] [PubMed]

46. Tsiatsiani, L.; Van Breusegem, F.; Gallois, P.; Zavialov, A.; Lam, E.; Bozhkov, P.V. Metacaspases. *Cell. Death Differ.* **2011**, *18*, 1279–1288. [CrossRef] [PubMed]

47. Levitsky, L.I.; Ivanov, M.V.; Lobas, A.A.; Bubis, J.A.; Tarasova, I.A.; Solovyeva, E.M.; Pridatchenko, M.L.; Gorshkov, M.V. IdentiPy: An Extensible Search Engine for Protein Identification in Shotgun Proteomics. *J. Proteome Res.* **2018**, *17*, 2249–2255. [CrossRef] [PubMed]

48. Kim, J.S.; Monroe, M.E.; Camp, D.G.; Smith, R.D.; Qian, W.J. In-source fragmentation and the sources of partially tryptic peptides in shotgun proteomics. *J. Proteome Res.* **2013**, *12*, 910–916. [CrossRef]

49. Alves, P.; Arnold, R.J.; Clemmer, D.E.; Li, Y.; Reilly, J.P.; Sheng, Q.; Tang, H.; Xun, Z.; Zeng, R.; Radivojac, P. Fast and accurate identification of semi-tryptic peptides in shotgun proteomics. *Bioinformatics* **2008**, *24*, 102–109. [CrossRef]

50. Carapito, C.; Kuhn, L.; Karim, L.; Rompais, M.; Rabilloud, T.; Schwenzer, H.; Sissler, M. Two proteomic methodologies for defining N-termini of mature human mitochondrial aminoacyl-tRNA synthetases. *Methods* **2017**, *113*, 111–119. [CrossRef] [PubMed]

51. Coll, N.S.; Epple, P.; Dangl, J.L. Programmed cell death in the plant immune system. *Cell Death Differ.* **2011**, *18*, 1247–1256. [CrossRef]

52. Richau, K.H.; Kaschani, F.; Verdoes, M.; Pansuriya, T.C.; Niessen, S.; Stuber, K.; Colby, T.; Overkleeft, H.S.; Bogyo, M.; Van der Hoorn, R.A. Subclassification and biochemical analysis of plant papain-like cysteine proteases displays subfamily-specific characteristics. *Plant Physiol.* **2012**, *158*, 1583–1599. [CrossRef]

53. Zou, Z.; Xie, G.; Yang, L. Papain-like cysteine protease encoding genes in rubber (Hevea brasiliensis): Comparative genomics, phylogenetic, and transcriptional profiling analysis. *Planta* **2017**, *246*, 999–1018. [CrossRef] [PubMed]

54. Golldack, D.; Vera, P.; Dietz, K.J. Expression of subtilisin-like serine proteases in Arabidopsis thaliana is cell-specific and responds to jasmonic acid and heavy metals with developmental differences. *Physiol. Plant* **2003**, *118*, 64–73. [CrossRef] [PubMed]

55. Martinez, D.E.; Borniego, M.L.; Battchikova, N.; Aro, E.M.; Tyystjärvi, E.; Guiamét, J.J. SASP, a Senescence-Associated Subtilisin Protease, is involved in reproductive development and determination of silique number in Arabidopsis. *J. Exp. Bot.* **2015**, *66*, 161–174. [CrossRef] [PubMed]

56. Neuteboom, L.W.; Veth-Tello, L.M.; Clijdesdale, O.R.; Hooykaas, P.J.J.; Van Der Zaal, B.J. A novel subtilisin-like protease gene from Arabidopsis thaliana is expressed at sites of lateral root emergence. *DNA Res.* **1999**, *6*, 13–19. [CrossRef] [PubMed]

57. Engineer, C.B.; Ghassemian, M.; Anderson, J.C.; Peck, S.C.; Hu, H.; Schroeder, J.I. Carbonic anhydrases, EPF2 and a novel protease mediate CO2control of stomatal development. *Nature* **2014**, *513*, 246–250. [CrossRef] [PubMed]

58. Book, A.J.; Yang, P.; Scalf, M.; Smith, L.M.; Vierstra, R.D. Tripeptidyl peptidase II. An oligomeric protease complex from Arabidopsis. *Plant Physiol.* **2005**, *138*, 1046–1057. [CrossRef] [PubMed]

59. Watanabe, N.; Lam, E. Arabidopsis metacaspase 2d is a positive mediator of cell death induced during biotic and abiotic stresses. *Plant J.* **2011**, *66*, 969–982. [CrossRef]

60. Petutschnig, E.K.; Stolze, M.; Lipka, U.; Kopischke, M.; Horlacher, J.; Valerius, O.; Rozhon, W.; Gust, A.A.; Kemmerling, B.; Poppenberger, B.; et al. A novel Arabidopsis CHITIN ELICITOR RECEPTOR KINASE 1 (CERK1) mutant with enhanced pathogen-induced cell death and altered receptor processing. *New Phytol.* **2014**, *204*, 955–967. [CrossRef] [PubMed]

61. Lin, R. Arabidopsis FHY3/FAR1 Gene Family and Distinct Roles of Its Members in Light Control of Arabidopsis Development. *Plant Physiol.* **2004**, *136*, 4010–4022. [CrossRef] [PubMed]

62. Tsiatsiani, L.; Timmerman, E.; De Bock, P.J.; Vercammen, D.; Stael, S.; van de Cotte, B.; Staes, A.; Goethals, M.; Beunens, T.; Van Damme, P.; et al. The Arabidopsis metacaspase9 degradome. *Plant Cell* **2013**, *25*, 2831–2847. [CrossRef] [PubMed]

63. Kersey, P.J.; Allen, J.E.; Armean, I.; Boddu, S.; Bolt, B.J.; Carvalho-Silva, D.; Christensen, M.; Davis, P.; Falin, L.J.; Grabmueller, C.; et al. Ensembl Genomes 2016: More genomes, more complexity. *Nucleic Acids Res.* **2016**, *44*, D574–D580. [CrossRef] [PubMed]

64. Eddy, S. Profile hidden Markov models. *Bioinformatics* **1998**, *14*, 755–763. [CrossRef] [PubMed]

65. Kuraku, S.; Zmasek, C.M.; Nishimura, O.; Katoh, K. aLeaves facilitates on-demand exploration of metazoan gene family trees on MAFFT sequence alignment server with enhanced interactivity. *Nucleic Acids Res.* **2013**, *41*, 22–28. [CrossRef] [PubMed]

66. Kumar, S.; Stecher, G.; Tamura, K. MEGA7: Molecular Evolutionary Genetics Analysis Version 7.0 for Bigger Datasets. *Mol. Biol. Evol.* **2017**, *33*, 1870–1874. [CrossRef] [PubMed]

67. Jenkyn, J.F.; King, J.E. Observations on the Origins of Septoria nodorum Infection of Winter Wheat. *Plant Pathol.* **1977**, *26*, 153–160. [CrossRef]

68. Browder, L.E. *Pathogenic Specialization in Cereal Rust Fungi, Especially Puccinia recondita f.sp. Tritici: Concepts, Methods of Study and Application*; United States Department of Agriculture (USDA): Washington, DC, USA, 1971; pp. 37–51.

69. Wiśniewski, J.R.; Zougman, A.; Nagaraj, N.; Mann, M. Universal sample preparation method for proteome analysis. *Nat. Methods* **2009**, *6*, 359. [CrossRef] [PubMed]

70. Kiran, K.; Rawal, H.C.; Dubey, H.; Jaswal, R.; Devanna, B.N.; Gupta, D.K.; Bhardwaj, S.C.; Prasad, P.; Pal, D.; Chhuneja, P.; et al. Draft genome of the wheat rust pathogen (Puccinia triticina) unravels genome-wide structural variations during evolution. *Genome Biol. Evol.* **2016**, *8*, 2702–2721. [CrossRef] [PubMed]

71. Syme, R.A.; Hane, J.K.; Friesen, T.L.; Oliver, R.P. Resequencing and Comparative Genomics of *Stagonospora nodorum*: Sectional Gene Absence and Effector Discovery. *Genes Genomes Genet.* **2013**, *3*, 959–969. [CrossRef] [PubMed]

72. Craig, R.; Beavis, R.C. TANDEM: Matching proteins with tandem mass spectra. *Bioinformatics* **2004**, *20*, 1466–1467. [CrossRef]

73. Ivanov, M.V.; Levitsky, L.I.; Lobas, A.A.; Panic, T.; Laskay, Ü.A.; Mitulovic, G.; Schmid, R.; Pridatchenko, M.L.; Tsybin, Y.O.; Gorshkov, M.V. Empirical multidimensional space for scoring peptide spectrum matches in shotgun proteomics. *J. Proteome Res.* **2014**, *13*, 1911–1920. [CrossRef] [PubMed]

74. Camacho, C.; Coulouris, G.; Avagyan, V.; Ma, N.; Papadopoulos, J.; Bealer, K.; Madden, T.L. BLAST+: Architecture and applications. *BMC Bioinf.* **2009**, *10*, 1–9. [CrossRef] [PubMed]

International Journal of
Molecular Sciences

MDPI

Article

Quantitative Proteomic Analysis of Castor (*Ricinus communis* L.) Seeds During Early Imbibition Provided Novel Insights into Cold Stress Response

Xiaoyu Wang [1,2,3,4,5,*], Min Li [6], Xuming Liu [1,2,3,4,5], Lixue Zhang [1,2,3,4,5], Qiong Duan [1,2,3,4,5] and Jixing Zhang [1,2,3,4,5,*]

1   College of Life Science, Inner Mongolia University for Nationalities, Tongliao 028000, China;
    liuxuminggenes@hotmail.com (X.L.); 15334940513@163.com (L.Z.); duanqiong0209@163.com (Q.D.)
2   Inner Mongolia Key Laboratory for Castor, Tongliao 028000, China
3   Inner Mongolia Industrial Engineering Research Center of Universities for Castor, Tongliao 028000, China
4   Inner Mongolia Collaborate Innovation Cultivate Center for Castor, Tongliao 028000, China
5   Horqin Plant Stress Biology Research Institute of Inner Mongolia University for Nationalities,
    Tongliao 028000, China
6   College of Agriculture, Inner Mongolia University for Nationalities, Tongliao 028000, China;
    lm@imun.edu.cn
*   Correspondence: xiaoyuwang1987@hotmail.com (X.W.); zhangjixing@imun.edu.cn (J.Z.);
    Tel.: +86-158-4851-9320 (X.W.); +86-159-2447-5588 (J.Z.)

Received: 19 December 2018; Accepted: 8 January 2019; Published: 16 January 2019

Abstract: Early planting is one of the strategies used to increase grain yield in temperate regions. However, poor cold tolerance in castor inhibits seed germination, resulting in lower seedling emergence and biomass. Here, the elite castor variety Tongbi 5 was used to identify the differential abundance protein species (DAPS) between cold stress (4 °C) and control conditions (30 °C) imbibed seeds. As a result, 127 DAPS were identified according to isobaric tag for relative and absolute quantification (iTRAQ) strategy. These DAPS were mainly involved in carbohydrate and energy metabolism, translation and posttranslational modification, stress response, lipid transport and metabolism, and signal transduction. Enzyme-linked immunosorbent assays (ELISA) demonstrated that the quantitative proteomics data collected here were reliable. This study provided some invaluable insights into the cold stress responses of early imbibed castor seeds: (1) up-accumulation of all DAPS involved in translation might confer cold tolerance by promoting protein synthesis; (2) stress-related proteins probably protect the cell against damage caused by cold stress; (3) up-accumulation of key DAPS associated with fatty acid biosynthesis might facilitate resistance or adaptation of imbibed castor seeds to cold stress by the increased content of unsaturated fatty acid (UFA). The data has been deposited to the ProteomeXchange with identifier PXD010043.

Keywords: *Ricinus communis* L.; cold stress; seed imbibition; iTRAQ; proteomics

1. Introduction

Cold stress is one of the major threats to plant growth, spatial distribution, agricultural productivity, and crop yield [1]. Most temperate plants, such as winter wheat, oats, and barley, can acquire cold acclimation and tolerate ice formation in their tissues; however, many important crops, such as rice, maize, and soybeans, are sensitive to cold stress and incapable of cold acclimation [2]. *Ricinus communis* L. (Euphorbiaceae) is an important non-edible oilseed crop originating in tropical regions but cultivated in many subtropical regions worldwide. The seed oil of castor bean is mainly used for pharmaceutical and industrial applications, as it is a rich source of ricinoleic acid, an unusual hydroxylated fatty acid [3]. Castor beans can be cultivated in some unfavorable environments, such as saline and drought conditions, where other crops would not grow and produce a good yield [4]. However, castor bean is sensitive to cold stress in temperate regions, where the temperature drops frequently during the early growing season. It has been reported that temperature below 20 °C can dramatically decrease seed germination capability [5]. Thus, a prime target for breeding efforts is to improve seed germination under cold stress. A combination of traditional and molecular assistant selection breeding is an effective strategy for generating stress-tolerant and widely adapted castor varieties. Therefore, it is imperative to understand the molecular response to cold stress and to identify some novel responsive genes or proteins in castor bean with strong potential for the improvement of cold tolerance by genetic engineering.

Seed germination, the first and important phase for plant propagation, starts by seed imbibition, leading to embryo transition from a state of quiescence in a dry seed to a state of highly active metabolism, and terminates with embryonic axis elongation [6]. Generally, seed germination can be divided into three phases: a rapid phase of water uptake (Phase I), followed by a plateau phase of water uptake (Phase II), in which grain morphology and structure obviously change and the radical and germ appear, and post-germination (Phase III), a rapid phase of water uptake with the initiation of growth [7,8]. During seed imbibition, phase I represents the initial seed germination stage, in which necessary structures and enzymes are present and storage substances, such as starch, proteins, and lipids, which provide energy and nutrition for seed germination, begin to be activated [8].

A handful of studies have proved that the transcript level of gene expression does not necessarily directly correlate well with abundance of corresponding protein species [9,10]. Since proteins were a direct effector of plant stress response, it was highly important to investigate changes in proteome level to identify potential protein markers whose abundance changes could be linked with changes in physiological indices under cold stress [11]. Early study showed that the abundance of cold-regulated/late embryogenesis-abundant (COR/LEA) proteins were enhanced by cold stress [12]. Several reactive oxygen species scavenging enzymes involved in metabolism of ascorbate-glutathione cycle were up-accumulated under cold stress [13]. Up-accumulation of chaperones, especially various heat shock proteins (HSPs), can play crucial roles in preventing protein misfolding caused by low temperature [13]. Recently, high-throughput proteomic technology was widely used to identify a set of differential abundance protein species (DAPS) associated with cellular responses to cold stress in diverse plants [1,14]. For example, 173 DAPS associated with carbohydrate and energy metabolism, translation and posttranslation modification, stress response, signal transduction etc. were detected in maize leaves after cold stress [1]. These studies can significantly contribute to our understanding of cold response. However, less proteome analyses were applied to elucidate the molecular mechanism of cold adaptation or resistance in seed germination compared to seedlings.

To date, several proteomics analyses identified a range of proteins related to germinating or developing castor seeds [15–20]. Several classes of seed reserve proteins such as 2S albumins, legumin-like and seed storage proteins, and proteins involved in plant defenses against biotic and abiotic stresses, were identified from developing castor seeds using two-dimensional electrophoresis (2-DE) [18]. Some important proteins involved in fatty acid metabolism, seed storage proteins, toxins, and allergens associated with developing castor oil seeds were identified by employing isotope coded protein label (ICPL) and isobaric tag for relative and absolute quantification (iTRAQ) technologies [20]. Fourteen proteins were identified in the endoplasmic reticulum of germinating castor seeds; ten of these proteins were concerned with roles in protein processing and storage, and lipid metabolism [15]. However, the proteomics analyses of early seed imbibition for castor under cold stress are rare. Recently, iTRAQ is more accurate and reliable for quantitation of protein species than traditional 2-DE analysis and is currently widely used for identifying cold-accumulated protein species in many plants [1,21].

In this study, we used iTRAQ-based proteomics to detect DAPS between cold-stressed imbibed seeds and unstressed control seeds. Possible biological functions and potential effects of these DAPS on cold tolerance were discussed. This analysis revealed complex changes at the proteomics level in early imbibed castor seeds under cold stress and provided new information concerning the plant responses to cold stress.

2. Results

2.1. Germination Analysis of Castor Beans During Imbibition

The changes in water uptake were assessed by measuring changes in the seed weight during imbibition at 30 °C and 4 °C. As shown in Figure 1A, seeds imbibed at 30 °C and 4 °C revealed a triphasic pattern during germination; however, the seeds imbibed at 30 °C could absorb more water than those at 4°C during the same period. Water uptake increased rapidly before 12 h (Phase I), followed by a plateau of seed imbibition from 12-30 h (Phase II) and rapid water uptake after 30 h (Phase III). The seed radical began to emerge during Phase II. However, in the seeds imbibed at 4 °C for 12 h, and then transferred to 30 °C, germination was retarded by 1 day compared to the control conditions. Only 21.5% of seeds germinated at day 3, in contrast to the imbibed seeds at a constant 30 °C, with 42.5% (Figure 1B). Thus, the imbibed seeds in the Phase I (12 h) were collected for further proteomic analysis.

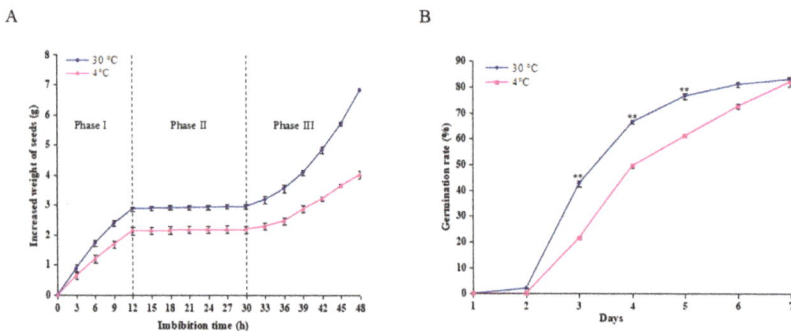

Figure 1. Effect of cold stress on seed imbibition and germination. (**A**) Triphasic pattern of water uptake under the cold stress (4 °C) and control conditions (30 °C). (**B**) Seed germination rate was calculated after the seeds imbibed at 4 °C and 30 °C for 12 h. For the cold stress treatment, the seeds were imbibed at 4 °C for 12 h, and then transferred to 30 °C for germination, and for the control, the seeds were allowed to germinate at a constant 30 °C only. Values represent the means ± SD from three fully independent biological replicates. * $p < 0.05$, ** $p < 0.01$ by Student's *t*-test.

2.2. Primary Data Analysis and Protein Identification

iTRAQ-based comparative proteome was used to identify the DAPS between cold stress (4 °C) and control conditions (30 °C) imbibed seeds. IPeak identified a total of 38863 spectra, 8280 peptides and 1670 proteins. In total, 74% proteins included at least two unique peptides. The mass of the identified proteins with 0–50, 51–100, and >100 kDa accounted for 64.4, 30.2, and 5.4% separately. 26 low molecular weight proteins (Mr > 10 kDa) and 91 high molecular weight proteins (Mr > 91 kDa) were identified using the iTRAQ strategy. The distribution of protein sequence coverage with 40–100, 30–40, 20–30, 10–20, and under 10% variations accounted for 8.7, 7.7, 13.5, 23.9 and 46.2%, respectively (Table S2). The mass spectrometry proteomic data of the present study have been deposited to the ProteomeXchange with identifier PXD010043 (http://proteomecentral.proteomexchange.org/cgi/GetDataset?ID=PXD010043).

2.3. Identification of DAPS by iTRAQ

A protein species was considered differentially accumulated when it exhibited a fold change >1.2 and *p* value <0.05. Based on these criteria, 127 DAPS were identified, of which, 109 were up-accumulated and 18 were down-accumulated under cold stress versus control conditions. Detailed information is provided in Table S3.

2.4. Bioinformatics Analysis of DAPS Identified by iTRAQ

Gene ontology (GO) annotations were carried out to identify the significantly enriched functional groups of DAPS. A total of 84 DAPS under cold stress versus control conditions were classified into 24 functional groups (Figure 2, Tables S1–S4), of which biological processes accounted for 14 GO terms (the most representative were "response to stimulus"), cellular components accounted for 4 GO terms (the most representative were "macromolecular complex"), and molecular functions accounted for 6 GO terms (the most representative were "binding").

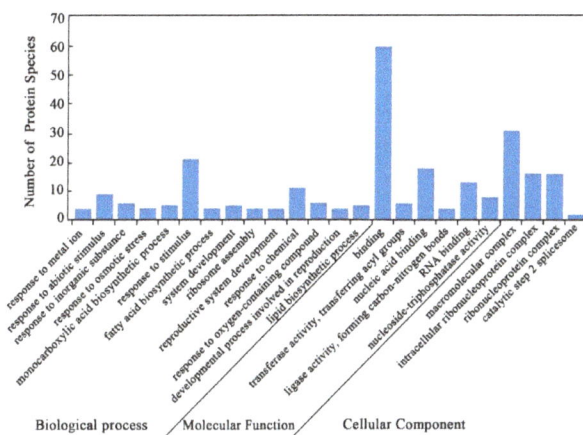

Figure 2. GO annotation of DAPS identified by the iTRAQ.

A total of 70 DAPS by iTRAQ were classified into 16 categories of Clusters of Orthologous Groups of proteins (COG), among which, "translation, ribosomal structure and biogenesis" represented the largest group (group J, 18 DAPS), followed by "posttranscriptional modification, protein turnover, chaperones" (group O, 9 DAPS) (Figure 3, Tables S2–S4).

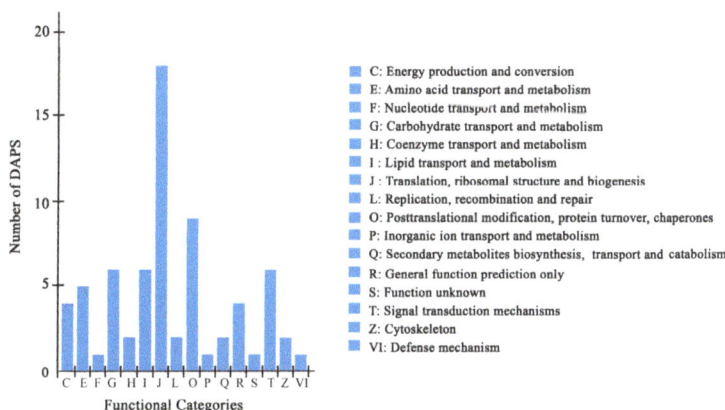

Figure 3. COG classification of DAPS identified by the iTRAQ.

To further explore the biological functions of these proteins, 56 DAPS were mapped to 16 pathways in the KEGG database (Tables S3 and S4). These annotated protein species were significantly enriched in the following pathways: "fatty acid biosynthesis", "biotin metabolism", "fatty acid metabolism", "cyanoamino acid metabolism", and "ribosome" (Table 1).

Table 1. Significantly enriched pathway annotation of DAPS identified by iTRAQ.

No.	Pathway	Number of DAPS	*p*-value
1	Fatty acid biosynthesis	5	0.016388
2	Biotin metabolism	3	0.017208
3	Fatty acid metabolism	6	0.031722
4	Cyanoamino acid metabolism	3	0.033034
5	Ribosome	12	0.044644

2.5. Confirmation of DAPS by ELISA

To validate the reliability of the DAPS as determined by iTRAQ, we used ELISA to assess the expression level of six DAPS. Four of the six DAPS were involved in fatty acid metabolism, including β-ketoacyl-acyl carrier protein synthase (KAS) II (KASII), KASI, biotin carboxylase (BC) subunit of Het-ACCase, and β-carboxyltransferase (β-CT) subunit of Het-ACCase, whereas the other two DAPS were involved in the pentose phosphate pathway: 6-phosphogluconolactonase (6PGL) and glucose-6-phosphate 1-dehydrogenase (G6PDH). As shown in Figure 4, the level of six DAPS was increased by cold stress, which showed a good correlation between the ELISA and iTRAQ-based datasets.

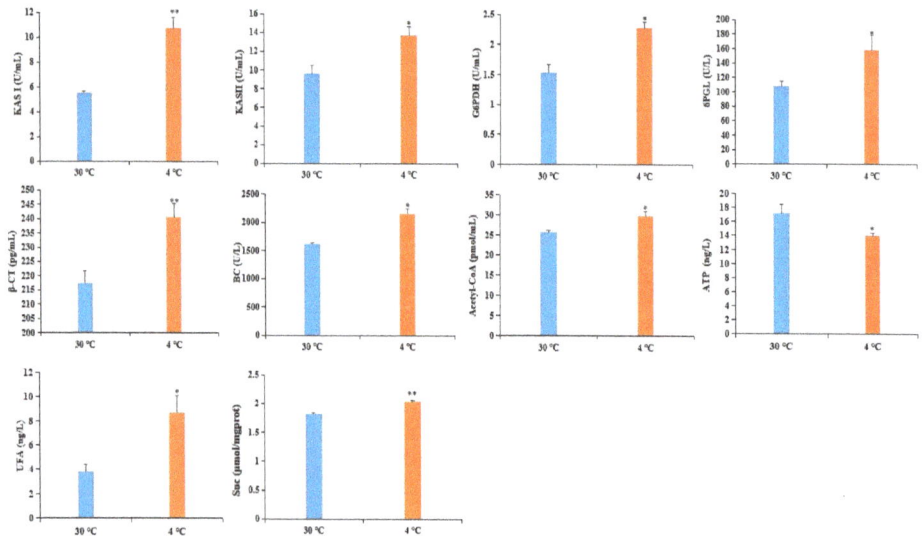

Figure 4. Validation of DAPS by ELISA. Values represent the means ±SD from three fully independent biological replicates. * $p < 0.05$, ** $p < 0.01$ by Student's *t*-test.

2.6. Transcriptional Analyses of the Corresponding Genes Encoding DAPS

To know the correlation between the abundance of DAPS and the transcript level of their corresponding genes, eight up-accumulated DAPS were selected for qRT-PCR analyses. The eight DAPS were dehydrin Xero (B9S696), glutathione peroxidase (GPX) (B9RCA6), late embryogenesis abundant protein D-34 (LEA D-34) (B9RTR0), LEA D-34 (B9S3Z7), SNF1-related protein kniase (SnRK1) α catalytic subunit (B9SVJ9), KAS II (Q41134), eukaryotic translation initiation factor 5A (IF5A) (B9STQ5), protein phosphatase 2c (PP2C) (B9SB19). The results showed that the eight selected DAPS can be clustered into two groups, i.e., group I, up-regulated at both transcript and protein level; Group II, no change at transcript level while up-accumulated at protein level (Table 2). This discrepancy between the mRNA and protein expression profiles indicated that the abundance of protein depends not only on the transcript level but also on transcript stability, post-transcriptional regulation, post-translational modifications, and protein degradation [9,10].

Table 2. Comparison of expression pattern at the mRNA and protein level of DAPS.

Protein ID	Description	iTRAQ Ratio	*p* Value	qPCR Ratio	*p* Value	Trends [a] iTRAQ	qPCR
B9S696	dehydrin Xero	1.45 ± 0.133	0.004	4.52 ± 0.009	0.000	+	+
B9RCA6	GPX	1.62 ± 0.136	0.001	2.30 ± 0.116	0.003	+	+
B9RTR0	LEA D-34	1.75 ± 0.138	0.000	5.34 ± 0.512	0.013	+	+
B9SB19	PP2C	1.21 ± 0.114	0.029	6.40 ± 0.952	0.029	+	+
Q41134	KASII	1.38 ± 0.207	0.023	2.76 ± 0.501	0.050	+	+
B9SVJ9	SnRK1 α catalytic subunit	1.32 ± 0.162	0.022	3.85 ± 0.530	0.032	+	+
B9S3Z7	LEA D-34	1.30 ± 0.150	0.021	1.16 ± 0.117	0.232	+	=
B9STQ5	IF5A	1.46 ± 0.279	0.032	0.42 ± 0.073	0.121	+	=

[a] +: up-regulated; =: not significantly changed.

3. Discussion

Castor beans, which possess a high economic value, are very sensitive to cold stress, especially at the germination stage. Transcriptome analysis has identified many differentially expressed genes (DEGs) mainly related to plant secondary metabolism in germinated seeds under cold stress [22]. However, the mechanisms underlying the effects of cold stress on early seed imbibition are largely unknown. This initial proteomics analysis of castor seeds during early imbibition identified several cold-accumulated protein species and unraveled a complex cellular network affected by cold stress. In this study, 127 DAPS were identified, among which, 109 were up-accumulated and 18 were down-accumulated under cold stress compared to control conditions. Unsurprisingly, these DAPS included some well-known stress-inducible proteins, such as the LEA, dehydrin. In addition, some cold-accumulated protein (methionine aminopeptidase) identified here have been verified in other plants based on iTRAQ strategy. Furthermore, this approach also identified some novel proteins that were not previously known to be associated with cold stress response. It has been reported that mature dry seeds can rapidly restart metabolic activity including protein synthesis after imbibing water. De novo protein synthesis was necessary for seed germination in rice [23]. Our results showed that "ribosome" was the significantly enriched pathway involved in early seed imbibition under cold stress. Dry seed also contained a myriad of "long lived mRNA", which was thought to be translated after imbibition [23]. It is interesting to further investigate if de novo transcription was required for germination of castor seeds by control-treated imbibed seeds versus dry seeds and translation of long-lived mRNAs was induced or regulated by cold stress during the germination of castor seeds. In brief, bioinformatics analysis revealed that 84 and 70 DAPS were annotated in 24 GO functional groups and 16 COG categories, respectively. In total, 56 DAPS were mapped into 16 KEGG pathways. The possible biological significance of some key DAPS and their relevant metabolic pathways in cold stress adaptation are discussed below.

3.1. DAPS Involved in Translation and Posttranslational Modification

Fifteen DAPS, including eleven ribosomal proteins (RPs) and four elongation factors, were up-accumulated in cold-treated imbibed seeds (Table 3, Table S3). RPs are essential for protein synthesis and play a critical role in metabolism, cell division and growth, and regulation of cold stress [21,24]. For example, three soybean ribosomal protein genes were induced by low temperature treatment [25]. Recently, there is increasing evidence that shows direct links between RPs and cold stress. Wang et al. reported that all the DAPS involved in mature ribosome assembly and translation processes were increased in maize leaves after a 12 h cold treatment [1]. Thus, up-accumulation of RPs in cold-treated imbibed castor seeds might be required for de novo transcription or participate as regulatory components in response to cold stress. Elongation factor Tu (EF-Tu), which is responsible for the elongation phase of protein synthesis, has been extensively studied in plant responses to various environmental challenges, such as cold and heat stresses [26]. In the present study, the abundance of EF-Tu was up-accumulated under cold stress, which agrees with the previous study that cold stress can increase the protein level of EF-Tu in rice [27,28]. The eukaryotic translation initiation factor 5A (eIF5A) promotes the first peptide bond formation at the onset of protein synthesis [29]. Plant eIF5A is involved in multiple biological processes, including protein synthesis regulation, translation elongations, mRNA turnover and decay, and abiotic stress responses [30–32]. For example, Wang et al. reported that transgenic yeast and poplar expressing TaeIF5A displayed elevated protein levels combined with improved abiotic stresses tolerance [33]. Elongation factor 1α promotes codon-directed binding of aminoacyl-tRNA (aa-tRNA) in the ribosome [34]. Transgenic plants with transgene AtEF1α were more tolerant to NaCl than the wild-type [35]. The increased accumulation of four elongation factors might confer cold tolerance by promoting protein synthesis or regulating physiological pathways. N-terminal Met excision (NME) is a process by which methionine aminopeptidase (MAP) specifically removes the first Met in most newly synthesized proteins. Overexpression of barley DNA-binding MAP in Arabidopsis exhibited stronger freezing tolerance compared to the wild type [36]. The abundance of MAP observed in this study showed increased accumulation in cold-treated imbibed seeds, which agrees with the recent report that cold stress increased the MAP level in petunia [21]. These results led us to speculate that the responsiveness of MAP to cold stress might be a common event that deserves further investigation.

Table 3. Information of DAPS in ribosome pathway.

Protein Accession	Fold Change	Accumulated	Description
B9SKD1	1.25	Up	60S ribosomal protein L3
B9RMF8	1.21	Up	Zn-dependent exopeptidases superfamily protein
B9S4D5	1.24	Up	40S ribosomal protein S26
B9SKG4	1.27	Up	40S ribosomal protein S11
B9SBM0	1.25	Up	60S ribosomal protein L9
B9SIV4	1.35	Up	60S ribosomal protein L7a
B9R982	1.21	Up	40S ribosomal protein S9
B9RG16	1.23	Up	40S ribosomal protein S27
B9SYV4	1.23	Up	60S ribosomal protein L21
B9RQ66	1.43	Up	60S ribosomal protein L28
B9SCT8	1.22	Up	60S ribosomal protein L10a
B9T040	1.5	Up	60S ribosomal protein L35

Posttranslational modifications play critical roles in the regulation of abiotic stresses, such as ubiquitination and phosphorylation [37]. The ubiquitination-proteasomal pathway has been implicated in diverse aspects of eukaryotic cellular regulation due to its ability to degrade intracellular protein [38,39]. Ubiquitin-activating enzymes catalyze the first step in the ubiquitination reaction, which activates ubiquitin and transfers the activated Ub to a ubiquitin-conjugating enzyme to form an E2-Ub thiolester [40]. The proteasome plays a fundamental role in retaining cellular homeostasis and is the major cellular proteolytic machinery responsible for the degradation of both normal and damaged proteins [41]. The increased accumulation of three protein species involved in the ubiquitin/26S proteasome system might contribute to potentially harmful polypeptide degradation (Table S3). Heat shock proteins (HSPs) are highly conserved proteins that are present in organisms, function as "molecular chaperones", promote the degradation of abnormal proteins and prevent the aggregation of denatured proteins [42,43]. Our results showed that the abundance of three heat-shock proteins was increased under cold stress (Table S3).

3.2. DAPS Involved in Stress Response

Cold stress can result in overproduction of reactive oxygen species (ROS). ROS can perturb cellular redox homeostasis and lead to oxidative damage to membrane lipids, nucleic acids, and proteins [21]. To relieve cellular damage by ROS, plants have developed ROS scavenging systems including antioxidants and antioxidant enzymes [44]. GPX is a ubiquitous enzyme in plant cells that use glutathione to reduce H_2O_2 and lipid hydroperoxides [45,46]. Thus, it was not surprising that the abundance of GPX was increased in cold-stressed imbibed seeds (Table S3). It has been reported that overexpression of GPX can enhance the growth of transgenic tobacco under cold and salt stresses [47]. LEA proteins are involved in many physiological processes, act as protectors of enzyme activities and stabilize membranes associated with anionic phospholipid vesicles at freezing temperature [48,49]. Consistent with previous findings, we identified four LEA proteins that showed increased accumulation in cold-treated lines compared to control lines (Table S3). Dehydrins, which are known as group 2 or D-11 family LEA proteins, can protect the cell against damage caused by stress. A high level of dehydrin transcripts or proteins is closely associated with cold tolerance in numerous plants such as Arabidopsis, rice and Rhododendron [50–52]. The increased accumulation of dehydrin may facilitate the resistance or adaptation of castor to cold stress (Table S3). Early responsive to dehydration (ERD) genes can be rapidly induced to counteract abiotic stresses, such as drought, low temperature or high salinity. Overexpression of VaERD15 in Arabidopsis resulted in higher cold tolerance and accumulation of antioxidants compared to wild type under cold stress [53]. The increased abundance of ERD was observed in cold-treated imbibed seeds (Table S3).

3.3. DAPS Involved in Carbohydrate and Energy Metabolism

Carbon metabolism provides the necessary energy for subsequent plant growth and development during germination. Simultaneously, carbon metabolism could be an effective connection with other metabolic processes [54]. The pentose phosphate pathway (PPP) is a central metabolic pathway including the irreversible oxidative pathway and reversible non-oxidative pathway, which is catalyzed by several different enzymes such as G6PDH, 6PGL and ribulose-phosphate 3-epimerase (RPE) [55–57]. G6PDH catalyzes the first and rate-limiting enzyme of PPP by converting glucose-6-phosphate to 6-phosphogluconolactone. 6PGL catalyzes the hydrolysis of 6-phosphogluconolactone, which was thought to occur spontaneously [57]. Numerous studies regarding G6PDH function in response to abiotic stresses have been performed. Overexpressing of PsG6PDH in transgenic tobacco resulted in enhanced cold tolerance [58]. In this study, the increased accumulation of G6PDH, 6PGL and RPE might provide more reducing equivalent NADPH for anabolic pathways including fatty acid synthesis, and carbon skeletons for the synthesis of acetyl-CoA, etc. (Table S3).

3.4. DAPS Involved in Lipid Transport and Metabolism

Fatty acids (FAs) are major components of cell or organelle membrane lipids. FAs are also precursors of messenger compounds such as jasmonic acid and phosphatidylinositol, which play key roles in certain signal transduction pathways, and are used as substrates for the synthesis of storage lipids that are important materials for seed germination and provide energy for humans [59–63]. During FA biosynthesis, acetyl-CoA carboxylase (ACCase) catalyzes the committed step of the de novo FA biosynthesis pathways by converting acetyl-CoA to malonyl-CoA. KAS is vital for carbon chain condensation and elongation from C4-C18. KASI has high activity when butyryl- to myristyl-ACP (C4:0-C14:0 ACP) is used as the substrate to produce hexanoyl- to palmitoyl-ACP (C6:0-C16:0 ACP), whereas KASII catalyzes the last condensation reaction of palmitoyl-ACPs to stearoyl-ACPs [64]. It is well known that alteration of the lipid composition of cell membrane is associated with cold tolerance [65]. The increased accumulation of four DAPS related to FA biosynthesis might adapt castor to cold stress by preventing membrane transition from liquid crystalline phase to gel phase (Table 4). This inference was also supported by our observation that the content of UFA was obviously increased in cold-treated imbibed seeds (Figure 4).

Table 4. Information of DAPS in fatty acid biosynthesis.

Protein Accession	Fold Change	Accumulated	Description
Q41134	1.38	Up	β-ketoacyl-acyl carrier protein synthase II
Q41135	1.45	Up	β-ketoacyl-acyl carrier protein synthase I
B9S1E2	1.26	Up	Biotin caboxylase (BC) subunit of Het-ACCase
B9TAH3	1.47	Up	β-carboxyltransferase (β-CT) subunit of Het-ACCase
B9RF47	1.38	Up	Short chain dehydrogenase

3.5. DAPS Involved in Signal Transduction

SnRK1 complex is a heterotrimeric complex composed of a α catalytic subunit that interacts with two other subunits [66]. SnRK1 α catalytic subunit triggers vast transcriptional and metabolic reprogramming and promotes tolerance to adverse conditions [67]. Several evidences have suggested that SnRK1 had the potential to regulate the carbohydrate metabolism of higher plants. For instance, antisense expression of the SnRK1 α catalytic subunit resulted in the reduction of sucrose synthase gene expression in leaves and tubers [68]. Our study showed that the increased level of SnRK1 α catalytic subunit might regulate sucrose metabolism. This inference was supported by our observation that the content of sucrose was increased in cold-treated imbibed seeds (Figure 4). Rosnoblet et al. showed that the amount of SnRK1 gamma subunit in the radical decreased during imbibition and was no longer detectable in the protruded radical [69]. However, conditions that block germination of imbibed seeds, including low water potential or ABA, maintain SnRK1 gamma subnuit expression [70]. Consistent with previous findings, up-accumulation of SnRK1 gamma subunit was observed in cold-treated imbibed seeds (Table S3). Reversible protein phosphorylation mediated by protein kinases and phosphatases is a central mechanism for modulating a body of cellular processes such as signaling transduction, cell division and development [71,72]. Serine/threonine protein phosphatases 2C (PP2C) have been suggested to play an important role in stress signaling [73]. Overexpression of ZmPP2C2 in tobacco enhanced cold tolerance [74]. Consistent with these findings, we identified two PP2Cs which was up-accumulated 1.21- and 1.42- fold in cold-treated lines compared to the control lines (Table S3). Purple acid phosphatase (PAP) represents a diverse group of acid phosphatases in animals, microorganisms, and plants that catalyze the hydrolysis of phosphate esters and anhydrides. It has been reported that the expression of the *GmPAP3* gene can be induced by abiotic stresses, such as salinity and drought. The GmPAP3 transgenic Arabidopsis displayed higher abiotic stress tolerance, which was probably related to alleviation of oxidative damage [75]. Our results showed that the abundance of purple acid phosphatase was increased in cold-treated imbibed seeds (Table S3).

3.6. A Proposed Metabolic Pathway for Ricinus Communis During Early Seed Imbibition in Response to Cold Stress

Based on the above findings and the data presented here, we proposed a view of resistance and adaptation strategies of early seed imbibition for castor under cold stress (Figure 5). Cold stress is partly decoded as an energy-deficiency signal, regardless of their site and mode of perception [76]. SnRK1 α subunit in plants can sense the energy deficit to trigger extensive transcriptional reprogramming that contribute to restoring homeostasis and elaborating long-term responses for adaptation [67]. In our study, cold stress resulted in energy deprivation (Figure 4). Upon sensing the energy deficit, up-accumulation of the SnRK1 α subunit in cold-treated imbibed seeds regulated the sucrose metabolism to produce more substrate for PPP. PPP is an effective connection with other metabolic processes.

The up-accumulation of rate-limiting enzymes (G6PDH and 6PGL) and RPE involved in PPP provided more acetyl-CoA for fatty acid synthesis (Figure 4). Chilling stress can impair membrane permeability by the transition of membrane lipids from liquid-crystalline phase to gel phase [77]. Tolerance to chilling stress is closely connected with the fatty acid desaturation of plant membrane lipids [78]. In this study, up-accumulation of some important DAPS significantly enriched in fatty acid biosynthesis might facilitate the resistance or adaptation of imbibed castor to cold stress by increasing the content of UFA (Figure 4).

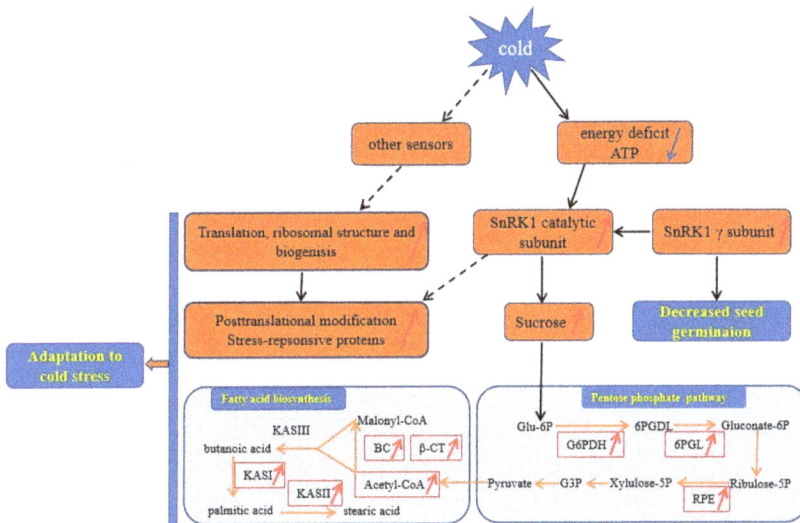

Figure 5. A proposed model for *Ricinus communis* during early seed imbibition in response to cold stress. The up-accumulated DAPS and increased content of acetyl-CoA and sucrose were indicated by the red arrow; the decreased content of ATP were indicated by blue arrow. Abbreviations: Glu-6P, Glucose 6-phosphate; 6PGDL, 6-phosphogluconolactone; G6PDH, glucose-6-phosphate 1-dehydrogenase; 6PGL, 6-phosphogluconolactonase; Gluconate-6P, 6-phosphogluconate; Ribulose-5P, Ribulose-5-phosphate; Xylulose-5P, Xylulose-5-phosphate; RPE, Ribulose-phosphate 3-epimerase; G3P, Glyceraldehyde 3-phosphate; KASI, β-ketoacyl-acyl carrier protein synthase I; KASII, β-ketoacyl-acyl carrier protein synthase II; KASIII, β-ketoacyl-acyl carrier protein synthase III. Solid lines denoted proven connections in plants, whereas broken lines represented connections that might exist in plants.

The absence of a SnRK1 gamma subunit (LeSNF4) in tomato contribute to the transition to the "germination/growth" mode rather than the maintenance of "maturation/dormancy" metabolic state. However, conditions that block germination of imbibed seeds, including ABA and FR light, can maintain LeSNF4 expression [70]. Therefore, the up-accumulation of SnRK1 gamma subunit in cold-imbibed seeds might be one of the reasons for the decreased seed germination ability. Energy deficiency is sensed by the SnRK1 that trigger the repression of genes involved in protein synthesis [79]. Up-accumulation of all mature ribosome assembly and translation in cold-imbibed seeds might be regulated by other sensors to confer cold tolerance by producing more important proteins, such as cold-responsive proteins. Chilling stress can cause dysfunction/denaturation of structural and functional protein. Some DAPS involved in the ubiquitin/26S proteasome system might be responsible for maintaining functional protein conformations and contributing to potentially harmful polypeptides degradation.

4. Materials and Methods

4.1. Plant Materials and Stress Treatment

The *ricinus communis* seeds (genotype Tongbi 5) used in this work were kindly supplied by the Tongliao Academy of Agricultural Sciences, China. Castor seeds were sown on 15 April and harvested on 1 October, 2017 in experimental field of Inner Mongolia University for Nationalities, China. This genotype was the elite variety with high seed yield potential (average up to 2316.03 kg/hm^2) and widely cultivated in China [80]. Briefly, 50 sterilized seeds of each biological replicate with similar size and weight were sown in Petri dishes (d = 12 cm) with filter papers soaked in 16 mL sterile distilled water at 30 °C or 4 °C for 7 days. The weight of the imbibed seeds was recorded every 3 h to calculate the water content of the seeds. The seeds were allowed to germinate at a constant 30 °C for optimal germination and at 4 °C for 12 h and then transferred to 30 °C for cold germination. Three biological replicates were conducted. Germination was determined when the tip of radical grew free of the seed coat [81].

4.2. Protein Extraction

Four Petri dishes of 50 sterilized seeds each (a total of 200 seeds) as one biological replicate were imbibed under 4 °C for 12 h along with the seeds imbibed at 30 °C for 12 h as the control. Two biological replicates were conducted for iTRAQ-based comparative proteomics analysis. Cold- and control-treated imbibed seeds were ground into powder with liquid nitrogen. The powder was slowly stirred into a beaker with 100 mL preboiling 2% (*v/v*) sodium dodecyl sulfate (SDS) solution and 2 mM phenylmethylsulfonyl fluoride (PMSF) (final concentration) for 3 min. The suspension was rapidly cooled, transferred to two new 50 mL tubes, sonicated for 30 min on ice and then centrifuged at 20,000× *g* for 15 min. The upper oil layer and lower precipitate were discarded, and the middle clear protein solution was dried by lyophilization and concentrated to 5 mL. The protein solution was mixed well with 30 mL 10% chilled trichloroacetic acid (TCA) acetone and incubated at −20 °C overnight. After centrifugation at 4 °C and 20,000× *g*, the supernatant was discarded. The precipitate was washed three times with chilled acetone. The pellet was air-dried and dissolved in lysis buffer (7 M urea, 2 M thiourea, 4% NP40, and 20 mM Tris-HCl, pH 8.0–8.5). The suspension was sonicated at 200 W for 5 min and centrifuged at 4 °C at 30,000× *g* for 15 min. The supernatant was then collected. The sample protein concentration was determined with the Bradford assay using BSA as the calibrant. The quality of the protein sample was measured by SDS-PAGE.

4.3. Protein Digestion and iTRAQ Labeling

A total of 100 µg protein for each sample was digested with Trypsin Gold (Promega, Madison, WI, USA) with the ratio of protein: trypsin = 20:1 at 37 °C for 16 h. Peptide samples were labeled with 8-plex iTRAQ reagents (Applied Biosystems, Foster City, CA, USA) according to the manufacturer's protocol. The control replicates (30 °C) were labeled with the iTRAQ tags 114 and 116, and the cold-treated sample replicates were labeled with the iTRAQ tags 117 and 121. The labeled peptide mixtures were pooled, dried by vacuum centrifugation and fractionated by strong cationic exchange (SCX) chromatography.

4.4. Fractionation by SCX

The Shimadzu LC-20AB HPLC Pump system (Kyoto, Japan) was used for SCX chromatography. The iTRAQ-labeled peptide mixtures were reconstituted with 4 mL buffer A (25 mM NaH$_2$PO$_4$ in 25% acetonitrile, pH 2.7) and loaded onto a 4.6×250 mm Ultremex SCX column containing 5 µm particles (Phenomenex). The peptides were eluted with a gradient of buffer A for 10 min, buffer B (25 mM NaH$_2$PO$_4$, 1 M KCl in 25% acetonitrile, pH 2.7) for 25 min, and 35–80% buffer B for 1 min. Elution was monitored by measuring absorbance at 214 nm, and fractions were collected every minute.

The eluted peptides were pooled as 15 fractions, desalted with a Strata X C18 column (Phenomenex) and vacuum-dried.

4.5. LC-ESI-MS/MS Analysis by Q-Exactive

Each fraction was re-dissolved in buffer A (5% acetonitrile, 0.1% formic acid) and centrifuged at $20,000 \times g$ for 10 min; the final concentration of peptides was approximately 0.5 µg/µL. Then, 8 µL supernatant was loaded on a LC-20AD nanoHPLC (Shimadzu, Kyoto, Japan) by the autosampler onto a C18 trap column, and the peptides were eluted onto an analytical C18 column (inner diameter, 75 µm) packed in-house. The samples were loaded at 8 µL/min for 4 min, and then the 40-min gradient was run at 300 nL/min starting from 2–35% buffer B (95% acetonitrile, 0.1% formic acid), followed by 5 min linear gradient to 80%, maintained at 80% buffer B for 4 min, and finally returned to 5% in 1 min.

Data acquisition was performed in the Q-Exactive (ThermoFinnigan, San Jose, CA, USA) mass spectrometer with a mass/charge (m/z) scanned range of 350–2000 Da. Intact peptides and ion fragments were detected in the Orbitrap at resolutions of 70,000 and 17,500, separately. Peptides were selected using the high-energy collision dissociation mode with a normalized collision energy setting of 27.0. MS/MS data were obtained using a data-dependent procedure to capture the 15 most abundant precursor ions. An electrospray voltage of 1.6 kV was applied, and the dynamic exclusion duration was set for 15 s.

4.6. Protein Identification and Quantification

Raw data files from the Orbitrap were merged and transformed into the MASCOT generic format (MGF) by Proteome Discoverer 1.3 (ThermoFisher Scientific, San Jose, CA, USA). The MS/MS data were searched by three search engines (MyriMatch v2.2.8634, X!Tandem v2015.04.01.1 and MS-GF+v2016.06.29) through IPeak against the Uniprot database of *Ricinus communis* with the following parameters: carbamidomethyl (C), iTRAQ 8 plex (N-term), and iTRAQ 8 plex (K) as fixed modification; oxidation (M) and iTRAQ 8 plex (Y) as variable modifications; full cleavage by trypsin with one missed cleavage permitted; peptide mass tolerance at ±10 ppm and fragment mass tolerance ±0.05 Da [82]. After MS/MS searching, IQuant was used for protein quantification with VSN normalization [83]. The false discovery rate (FDR) at both peptide spectrum matching (PSM) and protein levels was set ≤1%. Proteins with at least one unique peptide were used for quantification. T-tests were used to calculate a p-value for each protein. Only ratios with fold change > 1.2 and p-value < 0.05 were considered as significant.

4.7. Bioinformatics Analysis

Functional annotations of DAPS were conducted using GO (http://www.geneontology.org). The COG (http://www.ncbi.nlm.nih.gov/COG/) database was used for the functional classification of DAPS. The Kyoto Encyclopedia of Genes and Genomes (KEGG) (http://www.genome.jp/kegg/) was used to predict the main metabolic pathways and biochemical signal transduction pathways that the DAPS were involved in. A p-value < 0.05 was used as the threshold to determine the significant enrichments of GO and KEGG pathways.

4.8. Enzyme-Linked Immunosorbent Assays

Cold-and control-treated imbibed seeds were ground into powder with liquid nitrogen. 5 g powder was placed in a stoppered flask with 25 mL of 60% aqueous methanol solution and 20 mL of petroleum ether and shaken for 10 min. The homogenate was filtrated by filter paper and the filtrate was collected in the separatory funnel and released in the lower layer (60% aqueous methanol extract) after stratification. The extract was diluted to 30% of the final methanol concentration as the sample to be tested. The content of G6PDH, 6PGL, β-CT subunit of Het-ACCase, BC subunit of Het-ACCase, KASI, KASII, acetyl-CoA, ATP, and UFA were measured using reagent kits from the Jianglai Biotechnology Company Limited of Shanghai, China (Cat.No. JL-F22767, JL-F46429, JL-F46473,

JL-F46442, JL-F46461, JL-F46461, JL-F14042, JL13631, JL45670). Briefly, the purified antibody was used to coat microtiter plate wells to make a solid-phase antibody. 50 μL of different standard concentrations (kit available) was added to the coated wells. Blank and testing sample wells were set separately. 40 μL of sample dilution and 10 μL of testing samples were added to the testing sample wells and gently mixed. 100 μL of HRP-conjugate reagents were added to every well except the blank control well, and incubated for 60 min at 37 °C. After incubation, liquid in the well was discarded. The well was washed using a wash solution 5 times and dried. 50 μL chromogen solution A and 50 μL chromogen B were added to each well under dark incubation for 15 min at 37 °C. The reaction was terminated by the addition of a stop solution and color change was measured using a microplate reader at a wavelength of 450 nm.

Sucrose content was determined by an assay kit from Nanjing Jiancheng Bioengineering Institute, Jiangsu, China (A099-1). Briefly, a homogenate of cold-and control-treated imbibed seeds were prepared by the same method as ELISA. The content of sucrose was assayed by measuring the product of sucrose hydrolysis at 290 nm. Data are presented as means of three biological replicates ±SD. Results were considered statistically significant at p value < 0.05 using the Student's t-test.

4.9. RNA Extraction and qRT-PCR Analysis

Total RNA was extracted from the seeds imbibed at 4 °C and 30 °C for 12 h using a TaKaRa MiniBEST Plant RNA Extraction Kit according to the manufacturer's instructions (Code No. 9769). RNA integrity were measured using the Agilent 2100 Bioanalyzer (Agilent Technologies, Palo Alto, CA) with RNA integrity number (RIN) >7. Total RNA was treated with DNaseI to remove genomic DNA contamination. cDNA was synthesized, starting with 200 ng of total RNA and using UEIris RT-PCR system for First-Strand cDNA Synthesis (US Everbright, Suzhou, China). qRT-PCR was carried on qTOWER2.2 (Analytik Jena, Jena, Germany) using 2 × SYBR Green qPCR Master Mix (US Everbright, Suzhou, China) according to the manufacturer's instructions. qRT-PCR was performed with 2 μL of each cDNA diluted at 1:5. Three independent biologically replicated experiments were set up with three technical replicates per experiment. All expression data were normalized against the expression levels of the *RcActin* (XM_002522148) and *RcGAPDH* (XM_002513282) applied as internal standards. The relative levels of eight genes were analyzed using the $2^{-\Delta\Delta Ct}$ method [84]. The primers were designed according to the corresponding nucleotide sequences of castor bean in phytozome v10.2 (http://phytozome.jgj.doe.gov/pz/portal.html). Gene-specific and internal genes primers for qRT-PCR analysis are presented in Table S1. The values were means ±SD. Data were considered statistically significant at p value < 0.05 using the Student's t-test.

5. Conclusions

Imbibition is a critical process during seed germination. To identify the DAPS that contribute to seed germination under cold stress, the changes of proteomic in 12 h early imbibed seeds under cold stress were investigated in this study. A total of 127 DAPS were identified; these DAPS were involved in carbohydrate and energy metabolism, translation, ribosomal structure and biogenesis, posttranslational modification, protein turnover, chaperones, lipid transport and metabolism, signal transduction etc. These processes can work cooperatively to establish the beneficial equilibrium of physiological and cellular homeostasis. This approach identified new proteins that were not previously known to be associated with cold stress response. Future work will focus on improving cold tolerance via overexpression of β-ketoacyl-acyl carrier protein synthase I and II, and investigating the relationship between accumulation of SnRK1 gamma subunit and seed germination under cold stress. In summary, our proteomics analysis of early imbibed seeds not only increased our understanding of molecular mechanisms associated with castor seed germination under cold stress but also laid the theoretical foundation for the creation of a new cold-tolerant castor bean, and provided a reference for the improvement of cold-tolerant traits in other important economic crops.

Supplementary Materials: Supplementary materials can be found at http://www.mdpi.com/1422-0067/20/2/355/s1.

Author Contributions: X.W. and J.Z. conceived and designed the experiments; X.W. and X.L. performed the experiments; X.W., M.L., L.Z. and Q.D. analyzed the data; and X.W. wrote the paper.

Funding: This work was supported by the National Natural Science Foundation of China (Grant No. 31701467, 31760399), Inner Mongolia Key Laboratory of Castor Breeding Fund (Grant No. MDK2017037) and the Doctoral Program Fund of Inner Mongolia University for Nationalities (Grant No. BS398).

Conflicts of Interest: The authors declare no conflict of interest.

Abbreviations

iTRAQ	Isobaric tag for relative and absolute quantification
DAPS	Differential abundance protein species
ELISA	Enzyme-linked immunosorbent assays
DEGs	Differentially expressed genes
2-DE	Two-dimensional electrophoresis
GO	Gene ontology
COG	Clusters of Orthologous Groups of proteins
KAS	β-ketoacyl-acyl carrier protein synthase
6PGL	6-phosphogluconolactonase
G6PDH	Glucose-6-phosphate 1-dehydrogenase
GPX	Glutathione peroxidase
LEA	Late embryogenesis abundant protein
SnRK1	SNF1-related protein kniase
IF5A	Eukaryotic translation initiation factor 5A
PP2C	Protein phosphatase 2c

References

1. Wang, X.Y.; Shan, X.H.; Wu, Y.; Su, S.Z.; Li, S.P.; Liu, H.K.; Han, J.Y.; Xue, C.M.; Yuan, Y.P. iTRAQ-based quantitative proteomic analysis reveals new metabolic pathways responding to chilling stress in maize leaves. *J. Proteom.* **2016**, *146*, 14–24. [CrossRef] [PubMed]
2. Chinnusamy, V.; Zhu, J.H.; Zhu, J.K. Cold stress regulation of gene expression in plants. *Trends Plant Sci.* **2007**, *12*, 444–451. [CrossRef]
3. Scholz, V.; da Silva, J.N. Prospects and risks of the use of castor oil as a fuel. *Biomass Bioenergy* **2008**, *32*, 95–100. [CrossRef]
4. Lima Da Silva, N.; Maciel, M.; Batistella, C.; Filho, R. Optimization of biodiesel production from castor oil. *Appl. Biochem. Biotechnol.* **2006**, *130*, 405–414. [CrossRef]
5. Jiang, X.J.; Wen, X.D. The effect of temperature on the germination rates in castor bean. *Seed* **2008**, *27*, 67–69.
6. Wang, Z.F.; Wang, J.F.; Bao, Y.M.; Wu, Y.Y.; Zhang, H.S. Quantitative trait loci controlling rice seed germination under salt stress. *Euphytica* **2011**, *178*, 297–307. [CrossRef]
7. Dong, K.; Zhen, S.M.; Cheng, Z.W.; Cao, H.; Ge, P.; Yan, Y.M. Proteomic analysis reveals key proteins and phosphoproteins upon seed germination of wheat (*Triticum maestivum* L.). *Front. Plant Sci.* **2015**, *6*, 1017. [CrossRef] [PubMed]
8. Bewley, J.D. Seed germination and dormancy. *Plant Cell* **1997**, *9*, 1055–1066. [CrossRef]
9. Pradet-Balade, B.; Boulme, F.; Beug, H.; Mullner, E.W.; Garcia-Sanz, J.A. Translation control: Bridging the gap between genomics and proteomics? *Trends Biochem. Sci.* **2001**, *26*, 225–229. [CrossRef]
10. Greenbaum, D.; Colangelo, C.; Williams, K.; Gerstein, M. Comparing protein abundance and mRNA expression levels on a genomic scale. *Genome Biol.* **2003**, *4*, 117. [CrossRef]
11. Kosová, K.; Vítámvás, P.; Prášil, I.T.; Renaut, J. Plant proteome changes under abiotic stress—Contribution of proteomics studies to understanding plant stress response. *J. Proteom.* **2011**, *74*, 1301–1322. [CrossRef] [PubMed]
12. Cheng, L.; Gao, X.; Li, S.; Shi, M.; Javeed, H.; Jing, X.M.; Yang, G.X.; He, G.Y. Proteomic analysis of soybean (*Glycine max* (L.) Meer.) seeds during imbibition at chilling temperature. *Mol. Breed.* **2010**, *26*, 1–17. [CrossRef]

13. Kosmala, A.; Bocian, A.; Rapacz, M.; Jurczyk, B.; Zwierzykowski, Z. Identification of leaf proteins differentially accumulated during cold acclimation between *Festuca pratensis* plants with distinct levels of frost tolerance. *J. Exp. Bot.* **2009**, *60*, 3595–3609. [CrossRef] [PubMed]

14. Xu, Y.; Zeng, X.; Wu, J.; Zhang, F.Q.; Li, C.X.; Jiang, J.J.; Wang, Y.P.; Sun, W.C. iTRAQ-based quantitative proteome revealed metabolic changes in winter turnip rape (*Brassica rapa* L.) under cold stress. *Int. J. Mol. Sci.* **2018**, *19*, 3346. [CrossRef]

15. Maltman, D.J.; Simon, W.J.; Wheeler, C.H.; Dunn, M.J.; Wait, R.; Slabas, A.R. Proteomic analysis of the endoplasmic reticulum from developing and germinating seed of castor (*Ricinus communis*). *Electrophoresis* **2002**, *23*, 626–639. [CrossRef]

16. Maltman, D.J.; Gadd, S.M.; Simon, W.J.; Slabas, A.R. Differential proteomic analysis of the endoplasmic reticulum from developing and germinating seeds of castor (*Ricinus communis*) identifies seed protein precursors as significant components of the endoplasmic reticulum. *Proteomics* **2007**, *7*, 1513–1528. [CrossRef] [PubMed]

17. Houston, N.L.; Hajduch, M.; Thelen, J.J. Quantitative proteomics of seed filling in castor: Comparison with soybean and rapeseed reveals differences between photosynthetic and nonphotosynthetic seed metabolism. *Plant Physiol.* **2009**, *151*, 857–868. [CrossRef] [PubMed]

18. Campos, F.A.P.; Nogueira, F.C.S.; Cardoso, K.C.; Costa, G.C.L.; Del Bem, L.E.V.; Domont, G.B.; Da Silva, M.J.; Moreira, R.C.; Soares, A.A.; Juca, T.L. Proteome analysis of castor bean seeds. *Pure Appl. Chem.* **2010**, *82*, 259–267. [CrossRef]

19. Nogueira, F.C.; Palmisano, G.; Soares, E.L.; Shah, M.; Soares, A.A.; Roepstorff, P.; Campos, F.A.; Domont, G.B. Proteomic profile of the nucellus of castor bean (*Ricinus communis* L.) seeds during development. *J. Proteom.* **2012**, *75*, 1933–1939. [CrossRef]

20. Nogueira, F.C.S.; Palmisano, G.; Schwämmle, V.; Soares, E.L.; Soares, A.A.; Roepstorff, P.; Domont, G.B.; Campos, F.A.P. Isotope labeling-based quantitative proteomics of developing seeds of castor oil seed (*Ricinus communis* L.). *J. Proteome Res.* **2013**, *12*, 5012–5024. [CrossRef]

21. Zhang, W.; Zhang, H.L.; Ning, L.Y.; Li, B.; Bao, M.Z. Quantitative proteomic analysis provides novel insights into cold stress response in petunia seedlings. *Front. Plant Sci.* **2016**, *7*, 136. [CrossRef]

22. Wang, X.; Wang, L.J.; Yan, X.C.; Wang, L.; Tan, M.L.; Geng, X.X.; Wei, W.H. Transcriptome analysis of the germinated seeds identifies low-temperature responsive genes involved in germination process in *Ricinus communis. Acta Physiol. Plant* **2016**, *38*, 6. [CrossRef]

23. Sano, N.; Permana, H.; Kumada, R.; Shinozaki, Y.; Tanabata, T.; Yamada, T.; Hirasawa, T.; Kanekatsu, M. Proteomic analysis of embryonic proteins synthesized from long-lived mRNAs during germination of rice seeds. *Plant Cell Physiol.* **2012**, *53*, 687–698. [CrossRef]

24. Wang, J.; Lan, P.; Gao, H.; Zheng, L.; Li, W.; Schmidt, W. Expression changes of ribosomal proteins in phosphate and iron-deficient *Arabidopsis* roots predict stress-specific alterations in ribosome composition. *BMC Genom.* **2013**, *14*, 783. [CrossRef]

25. Kim, K.Y.; Park, S.W.; Chung, Y.S.; Chung, C.H.; Kim, J.I.; Lee, J.H. Molecular cloning of low-temperature-inducible ribosomal proteins from soybean. *J. Exp. Bot.* **2004**, *55*, 1153–1155. [CrossRef]

26. Bukovnik, U.; Fu, J.; Bennett, M.; Prasad, P.V.; Ristic, Z. Heat tolerance and expression of protein synthesis elongation factors, EF-Tu and EF-1α, in spring wheat. *Funct. Plant Biol.* **2009**, *36*, 234–241. [CrossRef]

27. Cui, S.; Huang, F.; Wang, J.; Ma, X.; Cheng, Y.; Liu, Y.J. A proteomic analysis of cold stress responses in rice seedlings. *Proteomics* **2005**, *5*, 3162–3172. [CrossRef]

28. Lee, D.G.; Ahsan, N.; Lee, S.H.; Kang, K.Y.; Bahk, J.D.; Lee, I.J.; Lee, B.H. A proteomic approach in analyzing heat-responsive proteins in rice leaves. *Proteomics* **2007**, *7*, 3369–3383. [CrossRef]

29. Li, A.L.; Li, H.Y.; Jin, B.F.; Ye, Q.N.; Zhou, T.; Yu, X.D.; Pan, X.; Man, J.H.; He, K.; Yu, M.; et al. A novel eIF5A complex functions as a regulator of p53 and p53-dependent apoptosis. *J. Biol. Chem.* **2004**, *279*, 49251–49258. [CrossRef]

30. Hopkins, M.T.; Lampi, Y.; Wang, T.W.; Liu, Z.; Thompson, J.E. Eukaryotic translation initiation factor 5A is involved in pathogen-induced cell death and development of disease symptoms in *Arabidopsis. Plant Physiol.* **2008**, *148*, 479–489. [CrossRef] [PubMed]

31. Wang, T.W.; Lu, L.; Zhang, C.G.; Taylor, C.; Thompson, J.E. Pleiotropic effects of suppressing deoxyhypusine synthase expression in *Arabidopsis thaliana. Plant Mol. Biol.* **2003**, *52*, 1223–1235. [CrossRef]

32. Xu, J.; Zhang, B.; Jiang, C.; Ming, F. RceIF5A, encoding an eukaryotic translation initiation factor 5A in *Rosa chinensis*, can enhance thermotolerance, oxidative and osmotic stress resistance of *Arabidopsis thaliana*. *Plant Mol. Biol.* **2011**, *75*, 167–178. [CrossRef] [PubMed]

33. Wang, L.Q.; Xu, C.X.; Wang, C.; Wang, Y.C. Characterization of a eukaryotic translation initiation factor 5A homolog from *Tamarix androssowii* involved in plant abiotic stress tolerance. *BMC Plant Biol.* **2012**, *12*, 118. [CrossRef]

34. Anand, M.; Chakraburtty, K.; Marton, M.J.; Hinnebusch, A.G.; Kinzy, T.G. Functional interactions between yeast translation eukaryotic elongation factor (eEF)1A and eEF3. *J. Biol. Chem.* **2003**, *278*, 6985–6991. [CrossRef] [PubMed]

35. Shin, D.; Moon, S.J.; Park, S.R.; Kim, B.G.; Byun, M.O. Elongation factor 1α from *A. thaliana* functions as molecular chaperone and confers resistance to salt stress in yeast and plants. *Plant Sci.* **2009**, *177*, 156–160. [CrossRef]

36. Jeong, H.J.; Shin, J.S.; Ok, S.H. Barley DNA-binding methionine aminopeptidase, which changes the localization from the nucleus to the cytoplasm by low temperature, is involved in freezing tolerance. *Plant Sci.* **2011**, *180*, 53–60. [CrossRef]

37. Mazzucotellin, E.; Mastrangelo, A.M.; Crosatti, C.; Guerra, D.; Stanca, A.M.; Cattivelli, L. Abiotic stress response in plants: When post-transcriptional and post-translational regulations control transcription. *Plant Sci.* **2008**, *174*, 420–431. [CrossRef]

38. Hershko, A.; Ciechanover, A. The ubiquitin system. *Annu. Rev. Biochem.* **1998**, *67*, 425–479. [CrossRef]

39. Callis, J.; Vierstra, R.D. Protein degradation in signaling. *Curr. Opin. Plant Biol.* **2000**, *3*, 381–386. [CrossRef]

40. Zhou, G.A.; Chang, R.Z.; Qiu, L.J. Overexpression of soybean ubiquitin-conjugating enzyme gene *GmUBC2* confers enhanced drought and salt tolerance through modulating abiotic stress-responsive gene expression in *Arabidopsis*. *Plant Mol. Biol.* **2010**, *72*, 357–367. [CrossRef]

41. Chondrogianni, N.; Tzavelas, C.; Pemberton, A.J.; Nezis, I.P.; Rivett, A.J.; Gonos, E.S. Overexpression of proteasome 5 subunit increases the amount of assembled proteasome and confers ameliorated response to oxidative stress and higher survival rates. *J. Biol. Chem.* **2005**, *280*, 11840–11850. [CrossRef] [PubMed]

42. Parsell, D.A.; Lindquist, S. The function of heat-shock proteins in stress tolerance: Degradation and reactivation of damaged proteins. *Annu. Rev. Genet.* **1993**, *27*, 437–496. [CrossRef]

43. Wang, W.X.; Vinocur, B.; Shoseyov, O.; Altman, A. Role of plant heat-shock proteins and molecular chaperones in the abiotic stress response. *Trends Plant Sci.* **2004**, *9*, 244–252. [CrossRef] [PubMed]

44. Blokhina, O.; Virolainen, E.; Fagerstedt, K. Antioxidants, oxidative damage and oxygen deprivation stress: A review. *Ann. Bot.* **2003**, *91*, 179–194. [CrossRef] [PubMed]

45. Milla, M.A.R.; Maurer, A.; Huete, A.R.; Gustafson, J.P. Glutathione peroxidase genes in *Arabidopsis* are ubiquitous and regulated by abiotic stresses through diverse signaling pathways. *Plant J.* **2003**, *36*, 602–615. [CrossRef]

46. Navrot, N.; Collin, V.; Gualberto, J.; Gelhaye, E.; Hirasawa, M.; Rey, P.; Knaff, D.B.; Issakidis, E.; Jacquot, J.P.; Rouhier, N. Plant glutathione peroxidases are functional peroxiredoxins distributed in several subcellular compartments and regulated during biotic and abiotic stresses. *Plant Physiol.* **2006**, *142*, 1364–1379. [CrossRef] [PubMed]

47. Roxas, V.P.; Lodhi, S.A.; Garrett, D.K.; Mahan, J.R.; Allen, R.D. Stress tolerance in transgenic tobacco seedlings that overexpress glutathione S-transferase/glutathione peroxidase. *Plant Cell Physiol.* **2000**, *41*, 1229. [CrossRef] [PubMed]

48. Kosová, K.; Vítámvás, P.; Prášil, I.T. The role of dehydrins in plant response to cold. *Biol. Plant.* **2007**, *51*, 601–617. [CrossRef]

49. Tolleter, D.; Hincha, D.K.; Macherel, D.A. Mitochondrial late embryogenesis abundant protein stabilizes model membranes in the dry state. *Biochim. Biophys. Acta* **2010**, *1798*, 1926–1933. [CrossRef]

50. Kawamura, Y.; Uemura, M. Mass spectrometric approach for identifying putative plasma membrane proteins of *Arabidopsis* leaves associated with cold acclimation. *Plant J.* **2003**, *36*, 141–154. [CrossRef]

51. Lee, S.C.; Lee, M.Y.; Kim, S.J.; Jun, S.H.; An, G.; Kim, S.R. Characterization of an abiotic stress-inducible dehydrin gene, *OsDhn1*, in rice (*Oryza sativa* L.). *Mol. Cells* **2005**, *19*, 212–218. [PubMed]

52. Peng, Y.H.; Reyes, J.L.; Wei, H.; Yang, Y.; Karlson, D.; Covarrubias, A.A.; Krebs, S.L.; Fessehaie, A.; Arora, R. RcDhn5, a cold acclimation–responsive dehydrin from *Rhododendron catawbiense* rescues enzyme activity from dehydration effects in vitro and enhances freezing tolerance in RcDhn5-overexpressing *Arabidopsis* plants. *Physiol. Plant.* **2008**, *134*, 583–597. [CrossRef]

53. Yu, D.D.; Zhang, L.H.; Zhao, K.; Niu, R.X.; Zhai, H.; Zhang, J.X. *VaERD15*, a transcription factor gene associated with cold-tolerance in Chinese Wild *Vitis amurensis*. *Front. Plant Sci.* **2017**, *8*, 297. [CrossRef] [PubMed]

54. Nicolás, G.; Aldasoro, J.J. Activity of the pentose phosphate pathway and changes in nicotinamide nucleotide content during germination of seeds of *Cicer arietinum* L. *J. Exp. Bot.* **1979**, *30*, 1163–1170. [CrossRef]

55. Dennis, D.T.; Blakeley, S.D. Carbohydrate metabolism. In *Biochemistry & Molecular Biology of Plants*; Buchanan, B.B., Gruissem, W., Eds.; American Society of Plant Physiologists: Rockville, MD, USA, 2000; pp. 652–654.

56. Xiong, Y.Q.; Defraia, C.; Williams, D.; Zhang, X.D.; Mou, Z.L. Characterization of *Arabidopsis* 6-phosphogluconolactonase T-DNA insertion mutants reveals an essential role for the oxidative section of the plastidic pentose phosphate pathway in plant growth and development. *Plant Cell Physiol.* **2009**, *50*, 1277–1291. [CrossRef] [PubMed]

57. Miclet, E.; Stoven, V.; Michels, P.A.; Opperdoes, F.R.; Lallemand, J.Y.; Duffieux, F. NMR spectroscopic analysis of the first two steps of the pentose-phosphate pathway elucidates the role of 6-phosphogluconolactonase. *J. Biol. Chem.* **2001**, *276*, 34840–34846. [CrossRef] [PubMed]

58. Lin, Y.Z.; Lin, S.Z.; Guo, H.; Zhang, Z.Y.; Chen, X.Y. Functional analysis of *PsG6PDH*, a cytosolic glucose-6-phosphate dehydrogenase gene from *Populus suaveolens*, and its contribution to cold tolerance improvement in tobacco plants. *Biotechnol. Lett.* **2013**, *35*, 1509. [CrossRef] [PubMed]

59. Bonaventure, G.; Salas, J.J.; Pollard, M.R.; Ohlrogge, J.B. Disruption of the *FATB* gene in *Arabidopsis* demonstrates an essential role of saturated fatty acids in plant growth. *Plant Cell* **2003**, *15*, 1020–1033. [CrossRef]

60. Zheng, H.Q.; Rowland, O.; Kunst, L. Disruptions of the *Arabidopsis* enoyl-CoA reductase gene reveal an essential role for very-long-chain fatty acid synthesis in cell expansion during plant morphogenesis. *Plant Cell* **2005**, *17*, 1467–1481. [CrossRef] [PubMed]

61. Roudier, F.; Gissot, L.; Beaudoin, F.; Haslam, R.; Michaelson, L.; Marion, J.; Molino, D.; Lima, A.; Bach, L.; Morin, H.; et al. Very-long-chain fatty acids are involved in polar auxin transport and developmental patterning in *Arabidopsis*. *Plant Cell* **2010**, *22*, 364–375. [CrossRef] [PubMed]

62. Durrett, T.P.; Benning, C.; Ohlrogge, J. Plant triacylglycerols as feedstocks for the production of biofuels. *Plant J.* **2008**, *54*, 593–607. [CrossRef]

63. Dyer, J.M.; Stymne, S.; Green, A.G.; Carlsson, A.S. High value oil from plants. *Plant J.* **2008**, *54*, 640–655. [CrossRef] [PubMed]

64. Gornicki, P.; Haselkorn, R. Wheat acetyl-CoA carboxylase. *Plant Mol. Biol.* **1993**, *22*, 547–552. [CrossRef] [PubMed]

65. Liu, X.Y.; Li, B.; Yang, J.H.; Sui, N.; Yang, X.M.; Meng, Q.W. Overexpression of tomato chloroplast omega-3 fatty acid desaturase gene alleviates the photoinhibition of photosystems 2 and 1 under chilling stress. *Photosynthesis* **2008**, *46*, 185–192. [CrossRef]

66. Hardie, D.G.; Carling, D.; Carlson, M. The AMP-activated/SNF1 protein kinase subfamily: Metabolic sensors of the eukaryotic cell? *Annu. Rev. Biochem.* **1998**, *67*, 821–855. [CrossRef]

67. Rodrigues, A.; Adamo, M.; Crozet, P.; Margalha, L.; Confraria, A.; Martinho, C.; Elias, A.; Rabissi, A.; Lumbreras, V.; González-Guzmán, M.; et al. ABI1 and PP2CA phosphatases are negative regulators of Snf1-related protein kinase1 signaling in *Arabidopsis*. *Plant Cell* **2013**, *25*, 3871–3884. [CrossRef] [PubMed]

68. Purcell, P.C.; Smith, A.M.; Halford, N.G. Antisense expression of a sucrose non-fermenting-1-related protein kinase sequence in potato results in decreased expression of sucrose synthase in tubers and loss of sucrose-inducibility of sucrose synthase transcripts in leaves. *Plant J.* **1998**, *14*, 195–202. [CrossRef]

69. Rosnoblet, C.; Aubry, C.; Leprince, O.; Vu, B.L.; Rogniaux, H.; Buitink, J. The regulatory gamma subunit SNF4b of the sucrose non-fermenting-related kinase complex is involved in longevity and stachyose accumulation during maturation of *Medicago truncatula* seeds. *Plant J.* **2007**, *51*, 47–59. [CrossRef]

70. Bradford, K.J.; Downie, A.B.; Gee, O.H.; Alvarado, V.; Yang, H.; Dahal, P. Abscisic acid and gibberellin differentially regulate expression of genes of the SNF1-related kinase complex in tomato seeds. *Plant Physiol.* **2003**, *132*, 1560–1576. [CrossRef]

71. Stone, J.M.; Walker, J.C. Plant protein kinase families and signal transduction. *Plant Physiol.* **1995**, *108*, 451–457. [CrossRef]

72. Schenk, P.W.; Snaar-Jagalska, B.E. Signal perception and transduction: The role of protein kinases. *Biochim. Biophys. Acta* **1999**, *1449*, 1–24. [CrossRef]

73. Yoshida, T.; Nishimura, N.; Kitahata, N.; Kuromori, T.; Ito, T.; Asami, T.; Shinozaki, K.; Hirayama, T. ABA-hypersensitive germination 3 encodes a protein phosphatase 2C (AtPP2CA) that strongly regulates abscisic acid signaling during germination among *Arabidopsis* protein phosphatase 2Cs. *Plant Physiol.* **2006**, *140*, 115–126. [CrossRef] [PubMed]

74. Hu, X.L.; Liu, L.X.; Xiao, B.L.; Li, D.P.; Xing, X.; Kong, X.P.; Li, D.Q. Enhanced tolerance to low temperature in tobacco by over-expression of a new maize protein phosphatase 2C, ZmPP2C2. *J. Plant Physiol.* **2010**, *167*, 1307–1315. [CrossRef]

75. Li, W.Y.F.; Shao, G.; Lam, H.M. Ectopic expression of *GmPAP3* alleviates oxidative damage caused by salinity and osmotic stresses. *New Phytol.* **2010**, *178*, 80–91. [CrossRef]

76. Baenagonzález, E.; Sheen, J. Convergent energy and stress signaling. *Trends Plant Sci.* **2008**, *13*, 474. [CrossRef] [PubMed]

77. Chinnusamy, V.; Zhu, J.K.; Sunkar, R. Gene regulation during cold stress acclimation in plants. *Methods Mol. Biol.* **2010**, *639*, 39–55. [PubMed]

78. Moon, B.Y.; Higashi, S.; Gombos, Z.; Murata, N. Unsaturation of the membrane lipids of chloroplasts stabilizes the photosynthetic machinery against low-temperature photoinhibition in TG tobacco plants. *Proc. Natl. Acad. Sci. USA* **1995**, *92*, 6219–6233. [CrossRef] [PubMed]

79. Baena-González, E.; Rolland, F.; Thevelein, J.M.; Sheen, J. A central integrator of transcription networks in plant stress and energy signalling. *Nature* **2007**, *448*, 938. [CrossRef]

80. Li, J.Q.; Zhu, G.L.; Li, J.X.; Tian, F.D.; Zhang, C.H.; Wu, G.L.; Wang, J.W. Breeding of new castor variety Tongbi 5. *Inn. Mong. Agric. Sci. Technol.* **2004**, *1*, 10–11.

81. Auld, D.L.; Bettis, B.L.; Crock, J.E.; Kephart, D. Planting date and temperature effects on germination and seed yield of Chickpea. *Agron. J.* **1988**, *80*, 909–914. [CrossRef]

82. Wen, B.; Du, C.; Li, G.; Ghali, F.; Jones, A.R.; Käll, L.; Xu, S.; Zhou, R.; Ren, Z.; Feng, Q.; et al. IPeak: An open source tool to combine results from multiple MS/MS search engines. *Proteomics* **2015**, *15*, 2916–2920. [CrossRef] [PubMed]

83. Wen, B.; Zhou, R.; Feng, Q.; Wang, Q.; Wang, J.; Liu, S. IQuant: An automated pipeline for quantitative proteomics based on isobaric tags. *Proteomics* **2014**, *14*, 2280–2285. [CrossRef] [PubMed]

84. Livak, K.J.; Schmittgen, T.D. Analysis of relative gene expression data using real-time quantitative PCR and the $2^{-\Delta\Delta Ct}$ method. *Methods* **2001**, *25*, 402–408. [CrossRef] [PubMed]

International Journal of
Molecular Sciences

MDPI

Article

Quantitative Proteomic Analysis of the Response to Cold Stress in Jojoba, a Tropical Woody Crop

Fei Gao [1,2,*], Pengju Ma [1], Yingxin Wu [1], Yijun Zhou [1] and Genfa Zhang [2,3,*]

[1] College of Life and Environmental Sciences, Minzu University of China, Beijing 100081, China; s161088@muc.edu.cn (P.M.); 16051039@muc.edu.cn (Y.W.); zhouyijun@muc.edu.cn (Y.Z.)
[2] Beijing Key Laboratory of Gene Resource and Molecular Development, Beijing 100875, China
[3] College of Life Sciences, Beijing Normal University, Beijing 100875, China
* Correspondence: gaofei@muc.edu.cn (F.G.); gfzh@bnu.edu.cn (G.Z.); Tel.: +86-10-6893-2633 (F.G.); +86-10-5880-9453 (G.Z.)

Received: 12 September 2018; Accepted: 3 January 2019; Published: 9 January 2019

Abstract: Jojoba (*Simmondsia chinensis*) is a semi-arid, oil-producing industrial crop that have been widely cultivated in tropical arid region. Low temperature is one of the major environmental stress that impair jojoba's growth, development and yield and limit introduction of jojoba in the vast temperate arid areas. To get insight into the molecular mechanisms of the cold stress response of jojoba, a combined physiological and quantitative proteomic analysis was conducted. Under cold stress, the photosynthesis was repressed, the level of malondialdehyde (MDA), relative electrolyte leakage (REL), soluble sugars, superoxide dismutase (SOD) and phenylalanine ammonia-lyase (PAL) were increased in jojoba leaves. Of the 2821 proteins whose abundance were determined, a total of 109 differentially accumulated proteins (DAPs) were found and quantitative real time PCR (qRT-PCR) analysis of the coding genes for 7 randomly selected DAPs were performed for validation. The identified DAPs were involved in various physiological processes. Functional classification analysis revealed that photosynthesis, adjustment of cytoskeleton and cell wall, lipid metabolism and transport, reactive oxygen species (ROS) scavenging and carbohydrate metabolism were closely associated with the cold stress response. Some cold-induced proteins, such as cold-regulated 47 (COR47), staurosporin and temperature sensitive 3-like a (STT3a), phytyl ester synthase 1 (PES1) and copper/zinc superoxide dismutase 1, might play important roles in cold acclimation in jojoba seedlings. Our work provided important data to understand the plant response to the cold stress in tropical woody crops.

Keywords: *Simmondsia chinensis*; cold stress; proteomics; leaf; iTRAQ

1. Introduction

Jojoba, *Simmondsia chinensis* (link) Schneider, also called wild hazel, deer nut, oat nut and coffeeberry, is an important and unique oil crop. The importance lies in the crushing oil of its seeds—jojoba oil has a wide range of commercial applications, including cosmetic formulations, food products and aerospace lubricants [1]. The composition and physical properties of the oil extracted from Jojoba seeds are similar to those of sperm oil and thus jojoba oil is a promising alternative to the threatened sperm whale oil [2]. Its uniqueness lies in two aspects, on the one hand, jojoba oil is a kind of vegetable oils with unique physical property and no other vegetable oil has physical properties comparable to jojoba oil. On the other hand, jojoba is a dryland crop and jojoba can be grown in deserts and various arid land areas without competing with common crops for farmland. Jojoba exhibit extremely high level of tolerance to drought and high temperature stresses and jojoba is proposed to have the ability to curb desert expansion around the world [3].

Jojoba is a desert shrub native to the semi-arid region of the Sonoran desert at the junction of Mexico and USA. Since the discovery of the fine properties of jojoba, has been successfully introduced into tropical and subtropical regions of many other countries, such as Australia, India, Egypt and China [4]. Although Jojoba has high tolerance to drought and high temperature, it is sensitive to cold stress. Hindered by the low tolerance to low temperature stress, jojoba is difficult to grow in temperate zones. Especially, although jojoba has been successfully introduced in parts of Yunnan and Sichuan province, China, many introduction studies in temperate regions of China like Henan province have failed [5]. It is necessary to analyze the physiological and biochemical response of jojoba to the cold stress and to investigate the response of jojoba to cold stress at the molecular level.

Low temperature is one of the key environmental cues that negatively affect plant growth and development and limit the geographic distribution area of plants. To understand the plant response to low temperature stress, researchers have conducted a number of physiological, biochemical and molecular biological studies [6]. Through these results, we learned that, upon perception of the low temperature signal in plants, the stress signal is transmitted downstream to activate many transcription factors mediating stress tolerance and modulate the expression levels of many cold-responsive genes, finally leading to adjustment of a large number of biological processes, including photosynthesis, signaling, transcription, metabolism, cell wall modification and stress response [7]. However, most of the studies on plant responses to cold stress were conducted in model plants and common crops such as Arabidopsis [8], rice [9] and wheat [10], no systematic analysis of the cold stress response in jojoba was reported by far, despite its importance as a unique semi-arid, oil-producing industrial crop.

Since proteins are the key players in the majority of cellular biological processes, proteomics techniques have been the powerful tools for detection of the quantitative alterations in protein abundance in plant response to environmental stress. The classical proteomics approach was two-dimensional gel electrophoresis (2-DE) coupled with mass spectrometry (MS) identification. With the rapid development of quantitative MS, the gel-based proteomic techniques are gradually giving way to some newly-developed technologies, for example, stable isotope labeled quantitative proteomics methods such as the isobaric tags for relative and absolute quantitation (iTRAQ) labeling technique. iTRAQ coupled to liquid chromatography-quadrupole mass spectrometry (LC-MS/MS) represents an efficient proteomic approach for the fast identification and accurate quantification of the high complexity protein mixture [11] and is currently being widely used for the quantitative comparative analysis of plant proteomes to various environmental stresses [12–15].

In the present study, the physiological and proteomic responses of jojoba to cold stress were investigated using iTRAQ-coupled LC-MS/MS technique. This study will reveal how leaf proteins and their related pathways were regulated for jojoba's response to cold stress, our study can also identify the candidate proteins which play key role in cold acclimation in jojoba seedlings, which should facilitate the understanding of the low temperature stress response in jojoba at the molecular level.

2. Results

2.1. Physiological Response of Jojoba Seedlings to Cold Stress

To investigate the physiological changes in jojoba leaves exposed to cold condition, the jojoba seedlings were treated with non-lethal cold treatment and several physiological and biochemical parameters were measured. Firstly, as expected, the physiological status of the jojoba was affected by cold stress and after cold treatment, the color of jojoba leaves changed from green to gray-green (Figure S1). The retarded growth typically induced by cold stress might be associated to the impaired photosynthesis in jojoba seedlings under cold stress conditions (Figure 1) and change of leaf color may result from the decreased chlorophyll content in jojoba leaves (Figure 2a).

Cold stress is expected to promote the membrane peroxidation, resulting in the elevated level of plasma membrane permeability. Malondialdehyde (MDA) and relative electrolyte leakage (REL) can be used to evaluate the plasma membrane lipid peroxidation and integrity, respectively. In the present study, both MDA content and REL level in jojoba leaves were increased significantly under cold stress (Figure 2b,c), indicating that the treatment regimen we used caused plasma membrane damage in jojoba leaf. Osmotic homeostasis may be disturbed under cold stress and we determined the cold stress induced changes in soluble sugars and proline in jojoba leaves (Figure 2d,e). The levels of soluble sugars and proline were increased significantly under cold stress. We speculated that the accumulation of soluble sugars and proline in jojoba leaves probably help to maintain osmotic balance during adaptation to the cold stress.

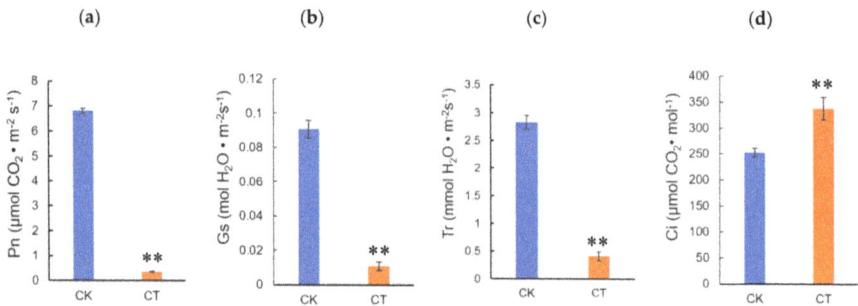

Figure 1. Cold stress-induced changes in photosynthesis related parameters in jojoba leaves from cold stress-treated group (CT) and control group (CK). (**a**) net photosynthesis rate (Pn); (**b**) stomatal conductance (Gs); (**c**) transpiration rate (Tr); (**d**) intercellular carbon dioxide concentration (Ci). Data were represented as means $\pm$ SD from five biological replicates (* $p < 0.05$, ** $p < 0.01$).

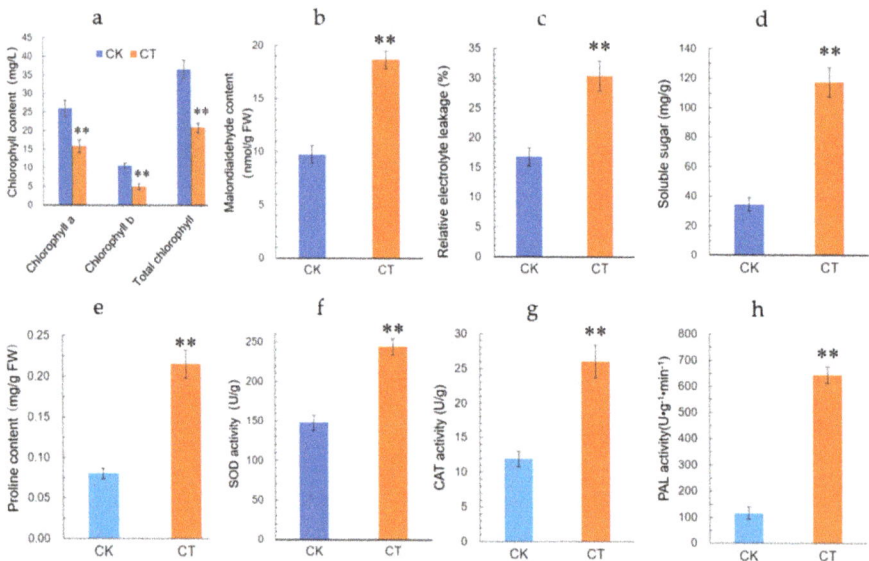

Figure 2. Cold stress-induced alterations of physiological parameters in jojoba leaves from cold stress-treated group (CT) and control group (CK). (**a**) chlorophyll content; (**b**) malondialdehyde (MDA); (**c**) relative electrolyte leakage (REL); (**d**) content of soluble sugars; (**e**) proline content; (**f**) activity of superoxide dismutase (SOD); (**g**) activity of catalase (CAT); (**h**) activity of phenylalanine ammonia-lyase (PAL). Data were represented as means $\pm$ SD from five biological replicates (* $p < 0.05$, ** $p < 0.01$).

Reactive oxygen species (ROS) always accumulated in stressed plants and the activities of antioxidant enzymes will be regulated correspondingly. In cold-stressed jojoba leaves, the activities of superoxide dismutase (SOD) and catalase (CAT) were up-regulated (Figure 2f,g) and these antioxidant enzymes probably contribute to ROS scavenging in cold-stressed jojoba leaves. In addition, the activity of phenylalanine ammonia-lyase (PAL), a key enzyme catalyzes the first metabolic step from primary metabolism to the secondary phenylpropanoid metabolism, was observed to up-regulated in cold-stressed jojoba leaves (Figure 2h).

2.2. iTRAQ Analysis and Identification of Differentially Accumulated Proteins

To investigate the proteomic changes associated with cold stress exposure in leaves of jojoba seedlings, iTRAQ analysis was conducted to identify the differentially accumulated proteins (DAPs) between the control and cold-treated plants. High-resolution LC–MS/MS was employed to detect and quantitate proteins in the jojoba leaves. The protein concentration of protein samples was measured by BCA method and the quality of each protein samples were evaluated by (polyacrylamide gel electrophoresis) SDS-PAGE analysis (Figure S2).

After labeling, the combined iTRAQ labeled peptides were fractionated by strong cation exchange (SCX) chromatography (Figure S3). The mass spectrometry proteomic data of the present study have been deposited in the PRIDE PRoteomics IDEntifications (PRIDE) database under the database identifier PXD007063.

A total of 23,422 unique peptides (FDR $\leq$ 0.01) were obtained and 2821 proteins were ultimately identified. The distribution of peptide number is shown in Figure 3a and all of the identified proteins having at least two peptides. The predicted molecular weights and isoelectric points (pIs) of the various identified proteins also showed high degrees of variation (Figure 3b,c), with molecular weights ranging from 10.3 to 254.6 kDa with a median of 37.8 kDa and pIs ranging from 3.95 to 12.06 with a median of 6.98. Moreover, most of the identified proteins have good peptide coverage (Figure 3d).

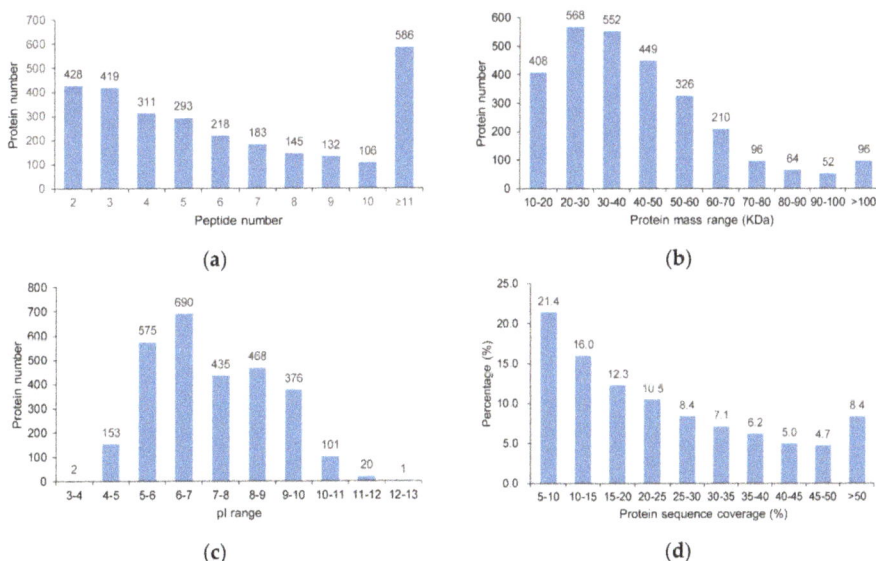

Figure 3. Characteristics of the identified unique proteins in jojoba leaf samples. (**a**) Unique peptide number distribution; (**b**) Protein mass distribution; (**c**) Protein isoelectric point distribution; (**d**) Peptide coverage of the identified proteins.

Of all the detected proteins, 2821 appear in each replicate of all samples and the relative quantifications of these proteins were used for further analyses. Statistical t-test analysis was used to identify the candidate proteins that are involved in the cold stress response in jojoba leaves. Of the 2821 proteins that were quantitated, a total of 109 unique proteins showed differential accumulation pattern (Table S2). Among these DAPs, 31 were up-regulated under cold stress, while the other 78 DAPs were down-regulated.

2.3. Functional Annotation and Classification of the Differentially Accumulated Proteins

To understand the biological roles of the DAPs in response to cold stress in jojoba leaves, we annotated the DAPs by the enrichment analysis in the Gene Ontology (GO) function term and the Kyoto Encyclopedia of Genes and Genomes (KEGG) pathway.

The amino acid sequences of the 109 DAPs were extracted from customized jojoba protein database based on their ID, then blastp algorithm was performed against the GO and KEGG databases. A total of 1215 GO terms and 38 KEGG terms were identified with a *p* value < 0.05. The top 3 categories of Biological Process terms were metabolic process, cellular process and single-organism process, the top 3 class of cellular component terms were cell, cell part and organelle and the top 3 categories of molecular function terms were catalytic activity, binding and structural molecule activity (Figure 4). The major KEGG pathways included biosynthesis of secondary metabolites (9 DAPs), Carbon fixation in photosynthetic organisms (6 DAPs), phagosome (7 DAPs), Ribosome (5 DAPs), oxidative phosphorylation (4 DAPs), protein processing in endoplasmic reticulum (3 DAPs) and phenylpropanoid biosynthesis (3 DAPs).

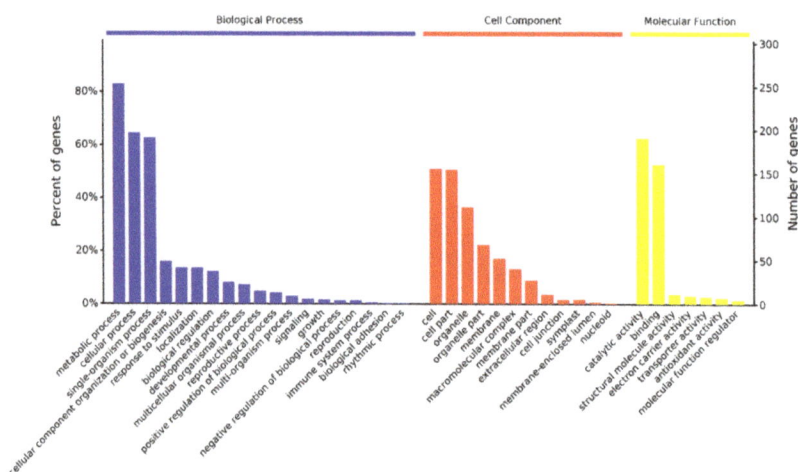

Figure 4. GO functional classification of the identified DAPs.

GO enrichment analyses of the DAPs using agriGO [16] revealed 57 enriched GO terms. These enriched GO terms were associated with various biological processes, including response to abiotic stimulus, phenylpropanoid biosynthetic process and carbohydrate metabolic process. The overrepresented GO cellular component terms included cytoplasm, chloroplast and ribosome. The top 3 enriched GO molecular function terms were catalytic activity, lyase activity and structural molecule activity.

Although GO and KEGG analysis can provide similar and overlapping results, integrating these results will help to reveal more accurately the biological processes represented by the DAPs and their biological significance. All DAPs were also annotated by aligning to Arabidopsis protein database (TAIR10) and Swiss-Prot database. Based on the annotation results, together with results of the GO and KEGG analyses, the DAPs were classified into 14 categories according to their putative biological functions, i.e., photosynthesis, cytoskeleton and cell wall, protein synthesis, folding and degradation, lipid metabolism and transport, stress response and defense, signal transduction, RNA splicing and transport, vesicle transport, carbohydrate metabolism, transmembrane transport, ROS scavenging, secondary metabolism and miscellaneous and unknown proteins (Figure 5 and Table S2). Their possible functions in cold stress signaling and response will be discussed later.

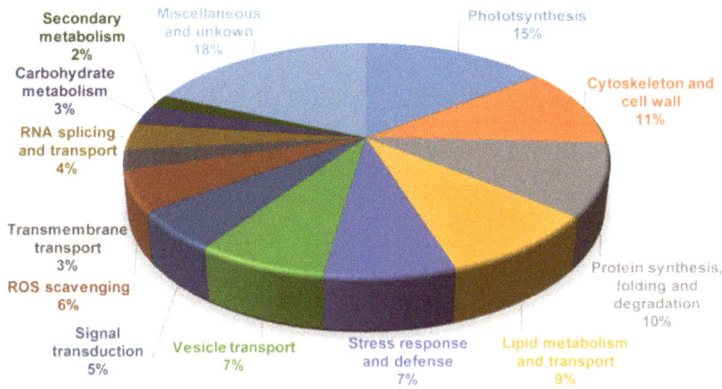

Figure 5. Functional categorization of the differentially accumulated proteins in jojoba leaves under cold stress.

2.4. Gene Expression Analysis of the Cold Stress Responsive Proteins

To validate the results of the quantitative proteomic analysis, 7 DAP coding genes were randomly selected for quantitative real time PCR (qRT-PCR) analysis (Figure 6). The expression levels of most of these genes exhibited the same trend with the protein abundance of the corresponding DAPs. However, the expression level of two genes (c89788_g1 and c75260_g1) showed the opposite change pattern with the abundance of their corresponding proteins. The discrepancy between the transcription level of the DAPs and the abundance of the corresponding proteins have been reported in previous studies [13] and this difference probably resulted from posttranslational modifications of proteins under cold stress, such as protein phosphorylation.

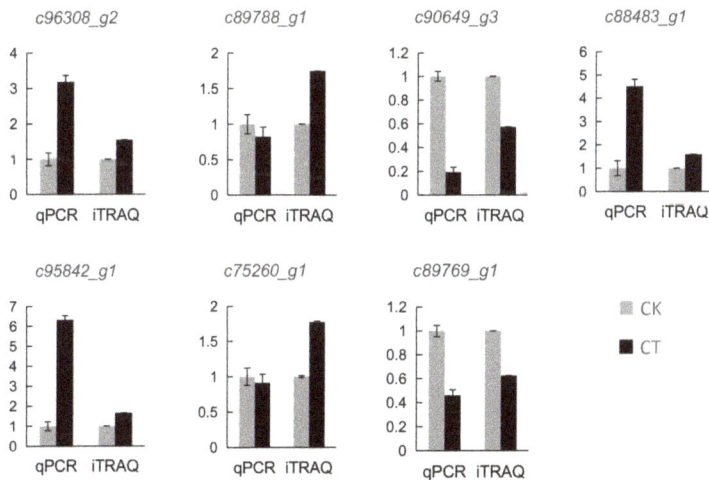

Figure 6. Gene expression analysis of the DAPs and comparison with the change pattern at the protein level revealed by iTRAQ analysis. Values were represented as means ± SD from three biological replicates. CK, control group; CT, cold-treated group.

2.5. Molecular Network Involved in Cold Stress Response in Jojoba Leaves

To reveal the interaction networks associated with the cold stress response in jojoba leaves, the protein-protein interaction (PPI) networks were constructed using the STRING protein-protein interaction database (Figure 7). Due to the lack of protein interaction data of jojoba and its closely related species, we used the homologous proteins in *Arabidopsis thaliana* to construct the protein interaction networks. For ease of understanding, the names of the DAPs were represented by the names or the locus numbers of the homologous proteins in Arabidopsis (http://www.arabidopsis.org) in the PPI map.

The largest network (Figure 7a) consists of 11 DAPs associated with proteins synthesis and folding, suggesting the cold stress significantly affected the protein synthesis in jojoba. The second largest network (Figure 7b) consisted of 4 proteins and most of them were related to the mitochondrial respiratory chain. The other subnetworks were associated with photosynthesis (Figure 7c), cell wall (Figure 7d) and transmembrane transport (Figure 7e). In sum, most of the cold stress-regulated biological processes identified via functional annotation and classification analyses were also highlighted in the PPI map.

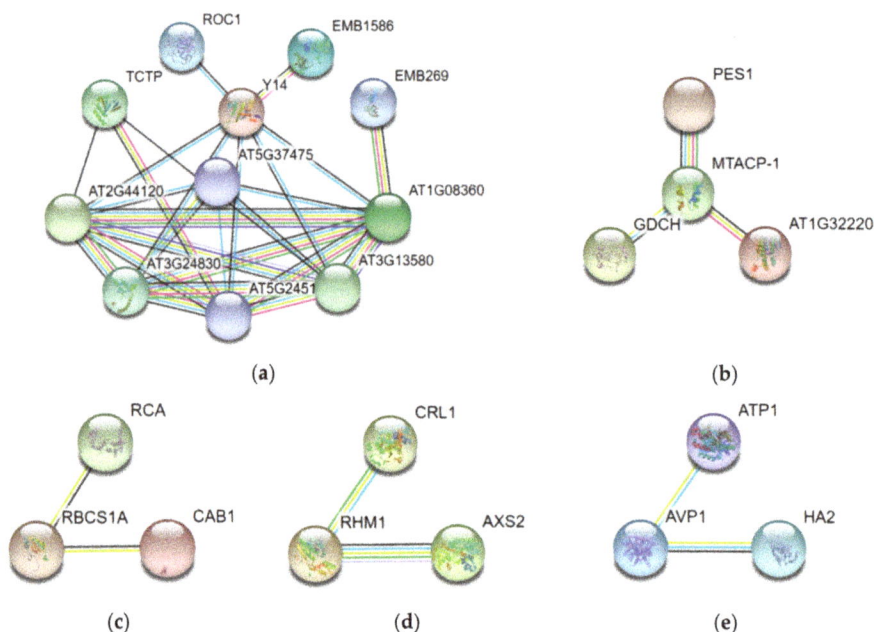

Figure 7. The protein-protein interaction (PPI) network of the differentially accumulated proteins (DAPs) in jojoba leaves under cold stress. Edge color represents protein-protein associations among different proteins: blue for known interactions from curated database, black for predicted interactions base on co-expression, pink for experimentally determined known interactions, green for predicted interactions base on gene neighborhood, red for predicted interactions base on gene fusions, dark blue for predicted interactions base on gene co-occurrence, yellow-green for predicted interactions base on text mining and light blue for predicted interactions base on protein homology. The largest network (**a**) was mainly associated with proteins synthesis and folding, the second largest network (**b**) was related to the mitochondrial respiratory chain. The other subnetworks were associated with photosynthesis (**c**), cell wall (**d**) and transmembrane transport (**e**).

3. Discussion

Our previous EST analysis had identified several candidate genes which may be involved in the water-deficient stress response in jojoba plants [17]. In the present proteomic analysis, a large number of cold stress-responsive proteins were identified in jojoba leaves. As expected, some well-known stress-inducible proteins were found, such as copper/zinc superoxide dismutase 1 (CSD1) and cold-regulated 47 (COR47). Some cold-responsive proteins reported previously in other plant species were also presented in the list of the DAPs, including ferredoxin 3 (FD3) and PHE ammonia lyase 1 (PAL1) [18]. These results support the reliability and robustness of the iTRAQ technology in investigating the plant response to environmental stress. Our data showed that several proteins were up-regulated significantly under cold stress and some of them might play important roles in the response to cold stress in jojoba. The possible biological functions of these DAPs in cold stress adaptation are further discussed below.

3.1. Proteins Involved in Stress Signal Transduction

In plant cells, perception of extracellular stimuli was mediated by the plasma membrane receptors and transduced by signaling pathway. In the present study, several components in Ca^{2+} and abscisic acid (ABA) signaling were regulated under cold stress, highlighting their pivotal roles in the jojoba's response to cold stress.

Ca^{2+} plays an essential role in plant cells in response to environmental stimuli as a second messenger and Ca^{2+} concentration has been found to increase in response to cold stress [19]. In our study, two calcium-binding proteins, i.e., calcium-binding EF-hand family protein and annexin 4, were found to be down-regulated under cold stress. Although calcium ion has been demonstrated to play an important role in the low temperature perception and signaling, in the present study, we can still find some components of calcium signaling pathway changes in abundance after 7 days of cold stress, indicating that the calcium ion signaling pathway may also be involved in low temperature adaptation in jojoba.

ABA signaling plays important role in stress response in plant and as expected, several DAPs involved ABA signaling were identified, including Serine/threonine-protein kinase GRIK2 (GRIK2), CPCK2 (chloroplast localized subunit of casein kinase 4) and annexin 4. SNF1-RELATED PROTEIN KINASE 1.1 (SnRK1.1) is a key component in abscisic acid-activated signaling pathway and Arabidopsis GRIK1 specifically activates SnRK1.1 by phosphorylation of its activation-loop [20]. Annexin 4 has been shown to play a vital role in abscisic acid signal transduction in Arabidopsis in a Ca^{2+}-dependent manner [21].

In addition, several DAPs involved synthesis and signaling of other phytohormones were also identified. Of them, ethylene-forming enzyme (EFE) is an enzyme involved in the ethylene biosynthesis and EXORDIUM like 5 (EXL5) is involved in brassinosteroid-dependent regulation of growth and development [22]. These data indicated that multiple phytohormone signaling pathways were adjusted in jojoba to adapt to the cold condition.

3.2. Proteins Involved in Photosynthesis and Carbohydrate Metabolism

Photosynthesis is greatly inhibited by low temperature in various plant species, especially for cold sensitive tropical crops such as jojoba. We observed that the physiological status of jojoba seedlings was affected by the cold stress treatment. The effect might relate to the decrease of the net photosynthesis as revealed by photosynthetic performance measurement (Figure 1).

Sixteen DAPs involved in photosynthesis were regulated by cold stress. In line with the impaired photosynthesis, most of the photosynthesis related DAPs were down-regulated in abundance, including three ribulose bisphosphate carboxylase small chain proteins and a Ribulose bisphosphate carboxylase/oxygenase activase. The only three up-regulated DAPs were FD3, NADP-malic enzyme 4 (NADP-ME4) and glyceraldehyde-3-phosphate dehydrogenase of plastid 1 (GAPCP-1). Ferredoxins

Int. J. Mol. Sci. **2019**, *20*, 243

are small, soluble iron-sulfur proteins that deliver electrons in many metabolic reactions and the chloroplast localized ferredoxins mainly function as electron transfer proteins to transfer reducing equivalents from photosystem I (PSI) to NADPH during linear electron flow (LEF). Ferredoxins were down or up-regulated under abiotic stress in several plant species such as rice and maize [18] and expression of a sweet pepper ferredoxin enhanced the tolerance to heat stress in *Arabidopsis thaliana* [23]. In the present study, the up-regulation of FD3 may help to prevent photo-oxidative damage under cold stress through cyclic electron flow (CEF) in jojoba leaves.

We observed significant increase in soluble sugars and proline level in jojoba leaves under cold stress. These molecules not only function as osmoprotectants but also protect the membrane via the interaction with the lipid biolayer and high levels of sugars inhibit photosynthesis in plant under cold stress [24]. Considered that the photosynthesis is greatly inhibited by cold stress, the accumulation of soluble sugars probably resulted from enhanced starch degradation, which was consistent with observations in tea plant [25]. As expected, many enzymes affecting sugar content were regulated under cold stress and these enzymes included a trehalose-6-phosphatase synthase s7, which is involved in trehalose biosynthesis [26] and two enzymes involved in starch synthesis, glucose-1-phosphate adenylyltransferase family protein (AGPase large subunit 3, APL3) and granule bound starch synthase 1) (GBSS1) (Table S2). ADP-Glucose pyrophosphorylase (AGP) catalyzes the rate limiting step in starch biosynthesis and AGPase large subunit 1 (APL1) and AGPase small subunit (APS1) are abundant in photosynthetic tissues and play the dominant role in leaves. APL3 is a large subunit isoform of AGP and present mainly in root [27]. The up-regulation of APL3 in cold-stressed jojoba leaves suggested that APL3 might also play a role in starch synthesis in leaves when the plants were exposed to cold stress. GBSS1 is the only starch synthase isoform required for amylose synthesis in chloroplast [28] and the up-regulation of GBSS1 in cold-stressed jojoba indicates that the proportion of amylose in starch may increase in cold-stressed jojoba leaves. In brief, our data showed that reorganization of starch metabolism was an essential process for jojoba to survive under low temperature conditions.

3.3. Proteins Involved in ROS Scavenging

Dysfunction of the photosynthetic apparatus under cold conditions exposes the plant to photoinhibition and can lead to elevated levels of ROS. In the present study, 6 DAPs involved in ROS scavenging were identified. Glutathione (GSH) plays key role in cell redox homeostasis. Three of these DAPs were involved in regulating GSH concentration, including GLYOXALASEI 6 (GLYI6) and two glutathione *S*-transferase family proteins. These results indicated that GSH metabolism was adjusted in jojoba leaf cells upon cold stress. Among these DAPs, glyoxalase Is (GLYIs) are one of the two groups of enzymes forming glyoxalase pathway and glyoxalase pathway has been shown to play an important role in stress tolerance. GLYI uses GSH as a cofactor for the detoxification of methylglyoxal (MG), high level of which is toxic to cells [29].

CSD1 were found to be up-regulated significantly in jojoba leaves under cold stress. In line with the change in protein abundance, the superoxide dismutase activity was elevated in cold stress jojoba leaves (Figure 2f). CSD1 encodes a cytosolic copper/zinc superoxide dismutase and its expression is negatively regulated by miR398. SODs catalyze the dismutation of superoxide into oxygen and hydrogen peroxide, constitute the first line of defense against ROS in cell. CSD1 has been observed to up-regulated under stressful conditions in many plant species [18,30] and transgenic plants that express CSD1 have shown enhanced tolerance to multiple stresses [30]. The up-regulation of CSD1 might contribute to cold acclimation in jojoba by repressing the elevation of ROS level.

3.4. Proteins Involved in Stress Response and Defense

Eight proteins involved in stress response and defense were identified as DAPs in the present study. Several proteins of these category, i.e., COR47, cystatin B, ARABIDOPSIS THALIANA KUNITZ TRYPSIN INHIBITOR 5 (ATKTI5), MLP-like protein 34 (MLP34), were frequently identified as differentially accumulated proteins under stressful conditions in previous proteomic studies in other plant species [18].

It is noteworthy that almost all DAPs involved in stress response and defense were down-regulated under cold stress in jojoba leaves and the only cold-induced DAPs in this category is COR47. Dehydrin (DHN) is a large family of proteins present in plants and DHNs are produced in response to environmental stresses. COR47 is one of DHNs that accumulate during the abiotic stress such as drought, salinity, freezing, or by treatment with ABA. COR47 is one the principal DHNs that accumulate under low temperature stress in *A. thaliana* and overexpression of COR47 improved cold stress tolerance of *A. thaliana* seedling [31]. COR47 is regarded as the marker genes of CBF/DREB pathway during cold acclimation, thus the up-regulation of COR47 under cold exposure probably play important role in cold adaptation of jojoba seedlings.

3.5. Proteins Involved in Cell Wall Modification and Osmotic Homeostasis

Cold stress was previously reported to affect the cell walls in pea seedlings [32] and cell wall modification is essential for plant acclimation to environmental stresses [33]. As expected, several enzymes involved in cell wall modification were differentially accumulated in cold-stressed jojoba leaves. Among these DAPs, pectin methylesterase CGR2 (CGR2) functions in the modification of cell walls via methylesterification of cell wall pectin [34]. Lignin is an important component of cell walls and several genes involved in lignin biosynthesis, i.e., PAL1, CCR (cinnamoyl CoA: NADP oxidoreductase)-LIKE 1 and a class III peroxidases, peroxidase 52, were identified as DAPs, indicating lignin synthesis was regulated under cold stress. Among these DAPs, PAL1 was up-regulated significantly and the PAL enzyme activity measurement validated the up-regulation of PAL at the protein level. PAL1 catalyzes the first step in the phenylpropanoid pathway [35] and another important enzyme in the phenylpropanoid pathway, flavonol synthase 1 (FLS1) was also up-regulated in cold-stressed jojoba leaves. In addition to act as the precursors for lignin synthesis, derivatives of phenylpropanoid pathway have various biological functions in plants, for example, some flavonoids function as protectants against oxidative stress induced by abiotic stress or pathogen attack [36].

Cold stress can induce inhibition of water uptake and indirectly resulted in osmotic stress in cells [37], thus, maintenance of the cell's osmotic potential under cold conditions is one of the major challenges for plant growth and development. In the present study, the increased soluble sugars in cold-stressed jojoba leaves can help the cell to lower water potential in cytoplasmic matric (Figure 2d). At the same time, several aquaporins and ion transporters were found to be differentially accumulated under cold stress and these DAPs included plasma membrane intrinsic protein 3 (PIP3), inorganic H pyrophosphatase family protein AVP1 (AVP1) and PLASMA MEMBRANE PROTON ATPASE 2 (PMA2). Of these DAPs, PMA2 probably contributes to the H^+-electrochemical potential difference across cytoplasma membrane that drives the active transport of nutrients by H^+-symport and PMA2 is also shown to be involved in cell expansion, by acidifying the apoplasm and thus, activating proteins involved in loosening the cell wall like expansins [38]. In the present study, the decreased PMA2 may be associated to the cell wall rigidification induced by cold stress [39]. AVP1 plays an important role in trans-tonoplast membrane proton gradient that is used to energize secondary transporters like vacuolar Na^+/H^+ antiporters and ectopic expression of an Arabidopsis AVP1 improves drought- and salt tolerance in cotton [40]. The up-regulation of AVP1 in cold-stressed jojoba leaves might contribute to the osmotic homeostasis of the cells.

Under the low temperature environment, the cell volume might become smaller. To cope with such stress, the cytoskeleton, cell wall and plasma membrane would change correspondingly. We found five cytoskeleton-related proteins differentially accumulated under cold stress. These DAPs included tubulin alpha-4 chain (TUA4), tubulin beta 8 (TUB8) and actin 1 (ACT1). The cytoskeleton-related DAPs probably participate in modulating the cytoskeleton organization in cold-stressed jojoba leaf cells.

3.6. Proteins Involved in Protein Synthesis, Folding and Degradation

Cold stress seriously affected the protein synthesis and folding and previous proteomic studies have identified many stress-responsive proteins associated with protein synthesis, folding and degradation in plants [18]. In jojoba, 11 DAPs involved in protein metabolism were identified, including 6 involved in translation, 3 in protein folding and 2 in protein modification. Extensive protein interactions were predicted among the DAPs in this category (Figure 7a).

Although most of the DAPs involved in protein metabolism were down-regulated, some of them were induced under cold stress. For example, 60S ribosomal protein L7-3 and L10a-1 and an eukaryotic translation initiation factor 3 (eIF-3) subunit were down-regulated, while 60S ribosomal protein L7-4 and L13a-2 were increased in abundance under cold stress. These data indicated that the translation apparatus was adjusted to adapt to cold environment.

Besides 60S ribosomal protein L7-4 and L13a-2, two cold-induced DAPs, i.e., signal peptidase complex catalytic subunit SEC11C (SEC11C) and staurosporin and temperature sensitive 3-like a (STT3A), were identified in jojoba leaves under cold stress. Signal peptidase is a group of enzymes help to remove signal peptides from nascent proteins Since no stress responsive signal peptidase have been reported by now, the biological significance of the up-regulation of SEC11C observed in the present study is still to be investigated.

Environmental stresses often lead to the accumulation of misfolded proteins in the endoplasmic reticulum (ER) lumen. Besides molecular chaperone and peptide disulfide isomerase, which help the refolding of the misfolded proteins, *N*-glycosylation in ER has been shown to regulate protein quality control. Protein *N*-glycosylation in ER is catalyzed by a multi-subunit enzyme, the oligosaccharyltransferase (OST) complex. One of the cold-induce DAPs in the present study, STT3A, is one of the catalytic subunits of OST. Previous studies showed that STT3a is required for recovery from the unfolded protein response and for cell during salt/osmotic stress recovery [41]. Our results indicated that cold stress induced STT3a possibly contribute to the cold acclimation of jojoba seedling by participated in protein folding and strengthening protein quality control in ER.

3.7. Proteins Involved in Lipid Metabolism and Transport

Lipids are involved in a wide variety of physiological processes in plant cells. For example, lipids form the cytoplasma membrane and membrane of organelles, protect tissues by forming leaf cuticle and wax; lipids participate in the photosynthetic capture of light in chloroplast and function as signaling molecules in stress signal transduction [42]. In the present study, 10 DAPs involved in fatty acid metabolism and transport were identified, indicating a significant change were taken place in lipid metabolism in cold-stressed jojoba leaves and such a phenomenon was reported in Arabidopsis and *Eutrema salsugineum* [42]. There of these DAPs, i.e., glycosylphosphatidylinositol-anchored lipid protein transfer 6 (LTPG6), fatty acid export 1 (FAX1) and ROSY1 are involved in lipid and sterol transfer. Among them, ROSY1 has been shown to be involved in the regulation of gravitropic response and basipetal auxin transport in roots [43] and the down-regulation of ROSY1 in jojoba leaves suggested its possible role in cold stress response. The remaining 8 DAPs in this category were mainly associated with lipid synthesis and catabolism. Among them, 3-DEOXY-D-ARABINO-HEPTULOSONATE-7-PHOSPHATE 2 (DAHP2) catalyzes the first step of the shikimate pathway, a key pathway for the synthesis of aromatic primary and secondary metabolites [44]. The up-regulation of DAHP2 in cold-stressed jojoba leaves indicated that more carbon flux was used to synthesize lipid and secondary metabolites. Mitochondrial acyl carrier protein 1 (MTACP-1), a member of the mitochondrial acyl carrier protein (ACP) family, is involved in fatty acid and lipoic acid synthesis in mitochondria [45]. Cytochrome B5 isoform B and E are membrane bound hemoproteins and are involved in oxidation-reduction process by functioning as electron carrier for some membrane bound oxygenases like fatty acid desaturases [46].

When plants were exposed to environmental stresses, the thylakoid membranes in chloroplasts are disintegrated and galactolipid are broken down, leading to the accumulation of phytol and free fatty acids, which are toxic to cells. Theses phytol and fatty acids can be converted into fatty acid phytyl esters and triacylglycerol by phytyl ester synthase (PES) [47]. PES1 was up-regulated in jojoba seedling under cold treatment, which would help to maintain the integrity of the photosynthetic membrane during cold stress. Thus, the up-regulation of PES1 probably play a vital role in cold adaptation to cold conditions for jojoba.

4. Materials and Methods

4.1. Plant Growth and Cold Stress Treatment

Jojoba seedlings were grown in commercial pots at 25/20 °C (d/n) under a photosynthetic photon flux density of 150 µmol m^{-2} s^{-1} with long-day conditions (16/8 h light/dark cycle) for 1 month. Thirty female Jojoba seedlings were divided into two groups randomly and all the seedling have similar height and 7–8 pairs of true leaves. The first group was still grown in tissue culture room for additional 7 days and served as the control group. The second group were transferred to the cold-stress conditions (150 µmol m^{-2} s^{-1} light intensity, 20/15 °C for the first two days and 15/10 °C d/n for the remaining five days). A growth chamber Pervical LT-36VL (Percival Scientific, Inc., Perry, IA, USA) was used for plant growth and cold treatment. After 7 days of treatment, the fifth and sixth pairs of leaves in the both groups were collected. Part of these leaves were used for biochemical and physiological parameter measurement and the remaining leaves were frozen immediately in liquid nitrogen and then stored at −70 °C for further RNA and protein extraction. Each sample were pooled from three individual plants and each group had at least three biological replicates.

4.2. Physiological and Biochemical Parameter Measurements

Measurements of photosynthesis, stomatal conductance, intercellular carbon dioxide concentration and transpiration rate were conducted using a portable gas analysis system, LI-COR 6400 (LICOR Inc., Lincoln, NE, USA). The chlorophyll contents were determined using a UV-vis spectrophotometer according to a previously described method [48]. MDA content and REL were measured using the previous described methods [49]. The content of soluble sugars was measured using the anthrone method [50]. Proline concentrations were measured in a UV-vis spectrophotometer by using the ninhydrin reaction method [51]. For enzyme activity measurement, five hundred milligram fresh leaves were homogenized with 5 mM phosphate buffer (pH 7.0) containing 1 mM EDTA and 2% PVPP at 4 °C. After centrifugation (12,000 g) for 15 min at 4 °C, the supernatant was collected and used for enzymes measurement. The activities of CAT and SOD were conducted according to the protocol provided by the manufacturer (Nanjing Jiancheng Institute of Biotechnology, Nanjing, China). PAL activity was assayed using the method of McCallum and Walker [52]. Data are presented as means ± standard deviation (SD) from five independent biological replicates.

4.3. Protein Extraction

Leaf samples from cold stress-treated seedlings (CT) and unstressed seedlings (CK) were ground into fine powder with liquid nitrogen. The power was suspended in an acetone solution containing 10% trichloroacetic acid and 65 mM Dithiothreitol (DTT). After thoroughly mixing by vortexing, proteins were precipitated at −20 °C for 1 h. Proteins were then harvested by centrifuging at 10,000 rpm (Eppendorf 5430R; Eppendorf Ltd., Hamburg, Germany) at 4 °C for 50 min. The resulting supernatant was removed and the protein pellet was washed three times with cold acetone and then dried by lyophilization. The pellet was then suspended in STD buffer (4% SDS, 1 mM DTT, 150 mM Tris-HCl, pH 8.0) and mixed thoroughly. After 5 min incubation in boiling water, the suspensions were dispersed by ultrasonication (80 w: 10 times for 10 s each, with 15 s intervals). After incubated in a boiling water bath for 5 min, the final protein solutions were collected by centrifugation. The protein concentration was measured using the bicinchoninic acid protein assay kit (Beyotime, Shanghai, China). In addition, the quality of the protein samples was further inspected by SDS-PAGE electrophoresis.

4.4. Protein Digestion and ITRAQ Labeling

Protein digestion was conducted using the FASP method [53]. The peptide content was determined by spectra density using UV absorption at 280 nm. The peptide mixture was labeled with the 8-plex iTRAQ reagent according to the manufacturer's recommendation (AB SCIEX) (113 to 115 tags for the three biological replicates of control group, 116 to 118 tags for the three biological replicates of cold stress-treated group) and vacuum dried.

4.5. Strong Cation Exchange (SCX) Chromatography and LC–MS/MS

All the labeled peptide samples were mixed together and then fractionated by strong cation exchange (SCX) chromatography using an AKTA Purifier system (GE Healthcare, Waukesha, WI, USA). The collected fractions (33 fractions) were finally combined into 15 pools and desalted on C18 Cartridges (66872-U, Sigma, St Louis, MO, USA).

Peptides separation was conducted with an automated Easy-nLC1000 system coupled to a Q-Exactive mass spectrometer (Thermo Finnigan, San Jose, CA, USA). The peptides were loaded onto a Thermos scientific EASY column (2 cm × 100 μm, 5 μm-C18) equilibrated with 95% Buffer A (Buffer A, 0.1% formic acid) and then the peptides were loaded and separated on a C18 column (75 mm × 250 mm, 3μm-C18) at a flow rate of 250 nL/min. The peptides were separated with an elution buffer B (0.1% formic acid in 84% acetonitrile) gradient as follows: 0–50% for 100 min, 50–100% for 8 min and 100% for 12 min. The Q-Exactive (Thermo Finnigan, San Jose, CA, USA) mass spectrometer was used to collect data in the positive ion mode according to the method described previously [13].

4.6. Protein Identification and Quantification

Raw MS/MS data were interpreted with Proteome Discoverer (version 1.4, Thermo Fisher Scientific, Waltham, MA, USA) using the search engine Mascot (Version 2.2, Matrix Science, London, UK) against the customized protein database of jojoba (47,062 sequences, translated from transcriptome sequencing of jojoba leaves and young fruits) and the decoy database. The Mascot parameters were set as follows: enzyme, trypsin; mass values, monoisotopic; peptide mass tolerance, ±20 ppm; MS/MS tolerance, 0.1 Da; max missed cleavages, 2; fixed modifications, Carbamidomethyl (C), iTRAQ8plex (N-term), iTRAQ8plex (K); variable modifications, Oxidation (M). The reverse of the target database was used as the decoy database. False discovery rate (FDR) of both proteins and peptides identification was not more than 1%. Protein identification was supported by a minimum of two unique peptide identification and an amino acid coverage $\geq$ 5%.

The intensity of the reporter ions from analyzed fragmentation spectrums was used for peptide quantification and the relative quantity was calculated using the Proteome Discoverer 1.4 software according to the user's guide. Each confident protein quantification required at least two unique peptide and the quantification values were rejected if not all quantification channels are present. The relative quantification of proteins in each group was based on the strength of the reporter ion. The average value of channels of control group was used as internal reference. Ratio was used to assess the fold changes in the abundance of the proteins identified in cold-treated group versus control group. Student's *t*-test was used to identify significant (p-value < 0.05) differences in means between cold-treated and control plants. To be identified as differentially accumulated protein (DAP), a protein must pass *t*-test with p-value < 0.05 and with a change ratio > 1.5 or < 0.667.

4.7. Function Annotation, Classification of the DAPs

Proteins were functionally annotated by using Blast2GO. GO enrichment analysis was conducted using the singular enrichment analysis (SEA) under agriGO [16] and the Arabidopsis thaliana (TAIR10) were used as backgrounds in combination with Fisher's test and the Yekutieli multiple-test with a threshold of FDR = 0.05. Subcellular localization information for proteins was collected from UniProt database and TAIR (www.arabidopsis.org). The biological functions of the DAPs were classified based on their gene ontology annotations and their annotation in KEGG (Kyoto Encyclopedia of Genes and Genomes) database (https://www.kegg.jp/). Protein-protein interaction networks were analyzed using the program STRING (http://string-db.org/), a database of known and predicted protein-protein interactions. The confidence score was set at the high level ($\geq$0.900).

4.8. Quantitative Real-Time PCR Analysis

Total RNA was extracted from jojoba leaves according to a previously described method [17]. qRT-PCR were performed according to previous descriptions [54]. The gene expressions of selected protein-coding genes were normalized against an internal reference gene, 18S rRNA (NCBI accession number: AF094562). The relative gene expression was determined using the $2^{-\Delta\Delta Ct}$ method [55]. Data are presented as means $\pm$ standard deviation (SD) from three independent biological replicates. A *p*-value < 0.05 was determined to be statistically significant. All primers used in the present study are listed in Table S1.

5. Conclusions

A combined physiological and quantitative proteomic analysis was performed to investigate the response to cold stress in jojoba, a semi-arid oil producing crop. Our results indicated that cold stress promote the membrane peroxidation and impair the membrane integrity and jojoba leaves respond to cold stress by reducing chlorophyll content and inhibiting photosynthesis, enhancing the ROS scavenging activities via up-regulating the abundance of antioxidant enzymes, adjusting the protein turnover and regulating the abundance of various defense and stress-response related proteins. Ca^{2+} and ABA signaling participated in the stress signaling transduction in the jojoba's response to the cold stress. Several cold stress induced proteins, including CSD1, COR47, STT3a and PES1 probably play essential roles in jojoba adaptation to the cold stress.

In future studies, we will collect more jojoba low temperature resistant and sensitive varieties, analyze the expression of the low temperature-responsive proteins identified in the present study in these varieties and further identify the protein markers closely associated with the low temperature tolerance in jojoba. We will conduct functional analysis of the low-temperature inducible proteins identified in the present study by expressing these proteins in *Escherichia coli*, yeast and/or *Arabidopsis thaliana* to evaluate their effect on low temperature tolerance of cells.

Supplementary Materials: The following are available online at http://www.mdpi.com/1422-0067/20/2/243/s1.

Author Contributions: P.M. and Y.W. performed the experiments and analyzed the data; F.G., Y.Z. and G.Z. conceived of the project, designed and the experiments; F.G. and Y.Z. wrote the manuscript. All the authors read and approved the final manuscript.

Funding: This work was supported by the National Natural Science Foundation of China (31670335, 31872672 and 31770363), Open fund from Beijing Key Laboratory of Gene Resource and Molecular Development (201706), National Innovation Training Program for Undergraduate (GCCX2018110014), Graduate Independent Research Projects (182115) and Double First-rate plan (Yldxxk201819).

Conflicts of Interest: The authors declare no conflict of interest.

References

1. Dyer, J.M.; Stymne, S.; Green, A.G.; Carlsson, A.S. High-value oils from plants. *Plant J.* **2008**, *54*, 640–655. [CrossRef] [PubMed]
2. Hill, K.; Hofer, R. Natural fats and oils. In *Sustainable Solutions for Modern Economies*; The Royal Society of Chemistry: London, UK, 2009; pp. 167–237. ISBN 978-1-84755-905-0.
3. Kar, A.; Rajaguru, S.N.; Singhvi, A.K.; Juyal, N. Chronostratigraphic evidence for episodes of desertification since the Last Glacial Epoch in the southern margin of Thar desert, India. In *Desertifcation in the Third Millennium*; Taylor & Francis: Thames, UK, 2003; pp. 123–128. ISBN 90-580-9571-1.

4. Abdel-Mageed, W.M.; Bayoumi, S.A.L.H.; Salama, A.A.R.; Salem-Bekhit, M.M.; Abd-Alrahman, S.H.; Sayed, H.M. Antioxidant lipoxygenase inhibitors from the leaf extracts of *Simmondsia chinensis*. *Asian Pac. J. Trop. Med.* **2014**, *7*, S521–S526. [CrossRef]

5. Zhang, G.F.; Beijing Normal University, Beijing, China. Personal communication, 2006.

6. Kumar, M.; Gho, Y.S.; Jung, K.H.; Kim, S.R. Genome-wide identification and analysis of genes, conserved between japonica and indica rice cultivars, that respond to low-temperature stress at the vegetative growth stage. *Front. Plant Sci.* **2017**, *8*, 1120. [CrossRef] [PubMed]

7. Barah, P.; Jayavelu, N.D.; Rasmussen, S.; Nielsen, H.B.; Mundy, J.; Bones, A.M. Genome-scale cold stress response regulatory networks in ten *Arabidopsis thaliana* ecotypes. *BMC Genom.* **2013**, *14*, 722. [CrossRef] [PubMed]

8. Chen, W.; Provart, N.J.; Glazebrook, J.; Katagiri, F.; Chang, H.S.; Eulgem, T.; Mauch, F.; Luan, S.; Zou, G.; Whitham, S.A.; et al. Expression profile matrix of Arabidopsis transcription factor genes suggests their putative functions in response to environmental stresses. *Plant Cell* **2002**, *14*, 559–574. [CrossRef] [PubMed]

9. Kim, C.Y.; Vo, K.T.X.; Nguyen, C.D.; Jeong, D.H.; Lee, S.K.; Kumar, M.; Kim, S.R.; Park, A.H.; Kim, J.K.; Jeon, J.S. Functional analysis of a cold-responsive rice WRKY gene, OsWRKY71. *Plant Biotechnol. Rep.* **2016**, *10*, 13–23. [CrossRef]

10. Winfield, M.O.; Lu, C.; Wilson, I.D.; Coghill, J.A.; Edwards, K.J. Plant responses to cold: Transcriptome analysis of wheat. *Plant Biotechnol. J.* **2010**, *8*, 749–771. [CrossRef] [PubMed]

11. Evans, C.; Noirel, J.; Ow, S.Y.; Salim, M.; Pereira-Medrano, A.G.; Couto, N.; Pandhal, J.; Smith, D.; Pham, T.K.; Karunakaran, E.; et al. An insight into iTRAQ: Where do we stand now? *Anal. Bioanal. Chem.* **2012**, *404*, 1011–1027. [CrossRef] [PubMed]

12. Zheng, B.B.; Fang, Y.N.; Pan, Z.Y.; Sun, L.; Deng, X.X.; Grosser, J.W.; Guo, W.W. iTRAQ-Based quantitative proteomics analysis revealed alterations of carbohydrate metabolism pathways and mitochondrial proteins in a male sterile cybrid pummelo. *J. Proteome Res.* **2014**, *13*, 2998–3015. [CrossRef] [PubMed]

13. Zhang, W.; Zhang, H.; Ning, L.; Li, B.; Bao, M. Quantitative proteomic analysis provides novel insights into cold stress responses in petunia seedlings. *Front. Plant Sci.* **2016**, *7*, 137. [CrossRef] [PubMed]

14. Li, W.; Zhao, F.; Fang, W.; Xie, D.; Hou, J.; Yang, X.; Zhao, Y.; Tang, Z.; Nie, L.; Lv, S. Identification of early salt stress responsive proteins in seedling roots of upland cotton (*Gossypium hirsutum* L.) employing iTRAQ-based proteomic technique. *Front. Plant Sci.* **2015**, *6*, 732. [CrossRef]

15. Xie, H.; Yang, D.H.; Yao, H.; Bai, G.; Zhang, Y.H.; Xiao, B.G. iTRAQ-based quantitative proteomic analysis reveals proteomic changes in leaves of cultivated tobacco (*Nicotiana tabacum*) in response to drought stress. *Biochem. Biophys. Res. Commun.* **2016**, *469*, 768–775. [CrossRef]

16. Du, Z.; Zhou, X.; Ling, Y.; Zhang, Z.H.; Su, Z. agriGO: A GO analysis toolkit for the agricultural community. *Nucleic Acids Res.* **2010**, *38*, W64. [CrossRef]

17. Geng, H.; Shi, L.; Li, W.; Zhang, B.; Chu, C.; Li, H.; Zhang, G. Gene expression of jojoba (*Simmondsia chinensis*) leaves exposed to drying. *Environ. Exp. Bot.* **2008**, *63*, 137–146. [CrossRef]

18. Mousavi, S.A.; Pouya, F.M.; Ghaffari, M.R.; Mirzaei, M.; Ghaffari, A.; Alikhani, M.; Ghareyazie, M.; Komatsu, S.; Haynes, P.A.; Salekdeh, G.H. PlantPReS: A database for plant proteome response to stress. *J. Proteom.* **2016**, *143*, 69–72. [CrossRef]

19. Tuteja, N.; Mahajan, S. Calcium signaling network in plants: An overview. *Plant Signal. Behav.* **2007**, *2*, 79–85. [CrossRef]

20. Shen, W.; Reyes, M.I.; Hanley-Bowdoin, L. Arabidopsis protein kinases GRIK1 and GRIK2 specifically activate SnRK1 by phosphorylating its activation loop. *Plant Physiol.* **2009**, *150*, 996–1005. [CrossRef]

21. Lee, S.; Lee, E.J.; Yang, E.J.; Lee, J.E.; Park, A.R.; Song, W.H.; Park, O.K. Proteomic identification of annexins, calcium-dependent membrane binding proteins that mediate osmotic stress and abscisic acid signal transduction in Arabidopsis. *Plant Cell* **2004**, *16*, 1378–1391. [CrossRef]

22. Schröder, F.; Lisso, J.; Lange, P.; Müssig, C. The extracellular EXO protein mediates cell expansion in Arabidopsis leaves. *BMC Plant Biol.* **2009**, *9*, 20. [CrossRef]

23. Lin, Y.H.; Huang, L.F.; Hase, T.; Huang, H.E.; Feng, T.Y. Expression of plant ferredoxin-like protein (PFLP) enhances tolerance to heat stress in *Arabidopsis thaliana*. *New Biotechnol.* **2015**, *32*, 235–242. [CrossRef]

24. Sami, F.; Yusuf, M.; Faizan, M.; Faraz, A.; Hayat, S. Role of sugars under abiotic stress. *Plant Physiol. Biochem.* **2016**, *109*, 54–61. [CrossRef]

25. Yue, C.; Cao, H.L.; Wang, L.; Zhou, Y.H.; Huang, Y.T.; Hao, X.Y.; Wang, Y.C.; Wang, B.; Yang, Y.J.; Wang, X.C. Effects of cold acclimation on sugar metabolism and sugar-related gene expression in tea plant during the winter season. *Plant Mol. Biol.* **2015**, *88*, 591–608. [CrossRef]

26. Eastmond, P.J.; Li, Y.; Graham, I.A. Is trehalose-6-phosphate a regulator of sugar metabolism in plants? *J. Exp. Bot.* **2003**, *54*, 533–537. [CrossRef]

27. Ventriglia, T.; Kuhn, M.L.; Ruiz, M.T.; Ribeiro-Pedro, M.; Valverde, F.; Ballicora, M.A.; Preiss, J.; Romero, J.M. Two Arabidopsis ADP-glucose pyrophosphorylase large subunits (APL1 and APL2) are catalytic. *Plant Physiol.* **2008**, *148*, 65–76. [CrossRef]

28. Seung, D.; Soyk, S.; Coiro, M.; Maier, B.A.; Eicke, S.; Zeeman, S.C. PROTEIN TARGETING TO STARCH is required for localising GRANULE-BOUND STARCH SYNTHASE to starch granules and for normal amylose synthesis in Arabidopsis. *PLoS Biol.* **2015**, *13*, e1002080. [CrossRef]

29. Yadav, S.K.; Singla-Pareek, S.L.; Sopory, S.K. An overview on the role of methylglyoxal and glyoxalases in plants. *Drug Metab. Drug Interact.* **2008**, *23*, 51–68. [CrossRef]

30. Prashanth, S.R.; Sadhasivam, V.; Parida, A. Over expression of cytosolic copper/zinc superoxide dismutase from a mangrove plant Avicennia marina in indica rice var Pusa Basmati-1 confers abiotic stress tolerance. *Transgen. Res.* **2008**, *17*, 281–291. [CrossRef]

31. Puhakainen, T.; Hess, M.W.; Mäkelä, P.; Svensson, J.; Heino, P.; Palva, E.T. Overexpression of multiple dehydrin genes enhances tolerance to freezing stress in Arabidopsis. *Plant Mol. Biol.* **2004**, *54*, 743–753. [CrossRef]

32. Weiser, R.L.; Wallner, S.J.; Waddell, J.W. Cell wall and extensin mRNA changes during cold acclimation of pea seedlings. *Plant Physiol.* **1990**, *93*, 1021–1026. [CrossRef]

33. Sasidharan, R.; Voesenek, L.A.C.J.; Pierik, R. Cell wall modifying proteins mediate plant acclimatization to biotic and abiotic stresses. *Crit. Rev. Plant Sci.* **2011**, *30*, 548–562. [CrossRef]

34. Kim, S.J.; Held, M.A.; Zemelis, S.; Wilkerson, C.; Brandizzi, F. CGR2 and CGR3 have critical overlapping roles in pectin methylesterification and plant growth in *Arabidopsis thaliana*. *Plant J.* **2015**, *82*, 208–220. [CrossRef]

35. Olsen, K.M.; Lea, U.S.; Slimestad, R.; Verheul, M.; Lillo, C. Differential expression of four Arabidopsis PAL genes; PAL1 and PAL2 have functional specialization in abiotic environmental-triggered flavonoid synthesis. *J. Plant Physiol.* **2008**, *165*, 149–1499. [CrossRef]

36. Falcone Ferreyra, M.L.; Rius, S.P.; Casati, P. Flavonoids: Biosynthesis, biological functions, and biotechnological applications. *Front. Plant Sci.* **2012**, *3*, 222. [CrossRef]

37. Chinnusamy, V.; Zhu, J.; Zhu, J.K. Cold stress regulation of gene expression in plants. *Trends Plant Sci.* **2007**, *12*, 444–451. [CrossRef]

38. Niczyj, M.; Champagne, A.; Alam, I.; Nader, J.; Boutry, M. Expression of a constitutively activated plasma membrane H$^+$-ATPase in *Nicotiana tabacum* BY-2 cells results in cell expansion. *Planta* **2016**, *244*, 1109–1124. [CrossRef]

39. Rajashekar, C.B.; Lafta, A. Cell-wall changes and cell tension in response to cold acclimation and exogenous abscisic acid in leaves and cell cultures. *Plant Physiol.* **1996**, *111*, 605–612. [CrossRef]

40. Pasapula, V.; Shen, G.; Kuppu, S.; Paez-Valencia, J.; Mendoza, M.; Hou, P.; Chen, J.; Qiu, X.; Zhu, L.; Zhang, X.; et al. Expression of an Arabidopsis vacuolar H$^+$-pyrophosphatase gene (AVP1) in cotton improves drought- and salt tolerance and increases fibre yield in the field conditions. *Plant Biotechnol. J.* **2011**, *9*, 88–99. [CrossRef]

41. Koiwa, H.; Li, F.; McCully, M.G.; Mendoza, I.; Koizumi, N.; Manabe, Y.; Nakagawa, Y.; Zhu, J.; Rus, A.; Pardo, J.M.; et al. The STT3a subunit isoform of the Arabidopsis oligosaccharyltransferase controls adaptive responses to salt/osmotic stress. *Plant Cell* **2003**, *15*, 2273–2284. [CrossRef]

42. Barrero-Sicilia, C.; Silvestre, S.; Haslam, R.P.; Michaelson, L.V. Lipid remodelling: Unravelling the response to cold stress in Arabidopsis and its extremophile relative *Eutrema salsugineum*. *Plant Sci.* **2017**, *263*, 194–200. [CrossRef]

43. Meyer, E.H.; Heazlewood, J.L.; Millar, A.H. Mitochondrial acyl carrier proteins in Arabidopsis thaliana are predominantly soluble matrix proteins and none can be confirmed as subunits of respiratory Complex I. *Plant Mol. Biol.* **2007**, *64*, 319–327. [CrossRef]

44. Langer, K.M.; Jones, C.R.; Jaworski, E.A.; Rushing, G.V.; Kim, J.Y.; Clark, D.G.; Colquhoun, T.A. PhDAHP1 is required for floral volatile benzenoid/phenylpropanoid biosynthesis in Petunia × hybrida cv 'Mitchell Diploid'. *Phytochemistry* **2014**, *103*, 22–31. [CrossRef]

45. Dalal, J.; Lewis, D.R.; Tietz, O.; Brown, E.M.; Brown, C.S.; Palme, K.; Muday, G.K.; Sederoff, H.W. ROSY1, a novel regulator of gravitropic response is a stigmasterol binding protein. *J. Plant Physiol.* **2016**, *196*, 28–40. [CrossRef]

46. Kumar, R.; Tran, L.S.; Neelakandan, A.K.; Nguyen, H.T. Higher plant cytochrome b5 polypeptides modulate fatty acid desaturation. *PLoS ONE.* **2012**, *7*, e31370. [CrossRef]

47. Lippold, F.; vom Dorp, K.; Abraham, M.; Hölzl, G.; Wewer, V.; Yilmaz, J.L.; Lager, I.; Montandon, C.; Besagni, C.; Kessler, F.; et al. Fatty acid phytyl ester synthesis in chloroplasts of Arabidopsis. *Plant Cell* **2012**, *24*, 2001–2014. [CrossRef]

48. Harris, E. Introduction to Chlamydomonas and Its Laboratory Use. In *The Chlamydomonas Sourcebook*; Elsevier LTD: Oxford, UK, 2008; Volume 1, pp. 235–252. ISBN 978-0-12370-873-1.

49. Wang, X.; Chen, S.; Zhang, H.; Shi, L.; Cao, F.; Guo, L.; Xie, Y.; Wang, T.; Yan, X.; Dai, S. Desiccation tolerance mechanism in resurrection fern-ally *Selaginella tamariscina* revealed by physiological and proteomic analysis. *J. Proteome Res.* **2010**, *9*, 6561–6577. [CrossRef]

50. Yemm, E.W.; Willis, A.J. The estimation of carbohydrates in plant extracts by anthrone. *Biochem. J.* **1954**, *57*, 508–514. [CrossRef]

51. Tang, R.; Zheng, J.; Jin, Z.; Zhang, D.; Huang, Y.; Chen, L. Possible correlation between high temperature-induced floret sterility and endogenous levels of IAA, GAs and ABA in rice (*Oryza sativa*, L.). *Plant Physiol. Biochem.* **2008**, *54*, 37–43. [CrossRef]

52. McCallum, J.A.; Walker, J.R.L. Phenolic biosynthesis during grain development in wheat changes in phenylalanine ammonia lyase activity and soluble phenolic content. *J. Cereal Sci.* **1990**, *11*, 35–49. [CrossRef]

53. Wisniewski, J.; Zougman, A.; Nagaraj, N.; Mann, M. Universal sample preparation method for proteome analysis. *Nat. Methods* **2009**, *6*, 359–362. [CrossRef]

54. Gao, F.; Wang, J.; Wei, S.; Li, Z.; Wang, N.; Li, H.; Feng, J.; Li, H.; Zhou, Y.; Zhang, F. Transcriptomic analysis of drought stress responses in *Ammopiptanthus mongolicus* leaves using the RNA-Seq technique. *PLoS ONE* **2015**, *10*, e0124382. [CrossRef]

55. Livak, K.J.; Schmittgen, T.D. Analysis of relative gene expression data using real-time quantitative PCR and the 2(-Delta Delta C(T)) method. *Methods* **2001**, *25*, 402–408. [CrossRef] [PubMed]

International Journal of
Molecular Sciences

MDPI

Article

Proteomics Analysis Reveals Non-Controlled Activation of Photosynthesis and Protein Synthesis in a Rice *npp1* Mutant under High Temperature and Elevated CO₂ Conditions

Takuya Inomata [1], **Marouane Baslam** [2], **Takahiro Masui** [1], **Tsutomu Koshu** [1],
Takeshi Takamatsu [1,2], **Kentaro Kaneko** [1], **Javier Pozueta-Romero** [3] **and Toshiaki Mitsui** [1,2,*]

[1]   Graduate School of Science and Technology, Niigata University, 2-8050 Ikarashi, Niigata 950-2181, Japan;
      f14m005a@mail.cc.niigata-u.ac.jp (T.I.); zoujoukikou@gmail.com (T.M.);
      tsutomu58koshu@hotmail.co.jp (T.K.); ttakamatsu@agr.niigata-u.ac.jp (T.T.); k-neko@gs.niigata-u.ac.jp (K.K.)
[2]   Department of Biochemistry, Niigata University, Niigata 950-218, Japan; mbaslam@gs.niigata-u.ac.jp
[3]   Instituto de Agrobiotecnología (CSIC, UPNA, Gobierno de Navarra), Mutiloako Etorbidea Zenbaki Gabe,
      31192 Mutiloabeti, Nafarroa, Spain; javier.pozueta@unavarra.es
*   Correspondence: t.mitsui@agr.niigata-u.ac.jp; Fax: +81-25-262-6641

Received: 15 August 2018; Accepted: 3 September 2018; Published: 7 September 2018

Abstract: Rice nucleotide pyrophosphatase/phosphodiesterase 1 (NPP1) catalyzes the hydrolytic breakdown of the pyrophosphate and phosphodiester bonds of a number of nucleotides including ADP-glucose and ATP. Under high temperature and elevated CO₂ conditions (HT + ECO₂), the *npp1* knockout rice mutant displayed rapid growth and high starch content phenotypes, indicating that *NPP1* exerts a negative effect on starch accumulation and growth. To gain further insight into the mechanisms involved in the *NPP1* downregulation induced starch overaccumulation, in this study we conducted photosynthesis, leaf proteomic, and chloroplast phosphoproteomic analyses of wild-type (WT) and *npp1* plants cultured under HT + ECO₂. Photosynthesis in *npp1* leaves was significantly higher than in WT. Additionally, *npp1* leaves accumulated higher levels of sucrose than WT. The proteomic analyses revealed upregulation of proteins related to carbohydrate metabolism and the protein synthesis system in *npp1* plants. Further, our data indicate the induction of 14-3-3 proteins in *npp1* plants. Our finding demonstrates a higher level of protein phosphorylation in *npp1* chloroplasts, which may play an important role in carbohydrate accumulation. Together, these results offer novel targets and provide additional insights into carbohydrate metabolism regulation under ambient and adverse conditions.

Keywords: chloroplast; elevated CO₂; heat stress; nucleotide pyrophosphatase/phosphodiesterase; (phospho)-proteomics; photosynthesis; protein phosphorylation; 14-3-3 proteins; *Oryza sativa* L.; starch; sucrose

---

1. Introduction

Changes in climate explain a major portion (32–39%) of yield variability, which correspondingly may translate into large fluctuations in global crop production [1]. Climate change that alters photosynthetic rates may modify physiological responses, plant growth rates (and overall productivity), and resource uses. Therefore, the potential for physiological functioning to evolve in response to climate change will be a key indicator of plant resilience in future environments [2,3]. As they are mostly immobile, plants must adapt to their biotic and abiotic environments limiting or promoting productivity and carbon biosequestration. Photosynthesis can be critical to mitigating the effects of changing climatic conditions. Although the relationship between stomatal conductance, CO₂ uptake, and photosynthesis fluctuates in nature, leaf photosynthesis is known to be highly correlated with stomatal conductance in rice [4,5]. Studies employing mutants deficient in a stomatal anion channel protein SLAC1 (Slow Anion Channel-associated 1) revealed that stomatal conductance is a major determinant of photosynthetic rate in higher plants [6,7]. Thus, terrestrial plants regulate CO_2 uptake by stomatal switch in response to environmental and biochemical stimuli. The assimilated carbon is further converted to starch in the plastid stroma or to sucrose in the cytosol via sugar nucleotides. In rice leaves, it has been shown that the assimilated carbon is partitioned into sucrose rather than starch [8,9]. Under varying environmental conditions, plant growth and development have to be counterbalanced for defense and stress adaption. Notably, proportional changes in protein composition in response to environmental changes are a major cellular response to requirements of homeostatic adjustment and metabolic remodeling.

Starch is synthesized in plant leaves during the day from photosynthetically fixed carbon. It is then mobilized and thus constitutes an important carbon reserve as well as an energy source for plants and provides the major nutritional value of food crops. Starch biosynthesis occurs in plastids and requires several key enzymes [10,11]. Starch, probably the most important metabolite of plant carbohydrate metabolism, is an insoluble polyglucan produced by starch synthase (SS) using ADP-glucose as the sugar donor nucleotide. Among the many enzymes participating in plant nucleotide catabolism, the pyrophosphatase/phosphodiesterase (NPP) family members, including *NPP1*, actively catalyze the hydrolytic breakdown of the pyrophosphate and phosphodiester bonds of ADP-glucose [12–14]. In rice, six *NPP* genes (*NPP1–6*) have been identified and characterized [14]. The *NPP* genes are located on chromosomes 3 (*NPP3*), 8 (*NPP1*), 9 (*NPP6*), and 12 (*NPP2, 4, 5*) in the rice genome. Although *NPP3* was exclusively localized in the endomembrane system [15], *NPP1, 2*, and 6 exhibited a dual localization in both the plastid and endomembrane systems [13,16,17], as has been found for other plastidial glycoproteins [18–22]. Whereas *NPP1* and *NPP6* recognized nucleotide sugars, *NPP2* did not recognize these compounds as substrates but preferentially hydrolyzed uridine diphosphate (UDP), ADP, and adenosine 5′-phosphosulfate (APS). *NPP1* best hydrolyzed ADP-glucose, ADP-ribose, and ATP, while ADP and ADP-glucose were the best substrates for *NPP6*. Rice *NPP* genes showed tissue- and stage-specific expression [14,15]. Thus, it is possible to suppose that rice NPPs are engaged in multiple biological functions through the influence of the turnover of nucleotides and nucleotide derivatives. *NPP1* is a major determinant of ADP-glucose pyrophosphatase (AGPPase) activity and can reduce the plastidic pool of ADP-glucose required for starch accumulation in rice [14]. Plant NPPs seem to play a crucial role in carbon flux by transporting carbon taken up from starch and from cell wall polysaccharide biosynthesis to other metabolic pathways in response to the physiological needs of the cell. We have previously shown that when the *npp1* null mutant is grown under different CO_2 concentration and temperature conditions, *NPP1* exerts a negative effect on plant growth and starch accumulation. This provided the first in vivo evidence for the role of *NPP1* in the control of growth and reserve pool of carbohydrate in rice under fluctuating climatic conditions [14]. Recently, the presence of NPPs in the plastidial compartment was further confirmed [13,15,16]. Moreover, we have tentatively identified complex-type and pauci-mannosidic-type oligosaccharide chains with β1,2-xylose and/or differential core α1,3-fucose residue(s) in rice *NPP1*, which provides strong evidence that the trans-Golgi compartments participate in the Golgi-to-plastid trafficking and targeting mechanism of *NPP* in rice [16]. Plants with a defective endoplasmic reticulum (ER) N-glycosylation pathway and N-glycan maturation are associated with diverse phenotypes and are more sensitive to environmental conditions [23]. Nonetheless, the regulatory mechanisms of high temperature and elevated CO_2-induced starch accumulation and growth stimulation in *npp1* leaves are unknown. Herein, we highlight that the loss-of-function of *NPP1* not only influences stomatal conductance, photosynthesis, and plants' storage of starch and sucrose but also impacts a set of proteins involved in carbohydrate metabolism and the protein synthetic system in rice leaves under high temperature and CO_2 conditions (HT + ECO_2).

2. Results and Discussion

2.1. Knocking out NPP1 Decreases Leaf Temperature and Enhances Stomatal Conductance under High Temperature and Elevated CO_2 Conditions

In order to examine phenotypic characteristics of the *npp1* mutant, 60-day old wild-type (WT) and *npp1* mutant rice plants were subjected to thermographic analysis in a spacious Biotron (length 4.5 × width 4.5 × height 1.8–2.2 m) with varying temperatures (28 and 33 °C) and CO_2 concentrations (40 and 160 Pa) under natural sunlight. As shown in Figure 1, the leaf temperatures of WT and *npp1* plants increased concomitantly with the CO_2 concentration irrespective of temperature conditions. Notably, the leaf temperature in *npp1* plants was apparently lower than in WT plants at all temperature and CO_2 conditions examined (Figure 1). We further examined the flag leaves' temperatures in WT and *npp1* under the different temperature (28 and 33 °C) and CO_2 conditions (40 to 200 Pa) under fully controlled environmental chamber conditions (width 64 × width 56 × height 108 cm) using a portable photosynthesis measuring system (Figure 2A). The leaf temperatures of both WT and *npp1* gradually increased as CO_2 concentrations increased, irrespective of temperature conditions, and reached saturation at 160 Pa (Figure 2A).

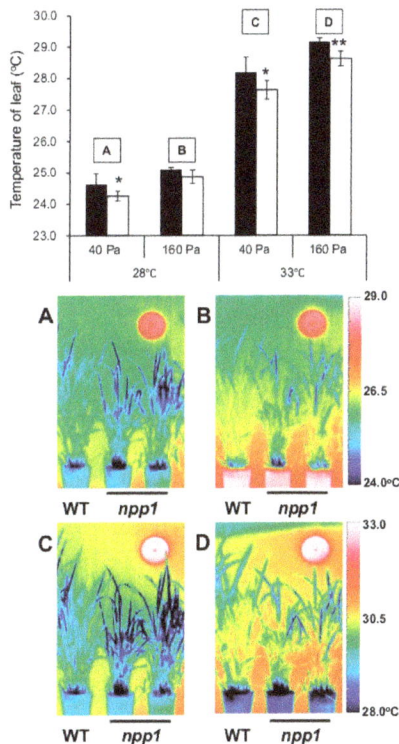

Figure 1. Leaf temperatures of WT and *npp1* mutant plants under different temperatures and CO_2 concentrations in Biotron's environmental conditions. WT (filled bars) and *npp1* (open bars) plants grown for 60 days under normal conditions (14 h light/10 h dark: 28/23 °C, 40 Pa CO_2) after germination were subjected to thermographic analysis of their leaf temperatures under different temperature and CO_2 conditions: (**A**) 28 °C, 40 Pa, (**B**) 28 °C, 160 Pa, (**C**) 33 °C, 40 Pa, and (**D**) 33 °C, 160 Pa. Values in upper panel show the means ± standard deviation (s.d.) (*n* = 5). Asterisks indicate significant differences by Student's *t*-test (*, $p > 0.05$; **, $p > 0.01$). Lower panels represent the thermographic images. WT: wild-type.

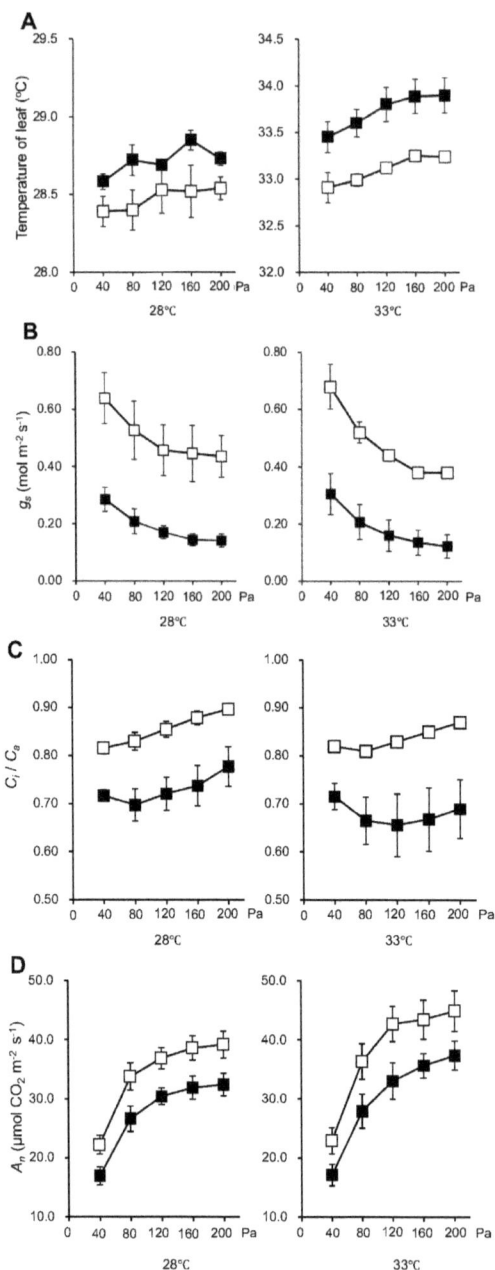

Figure 2. Changes in temperatures, g_s, C_i/C_a, and A_n of WT and *npp1* mutant plants under different temperatures and CO_2 concentrations in a controlled growth chamber. Top leaves of main culm of WT (■) and *npp1* (□) plants grown for 60 days under normal conditions (14 h light/10 h dark: 28/23 °C, 40 Pa CO_2) were subjected to measurements of temperature (**A**), stomatal conductance to water (g_s) (**B**), ratios of internal [CO_2] to ambient [CO_2] (C_i/C_a) (**C**), and photosynthetic rates (A_n) (**D**) under the following conditions: temperature, 28 and 33 °C; CO_2, 40 to 200 Pa; relative humidity, 50%; light, 1000 μmol·photons·m^{-2}·s^{-1}. Values show the means ± s.d. ($n = 3$).

The reduction of leaf temperature in *npp1* strongly suggested that knocking out *NPP1* increases the transpiration rate of the plant. This inference was corroborated by the analysis of the stomatal conductance (g_s) under varying temperatures and CO_2 concentrations. As shown in Figure 2B, elevated CO_2 concentrations caused partial stomatal closure in the two genotypes. No significant differences were observed in the g_s values between the two temperature regimes. Notably, the g_s values in *npp1* plants were higher than in WT plants under all temperatures and CO_2 concentrations.

Previous reports have shown that high CO_2 decreases g_s in a temperature-independent manner [6,24–28]. Changes in g_s value in the *npp1* mutant under different temperature and CO_2 conditions were similar to those in WT, indicating that the stomatal regulatory mechanisms of *npp1* work normally.

2.2. Knocking out NPP1 Enhances Photosynthetic Capacity under High Temperature and Elevated CO2 Conditions

The high g_s values observed in the *npp1* mutant suggested that the lack of *NPP1* expression enhances CO_2 diffusion. In line with this presumption we found that, irrespective of the temperature regime, the intercellular-to-atmospheric CO_2 mole fraction (C_i/C_a) in *npp1* was higher than in WT plants at all atmospheric CO_2 concentrations (Figure 2C). Consistently, *npp1* plants had higher net rates of CO_2 assimilation (A_n) than WT plants at all atmospheric CO_2 levels in the two temperature regimes (Figure 2D).

Whether enhanced photosynthesis in *npp1* plants is solely the consequence of enhanced CO_2 diffusion was investigated by measuring A_n under varying C_i. As shown in Figure 3, these analyses revealed that the *npp1* mutant showed higher A_n values than WT plants at all C_i levels. In rice plants, sucrose is the primary transport sugar and plays a central role in plant growth and development. As expected, levels of sucrose in *npp1* mutants were found to be significantly higher compared with their control WT in any incubation conditions (Figure 4). Previously, the *npp1* mutants have been shown to accumulate starch under high CO_2 concentration conditions [14]. Thus, it is evident that *NPP1* exerts a negative effect on carbohydrate accumulation under HT + ECO_2. In these conditions, the *npp1* mutants increase photosynthesis and the Calvin-Benson cycle (as detailed below), the latter via increased accumulation of Rubisco activase, enabling *npp1* plants to adjust downstream reactions such as sucrose biosynthesis and phloem loading to the increased leaf assimilation. These results indicate that the stimulation of photosynthesis in the *npp1* mutant was caused by both stomatal conductance and non-stomatal conductance related causes. Stomata opening and closure occurs via changes in the turgor pressure of guard cells, and the physiological event is regulated by ion and sugar movements in the guard cells [6,29,30]. Increased CO_2 concentrations enhance anion channel activity (which has been proposed to be a means of mediating the efflux of osmoregulatory anions from guard cells) [31], causing stomatal closure in rice leaves (Figure 2B, [32]). In *npp1* guard cells, the accumulation of sucrose probably causes stomata opening, further activating the photosynthesis of *npp1* leaves under the elevated CO_2. The *NPP1* enzyme, which hydrolyzes sugar nucleotides and ATP, is localized and functions in the chloroplast of rice, therefore, the contents of ATP in chloroplasts should be increased. The high level of ATP in *npp1* chloroplasts would boost the turnover of the Calvin cycle and actively convert CO_2 to starch and sucrose. In fact, in *npp1* null mutants grown under different temperature conditions and CO_2 concentrations, a negative effect on the reserve pool of carbohydrates is seen. In this case, *NPP* activity may occupy a central position in carbon metabolism and its metabolic output to provide products such as the amino acids and lipids necessary for increasing plant biomass.

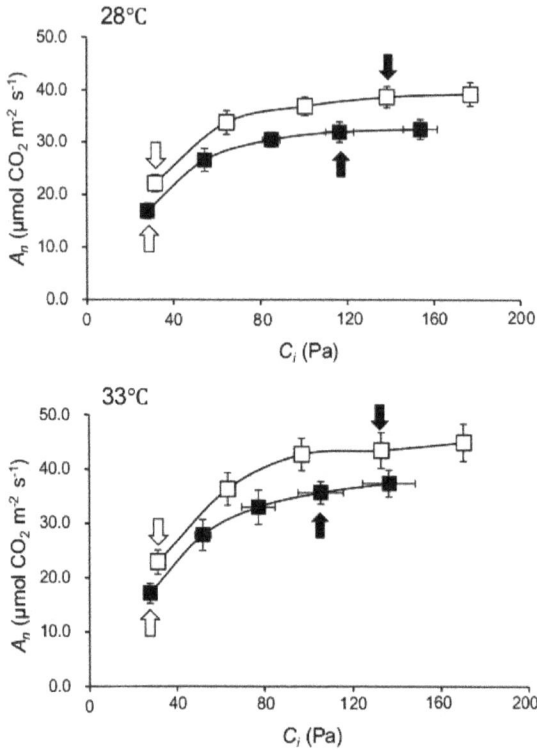

Figure 3. Photosynthetic rates at different intracellular [CO_2] (A_n /C_i curve) of leaves from WT (■) and *npp1* (□) plants. The results of Figure 2 were used to draw an A_n/C_i curve. Open and filled arrows represent the data obtained at C_a of 40 and 160 Pa, respectively.

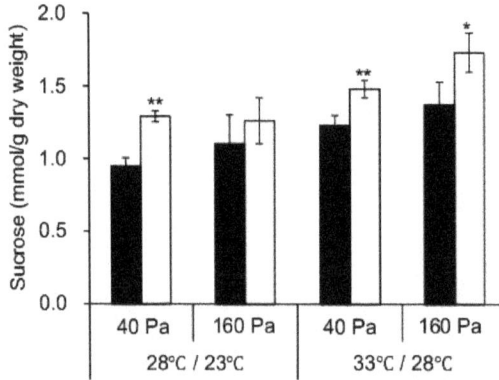

Figure 4. Sucrose accumulation in leaves of WT and *npp1* mutant plants. WT (filled bars) and *npp1* (open bars) plants grown for 60 days under normal conditions (28/23 °C, 40 Pa CO_2) were further incubated under different temperatures and CO_2 concentrations. At the end of a light cycle, the leaves of the WT and the *npp1* mutant plants were subjected to sucrose assays. Values show the means ± s.d. (n = 3~5). Asterisks indicate significant differences by Student's *t*-test (*, $p > 0.05$; **, $p > 0.01$).

2.3. Proteomic Characterization of npp1 Leaves under High Temperature and Elevated CO_2 Conditions

To characterize changes in the proteome of *npp1* leaves under HT + ECO_2, we carried out a quantitative proteomic analysis. Proteins extracted from leaves of WT and *npp1* plants grown under normal (28/23 °C and 40 Pa CO_2) and HT + ECO_2 (33/28 °C and 160 Pa CO_2) conditions were labeled by iTRAQ (isobaric tag for relative and absolute quantitation), followed by tandem mass spectrometry (MS/MS) analysis. Using this approach, 103 differentially expressed proteins were successfully identified among 1701 detected proteins in total. The general trend indicates that the response of the *npp1* mutant to the HT + ECO_2 treatment is due, at least partly, to changes in the expression of proteins from the following groups: photosynthesis, carbohydrate metabolism, protein synthesis, and signaling.

2.4. Photosynthesis and Carbohydrate Metabolism

Various proteins associated with photosynthesis and carbohydrate metabolism were upregulated in HT + ECO_2 grown *npp1* plants (Figure 5A) compared with WT plants (Figure 5B). The HT + ECO_2 treatment promoted the expression of Rubisco activase, a protein that acts as Rubisco's catalytic chaperone [33]. This result is consistent with the enhanced photosynthetic carbon assimilation and also a rise in electron transport capacity. Growth under HT + ECO_2 also upregulated the expression of Calvin-Benson enzymes (e.g., fructose-bisphosphate aldolase (FBPA), phosphoglycerate kinase (PGK), and phosphoribulokinase (PRK)). The data presented thus indicate that the HT + ECO_2 enhancement of photosynthesis in *npp1* is the result of enhanced enzymatic activities involved in CO_2 fixation.

The starch synthesis-related enzymes, G1P adenylyltransferase (AGPase), sucrose synthase (SuSy), and 4-α-glucanotransferase protein (DPE2), exhibited a clear upregulation under HT + ECO_2. AGPase is produced from ATP and glucose 1-phosphate to the ADP-glucose necessary for starch biosynthesis. SuSy is responsible for the conversion of sucrose and a nucleoside diphosphate into the corresponding nucleoside diphosphate glucose and fructose, however, this sucrolytic protein may participate in the direct conversion of sucrose into ADP-glucose linked to starch biosynthesis. The sucrose and starch metabolic pathways are tightly interconnected by means of cytosolic ADP-glucose producing enzymes such as SuSy and by the action of an ADP-glucose translocator located on the chloroplast envelope membranes [10,34]. Furthermore, the levels of cytosolic glucanotransferase DPE2 involved in the conversion of starch into sucrose [35] were high in *npp1* leaves under HT + ECO_2 conditions. The high expression of AGPase, SuSy, and DPE2 induced under HT + ECO_2 conditions would be crucial factors for carbohydrate accumulation.

2.5. Protein Synthesis System

The marked upregulation of a large set of proteins related to the protein synthesis system when knocking out *npp1* was also produced by the enhancement of temperature and CO_2 (Figure 5C), although such an increase was not observed in WT leaves (Figure 5D). Ribosome components such as 60S, 50S, 40S, and 30S ribosomal proteins were identified as being upregulated. In addition, elongation factor 1 (EF-Iα and β), 2, T (EF-Ts and -Tu), and four rice eukaryotic initiation factor proteins (eIF 3A, eIF 3F, eIF4A, and eIF4F) were increased in *npp1* under HT + ECO_2. EF-Iα controls GTP-binding proteins responsible for cytoskeletal tubulin, which is positively correlated with starch accumulation [36]. Early studies showed that EF-Iα can influence the organization of cytoskeletal components surrounding protein bodies around the endoplasmic reticulum membranes, which appear to reorganize coincident with the actively accumulating storage proteins in the endosperm [37]. eIF participates in most translation initiation processes and plays important roles in the growth and development (and organ size) of *Arabidopsis* and rice [38–40].

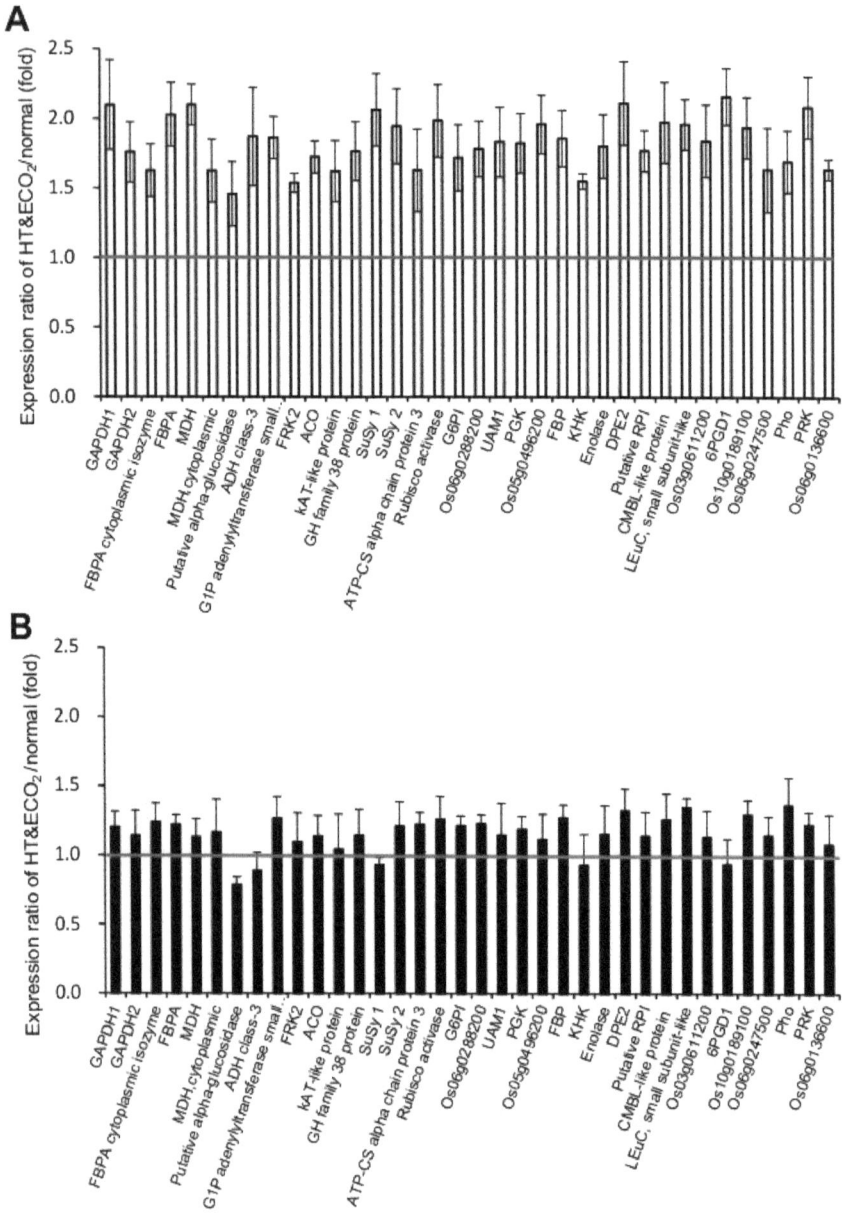

Figure 5. *Cont.*

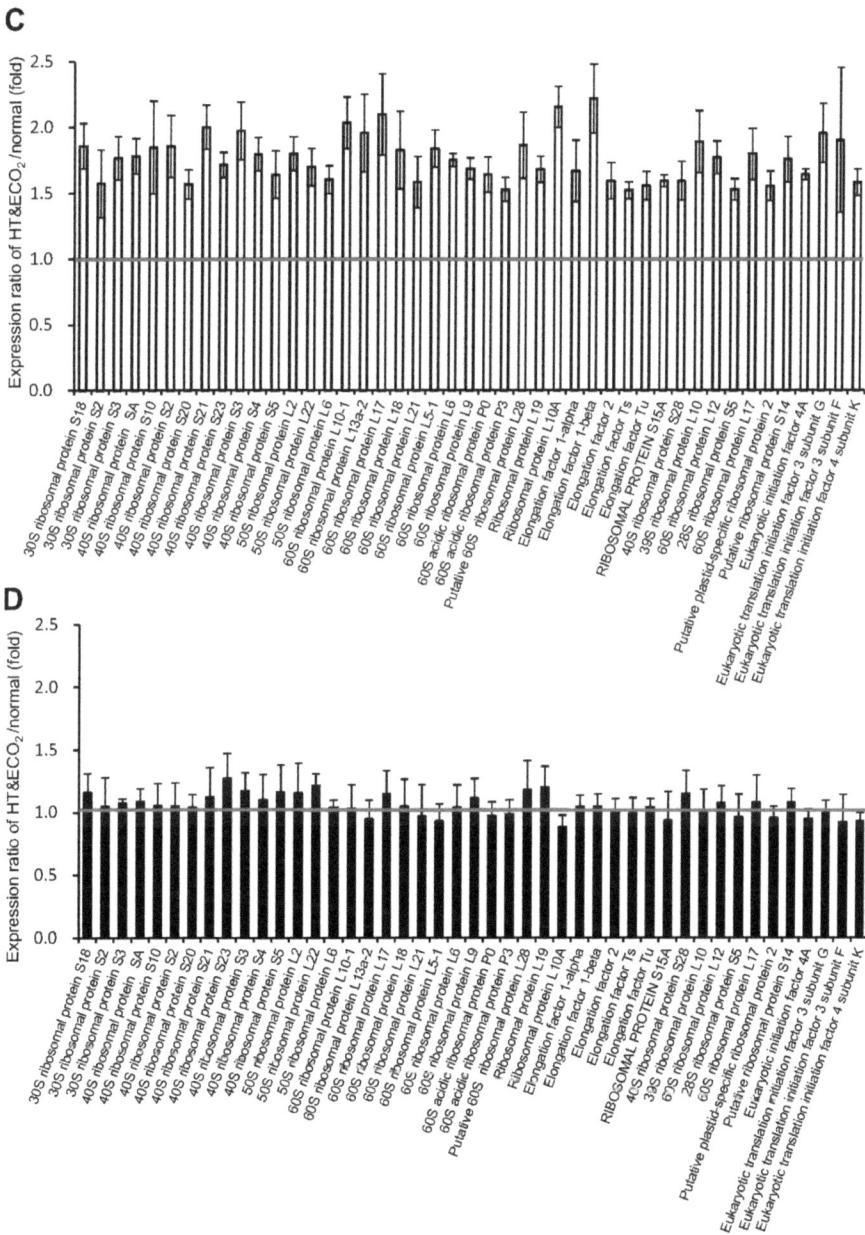

Figure 5. Changes in the expression of carbohydrate- and protein synthesis-related proteins in leaves of WT and *npp1* mutant plants under high temperature and elevated CO_2 concentrations. The leaves of WT (**B,D**) and *npp1* (**A,C**) plants were incubated under normal (28/23 °C, 40 Pa CO_2) and HT + ECO_2 (33/28 °C, 160 Pa CO_2) conditions, and then subjected to a proteomic analysis with Isobaric tags for relative and absolute quantitation (iTRAQ) labeling. Values show the means ± s.d. (*n* = 3). The red line shows the ratio between HT+ECO_2/normal condition mean equal to 1.

2.6. Signaling

Several 14-3-3 proteins exhibited significant changes in *npp1* mutants under HT + ECO$_2$ conditions (Figure 6) and were thus deemed to play important roles in starch biosynthesis in rice leaves. In this study, the upregulation of three 14-3-3 isoforms (14-3-3 GF14C, 14-3-3 GF14E, and 14-3-3 GF14F) was detected, suggesting changes in the phosphorylation status in *npp1* mutants. Recently, the ectopic overexpression of the cassava 14-3-3 gene in *Arabidopsis* showed an increase in starch content in the leaves from transgenic plants [36]. A striking feature of the 14-3-3 proteins is their ability to bind a multitude of functionally diverse signaling proteins, including kinases, phosphatases, and transmembrane receptors [41], and 14-3-3 proteins play roles in regulating plant development and stress responses by effecting direct protein-protein interactions [42–44]. The interactions with 14-3-3s are subject to environmental control through signaling pathways that impact on 14-3-3 binding sites. The phosphoserine/threonine-binding 14-3-3 proteins participate in environmentally responsive phosphorylation-linked regulatory functions in plants and are potentially involved in starch regulation [45]. Previous studies have identified the protein-protein interactions between 14-3-3 and key enzymes of primary metabolism (e.g., sucrose-phosphate synthase and glyceraldehyde-3-phosphate dehydrogenase) and showed the central role of 14-3-3s as regulators of enzymes of cytosolic metabolism and ion pumps [46–49]. A molecular genetic analysis of 14-3-3 isoforms using overexpressed and knockout plants with studies of protein-protein interactions revealed alterations in the level of metabolic intermediates of glycolysis, tricarboxylic acid (TCA), and biosynthesis of aromatic compounds [50]. Hence, we may speculate that 14-3-3 may play a role in the accumulation of starch in *npp1* mutants by interacting with metabolic enzymes and thus appears to maintain those enzymes in an active state in the cell. Notably, 14-3-3 proteins have been localized to the chloroplast stroma and the stromal side of thylakoid membranes [51], thereby implicating a potential role in starch regulation.

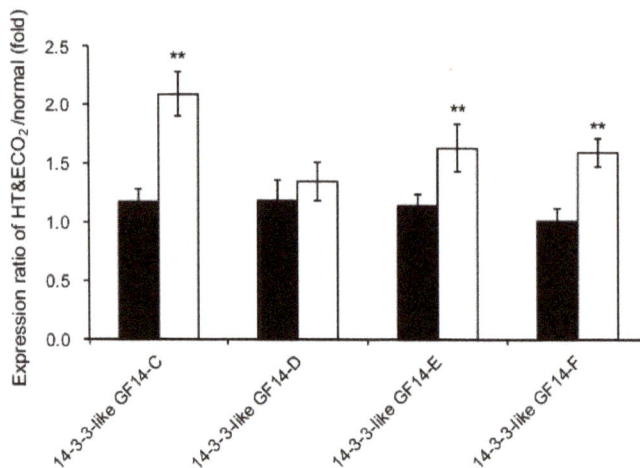

Figure 6. Changes in expression of 14-3-3 proteins in leaves of WT and *npp1* mutant plants under high temperature and elevated CO$_2$ concentrations. Details of incubation conditions via a proteomic analysis were described in Figure 5. The leaves of WT (filled bars) and *npp1* (open bars) plants were incubated under normal and HT + ECO$_2$ conditions. Values show the means ± s.d. (*n* = 3). Asterisks indicate significant differences by Student's *t*-test (**, *p* > 0.01).

Photosynthesis, sugar assays, and quantitative proteomic analyses of leaves of knockout *npp1* revealed that the mutant plants always become highly active regardless of normal or HT and ECO$_2$ conditions (Figures 3–5). The higher A_n in *npp1* mutants could be the consequence of various factors: (i) higher CO_2 diffusion; (ii) more efficient conversion of light energy into ATP; and (iii) the upregulation of photosynthetic enzymes (Rubisco activase and Calvin-Benson enzymes). We consider that an activation effector would be the high level of ATP in chloroplasts, since *NPP1* preferentially hydrolyzes ATP [14]. Of note, a series of 14-3-3 proteins were upregulated in *npp1* mutants (Figure 6), suggesting that the protein phosphorylation status is possibly changed by the disruption of the *NPP1* gene. Phosphorylation of proteins in chloroplasts plays a major role in regulating both the light and dark reactions of photosynthesis. Light-harvesting chlorophyll a/b binding proteins [52–54], Rubisco [55], Rubisco activase [56], phosphoglycerate kinase [57], Glyceraldehyde 3-phosphate dehydrogenase (GAPDH) [58], and transketolase [59] were shown to be phosphorylated and the enzyme functions were regulated. In addition, sigma factor [60] and 24/28 RNA binding proteins [61] were also phosphorylated.

We analyzed the phosphorylation state of WT and *npp1* chloroplast proteins under HT and ECO_2 conditions by employing TiO_2 chromatography and mass spectrometry. The phosphopeptides derived from chlorophyll a/b binding protein, Rubisco large chain, pyruvate phosphate dikinase, and some other plastidial enzymes and proteins were detected with good reproducibility. The level of protein phosphorylation in *npp1* chloroplasts under HT and ECO_2 was higher than that of *npp1* under the normal condition or the WT chloroplasts (Table 1). The chlorophyll a/b binding protein is the main component of the light-harvesting complex (LHC), which is a light receptor that captures and delivers excitation energy to photosystems PSI and PSII. The reversible phosphorylation of light harvesting chlorophyll a/b binding proteins (LHCII) has been observed, which regulates state transitions for balancing the excitation energy between PSI and PSII [54,62]. As shown in Table 1, several phosphorylation sites of chlorophyll a/b binding proteins in the *npp1* mutant were more highly phosphorylated in comparison with the WT chloroplasts. In addition, the phosphorylation level of chlorophyll a/b binding proteins in both *npp1* and WT were increased under HT and ECO_2 conditions. In our phosphoproteomic analysis, two phosphopeptides plastid movement impaired1 (PMI1) and plastid transcriptionally active 16 (pTac16) were found to be more abundant under ECO_2 + HT condition in *npp1* mutant plants. The PMI1 is a plant-specific C2-domain protein that plays a role in organelle movement and positioning [63]. The movements of *npp1* organelles (i.e chloroplasts) within the cell, which become appropriately positioned under ECO_2 + HT, could be a fundamental cellular activity to accomplish their functions and adapt to environmental stress. The pTac16 is a plastid membrane-attached multimeric protein complex involved in plastid transcription and translation. The knockout lines *ptac* seedlings developed white or yellow cotyledons, failed to accumulate chlorophyll even under low light intensities, impaired plastid structure [64], and downregulated levels of plastid-encoded polymerase (PEP) responsible exclusively for the expression of chloroplast-encoded photosynthetic genes [65,66]. The upregulation of pTac 16 in *npp1* mutant plants under HT + ECO_2 could at least partly explain the enhancement of photosynthesis-related genes and activity, and plastid metabolism. Hence, we consider that the high phosphorylation status in *npp1* under HT and ECO_2 could be related to the activation of growth and carbohydrate accumulation of the seedlings.

Table 1. Phosphopeptide detection in WT and *npp1* chloroplasts with or without HT and ECO_2 treatment. n.d.: not detected.

Description	Accession	WT Control	WT HT&ECO₂	npp1 Control	npp1 HT&ECO₂
Chlorophyll a/b-binding protein	Q6Z411	$(1.15 \pm 0.78) \times 10^8$	$(2.43 \pm 2.06) \times 10^8$	$(1.05 \pm 0.53) \times 10^9$	$(7.43 \pm 4.27) \times 10^8$
Chlorophyll a-b binding protein 2	P12331	n.d.	$(4.84 \pm 2.39) \times 10^6$	n.d.	$(4.29 \pm 4.18) \times 10^6$
Chlorophyll a-b binding protein	Q7XV11	$(1.71 \pm 1.01) \times 10^6$	$(3.83 \pm 0.95) \times 10^6$	$(1.10 \pm 0.44) \times 10^7$	$(9.44 \pm 4.78) \times 10^6$
Ribulose bisphosphate carboxylase large chain	P0C512	$(1.08 \pm 0.97) \times 10^7$	n.d.	$(1.52 \pm 0.29) \times 10^7$	$(1.04 \pm 0.78) \times 10^7$
ATP synthase subunit beta	P12085	$(3.65 \pm 2.51) \times 10^6$	n.d.	n.d.	n.d.
PLASTID TRANSCRIPTIONALLY ACTIVE 16	Q0DJF9	$(2.55 \pm 1.90) \times 10^7$	$(2.05 \pm 1.54) \times 10^7$	$(0.81 \pm 1.17) \times 10^7$	$(5.19 \pm 7.87) \times 10^7$
protein CURVATURE THYLAKOID 1A	Q5Z6P4	$(8.76 \pm 1.51) \times 10^7$	$(3.11 \pm 0.67) \times 10^7$	$(2.55 \pm 0.04) \times 10^7$	$(6.79 \pm 4.23) \times 10^7$
PLASTID MOVEMENT IMPAIRED1	Q0IZR7	n.d.	n.d.	n.d.	$(4.16 \pm 1.49) \times 10^6$
Pyruvate, phosphate dikinase 1	Q6AVA8	$(3.35 \pm 2.78) \times 10^5$	$(4.62 \pm 0.02) \times 10^5$	$(1.35 \pm 1.24) \times 10^6$	n.d.

3. Material and Methods

3.1. Plant Material and Growth Condition

The rice variety used in this study was *Oryza sativa* L. cv. Nipponbare. The *Tos17*-inserted line of *NPP1* (ND8012) was obtained from the National Institute of Agrobiological Sciences (NIAS, Tsukuba, Japan; [67]), and the *npp1-1-1* line with a single copy of *Tos17* inserted into the *NPP1* gene was established previously. WT and *npp1* mutant plants were grown and harvested at Niigata University paddy field (Niigata, Japan).

WT and *npp1* mutant seeds were grown in a commercial soil (Kumiai Gousei Baido 3, JA, Tokyo, Japan) in plastic pots and incubated in the growth chamber (CFH-415, Tomy Seiko, Tokyo, Japan) at 28 °C (14 h day)/23 °C (10 h night) cycles with fluorescent lighting (300 $\mu mol \cdot m^{-2} \cdot s^{-1}$). Seeds and plant samples were stored at 4 °C before analysis.

3.2. Thermal Imaging

Thermal images of WT and *npp1* mutant plants were obtained using the InfRec (NEC) thermal video system. Plants grown for 60 days on soil (Kumiai Gousei Baido 3) were transferred to Biotron LPH-1.5PH-NCII (length 4.5 × width 4.5 × height 1.8–2.2 m, Nihon-ika, Osaka, Japan) and incubated under four different conditions (28 °C, 40 Pa CO_2; 28 °C, 160 Pa CO_2; 33 °C, 40 Pa CO_2; 33 °C/160 Pa CO_2) at 70% relative humidity under natural light.

3.3. Gas Exchange Measurements

Photosynthetic rate (A_n), leaf conductance (g_s), and intercellular CO_2 concentrations [CO_2] (C_i) were measured with a portable photosynthesis LI-6400XL system (LI-6400-20, LiCor Biosciences, Lincoln, NE, USA). Gas exchange of WT and *npp1* leaves was recorded in the central segment of top leaves attached to the main culm between 3 and 8 h after the start of the photoperiod. Leaf cuvette conditions were set as follows: block temperature was set at ambient (growth chamber; 28 °C) and high (33 °C) temperatures; [CO_2] was set at 400 to 2000 $\mu mol \cdot mol^{-1}$; relative humidity was maintained equal to that in the growth chamber; and Photosynthetically Active Radiation (PAR) was set at 1200 $\mu mol \cdot m^{-2} \cdot s^{-1}$, resulting in light-saturated photosynthesis and no decline as a result of photorespiration.

3.4. Assay

Sucrose content was measured according to the methods described previously [18,68]. Chlorophyll extracted from the isolated chloroplasts with 80% (*v/v*) acetone was assayed by the method described by Porra et al. [69]. Protein concentration was determined by the Pierce 660 nm Protein Assay Kit (Thermo Fisher Scientific, Waltham, MA, USA) using bovine serum albumin (BSA) as a standard.

3.5. Analysis of Leaf Proteome

Two hundred milligrams of leaves of WT and *npp1* grown for 7 days under normal (28 °C, 14 h day/23 °C, 10 h night, 40 Pa CO_2) and elevated temperature and CO_2 (33 °C, 14 h day/28 °C, 10 h night, 160 Pa CO_2) conditions were ground in liquid nitrogen and suspended in 7 M urea, 2 M thiourea, 3% (*w/v*) CHAPS (3-((3-cholamidopropyl) dimethylammonio)-1-propanesulfonate), 1% (*v/v*) Triton X-100, and 10 mM dithiothreitol. After centrifugation at 10,000× *g* at 4 °C for 5 min, the supernatants were mixed with 1/10 volume of 100% (*w/v*) TCA, incubated on ice for 15 min, and then centrifuged at 10,000× *g* at 4 °C for 15 min. The resulting precipitates were washed three times with ice-cold acetone and resuspended in 8 M urea.

The procedure of quantitative shotgun proteomic analysis was the same as previously described [68]. The protein samples (50 µg) were thoroughly digested with endoproteinase Lys-C and trypsin at 37 °C for 12 h. iTRAQ labeling of peptides was carried out with 4-plex iTRAQ tags according to the manufacturer's protocol (AB Sciex, Framingham, MA, USA), and the resultant 4-iTRAQ-labeled peptide samples were mixed. iTRAQ analysis was performed by employing a DiNa-A-LTQ-Orbitrap-XL system operated with Xcalibur 2.0 software (Thermo Fisher Scientific). Proteins were identified with Proteome Discoverer v. 1.4 software, and the SEQUEST HT (Thermo Fisher Scientific) and MsAmanda [70] search tool using the UniProt (http://www.uniprot.org/) *O. sativa* subsp. japonica database (63,535 proteins) with the following parameters: enzyme, trypsin; maximum missed cleavages site, 2; peptide charge, 2+ or 3+; MS tolerance, 5 ppm; MS/MS tolerance, ±0.5 Da; dynamic modification, carboxymethylation (C), oxidation (H, M, W), iTRAQ 4-plex (K, Y, N-terminus). False discovery rates were <1%.

The mass spectrometry proteomics data (JPST000338) have been deposited in the jPOST repository (https://repository.jpostdb.org/).

3.6. Analysis of Chloroplast Phosphoproteome

The procedure of chloroplast isolation was essentially identical to the method described earlier [16,71]. Rice seeds were germinated and grown for 7 days under normal (28 °C, 14 h day/23 °C, 10 h night, 40 Pa CO_2) and elevated temperature and CO_2 (33 °C, 14 h day/28 °C, 10 h night, 160 Pa CO_2) conditions. Thirty grams of leaves were homogenized with an equal volume of solution A mixture consisting of 50 mM HEPES-KOH pH 7.5, 0.33 M sorbitol, 5 mM $MgCl_2$, 5 mM $MnCl_2$ and 5 mM EDTA (ethylenediaminetetraacetic acid), 50 mM sodium ascorbate, and then the homogenates were passed through four layers of gauze and four layers of Miracloth (Merck, Darmstadt, Germany). The filtrate was layered onto an 80% (*v/v*) Percoll (Sigma, St.Louis, USA) cushion containing solution A, and centrifuged at 2000× *g* at 4 °C for 4 min. The crude chloroplasts on the Percoll surface were diluted with more than twice the volume of solution A, then layered onto a discontinuous density gradient consisting of 40% and 80% Percoll solutions. The gradient was centrifuged at 4000× *g* at 4 °C for 10 min. Intact chloroplasts enriched around the 40%/80% Percoll interface were collected and subjected again to the Percoll gradient centrifugation. Intact chloroplasts were diluted with five times the volume of solution A, and centrifuged at 2000× *g* at 4 °C for 4 min, followed by chlorophyll and protein extractions.

Analyses of phosphoproteins were carried out according to the method described by Fukuda et al. [72]. Intact chloroplasts were suspended in 7 M urea, 2 M thiourea, 3% (*w/v*) CHAPS, 1% (*v/v*) Triton X-100, and 10 mM dithiothreitol. After centrifugation at 10,000× *g* at 4 °C for 5 min, the supernatants were mixed with 1/10 volume of 100% (*w/v*) TCA, incubated on ice for 15 min, and then centrifuged at 10,000× *g* at 4 °C for 15 min. The resulting precipitates were washed three times with ice-cold acetone and resuspended in 8 M urea. The protein preparations (100 μg) were digested with 2% (*w/w*) endoproteinase Lys-C and trypsin in 25 mM NH_4HCO_3 and 0.8 M urea at 37 °C for 12 h. The reaction mixtures were dried on a Centrifugal Concentrator (CC-105, Tomy, Japan) and then dissolved in buffer A consisting of 60% (*v/v*) acetonitrile (ACN), 5% (*v/v*) glycerol, 0.1% (*v/v*) Trifluoroacetic acid (TFA). A MonoSpin TiO column (1000 μL: GL Science, Tokyo, Japan) was pre-equilibrated with buffer B consisting of 80% (*v/v*) ACN, 0.1% (*v/v*) TFA, followed by buffer A, and the samples were centrifuged at 3000× *g* for 1 min each. The obtained peptide samples were applied to the tip column and centrifuged at 3000× *g* for 5 min. The flow-through fraction was applied to the tip column again. Subsequently, the tip column was washed three times with buffer A and centrifuged at 3000× *g* for 1 min. The binding phosphopeptides were eluted with 100 μL of 5% NH_4OH with centrifugation at 1000× *g* for 5 min. After elution with 5% NH_4OH, the tightly binding phosphopeptides were eluted with 100 μL of 1 M (*w/v*) bis-Tris propane with centrifugation at 1000× *g* for 1 min. The phosphopeptides eluted with 5% NH_4OH solution were dried on a Centrifugal Concentrator and then dissolved in 5% (*v/v*) formic acid (FA). The phosphopeptides eluted with 1 M bis-Tris propane were acidified with 900 μL of 5% FA and then desalted using a MonoSpin C18 column (GL Sciences, Tokyo, Japan).

Int. J. Mol. Sci. **2018**, *19*, 2655

Each phospho-peptide fraction was loaded on a HiQ sil C-18 W-3 trap column with buffer C consisting of 0.1% (*v/v*) FA and 2% (*v/v*) ACN using a DiNa-A system (KYA Tech., Tokyo, Japan). A linear gradient from 0% to 33% buffer D consisting of 0.1% FA and 80% ACN for 600 min, 33% to 100% D for 10 min and back to 0% D in 15 min was applied, and peptides eluted from the HiQ sil C-18 W-3 column were directly loaded on a MonoCap C18 High Resolution 2000 separation column. The separated peptides were introduced into a LTQ-Orbitrap XL mass spectrometer (Thermo Fisher Scientific, Waltham, MA, USA) with a flow rate of 300 nL·min^{-1} and an ionization voltage of 1.7–2.5 kV. The mass range selected for MS scan was set to 350–1600 *m/z* and the top five peaks were subjected to MS/MS analysis. The full MS scan was detected in the Orbitrap, and the MS/MS scans were detected in the linear ion trap. The normalized collision energy for MS/MS was set to 35 eV for collision-induced dissociation (CID).

Operation of protein identification with software and database was carried out as described above [68]. The phosphorylation of S/T/Y and oxidation of H/M/W residues were set as dynamic modifications. False discovery rates were <1%.

The chloroplast phosphoproteome data (JPST000462) have been deposited in the jPOST repository (https://repository.jpostdb.org/).

4. Conclusions

Previous investigations have revealed that *NPP1* exerts a negative effect on starch accumulation and growth. *NPP1* localizes to the chloroplasts and degrades a number of nucleotides including ADP-glucose and ATP, thus, it is possibly a kind of room (for example, chloroplast stroma) cleaning enzyme. The molecular physiological phenotype of the *npp1* mutant was further analyzed in the present study. Lower temperatures and changes in the transpiration rates of *npp1* leaves were observed, indicating that the disruption of the *NPP1* gene caused the stomatal opening of rice leaves. Furthermore, the analysis of the A_n/C_i curve indicated that the enhancement of photosynthesis in *npp1* resulted from multiple causes in addition to the stomatal conductance. The proteome of carbohydrate metabolism and protein synthesizing system in *npp1* leaves was strongly upregulated by HT and ECO$_2$. Furthermore, the protein phosphorylation status in *npp1* chloroplasts was significantly higher than in WT chloroplasts. An increase in ATP in chloroplasts might be a key stimulus, because *NPP1* preferentially hydrolyzes ATP. Judging from the overall results, we consider that the remarkable enhancement of plant growth and carbohydrate accumulation in *npp1* mutant plants under HT and ECO$_2$ conditions was a consequence of the non-controlled activation of photosynthesis and protein synthesis by loss of function of a fine-tuning enzyme *NPP1*.

Author Contributions: T.M. (Toshiaki Mitsui) conceived and designed the project. T.I., T.M. (Takahiro Masui), K.K., T.K., and T.T. (Toshiaki Mitsui) performed the experiments. T.M. (Toshiaki Mitsui) and M.B. supervised and interpreted the data. T.M., M.B., and J.P.-R. wrote the manuscript. All authors discussed the results and implications and commented on the manuscript at all stages.

Funding: This research was supported by Grants-in-Aid for Scientific Research (A) (15H02486) and Strategic International Collaborative Research Program by the Japan Science and Technology Agency (JST SICORP).

Conflicts of Interest: The authors declare that the research was conducted in the absence of any commercial or financial relationships that could be construed as potential conflicts of interest.

References

1. Ray, D.K.; Gerber, J.S.; MacDonald, G.K.; West, P.C. Climate variation explains a third of global crop yield variability. *Nat. Commun.* **2015**, *6*. [CrossRef] [PubMed]
2. Becklin, K.M.; Anderson, J.T.; Gerhart, L.M.; Wadgymar, S.M.; Wessinger, C.A.; Ward, J.K. Examining plant physiological responses to climate change through an evolutionary Lens. *Plant Physiol.* **2016**, *172*, 635–649. [CrossRef] [PubMed]
3. Becklin, K.M.; Walker, S.M.; Way, D.A.; Ward, J.K. CO$_2$ studies remain key to understanding a future world. *New Phytol.* **2017**, *214*, 34–40. [CrossRef] [PubMed]

4. Hirasawa, T.; IIda, Y.; Ishihara, K. Effect of leaf water potential and air humidity on photosynthetic rate and diffusive conductance in rice plants. *Jpn. J. Crop Sci.* **1988**, *57*, 112–118. [CrossRef]

5. Ishihara, K.; Saitoh, K. Diurnal courses of photosynthesis, transpiration, and diffusive conductance in the single-leaf of the rice plants grown in the paddy field under submerged condition. *Jpn. J. Crop Sci.* **1987**, *56*, 8–17. [CrossRef]

6. Negi, J.; Matsuda, O.; Nagasawa, T.; Oba, Y.; Takahashi, H.; Kawai-Yamada, M.; Uchimiya, H.; Hashimoto, M.; Iba, K. CO_2 regulator SLAC1 and its homologues are essential for anion homeostasis in plant cells. *Nature* **2008**, *452*, 483–486. [CrossRef] [PubMed]

7. Kusumi, K.; Chono, Y.; Shimada, H.; Gotoh, E.; Tsuyama, M.; Iba, K. Chloroplast biogenesis during the early stage of leaf development in rice. *Plant Biotechnol.* **2010**, *27*, 85–90. [CrossRef]

8. Ishimaru, K. Identification of a locus increasing rice yield and physiological analysis of its function. *Plant Physiol.* **2003**, *133*, 1083–1090. [CrossRef] [PubMed]

9. Scofield, G.N.; Hirose, T.; Aoki, N.; Furbank, R.T. Involvement of the sucrose transporter, OsSUT1, in the long-distance pathway for assimilate transport in rice. *J. Exp. Bot.* **2007**, *58*, 3155–3169. [CrossRef] [PubMed]

10. Bahaji, A.; Li, J.; Sánchez-López, Á.M.; Baroja-Fernández, E.; Muñoz, F.J.; Ovecka, M.; Almagro, G.; Montero, M.; Ezquer, I.; Etxeberria, E.; et al. Starch biosynthesis, its regulation and biotechnological approaches to improve crop yields. *Biotechnol. Adv.* **2014**, *32*, 87–106. [CrossRef] [PubMed]

11. Ball, S.G.; Morell, M.K. From bacterial glycogen to starch: Understanding the Biogenesis of the Plant Starch Granule. *Annu. Rev. Plant Biol.* **2003**, *54*, 207–233. [CrossRef] [PubMed]

12. Rodriguez-López, M.; Baroja-Fernández, E.; Zandueta-Criado, A.; Pozueta-Romero, J. Adenosine diphosphate glucose pyrophosphatase: A plastidial phosphodiesterase that prevents starch biosynthesis. *Proc. Natl. Acad. Sci. USA* **2000**, *97*, 8705–8710. [CrossRef] [PubMed]

13. Nanjo, Y.; Oka, H.; Ikarashi, N.; Kaneko, K.; Kitajima, A.; Mitsui, T.; Muñoz, F.J.; Rodríguez-López, M.; Baroja-Fernández, E.; Pozueta-Romero, J. Rice plastidial *N*-glycosylated nucleotide pyrophosphatase/phosphodiesterase is transported from the ER-golgi to the chloroplast through the secretory pathway. *Plant Cell* **2006**, *18*, 2582–2592. [CrossRef] [PubMed]

14. Kaneko, K.; Inomata, T.; Masui, T.; Koshu, T.; Umezawa, Y.; Itoh, K.; Pozueta-romero, J.; Mitsui, T. Nucleotide Pyrophosphatase/Phosphodiesterase 1 exerts a negative effect on starch accumulation and growth in rice seedlings under high temperature and CO_2 concentration conditions. *Plant Cell Physiol.* **2014**, *55*, 320–332. [CrossRef] [PubMed]

15. Kaneko, K.; Yamada, C.; Yanagida, A.; Koshu, T.; Umezawa, Y.; Itoh, K.; Hori, H.; Mitsui, T. Differential localizations and functions of rice nucleotide pyrophosphatase/phosphodiesterase isozymes 1 and 3. *Plant Biotechnol.* **2011**, *28*, 69–76. [CrossRef]

16. Kaneko, K.; Takamatsu, T.; Inomata, T.; Oikawa, K.; Itoh, K.; Hirose, K.; Amano, M.; Nishimura, S.-I.; Toyooka, K.; Matsuoka, K.; et al. *N*-glycomic and microscopic subcellular localization analyses of *NPP1*, *2* and *6* strongly indicate that trans-Golgi compartments participate in the Golgi-to-plastid traffic of nucleotide pyrophosphatase/phosphodiesterases in rice. *Plant Cell Physiol.* **2016**, *57*, 1610–1628. [CrossRef] [PubMed]

17. Baslam, M.; Oikawa, K.; Kitajima-koga, A.; Kaneko, K.; Mitsui, T. Golgi-to-plastid trafficking of proteins through secretory pathway: Insights into vesicle-mediated import toward the plastids. *Plant Signal. Behav.* **2016**, *11*, e1221558. [CrossRef] [PubMed]

18. Asatsuma, S.; Sawada, C.; Itoh, K.; Okito, M.; Kitajima, A.; Mitsui, T. Involvement of α-amylase I-1 in starch degradation in rice chloroplasts. *Plant Cell Physiol.* **2005**, *46*, 858–869. [CrossRef] [PubMed]

19. Villarejo, A.; Burén, S.; Larsson, S.; Déjardin, A.; Monné, M.; Rudhe, C.; Karlsson, J.; Jansson, S.; Lerouge, P.; Rolland, N.; et al. Evidence for a protein transported through the secretory pathway en route to the higher plant chloroplast. *Nat. Cell Biol.* **2005**, *7*, 1224–1231. [CrossRef] [PubMed]

20. Kitajima, A.; Asatsuma, S.; Okada, H.; Hamada, Y.; Kaneko, K.; Nanjo, Y.; Kawagoe, Y.; Toyooka, K.; Matsuoka, K.; Takeuchi, M.; et al. The rice α-amylase glycoprotein is targeted from the Golgi apparatus through the secretory pathway to the plastids. *Plant Cell* **2009**, *21*, 2844–2858. [CrossRef] [PubMed]

21. Burén, S.; Ortega-Villasante, C.; Blanco-Rivero, A.; Martínez-Bernardini, A.; Shutova, T.; Shevela, D.; Messinger, J.; Bako, L.; Villarejo, A.; Samuelsson, G. Importance of post-translational modifications for functionality of a chloroplast-localized carbonic anhydrase (CAH1) in Arabidopsis thaliana. *PLoS ONE* **2011**, *6*, e21021. [CrossRef] [PubMed]

22. Shiraya, T.; Mori, T.; Maruyama, T.; Sasaki, M.; Takamatsu, T.; Oikawa, K.; Itoh, K.; Kaneko, K.; Ichikawa, H.; Mitsui, T. Golgi/plastid-type manganese superoxide dismutase involved in heat-stress tolerance during grain filling of rice. *Plant Biotechnol. J.* **2015**, *13*, 1251–1263. [CrossRef] [PubMed]

23. Kang, J.S.; Frank, J.; Kang, C.H.; Kajiura, H.; Vikram, M.; Ueda, A.; Kim, S.; Bahk, J.D.; Triplett, B.; Fujiyama, K.; et al. Salt tolerance of *Arabidopsis thaliana* requires maturation of *N*-glycosylated proteins in the Golgi apparatus. *Proc. Natl. Acad. Sci. USA* **2008**, *105*, 5933–5938. [CrossRef] [PubMed]

24. Teskey, R.O.; Fites, J.A.; Samuelson, L.J.; Bongarten, B.C. Stomatal and nonstomatal limitations to net photosynthesis in *Pinus taeda* L. under different environmental conditions. *Tree Physiol.* **1986**, *2*, 131–142. [CrossRef] [PubMed]

25. Sage, R.F.; Sharkey, T.D. The effect of temperature on the occurrence of O_2 and CO_2 insensitive photosynthesis in field grown plants. *Plant Physiol.* **1987**, *84*, 658–664. [CrossRef] [PubMed]

26. Cerasoli, S.; Wertin, T.; McGuire, M.A.; Rodrigues, A.; Aubrey, D.P.; Pereira, J.S.; Teskey, R.O. Poplar saplings exposed to recurring temperature shifts of different amplitude exhibit differences in leaf gas exchange and growth despite equal mean temperature. *AoB Plants* **2014**, *6*, plu018. [CrossRef] [PubMed]

27. Von Caemmerer, S.; Evans, J.R. Temperature responses of mesophyll conductance differ greatly between species. *Plant. Cell Environ.* **2015**, *38*, 629–637. [CrossRef] [PubMed]

28. Uprety, D.C.; Dwivedi, N.; Jain, V.; Mohan, R. Effect of elevated carbon dioxide concentration on the stomatal parameters of rice cultivars. *Photosynthetica* **2002**, *40*, 315–319. [CrossRef]

29. Shimazaki, K.; Iino, M.; Zeiger, E. Blue light-dependent proton extrusion by guard-cell protoplasts of *Vicia faba*. *Nature* **1986**, *319*, 324–326. [CrossRef]

30. Talbott, L.D.; Zeiger, E. The role of sucrose in guard cell osmoregulation. *J. Exp. Bot.* **1998**, *49*, 329–337. [CrossRef]

31. Schroeder, J.I.; Hagiwara, S. Cytosolic calcium regulates ion channels in the plasma membrane of *Vicia faba* guard cells. *Nature* **1989**, *338*, 427–430. [CrossRef]

32. Kusumi, K.; Hirotsuka, S.; Kumamaru, T.; Iba, K. Increased leaf photosynthesis caused by elevated stomatal conductance in a rice mutant deficient in SLAC1, a guard cell anion channel protein. *J. Exp. Bot.* **2012**, *63*, 5635–5644. [CrossRef] [PubMed]

33. Portis, A. Rubisco Activase—Rubisco's catalytic chaperone. *Photosynth. Res.* **2003**, *75*, 11–27. [CrossRef] [PubMed]

34. Li, J.; Baroja-Fernandez, E.; Bahaji, A.; Munoz, F.J.; Ovecka, M.; Montero, M.; Sesma, M.T.; Alonso-Casajus, N.; Almagro, G.; Sanchez-Lopez, A.M.; et al. Enhancing sucrose synthase activity results in increased levels of starch and ADP-glucose in maize (*Zea mays* L.) seed endosperms. *Plant Cell Physiol.* **2013**, *54*, 282–294. [CrossRef] [PubMed]

35. Lu, Y.; Gehan, J.P.; Sharkey, T.D. Daylength and circadian effects on starch degradation and maltose metabolism. *Plant Physiol.* **2005**, *138*, 2280–2291. [CrossRef] [PubMed]

36. Wang, X.; Chang, L.; Tong, Z.; Wang, D.; Yin, Q.; Wang, D.; Jin, X.; Yang, Q.; Wang, L.; Sun, Y.; et al. Proteomics profiling reveals carbohydrate metabolic enzymes and 14-3-3 proteins play important roles for starch accumulation during cassava root tuberization. *Sci. Rep.* **2016**, *6*, 19643. [CrossRef] [PubMed]

37. Clore, A.M.; Dannenhoffer, J.M.; Larkins, B.A. EF-1α is associated with a cytoskeletal network surrounding protein bodies in maize endosperm cells. *Plant Cell* **1996**, *8*, 2003–2014. [CrossRef] [PubMed]

38. Yahalom, A.; Kim, T.-H.; Roy, B.; Singer, R.; Von Arnim, A.G.; Chamovitz, D.A. *Arabidopsis* eIF3e is regulated by the COP9 signalosome and has an impact on development and protein translation. *Plant J.* **2007**, *53*, 300–311. [CrossRef] [PubMed]

39. Li, Q.; Deng, Z.; Gong, C.; Wang, T. The rice eukaryotic translation initiation factor 3 subunit f (OseIF3f) is involved in microgametogenesis. *Front. Plant Sci.* **2016**, *7*, 532. [CrossRef] [PubMed]

40. Kim, T.-H.; Kim, B.-H.; Yahalom, A.; Chamovitz, D.A.; von Arnim, A.G. Translational regulation via 5′ mRNA leader sequences revealed by mutational analysis of the *Arabidopsis translation* initiation factor subunit eIF3h. *Plant Cell* **2004**, *16*, 3341–3356. [CrossRef] [PubMed]

41. Fu, H.; Subramanian, R.R.; Masters, S.C. 14-3-3 proteins: structure, function, and regulation. *Annu. Rev. Pharmacol. Toxicol.* **2000**, *40*, 617–647. [CrossRef] [PubMed]

42. Roberts, M.R.; Salinas, J.; Collinge, D.B. 14-3-3 proteins and the response to abiotic and biotic stress. *Plant Mol. Biol.* **2002**, *50*, 1031–1039. [CrossRef] [PubMed]

43. Wilson, R.S.; Swatek, K.N.; Thelen, J.J. Regulation of the regulators: Post-Translational Modifications, subcellular, and spatiotemporal distribution of plant 14-3-3 proteins. *Front. Plant Sci.* **2016**, *7*, 611. [CrossRef] [PubMed]

44. Xu, W.; Jia, L.; Shi, W.; Liang, J.; Zhang, J. Smart role of plant 14-3-3 proteins in response to phosphate deficiency. *Plant Signal. Behav.* **2012**, *7*, 1047–1048. [CrossRef] [PubMed]

45. Sehnke, P.C.; Chung, H.J.; Wu, K.; Ferl, R.J. Regulation of starch accumulation by granule-associated plant 14-3-3 proteins. *Proc. Natl. Acad. Sci. USA* **2001**, *98*, 765–770. [CrossRef] [PubMed]

46. Baunsgaard, L.; Fuglsang, A.T.; Jahn, T.; Korthout, H.A.; de Boer, A.H.; Palmgren, M.G. The 14-3-3 proteins associate with the plant plasma membrane H$^+$-ATPase to generate a fusicoccin binding complex and a fusicoccin responsive system. *Plant J.* **1998**, *13*, 661–671. [CrossRef] [PubMed]

47. Cotelle, V.; Meek, S.E.; Provan, F.; Milne, F.C.; Morrice, N.; MacKintosh, C. 14-3-3s regulate global cleavage of their diverse binding partners in sugar-starved Arabidopsis cells. *EMBO J.* **2000**, *19*, 2869–2876. [CrossRef] [PubMed]

48. Moorhead, G.; Douglas, P.; Cotelle, V.; Harthill, J.; Morrice, N.; Meek, S.; Deiting, U.; Stitt, M.; Scarabel, M.; Aitken, A.; et al. Phosphorylation-dependent interactions between enzymes of plant metabolism and 14-3-3 proteins. *Plant J.* **1999**, *18*, 1–12. [CrossRef] [PubMed]

49. Toroser, D.; Athwal, G.S.; Huber, S.C. Site-specific regulatory interaction between spinach leaf sucrose-phosphate synthase and 14-3-3 proteins. *FEBS Lett.* **1998**, *435*, 110–114. [CrossRef]

50. Diaz, C.; Kusano, M.; Sulpice, R.; Araki, M.; Redestig, H.; Saito, K.; Stitt, M.; Shin, R. Determining novel functions of Arabidopsis 14-3-3 proteins in central metabolic processes. *BMC Syst. Biol.* **2011**, *5*, 192. [CrossRef] [PubMed]

51. Sehnke, P.C.; Henry, R.; Cline, K.; Ferl, R.J. Interaction of a plant 14-3-3 protein with the signal peptide of a thylakoid-targeted chloroplast precursor protein and the presence of 14-3-3 isoforms in the chloroplast stroma. *Plant Physiol.* **2000**, *122*, 235–242. [CrossRef] [PubMed]

52. Bellafiore, S.; Barneche, F.; Peltier, G.; Rochaix, J.-D. State transitions and light adaptation require chloroplast thylakoid protein kinase STN7. *Nature* **2005**, *433*, 892–895. [CrossRef] [PubMed]

53. Bonardi, V.; Pesaresi, P.; Becker, T.; Schleiff, E.; Wagner, R.; Pfannschmidt, T.; Jahns, P.; Leister, D. Photosystem II core phosphorylation and photosynthetic acclimation require two different protein kinases. *Nature* **2005**, *437*, 1179–1182. [CrossRef] [PubMed]

54. Pietrzykowska, M.; Suorsa, M.; Semchonok, D.A.; Tikkanen, M.; Boekema, E.J.; Aro, E.-M.; Jansson, S. The light-harvesting chlorophyll a/b binding proteins Lhcb1 and Lhcb2 play complementary roles during state transitions in Arabidopsis. *Plant Cell* **2014**, *26*, 3646–3660. [CrossRef] [PubMed]

55. Chen, X.; Zhang, W.; Zhang, B.; Zhou, J.; Wang, Y.; Yang, Q.; Ke, Y.; He, H. Phosphoproteins regulated by heat stress in rice leaves. *Proteome Sci.* **2011**, *9*, 37. [CrossRef] [PubMed]

56. Lemeille, S.; Turkina, M.V.; Vener, A.V.; Rochaix, J.-D. Stt7-dependent phosphorylation during state transitions in the green alga *Chlamydomonas reinhardtii*. *Mol. Cell. Proteom.* **2010**, *9*, 1281–1295. [CrossRef] [PubMed]

57. Facette, M.R.; Shen, Z.; Björnsdóttir, F.R.; Briggs, S.P.; Smith, L.G. Parallel proteomic and phosphoproteomic analyses of successive stages of maize leaf development. *Plant Cell* **2013**, *25*, 2798–2812. [CrossRef] [PubMed]

58. Baginsky, S. Protein phosphorylation in chloroplasts—A survey of phosphorylation targets. *J. Exp. Bot.* **2016**, *67*, 3873–3882. [CrossRef] [PubMed]

59. Rocha, A.G.; Mehlmer, N.; Stael, S.; Mair, A.; Parvin, N.; Chigri, F.; Teige, M.; Vothknecht, U.C. Phosphorylation of Arabidopsis transketolase at Ser 428 provides a potential paradigm for the metabolic control of chloroplast carbon metabolism. *Biochem. J.* **2014**, *458*, 313–322. [CrossRef] [PubMed]

60. Shimizu, M.; Kato, H.; Ogawa, T.; Kurachi, A.; Nakagawa, Y.; Kobayashi, H. Sigma factor phosphorylation in the photosynthetic control of photosystem stoichiometry. *Proc. Natl. Acad. Sci. USA* **2010**, *107*, 10760–10764. [CrossRef] [PubMed]

61. Vargas-Suárez, M.; Castro-Sánchez, A.; Toledo-Ortiz, G.; González de la Vara, L.E.; García, E.; Loza-Tavera, H. Protein phosphorylation regulates in vitro spinach chloroplast petD mRNA 3′-untranslated region stability, processing, and degradation. *Biochimie* **2013**, *95*, 400–409. [CrossRef] [PubMed]

62. Tikkanen, M.; Aro, E.-M. Integrative regulatory network of plant thylakoid energy transduction. *Trends Plant Sci.* **2014**, *19*, 10–17. [CrossRef] [PubMed]

63. Suetsugu, N.; Higa, T.; Kong, S.-G.; Wada, M. Plastid Movement Impaired1 and Plastid Movement Impaired1-Related1 mediate photorelocation movements of both chloroplasts and nuclei. *Plant Physiol.* **2015**, *169*, 1155–1167. [CrossRef] [PubMed]

64. Pfalz, J.; Liere, K.; Kandlbinder, A.; Dietz, K.-J.; Oelmüller, R. pTAC2, -6, and -12 are components of the transcriptionally active plastid chromosome that are required for plastid gene expression. *Plant Cell* **2006**, *18*, 176–197. [CrossRef] [PubMed]

65. Gao, Z.-P.; Chen, G.-X.; Yang, Z.-N. Regulatory role of *Arabidopsis* pTAC14 in chloroplast development and plastid gene expression. *Plant Signal. Behav.* **2012**, *7*, 1354–1356. [CrossRef] [PubMed]

66. Gao, Z.-P.; Yu, Q.-B.; Zhao, T.-T.; Ma, Q.; Chen, G.-X.; Yang, Z.-N. A functional component of the transcriptionally active chromosome complex, *Arabidopsis* pTAC14, interacts with pTAC12/HEMERA and regulates plastid gene expression. *Plant Physiol.* **2011**, *157*, 1733–1745. [CrossRef] [PubMed]

67. Miyao, A.; Tanaka, K.; Murata, K.; Sawaki, H.; Takeda, S.; Abe, K.; Shinozuka, Y.; Onosato, K.; Hirochika, H. Target site specificity of the Tos17 retrotransposon shows a preference for insertion within genes and against insertion in retrotransposon-rich regions of the genome. *Plant Cell* **2003**, *15*, 1771–1780. [CrossRef] [PubMed]

68. Kaneko, K.; Sasaki, M.; Kuribayashi, N.; Suzuki, H.; Sasuga, Y.; Shiraya, T.; Inomata, T.; Itoh, K.; Baslam, M.; Mitsui, T. Proteomic and glycomic characterization of rice chalky grains produced under moderate and high-temperature conditions in field system. *Rice* **2016**, *9*, 26. [CrossRef] [PubMed]

69. Porra, R.J.; Thompson, W.A.; Kriedemann, P.E. Determination of accurate extinction coefficients and simultaneous equations for assaying chlorophylls a and b extracted with four different solvents: Verification of the concentration of chlorophyll standards by atomic absorption spectroscopy. *Biochim. Biophys. Acta Bioenerg.* **1989**, *975*, 384–394. [CrossRef]

70. Dorfer, V.; Pichler, P.; Stranzl, T.; Stadlmann, J.; Taus, T.; Winkler, S.; Mechtler, K. MS Amanda, a universal identification algorithm optimized for high accuracy tandem mass spectra. *J. Proteome Res.* **2014**, *13*, 3679–3684. [CrossRef] [PubMed]

71. Takamatsu, T.; Baslam, M.; Inomata, T.; Oikawa, K.; Itoh, K.; Ohnishi, T.; Kinoshita, T.; Mitsui, T. Optimized method of extracting rice chloroplast DNA for high-quality plastome resequencing and *de novo* assembly. *Front. Plant Sci.* **2018**, *9*, 266. [CrossRef] [PubMed]

72. Fukuda, I.; Hirabayashi-Ishioka, Y.; Sakikawa, I.; Ota, T.; Yokoyama, M.; Uchiumi, T.; Morita, A. Optimization of enrichment conditions on TiO_2 chromatography using glycerol as an additive reagent for effective phosphoproteomic analysis. *J. Proteome Res.* **2013**, *12*, 5587–5597. [CrossRef] [PubMed]

International Journal of
Molecular Sciences

MDPI

Article

Quantitative Proteomics Analysis of Lettuce (*Lactuca sativa* L.) Reveals Molecular Basis-Associated Auxin and Photosynthesis with Bolting Induced by High Temperature

Jing-Hong Hao [1,†], Li-Li Zhang [1,†], Pan-Pan Li [1], Yan-Chuan Sun [1], Jian-Ke Li [2], Xiao-Xiao Qin [1], Lu Wang [1], Zheng-Yang Qi [1], Shuang Xiao [1], Ying-Yan Han [1], Chao-Jie Liu [1] and Shuang-Xi Fan [1,*]

[1]   Beijing Key Laboratory of New Technology in Agricultural Application, National Demonstration Center for Experimental Plant Production Education, Plant Science and Technology College, Beijing University of Agriculture, Beijing 102206, China; haojinghong2013@126.com (J.-H.H.); zll0224@126.com (L.-L.Z.); plee0616@163.com (P.-P.L.); syc546421184@163.com (Y.-C.S.); vipqindada@163.com (X.-X.Q.); 18810986422@163.com (L.W.); 15100258302@163.com (Z.-Y.Q.); jiumingxs@sina.com (S.X.); hyybac@126.com (Y.-Y.H.); cliu@bua.edu.cn (C.-J.L.)

[2]   Institute of Apicultural Research, Chinese Academy of Agricultural Science, No. 1 Beigou Xiangshan, Beijing 100093, China; apislijk@126.com

*   Correspondence: fsx20@bua.edu.cn; Tel./Fax: +86-10-8079-7238

†   These authors contributed equally to this work.

Received: 16 July 2018; Accepted: 26 September 2018; Published: 28 September 2018

Abstract: Bolting is a key process in the growth and development of lettuce (*Lactuca sativa* L.). A high temperature can induce early bolting, which decreases both the quality and production of lettuce. However, knowledge of underlying lettuce bolting is still lacking. To better understand the molecular basis of bolting, a comparative proteomics analysis was conducted on lettuce stems, during the bolting period induced by a high temperature (33 °C) and a control temperature (20 °C) using iTRAQ-based proteomics, phenotypic measures, and biological verifications using qRT-PCR and Western blot. The high temperature induced lettuce bolting, while the control temperature did not. Of the 5454 identified proteins, 619 proteins presented differential abundance induced by high-temperature relative to the control group, of which 345 had an increased abundance and 274 had a decreased abundance. Proteins with an abundance level change were mainly enriched in pathways associated with photosynthesis and tryptophan metabolism involved in auxin (IAA) biosynthesis. Moreover, among the proteins with differential abundance, proteins associated with photosynthesis and tryptophan metabolism were increased. These findings indicate that a high temperature enhances the function of photosynthesis and IAA biosynthesis to promote the process of bolting, which is in line with the physiology and transcription level of IAA metabolism. Our data provide a first comprehensive dataset for gaining novel understanding of the molecular basis underlying lettuce bolting induced by high temperature. It is potentially important for further functional analysis and genetic manipulation for molecular breeding to breed new cultivars of lettuce to restrain early bolting, which is vital for improving vegetable quality.

Keywords: lettuce; bolting; proteome; high temperature; iTRAQ

1. Introduction

Bolting is a clear characteristic of the transition from vegetative to reproductive growth in blade root vegetable plants. It is defined as the accelerated and sustained rapid elongation of the stem flower bud differentiation, after which flowering begins. Early bolting refers to the phenomenon of bolting before the formation of the plant product, and seriously affects crop yield and quality. When early bolting occurs, the main characteristics of leaf vegetable performance include the following: premature differentiation of flower buds, reduction in the number of leaves, textural decline, development of a bitter flavor, and the inability to form a compact leaf ball. This process usually causes huge economic losses for producers, and thus strategies to prevent early bolting are urgently needed.

Bolting is regulated by diverse environmental and endogenous factors, such as the temperature, light signals, day length, developmental stage and plant hormones [1]. It consists of a series of physiological and biochemical reactions in plant cells. Researchers have found that the soluble protein, soluble sugar, free amino acid, peroxisome (POD), and vitamin C (Vc) contents are likely germane to the initiation of bolting in mustard and cabbage [2]. Gibberellins (GAs) are phytohormones that regulate many aspects of plant growth and development, including bolting [3]. A puzzling and controversial phenomenon is that ethylene delays bolting in wild-type *Arabidopsis*, yet both the constitutive triple response mutant (*ctr1-1*) and ethylene-insensitive mutants (*etr1-1* and *ein2-5*) exhibit a similar delayed bolting phenotype [4].

In recent studies, several genes and proteins have been implicated in a bolting control according to the isolation of loss-of-function mutants or analysis of transgenic plants. It was found that sugar beet contains a large *CONSTANS*-like gene family, independent of the early-bolting (*B*) gene locus [5]. It was also speculated that the *BrVHA-E1*, *BrSAMS*, *BrrbcL*, and *BrTUA6* genes might be involved in regulating the flower differentiation and bolting of *Brassica rapa* according to their significantly different expression levels [6]. S-adenosylmethionine synthetase (SAMS) is a precursor of ethylene [7]. Zhu et al. found that the serine-62 (Ser-62) phosphorylation of Ethylene Response Factor110 (ERF110) is involved regulating the bolting time of Arabidopsis [8]. Meanwhile, liu et al. found that ETH promotes flowering of Pineapple. It can be speculated that there is a connection between SAMS and bolting, but the specific mechanism of action is not clear. Research by Strompen et al. shows that VHA-E1 is involved in regulating early embryogenesis in Arabidopsis [9]. Zhang et al. speculated that BrVHA-E1 is only associated with the differentiation of flower buds [6]. Microtubules exist widely in the cytoskeleton, and are important for cellular morphology, cell division, transportation, energy transfer, and signal transduction, and α-tubulin (*TUA*) is the basic unit of microtubules. The expression of three *TUAs* was determined in *Brassica rapa*, and the results showed that *BrTUAs* were involved in the regulation of bolting in *Brassica rapa*; *TUAs* affect the speed of shaft elongation [10]. Rubisco is the key regulatory enzyme catalyzing CO_2 fixation and ribulose diphosphate oxygenase reaction dual function, which determines net photosynthesis [11]. The holoenzyme of Rubisco is composed of eight small subunits encoded by a nuclear multigene family (rbcS), and eight large subunits encoded by a single gene (rbcL) in the multicopy chloroplast genome. Sucrose has been found to directly regulate the expression of the flower regulator LFY [12]. The chloroplast *rbcL* gene encoded Ribulose-1,5-bisphosphate carboxylase oxygenase (RuBisCO), which is the key enzyme involved in the calvin cycle of carbon assimilation during photosynthesis [13]. *BrrbcL* was up-regulated after color change. It boosts the synthesis of photosynthetic products, leading to an increased C/N ratio, and the subsequent occurrence of bolting and flowering in the plant [6]. ZCE1 is a *cis-cinnamic acid* (*Zusammen-cinnamic acid*)-*Enhanced* gene, which encodes a member of the major latex protein-like (*MLPL*) gene family. The zce1 mutant produced by the RNA-interference technique shows an earlier bolting phenotype in *Arabidopsis*, indicating that ZCE1 plays a role in delaying bolting [14]. From the perspective of previous researches, many physiological metabolisms and some genes or proteins are related to bolting. However, it is possible that other metabolisms may take part in bolting but are still unknown.

Lettuce (*Lactuca sativa* L.) has economic importance because it is a globally consumed, popular leafy vegetable. This vegetable originated along the Mediterranean coast, and has an optimum growth temperature of 15–20 °C; hence, it is sensitive to high temperatures. When temperatures exceed 30 °C, lettuces undergo early bolting, which lead to decreased nutritional quality, reduced commercial value, and significant losses in productivity and economic benefits. Little is known about bolting in lettuce. It was reported that a high temperature induced bolting in lettuce and that GAs played an important role in this process. *LsGA3ox1* is a gene that is possibly responsible for the

Int. J. Mol. Sci. **2018**, *19*, 2967

increased GA1, but the mechanism of GA metabolism and/or action may differ among cultivars with different bolting characteristics [3]. *SUPPRESSOR OF OVEREXPRESSION OF CO 1 (SOC1)* encodes a MADS-box protein that integrates multiple flowering signals derived from photoperiod, temperature, hormone and age-related pathways [15]. *SOC1* interacts with multiple MADS-box proteins, including *FRUITFULL (FUL)*, *AP1* and *AGAMOUS LIKE24 (AGL24)*, and regulates several flowering genes, e.g. by directly binding to their regulatory sequences [16]. Han and colleagues from our laboratory [17] reported that, although GA regulates bolting in lettuce, it may be the MADS-box genes instead, which play a major role in differing the bolting resistance between a bolting resistant line and a bolting sensitive line. A total of 12 MADS-box transcription factors were dramatically induced in lettuce during bolting such as putative *LsSOC1*, *LsAP1*, *LsFUL* and *LsAGL24*.

However, it is unknown whether other hormones and metabolism or main genes/proteins functions in bolting, and the molecular mechanism remains elusive.

Proteomics is becoming an increasingly important tool to reveal molecular mechanisms at the overall level of protein expression because proteins are directly linked to cellular functions. The mRNA only shows changes at the transcriptional level and cannot fully represent the true level of protein expression. Therefore, proteins and their functions must be investigated to study genetic features. Recently, proteomics has been widely used in the exploration of the resistance mechanism or development characters of plants, such as heat resistance [18], cold resistance and salt resistance [19]. Applying the proteomics approach, the bolting pattern in several plant species has been analyzed, such as *Brassica rapa* [5], *Lactuca sativa* [17], and *Arabidopsis thaliana* [8,14]. Furthermore, the proteomics analysis of the molecular basis of bolting in lettuce (a non-vernalization plant) induced by a high temperature was reported by Han and colleagues from our laboratory [17]. Due to the technological limitation of two-dimensional electrophoresis-based proteomics, only 30 proteins with differential abundance were identified in lettuce in the previous work from our laboratory [17]. The technological advances in the resolution and accuracy of mass spectrometry and efficient labeling quantification of protein abundance levels, allow for a greater depth of proteome coverage. The isobaric tagging for relative and absolute quantification (iTRAQ) was performed. This provides an opportunity for gaining new insight into the molecular basis that drives the bolting of lettuce induced by high temperatures. Here, we revealed potential mechanisms involved in the regulation of bolting by proteomics. The results were verified using Western blot, real-time quantitative fluorescence PCR, and physiological analyses. Finally, a possible pathway map of bolting in lettuce induced by high temperatures was proposed. Our data may potentially be important for resolving the challenging problem of early bolting in vegetable cultivation, the genetic manipulation of lettuce and other bolting plants.

2. Results

2.1. The Morphological and Physiological Changes of Lettuce Stems During Bolting Induced by High Temperatures

High temperature treatment promoted stem elongation. On Day 8, the length of the stem had increased, with a significant difference ($p < 0.05$) in the high-temperature group compared to the control group. With a longer treatment time, the stem elongation rate accelerated significantly. On Day 16, a strongly significant increase ($p < 0.01$) of 92.9% in the stem length was observed in the high-temperature group, relative to the control group. Afterwards, the stem increase trend was more significant (Figure 1A). The leaves of the high-temperature group turned yellow and withered on Day 40, whereas the leaves of the control group grew well (Figure 1B). A significant change was observed at the stem tip (Figure 2). As shown in Figure 2, the stem tip in the control group remained conical throughout the entire experimental/observational period. The growing point of the stem tip in the high-temperature group remained conical until Day 8. Subsequently, the growing point became larger and less prominent. On Day 24 of the treatment group, the growing point was completely flattened, and the basal inflorescence was entirely raised. On Day 40, part of the phyllary was differentiated at the base of the inflorescence. Combining the change of stem elongation with the progress of flower bud differentiation under a high temperature, it was concluded that obvious bolting had occurred from Day 8 to Day 40 after treatment with a high temperature. The plants showed obvious bolting on Day 32 under a high temperature, thus Day 32 was selected for sampling for the proteome and physiology analyses.

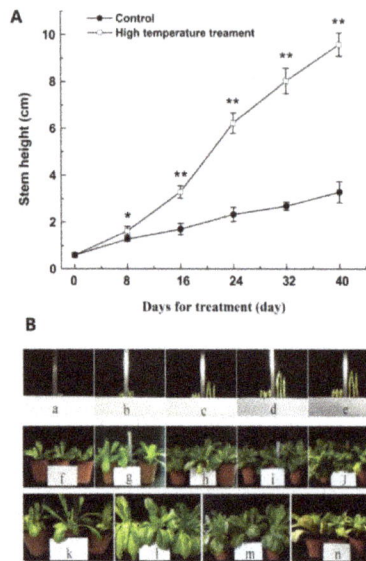

Figure 1. Change of stem height in lettuce after high temperature treatment of 33/25 °C. (**A**) Changes of stem length under 20/13 °C (day/night) (control) and 33/25 °C (day/night) treatment for 40 days. The data (mean ± SD) are the means of three replicates with standard errors shown by vertical bars, $n = 9$. * and ** indicate significant difference at $p < 0.05$ and $p < 0.01$ by Tukey's test, respectively. (**B**) The phenotypes of lettuce under 20/13 °C (day/night) (control) and 33/25 °C (day/night) treatment for 40 days. (a–e) Stem growth after different temperature treatment for 0, 8, 24, 32 and 40 days (control on left and high temperature treatment on right). Representative images of plants under control (g–j) and high temperature treatment (k–n) for 8, 24, 32 and 40 days, and (f) plant growth at Day 0.

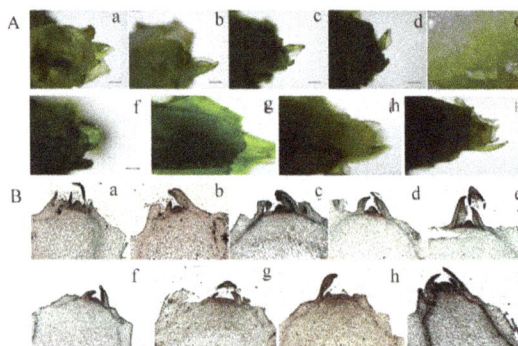

Figure 2. Change of flower bud differentiation of lettuce after high temperature treatment of 33/25 °C. (**A**) Representative images of stem tips under 20/13 °C (day/night) (control) and 33/25 °C (day/night) treatment for 40 days. Representative images of stem tip growth under control (b–e) and high temperature treatment (f–i) for 8, 24, 32 and 40 days, and (a) stem tip growth at Day 0. (**B**) The progress of flower bud differentiation. Representative images of morphology of flower bud under control (b–e) and high temperature treatment (f–i) for 8, 24, 32 and 40 days, and (a) stem tip growth at Day 0.

From the stem tissues of lettuces in control and high-temperature groups, the contents of six endogenous hormones, namely gibberellins (GA_{1+3}), zeatin (ZR), brassinosteroid (BR), jasmonic acid methyl ester (JA-ME), auxin (IAA), and abscisic acid (ABA), were examined. As compared to the control group, the contents of ABA, GA_{1+3}, ZR, IAA, JA-ME, and BR were significantly higher in the high-temperature group, and there was a stronger significant difference ($p < 0.01$) in IAA than in ABA, GA_{1+3}, JA-ME, and BR ($p < 0.05$), with increases of 44.9%, 17.8%, 23.7%, 33.3%, and 34.4%, respectively (Figure 3A).

To determine whether IAA mediates bolting in lettuce, we next explored the effect of exogenous IAA on the bolting of lettuce. As shown in Figure 3B, IAA could promote bolting. In the IAA treated plants, lettuce was obviously extended on Day 5, and the stem length was increased by 32.6% with significant difference ($p < 0.05$). The largest increase ratio was 138.3% on Day 30 compared to mock-treated plants, suggesting that exogenous IAA accelerates bolting.

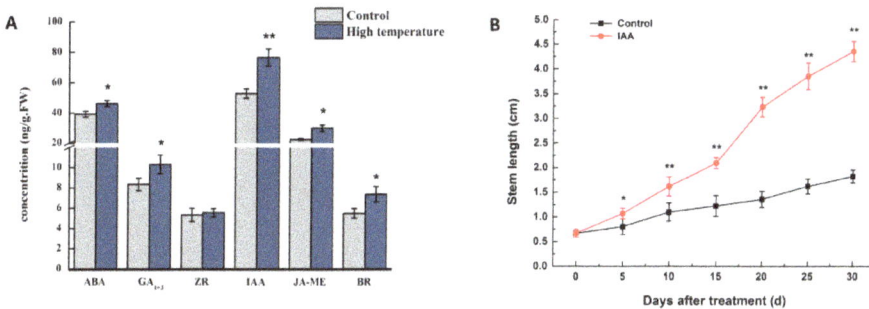

Figure 3. Physiological measurements of lettuce after high temperature treatment of 33/25 °C. (**A**) Content of endogenous hormones (auxin (IAA), gibberellins (GA_{1+3}), zeatin (ZR), jasmonic acid methyl ester (JA-ME), abscisic acid (ABA), and brassinosteroid (BR)) in the stems of lettuce in the condition of 20/13 °C (day/night) (control) and 33/25 °C (day/night) treatment for 32 days. (**B**) Stem elongation of lettuce after exogenous 40 mg/L auxin treatment. * and ** indicate significant difference at $p < 0.05$ and $p < 0.01$ by Tukey's test, respectively.

2.2. Identification of Differential Abundance Proteins Using iTRAQ in Lettuce Stems During Bolting Induced by High Temperature

By means of the iTRAQ-labeled proteomics approach, 5454 proteins were identified in the lettuce stems, as shown in Table S1. The mass spectrometry proteomics data have been deposited in the ProteomeXchange Consortium (http://www.proteomexchange.org) via the PRIDE database, with the dataset identifier, PXD008610. A total of 619 proteins changed significantly in abundance, and 345 of these proteins had increased abundance (red section in Figure 4) while 274 had decreased abundance (yellow section in Figure 4). Detailed information on proteins with differential abundance is shown in Table S2, and the spectra of these proteins, with one unique peptide, are shown in Table S3. Among these, the increased abundance proteins with the highest fold change were hypothetical protein Ccrd_011733 (3.22) and bet v i domain-containing protein (2.51). For the decreased abundance proteins, these were 3-n-debenzoyl-2-deoxytaxol n-benzoyltransferase (0.38), and protein light-dependent short hypocotyls 5-like (0.39). All peptide match information including m/z, score, delta, PTM. expect value, PSMs, PEP, Charge and RT are shown in Table S4.

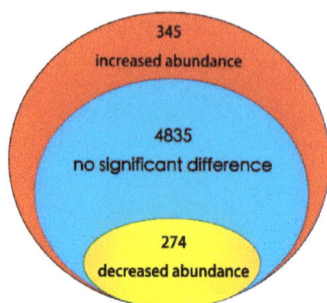

Figure 4. The distribution of proteins with differential abundance.

2.3. Functional Classification and Metabolic Pathways of Differential Abundance Proteins

We aimed to study the mechanisms by which proteins modulate lettuce bolting induced by high temperatures. Based on the BLAST alignment, Gene Ontology (GO) classification, and literature [20], the identified proteins were classified into 13 functional categories. To identify the significant changes in biological process (BP), molecular function (MF), and cellular component (CC) between the control and high-temperature treatment groups, GO annotation was performed using the Trinotate through BLAST search against the well-annotated protein sequences (SwissProt). Of the total differential proteins, 1796 GO terms were annotated (Figure 5). At level 2, the proteins with differential abundance in bolting under a high temperature in the BP category were annotated with the following terms: Metabolic process (36.49%), cellular process (28.83%), response to stimulus (7.36%), localization (5.86%), biological regulation (4.65%), etc. Similarly, the catalytic activity (50.15%), binding (38.30%), structural molecule activity (4.45%), transporter activity (3.29%), and other (3.87%) terms were annotated in the MF category. In the CC category, cell (21.96%), cell part (21.85%), membrane (16.11%) and organelle (12.91%) terms were annotated. Next, the proteins with increased and decreased abundance were utilized for the GO term enrichment analysis (Figure 5D). Results show that the main GO enrichment functions of increased abundance proteins were: Plastid (43), chloroplast (37), thylakoid (29), cell periphery (26), organelle subcompartment (23), plastid part (23), and chloroplast part (23). The main GO enrichment functions of decreased abundance proteins were carbohydrate metabolic process (28), cell periphery (27), hydrolase activity, acting on glycosyl bonds (20), plastid (13), and chloroplast (12). Overall, most proteins with differential abundance taking part in thylakoid, thylakoid part, photosynthetic membrane, plastid thylakoid, photosynthesis, etc.

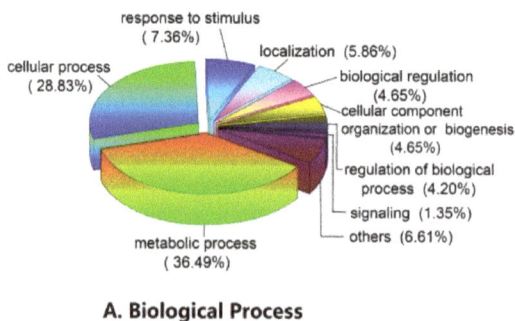

A. Biological Process

Figure 5. *Cont.*

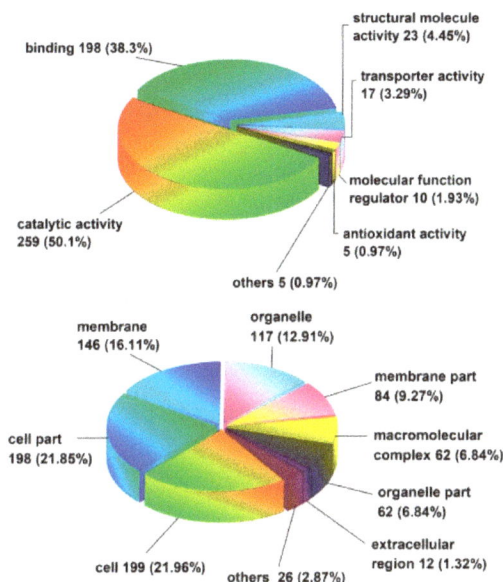

C. Celluclar Component

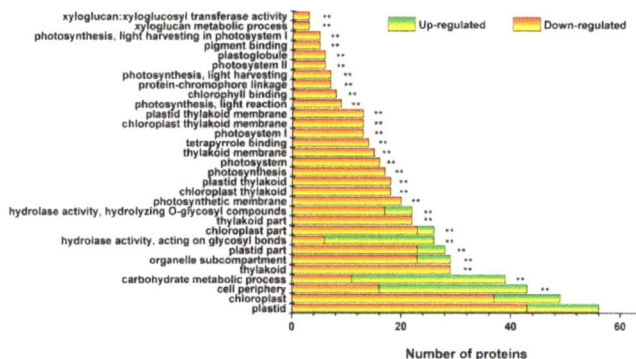

D. GO Enrichment

Figure 5. ClueGO and GO enrichment analysis of proteins with differential abundance: (**A**) Biological Process; (**B**) Molecular Function; (**C**) Cellular Component; and (**D**) GO enrichment. ** indicate significant difference at $p < 0.01$ by Tukey's test, respectively.

To analyze and identify the major metabolic and signal transduction pathways of the proteins with differential abundance, the KEGG database [21] was used. After annotation, the KO number of proteins that were homologous/similar to that of the related KEGG pathway via sequence alignment was determined, and 189 KEGG signaling/metabolic pathways associated with 467 proteins were extracted. Related plants accounted for 76.90% of these proteins. As shown in Figure 6A,B, the metabolic pathways observed in over half of the proteins were: Ribosome (15.03%), photosynthesis (9.80%), phenylpropanoid biosynthesis (7.19%), pyruvate metabolism (5.23%), photosynthesis-antenna proteins (5.23%), starch and sucrose metabolism (4.58%), carbon fixation in photosynthetic organisms (4.58%), protein processing in endoplasmic reticulum (4.58%), spliceosome (4.58%), oxidative phosphorylation (3.92%), glycolysis/gluconeogenesis (3.92%), Cysteine and methionine metabolism (3.92%), Glutathione metabolism (3.92%), and necroptosis (3.92%). Several other significant metabolic pathways were observed as well, such as plant–pathogen interaction (3.27%), peroxisome (2.61%), tryptophan metabolism (2.61%), plant hormone signal transduction (2.61%), and metabolism of xenobiotics by cytochrome P450 (1.96%). To better understand the key metabolic pathways involved in the bolting of lettuce, all differential proteins were successfully enriched and aligned with 14 KEGG pathways (Figure 6C). As shown in Figure 6C, we found that the significantly enriched metabolic pathways of increased abundance proteins were photosynthesis (15), ribosome (10), photosynthesis antenna proteins (8), phenylpropanoid biosynthesis (6), ascorbate and aldarate metabolism (5), tryptophan metabolism (4), mineral absorption (3) and sesquiterpenoid and triterpenoid biosynthesis (3). The significantly enriched metabolic pathways of decreased abundance proteins were ribosome (13), phenylpropanoid biosynthesis (5) and ascorbate and aldarate metabolism (2). Furthermore, all the proteins with differential abundance that take part in photosynthesis, photosynthesis-antenna proteins, tryptophan metabolism, mineral absorption, and sesquiterpenoid and triterpenoid biosynthesis had increased abundance.

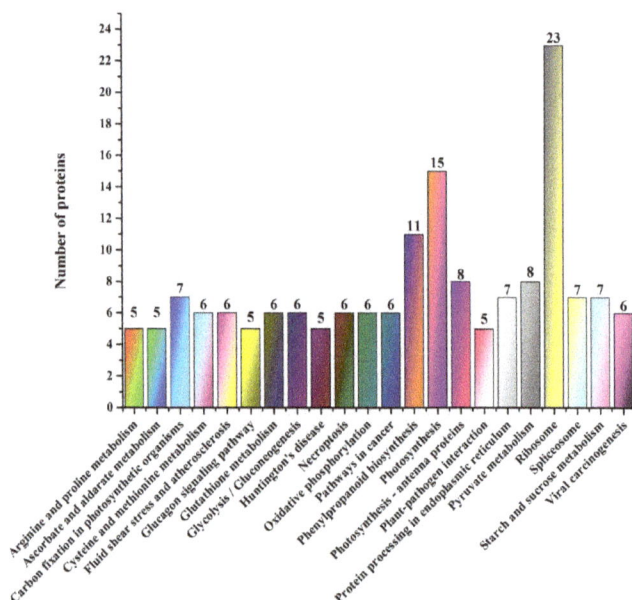

A. KEGG Pathway

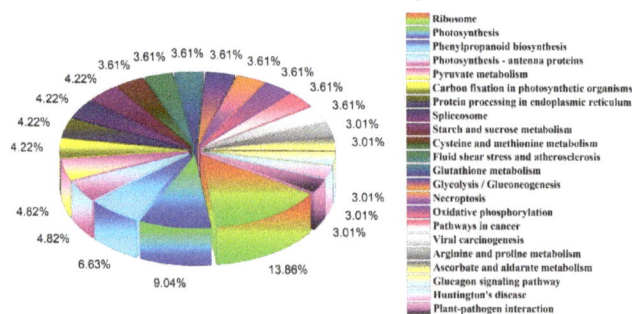

B. The percentage of Matebolic pathway

Figure 6. *Cont.*

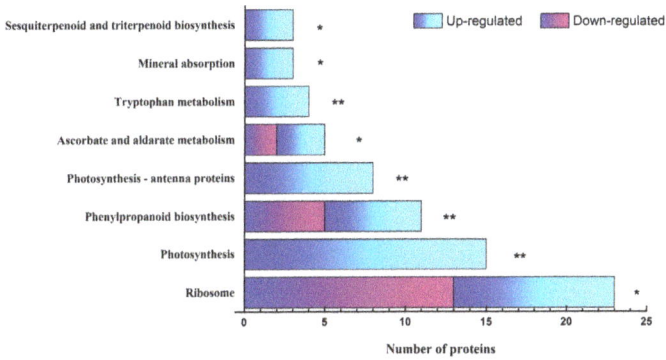

C. KEGG Enrichment

Figure 6. Metabolic analysis of KEGG pathway (matched number more than four) and KEGG Enrichment: (**A**) KEGG Pathway; (**B**) the percentage of Metabolic Pathway; and (**C**) KEGG Enrichment. * and ** indicate significant difference at $p < 0.05$ and $p < 0.01$ by Tukey's test, respectively.

In the last step of the biosynthesis of indoleacetate in tryptophan metabolism (auxin biosynthesis) (Figure 7A), four proteins with increased abundance were identified: aldehyde dehydrogenase family 2 member mitochondrial (ALDH), amidase family protein, catalase (KatE), and Acetyl-CoA c-acetyltransferase. In the plant hormone signal transduction pathway (Figure 7B), four proteins with differential abundance involved three hormone signal transduction pathways: GRP in gibberellin signal transduction, AHP in cytokinine signal transduction, and NPR1 and PR-1 in salicylic acid signal transduction. Among these pathways, the proteins with increased abundance were GRP and AHP and the proteins with decreased abundance were NPR1 and PR-1.

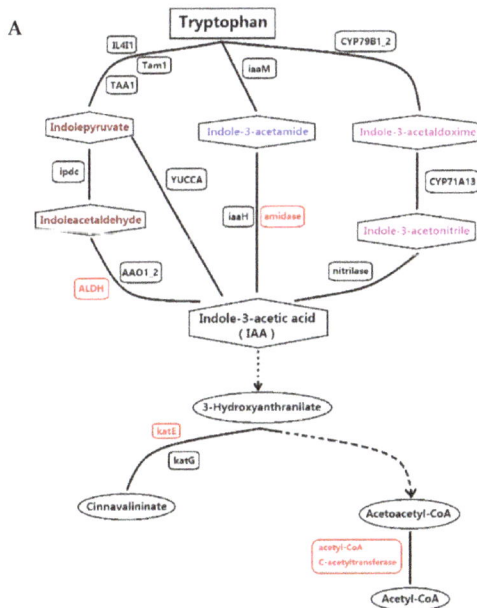

Figure 7. *Cont.*

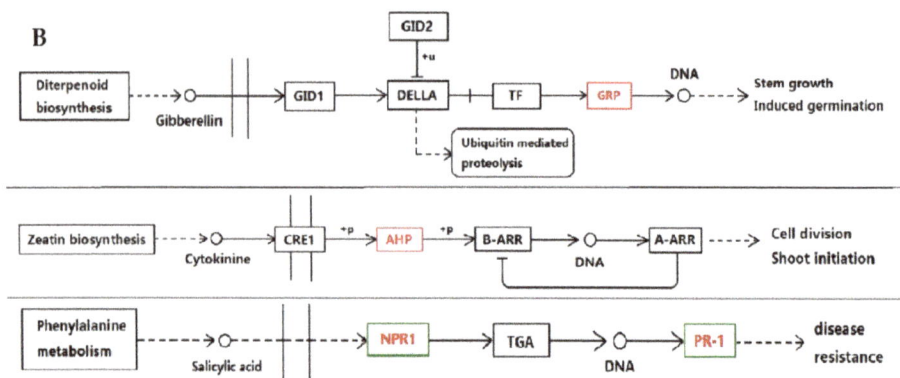

Figure 7. Maps of tryptophan metabolism and plant hormone signal transduction pathways: (**A**) tryptophan metabolism (auxin biosynthesis pathway); and (**B**) plant hormone signal transduction pathway. The red rectangles represent increased abundance proteins, and the green rectangles represent decreased abundance proteins. Definition of proteins indicated in Figure 7A: ALDH, aldehyde dehydrogenase family 2 member mitochondrial; KatE, catalase. Definition of proteins indicated in Figure 7B: GRP, gibberellin regulated protein; AHP, histidine-containing phosphotransfer protein; NPR1, regulatory protein NPR1; PR-1, pathogenesis-related protein 1. The dotted line means many steps of process; The double line means cytomembrane; +u means ubiquitination; +p means Phosphorylation; Vertical lines mean dissociation; The dotted line with a circle means chemical molecule undergoing several steps to combine with next substance.

In photosynthesis (Figure 8A) and photosynthesis-antenna proteins (Figure 8B) pathways, there were 14 and 7 proteins with differential abundance, respectively, all of which had increased abundance. In the photosynthesis pathway, the proteins were associated with photosystem I, photosystem II, cytochrome b6/f complex, photosynthetic electron transport, and f-type ATPase (Figure 8A). In the photosynthesis-antenna proteins pathway, the proteins with differential abundance were implicated in allophycocyanin (AP), allophycocyanin (PC)/phycoerythrocyanin (PEC), phycoerythrin (PE), and the light-harvesting chlorophyll protein complex (LHC) (Figure 8B).

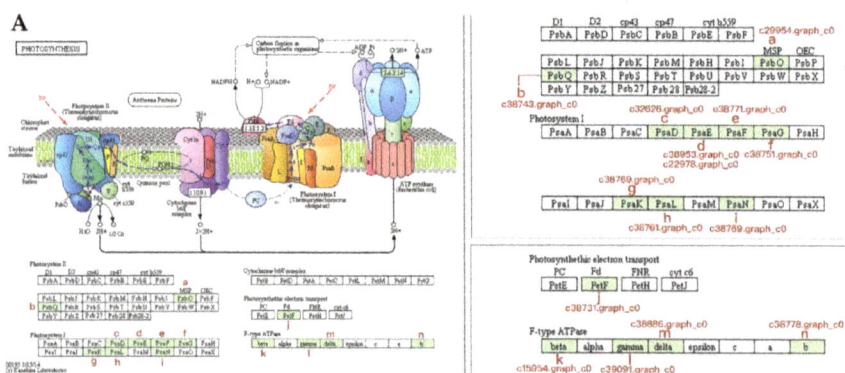

Figure 8. *Cont.*

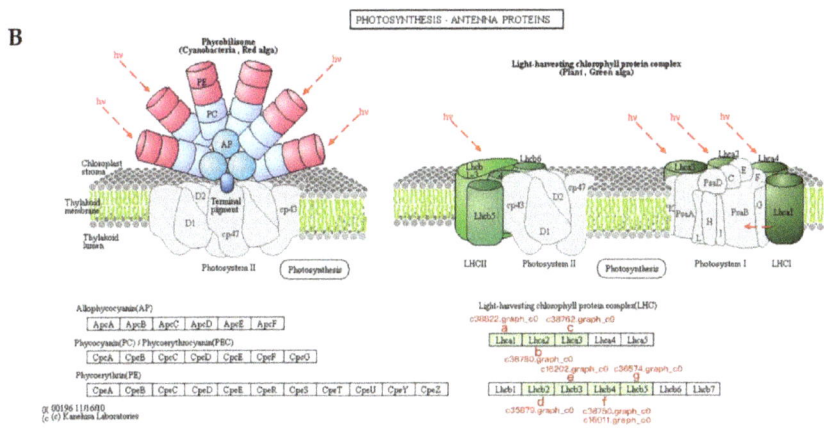

Figure 8. Maps of photosynthesis and photosynthesis-antenna proteins pathways: (**A**) photosynthesis pathway; and (**B**) photosynthesis-antenna proteins pathway. (**A**) The green background rectangles on the left side of the graph represent the identified proteins with differential abundance. The right side of the graph is the enlarged image of the important part, and the proteins with differential abundance are labeled with accession no. (**B**) The green background rectangles represent the identified proteins with differential abundance and labeled with accession no. Definition of proteins indicated in Figure 8A: a: PsbO 1, photosystem II oxygen-evolving enhancer protein 1; b: PsbQ 3, photosystem II oxygen-evolving enhancer protein 3; c: PsaD, photosystem I subunit II; d: PsaE, photosystem I reaction center subunit IV; e: PsaF, photosystem I reaction center subunit III; f: PsaG, photosystem I subunit V; g: PsaK, photosystem I subunit X; h: PsaL, photosystem I subunit XI; i: PsaN, photosystem I subunit PsaN; j: PetF, ferredoxin; k: beta (β), F-type H$^+$-transporting ATPase subunit beta; l: gamma (γ), F-type H$^+$-transporting ATPase subunit gamma; m: delta (δ), F-type H$^+$-transporting ATPase subunit delta; n: b, F-type H$^+$-transporting ATPase subunit b. Definition of proteins indicated in Figure 8B: Lhca1–3, light-harvesting complex I chlorophyll a/b binding protein 1–3; Lhcb2–5, light-harvesting complex II chlorophyll a/ binding protein 2–5. alpha (α), F-type H$^+$-transporting ATPase subunit alpha; D1: protein subunit D1; D2: protein subunit D2; Arrow: light.

2.4. Hierarchical Clustering of Protein Profiles

To identify the proteins with similar expression patterns, hierarchical clustering was performed. An uncentered correlation was used to define the similarity. The hierarchical clusters were assembled using the average linkage clustering method. The proteins with differential abundance were classified into six main clusters (Figure 9). Compared to the control group, the proteins in Clusters 1–3 of the high-temperature group have increased abundance and the proteins in Clusters 4–6 have decreased abundance. Cluster 1 had the highest increased abundance and included 37 proteins; Cluster 6 had the highest decreased abundance and included 94 proteins. The number of proteins in Clusters 2 and 5 accounted for approximately half of the total number of proteins with differential abundance and consisted of 198 and 214 proteins, respectively, whereas Clusters 3 and 4 consisted of 39 and 37 proteins, respectively. To better visualize the protein clusters during bolting under a high temperature, the significant proteins with differential abundance were also sorted and analyzed by clusters in the form as shown on the right side of Figure 9. After performing a comparison and analysis of the six images, we found that the major biological processes involved in each cluster were as follows: Cluster 1: catalytic activity (13), metabolic process (8), membrane (8), cell (7), cell part (7), cellular process (6) and binding (6); Cluster 2: metabolic process (87), catalytic activity (82), cell (76), cell part (76), cellular process (68), binding (67) and organelle (45); Cluster 3: catalytic activity (24), metabolic process (21), binding (11), cellular process (10), cell (7) and cell part (7); Cluster 4: metabolic process (14), cellular process (14), membrane (13), cell (13), cell part (13) and binding (12); Cluster 5: catalytic activity (94), metabolic process (81), binding

(74), cellular process (65), cell (64) and cell part (63); and Cluster 6: catalytic activity (35), cell (32), cell part (32), metabolic process (30), cellular process (27), membrane (26) and binding (25).

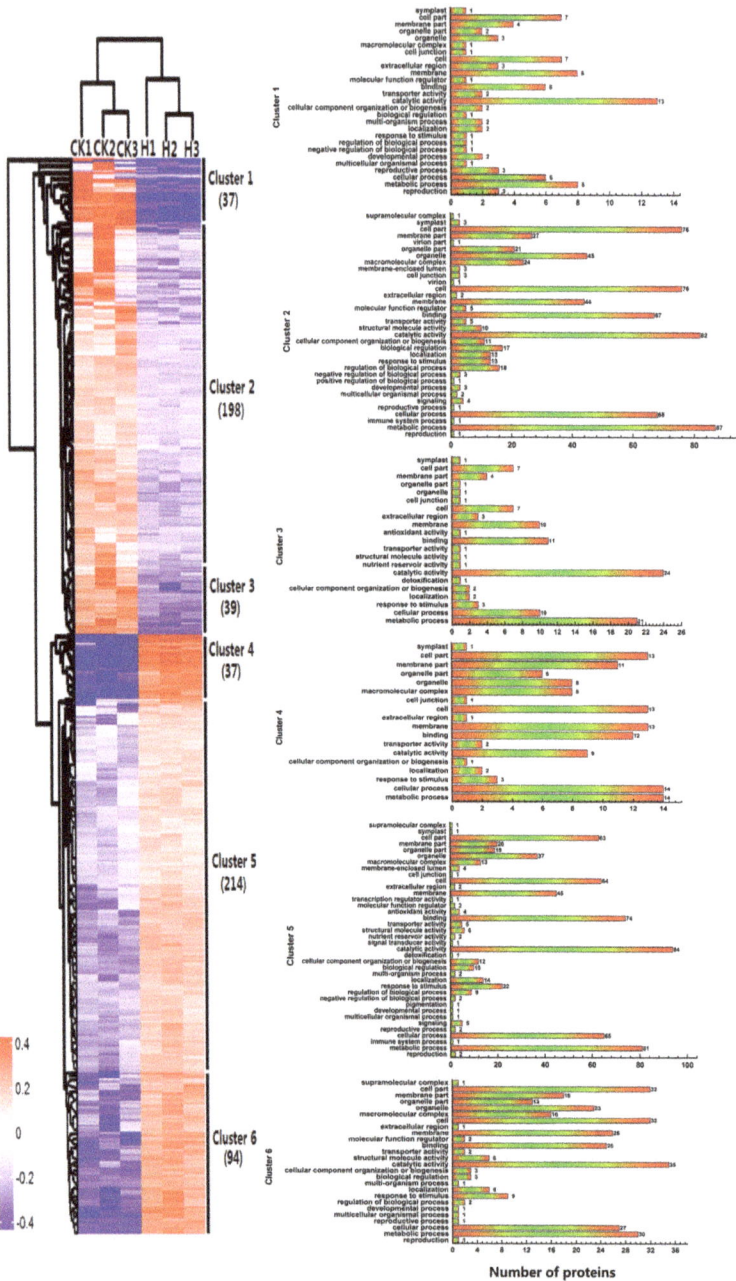

Figure 9. Hierarchical clustering analysis of the differential proteins in lettuce stem under high temperature. The left graph represents hierarchical clustering of proteins with differential abundance. The right graphs represent protein functional classification of the clusters.

2.5. Expression Levels of Genes Encoding Some Identified Proteins

To understand the relationship between the abundance of a protein and the level of its gene transcripts, we measured the expression profiles of genes encoding 13 selected key node proteins. These proteins are primarily observed in five major functional groups: Phytohormone metabolism (GA, IAA, and ETH), signal transduction, oxido-reduction, ubiquitin degradation, and protein kinase. The mRNA expression trend showed that eight proteins (ADF2MC4, ADF2MM, GSTL3L, PD, ACO1, STPK, NPR1 and PLRLSTPKRIX1) were consistent with the protein abundance (Figure 10). However, the expression levels of five proteins (EIX1, AFP, AACT, CYP71A22 and GSTL3) did not conform to at the mRNA and protein level, which might have been caused by the presence of post-translational modifications.

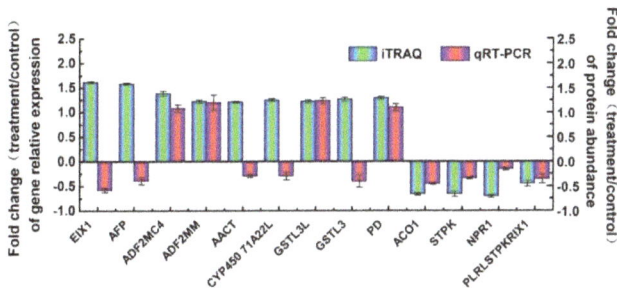

Figure 10. Correlation of mRNA level and protein abundance by iTRAQ. The fold-change of treatment/control in transcript level using the qRT-PCR approach of 19 candidate genes involved in the identified proteins with differential abundance and the protein abundance level by iTRAQ is shown in the figure. The positive number indicates increased abundance, and the negative number indicates decreased abundance. Each histogram represents the mean value of three biological replicates, and the vertical bars indicate the standard error (n = 3). Definition of 19 candidate genes involved in the identified proteins with differential abundance: EIX1, esterase isoform x1; AFP, amidase family protein; ADF2MC4, aldehyde dehydrogenase family 2 member c4; ADF2MM, aldehyde dehydrogenase family 2 member mitochondrial; AACT, acetyl-CoA c-acetyltransferase 3; CYP 71A22L, cytochrome p450 71a22-like; GSTL3L, glutathione s-transferase l3-like; GSTL3, glutathione s-transferase l3; PD, phytanoyl-dioxygenase; ACO1,1-aminocyclopropane-1-carboxylate oxidase 1; STPK, serine threonine-protein kinase; NPR1, regulatory protein NPR1; PLRLSTPKRIX1, probable lrr receptor-like serine threonine-protein kinase rfk1 isoform x1.

3. Discussion

3.1. Proteins Implicated in Hormone Metabolism During Bolting in Lettuce Under a High Temperature

Flower bud differentiation, bolting, and flowering in plants are complex processes. When plants enter the reproductive stage from vegetative growth, nutrients are gradually redistributed to the reproductive organs. This process is the result of the combined effects of various factors. During the whole plant growth and development, including bolting, phytohormones play a critical regulatory role in physiological metabolism and morphogenesis [22]. IAA was the first plant hormone to be discovered. The famous "acid growth theory" shows that, after growing cells are treated with IAA, the pH of the cell wall is reduced, and cell wall relaxation is increased, leading to increased cell elongation. The researchers of this theory found that the primary transcript effect, associated with elevated steady-state auxin concentrations, on elongating root cells, is the up-regulation of cell wall remodeling factors, notably expansins, whereas plant hormone signaling pathways maintain remarkable homeostasis [23]. The *SAUR19* subfamily of *SMALL AUXIN UP RNA* genes promotes cell expansion [24], and the NPH4/auxin response factors ARF 7 and ARF19 promote leaf expansion and auxin-induced lateral root formation [25]. IAA promotes the elongation of stems, whereas auxin deficiencies lead to the inhibition of stem elongation [26,27]. In this study, we found that the following four proteins in the tryptophan metabolism pathway had increased abundance in the last step of the synthesis of indole acetic acid: The aldehyde dehydrogenase family 2 member mitochondrial (ALDH), amidase family protein, catalase (KatE), and Acetyl-CoA c-acetyltransferase. These results suggested that the regulation of IAA for bolting in lettuce may be related to the increased abundance of the above four enzymes, and the specific performance of these proteins affect the biosynthesis of IAA. The induced gene expression was consistent with the increased IAA content in stems subjected to a high temperature (Figure 3). In combination with the finding that exogenous IAA-accelerated bolting (Figure 3), this further suggests that IAA was closely related to bolting. However, applications of IAA and IAA inhibitors in spinach plants had no clear effect on flower bud development and bolting in either treatment. The possible reason for this is that the principle of IAA action is different among different species [28]. Hence, to determine the function of IAA, future study will focus on the synergistic effect with other hormones on bolting in lettuce.

GAs is a large family of plant hormones that promote the relaxation of cell walls and cell elongation resulting in stem elongation, which plays an important role in the bolting of vegetables [29]. The three types of enzymes involved in the synthesis of biologically active GA from GA precursors are terpene synthases (TPSs), cytochrome P450 monooxygenases (P450s), and 2-oxoglutarate-dependent dioxygenases (2ODDs) [30]. In the current study, it was found that the GA-regulated protein 1-like (GRP1L) and 8 cytochrome P450 monooxygenases had increased abundance. Among them, P450s are key enzymes in the synthesis of GA, and they catalyze a number of oxidation steps in the middle part of the pathway [31]. However, needs to be the very specific P450s, and the identified P450s in this study were not the types involved in GA biosynthesis. The increased abundance of GA-regulated proteins indicated that the change of GA signal transduction might be consistent with the bolting. Researchers who found that exogenous GA caused a severe elongation of the lettuce stem [32], and promoted bolting in both a bolting-resistant lettuce line and a bolting-sensitive line [17]. GA could promote bud differentiation in the rape flower [33], the AtGA20ox1 could greatly contribute to internode and filament elongation in *Arabidopsis* [34], and the application of GA$_3$ to spinach plants rapidly induced bolting [28]. T better determine the function of P450s in GA synthesis and its relationship with bolting, the type of P450s that are responsible for this needs to be determined, and a transgenosis analysis needs to be carried out in the future. Additionally, whether the GA signal transduction affects plant bolting may be a new research focus.

3.2. Proteins Related to Phosphorylation During Bolting in Lettuce Under a High Temperature

Protein phosphorylation is an important mechanism for the regulation of cellular responses to various external signals [35], and it regulates the basic processes of cell metabolism, such as cell division, differentiation, and growth development. In addition, protein phosphorylation plays an important role in hormone regulation as well as biological and abiotic stress responses [36]. According to the amino acid residue types of substrate proteins, protein kinases can be divided into the categories serine/threonine protein kinases (STPK), tyrosine protein kinases (TPK), and histidine protein kinases (HPK), tryptophan protein kinases and aspartate aminoacyl/glutamyl protein kinases. Among these, STPK is one of the most important protein kinases and regulates many cell life activities [37]. The receptor serine/threonine kinases (RSTK), which is a type of single transmembrane protein receptor, shows that STPK activity in the cell and always exerts a normal physiological function as a heterodimer. This protein kinase may primarily cause the phosphorylation of serine or threonine downstream signaling proteins, and it can pass extracellular signals into cells, and can achieve a variety of biological functions through its influence on gene transcription. Recently, many studies about the phosphorylated protein functions in some species have been reported, including peanuts [38], alfalfa [39], and *Arabidopsis* [40], have been reported. In this study, we found that four RSTKs in lettuce stems during bolting under a high temperature had decreased abundance, and one STPK had increased abundance, which affected the phosphorylation of serine or threonine in downstream signaling proteins, ans might further impact the bolting induced by high temperature.

3.3. Defense Proteins Play an Important Role in Defense Reaction and Hormone Metabolism During Bolting in Lettuce Under a High Temperature

Glutathione S-transferase (GST) is a common enzyme family found in bacteria, fungi, plants and animals, and has many biological functions. As a type of isozyme, GSTs are encoded by a large and complex gene family [41]. This kind of enzyme can catalyze the conjugate reaction of reduced glutathione and toxic, alien or oxidation products [42], protects cells from oxidative damage [43], and plays an important role in plant resistance to abiotic stresses, including chilling, drought, and high salt stress [44]. In this study, two of the GST family proteins, namely, glutathione *S*-transferase l3 (1.26) and glutathione *S*-transferase l3-like (1.22), were detected in bolting lettuce stems. Thus, GSTs are also found to be involved in the bolting process of lettuce under high temperature.

Ascorbic acid (AsA) is a common small-molecule antioxidant in higher plants that plays an important role in the resistance of plant cells to oxidative stress. AsA is an important and required cofactor of metabolic enzymes in the synthesis of secondary metabolites, such as ethylene, GA, and anthocyanin, and it also regulates cell growth. During the progress of bolting in *Arabidopsis*, the AsA content was greatly reduced in parallel with an increased expression of OgLEAFY, and the gene encoded a key transcription factor that integrates different flowering-inducing pathways [45]. In the current study, changes in the expression of AsA metabolism-related proteins likely lead to the increase of endogenous AsA content in lettuce stems, thereby indirectly promoting the synthesis of plant hormones, as well as initiating a protective function.

3.4. Proteins Associated with Photosynthesis During Bolting in Lettuce

Plant growth and development must be coordinated with metabolism, notably with the efficiency of photosynthesis and the uptake of nutrients. This coordination requires local connections between the hormone response and metabolic state, as well as long-distance connections between shoot and root tissues. In studies on Chinese cabbage [46], 19 expression sequence tags (ESTs) associated with bolting or flowering were isolated and cloned, and the blast results indicated that 15 of them were involved in the synthesis of anthocyanins, photosynthesis, and signal transduction. During the bolting process of Chinese cabbage, it was confirmed that photosynthesis and abiotic signal response genes, such as BrPIF4, BrPIF5, and BrCOLs, were highly expressed in the outer leaves [47]. In our study, the photosynthesis-related proteins have significant increased abundance in the bolting process under a high temperature, indicating that photosynthesis may be significantly enhanced. While our photosynthesis analysis was on the lettuce stem, these findings are similar to previous research on lettuce leaves. Some lettuce leaves will likely be injured under a high temperature, as a consequence of tissue necrosis, so the enhancement of stem photosynthesis might be a way to supplement affected leaf photosynthesis. This could be an explanation for the accumulation of a large amount of organic matter (sugar) in the stem, which is needed for the growth of the stem and bolting. The subsequent identification will be further confirmed by the determination of stem photosynthesis. Previous research shows that a relationship exists between photosynthesis and phytohormones in plants. Exogenous GA enhanced the expression of many key photosynthetic genes, such as GID1, RGA, GID2, and MYBGa, which is in agreement with the observed increase in the measurements of photosynthesis [48]. Huerta found that an extensive up-regulation of genes involved in photosynthesis and carbon utilization, and down-regulation of those involved in protein synthesis and ribosome biogenesis, were shown for the first time in plants with a higher GA content [49]. The maximum IAA content and ethylene evolution was noted when the upper leaves were removed, the photosynthetic rate and photosynthetic water-use efficiency showed a reverse trend, and the application of IAA could recover the photosynthesis and stomatal conductance of 50% of upper leaf removal plants [50]. Combining the changes in the hormone pathway, it was speculated that the increased abundance proteins involved in GA and IAA metabolism increased the content of endogenous GA and IAA, thereby enhancing photosynthesis and benefitting plant bolting. Comprehensively, it was suggested that the drastic changes in photosynthesis, carbon metabolism, ribosome biogenesis, glycolysis/gluconeogenesis, phenylpropanoid biosynthesis, tryptophan metabolism may play an important role in the inducement of bolting, and the enhanced function of photosynthesis and tryptophan metabolism may locate the important place. Reports have indicated that the tryptophan-dependent IAA synthesis pathway is an important route for IAA synthesis in plants [51]. Similarly, the elevated gene expression implicated in tryptophan metabolism matched the higher levels of IAA content observed in stems of plants subjected to high temperature (Figure 3).

3.5. Proteins Associated with Expansin During Bolting in Lettuce

Expansins were first discovered in cucumbers [52] Expansins are a class of cell wall proteins that uniquely induce pH-dependent cell wall elongation and alleviate wall pressure [53]. As an important part of plant cell wall, expandable protein makes cell wall component loose and cell stretch by means of enzyme catalysis [54]. Wittwer and Bukovac reported that the application of GA induced bolting and earlier flowering in lettuce [55]. Cosgrove found that gibberellin had similar effects on stem elongation and expansins [54]. In this study, we found that one expansin protein (expansin-like b1) had increased abundance in the lettuce stem during bolting induced by high temperature, which was significantly different from the control group. At the same time, a beta expansin precursor had decreased abundance. Therefore, expansin proteins might be related to bolting induced by high temperature in lettuce. Thus, we will focus on the relationship between expansin proteins and bolting in future study.

4. Materials and Methods

4.1. Plant Materials and Treatment

Seeds of GB-30 lettuce (*Lactuca sativa* L.), a variety that bolts easily, were numbered and conserved in our laboratory, sown in a sand/soil/peat (1:1:1 v/v) mixture, and grown in the Beijing University of Agriculture Experimental Station of Beijing under standard greenhouse conditions (14 h light; 300–1300 $\mu mol/(m^2 s)$; 20 ± 2 °C during the day; 13 ± 2 °C at night; 10 h dark; and 50–70% relative humidity). The seedlings were transplanted into 10 cm pots at the trefoil stage. When the lettuce plants developed the sixth true leaf, they were moved to a growth chamber under the following condition: Temperatures of 20/13 °C (day/night), a 14/10 h photoperiod, and 60% relative humidity for two days of acclimatization. After that, the plants were divided into two groups. The control group (group CK) was kept under the standard greenhouse conditions as described above. The other group (group H) was moved to another growth chamber and treated with high temperatures of 33 and 25 °C during the day and night, respectively. The other environmental conditions were unchanged. The stem lengths (in cm) of the control and treatment plants were measured every eight days using a ruler. At the same time, the blossom buds were observed by the stereoscopic microscope and paraffin methods [56] to define the progress of flower bud differentiation. After 32 days, stem samples from the control and treatment plants were collected, frozen in liquid nitrogen, and stored at −80 °C for further measurements on endogenous hormones and proteome analysis.

4.2. Endogenous Hormone Measurement

The endogenous hormone measurements were performed using enzyme-linked immune sorbent assays (ELISA), as previously described [57]. Standard auxin (IAA), gibberellin (GA_{1+3}), zeatin (ZR), abscisic acid (ABA), Jasmonic acid methyl ester (JA-ME), and brassinolide (BR) (Sangon Biotech Co. Ltd., Shanghai, China) were used for calibration. The unit of endogenous hormone content was ng/g FW.

4.3. Exogenous Auxin (IAA) Treatment

After observing the effects of exogenous auxin at different concentrations (10, 40, 70, and 100 mg/L), 40 mg/L was chosen for the formal exogenous auxin treatment of bolting in the preparatory experiment. Plants at the sixth true leaf stage with uniform growth were selected and sprayed with 40 mg/L auxin. Water was used as a control. Twelve plants were used for each treatment, and the stem length (in cm) was measured using a ruler, every five days from the start of the treatment.

4.4. Protein Extraction

Approximately 2.5 g of each sample was ground into fine powder in liquid nitrogen. The powder was resuspended in 30 mL of 10% (*w*/*v*) trichloroacetic acid (TCA)/acetone (65 mM dithiothreitol (DTT)) in a 50 mL tube. The mixture was stored overnight (minimum duration) at −20 °C for precipitation. After centrifugation at 10,000 rpm for 30 min at 4 °C, the supernatant was discarded. Subsequently, 40 mL pre-cooling acetone was added and centrifuged at 7000 rpm for 15 min. This supernatant was also discarded and the pellet was washed three times with acetone. Afterwards, 200 uL lysis buffer (SDT buffer (4% (*v*/*v*) SDS, 100 mM Tris-HCl, 1 mM DTT, pH 7.6) was added to the precipitate, and placed on ice for 20 min after ultrasonic treatment for 30 min. After centrifugation at 12,000 rpm for 10 min at 4 °C, the supernatant was extracted. The precipitate was vacuum-dried. The total protein in the supernatant was quantified using the BCA Protein Assay Kit (Bio-Rad, Hercules, CA, USA).

4.5. Protein Digestion and iTRAQ Labeling

Protein digestion was performed according to the FASP procedure described by Wiśniewski and colleagues [58], and the resulting peptide mixture was labeled using the 8-plex iTRAQ reagent according to the manufacturer's instructions (Applied Biosystems, Foster City, CA, USA). In brief, each sample of 200 μg of protein was incorporated into a 30 μL SDT buffer (4% (*v*/*v*) SDS, 100 mM DTT, 150 mM Tris-HCl pH 8.0). The detergent, DTT, and other low-molecular-weight components were removed using a UA buffer (8 M Urea, 150 mM Tris-HCl pH 8.0) and repeated ultrafiltration (Microcon units, 30 kD). Then, 100 μL 0.05 M iodoacetamide in UA buffer was added to block reduced cysteine residues, and the samples were incubated in the dark for 20 min. The filters were washed three times with 100 μL UA buffer and, subsequently, twice with 100 μL DS buffer (50 mM triethylammoniumbicarbonate at pH 8.5). Finally, the protein suspensions were digested at 37 °C overnight using 2 μg trypsin (Promega, Madison, WI, USA) in 40 μL DS buffer, and the resulting peptides were collected as a filtrate. The peptide content was estimated by UV light spectral density at 280 nm using an extinction coefficient of 1.1 of 0.1% (g/L) solution, which was calculated based on the frequency of tryptophan and tyrosine in vertebrate proteins. For labeling, each iTRAQ reagent was dissolved in 70 μL of ethanol and added to the respective peptide mixture. The experiment was performed in three independent biological replicates, and each independent biological replication consisted of a pool of three plants. The three independent biological replications of the control were labeled as (CK1)-113, (CK2)-114, and (CK3)-115, and the three independent biological replications of the treatment were labeled as (H1)-116, (H2)-117, and (H3)-118.

4.6. Peptide Fractionation with Strong Cation Exchange (SCX) Chromatography

The iTRAQ labeled peptides were fractionated by SCX chromatography using the AKTA Purifier system (GE Healthcare, Chicago, IL, USA). The dried peptide mixture was reconstituted and acidified with 2 mL buffer A (10 mM KH_2PO_4 in 25% (*v*/*v*) of ACN, pH 2.7) and loaded onto a PolySULFOETHYL 4.6 × 100 mm column (5 μm, 200 Å, PolyLC Inc, Columbia, MD, USA). The peptides were eluted with buffer B (500 mM KCl, 10 mM KH_2PO_4 in 25% (*v*/*v*) of ACN, pH 2.7) at a flow rate of 1 ml/min with the following gradients of 0–8% buffer B (500 mM KCl, 10 mM KH2PO4 in 25% of ACN, pH 3.0) for 22 min, 8–52% buffer B from 22–47 min, 52–100% buffer B from 47–50 min, 100% buffer B from 50–58 min, and then, after 58 min, buffer B was reset to 0%. The elution was monitored by absorbance at 214 nm, and fractions were collected every minute. The collected fractions were combined into 15 fractions and desalted on C18 Cartridges (Empore™ SPE Cartridges C18 (standard density), bed I.D 7 mm, volume 3 mL, Sigma, St. Louis, MI, USA). Each fraction was concentrated by vacuum centrifugation and reconstituted in 40 µL of 0.1% (*v/v*) acetic acid. All samples were stored at −80 °C until LC-MS/MS analysis.

4.7. Liquid Chromatography (LC)-Electrospray Ionization (ESI) Tandem MS (MS/MS) Analysis

Experiments were performed on a Q-Exactive mass spectrometer coupled with Easy nLC (Proxeon Biosystems, now Thermo Fisher Scientific). Into each fraction, 10 µL was injected for nano LC-MS/MS analysis. The peptide mixture (1–2 µg) was loaded onto a C18–reversed phase column (Thermo Scientific Easy Column, 10 cm long, 75 µm inner diameter, 3 µm resin) in buffer A (0.1% (*v/v*) formic acid) and separated with a linear gradient of buffer B (80% (*v/v*) acetonitrile and 0.1% (*v/v*) Formic acid) at a flow rate of 250 nL/min, controlled by IntelliFlow technology for 140 min. MS data were acquired using a data-dependent top 10 method, which dynamically chooses the most abundant precursor ions from the survey scan (300–1800 *m/z*) for HCD fragmentation. The determination of the target value is based on predictive Automatic Gain Control (pAGC). The dynamic exclusion duration was 60 s. Survey scans were acquired at a resolution of 70,000 at *m/z* 200, and the resolution for HCD spectra was set to 17,500 at *m/z* 200. Normalized collision energy was 30 eV, and the underfill ratio, which specifies the minimum percentage of the target value likely to be reached at the maximum fill time, was defined as 0.1%. The instrument was run with peptide recognition mode enabled.

4.8. Database Search and Protein Quantification

MS/MS spectra were searched using the MASCOT engine (Matrix Science, London, UK; version 2.2) embedded into Proteome Discoverer 1.3 (Thermo Electron, San Jose, CA, USA), against Lactuca.Unigene.pep.fasta (lettuce protein database translated from transcriptome, created by our laboratory). For protein identification, the following options were used: Peptide mass tolerance = 20 ppm, MS/MS tolerance = 0.1 Da, Enzyme = Trypsin, Max Missed cleavage = 2, Fixed modification: Carbamidomethyl (C), iTRAQ 8-plex (K), iTRAQ 8-plex (N-term), Variable modification: Oxidation (M), iTRAQ8plex (Y). Each confident protein identification and quantification required at least one unique peptide, and the proteins with one unique peptide were supplied high quality spectra in Table S3. The false discovery rate (FDR) of identified proteins was ≤0.01.

The relative quantification of proteins was based on the strength of the reporter ion, which reflects the relative abundance of peptides. The fold-change was obtained according to different comparison groups (control and high-temperature treatment), through the reporter ion ratio labeled with different isotopes, as described above. In the identified proteins, the fold-changes of >1.20 or <0.84, and the *p*-values of <0.05 using one sample t test, were considered as significant.

4.9. Bioinformatics Analysis of Proteins

Functional category analysis was performed with Blast2GO software (http://www.geneontology.org) [59]. The online Kyoto Encyclopedia of Genes and Genomes (KEGG) database (http://www.genome.jp/kegg/) was used to retrieve their KEGG Orthology (KO) and the data were subsequently mapped on pathways in KEGG [60]. The corresponding KEGG pathways were extracted. To further explore the impact of proteins with differential abundance in the cell physiological processes of cells and discover internal relations between proteins with differential abundance, enrichment analysis was performed. GO enrichment on three ontologies (biological process, molecular function, and cellular component), and KEGG pathway enrichment analyses were applied based on the Fisher's exact test, considering the whole quantified protein annotations as the background dataset. The Benjamini–Hochberg correction for multiple testing was further applied to adjust the derived *p*-values. Only functional categories and pathways with *p*-values <0.05 were considered as significant. The studied protein-relative abundance data were used to perform hierarchical clustering analysis. For this purpose, Cluster 3.0 (http://bonsai.hgc.jp/~mdehoon/software/cluster/software.htm) and the Java Treeview software (http://jtreeview.sourceforge.net) were used. The Euclidean distance algorithm for similarity measurement and the average linkage clustering algorithm for clustering were selected when performing hierarchical clustering.

The workflow in the proteome is shown in Figure S1.

4.10. Total RNA Extraction and Real-Time PCR

The total RNA was extracted using the RNA pure Total RNA Kit (Aidlab Biotech, Beijing, China) according to the manufacturer's instructions. The RNA samples were reversely transcribed into cDNAs using TransScript First-Strand cDNA Synthesis SuperMix (TransGen Biotech, Beijing, China). The procedure was as follows: RNA (2 µg) mixed with 1 µL Oligod (T) 18 (0.5 µg/ µL), 2 × TS Reaction Mix (10 µL) and, TransScript RT/RI Enzyme Mix (1 µL) with an additional 20 µL of RNase-free Water to. The mixture was mixed gently and incubated at 42 °C for 15 min. The reaction was terminated by incubation at 85 °C for 5 s, and the cDNAs of the product were stored at −20 °C. The cDNA samples were used as a template, then mixed with 200 nmol primer and SYBR Green PCR Real Master Mix (Takara, Kusatsu, Japan) for real-time PCR analysis using Bio-Rad CFX 96 real-time PCR instruments and CFX manager software ver 3.0 (Bio-Rad laboratories, California, USA). The temperature procedure was as follows: 94 °C for 3 min, 32 cycles of 94 °C for 30 s, 57 °C for 30 s, and 72 °C for 20 s. The fluorescence signal was collected during the elongation of every cycle at 72 °C. The 18S was used as an internal standard for normalization. The primers used in qRT-PCR are listed in Table S5.

4.11. Statistical Analysis

All tests were performed in three replicates. For the measurement of stem length, each biological replicate had nine samples from nine plants. For the observation of flower bud differentiation, each biological replicate had five samples from five plants. For physiology and proteome analyses, three different stems were pooled together as one biological sample, and this was done three times to produce three independent biological replicates (of three pooled stems) for both physiology and proteome analysis. The presented data represent the means ± SD of three replications, and were statistically analyzed using an analysis of variance (ANOVA) by SPSS 10.0 (International Business Machine, Chicago, IL, USA). Tukey's test was used to identify significant differences among groups ($p < 0.05$, $p < 0.01$). Figures representing the physiological parameters were drawn using Origin Pro 8.0 SR4 (Origin Lab, Northampton, MA, USA) and Microsoft Office PowerPoint 2007. Western blot immunoreactive protein bands were quantified by densitometry using ImageLab 3.0 software (Bio-Rad, Hercules, CA, USA).

5. Conclusions

The bolting of plants is usually induced by a wide variety of biological changes. Here, we observed the phenotypic changes of lettuce bolting induced by a high temperature of 33 °C, which are in sharp contrast to lettuce subjected to the normal growth temperature of 20 °C, under which condition no bolting occurred. The activity of enzymes involved in protein synthesis and defense systems was functionally enhanced. Subsequently, the functional classes associated with reproductive growth, such as hormone metabolism, were highly activated, especially for IAA. These findings by proteomics were in agreement with the validation in the physiology and gene expression of IAA metabolism. Furthermore, the functions of IAA may promote photosynthesis. Thus, the synergistic reactions of these metabolisms led to the synthesis of expansin and cell cycle proteins, eventually resulting in plant bolting.

Supplementary Materials: The supplementary materials are available online at http://www.mdpi.com/1422-0067/19/10/2967/s1.

Author Contributions: J.-H.H., L.-L.Z. and S.-X.F. conceived and designed the experiments. J.-H.H., L.-L.Z., Y.-C.S. and P.-P.L. performed the experiments. S.X., Z.-Y.Q. and L.W. performed the data analysis and prepared the figures and tables. J.-K.L. advised on the analysis and interpretation of protein data. Y.-Y.H., C.-J.L., P.-P.L. and

Int. J. Mol. Sci. **2018**, *19*, 2967

X.-X.Q. prepared the figures and tables. J.-H.H. and L.-L.Z. wrote the paper. All authors read and approved the final manuscript.

Funding: This work was financially supported by the 2018 Joint Funding Project of Beijing Natural Science Foundation-the Municipal Education Commission (KZ201810020027), the National Natural Science Foundation of China (Grant No. 31372057), the Education Commission of Beijing (PXM2013_014207_0000_73), the 2017 Agriculture Education Teaching Reform Project of Beijing University of Agriculture (BUA2017JG030), 2017 Research Fund for Academic Degree & Graduate Education of Beijing University of Agriculture, and Earmarked Fund for Beijing Leaf Vegetables Innovation. Team of Modern Agro-industry Technology Research System (BAIC08).

Acknowledgments: We are grateful to Rui Yan and Wenfeng Sun from Shanghai Applied Protein Technology for proteomic and bioinformatics analysis.

Conflicts of Interest: The authors declare no conflict of interest.

References

1. Pajoro, A.; Biewers, S.; Dougali, E.; Valentim, F.L.; Mendes, M.A.; Porri, A. The (r)evolution of gene regulatory networks controlling Arabidopsis plant reproduction; a two decades history. *J. Exp. Bot.* **2014**, *65*, 4731–4745. [CrossRef] [PubMed]

2. Tang, Q.L.; Song, M.; Wang, X.J.; Wang, Z.M.; Ren, X.S. Study on physiological and biochemical changes during startup bolting of mustard and cabbage induced by temperature and photoperiod. *J. Southwest Univ.* **2009**, *31*, 52–57.

3. Fukuda, M.; Matsuo, S.; Kikuchi, K.; Mitsuhashi, W.; Toyomasu, T.; Honda, I. Gibberellin metabolism during stem elongation stimulated by high temperature in lettuce. *Acta Hortic.* **2012**, *932*, 259–264. [CrossRef]

4. Achard, P.; Baghour, M.; Chapple, A.; Hedden, P.; Van der Straeten, D.; Genschik, P.; Moritz, T.; Harberd, N.P. The plant stress hormone ethylene controls floral transition via DELLA-dependent regulation ot floral meristem-identity genes. *Proc. Natl. Acad. Sci. USA* **2007**, *104*, 6484–6489. [CrossRef] [PubMed]

5. Chia, T.Y.; Müller, A.; Jung, C.; Mutasa-Göttgens, E.S. Sugar beet contains a large *CONSTANS-LIKE* gene family including a CO homologue that is independent of the early-bolting (B) gene locus. *J. Exp. Bot.* **2008**, *59*, 2735–2748. [CrossRef] [PubMed]

6. Zhang, Y.W.; Guo, M.H.; Tang, X.B.; Jin, D.; Fang, Z.Y. Proteomic and gene expression analyses during bolting-related leaf color change in Brassica rapa. *Genet. Mol. Res.* **2016**, *15*, 1–11. [CrossRef] [PubMed]

7. Guo, Z.; Tan, J.; Zhuo, C.; Wang, C.Y.; Xiang, B.; Wang, Z.Y. Abscisic acid, H_2O_2 and nitric oxide interactions mediated cold-induced S-adenosylmethionine synthetase in Medicago sativa subsp. falcata that confers cold tolerance through up-regulating polyamine oxidation. *Plant Biotechnol. J.* **2014**, *12*, 601–612. [CrossRef] [PubMed]

8. Zhu, L.; Liu, D.; Li, Y.; Li, N. Functional phosphoproteomic analysis reveals that a Serine-62-Phosphorylated isoform of ethylene response factor110 is involved in Arabidopsis bolting. *Plant Physiol.* **2013**, *161*, 904–917. [CrossRef] [PubMed]

9. Strompen, G.; Dettmer, J.; Stierhof, Y.D.; Schumacher, K.; Jurgens, G.; Mayer, U. Arabidopsis vacuolar H-ATPase subunit E isoform 1 is required for Golgi organization and vacuole function in embryogenesis. *Plant J.* **2010**, *41*, 125–132. [CrossRef] [PubMed]

10. Zhang, Y.W.; Jin, D.; Xu, C.; Zhang, L.; Guo, M.H.; Fang, Z.Y. Regulation of bolting and identification of the α-tubulin gene family in *Brassica rapa* L. ssp pekinensis. *Genet. Mol. Res.* **2015**, *15*, 1–13. [CrossRef] [PubMed]

11. Andersson, I.; Backlund, A. Structure and function of Rubisco. *Plant Physiol. Biochem.* **2008**, *46*, 275–291. [CrossRef] [PubMed]

12. Ohto, M.; Onai, K.; Furukawa, Y.; Aoki, E.; Araki, T.; Nakamura, K. Effects of sugar on vegetative development and floral transition in *Arabidopsis*. *Plant Physiol.* **2001**, *127*, 252–261. [CrossRef] [PubMed]

13. Spreitzer, R.J.; Salvucci, M.E. Rubisco: Structure, regulatory interactions, and possibilities for a better enzyme. *Annu. Rev. Plant Biol.* **2002**, *53*, 449–475. [CrossRef] [PubMed]

14. Guo, D.; Wong, W.S.; Xu, W.Z.; Sun, F.F.; Qing, D.J.; Li, N. Cis-cinnamic acid-enhanced 1 gene plays a role in regulation of Arabidopsis bolting. *Plant Mol. Biol.* **2011**, *75*, 481–495. [CrossRef] [PubMed]

15. Lee, J.; Lee, I. Regulation and function of SOC1, a flowering pathway integrator. *J. Exp. Bot.* **2010**, *61*, 2247–2254. [CrossRef] [PubMed]

16. Balanza, V.; Martinez-Fernandez, I.; Ferrandiz, C. Sequential action of FRUITFULL as a modulator of the activity of the floral regulators SVP and SOC1. *J. Exp. Bot.* **2014**, *65*, 1193–1203. [CrossRef] [PubMed]

17. Han, Y.Y.; Chen, Z.J.; Lv, S.S.; Ning, K.; Ji, X.L.; Liu, X.Y.; Wang, Q.; Liu, R.Y.; Fan, S.X.; Zhang, X.L. MADS-Box genes and gibberellins regulate bolting in Lettuce (*Lactuca sativa* L.). *Front. Plant Sci.* **2016**, *7*, 1889–1903. [CrossRef] [PubMed]

18. Rollins, J.A.; Habte, E.; Templer, S.E.; Colby, T.; Schmidt, J.; von Korff, M. Leaf proteome alterations in the context of physiological and morphological responses to drought and heat stress in barley (*Hordeum vulgare* L.). *J. Exp. Bot.* **2013**, *64*, 3201–3212. [CrossRef] [PubMed]

19. Hwang, I.; Jung, H.J.; Park, J.I.; Yang, T.J.; Nou, I.S. Transcriptome analysis of newly classified bZIP transcription factor of Brassica rapa in cold stress response. *Genomics* **2014**, *104*, 194–202. [CrossRef] [PubMed]

20. Li, L.Q.; Zhang, Y.; Zou, L.Y.; Li, C.Q.; Yu, B.; Zheng, X.Q.; Zho, Y. An Ensemble Classifier for Eukaryotic Protein Subcellular Location Prediction Using Gene Ontology Categories and Amino Acid Hydrophobicity. *PLoS ONE* **2012**, *7*, e31057. [CrossRef] [PubMed]

21. Kanehisa, M.; Goto, S.; Sato, Y.; Furumichi, M.; Tanabe, M. KEGG for integration and interpretation of large-scale molecular data sets. *Nucleic Acids Res.* **2012**, *40*, D109–D114. [CrossRef] [PubMed]

22. Wolters, H.; Jürgens, G. Survival of the flexible: Hormonal growth control and adaptation in plant development. *Nat. Rev. Genet.* **2009**, *10*, 305–317. [CrossRef] [PubMed]

23. Pacheco-Villalobos, D.; Díaz-Moreno, S.M.; van der Schuren, A.; Tamaki, T.; Kang, Y.H.; Gujas, B.; Novak, O.; Jaspert, N.; Li, Z.; Wolf, S.; et al. The effects of high steady state auxin levels on root cell elongation in Brachypodium. *Plant Cell* **2016**, *28*, 1–14. [CrossRef] [PubMed]

24. Spartz, A.K.; Sang, H.L.; Wenger, J.P.; Gonzalez, N.; Itoh, H.; Inzé, D.; Peer, W.A.; Murphy, A.S.; Overvoorde, P.J.; Gray, W.M. The SAUR19 subfamily of SMALL AUXIN UP RNA genes promote cell expansion. *Plant J.* **2012**, *70*, 978–990. [CrossRef] [PubMed]

25. Wilmoth, J.C.; Wang, S.; Tiwari, S.B.; Joshi, A.D.; Hagen, G.; Guilfoyle, T.J.; Alonso, J.M.; Ecker, J.R.; Reed, J.W. NPH4/ARF7 and ARF19 promote leaf expansion and auxin-induced lateral root formation. *Plant J.* **2005**, *43*, 21–30. [CrossRef] [PubMed]

26. Yin, C.X.; Gan, L.J.; Ng, D.; Zhou, X.; Xia, K. Decreased paniclederived indole-3-acetic acid reduces gibberellin A1 level in the uppermost internode, causing panicle enclosure in male sterile rice Zhenshan 97A. *J. Exp. Bot.* **2007**, *58*, 2441–2449. [CrossRef] [PubMed]

27. Mockaitis, K.; Estelle, M. Auxin receptors and plant development: A new signaling paradigm. *Annu. Rev. Cell Dev. Biol.* **2008**, *24*, 55–80. [CrossRef] [PubMed]

28. Fukuda, N.; Kondo, M.; Nishimura, S.; Koshioka, M.; Tanakadate, S.; Ito, A.; Mander, L. The role of phytohormones in flowering and bolting of Spinach (*Spinacia oleracea* L.) under mid-night lighting. *Acta Hortic.* **2006**, *711*, 247–254. [CrossRef]

29. Richards, D.E.; King, K.E.; Ait-Ali, T.; Harberd, N.P. HOW GIBBERELLIN REGULATES PLANT GROWTH AND DEVELOPMENT: A molecular genetic analysis of gibberellin signaling. *Annu. Rev. Plant Physiol. Plant Mol. Biol.* **2001**, *52*, 67–88. [CrossRef] [PubMed]

30. Kasahara, H.; Hanada, A.; Kuzuyama, T.; Takagi, M.; Kamiya, Y.; Yamaguchi, S. Contribution of the mevalonate and methylerythritol phosphate pathways to the biosynthesis of gibberellins in *Arabidopsis*. *J. Biol. Chem.* **2002**, *277*, 45188–45194. [CrossRef] [PubMed]

31. Davidson, S.E.; Reid, J.B.; Helliwell, C.A. Cytochromes P450 in gibberellin biosynthesis. *Phytochem. Rev.* **2006**, *5*, 405–419. [CrossRef]

32. Nothmann, J. Effect of growth regulator treatments on heading, bolting, spiralled leaf formation and yield performance of cos lettuce (L. var. roman). *J. Hortic. Sci.* **1973**, *48*, 379–386. [CrossRef]

33. Jiang, X.M.; Dan, L.; Wang, F.J. Effects of exogenous GA3 on flower bud differentiation and flower-ball development of Broccoli (Brassica oleracea var. italica). *Plant Physiol. Commun.* **2008**, *44*, 639–642.

34. Rieu, I.; Ruiz-Rivero, O.; Fernandez-Garcia, N.; Griffiths, J.; Powers, S.J.; Gong, F.; Linhartova, T.; Eriksson, S.; Nilsson, O.; Thomas, S.G.; et al. The gibberellin biosynthetic genes AtGA20ox1 and AtGA20ox2 act, partially redundantly, to promote growth and development throughout the Arabidopsis life cycle. *Plant J.* **2007**, *53*, 488–504. [CrossRef] [PubMed]

35. Henderson, H.; Macleod, G.; Hrabchak, C.; Varmuza, S. New candidate targets of protein phosphatase-1c-gamma-2 in mouse testis revealed by a differential phosphoproteome analysis. *Int. J. Androl.* **2011**, *34*, 339–351. [CrossRef] [PubMed]

36. Liu, Y.H.; Li, H.Y.; Shi, Y.S.; Song, Y.C.; Wang, T.Y.; Li, Y. A maize early responsive to dehydration gene, *ZmERD4*, provides enhanced drought and salt tolerance in Arabidopsis. *Plant Mol. Biol. Rep.* **2009**, *27*, 542–548. [CrossRef]

37. Hardie, D.G. Plant Protein Serine/Threonine Kinases: Classification and Functions. *Annu. Rev. Plant Biol.* **1999**, *50*, 97–131. [CrossRef] [PubMed]

38. Rudrabhatla, P.; Rajasekharan, R. Mutational analysis of stress-responsive peanut dual specificity kinase: Identification of tyrosine residues involved in regulation of protein kinase activity. *J. Biol. Chem.* **2003**, *278*, 17328–17335. [CrossRef] [PubMed]

39. Chinchilla, D.; Merchan, F.; Megias, M.; Kondorosi, A.; Sousa, C.; Crespi, M. Ankyrin protein kinases: A novel type of plant kinase gene whose expression is induced by osmotic stress. *Plant Mol. Biol.* **2003**, *51*, 555–566. [CrossRef] [PubMed]

40. Rudrabhatla, P.; Reddy, M.M.; Rajasekharan, R. Genome-wide analysis and experimentation of plant serine/threonine/tyrosine-specific protein kinases. *Plant Mol. Biol.* **2006**, *60*, 293–319. [CrossRef] [PubMed]

41. Dixon, D.P.; Skipsey, M.; Edwards, R. Roles for glutathione transferases in plant secondary metabolism. *Phytochemistry* **2010**, *71*, 338–350. [CrossRef] [PubMed]

42. Møldrup, M.E.; Geu-Flores, F.; Olsen, C.E.; Halkier, B.A. Modulation of sulfur metabolism enables efficient glucosinolate engineering. *BMC Biotechnol.* **2011**, *11*, 12. [CrossRef] [PubMed]

43. Bartling, D.; Radzio, R.; Steiner, U.; Weiler, E.W. A glutathione S-transferase with glutathione-peroxidase activity from Arabidopsis thaliana; molecular cloning and functional characterization. *Eur. J. Biochem.* **1993**, *216*, 579–586. [CrossRef] [PubMed]

44. Xu, F.X.; Lagudah, E.S.; Moose, S.P.; Riechers, D.E. Tandemly duplicated safener-induced glutathione S-transferase genes from Triticum tauscii contribute to genome- and organ-specific expression in hexaploid wheat. *Plant Physiol.* **2002**, *130*, 362–373. [CrossRef] [PubMed]

45. Shen, C.H.; Krishnamurthy, R.; Yeh, K.W. Decreased L-ascorbate content mediating bolting is mainly regulated by the galacturonate pathway in oncidium. *Plant Cell Physiol.* **2009**, *50*, 935–946. [CrossRef] [PubMed]

46. Xiao, X.F.; Lei, J.J.; Cao, B.H.; Chen, G.J.; Chen, C.M. cDNA-AFLP analysis on bolting or flowering of flowering Chinese cabbage and molecular characteristics of BrcuDFR-like /BrcuAXS gene. *Mol. Biol. Rep.* **2012**, *39*, 7525–7531. [CrossRef] [PubMed]

47. Dong, H.P.; Peng, J.L.; Bao, Z.L.; Meng, X.D.; Bonasera, J.M.; Chen, G.Y.; Beer, S.V.; Dong, H.S. Downstream divergence of the ethylene signaling pathway for Harpin-stimulated Arabidopsis growth and insect defense. *Plant Physiol.* **2004**, *136*, 3628–3638. [CrossRef] [PubMed]

48. Xie, J.B.; Tian, J.X.; Du, Q.Z.; Chen, J.H.; Li, Y.; Yang, X.H.; Li, B.L.; Zhang, D.Q. Association genetics and transcriptome analysis reveal a gibberellin-responsive pathway involved in regulating photosynthesis. *J. Exp. Bot.* **2016**, *67*, 3325–3338. [CrossRef] [PubMed]

49. Huerta, L.; Forment, J.; Gadea, J.; Fagoaga, C.; Peña, L.; Perez-Amador, M.A.; García-Martínez, J.L. Gene expression analysis in citrus reveals the role of gibberellins on photosynthesis and stress. *Plant Cell Environ.* **2008**, *31*, 1620–1633. [CrossRef] [PubMed]

50. Khan, N.A.; Khan, M.; Ansari, H.R. Samiullah. Auxin and defoliation effects on photosynthesis and ethylene evolution in mustard. *Sci. Hortic.* **2002**, *96*, 43–51. [CrossRef]

51. Vanneste, S.; Friml, J. Auxin: A trigger for change in plant development. *Cell* **2009**, *136*, 1005–1016. [CrossRef] [PubMed]

52. Mcqueen-Mason, S.; Durachko, D.M.; Cosgrove, D.J. Two endogenous proteins that induce cell wall extension in plants. *Plant Cell* **1992**, *4*, 1425–1433. [CrossRef] [PubMed]

53. Daniel, J.C.; Li, L.C.; Cho, H.T.; Susanne, H.B.; Richard, C.M.; Douglas, B. The growing world of expansins. *Plant Cell Physiol.* **2002**, *43*, 1436–1444.

54. Cosgrove, D.J. Loosening of plant cell walls by expansins. *Nature* **2000**, *407*, 321–326. [CrossRef] [PubMed]

55. Wittwer, S.H.; Bukovac, M.J.; Sell, H.M.; Weller, L.E. Some Effects of Gibberellin on Flowering and Fruit Setting. *Plant Physiol.* **1957**, *32*, 39–41. [CrossRef] [PubMed]

56. Zhang, Y.M.; Ma, H.L.; Calderón-Urrea, A.; Tian, C.X.; Bai, X.M.; Wei, J.M. Anatomical changes to protect organelle integrity account for tolerance to alkali and salt stresses in Melilotus officinalis. *Plant Soil* **2016**, *406*, 327–340. [CrossRef]

57. Swaczynová, J.; Novák, O.; Hauserová, E.; Fuksová, K.; Šíša, M.; Kohout, L.; Strnad, M. New techniques for the estimation of naturally occurring brassinosteroids. *J. Plant Growth Regul.* **2007**, *26*, 1–14. [CrossRef]

58. Wiśniewski, J.R.; Zougman, A.; Nagaraj, N.; Mann, M. Universal sample preparation method for proteome analysis. *Nat. Methods* **2009**, *6*, 359–362. [CrossRef] [PubMed]

59. Conesa, A.; Götz, S.; García-Gómez, J.M.; Terol, J.; Talón, M.; Robles, M. Blast2GO: A universal tool for annotation, visualization and analysis in functional genomics research. *Bioinformatics* **2005**, *21*, 3674–3676. [CrossRef] [PubMed]

60. Moriya, Y.; Itoh, M.; Okuda, S.; Yoshizawa, A.C.; Kanehisa, M. KAAS: An automatic genome annotation and pathway reconstruction server. *Nucleic Acids Res.* **2007**, *35*, 182–185. [CrossRef] [PubMed]

International Journal of
Molecular Sciences

MDPI

Article

Comparative Proteomic and Physiological Analyses of Two Divergent Maize Inbred Lines Provide More Insights into Drought-Stress Tolerance Mechanisms

Tinashe Zenda [1,2], Songtao Liu [1,2], Xuan Wang [1,2], Hongyu Jin [1,2], Guo Liu [1,2] and Huijun Duan [1,2,*]

1   Department of Crop Genetics and Breeding, College of Agronomy, Hebei Agricultural University, Baoding 071001, China; tzenda@hebau.edu.cn (T.Z.); m15028293845@163.com (S.L.); 15733289921@163.com (X.W.); m15633790536@163.com (H.J.); m15612245597@163.com (G.L.)
2   North China Key Laboratory for Crop Germplasm Resources of the Education Ministry, Hebei Agricultural University, Baoding 071001, China
*   Correspondence: hjduan@hebau.edu.cn; Tel.: +86-1393-1279-716

Received: 6 September 2018; Accepted: 15 October 2018; Published: 18 October 2018

Abstract: Drought stress is the major abiotic factor threatening maize (*Zea mays* L.) yield globally. Therefore, revealing the molecular mechanisms fundamental to drought tolerance in maize becomes imperative. Herein, we conducted a comprehensive comparative analysis of two maize inbred lines contrasting in drought stress tolerance based on their physiological and proteomic responses at the seedling stage. Our observations showed that divergent stress tolerance mechanisms exist between the two inbred-lines at physiological and proteomic levels, with YE8112 being comparatively more tolerant than MO17 owing to its maintenance of higher relative leaf water and proline contents, greater increase in peroxidase (POD) activity, along with decreased level of lipid peroxidation under stressed conditions. Using an iTRAQ (isobaric tags for relative and absolute quantification)-based method, we identified a total of 721 differentially abundant proteins (DAPs). Amongst these, we fished out five essential sets of drought responsive DAPs, including 13 DAPs specific to YE8112, 107 specific DAPs shared between drought-sensitive and drought-tolerant lines after drought treatment (SD_TD), three DAPs of YE8112 also regulated in SD_TD, 84 DAPs unique to MO17, and five overlapping DAPs between the two inbred lines. The most significantly enriched DAPs in YE8112 were associated with the photosynthesis antenna proteins pathway, whilst those in MO17 were related to C5-branched dibasic acid metabolism and RNA transport pathways. The changes in protein abundance were consistent with the observed physiological characterizations of the two inbred lines. Further, quantitative real-time polymerase chain reaction (qRT-PCR) analysis results confirmed the iTRAQ sequencing data. The higher drought tolerance of YE8112 was attributed to: activation of photosynthesis proteins involved in balancing light capture and utilization; enhanced lipid-metabolism; development of abiotic and biotic cross-tolerance mechanisms; increased cellular detoxification capacity; activation of chaperones that stabilize other proteins against drought-induced denaturation; and reduced synthesis of redundant proteins to help save energy to battle drought stress. These findings provide further insights into the molecular signatures underpinning maize drought stress tolerance.

Keywords: proteome profiling; iTRAQ; differentially abundant proteins (DAPs); drought stress; physiological responses; *Zea mays* L.

---

1. Introduction

Maize (*Zea mays* L.) is one of the world's most agro-economically important crops because of its raw material use in the food, feed, and biofuel production for humans and animals [1–3].

However, it is under severe threat from various abiotic stresses including drought, salinity, cold, heat, and flooding [4–8]. Among these, drought or moisture deficit is the most serious environmental factor posing a substantial menace to maize production worldwide, especially under rain-fed conditions [9–11].

The crop is susceptible to drought at various growth stages, including seedling, pre-flowering and grain-filling [4]. In particular, drought stress can affect plant growth at the seedling stage [12]. In arid and semi-arid regions such as Hebei Province in Northern China, maize often undergo drought stress in spring and early summer when water deficits threaten germination and seedling growth [3,13]. Although maize seedlings require less water compared to later vegetative and reproductive stages, moisture stress at seedling stage influences their adaptation at the early crop establishment phase and their grain yield potential, due to premature flowering and a longer anthesis-silk interval [14,15]. Revealing the mechanism of maize drought response at the seedling stage and improving early crop establishment in regions where drought occurs during the early crop development phase therefore become priority goals of the maize drought-tolerant breeding program [3].

Scientific research has made tremendous progress in unravelling maize drought stress response mechanisms at the vegetative and reproductive stages [16]. Despite this, however, and the existence of several reports on drought tolerance analyses between inbred lines at the seedling stage [4,17,18], our understanding of seedling drought stress response mechanisms and genes involved still remain unclear. Several reports have focused on physiological and biochemical [19–21], as well as large-scale transcriptomic analyses [1,3,22–24]. However, transcriptome profiling has limitations because mRNA levels are not always correlated to those of corresponding proteins due to post-transcriptional and post-translational modifications [5,25,26].

Elucidating the molecular changes at protein level has become extremely important for studying drought stress responses in plants. Since proteins are directly involved in plant stress responses, proteomic studies can eventually contribute to dissecting the possible relationships between protein changes and plant stress tolerance [27,28]. This, therefore, provides new insights into plant responses to drought stress at the protein level [10,29,30]. High-throughput proteomics has become a powerful tool for performing large-scale studies and comprehensive identification of drought responsive proteins in plants [31–35]. The iTRAQ (isobaric tags for relative and absolute quantification) analysis method is a second generation proteomic technique that provides a gel-free shortgun quantitative analysis. It utilizes isobaric reagents to label tryptic peptides and monitor relative changes in protein and PMT (peptide mass tolerance) abundance, and it allows for up to eight samples [36]. Thus, the method especially facilitates the analysis of time courses of plant stress responses or biological replicates in a single experiment, and the technique has become increasingly popular in plant stress response studies [37].

Here, in order to study maize drought stress responses at the protein level, we have also employed an iTRAQ-based quantitative strategy to perform proteome profiling of two contrasting maize inbred lines (drought-tolerant YE8112 and drought-sensitive MO17) at the seedling stage. We conducted a comparative proteomic analysis of these two lines' leaves after a seven-day moisture-deficit exposure period. In addition, we evaluated some physiological responses of these two inbred lines under drought stress, and the results of this study provide further insights into the drought stress tolerance signatures in maize.

2. Results

2.1. Phenotypic and Physiological Differences between YE8112 and MO17 in Response to Drought Stress

To validate the previous observations that MO17 is drought-sensitive [38] and YE8112 drought tolerant [39] and to investigate the molecular mechanisms underlying YE8112 drought tolerance, seedlings at the three-leaf stage were treated with or without moisture deficit stress for 7 days in a greenhouse environment. Several drought-induced phenotypic responses were then observed.

As expected, no significant phenotypic differences were observed between the two lines under water-sufficient conditions, as they both maintained intact plant architecture (Figure 1A). However, post drought exposure; there were significant differences in the performances of the two lines. The leaves of MO17 were distinctly shriveled up (Figure 1B), whilst YE8112 seedlings displayed little phenotypic change by maintaining fully expanded green leaves and intact plant architecture (Figure 1C).

Drought stress significantly ($p < 0.05$) decreased the leaf relative water content (RWC) from day 1 in MO17, and from day 3 in YE8112 (Figure 1D). This shows that, upon exposure to drought stress, the sensitive line MO17 lost leaf water significantly quicker than tolerant line YE8112. Moreover, the RWC of YE8112 was higher than that of MO17 in water-deficit conditions (Figure 1D); these results corresponding to our visual observation. Further, the RWC change in the sensitive line MO17 was evidently higher than that of the tolerant line (Figure 1D), which indicates that the tolerant line YE8112 had higher water retention capacity than sensitive line MO17. The POD activity showed an increasing trend, in pace with increasing number of treatment days (Figure 1E). This indicates that certain drought stress intensity could result in increased production and activity of antioxidant enzymes and protective osmolytes in maize seedlings leaves. The proline content was significantly ($p < 0.05$) increased in both MO17 and YE8112 upon drought stress exposure, commencing from day 1 in both inbred lines (Figure 1F). Additionally, the proline content was generally higher in YE8112 than in MO17 at most time points under stress conditions (Figure 1F). Results on leaf malondialdehyde (MDA) content showed that overall; it was significantly higher in MO17 than in YE8112 under both stressed and non-stressed conditions. In both inbred lines, MDA content showed an increasing trend, until the third day, and then declined significantly thereafter (Figure 1G). From the fifth day onwards, MDA content exhibited a gradual decline or a uniform level in MO17 and YE8112, respectively (Figure 1G). This may suggest that with the increase of stress exposure period, leaf cell membranes are severely injured, ultimately leading to membrane lipid release and destruction of membrane structures. Trypan blue staining results indicated that under control conditions, leaf cells of both inbred lines remained intact and viable, hence, unstained (Figure 2A,B). However, post drought exposure, sensitive line MO17 had lower active cells and cell membranes were significantly damaged (Figure 2C). In contrast, tolerant line YE8112 still had more active cells (Figure 2D).

2.2. Inventory of Maize Seedling Leaf Proteins Identified by iTRAQ

Using the Mascot software, 172,775 spectra were matched with known spectra, and 19,678 peptides, 12,054 unique peptides, and 3785 proteins were identified. Amongst these 3785 identified proteins (Table S1), 100 (2.65%) were <10 kDa, 3301 (87.21%) were 10–70 kDa, 259 (6.84%) were 70–100 kDa, and 125 (3.30%) were >100 kDa in weight (Figure S1A). In addition, 2084 (55.06%) proteins were detected based on at least two unique peptides whilst the remaining 1701 (44.94%) proteins had only one identified unique peptide (Figure S1B). Protein sequence coverage was generally below 25% (Figure S1C). Proteins with at least one unique peptide were used for a subsequent analysis of differentially abundant proteins (DAPs). The distribution of the peptide lengths defining each protein showed that over 85% of the peptides had lengths between 5 and 20 amino acids, with 9–11 and 11–13 amino acids being modal lengths (Figure S1D).

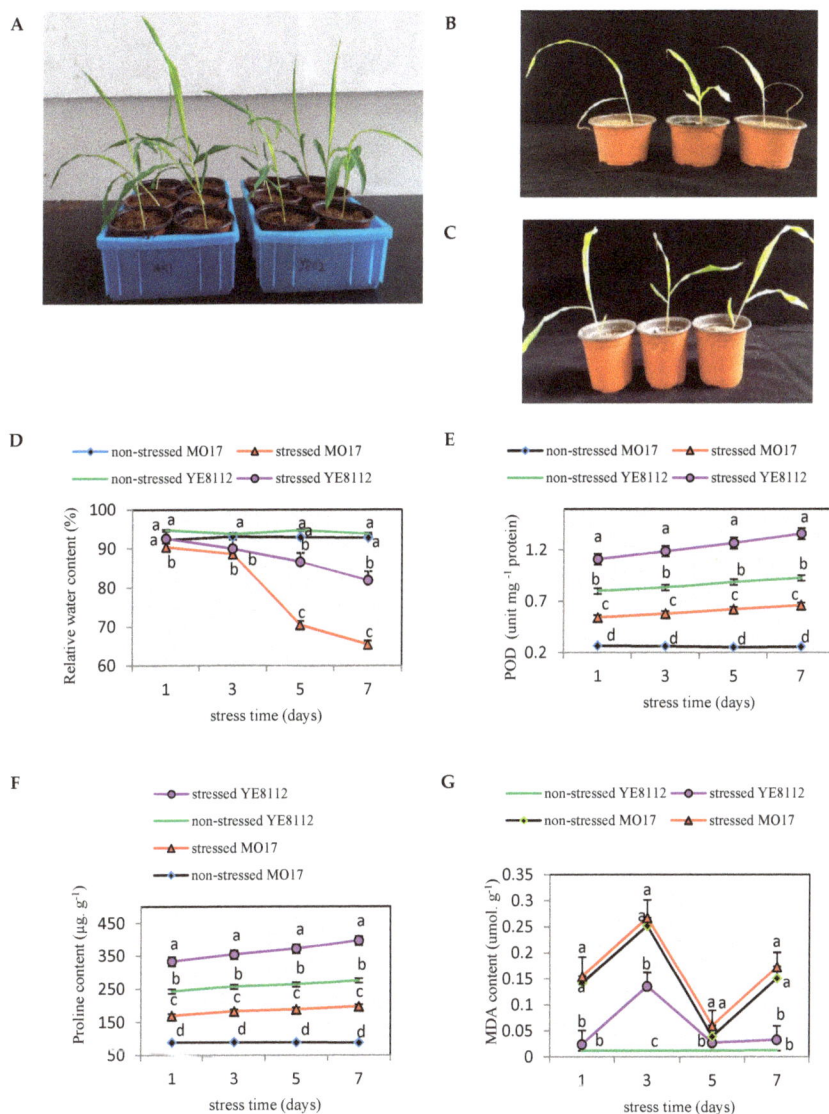

Figure 1. Phenotypic (**A–C**) and physiological (**D–G**) responses of two maize inbred lines to drought stress. Phenotypic displays presented here are for three-leaf-stage seedlings after 7 days of moisture deficit treatment. (**A**) MO17 and YE8112 inbred lines under non-stressed (water-sufficient) conditions; (**B**) sensitive line MO17 drought stressed; (**C**) tolerant line YE8112 drought stressed; (**D–G**) physiological changes were measured in leaf tissues at different stress exposure periods/time points (1, 3, 5, and 7 days); (**D**) leaf relative water content, (**E**) peroxidase (POD) enzyme activity, (**F**) proline content and (**G**) level of lipid peroxidation (MDA (malondialdehyde) content). Data are presented as the mean ± standard error (*n* = 3). Different letters above line graphs show significant difference among treatments at a given day of treatment ($p \leq 0.05$).

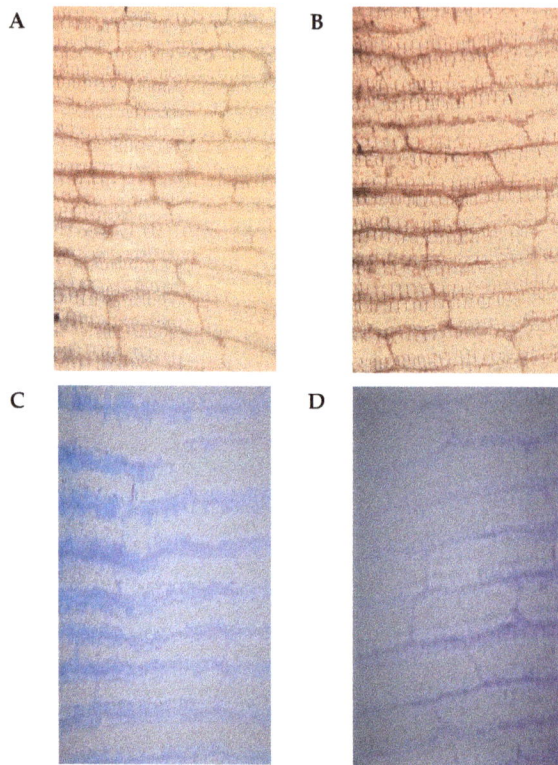

Figure 2. Results of trypan blue staining of leaves. (**A**) Non-stressed sensitive inbred line MO17, (**B**) non-stressed tolerant line YE8112, (**C**) drought-stressed MO17, and (**D**) drought stressed YE8112, seven days post drought exposure. Scale bars = 200 μm.

2.3. Analysis of Diffentially Abundant Proteins (DAPs) Observed in Different Experimental Comparisons

Comparative proteomic analysis was used to investigate the changes of protein profiles in leaves of YE8112 (drought-tolerant, T) and MO17 (drought-sensitive, S) inbred lines under drought stress conditions. A pairwise comparison of before and after treatments (drought, D, and control, C) was performed in YE8112 (TD_TC) and MO17 (SD_SC) individually. In addition, a comparative study on the drought stress proteome was performed between the tolerant and sensitive lines, under drought (SD_TD) and under water-sufficient (control) (SC_TC) conditions, giving four comparison groups (Table 1). Before drought treatment, a total of 258 differentially abundant proteins were identified between the tolerant and sensitive lines (SC_TC). Of these DAPs, 119 had higher accumulation levels in the tolerant line compared to the sensitive line (Table 1). After drought treatment, we found 269 DAPs between the tolerant and sensitive lines (SD_TD). Of these DAPs, 116 had higher expression levels in the tolerant line compared to the sensitive line (Table 1). In the tolerant line, 37 proteins (Table S2) showed differential abundance before and after drought treatment (TD_TC); 11 of these DAPs were up-regulated (Table 1). In the sensitive line, we observed 157 DAPs (Table S3) before and after drought treatment (SD_SC); 65 of these DAPs were up-regulated whilst 92 were down-regulated (Table 1). In total, 721 DAPs were found among the four comparison groups (Table 1, Figure 3).

Table 1. Number of differentially abundant proteins (DAPs) identified in each comparison group.

Comparisons [1]	Up-Regulated [2]	Down-Regulated [3]	Total [4]
SD_SC	65	92	157
TD_TC	11	26	37
SD_TD	116	153	269
SC_TC	119	139	258

[1] Comparisons, differential comparison groups; SD, sensitive inbred line (MO17) under drought treatment conditions; SC, sensitive inbred line under well-watered (control) conditions; TD, tolerant inbred line (YE8112) under drought conditions; TC, tolerant inbred line under control conditions; [2] up-regulated: increased differential abundant protein; [3] down-regulated: reduced differential abundant protein; [4] Total: total of all the differentially abundant proteins in a comparison group. An underscore between two line-treatment combinations implies comparison of those combinations.

With reference to Figure 3, the combinations of the four comparisons reflect the impact of lines or treatment. Some of the combinations are more important than others in respect of drought tolerance. Area I represents specific DAPs of TD_TC, that is, the specific drought responsive DAPs of the drought tolerant line YE8112. Of these 13 DAPs, five were up-regulated and eight were down-regulated (Table 2). For comparative analysis, Table 3 shows the 84 drought responsive DAPs unique to SD_SC (labeled V in Figure 3); of which 35 were up-regulated and 49 down-regulated. Area II represents specific DAPs of SD_TD, that is, specific DAPs shared between the drought sensitive and drought tolerant lines after drought treatment. For detailed analysis of these 107 specific DAPs of SD_TD, please refer to Figure 4 and Table S4.

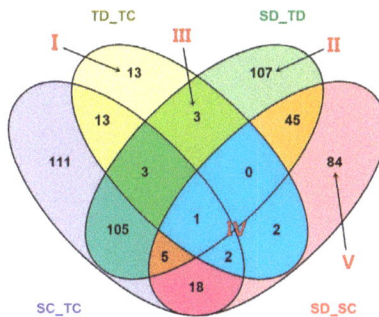

Figure 3. Venn diagram analysis of differentially abundant proteins (DAPs) identified in the four experimental comparisons. The overlapping regions of the Venns indicate the DAPs shared between/among corresponding groups. Area I represents 13 drought responsive DAPs specific to TD_TC; Area II represents 107 DAPs exclusive to SD_TD; Area III shows the 3 DAPs specifically shared between TD_TC and SD_TD; Area IV shows the five overlapping DAPs within line (shared between TD_TC and SD_SC); Area V shows 84 DAPs exclusive to SD_SC comparison.

Area III represents the three specifically shared DAPs between TD_TC and SD_TD, that is, drought responsive DAPs of the tolerant line that were also differentially expressed between the tolerant and sensitive lines after drought treatment. Of these three DAPs, all were up-regulated in the TD_TC comparison, but all down-regulated in the SD_TD comparison (Table 4). Area IV represents the five DAPs shared by TD_TC and SD_SC, that is, the common (overlapping) drought responsive DAPs within line. Of these five common drought responsive DAPs, all were down-regulated in tolerant line YE8112; whereas three were up-regulated and two down-regulated in sensitive line MO17 (Table 5).

An analysis of the $\log_2$ fold-changes of the significant differentially abundant proteins revealed that, in response to drought stress, DAPs in MO17 had significantly higher fold changes than DAPs in drought tolerant line YE8112 (Figure 4, Figures S2 and S3).

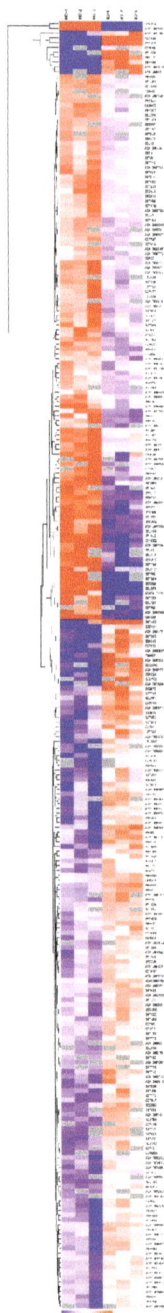

Figure 4. Clustering analysis of differentially abundant proteins (DAPs) in SD_TD comparison. Each row represents a protein significantly abundantly expressed. First three columns refer to technical replicates (MD1–3) for MO17 drought stressed, whilst the last three columns (8D1–3) refer to replicates for YE8112 drought stressed. The scale bar on the X-axis indicates the logarithmic value (log 2) expression of the DAPs, up-regulated (red) and down-regulated (blue).

Table 2. Drought-responsive maize seedling leaf proteins observed specifically in tolerant line YE8112.

No.	Protein ID [1]	Gene Name/ID [2]	Description [3]	Coverage (%) [4]	Peptide Fragments [5]	Fold Change [6]	*p* Value [7]	Pathways [8]
1	C0HJ06	541618	Uncharacterized protein [9]	22.4	1	1.37	0.0109	MAPK signaling pathway/Plant hormone signaling
2	Q41746	Lhcb5-1	Chlorophyll a-b binding protein, chloroplastic	55.8	10	1.24	0.0131	X3
3	C0HGH7	100193714	Universal stress family protein	20.4	3	1.23	0.0430	
4	A0A1D6GAZ6	ZEAMMB73_Zm00001c012677	Glycerophosphodiester phosphodiesterase GDPD5	16.8	5	1.22	0.0136	Glycerophospholipid metabolism
5	C0P948	Zm00001d024886	Uncharacterized protein	55.9	20	1.21	0.0350	
6	A0A1D6PQ00	100286059	U2 snRNP auxiliary factor large subunit	9.4	2	0.83	0.0171	Spliceosome
7	A0A1D6IU11	100383306	Ubiquitin carboxyl-terminal hydrolase 13	2.7	3	0.83	0.0217	
8	A0A1D6MJP2	ZEAMMB73_Zm00001d039613	Uncharacterized protein	19.9	4	0.82	0.0111	
9	B4FTP2	ZEAMMB73_Zm00001d021334	Thioredoxin-like protein CDSP32 chloroplastic	23.7	6	0.81	0.0246	
10	B4F845	100191245	Uncharacterized protein	3.0	1	0.81	0.0027	
11	H9BG22	101027254	Alpha-dioxygenase	4.4	3	0.80	0.0162	alpha-linolenic acid metabolism
12	Q5GJ59	TPS7	Terpene synthase 7	14.8	5	0.78	0.0179	
13	C0PHF6	100383595	AAA-ATPase ASD mitochondrial	10.6	5	0.55	0.0487	

[1] Protein ID, unique protein identifying number in the UniProt database; [2] Gene name/ID; name or ID number of the corresponding gene of the identified differentially abundant protein as searched against the maize sequence database Gramene (http://ensemble.gramene.org/Zeamays); [3] Description, annotated biological functions based on Gene Ontology (GO) analysis; [4] Coverage (%), sequence coverage is calculated as the number of amino acids in the peptide fragments observed divided by the protein amino acid length; [5] Peptides fragments, refer to the number of matched peptide fragments generated by trypsin digestion; [6] Fold change, is expressed as the ratio of intensities of up-regulated or down-regulated proteins between drought stress treatments and control (well-watered conditions); All the fold change figures below 1 represents that the proteins were down-regulated. All the figures above 1 means the proteins were up-regulated; [7] *p* value, statistical level (using Student's *t*-test) below <0.05, at which protein differential expression was accepted as significant; [8] Pathways, metabolic Kyoto Encyclopedia of Genes and Genomes (KEGG) pathways in which the identified protein was found to be significantly enriched; [9] uncharacterized protein, a protein without any functional annotations ascribed to it at the present.

Int. J. Mol. Sci. **2018**, 19, 3225

Table 3. Drought-responsive maize seedling leaf proteins observed specifically in sensitive line MO17.

No.	Protein ID [1]	Gene Name/ID [2]	Description [3]	Coverage (%) [4]	Peptide Fragments [5]	Fold Change [6]	p Value [7]	Pathways [8]
1	B4FV94	Zm00001d032197	Chlorophyll a-b binding protein, chloroplastic	49.8	7	1.66	0.0326	Photosynthesis-antenna proteins
2	B4FCG6	Zm00001d004386	9 Uncharacterized protein	9.0	1	1.48	0.0036	
3	B4FTN5	100273215	Metal-dependent protein hydrolase	5.7	1	1.45	0.0459	
4	B8A3B7	Zm00001d043059	Uncharacterized protein	20.8	3	1.33	0.0278	
5	C0P6L9	Zm00001d053377	Uncharacterized protein	40.2	7	1.33	0.0011	Ribosome
6	B4FLE3	100282216	HSP20-like chaperones superfamily protein	33.0	4	1.32	0.0484	
7	B6U3Z0	Zm00001d053377	50S ribosomal protein L21	42.5	7	1.31	0.0149	Ribosome
8	K7TP80	Zm00001d024014	Zinc finger (C3HC4-type RING finger) family protein	36.0	14	1.31	0.0028	
9	A0A1D6JW44	Zm00001d028428	Calcium-binding EF-hand family protein	9.0	1	1.30	0.0014	
10	A0A097PND9	Zm00001d015195	AT5G11810-like protein (Fragment)	6.9	5	1.29	0.0038	
11	B4FE30	100193174	10 kDa chaperonin	45.9		1.29	0.0024	Carbon metabolism, Pyruvate metabolism, Cysteine and methionine metabolism
12	B4FZU8	100274264	Malate dehydrogenase	56.8	12	1.28	0.0150	
13	Q4A1J8	cc3	Cysteine proteinase inhibitor	11.3	1	1.28	0.0293	
14	A0A1X7YHJ3	Zm00001d000282	Photosystem II CP47 reaction center protein	46.9	16	1.28	0.0062	Photosynthesis
15	B4FWP6	Zm00001d039452	Uncharacterized protein	9.9	4	1.27	0.0374	Spliceosome
16	B4FTL2	Zm00001d044931	Protein TIC 22 chloroplastic	9.3	2	1.27	0.0001	
17	C0P8X5	100294068	Electron transfer flavoprotein subunit beta mitochondrial	14.9	1	1.25	0.0020	
18	A0A1D6HE45	ZEAMMB73_Zm00001d017330	ATP-dependent Clp protease proteolytic subunit	33.7	5	1.25	0.0218	
19	Q2XX37	plt2	Non-specific lipid-transfer protein	46.2	4	1.25	0.0435	
20	A0A1D6JYF7	103634473	Kinesin-like protein	3.1	1	1.24	0.0409	
21	A0A1D6E501	ZEAMMB73_Zm00001d002880	3-isopropylmalate dehydrogenase	50.1	12	1.24	0.0449	Oxocarboxylic acid metabolism, C5-Branched dibasic acid metabolism, Biosynthesis of amino acids
22	A0A1D6L0Y0	ZEAMMB73_Zm00001d033634	Uncharacterized protein	7.6	1	1.24	0.0111	
23	A0A096PRE6	100282938	Fibrillin1	31.4	9	1.23	0.0421	
24	K7UWX4	ZEAMMB73_Zm00001d051062	GrpE protein homolog	44.2	11	1.23	0.0083	
25	B4FMA5	100217267	Chaperone DnaJ-domain superfamily protein	14.6	2	1.23	0.0378	
26	B7ZZT1	Zm00001d027326	Uncharacterized protein	6.5	1	1.22	0.0039	
27	B8A045	100279815	Phospholipase D	2.9	2	1.22	0.0211	Endocytosis, Ether lipid metabolism, Glycero phospholipid metabolism
28	B6TGF1	Zm00001d009640	Malate dehydrogenase 2 mitochond.	72.4	14	1.22	0.0092	Carbon metabolism, Pyruvate metabolism, Cysteine and methionine metabolism, Carbon fixation in photosynthetic organisms

Table 3. *Cont.*

No.	Protein ID [1]	Gene Name/ID [2]	Description [3]	Coverage (%) [4]	Peptide Fragments [5]	Fold Change [6]	*p* Value [7]	Pathways [8]
29	A0A1D6FJ49	ZEAMMB73_Zm00001d009189	TPR repeat	6.0	1	1.22	0.0283	RNA transport, RNA degradation, mRNA surveillance pathway
30	B6UHD9	Zm00001d021715	Peptide chain release factor 2	8.0	2	1.22	0.0374	
31	B6TDF7	100282980	Plastid-specific 30S ribosomal protein 2	45.4	9	1.21	0.0014	Oxidative phosphorylation
32	Q1KKB7	nad1	NADH-ubiquinone oxidoreductase chain 1	5.9	1	1.21	0.0355	Photosynthesis
33	A0A059Q7D4	psbD	Photosystem II D2 protein	25.2	7	1.20	0.0058	
34	C4J3Q4	100277436	YCF37-like protein	17.7	2	1.20	0.0017	
35	B4FTK9	100282281	Alpha/beta-Hydrolases superfamily protein	33.6	6	1.20	0.0421	
36	B6TBW4	100282838	ERBB-3 BINDING PROTEIN 1	30.5	10	0.83	0.0174	
37	A0A1D6DVJ8	ZEAMMB73_Zm00001d012006	H(+)-ATPase 5	34.6	18	0.83	0.0322	Oxidative phosphorylation
38	A0A1D6DYT2	100383668	Signal recognition particle 14 kDa protein	11.3	1	0.83	0.0172	Protein export
39	B6T346	100279524	THO complex subunit 4	14.2	3	0.83	0.0390	mRNA surveillance pathway, RNA transport
40	A0A1D6GKY6	100192032	Uncharacterized protein	4.9	1	0.83	0.0411	
41	B6SJ21	100280585	Guanine nucleotide-binding protein beta subunit-like protein	59.3	13	0.83	0.0232	
42	C0PJ72	Zm00001d017459	Uncharacterized protein	8.3	1	0.82	0.0201	Valine, leucine and isoleucine degradation
43	C0HJ59	100381692	Uncharacterized protein	13.3	5	0.82	0.0181	
44	A0A1D6M4E1	ZEAMMB73_Zm00001d138192	Glutathione transferase41	8.6	1	0.82	0.0025	Glutathione metabolism
45	A0A1D6GES6	103625778	DNA gyrase subunit A chloroplastic/mitochondrial	1.9	1	0.81	0.0372	
46	B6TIL4	Zm00001d048954	GDP-mannose 3,5-epimerase 2	20.5	6	0.81	0.0265	Amino sugar and nucleotide sugar metabolism, Ascorbate and aldarate metabolism
47	B6T3J2	100282096	Eukaryotic translation initiation factor 2 beta subunit	12.9	3	0.81	0.0206	RNA transport
48	A0A1D6F8L4	100194138	Coatomer subunit gamma	7.2	4	0.81	0.0316	
49	C0PI69	Zm00001d040286	Uncharacterized protein	18.5	2	0.81	0.0092	
50	A0A0B4J3C2	ZEAMMB73_Zm00001d037873	Elongation factor 1-alpha	42.1	15	0.81	0.0463	RNA transport
51	B4FEV5	Zm00001d031689	Uncharacterized protein	13.8	1	0.81	0.0400	Plant-pathogen interaction
52	P26566	rpl20	50S ribosomal protein L20, chloroplastic	20.2	3	0.81	0.0476	Ribosome
53	A0A1D6KBW7	ZEAMMB73_Zm00001d030317	Hsp20/alpha crystallin family protein	17.8	2	0.81	0.0098	
54	A0A1D6ICZ3	542526	Calcium dependent protein kinase8	7.0	3	0.80	0.0465	Plant-pathogen interaction
55	B4FAJ4	Zm00001d008759	Uncharacterized protein	2.8	1	0.80	0.0260	Peroxisome
56	B6T9T5	N/A	Uncharacterized protein	4.3	1	0.80	0.0002	
57	Q9M7E2	Zm00001d036904	Elongation factor 1-alpha	30.7	10	0.80	0.0134	RNA transport
58	B7ZZ42	103650526	Heat shock 70 kDa protein 3	58.6	30	0.80	0.0076	Spliceosome, Endocytosis, Protein processing in endoplasmic reticulum
59	A0A1D6N9X4	103651144	Insulin-degrading enzyme-like 1 peroxisomal	3.5	3	0.79	0.0149	
60	A0A1D6IHP2	103633334	ARM repeat superfamily protein	6.5	5	0.79	0.0161	

Table 3. *Cont.*

No.	Protein ID [1]	Gene Name/ID [2]	Description [3]	Coverage (%) [4]	Peptide Fragments [5]	Fold Change [6]	p Value [7]	Pathways [8]
61	B4FLV6	100286322	Protein translation factor SUI1	20.0	3	0.79	0.0269	RNA transport
62	B4FQM2	100282190	Pyrophosphate–fructose 6-phosphate 1-phosphotransferase subunit beta	6.7	2	0.79	0.0123	Fructose and mannose metabolism, Pentose phosphate pathway; Glycolysis/Gluconeogenesis
63	B6TY02	Zm00001d017866	Aspartic proteinase nepenthesin-1	5.6	2	0.78	0.0276	
64	A0A1D6PW61	100191474	DNA topoisomerase 1 beta	3.1	1	0.78	0.0189	
65	B6SR37	Zm00001d011799	Uncharacterized protein	17.3	2	0.78	0.0070	
66	A0A1D6JQY8	100192907	Uroporphyrinogen-III synthase	2.8	1	0.78	0.0294	Porphyrin and chlorophyll metabolism
67	A0A1D6IIC2	ZEAMMB73_Zm00001d021999	Nuclear transport factor 2 (NTF2) family protein	5.6	1	0.77	0.0092	
68	B6U4J6	Zm00001d045774	Embryogenesis transmembrane protein	4.5	1	0.77	0.0258	
69	C0P626	Zm00001d011454	Carbonic anhydrase	74.3	13	0.77	0.0272	Nitrogen metabolism
70	Q9M7E3	Zm00001d009868	Elongation factor 1-alpha	37.8	13	0.76	0.0045	RNA transport
71	B6SI29	1005011869	Histone H2A	29.3	4	0.76	0.0326	
72	B4FLA6	100194327	Histone H2A	28.9	3	0.76	0.0406	
73	A0A1D6JVL9	ZEAMMB73_Zm00001d028377	Small nuclear ribonucleoprotein Sm D3	21.7	2	0.75	0.0384	Spliceosome
74	B6SLI1	100282946	40S ribosomal protein S30	16.1	1	0.74	0.0110	Ribosome
75	A0A1D6LBT4	100279572	Protein prenyltransferase superfamily protein	7.0	1	0.72	0.0475	
76	A0A1D6P0E7	ZEAMMB73_Zm00001d046001	Triose phosphate/ phosphate translocator TPT chloroplastic	22.1	2	0.72	0.0111	
77	B4FFS7	Zm00001d036233	Uncharacterized protein	7.8	1	0.71	0.0047	
78	A0A1D6FPL0	100382596	Fructose-16-bisphosphatase cytosolic	21.2	8	0.70	0.0179	Fructose and mannose metabolism, Pentose phosphate pathway Carbon metabolism, Glycolysis/Gluconeogenesis.
79	Q8LLS4	Pgk-1	Phosphoglycerate kinase (Fragment)	32.2	9	0.69	0.0440	
80	A0A1D6K8W1	ZEAMMB73_Zm00001d030005	Dynamin-related protein 1E	2.7	1	0.68	0.0411	
81	A0A1D6QSH1	100383873	Cullin-associated NEDD8-dissociated protein 1	3.6	3	0.65	0.0139	
82	B6TNP4	Zm00001d034479	Histone H1	41.0	11	0.65	0.0485	
83	A0A1D6MEZ2	ZEAMMB73_Zm00001d039282	Serine/threonine-protein kinase AGC1-5	1.4	1	0.55	0.0120	
84	E7DDW6	Zm00001d026630	Clathrin light chain 2	23.0	4	0.52	0.0203	

[1] Protein ID, unique protein identifying number in the UniProt database; [2] Gene name/ID; name or ID number of the corresponding gene of the identified differentially abundant protein as searched against the maize sequence database Gramene (http://ensemble.gramene.org/Zeamays); [3] Description, annotated biological functions based on Gene Ontology (GO) analysis; [4] Coverage (%), sequence coverage is calculated as the number of amino acids in the peptide fragments observed divided by the protein amino acid length; [5] Peptides fragments, refer to the number of matched peptide fragments generated by trypsin digestion; [6] Fold change, is expressed as the ratio of intensities of up-regulated or down-regulated proteins between drought stress treatments and control (well-watered conditions); All the fold change figures below 1 represents that the proteins were down-regulated. All the figures above 1 means the proteins were up-regulated; [7] p value, statistical level (using Student's t-test) below <0.05, at which protein differential expression was accepted as significant; [8] Pathways, metabolic pathways in which the identified protein was found to be significantly enriched; [9] uncharacterized protein, a protein without any functional annotations ascribed to it at the present.

Table 4. Drought responsive DAPs of the tolerant line that were also differentially expressed between the tolerant and sensitive lines after drought treatment.

No.	Protein ID[1]	Gene Name/ID[2]	Description[3]	Coverage (%)[4]	Peptide Fragments[5]	YE8112 Fold Change[6]	p Value[7]	SD_TD Fold Change[8]	p Value[7]	Pathways[9]
1	B6SQW8	Zm00001d024893	Uncharacterized protein	27.2	3	1.59	0.0155	0.53	0.0093	No significant enrichment
2	B4FKG5	542304	Abscisic acid stress ripening 1	47.1	4	1.34	0.0096	0.60	0.0325	No significant enrichment
3	A0A1D6HWS1	100282063	Dirigent protein	34.3	4	1.29	0.0207	0.67	0.0118	Not significant enrichment

[1] Protein ID, unique protein identifying number in the UniProt database; [2] Gene name/ID; name or ID number of the corresponding gene of the identified differentially abundant protein as searched against the maize sequence database Gramene (http://ensemble.gramene.org/Zeamays); [3] Description, annotated biological functions based on Gene Ontology (GO) analysis; [4] Coverage (%), sequence coverage is calculated as the number of amino acids in the peptide fragments observed divided by the protein amino acid length; [5] Peptides fragments, refer to the number of matched peptide fragments generated by trypsin digestion; [6] YE8112 fold change, is expressed as the ratio of intensities of up-regulated or down-regulated proteins between drought stress and control (well-watered) conditions; [7] p value, statistical level (using Student's t-test) below <0.05, at which protein differential expression was accepted as significant; [8] SD_TD fold change, is the ratio of intensities of up-regulated or down-regulated proteins between drought stressed sensitive line and drought stressed tolerant line; All the fold change figures below 1 represents that the proteins were down-regulated. All the figures above 1 means the proteins were up-regulated; [9] Pathways, metabolic pathways in which the identified protein was found to be significantly enriched.

Table 5. Common (overlapping) drought-responsive seedling leaf DAPs between MO17 and YE8112.

No.	Protein ID	Gene Name/ID	Description	Coverage (%)	Peptide Fragments	YE8112		MO17		Pathways
						Fold Change	p Value	Fold Change	p Value	
1	B6ITD62	100282951	Membrane steroid-binding protein 1	35.8	5	0.81	0.0223	1.50	0.0142	
2	A0A1D6GZE2	100272744	Ribose-phosphate pyrophosphokinase	5.4	1	0.82	0.0078	0.82	0.0068	Purine metabolism/Carbon metabolism/Pentose phosphate pathway
3	C4J0F8	Zm00001d038865	Uncharacterized protein	32.5	4	0.80	0.0090	0.81	0.0465	
4	C0PHL2	Zm00001d018627	Monosaccharide transporter1	3.8	1	0.79	0.0051	1.69	0.0495	
5	C0HDZ4	Zm00001d009082	SAM-dependent methyltransferase superfamily protein	14.1	2	0.73	0.0218	1.52	0.0245	Ribosome

For full description of the column items, please refer to Tables 2–4 captions above.

2.4. Gene Ontology (GO) Annotation and Functional Classification of the Drought Responsive DAPs

We performed gene ontology (GO) annotation to assign GO terms to the DAPs using Blast2GO web-based program (https://www.blast2go.com/). Further, GO functional classification of the GO-term-assigned-DAPs into biological processes (BP), molecular functions (MF), and cellular component (CC) categories was carried out. For the tolerant inbred line YE8112-specific DAPs (Area I of Figure 4), GO:0010196 (non-photochemical quenching), GO:1990066 (energy quenching), GO:0010155 (regulation of proton transport), GO:0009644 (response to high light intensity), and GO:0009743 (response to carbohydrates) were the most significantly enriched terms in the BP category; GO:0010333 (terpene synthase activity), GO:0003937 (IMP cyclohydrolase activity) and GO:0004126 (cytidine deaminase activity) were significant in the MF category; whereas GO:0009503 (thylakoid light-harvesting complex), GO:0030076 (light-harvesting complex), GO:0009783 (photosystem II antenna complex), GO:0098807 (chloroplast thylakoid membrane protein complex), and GO:0009517 (PSII associated light-harvesting complex II) were significant GO terms in the CC function (Table S5; Figure S4A).

In the SD_TD comparison (Area II of Figure 4), GO:0065004 (protein-DNA complex assembly), GO:0006323 (DNA packaging), GO:0006325 (chromatin organization) and GO:0006334 (nucleosome assembly) were the most significant terms in BP category; whilst GO:0046982 (protein heterodimerization activity), GO:0046983 (protein dimerization activity) and GO:0004473 (malate dehydrogenase (decarboxylating) (NADP+) activity) were the most significantly enriched under the MF category (Table S6; Figure S4B). Among the significant GO terms in the sensitive line MO17 (SC_SD) were GO:0051276 (chromosome organization), GO:0007059 (chromosome segregation) and GO:0006338 (chromatin remodeling) in the BP category; GO:0008135 (translation factor activity, RNA binding), GO:0003676 (nucleic acid binding), and GO:0003924 (GTPase activity) in the MF category; GO:0005694 (chromosome), GO:0000785 (chromatin) and GO:0044427 (chromosomal part) in the CC functions (Table S7; Figure S4C).

The significantly enriched GO terms in each of the three comparison groups (TC_TD, SD_TD, SC_SD) were mapped to the top 20 biological functions. Among the tolerant line YE8112 (TC_TD) -specific DAPs, metabolic process (46.86%), cellular process (36.23%) and response to stimuli (7.69%) were the most popular BP functions; catalytic activity (48.0%) and binding (40.47%) most prominent in MF category; whilst cells and cell parts (47.0%), organelles (22.31%), organelle parts (5.88%), membrane (17.01%), and membrane parts (7.75%) were the popular locations for the DAPs under CC functions (Figure 5A). In the Area II (SD_TD) DAPs, metabolic process (50%), cellular process (35%), and response to stimuli (15%) in BP category; catalytic activity (55%) and binding (43%) in MF category; cell (55%), cell part (45%), and organelle (50%) in CC functional category were prominent (Figure 5B). Among the sensitive line MO17 (SD_SC)-specific DAPs, metabolic process (48.03%), cellular process (37.41%) and cellular component organization (10.22%) were the most common biological processes; catalytic activity (46.37%), binding (48.19%), and structural molecule activity (5%) in the MF category; whereas cells and cell parts (44%), organelles and organelle parts (20%) and membrane (23%) were prominent in CC functions category (Figure 5C).

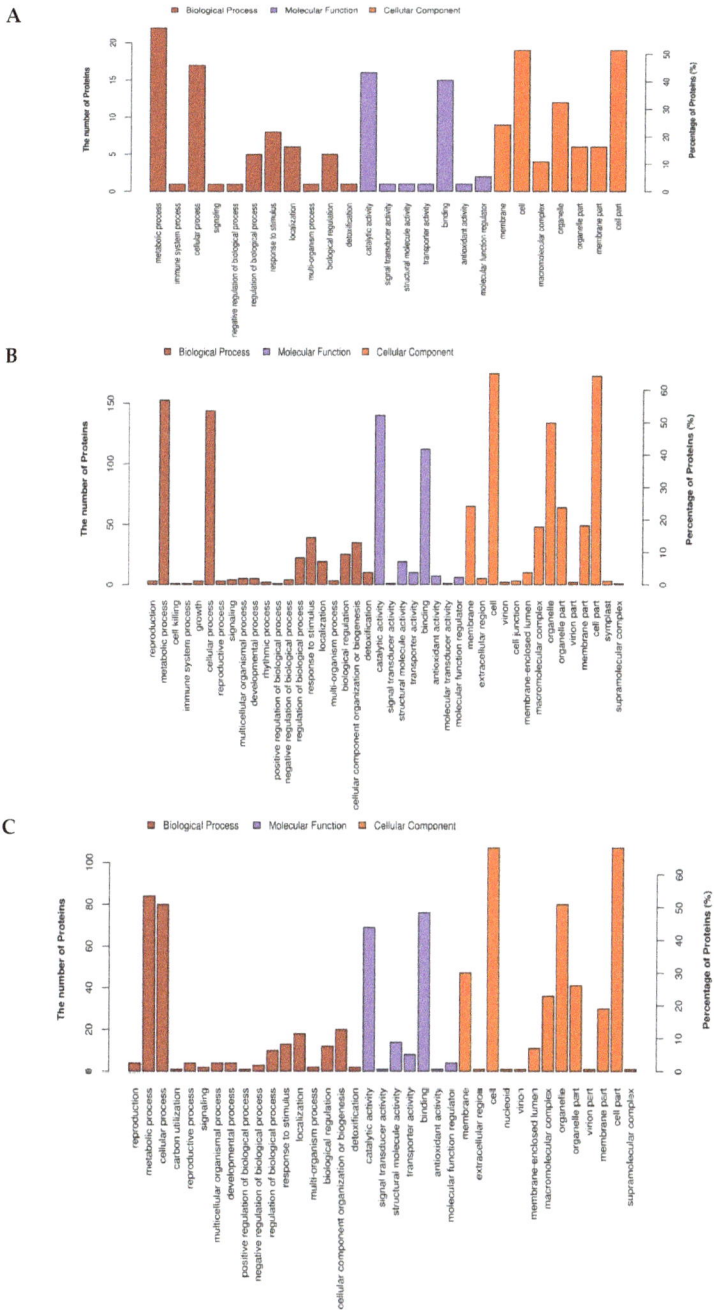

Figure 5. Gene ontology (GO) functional classification of drought responsive proteins. (**A**) YE8112 specific DAPs; (**B**) SD_TD specific DAPs; and (**C**) MO17 specific DAPs. *Y*-axis represents the number (and%) of proteins in each function; *X*-axis displays the protein functions, categorized into three broad biological functional groups.

2.5. KEGG Pathway Enrichment Analysis

To further analyze the functional consequences of the drought-responsive DAPs, we mapped them to the Kyoto Encyclopedia of Genes and Genomes (KEGG, available online: https://www.genome. jp/kegg/; accessed on 16 March 2018) database and the DAPs were assigned to various biological pathways. Additionally, significant KEGG pathway enrichment analysis was performed using the hypergeometric test. Proline metabolism (two proteins), photosynthesis antenna proteins (2) and biosynthesis of amino acids (2) were the top three enriched pathways in YE8112 (Figure 6A). However, the composition of the enriched KEGG pathways in MO17 differed significantly, with RNA transport (nine proteins), ribosome (9), carbon metabolism (7), biosynthesis of amino acids (6), and carbon fixation in photosynthetic organisms (6) being the top most enriched pathways (Figure 6B). These results show that more proteins were observed in the enriched pathways of MO17 than YE8112 and that the two inbred lines diverge significantly in pathway responses to drought stress. Using a hypergeometric test, KEGG pathways that had a p value < 0.05 were considered to be significantly affected by drought stress. We observed that only one pathway (photosynthesis antenna proteins) was considerably enriched (0.06) among the YE8112 pathways (Figure 6C), whist two KEGG pathways, RNA transport (0.16) and C5-branded dibasic acid metabolism (0.33), were significantly enriched among MO17 pathways (Figure 6D).

2.6. Protein-Protein Interactions (PPI)

Plant cell and tissue proteins do not act as individual molecules, but, rather, play coordinated and interlinked roles in the context of networks [30]. To determine how maize leaf cells' drought stress signals are transmitted through protein-protein interactions to affect specific cellular functions, the identified YE8112 and MO17 DAPs were further analyzed using the String 10.5 database. Three groups of interacting proteins were identified in YE8112 (Figure 7A). The first and largest network comprise Adenosylhomocysteinase (Zm 19562); hypothetical protein LOC100194360 (AC 199526.5_FGP002); 5-methyltetrahydropteroyltriglutamate-homocysteine methyltransferase (Zm 45026); O-succinylhomoserine sulfhydrylase (GRMZM2G450498_P01); Adenosylhomocysteinase (GRMZM2G111909_P01); uncharacterized protein (Zm 24266); and Glutamate synthase 2 (NADH) (GRMZM2G375064_P01). These proteins are crucial in amino acid metabolism, maintaining antioxidant defense and epigenetic regulation (DNA methylation and histone modifications). The second group was constituted by (Zm 24266)—(GRMZM2G375064_P01)—electron transporter/thiol-disulfide exchange intermediate (GRMZM5G869196_P01) linkage. These proteins are involved in amino acid metabolism, energy metabolism (NADPH production), electron transport and stress signaling, and maintaining redox homeostasis.

The third interaction network involved (Zm 24266)—hypothetical protein LOC 100274507 (AR4)—(GRMZM2G375064_P01)—Arginase 1 mitochondrial-like (GRMZM5G831308_P01). These proteins interact in energy (NADH) production and polyamines and proline synthesis. In addition, four protein pairs (including AY110562—GRMZM5G831308_P01, Zm 5448—AC 199526.5_FGP002, GRMZM5G869196_P01—GRMZM5G864335_P01, and GRMZM5G869196_P01—Zm 118187) were observed (Figure 7A). Meanwhile, separate protein interaction networks were predicted for MO17, including a large and complex network, several small networks, and protein pairs (Figure 7B).

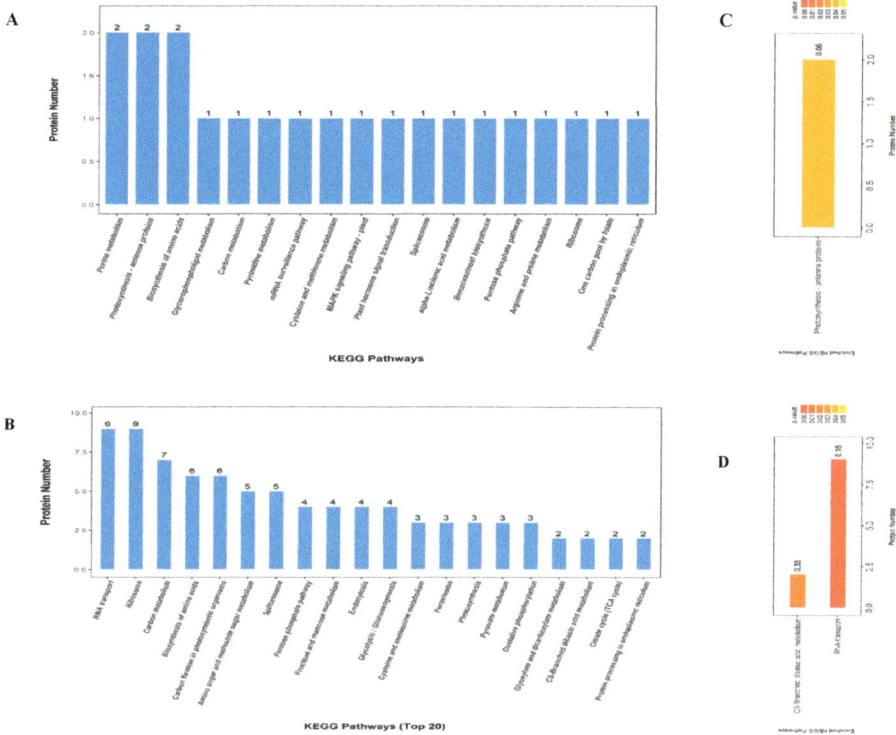

Figure 6. KEGG pathway enrichment analysis of the DAPs. (**A**) TD_TC comparison; (**B**) SD_SC comparison. The whole number above the bar (blue) graph represents number of DAPs enriched in the corresponding pathway. (**C**) Most significantly enriched pathway in TD_TC. (**D**) Most significantly enriched pathways in SD_SC based on the hypergeometric test. The significance of the enrichment of the KEGG path is based on the Student's *t*-test, $p < 0.05$. The color gradient represents the size of the p value; the color is from orange to red, and the nearer red represents the smaller the p value, the higher the significant level of enrichment of the corresponding KEGG pathway. The label above the bar graph shows the enrichment factor (rich factor ≤ 1), and the enrichment factor indicates the number of differentially abundant proteins participating in a KEGG pathway as a proportion of proteins involved in the pathway in all identified proteins.

2.7. Quantitative Real-Time RT-PCR (qRT-PCR) Analysis

To confirm our findings based on iTRAQ sequencing data, we conducted a supporting experiment by using quantitative real-time PCR (qRT-PCR). We made the selection of genes based on the following criteria: highly differentiated in response to drought stress and reported to be potentially associated with drought tolerance. A sample of 30 genes (Table S8) was selected from the drought responsive DAPs from different groups (labeled Areas I–IV of Figure 3). Results of the qRT-PCR analysis confirmed our findings based on iTRAQ seq data. In particular, the patterns of iTRAQ seq expression on all 30 genes were replicated by the qRT-PCR approach (Figure 8A–D; Table S9). A correlation coefficient (R^2) (of the fold changes between qRT-PCR and iTRAQ seq) of 83.51% was obtained (Figure S5), endorsing that our iTRAQ sequencing data was reliable.

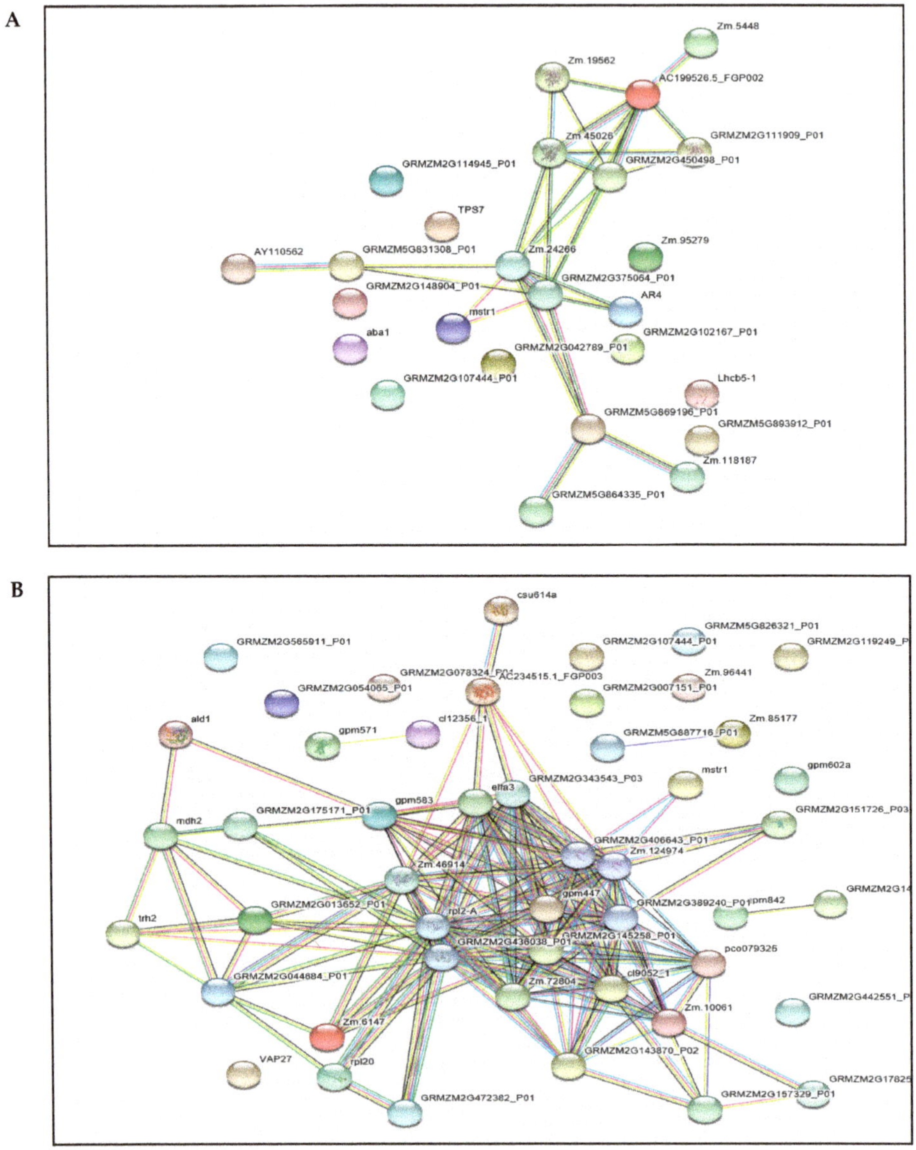

Figure 7. Protein interaction network consisting of DAPs identified in drought stressed maize seedling leaves of (**A**) YE8112 (**B**) MO17. The network was constructed using the String program (http:// www.string-db.org/) with a confidence score higher than 0.5. Nodes represent proteins, and the line thickness represents the strength of the supporting data.

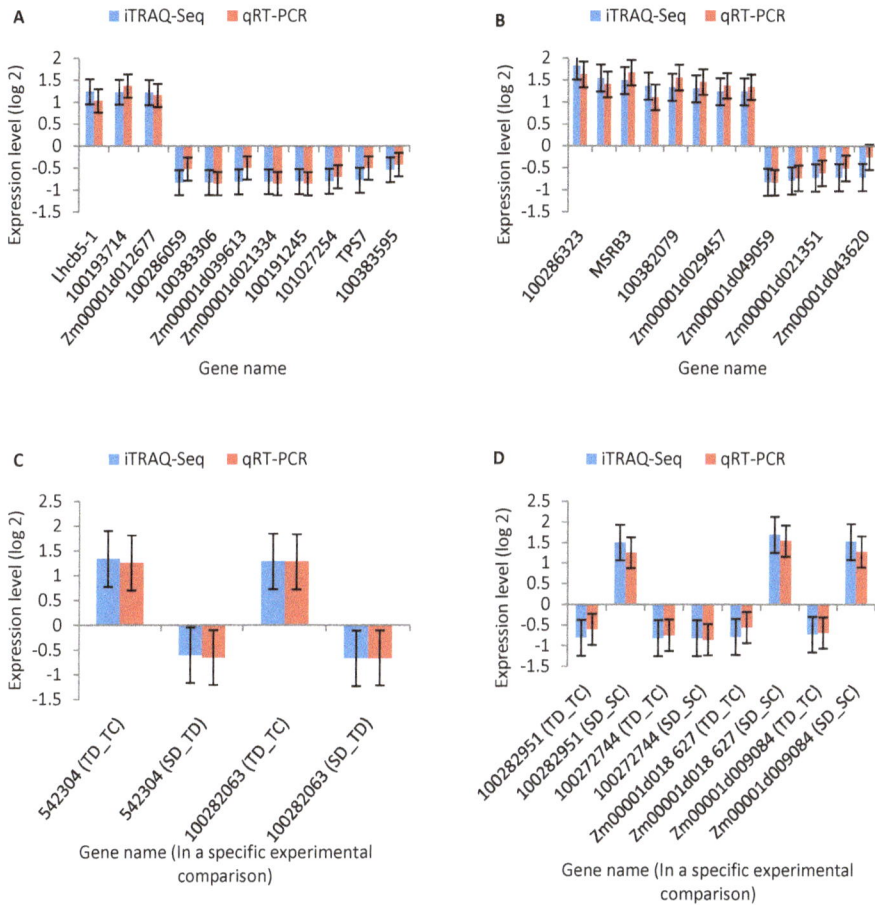

Figure 8. Confirmation of iTRAQ-seq results by quantitative real-time PCR (qRT-PCR). Quantitative RT-PCR analysis of the expression patterns of the maize seedling leaf genes encoded by differentially abundant proteins (DAPs) from different comparisons: (**A**) DAPs specific to TD_TC; (**B**) DAPs specific to SD_TD; (**C**) DAPs shared between TD_TC and SD_TD; and (**D**) Common DAPs shared between TD_TC and SD_SC. The *y*-axis represents qPCR relative expression levels (log$_2$-fold change) and fold-change of the iTRAQ-seq data. All genes with negative values of expression level means that they were down-regulated in response to drought stress. Maize gene *GAPDH* (accession no. X07156) was used as the internal reference. Error bars represent the SE (*n* = 3).

3. Discussion

Drought stress is the most serious environmental stress posing a severe threat to maize production worldwide [9–11,16,40]. In response to drought stress, plants evolve complex adaptive mechanisms at the physiological, biochemical, and molecular levels [41–45]. To gain in-depth understanding of the determinants underpinning drought tolerance in maize, herein, we have performed proteomic analysis of two contrasting maize inbred lines (drought-tolerant YE8112 and drought-sensitive MO17) after a seven-day moisture-deficit exposure period at the seedling stage. Further, we compared some physiological responses of these two inbred lines under drought stress conditions, and our findings provide further insights into the drought stress tolerance signatures in maize.

3.1. Inbred Lines YE8112 and MO17 Showed Significant Differences in Physiological Response to Drought Stress

Our experimental findings on physiological indices showed that the two maize inbred lines performed differently under drought stressed conditions. Relative water content (RWC) decreased significantly in leaf tissues of both inbred lines under drought stress conditions, and at most stress exposure time periods (days). It was generally higher in YE8112 seedlings both under non-stress and drought stress conditions (Figure 1D). We suggest that the high relative water content could help the tolerant inbred line YE8112 to perform physio-biochemical processes more efficiently under drought stress environment than the sensitive line MO17. Similarly, Moussa and Abdel-Aziz [46] observed RWC to be significantly higher in the tolerant maize genotype Giza 2 than sensitive genotype Trihybrid 321 under both control and water stress conditions.

Upon exposure to abiotic stresses, tolerant cells activate their enzymatic antioxidant system, which then starts quenching the ROS (reactive oxygen species) and protecting the cell [47]. Peroxidases (POD) and superoxide dismutase (SOD) constitute the first line of defense via detoxification of superoxide radicals, by acting as H_2O_2 scavenging enzymes [48]. Our investigation of the POD and proline contents of the two lines revealed that tolerant YE8112 seedlings always accumulated greater amounts of the antioxidant enzyme and protective osmolyte than sensitive MO17 seedlings under drought conditions (Figure 1E,F). The POD activity was enhanced continuously with increasing drought stress exposure period (days) in both inbred lines. However, the percent enhancement was significantly greater in tolerant line YE8112 than sensitive line MO17 (Figure 1E). It has been recognized that improved POD activity under stress conditions protects plant cells from oxidative damage emanating from reactive oxygen species (ROS) generated under such conditions [46]. In the current study, the tolerant inbred line YE8112 had greater POD activity than sensitive line MO17, which may infer better ROS quenching capacity of YE8112 than MO17. Moreover, higher proline content in YE8112 leaf tissues may explain the improved cell homeostasis in YE8112 than MO17 [49]. Higher proline content in the cells has been revealed to lower cell water potential, consequently promoting increased water absorption into those cells, thereby providing cells with immediate short-term cushion against the effects of water deficit [50].

In the present investigation, tolerant maize inbred line YE8112 maintained a higher cell membrane stability index under drought-stress conditions (Figure 2). Contrastingly, the lower membrane stability index in MO17 reflects the extent of lipid peroxidation, which in turn is a consequence of higher oxidative stress due to water stress conditions [48]. The MDA content was significantly higher in MO17 than in YE8112, both under non-stress and drought-stress conditions (Figure 1G). The rise in MDA content in both inbred lines under stress conditions suggests that drought stress could induce membrane lipid peroxidation and membrane injury by means of ROS [47,51]. In the current study, the tolerant line YE8112 had lower MDA content values than the sensitive line MO17, indicating that YE8112 cells had a better ROS quenching ability than MO17 cells, hence improved cell membrane stability. Previously, it has been revealed that higher cell membrane stability and improved cell water preservation capacity of the tolerant maize lines help them better endure moisture deficit as compared to (the low membrane stability and poor water retention capacity of) the sensitive lines [3,46]. Further, the iTRAQ analysis showed that the two genotypes' responses to drought stress were quite different. After drought stress treatment and at the standard fold change of ≥ 2 and false discovery rates (FDR) <0.001, drought-tolerant YE8112 had relatively lower DAPs than drought-sensitive MO17 (Figure 4). Under drought conditions and compared to inbred line MO17, tolerant line YE8112 maintained higher leaf RWC (Figure 1D), consequently leading to relatively lower stress at the cellular level. This has been further confirmed by trypan blue staining (Figure 2). Thus, YE8112 had a more limited proteome response. A series of reports on maize seedling-stage abiotic stress analyses between different inbred lines exist [2,4]. In particular, Li et al. [2] found relatively large number of differentially expressed genes (DEGs) in freezing-sensitive inbred line Hei8834 than freezing-tolerant line KR701 after freezing treatment. Similarly, Zheng et al. [4] realized greater proportion of DEGs in drought-sensitive Ye478 than drought-tolerant Han21 after drought treatment. From the above analysis, we could confidently conclude that in our study, there was high consistence between proteome profiling data and the phenotypic and physiological characterization of the two inbred lines.

Thus, from these findings, it can be inferred that the stress tolerance mechanism exists at seedling stage of maize inbred lines. The YE8112 inbred line is comparatively tolerant to drought stress owing to its maintenance of higher RWC and proline contents under both non-stressed and stressed conditions, higher increase in POD enzyme activity, along with decreased level of lipid peroxidation (MDA content). The higher membrane stability index and high water retention capacity might have also imparted drought stress tolerance in YE8112.

3.2. Drought Responsive DAPs Observed in the Tolerant Inbred Line YE8112

3.2.1. Photosynthesis (Photosystem II) Related Proteins Are the Major Drought Tolerance Signature in YE8112

Among the up-regulated DAPs observed in YE8112 were chlorophyll a-b binding proteins (Q41746 encoding *Lhcb5-1*; and B4FL55 encoding *542320/Lhcb5-2*). It has been noted that drought stress induced mismatch between photosynthetic light capture and utilization limits the overall plant cell photosynthetic efficiency [48]. The inhibition in photosynthesis activity results from the cell-damaging ROS that are generated in the PSII reaction center of the thylakoid membranes when cells exude excess light energy [52]. In response, plants activate the proteins involved in balancing photosynthesis light capture and utilization and non-photochemical quenching. In the current study, chlorophyll a-b binding proteins (Q41746 and B4FL55) were significantly up-regulated in response to drought stress. In addition, GO annotation analysis (See Section 2.4 above; Figure S4A) showed that under the biological process and cell component categories, the terms related to non-photochemical quenching, energy quenching, response to high light intensity, PSII antenna complex, PSII associated light harvesting complex II, and thylakoid light-harvesting complex were dominating and most significantly enriched. Furthermore, 'photosynthesis antenna proteins' KEGG pathway was the most significantly enriched in YE8112 (Figure 6A). Thus, these genes (*Lhcb5-1*, *542320/Lhcb5-2*) play pivotal roles in PSII associated light-harvesting complex and cysteine biosynthesis process [53,54]. This appears to be the tolerant inbred line YE8112's major molecular signature in drought stress tolerance.

3.2.2. Up-Regulation of Lipid-Metabolism Related Proteins Could Contribute to Increased Signaling and Water Conservation in the Cell

Lipid metabolism related proteins (Q2XX23, nsLTPs; A0A1D6GAZ6, GDPD) were up-regulated in response to drought stress (Table 2; Table S2). Several potential biological functions of nsLTPs have been proposed, including their (nsLTPs) involvement in long-distance signaling that possibly is implicated in plant defense against pathogens [55], and the formation of protective hydrophobic layer on the surfaces of plant aerial organs [56]. In barley (*Hordeum vulgaries* L.) and *Zea mays* L. leaves, nsLTPs, working in synergy with thionins, were identified as potent inhibitors of bacterial and fungal plant pathogens [57]. This may indicate that plants may have developed cross-tolerance mechanisms to cope with abiotic and biotic stresses [5,58]. The GDPD (Glycerophosphodiester phosphodiesterase) which hydrolyzes glycerophosphodiesters into sn-glycerol-3-phosphate (G-3-P) and the corresponding alcohols, plays a crucial role in lipid metabolism in both prokaryotes and eukaryotes [59]. Cheng et al. [60], studying on *Arabidopsis thaliana*, suggested that the GDPD-mediated lipid metabolic pathway may be involved in release of inorganic phosphate from phospholipids during phosphate starvation. Here, we also submit that the enhancement (up-regulation) of lipid-metabolism related proteins could contribute to increased signaling and water conservation in the cell through formation of hydrophobic layer on leaf surface (which enables the leaves of stressed maize to normal growth under stress), and thus, is an indispensable adaptive response to drought stress in maize seedlings.

3.2.3. Enhancement of Molecular Chaperons Is a Vital Strategy for Drought Stress Tolerance in YE8112

To confront protein inactivation or denaturation resulting from drought stress, plants activate protective mechanisms that include chaperones and chaperone-like proteins, osmolytes or compatible solutes [61]. Here, abscisic acid stress ripening 1 (ASR1) protein was up-regulated in response to drought stress. Previously, the combined effort of tomato *ASR1* gene analogue (*S1ASR1*) and osmolyte glycine-betaine has been shown to stabilize other proteins against heat and cold stress induced denaturation, thereby protecting those proteins under such conditions [62]. Kalifa et al. [63] had observed that overexpression of the water and salt stress-regulated *Asr1* gene confers an increased salt tolerance. Earlier, they had concluded that steady-state cellular levels of tomato ASR1 mRNA and protein are transiently increased following exposure of plants to poly (ethylene glycol), NaCl or abscisic acid [64]. Universal stress proteins (USP) are widely spread proteins in nature, belonging to the PF00582 superfamily (COG0589) and are suggested to function in nucleotide binding and signal transduction [65]. In stress conditions such as heat shock, nutrient starvation, the presence of oxidants, DNA-damaging agents, or other stress agents which may arrest cell growth, USPs are overproduced and through a variety of mechanisms aid the organism survive such uncomfortable condition [66]. Furthermore, HSP protein (B4FVB8), alpha/beta hydrolase superfamily protein (B6T6N9) and Clp B3 protein chloroplastic (A0A1D6HFD3) were up-regulated in YE8112 (in the SD_TD comparison) in response to drought (Table S4; Figure 4). Alpha/beta hydrolase (ABH) functions as chaperons and hormone precursors in the stress response process, by way of its fold acting as bona fide ligand receptor in the strigolactone, karrin-smoke receptor, and gibberellin response pathways [67]. Chaperon protein Clp B3 chloroplastic confers thermo-tolerance to chloroplasts during heat stress in Arabidopsis [68]. From these reports, we can conclude that up-regulation of chaperons and USP genes is an important strategy to tolerate drought in maize seedlings.

3.2.4. Proteins/Enzymes Involved in Cellular Detoxification under Drought Stress

Plant stress response process is a complex phenomenon, involving stress signals perception, cell homeostasis adjustment, DNA cell cycle check points arresting, and damage-induced DNA repair processes [9]. In addition, mitogen-activated protein kinase (MAPK) cascades, calcium-regulated proteins, ROS, and transcriptional factors cross-talk are active in stress signaling and defense response and acclimation pathways, rendering the whole network intricate [69]. Generally, ROS perturb cellular redox homeostasis resulting in oxidative damage to many mitochondrial cellular components along with over-reduction of electron transport chain components in the mitochondria, plastids and several detoxification reaction centers. This also results in an imbalance between ROS and the antioxidative defense system [70]. It is critical that proteins involved in redox homeostasis be instituted for fine regulation of the steady state and responsive signaling levels of ROS in order to avoid injury and maintain an appropriate level by which different developmental and environmental signals can be perceived and transmitted [30,71]. Here, we observed that glutathione transferase (B4G1V3), thioredoxin-like protein (A0A1D6K5D2) and ferredoxin-oxidoreductase (COP472) were up-regulated in response to drought (see SD_TD comparison, Figure 4; Table S4).

Glutathione transferases (GSTs) are key cellular detoxification enzymes involved in scavenging of excessive amounts of ROS generated in plant tissues under oxidative stress conditions, and thus, protect plants from oxidative damage [72,73]. They also participate in the signal transduction pathways, cellular responses to auxins and cytokinins, as well as metabolic turnover of cinnamic acid and anthocyanins [74,75]. GSTs have also been up-regulated in response to aluminum toxicity [76]. Ferrodoxin oxido-reductase is vital in oxidation-reduction, electron transfer and signaling processes, as well as catalyzing light dependent photosynthesis [77,78]. Thioredoxins (TRXs) are involved in the protection against oxidative stress as electron donors for thioredoxin peroxidases, which detoxify hydrogen peroxide and alkyl hydroperoxides [79]. Potato plants lacking the CDSP32 plastidic thioredoxin exhibited overoxidation of the BAS1 2-cysteine peroxiredoxin and increased lipid peroxidation in thylakoids under photooxidative stress [79]. Thus, the up-regulation of these antioxidant enzymes herein aids in countering the ROS effects, thereby protecting cells from oxidative damage. Overall, we can suggest that YE8112 endured drought stress better than MO17 because of its enhanced activation of proteins involved in detoxification signaling, response to stress and oxidation-reduction.

However, in the TD_TC comparison, we observed that five proteins involved in stress oxidation-reduction (B6TD62, membrane steroid binding protein; B4FQR3, Aldose reductase; Q84TC2, DIBOA-glucoside dioxygenase BX6; B4FTP2; and H9BG22) and ribosome biogenesis (A0A1D6PT84 and C4J0F8) were differentially down-regulated (Table S2). The down-regulation of these stress redox homeostasis proteins in TD_TC implies the complexity of the cell redox system in stress response. Further, the repression of proteins involved in ribosome biogenesis in leaves of YE8112 may, on one hand, simply indicate the drastic effect of drought on stress-defense protein biosynthesis [80]. However, on the other hand, here, we suggest that the down-regulation of proteins involved in ribosome biosynthesis is an indication that, under drought stress, the tolerant line YE8112 had the ability to reduce the synthesis of redundant proteins, which may help the plant save energy to battle that stress [5,81].

3.2.5. Proteins Related to 'Response to Stimuli' under Drought Stress

Several DAPs were enriched in 'response to stimuli' under the biological processes (BP) category of the GO functional classification in the tolerant line YE8112 (Table S5). Among the up-regulated DAPs in this function were two uncharacterized proteins (C0HJ06, B6UFE3), two chlorophyll a-b binding proteins (Q41746, B4FL55); Abscic acid stress ripening 1 (B4FKG5), and a universal stress protein (C0HGH7) (Table 2). Additionally, in the SD_TD comparison, cytokinin riboside 5′–monophosphate phosphoribohydrolase protein (A0A1D6NKY3) (*LOG*) was up-regulated in response to drought stress (Figure 4; Table S4). The *LOG* enzyme is involved in cytokinin activation [82]. Cytokinin is a multifaceted phytohormone that plays crucial roles in diverse aspects of plant growth and development, including leaf senescence, apical dominance, lateral root formation, stress signaling and tolerance [83]. Cytokinin signaling cascades are evolutionarily related to the two-component systems that participate in environmental-stimuli-triggered signal transduction [84]. Taken collectively, we can conclude that cytokinin metabolism and signaling; in cross-link with photosynthesis proteins and some chaperons constitute a vital drought response cascade in YE8112.

However, six proteins (A0A1D6IUI1, ubiquitin carboxyl-terminal hydrolase 13; H9BG22, alpha-dioxygenase; A0A1D6PQ00, U2 snRNP auxiliary factor large subunit; B4FTP2, thioredoxin like protein CDSP32; Q5GJ59, terpene synthase 7; COPHF6, AAA-ATPase ASD mitochondrial) were down-regulated in response to drought in the TC_TD group (Table 2). The ubiquitin-dependent proteolytic pathway degrades most proteins and is the primary proteolysis mechanism in eukaryotic cells [85]. Whereas ubiquitin regulates the degradation of proteins, deubiquitinating enzymes (deubiquitinases) play the antagonistic role, therefore reversing the fate of the proteins [86]. Here, the down-regulation of ubiquitin carboxyl-terminal hydrolase 13 implies that cells suppress the proteins and enzymes involved in protein ubiqutination in order to protect themselves against unnecessary protein degradation under drought stress. Alpha-dioxygenase (α-DOX) catalyzes the primary oxygenation of fatty acids into oxylipins, which are important in plant signaling pathways. It has been shown to be up-regulated in response to different abiotic stresses including drought, salt, cold, and heavy metal; and may also be involved in the leaf senescence process [87]. Here we suggest that the down-regulation of α-DOX may be a way to retard leaf senescence in stressed maize seedlings, thereby improving drought tolerance.

Terpenes constitute a large class of secondary metabolites that serve multiple roles in the interactions between plants and their environment, including biotic and abiotic stress responses [88]. They are involved in environmental stimuli perception, stress, and phytohormone signaling [89,90]. In addition, MAPK cascade (signal transduction mechanism) plays an important role in activation and de-activation of enzymes through phosphorylation/de-phosphorylation, which allows for fast and specific signal transduction and amplification of external stimuli [91]. Previous studies [92–94] have revealed the role of MAPK cascade in intracellular pathogen immunity and abiotic stress signaling. However, in the current study, MAPK (A0A1D6GZE2) and terpene proteins were down-regulated reflecting the importance and complexity of the cell redox system, signaling, and abiotic-biotic stress cross talks in drought response. Furthermore, splicing is an essential process in eukaryotic gene expression, and the precise excision of introns from premRNA requires a dynamically assembled RNA protein complex (spliceosome). U2 snRNP is one such essential splicing factor that participates in intron and exon definition [95]. Thus, here, the down-regulation of U2 snRNP may imply that mRNA processing is negatively hampered by drought stress.

3.2.6. Key Epigenetic Regulation Mechanisms of the Tolerant Line YE8112

Plants also cope with abiotic stresses by prompt and harmonized changes at transcriptional and post-transcriptional levels, including the epigenetic mechanisms [96]. DNA methylation is essential for stress memory and adaptation in plants [97]. Abiotic or biotic factors can influence gene expression regulation via DNA methylation [98]. In chick pea (*Cicer arietinum* L.) leaf tissues, drought stress triggered DNA hyper-methylation [99]. Combined drought and salinity stresses triggered a shift from C3 to CAM photosynthesis mode in *Mesembryanthemum crystallinum* L. plants, as a result of DNA CpHpG-hypermethylation [100]. In the current study, proteins involved in S-adenosyl-methionine (SAM) dependent methyltransferase (MTases) activity (A0A1D6NE76 and C0HDZ4) were differentially expressed in response to drought stress (Table S2). SAM serves as methyl donor for SAM-dependent methyltransferases (MTases). The resultant transmethylation of biomolecules constitutes a significant biochemical mechanism in epigenetic regulation, cellular signaling, and metabolite degradation [101]. The DEP C0HDZ4 encode the maize gene ZEAMMB73_Zm0001d009084 and is important for DNA methylation. Thus, here, YE8112 induced dynamic DNA methylation alterations as part of a complex drought-stress response network, with bias towards down-regulation of SAM-D-MTase. Furthermore, acetyltransferase (B6UHR7) was up-regulated in YE8112 (see the SD_TD comparison, Table S4). Histone acetyltransferases (HATs) play an important role in eukaryotic transcriptional activation in the epigenetic regulation process [102]. Thus, the key epigenetic regulation mechanisms in YE8112 were DNA methylation (via down-regulation of overlapping protein A0A1D6NE76) and enhanced histone acetylation through up-regulation of HATs related proteins.

3.3. Drought Responsive DAPs Observed in Sensitive Inbred-Line MO17

The iTRAQ analysis identified a higher number of DAPs in MO17 than in YE8112 in response to drought stress (compare Tables 2 and 3). Variation in abundance of the DAPs in response to drought stress implies specific sensitivity or adaptation of these two maize lines [30]; the two inbred line plants detected the extent of the same drought stress conditions differentially. Drought tolerant-line YE8112 might have perceived the prevailing drought conditions as mild and then modulated fewer DAPs, whilst sensitive-line MO17 perceived the same conditions as severe and modulated more abundant DAPs in response.

3.3.1. Enhanced Expression of Heat Shock Proteins (HSP20-Like Chaperons) and 50S Ribosomal Proteins Constitutes a Critical Defensive Response in MO17

Among the dominating up-regulated DAPs in MO17, we observed heat shock proteins (HSP 20-like chaperons superfamily), chaperon DNA-J domain superfamily proteins and ribosomal proteins (50S Ribosomal protein L20) (Table 3). Molecular chaperons facilitate the stabilization of other macromolecular structures, including other proteins, under stress conditions [80]. Precisely, heat shock proteins (HSPs) are vital in protecting plants against stress by preserving other proteins in their functional confirmations [103]. HSPs have been greatly accumulated in alfalfa (*Medicago sativa* L.) leaves in response to salinity stress [104]. As anticipated, the increased accumulation (up-regulation) of HSPs could be regarded as a crucial defensive response of MO17 against drought stress. Additionally, ribosomal proteins (40S, 50S, and 60S) are an integral component of stress-defense protein biosynthesis machinery [105], hence were up-regulated under drought stress. Similarly, Ziogas et al. [106] found out that the 40S and 60S ribosomal proteins were up-regulated in citrus response to PEG-induced osmotic stress.

3.3.2. Up-Regulation of Cell Detoxification and Photosynthesis Related Proteins May Contribute to Enhanced Drought Stress Tolerance in MO17

Superoxide dismutase protein (B4F925), together with the photosynthesis related proteins: chlorophyll a-b binding protein (B4FV94), oxygen evolving enhancer protein (B6SUJ9), photosystem II CP47 reaction center protein (A0A1X7YHJ3), and pyruvate phosphate dikinase proteins were up-regulated in response to drought stress (Table 3). Enhanced antioxidant enzyme activity is a part of an array of complex detoxification and defense mechanisms to protect cells from the oxidative damage by excessive ROS [9]. Enhanced accumulation of SOD proteins suggests that the activation of enzymatic antioxidant systems is a crucial protective mechanism for drought stressed MO17. The SOD and oxygen evolving enhancer proteins may increase drought tolerance by playing a role in cellular detoxification and protecting cells from oxygen toxicity [80,85]. Photosystem II proteins, together with other auxiliary proteins, enzymes, or components of thylakoid protein trafficking/targeting systems, are directly or indirectly involved in de novo assembly and/or the repair and reassembly cycle of PSII [107,108]. Pyruvate phosphate dikinase (PPDK) is one of the most important enzymes in C_4 photosynthesis, catalyzing the reversible phosphorylation of pyruvate to phospho*enol*pyruvate, thus, the most crucial rate-limiting C_4 cycle enzyme [109,110]. Taken collectively, the above results indicated that the up-regulation of cell detoxification and photosynthesis enhancing proteins constitute a vital drought stress response strategy in the sensitive maize inbred-line MO17.

3.3.3. Glutathione Transferases and Ca^{2+}-Dependent Kinases Negatively Influenced by Short Term Drought Stress

Among the down-regulated DAPs in MO17 were those associated with signaling recognition, especially glutathione transferases (GSTs; A0A1D6M4E1) and calcium dependent protein kinase (A0A1D6ICZ3) (Table 3). The GSTs are key participants in plant growth and development, shoot regeneration processes, and adaptability to adverse environmental stimuli [72]. Crucially, GSTs are major cellular detoxification enzymes protecting plants from oxidative damage [73]. Calcium-dependent protein kinases (CDPKs) represent potential Ca^{2+} decoders to translate developmental and environmental stress cues [111,112]. However, the down-regulation of DAPs regulating these enzymes herein implies that short-term drought stress negatively influenced the signal transduction processes involving these enzymes.

3.3.4. Key Epigenetic Regulation Mechanisms of the Sensitive Line MO17

In addition to the DNA methylation related protein A0A1D6NE76 (overlapping between the two inbred lines; down-regulated in YE8112, but up-regulated in MO17), we also observed proteins associated with histones (histones H2A and H1) to be down-regulated in response to drought stress (Table 3). Histone modification is the key epigenetic regulation mechanisms in plants and eukaryotic cells [113]. Phosphorylation of H2A histones functions in DNA double strand breaks (DSBs) repair [114]. Thus, whilst DNA methylation (through down-regulation of related proteins) and histone acetylation were dominant epigenetic regulation mechanisms in YE8112, DNA methylation (via up-regulation of related proteins) and histone modification (probably phosphorylation; via down-regulation of H2A and H1 proteins) were preferred in MO17 in response to drought stress.

3.4. Overlapping Drought Responsive Proteins Between YE8112 and MO17 under Drought Conditions

Venn diagram (Figure 3) analysis showed that only five significant DAPs were common between TD_TC and SD_SC. All the 5 proteins (Table 5) were down-regulated in tolerant line

YE8112 in response to drought treatment. Comparably, among these five common proteins, two (ribose-phosphate pyrophosphokinase and uncharacterized protein C4J0F8) were down-regulated, whilst the other 3 (membrane steroid binding protein 1, monosaccharide transporter 1, SAM-dependent methyltransferase superfamily protein) were up-regulated in sensitive line MO17 in response to drought treatment. Moreover, the two common down-regulated proteins showed similar fold changes in both inbred lines under drought stress (Table 5). In *Arabidopsis thaliana*, membrane steroid binding protein 1 (MSBP1) is involved in inhibition of cell elongation [113]. Additionally, Yang et al. [114] realized that the inhibitory effects by 1-N-naphthylphthalamic acid (NPA), an inhibitor of polar auxin transport, are suppressed under the MSBP1 overexpression, suggesting the positive effects of MSBP1 on polar auxin transport. They concluded that MSBP1stimulates tropism by regulating vesicle trafficking and auxin redistribution in Arabidopsis seedling roots. Here, we suggest that maize seedlings endure drought stress by down-regulating MSBP1 in tolerant line YE8112, but up-regulating (overexpression) it in sensitive line MO17, as a way to enhance cell elongation and growth under stress. Ribose-phosphate pyrophosphokinase (PRPP synthetase) catalyzes the nucleotide biosynthesis process. PRPP is an essential substrate for purine and pyrimidine nucleotides, both in the de novo synthesis and in the salvage pathway [115]. In the current study, therefore, the down-regulation of the PRPP synthetase enzyme in both inbred lines under drought stress is consistent with the inhibition of nucleotide biosynthesis as a general feature of abiotic stresses. Moreover, our observation that an uncharacterized protein C4J0F8 was down-regulated, and at the same fold change in both lines, suggests that the protein has a common function in the two maize inbred lines' drought stress responses. This could serve as a targeted protein for further elucidation in our future studies.

Monosaccharide transporters (MSTs) are integral membrane proteins whose trans-membrane-spanning domains interact to form a central pore that shuttles soluble monosaccharides across hydrophobic membranes [116]. Expression of plant MST genes is also regulated by environmental stimuli such as pathogen infection (AtSTP4) [117] or wounding (AtSTP3 and AtSTP4) [118]. The MSTs catalyze monosaccharide import into classic sinks such as root tips and anthers, and, most importantly, help to meet the increased carbohydrate demand of cells responding to environmental stress [117]. Based on these discussions, we herein suggest MSTs to play an important adaptive role in the supply of carbohydrates to rapidly growing or metabolically hyperactive cells or tissues fighting drought stress, especially in sensitive line MO17, whilst down-regulation in tolerant line YE8112 may imply genotype diversity and the negative effects of drought stress on carbohydrates translocation in YE8112. The SAM synthetase gene is expressed in all living cells, and its product, Sadenosyl-L-methionine, is the major methyl donor in all cells [119]. Previously, the expression of SAM synthetase in soybean root was shown to be decreased upon exposure to drought stress [120]. Here, we state that, on one hand, the down-regulation of this enzyme in tolerant inbred YE8112 is consistent with the inhibition of photosynthetic activity as a general feature of abiotic stresses. On the other hand, this observation may imply SAM-dependent methyltransferase (SAM-D-Mtases) protein's variability in epigenetic mechanism (DNA methylation) regulation, as determined by genotypic differences, considering that the same protein was up-regulated in sensitive line MO17 in response to drought stress.

3.5. Significantly Enriched Metabolic Pathways of DAPs under Drought Stress

Metabolic adaptation of plants exposed to different stress requires sophisticated metabolic reorganization of multiple metabolic pathways [80], hence, we employed KEGG pathway enrichment analysis to identify key pathways related to drought stress response in maize seedlings. Photosynthesis antenna proteins pathway was the most significantly enriched, followed by proline metabolism and biosynthesis of amino acids pathways (Figure 6A). Photosynthesis of C_4 plants is highly sensitive to drought stress [121,122]. Chloroplasts, particularly the thylakoid membranes—PSII reaction centers, are one of the organelles most influenced by drought stress [54,123]. In the current study, the protein (B4FL55) encoding the *Lhcb5-2* gene and protein (Q41746) encoding *Lhcb5-1* gene were up-regulated in both inbred lines and significantly enriched in the photosynthesis (antenna protein) pathway

(Tables 2 and 3; Figure 6A). These proteins are a part of the light harvesting complexes (LHCs) and the electron transport components of the photosystem II (PSII) of the plant photosynthesis machinery [124]. They act as peripheral antenna systems enabling more efficient absorption of light energy [125]. Further, Lhch5-1 is involved in the intracellular non-photochemical quenching and the cysteine biosynthesis processes [124]. Previously, Zhao et al. [85] observed photosynthesis as the top signaling pathway affected by drought stress in maize, with chlorophyll a-b binding protein being up-regulated in an ABA-dependent manner. Remarkably, Dudhate et al. [126] also observed photosynthesis pathway to be highly enriched in pearl millet in response to drought stress. Taken together, these proteins play critical roles in light capture and utilization balancing to avoid photoinhibition (photodamage or photoinactivation) of the PSII due to excess light, as well as electron transport system, thus their involvement in photosynthesis pathway in tolerant line YE8112.

Comparatively, drought sensitive line MO17 showed two significantly enriched pathways, C5-branched dibasic acid metabolism (C5-BDAM) and RNA transport (Figure 6D). In *Physcomitrella patens* L., the C5-BDAM pathway has been observed critical in protoplast reprogramming to stem cells during the process of cell division [127]. In a stage-specific analysis, C5-BDAM pathway is specifically enriched from 24 h to 48 h during the process (a stage of stem cell re-entering cell cycle). Together with other pathways such as pentose phosphate pathway and leucine and isoleucine biosynthesis, C5-BDAM is closely associated with cell fate transition during protoplast reprogramming into stem cells [127]. Transport of RNAs from the nucleus to the cytoplasm, as ribonucleoprotein complexes (RNPs), is functionally coupled to gene expression processes such as splicing and translation [128]. Here, we suggest that translation and post translational processes are altered by drought stress as the cells modulate gene expressions related to stress tolerance, more prominently in sensitive line MO17. Similarly, Zhao et al. [85] observed RNA transport pathway to be significantly enriched in maize leaves in response to drought. For the pictorial view of the two most significantly enriched pathways described herein, please refer to Figure S6.

3.6. Function-Unknown Proteins Identified Under Drought Stress Conditions

We identified proteins with known critical roles in drought stress responses, together with unknown or predicted proteins that may have important functions in the regulatory network for drought stress. Of the 37 DAPs identified in tolerant-line YE8112, seven were of unknown functions, including four (B6SQW8, C0HJ06, B6UFE3, and C0P948) up-regulated and three (A0A1D6MJP2, B4F845, and C4J0F8) down-regulated. Interestingly, protein B6SQW8 was the most significantly expressed in tolerant line (TD_TC) (Table S2). Additionally, out of the 157 DAPs identified in sensitive-line (SD_SC), thirty were uncharacterized proteins, including 12 up-regulated and 18 down-regulated (Table S3). Moreover, one unknown protein (C4J0F8) was observed to overlap and exhibited a similar expression pattern (down-regulation) under drought stress, suggesting it has a common stress response function in the two inbred lines. These stress-responsive proteins with predicted functions may confer drought tolerance. Therefore, further studies of these proteins will help elucidate the molecular mechanisms underlying drought stress responses of maize lines differing in drought tolerance.

3.7. Protein-Protein Interaction (PPI) Analysis

Proteins in the cell are usually found as complexes, and biological processes within the cell are controlled by interactions between various proteins [119]. Therefore, identifying potential protein partners and studying protein–protein interactions becomes imperative for drought stress response research. Here, we used String 10.5 database analysis to determine how the identified differentially abundant proteins interact with others in networks to effect specific cellular functions. Some of the drought responsive proteins were predicted to interact with each other and hold central positions in certain PPI networks whereas some nodes showed no direct connections (Figure 7). The linkages created by these identified proteins in interaction networks can provide deeper insights into their relative importance in biological processes. 'Protein hubs' (connected to various other proteins) such as uncharacterized protein (Zm45026) in YE8112 and elongation factor 1-alpha (GRMZM2G343543_P03) in MO17; and 'bottlenecks' (key connectors of sub-networks), such as electron transporter/thiol-disulfide exchange intermediate (GRMZM5G869196_P01) in YE8112 and hypothetical protein LOC100383576 (AC234515.1_FGP003) in MO17 represent central points for communication co-ordination within the interaction network and tend to play critical roles in drought stress responses.

Analysis of PPI networks in tolerant inbred line YE8112 (Figure 7A) revealed that the interaction constituted by proteins involved in stress signaling, maintaining antioxidant defense, electron transport, and amino acid (protein) metabolism occupied a central position and may play a critical role in maize seedling drought stress responses. In addition, another protein interaction made up of proteins involved in energy metabolism, amino acid metabolism, maintaining redox homeostasis, and epigenetic regulation was prominent in YE8112. Moreover, a smaller connection had proteins involved in energy (NADH) metabolism and secondary metabolite (polyamines and proline) synthesis. These observations confirm the importance of these metabolic processes in drought stress response as revealed previously [71,80,85,103,105].

Most hub proteins in the larger complex and small networks in MO17 such as elfa3 (elongation factor 1-alpha), 50S Ribosomal protein L2 (rpl2-A), plastid specific 30S ribosomal protein 2 (GRMZM2G143870) and GRMZM2G343543_P03 were involved in protein biosynthesis and de-ubiquitination, suggesting these processes are critical drought responses [105,106], in sensitive line MO17. Furthermore, the several nodes that are not connected with other proteins within the interaction networks (for example Lhcb1 and aba1 in YE8112, and GRMZM5G826321_P01 in MO17) showed that those proteins did not interact with others based on the String database analysis [30]. However, these proteins may play indirect roles in maize seedling responses to drought stress.

3.8. Proposed Models of Drought Stress Tolerance in Maize Seedlings

Based on the annotated biological functions and the relevant published literature on the key drought responsive/related proteins or genes identified in the current study, we have developed models for drought stress tolerance in maize as shown below (Figure 9).

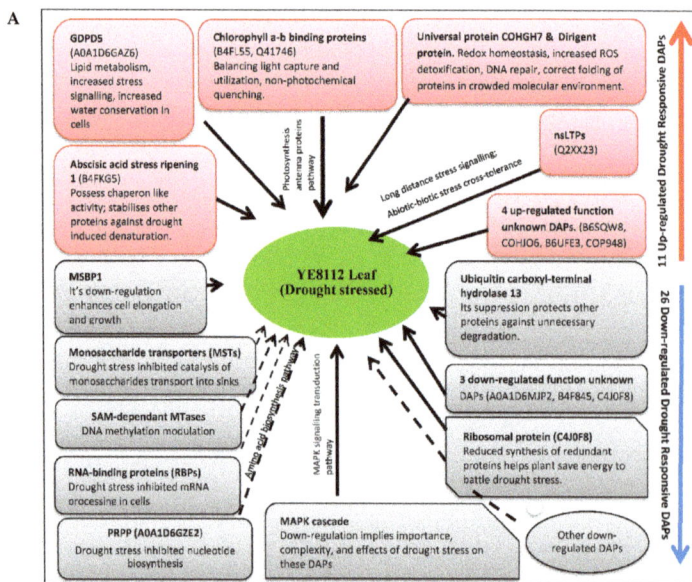

Figure 9. *Cont.*

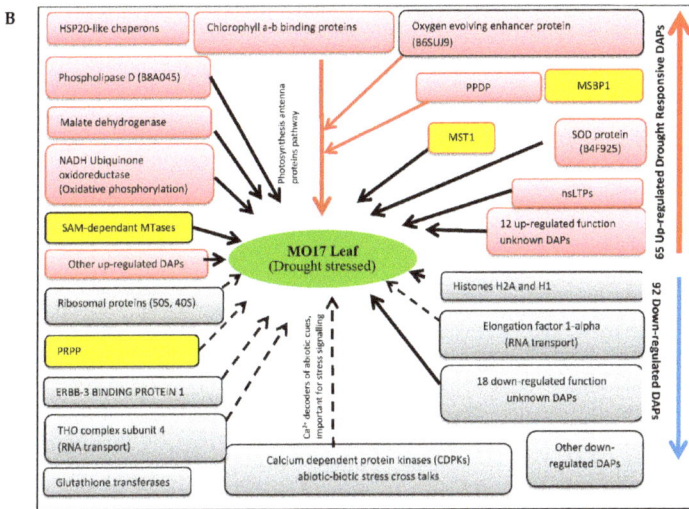

Figure 9. Molecular models of drought tolerance in maize seedling leaves of: (**A**) tolerant inbred line YE8112 and (**B**) sensitive line MO17. Red nodes (rectangles/circles) signify up-regulated DAPs; gray nodes signify down-regulated DAPs; yellow nodes in MO17 model (Figure 9B) represents overlapping DAPs also observed in YE8112. Dotted black connectors/arrows imply drought stress imposed negative effects on respective proteins or pathways; compound type black connectors imply desirable drought stress response outcomes on respective proteins. Note: nsLTPs, non-specific lipid transfer proteins; GDPD5, Glycerophosphodiester phosphodiesterase 5; MSBP1, membrane steroid binding protein 1; MAPK, mitogen-activated protein kinases; PRPP, Ribose-phosphate pyrophosphokinase; MST1, monosaccharide transporter 1; PPDP, pyruvate phosphate dikinase proteins.

4. Materials and Methods

4.1. Plant Materials and Drought Stress Treatment

Two maize (*Zea mays* L.) inbred lines (ILs) with contrasting drought sensitivity (tolerant YE8112 and sensitive MO17) were used in this experiment. Seeds of the two inbred lines were provided by the North China Key Laboratory for Crop Germplasm Resources of Education Ministry, Hebei Agricultural University, China. In selecting the two ILs, we employed our lab screening on seedling survival rates of dozens of maize inbred lines under drought stress treatment; this finding was supported with previous experiments [38,39]. Seeds were surface sterilized in 10% hydrogen peroxide for 5 min, followed by washing three-times with sterile water. Then, the seeds were germinated by laying them between two layers of damp filter paper at 28 °C for 24 h according to the procedures of Lei et al. [10]. Germinated seeds were placed in the same size PVC pots with uniform soil and grown under greenhouse controlled conditions (light/dark cycles: 14/10; 28/22 °C; 60 ± 5% relative humidity) at Hebei Agricultural University, Baoding, China. Maize seedlings were grown under normal conditions until the three leaves were fully expanded. Then, both the tolerant and sensitive inbred lines were exposed to drought conditions for a 7-day period. For both tolerant and sensitive lines, a half of the plants were subjected to drought by withholding irrigation to 50% soil moisture content (which was detected using a TZS-1 soil moisture meter, Zhejiang Top Cloud-Agri Technology Co., Ltd., Hanzhou; China) and the rest of the plants were grown under well-watered condition (control). Flag leaves from the control and drought stress treated plants collected after 1, 3, 5, and 7 days of treatment (for physiological analyses), and collected once at 7 days post treatment exposure (for proteomic analysis) were immediately frozen in liquid nitrogen and stored at −80 °C prior to respective analyses. Each treatment was replicated three times.

4.2. Phenotypic and Physiological Characterizations

Phenotypic and physiological characterizations were measured for the YE8112 and MO17 seedlings under well-watered and drought-stress conditions. Relative water content (RWC) was estimated according to Galmés et al. [129]. Trypan blue staining of the leaves of both inbred lines under water-deficit conditions was also conducted [130,131]. The leaf peroxidase (POD) activity was estimated by the guaiacol method [132]. The level of lipid peroxidation (MDA content) in the leaves was measured by thiobarbituric acid (TBA) method [133]. The osmolytes proline content was determined using ninhydrin as per the protocol of Bates et al. [134].

4.3. Protein Extraction

Total proteins were extracted from the non-stressed and stressed leaf tissues of two maize inbred lines with three biological replicates (each containing 500 mg maize leaves) using the cold acetone method as described in previous reports [30,135]. In brief, samples were ground to a powder in liquid nitrogen and lysed with 2 mL lysis buffer containing 8 M urea, 2% SDS, and 1× Protease Inhibitor Cocktail (Thermo Fisher Scientific, Shanghai, China). Then, the solution was kept on ice for 30 min prior to centrifugation at 11,500 rpm (18,000× g) for 15 min at 4 °C. The supernatant was then transferred into a new tube and precipitated with 10% TCA/90% acetone, followed by incubation at −20 °C overnight. Pellets were washed thrice with acetone. Finally, the precipitate was dissolved in 8 M urea under ultrasound irradiation. Total protein concentrations of the extracts were determined using a Coomassie Bradford Protein Assay Kit (23200, Thermo Fisher Scientific, Shanghai, China), with bovine serum albumin (BSA) as standard, according to the manufacturer's instructions. The absorbance was determined at 562 nm using an xMark microplate absorbance spectrophotometer (Bio-Rad Laboratories Inc., Hercules, CA, USA), and protein extracts quality was examined with SDS-PAGE (tricine-sodium dodecyl sulfate polyacrylamide gel electrophoresis) [136].

4.4. Protein Digestion and iTRAQ (Isobaric Tags for Relative and Absolute Quantification) Labeling

For each sample, the solution was transferred to a new tube and adjusted to 100 µL using 8 M urea, mixed with 11 µL 1 M DTT, and incubated at 37 °C for 1 h followed by centrifugation at 4 °C at 14,000× g for 10 min. The supernatant was incubated in a dark room for 20 min after the addition of 120 µL 55 mM iodoacetamide. This followed washing of the supernatant using 100 µL mM TEAB (triethylammonium bicarbonate) and centrifugation at 14,000× g for 10 min at 4 °C, followed by discarding the eluate. This washing step was repeated thrice before trypsin digestion. Total proteins were digested using trypsin (Promega, Madison, WI, USA) at a ratio of protein:trypsin = 30:1 at 37 °C overnight (16 h). The peptides were dried in a centrifugal vacuum concentrator and reconstituted in 0.5 M TEAB. Detailed protein digestion procedures are contained in a previous report [80].

Protein iTRAQ labeling was conducted by Applied Protein Technology Co., Limited (Shanghai, China) using an iTRAQ Reagents 8-plex kit (AB Sciex, Foster City, CA, USA) according to the manufacturer's protocol. In brief, one unit of iTRAQ reagent (defined as the amount of reagent required to label 100 µg of protein) was thawed and reconstituted in 70 µL isopropanol. The control replicates were labeled with iTRAQ tag 115 for the drought-sensitive inbred line (MO17) and tag 117 for drought-tolerant inbred line (YE8112). The drought treated replicates were labeled with tags 114 and 116 for drought-sensitive and drought–tolerant lines, respectively. Three technical replicates were performed.

4.5. Strong Cation Exchange (SCX) and LC-MS/MS Analysis

Sample fractionation was conducted before LC-MS/MS (liquid chromatography-tandem mass spectrometry) analysis as described in previous report by Ross et al. [36] with some modifications. Briefly, the iTRAQ labeled peptide mixtures were separated by strong cation exchange (SCX) chromatography on an Agilent 1100 HPLC system (Agilent Technologies, Waldbronn, Germany) using a PolySulfoethyl A column (4.6 × 100 mm^2, 5 μm, 300 Å; PolyLC, Columbia, MD, USA) as per the manufacturer's guidelines. The sample was dissolved in 4 mL of SCX loading buffer (25% *v/v* ACN, 10 mM KH$_2$PO$_4$, pH 3, with phosphoric acid), loaded and washed isocratically for 20 min at 0.5 mL/min to remove excess reagent. The retained peptides were eluted with a linear gradient of 0–500 mM KCl (25% *v/v* ACN, 10 mM KH$_2$PO$_4$, pH 3) over 15 min at a flow rate of 1 mL/min, with fractions collected at 1 min intervals. The elution was monitored by measuring absorbance at 214 nm, and the eluted peptides were pooled into 10 fractions.

Each SCX fraction was subjected to reverse phase nanoflow HPLC separation and quadruple time-of-flight (QSTAR XL) mass spectrometry analysis. Protocols for the analysis of reverse phase nanoflow HPLC and tandem mass spectrometry have been explicitly described in previous reports [80,137]. In short, peptides were subjected to nano electrospray ionization followed by tandem mass spectrometry (MS/MS). The mass spectrometry was analyzed by Q-Exactive mass spectrometer (Thermo Fisher Scientific, Shanghai, China) after the sample had been analyzed by chromatography. The MS spectra with a mass range of 300–1800 *m/z* were acquired at a resolving power of 120 K, the primary mass spectrometry resolution of 70,000 at 200 *m/z*, AGC (automatic gain control) target of 1e6, maximum IT of 50 ms, and dynamic exclusion time (active exclusion) of 60.0 s The mass charge ratio of polypeptides and polypeptide fragments were set according to the following parameters: 20 fragments (MS2 scan) were collected after each scan (full scan), MS2 activation type was HCD, isolation window 2 *m/z*, two-grade mass spectrometry resolution of 17,500 at 200 *m/z*, the normalized collision energy of 30 eV, underfill of 0.1%. The electrospray voltage applied was 1.5 kV. Maximum ion injection times for the MS and MS/MS were 50 and 100 ms, respectively.

4.6. Protein Identification and Quantification

All of the mass spectrometry data from the LC-MS/MS raw files were obtained using Mascot software version 2.2 (Matrix Science, London, UK) and converted into MGF files using Proteome Discovery 1.4 (Thermo Fisher Scientific Inc., Waltham, MA, USA). For protein identification, MGF data files from the LC-MS/MS were searched against the Uniprot database (available online: https: //www.uniprot.org; accessed on 12 January 2018; uniprot_*Zea mays*_132339_20180112.FASTA; 76,417 sequences) using Mascot search engine. The search parameters were set as follows: trypsin as the cleavage enzyme; two maximum missed cleavages allowed; fragment mass tolerance was set at ±0.1 Da; and peptide mass tolerance was set at ±20 ppm; monoisotopic as the mass values; iTRAQ 8 plex (Y) and Qxidation (M) as variable modifications; and Carbamidomethyl (C), iTRAQ 8 plex (N-term) and iTRAQ 8 plex (K) selected as fixed modifications. Only peptides with a false discovery rate (FDR) estimation ≤1% and a 95% confidence interval were counted as being successfully identified.

As described in a previous study [137], protein relative quantification was dependent on the reporter ions ratios, from which relative peptides abundance can be estimated. Only proteins that were present in all the samples were considered for quantification; shared peptides were omitted. Reporter ion ratios determination used the peak intensities of the reporter ions, with control-treated YE8112 sample serving as reference. Further normalization of the final protein quantification ratios was conducted using the median average of those ratios. The unique peptide ratios' median represented the protein ratio. Student's *t*-test was used to analyze the differentially abundant proteins (DAPs), with proteins exhibiting fold-changes >1.2 or <0.83 (*p* < 0.05) considered to be statistically significant DAPs [138].

4.7. DAPs Functional Classification, Pathway Enrichment, and Hierarchal Clustering Analysis

The successfully identified DAPs were used as queries to search the Interpro (https://www.ebi.ac. uk/interpro/) and Pfam (http://pfam.xfam.org/); Gene Ontology (GO) (http://www.geneontology. org/) and the KEGG (http://www.genome.jp/kegg/) databases. The corresponding gene sequences of the DAPs were obtained by searching the maize sequence database Gramene (http://ensemble. gramene.org/Zea_mays/). GO analysis [139] was used for functional annotation and classification of the DAPs identified to describe the biological processes, cellular component, and molecular functions involved in the response to drought stress. Additionally, GO (protein) terms were assigned to each DAP based on BLASTX similarity (E-value < 1.0×10^5) and known GO annotations, using the Blast2GO tool (available online: https://www.blast2go.com; accessed on 6 February 2018) [140]. The DAPs were assigned to various biological pathways using the KEGG pathway analysis. Further, significant KEGG pathway enrichment analysis was performed using the hypergeometric test, with Q (Bonferroni-corrected *p*-value) less than 0.05 defined as statistically significant. A protein interaction network was constructed using the String program (version 10.5) (http://www.string-db.org/).

4.8. RNA Extraction, cDNA Synthesis, and RT-qPCR Analysis

Total RNA was isolated from non-stressed and stressed seedling leaves of the two inbred lines (YE8112 and MO17) and prepared for qRT-PCR analysis using the Omini Plant RNA Kit (DNase I) (CWBIO, Beijing, China) based on the manufacturer's instructions. For cDNA synthesis, 1 µg of total RNA was reverse-transcribed in a total volume of 20 µL, using HiFiscript cDNA Synthesis Kit (CWBIO, Beijing, China) according to the manufacturer's instructions. Thirty DAPs were selected and gene-specific primers (Table S7) designed for qRT-PCR using Primer Premier 5 Designer software. qRT-PCR was conducted with a C1000 (CFX96 Real-Time System) Thermal Cycler (Bio-Rad) using 2× Fast Super EvaGreen® qPCR Mastermix (US Everbright Inc., Suzhou, China). Each qRT-PCR reaction mixture comprised 1 µL of template cDNA, 1 µl of forward primer (50 pmol), 1 µL of reverse primer (50 pmol), and 10 µL of 2× Fast Super EvaGreen® qPCR Mastermix (US Everbright Inc., Suzhou, China) in a total reaction volume of 20 µL. The amplification program was set as follows: 95 °C for 2 min followed by 40 cycles of 95 °C for 10 s and 55 °C for 30 s [80]. A steady and constitutively expressed maize gene *GAPDH* (accession no. X07156) was used as the internal reference gene, together with the forward primer (GAPDH-F: 5′-ACTGTGGATGTCTCGGTTGTTG-3′) and reverse primer (GAPDH-R: 5′-CCTCGGAAGCAGCCTTAATAGC-3′). Each sample had three technical replicates [2]. The relative mRNA abundance was calculated according to the $2^{-\Delta\Delta CT}$ method [141].

4.9. Statistical Data Analysis of Physiological Changes

Physiological data analysis was performed using the SPSS statistical software package (version 19.0; SPSS Institute Ltd., Armonk, NY, USA), and the significance of differences were tested by Fisher's protected least significant differences (PLSD) test with a *p*-value ≤ 0.05 set as statistically significant.

5. Conclusions

In the present study, we conducted a comprehensive comparative analysis of two maize inbred lines contrasting in drought stress tolerance based on their physiological and proteomic responses. Our results have shown that divergent stress tolerance mechanisms exist between the two lines at the seedling stage. Both qualitative and quantitative differences, at physiological and proteomic levels, showed that YE8112 is comparatively more tolerant to drought stress than MO17 owing to its maintenance of higher RWC and proline contents, higher increase in POD enzyme activity, along with decreased level of lipid peroxidation under stressed conditions. Using an iTRAQ-based method, we obtained a total of 721 differentially abundant proteins (DAPs). Amongst these, we identified five essential sets of drought responsive DAPs, including 13 DAPs specific to YE8112, 107 DAPs specific to TD_SD comparison, three DAPs of YE8112 also regulated in TD_SD, 84 DAPs unique to MO17, and five overlapping DAPs between the two inbred lines. The most significantly enriched proteins in YE8112 were associated with the photosynthesis antenna proteins pathway, whilst those in MO17 were related to C5-branched dibasic acid metabolism and RNA transport pathways. The changes in protein abundance were consistent with the observed physiological characterizations of the two inbred lines. Further, our qRT-PCR analysis results confirmed the iTRAQ sequencing based findings. We have clarified the two maize inbred lines' strategies to tolerate drought stress and elucidated the fundamental molecular networks associated. Based on our findings, and relevant literature cited herein this report, we have proposed molecular models of drought tolerance in the two inbred line seedling leaves as provided in Figure 9. The higher drought stress tolerance of YE8112 may be attributed to: (a) Activation of photosynthesis (PSII) proteins involved in balancing light capture and utilization, and improving non-photochemical quenching; (b) Enhancement of lipid-metabolism related proteins, contributing to increased stress signaling and water conservation in the cell. Furthermore, plants may have developed abiotic-biotic stress cross-tolerance mechanisms; (c) Stimulation of chaperons such as ASR1 protein in order to stabilize a number of other proteins against drought-induced denaturation; (d) Increased cell ROS detoxification capacity; (e) Reduced synthesis of redundant proteins to help the plant save energy to battle drought stress; and (f) Suppression of protein ubiquitnation in order to protect proteins against unnecessary degradation under drought stress, and thus, reversing the fate of those proteins. These results provide more insights into the physiological and molecular mechanisms underpinning drought stress tolerance in maize seedlings.

Supplementary Materials: Supplementary materials can be found at http://www.mdpi.com/1422-0067/19/10/3225/s1.

Author Contributions: Conceptualization, T.Z. and H.D.; Data curation, T.Z. and H.D.; Formal analysis, T.Z., S.L., X.W., H.J., G.L., and H.D.; Funding acquisition, H.D.; Investigation, T.Z., S.L., X.W., H.J., and G.L.; Methodology, T.Z. and X.W.; Project administration, H.D.; Resources, H.D.; Software, H.J.; Supervision, H.D.; Validation, T.Z., S.L., X.W., H.J., and H.D.; Visualization, G.L.; Writing—original draft, T.Z. and S.L.; Writing—review & editing, T.Z., S.L., X.W., H.J., G.L., and H.D.

Funding: This research was supported by the National Key Research and Development Project of China (Selection and Efficient Combination Model of Wheat and Maize Water Saving, High Yield and High Quality Varieties) (Grant No. 2017YFD0300901).

Conflicts of Interest: The authors declare that they have no conflict of interest. Additionally, the founding sponsors had no role in the design of the study; in the collection, analyses, or interpretation of data; in the writing of the manuscript, and in the decision to publish the results.

Abbreviations

ASR1	Abscisic acid stress ripening 1
CDPKs	Calcium dependent protein kinases
DAPs	Differentially abundant proteins
GDPD	Glycerophosphodiester phosphodiesterase
GO	Gene ontology
GST	Glutathione-S-transferase
HSPs	Heat shock proteins
iTRAQ	Isobaric tags for relative and absolute quantification
KEGG	Kyoto Encyclopedia of Genes and Genomes
LC-MS/MS	Liquid chromatography-tandem mass spectrometry
MAPK	Mitogen-activated protein kinases
MDA	Malondialdehyde
MSBP1	Membrane steroid binding protein 1
MSTs	Monosaccharide transporters
nsLTPs	Non-specific lipid transfer proteins
PPI	Protein-protein interaction
POD	Peroxidases
PRPP	Ribose-phosphate pyrophosphokinase
qRT-PCR	Quantitative real-time polymerase chain reaction
RBPs	RNA binding proteins
ROS	Reactive oxygen species
SAM	S-adenosyl-methionine
SAM-D-Mtases	SAM-dependent methyltransferases
SOD	Superoxide dismutase
TRX	Thioredoxin

References

1. Campos, H.; Cooper, M.; Habben, J.E.; Edmeades, G.O.; Schussler, J.R. Improving drought tolerance in maize: A view from industry. *Field Crops Res.* **2004**, *90*, 19–34. [CrossRef]
2. Li, G.K.; Gao, J.; Peng, H.; Shen, Y.O.; Ding, H.P.; Zhang, Z.M.; Pan, G.T.; Lin, H.J. Proteomic changes in maize as a response to heavy metal (lead) stress revealed by iTRAQ quantitative proteomics. *Genet. Mol. Res.* **2016**, *15*, 1–14. [CrossRef] [PubMed]
3. Min, H.; Chen, C.; Wei, S.; Shang, X.; Sun, M.; Xia, R.; Liu, X.; Hao, D.; Chen, H.; Xie, Q. Identification of drought tolerant mechanisms in maize seedlings based on transcriptome analysis of recombination inbred lines. *Front Plant Sci.* **2016**, *7*, 1080. [CrossRef] [PubMed]
4. Zheng, J.; Fu, J.; Gou, M.; Huai, J.; Liu, Y.; Jian, M.; Huang, Q.; Guo, X.; Dong, Z.; Wang, H.; et al. Genome-wide transcriptome analysis of two maize inbred lines under drought stress. *Plant Mol. Biol.* **2010**, *72*, 407–421. [CrossRef] [PubMed]
5. Cui, D.; Wu, D.; Liu, J.; Li, D.; Xu, C.; Li, S.; Li, P.; Zhang, H.; Liu, X.; Jiang, C.; et al. Proteomic analysis of seedling roots of two maize inbred lines that differ significantly in the salt stress response. *PLoS ONE* **2015**, *10*, e0116697. [CrossRef] [PubMed]
6. Wang, X.; Shan, X.; Wu, Y.; Su, S.; Li, S.; Liu, H.; Han, J.; Xue, C.; Yuan, Y. iTRAQ-based quantitative proteomic analysis reveals new metabolic pathways responding to chilling stress in maize seedlings. *J. Proteom.* **2016**, *146*, 14–24. [CrossRef] [PubMed]
7. Chen, J.; Xu, W.; Velten, J.; Xin, Z.; Stout, J. Characterization of maize inbred lines for drought and heat tolerance. *J. Soil Water Conserv.* **2012**, *67*, 354–364. [CrossRef]
8. Kaul, J.; Kumar, R.; Kumar, R.S.; Dass, S.; Bhat, B.; Kamboj, O.P.; Nara, S.; Yadav, A.K. Response of maize (*Zea mays* L.) hybrids to excess soil moisture stress at different growth stages. *Res. Crops* **2013**, *4*, 439–443.
9. Farooq, M.; Wahid, A.; Kobayashi, N.; Fujita, D.; Basra, S.M.A. Plant drought stress: Effects, mechanisms and management. *Agron. Sustain. Dev.* **2009**, *29*, 185–212. [CrossRef]
10. Lei, L.; Shi, J.; Chen, J.; Zhang, M.; Sun, S.; Xie, S.; Li, X.; Zeng, B.; Peng, L.; Hauck, A.; et al. Ribosome profiling reveals dynamic translational landscape in maize seedlings under drought stress. *Plant J.* **2015**, *84*, 1206–1218. [CrossRef] [PubMed]
11. Raos, G.J.N.; Reddy, J.N.; Variar, M.; Mahender, A. Molecular breeding to improve plant breeding to improve plant resistance to abiotic stresses. *Adv. Plant Breed. Strateg.* **2016**, *2*, 283–326. [CrossRef]
12. Ahmadi, A.; Emam, Y.; Pessarakli, M. Biochemical changes in maize seedlings exposed to drought stress conditions at different nitrogen levels. *J. Plant Nutr.* **2010**, *33*, 541–556. [CrossRef]
13. Yin, X.G.; Jørgen, E.O.; Wang, M.; Kersebaum, K.-C.; Chen, H.; Baby, S.; Öztürk, I.; Chen, F. Adapting maize production to drought in the northeast farming region of China. *Eur. J. Agron.* **2016**, *77*, 47–58. [CrossRef]
14. Maiti, R.K.; Maiti, L.E.; Sonia, M.; Maiti, A.M.; Maiti, M.; Maiti, H. Genotypic variability in maize (*Zea mays* L.) for resistance to drought and salinity at the seedling stage. *J. Plant Physiol.* **1996**, *148*, 741–744. [CrossRef]
15. Cao, L.Z.X.; Wj, B.X.P. Discuss on evaluating method to Drought-resistance of Maize in seedling stage. *J. Maize Sci.* **2004**, *12*, 73–75.
16. Miao, Z.; Han, Z.; Zhang, T.; Chen, S.; Ma, C. A systems approach to spatio-temporal understanding of the drought stress response in maize. *Sci. Rep.* **2017**, *7*, 1–14. [CrossRef] [PubMed]
17. Liu, X.; Li, X.; Li, W.; Li, M.; Liu, X. Analysis on difference for drought responses of maize inbred lines at seedling stage. *J. Maize Sci.* **2004**, *12*, 63–65.
18. Wu, B.; Li, X.; Xiao, M. Genetic variation in fifty-three maize inbred lines in relation to drought tolerance at seedling stage. *Sci. Agric. Sin.* **2007**, *40*, 665–676.
19. Oliver, S.N.; Dennis, E.S.; Dolferus, R. ABA regulates apoplastic sugar transport and is a potential signal for cold-induced pollen sterility in rice. *Plant Cell Physiol.* **2007**, *48*, 1319–1330. [CrossRef] [PubMed]
20. Jogaiah, S.; Govind, S.R.; Tran, L.S. Systems biology-based approaches toward understanding drought tolerance in food crops. *Crit. Rev. Biotechnol.* **2013**, *33*, 23–39. [CrossRef] [PubMed]
21. Luan, M.; Xu, M.; Lu, Y.; Zhang, L.; Fan, Y.; Wang, L. Expression of zma-miR169 miRNAs and their target *ZmNF-YA* genes in response to abiotic stress in maize leaves. *Gene* **2015**, *555*, 178–185. [CrossRef] [PubMed]
22. Wang, G.; Zhu, Q.G.; Meng, Q.W.; Wu, C.A. Transcript profiling during salt stress of young cotton (*Gossypium hirsutum*) seedlings via Solexa sequencing. *Acta Physiol. Plant.* **2012**, *34*, 107–115. [CrossRef]

23. Chen, Y.; Liu, Z.H.; Feng, L.; Zheng, Y.; Li, D.D.; Li, X.B. Genome-wide functional analysis of cotton (*Gossypium hirsutum*) in response to drought. *PLoS ONE* **2013**, *8*, e80879. [CrossRef] [PubMed]

24. Chan, Z.L. Expression profiling of ABA pathway transcripts indicates crosstalk between abiotic and biotic stress responses in *Arabidopsis*. *Genomics* **2012**, *100*, 110–115. [CrossRef] [PubMed]

25. Canovas, F.M.; Dumas-Gaudot, E.; Recorbet, G.; Jorrin, J.; Mock, H.P.; Rossignol, M. Plant proteome analysis. *Proteomics* **2004**, *4*, 285–298. [CrossRef] [PubMed]

26. Zhao, Q.; Zhang, H.; Wang, T.; Chen, S.; Dai, S. Proteomics-based investigation of salt-responsive mechanisms in plant roots. *J. Proteom.* **2013**, *82*, 230–253. [CrossRef] [PubMed]

27. Komatsu, S.; Hiraga, S.; Yanagawa, Y. Proteomics techniques for the development of flood tolerant crops. *J. Proteome Res.* **2012**, *11*, 68–78. [CrossRef] [PubMed]

28. Wu, S.; Ning, F.; Zhang, Q.; Wu, X.; Wang, W. Enhancing omics research of crop responses to drought under field conditions. *Front. Plant Sci.* **2017**, *8*, 1–5. [CrossRef] [PubMed]

29. Kosova, K.; Vitamvas, P.; Prasil, I.T.; Renaut, J. Plant proteome changes under abiotic stress-contribution of proteomics studies to understanding plant stress response. *J. Proteom.* **2011**, *74*, 1301–1322. [CrossRef] [PubMed]

30. Yang, L.; Jiang, T.; Fountain, J.C.; Scully, B.T.; Lee, R.D.; Kemerait, R.C.; Chen, S.; Guo, B. Protein profiles reveal diverse responsive signaling pathways in kernels of two maize inbred lines with contrasting drought sensitivity. *Int. J. Mol. Sci.* **2014**, *15*, 18892–18918. [CrossRef] [PubMed]

31. Hu, X.; Wu, L.; Zhao, F.; Zhang, D.; Li, N.; Zhu, G.; Li, C.; Wang, W. Phosphoproteomic analysis of the response of maize leaves to drought, heat and their combination stress. *Front. Plant Sci.* **2015**, *6*, 298. [CrossRef] [PubMed]

32. Benešová, M.; Holá, D.; Fischer, L.; Jedelský, P.L.; Hnilička, F.; Wilhelmová, N.; Rothová, O.; Kočová, M.; Procházková, D.; Honnerová, J.; et al. The physiology and proteomics of drought tolerance in maize: Early stomatal closure as a cause of lower tolerance to short-term dehydration? *PLoS ONE* **2012**, *7*, e38017. [CrossRef] [PubMed]

33. Gong, F.; Hu, X.; Wang, W. Proteomic analysis of crop plants under abiotic stress conditions: Where to focus our research? *Front. Plant Sci.* **2015**, *6*, 418. [CrossRef] [PubMed]

34. Dong, M.; Gu, J.; Zhang, L.; Chen, P.; Liu, T.; Deng, J.; Lu, H.; Han, L.; Zhao, B. Comparative proteomics analysis of superior and inferior spikelets in hybrid rice during grain filling and response of inferior spikelets to drought stress using isobaric tags for relative and absolute quantification. *J. Proteom.* **2014**, *109*, 382–399. [CrossRef] [PubMed]

35. Zhao, Y.; Du, H.; Wang, Z.; Huang, B. Identification of proteins associated with water-deficit tolerance in C4 perennial grass species, *Cynodon dactylonxCynodon transvaalensis* and *Cynodon dactylon*. *Physiol. Plant.* **2011**, *141*, 40–55. [CrossRef] [PubMed]

36. Ross, P.L.; Huang, Y.N.; Marchese, J.N.; Williamson, B.; Parker, K.; Hattan, S.; Khainovski, N.; Pillai, S.; Dey, S.; Daniels, S.; et al. Multiplexed protein quantitation in *Saccharomyces cerevisiae* using amine-reactive isobaric tagging reagents. *Mol. Cell. Proteom.* **2004**, *3*, 1154–1169. [CrossRef] [PubMed]

37. Wu, X.; Wang, W. Increasing confidence of proteomics data regarding the identification of stress-responsive proteins in crop plants. *Front. Plant Sci.* **2016**, *7*, 702. [CrossRef] [PubMed]

38. Qi, S.; Li, X.; Zhang, D.; Li, X.; Hao, Z.; Weng, J.; Li, M.; Zhang, S. Trends in drought tolerance in Chinese maize cultivars from the 1950s to the 2000s. *Field Crops Res.* **2017**, *201*, 175–183. [CrossRef]

39. Zhang, Z.; Tang, W.; Tao, Y.; Zheng, Y. cDNA microarray analysis of early response to submerging stress in Zea mays roots. *Russ. J. Plant Physiol.* **2005**, *52*, 43–49. [CrossRef]

40. Edmeades, G.O. *Progress in Achieving and Delivering Drought Tolerance in Maize—An Update*; ISAA: Ithaca, NY, USA, 2013; pp. 1–39.

41. Zhu, J.K. Abiotic stress signaling and responses in plants. *Cell* **2016**, *167*, 313–324. [CrossRef] [PubMed]

42. Wang, X.; Komatsu, S. Plant subcellular proteomics: Application for exploring optimal cell function in soybean. *J. Proteom.* **2016**, *143*, 45–56. [CrossRef] [PubMed]

43. Gill, S.S.; Tuteja, N. Reactive oxygen species and antioxidant machinery in abiotic stress tolerance in crop plants. *Plant Physiol. Biochem.* **2010**, *48*, 909–930. [CrossRef] [PubMed]

44. Islam, M.; Begum, M.C.; Kabir, A.H.; Alam, M.F. Molecular and biochemical mechanisms associated with differential responses to drought tolerance in wheat (*Triticum aestivum* L.). *J. Plant Interact.* **2015**, *10*, 195–201. [CrossRef]

45. Kamies, R.; Farrant, J.M.; Tadele, Z.; Cannarozzi, G.; Rafudeen, M.S. A Proteomic Approach to Investigate the Drought Response in the Orphan Crop Eragrostis tef. *Proteome* **2017**, *5*, 32. [CrossRef] [PubMed]

46. Moussa, H.R.; Abdel-Aziz, S.M. Comparative response of drought tolerant and drought sensitive maize genotypes to water stress. *Aust. J. Crop Sci.* **2008**, *1*, 31–36.

47. Miller, G.; Suzuki, N.; Ciftciyilmaz, S.; Mittler, R. Reactive oxygen species homeostasis and signalling during drought and salinity stresses. *Plant Cell Environ.* **2010**, *33*, 453–467. [CrossRef] [PubMed]

48. Sharma, P.; Jha, A.B.; Dubey, R.S.; Pessarakli, M. Reactive Oxygen Species, Oxidative Damage, and Antioxidative Defense Mechanism in Plants under Stressful Conditions. *J. Botany* **2012**, *10*, 26. [CrossRef]

49. Mattioni, C.; Lacerenza, N.G.; Troccoli, A.; De Leonardis, A.M.; Di Fonzo, N. Water and salt stress-induced alterations in proline metabolism of *Triticum durum* seedlings. *Physiol. Plant.* **1997**, *101*, 787–792. [CrossRef]

50. Kumar, S.G.; Matta Reddy, A.; Sudhakar, C. NaCl effects on proline metabolism in two high yielding genotypes of mulberry (*Morus alba* L.) with contrasting salt tolerance. *Plant Sci.* **2003**, *165*, 1245–1251. [CrossRef]

51. Sairam, R.K.; Roa, K.V.; Srivastava, G.C. Differential response of wheat cultivar genotypes to long term salinity stress in relation oxidative stress, antioxidant activity, and osmolyte concentration. *Plant Sci.* **2000**, *163*, 1037–1048. [CrossRef]

52. Foyer, C.H.; Harbinson, J. Oxygen metabolism and the regulation of photosynthetic electron transport. In *Causes of Photooxidative Stresses and Amelioration of Defense Systems in Plants*; Foyer, C.H., Mullineaux, P., Eds.; CRC Press: Boca Raton, FL, USA, 1994; pp. 1–42.

53. Luca, D.; Stefano, C.; Mauro, B.; David, P.; Karekl, Ž.; Niyogi, K.K.; Fleming, G.R.; Zigmantas, D.; Bassi, R. Two mechanisms for dissipation of excess light in monomeric and trimeric light-harvesting complexes. *Nat. Plants* **2010**, *3*, 17033. [CrossRef]

54. Tai, F.J.; Yuan, Z.L.; Wu, X.L.; Zhao, P.F.; Hu, X.L.; Wang, W. Identification of membrane proteins in maize leaves, altered in expression under drought stress through polyethylene glycol treatment. *Plant Omics J.* **2011**, *4*, 250–256.

55. Edstam, M.M.; Viitanen, L.; Salminen, T.A.; Edqvist, J. Evolutionary history of the non-specific lipid transfer proteins. *Mol. Plant* **2011**, *4*, 947–964. [CrossRef] [PubMed]

56. Lee, S.B.; Go, Y.S.; Bae, H.J.; Park, J.H.; Cho, S.H.; Cho, H.J.; Lee, D.S.; Park, O.K.; Hwang, I.; Suh, M.C. Disruption of glycosylphosphatidylinositol-anchored lipid transfer protein gene altered cuticular lipid composition, increased plastoglobules and enhanced susceptibility to infection by the fungal pathogen. *Alternaria brassicicola. Plant Physiol.* **2009**, *150*, 42–54. [CrossRef] [PubMed]

57. Molina, A.; Segura, A.; García-Olmedo, F. Lipid transfer proteins (nsLTPs) from barley and maize leaves are potent inhibitors of bacterial and fungal plant pathogens. *FEBS Lett.* **1993**, *316*, 119–122. [CrossRef]

58. Tuteja, N. Mechanisms of high salinity tolerance in plants. *Methods Enzymol.* **2007**, *428*, 419–438. [CrossRef] [PubMed]

59. Nakamura, Y.; Koizumi, R.; Shui, G.; Shimojima, M.; Wenk, M.R.; Ito, T.; Ohta, H. Arabidopsis lipins mediate eukaryotic pathway of lipid metabolism and cope critically with phosphate starvation. *Proc. Natl. Acad. Sci. USA* **2009**, *106*, 20978–20983. [CrossRef] [PubMed]

60. Cheng, Y.; Zhou, W.; El Sheery, N.I.; Peters, C.; Li, M.; Wang, X.; Huang, J. Characterization of the Arabidopsis glycerophosphodiester phosphodiesterase (GDPD) family reveals a role of the plastid-localized AtGDPD1 in maintaining cellular phosphate homeostasis under phosphate starvation. *Plant J.* **2011**, *66*, 781–795. [CrossRef] [PubMed]

61. Rom, S.; Gilad, A.; Kalifa, Y.; Konrad, Z.; Karpasas, M.M.; Goldgur, Y.; Bar-Zvi, D. Mapping the DNA- and zinc-binding domains of ASR1 (abscisic acid stress ripening), an abiotic-stress regulated plant specific protein. *Biochimie* **2006**, *88*, 621–628. [CrossRef] [PubMed]

62. Konrad, Z.; Bar-Zvi, D. Synergism between the chaperone-like activity of the stress regulated ASR1 protein and the osmolyte glycine-betaine. *Planta* **2008**, *227*, 1213–1219. [CrossRef] [PubMed]

63. Kalifa, Y.; Perlson, E.; Gilad, A.; Konrad, Z.; Scolnik, P.A.; Bar-Zvi, D. Over expression of the water and salt stress-regulated *Asr1* gene confers an increased salt tolerance. *Plant Cell Environ.* **2004**, *27*, 1459–1468. [CrossRef]

64. Kalifa, Y.; Gilad, A.; Konrad, Z.; Zaccai, M.; Scolnik, P.A.; Bar-Zvi, D. The water- and salt-stress-regulated *Asr1* (abscisic acid stress ripening) gene encodes a zinc-dependent DNA-binding protein. *Biochem. J.* **2004**, *381 Pt 2*, 373–378. [CrossRef]

65. Isokpehi, R.D.; Simmons, S.S.; Cohly, H.H.P.; Ekunwe, S.I.; Begonia, G.B.; Ayensu, W.K. Identification of drought-responsive universal stress proteins in Viridiplantae. *Bioinform. Biol. Insights* **2011**, *5*, 41–58. [CrossRef] [PubMed]

66. Tkaczuk, K.L.; Shumilin, I.A.; Chruszcz, M.; Evdokimova, E.; Savchenko, A.; Minor, W. Structural and functional insight into the universal stress protein family. *Evol. Appl.* **2013**, *6*, 434–449. [CrossRef] [PubMed]

67. Mindrebo, J.T.; Nartey, C.M.; Seto, Y.; Burkart, M.D.; Noel, J.P. Unveiling the functional diversity of the alpha/beta hydrolase superfamily in the plant kingdom. *Curr. Opin. Struct. Biol.* **2016**, *41*, 233–246. [CrossRef] [PubMed]

68. Lee, U.; Rioflorido, I.; Hong, S.W.; Larkindale, J.; Waters, E.R.; Vierling, E. The Arabidopsis ClpB/Hsp100 family of proteins: Chaperones for stress and chloroplast development. *Plant J.* **2007**, *49*, 115–127. [CrossRef] [PubMed]

69. Chen, Z.Y.; Brown, R.L.; Damann, K.E.; Cleveland, T.E. Identification of unique or elevated levels of kernel prpteins in aflatoxin-resistant maize genotypes through proteome analysis. *Phytopathology* **2002**, *92*, 1084–1094. [CrossRef] [PubMed]

70. Cruz de Carvalho, M.H. Drought stress and reactive oxygen species:Production, scavenging and signaling. *Plant Signal. Behav.* **2008**, *3*, 156–165. [CrossRef] [PubMed]

71. Wan, X.Y.; Liu, J.Y. Comparative proteomics analysis reveals an intimate protein network provoked by hydrogen peroxide stress in rice seedling leaves. *Mol. Cell. Proteom.* **2008**, *7*, 1469–1488. [CrossRef] [PubMed]

72. Gong, H.; Jiao, Y.; Hu, W.W.; Pua, E.C. Expression of glutathione-*S*-transferase and its role in plant growth and development in vivo and shoot morphogenesis in vitro. *Plant Mol. Biol.* **2005**, *57*, 53–66. [CrossRef] [PubMed]

73. Sytykiewicz, H.; Chrzanowski, G.; Czerniewicz, P.; Sprawka, I.; Łukasik, I.; Goławska, S.; Sempruch, C. Expression profiling of selected glutathione transferase genes in *Zea mays* (L.) seedlings infested with cereal aphids. *PLoS ONE* **2014**, *9*, e111863. [CrossRef] [PubMed]

74. Cummins, I.; Wortley, D.J.; Sabbadin, F.; He, Z.; Coxon, C.R.; Straker, H.E.; Sellars, J.D.; Knight, K.; Edwards, L.; Hughes, D.; et al. Key role for a glutathione transferase in multiple-herbicide resistance in grass weeds. *Proc. Natl. Acad. Sci. USA* **2013**, *110*, 5812–5817. [CrossRef] [PubMed]

75. Lan, T.; Yang, Z.L.; Yang, X.; Liu, Y.J.; Wang, X.R.; Zeng, Q.Y. Extensive functional diversification of the *Populus* glutathione *S*-transferase supergene family. *Plant Cell* **2009**, *21*, 3749–3766. [CrossRef] [PubMed]

76. Cançado, G.M.A.; De Rosa, V.E.; Fernandez, J.H.; Maron, L.G.; Jorge, R.A.; Menossi, M. Glutathione S-transferase and aluminum toxicity in maize. *Funct. Plant Biol.* **2005**, *32*, 1045–1055. [CrossRef]

77. Buchanan, B.; Schurmann, P.; Wolosiuk, R.; Jacquot, J. The ferredoxin/thioredoxin system: From discovery to molecular structures and beyond. *Discov. Photosynth.* **2002**, *73*, 215–222. [CrossRef] [PubMed]

78. Balmer, Y.; Vensel, W.H.; Cai, N.; Manieri, W.; Schürmann, P.; Hurkman, W.J.; Buchanan, B.B. A complete ferredoxin_thioredoxin system regulates fundamental processes in amyloplasts. *Proc. Natl. Acad. Sci. USA* **2006**, *103*, 2988–2993. [CrossRef] [PubMed]

79. Broin, M.; Rey, P. Potato plants lacking the CDSP32 plastidic thioredoxin exhibit overoxidation of the BAS1 2-cysteine peroxiredoxin and increased lipid peroxidation in thylakoids under photooxidative stress. *Plant Physiol.* **2003**, *132*, 1335–1343. [CrossRef] [PubMed]

80. Zhang, C.; Shi, S. Physiological and Proteomic Responses of Contrasting Alfalfa (*Medicago sativa* L.) Varieties to PEG-Induced Osmotic Stress. *Front. Plant Sci.* **2018**, *9*, 1–21. [CrossRef] [PubMed]

81. Wahid, A.; Gelani, S.; Ashraf, M.; Foolad, M.R. Heat tolerance in plants: An overview. *Environ. Exp. Bot.* **2007**, *61*, 199–223. [CrossRef]

82. Kuroha, T.; Tokunaga, H.; Kojima, M.; Ueda, N.; Ishida, T.; Nagawa, S.; Fukuda, H.; Sugimoto, K.; Sakakibara, H. Functional analyses of LONELY GUY cytokinin-activating enzymes reveal the importance of the direct activation pathway in Arabidopsis. *Plant Cell* **2009**, *21*, 3152–3169. [CrossRef] [PubMed]

83. Argueso, C.T.; Ferreira, F.J.; Kieber, J.J. Environmental perception avenues: The interaction of cytokinin and environmental response pathways. *Plant Cell Environ.* **2009**, *32*, 1147–1160. [CrossRef] [PubMed]

84. Pavlů, J.; Novák, J.; Koukalová, V.; Luklová, M.; Brzobohatý, B.; Černý, M. Cytokinin at the crossroads of abiotic stress signalling pathways. *Int. J. Mol. Sci.* **2018**, *19*, 2450. [CrossRef] [PubMed]

85. Zhao, Y.; Wang, Y.; Yang, H.; Wang, W.; Wu, J.; Hu, X. Quantitative proteomic analyses identify aba-related proteins and signal pathways in maize leaves under drought conditions. *Front. Plant Sci.* **2016**, *7*, 1–23. [CrossRef] [PubMed]

86. Zhang, Z.; Li, J.; Liu, H.; Chong, K.; Xu, Y. Roles of ubiquitination-mediated protein degradation in plant responses to abiotic stresses. *Environ. Exp. Bot.* **2015**, *114*, 92–103. [CrossRef]

87. Bannenberg, G.; Mart'nez, M.; Rodrı'guez, M.J.; Lo'pez, M.A.; Ponce de León, I.; Hamberg, M.; Castresana, C. Functional analysis of a-DOX2, an active a-Dioxygenase critical for normal development in tomato plants. *Plant Physiol.* **2009**, *151*, 1421–1432. [CrossRef] [PubMed]

88. Mazid, M.; Khan, T.A.; Mohammad, F. Role of secondary metabolites in defense mechanisms of plants. *Biol. Med.* **2011**, *3*, 232–249.

89. Falara, V.; Akhtar, T.A.; Nguyen, T.T.; Spyropoulou, E.A.; Bleeker, P.M.; Schauvinhold, I.; Matsuba, Y.; Bonini, M.E.; Schilmiller, A.L.; Last, R.L.; et al. The tomato terpene synthase gene family. *Plant Physiol.* **2011**, *157*, 770–789. [CrossRef] [PubMed]

90. Jia, Q.; Köllner, T.G.; Gershenzon, J.; Chen, F. MTPSLs: New terpene synthases in nonseed plants. *Trends Plant Sci.* **2018**, *23*, 121–128. [CrossRef] [PubMed]

91. Jonak, C.; Heberle-Bors, E.; Hirt, H. MAP kinases: Universal multipurpose signalling tools. *Plant Mol. Biol.* **1994**, *24*, 407–416. [CrossRef] [PubMed]

92. Morris, P.C. MAP kinase signaltransduction pathways in plants. *New Phytol.* **2001**, *151*, 67–89. [CrossRef]

93. Xu, J.; Li, Y.; Wang, Y.; Liu, H.; Lei, L.; Yang, H.; Liu, G.; Ren, D. Activation of MAPK kinase 9 induces ethylene and camalexin biosynthesis and enhances sensitivity to salt stress in *Arabidopsis*. *J. Biol. Chem.* **2008**, *283*, 26996–27006. [CrossRef] [PubMed]

94. Rasmussen, M.W.; Milena, R.; Morten, P.; John, M. MAP kinase cascades in arabidopsis innate immunity. *Front. Plant Sci.* **2012**, *3*, 1–6. [CrossRef] [PubMed]

95. Wang, B.B.; Brendel, V. Molecular characterization and phylogeny of U2AF homologs in plants. *Plant Physiol.* **2006**, *140*, 624–636. [CrossRef] [PubMed]

96. Kumar, S.; Singh, A. Epigenetic regulation of abiotic stress tolerance in plants. *Adv. Plants Agric. Res.* **2016**, *5*, 15406. [CrossRef]

97. Yaish, M.W. DNA methylation-associated epigenetic regulation changes in stress tolerance of plants. In *Molecular Stress Physiology of Plants*; Rout, G.R., Das, A.B., Eds.; Springer: New Dehli, India, 2013; pp. 427–439.

98. Chinnusamy, V.; Zhu, J.K. Epigenetic regulation of stress responses in plants. *Curr. Opin. Plant Biol.* **2009**, *12*, 133–139. [CrossRef] [PubMed]

99. Labra, M.; Ghiani, A.; Citterio, S.; Sgorbati, S.; Sala, F.; Vannini, C.; Ruffini-Castiglione, M.; Bracale, M. Analysis of cytosine methylation pattern in response to water deficit in pea root tips. *Plant Biol.* **2002**, *4*, 694–699. [CrossRef]

100. Dyachenko, O.V.; Zakharchenko, N.S.; Shevchuk, T.V.; Bohnert, H.J.; Cushman, J.C.; Buryanov, Y.I. Effect of hypermethylation of CCWGG sequences in DNA of *Mesembryanthemum crystallinum* plants on their adaptation to salt stress. *Biochemistry* **2006**, *71*, 461–465. [CrossRef] [PubMed]

101. Kozbial, P.Z.; Mushegian, A.R. Natural history of S-adenosylmethionine-binding proteins. *BMC Struct. Biol.* **2005**, *5*, 19. [CrossRef] [PubMed]

102. Liu, X.; Luo, M.; Zhang, W.; Zhao, J.; Zhang, J.; Wu, K.; Tian, L.; Duan, J. Histone acetyltransferases in rice (*Oryza sativa* L.): Phylogenetic analysis, subcellular localization and expression. *BMC Plant Biol.* **2012**, *12*, 145. [CrossRef] [PubMed]

103. Wang, W.; Vinocur, B.; Shoseyov, O.; Altman, A. Role of plant heat-shock proteins and molecular chaperons in the abiotic stress response. *Trends Plant Sci.* **2004**, *9*, 244–252. [CrossRef] [PubMed]

104. Ma, Q.-L.; Kang, J.-M.; Long, R.-C.; Cui, Y.-J.; Zhang, T.-J.; Xiong, J.-B.; Yang, Q.-C.; Sun, Y. Proteomic analysis of salt and osmotic-drought stress in alfalfa seedlings. *J. Integr. Agric.* **2016**, *15*, 2266–2278. [CrossRef]

105. De la Cruz, J.; Karbstein, K.; Woolford, J.L. Functions of ribosomal proteins in assembly of eukaryotic ribosomes in vivo. *Annu. Rev. Biochem.* **2015**, *84*, 93–129. [CrossRef] [PubMed]

106. Ziogas, V.; Tanou, G.; Belghazi, M.; Filippou, P.; Fotopoulos, V.; Grigorios, D.; Molassiotis, A.A. Roles of sodium hydrosulfide and sodium nitroprusside as priming molecules during drought acclimation in citrus plants. *Plant Mol. Biol.* **2015**, *89*, 433–450. [CrossRef] [PubMed]

107. Lu, Y. Identification and roles of Photosystem II assembly, stability, and repair factors in Arabidopsis. *Front. Plant Sci.* **2016**, *7*, 1–27. [CrossRef] [PubMed]

108. Theis, J.; Schroda, M. Revisiting the photosystem II repair cycle. *Plant Signal. Behav.* **2016**, *11*, e1218587. [CrossRef] [PubMed]

109. Edwards, G.E.; Nakamoto, H.; Burnell, J.N.; Hatch, M.D. Pyruvate, Pi dikinase and NADP-malate dehydrogenase in C_4 photosynthesis: Properties and mechanism of light/dark regulation. *Annu. Rev. Plant Physiol.* **1985**, *36*, 255–286. [CrossRef]

110. Jiang, H.-X.; Yang, L.-T.; Qi, Y.-P.; Lu, Y.-B.; Huang, Z.-R.; Chen, L.-S. Root iTRAQ protein profile analysis of two citrus species differing in aluminium-tolerance in response to long-term aluminium-toxicity. *BMC Genom.* **2015**, *16*, 949. [CrossRef] [PubMed]

111. Liese, A.; Romeis, T. Biochemical regulation of in vivo function of plant calcium-dependent protein kinases (CDPK). *Biochim. Biophys. Acta* **2013**, *1833*, 1582–1589. [CrossRef] [PubMed]

112. Liu, W.; Li, W.; He, Q.L.; Daud, M.K.; Chen, J.; Zhu, S. Genome-wide survey and expression analysis of calcium-dependent protein kinase in *Gossypium raimondii*. *PLoS ONE* **2014**, *9*, e98189. [CrossRef] [PubMed]

113. Yang, X.H.; Xu, Z.H.; Xue, H.W. Arabidopsis membrane steroid binding protein 1 is involved in inhibition of cell elongation. *Plant Cell* **2005**, *17*, 116–131. [CrossRef] [PubMed]

114. Yang, X.; Song, L.; Xue, H.W. Membrane steroid binding protein 1 (MSBP1) stimulates tropism by regulating vesicle trafficking and auxin redistribution. *Mol. Plant* **2008**, *1*, 1077–1078. [CrossRef] [PubMed]

115. Ashihara, H. Biosynthesis of 5-phosphoribosyl-1-pyrophosphate in plants: A review. *Eur. Chem. Bull.* **2016**, *5*, 314–323.

116. Büttner, M.; Sauer, N. Monosaccharide transporters in plants: Structure, function and physiology. *Biochim. Biophys. Acta* **2000**, *1465*, 263–274. [CrossRef]

117. Truernit, E.; Schmid, J.; Epple, P.; Illig, J.; Sauer, N. The sink-specific and stress-regulated Arabidopsis STP4 gene: Enhanced expression of a gene encoding a monosaccharide transporter by wounding, elicitors, and pathogen challenge. *Plant Cell* **1996**, *8*, 2169–2182. [CrossRef] [PubMed]

118. Büttner, M.; Truernit, E.; Baier, K.; Scholz-Starke, J.; Sontheim, M.; Lauterbach, C.; Huss, V.A.R.; Sauer, N. AtSTP3, a green leaf-specific, low affinity monosaccharide-H+ symporter of Arabidopsis thaliana. *Plant Cell Environ.* **2000**, *23*, 175–184. [CrossRef]

119. Nouri, M.Z.; Toorchi, M.; Komatsu, S. Proteomics approach for identifying abiotic stress responsive proteins in soybean. Soybean Aleksandra Sudarić. *IntechOpen* **2011**, *276*, 514. [CrossRef]

120. Alam, I.; Sharmin, S.A.; Kim, K.H.; Yang, J.K.; Choi, M.S.; Lee, B.H. Proteome analysis of soybean roots subjected to short-term drought stress. *Plant Soil* **2010**, *333*, 491–505. [CrossRef]

121. Ghannoum, O. C4 photosynthesis and water stress. *Ann. Bot.* **2009**, *103*, 635–644. [CrossRef] [PubMed]

122. Zhang, X.B.; Lei, L.; Lai, J.S.; Zhao, H.M.; Song, W.B. Effects of drought stress and water recovery on physiological responses and gene expression in maize seedlings. *BMC Plant Biol.* **2018**, *18*, 1–16. [CrossRef] [PubMed]

123. Zhao, F.; Zhang, D.; Zhao, Y.; Wang, W.; Yang, H.; Tai, F.; Li, C.; Hu, X. The difference of physiological and proteomic changes in maize leaves adaptation to drought, heat, and combined both stresses. *Front. Plant Sci.* **2016**, *7*, 1–19. [CrossRef] [PubMed]

124. Murata, N.; Allakhverdiev, S.I.; Nishiyama, Y. The mechanism of photoinhibition in vivo: Re-evaluation of the roles of catalase, α-tocopherol, non-photochemical quenching, and electron transport. *Biochim. Biophys. Acta* **2012**, *1817*, 1127–1133. [CrossRef] [PubMed]

125. Melkozernov, A.N.; Blankenship, R.E. Photosynthetic functions of chlorophylls. In *Chlorophylls and Bacteriochlorophylls*; Grimm, B., Porra, R.J., Rüdiger, W., Scheer, H., Eds.; Springer: Dordrecht, The Netherlands, 2006; pp. 397–412.

126. Dudhate, A.; Shinde, H.; Tsugama, D.; Liu, S.; Takano, T. Transcriptomic analysis reveals the differentially expressed genes and pathways involved in drought tolerance in pearl millet [*Pennisetum glaucum* (L.) R. Br]. *PLoS ONE* **2018**, *13*, e0195908. [CrossRef] [PubMed]

127. Xiao, L.; Zhang, L.; Yang, G.; Zhu, H.; He, Y. Transcriptome of Protoplasts Reprogrammed into Stem Cells in *Physcomitrella patens*. *PLoS ONE* **2012**, *7*, e35961. [CrossRef] [PubMed]

128. Izaurralde, E.; Mattaj, I.W. Transport of RNA between nucleus and cytoplasm. *Semin. Cell Biol.* **1992**, *3*, 279–288. [CrossRef]

129. Galmés, J.; Flexas, J.; Savé, R.; Medrano, H. Water relations and stomatal characteristics of Mediterranean plants with different growth forms and leaf habits: Responses to water stress and recovery. *Plant Soil* **2007**, *290*, 139–155. [CrossRef]

130. Keogh, R.C.; Deverall, B.J.; Mcleod, S. Comparison of histological and physiological responses to phakopspora pachyrhizi in pesistant and susceptible soybean. *Great Br.* **1980**, *74*, 329–333. [CrossRef]

131. Lade, H.; Kadam, A.; Paul, D.; Govindwar, S. A low-cost wheat bran medium for biodegradation of the benzidine-based carainogenic dye trypan blue using a micronial consortium. *Environ. Res. Public Health* **2015**, *12*, 3480–3505. [CrossRef] [PubMed]

132. Han, L.B.; Song, G.L.; Zhang, X. Preliminary observation of physiological responses of three turfgrass species to traffic stress. *HortTechnology* **2008**, *18*, 139–143.

133. Dhindsa, R.S.; Plumb-Dhindsa, P.; Thorpe, T.A. Leaf senescence: Correlated with increased leaves of membrane permeability and lipid peroxidation, and decreased levels of superoxide dismutase and catalase. *J. Exp. Bot.* **1981**, *32*, 93–101. [CrossRef]

134. Bates, T.S.; Waldren, R.P.; Teare, I.D. Rapid determination of free proline for water-stress studies. *Plant Soil* **1973**, *39*, 205–207. [CrossRef]

135. Wisniewski, J.R.; Zougman, A.; Nagaraj, N.; Mann, M. Universal sample preparation method for proteome analysis. *Nat. Methods* **2009**, *6*, 359–362. [CrossRef] [PubMed]

136. Swägger, H. Tricine-SDS-PAGE. *Nat. Protoc.* **2006**, *1*, 16–22. [CrossRef] [PubMed]

137. Luo, M.; Zhao, Y.; Wang, Y.; Shi, Z.; Zhang, P.; Zhang, Y.; Song, W.; Zhao, J. Comparative proteomics of contrasting maize genotypes provides insights into salt-stress tolerance mechanisms. *J. Proteome Res.* **2018**, *17*, 141–153. [CrossRef] [PubMed]

138. Song, H.; Wang, H.Y.; Zhang, T. Comprehensive and quantitative proteomic analysis of metamorphosis-related proteins in the veined rapa whelk. *Int. J. Mol. Sci.* **2016**, *17*, 924. [CrossRef] [PubMed]

139. Ashburner, M.; Ball, C.A.; Blake, J.A.; Botstein, D.; Butler, H.; Cherry, J.M.; Davis, A.P.; Dolinski, K.; Dwight, S.S.; Eppig, J.T.; et al. Gene ontology: Tool for the unification of biology. *Nat. Genet.* **2000**, *25*, 25–29. [CrossRef] [PubMed]

140. Gotz, S.; García-Gómez, J.M.; Terol, J.; Williams, T.D.; Nagaraj, S.H.; Nueda, M.J.; Robles, M.; Talón, M.; Dopazo, J.; Conesa, A. High-throughput functional annotation and data mining with the Blast2GO suite. *Nucleic Acids Res.* **2008**, *36*, 3420–3435. [CrossRef] [PubMed]

141. Livak, K.J.; Schmittgen, T.D. Analysis of relative gene expression data using real-time quantitative PCR and the $2^{-\Delta\Delta CT}$ method. *Methods* **2001**, *25*, 402–408. [CrossRef] [PubMed]

International Journal of
Molecular Sciences

MDPI

Article

Comparative Proteomic Analysis Reveals Elevated Capacity for Photosynthesis in Polyphenol Oxidase Expression-Silenced *Clematis terniflora* DC. Leaves

Xi Chen [1,†], Bingxian Yang [1,†], Wei Huang [1], Tantan Wang [1], Yaohan Li [1], Zhuoheng Zhong [1], Lin Yang [2], Shouxin Li [3] and Jingkui Tian [1,4,*]

[1]   College of Biomedical Engineering & Instrument Science, Zhejiang University, Zheda Road 38, Hangzhou 310027, China; 21615052@zju.edu.cn (X.C.); xianyb@zju.edu.cn (B.Y.); 21615051@zju.edu.cn (W.H.); 21615085@zju.edu.cn (T.W.); 21815031@zju.edu.cn (Y.L.); zhongzhh@zju.edu.cn (Z.Z.)
[2]   Zhuhai Weilan Pharmaceutical Co., Ltd., Zhuhai 519030, China; LinYangzhuhai@163.com
[3]   Changshu Qiushi Technology Co., Ltd., Suzhou 215500, China; 0016837@zju.edu.cn
[4]   Zhejiang-Malaysia Joint Research Center for Traditional Medicine, Zhejiang University, Hangzhou 310027, China
*   Correspondence: tjk@zju.edu.cn; Tel.: +86-571-88273823; Fax: +86-571-87951676
†   These authors contributed equally to this work.

Received: 21 September 2018; Accepted: 2 December 2018; Published: 5 December 2018

Abstract: Polyphenol oxidase (PPO) catalyzes the o-hydroxylation of monophenols and oxidation of o-diphenols to quinones. Although the effects of PPO on plant physiology were recently proposed, little has been done to explore the inherent molecular mechanisms. To explore the in vivo physiological functions of PPO, a model with decreased PPO expression and enzymatic activity was constructed on *Clematis terniflora* DC. using virus-induced gene silencing (VIGS) technology. Proteomics was performed to identify the differentially expressed proteins (DEPs) in the model (VC) and empty vector-carrying plants (VV) untreated or exposed to high levels of UV-B and dark (HUV-B+D). Following integration, it was concluded that the DEPs mainly functioned in photosynthesis, glycolysis, and redox in the PPO silence plants. Mapman analysis showed that the DEPs were mainly involved in light reaction and Calvin cycle in photosynthesis. Further analysis illustrated that the expression level of adenosine triphosphate (ATP) synthase, the content of chlorophyll, and the photosynthesis rate were increased in VC plants compared to VV plants pre- and post HUV-B+D. These results indicate that the silence of PPO elevated the plant photosynthesis by activating the glycolysis process, regulating Calvin cycle and providing ATP for energy metabolism. This study provides a prospective approach for increasing crop yield in agricultural production.

Keywords: *Clematis terniflora* DC.; polyphenol oxidase; virus induced gene silencing; photosynthesis; glycolysis

---

1. Introduction

Polyphenol oxidase (PPO) is an oxidoreductase that catalyzes the oxidation of monophenols and/or o-diphenols to o-quinones, which form brown melanin pigments in fruits and vegetables by covalently modifying and cross-linking proteins [1,2]. However, PPO is not only involved in the formation of pigments, but also plays a crucial role in the biosynthesis of secondary metabolites such as aurones [3] and betalins [4]. PPO solitarily catalyzes the hydroxylation and oxidative cyclization of chalcones, leading to the formation of aurone [3]. In addition, PPO, a tyrosinase can also hydroxylate tyramine into dopamine, which in the presence of betalamic acid yields dopamine-betaxanthin that can be further oxidized to yield 2-des-carboxy-betanidin [4]. Furthermore, PPO also has roles in plant defense against insects and pathogens. In transgenic tomato plants, PPO overexpression greatly increases resistance to *Pseudomonas syringae* [5], and the manipulation of PPO activity provides simultaneous resistance to both disease and insect pests [6]. A recent study on strawberries illustrated that PPO overexpression delays the fungal infection [7]. PPOs may perform different functions in diverse plant species and possibly have multiple roles in plants for the huge PPO gene families.

Potential roles for PPO in plants have also been suggested for adaption to abiotic stresses. A study in olive trees showed oxidation of phenolic is inhibited by decreased PPO, suggesting there is an association between decreased PPO activity and improved antioxidant capacity [8]. The role of PPO in regulation of cell death in walnuts was explored, where silencing of PPO resulted in increased tyramine, which elicited cell death in walnuts [9]. Furthermore, a recent study also demonstrated the potential of PPO in bioremediation and food/drug industries as it significantly reduces the phenol content in an artificial solution [10]. As well as evidence supporting that PPOs play a role in plant defense against biotic stressors, several independent lines of evidence identified PPO with a chloroplastic location, linking PPO with photosynthesis. However, although an interaction between photosynthesis and PPO activity has been presented more than once [11–13], evidence either for or against direct involvement has been ambiguous, and the absence of chloroplastic substrates that were identified remains an issue.

Virus-induced gene silencing (VIGS) takes advantage of plant RNAi-mediated antiviral defense and has been used widely in plants to analyze gene function [14]. Lee et al. [15] used VIGS to analyze the susceptibility and resistance functional wheat genes involved in *Zymoseptoria tritici*. Using the barley stripe mosaic virus—VIGS technique, Zhao et al. [16] indicated that the TNBL1 gene is an important gene positively involved in wheat defense response to barley yellow dwarf virus infection. Groszyk et al. [17] analyzed the Bx1 gene ortholog in rye using VIGS and found it to be functionally involved in benzoxazinoid biosynthesis. Recently, VIGS has been a powerful alternative technology for determining the unknown functions of genes and a combination of VIGS and omics is evolving as a competitive strategy for gene function analysis. VIGS and proteomic analysis of the resistance of cotton to *Veticillium dahia* revealed that gossypol, brassinosteroids, and jasmonic acid contribute to this process [18]. The role of the tomato TAGL1 gene in regulating the accumulation of fruit metabolites was investigated using VIGS and metabolomics analyses [19]. An improved virus-induced gene silencing approach was used to elucidate the role of highly homologous *Nicotiana benthamiana* ubiquitin enzymes gene family members in plant immunity [20]. In addition, VIGS technology as a powerful investigation tool is still on the way with its application potentials remaining to be fully developed.

Clematis terniflora DC. is a Chinese folk medicinal resource with important pharmaceutical value in the treatment of inflammatory symptoms in the respiratory and urinary systems [21]. Studies have shown that the extracts from the leaves and stems of *C. terniflora* have anti-inflammatory, anti-tumor, and anti-nociceptive effects [22–25]. Furthermore, omics technologies were used on *C. terniflora* to prospectively understand the inherent mechanism underlying its medicinal quality [24,26]. Considering the role PPO played in stress resistance of plant and biosynthesis of secondary metabolites, regulating PPO activity or gene expression can be beneficial to the yields and quality of medicinal plants. Therefore, the potential relationships between photosynthesis and PPO activity are highly relevant to the improvement of yields and quality [27]. To have insight into this relationship, a comparative proteomic analysis was performed on the leaves of *C. terniflora* DC. with down-regulated PPO activity by VIGS in this study. High-intensity UV-B and dark incubation (HUV-B+D), which are helpful stressors for medicinal plants [28], were used for an integration study.

2. Results

2.1. Cloning, Sequence Analysis, and Phylogenetic Tree Analysis of CtPPO

Following the identification of CtPPO transcript in *C. terniflora* transcriptome data, 1681 bp of PPO cDNA was cloned from *C. terniflora* leaves using RT-PCR and 5′ RACE with the complementary application of genome-walking technologies (Figure S2a). Sequence analysis of CtPPO cDNA revealed a 166-bp

5′-untranslated region (UTR) located upstream of a start codon and an open reading frame (ORF) of 1681 bp encoding 586 amino acids with a calculated molecular mass of 66.11 kDa and an isoelectric point of 7.18.

Blast analysis of the predicted CtPPO amino acid sequence revealed 62% similarity with *Nelumbo nucifera*. To further understand the evolutionary relationships between CtPPO and PPO in other plants, a phylogenetic tree was constructed based on amino acid sequence. Phylogenetic analysis showed CtPPO has a close evolutionary relationship with the PPO gene in *N. nucifera*, but not other plant species (Figure S2b).

2.2. Virus-Induced Gene Silencing of CtPPO in C. terniflora DC

VIGS provides an alternative approach for functional analysis of genes. Here, to determine the role of CtPPO, we silenced CtPPO using tobacco rattle virus (TRV)-mediated gene silencing. We constructed two modified TRV vectors: TRV-CtPPO and TRV-PDS. To determine the extent of silencing, we performed qRT-PCR to analysis the transcript levels of the CtPPO gene in inoculated plants. RNA extracted from *C. terniflora* agro-infiltrated with empty vector TRV or TRV-CtPPO or control leaves 20 days post-infection were used for qRT-PCR. As shown in Figure 1a, CtPPO expression levels were reduced in plants infiltrated with TRV-CtPPO compared to the control plants (VIGS-vector and control). To further confirm CtPPO expression, we measured CtPPO activity in control, TRV-vector and TRV-CtPPO plants. In VIGS-CtPPO (VC), PPO activity was significantly lower than in VIGS-vector (VV) and control, illustrating that a successful VC plant model was constructed (Figure 1b). Despite these obvious changes in CtPPO in plants, however, there were no macroscopic phenotypic differences in the VC, VV, and control leaves at the starting point (Figure 1c).

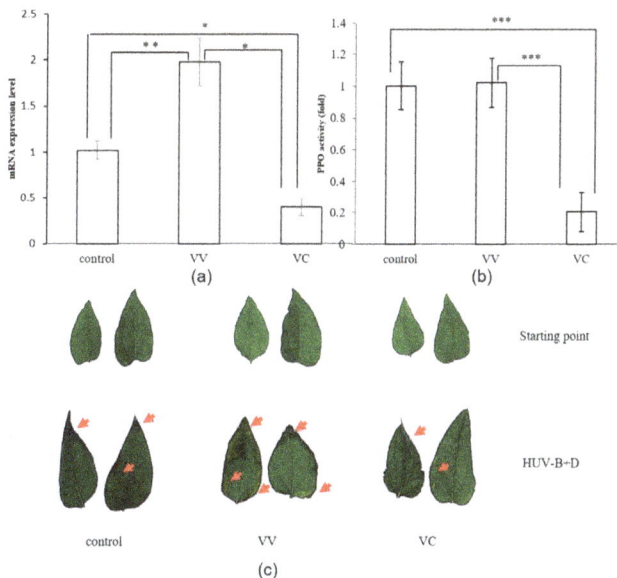

Figure 1. (**a**) Expression levels of CtPPO gene mRNA in leaves of *C. terniflora* in control, VIGS-vector, and VIGS-CtPPO plants. Data are shown as mean ± SD of the independent biological replicates. Asterisks indicate significant changes as measured by Student's *t*-test (* $p < 0.05$, ** $p < 0.01$, and *** $p < 0.001$); (**b**) PPO activity was assayed in *C. terniflora* leaves of control, VIGS- vector and VIGS-CtPPO plants. Data are shown as mean ± SD of independent biological replicates. Asterisks indicate significant changes as measured by Student's *t*-test (* $p < 0.05$, ** $p < 0.01$, and *** $p < 0.001$); (**c**) Morphology of control, VIGS-vector, and VIGS-CtPPO *C. terniflora* plants before and after HUV-B+D treatment. Photograph showing *C. terniflora* growth under different treatments. The damaged areas were marked with red arrows.

2.3. Effects of VIGS-CtPPO and VIGS-Vector on Leaf Proteins in C. terniflora DC.

To investigate the effects of CtPPO on *C. terniflora* leaf proteins, a gel-free/label-free proteomic technique was used. Proteins extracted from VC- and VV-infected *C. terniflora* leaves were reduced, alkylated, digested, and analyzed by nano-LC-MS/MS. The obtained proteomic data were analyzed using the UniProtKB/Swiss-Prot database and protein content was estimated by mol%. The functions of identified proteins were predicted based on comparisons with functional annotations of the *Arabidopsis* genome and classified using MapMan bin codes. The results of the functional analyses demonstrated that the proteins with increased expression were mainly enriched in photosynthesis, glycolysis, redox, protein metabolism, and secondary metabolism, while the proteins with decreased expression were mainly enriched in protein metabolism and photosynthesis. Notably, expression of proteins related to secondary metabolism, development, and glycolysis were increased in the VC plants, whereas the expression of proteins related to N-metabolism, amino acid metabolism, and cell were decreased (Figure 2a; Tables S8 and S9).

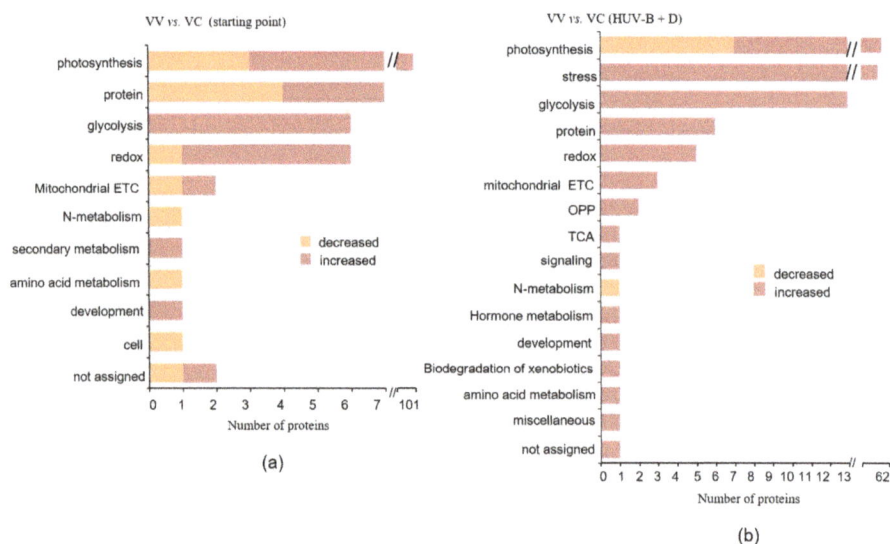

Figure 2. (**a**,**b**) Functional distribution of proteins identified in VIGS-vector and VIGS-CtPPO *C. terniflora* DC. leaves. The leaves were collected at the (**a**) starting point and (**b**) after HUV-B treatment. Proteins were extracted, reduced, alkylated, digested, and analyzed by nanoLC-MS/MS. Protein content is reported as mol %. Protein functions were predicted and categorized using MapMan bin codes. Abbreviations: cell, cell division/organization/cycle; TCA, tricarboxylic acid; ETC, electron transport chains; OPP, oxidative pentose phosphate.

2.4. Effects of High-Level UV-B and Dark Treatment on Proteins in VIGS-CtPPO and VIGS-Vector C. terniflora DC. Leaves

HUV-B+D dramatically impacts plant physiology [24]. To further investigate the effects of CtPPO on *C. terniflora*, we exposed VC, VV, and control plants to HUV-B+D. Subsequent phenotypic analysis revealed the leaves to be wilted with burned patches and a high degree of crispation in control plants. The leaves of VV plants were similar to the control plants, while the leaves from the VC plants appeared significantly less damaged than in the control and VV plants (Figure 1c). To determine the expression profiles of leaf proteins in VC and VV after HUV-B+D, a gel-free/label-free proteomic approach was used. Proteins extracted from VC and VV plant leaves after HUV-B+D were reduced, alkylated, digested, and analyzed by nano-LC-MS/MS. Obtained proteomic data were analyzed using the UniProtKB/Swiss-Prot database and the protein content was estimated by mol%. The functions of identified proteins were predicted based on comparisons with functional annotations of the *Arabidopsis* genome and classified using MapMan bin codes. The results of the functional analyses demonstrated that HUV-B+D increased expression of proteins involved in most of the functional classes, including photosynthesis, stress, glycolysis, protein metabolism, and redox, in VC plants compared to VV plants. However, expression of proteins involved in N-metabolism was decreased in VC plants compared to VV plants (Figure 2b).

2.5. Integrated Analysis of Proteins in VIGS-CtPPO and VIGS-Vector C. terniflora Differentially Expressed after HUV-B+D

The proteins identified in leaves collected from VC and VV plants at the starting point were 676 and 662, respectively. Among the identified proteins, 480 and 467 proteins were commonly expressed in VV and VC plants, respectively, at the starting point and after HUV-B+D. Alternatively, there were 443 and 455 proteins commonly expressed at the starting point and HUV-B+D in VV and VC plants, respectively (Figure S3).

For proteins differentially expressed in response to HUV-B+D in VV and VC plants, further integration analysis was performed. Figure 3 shows that the number of proteins involved in photosynthesis displaying increased expression was decreased in VC plants compared with that in VV plants. The number of increased proteins related to stress, OPP, secondary metabolism, and transport was increased in VC plants compared to VV plants. The number of decreased proteins associated with stress, glycolysis, amino acid metabolism, and carbon-one(C1)-metabolism was decreased in VC plants compared to VV plants. Furthermore, proteins involved in glycolysis, amino acid metabolism, cell, mitochondrial ETC, TCA, and signaling displaying increased expression were only detected in VC plants (Figure 3, Tables S6 and S7).

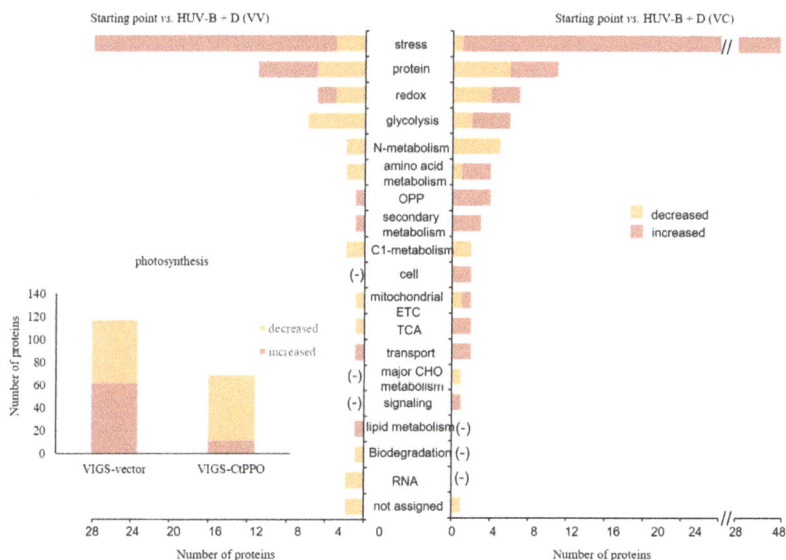

Figure 3. Functional distribution of proteins identified in VIGS-vector and VIGS-CtPPO *C. terniflora* DC. leaves. Leaves were collected at the starting point and after HUV-B+D treatment. Proteins were extracted, reduced, alkylated, digested, and analyzed by nanoLC-MS/MS. Protein content is reported as mol %. Protein functions were predicted and categorized using MapMan bin codes. Abbreviations: cell, cell division/organization/cycle; TCA, tricarboxylic acid; ETC, electron transport chains; CHO, carbohydrate; OPP, oxidative pentose phosphate; (-), not determined.

2.6. MapMan Analysis of the VIGS-CtPPO and VIGS-Vector C. terniflora Leaf Proteomic Data at the Starting Point and after HUV-B+D

To examine changes in the levels of the identified proteins in-depth in VV and VC plants, the DEPs were analyzed using MapMan software. The analysis identified the main functional categories of the proteins displaying significant changes in expression as involved in the light reaction, photorespiration, and Calvin cycle. HUV-B+D conditions induced dramatic decrease in Calvin cycle and light reaction related proteins in VV, however, very interestingly, it was significantly rescued by the silencing of CtPPO in VC. Further analysis showed that the silencing of CtPPO led to the increase of Calvin cycle related proteins in VC than VV at starting point (Figure 4a,b). After exploring HUV-B+D, the silencing of CtPPO gave a further increase of Calvin cycle related proteins. Additionally, it also resulted in the synchronously increase on light reaction and photorespiration related proteins in VC than VV at HUV-B+D (Figure 4c,d).

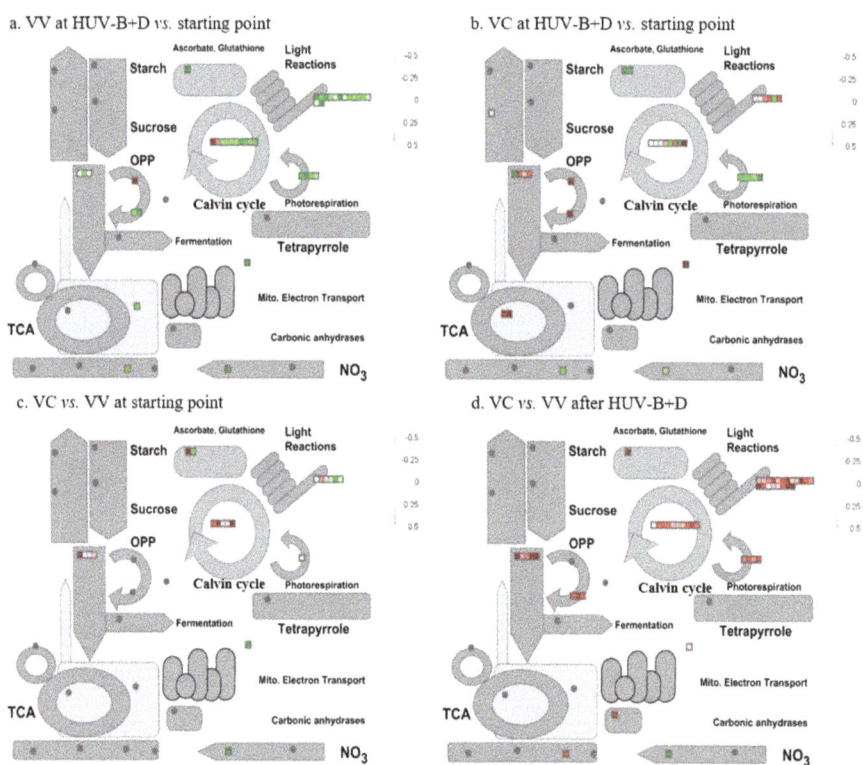

Figure 4. Metabolic pathways of proteins identified in VIGS-vector and VIGS-CtPPO *C. terniflora* DC. Leaves were collected at the starting point and after HUV-B+D treatment. The proteins were grouped into functional categories related to primary metabolism and changes in abundance were visualized using MapMan software. Each square and color indicates the fold-change of a differentially expressed protein. Green and red indicate a decrease and increase, respectively, in fold change compared with the corresponding group: (**a**) comparison of proteins in VIGS-vector leaves after HUV-B+D treatment at starting point, (**b**) comparison of proteins in VIGS-CtPPO leaves after HUV-B+D treatment at starting point, (**c**) comparison of proteins at starting point in VIGS-CtPPO and VIGS-vector plants, and (**d**) comparison of proteins after HUV-B+D treatment in VIGS-CtPPO and VIGS-vector plants. Abbreviations: OPP, oxidative pentose phosphate; CHO, carbohydrate; TCA, tricarboxylic acid cycle.

Following this preliminary analysis, we further studied the differences in enzymes in the light reaction and Calvin cycle. The proteins involved in the light reaction and Calvin cycle was mostly decreased in VV but there was no obvious change in VC after HUV-B+D treatment. At starting point, 1 and 3 proteins in photosystem I and II of light reaction was increased in VC compared to VV. In Calvin cycle, 3 and 2 proteins involved in carboxylation and reduction was increased in VC compared with VV. However, after HUV-B+D treatment, 19 proteins related to NADP$^+$-dependent aldehyde reductase, ATP synthase, and glycerol phosphate dehydrogenase in light reactions was increased and 11 proteins related to the enzymes in reduction and regeneration of Calvin cycle was increased in comparison of VC and VV (Figure S4).

2.7. Expression Profile Analysis of Glycolysis-Related Proteins in C. terniflora DC. Leaves

Glycolysis related proteins were significantly changed by the silencing of CtPPO. To further characterize the relationship between glycolysis and CtPPO, expression profile analysis of glycolysis-related proteins in *C. terniclora* leaves was conducted. The abundances of glyceraldehyde-3-phosphate dehydrogenase (GAPDH), UTP-glucose-1-phosphate uridylytransferase (GPUT), phosphoglycerate kinase (PGK), phosphoenolpyruvate carboxylase (PEPC), and enolase were increased in VC plants compared to VV plants at the starting point. Moreover, after HUV-B+D, the abundances of GPUT, fructose-1,6-bisphosphate aldolase (FBA), GAPDH, PGK, BPGM, enolase, and PEPC were all increased in VC plants compared to VV plants (Figure 5).

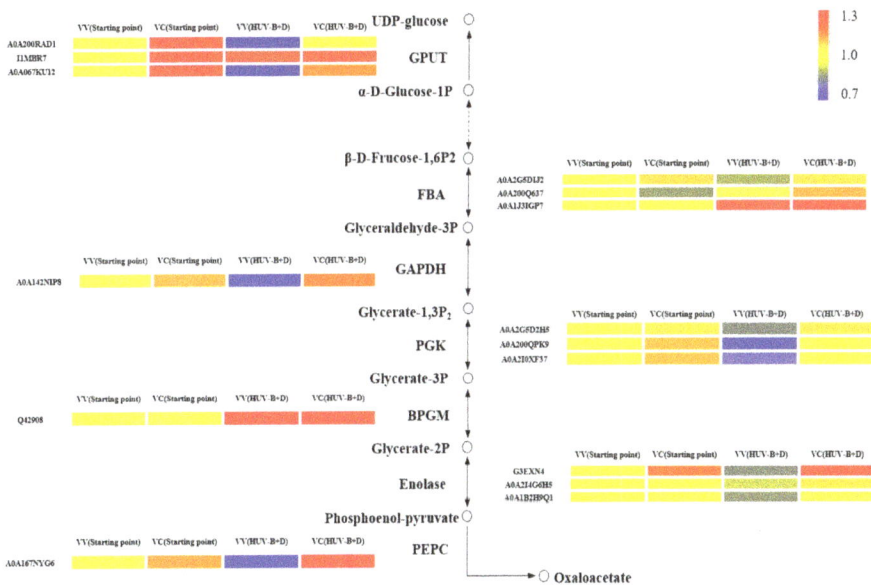

Figure 5. Expression profile analysis of glycolysis-associated proteins. Each square and color indicates the fold-change for a differentially expressed protein. Blue and red indicate a decrease and increase, respectively, in fold-change compared with the corresponding group: VV(SP) proteins in VIGS-vector plants at starting point, VC(SP) proteins in VIGS-CtPPO plants at starting point, VV(HUV-B+D) proteins in VIGS-vector plants after HUV-B+D treatment, and VC(HUV-B+D) proteins in VIGS-CtPPO plants after HUV-B+D treatment. FBA: fructose-1,6-bisphosphate aldolase; GAPDH: glyceraldehyde-3-phosphate dehydrogenase; PEPC: phosphoenolpyruvate carboxylase; circle: metabolite; arrow: direction; solid line: single step; dotted line: multiple step.

2.8. Effects of VIGS-CtPPO on Photosynthesis Characteristics in C. terniflora DC. Leaves

To confirm the relationship between CtPPO and photosynthesis, the chlorophyll content was measured first. The results showed that the chlorophyll a, chlorophyll b, and total chlorophyll contents were all higher in VC plants than in VV and control plants (Figure 6). Then the photosynthesis characteristics were measured by using a Li-6400 Portable Photosynthesis System.

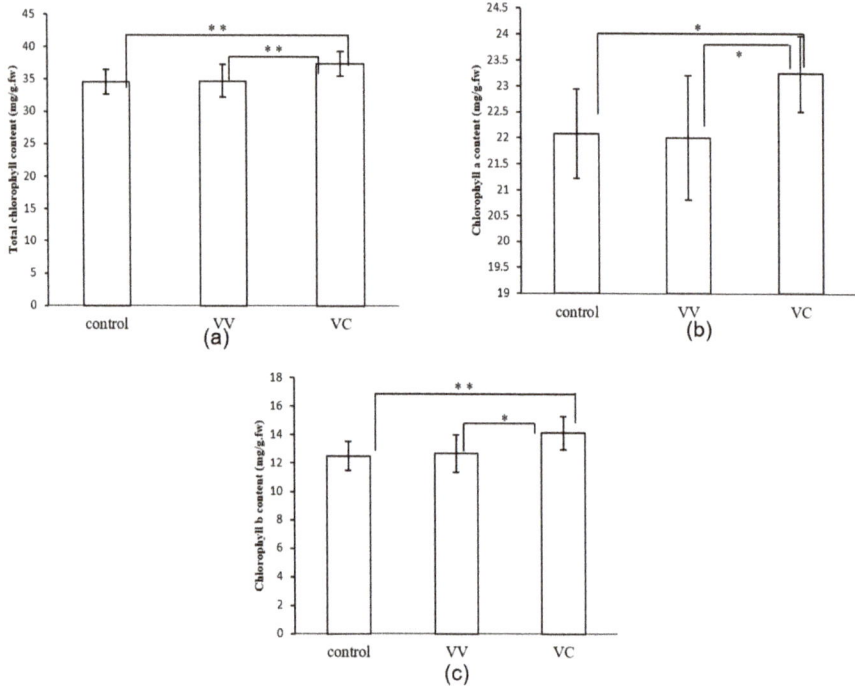

Figure 6. Chlorophyll content in in control, VIGS- vector and VIGS-CtPPO *C. terniflora* leaves. (**a**) Total chlorophyll content, (**b**) chlorophyll a content, and (**c**) chlorophyll b content. Data are shown as mean ± SD for independent biological replicates. Asterisks indicate significant changes as measured by Student's *t*-test (* $p < 0.05$, ** $p < 0.01$). fw: fresh weight.

The photosynthesis rate, intercellular CO_2, stomatal conductance, and transpiration rate were further measured in the study. The photosynthesis rate, transpiration rate, and stomatal conductance in VC plants were higher than in VV plants, while the intercellular CO_2 was lower than that in VV (Figure 7). After HUV-B+D, the damage of photosynthesis rate in VC plants was significantly lower than in VV and control plants, while the stomatal conductance and transpiration rate were low in control, VV, and VC plants. The intercellular CO_2 was still lower in VC and control plants than in VV plants, but both were higher than that at the starting point (Figure S5). The photosynthesis rate was dramatically decreased in by HUV-B+D in *C. terniflora*, however, it seems that the silencing of CtPPO has a mitigating effect to the damage degree in VC (Figure 7 and Figure S5).

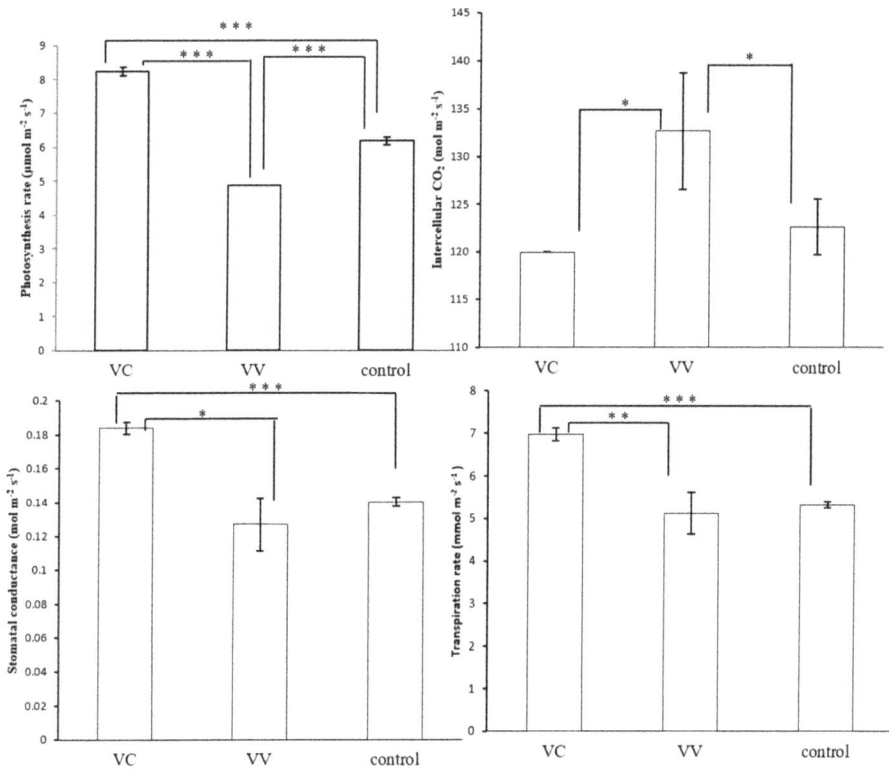

Figure 7. Analysis of photosynthesis characteristics in control, VIGS-vector, and VIGS-CtPPO at starting point. The photosynthesis rate, intercellular CO_2, stomatal conductance, and transpiration rate were measured using an open gas-exchange system. Data are shown as mean $\pm$ SD of independent biological replicates. Asterisks indicate significant changes as measured by Student's t t-test (* $p < 0.05$, ** $p < 0.01$, and *** $p < 0.001$).

2.9. Effects of VIGS-CtPPO on ATP Synthase in C. terniflora DC. Leaves

To confirm the relationship between CtPPO and energy metabolism, qRT-PCR-based analysis of ATP synthase was performed. Leaves of VV and VC were collected at starting point and after HUV-B+D. At the starting point, the expression of ATP synthase was increased without statistical significance in VC compared with control and VV (Figure 8). However, after HUV-B+D, ATP synthase was increased by 2-folds in VC plants compared with VV and control plants (Figure 8). These results indicate that the silencing of CtPPO might activate the energy metabolism by up-regulating the ATP synthase in response to HUV-B+D.

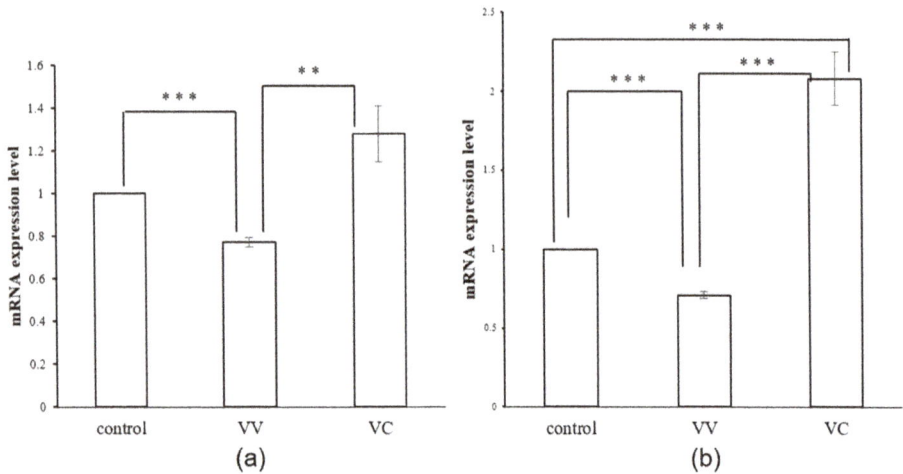

Figure 8. Levels of chloroplast ATP synthase mRNA in control, VIGS-vector, and VIGS-CtPPO *C. terniflora* leaves. (**a**,**b**) *C. terniflora* leaves at the (**a**) starting point and (**b**) after HUV-B+D treatment. Data are shown as mean ± SD of independent biological replicates. Asterisks indicate significant changes as measured by Student's *t*-test (** $p < 0.01$, and *** $p < 0.001$).

3. Discussion

3.1. Silencing of PPO Promoted the Light Reaction in C. terniflora

The photosynthetic process in plants can operate through linear or cyclic electron flow, which involves three major complexes of the electron transfer chain: photosystem II (PSII), photosystem I (PSI), and cytochrome b6f complex [29,30]. Light reactions are an important step in photosynthesis and can transfer light energy to ATP and NADPH [31]. The initial step in this process is the absorption of light energy by the chlorophyll molecule. Ferredoxins were extensively employed as electron shuttles by anaerobes long before the advent of oxygenic photosynthesis [32], while the ferredoxin-NADP$^+$ reductase catalyzes an electron-hydride exchange between reduced ferredoxin and NADP$^+$ to yield NADPH [33]. The activation state of chloroplast ATP synthase is regulated by proton-motive forces generated by photosynthetic electron transfer reactions and reduction of disulphides in the γ subunit by TRX [34,35]. A lower ATP content resulting from the loss of ATP synthesis can inhibit the synthesis of ribulose biphosphate and influence the photosynthetic assimilation of CO_2 [36]. In our study, the amounts of chlorophyll a and chlorophyll b were increased in VC plants compared with the VV and control plants, indicating the silencing of CtPPO could upregulate the PSs system in photosynthesis. In addition, the expression of ATP synthase, and ferredoxin-NADP$^+$ reductase were decreased in response to HUV-B+D (Table S6), indicating the inhibition effect of HUV-B+D on *C. terniflora*. However, the expression of ATP synthase (Figure 8) and ferredoxin-NADP$^+$ reductase (Table 1) were dramatically increased in VC plants compared with VV plants after HUV-B+D. These results indicate that VIGS of CtPPO had a positive effect on upregulating the light reactions to provide ATP and NADPH for plants under stress.

Table 1. Differential Expressed Proteins in *Clematis terniflora* DC. between VV and VC after HUV-B+D treatment (1.55 > fold change > 1.27).

No.	Protein ID [a]	Description	Abundance		FC	*p*-Value	Annotation [b]
			VV	VC			
	Increased						
1	F6HCT7	Uncharacterized protein	0.01	0.02	1.55	0.01	stress
2	M5W912	Ferredoxin-NADP reductase	0.04	0.06	1.51	0.00	PS
3	A0A2G5CER3	Uncharacterized protein (Fragment)	0.01	0.01	1.50	0.03	signaling
4	A0A200RAD1	UTP-glucose-1-phosphate uridylyltransferase	0.02	0.03	1.48	0.03	glycolysis
5	I6P9I5	Cytosolic ascorbate peroxidase (Fragment)	0.03	0.04	1.47	0.02	redox
6	J3MAL0	Ferredoxin-NADP reductase	0.04	0.06	1.47	0.00	PS
7	M4CGU3	Uncharacterized protein	0.02	0.02	1.45	0.02	PS
8	Q42908	2,3-bisphosphoglycerate-independent phosphoglycerate mutase	0.01	0.02	1.42	0.04	glycolysis
9	A0A1J3H2G2	Ubiquitin-NEDD8-like protein RUB1 (Fragment)	0.27	0.38	1.40	0.01	protein
10	A0A2G5F096	Uncharacterized protein	0.02	0.03	1.40	0.05	development
11	C5YTC0	Uncharacterized protein	0.02	0.03	1.39	0.00	not assigned
12	A0A067L6G5	Uncharacterized protein	0.02	0.03	1.39	0.00	amino acid metabolism
13	W9RXI1	Glycerate dehydrogenase	0.08	0.11	1.38	0.00	PS
14	Q1EP00	Chlorophyll a-b binding protein, chloroplastic	0.05	0.07	1.37	0.03	PS
15	W9SCQ6	Ferredoxin-NADP reductase	0.05	0.07	1.36	0.00	PS
16	A0A2G5DAJ3	Carbonic anhydrase	0.03	0.04	1.36	0.04	TCA
17	W8TP69	Glycerate dehydrogenase-like protein	0.04	0.06	1.36	0.02	PS
18	A0A0K9P513	Phosphoglycerate kinase	0.08	0.10	1.35	0.01	PS
19	A0A0K9Q3W1	70 kDa heat shock protein	0.06	0.07	1.35	0.01	stress
20	A0A1J6I7J0	2-cys peroxiredoxin bas1, chloroplastic	0.05	0.07	1.35	0.01	redox
21	C0PRV0	Lactoylglutathione lyase	0.03	0.04	1.33	0.02	Biodegradation of Xenobiotics
22	Q19U04	NADH-dependent hydroxypyruvate reductase (Fragment)	0.11	0.15	1.33	0.00	PS
23	D2XUU3	Chloroplast managanese stabilizing protein (Fragment)	0.17	0.22	1.32	0.04	PS
24	A0A0A0KBL8	Uncharacterized protein	0.03	0.04	1.32	0.00	PS
25	A5BVF4	Uncharacterized protein	0.16	0.21	1.32	0.02	PS
26	A0A200QG47	Aminotransferase	0.05	0.06	1.32	0.02	PS
27	W1P8B5	Uncharacterized protein	0.01	0.01	1.31	0.05	OPP
28	A0A0D3B1C7	Uncharacterized protein	0.02	0.02	1.30	0.02	OPP
29	K7KB09	Uncharacterized protein	0.03	0.04	1.30	0.01	hormone metabolism
30	K4BW79	2-methylene-furan-3-one reductase	0.06	0.07	1.30	0.00	misc
31	A0A2H5NQP8	Uncharacterized protein	0.08	0.10	1.30	0.01	PS
32	A0A0D2Q3K9	Uncharacterized protein	0.04	0.05	1.30	0.00	stress
33	A0A200PYZ1	ATPase	0.08	0.11	1.29	0.01	PS
34	A1BQW9	Transketolase (Fragment)	0.05	0.06	1.29	0.00	PS
35	A0A251VGE5	Putative photosystem I PsaA/PsaB	0.05	0.06	1.29	0.03	PS
36	A0A1D8H339	2-Cys peroxiredoxin	0.07	0.09	1.28	0.01	redox
37	S8EAM3	Heat shock protein hsp70 (Fragment)	0.03	0.03	1.28	0.00	stress

[a] Protein ID, according to UniProtKB/Swiss-Prot database. [b] Function, protein function categorized using MapMan bin codes. FC, fold change; PS, photosynthesis; TCA, tricarboxylic acid; OPP, oxidative pentose phosphate; misc: miscellaneous.

3.2. Silencing of PPO Activated the Calvin Cycle in C. terniflora

The Calvin cycle is an essential carbon fixation process in photosynthesis that converts carbon dioxide to glucose, accompanied by reduction reactions and ribulose 1,5-bisphosphate (RuBP) regeneration [37]. The Calvin cycle is redox-activated process, and the redox homeostasis is mediated by reducing power from photosynthetic electron transport to ferredoxin (Fd) and NADPH, via Fd-thioredoxin (TRX) reductase (FTR) and NADPH-dependent TRX reductase (NTRC) [38]. FTR and multiple TRXs consist the Fd-TRX system, while the NTRC contains a complete TRX system in a single polypeptide. Both systems have an impact on activation of Calvin cycle-related enzymes, including fructose-1,6-bisphosphatase (FBPase), sedoheptulose-1,7-bisphosphatase (SBPase), phosphoribulokinase (PRK), and GAPDH [39–41]. In our current study, Calvin cycle-related PGK, GAPDH, PRK, TK, and RuBP were decreased in response to HUV-B+D, indicating HUV-B+D significantly inhibited the Calvin cycle. However, the drop in CtPPO activity led to a dramatic increase in Calvin cycle-related proteins in *C. terniflora* at the starting point and after HUV-B+D. The redox-related proteins were further analyzed and TRX was found to be increased in VC plants compared with VV plants at the starting point (Table 2) and Fd-NADP reductase was increased in VC plants compared to VV plants after HUV-B+D (Table 1). As PPO is located in chloroplast [42,43], these results indicate VIGS of CtPPO can protect the Calvin cycle by activating the FTR and NTRC system.

Table 2. Differential Expressed Proteins in *Clematis terniflora* DC. between VV and VC at starting point (fold change > 1.40).

No.	Protein ID [a]	Description	Abundance		FC	*p*-Value	Annotation [b]
			VV	VC			
	Increased						
1	A0A151U9E4	Uncharacterized protein	0.02	0.03	1.92	0.01	redox
2	Q8M9K2	Ribulose-bisphosphate carboxylase (Fragment)	0.79	1.46	1.84	0.00	PS
3	K8ECB3	Thioredoxin	0.04	0.08	1.82	0.00	redox
4	A0A067KU12	UTP-glucose-1-phosphate uridylyltransferase	0.03	0.04	1.57	0.02	glycolysis
5	A0A068TPY5	Uncharacterized protein	0.03	0.05	1.53	0.01	PS
6	A0A2G3DEA2	3-oxo-Delta(4,5)-steroid 5-beta-reductase	0.03	0.04	1.45	0.02	development
7	A1X444	Ribulose-1,5-bisphosphate carboxylase/oxygenase large	0.27	0.39	1.44	0.00	PS
8	A0A1U8LIR6	2-methyl-6-phytyl-1,4-hydroquinone methyltransferase	0.02	0.02	1.43	0.01	secondary metabolism
9	A0A200RAD1	UTP-glucose-1-phosphate uridylyltransferase	0.03	0.04	1.42	0.04	glycolysis
10	A0A2G5DW13	Uncharacterized protein	0.04	0.06	1.42	0.01	redox
11	A0A067KC46	Ribulose bisphosphate carboxylase small chain	0.06	0.09	1.41	0.02	PS
12	W9QII5	Peroxiredoxin Q	0.03	0.04	1.41	0.02	redox
13	A0A061EH79	Ribulose bisphosphate carboxylase small chain	0.05	0.07	1.41	0.00	PS

[a] Protein ID, according to UniProtKB/Swiss-Prot database. [b] Function, protein function categorized using MapMan bin codes. FC, fold change; PS, photosynthesis.

Chlorophyll biosynthetic enzymes glutamyl-tRNA reductase, Mg-protoporphyrin IX monomethylester cyclase, and plastidic 2-Cys PRXs are direct substrates of NTRC in *Arabidopsis* [44]. As a result, NTRC protects the formation of chlorophyll in *Arabidopsis* [45]. The NTRC, together with 2-Cys PRXs, forms a two component peroxide detoxifying system that acts as a reductant under stress conditions [44]. Overexpression of NTRC raises the CO_2 fixation rate and lowers non-photochemical quenching by enhancing the activation of TRX-regulated enzymes and chloroplast ATP synthase in the Calvin cycle [46]. In this study, the amounts of total chlorophyll, chlorophyll a, and chlorophyll b were increased in VC plants compared with VV and control plants (Figure 6). The amount of 2-Cys peroxiredoxin was decreased in VV plants in response to HUV-B+D (Table S6), however, it was increased in VC plants compared with VV plants after HUV-B+D (Table 1). Furthermore, chloroplast ATP synthase was increased in VC plants compared with VV and control plants at the starting point and were increased more significantly in VC plants than in VV and control plants following HUV-B+D (Figure 8). The intercellular CO_2 was significantly lower in VC than VV and control (Figure 7) which illustrated the silencing of CtPPO increased the utilization of carbon resource. These results help to increase understanding that CtPPO might influence the carbon fixation by regulating NTRC system.

3.3. Silencing of PPO Enhanced the glycolysis in C. terniflora

Glycolysis is a central metabolic pathway in plants that oxidizes hexoses to provide ATP, reduces power and pyruvate, and produces precursors for anabolism [47]. Studies have proved that the increase of PGK, and BPGM in *Arabidopsis thaliana* provide energy supply by ATP production [48,49]. Arabidopsis double mutants lacking BPGM enzyme activity exhibited defects in blue light, low CO_2, and abscisic acid-regulated stomatal movements [49], which are responsible for all gaseous diffusion and can control photosynthetic CO_2 uptake to influence photosynthesis [50]. The increased glycolytic proteins in root of oat enhanced ATP production and promoted adaptation to anaerobic conditions [51]. It has been suggested that the stomatal red-light response signal may be provided by the redox state of photosynthetic electron transport chain components, such as the redox state of plastoquinone and production of ATP [52]

In our study, the increases in GPUT, FBA, GAPDH, PGK, Enolase, and PEPC in CtPPO-silenced plants demonstrated the activation of glycolysis, which could provide more ATP for energy metabolism in *C. terniflora*. Further analysis also proved the enhanced stomatal conductance by the silencing of CtPPO in VC (Figure 7). The interaction of glyceraldehyde-3-phosphate dehydrogenase and phospholipase D might provide a direct connection between signal transduction, energy metabolism, and growth control in plant response to stress conditions [53]. Yasmeen's study suggest that Cu nanoparticles might enhance the tolerance of wheat to drought and salinity by increasing the glycolysis related protein abundance [54]. These results indicate the decrease in CtPPO activity in *C. terniflora* might elevate its stress tolerance by activating glycolysis metabolism.

3.4. Artificial Interference with PPO Activity Has Potential Applications in Agricultural Production

PPOs catalyze the oxidation of phenols to quinones that subsequently react with amino acids or proteins to form brown and black pigments, greatly reducing the appearance of quality of wheat products [55]. The role of plant PPO in postharvest browning has been the primary focus of research [56]. Although there is no evidence that high PPO activity is associated with depressed nutritional value, the darkened color still negatively affects consumer choice [57]. Studies on wheat concluded that the development of wheat cultivars with low grain PPO activity is one of the main objectives in wheat breeding programs [58]. Photosynthesis is the major trait for any further increase in the yield potential of crops [59]. Various genetic engineering approaches for enhancing C3 plant photosynthesis have been consistently proposed over the past years to improve the crop productivity [60–62]. Our study clearly suggests the potential for suppression of PPO activity on photosynthesis in *C. terniflora*, which indicates the feasibility of artificially mediating photosynthesis through the control of PPO gene expression. Nevertheless, it is undeniable that PPO has positive effects on plant defense in response to abiotic or biotic stresses as many PPO genes have been shown to be upregulated by wounding, pathogens, and hormones [63]. However, latent PPO enzymes can be activated by interactions with their substrates [64], occurring in great measure when plants are under stress. All in all, our current work provides a prospective approach for increasing crop yield in agricultural production.

4. Material and Methods

4.1. Plant Materials and Growth Conditions

C. terniflora DC. seeds sprouted in incubators and the sprouts were sown into seedbeds. The seedlings were then transplanted into pots and placed in a greenhouse, which was controlled at 28–30 °C, 70–80% relative humidity, and 160 µmol m^{-2} s^{-1} of white light irradiance. In the garden, conditions ranging from the soil to microclimate were equivalent among all plant samples. After six weeks, the plants were used for experiments [28].

4.2. High Level UVB and Dark Treatment

For HUV-B+D, 6-week-old plants were exposed to 104.4 kJ m^{-2} d^{-1} of UV-B irradiation at conditions of 25–30 °C and 80% relative humidity in a cabinet. Plants before irradiated were regarded as starting point samples. Intensity of UV-B irradiation was determined by a UV Light Meter (Beijing Normal University, Beijing, China). After irradiated by HUV-B for 5 h, plants were incubated in dark for 48 h. Leaves in the basal 10 to 60 cm of each experimental plant were collected as simples for further proteomic analyses. Five leaves were collected for each replicate and 3 independent biological replicates were assessed [28].

4.3. RNA Extraction and Cloning of CtPPO Gene

An RNeasy Plant Mini kit (Qiagen, Hilden, Germany) was used to obtain total RNA. Quantity and quality of obtained RNA were determined using an Agilent 2100 Bioanalyzer (Agilent Technologies, Palo Alto, CA, USA). The extracted total RNA was used as template to synthesize first-strand complementary DNA (cDNA) using an Oligo (dT) and OneScript^{TM} Reverse Transcriptase OneScrpt^{TM} cDNA Synthesis Kit (Applied Biological Materials Inc., Vancouver, BC, Canada). The first-strand cDNA was utilized as template for PCR using the primers listed in Table S5, which was designed based on the single EST sequence of a suspected CtPPO gene in the transcriptome of *C. terniflora* and amplification was performed as described [28]. The amplified PCR fragments, which were approximately 264 bp in length, were isolated, inserted into the pMD^{TM}18-T Vector (pMD^{TM}18-T Vector Cloning Kit, Takara, Kyoto, Japan), and sequenced. The complete coding sequence of CtPPO was obtained using 5'- and 3'-rapid amplification of cDNA ends (RACE) with internal specific primers (Table S3) using the SMARTer RACE cDNA Amplification Kit (Takara) according to the manufacturer's instructions.

The downstream sequence of CtPPO was obtained using a PCR-based genome walking approach with the Genome Walking Kit (Takara). Three designed primers (gene specific primer (GSP); Table S2) and 4 internal short degenerate arbitrarily primed (AP) primers included in the kit (AP1, AP2, AP3, and AP4) were used to perform genome-walking PCR. The PCR conditions and designed primer Tm were based on the manufacturer's instructions. Three-step nested PCR was performed to increase specificity. The nested PCR products were excised, purified, and sequenced.

4.4. Construction of CtPPO Virus-Derived Vectors

A 439-bp fragment of the obtained CtPPO gene was amplified using Permix TaqTM (Ex TaqTM Version 2.0 plus dry; Takara) and inserted into a pMD^{TM}18-T vector (pMD^{TM}18-T Vector Cloning Kit, Takara) using primers TXPPO-F and TXPPO-R (Table S4) containing XbaI and BamHI restriction enzyme sites. The CtPPO sequence was harvested from the pMD^{TM}18-T vector using the EcoRI restriction enzyme and inserted into the EcoRI site between the CaMV 35S promoter (2 × 35S) and NOS terminator of the pTRV2 vector. The orientation of the plasmid with pTRV2-CtPPO (Figure S1) was verified by sequencing. Assembly of the pTRV vector was performed as described [65]. The plasmids were sequenced to verify correct insertion of the fragment and transformed into *Agrobacterium tumefaciens* GV3101.

4.5. Virus-Induced Gene Silencing

VIGS was performed according to the method described by Salim et al. [66]. The GV3101 strains of *A. tumefaciens* carrying pTRV1, pTRV2, and pTRV2-PPO were stored at −80 °C. Agrobacteria were cultured in 300 mL of Luria-Bertani liquid medium containing 10 mM 2-(N-morpholino) ethanesulfonic acid (MES), 50 µg mL^{-1} kanamycin, and 20 µM acetosyringone. Then the cultures were centrifuged at the speed of 5000× *g* for 10 min and the resulting bacterial pellets were resuspended in 5 mL infiltration buffer containing 10 mM MgCl$_2$, 10 mM MES and 200 µM acetosyringone, and incubated at 28 °C with shaking for 3 h. The mixed suspensions were used for surface infiltration of three-week-old *Nicotiana benthamiana* seedlings using a needleless syringe and 4–6 leaves were infected. The PDS (phytoene desaturase) gene was used as a marker for the evaluation of the efficiency of VIGS in CtPPO genes silencing in *N. benthamiana*. The VIGS *N. benthamiana* plants were grown in a greenhouse and the TRV virus was detected by PCR to confirm infection of the *N. benthamiana* leaves. The successfully infected *N. benthamiana* leaves were used for the abrasion inoculation of *C. terniflora* leaves as described below.

Virus-infected *N. benthamiana* leaves (1 g) were ground with 2 mL of 10 mM potassium phosphate buffer (pH 7) containing 1–2% *w/v* Celite 545 AW abrasive (Sigma-Aldrich, Milwaukee, WI, USA) using a mortar and pestle. The resulting homogenate was used for inoculation by the abrasion method on all tender leaves of 3-week-old *C. terniflora* plants. Two groups of plants, one containing the pTRV virus vector and the other the pTRV-PPO virus vector, were prepared and each was inoculated, where six plants were used as biological replicates.

4.6. Protein Extraction, Enrichment, and Digestion for Proteomics Analysis

The tissues were ground in liquid nitrogen. For protein extraction, the phenol extraction method was performed according to a previously published protocol [67]. Briefly, 6 g of frozen powder was suspended in 15 mL of homogenization medium containing 50% (*v/v*) phenol, 100 mM KCl, 50 mM EDTA, 0.2% (*v/v*) 2-mercaptoethanol, 1 mM PMSF, 700 mM sucrose, and 500 mM Tris-HCl at pH 7.5, shaken at 37 °C for 15 min, and then centrifuged at the speed of 2500× *g* at 4 °C for 20 min. Four volumes of precipitation solution (0.1 M ammonium acetate in 100% methanol) were added in to the phenolic phase collected to precipitate the proteins. The precipitate was collected by centrifugation for 10 min at the speed of 2500× *g*. The supernatant was discarded and washed with precipitation solution. Precipitation was allowed to occur at −20 °C for 3 h and the precipitate was obtained by centrifugation, which was repeated twice. The collected proteins were dissolved in 0.5 mL of 50 mM ammonium bicarbonate (ABC). Protein concentrations were determined using the Bradford assay with bovine serum albumin as the standard.

The protein (100 µg) and an additional 50 mM ABC were transferred into a new tube to create a final volume of 100 µL. The protein was reduced with 10 mM dithiothreitol for 1 h at 55 °C and alkylated with 50 mM iodoacetamide at 37 °C for 30 min in darkness, then digested with sequence-grade modified trypsin (Promega, Fitchburg, WI, USA). Trypsin and 1% lysyl endopeptidase were added to the protein and then the solutions were incubated in a water bath at 37 °C for 4 h and another 1% enzyme was added. The solutions were incubated at 37 °C for 12 h, centrifuged at 10,000× *g*, and then dried in a vacuum. Trypsinized peptides in 200 µL of 0.1% trifluoroacetic acid (TFA) and 0.5% acetonitrile were loaded onto activated and balanced C18 SPE columns (Sep-Pak C18; Waters, Milford, MA, USA). After washing the column with 200 µL of 0.1% TFA and 0.5% acetonitrile, 300 µL of 0.1% TFA and 60 % acetonitrile were added and then the elution solution was collected and dried in a vacuum. Obtained supernatant was collected for subsequent analysis.

4.7. Nano-HPLC-MS/MS Analysis

The peptides were resuspended in 100 µL of solvent A (water with 0.1% formic acid; B: ACN with 0.1% formic acid), separated by nano-UPLC, and analyzed by online electrospray tandem MS, which were performed on a Nano Aquity UPLC system (Waters Corporation, Milford, MA, USA) connected to a quadrupole-Orbitrap mass spectrometer (Q-Exactive; Thermo Fisher Scientific, Bremen, Germany) equipped with an online nano-electrospray ion source. Peptide samples (2 μL) were loaded onto the trap column (Thermo Scientific Acclaim PepMap C18, 100 μm × 2 cm) with a flow of 10 μL/min for 3 min and subsequently separated on an analytical column (Acclaim PepMap C18, 75 μm × 25 cm) with a linear gradient of 1 to 30% solvent B (acetonitrile with 0.1% formic acid) in 95 min. The column was re-equilibrated at the initial conditions for 15 min. The column flow rate was maintained at 300 nL/min. The electrospray voltage of 2.0 kV versus the inlet of the mass spectrometer was used.

The Q-Exactive mass spectrometer was set at the data-dependent mode allowing the machine switch between MS and MS/MS acquisition automatically. Survey full-scan MS spectra (m/z 350–1600) were acquired with a mass resolution of 70 K followed by fifteen sequential high-energy collisional dissociation MS/MS scans with a resolution of 17.5 K. One microscan was recorded using a dynamic exclusion of 30 s in all cases. The MS/MS fixed first mass was set at 100.

4.8. Protein Identification Based on Mass Spectrometry Data

Tandem mass spectra were extracted by Proteome Discoverer software (Thermo Fisher Scientific, version 1.4.0.288, Bremen, Germany). Charge state deconvolution and deisotoping were not performed. All MS/MS samples were analyzed using Mascot (Matrix Science, London, UK; version 2.3). Mascot was set up to search the Uniprot database (Taxonomy: Viridiplantae, 5716577 entries, https://www.uniprot.org/) assuming the digestion enzyme trypsin. Mascot was searched with a fragment ion mass tolerance of 0.050 Da and a parent ion tolerance of 10.0 PPM. Carbamidomethyl of cysteine was specified in Mascot as fixed modifications. Oxidation of methionine was specified in Mascot as a variable modification. Minimum peptide length was set to six amino acids the false discovery rate was set to 0.01 for peptide identifications. Identified peptides shared between two proteins were combined and reported as one protein group. The minimum requirement for the identification of a protein was a minimum of 2 matched peptides and a p-value < 0.05. Spectral counting was used to estimate relative protein abundance [68]. The mass spectrometry proteomics data have been deposited to the ProteomeXchange Consortium via the PRIDE [69] partner repository with the dataset identifier PXD011439.

4.9. Functional Annotation

Protein functions were categorized using MapMan bin codes (available online: http://mapman.gabipd.org/) [70]. Annotations were transferred to the *Arabidopsis* genome with consideration for orthologous genes to predict functions of identified proteins from *C. terniflora*. Pathway mapping of identified proteins was performed using the Kyoto Encyclopedia of Genes and Genomes (KEGG) database (available online: http://www.genome.jp/kegg/) [71].

4.10. Cluster Analysis of Protein Abundance

Protein abundance ratios were used for cluster analysis, performed on Cluster 3.0 software version 3.0 (available online: http://bonsai.hgc.jp/~{}mdehoon/software/cluster/) [72].

4.11. Phylogenetic Analysis

The amino acid sequences of CtPPO were aligned using CLUSTALW (available online: http://www.clustal.org/) [73] and a neighbor-joining phylogenetic tree was computed with MEGA6 (available online: http://www.clustal.org/) using 1000 bootstrapped replicates [74].

4.12. qRT-PCR

Total RNA of leaves was extracted using an RNA isolation kit (Huayueyang, Beijing, China) and served as the template for reverse transcription using 5× All-In-One RT MasterMix with an AccuRT Genomic DNA Removal Kit (Applied Biological Materials, Richmond, BC, Canada) to obtain cDNA. Next, qRT-PCR was performed on a Bio-Rad IQ2 Multicolor Real-Time PCR Detection System (Bio-Rad, Hercules, CA, USA) with EvaGreen 2× qPCR Master Mix-iCycler (Applied Biological Materials) as the fluorescent dye. GAPDH, a housekeeping gene, was used as a standard for relative quantification of target genes. The gene-specific primers are listed in Additional file 1: Table S1. Three biological replicates were assessed and the relative expression levels were calculated using the $2^{-\Delta\Delta Ct}$ method [26].

4.13. PPO Enzymatic Activity

Enzymatic activity of PPO was assayed as described by Marko et al. [75]. Frozen leaf (0.5 g) was homogenized with 1 mL of 50 mM sodium phosphate buffer (pH 7.5) containing 1% (w/v) polyvinyl polypyrrolidone and 0.1 mM EDTA using a pre-chilled mortar and pestle. After extraction in an end-over-end shaker at 4 °C for 20 mins, the mixture was centrifuged at $16000\times g$ for 20 min at 4 °C and the resulting supernatant was used in enzyme extraction.

Then, 800 µL of 0.1 M sodium phosphate buffer (pH 6.0) containing 50 mM (+)-catechin (Carl Roth, Karlsruhe, Germany) was mixed with 150 µL of enzyme extraction sample and incubated for 15 min at 37 °C. The reaction was terminated by adding 150 µL of 6 N HCl and the absorbance at 420 nm was recorded and used in further calculations. Protein concentration was determined according to Bradford using bovine serum albumin as the reference protein. Specific PPO activity was calculated as units (U) per mg sample protein with 1 U corresponding to an absorbance change of 0.01/min. Since levels of PPO activity in the control plants were somewhat variable between independent experiments, data were normalized to the specific PPO activity in leaves of healthy wild-type controls and the results are shown as fold induction (with 1 corresponding to a specific activity of 3.2 to 23 U/mg).

4.14. Analysis of Photosynthesis Characteristics

An Li-6400 Portable Photosynthesis System (Li-Cor, Lincoln, NE, USA) equipped with a red/blue LED light source were used to measure the photosynthesis characteristics of the leaves, and was operated using air from a large volume with a stable CO_2 partial pressure. All measurements were carried out at a photon flux density of 500 µmol m^{-2} s^{-1} at 28 °C and made after stable reading was achieved. The photosynthesis rate (µmol m^{-2} s^{-1}), stomatal conductance (mol m^{-2} s^{-1}), and transpiration rate (mmol m^{-2} s^{-1}) of each leaf were recorded. Each experiment was repeated for different plants three times and three independent biological replicates were assessed.

4.15. Analysis of Chlorophyll Content

C. terniflora leaves (0.2 g) were cut up and blended with 10 mL of 95% alcohol in a tube. The tube was covered and the leaves soaked in the dark until they had turned completely white. The absorbances of the extracting solutions were measured at wavelengths of 665, 649, and 470 nm separately. The chlorophyll content was calculated using Chlorophyll a (Ca) = 13.95 D_{665} − 6.88 D_{649}, Chlorophyll b (Cb) = 24.96 D_{649} − 7.32 D_{665}, Carotenoid = (1000 D_{470} − 2.05 Ca − 114 Cb)/245, and Total chlorophyll = Ca + Cb.

4.16. Statistical Analysis

Statistical significance comparing two groups was determined using Student's *t*-test and multiple groups using one-way ANOVA followed by Tukey's test. Statistical evaluation was performed on SPSS statistical software version 22.0 (IBM, Armonk, NY, USA). A *p*-value <0.05 was considered statistically significant.

4.17. Accession Codes

The nucleotide sequence of CtPPO has been deposited in the GenBank™ database under the accession number MK070494.

5. Conclusions

PPO is an oxidoreductase that plays a crucial role in the biosynthesis of secondary metabolites. *C. terniflora* is a Chinese folk medical resource with potential pharmaceutical value for the treatment of various diseases [21]. High-intensity UV-B and dark incubation (HUV-B+D), which are helpful stressors for medicinal plants [28]. Omics technologies were used on *C. terniflora* to prospectively understand the inherent mechanism underlying its medicinal quality [24,26]. To clarify the in vivo physiological functions of PPO, we performed a comparative proteomic analysis on the leaves of *C. terniflora* with down-regulated PPO activity by VIGS. The main findings are as follows: (i) The differentially expressed proteins were mainly functioned in photosynthesis, glycolysis and redox in the VC plants; (ii) the differentially expressed proteins related to photosynthesis were mainly involved in light reaction and Calvin cycle; (iii) the expression level of adenosine triphosphate (ATP) synthase, the content of chlorophyll, and the photosynthesis rate were increased in VC plants compared to VV plants pre- and post HUV-B+D. Taken together, these results indicate that the silence of PPO can activate the glycolysis process, regulate Calvin cycle and provide ATP for energy metabolism to elevate the plant photosynthesis. And this study provides a prospective approach for increasing crop yield in agricultural production.

Supplementary Materials: The following are available online at http://www.mdpi.com/1422-0067/19/12/3897/s1.

Author Contributions: Conceptualization, X.C., B.Y. and J.T.; methodology, X.C. and B.Y.; software, W.H. and T.W.; validation, L.Y. and S.L.; formal analysis, W.H.; investigation, X.C. and Y.L.; resources, B.Y.; data curation, T.W.; writing—original draft preparation, X.C. and B.Y.; writing—review and editing, X.C., B.Y. and Z.Z.; visualization, Y.L.; supervision, B.Y.; project administration, X.C.; funding acquisition, B.Y and J.T.

Funding: This research was funded by National Science and Technology Major Project of China, grant number No. 2017ZX09301007; National Natural Science Foundation of China, grant number No. 81703634; and Zhejiang Provincial Science and Technology Planning Project, grant number No. 2016C04005.

Conflicts of Interest: The authors declare no conflict of interest.

Abbreviations

3-PGA	3-Phosphoglycerate
ABA	Abscisic acid
ABC	Ammonium bicarbonate
ACN	Acetonitrile
ATP	Adenosine triphosphate
BPGM	Phosphoglycerate mutase
C. terniflora	*Clematis terniflora*
CtPPO	PPO gene in *C. terniflora*
DEPs	Differentially expressed proteins
ETC	Electron transport chain
FBA	Fructose-1,6-bisphosphate aldolase
FBPase	Fructose-1,6-bisphosphatase
Fd	Ferredoxin
FTR	Fd-TRX reductase
GAP	Glyceraldehyde 3-phosphate
GAPDH	Glyceraldehyde-3-phosphate dehydrogenase
GPI	Glucose-6-phosphate isomerase
GPUT	UTP-glucose-1-phosphate uridylytransferase
HCD	High energy collisional dissociation
HUV-B+D	High level of UV-B and dark
KEGG	Kyoto Encyclopedia of Genes and Genomes
MES	2-(N-morpholino) ethanesulfonic acid
NTRC	NADPH-dependent TRX reductase

Int. J. Mol. Sci. **2018**, *19*, 3897

ORF	Open reading frame
PDS	Phytoene desaturase
PEP	Phosphoenolpyruvate
PEPC	Phosphoenolpyruvate carboxylase
PGK	Phosphoglycerate kinase
PPO	Polyphenol oxidase
PRK	Phosphoribulokinase
PS I	Photosystem I
PS II	Photosystem II
PS	Photosystem
RACE	5′ Rapid amplification of cDNA ends
Rubisco	Ribulose-1,5-bisphosphate carboxylase/oxygenase
RuBP	Ribulose 1,5-bisphosphate
SBPase	Sedoheptulose-1,7-bisphosphatase
TCA	Tricarboxylic acid
TFA	Trifluoroacetic acid
TK	Transketolase
TRV	Tobacco rattle virus
TRX	Thioredoxin
UTR	Untranslated Region
VIGS	Virus-induced gene silencing
VC	VIGS-CtPPO
VV	VIGS-vector

References

1. Hart, E.H.; Onime, L.A.; Davies, T.E.; Morphew, R.M.; Kingston-Smith, A.H. The effects of PPO activity on the proteome of ingested red clover and implications for improving the nutrition of grazing cattle. *J. Proteom.* **2016**, *141*, 67–76. [CrossRef]
2. Kroll, J.; Rawel, H.M. Reactions of Plant Phenols with Myoglobin: Influence of Chemical Structure of the Phenolic Compounds. *J. Food Sci.* **2010**, *66*, 48–58. [CrossRef]
3. Nakayama, T.; Yonekura-Sakakibara, K.; Sato, T.; Kikuchi, S.; Fukui, Y.; Fukuchi-Mizutani, M.; Ueda, T.; Nakao, M.; Tanaka, Y.; Kusumi, T.; et al. Aureusidin synthase: A polyphenol oxidase homolog responsible for flower coloration. *Science* **2000**, *290*, 1163–1166. [CrossRef] [PubMed]
4. Gandia-Herrero, F.; Escribano, J.; Garcia-Carmona, F. Characterization of the monophenolase activity of tyrosinase on betaxanthins: The tyramine-betaxanthin/dopamine-betaxanthin pair. *Planta* **2005**, *222*, 307–318. [CrossRef] [PubMed]
5. Li, L.; Steffens, J.C. Overexpression of polyphenol oxidase in transgenic tomato plants results in enhanced bacterial disease resistance. *Planta* **2002**, *215*, 239–247. [CrossRef]
6. Thipyapong, P.; Mahanil, S.; Bhonwong, A.; Attajarusit, J.; Stout, M.J.; Steffens, J.C. Increasing Resistance of Tomato to Lepidopteran Insects by Overexpression of Polyphenol Oxidase. *Acta Hortic.* **2006**, *22*, 29–38. [CrossRef]
7. Jia, H.; Zhao, P.; Wang, B.; Tariq, P.; Zhao, F.; Zhao, M.; Wang, Q.; Yang, T.; Fang, J. Overexpression of Polyphenol Oxidase Gene in Strawberry Fruit Delays the Fungus Infection Process. *Plant Mol. Biol. Report.* **2016**, *34*, 592–606. [CrossRef]
8. Sofo, A.; Dichio, B.; Xiloyannis, C.; Masia, A. Antioxidant defences in olive trees during drought stress: Changes in activity of some antioxidant enzymes. *Funct. Plant Biol.* **2005**, *32*, 351–364. [CrossRef]
9. Araji, S.; Grammer, T.A.; Gertzen, R.; Anderson, S.D.; Mikulic-Petkovsek, M.; Veberic, R.; Phu, M.L.; Solar, A.; Leslie, C.A.; Dandekar, A.M.; et al. Novel roles for the polyphenol oxidase enzyme in secondary metabolism and the regulation of cell death in walnut. *Plant Physiol.* **2014**, *164*, 1191–1203. [CrossRef]
10. Murniati, A.; Buchari, B.; Gandasasmita, S.; Nurachman, Z.; Nurhanifah, N. Characterization of Polyphenol Oxidase Application as Phenol Removal in Extracts of Rejected White Oyster Mushrooms (*Pleurotus ostreatus*). *Orient. J. Chem.* **2018**, *34*, 1457–1468. [CrossRef]

11. Sommer, A.; Ne'Eman, E.; Koussevitzky, S.; Hunt, M.D.; Steffens, J.C.; Mayer, A.M.; Harel, E. The Inhibition by Cu2+ of the Import of Polyphenol Oxidase into Chloroplasts. *Photosynth. Light Biosph.* **1995**, *3*, 827–830.
12. Koussevitzky, S.; Ne'Eman, E.; Harel, E. Import of polyphenol oxidase by chloroplasts is enhanced by methyl jasmonate. *Planta* **2004**, *219*, 412–419. [CrossRef] [PubMed]
13. Mayer, A.M. Polyphenol oxidases in plants and fungi: Going places? A review. *Phytochemistry* **2006**, *67*, 2318–2331. [CrossRef] [PubMed]
14. Velásquez, A.C.; Chakravarthy, S.; Martin, G.B. Virus-induced gene silencing (VIGS) in *Nicotiana benthamiana* and tomato. *J. Vis. Exp.* **2009**, *28*, 1292.
15. Lee, W.S.; Rudd, J.J.; Kanyuka, K. Virus induced gene silencing (VIGS) for functional analysis of wheat genes involved in *Zymoseptoria tritici* susceptibility and resistance. *Fungal Genet. Biol.* **2015**, *79*, 84–88. [CrossRef]
16. Zhao, D.; Zhao, J.R.; Huang, X.; Li, N.; Liu, Y.; Huang, Z.J.; Zhang, Z.Y. Functional Analysis of TNBL1 Gene in Wheat Defense Response to Barley yellow dwarf virus Using BSMV-VIGS Technique. *Acta Agron. Sin.* **2011**, *37*, 2106–2110. [CrossRef]
17. Groszyk, J.; Kowalczyk, M.; Yanushevska, Y.; Stochmal, A.; Rakoczy-Trojanowska, M.; Orczyk, W. Identification and VIGS-based characterization of Bx1 ortholog in rye (*Secale cereale* L.). *PLoS ONE* **2017**, *12*, e0171506. [CrossRef] [PubMed]
18. Gao, W.; Long, L.; Zhu, L.F.; Rong, Y.M.; Lu, X.H.; Sun, M.Y.; Zhang, L.; Tian, J.K. Proteomic and Virus-induced Gene Silencing (VIGS) Analyses Reveal That Gossypol, Brassinosteroids, and Jasmonic acid Contribute to the Resistance of Cotton to *Verticillium dahliae*. *Mol. Cell. Proteom.* **2013**, *12*, 3690–3703. [CrossRef] [PubMed]
19. Zhao, X.D.; Yuan, X.Y.; Chen, S.; Meng, L.; Fu, D. Role of the tomato TAGL1 gene in regulating fruit metabolites elucidated using RNA sequence and metabolomics analyses. *PLoS ONE* **2018**, *13*, e0199083. [CrossRef] [PubMed]
20. Zhou, B.J.; Zeng, L. Elucidating the role of highly homologous Nicotiana benthamiana ubiquitin E2 gene family members in plant immunity through an improved virus-induced gene silencing approach. *Plant Methods* **2017**, *13*, 59. [CrossRef]
21. Chen, R.Z.; Cui, L.; Guo, Y.J.; Rong, Y.M.; Lu, X.H.; Sun, M.Y.; Zhang, L.; Tian, J.K. In vivo study of four preparative extracts of *Clematis terniflora*, DC. for antinociceptive activity and anti-inflammatory activity in rat model of carrageenan-induced chronic non-bacterial prostatitis. *J. Ethnopharmacol.* **2011**, *134*, 1018–1023. [CrossRef] [PubMed]
22. Zhang, L.J.; Huang, H.T.; Huang, S.Y.; Lin, Z.H.; Shen, C.C.; Tsai, W.J.; Kuo, Y.H. Antioxidant and Anti-Inflammatory Phenolic Glycosides from *Clematis tashiroi*. *J. Nat. Prod.* **2015**, *78*, 1586–1592. [CrossRef] [PubMed]
23. Wang, L.L.; Huang, Y.H.; Yan, C.Y.; Wei, X.D.; Hou, J.Q.; Pu, J.X.; Lv, J.X. N-acetylcysteine Ameliorates Prostatitis via miR-141 Regulating Keap1/Nrf2 Signaling. *Inflammation* **2016**, *39*, 938–947. [CrossRef] [PubMed]
24. Yang, B.X.; Wang, X.; Gao, C.X.; Chen, M.; Guan, Q.J.; Tian, J.K.; Komatsu, S. Proteomic and Metabolomic Analyses of Leaf from *Clematis terniflora* DC. Exposed to High-Level Ultraviolet-B Irradiation with Dark Treatment. *J. Proteome Res.* **2015**, *15*, 2643–2657. [CrossRef] [PubMed]
25. Liu, X.B.; Yang, B.X.; Zhang, L.; Lu, Y.Z.; Gong, M.H.; Tian, J.K. An in vivo, and in vitro, assessment of the anti-inflammatory, antinociceptive, and immunomodulatory activities of *Clematis terniflora*, DC. extract, participation of aurantiamide acetate. *J. Ethnopharmacol.* **2015**, *169*, 287–294. [CrossRef] [PubMed]
26. Gao, C.X.; Yang, B.X.; Zhang, D.D.; Chen, M.; Tian, J.K. Enhanced metabolic process to indole alkaloids in *Clematis terniflora*, DC. after exposure to high level of UV-B irradiation followed by the dark. *BMC Plant Biol.* **2016**, *16*, 231. [CrossRef]
27. Boeckx, T.; Winters, A.L.; Webb, K.J.; Kingston-Smith, A.H. Polyphenol oxidase in leaves: Is there any significance to the chloroplastic localization? *J. Exp. Bot.* **2015**, *66*, 3571–3579. [CrossRef]
28. Yang, B.X.; Guan, Q.J.; Tian, J.K.; Komatsu, S. Transcriptomic and proteomic analyses of leaves from *Clematis terniflora*, DC. under high level of ultraviolet-B irradiation followed by dark treatment. *J. Proteom.* **2016**, *150*, 323–340. [CrossRef]
29. Joliot, P.; Joliot, A. Cyclic electron flow in C3 plants. *BBA Bioenerg.* **2015**, *1757*, 362–368. [CrossRef]
30. Arnon, D.I.; Whatley, F.R.; Allen, M.B. Vitamin K as a cofactor of photosynthetic phosphorylation. *BBA Biochim. Biophys. Acta* **1955**, *16*, 607–608. [CrossRef]

31. Grondelle, R.V.; Zuber, H. The light reactions of photosynthesis. *Proc. Natl. Acad. Sci. USA* **1971**, *68*, 2883–2892. [CrossRef]

32. Karlusich, J.J.P.; Carrillo, N. Evolution of the acceptor side of photosystem I: Ferredoxin, flavodoxin, and ferredoxin-NADP +, oxidoreductase. *Photosynth. Res.* **2017**, *134*, 235–250. [CrossRef] [PubMed]

33. Aliverti, A.; Pandini, V.; Pennati, A.; Rosa, M.; Zanetti, G. Structural and functional diversity of ferredoxin-NADP(+) reductases. *Arch. Biochem. Biophys.* **2008**, *474*, 283–291. [CrossRef] [PubMed]

34. Hisabori, T.; Sunamura, E.I.; Kim, Y.; Konno, H. The Chloroplast ATP Synthase Features the Characteristic Redox Regulation Machinery. *Antioxid. Redox Signal.* **2013**, *19*, 1846–1854. [CrossRef] [PubMed]

35. Kohzuma, K.; Bosco, C.D.; Meurer, J.; Kramer, D.M. Light- and Metabolism-related Regulation of the Chloroplast ATP Synthase Has Distinct Mechanisms and Functions. *J. Biol. Chem.* **2013**, *288*, 13156–13163. [CrossRef] [PubMed]

36. Tezara, W.; Mitchell, V.J.; Driscoll, S.D.; Lawlor, D.W. Water stress inhibits plant photosynthesis by decreasing coupling factor and ATP. *Nature* **1999**, *401*, 914–917. [CrossRef]

37. Shu, S.; Chen, L.; Lu, W.; Sun, J.; Guo, S.; Yuan, Y.; Li, J. Effects of exogenous spermidine on photosynthetic capacity and expression of Calvin cycle genes in salt-stressed cucumber seedlings. *J. Plant Res.* **2014**, *127*, 763–773. [CrossRef]

38. Hanke, G.; Mulo, P. Plant type ferredoxins and ferredoxin-dependent metabolism. *Plant Cell Environ.* **2013**, *36*, 1071–1084. [CrossRef]

39. Brandes, H.K.; Hartman, F.C.; Lu, T.Y.S.; Larimer, F.W. Efficient Expression of the Gene for Spinach Phosphoribulokinase in Pichia pastoris and Utilization of the Recombinant Enzyme to Explore the Role of Regulatory Cysteinyl Residues by Site-directed Mutagenesis. *J. Biol. Chem.* **1996**, *271*, 6490–6496. [CrossRef]

40. Nikkanen, L.; Rintamäki, E. Thioredoxin-dependent regulatory networks in chloroplasts under fluctuating light conditions. *Philos. Trans. R Soc. Lond.* **2014**, *369*, 20130224. [CrossRef]

41. Thormählen, I.; Meitzel, T.; Groysman, J.; Öchsner, A.B.; von Roepenach-Lahaye, E.; Naranjo, B.; Cejudo, F.J.; Geigenberger, P. Thioredoxin f1 and NADPH-Dependent Thioredoxin Reductase C Have Overlapping Functions in Regulating Photosynthetic Metabolism and Plant Growth in Response to Varying Light Conditions. *Plant Physiol.* **2015**, *169*, 1766–1786. [CrossRef] [PubMed]

42. Sommer, A.; Ne'Eman, E.; Steffens, J.C.; Mayer, A.M.; Harel, E. Import, targeting, and processing of a plant polyphenol oxidase. *Plant Physiol.* **1994**, *105*, 1301–1311. [CrossRef] [PubMed]

43. Qi, J.; Li, G.Q.; Dong, Z.; Zhou, W. Transformation of tobacco plants by Yali PPO-GFP fusion gene and observation of subcellular localization. *Am. J. Transl. Res.* **2016**, *8*, 698. [PubMed]

44. Richter, A.S.; Peter, E.; Rothbart, M.; Schlicke, H.; Toivola, J.; Rintamäki, E.; Grimm, B. Posttranslational influence of NADPH-dependent thioredoxin reductase C on enzymes in tetrapyrrole synthesis. *Plant Physiol.* **2013**, *162*, 63–73. [CrossRef] [PubMed]

45. Lepisto, A.; Kangasjarvi, S.; Luomala, E.M.; Brader, G.; Sipari, N.; Keränen, M.; Keinänen, M.; Rintamäki, E. Chloroplast NADPH-Thioredoxin Reductase Interacts with Photoperiodic Development in Arabidopsis. *Plant Physiol.* **2009**, *149*, 1261–1276. [CrossRef] [PubMed]

46. Nikkanen, L.; Toivola, J.; Rintamäki, E. Crosstalk between chloroplast thioredoxin systems in regulation of photosynthesis. *Plant Cell Environ.* **2016**, *39*, 1691–1705. [CrossRef] [PubMed]

47. Lunt, S.Y.; Vander Heiden, M.G. Aerobic glycolysis: Meeting the metabolic requirements of cell proliferation. *Annu. Rev. Cell Dev. Biol.* **2011**, *27*, 441–464. [CrossRef]

48. Rosa-Téllez, S.; Anoman, A.D.; Flores-Tornero, M.; Toujani, W.; Alseek, S.; Femie, A.R.; Nebauer, S.G.; Muñoz-Bertomeu, J.; Segura, J.; Ros, R. Phosphoglycerate Kinases Are Co-Regulated to Adjust Metabolism and to Optimize Growth. *Plant Physiol.* **2017**, *176*, 1182–1198. [CrossRef]

49. Zhao, Z.; Assmann, S.M. The glycolytic enzyme, phosphoglycerate mutase, has critical roles in stomatal movement, vegetative growth, and pollen production in *Arabidopsis thaliana*. *J. Exp. Bot.* **2011**, *62*, 5179–5189. [CrossRef]

50. Lawson, T.; Blatt, M.R. Stomatal size, speed, and responsiveness impact on photosynthesis and water use efficiency. *Plant Physiol.* **2014**, *164*, 1556–1570. [CrossRef]

51. Bai, J.; Liu, J.; Jiao, W.; Sa, R.; Zhang, N.; Jia, R. Proteomic analysis of salt-responsive proteins in oat roots (*Avena sativa* L.). *J. Sci. Food Agric.* **2016**, *96*, 3867–3875. [CrossRef] [PubMed]

52. Azoulay-Shemer, T.; Palomares, A.; Bagheri, A.; Israelsson-Nordstrom, M.; Engineer, C.B.; Bargmann, B.O.; Stephan, A.B.; Schroeder, J.I. Guard cell photosynthesis is critical for stomatal turgor production, yet does not directly mediate CO_2- and ABA-induced stomatal closing. *Plant J.* **2015**, *83*, 567–581. [CrossRef]

53. Guo, L.; Devaiah, S.P.; Narasimhan, R.; Pan, X.; Zhang, Y.; Zhang, W.; Wang, X. Cytosolic glyceraldehyde-3-phosphate dehydrogenases interact with phospholipase Dδ to transduce hydrogen peroxide signals in the *Arabidopsis* response to stress. *Plant Cell* **2012**, *24*, 2200–2212. [CrossRef] [PubMed]

54. Yasmeen, F.; Raja, N.I.; Razzaq, A.; Komatsu, S. Proteomic and physiological analyses of wheat seeds exposed to copper and iron nanoparticles. *BBA Proteins Proteom.* **2017**, *1865*, 28–42. [CrossRef] [PubMed]

55. Feillet, P.; Autran, J.C.; Icard-Vernière, C. Pasta brownness: An assessment. *J. Cereal Sci.* **2000**, *32*, 215–233. [CrossRef]

56. Mesquita, V.L.V.; Queiroz, C. *Enzymatic Browning Biochemistry of Foods*; Elsevier Inc.: Amsterdam, The Netherlands, 2013.

57. Simeone, R.; Pasqualone, A.; Clodoveo, M.L.; Blanco, A. Genetic mapping of polyphenol oxidase in tetraploid wheat. *Cell. Mol. Biol. Lett.* **2002**, *7*, 763–769. [PubMed]

58. Sun, J.Z.; Zhao, J.T.; Liu, D.C.; Yang, W.L.; Luo, G.B.; Zhang, L.Y.; Zhang, X.Q.; Zhang, A.M. Modification to the Test Method of Polyphenol Oxidase(PPO) Activity in Wheat Seeds and Its Usage in Breeding Programs. *J. Triticeae Crops* **2012**, *3*, 13.

59. Long, S.P.; Zhu, X.; Naidu, S.L.; Ort, D.R. Can improvement in photosynthesis increase crop yields? *Plant Cell Environ.* **2006**, *29*, 315–330. [CrossRef]

60. Parry, M.A.; Reynolds, M.; Salvucci, M.E.; Raines, C.; Andralojc, P.J.; Zhu, X.G.; Price, G.D.; Condon, A.G.; Furbank, R.T. Raising yield potential of wheat. II. Increasing photosynthetic capacity and efficiency. *J. Exp. Bot.* **2011**, *62*, 453–467. [CrossRef]

61. Lin, M.T.; Occhialini, A.; Andralojc, P.J.; Devonshire, J.; Hines, K.M.; Parry, M.A.; Hanson, M.R. β-Carboxysomal proteins assemble into highly organized structures in *Nicotiana* chloroplasts. *Plant J. Cell Mol. Biol.* **2014**, *79*, 1–12. [CrossRef] [PubMed]

62. Kromdijk, J.; Głowacka, K.; Leonelli, L.; Gabilly, S.T.; Iwai, M.; Niyogi, K.K.; Long, S.P. Improving photosynthesis and crop productivity by accelerating recovery from photoprotection. *Science* **2016**, *354*, 857–861. [CrossRef] [PubMed]

63. Constabel, C.P.; Ryan, C.A. A survey of wound- and methyl jasmonate-induced leaf polyphenol oxidase in crop plants. *Phytochemistry* **1998**, *47*, 507–511. [CrossRef]

64. Winters, A.L.; Minchin, F.R.; Michaelson-Yeates, T.P.; Lee, M.R.; Morris, P. Latent and active polyphenol oxidase (PPO) in red clover (*Trifolium pratense*) and use of a low PPO mutant to study the role of PPO in proteolysis reduction. *J. Agric. Food Chem.* **2008**, *56*, 2817–2824. [CrossRef] [PubMed]

65. Becker, A. Virus-Induced Gene Silencing. *Methods Mol. Biol.* **2011**, *236*, 287–294.

66. Salim, V.; Yu, F.; Altarejos, J.; Luca, V.D. Virus-induced gene silencing identifies *Catharanthus roseus* 7-deoxyloganic acid-7-hydroxylase, a step in iridoid and monoterpene indole alkaloid biosynthesis. *Plant J.* **2013**, *76*, 754–765. [CrossRef] [PubMed]

67. Isaacson, T.; Damasceno, C.M.B.; Saravanan, R.S.; He, Y.; Catalá, C.; Saladié, M.; Rose, J.K. Sample extraction techniques for enhanced proteomic analysis of plant tissues. *Nat. Protoc.* **2006**, *1*, 769. [CrossRef] [PubMed]

68. Carvalho, P.C.; Fischer, J.S.G.; Xu, T.; Yates, J.R.; Barbosa, V.C. PatternLab: From mass spectra to label-free differential shotgun proteomics. *Curr. Protoc. Bioinform.* **2012**, *13*, 13–19.

69. Vizcaíno, J.A.; Csordas, A.; del-Toro, N.; Dianes, J.A.; Lavidas, I.; Mayer, G.; Perez-Riverol, Y.; Reisinger, F.; Ternent, T.; Xu, Q.W.; et al. 2016 update of the PRIDE database and related tools. *Nucleic Acids Res.* **2016**, *44*, D447–D456. [PubMed]

70. Usadel, B.; Nagel, A.; Thimm, O.; Redestig, H.; Blaesing, O.E.; Palacios-Rojas, N.; Selbig, J.; Hannemann, J.; Piques, M.C.; Steinhauser, D.; et al. Extension of the visualization tool MapMan to allow statistical analysis of arrays, display of corresponding genes, and comparison with known responses. *Plant Physiol.* **2005**, *138*, 1195–1204. [CrossRef]

71. Kanehisa, M.; Goto, S. KEGG: Kyoto Encyclopaedia of Genes and Genomes. *Nucleic Acids Res.* **2000**, *28*, 27–30. [CrossRef]

72. Hoon, M.J.L.D.; Imoto, S.; Nolan, J.; Miyano, S. Open source clustering software. *Bioinformatics* **2004**, *20*, 1453. [CrossRef] [PubMed]

73. Larkin, M.A.; Blackshields, G.; Brown, N.P.; Chenna, R.; Mcgettigan, P.A.; Mcwilliam, H.; Valentin, F.; Wallace, I.M.; Wilm, A.; Lopez, R. Clustal W and Clustal X version 2.0. *Bioinformatics* **2007**, *23*, 2947–2948. [CrossRef] [PubMed]
74. Kumar, S.; Nei, M.; Dudley, J.; Tamura, K. MEGA: A biologist-centric software for evolutionary analysis of DNA and protein sequences. *Brief. Bioinform.* **2008**, *9*, 299–306. [CrossRef] [PubMed]
75. Bosch, M.; Berger, S.; Schaller, A.; Stintzi, A. Jasmonate-dependent induction of polyphenol oxidase activity in tomato foliage is important for defense against *Spodoptera exiguabut* not against *Manduca sexta*. *BMC Plant Biol.* **2014**, *14*, 257. [CrossRef] [PubMed]

International Journal of
Molecular Sciences

MDPI

Article

Protein Phosphatase (*PP2C9*) Induces Protein Expression Differentially to Mediate Nitrogen Utilization Efficiency in Rice under Nitrogen-Deficient Condition

Muhammad Waqas [1,2,*], **Shizhong Feng** [3], **Hira Amjad** [2,3], **Puleng Letuma** [2,3], **Wenshan Zhan** [3], **Zhong Li** [1,2], **Changxun Fang** [1,2,3], **Yasir Arafat** [2,3], **Muhammad Umar Khan** [2,3], **Muhammad Tayyab** [1,2] and **Wenxiong Lin** [1,2,3,*]

1   Key Laboratory for Genetics, Breeding and Multiple Utilization of Crops, Ministry of Education/College of Crop Sciences, Fujian Agriculture and Forestry University, Fuzhou 350002, China; lizhong021@126.com (Z.L.); fcx007@fafu.edu.cn (C.F.); tyb.pk@hotmail.com (M.T.)
2   Fujian Provincial Key Laboratory of Agroecological Processing and Safety Monitoring, College of Life Sciences, Fujian Agriculture and Forestry University, Fuzhou 350002, China; heraamahmood@yahoo.com (H.A.); pulengletuma@yahoo.com (P.L.); arafat_pep@yahoo.com (Y.A.); umar.khan018@yahoo.com (M.U.K.)
3   Key Laboratory of Crop Ecology and Molecular Physiology (Fujian Agriculture and Forestry University), Fujian Province University, Fuzhou 350002, China; M18120821563@163.com (S.F.); 1170525027@fafu.edu.cn (W.Z.)
*   Correspondence: waqasjutt_19@yahoo.com (M.W.); lwx@fafu.edu.cn (W.L.)

Received: 18 July 2018; Accepted: 10 September 2018; Published: 19 September 2018

Abstract: Nitrogen (N) is an essential element usually limiting in plant growth and a basic factor for increasing the input cost in agriculture. To ensure the food security and environmental sustainability it is urgently required to manage the N fertilizer. The identification or development of genotypes with high nitrogen utilization efficiency (NUE) which can grow efficiently and sustain yield in low N conditions is a possible solution. In this study, two isogenic rice genotypes i.e., wild-type rice kitaake and its transgenic line *PP2C9TL* overexpressed protein phosphatase gene (*PP2C9*) were used for comparative proteomics analysis at control and low level of N to identify specific proteins and encoding genes related to high NUE. 2D gel electrophoresis was used to perform the differential proteome analysis. In the leaf proteome, 30 protein spots were differentially expressed between the two isogenic lines under low N level which were involved in the process of energy, photosynthesis, N metabolism, signaling, and defense mechanisms. In addition, we have found that protein phosphatase enhances nitrate reductase activation by downregulation of SnRK1 and 14-3-3 proteins. Furthermore, we showed that *PP2C9TL* exhibits higher NUE than *WT* due to higher activity of nitrate reductase. This study provides new insights on the rice proteome which would be useful in the development of new strategies to increase NUE in cereal crops.

Keywords: N utilization efficiency; proteomics; 2D; protein phosphatase; rice isogenic line; SnRK1; 14-3-3

1. Introduction

The present capacity to provide food for the increasing global population is due to the green revolution, which is based on the adoption of semidwarf cereals with high yield. However, to achieve an increase in crop production depends on the application of nitrogen (N) fertilizers [1–3]. Application of N fertilizer has become a key factor to improve the crop productivity. Unfortunately, the extensive use of N fertilizers is causing harm to the soil as well as water bodies. Nitrogen leached from agricultural lands, in the form of nitrate causing eutrophication in rivers, lakes, and oceans, is decreasing aquatic diversity and damaging drinking water [4]. Therefore, it is urgently required to limit or reduce the application of fertilizers without affecting crop production [5,6]. Nitrogen (N) is an essential plant macronutrient as it is the fundamental element of plant components such as nucleic acids, amino acids, proteins, enzymes, chlorophyll, and various hormones. Availability of N plays a vital role in plant growth, senescence, flowering time, photosynthesis, and translocation of photosynthates [7]. Worldwide consumption of N-based fertilizers is approximately 119.40 million tons with an annual growth of 1.4% [8]. Asia uses 62.1% of the total nitrogenous fertilizers and China alone shares 18% of the Asian N consumption [8]. However, it is alarming that major cereal crops like wheat, rice, and maize only utilize 30–40% of the applied N. The remaining unutilized 60–70% of applied N causes severe health and environmental risks [9].

Improving the N use efficiency (NUE) in crops would be helpful to increase crop production without any penalty to the environment. It is estimated that 1.1 billion USD can be saved by increasing 1% NUE and can also help reduce environmental pollution [10]. Several approaches from agronomic methods to transgenic efforts have been tried to solve this issue such as the split application of N, use of nitrification inhibitors, and the slow release of fertilizers [11,12]. Conventional procedures such as selective breeding improve grain yield heritability [13]. However, it cannot explain the genetic basis for the improvements of complex quantitative traits like NUE [14]. Scientists have developed gene-overexpressed mutants to increase the biomass and plant N contents attempting to enhance NUE in crop production [15–17]. The overexpression of high-affinity ammonium transport system (HATS) like NRT2.1 increases the nitrate influx, but there was no improvement in the phenotypic NUE [18,19]. However, the overexpression of nitrite reductase (NiR) and nitrate reductase (NR) in tobacco and Arabidopsis decreases nitrate levels in plant tissues but fail to improve the biomass or grain yield [12,20]. The ectopic expression of cytoplasmic glutamine synthetase1 (GS1) [21] and glutamine synthetase 2 (GS2) [22,23] have some positive effects on plant biomass and grain yield. Studies suggest that targeted manipulation of one gene or just one component of the N signaling pathway may not be enough to significantly improve overall NUE because it is a quantitative trait controlled by many genes and interacting regulatory pathways that are linked with the actual NUE of a crop plant [24].

Many scientists have tried to exploit the regulatory pathways involved in the transport of nitrate in plant organs. The reversible proteins phosphorylation by protein phosphatases is an essential mechanism to regulate the different biological processes. In plants, PP2Cs is a major phosphatase-encoding gene family that has emerged as a key regulator of stress signaling [25,26]. Protein phosphatases (PP2Cs) negatively regulate abscisic acid (ABA) signaling. In Arabidopsis, PP2C proteins such as ABA-insensitive 1 (ABI1), ABI2, and Hypersensitive to ABA 1 (HAB1) have been identified to regulate ABA-induced signaling under biotic and abiotic stresses by interacting with SnRK2s and PYR/PYL/RCARs [27–31]. The activity of nitrate transporters (NPF6.3) is regulated by CBL9 (Calcineurin B like protein 9) and CIPK23 complex (CBL interacting protein kinase 23), CBL9 phosphorylates the CIPK23 activating the protein complex. The activated CIPK23 inhibits the activity of NPF6.3 [32]. However, protein phosphatase (PP2C9s) enhances the NPF6.3-dependent nitrate sensing by dephosphorylating the CIPK23 and CBL9 complex, nitrate signaling, and nitrate transport [33]. Recently, scientists have started to focus on the critical role of protein phosphates in NUE. In our preliminary study, we have found that the protein phosphatase (PP2C9) was closely related with the improvement of the NUE in rice by enhancing N uptake and assimilation, however, the appropriate information still lacks, especially regarding the PP2C9 regulatory mechanism for high NUE in rice plants.

In order to exploit the regulatory role of protein phosphatase (PP2C9) regarding NUE, we used the transgenic japonica rice line overexpressing the protein phosphatase (PP2C9) for differential proteomics analysis to identify the genes and signaling pathways mediated by PP2C9 under nitrogen limiting conditions. Although the plant organs contain the same complement of the genome, the expression of genes and proteins accumulation varies widely. Proteomic studies can overcome the limitations of post-translational modifications that occur during DNA/RNA transcription and expression processes, and provide proper information about plant biological functions at a particular time course. Rice (*Oryza sativa* L.) a staple food for half of the population worldwide and of immense agricultural importance, has been a popular research subject for agriculture scientists [34,35]. In China, by 2030, rice demand will have increased up to 14%, to fulfill the increasing demand farmers are extensively using N fertilizers [36,37]. The amount of N fertilizer used (209 kg ha^{-1}) for rice production in China is 90% more than the global use [38], and this makes China the world leading N fertilizer consumer with low N utilization efficiency [39,40] and high environmental risk. Considering all these issues, we have developed isogenic lines including the transgenic rice line overexpressing protein phosphatase (*PP2C9TL*) and its wild-type (kitaake); we used these materials to investigate the underlying mechanism associated with NUE in rice exposed to a limited nitrogen supply condition through differential proteomics. We have identified that the protein phosphatase (PP2C9) gene, which functions to significantly improve the NUE by enhancing N uptake and assimilation by regulating nitrate reductase activation via dephosphorylation of SnRK1 and 14-3-3 proteins.

2. Results

2.1. Physiological Performance of PP2C9TL and WT

The present study shows that the overexpression of protein phosphatase (*PP2C9*) significantly improves rice plant performance under N deficient conditions. We measured the dry matter of leaf from the tillering (T) to maturity stage of rice plants which were sampled in the time courses at 5 days after flowering (DAF), 10 DAF,15 DAF, 20 DAF, 25 DAF, and 30 DAF, based on our initial findings [41]. As compared to *WT*, in *PP2C9TL* the dry matter of leaf significantly increased up to 10 DAF, after that it slowly started to decrease from 15 DAF to 20 DAF and finally, a significant decline was observed from 20 DAF up to 30 DAF under control and N deficient conditions, as shown in (Figure 1A, Figures S1 and S2). In *WT*, the dry leaf matter increased up to 10 DAF and then insignificantly decreased from 15 DAF to 30 DAF. We found that *PP2C9TL* efficiently tolerated N stress and produced the almost same amount of dry matter at low N as the *WT* produced at the control level of N, whereas, low N stress affected the *WT* plants' growth and decreased the leaf biomass as shown in Figure 1A. Similarly, higher chlorophyll content was observed in *PP2C9TL* than *WT*. We used the SPAD meter to measure the chlorophyll contents. The chlorophyll content increased from T to 10 DAF and decreased at 30 DAF in both *WT* and *PP2C9TL*, as shown in (Figure 1B, Figures S1 and S2). The physiological indices showed that *PP2C9TL* could effectively increase dry leaf matter and photoassimilates, which could then be transported to grain to increase the yield.

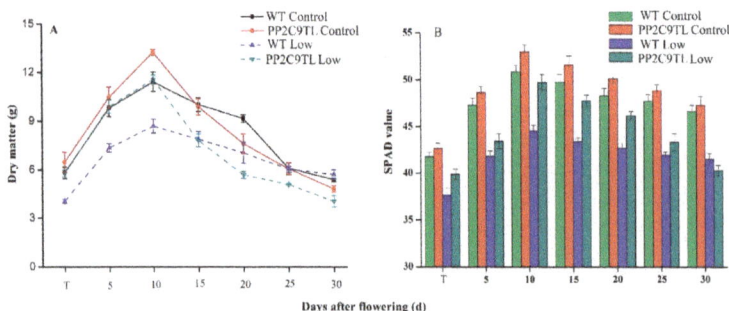

Figure 1. (**A**) The leaf dry matter (g) of *PP2C9TL* and *WT* at tillering (T), 5, 10, 15, 20, 25, and 30 DAF. (**B**) The chlorophyll contents of *PP2C9TL* and *WT* at tillering (T), 5, 10, 15, 20, 25, and 30 DAF.

2.2. NUE and Yield Performance of PP2C9TL and WT

In order to estimate the nitrogen uptake and NUE, we have calculated the leaf N content in *PP2C9TL* and *WT* in control and low N conditions. The significant differences in N content of the leaf were found at 10 DAF between the two genotypes under low N conditions. The *PP2C9TL* showed efficient N uptake compared to *WT* (Figure 2A). The highest difference in leaf N contents between the two genotypes was found at the heading stage (10 DAF). The grain yields across different levels of N are shown in Figure 2B. The difference between yields of *PP2C9TL* and *WT* is higher without N and with a low level of N, whereas, the yield differences decreased with the increase in N level. *PP2C9TL* produced a higher yield (58 g/pot) than *WT* (43 g/pot) at low N condition. The yield of *PP2C9TL* in low N level is almost equal to *WT* yield at the control level of N, these results indicate that *PP2C9TL* produced a high yield by consuming less N. The differences in yield at given levels of N confirmed the genetic variation for NUE between the two genotypes. Physiological attributes showed significant differences at 10 DAF between the two genotypes, clearly the *PP2C9TL* performed better than the *WT* so, to investigate the molecular pathway involved for higher NUE in *PP2C9TL* at low N, we designed differential proteomic experiments at 10 DAF.

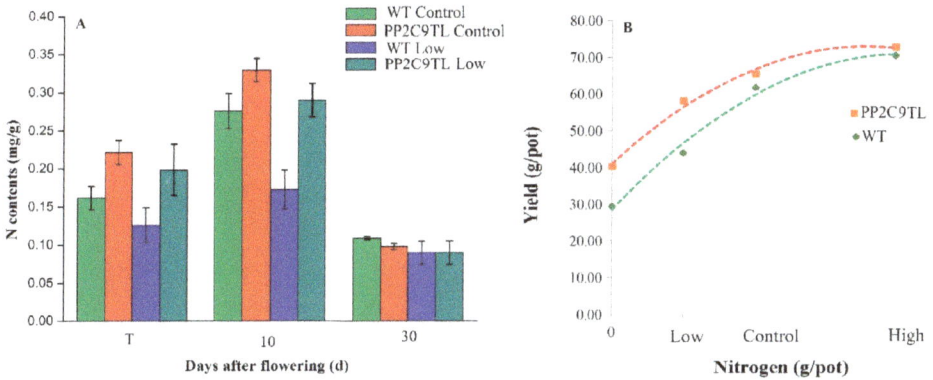

Figure 2. (**A**) The leaf N contents of *PP2C9TL* and *WT* at tillering T, 10 DAF, and 30 DAF; (**B**) the NUE of *WT* and *PP2C9TL* at low, control, and high levels of N.

2.3. Leaf Proteome Analysis of the Two Isogenic Lines under N Deficient Conditions

Differential proteomics of leaves from both *PP2C9TL* and its wild-type, *Kitaake japonica* rice, in low nitrogen conditions, lead to the identification of proteins involved in regulation of N uptake. Representative gels were shown in Figure 3, Figures S3 and S4. A total of 30 protein spots were found to be differentially expressed between the pH ranges of 4 to 7 (Figure 4, Figures S5 and S6). Imagemaster 5.0 was used for expression abundance of protein spots based on their relative volume. However, both genotypes showed differential protein expression under the N deficient condition.

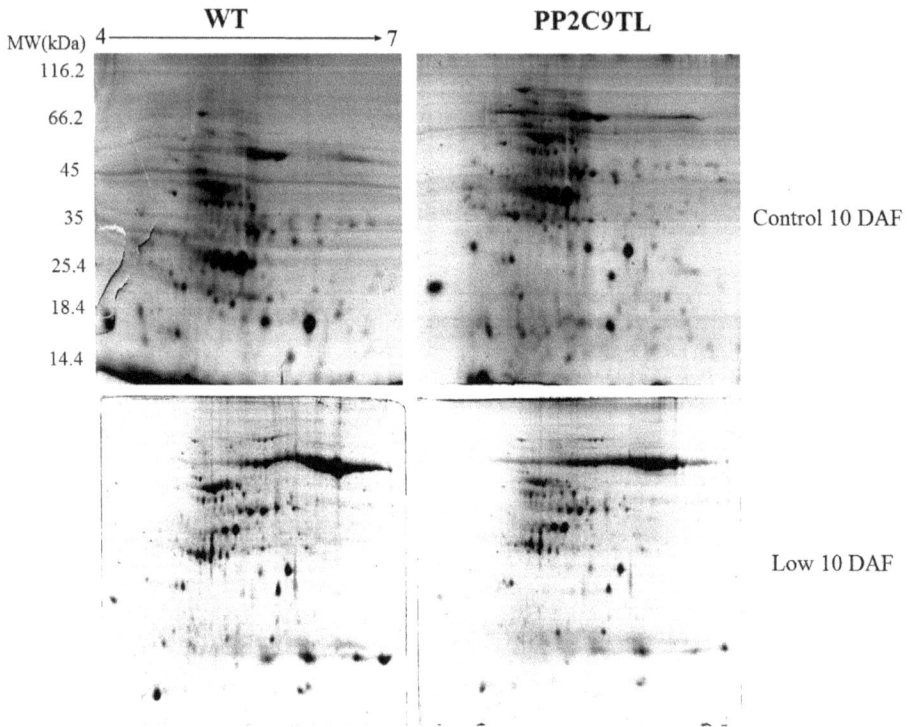

Figure 3. Representative 2-DE gel electrophoresis images of leaf proteins from the *WT* and *PP2C9TL* at 10 DAF under control and low N conditions. The MW (kilodaltons) and pI of the proteins are shown on the left and at the top, respectively.

Figure 4. A representative 2-D gel electrophoresis image showing total differentially expressed protein spots in the leaf under low N condition.

2.4. Functional Characterization of the Identified Proteins

The 30 proteins showed differential expression in the leaf (Table 1). Among these identified proteins, 14 (47%) were upregulated and 16 (53%) were downregulated. Web gene ontology annotation plot (WEGO) analysis was carried out to identify the biological function and cellular location of these proteins. These identified proteins are divided into nine groups according to their molecular and biological function: energy (23.33%), carbohydrate metabolism (16.67%), defense (13%), signaling (10%), transcription (10%), nitrogen metabolism (6.67%), cell growth and division (6.67%), and protein folding and storage (3.33%), as shown in Figure 5.

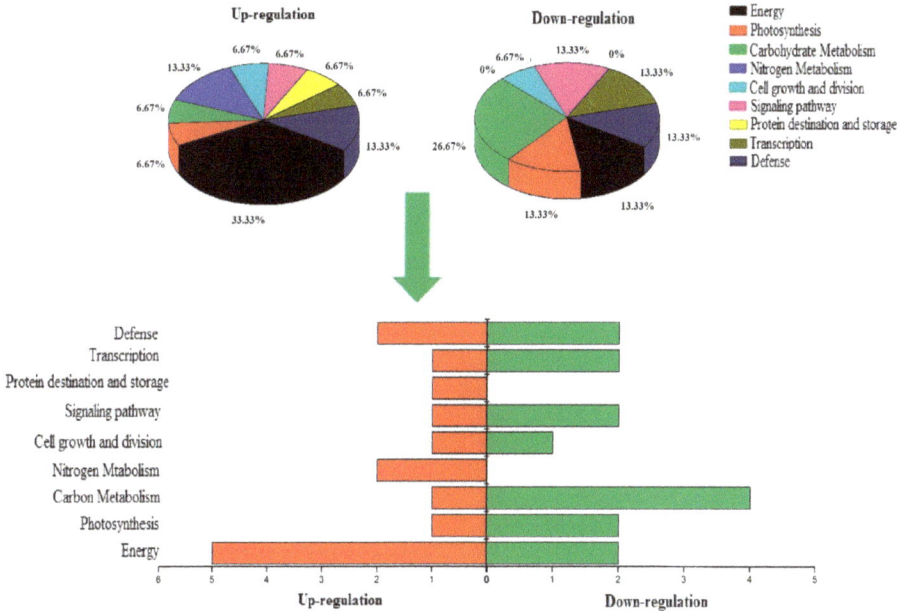

Figure 5. Pie graph percentage distribution display of the identified proteins described in (Table 1) based on biological function along with sidewise bar graph depicting up-and down-regulating proteins.

Table 1. Differential expression of leaf proteins between *WT* and *PP2C9TL* at 10 DAF under low N conditions.

SN [a]	AN [b]	Protein Name	Functional Characterization	Subcellular Location	Score [c]	MW (kDa)/pI [d]	MP [e]	CT [f]
Energy Metabolism								
3	A6N1P5.1	Ribulose-1,5-bisphosphate carboxylase/oxygenase	Energy (photosynthesis)	Chloroplast	69	30.36/5.22	12	UR
10	P93431-1	RuBisCO	Energy (photosynthesis)	Chloroplast	45	51.42/5.16	4	UR
20	Q6ZC90	ATP synthase	Energy	Mitochondria	46	27.3/6.93	4	UR
7	Q8S6Z1.1	ATP synthase subunit alpha	ATP energy	Mitochondria	136	29.30/5.33	9	UR
2	A3C6G9.1	Glycine cleavage system H protein	Energy	Mitochondria	108	17.3/5.03	11	UR
24	A1YQK1	Malate dehydrogenase	TCA cycle	Cytoplasm	155	36.5/6.33	6	DR
Photosynthesis								
26	Q9FYX8	Phosphoenolpyruvate carboxylase	Photosynthesis	Cytosol	98	105.2/5.73	8	DR
30	E9KIQ8	Photosystem II CP47	Photosynthesis	Chloroplast	89	53.24/5.67	7	UR
22	Q652K1	Stay-green protein	Photosynthesis	Chloroplast	234	121.32/4.66	13	DR
Carbohydrate Metabolism								
1	Q6Z8F4.1	Phosphoribulokinase	Calvin cycle	Chloroplast	43	44.83/6.02	2	DR
5	Q53P94.1	Fructose-bisphosphate aldolase	Glycolysis	Chloroplast	138	14.24/7.50	4	DR
9	Q6EQ16.1	ATP-dependent 6-phosphofructokinase	Glycolysis	Chloroplast	121	52.7/7.65	10	DR
14	Q0INM3.1	Beta-galactosidase 15	Carbohydrate metabolism	Cytosol	269	100.8/5.95	16	DR
6	Q84JG8	Chloroplast sedoheptulose-1,7-bisphosphatase	Calvin cycle	Chloroplast	75	42.21/5.84	6	UR
Nitrogen Metabolism								
23	Q6ZHH7	Nitrate reductase	Nitrogen metabolism	Chloroplast	271	97.79/534	13	UR
19	P14655	Glutamine synthetase	Nitrogen metabolism	Chloroplast	67	47.08/5.96	7	UR
Cell Growth and Division								
18	Q10PE7	Deoxymugineic acid synthase 1	Growth	Cytosol	129	33.6/7.541	13	UR
13	Q7XE16	Cell division cycle 48	Cell division	Cell membrane	134	88.8/5.23	15	DR
Signaling								
15	Q6967	14-3-3 protein GF 14-B	Defense signaling	Nucleus	114	18.27/4.46	21	DR
27	Q18PR8	Beta subunit 2 of SnRK1	signaling	Membrane	46	32.76/5.44	9	DR
25	Q6K1U4	Probable protein phosphatase 2C 16	Phosphatase	Nucleus	94	56.21/4.17	19	UR
Transcription								
11	Q6IER3	WRKY 8	Transcription	Nucleus	357	30.36/7.07	20	DR
8	E5RQA1	GHD 7	Transcription	Nucleus	112	26.83/7.23	14	DR
21	Q84MM9	Monoculm protein 1	Transcription	Nucleus	123	48.5/7.54	18	UR
Storage and Structural Protein								
29	Q8H903	Chaperonin 60 kDa protein	Protein synthesis, folding	Cytosol	336	60.8/5.78	12	DR
Defense								
17	Q943K7	HSP 70	Defense	Nucleus	90	71.23/5.21	7	UR
28	Q5QT28	Remorin 1 protein/Hsp20	Defense	Nucleus	44	22.4/537	6	UR
16	Q9FE01	L-ascorbate peroxidase	Defense	Chloroplast	75	28.49/5.36	14	DR
4	Q6YUZ0	Thaumatin-like protein	Defense	Cytoplasm	97	18.81/4.88	10	DR

Note: [a] Protein spot numbers correspond to 2-DE gels shown in Figure 4. [b] Accession number according to the UniProt database. [c] Score, ions score of identified proteins using rice genome sequence databases. [d] Theoretical MW (kDa) and pI values. MW, molecular weight; pI, isoelectric point. [e] M.P., number of query matched peptides. [f] Changes in the protein spots between PP2C9TL and WT under low N conditions; UR: protein spots upregulated at grain-filling stage (10 DAF), DR: protein spots downregulated at grain-filling stage (10 DAF).

2.5. Subcellular Characterization of the Identified Proteins

Almost all the organelle functions were affected by low N as the identified proteins belong to different organelles. The expressed proteins were mostly associated to chloroplast (40%) followed by nucleus (20%), cytosol (10%), mitochondria (10%), cell membrane (10%), and cytoplasm (7%) (Figure 6).

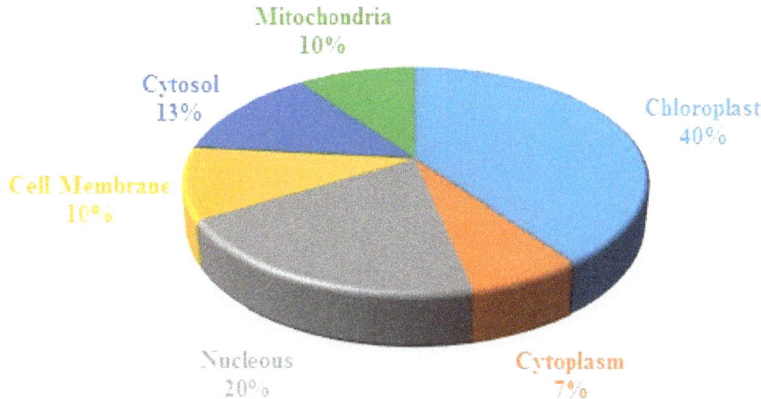

Figure 6. Subcellular indexing of the identified proteins.

2.6. Potential Molecular Pathway Based on Differentially Expressed Proteins

The differential expression of proteins showed that low N influenced different molecular pathways involved in photosynthesis, Calvin cycle, glycolysis, N metabolism, and defense mechanism. The *PP2C9TL* efficiently tolerate the N stress than *WT*. In *PP2C9TL*, the RuBisCO (Ribulose-1, 5-bisphosphate carboxylase/oxygenase), Photosystem II, and oxygen evolving enhancer proteins (OOEs) were upregulated compared to *WT*, and the stay green protein was downregulated. The photosystem II and RuBisCO helps *PP2C9TL* to maintain photosynthesis during N deficiency better than the *WT*. RuBisCO plays an important role in the Calvin cycle to generate energy in *PP2C9TL* and is interlinked with the glycolysis and TCA cycle, as shown in Figure 7. However, in *PP2C9TL* the overexpressed PP2C9 gene takes part in the activation of nitrate reductase (NR) via dephosphorylation of 14-3-3 and SnRK, as shown in Figure 8. The differential expression showed that 14-3-3 and SnRK are downregulated in *PP2C9TL*. The higher activity of NR in *PP2C9TL* generates more nitric oxide which increases N uptake by enhancing the development of lateral roots.

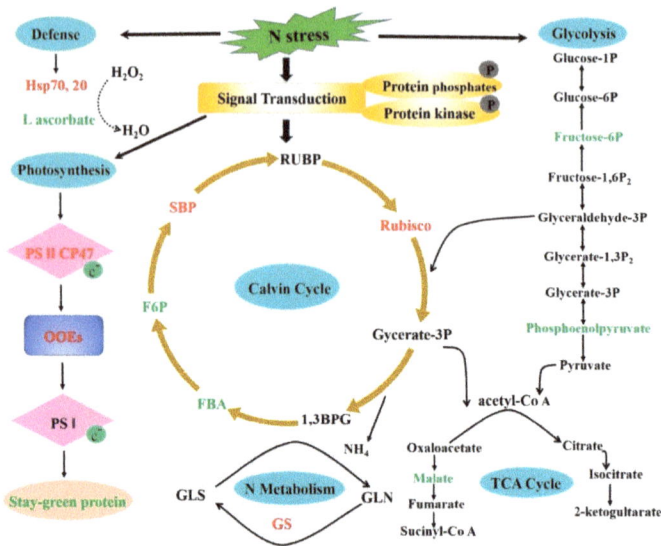

Figure 7. Potential molecular pathway based on differentially expressed proteins in *PP2C9TL* under N deficient conditions. The red color text shows the upregulation and green color text shows downregulation of proteins. Note RUBP (Ribulose 1, 5-bisphosphate), RuBisCO (Ribulose-1,5-bisphosphate carboxylase/oxygenase), 1,3 BPG (1,3-Bisphosphoglycerate), FBA (Fructose-bisphosphate aldolase), F6P (Fructose-bisphosphate aldolase), SBP (sedoheptulose-1,7-bisphosphatase), PSII (Photosystem II), OOEs (oxygen evolving enhancer proteins), PSI (Photosystem I), GLS (Glutaminase), and GS (Glutamate synthase).

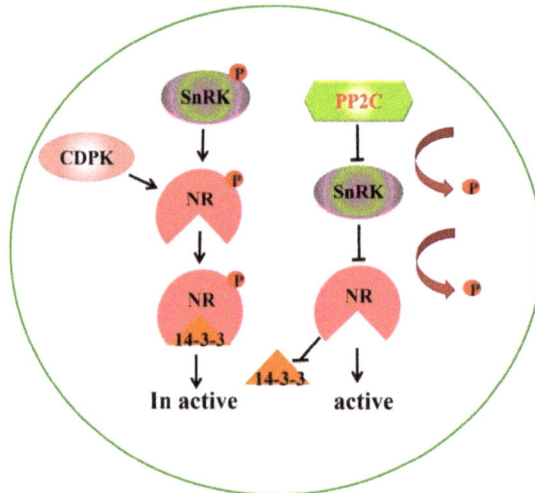

Figure 8. *PP2C9* role for nitrate reductase activation by dephosphorylation of SnRK. 14-3-3 proteins SnRK (SNF1-related kinase) and CDPK (calcium-dependent protein kinases) can phosphorylate NR in crop plants. After phosphorylation of NR, it binds with the 14-3-3 protein to form a complex which is inactive or has low activity. *PP2C9* interacts with 14-3-3 and SnRK and dephosphorylates them to activate NR.

2.7. Western Blotting of the Important Differentially Expressed Protein

The 14-3-3 protein is a highly conserved protein in crop plants and is directly involved in the NR inactivation mechanism. So Western blotting was carried out to verify the *PP2C9TL* and the differential expression of 14-3-3 protein detected by 2DE gels using primary mouse antisera and the HRP-anti-mouse IgG as the secondary antibodies against Flag-tag PP2C9 and 14-3-3. The Western blot results confirmed the downregulation of 14-3-3 in *PP2C9TL*, whereas, the presence of Flag-tag verified the enhanced expression of *PP2C9* in *PP2C9TL*, as shown in Figure 9A.

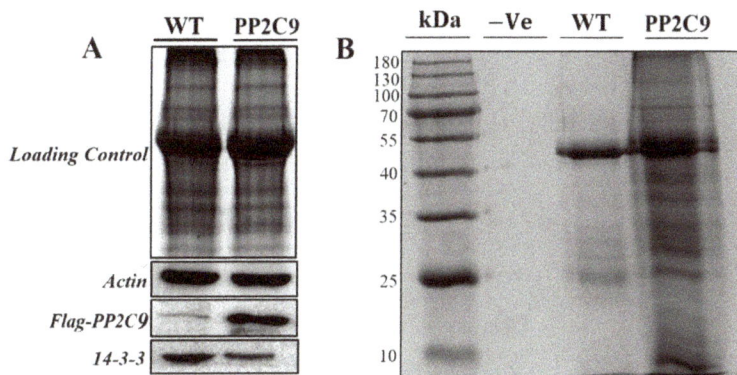

Figure 9. (**A**)Western blotting of differentially expressed proteins in *PP2C9TL* and *WT* at 10 DAF stage. The top photo shows the SDS-PAGE separation of two protein samples used as loading control. The bottom photo shows gel transferred onto nitrocellulose membrane for western blotting to detect actin, anti-Flag-*PP2C9*, and 14-3-3 protein for *WT* and *PP2C9TL* at the 10 DAF stage. (**B**) Co-immunoprecipitation (Co-IP) assays of *PP2C9* associated protein analyzed by SDS-PAGE. Protein molecular weight marker, negative control without antibodies, *WT* protein, and the third is the *PP2C9TL* protein. 2.8. Identification of PP2C9 Associated Protein by Co-Immunoprecipitation (CO-IP) and Mass Spectrometry.

In order to find the PP2C9-associated protein interactions and to pull out all protein complexes among *PP2C9TL* and *WT*, we carried out the immunoprecipitation assay by using the (rabbit antisera Flag-tag antibodies). The immunoprecipitation assay results showed that *PP2C9TL* had obvious differences in bands compared with *WT*. These differential protein bands were located at 20 to 80 kDa as shown in Figure 9B. Protein bands were excised and exposed to digestion and analysis followed by tandem mass spectrometry (MS/MS). The identified proteins by (MS/MS) analysis were shown in Table 2. The Co-IP validates the results obtained by comparative proteomics as most of the identified proteins were similar with 2DE proteins and involved in photosynthesis, energy, N metabolism, signaling cascades, and defense mechanisms. So these identified differential proteins from Co-IP and 2DE were responsible for higher NUE of *PP2C9TL* than *WT*.

Table 2. LC-MS/MS analysis of the Co-immunoprecipitation of *PP2C9TL*.

AN [a]	Protein Name [b]	Functional Characterization	Subcellular Location	Score [c]	MW (kDa)/pI [d]
Energy Metabolism					
LOC_Os12g10580.1	Ribulose bisphosphate carboxylase large chain	Energy (photosynthesis)	Chloroplast	371.7	56/8.9
LOC_Os10g21268.1	Ribulose bisphosphate carboxylase large chain	Energy (photosynthesis)	Chloroplast	683.23	53.7/7.03
LOC_Os06g39740.1	ATP synthase subunit beta	Energy	Mitochondria	576.28	54.2/5.5
LOC_Os06g45120.1	ATP synthase	ATP energy	Mitochondria	178.54	68.4/5.34
LOC_Os10g17280.1	ATP synthase gamma chain	Energy	Mitochondria	16.83	35.2/7.03
LOC_Os04g16740.1	ATP synthase subunit alpha	ATP energy	Mitochondria	279.38	55.6/6.25
LOC_Os03g17070.1	ATP synthase B chain chloroplast precursor	ATP energy	Mitochondria	39.69	22.7/5.85
LOC_Os10g37180.1	Glycine cleavage system H protein	Energy	Mitochondria	50.25	17.4/5.03
LOC_Os03g56280.1	Malate dehydrogenase	TCA cycle	Cytoplasm	146.16	37/7.94
Photosynthesis					
LOC_Os08g27840.1	Phosphoenolpyruvate carboxylase	Photosynthesis	Cytosol	260.03	110/5.8
LOC_Os04g16874.1	Photosystem II 44 kDa reaction center protein	Photosynthesis	Chloroplast	94.42	44.7/6.6
LOC_Os03g21560.1	Photosystem II 11 kD protein	Photosynthesis	Chloroplast	21.05	17.6/9.91
LOC_Os07g05360.1	Photosystem II 10 kDa polypeptide, chloroplast	Photosynthesis	Chloroplast	22.11	14.2/9.81
LOC_Os02g24634.1	Photosystem II D2 protein	Photosynthesis	Chloroplast	64.09	39.6/5.4
LOC_Os06g39708.1	Photosystem II P680 chlorophyll a Apoprotein	Photosynthesis	Chloroplast	161.07	56.2/6.64
Carbohydrate Metabolism					
LOC_Os02g47020.1	Phosphoribulokinase	Calvin cycle	Chloroplast	266.6	44.8/6.02
LOC_Os04g16680.1	Fructose-1,6-bisphosphatase	Glycolysis	Chloroplast	256.8	42.2/6.09
LOC_Os11g07020.1	Fructose-bisphosphate aldolase isozyme	Glycolysis	Chloroplast	403.24	42/6.8
LOC_Os01g64660.2	Fructose-1,6-bisphosphatase	Carbohydrate metabolism	Cytosol	127.72	37/5.77
LOC_Os08g03290.1	Glyceraldehyde-3-phosphate dehydrogenase	Carbohydrate metabolism	Cytosol	97.76	36.4/7.11
Nitrogen Metabolism					
LOC_Os02g50240.1	Glutamine synthetase	Nitrogen metabolism	Chloroplast	134.89	39.2/5.73
LOC_Os05g04220.1	Nitrogen regulatory protein P-II	Nitrogen metabolism	Chloroplast	5.02	22.7/9.91
LOC_Os02g52730.1	Ferredoxin-nitrite reductase	Nitrogen metabolism	Chloroplast	3.35	72.4/8.29
LOC_Os04g01590.1	Arginase	Nitrogen metabolism	Mitochondria	14.54	36.9/5.3
LOC_Os09g28050.1	Asparate aminotransferase	Nitrogen metabolism	Chloroplast	19.62	50.6/6

Table 2. *Cont.*

AN [a]	Protein Name [b]	Functional Characterization	Subcellular Location	Score [c]	MW (kDa)/pI [d]
		Cell Growth and Division			
LOC_Os10g30580.1	Cell division	Growth	Cytosol	42.5	89.8/5.21
LOC_Os02g58790.1	Cell division inhibitor	Growth	Cytosol	6.35	38.8/9.16
		Signaling			
LOC_Os03g50290.1	14-3-3 protein	Defense	Nucleus	112.53	29.2/4.88
LOC_Os04g38870.3	14-3-3 protein	Signaling	Membrane	88.34	29.8/4.81
LOC_Os08g33370.2	14-3-3 protein	Phosphatase	Nucleus	74.19	28.8/4.84
LOC_Os05g11550.1	Serine/threonine protein phosphatase 5	Growth	Cytosol	5.86	54.4/6.02
LOC_Os07g32380.1	Protein phosphatase 2C	Phosphatase	Nucleus	4.9	20/8.18
LOC_Os09g06230.1	Serine/threonine-protein kinase 16	Phosphatase	Nucleus	3.21	25.5/4.7
LOC_Os04g56450.1	SnRK1-interacting protein 1	Signaling	Membrane	1.89	30.6/5.15
LOC_Os09g33790.1	Proten phosphatase 2C	Signaling	Membrane	2.95	109/7.15
LOC_Os02g38300.1	SNf7 domain-containing protein	Signaling	Membrane	2.69	25.5/4.7
		Transcription			
LOC_Os03g55164.1	WRKY4	Transcription	Nucleus	2.47	14.8/7.28
		Storage and structural protein			
LOC_Os06g09679.2	Chaperonin	Protein synthesis, folding	Cytosol	81.66	26.3/8.02
LOC_Os10g41710.1	Chaperonin, putative expressed	Protein synthesis, folding	Cytosol	5.18	21.1/8.92
LOC_Os09g26730.1	Chaperonin	Protein synthesis, folding	Cytosol	70.68	69.1/6.37
		Defense			
LOC_Os10g07210.1	Hsp20/alp:a crystallin family protein	Defense	Nucleus	2.26	18.6/8.2
LOC_Os06g37150.1	L-ascorbate oxidase precursor	Defense	Chloroplast	2.9	27.4/7.65
LOC_Os07g49400.2	Cytosolic Ascorbate Peroxidase	Defense	Chloroplast	68.37	25.2/5.71
LOC_Os12g43440.1	Thaumatin putative,	Defense	Cytoplasm	9.64	22.4/5.5
LOC_Os07g47510.1	Stress-related protein,	Defense	Nucleus	2.38	27.4/7.65
LOC_Os09g29200.1	Glutathione S-transferase,	Defense	Cytoplasm	134.49	25.2/5.71
LOC_Os04g45070.1	Remorin	Defense	Nucleus	45.89	22.4/5.5

3. Discussion

3.1. Regulatory Role of PP2C9 for Higher NUE in PP2C9TL under N Deficient Conditions

The reversible proteins phosphorylation by protein phosphatases is an essential mechanism to regulate the different biological processes. In plants, PP2C9s are a major phosphatase-encoding gene family and have emerged as key regulators of stress signaling. To gain a more comprehensive understanding of the function of PP2C9 in response to N stress, we devised a comparative proteomics strategy, with this approach, we identified three regulatory proteins the PP2C9, 14-3-3 family, and SnRK. PP2C9 deals with phosphorylation/dephosphorylation of several proteins, so it is quite possible to alter the signaling pathways. The proteins involved in signaling (14-3-3 Spot 15 and Beta subunit 2 of SnRK1 Spot 27) were downregulated in *PP2C9TL* compared with that in *WT*, whereas the protein phosphatase 2C 16 (Spot 25) was upregulated in *PP2C9TL*. The upregulation of protein phosphatase 2C 16 is not surprising as it is overexpressed in *PP2C9TL*. The downregulation of SnRK1 was also observed in Arabidopsis during N and K stress [42]. 14-3-3 was reported to be a key regulator of nitrogen and carbon metabolism through interaction with multiple signal transduction pathways [43]. In Arabidopsis, the overexpression of 14-3-3 decreased the sugar- and N-based compounds and also reduced the levels of malate and citrate, which are the intermediate compounds of the TCA cycle [44]. In *PP2C9TL*, the 14-3-3 protein was downregulated, which might be due to the overexpressed *PP2C9* gene which dephosphorylates the 14-3-3, the downregulation of 14-3-3 was confirmed by the Western blot results. *PP2C9* interacts with 14-3-3 and SnRK1 and dephosphorylates them to activate NR. SnRK1 and CDPK (calcium-dependent protein kinases) can phosphorylate NR in crop plants [45–47]. Phosphorylation itself does not alter NR activity. However, 14-3-3 proteins and cations, such as Mg^{2+}, are present, and this complex of phosphorylated NR shows low activity, whereas *PP2C9* reactivates NR by dephosphorylation [48,49]. Our results suggest that the 14-3-3s regulate the C and N metabolism through interaction with SnRK1 and PP2C9. NR activity is most important for nitrogen uptake and assimilation is the most probable reason for higher NUE of *PP2C9TL* in low nitrogen.

3.2. The Physiological Basis for NUE and Grain Yield

NUE of crop plants roughly depends upon two factors. First, how efficiently a plant absorbs N from the soil (N uptake efficiency) and second, how efficiency can the plant utilize this N to produce a high grain yield [50]. Whatever the N level, the high vegetative growth at flowering stage was favorable for higher N uptake efficiency. The N uptake at flowering was significantly correlated with grain yield at both a low and high level of N, especially at low N where it determines seed number [20,51]. Grain yield depends upon complex biochemical processes, especially C and N interactions, which lead to the production of dry matter [52]. In this study, *PP2C9TL* have higher vegetative growth at the flowering stage. *PP2C9TL* increased leaf dry matter and N contents at 10 DAF and thereafter, consequently decreased until 30 DAF. This might occur because the proteins, fats, carbohydrates, minerals, and vitamins are transported and accumulated into the grain from vegetative parts that ultimately reduce the dry weight of plants. The relationship between source–sink is the key factor for the grain yield of cereal crops, the plant leaves are the primary source of photoassimilates, whereas, the grains are the primary sink [53]. Our results are consistent with previous findings that a positive correlation was observed between the dry matter accumulations at the heading stage which is negatively correlated at the grain filling stage [54]. Arginine is a major storage form of N in plant organs and is well-known for N transport [55]. The *PP2C9TL* COIP results found arginase and aspartate aminotransferase enzymes suggesting the higher metabolism of arginine and aspartate which ensures efficient transport and storage of N in *PP2C9TL*. PP2C9 regulates N and C metabolism by activation of NR, sucrose phosphate synthase (SPS), and inactivation of PEPase [52]. Thus, it might be possible that PP2C9 together with SnRK are involved in the regulation of N and C metabolism-related enzymes through phosphorylation/dephosphorylation. *PP2C9TL* has a higher photosynthetic rate because photosynthesis is directly related with leaf N and chlorophyll contents which were higher in the *PP2C9TL* compared to *WT*. We further demonstrated that in *PP2C9TL* a high NUE and N uptake efficiency were observed at low level of N which decreases with the increase of N. The yield differences between two genotypes was decreased with the increase in N level, this might be due to surplus N availability which ultimately slow down or stop the increase in yield and reduces NUE. Extensive N application usually resulted in unlimited N absorption, lodging and yield reduction [56]. The NUE is also govern by the rice genotypes as the yield differences at same level of N depicts that two genotypes exhibit the genetic variability for NUE, this genetic variability could be helpful for crop breeders to develop the new rice genotypes with higher NUE.

3.3. Proteins Expression Involved in Energy of PP2C9TL and WT Genotypes under N Deficient Conditions

As expected the proteins differential expressed under low N condition were mostly belonged to energy. RuBisCO (Ribulose-1,5-bisphosphate carboxylase/oxygenase Spot 3 and Spot 10) was downregulated under low N conditions in *WT* as compared to *PP2C9TL*. In C3 plants, about 12% of total N is consists of RuBisCO during vegetative growth so it has major significance for NUE. RuBisCO has a dual catalytic function as carboxylase for carbon dioxide assimilation and as oxygenase to trigger the photo-respiratory pathway in plants [57]. Under the low concentration of N, availability of RuBisCO was also decreased in cereals [58] and in *Arabidopsis* [59]. In *WT* RuBisCO is downregulated suggests that extensive degradation of photosynthetic apparatus [59]. However in *PP2C9TL*, this degradation is limited under low N that might be due to high uptake of N. Nitrogen influx into leaves directly determined the RuBisCO synthesis, fluctuations in RuBisCO synthesis correlate well with those for N influx throughout the leaf lifetime in rice [60]. ATP synthase and the alpha subunit of ATP synthase (Spot 20, Spot 7), mainly produced in the mitochondria or chloroplast membranes, catalyze the synthesis of ATP from ADP during energy producing biochemical cycles. In the electron transport chain of photosynthesis, the ATP synthase complex takes part in photophosphorylation of ADP to ATP providing energy to the Calvin cycle for subsequent biosynthesis [61]. In our study, under the low N condition, the concentration of this (spot 22) was higher in *PP2C9TL* than *WT*. On the other hand, the spot (7) intensity was lower in *PP2C9TL* than *WT*. The ATP synthase protein complex takes part in maintaining the function of the chloroplast during stress conditions [62]. The increased expression of ATP synthase protein in *PP2C9TL* plants provides tolerance against N deficiency; however, protein phosphatases take part in reversible phosphorylation of proteins to activate/deactivate them in different signaling pathways. In cereal crops, differential expression of ATP synthase has been observed [63]. Glycine cleavage system H protein (Spot 2), also named glycine dehydrogenase (GDC), is a multiprotein complex that is found in all living organisms. GDC is required for photorespiration in C3 plants, and it takes glycine, which is produced as a byproduct of the Calvin cycle, and converts it to serine through an interaction with serine hydroxyl methyltransferase (SHMT). The glycine cleavage system generates ammonia which is later assimilated by glutamine synthetase through the glutamine oxoglutarate aminotransferase cycle at the expense of one ATP and one NADPH. The intensity of this protein was higher in *PP2C9TL* than *WT*. Malate dehydrogenase (Spot 24) catalyzes the terminal step of the citric acid cycle and converts malate to oxaloacetic acid (OAA) by generating NADH. The malate dehydrogenase was downregulated in *PP2C9TL* and *WT* under low N treatment. In Arabidopsis and cereal crops, a slight upregulation of malate dehydrogenase was observed with an increase in nitrate level [63–65].

3.4. Proteins Expression Involved in Photosynthesis of PP2C9TL and WT Genotypes under N Deficient Conditions

Phosphoenolpyruvate carboxylase or PEPC primarily fixes and assimilates photosynthetic CO_2 in C4 and crassulacean acid metabolism (CAM) plants, whereas, PEPC replenishes the TCA cycle with intermediates in the C3 plant [66,67]. PEPC was reported to be phosphorylated by the Ser/Thr kinase and PEPC kinase (PPCK), whereas, protein phosphatase 2A dephosphorylates the PEPC [68]. The PEPC (Spot 26) was found to be downregulated in the *PP2C9TL* under low N. The downregulated expression of PEPC protein in *PP2C9TL* might be due to the overexpression of *PP2C9* which ultimately deactivates the PEPC by dephosphorylation. Photosystem II CP47 reaction center protein is the constitutive transmembrane antenna proteins which interact with chlorophyll a and beta-carotene to pass the excitation energy on to the reaction center. It takes part in photosynthesis by sustaining the stability of PSII [69,70]. This protein was increased in *PP2C9TL* under low N (Spot 30). The degradation of oxygen-evolving complex (OEC) proteins was observed during low N, in response, OEC releases oxygen evolving enhancer proteins (OEEs) as a degradation product which enables the plant to adapt and survive in unfavorable conditions [71]. Oxygen-evolving enhancer proteins (OEEs) were bound to photosystem II (PSII) at the lumen side of the thylakoid membrane. The electrons generated during these reactions were transferred to photosystem I through electron transport chain which are ultimately used during NADP reduction. Electron transport from PSII to PSI, together with RuBisCO and carbon metabolism enzymes, are crucial to the photosynthetic rate [72]. In *PP2C9TL*, an increase in the OEEs helps PSII to maintain its ability to assimilate maximum photoassimilates under low N levels as reported by previous studies [73]. The stay-green protein (Spot 22) located in the thylakoid membranes triggers chlorophyll degradation during natural and dark-induced leaf senescence [74]. The stay-green protein was found to be downregulated in *PP2C9TL* relative to *WT* in low nitrogen. The downregulation of the stay-green protein shows the high tolerance ability of *PP2C9TL* in low N conditions because overexpression of stay-green protein reduced the number of lamellae in the grana thylakoids and increases chlorophyll breakdown in rice [75].

3.5. Proteins Expression Involved in Nitrogen Metabolism of PP2C9TL and WT Genotypes under N Deficient Conditions

In crop plants, N metabolism is most important to their nutritional availability. Two of the most important N metabolism proteins showed an upregulated expression pattern in *PP2C9TL* relative to *WT*. Nitrate reductase (NR) (Spot 23) is the most important enzyme for N metabolism and directly accounts for N assimilation rate in plants, NR is used as the main product for organ development and plant growth [76]. However, *PP2C9TL* showed an increase in expression of NR protein. We found *PP2C9TL* efficiently increased the uptake and assimilation of N under low N levels compared to the *WT*. Moreover, nitrate reductase is activated by protein phosphatase by dephosphorylation at the post-translational level. [77,78]. *PP2C9TL* overexpressed the protein phosphatase that could dephosphorylate NR to keep active on nitrate availability. Glutamine synthetase (GS) catalyzes ammonia produced by photorespiration, protein catabolism, nitrate, and ammonia metabolism and takes part in N assimilation and transportation [79,80]. Glutamine synthetase (GS) protein (Spot 19) was upregulated in *PP2C9TL* relative to *WT* during low nitrogen. This upregulated GS expression in *PP2C9TL* provides sufficient GS level to incorporate ammonia into organic compounds efficiently [81]. In the *WT* genotype, GS expression was downregulated in low nitrogen showing that N-deprived plants are undergoing more stress.

3.6. Proteins Expression Involved in Defense and Protein Folding of PP2C9TL and WT Genotypes under N Deficient Conditions

N stress caused upregulation of (HSP 70 Spot 17 and remorin 1 protein/Hsp20 Spot 28) in *PP2C9TL* whereas, (L-ascorbate peroxidase Spot 16 and thaumatin-like protein Spot 4) was downregulated in *PP2C9TL*. The upregulation of HSP is the most conserved adaptive strategy of plants in response to stress [82–85]. Yang et al. [86] reported an upregulation of HSP 70 in *Phaeodactylum tricornutum* under N stress. Ascorbate peroxidase (APX) exists in isoforms in different subcellular organelles such as peroxisome, chloroplasts, cytosol, and mitochondria and helps plants detoxify H_2O_2 by converting it into H_2O [87,88]. Chaperonin 60 kDa protein (Spot 29), was upregulated in *PP2C9TL* under low N, these chaperonins which help and guide cells in the correct folding of proteins both in normal conditions [89] and under stress [90]. The upregulation of chaperonin 59.7 kDa protein was also reported in microbes in the low N condition [91].

4. Materials and Methods

4.1. Plant Materials and Growth Conditions

These research experiments were carried out at the greenhouse of the experimental farm of Fujian Agriculture and Forestry University, Fuzhou, Fujian province, China during the rice growing season (April to October 2017). The region has a humid subtropical climate with mean temperatures ranging from 15 to 34 °C in 2017. Two isogenic rice lines of the wild-type, Kitaake (*Oryza sativa* L. ssp. *Japonica*) and its counterpart, the *PP2C9* transgenic line (*PP2C9TL*), in which the *PP2C9* 9 gene (LOC8058897) has been transformed from sorghum bicolor, overexpressing the protein phosphatase 2C 9 protein (XP_002456624.1) were used as research materials. Rice seeds were soaked in distilled water at 25 °C for 24 h and later incubated at 37 °C for 48 h. The germinated seeds of similar size were sown to obtain uniform seedlings which were transplanted to plastic buckets sized in 0.3 m top and 0.23 m bottom diameter with four hills per pot. The sandy loam soil was mixed well to make it uniform before being used for potting with available N (190.6 mg kg^{-1}), phosphorus (126.6 mg kg^{-1}), and potassium (201.6 mg kg^{-1}). The recommended level of N in Fuzhou (225 kg ha^{-1}) and half of this amount (112.5 kg ha^{-1}) were set as the treatment and control, respectively. Phosphorus was used as the base fertilizer and potassium as the top dressing, at the rate of 112.5 kg ha^{-1} (P$_2$O$_5$), and 180 kg ha^{-1} (K$_2$O) converted to the amounts per barrel, respectively.

4.2. Transgenic Line Generation

The full length ORF of PP2C9 (GenBank accession no. LOC8058897) was cloned into the p3301 vector under the control of cauliflower mosaic virus promoter (CaMV 35S) including a FLAG Octapeptide tag. The agrobacterium-mediated transformation of rice immature embryos from mature seeds was used following the protocol of a past paper [92]. Transformed cells obtained from these tissues were selected on the basis of hygromycin resistance, and transgenic plants were finally regenerated. The transgenic plants were screened and confirmed with a Flag-tag assay.

4.3. Physiological Measurement

The leaves were sampled from both genotypes at tillering, 5, 10, 15, 20, 25, and 30 days after flowering (DAF). To collect samples, four plants were collected from plastic buckets and the leaves were water-washed and separated. Collected leaf samples were dried in an oven at 105 °C for 30 min, and afterward, dried at 80 °C for 48 h until constant weight. Dry weight was recorded as described by Yoshida [93]. Chlorophyll content from the leaves was measured according to Xiong et al. [94], by using a SPAD-502 chlorophyll meter (Konica-Minolta, Osaka, Japan). The readings were taken in the morning between 9:00 and 10:30 a.m. from flag leaves in four replications. The leaf samples were digested with H$_2$SO$_4$-H$_2$O$_2$ and the leaf N concentration was determined using the Kjeldahl method according to Sáez-Plaza et al. [95]. Sampling was done in three replications and SPSS software was used for the analysis of variance (ANOVA) by the least significant difference at $p < 0.05$ (LSD$_{0.05}$).

4.4. Protein Sample Preparation

Leaf samples were collected at tillering, 10 DAF, and 30 DAF from *WT* and *PP2C9TL* plants under control and low N conditions. Leaf samples were washed, freeze-dried with liquid nitrogen, and saved at −80 °C. Protein extraction from leaves was performed by following the protocol of Li et al. [96]. Concisely, 5 g of frozen leaf samples at −80 °C was ground to powder along with polyvinylpyrrolidone (PVP) and liquid nitrogen by using mortar and pestle. The obtained fine powder was mixed in precooled acetone containing 10% TCA (Trichloroacetic acid) and 0.07% β-mercaptoethanol and kept at −20 °C overnight. The sample was centrifuged at 16,000 g for 30 min at 4 °C. The supernatant was discarded and obtained pellet was dissolved in 20mL precooled 100% acetone containing 0.07% β-mercaptoethanol, slightly vortexed and kept at −20 °C for 3 h., a transparent supernatant was obtained by repeating this step many times. The precipitate was freeze-dried under vacuum to powder.

The obtained powder was dissolved in a buffer (pH 8.0) containing 8 mol/L thiourea, 4% CHAPS (3-[(3-cholamidopropyl) dimethylammonio]-1-propanesulfonate), 40 mmol/L Tris, and 65mmol/L DTT (dithiothreitol). Bradford method is used to measure the protein concentration by using BSA (Bovine Serum Albumin) as a standard [97].

4.5. Two-Dimensional Gel Electrophoresis

Based on the physiological response information of the two rice genotypes exposed to the nitrogen treatment at the pot, 2DE of proteins was performed to separate the leaf proteins at 10 DAF using the isoelectric focusing (IEF) strip gels (17 cm, pH 4–7; Bio-Rad, Hercules, CA, USA) for the first dimension. For IEF, an immobiline dry strip gel was rehydrated at 20 °C for 14 h and 1.3 mg protein was loaded in each strip. Protean IEF apparatus (Bio-Rad, USA) was used to do IEF. The voltages applied were at a gradient of 500 V for 1 h; gradient of 1000 V for 2 h; gradient of 8000 V for 3 h; held at 8000 V for 3 h; and then at a gradient of 1000 V for 24 h, as mentioned in [45]. After completing the IEF, the strips were further exposed to an equilibration buffer. Equilibration buffer I (65 mM DTT) was used for 15 min shaking. Iodoacetamide (2.5% (*w/v*) was used and kept shaking for 15 min. The second-dimensional separation was carried out on SDS-PAGE containing 12% (*v/v*) polyacrylamide gels in 2D-Electrophoresis SDS-PAGE Apparatus (GE) at 15 mA current per gel until completion of electrophoresis. The gels were stained with Colloidal Coomassie Blue G-250 for 12 h followed by destaining. GE Image scanner III was used to scan the obtained protein gels and Imagemaster 5.0 software was used to identify comparative protein spots.

4.6. In-Gel Protein Digestion

Differential protein spots were cut from the gel and washed with deionized water and rehydrated with acetonitrile and ammonium bicarbonate as mentioned in [45]. The gel was destained twice with 100 μL of acetonitrile (50%)/100 mM ammonium bicarbonate (50%) for 10 min and then dehydrated with 100% acetonitrile. Lastly, samples were digested with 20 μL of trypsin (12.5 μg mL^{-1} 50 mM ammonium bicarbonate) for 30 min on the ice and then incubated at 37 °C overnight. The 20% acetonitrile and 1% formic acid were added to the gel and centrifuged to obtain the supernatant which later used for LC-ESI-MS/MS analysis.

4.7. LC-ESI-MS/MS Analysis and Protein Identification

The LC-ESI-MS/MS analysis was performed by the protocol of Li, Li, Muhammad, Lin, Azeem, Zhao, Lin, Chen, Fang, and Letuma [96]. High performance liquid chromatography: Thermo Scientific Accera System; Chromatographic Column: Bio Basic C18 Column (100 × 0.18 mm, the particle size: 5 μm). Loading quantity of sample: 10 uL. Mobile phase: Solvent A was 0.1% HCO_2H mixed in water, and Solvent B was 0.1% HCO_2H mixed in CH_3CN. Gradient was held at 2% solvent B for 2 min and increased linearly up to 90% Solvent B for 60 min. The peptides were eluted from a C18 column at a flow rate of 160 μL min^{-1} and then electro-sprayed directly into an LTQ mass spectrometer using a spray voltage of 3.5 kV and a constant capillary temperature of 275 °C. Data retrieved from data-dependent MS/MS scanning mode. Mass spectrometry analysis of the raw data obtained in Proteome Discoverer 1.2 relative quantitative analysis software and database retrieval was performed through UniProt database (http://www.uiprot.org/) using the *Oryza sativa* Fasta protein libraries 2.6 Software Analysis.

4.8. Confirmation of Important Protein 14-3-3 by Western Blotting Analysis

To perform Western blotting, the primary antibody for 14-3-3 (Catalog No. 51-0700) and Flag DYKDDDDK Tag Monoclonal Antibody (FG4R) (Catalog No. MA1-91878) were obtained from (Thermo Fisher Scientific, Waltham, MA, USA). The proteins for western blotting were transferred to tubes containing 300 μL SDS loading buffer and boiled at 95 °C for 10 min. Each sample was separated by SDS-PAGE and blotted on nitrocellulose membranes. The membranes were blocked with 5%

bovine serum albumin (BSA) in phosphate buffered saline with tween 20 (PBST), after that, incubated with primary antibodies (rabbit antisera to 14-3-3 or Flag) with dilutions following manufactures instructions. Horseradish peroxidase (HRP) conjugated goat anti-rabbit antibodies were used as the secondary antibodies. Antibody-tagged protein spots were detected by 3, 3′-diaminobenzidine (DAB).

4.9. Co-Immunoprecipation (Co-IP) Assay

The plant cell lysate in 1 ml of western and IP buffer [20 mM Tris (pH 7.5), 150 mM NaCl, 1% Triton X-100] were incubated with the lab-preserved rabbit antisera polyclonal Flag-tag antibody (Thermo Fisher Scientific, USA) dilutions following the manufacturer's instructions for overnight at 4 °C. The next day samples incubated together with protein G Agarose (PGS) beads for 3 to 5 h of slowly shaking at 4 °C then centrifuged ($1500\times g$) to collect proteins, and were then washed with PBS buffer. After 5 times centrifugation and washing, the protein sample was transferred to a new Eppendorf tube and SDS-PAGE was performed, as mentioned above.

4.10. Statistical Analysis

Functional characterization of differential proteins was done by using, Web gene ontology annotation plot (WEGO), and Kyoto encyclopedia of genes and genomes (KEGG) by Ye et al. [98]. Origin 8.0 was used for graphical analysis [99] and SPSS was used for statistical analysis [100].

5. Conclusions

Nitrogen (N)-based fertilizer is a key factor for crop productivity. To maximize yields, farmers extensively use the nitrogenous fertilizers, however, a limited amount of N is utilized by crop plants and the remaining amount causes severe environmental pollution. The development of genotypes with high nitrogen utilization efficiency (NUE) which can grow efficiently and sustain yields in low N conditions are imperative for sustainable agriculture. In the present study, two isogenic lines wild-type, kitaake, and *PP2C9TL* overexpressing the protein phosphatase gene were used for comparative proteomics to identify candidate genes related with high NUE. This study confirmed the differential expression of the protein in the two isogenic lines. In the leaf proteome, 30 protein spots were differentially expressed between two genotypes under low N level, and mostly, the proteins were involved in energy and metabolism. This provides evidence about the connection between N and carbon metabolism. However, the 14-3-3 protein was reported to reduce N- and C-containing compounds and negatively affect the TCA cycle in low N conditions. In this study, the *PP2C9TL* overexpressed *PP2C9* gene downregulated 14-3-3 to enhance NR activity. The proteomics analysis confirmed that *PP2C9TL* exhibits a higher energy level (TCA cycle) than *WT* due to the downregulation of 14-3-3 protein. Physiological and proteomics studies showed that overexpression of protein phosphatase in *PP2C9TL* significantly increased the NUE. Physiological study coupled with proteomic analysis provides a potential basis for genetic-based selection of N-related genotypes to increase N efficiency in cereal crops.

Supplementary Materials: Supplementary materials can be found at http://www.mdpi.com/1422-0067/19/9/2827/s1. Figure S1 Phenotypic response of *WT* and *PP2C9TL* under different levels of N at heading stage. Figure S2: Plant height of *WT* and *PP2C9TL* under different levels of N at Tillering, Heading, and Maturity stage. Figure S3: Representative 2-DE gel electrophoresis images from the leaf total proteins of WT and PP2C9TL, sampled at 10 DAF under control N treatment. Figure S4: Representative 2-DE gel electrophoresis images from the leaf total proteins of WT and PP2C9TL, sampled at 10 DAF under low N treatment. Figure S5: Representative 2-DE gel electrophoresis images of differentially expressed proteins of WT and PP2C9TL leaves, sampled at 10 DAF under low N treatment. Figure S6: Representative 2-DE gel electrophoresis images of differentially expressed proteins of WT and PP2C9TL leaves, sampled at 10 DAF under control N treatment.

Author Contributions: M.W., S.F., H.A., Z.L., and W.Z. carried out the experiments. W.M. analyzed the results and wrote the manuscripts. W.L. designed experiments and corrected the manuscript. C.F., Y.A., M.U.K., P.L., and M.T. contributed scientific advice, critical reading, and editing of the manuscript. All authors read and approved the manuscript.

Funding: This work was sponsored by the National Key Research and Development program of China (2016YFD0300508), the National Natural Science Foundation of China (No. 31271670, 31401306), Fujian-Taiwan Joint Innovative Centre for Germplasm Resources and Cultivation of Crop (grant No. 2015-75, Fujian 2011 program, China), and the Central leading local science and technology development special fund project (2017L3010008).

Conflicts of Interest: The authors declare no conflicts of interest.

References

1. Amaral, L.R.; Molin, J.P.; Schepers, J.S. Algorithm for variable-rate nitrogen application in sugarcane based on active crop canopy sensor. *Agron. J.* **2015**, *107*, 1513–1523. [CrossRef]

2. Cao, Q.; Miao, Y.; Shen, J.; Yu, W.; Yuan, F.; Cheng, S.; Huang, S.; Wang, H.; Yang, W.; Liu, F. Improving in-season estimation of rice yield potential and responsiveness to topdressing nitrogen application with crop circle active crop canopy sensor. *Precis. Agric.* **2016**, *17*, 136–154. [CrossRef]

3. Khush, G.S. What it will take to feed 5.0 billion rice consumers in 2030. *Plant Mol. Biol.* **2005**, *59*, 1–6. [CrossRef] [PubMed]

4. Good, A.G.; Beatty, P.H. Fertilizing nature: A tragedy of excess in the commons. *PLoS Biol.* **2011**, *9*, e1001124. [CrossRef] [PubMed]

5. Tilman, D.; Clark, M. Food, agriculture & the environment: Can we feed the world & save the earth? *Daedalus* **2015**, *144*, 8–23.

6. Tilman, D.; Fargione, J.; Wolff, B.; D'antonio, C.; Dobson, A.; Howarth, R.; Schindler, D.; Schlesinger, W.H.; Simberloff, D.; Swackhamer, D. Forecasting agriculturally driven global environmental change. *Science* **2001**, *292*, 281–284. [CrossRef] [PubMed]

7. Prinsi, B.; Negri, A.S.; Pesaresi, P.; Cocucci, M.; Espen, L. Evaluation of protein pattern changes in roots and leaves of *Zea mays* plants in response to nitrate availability by two-dimensional gel electrophoresis analysis. *BMC Plant Biol.* **2009**, *9*, 113–129. [CrossRef] [PubMed]

8. FAO. *World Fertilizer Trends and Outlook to 2018*; FAO: Rome, Italy, 2018; p. 66.

9. Hakeem, K.R.; Ahmad, A.; Iqbal, M.; Gucel, S.; Ozturk, M. Nitrogen-efficient rice cultivars can reduce nitrate pollution. *Environ. Sci. Pollut. Res.* **2011**, *18*, 1184–1193. [CrossRef] [PubMed]

10. Kant, S.; Bi, Y.-M.; Rothstein, S.J. Understanding plant response to nitrogen limitation for the improvement of crop nitrogen use efficiency. *J. Exp. Bot.* **2010**, *62*, 1499–1509. [CrossRef] [PubMed]

11. Huang, S.; Zhao, C.; Zhang, Y.; Wang, C. Nitrogen use efficiency in rice. In *Nitrogen in Agriculture-Updates*; InTech: Christchurch, New Zealand, 2018.

12. Pathak, R.R.; Ahmad, A.; Lochab, S.; Raghuram, N. Molecular physiology of plant nitrogen use efficiency and biotechnological options for its enhancement. *Curr. Sci.* **2008**, 1394–1403.

13. Waqas, M.; Faheem, M.; Khan, A.S.; Shehzad, M.; Ansari, M.A.A. Estimation of heritability and genetic advance for some yield traits in eight f2 populations of wheat (*Triticum aestivum* L.). *Sci. Lett.* **2014**, *2*, 43–47.

14. Masclaux-Daubresse, C.; Daniel-Vedele, F.; Dechorgnat, J.; Chardon, F.; Gaufichon, L.; Suzuki, A. Nitrogen uptake, assimilation and remobilization in plants: Challenges for sustainable and productive agriculture. *Ann. Bot.* **2010**, *105*, 1141–1157. [CrossRef] [PubMed]

15. Konishi, M.; Yanagisawa, S. The regulatory region controlling the nitrate-responsive expression of a nitrate reductase gene, nia1, in *arabidopsis*. *Plant Cell Physiol.* **2011**, *52*, 824–836. [CrossRef] [PubMed]

16. Bloom, A.J.; Burger, M.; Asensio, J.S.R.; Cousins, A.B. Carbon dioxide enrichment inhibits nitrate assimilation in wheat and *arabidopsis*. *Science* **2010**, *328*, 899–903. [CrossRef] [PubMed]

17. McAllister, C.H.; Beatty, P.H.; Good, A.G. Engineering nitrogen use efficient crop plants: The current status. *Plant Biotechnol. J.* **2012**, *10*, 1011–1025. [CrossRef] [PubMed]

18. Fraisier, V.; Gojon, A.; Tillard, P.; Daniel-Vedele, F. Constitutive expression of a putative high-affinity nitrate transporter in *Nicotiana plumbaginifolia*: Evidence for post-transcriptional regulation by a reduced nitrogen source. *Plant. J.* **2000**, *23*, 489–496. [CrossRef] [PubMed]

19. Laugier, E.; Bouguyon, E.; Mauriès, A.; Tillard, P.; Gojon, A.; Lejay, L. Regulation of high-affinity nitrate uptake in roots of arabidopsis depends predominantly on posttranscriptional control of the nrt2. 1/nar2. 1 transport system. *Plant Physiol.* **2012**, *158*, 1067–1078. [CrossRef] [PubMed]

20. Han, M.; Okamoto, M.; Beatty, P.H.; Rothstein, S.J.; Good, A.G. The genetics of nitrogen use efficiency in crop plants. *Annu. Rev. Genet.* **2015**, *49*, 269–289. [CrossRef] [PubMed]

21. Hoshida, H.; Tanaka, Y.; Hibino, T.; Hayashi, Y.; Tanaka, A.; Takabe, T.; Takabe, T. Enhanced tolerance to salt stress in transgenic rice that overexpresses chloroplast glutamine synthetase. *Plant Mol. Biol.* **2000**, *43*, 103–111. [CrossRef] [PubMed]

22. Habash, D.; Massiah, A.; Rong, H.; Wallsgrove, R.; Leigh, R. The role of cytosolic glutamine synthetase in wheat. *Ann. Appl. Biol.* **2001**, *138*, 83–89. [CrossRef]

23. Konishi, N.; Saito, M.; Imagawa, F.; Kanno, K.; Yamaya, T.; Kojima, S. Cytosolic glutamine synthetase isozymes play redundant roles in ammonium assimilation under low-ammonium conditions in roots of *Arabidopsis thaliana*. *Plant Cell Physiol.* **2018**, *59*, 601–613. [CrossRef] [PubMed]

24. Moose, S.; Below, F.E. Biotechnology approaches to improving maize nitrogen use efficiency. In *Molecular Genetic Approaches to Maize Improvement*; Springer: New York, NY, USA, 2009; pp. 65–77.

25. Moorhead, G.B.G.; Trinkle-Mulcahy, L.; Ulke-Lemée, A. Emerging roles of nuclear protein phosphatases. *Nat. Rev. Mol. Cell Biol.* **2007**, *8*, 234–244. [CrossRef] [PubMed]

26. Singh, A.; Pandey, A.; Srivastava, A.K.; Tran, L.-S.P.; Pandey, G.K. Plant protein phosphatases 2c: From genomic diversity to functional multiplicity and importance in stress management. *Crit. Rev. Biotechnol.* **2016**, *36*, 1023–1035. [CrossRef] [PubMed]

27. de Torres-Zabala, M.; Truman, W.; Bennett, M.H.; Lafforgue, G.; Mansfield, J.W.; Egea, P.R.; Bögre, L.; Grant, M. *Pseudomonas syringae* pv. *Tomato* hijacks the *arabidopsis* abscisic acid signalling pathway to cause disease. *EMBO J.* **2007**, *26*, 1434–1443. [PubMed]

28. Cai, Z.; Liu, J.; Wang, H.; Yang, C.; Chen, Y.; Li, Y.; Pan, S.; Dong, R.; Tang, G.; de Dios Barajas-Lopez, J. Gsk3-like kinases positively modulate abscisic acid signaling through phosphorylating subgroup iii snrk2s in *arabidopsis*. *Proc. Natl. Acad. Sci. USA* **2014**, *111*, 9651–9656. [CrossRef] [PubMed]

29. Castillo, M.-C.; Lozano-Juste, J.; González-Guzmán, M.; Rodriguez, L.; Rodriguez, P.L.; León, J. Inactivation of pyr/pyl/rcar aba receptors by tyrosine nitration may enable rapid inhibition of aba signaling by nitric oxide in plants. *Sci. Signal.* **2015**, *8*, ra89. [CrossRef] [PubMed]

30. Wang, K.; He, J.; Zhao, Y.; Wu, T.; Zhou, X.; Ding, Y.; Kong, L.; Wang, X.; Wang, Y.; Li, J. Ear1 negatively regulates aba signaling by enhancing 2c protein phosphatase activity. *Plant Cell* **2018**, *30*, 815–834. [CrossRef] [PubMed]

31. Krzywińska, E.; Bucholc, M.; Kulik, A.; Ciesielski, A.; Lichocka, M.; Dębski, J.; Ludwików, A.; Dadlez, M.; Rodriguez, P.L.; Dobrowolska, G. Phosphatase abi1 and okadaic acid-sensitive phosphoprotein phosphatases inhibit salt stress-activated snrk2. 4 kinase. *BMC Plant Biol.* **2016**, *16*, 136–147. [CrossRef] [PubMed]

32. Ho, C.-H.; Lin, S.-H.; Hu, H.-C.; Tsay, Y.-F. Chl1 functions as a nitrate sensor in plants. *Cell* **2009**, *138*, 1184–1194. [CrossRef] [PubMed]

33. Léran, S.; Edel, K.H.; Pervent, M.; Hashimoto, K.; Corratgé-Faillie, C.; Offenborn, J.N.; Tillard, P.; Gojon, A.; Kudla, J.; Lacombe, B. Nitrate sensing and uptake in *arabidopsis* are enhanced by abi2, a phosphatase inactivated by the stress hormone abscisic acid. *Sci. Signal.* **2015**, *8*, ra43. [CrossRef] [PubMed]

34. Zhang, Z.; Chen, J.; Lin, S.; Li, Z.; Cheng, R.; Fang, C.; Chen, H.; Lin, W. Proteomic and phosphoproteomic determination of aba's effects on grain-filling of *Oryza sativa* L. Inferior spikelets. *Plant. Sci.* **2012**, *185*, 259–273. [CrossRef] [PubMed]

35. Zhang, Z.; Zhao, H.; Tang, J.; Li, Z.; Li, Z.; Chen, D.; Lin, W. A proteomic study on molecular mechanism of poor grain-filling of rice (*Oryza sativa* L.) inferior spikelets. *PLoS ONE* **2014**, *9*, e89140. [CrossRef] [PubMed]

36. West, P.C.; Gerber, J.S.; Engstrom, P.M.; Mueller, N.D.; Brauman, K.A.; Carlson, K.M.; Cassidy, E.S.; Johnston, M.; MacDonald, G.K.; Ray, D.K. Leverage points for improving global food security and the environment. *Science* **2014**, *345*, 325–328. [CrossRef] [PubMed]

37. Ju, X.-T.; Xing, G.-X.; Chen, X.-P.; Zhang, S.-L.; Zhang, L.-J.; Liu, X.-J.; Cui, Z.-L.; Yin, B.; Christie, P.; Zhu, Z.-L. Reducing environmental risk by improving n management in intensive chinese agricultural systems. *Proc. Natl. Acad. Sci. USA* **2009**, *106*, 3041–3046. [CrossRef] [PubMed]

38. Chen, X.; Cui, Z.; Fan, M.; Vitousek, P.; Zhao, M.; Ma, W.; Wang, Z.; Zhang, W.; Yan, X.; Yang, J. Producing more grain with lower environmental costs. *Nature* **2014**, *514*, 486–489. [CrossRef] [PubMed]

39. Peng, S.; Buresh, R.J.; Huang, J.; Yang, J.; Zou, Y.; Zhong, X.; Wang, G.; Zhang, F. Strategies for overcoming low agronomic nitrogen use efficiency in irrigated rice systems in china. *Field Crops Res.* **2006**, *96*, 37–47. [CrossRef]

40. Xu, G.; Fan, X.; Miller, A.J. Plant nitrogen assimilation and use efficiency. *Annu. Rev. Plant Biol.* **2012**, *63*, 153–182. [CrossRef] [PubMed]

41. Zhang, Z.; Zhang, Y.; Liu, X.; Li, Z.; Lin, W. The use of comparative quantitative proteomics analysis in rice grain-filling in determining response to moderate soil drying stress. *Plant Growth Regul.* **2017**, *82*, 219–232. [CrossRef]

42. Shin, R.; Alvarez, S.; Burch, A.Y.; Jez, J.M.; Schachtman, D.P. Phosphoproteomic identification of targets of the *arabidopsis* sucrose nonfermenting-like kinase snrk2. 8 reveals a connection to metabolic processes. *Proc. Natl. Acad. Sci. USA* **2007**, *104*, 6460–6465. [CrossRef] [PubMed]

43. Comparot, S.; Lingiah, G.; Martin, T. Function and specificity of 14-3-3 proteins in the regulation of carbohydrate and nitrogen metabolism. *J. Exp. Bot.* **2003**, *54*, 595–604. [CrossRef] [PubMed]

44. Diaz, C.; Kusano, M.; Sulpice, R.; Araki, M.; Redestig, H.; Saito, K.; Stitt, M.; Shin, R. Determining novel functions of *arabidopsis* 14-3-3 proteins in central metabolic processes. *BMC Syst. Biol.* **2011**, *5*, 192–203. [CrossRef] [PubMed]

45. Robertlee, J.; Kobayashi, K.; Suzuki, M.; Muranaka, T. Akin10, a representative *arabidopsis* snf1-related protein kinase 1 (snrk1), phosphorylates and downregulates plant hmg-coa reductase. *FEBS Lett.* **2017**, *591*, 1159–1166. [CrossRef] [PubMed]

46. Wang, F.; Ren, G.; Li, F.; Wang, B.; Yang, Y.; Ma, X.; Niu, Y.; Ye, Y.; Chen, X.; Fan, S. Overexpression of gmsnrk1, a soybean sucrose non-fermenting-1 related protein kinase 1 gene, results in directional alteration of carbohydrate metabolism in transgenic *arabidopsis*. *Biotechnol. Biotechnol. Equip.* **2018**, *32*, 1–11. [CrossRef]

47. Ross, F.A.; MacKintosh, C.; Hardie, D.G. Amp-activated protein kinase: A cellular energy sensor that comes in 12 flavours. *FEBS J.* **2016**, *283*, 2987–3001. [CrossRef] [PubMed]

48. Xu, H.; Zhao, X.; Guo, C.; Chen, L.; Li, K. Spinach 14-3-3 protein interacts with the plasma membrane h+-atpase and nitrate reductase in response to excess nitrate stress. *Plant Physiol. Biochem.* **2016**, *106*, 187–197. [CrossRef] [PubMed]

49. Bachmann, M.; Huber, J.L.; Athwal, G.S.; Wu, K.; Ferl, R.J.; Huber, S.C. 14-3-3 proteins associate with the regulatory phosphorylation site of spinach leaf nitrate reductase in an isoform-specific manner and reduce dephosphorylation of ser-543 by endogenous protein phosphatases. *FEBS Lett.* **1996**, *398*, 26–30. [CrossRef]

50. Moll, R.H.; Kamprath, E.J.; Jackson, W.A. Analysis and interpretation of factors which contribute to efficiency of nitrogen utilization. *Agron. J.* **1982**, *74*, 562–564. [CrossRef]

51. Gallais, A.; Hirel, B. An approach to the genetics of nitrogen use efficiency in maize. *J. Exp. Bot.* **2004**, *55*, 295–306. [CrossRef] [PubMed]

52. Liu, R.; Zhang, H.; Zhao, P.; Zhang, Z.; Liang, W.; Tian, Z.; Zheng, Y. Mining of candidate maize genes for nitrogen use efficiency by integrating gene expression and QTL data. *Plant Mol. Biol. Rep.* **2012**, *30*, 297–308. [CrossRef]

53. Basu, S.; Ramegowda, V.; Kumar, A.; Pereira, A. Plant adaptation to drought stress. *F1000Res* **2016**, *5*, 1554–1563. [CrossRef] [PubMed]

54. Foulkes, M.J.; Hawkesford, M.J.; Barraclough, P.B.; Holdsworth, M.J.; Kerr, S.; Kightley, S.; Shewry, P.R. Identifying traits to improve the nitrogen economy of wheat: Recent advances and future prospects. *Field Crops Res.* **2009**, *114*, 329–342. [CrossRef]

55. Rennenberg, H.; Wildhagen, H.; Ehlting, B. Nitrogen nutrition of poplar trees. *Plant. Biol.* **2010**, *12*, 275–291. [CrossRef] [PubMed]

56. Liu, X.; Ju, X.; Zhang, F.; Pan, J.; Christie, P. Nitrogen dynamics and budgets in a winter wheat–maize cropping system in the north china plain. *Field Crops Res.* **2003**, *83*, 111–124. [CrossRef]

57. Wang, J.-G.; Liu, H.; Sun, H.; Hua, W.; Wang, H.; Liu, X.; Wei, B. One-pot synthesis of nitrogen-doped ordered mesoporous carbon spheres for high-rate and long-cycle life supercapacitors. *Carbon* **2018**, *127*, 85–92. [CrossRef]

58. Hakeem, K.R.; Chandna, R.; Ahmad, A.; Qureshi, M.I.; Iqbal, M. Proteomic analysis for low and high nitrogen-responsive proteins in the leaves of rice genotypes grown at three nitrogen levels. *Appl. Biochem. Biotechnol.* **2012**, *168*, 834–850. [CrossRef] [PubMed]

59. Wang, X.; Bian, Y.; Cheng, K.; Zou, H.; Sun, S.S.-M.; He, J.-X. A comprehensive differential proteomic study of nitrate deprivation in *arabidopsis* reveals complex regulatory networks of plant nitrogen responses. *J. Proteome Res.* **2012**, *11*, 2301–2315. [CrossRef] [PubMed]

60. Imai, K.; Suzuki, Y.; Mae, T.; Makino, A. Changes in the synthesis of RuBisCO in rice leaves in relation to senescence and n influx. *Ann. Bot.* **2008**, *101*, 135–144. [CrossRef] [PubMed]

61. Koyama, K.; Kang, S.-S.; Huang, W.; Yanagawa, Y.; Takahashi, Y.; Nagano, M. Aging-related changes in in vitro-matured bovine oocytes: Oxidative stress, mitochondrial activity and atp content after nuclear maturation. *J. Reprod. Dev.* **2014**, *60*, 136–142. [CrossRef] [PubMed]

62. Khueychai, S.; Jangpromma, N.; Daduang, S.; Jaisil, P.; Lomthaisong, K.; Dhiravisit, A.; Klaynongsruang, S. Comparative proteomic analysis of leaves, leaf sheaths, and roots of drought-contrasting sugarcane cultivars in response to drought stress. *Acta Physiol. Plant.* **2015**, *37*, 88–103. [CrossRef]

63. Nazir, M.; Pandey, R.; Siddiqi, T.O.; Ibrahim, M.M.; Qureshi, M.I.; Abraham, G.; Vengavasi, K.; Ahmad, A. Nitrogen-deficiency stress induces protein expression differentially in low-n tolerant and low-n sensitive maize genotypes. *Front. Plant Sci.* **2016**, *7*, 298. [CrossRef] [PubMed]

64. Wang, B.; Ratajczak, R.; Zhang, J. Activity, amount and subunit composition of vacuolar-type h+-atpase and h+-ppase in wheat roots under severe nacl stress. *J. Plant Physiol.* **2000**, *157*, 109–116. [CrossRef]

65. Asaoka, M.; Segami, S.; Ferjani, A.; Maeshima, M. Contribution of ppi-hydrolyzing function of vacuolar h+-pyrophosphatase in vegetative growth of *arabidopsis*: Evidenced by expression of uncoupling mutated enzymes. *Front. Plant Sci.* **2016**, *7*, 415. [CrossRef] [PubMed]

66. Suzuki, S.; Murai, N.; Kasaoka, K.; Hiyoshi, T.; Imaseki, H.; Burnell, J.N.; Arai, M. Carbon metabolism in transgenic rice plants that express phosphoenolpyruvate carboxylase and/or phosphoenolpyruvate carboxykinase. *Plant. Sci.* **2006**, *170*, 1010–1019. [CrossRef]

67. Melzer, E.; O'Leary, M.H. Anapleurotic CO_2 fixation by phosphoenolpyruvate carboxylase in c3 plants. *Plant Physiol.* **1987**, *84*, 58–60. [CrossRef] [PubMed]

68. Shi, J.; Yi, K.; Liu, Y.; Xie, L.; Zhou, Z.; Chen, Y.; Hu, Z.; Zheng, T.; Liu, R.; Chen, Y. Phosphoenolpyruvate carboxylase in *arabidopsis* leaves plays a crucial role in carbon and nitrogen metabolism. *Plant Physiol.* **2015**, *167*, 671–681. [CrossRef] [PubMed]

69. Moghaddam, S.S.; Ibrahim, R.; Damalas, C.A.; Noorhosseini, S.A. Effects of gamma stress and carbon dioxide on eight bioactive flavonoids and photosynthetic efficiency in *Centella asiatica*. *J. Plant Growth Regul.* **2017**, *36*, 957–969. [CrossRef]

70. Yang, Z.; Yuan, C.; Han, W.; Li, Y.; Xiao, F. Effects of low irradiation on photosynthesis and antioxidant enzyme activities in cucumber during ripening stage. *Photosynthetica* **2016**, *54*, 251–258. [CrossRef]

71. Campos, H.; Trejo, C.; Peña-Valdivia, C.B.; García-Nava, R.; Conde-Martínez, F.V.; Cruz-Ortega, M.R. Stomatal and non-stomatal limitations of bell pepper (*Capsicum annuum* L.) plants under water stress and re-watering: Delayed restoration of photosynthesis during recovery. *Environ. Exp. Bot.* **2014**, *98*, 56–64. [CrossRef]

72. Medrano, H.; Escalona, J.M.; Bota, J.; Gulías, J.; Flexas, J. Regulation of photosynthesis of c3 plants in response to progressive drought: Stomatal conductance as a reference parameter. *Ann. Bot.* **2002**, *89*, 895–905. [CrossRef] [PubMed]

73. Jiang, S.-S.; Liang, X.-N.; Li, X.; Wang, S.-L.; Lv, D.-W.; Ma, C.-Y.; Li, X. H.; Ma, W.-J.; Yan, Y.-M. Wheat drought-responsive grain proteome analysis by linear and nonlinear 2-de and maldi-tof mass spectrometry. *Int. J. Mol. Sci.* **2012**, *13*, 16065–16083. [CrossRef] [PubMed]

74. Sato, T.; Shimoda, Y.; Matsuda, K.; Tanaka, A.; Ito, H. Mg-dechelation of chlorophyll a by stay-green activates chlorophyll b degradation through expressing non-yellow coloring 1 in *Arabidopsis thaliana*. *J. Plant Physiol.* **2018**, *222*, 94–102. [CrossRef] [PubMed]

75. Jiang, H.; Li, M.; Liang, N.; Yan, H.; Wei, Y.; Xu, X.; Liu, J.; Xu, Z.; Chen, F.; Wu, G. Molecular cloning and function analysis of the stay green gene in rice. *Plant. J.* **2007**, *52*, 197–209. [CrossRef] [PubMed]

76. Rosales, E.P.; Iannone, M.F.; Groppa, M.D.; Benavides, M.P. Nitric oxide inhibits nitrate reductase activity in wheat leaves. *Plant Physiol. Biochem.* **2011**, *49*, 124–130. [CrossRef] [PubMed]

77. Kim, J.Y.; Park, B.S.; Park, S.W.; Lee, H.Y.; Song, J.T.; Seo, H.S. Nitrate reductases are relocalized to the nucleus by atsiz1 and their levels are negatively regulated by cop1 and ammonium. *Int. J. Mol. Sci.* **2018**, *19*, 1202–1216. [CrossRef] [PubMed]

78. Su, W.; Mertens, J.A.; Kanamaru, K.; Campbell, W.H.; Crawford, N.M. Analysis of wild-type and mutant plant nitrate reductase expressed in the methylotrophic yeast *Pichia pastoris*. *Plant Physiol.* **1997**, *115*, 1135–1143. [CrossRef] [PubMed]

79. Lea, P.J.; Miflin, B.J. Nitrogen assimilation and its relevance to crop improvement. *Annu. Plant. Rev.* **2018**, *50*, 1–40.

80. Zhang, H.; Liang, C.; Aoki, N.; Kawai, K.; Takane, K.-I.; Ohsugi, R. Introduction of a fungal nadp (h)-dependent glutamate dehydrogenase (gdha) improves growth, grain weight and salt resistance by enhancing the nitrogen uptake efficiency in forage rice. *Plant Prod. Sci.* **2016**, *19*, 267–278. [CrossRef]

81. Wallsgrove, R.M.; Turner, J.C.; Hall, N.P.; Kendall, A.C.; Bright, S.W.J. Barley mutants lacking chloroplast glutamine synthetase—Biochemical and genetic analysis. *Plant Physiol.* **1987**, *83*, 155–158. [CrossRef] [PubMed]

82. Gazanchian, A.; Hajheidari, M.; Sima, N.K.; Salekdeh, G.H. Proteome response of *Elymus elongatum* to severe water stress and recovery. *J. Exp. Bot.* **2007**, *58*, 291–300. [CrossRef] [PubMed]

83. Zhang, M.; Li, G.; Huang, W.; Bi, T.; Chen, G.; Tang, Z.; Su, W.; Sun, W. Proteomic study of *Carissa spinarum* in response to combined heat and drought stress. *Proteomics* **2010**, *10*, 3117–3129. [CrossRef] [PubMed]

84. Wang, X.; Dinler, B.S.; Vignjevic, M.; Jacobsen, S.; Wollenweber, B. Physiological and proteome studies of responses to heat stress during grain filling in contrasting wheat cultivars. *Plant Sci.* **2015**, *230*, 33–50. [CrossRef] [PubMed]

85. Kosová, K.; Vítámvás, P.; Prášil, I.T. Proteomics of stress responses in wheat and barley—Search for potential protein markers of stress tolerance. *Front. Plant Sci.* **2014**, *5*, 711. [CrossRef] [PubMed]

86. Yang, Z.-K.; Ma, Y.-H.; Zheng, J.-W.; Yang, W.-D.; Liu, J.-S.; Li, H.-Y. Proteomics to reveal metabolic network shifts towards lipid accumulation following nitrogen deprivation in the diatom *Phaeodactylum tricornutum*. *J. Appl. Phycol.* **2014**, *26*, 73–82. [CrossRef] [PubMed]

87. Dietz, K.-J. Thiol-based peroxidases and ascorbate peroxidases: Why plants rely on multiple peroxidase systems in the photosynthesizing chloroplast? *Mol. Cells* **2016**, *39*, 20–25. [PubMed]

88. Turner, D.D.; Lad, L.; Kwon, H.; Basran, J.; Sharp, K.H.; Moody, P.C.E.; Raven, E.L. The role of ala134 in controlling substrate binding and reactivity in ascorbate peroxidase. *J. Inorg. Biochem.* **2017**, *180*, 230–234. [CrossRef] [PubMed]

89. Motojima, F.; Fujii, K.; Yoshida, M. Chaperonin facilitates protein folding by avoiding polypeptide collapse. *bioRxiv* **2017**. [CrossRef]

90. Rao, T.; Lund, P.A. Differential expression of the multiple chaperonins of *Mycobacterium smegmatis*. *FEMS Microbiol. Lett.* **2010**, *310*, 24–31. [CrossRef] [PubMed]

91. Rajaram, H.; Chaurasia, A.K.; Potnis, A.A. Multiple chaperonins in cyanobacteria: Why one is not enough! In *Prokaryotic Chaperonins*; Springer: New York, NY, USA, 2017; pp. 93–109.

92. Hiei, Y.; Komari, T. *Agrobacterium*-mediated transformation of rice using immature embryos or calli induced from mature seed. *Nat. Protoc.* **2008**, *3*, 824–834. [CrossRef] [PubMed]

93. Yoshida, S. Routine procedure for growing rice plants in culture solution. *Lab. Man. Physiol. Stud. Rice* **1976**, 61–66.

94. Xiong, D.; Chen, J.; Yu, T.; Gao, W.; Ling, X.; Li, Y.; Peng, S.; Huang, J. Spad-based leaf nitrogen estimation is impacted by environmental factors and crop leaf characteristics. *Sci. Rep.* **2015**, *5*, 13389–13399. [CrossRef] [PubMed]

95. Sáez-Plaza, P.; Navas, M.J.; Wybraniec, S.; Michałowski, T.; Asuero, A.G. An overview of the kjeldahl method of nitrogen determination. Part ii. Sample preparation, working scale, instrumental finish, and quality control. *Crit. Rev. Anal. Chem.* **2013**, *43*, 224–272. [CrossRef]

96. Li, Z.; Li, Z.; Muhammad, W.; Lin, M.; Azeem, S.; Zhao, H.; Lin, S.; Chen, T.; Fang, C.; Letuma, P. Proteomic analysis of positive influence of alternate wetting and moderate soil drying on the process of rice grain filling. *Plant Growth Regul.* **2018**, *84*, 533–548. [CrossRef]

97. Bradford, M.M. A rapid and sensitive method for the quantitation of microgram quantities of protein utilizing the principle of protein-dye binding. *Anal. Biochem.* **1976**, *72*, 248–254. [CrossRef]

98. Ye, J.; Fang, L.; Zheng, H.; Zhang, Y.; Chen, J.; Zhang, Z.; Wang, J.; Li, S.; Li, R.; Bolund, L. Wego: A web tool for plotting go annotations. *Nucleic Acids Res.* **2006**, *34*, W293–W297. [CrossRef] [PubMed]

99. Deschenes, L.A.; Vanden Bout, D.A. Origin 6.0: Scientific data analysis and graphing software origin lab corporation (formerly Microcal software, Inc.). *J. Am. Chem. Soc.* **2000**, *122*, 9567–9568. [CrossRef]

100. Barrett, K.C.; Morgan, G.A.; Leech, N.L.; Gloeckner, G.W. *IBM SPSS for Introductory Statistics: Use and Interpretation*; Routledge: Abingdon, UK, 2012.

International Journal of
Molecular Sciences

MDPI

Article

New Insights on *Arabidopsis thaliana* Root Adaption to Ammonium Nutrition by the Use of a Quantitative Proteomic Approach

Inmaculada Coleto [1], Izargi Vega-Mas [1], Gaetan Glauser [2], María Begoña González-Moro [1], Daniel Marino [1,3,*] and Idoia Ariz [4,*]

[1] Department of Plant Biology and Ecology, University of the Basque Country (UPV/EHU), Apdo. 644,
E-48080 Bilbao, Spain; inmaculada.coleto@ehu.eus (I.C.); izargiaida.vega@ehu.eus (I.V.-M.);
mariabegona.gonzalez@ehu.eus (M.B.G.-M.)
[2] Neuchâtel Platform of Analytical Chemistry, University of Neuchâtel, Avenue de Bellevaux 51,
2000 Neuchâtel, Switzerland; gaetan.glauser@unine.ch
[3] Ikerbasque, Basque Foundation for Science, E-48011 Bilbao, Spain
[4] Departamento de Biología Ambiental. Facultad de Ciencias, Universidad de Navarra, C/Irunlarrea 1,
31008 Pamplona, Spain
* Correspondence: daniel.marino@ehu.eus (D.M.); iariza@unav.es (I.A.);
Tel.: +34-946-017-957 (D.M.); +34-948-425-600 (I.A.)

Received: 22 January 2019; Accepted: 5 February 2019; Published: 14 February 2019

Abstract: Nitrogen is an essential element for plant nutrition. Nitrate and ammonium are the two major inorganic nitrogen forms available for plant growth. Plant preference for one or the other form depends on the interplay between plant genetic background and environmental variables. Ammonium-based fertilization has been shown less environmentally harmful compared to nitrate fertilization, because of reducing, among others, nitrate leaching and nitrous oxide emissions. However, ammonium nutrition may become a stressful situation for a wide range of plant species when the ion is present at high concentrations. Although studied for long time, there is still an important lack of knowledge to explain plant tolerance or sensitivity towards ammonium nutrition. In this context, we performed a comparative proteomic study in roots of *Arabidopsis thaliana* plants grown under exclusive ammonium or nitrate supply. We identified and quantified 68 proteins with differential abundance between both conditions. These proteins revealed new potential important players on root response to ammonium nutrition, such as H^+-consuming metabolic pathways to regulate pH homeostasis and specific secondary metabolic pathways like brassinosteroid and glucosinolate biosynthetic pathways.

Keywords: ammonium; *Arabidopsis thaliana*; carbon metabolism; nitrogen metabolism; nitrate; proteomics; root; secondary metabolism

1. Introduction

Nitrogen (N), despite being one of the essential macronutrients for plant development, is often a limiting element in agricultural soils. The two major inorganic N sources for plants in soils are nitrate (NO_3^-) and ammonium (NH_4^+). The first one is an anion (oxidation state of N, +5) and the second one, a cation (oxidation state of N, −3), thus, both N sources differ extremely in their chemical properties [1]. In the soil, both N sources are present. However, their relative abundance is reliant on its interaction with the microbiological and physicochemical characteristics of the soil, and plant preference for one or another form depends on the interplay between plant species and environmental variables such as soil properties or light [2]. The use of NO_3^- as fertilizer, as well as the high nitrification rates commonly observed in agricultural soils when urea or NH_4^+ are applied, have made that crop species are mostly adapted to nitrate nutrition. However, the increasing use of ammonium-based fertilizers formulated with nitrification inhibitors, which have been proven useful to mitigate the effect of agriculture on the environment [3,4], demands a deeper study of plants N source preference. Ammonium nutrition may represent a stressful situation for a wide range of plant species when it is applied at high concentrations. Growth reduction is the most common symptom of ammonium stress [5,6]. The toxicity degree is dependent on genetic features (inter- and intraspecies) and on chemical traits, such as external NH_4^+ concentration and pH [7–9]. Indeed, pH is known to play a key role in plants response to ammonium nutrition; importantly, plants adapted to acidic conditions have sometimes been reported to be tolerant to ammonium stress [10,11].

NO_3^-, with net negative charge, is co-transported with two protons, whereas NH_4^+, with net positive charge, is mainly transported through electrogenic transport via ammonium transporters (AMTs) or cation channel [12,13]. This uptake difference in terms of charge balance can significantly influence the uptake of other mineral nutrients and also cell metabolic homeostasis [14,15]. Moreover, NO_3^- and NH_4^+ assimilation is also different in terms of H^+ balance, reductants consumption and redox balance [16]. For the synthesis of one glutamate molecule, NO_3^- assimilation produces OH^- whereas NH_4^+ assimilation produces two H^+ [17,18]. Thus, the balance of H^+ production and consumption must be accurately controlled according to the N form absorbed to maintain the cytoplasmic pH and a favorable electrochemical gradient across cell membrane [19]. This control is exerted through the so-called "biophysical pH-stat" and "biochemical pH-stat". The "biophysical pH-stat" is based on the buffering capacity of HPO_4^{2-} and in the action of H^+-pumps (e.g., ATPases). The "biochemical pH-stat" is based on the activation of H^+-consuming metabolic pathways [20].

Although ammonium nutrition at high concentration (millimolar range, mM) is mostly reported as an unwanted situation that may affect crops yield, the metabolic adaptation to the presence of NH_4^+ as main N source, may entail benefits for plants. For instance, since ammonium nutrition is known to stimulate N assimilation machinery, an increase in protein yield has been reported in grains of wheat grown with NH_4^+ as N source [21]. Similarly, an increase in leaf glucosinolate content has been shown in ammonium-fed Arabidopsis and Brassica crops [15,22]. Besides, some works have reported a higher tolerance to abiotic stresses, such as drought and salinity, in different ammonium-fed plant species [23–26]. Moreover, considering the constant rising of atmospheric CO_2 concentrations, some authors have argued that C3 plants growing under ammonium nutrition responded more positively to elevated CO_2 than such plants growing under nitrate nutrition [1,27]. However, this is controversial since other works have not observed this effect [28,29].

Although plant response upon ammonium nutrition has been extensively studied, the molecular mechanisms governing the responses that lead plants to adapt their metabolism to tolerate this situation remain largely unknown. In this context, to find new actors, mechanisms and processes associated with plants ammonium response, we have performed a quantitative proteomic study in the root of Arabidopsis plants grown under a non-toxic ammonium condition, using nitrate nutrition as control. This approach provided some new clues for future research related to metabolic pathways and signaling processes involved in root adaptation to ammonium nutrition, such as the induction of secondary metabolism and the putative association between the gamma-aminobutyric acid (GABA) shunt, malate, and enzymes participating in biochemical pH-stat to regulate H^+ balance.

2. Results and Discussion

2.1. Physiological Response of Arabidopsis Roots under Ammonium Nutrition

Most plant species are sensitive to long-term ammonium nutrition at high concentration [5,6]. In this work, despite the different N source supplied (nitrate vs. ammonium), plants showed similar total and root biomass (Table 1). Indeed, a relief from toxicity symptoms has often been observed when nutrient solutions are pH-buffered or when medium pH increases because this counteracts the medium acidification derived from ammonium nutrition [9,11,30]. Thus, growing the plant in a buffered medium (pH 6.5), we achieved a condition to study ammonium tolerance in Arabidopsis. Since plant root is the first organ that senses and responds to nutritional conditions, understanding how root adapts to non-toxic NH_4^+ nutrition is an important step to design practices for mineral N nutrition management in plants. In general, it is considered that the deleterious effect of ammonium nutrition at high concentrations in plants is a consequence of the excessive NH_4^+ uptake and accumulation in tissues, and plants with an enhanced synthesis of N-reduced compounds are more tolerant to ammonium nutrition [31,32]. To further ascertain whether ammonium-fed plants were suffering or not stress, we quantified internal NH_4^+ content, free amino acid and total soluble protein contents in roots. In this work, roots of ammonium-fed plants showed an increase of amino acid and soluble protein contents, whereas only a slight increase of internal NH_4^+ content compared with those grown with NO_3^-. These results indicate that plants were not facing a stressful situation (Table 1).

Table 1. Biomass, and NH_4^+, amino acids and soluble protein contents in 21 days-grown *Arabidopsis thaliana* plants (9 days in agar plates plus 12 days in 24-well plates) with nitrate or ammonium as sole N source.

Parameter	Nitrate	Ammonium
Total plant biomass (mg·FW·plant^{-1})	24.88 ± 1.48	24.41 ± 2.79
Shoot biomass (mg·FW·plant^{-1})	16.61 ± 0.69	**14.35 ± 0.31**
Root biomass (mg·FW·plant^{-1})	8.27 ± 0.82	10.06 ± 1.89
Root NH_4^+ content (nmol·mg^{-1} FW)	0.41 ± 0.03	**0.57 ± 0.05**
Root total free amino acids (nmol·Gln·mg^{-1}·FW)	4.02 ± 1.5	**11.57 ± 0.89**
Root total soluble protein (µg·mg^{-1}·FW)	4.1 ± 0.39	**9.13 ± 1.19**

Values represent mean $\pm$ SE (n = 20, for biomass and n = 4, for ammonium, amino acids and protein values). Significant differences between treatments are highlighted in bold text (Student *t*-test; $p < 0.05$). FW: Fresh weight.

In order to identify root metabolic pathways differentially regulated in both N conditions that could be "targets" to further research in ammonium-fed plants, we performed a comparative proteomic analysis.

2.2. Overview of Proteomic Analysis in Arabidopsis Roots Grown under Exclusive Nitrate or Ammonium Supply

A quantitative proteomic analysis, with isobaric tags for relative and absolute quantitation (iTRAQ), was used to analyze relative abundance of proteins in four independent pools of Arabidopsis roots per treatment (1 pool = 120 individual plant roots). Peptides of six or more amino acids in length, and with a maximum of two missed cleavages were exclusively considered for the analyses. For protein quantification, only proteins identified in at least three out of four samples per treatment and with two or more unique peptides identified were considered. Following these criteria (detailed information in Materials and Methods section), we identified 4469 proteins and quantified 799, out of them 68 proteins were differentially abundant ($p \leq 0.05$) in both N conditions (Table 2 and Supplementary Dataset S1). Among these 68 proteins, 31 showed a higher abundance in roots of ammonium-grown plants, whereas 37 showed a higher abundance in roots of nitrate-grown ones. Functional classification of differentially abundant proteins according to MapMan software analysis [33] revealed that a significant number of the differentially regulated proteins were associated with categories related to primary carbon (C) metabolism, in particular, to organic acid transformation, photorespiration, glycolysis, gluconeogenesis, carbohydrate metabolism, and amino acid metabolism (Table 2 and Supplementary Dataset S2). Importantly, most of these proteins showed higher abundance in root of ammonium-fed plants (Table 2). Differentially abundant proteins were also included in categories such as protein turnover (synthesis/degradation), signaling, abiotic stress, and redox response, among others. In addition, a number of proteins was related with transport processes, notably H^+ transport, which is a key aspect when leading to pH homeostasis control under ammonium nutrition (Table 2). Despite the different H^+ balance driven by distinct N forms used as N source, cytoplasmic pH stays mostly unchanged because of the pH-stat mechanisms [34]. These mechanisms for pH regulation are those related mainly to the biophysical pH-stat, mainly constituted by H^+ pumps, H^+ inclusion in vacuoles and H^+ release in the rhizosphere, and biochemical pH-stat [34]. Curiously, in this study, two proteins related to H^+ pumps were downregulated under ammonium nutrition, a P-type ATPase from the superfamily of cation-transporting ATPases (ATPase 2; P19456) and a V-type proton ATPase subunit E3 (P0CAN7) (Table 2). Consistent with these results, Marino et al. reported the lesser abundance of the proton pump-interactor 1 (O23144), which stimulates plasmatic membrane H^+-ATPase activity in vitro conditions, in the leaves of Arabidopsis grown with ammonium as N source [15,35]. Furthermore, transcriptomic studies in ammonium-fed plants also showed downregulation of genes associated to H^+ transport in vacuole and plasma membrane such as the vacuolar cation/proton exchanger 3-like gene (Solyc06g006110.2.1) in tomato, and the H^+-transporting plasma membrane ATPases (AT3G60330; AT4G30190) in Arabidopsis [36,37]. In addition, sorghum roots exposed to ammonium concentrations above 1 mM also showed decreased H^+-ATPase gene expression and activity [38]. It has been suggested that the biophysical pH-stat would be regulating pH homeostasis upon transitory pH variations; in contrast, it would not be effective upon long-term intracellular pH alterations [39]. Future studies about the relationship between ammonium uptake and homeostasis, and the "biophysical pH-stat" mechanisms, will be essential to further understand the potential involvement of H^+ pumps and specifically ATPases on pH regulation associated with ammonium nutrition.

Table 2. Functional classification of proteins showing differential abundance in roots of nitrate- and ammonium-cultured *A. thaliana* plants.

Protein Description	TAIR ID	Uniprot ID	*p*-Value	Fold Change NO_3^-/NH_4^+
Organic acid transformation				
Dihydrolipoyllysine-residue acetyltransferase component 5 of pyruvate dehydrogenase complex	AT1G34430	Q9C8P0	0.010	5.27
Dihydrolipoyl dehydrogenase 1	AT1G48030	Q9M5K3	0.019	0.30
ATP-citrate synthase alpha chain protein 3	AT1G09430	O80526	0.001	0.11
Photorespiration				
Serine hydroxymethyltransferase 3, chloroplastic	AT4G32520	Q94JQ3	0.011	0.19
Carbohydrate metabolism				
ADP-glucose pyrophosphorylase family protein	AT1G74910	F4HXD1	0.008	0.13
Glycolysis				
Pyruvate kinase	AT5G56350	Q9FM97	0.009	0.21
Gluconeogenese/Glyoxylate cycle				
Phosphoenolpyruvate carboxykinase (ATP)	AT4G37870	Q9T074	0.010	0.14
Mitochondrial electron transport				
Gamma carbonic anhydrase-like 1	AT5G63510	F4KAG8	0.003	0.17
Amino acid metabolism				
Glutamate decarboxylase 2	AT1G65960	Q42472	0.027	0.24
Aspartate semialdehyde dehydrogenase	AT1G14810	Q8VYI4	0.040	5.29
Methylmalonate-semialdehyde dehydrogenase (acylating)	AT2G14170	A8MQR6	0.048	4.36
Secondary metabolism				
3-isopropylmalate dehydratase large subunit 1	AT4G13430	Q94AR8	0.003	5.48
Methylthioalkylmalate synthase 3	AT5G23020	Q9FN52	0.003	3.25
Acetyl-CoA acetyltransferase	AT5G48230	Q854Y1-2	0.049	0.30
Betaine aldehyde dehydrogenase 1	AT1G74920	F4HXD2	0.037	0.26
Chalcone synthase	AT5G13930	P13114	0.039	0.26

Table 2. *Cont.*

Protein Description	TAIR ID	Uniprot ID	p-Value	Fold Change NO_3^-/NH_4^+
Hormone metabolism				
Delta(24)-sterol reductase	AT3G19820	Q39085	0.037	0.31
Co-factor and vitamin metabolism				
Nicotinate-nucleotide pyrophosphorylase (carboxylating)	AT2G01350	F41NA0	0.015	0.10
Cell wall synthesis				
UDP-glucuronic acid decarboxylase 3	AT5G59290	F4KHU8	0.029	2.93
Tetrapyrrole synthesis				
Protoporphyrinogen oxidase 2	AT5G14220	Q8S9J1-2	0.003	0.19
Abiotic stress				
Endoplasmin homolog	AT4G24190	F4JQ55	0.031	3.16
DnaJ protein ERDJ38	AT3G62600	Q9LZK5	0.002	6.67
Probable methyltransferase PMt24	AT1G29470	Q6NPR7	0.004	5.51
Germin-like protein subfamily T member 1	AT1G18970	P92995	0.031	3.55
Redox response				
Protein disulfide isomerase-like 1-2	AT1G21750	F4HZN9	0.036	4.41
Protein disulfide-isomerase	AT1G77510	Q9SRG3	0.002	8.12
Protein disulfide isomerase-like 1-6	AT3G16110	Q66GQ3	0.025	0.21
Thioredoxin reductase 2	AT2G17420	Q39242	0.013	0.28
Nucleotide metabolism				
Nudix hydrolase 3	AT1G79690	Q8L831	0.018	0.26
RNA processing				
Polyadenylate-binding protein 2	AT4G34110	P42731	0.048	0.39
Polyadenylate-binding protein 4	AT2G23350	O22173	0.006	0.18
Reactive intermediate deaminase A	AT3G20390	Q94JQ4	0.002	11.07
Protein synthesis				
30S ribosomal protein S3	ATCG00800	P56798	0.043	2.82
40S ribosomal protein S3-3	AT5G35530	Q9FJA6	0.004	7.52
Elongation factor Tu (mitochondrial)	AT4G02930	Q9ZT91	0.034	3.43
Elongation factor Tu (chloroplastic)	AT4G20360	P17745	0.026	5.32
Protein degradation				
26S proteasome non-ATPase regulatory subunit 14 homolog	AT5G23540	Q9LT08	0.012	0.27
Proteasome subunit beta type-4	AT1G56450	Q7DLR9	0.035	0.27
Protein targeting				
Nuclear pore complex protein NUP155	AT1G14850	F4HXV6	0.033	2.99
ADP-ribosylation factor 2-A	AT3G62290	Q9M1P5	0.031	4.87
Signaling				
Rho GDP-dissociation inhibitor 1	AT3G07880	Q9SFC6	0.021	5.59
Dynamin-related protein 1A	AT5G42080	P42697	0.003	5.79
GTP-binding nuclear protein Ran-1	AT5G20010	P41916	0.020	4.92
14-3-3-like protein GF14 chi	AT4G09000	P42643	0.032	0.22
Cell vesicle transport				
Tubulin beta 6-chain (Cell organization)	AT5G12250	P29514	0.020	0.28
AP-4 complex subunit epsilon	AT1G31730	Q8L7A9	0.030	3.63
Coatomer subunit beta-1	AT4G31480	Q95V21	0.007	8.98
Golgin candidate 5	AT1G79830	F4HQB9	0.015	5.49
COG complex component-related protein	AT5G51430	Q9FGN0	0.023	6.23
Transport				
V-type proton ATPase subunit E3	AT1G64200	P0CAN7	0.044	2.89
ATPase 2, plasma membrane-type	AT4G30190	P19456	0.046	3.26
Mitochondrial dicarboxylate/tricarboxylate transporter DTC	AT5G19760	Q9C5M0	0.010	3.54
ABC transporter A family member 2	AT3G47730	Q84K47	0.010	3.82
ABC transporter F family member 1	AT5G60790	Q9FJH5	0.009	0.15

Table 2. *Cont.*

Protein Description	TAIR ID	Uniprot ID	*p*-Value	Fold Change NO_3^-/NH_4^+
Miscellaneous				
AT3g23600/MDB19_9	ATG23600	Q9LUG8	0.012	4.83
Dolichyl-diphosphooligosaccharide-protein glycosyltransferase 48 kDa subunit	AT5G66680	Q944K2	0.009	0.17
Methylesterase 3	AT2G23610	O80477	0.0369	2.93
Glutathione S-transferase L3	AT5G02790	Q9LZO6	0.0218	3.61
Peroxidase 34	AT3G49120	Q9SMU8	0.0070	0.16
Peroxidase 30	AT3G21770	Q9LSY7	0.0165	4.46
AT4g13180/F17N18-70	AT4G13180	Q9SVQ9	0.0322	0.22
Not assigned ontology				
NADH dehydrogenase (ubiquinone) iron-sulfur protein 2	ATMG00510	P93306	0.0167	1.78
tRNA (guanine-N(7)-)-methyltransferase non-catalytic subunit	AT1G03110	Q93WD7	0.0376	0.31
Pheromone receptor, putative (AR401)	AT1G66680	Q9C9M1	0.0143	4.38
WD40 domain-containing protein	AT5G24710	F4K1H8	0.0265	0.29
Protein EMBRYO DEFECTIVE 2734	AT5G19820	Q93V68	0.0208	0.20
Calcium-dependent lipid-binding family protein	AT1G48090	F4HWS2	0.0156	0.29
Metal-dependent protein hydrolase	AT5G41970	F4K000	0.0349	3.09

Gene ontology (GO) enrichment analysis for cellular component (Figure 1A and Supplementary Dataset S3) and biological process (Figure 1B and Supplementary Dataset S3) was performed with BioMaps tool of VirtualPlant 1.3 [40]. Regarding cellular locations, almost every compartment was enriched, the vacuole being the cellular component showing the highest fold enrichment, followed by the endoplasmic reticulum and the cell wall (Figure 1A). Regarding biological processes, the GO enrichment analysis highlighted "glucosinolate biosynthetic process" as the category with the highest fold enrichment, followed by "response to inorganic substance" and "sulfur compound biosynthetic process" (Figure 1B).

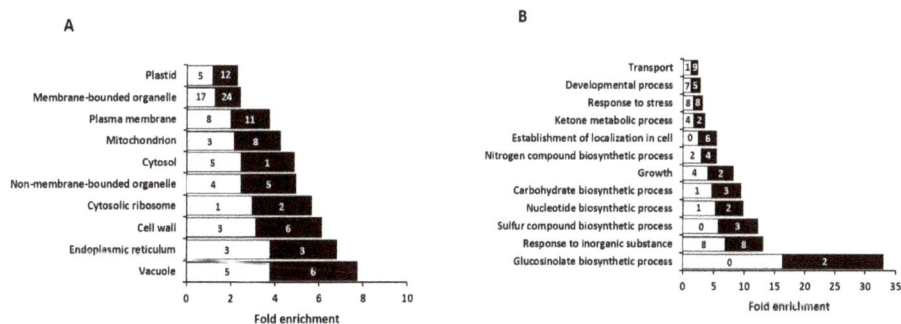

Figure 1. (**A**) Enriched categories for cellular component and (**B**) biological process of differentially abundant proteins in roots of *A. thaliana* plants cultured with ammonium ($p \leq 0.05$). Number of proteins upregulated (white) and downregulated (black) by ammonium relative to nitrate, is indicated inside the bars.

2.3. Glucosinolate Biosynthesis is Modulated by Ammonium or Nitrate as N Source

As stated, the GO enrichment analysis highlighted the regulation of the glucosinolate (GLS) biosynthetic process by the N source provided (Figure 1b). GLS are abundant sulphur-containing secondary metabolites found almost exclusively in the Brassicaceae family, which are classified in function of their precursor amino acids. Indolic GLS are derived from Trp, aromatic GLS are derived from Phe or Tyr and aliphatic GLS are derived from Ala, Ile, Leu, Met, or Val. Arabidopsis Col-0 produces up to 40 different GLS that are mainly derived from Met and Trp [41]. The classical function of GLS is plant defense from insect and pathogen attack. Indeed, herbivore triggers GLS degradation and the generated degradation products are toxic for the pathogen [42]. Besides, although this aspect has been studied to a much lesser extent, GLS seem to be related with the response of Brassica plants to abiotic stresses such as salinity and water deficit [43,44]. Because GLS synthesis is linked with S and N metabolism, N availability can influence their accumulation in different Brassica crops [45,46]. Regarding the effect of ammonium nutrition, GLS synthesis induction has been reported in the leaves of Brassicaceae plants such as Arabidopsis, broccoli, and oilseed rape [15,22,47]. However, in the present study, the two differentially abundant proteins associated with "glucosinolate biosynthetic process", identified and quantified in roots, were both downregulated in ammonium—relative to nitrate-fed roots; therefore, suggesting a different behavior of root tissue with respect to leaf tissue. These two downregulated proteins are 3-isopropylmalate dehydratase large subunit 1 (Q94AR8) and methylthioalkylmalate synthase 3 (Q9FN52), which are involved in side-chain methionine elongation, the precursor for aliphatic GLS biosynthesis (Table 2 and Supplementary Dataset 1). To assess whether the effect of N source on GLS metabolism proteins was also reflected in the content of GLS, individual GLS were quantified. In accordance with the downregulation observed in GLS-metabolic process, total GLS content was lower in ammonium-fed roots compared to nitrate-fed ones (Table 3). This decrease was mainly due to the contribution of aliphatic GLS; specifically, to glucohirsutin, 7-methylthioheptyl-GS and 8-methylthiooctyl-GS that were indeed the most abundant aliphatic GLS. Indolic GLS content was similar in both nutritional conditions and no aromatic GLS was detected (Table 3). Overall, it is clear that N source affects GLS synthesis; however, it remains to be elucidated how its differential regulation in shoots and roots takes place and whether GLS long-distance transport systems are involved in this organ-dependent regulation. This will be helpful for generating plants with increased GLS synthesis, which is desirable to promote natural plant defense and Brassica crops nutritional value. Indeed, GLS derivatives, in particular sulphoraphane, that is produced from glucoraphanin hydrolysis, have been associated with health-promoting activities [48].

Table 3. Individual glucosinolate content (ng mg^{-1} FW) in roots of *A. thaliana* plants grown with nitrate or ammonium as sole N source.

Aliphatic Glucosinolates	Nitrate	Ammonium
Glucoraphanin (4MSOB)	25.61 ± 4.13	**64.12 ± 8.76**
Glucoalyssin (5MSOP)	3.59 ± 0.32	**6.04 ± 0.52**
Glucoiberin (3MSOP)	1.88 ± 0.27	**4.50 ± 0.55**
Glucoerucin (4MTB)	2.20 ± 0.52	**4.25 ± 0.38**
Glucoberteroin (5MTP)	0.87 ± 0.08	**1.50 ± 0.08**
Glucoibarin (7MSOH)	54.17 ± 4.76	44.56 ± 3.01
Glucohirsutin (8MSOO)	610.00 ± 57.62	**452.58 ± 33.60**
C6-aliphatic GLS A ($C_{13}H_{24}NO_9S_2$)	0.57 ± 0.07	0.80 ± 0.10
7-Methylthioheptyl-GS ($C_{15}H_{28}NO_9S_3$)	125.12 ± 11.28	**92.59 ± 4.57**
8-Methylthiooctyl-GS ($C_{16}H_{31}NO_9S_3$)	1090.72 ± 58.47	**786.08 ± 33.54**
Total Aliphatic	1914.67 ± 125.02	**1457.04 ± 69.85**
Indolic Glucosinolates	**Nitrate**	**Ammonium**
Glucobrassicin (I3M)	96.97 ± 8.25	97.54 ± 4.64
Neoglucobrassicin (IMOI3M)	268.83 ± 25.97	214.38 ± 16.71
Hydroxyglucobrassicin (4OHI3M)	12.39 ± 1.00	11.73 ± 0.37
Methoxyglucobrassicin (4MOI3M)	21.90 ± 2.62	20.87 ± 2.58
Total Indolic	400.10 ± 36.42	344.52 ± 22.24
Total Glucosinolates	2337.88 ± 158.74	**1818.12 ± 90.34**

Values represent mean ± SE (*n* = 4). Significant differences among treatments are highlighted in bold text (Student *t*-test, *p* < 0.05).

2.4. Ammonium Nutrition and Secondary Metabolism in Arabidopsis Roots: Brassinosteroids and Hormonal Signaling Pathways

Regulation of secondary metabolism has been reported in several species exposed to ammonium stress such as in tomato [36] or in Arabidopsis [49]. In the present study, quantitative proteomic analysis revealed that, besides glucosinolate biosynthesis, alternative secondary metabolic routes were also influenced by the N source. Ammonium-fed roots showed increased abundance of the ATP-citrate synthase alpha chain protein 3 (O80526), the subunit A of the heteromeric enzyme complex ATP-citrate lyase (ACL) in charge of acetyl-CoA synthesis (Figure 2). On one hand, acetyl-CoA is the central precursor of flavonoids and indeed a chalcone synthase (CHS; P13114) was more abundant in ammonium than in nitrate nutrition (Table 2 and Figure 2). On the other hand, acetyl-CoA can be condensed to acetoacetyl-CoA, by the action of acetoacetyl-CoA thiolase (AACT; Q854Y1-2), which also was more abundant in ammonium-fed roots. Acetoacetyl-CoA leads to the synthesis of early mevalonate-mediated isoprenoids [50] (Figure 2). AACTs are involved in Step 1 (1 of 3) of the sub-pathway that synthesizes mevalonate (MVA) [51]. Phosphomevalonate (MVAP), generated in cytosol by phosphorylation of MVA, enters peroxisome and after a couple of reactions, isopentenyl diphosphate (IPP) and its isomer, dimethylallyl diphosphate (DMAPP), the direct precursors of the entire class of isoprenoids derived from mevalonate, are produced [51]. IPP and DMAPP return to the cytosol and by the hydrolysis of the terminal phosphate bond, mediated by cytosolic phosphohydrolases such as Nudix hydrolase 3 (NUDX3; Q8L831), can be transformed into isopentenyl phosphate (IP) and dimethylallyl phosphate (DMAP), respectively [51,52]. NUDX3 was also more abundant in roots of ammonium-fed plants with respect to nitrate-fed ones (Figure 2). Thus, the increased abundance of these three proteins in ammonium-fed roots, participating in the early mevalonate-mediated isoprenoid biosynthesis pathway, suggests that this metabolic route may be induced and strongly modulated at such nutrition conditions. This pathway may lead to the synthesis of brassinosteroids (BRs). Indeed, the protein delta(24)-sterol reductase (Q39085), involved in the conversion of the early BR precursor 24-methylenecholesterol to campesterol, also showed increased abundance in ammonium nutrition (Table 2 and Figure 2). Interestingly, BRs have been recently related to the regulation of the AMT1-type ammonium transport proteins in Arabidopsis and rice [53,54]. In rice, BRs induce the gene expression of OsAMT1;1 and OsAMT1;2 ammonium transporters [54]. Furthermore, the authors identified ABI3/VP1-Like 1 (RAVL1), a regulator of BRs homeostasis, as a direct regulator of OsAMT1;2, overall showing an important link between BRs and the transcriptional regulation of ammonium uptake [54]. In contrast, in Arabidopsis, it appears that BRs will be acting as negative regulators of AMT1 transporters [53]. Overall, it seems that BR-mediated regulatory circuits are somehow connected with ammonium uptake and signaling in a species-dependent manner. Future works will be essential to shed further light on the complex interaction between hormonal signaling pathways and nutrient uptake, notably in the context of BRs–ammonium relationship.

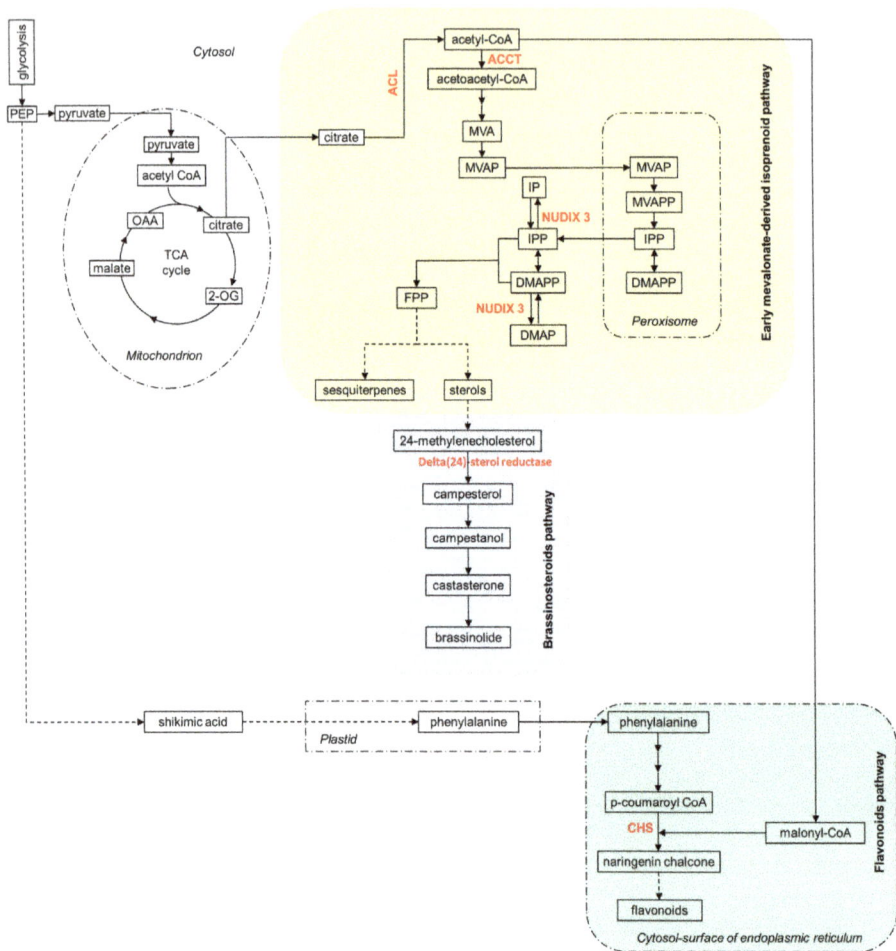

Figure 2. Induction of secondary metabolism in Arabidopsis roots by ammonium nutrition. Proteins with higher abundance in ammonium nutrition are highlighted in red bold text. Dotted arrows indicate non-detailed metabolic steps/transformations in a metabolic pathway. Abbreviations: 2-oxoglutarate (2-OG); acetyl-CoA acetyltransferase (ACCT); ATP-citrate lyase (ACL); chalcone synthase (CHS); dimethylallyl diphosphate (DMAPP); dimethylallyl phosphate (DMAP); farnesyl diphosphate (FPP); isopentenyl diphosphate (IPP); isopentenyl phosphate (IP); nudix hydrolase and a dipeptidyl peptidase III (NUDIX 3); mevalonate (MVA); oxalacetate (OAA); phosphoenolpyruvate (PEP); phosphomevalonate (MVAP).

2.5. C/N Metabolism Modulation in Ammonium-Fed Plants May Be Driven by Alternative C Provision Routes to Tricarboxylic Acid (TCA) Cycle while Contributing to H+ Balance

The C/N balance in plants is regulated by the availability of C skeletons, energy, and reductants for the N assimilatory pathways [55]. One of the known consequences of ammonium nutrition is the induction of ammonium assimilation machinery, which demands high energy and carbon consumption. In this study, several proteins associated with C metabolism were more abundant in roots of Arabidopsis plants grown under a non-toxic ammonium condition (Table 2 and Figure 3). In this line, previous studies reported that ammonium accumulation in roots triggering ammonium toxicity may be partially mitigated by the provision of extra C [29,56]. Interestingly, supplementary C would be not only serving N assimilation but also improving cell ion balance and managing respiration rates and ATP availability [56]. Indeed, although the respiratory cost of ammonium assimilation is not as much of that of nitrate, the overall effects of ammonium stress have been associated with the increased capacity of respiratory bypass pathways [57,58]. Specifically, the capacity of alternative oxidase (AOX) is substantially elevated in plants grown on ammonium [57,59]. AOX together with pyruvate kinase (PK; Q9FM97), protein more abundant in ammonium-fed roots of this study (Table 2 and Figure 3), have been described as a H$^+$-sink unit of the revised biochemical pH-stat mechanism under aerobic conditions [60,61]. This is a key aspect, since the control of pH homeostasis is critical for the plant to face ammonium stress [9,30,62].

As previously mentioned, this proteomic study also revealed a number of proteins associated with secondary metabolism (Table 2 and Figure 2) and several authors have associated the production of secondary metabolites with cytoplasmic acidification [61,63]. Furthermore, Sakano suggested that AOX activation may also be deeply involved in the oxidation of excess reducing equivalents produced during the synthesis of secondary metabolites [61].

Besides AOX and PK, this study also showed increased abundance of other C metabolism-related proteins whose enzymatic activity consumes H$^+$ such as phosphoenolpyruvate carboxykinase, (PEPCK; Q9T074) and glutamate decarboxylase (GAD; Q42472; Figure 3). We determined the enzyme activity of PEPCK and, in agreement with the proteomics results, it was also significantly higher in ammonium-fed plants (Figure 3). Malate dehydrogenase activity, which converts malate to the PEPCK substrate, oxaloacetate (OAA), was also increased under ammonium nutrition (Figure 3). The role of PEPCK in the metabolism of ammonium-fed plants regulating pH, by consuming H$^+$ via malate decarboxylation to pyruvate by the sequential action of MDH, PEPCK, and PK has been previously suggested, notably, in the more active tissues in the N metabolism, such as the pericycle [18,64,65]. Furthermore, PEPCK abundance and activity also increased in cucumber plants exposed to ammonium and acidification conditions [66].

In roots of plants grown under ammonium nutrition, TCA cycle usually functions in an open-mode, since almost all the 2-oxoglutarate (2-OG) generated is diverted into amino acids synthesis [67]. Thus, to replenish pyruvate pool to ensure the supply of 2-OG, the anaplerotic pathways associated to TCA cycle have been suggested to bear a predominant role. Importantly, MDH and PEPCK, apart from their role in the "biochemical pH-stat", are a part of these anaplerotic pathways together with malic enzyme (ME) and phosphoenolpyruvate carboxylase (PEPC). Overall, these anaplerotic routes were enhanced in ammonium-fed roots compared to the nitrate-fed ones (Figure 3).

The increased 2OG production by isocitrate dehydrogenase (ICDH) (Figure 3) may induce Glu production and its derivate amino acids (Table 1). For instance, GAD, in charge of GABA synthesis and whose abundance is increased in ammonium-fed plants (Table 2 and Figure 3) can be induced and stimulated by increases in cytosolic Ca^{2+} (via Ca^{2+}/CaM) or H$^+$ concentrations and thus, it has also been related to cell pH regulation [19,68]. GABA concentration is influenced by different environmental changes, inter alia, N form. Indeed, GABA content has already been shown to increase in ammonium-grown Arabidopsis plants relative to nitrate-fed plants [69]. Additionally, GABA and malate appear to be tightly connected and to participate, among others, in TCA cycle regulation and in the regulation of electrical potential across membranes acting on aluminum-activated anion transporters (AMLTs) [70,71]. Thus, a proper C and N metabolic adaptation in roots coordinated with NH$_4$$^+$ uptake, transport and storage appears essential in order to maintain cell pH, reductant, and electrochemical homeostasis upon plants growth under ammonium nutrition.

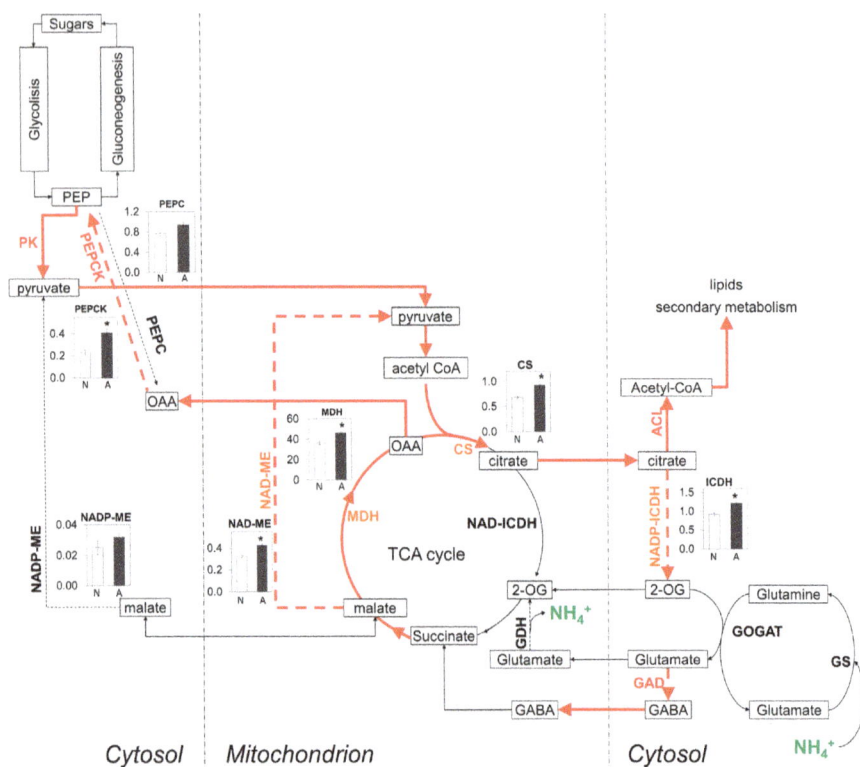

Figure 3. C anaplerotic routes (dotted arrows) in ammonium-fed roots of *A. thaliana* plants. Ammonium (NH_4^+) release or incorporation to metabolic pathways is highlighted in green bold text. Induced routes and proteins with higher abundance in ammonium relative to nitrate nutrition are highlighted in red (bold lines and text). Increased activity of anaplerotic and TCA-cycle enzymes are shown in orange bold text and emphasized with an asterisk (*) in the activity graphs. Graphs represent enzyme activity of Arabidopsis roots fed with ammonium (grey bars) or with nitrate (white bars). Enzyme activities are expressed as: CS (nmol CoA mg^{-1} FW min^{-1}); ICDH (nmol NADP mg^{-1} FW min^{-1}); MDH (nmol NADP mg^{-1} FW min^{-1}); NAD-ME (nmol NAD mg^{-1} FW min^{-1}); NADP-ME (nmol NADP mg^{-1} FW min^{-1}); PEPC (nmol NADH mg^{-1} FW min^{-1}); PEPCK (nmol NADH mg^{-1} FW min^{-1}). Data shown in graphics represent mean values $\pm$ SE (n = 4). Asterisk (*) indicates significant N source effect (*t*-test, $p < 0.05$). Abbreviations: ATP-citrate lyase (ACL); citrate synthase (CS); gamma-amniobutyric acid (GABA). glutamate decarboxylase (GAD); glutamate dehydrogenase (GDH); glutamate synthase (GOGAT); glutamine synthetase (GS); NAD-isocitrate dehydrogenase (NAD-ICDH); NADP-isocitrate dehydrogenase (NADP-ICDH); malate dehydrogenase (MDH); NADP-malic enzyme (NADP-ME); NAD-malic enzyme (NAD-ME); phosphoenolpyruvate (PEP); phosphoenolpyruvate carboxylase (PEPC); phosphoenolpyruvate carboxykinase (PEPCK); pyruvate kinase (PK).

Finally, the present study provides new hints of metabolic pathways and signaling processes that can be involved in root adaptation to ammonium nutrition, a process that although thoroughly studied continues being still poorly understood. Among the novel points that arise from this study, the connection between ammonium nutrition and secondary metabolism and the putative association between GABA shunt and TCA cycle associated enzymes to regulate H^+ balance and plasma membrane electrical potential deserve special attention in future research.

3. Materials and Methods

3.1. Plant Culture and Experimental Design

Plants used in this study were cultured as described in [15]. Briefly, seeds of *Arabidopsis thaliana* ecotype Col-0 were sterilized and cultured with a modified Murashige and Skoog solution (2.25 mM $CaCl_2$, 1.25 mM KH_2PO_4, 0.75 mM $MgSO_4$, 5 mM KCl, 0.085 mM Na_2EDTA, 5 μM KI, 0.1 μM $CuSO_4$, 100 μM $MnSO_4$, 100 μM H_3BO_3, 0.1 μM $CoCl_2$, 100 μM $FeSO_4$, 30 μM $ZnSO_4$, and 0.1 μM Na_2MoO_4) supplemented with 0.5% sucrose and 20.5 mM MES (pH 6.5) [72]. Nitrogen source was added at a concentration of 2 mM as 1 mM $(NH_4)_2SO_4$ for ammonium-based nutrition or 1 mM $Ca(NO_3)_2$ for nitrate nutrition. To equilibrate the Ca^{2+} supplied together with the NO_3^-, NH_4^+-fed plants were supplemented with 1 mM $CaSO_4$. The experimental design of this study, in terms of N concentration, pH, volume, and renew frequency of nutrient solution, was selected according to results obtained in previous nutritional studies with Arabidopsis plants [9,15].

Plants were stratified at 4 °C for four days in the dark and then moved into a growth chamber under the following controlled conditions: 14 h, 200 μmol·m^{-2}·s^{-1} light intensity, 60% relative humidity, and 22 °C (day conditions); 10 h, 70% relative humidity, and 18 °C (night conditions).

First, plants were germinated and cultured during 9 days in 0.6% agar Petri dishes with the nutrient solution described above. After this time, seedlings were transferred to sterile 24-well plates containing 1 mL of the same nutrient solution used for seed germination without agar (one plant per well). Then, plates were kept under continuous shaking (120 rpm) for 12 additional days and the liquid nutrient solution was renewed on days 5 and 9. Four independent experiments were carried out, each one with 10, 24-well plates. Each plate contained 12 plants per treatment. When harvesting, shoots and roots of plants within each plate and treatment were pooled separately, dried with paper towels and the biomass was recorded. For proteomic and metabolic analysis, all the roots within each experiment and treatment (120 plants) were pooled together, immediately frozen in liquid nitrogen and stored at −80 °C.

3.2. Ammonium and Total Free Amino Acid Quantification

Root extracts for ammonium and total free amino acids quantification were obtained by adding 20 μL of ultrapure water per milligram of tissue. The homogenates were incubated at 80 °C during 5 min and centrifuged at 16,000g and 4 °C for 20 min and then, supernatants were recovered.

Ammonium content was determined following the phenol hypochlorite method [73]. For total free amino acids quantification, the ninhydrin method was followed using glutamine as a standard for the calibration curve [74].

3.3. Soluble Protein Quantification and Enzyme Activities Determination

For soluble protein and enzyme activities determination, frozen root powder was homogenized with extraction buffer (10 μL per milligram of tissue). Extraction buffer was composed by 0.1% Triton X-100, 10% glycerol, 0.5% polyvinylpolypyrrolidone, 50 mM HEPES pH 7.5, 10 mM $MgCl_2$, 1 mM EDTA, 1 mM EGTA, 10 mM dithiothreitol, 1 mM phenylmethylsulfonyl fluoride, 1 mM ε-aminocaproic acid and 10 μM leupeptin. Homogenates were then centrifuged at 16,000g for 20 min at 4 °C and the supernatants recovered. Soluble protein content was determined by a dye binding protein assay (Bio-Rad Bradford Protein assay) using BSA as a standard for the calibration curve. Phosphoenolpyruvate carboxylase (PEPC), NAD-dependent malic (NAD-ME), NADP-dependent isocitrate dehydrogenase (ICDH), and malate dehydrogenase (MDH) were assayed as described in [9]. The following reaction buffers were used: for PEPC activity assay (100 mM Tricine-KOH (pH 8), 5 mM $MgCl_2$, 5 mM NaF, 0.25 mM NADH, 6.4 U of malate dehydrogenase mL^{-1}, 2 mM $NaHCO_3$ and 3 mM phosphoenolpyruvate); for NAD-ME activity assay (50 mM HEPES-KOH (pH 8), 0.2 mM EDTA-Na_2, 5 mM DTT, 2 mM NAD, 5 mM malate, 25 μM NADH, 0.1 mM Acetyl CoA and 4 mM $MnCl_2$); for NADP-dependent malic enzyme (NADP-ME) activity assay (100 mM Tris-HCl (pH 7), 10 mM $MgCl_2$,

0.5 mM NADP and 10 mM malate); for ICDH assay (100 mM Tricine-KOH (pH 8), 0.25 mM NADP, 5 mM $MgCl_2$ and 5 mM isocitrate); for MDH assay (100 mM HEPES-KOH (pH 7.5), 5 mM $MgSO_4$, 0.2 mM NADH and 2 mM oxaloacetate). Phosphoenolpyruvate carboxykinase (PEPCK) was assayed using a reaction buffer composed by 100 mM HEPES-KOH (pH 6.8), 25 mM DTT, 100 mM KCl, 90 mM $KHCO_3^-$, 1 mM ADP, 6 mM $MnCl_2$, 0.2 mM NADH, 7 U of malate dehydrogenase mL^{-1} and 6 mM phosphoenolpyruvate, as described in [75]. Enzyme activities were assayed by spectrophotometry at 340 nm, monitoring evolution (formation or extinction) of NAD(P)H at 30 °C for 20 min. In the case of MDH, 10-fold diluted extracts were used. In citrate synthase (CS) activity assay, the extraction buffer was the same as that described above, except for DTT, which was not added. Protein extracts were incubated at 30 °C for 20 min with a reaction buffer (100 mM Tris-HCl (pH 8), 1 mM oxaloacetate, 0.25 mM acetyl coenzyme A and 0.1 mM 5,5'-dithiobis (2-nitrobenzoic acid), DTNB). CS activity was measured by spectrophotometry at 412nm, monitoring the absorbance originated by the formation of 2-nitro-5-thiobenzoic acid (TNB) [76]. All reagents were purchased from Sigma-Aldrich (St. Louis, MO, USA).

3.4. Glucosinolate Determination

For glucosinolate determination around 25 mg of frozen root powder were extracted by adding 1 mL of MeOH:water (70:30). The mixtures were homogenized in a Tissue Lyser (Retsch MM 400, Haan, Germany) and incubated for 15 min at 80 °C to inactivate myrosinase. Then, homogenates were centrifuged for 20 min at 16,000g. Glucosinolates were determined from supernatants by ultra-high performance liquid chromatography-quadrupole time-of-flight mass spectrometry (UHPLC/Q-TOF-MS) analyses using an Acquity UPLC from Waters (Milford, MA, USA) interfaced to a Synapt G2 QTOF from Waters (Milford, MA, USA)with electrospray ionization as described previously [77]. Glucosinolates were quantified using glucoraphanin and glucobrassicin as standards (Phytolab, Vestenbergsgreuth, Germany).

3.5. Proteomic Analysis

3.5.1. Sample Preparation and Labeling for Proteomic Analysis

Proteins were extracted from 50 mg of root fresh weight (FW) homogenized in 0.5 mL of an extraction buffer composed by 7 M urea, 2 M thiourea, 4% CHAPS, 2% Triton X-100, 50 mM DTT, and 0.5% plant protease inhibitor and phosphatase inhibitors cocktails (Sigma-Aldrich, St. Louis, MO, USA). Then, homogenates were centrifuged for 15 min at 10,000g and 4 °C and total protein precipitated from 200 μL of supernatant with methanol and chloroform (600 μL methanol, 15 μL chloroform, and 450 μL ultrapure water). Mixtures were spun (in a vortex) and centrifuged for 1 min at 14,000g. The aqueous phase was then removed, an additional 450 μL of methanol added, and the centrifugation step was repeated. After discarding the methanol phase, protein pellets were dried in a vacuum centrifuge and resuspended into 7 M urea, 2 M thiourea, and 4% CHAPS. Global experiments were carried out with four independent biological samples in each experimental condition. Each sample corresponded to a pool of 120 plants. Protein extracts (150 μg) were precipitated with methanol/choloroform, and pellets dissolved in 7 M urea, 2 M thiourea, 4% (*v/v*) CHAPS. Protein was quantified with the Bradford assay kit (Bio-Rad, Hercules, CA; USA). A shotgun comparative proteomic analysis of total root extracts using an iTRAQ 8-plex experiment was performed [78]. iTRAQ labeling of each sample was made according to the manufacturer's protocol (Sciex, Framingham, MA, USA). Total protein (100 μg) from each sample was reduced with 50 mM tris (2-carboxyethyl) phosphine (TCEP) at 60 °C for 1 h. Cysteine residues were alkylated with 200 mM methylmethanethiosulfonate (MMTS) at room temperature for 15 min. Trypsin (Promega, Fitchburg, WI, USA), 1:20, *w/w*, was used for protein enzymatic cleavage at 37 °C for 16 h. Each root tryptic digest was labeled by incubation (1 h) according to the manufacturer's instructions with one isobaric amine-reactive tags, as follows: Tag113, ammonium media-1; Tag114, ammonium media-2; Tag115, ammonium media-3; Tag116, ammonium media-4;

Tag117, nitrate media-1; Tag118, nitrate media-2; Tag119, nitrate media-3; Tag121, nitrate media-4. Then, every set of labeled samples was independently pooled and evaporated until <40 µL by vacuum centrifugation. Unless otherwise stated all reagents were purchased from Sigma-Aldrich (St. Louis, MO, USA).

3.5.2. Peptide Fractionation

The peptide pool was injected to an Ettan LC system with a X-Terra RP18 pre-column (2.1 × 20 mm) and a high pH stable X-Terra RP18 column (C18; 2.1 mm × 150 mm; 3.5 µm) (Waters, Milford, MA, USA) at a flow rate of 40 µL/min, increasing in this way the proteome coverage. Elution of peptides was made with a mobile phase B of 5–65% linear gradient over 35 min (A, 5 mM ammonium bicarbonate in water at pH 9.8; B, 5 mM ammonium bicarbonate in acetonitrile at pH 9.8). Eleven fractions were collected, evaporated under vacuum and reconstituted into 20 µL of 2% acetonitrile, 0.1% formic acid, 98% MilliQ water previous to mass spectrometric analysis.

3.5.3. Triple-TOF 5600 Mass Spectrometry (MS) Analysis

The split of peptides was made by reverse phase chromatography using an Eksigent nanoLC ultra 2D pump fitted with a 75 µm ID column (Eksigent 0.075 mm × 25 cm). Samples were desalted and concentrated with a 0.5 cm length 300 µm ID pre-column, which was packed with the same chemistry as the separating column. Mobile phases were 100% water 0.1% formic acid (buffer A) and 100% Acetonitrile, 0.1% formic acid (buffer B). The column gradient (237 min) used was a two-step gradient, first from 5% B to 25% B in 180 min and second, from 25% B to 40% B in 30 min. Column was equilibrated in 95% B for 10 min and 5% B for 15 min. Along the entire process, the pre-column was in line with column, and flow during the gradient was maintained at 300 nL/min. The separated peptides eluted from the column were analyzed using an AB Sciex 5600 TripleTOF™ system (Sciex, Framingham, MA, USA). Information data was acquired upon a survey scan (mass range from 350 m/z up to 1250 m/z; scan time: 250 ms). Top 35 peaks were selected for fragmentation. Minimum accumulation time for MS/MS was set to 100 ms (3.8 s of total cycle time). Product ions were scanned in a mass range from 100 m/z up to 1700 m/z and excluded for further fragmentation during 15 s.

3.5.4. Data Analysis

Analyses of raw data (.wiff, Sciex) were performed with MaxQuant software [79]. Peak list was generated with the default Sciex Q-TOF instrument parameters except for the main search peptide tolerance that was set to 0.01 Da, and the MS/MS match tolerance that was increased up to 50 ppm. Minimum peptide length was set to six amino acids. Two databases were used. A contaminant database (.fasta) was first used to filter out contaminants. Peak lists were searched against the TAIR10 *A. thaliana* database (www.arabidopsis.org), and Andromeda was used as a search engine [80]. The search parameters allowed for methionine oxidation and cysteine modification by MMTS. Reporter ion intensities were bias corrected for the overlapping isotope contributions from the iTRAQ tags according to the certificate of analysis provided by the reagent manufacturer (Sciex, Framingham, MA, USA). The maximum false discovery rates (FDR) were set to 1% at the protein and peptide levels. Analyses were limited to peptides of six or more amino acids in length, and considering a maximum of two missed cleavages. Proteins identified by site (identification based only on a modification), reverse proteins (identified by decoy database) and potential contaminants were filtered out. Only proteins with more than one missing value was accepted (i.e., protein identified in three out of the four replicates) and was rescued by replacing it with the mean of the rest of the in-group samples. Data were normalized and transformed for later comparison using quantile normalization and log2 transformation, respectively. The Limma Bioconductor software package in R was used for ANOVA analyses. Significant and differential data were selected by a p value < 0.05.

3.5.5. Functional Classification and Gene Ontology Enrichment Analysis

Functional classification of the differentially abundant proteins was carried out according to MapMan software (http://mapman.gabipd.org/es/mapman, version 3.6.0) [33]. Gene ontology (GO) enrichment analysis and visualization for cellular component and biological process were performed with BioMaps tool of VirtualPlant 1.3 using the *A. thaliana* Columbia tair10 genome as background population [40]. Over-representation was calculated with Fisher's exact test, with a cut-off value of $p \leq 0.05$.

3.6. Statistical Analyses

Proteomics data statistical analysis is described in the above section.

For biomass and metabolic data, statistical analysis was carried out using IBM SPSS 22.0 software (IBM Corp., Armonk, NY, USA). The significance of the results was assessed using independent samples Student *t*-test with a *p* value < 0.05.

Supplementary Materials: Supplementary materials can be found at http://www.mdpi.com/1422-0067/20/4/814/s1.

Author Contributions: Conceptualization, D.M. and I.A.; Formal analysis, I.C., I.V.-M., G.G., M.B.G.-M., D.M., and I.A.; Funding acquisition, D.M.; Investigation, I.C., I.V.-M., G.G., D.M., and I.A.; Writing—original draft, I.C., D.M., and I.A.; Writing—review & editing, I.C., I.V.-M., G.G., M.B.G.-M., D.M., and I.A.

Funding: This research was financially supported by the Basque Government (IT932-16), the Spanish Ministry of Economy and Competitiveness (BIO2014-56271-R and BIO2017-84035-R co-funded by FEDER). I.A. was supported by a Juan de la Cierva postdoctoral contract (IJCI-2015-26002) from the Spanish Ministry of Economy and Competitiveness. I.C. was supported by a postdoctoral fellowship from the University of the Basque Country UPV/EHU.

Acknowledgments: Proteomic analysis (triple-TOF 5600 MS analysis) was performed at the Proteomics Unit Facility, Navarrabiomed, Fundación Miguel Servet, Instituto de Investigación Sanitaria de Navarra (IdiSNA), Spain. The Proteomics Unit of Navarrabiomed is a member of Proteored, PRB3-ISCIII.

Conflicts of Interest: The authors declare no conflict of interest. The funders had no role in the design of the study; in the collection, analyses, or interpretation of data; in the writing of the manuscript, or in the decision to publish the results.

Abbreviations

2-OG	2-Oxoglutarate
AACT	Acetoacetyl-CoA thiolase
ACCT	Acetyl-CoA acetyltransferase
ACL	ATP-citrate lyase
AMT	Ammonium transporter
AOX	Alternative oxidase
ATP	Adenosine triphosphate
BR	Brassinosteroid
CHS	Chalcone synthase
CoA	Coenzyme A
CS	Citrate synthase
DMAP	Dimethylallyl phosphate
DMAPP	Dimethylallyl diphosphate
FDR	False discovery rate
FPP	Farnesyl diphosphate
FW	Fresh weight
GAD	Glutamate decarboxylase
GABA	Gamma-aminobutyric acid
GDH	Glutamate dehydrogenase
GLS	Glucosinolate(s)
GO	Gene ontology

GOGAT	Glutamate synthase
GS	Glutamine synthetase
ICDH	Isocitrate dehydrogenase
IP	Isopentenyl phosphate
IPP	Isopentenyl diphosphate
iTRAQ	Isobaric tags for relative and absolute quantitation
MDH	Malate dehydrogenase
ME	Malic enzyme
MVA	Mevalonate
MVAP	Phosphomevalonate
NAD(H)	Nicotinamide adenine dinucleotide (reduced)
NADP(H)	Nicotinamide adenine dinucleotide phosphate (reduced)
NUDIX 3	Nudix hydrolase and a dipeptidyl peptidase III
OAA	Oxaloacetate
PEP	Phosphoenolpyruvate
PEPCK	Phosphoenolpyruvate carboxykinase
PEPC	Phosphoenolpyruvate carboxylase
PK	Pyruvate kinase
TCA	Tricarboxylic acid

References

1. Bloom, A.J. The increasing importance of distinguishing among plant nitrogen sources. *Curr. Opin. Plant Biol.* **2015**, *25*, 10–16. [CrossRef] [PubMed]

2. Britto, D.T.; Kronzucker, H.J. Ecological significance and complexity of N-source preference in plants. *Ann. Bot.* **2013**, *112*, 957–963. [CrossRef]

3. Huérfano, X.; Fuertes-Mendizábal, T.; Duñabeitia, M.K.; González-Murua, C.; Estavillo, J.M.; Menéndez, S. Splitting the application of 3,4-dimethylpyrazole phosphate (DMPP): Influence on greenhouse gases emissions and wheat yield and quality under humid Mediterranean conditions. *Eur. J. Agron.* **2015**, *64*, 47–57. [CrossRef]

4. Torralbo, F.; Menéndez, S.; Barrena, I.; Estavillo, J.M.; Marino, D.; González-Murua, C. Dimethyl pyrazol-based nitrification inhibitors effect on nitrifying and denitrifying bacteria to mitigate N_2O emission. *Sci. Rep.* **2017**, *7*, 13810. [CrossRef] [PubMed]

5. Britto, D.T.; Kronzucker, H.J. NH_4^+ toxicity in higher plants: A critical review. *J. Plant Physiol.* **2002**, *159*, 567–584. [CrossRef]

6. Esteban, R.; Ariz, I.; Cruz, C.; Moran, J.F. Review: Mechanisms of ammonium toxicity and the quest for tolerance. *Plant Sci.* **2016**, *248*, 92–101. [CrossRef] [PubMed]

7. Cruz, C.; Domínguez-Valdivia, M.D.; Aparicio-Tejo, P.M.; Lamsfus, C.; Bio, A.; Martins-Loução, M.A.; Moran, J.F. Intra-specific variation in pea responses to ammonium nutrition leads to different degrees of tolerance. *Environ. Exp. Bot.* **2011**, *70*, 233–243. [CrossRef]

8. Sarasketa, A.; González-Moro, M.B.; González-Murua, C.; Marino, D. Exploring ammonium tolerance in a large panel of *Arabidopsis thaliana* natural accessions. *J. Exp. Bot.* **2014**, *65*, 6023–6033. [CrossRef]

9. Sarasketa, A.; González-Moro, M.B.; González-Murua, C.; Marino, D. Nitrogen source and external medium pH interaction differentially affects root and shoot metabolism in Arabidopsis. *Front. Plant Sci.* **2016**, *7*, 29. [CrossRef]

10. Zhu, Y.; Lian, J.; Zeng, H.; Gan, L.; Di, T.; Shen, Q.; Xu, G. Involvement of plasma membrane H+-ATPase in adaption of rice to ammonium nutrient. *Rice Sci.* **2011**, *18*, 335–342. [CrossRef]

11. Wang, F.; Gao, J.; Tian, Z.; Liu, Y.; Abid, M.; Jiang, D.; Cao, W.; Dai, T. Adaptation to rhizosphere acidification is a necessary prerequisite for wheat (*Triticum aestivum* L.) seedling resistance to ammonium stress. *Plant Physiol. Biochem.* **2016**, *108*, 447–455. [CrossRef] [PubMed]

12. Mayer, M.; Dynowski, M.; Ludewig, U. Ammonium ion transport by the AMT/Rh homologue LeAMT1;1. *Biochem. J.* **2006**, *396*, 431–437. [CrossRef] [PubMed]

13. Marschner, H. *Mineral Nutrition of Higher Plants*; Elsevier: London, UK, 2012; ISBN 9780124735422.

14. Szczerba, M.W.; Britto, D.T.; Ali, S.A.; Balkos, K.D.; Kronzucker, H.J. NH$_4$$^+$-stimulated and -inhibited components of K$^+$ transport in rice (*Oryza sativa* L.). *J. Exp. Bot.* **2008**, *59*, 3415–3423. [CrossRef] [PubMed]

15. Marino, D.; Ariz, I.; Lasa, B.; Santamaría, E.; Fernández-Irigoyen, J.; González-Murua, C.; Aparicio Tejo, P.M. Quantitative proteomics reveals the importance of nitrogen source to control glucosinolate metabolism in *Arabidopsis thaliana* and *Brassica oleracea*. *J. Exp. Bot.* **2016**, *67*. [CrossRef] [PubMed]

16. Podgórska, A.; Burian, M.; Rychter, A.M.; Rasmusson, A.G.; Szal, B. Short-term ammonium supply induces cellular defence to prevent oxidative stress in *Arabidopsis* leaves. *Physiol. Plant.* **2017**, *160*, 65–83. [CrossRef] [PubMed]

17. Gerendás, J.; Abbadi, J.; Sattelmacher, B. Potassium efficiency of safflower (*Carthamus tinctorius* L.) and sunflower (*Helianthus annuus* L.). *J. Plant Nutr. Soil Sci.* **2008**, *171*, 431–439. [CrossRef]

18. Walker, R.; Chen, Z.-H. Phosphoenolpyruvate carboxykinase: Structure, function and regulation. *Adv. Bot. Res.* **2002**, *38*, 93–189. [CrossRef]

19. Shavrukov, Y.; Hirai, Y. Good and bad protons: genetic aspects of acidity stress responses in plants. *J. Exp. Bot.* **2016**, *67*, 15–30. [CrossRef]

20. Raven, J.A. Biochemical disposal of excess H$^+$ in growing plants? *New Phytol.* **1986**, *104*, 175–206. [CrossRef]

21. Fuertes-Mendizábal, T.; González-Torralba, J.; Arregui, L.M.; González-Murua, C.; González-Moro, M.B.; Estavillo, J.M. Ammonium as sole N source improves grain quality in wheat. *J. Sci. Food Agric.* **2013**, *93*, 2162–2171. [CrossRef]

22. Coleto, I.; de la Peña, M.; Rodríguez-Escalante, J.; Bejarano, I.; Glauser, G.; Aparicio-Tejo, P.M.; González-Moro, M.B.; Marino, D. Leaves play a central role in the adaptation of nitrogen and sulfur metabolism to ammonium nutrition in oilseed rape (*Brassica napus*). *BMC Plant Biol.* **2017**, *17*, 157. [CrossRef]

23. Gao, Y.; Li, Y.; Yang, X.; Li, H.; Shen, Q.; Guo, S. Ammonium nutrition increases water absorption in rice seedlings (*Oryza sativa* L.) under water stress. *Plant Soil* **2010**, *331*, 193–201. [CrossRef]

24. Fernández-Crespo, E.; Camañes, G.; García-Agustín, P. Ammonium enhances resistance to salinity stress in citrus plants. *J. Plant Physiol.* **2012**, *169*, 1183–1191. [CrossRef]

25. Fernández-Crespo, E.; Gómez-Pastor, R.; Scalschi, L.; Llorens, E.; Camañes, G.; García-Agustín, P. NH$_4$$^+$ induces antioxidant cellular machinery and provides resistance to salt stress in citrus plants. *Trees* **2014**, *28*, 1693–1704. [CrossRef]

26. Hessini, K.; Hamed, K. Ben; Gandour, M.; Mejri, M.; Abdelly, C.; Cruz, C. Ammonium nutrition in the halophyte *Spartina alterniflora* under salt stress: evidence for a priming effect of ammonium? *Plant Soil* **2013**, *370*, 163–173. [CrossRef]

27. Bloom, A.J.; Burger, M.; Asensio, J.S.R.; Cousins, A.B. Carbon dioxide enrichment inhibits nitrate assimilation in wheat and arabidopsis. *Science* **2010**, *328*, 899–903. [CrossRef]

28. Andrews, M.; Condron, L.M.; Kemp, P.D.; Topping, J.F.; Lindsey, K.; Hodge, S.; Raven, J.A. Elevated CO$_2$ effects on nitrogen assimilation and growth of C3 vascular plants are similar regardless of N-form assimilated. *J. Exp. Bot.* **2019**, *70*, 683–690. [CrossRef]

29. Vega-Mas, I.; Marino, D.; Sánchez-Zabala, J.; González-Murua, C.; Estavillo, J.M.; González-Moro, M.B. CO$_2$ enrichment modulates ammonium nutrition in tomato adjusting carbon and nitrogen metabolism to stomatal conductance. *Plant Sci.* **2015**, *241*, 32–44. [CrossRef]

30. Chaillou, S.; Vessey, J.K.; Morot-gaudry, J.F.; Raper, C.D.; Henry, L.T.; Boutin, J.P. Expression of characteristics of ammonium nutrition as affected by ph of the root medium. *J. Exp. Bot.* **1991**, *42*, 189–196. [CrossRef]

31. Britto, D.T.; Siddiqi, M.Y.; Glass, A.D.M.; Kronzucker, H.J. Futile transmembrane NH$_4$$^+$ cycling: A cellular hypothesis to explain ammonium toxicity in plants. *Proc. Natl. Acad. Sci. USA* **2001**, *98*, 4255–4258. [CrossRef]

32. Cruz, C.; Bio, A.F.M.; Domínguez-Valdivia, M.D.; Aparicio-Tejo, P.M.; Lamsfus, C.; Martins-Loução, M.A. How does glutamine synthetase activity determine plant tolerance to ammonium? *Planta* **2006**, *223*, 1068–1080. [CrossRef]

33. Usadel, B.; Nagel, A.; Thimm, O.; Redestig, H.; Blaesing, O.E.; Palacios-Rojas, N.; Selbig, J.; Hannemann, J.; Piques, M.C.; Steinhauser, D.; et al. Extension of the visualization tool MapMan to allow statistical analysis of arrays, display of corresponding genes, and comparison with known responses. *Plant Physiol.* **2005**, *138*, 1195–1204. [CrossRef]

34. Schubert, S.; Yan, F. Nitrate and ammonium nutrition of plants: Effects on acid/base balance and adaptation of root cell plasmalemma H$^+$ ATPase. *J. Plant Nutr. Soil Sci.* **1997**, *160*, 275–281. [CrossRef]

35. Anzi, C.; Pelucchi, P.; Vazzola, V.; Murgia, I.; Gomarasca, S.; Beretta Piccoli, M.; Morandini, P. The proton pump interactor (Ppi) gene family of *Arabidopsis thaliana*: expression pattern of Ppi1 and characterisation of knockout mutants for Ppi1 and 2. *Plant Biol.* **2008**, *10*, 237–249. [CrossRef]

36. Vega-Mas, I.; Pérez-Delgado, C.M.; Marino, D.; Fuertes-Mendizábal, T.; González-Murua, C.; Márquez, A.J.; Betti, M.; Estavillo, J.M.; Gonzáez-Moro, M.B. Elevated CO_2 induces root defensive mechanisms in tomato plants when dealing with ammonium toxicity. *Plant Cell Physiol.* **2017**, *58*, 2112–2125. [CrossRef]

37. Patterson, K.; Cakmak, T.; Cooper, A.; Lager, I.; Rasmusson, A.G.; Escobar, M.A. Distinct signalling pathways and transcriptome response signatures differentiate ammonium- and nitrate-supplied plants. *Plant Cell Environ.* **2010**, *33*, 1486–1501. [CrossRef]

38. Zeng, H.; Di, T.; Zhu, Y.; Subbarao, G.V. Transcriptional response of plasma membrane H^+-ATPase genes to ammonium nutrition and its functional link to the release of biological nitrification inhibitors from sorghum roots. *Plant Soil* **2016**, *398*, 301–312. [CrossRef]

39. Felle, H.H. pH Regulation in anoxic plants. *Ann. Bot.* **2005**, *96*, 519. [CrossRef]

40. Katari, M.S.; Nowicki, S.D.; Aceituno, F.F.; Nero, D.; Kelfer, J.; Thompson, L.P.; Cabello, J.M.; Davidson, R.S.; Goldberg, A.P.; Shasha, D.E.; et al. VirtualPlant: A software platform to support systems biology research. *Plant Physiol.* **2010**, *152*, 500–515. [CrossRef]

41. Sønderby, I.E.; Geu-Flores, F.; Halkier, B.A. Biosynthesis of glucosinolates—Gene discovery and beyond. *Trends Plant Sci.* **2010**, *15*, 283–290. [CrossRef]

42. Wittstock, U.; Burow, M. Glucosinolate breakdown in Arabidopsis: mechanism, regulation and biological significance. *Arab. B.* **2010**, *8*, e0134. [CrossRef]

43. Qasim, M.; Ashraf, M.; Ashraf, M.Y.; Rehman, S.-U.; Rha, E.S. Salt-induced changes in two canola cultivars differing in salt tolerance. *Biol. Plant.* **2003**, *46*, 629–632. [CrossRef]

44. Zhang, H.; Schonhof, I.; Krumbein, A.; Gutezeit, B.; Li, L.; Stützel, H.; Schreiner, M. Water supply and growing season influence glucosinolate concentration and composition in turnip root (*Brassica rapa* ssp.rapifera L.). *J. Plant Nutr. Soil Sci.* **2008**, *171*, 255–265. [CrossRef]

45. Schonhof, I.; Blankenburg, D.; Müller, S.; Krumbein, A. Sulfur and nitrogen supply influence growth, product appearance, and glucosinolate concentration of broccoli. *J. Plant Nutr. Soil Sci.* **2007**, *170*, 65–72. [CrossRef]

46. Omirou, M.D.; Papadopoulou, K.K.; Papastylianou, I.; Constantinou, M.; Karpouzas, D.G.; Asimakopoulos, I.; Ehaliotis, C. Impact of nitrogen and sulfur fertilization on the composition of glucosinolates in relation to sulfur assimilation in different plant organs of Broccoli. *J. Agric. Food Chem.* **2009**, *57*, 9408–9417. [CrossRef]

47. La, G.-X.; Yang, T.-G.; Fang, P.; Guo, H.-X.; Hao, X.; Huang, S.-M. Effect of NH_4^+/NO_3^- ratios on the growth and bolting stem glucosinolate content of Chinese kale (*Brassica alboglabra* L.H. Bailey). *Aust. J. Crop Sci.* **2013**, *7*, 618–624.

48. Traka, M.H.; Mithen, R.F. Plant science and human nutrition: Challenges in assessing health-promoting properties of phytochemicals. *Plant Cell* **2011**, *23*, 2483–2497. [CrossRef]

49. Menz, J.; Range, T.; Trini, J.; Ludewig, U.; Neuhäuser, B. Molecular basis of differential nitrogen use efficiencies and nitrogen source preferences in contrasting Arabidopsis accessions. *Sci. Rep.* **2018**, *8*, 3373. [CrossRef]

50. Fatland, B.L.; Nikolau, B.J.; Wurtele, E.S. Reverse genetic characterization of cytosolic acetyl-CoA generation by ATP-citrate lyase in Arabidopsis. *Plant Cell* **2005**, *17*, 182–203. [CrossRef]

51. Henry, L.K.; Thomas, S.T.; Widhalm, J.R.; Lynch, J.H.; Davis, T.C.; Kessler, S.A.; Bohlmann, J.; Noel, J.P.; Dudareva, N. Contribution of isopentenyl phosphate to plant terpenoid metabolism. *Nat. Plants* **2018**, *4*, 721–729. [CrossRef]

52. Karačić, Z.; Vukelić, B.; Ho, G.H.; Jozić, I.; Sučec, I.; Salopek-Sondi, B.; Kozlović, M.; Brenner, S.E.; Ludwig-Müller, J.; Abramić, M. A novel plant enzyme with dual activity: an atypical Nudix hydrolase and a dipeptidyl peptidase III. *Biol. Chem.* **2017**, *398*, 101–112. [CrossRef]

53. Zhao, B.T.; Zhu, X.F.; Jung, J.H.; Xuan, Y.H. Effect of brassinosteroids on ammonium uptake via regulation of ammonium transporter and N-metabolism genes in Arabidopsis. *Biol. Plant.* **2016**, *60*, 563–571. [CrossRef]

54. Xuan, Y.H.; Duan, F.Y.; Je, B. Il; Kim, C.M.; Li, T.Y.; Liu, J.M.; Park, S.J.; Cho, J.H.; Kim, T.H.; von Wiren, N.; et al. Related to *ABI3/VP1-Like 1* (*RAVL1*) regulates brassinosteroid-mediated activation of *AMT1;2* in rice (Oryza sativa). *J. Exp. Bot.* **2016**, *68*, erw442. [CrossRef]

55. Nunes-Nesi, A.; Fernie, A.R.; Stitt, M. Metabolic and signaling aspects underpinning the regulation of plant carbon nitrogen interactions. *Mol. Plant* **2010**, *3*, 973–996. [CrossRef]

56. Ariz, I.; Artola, E.; Asensio, A.C.; Cruchaga, S.; Aparicio-Tejo, P.M.; Moran, J.F. High irradiance increases NH$_4^+$ tolerance in *Pisum sativum*: Higher carbon and energy availability improve ion balance but not N assimilation. *J. Plant Physiol.* **2011**, *168*, 1009–1015. [CrossRef]

57. Escobar, M.A.; Geisler, D.A.; Rasmusson, A.G. Reorganization of the alternative pathways of the Arabidopsis respiratory chain by nitrogen supply: Opposing effects of ammonium and nitrate. *Plant J.* **2006**, *45*, 775–788. [CrossRef]

58. Hachiya, T.; Noguchi, K. Integrative response of plant mitochondrial electron transport chain to nitrogen source. *Plant Cell Rep.* **2011**, *30*, 195–204. [CrossRef]

59. Frechilla, S.; Lasa, B.; Aleu, M.; Juanarena, N.; Lamsfus, C.; Aparicio-Tejo, P.M. Short-term ammonium supply stimulates glutamate dehydrogenase activity and alternative pathway respiration in roots of pea plants. *J. Plant Physiol.* **2002**, *159*, 811–818. [CrossRef]

60. Sakano, K. Revision of biochemical pH-stat: Involvement of alternative pathway metabolisms. *Plant Cell Physiol.* **1998**, *39*, 467–473. [CrossRef]

61. Sakano, K. Metabolic regulation of pH in plant cells: role of cytoplasmic pH in defense reaction and secondary metabolism. *Int. Rev. Cytol.* **2001**, *206*, 1–44.

62. Hachiya, T.; Watanabe, C.K.; Boom, C.; Tholen, D.; Takahara, K.; Kawai-Yamada, M.; Uchimiya, H.; Uesono, Y.; Terashima, I.; Noguchi, K. Ammonium-dependent respiratory increase is dependent on the cytochrome pathway in *Arabidopsis thaliana* shoots. *Plant Cell Environ.* **2010**, *33*, 1888–1897. [CrossRef]

63. Hagendoorn, M.J.M.; Wagner, A.M.; Segers, C.; Van Der Plas, L.H.W.; Oostdam, A.; Van Walraven, H.S. Cytoplasmic acidification and secondary metabolite production in different plant cell suspensions (A comparative study). *Plant Physiol.* **1994**, *106*, 723–730. [CrossRef]

64. Walker, R.P.; Chen, Z.; Johnson, K.E.; Famiani, F.; Tecsi, L.; Leegood, R.C. Using immunohistochemistry to study plant metabolism: the examples of its use in the localization of amino acids in plant tissues, and of phosphoenolpyruvate carboxykinase and its possible role in pH regulation. *J. Exp. Bot.* **2001**, *52*, 565–576. [CrossRef]

65. Walker, R.P.; Benincasa, P.; Battistelli, A.; Moscatello, S.; Técsi, L.; Leegood, R.C.; Famiani, F. Gluconeogenesis and nitrogen metabolism in maize. *Plant Physiol. Biochem.* **2018**, *130*, 324–333. [CrossRef]

66. Chen, Z.-H.; Walker, R.P.; Técsi, L.I.; Lea, P.J.; Leegood, R.C. Phosphoenolpyruvate carboxykinase in cucumber plants is increased both by ammonium and by acidification, and is present in the phloem. *Planta* **2004**, *219*, 48–58. [CrossRef]

67. Sweetlove, L.J.; Beard, K.F.M.; Nunes-Nesi, A.; Fernie, A.R.; Ratcliffe, R.G. Not just a circle: flux modes in the plant TCA cycle. *Trends Plant Sci.* **2010**, *15*, 462–470. [CrossRef]

68. Kinnersley, A.M.; Turano, F.J. Gamma aminobutyric acid (GABA) and plant responses to stress. *CRC. Crit. Rev. Plant Sci.* **2000**, *19*, 479–509. [CrossRef]

69. Turano, F.J.; Fang, T.K. Characterization of two glutamate decarboxylase cDNA clones from Arabidopsis. *Plant Physiol.* **1998**, *117*, 1411–1421. [CrossRef]

70. Ramesh, S.A.; Tyerman, S.D.; Xu, B.; Bose, J.; Kaur, S.; Conn, V.; Domingos, P.; Ullah, S.; Wege, S.; Shabala, S.; et al. GABA signalling modulates plant growth by directly regulating the activity of plant-specific anion transporters. *Nat. Commun.* **2015**, *6*, 7879. [CrossRef]

71. Gilliham, M.; Tyerman, S.D. Linking Metabolism to Membrane Signaling: The GABA–Malate Connection. *Trends Plant Sci.* **2016**, *21*, 295–301. [CrossRef]

72. Hachiya, T.; Watanabe, C.K.; Fujimoto, M.; Ishikawa, T.; Takahara, K.; Kawai-Yamada, M.; Uchimiya, H.; Uesono, Y.; Terashima, I.; Noguchi, K. Nitrate addition alleviates ammonium toxicity without lessening ammonium accumulation, organic acid depletion and inorganic cation depletion in *Arabidopsis thaliana* shoots. *Plant Cell Physiol.* **2012**, *53*, 577–591. [CrossRef]

73. Solórzano, L. Determination of ammonia in natural waters by the phenolhypochlorite method. *Limnol. Oceanogr.* **1969**, *14*, 799–801. [CrossRef]

74. Yemm, E.W.; Cocking, E.C.; Ricketts, R.E. The determination of amino-acids with ninhydrin. *Analyst* **1955**, *80*, 209. [CrossRef]

75. Walker, R.P.; Chen, Z.-H.; Técsi, L.I.; Famiani, F.; Lea, P.J.; Leegood, R.C. Phosphoenolpyruvate carboxykinase plays a role in interactions of carbon and nitrogen metabolism during grape seed development. *Planta* **1999**, *210*, 9–18. [CrossRef]

76. Srere, P.A. [1] Citrate synthase: [EC 4.1.3.7. Citrate oxaloacetate-lyase (CoA-acetylating)]. *Methods Enzymol.* **1969**, *13*, 3–11. [CrossRef]

77. Glauser, G.; Schweizer, F.; Turlings, T.C.J.; Reymond, P. Rapid profiling of intact glucosinolates in Arabidopsis leaves by UHPLC-QTOFMS using a charged surface hybrid column. *Phytochem. Anal.* **2012**, *23*, 520–528. [CrossRef]

78. Unwin, R.D.; Griffiths, J.R.; Whetton, A.D. Simultaneous analysis of relative protein expression levels across multiple samples using iTRAQ isobaric tags with 2D nano LC–MS/MS. *Nat. Protoc.* **2010**, *5*, 1574–1582. [CrossRef]

79. Cox, J.; Mann, M. MaxQuant enables high peptide identification rates, individualized p.p.b.-range mass accuracies and proteome-wide protein quantification. *Nat. Biotechnol.* **2008**, *26*, 1367–1372. [CrossRef]

80. Cox, J.; Neuhauser, N.; Michalski, A.; Scheltema, R.A.; Olsen, J.V.; Mann, M. Andromeda: A peptide search engine integrated into the MaxQuant environment. *J. Proteome Res.* **2011**, *10*, 1794–1805. [CrossRef]

International Journal of
Molecular Sciences

MDPI

Article

Changes in the Proteome of *Medicago sativa* Leaves in Response to Long-Term Cadmium Exposure Using a Cell-Wall Targeted Approach

Annelie Gutsch [1,2], Salha Zouaghi [1], Jenny Renaut [1], Ann Cuypers [2], Jean-Francois Hausman [1] and Kjell Sergeant [1,*]

[1] Environmental Research and Innovation, Luxembourg Institute of Science and Technology, 5, avenue des Hauts-Fourneaux, Esch-sur-Alzette, 4362 Luxembourg, Luxembourg; annelie.gutsch@list.lu (A.G.); salhazouaghi@gmail.com (S.Z.); jenny.renaut@list.lu (J.R.); jean-francois.hausman@list.lu (J.-F.H.)
[2] Agoralaan building D, Hasselt University, Campus Diepenbeek, Centre for Environmental Science, 3590 Diepenbeek, Belgium; ann.cuypers@uhasselt.be
* Correspondence: kjell.sergeant@list.lu; Tel.: +352-275-888-1

Received: 4 August 2018; Accepted: 21 August 2018; Published: 24 August 2018

Abstract: Accumulation of cadmium (Cd) shows a serious problem for the environment and poses a threat to plants. Plants employing various cellular and molecular mechanisms to limit Cd toxicity and alterations of the cell wall structure were observed upon Cd exposure. This study focuses on changes in the cell wall protein-enriched subproteome of alfalfa (*Medicago sativa*) leaves during long-term Cd exposure. Plants grew on Cd-contaminated soil (10 mg/kg dry weight (DW)) for an entire season. A targeted approach was used to sequentially extract cell wall protein-enriched fractions from the leaves and quantitative analyses were conducted with two-dimensional difference gel electrophoresis (2D DIGE) followed by protein identification with matrix-assisted laser desorption/ionization (MALDI) time-of-flight/time of flight (TOF/TOF) mass spectrometry. In 212 spots that showed a significant change in intensity upon Cd exposure a single protein was identified. Of these, 163 proteins are predicted to be secreted and involved in various physiological processes. Proteins of other subcellular localization were mainly chloroplastic and decreased in response to Cd, which confirms the Cd-induced disturbance of the photosynthesis. The observed changes indicate an active defence response against a Cd-induced oxidative burst and a restructuring of the cell wall, which is, however, different to what is observed in *M. sativa* stems and will be discussed.

Keywords: *Medicago sativa*; leaf cell wall proteome; cadmium; quantitative proteomics; 2D DIGE

1. Introduction

Pollution of soil, water and air is one of the serious issues of recent decades. Amongst others, contamination with heavy metals is of great concern due to their stability in the ecosystem. Contaminated sites are inaccessible for humans in the context of urbanization, biomass- and food-production, which poses a major problem and exacerbates the already limited availability of soil. Cadmium (Cd) is one of the most common pollutants in the environment with a high degree of genotoxicity [1]. Plants exposed to Cd suffer from an impairment of physiological and biochemical processes. They show limited growth and chlorosis and Cd leads to oxidative stress by generating reactive oxygen species (ROS) [2]. Cadmium interferes with photosynthesis by reducing the chlorophyll content, depressing the photosynthetic rate and induces direct damage to photosynthetic enzymes in a concentration- and time-dependent manner. Thereby, it was shown that Cd interferes more profoundly with the activity of photosystem II than photosystem I [3,4]. Cadmium can displace calcium (Ca) in photosystem II, thus inhibiting the formation of a functional complex and preventing photoactivation [5].

The plant cell wall is a dynamic cell-surrounding structure, which provides mechanical support and rigidity. It consists of cellulose, hemicellulose, pectin, as well as phenolic compounds. Proteins responsible for intercellular communication and interaction between the cell and the environment are imbedded in the cell wall. Those proteins make about 10% of the cell wall mass and their tightly regulated enzymatic reactions can alter the cell wall structure and properties [6,7], not only during plant development, but also during plant defence responses to biotic and abiotic stress [8,9]. Pectin methylesterase (PME), a cell wall protein, de-esterifies the pectic polysaccharide homogalacturonan (HG) creating binding sites for Ca^{2+}. Bound Ca mediates the bridging between two HG molecules to form a stable gel (egg-box structure) [10,11]. In the presence of Cd, PME showed an enhanced activity and the degree of low-methylesterified pectin in the cell wall increases concurrently with the deposition of Cd. By having the same charge, Cd^{2+} can bind pectin and displace Ca^{2+} as the cross-linking ion in the egg-box structure [12–14]. Additionally, Cd exposure has been shown to enhance lignification of the cell wall through an increased activity of cell wall-bound peroxidases, which causes cell wall stiffening and growth inhibition [15,16]. Such Cd-induced alterations of the cell wall structure indicate that the cell wall is part of the defence mechanisms set-up by the plant and that those structural changes limit further translocation of Cd, thus, keeping cytosolic Cd concentrations low.

The plant cell wall proteome has been studied in different species including dicots and monocots. To date, the *Arabidopsis thaliana* cell wall proteome is the most comprehensive [17]. Yet, the leaf apoplastic proteome including cell wall proteins remains much less studied [18] and information about cell wall proteins that change in abundance due to a treatment is underrepresented in the current scientific literature [19,20]. However, comparative cell wall proteome studies in leaves already provided information on how the cell wall proteome changes when exposed to various stresses [21,22]. To understand the mechanisms that take place in the cell wall during exposure to environmental constrains, it is important to unravel the cell wall proteome, its involvement in stress detection and response as well as its role in maintaining cell wall integrity.

Medicago sativa, commonly known as alfalfa, is an important forage legume and often used for research on cell wall development and stress adaptation [23,24]. Contrary to most research, in the present study *M. sativa* plants were exposed to realistic Cd concentrations for a long-term period, which makes the here-obtained results relevant for agricultural practices. Relative quantitative changes of the cell wall protein-enriched subproteome from leaves were investigated using 2D DIGE, which not only enables relative quantification but also visualizes different protein isoforms and modified proteins caused by Cd exposure [25]. A protocol for the enrichment of cell wall proteins was recently developed for *M. sativa* stems [26] and used in the current study on *M. sativa* leaves. The number of cytosolic contaminants in the different cell wall protein-enriched fractions remain low, which facilitates an accurate understanding of the leaf cell wall proteome. Although *M. sativa* proteins can be identified based on homology with *M. truncatula* proteins, as performed in a recent study [27], the combination of a search against the NCBI database and the *M. sativa* nucleotide database enlarges the number of identified proteins and the sequence coverage of the identified proteins, giving more comprehensive results. To our knowledge, this is the first study of the cell wall proteome of *M. sativa* leaves after long-term exposure to Cd.

2. Results

Cell wall protein-enriched fractions were obtained by subsequently using three different buffers of increasing ionic strength containing $CaCl_2$, ethylene glycol-*bis*[β-aminoethyl ether]-*N,N,N',N'*-tetraacetic acid (EGTA) or LiCl, to extract proteins with various wall-binding affinities. Using a targeted extraction protocol, the contamination with cytosolic proteins is low. However, several proteins involved in photosynthesis were identified and quantitative changes in these proteins are consistent throughout the fractions and replicates. As photosynthetic proteins are highly abundant in leaves and as Cd affects photosynthesis, they are included in the results and discussion.

A principle component analysis (PCA) on the gel-based spot intensity data analysed with the SameSpots software (TotalLab) revealed a clear distinction between control and Cd-exposed samples in the three cell wall protein-enriched fractions (Figure 1).

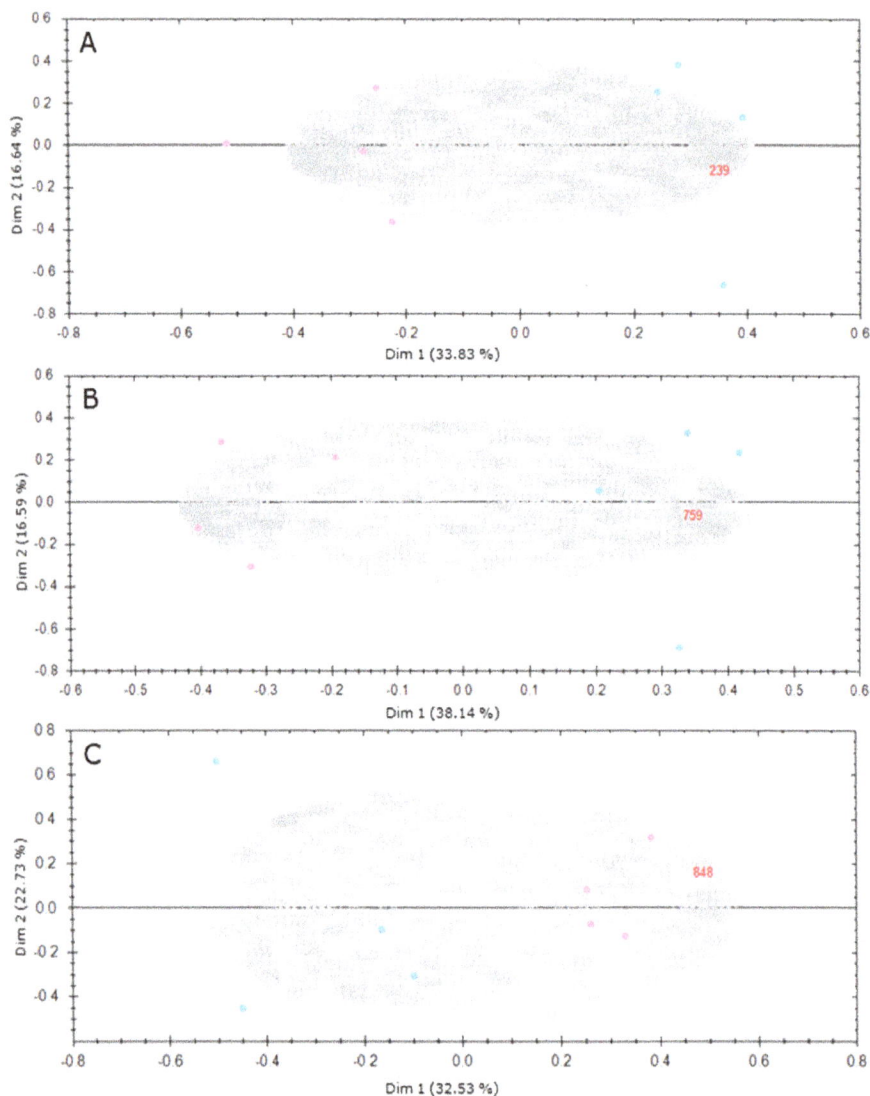

Figure 1. PCA analysis of the gel-based spot intensity data from the three cell wall protein-enriched fractions. (**A**) $CaCl_2$; (**B**) EGTA; (**C**) LiCl. Statistical analysis was done with SameSpots software (TotalLab). Blue dots represent the four biological replicates of the control. Pink dots represent the four biological replicates of Cd-exposed samples. Grey and red numbers correspond to spot numbers considered for the statistical analysis.

Int. J. Mol. Sci. **2018**, *19*, 2498

After manual and statistical evaluation of all detected spots, 306 spots showed a significant abundance change in response to Cd exposure (fold-change $\geq$ 1.2, ANOVA *p*-value $\leq$ 0.05) and were picked for identification. All mass spectra (MS) and MS/MS data that resulted in the protein identifications using the MASCOT server are provided in Table S1. Out of the total number of significantly changed spots, in 212 a single protein was identified and those were considered for biological interpretation. Based on the prediction for the subcellular localization with TargetP, 163 (76.9%) of these proteins are predicted to be secreted. Thirty seven are targeted to the chloroplast and 12 do not have a predicted subcellular target site. These predictions are mostly coherent with those from DeepLoc (Table S2), whereby DeepLoc distinguished the different locations after a protein has entered the secretory pathway as the most important change compared to TargetP. However, ongoing research in our lab indicates that DeepLoc predictions are not always reliable and vacuolar proteins are designates as extracellular and vice versa. Therefore, we based ourselves on TargetP in the results and in the discussion. In the CaCl$_2$ fraction, 65 spots gave a significant identification of a single protein, of which 22 are of lower abundance and 43 are of higher abundance in response to Cd-exposure. Most proteins were identified in the EGTA fraction. Here, 93 proteins were found to increase (55 proteins) or decrease (38 proteins) in abundance. In the LiCl fraction a total number of 54 proteins were identified, of which 19 decreased and 35 increased in abundance. All proteins were clustered according to their predicted biological function to gain better insight on their physiological role and how this can be related to the plant's response during Cd exposure (Figure 2). A complete list of all spots containing a single protein identification and, therefore, considered for biological interpretation is provided in Table S2, including statistical values obtained by the SameSpots software (TotalLab, Newcastle upon Tyne, UK) and their biological function. This information is summarised in Table 1. In several spots the same nominal protein was identified, however, the observation that it was identified at a different pI and/or molecular weight indicates that this concerns proteoforms. An example of this is a C-terminal truncation of eight amino acids from chitinase (e.g., the spots EGTA 1225 and 1279) that is probably determining for its subcellular location. The observation of a semi-tryptic peptide, corresponding to a cleavage in the middle of the papain family cysteine protease active domain in the spot EGTA 1971 (Table S1) indicates that the degradation of this protein increases in Cd-exposed plants. In the spot LiCl 1250 the same protein was identified but this time with a semi-tryptic peptide corresponding to the start of the active domain, after removal of the N-terminal inhibitor domain. These observations are confirmed by the position of the spots on the gel (Figure S1). Only by using a protein-based method, gel-based or gel-free, such proof of post-translational events can be obtained.

A large part of the higher abundant proteins are involved in plant defence (Figure 2). Another class of proteins with higher abundance upon Cd exposure have a designated function in oxidation-reduction processes. This classification includes different peroxidase isoforms present in all three fractions, plus a plastocyanin-like domain protein identified in the EGTA fraction (Table 1). Likewise, some proteins involved in carbohydrate metabolic processes and proteolysis are found to be of higher abundance. A minor part has a nutrient reserve function (rhicadhesin receptor, auxin-binding protein ABP19a) or is involved in photosynthesis (photosystem I reaction centre subunit II) (Table 1).

The functions assigned to proteins with a decreasing abundance are more diverse in the CaCl$_2$ and EGTA fraction (Figure 2). Proteins identified in the LiCl fraction were only classified in cell wall modification (47%) or nutrient reserve (53%) (Figure 2). In those 19 spots only three different proteins were identified, namely auxin-binding protein ABP19a, stem 28 kDa glycoprotein, and polygalacturonase non-catalytic protein (Table S2). Most proteins of lower abundance in the CaCl$_2$ fraction are involved in photosynthesis (oxygen-evolving enhancer protein, ribulose-1,5-bisphosphate carboxylase/oxygenase (RuBisCO) small chain and PS II oxygen-evolving enhancer protein) (Table S2). Photosynthetic proteins are highly abundant in leaves. As expected, the highest percentage of proteins that are not predicted to be cell wall localized is found in the CaCl$_2$ fraction. It must be noted that no spot corresponding to RuBisCO large chain, the main component of the leaf proteome, is present. This protein makes up more than 50% of the total leaf protein content and masking of lower abundant proteins by RubisCO large chain seems to be avoided using the here applied protocol for cell wall protein enrichment. Of the identified proteins, only the function of the plant/F18G18-200 protein, containing a DUF642 conserved domain, identified in the $CaCl_2$ and EGTA fraction remains unknown. The protein shows a decreased abundance in response to Cd and a secretion signal peptide was detected.

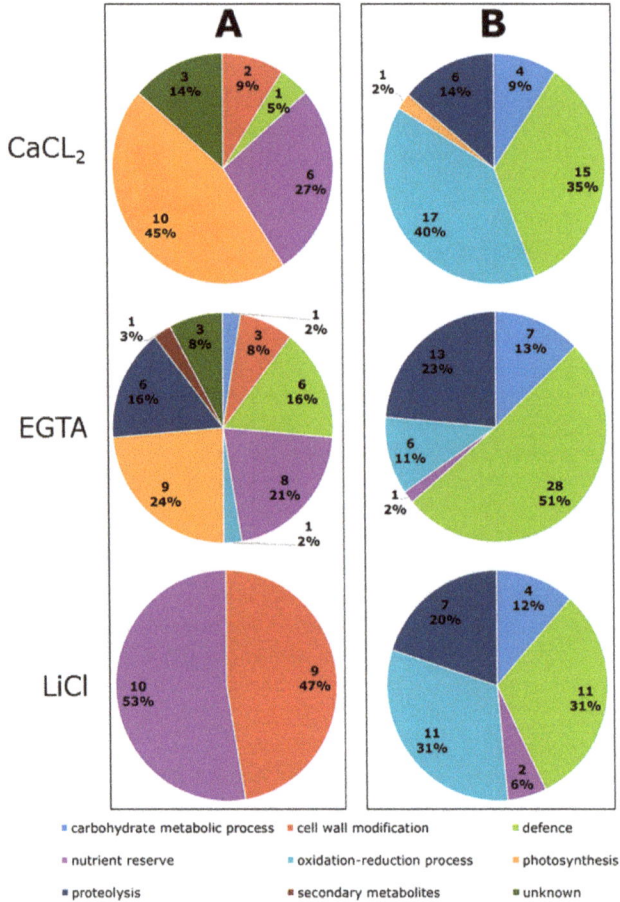

Figure 2. Functional classification of proteins present in cell wall protein-enriched fractions from *M. sativa* leaves after long-term exposure to Cd. *M. sativa* plants were exposed to Cd (10 mg/kg soil DW) for an entire season. Quantitative analysis based on four replicates were done with 2D DIGE comparing Cd-exposed samples to control samples and identified proteins clustered according to their predicted function using Blast2Go. (**A**) Functional classes of lower abundant proteins; and (**B**) functional classes of higher abundant proteins.

Table 1. Summary of all identified proteins in cell wall protein-enriched fractions from *M. sativa* leaves, which changed significantly in abundance after long-term Cd exposure. The table is based on all identifications provided in Table S2. The targeted location was predicted with TargetP. C: chloroplast; S: secretory pathway; /: any other location.

Protein Identification	NCBI Identification *	Nr. of Spots Wherein the Protein Was Identified	TargetP
Lower Abundant in Cd-Exposed Plants			
Carbohydrate Metabolic Process			
Sedoheptulose-1,7-bisphosphatase	gi | 357461143	1	C
Cell wall modification			
Pectinesterase/pectinesterase inhibitor	gi | 357504799	3	S
Polygalacturonase non-catalytic protein	gi | 922335979	10	S
Polygalacturonase-inhibiting protein 1	gi | 374634428	1	/
Defence			
Cystatin	gi | 74058377	1	/
Nod factor-binding lectin-nucleotide phosphohydrolase	gi | 357508587	4	S
Pathogenesis-related thaumatin family protein	gi | 922367846	1	S
CAP, cysteine-rich secretory protein, antigen 5	gi | 357446161	1	S
Nutrient reserve			
Auxin-binding protein ABP19a	gi | 357513969	11	S
Germin-like protein subfamily 3 member 1	gi | 502156424	1	S
Stem 28 kDa glycoprotein	gi | 357513539	12	S
Oxidation-reduction process			
1-cys peroxiredoxin PER1	gi | 922395795	1	C
Photosynthesis			
Chlorophyll a-b binding protein 2	gi | 3293555	1	C
Oxygen-evolving enhancer protein	gi | 922331371	6	C
Ribose-5-phosphate isomerase A	gi | 357512271	4	C
Ribulose bisphosphate carboxylase small chain	gi | 3914601	5	C
Photosystem I reaction centre subunit IV A	gi | 922402507	1	C
Photosystem II oxygen-evolving enhancer protein	gi | 922336891	1	C
Proteolysis			
Eukaryotic aspartyl protease family protein	gi | 922379288	6	S
Secondary metabolites			
Lactoylglutathione lyase-like protein	gi | 922388614	1	/
Unknown			
Plant/F18G18-200 protein	gi | 922395263	6	S
Higher Abundant in Cd-Exposed Plants			
Carbohydrate metabolic process			
Glucan endo-1,3-β-glucosidase	gi | 357474061	11	S
Glycoside hydrolase, family 17	gi | 87240471	1	/
Glycoside hydrolase family 18 protein	gi | 357454031	3	S
Defence			
Allergen Pru protein, putative	gi | 922401927	7	S
Chitinase (Class Ib)/Hevein	gi | 922329699	6	S
Chitinase/Hevein/PR-4/Wheatwin2	gi | 357447233	10	S
Chitinase	gi | 357443753	4	S
Class I chitinase	gi | 1800141	5	S
Disease resistance response protein	gi | 922325015	2	S or/
Pathogenesis-related protein 1	gi | 548592	1	S
Pathogenesis-related thaumatin family protein	gi | 922338021	4	S
Plant basic secretory protein (BSP) family protein	gi | 922407517	2	S
Pre-hevein-like protein	gi | 7381205	1	/
Stromal 70 kDa heat shock-related protein	gi | 821595433	1	C
CAP, cysteine-rich secretory protein, antigen 5	gi | 357446161	7	S
Nutrient reserve			
Auxin-binding protein ABP19a	gi | 357513969	1	S
Rhicadhesin receptor	gi | 357511665	2	S
Oxidation-reduction process			
Anionic peroxidase swpb3 protein	gi | 922380311	1	S
Class III peroxidase	gi | 357476371	10	S
Peroxidase	gi | 537317	7	S
Peroxidase family protein	gi | 357448431	1	S
Peroxidase1b	gi | 971560	3	S
Peroxidase2	gi | 13992528	10	S
Plastocyanin-like domain protein	gi | 922335020	2	S
Photosynthesis			
Photosystem I reaction centre subunit II	gi | 357480841	1	C
Proteolysis			
Carboxyl-terminal peptidase	gi | 922336331	2	S
Eukaryotic aspartyl protease family protein	gi | 922327497	14	C
Papain family cysteine protease	gi | 357437715	2	S
Polyubiquitin	gi | 695063425	6	/
Subtilisin-like serine protease	gi | 922333118	1	S

* The given NCBI identification is representative, more than one identifier was assigned to the same protein. A complete protein identification list including all NCBI identification numbers is provided in Table S2.

Generally, if the same protein was identified in different spots, their changes in abundance were consistent with a comparable fold-change. However, spots containing auxin-binding protein ABP19a changed in different directions in the LiCl fraction (Table S2, spot 181 down and spot 1516 up) and were found to be less abundant in the $CaCl_2$ fraction. Although the abundance of distinct isoforms of this protein is differently influenced by Cd exposure, the spectra of these proteins are identical and no hint was found to explain the observed dissimilarity. On the other side, of all spots containing CAP (cysteine-rich secretory protein, antigen 5) in the $CaCl_2$ fraction (Table S2, spot 3105 and spots 3028, 3030, 3031, 3034, 3037, 3081, and 3095), only spot 3105 has a decreased abundance. All of these spots contain a CAP most homologous to *Medicaco truncatula* CAP gi: 357446161. Nonetheless, only in spot 3028 the N-terminal peptide, after removal of the signal peptide, is QDSQADYVNAHNEAR, corresponding to contig 56,806 of *M. sativa*. In the other spots QDSQADYVNAHNDAR is identified as an N-terminal peptide corresponding to the contigs 111,668 and 1437 (Table S1). When calculating the pI of the translated contigs, the position on the gel confirms that spot 3028 is more acidic than the others (Figure 3).

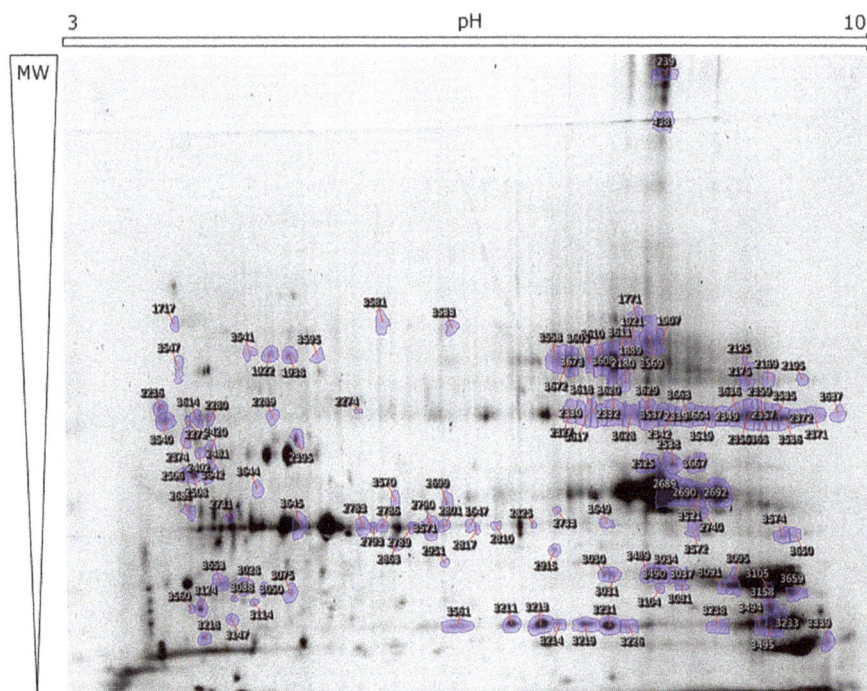

Figure 3. Illustration of 2D DIGE from the cell wall protein-enriched $CaCl_2$ fraction from leaves of *M. sativa*. Extractions were performed in four replicates. Proteins were pre-labelled with different CyDye to enable relative quantitative protein analysis. Labelled samples were loaded onto Immobiline™ DryStrip 3–10 NL, 24 cm (GE Healthcare) followed by migration on HPE™ Large Gel NF-12.5% (Serva Electrophoresis GmbH). Indicated spots were selected for picking based on statistical parameters calculated by the SameSpots software (TotalLab). Images of 2D DIGE from EGTA and LiCl fractions are provided in Figure S1. MW = molecular weight.

A number of spots in the EGTA fraction contain eukaryotic aspartyl protease family proteins. Some of these spots are of lower intensity after Cd-exposure (spot 956, 961, 962, 964, 978, and 983), while others are of higher intensity (904, 911, 917, 948, 955, 2239, and 2240, Table S2 and Figure S1).

Based on the spectra and the matched peptide sequences (Table S1), it was found that aspartyl protease proteins with a decreasing abundance are predicted to be secreted, while those that are of increasing abundance have a chloroplast transit peptide. The distinction can also be observed on the gel images as the chloroplastic aspartyl proteases are acidic and thus cluster on the left side of the gel (Figure S1).

The secreted aspartyl proteases (EGTA fraction spot 956, 961, 962, 964, 978, and 983) can be divided into three groups according to their location on the gel: (1) 978, 983; (2) 956; and (3) 961, 962, 964 (Figure S1). The *M. sativa* contigs corresponding to those spots were blasted against the NCBI database and split into two different groups. Spots 978 and 983 (corresponding to contig 4015) show homology to the *M. truncatula* sequence XP_003594399.1, while the contigs corresponding to the remaining spots show the highest homology to the *M. truncatula* sequence XP_013459881.1. Those *M. truncatula* sequences only have a sequence identity of 37%. However, both proteins carry the same conserved domain (pepsin_retropepsin_like superfamily) and no functional differences could be found in the literature.

The larger groups of spots with the same functional annotation were dissected, for instance, in 29 spots different chitinase proteins were identified. These can be divided in five groups based on the most homologous *M. truncatula* protein. This grouping is in agreement with the calculated pI and molecular weight of the different spots. In the spots 1279, 2259, 1225, and 2249 of the EGTA fraction, a class I chitinase (gi:1800141) was identified. Surprisingly, the identified protein lacks the eight C-terminal amino acids as shown by the identification of the peptide at m/z 2683.19 (Table S1). Similar observations have been conducted before [28] and may relate to the fact that the protein is actually vacuolar and only removal of the C-terminal octopeptide allows the protein to be secreted [29]. Of the chitinase-containing spots, those spots that are matched to the gi:357443753 (spot 1399, 1416, 1419, and 2508) have the highest fold change (Table S2). No clear functional differences between the chitinases could be found during this analysis.

In addition to sequence variants, signal sequences and the cleavage of activation/inhibition sequences, other post-translational modifications were also observed during MS analysis. Among these, α-β didehydrophenylalanine, as a potentially structure-determining modification, was identified in the β-subunit of polygalacturonase [25].

3. Discussion

The current study shows the impact of long-term Cd exposure on the *M. sativa* leaf subproteome, enriched in cell wall proteins. Identified proteins were clustered according to their biological function using Blast2Go (Figure 2) and those assignments as well as the number of proteins showing an increased or decreased abundance upon Cd exposure are comparable with results from a similar study on *M. sativa* stems [30]. In both studies, results confirm that the enrichment obtained with the used protocol is better than that obtained with comparable protocols, but not 100% and several non-apoplastic proteins, mainly chloroplast targeted, were identified most prominent in the CaCl$_2$ fraction (Figure 2) [30]. Those proteins will also be discussed as they are highly abundant in leaves and coherent, significant changes in abundance were observed between replicates and fractions in response to Cd exposure.

After planting, strong growth inhibition of Cd-exposed plants was observed at the end of the first growth cycle and coincided with the visual observation of leaf senescence. These phenotypical differences between Cd-exposed and control plants disappeared during the second growth cycle and no differences in appearance were observed at a later maturation state. Overall, long-term Cd exposure did not have any morphological impact on the leaves of *M. sativa*. Leaves from control and Cd-exposed plants had a rather heterogeneous appearance, due to which no leaf was representative for any of the conditions. Furthermore to attain the amount of leaf material needed, four to five grams for each replicate, all leaves from young to old were sampled. The visual observation of limited impact of long-term Cd exposure is supported by the fact that the average plant biomass taken from five replicates at the end of the experiment was not significantly different between Cd-exposed (120.32 $\pm$ 4.67 g) and control plants (110.65 ± 3.78 g) [30]. Although long-term experiments as presented here are rather scarce, the high sensitivity of initial growth stages followed by a low impact of the applied stress in adult plants was observed before for other plant species [15,31,32]. Obviously, plants do acquire a new steady-state when they are subjected to a constant severe but non-lethal stress [33].

3.1. Cd-Induced Degradation of Photosynthetic Proteins

Throughout all three cell wall protein-enriched fractions a decreasing abundance of photosynthetic proteins was revealed (Table 1). Among others, RuBisCO small chain was less abundant in Cd-exposed plants as well as subunits of photosystem I and II. The Cd-induced decreased abundance of photosynthetic proteins was recently reported in *M. sativa* stems after long-term Cd exposure when analysing cell wall protein-enriched fractions using 2D DIGE [30]. Likewise as in leaves, RuBisCO small chain, as well as a subunit of photosystem I, were less abundant in *M. sativa* stems when exposed to Cd. The fact that Cd is disrupting photosynthesis and that it causes oxidative stress in plants by ROS production is well known [2,34]. Reactive oxygen species are able to oxidise proteins, destining the oxidised protein for degradation, which goes together with an increasing protease activity [35,36]. Here, we observed that Cd induces a higher abundance of chloroplastic aspartyl protease isoforms in the EGTA fraction. The protein was identified before in leaf tissue of Cd-exposed *A. thaliana* [37] and its homology with cnd41-like proteins implies its function in senescence and nutrient recovery. Aspartyl proteases are implicated in degradation or processing of proteins and their occurrence during stress responses is established. They might have a crucial function in protein turn-over to prevent the accumulation of deactivated proteins, thereby increasing the pool of available amino acids, needed for the synthesis of defence-related proteins [38]. They are involved in RuBisCO degradation and degradation of chloroplastic proteins [39–41]. An increased abundance of this protease in *M. sativa* leaves correlates with the observed decreased abundance of photosynthetic proteins. Oxygen radicals, which appear during Cd exposure cause enhanced degradation of proteins involved in photosynthesis and impair the physiological processes in leaves. The increased abundance of a 70 kDa stromal heat shock protein ($CaCl_2$ spot 2402) likewise indicates that there is an increased need for protein refolding in the chloroplast.

A second group of spots in the EGTA fraction contains isoforms of aspartyl protease, which are of lower abundance in Cd-exposed plants. The proteins identified in these spots are predicted to be secreted (Table S2). Although no unambiguous function is established for secreted aspartyl proteases, they have been identified before, for instance in the pollen cell wall [42]. Functional studies have implicated them in the defence against biotic stresses [43] and the *Arabidopsis* homolog AED1 was recently proposed to be part of a homeostatic feedback mechanism regulating the systemic acquired resistance (SAR) response [44]. This link between SAR, a salicylic acid-regulated process, and the here-observed limited effect of long-term Cd exposure confirms previous studies, wherein salicylic acid application is shown to alleviate Cd-induced growth inhibition [45].

The only photosynthetic protein found with an increasing abundance is the photosystem I reaction centre subunit II ($CaCl_2$ fraction spot 3075, Table S2). The position of this spot is, however, at too low a pI and molecular weight, providing a further indication for increasing protein degradation in the chloroplast.

3.2. Cd Influences the Abundance of Proteins Related to the Cell Wall Structure

Pectinesterase/pectinesterase inhibitor proteins are identified in three spots (Table S2, $CaCl_2$ fraction spot 2175 and EGTA fraction spot 989, 1007). The abundance of these proteins decreases in Cd-exposed plants. PMEs and their inhibitor are expressed as a single polypeptide and get subsequently processed by cleavage between the inhibitor and active domain. Our MS data do not confirm that the inhibitor domain is cut from the PME domain. However, the position on the 2D gel matches with what is predicted for the active protein and are at the same position as those found in *M. sativa* stems [30]. In the latter study, cleavage between the inhibitor and the active domain was confirmed based on MS data, indicating that in leaves the inhibitor domain was also cleaved and thus the protein activated. However, the abundance of PME increased in *M. sativa* stems in response to long-term Cd exposure [30].

PME catalyses the demethylesterification of HG in plant cell walls. Those demethylated, acidic HG molecules form bridges between each other mediated by Ca^{2+} ions (egg-box structure) and confer rigidity to the cell wall [10,46]. Cadmium ions have the same charge (Cd^{2+}) and can displace Ca^{2+} as cross linker in the pectin egg-box structure. During Cd exposure a higher abundance of PME was found in several studies and an enhanced activity demonstrated [14,16]. Changes in HG pattern in response to Cd have been investigated together with a preferential allocation of Cd in the cell wall [12,47,48]. Thus, the cell wall structure changes during Cd exposure and the sequestration of Cd in the cell wall protects the cytosol. In a recent study on the cell wall composition in *M. sativa* stems upon long-term Cd exposure, the most significant changes appeared in the pectin fraction towards a higher abundance of HG upon Cd exposure and an increased activity of PME in response to Cd was determined (Gutsch et al., submitted). This suggests a high demethylation degree, which creates binding sites for Cd and immobilize it in the cell wall. Opposite to what is observed in *M. sativa* stems [30], Cd exposure led to a decreasing abundance of PME in leaves, which would lead to a low demethylation degree of HG. The methylation degree of pectin has an influence on the cell wall structure and, furthermore, limits the accessibility for pectin-degrading enzymes, such as polygalacturonases [49]. In response to long-term Cd exposure, the structural changes in the pectin network of the leaf cell wall seem to be different from those anticipated in the stems of *M. sativa* in response to Cd [30] and an organ-specific influence of Cd on the cell wall can be assumed. Together with so-far-unpublished data (Gutsch et al., submitted), the structural changes, which are observed in the stems of *M. sativa*, promote the creation of binding sites for Cd in the stem cell wall as a direct response to the applied stress. In this matter, the leaf cell wall plays probably a minor role in the retention of Cd but more data on the structural changes would be needed to draw a more comprehensive conclusion.

Polygalacturonase-inhibiting protein 1 and polygalacturonase non-catalytic protein or β-subunit are involved in pectin degradation. Conflicting data on the function of the latter has been obtained. Overexpression of the polygalacturonase non-catalytic β-subunit in rice resulted in a decreased pectin content and a higher susceptibility to abiotic stress due to lower cell adhesion [50]. On the other hand, during fruit ripening in tomato, it limits the extent of pectin solubilisation and depolymerization [51]. In the present study, a lower abundance occurred in the *M. sativa* leaf subproteome enriched in cell wall proteins (Table 1), as was previously reported in *M. sativa* stems [30]. However, a quantitative change in the abundance of β-subunit of polygalacturonase is not necessarily correlated with a change in polygalacturonase activity [52]. The functional linkage between these two polygalacturonase subunits remains to be solved as contradictory data exist. Furthermore, polygalacturonase inhibitor showed a decreasing abundance in our study. It can be speculated that this decreased abundance is a cross reaction to the lower abundance of polygalacturonase non-catalytic protein in order to keep the polygalacturonase activity in the cell wall stable. So far, polygalacturonase inhibitor has only been described during the defence against pathogen attacks by inhibiting the fungal polygalacturonase [53,54]. Those proteins have a specific binding site to interact with pectin in the plant cell wall, which is furthermore influenced by methylesterification patterns in the pectin network [55]. Their function in the alteration of the plant cell wall structure in response to abiotic stress remains unclear.

Various peroxidase isoforms were highly abundant in the leaves of Cd-exposed *M. sativa* (Table 1). Previous studies on *M. sativa* stems [30] and poplar leaves gave similar results [56]. Peroxidases are involved in oxidation-reduction and lignification processes in the cell wall [57]. Using H_2O_2 molecules as a co-substrate, peroxidases catalyse the oxidation of monolignols, which then cross-link to form lignin [58] and increasing peroxidase activity was positively correlated with the degree of cell wall lignification [59]. Cadmium enhances H_2O_2 accumulation in plants and increased peroxidase abundance and activity, leading to cell wall lignification, cell wall stiffening and growth impairment [15,

Int. J. Mol. Sci. **2018**, *19*, 2498

60,61]. Oxidative stress might also be responsible for the accumulation of carboxyl-terminal peptidase (Table 1). The protein contains two DUF domains (239 and 4409). The *Arabidopsis* homolog was found to be responsive to ROS and confers enhanced tolerance to oxidative stress [62]. The same protein was found to have a role in cell wall modification, influencing nutrient transport by modification of root endodermal barriers [63].

3.3. Enhanced Accumulation of Defence Proteins as a Response to Cd

Different defence related proteins were identified in *M. sativa* leaves (Table 1). Their increased accumulation due to Cd exposure underline the strong defence response of the plant, which had been reported before when plants were exposed to heavy metals [30,37,56,64]. Most prevalent, we identified different chitinase isoforms throughout all three fractions (Table S2), which is consistent with previous findings in *M. sativa* stems [30], where several chitinase isoforms increased in abundance upon long-term Cd-exposure. A study on different plant species exposed to different metals indicates that a metal-specific chitinase-expression profile may exist [65]. Although such metal-specific functions of chitinases remain uncertain, an increased abundance of chitinases has been proposed as a marker for the induction of a SAR-response. Despite the small impact of long-term Cd exposure on biomass production [30], it appears to have a significant negative impact on the presence of proteins classified as having a nutrient reserve function (Figure 2), which was also reported in *M. sativa* stems [30]. This may indicate that the plants are capable of maintaining growth under the conditions used in this study, but are not capable of establishing reserves. It must, however, be mentioned that multiple functions are attributed to these nutrient reserve proteins and that the observed decreased abundance may have other consequences [66].

4. Materials and Methods

4.1. Plant Material and Treatment

M. sativa (cv Giulia) seeds were inoculated with *Sinorhizobium meliloti* and sown in May 2015 on Cd-polluted soil (10 mg/kg soil DW added as $CdSO_4$) and uncontaminated soil. The used soil was prepared as one batch (2/3 potting soil mixed with 1/3 sand (w/w)) before splitting in two conditions. For each condition 12 times 12 pots were planted. The plants were kept in the greenhouse until flowering stage was reached (July) and a first cut was done as during agricultural cultivation of *M. sativa*. Subsequently, plants were kept outside to avoid insect infestation. After a re-growing period till the pre-flowering stage was reached, plants were put back into the greenhouse for one more week before sampling was done on the 10th of September. No temperature or day cycle control was done during the experiment and no fertilizer was applied. A pool of leaves was sampled in four replicates for each condition and directly frozen in liquid nitrogen. All samples were kept at −80 °C until further use.

4.2. Cell Wall Protein Enrichment

Cell wall protein-enriched fractions were obtained as described elsewhere [25] by using subsequently three different buffers with increasing ionic strength to extract also tightly bound proteins. Following the extraction, all three protein fractions ($CaCl_2$, ethylene glycol-*bis*(β-aminoethyl ether)-*N,N,N′,N′*-tetraacetic acid [EGTA], LiCl) were concentrated with Amicon Ultra 15 10 K (Millipore, Burlington, MA, USA) by centrifugation ($4700 \times g$, 4 °C) to an approximate volume of 200 µL. Subsequently, the ReadyPrep 2D Cleanup kit (Bio-Rad, Hercules, CA, USA) was used to desalt the samples following the manufacturer's instruction. Cleaned samples were solubilized in labelling buffer (7 M urea, 2 M thiourea, 2% w/v 3-[(3-Cholamidopropyl)dimethylammonio]-1-propanesulfonate (CHAPS), 30 mM Tris) and the protein concentrations determined using the Bradford method (Bradford reagent, Bio-Rad).

4.3. Quantitative Protein Analysis and Identification

A 2D DIGE was undertaken to compare protein abundances between conditions in the three different fractions. Therefore, 50 µg of protein from each sample were labelled with either Cy3 or Cy5 and a dye swap was applied to avoid a possible effect of preferential labelling. An internal standard composed of 25 µg protein from each replicate (four biological replicates from control and cadmium, respectively) was labelled with Cy2 for each fraction ($CaCl_2$, EGTA, LiCl). Labelled samples were mixed, 9 µL Servalyte pH 3–10 (Serva Electrophoresis GmbH, Heidelberg, Germany) and 2.7 µL Destreak Reagent (GE Healthcare, Chicago, IL, USA) were added. The volumes were adjusted with lysis buffer (7 M urea, 2 M thiourea, 4% *w/v* CHAPS) to 450 µL. Samples were loaded onto Immobiline™ DryStrip 3–10 NL, 24 cm (GE Healthcare, Chicago, IL, USA) overnight, followed by isoelectric focusing (IEF): (1) constant 100 V for 4 h; (2) linear gradient up to 1000 V for 4 h; (3) constant 1000 V for 5 h; (4) linear gradient up to 10,000 V for 6 h; and (5) constant 10,000 V until a total of 80,000 volt hours were reached. IEF-strips were equilibrated in equilibration buffer (Serva Electrophoresis GmbH) according to the manufacturer's instructions. 2D HPE™ Large Gels NF-12.5% (Serva Electrophoresis GmbH) were used for the second dimension and were run on a HPE tower system following the manufacturer's instruction. After fixation (15% ethanol *v/v*, 1% *w/v* citric acid), three images from each of the gels were acquired using different wavelengths for the different labelling dyes (Cy2 488 nm, Cy3 532 nm, Cy5 642 nm) (Typhoon FLA 9500, GE Healthcare). SameSpots software v4.5 (TotalLab, Newcastle upon Tyne, UK) was used for the relative quantitative image analysis. Since the same internal standard is run on each gel of a fraction, alignment and normalisation with the internal standard allows comparison of the spots between repetitions. All statistical analyses were automatically done by the software. A spot was chosen for protein identification if a treatment effect was reported (fold-change $\geq$ 1.2, ANOVA *p*-value $\leq$ 0.05, Table S3 for spot volumes), if the spot was matched on all replicates and after manual validation.

Selected spots (Figure 3 and Figure S1) were picked with an Ettan Spot Picker (GE Healthcare) and digested prior to MS/MS analyses as described before [25]. Mass spectra were acquired with 5800 MALDI TOF/TOF (AB Sciex, Framingham, MA, USA). The ten most abundant peaks were automatically selected for fragmentation and spectra submitted to an in-house MASCOT server (Matrix Science, Available online: www.matrixscience.com) for database-dependent identifications. A first search was performed against the NCBInr database limited to *Viridiplantae* (3,334,509 sequences) and a second one against *M. sativa* sequences downloaded from the Samuel Roberts Noble website (Available online: plantgrn.noble.org/AGED (675,756 sequences, 304,231,576 residues)) [67]. The search parameters were as follows: mass tolerance 100 ppm, fragment mass tolerance 0.5 Da, cysteine carbamidomethylation as fixed modification, methionine oxidation, double oxidation of tryptophan, and tryptophan to kynurenine as variable modification. When at least two peptides passed the MASCOT-calculated 0.05 threshold score of 40, proteins were considered as identified. Additionally, if a high-quality spectrum was not matched to a protein, the interpretation was done manually and search parameters adjusted (semitryptic, single amino acid change, and post-translational modification) to increase the sequence coverage of identified proteins. After manual validation of the identifications, the subcellular location was predicted with the TargetP server using the standard search parameters (Available online: http://www.cbs.dtu.dk/services/TargetP) [68]. Only proteins with a predicted signal peptide have been considered as cell wall proteins as done in current literature [17,69]. To validate the predicted subcellular location, a second location prediction was undertaken using DeepLoc [70]. In some cases predictions were corrected after literature research. The Blast2Go software was used to gather information about the biological function of the identified proteins as well as the available literature.

5. Conclusions

M. sativa plants were exposed to Cd (10 mg/kg soil DW) in a long-term experiment and the leaf cell wall protein-enriched subproteome was analysed. In total, 212 identified proteins changed significantly in response to Cd and a major part of these identified proteins is involved in defence responses, underlining the importance of the general defence machinery in response to Cd and linking the observations in this study with knowledge on the SAR response. Cell wall proteins related to oxidation-reduction processes are highly abundant in Cd-exposed plants and might counteract the Cd-induced oxidative burst in the plant. Germin-like proteins, although classified as nutrient reserve, may also contribute. Interestingly, Cd provokes tissue-specific alterations in the pectin network of the cell wall in *M. sativa* leaves and stems [30]. The leaf cell wall seems to be less involved in the assumed cell wall-promoted binding of Cd as a protective mechanism.

About 18% of the identified proteins are targeted to the chloroplast and their relative abundance decreases upon Cd exposure concomitantly with an increase in chloroplastic, proteolytic proteins. Therefore, the increased protein degradation in the chloroplast confirms interference of Cd with the photosynthetic activity of plants. Nonetheless, Cd-exposed plants showed no difference in biomass production or in the growth at the moment of sampling in comparison to control plants, which suggests that the plants established a new metabolic steady-state during the long-term stress exposure. The important decrease in proteins with nutrient reserve function, however, indicates that the plants are weakened and may perform worse than control plants when exposed to secondary stresses.

Supplementary Materials: Supplementary Materials can be found at http://www.mdpi.com/1422-0067/19/9/2498/s1. Table S1. Complete MASCOT protein identification data of picked spots (xls 819 KB). Table S2. The table contains all identified proteins from each fraction and their abundance change due to Cd exposure. Fold change and *p*-values were obtained using the SameSpots software (TotalLab). Functional classification of each protein was determined using Blast2Go. Subcellular localization was predicted with TargetP and DeepLoc (docx 37 KB). Figure S1. 2D DIGE images from the cell wall protein-enriched EGTA and LiCl fractions of *M. sativa* leaves (docx. 2.118 KB). Table S3. All spots and corresponding volumes (relative and normalized) are given as detected with SameSpots software (TotalLab) including statistical parameters automatically calculated by the software. Spots that were selected for identification are indicated (xlsx 536 KB).

Author Contributions: A.G., K.S., J.-F.H. and A.C. designed the experiment. S.Z. performed all the experimental work and data analysis and K.S. contributed to the protein identification and quantification. J.R. supplied and managed the technical equipment. A.G. and K.S. performed the data interpretation and drafted the manuscript which was critically revised and approved for submission by all the other authors.

Funding: This work was supported by the FNR and FWO and is part of the bilateral project CadWALL (FNR/FWO project INTER/FWO/13/14).

Acknowledgments: We specially thank Sébastien Planchon for his technical support during the 2D DIGE experiment and the following quantitative analysis.

Conflicts of Interest: The authors declare no conflict of interest. The funders had no role in the design of the study; in the collection, analyses, or interpretation of data; in the writing of the manuscript, or in the decision to publish the results.

Abbreviations

2D DIGE	Two-dimensional difference gel electrophoresis
Ca	Calcium
Cd	Cadmium
DW	Dry weight
HG	Homogalacturonan
kDa	Kilo Dalton
MALDI	Matrix-assisted Laser Desorption/Ionization
MS	Mass spectrum
MW	Molecular weight
PCA	Principal component analysis
PME	Pectin methylesterase
ROS	Reactive oxygen species
RuBisCO	Ribulose bisphosphate carboxylase/oxidase
SAR	Systemic acquired resistance
TOF	Time of flight

Int. J. Mol. Sci. **2018**, *19*, 2498

References

1. Kovalchuk, O.; Titov, V.; Hohn, B.; Kovalchuk, I. A sensitive transgenic plant system to detect toxic inorganic compounds in the environment. *Nat. Biotechnol.* **2001**, *19*, 568–572. [CrossRef] [PubMed]

2. Di Toppi, L.S.; Gabbrielli, R. Response to cadmium in higher plants. *Environ. Exp. Bot.* **1999**, *41*, 105–130. [CrossRef]

3. Chugh, L.K.; Sawhney, S.K. Photosynthetic activities of *Pisum sativum* seedlings grown in presence of cadmium. *Plant Physiol. Biochem.* **1999**, *37*, 297–303. [CrossRef]

4. Wang, Y.; Jiang, X.; Li, K.; Wu, M.; Zhang, R.; Zhang, L.; Chen, G. Photosynthetic responses of *Oryza sativa* L. seedlings to cadmium stress: Physiological, biochemical and ultrastructural analyses. *Biometals* **2014**, *27*, 389–401. [CrossRef] [PubMed]

5. Faller, P.; Kienzler, K.; Krieger-Liszkay, A. Mechanism of Cd^{2+} toxicity: Cd^{2+} inhibits photoactivation of Photosystem II by competitive binding to the essential Ca^{2+} site. *Biochim. Biophys. Acta Bioenerg.* **2005**, *1706*, 158–164. [CrossRef] [PubMed]

6. Cassab, G.I. Plant Cell Wall Proteins. *Annu. Rev. Plant Physiol. Plant Mol. Biol.* **1998**, *49*, 281–309. [CrossRef] [PubMed]

7. Fry, S.C.; Smith, R.C.; Renwick, K.F.; Martin, D.J.; Hodge, S.K.; Matthews, K.J. Xyloglucan endotransglycosylase, a new wall-loosening enzyme activity from plants. *Biochem. J.* **1992**, *282*, 821–828. [CrossRef] [PubMed]

8. Qin, Q.; Bergmann, C.W.; Rose, J.K.C.; Saladie, M.; Kolli, V.S.K.; Albersheim, P.; Darvill, A.G.; York, W.S. Characterization of a tomato protein that inhibits a xyloglucan-specific endoglucanase. *Plant J.* **2003**, *34*, 327–338. [CrossRef] [PubMed]

9. Bradley, D.J.; Kjellbom, P.; Lamb, C.J. Elicitor- and wound-induced oxidative cross-linking of a proline-rich plant cell wall protein: A novel, rapid defense response. *Cell* **1992**, *70*, 21–30. [CrossRef]

10. Caffall, K.H.; Mohnen, D. The structure, function, and biosynthesis of plant cell wall pectic polysaccharides. *Carbohydr. Res.* **2009**, *344*, 1879–1900. [CrossRef] [PubMed]

11. Krzesłowska, M. The cell wall in plant cell response to trace metals: Polysaccharide remodeling and its role in defense strategy. *Acta Physiol. Plant* **2011**, *33*, 35–51. [CrossRef]

12. Vollenweider, P.; Cosio, C.; Günthardt-Goerg, M.S.; Keller, C. Localization and effects of cadmium in leaves of a cadmium-tolerant willow (*Salix viminalis* L.). *Environ. Exp. Bot.* **2006**, *58*, 25–40. [CrossRef]

13. Parrotta, L.; Guerriero, G.; Sergeant, K.; Cai, G.; Hausman, J.-F. Target or barrier? The cell wall of early- and later-diverging plants vs. cadmium toxicity: Differences in the response mechanisms. *Front. Plant Sci.* **2015**, *6*, 133. [CrossRef] [PubMed]

14. Douchiche, O.; Soret-Morvan, O.; Chaïbi, W.; Morvan, C.; Paynel, F. Characteristics of cadmium tolerance in "Hermes" flax seedlings: Contribution of cell walls. *Chemosphere* **2010**, *81*, 1430–1436. [CrossRef] [PubMed]

15. Chaoui, A.; El Ferjani, E. Effects of cadmium and copper on antioxidant capacities, lignification and auxin degradation in leaves of pea (*Pisum sativum* L.) seedlings. *C. R. Biol.* **2005**, *328*, 23–31. [CrossRef] [PubMed]

16. Paynel, F.; Schaumann, A.; Arkoun, M.; Douchiche, O.; Morvan, C. Temporal regulation of cell-wall pectin methylesterase and peroxidase isoforms in cadmium-treated flax hypocotyl. *Ann. Bot.* **2009**, *104*, 1363–1372. [CrossRef] [PubMed]

17. Albenne, C.; Canut, H.; Jamet, E. Plant cell wall proteomics: The leadership of *Arabidopsis thaliana*. *Front. Plant Sci.* **2013**, *4*, 1–17. [CrossRef] [PubMed]

18. Haslam, R.P.; Downie, A.L.; Raveton, M.; Gallardo, K.; Job, D.; Pallett, K.E.; John, P.; Parry, M.A.J.; Coleman, J.O.D. The assessment of enriched apoplastic extracts using proteomic approaches. *Ann. Appl. Biol.* **2003**, *143*, 81–91. [CrossRef]

19. Ndimba, B.K.; Chivasa, S.; Hamilton, J.M.; Simon, W.J.; Slabas, A.R. Proteomic analysis of changes in the extracellular matrix of *Arabidopsis* cell suspension cultures induced by fungal elicitors. *Proteomics* **2003**, *3*, 1047–1059. [CrossRef] [PubMed]

20. Tran, H.T.; Plaxton, W.C. Proteomic analysis of alterations in the secretome of *Arabidopsis thaliana* suspension cells subjected to nutritional phosphate deficiency. *Proteomics* **2008**, *8*, 4317–4326. [CrossRef] [PubMed]

21. Meng, X.; Song, T.; Fan, H.; Yu, Y.; Cui, N.; Zhao, J.; Meng, K. A comparative cell wall proteomic analysis of cucumber leaves under *Sphaerotheca fuliginea* stress. *Acta Physiol. Plant.* **2016**, *38*, 260. [CrossRef]

22. Dani, V.; Simon, W.J.; Duranti, M.; Croy, R.R.D. Changes in the tobacco leaf apoplast proteome in response to salt stress. *Proteomics* **2005**, *5*, 737–745. [CrossRef] [PubMed]

23. Printz, B.; Guerriero, G.; Sergeant, K.; Audinot, J.-N.; Guignard, C.; Renaut, J.; Lutts, S.; Hausman, J.-F. Combining-Omics to Unravel the Impact of Copper Nutrition on Alfalfa (*Medicago sativa*) Stem Metabolism. *Plant Cell Physiol.* **2016**, *57*, 407–422. [CrossRef] [PubMed]

24. Verdonk, J.C.; Hatfield, R.D.; Sullivan, M.L. Proteomic analysis of cell walls of two developmental stages of alfalfa stems. *Front. Plant Sci.* **2012**, *3*, 279. [CrossRef] [PubMed]

25. Sergeant, K.; Printz, B.; Gutsch, A.; Behr, M.; Renaut, J.; Hausman, J.-F. Didehydrophenylalanine, an abundant modification in the β subunit of plant polygalacturonases. *PLoS ONE* **2017**, *12*, e0171990. [CrossRef] [PubMed]

26. Printz, B.; Morais, R.D.S.; Wienkoop, S.; Sergeant, K.; Lutts, S.; Hausman, J.-F.; Renaut, J. An improved protocol to study the plant cell wall proteome. *Front. Plant Sci.* **2015**, *6*, 237. [CrossRef] [PubMed]

27. Xiong, J.; Sun, Y.; Yang, Q.; Tian, H.; Zhang, H.; Liu, Y.; Chen, M. Proteomic analysis of early salt stress responsive proteins in alfalfa roots and shoots. *Proteome Sci.* **2017**, *15*, 19. [CrossRef] [PubMed]

28. Neuhaus, J.-M.; Pietrzak, M.; Boller, T. Mutation analysis of the C-terminal vacuolar targeting peptide of tobacco chitinase: Low specificity of the sorting system, and gradual transition between intracellular retention and secretion into the extracellular space. *Plant J.* **1994**, *5*, 45–54. [CrossRef] [PubMed]

29. Stigliano, E.; Di Sansebastiano, G.P.; Neuhaus, J.M. Contribution of chitinase A's C-terminal vacuolar sorting determinant to the study of soluble protein compartmentation. *Int. J. Mol. Sci.* **2014**, *15*, 11030–11039. [CrossRef] [PubMed]

30. Gutsch, A.; Keunen, E.; Guerriero, G.; Renaut, J.; Cuypers, A.; Hausman, J.-F.; Sergeant, K. Long-term cadmium exposure influences the abundance of proteins that impact the cell wall structure in *Medicago sativa* stems. *Plant Biol. J.* **2018**. [CrossRef] [PubMed]

31. Peralta-Videa, J.R.; De la Rosa, G.; Gonzalez, J.H.; Gardea-Torresdey, J.L. Effects of the growth stage on the heavy metal tolerance of alfalfa plants. *Adv. Environ. Res.* **2004**, *8*, 679–685. [CrossRef]

32. Wang, C.Q.; Song, H. Calcium protects *Trifolium repens* L. seedlings against cadmium stress. *Plant Cell Rep.* **2009**, *28*, 1341–1349. [CrossRef] [PubMed]

33. Gratão, P.L.; Monteiro, C.C.; Antunes, A.M.; Peres, L.E.P.; Azevedo, R.A. Acquired tolerance of tomato (*Lycopersicon esculentum* cv. Micro-Tom) plants to cadmium-induced stress. *Ann. Appl. Biol.* **2008**, *153*, 321–333. [CrossRef]

34. Schützendübel, A.; Polle, A. Plant responses to abiotic stresses: Heavy metal-induced oxidative stress and protection by mycorrhization. *J. Exp. Bot.* **2002**, *53*, 1351–1365. [CrossRef] [PubMed]

35. Romero-Puertas, M.C.; Palma, J.M.; Gómez, M.; Río, L.A.D.E.L.; Sandalio, L.M. Cadmium causes the oxidative modification of proteins in pea plants. *Plant Cell Environ.* **2002**, *25*, 677–686. [CrossRef]

36. Sandalio, L.M.; Dalurzo, H.C.; Gómez, M.; Romero-Puertas, M.C.; del Río, L.A. Cadmium-induced changes in the growth and oxidative metabolism of pea plants. *J. Exp. Bot.* **2001**, *52*, 2115–2126. [CrossRef] [PubMed]

37. Semane, B.; Dupae, J.; Cuypers, A.; Noben, J.P.; Tuomainen, M.; Tervahauta, A.; Kärenlampi, S.; van Belleghem, F.; Smeets, K.; Vangronsveld, J. Leaf proteome responses of *Arabidopsis thaliana* exposed to mild cadmium stress. *J. Plant Physiol.* **2010**, *167*, 247–254. [CrossRef] [PubMed]

38. Simões, I.; Faro, C. Structure and function of plant aspartic proteinases. *Eur. J. Biochem.* **2004**, *271*, 2067–2075. [CrossRef] [PubMed]

39. Bhalerao, R.; Keskitalo, J.; Sterky, F.; Erlandsson, R.; Björkbacka, H.; Birve, S.J.; Karlsson, J.; Gardeström, P.; Gustafsson, P.; Lundeberg, J.; et al. Gene expression in autumn leaves. *Plant Physiol.* **2003**, *131*, 430–442. [CrossRef] [PubMed]

40. Kato, Y.; Murakami, S.; Yamamoto, Y.; Chatani, H.; Kondo, Y.; Nakano, T.; Yokota, A.; Sato, F. The DNA-binding protease, CND41, and the degradation of ribulose-1,5-bisphosphate carboxylase/oxygenase in senescent leaves of tobacco. *Planta* **2004**, *220*, 97–104. [CrossRef] [PubMed]

41. Parrott, D.L.; McInnerney, K.; Feller, U.; Fischer, A.M. Steam-girdling of barley (*Hordeum vulgare*) leaves leads to carbohydrate accumulation and accelerated leaf senescence, facilitating transcriptomic analysis of senescence-associated genes. *New Phytol.* **2007**, *176*, 56–69. [CrossRef] [PubMed]

42. Radlowski, M.; Kalinowski, A.; Adamczyk, J.; Krolikowski, Z.; Bartkowiak, S. Proteolytic activity in the maize pollen wall. *Physiol. Plant.* **1996**, *98*, 172–178. [CrossRef]

43. Xia, Y.; Suzuki, H.; Borevitz, J.; Blount, J.; Guo, Z.; Patel, K.; Dixon, R.A.; Lamb, C. An extracellular aspartic protease functions in *Arabidopsis* disease resistance signaling. *EMBO J.* **2004**, *23*, 980–988. [CrossRef] [PubMed]

44. Breitenbach, H.H.; Wenig, M.; Wittek, F.; Jorda, L.; Maldonado-Alconada, A.M.; Sarioglu, H.; Colby, T.; Knappe, C.; Bichlmeier, M.; Pabst, E.; et al. Contrasting roles of the apoplastic aspartyl protease apoplastic, enhanced disease susceptibility1-dependent1 and legume lectin-like protein1 in *Arabidopsis* systemic acquired resistance. *Plant Physiol.* **2014**, *165*, 791–809. [CrossRef] [PubMed]

45. Liu, Z.; Ding, Y.; Wang, F.; Ye, Y.; Zhu, C. Role of salicylic acid in resistance to cadmium stress in plants. *Plant Cell Rep.* **2016**, *35*, 719–731. [CrossRef] [PubMed]

46. Micheli, F. Pectin methylesterases: Cell wall enzymes with important roles in plant physiology. *Trends Plant Sci.* **2001**, *6*, 414–419. [CrossRef]

47. Ramos, I.; Esteban, E.; Lucena, J.J.; Gárate, A. Cadmium uptake and subcellular distribution in plants of *Lactuca* sp. Cd-Mn interaction. *Plant Sci.* **2002**, *162*, 761–767. [CrossRef]

48. Douchiche, O.; Driouich, A.; Morvan, C. Spatial regulation of cell-wall structure in response to heavy metal stress: Cadmium-induced alteration of the methyl-esterification pattern of homogalacturonans. *Ann. Bot.* **2010**, *105*, 481–491. [CrossRef] [PubMed]

49. Wolf, S.; Mouille, G.; Pelloux, J. Homogalacturonan methyl-esterification and plant development. *Mol. Plant* **2009**, *2*, 851–860. [CrossRef] [PubMed]

50. Liu, H.H.; Ma, Y.; Chen, N.; Guo, S.; Liu, H.H.; Guo, X.; Chong, K.; Xu, Y. Overexpression of stress-inducible *OsBURP16*, the β subunit of polygalacturonase 1, decreases pectin content and cell adhesion and increases abiotic stress sensitivity in rice. *Plant Cell Environ.* **2014**, *37*, 1144–1158. [CrossRef] [PubMed]

51. Watson, C.F.; Zheng, L.; DellaPenna, D. Reduction of tomato polygalacturonase β subunit expression affects pectin solubilization and degradation during fruit ripening. *Plant Cell* **1994**, *6*, 1623–1634. [CrossRef] [PubMed]

52. Tucker, G.A.; Robertson, N.G.; Grierson, D. The conversion of tomato-fruit polygalacturonase isoenzyme 2 into isoenzyme 1 in vitro. *Eur. J. Biochem.* **1981**, *115*, 87–90. [CrossRef] [PubMed]

53. Favaron, F.; D'Ovidio, R.; Alghisi, P. Purification and molecular charecterisation of a soybean polygalacturonase-inhibiting protein. *Planta* **1994**, *195*, 80–87. [CrossRef] [PubMed]

54. De Lorenzo, G.; Ferrari, S. Polygalacturonase-inhibiting proteins in defense against phytopathogenic fungi. *Curr. Opin. Plant Biol.* **2002**, *5*, 295–299. [CrossRef]

55. Spadoni, S.; Zabotina, O.; di Matteo, A.; Mikkelsen, J.D.; Cervone, F.; de Lorenzo, G.; Mattei, B.; Bellincampi, D. Polygalacturonase-inhibiting protein interacts with pectin through a binding site formed by four clustered residues of arginine and lysine. *Plant Physiol.* **2006**, *141*, 557–564. [CrossRef] [PubMed]

56. Kieffer, P.; Schröder, P.; Dommes, J.; Hoffmann, L.; Renaut, J.; Hausman, J.F. Proteomic and enzymatic response of poplar to cadmium stress. *J. Proteom.* **2009**, *72*, 379–396. [CrossRef]

57. Loix, C.; Huybrechts, M.; Vangronsveld, J.; Gielen, M.; Keunen, E.; Cuypers, A. Reciprocal interactions between cadmium-induced cell wall responses and oxidative stress in plants. *Front. Plant Sci.* **2017**, *8*, 1867. [CrossRef] [PubMed]

58. Passardi, F.; Penel, C.; Dunand, C. Performing the paradoxical: How plant peroxidases modify the cell wall. *Trends Plant Sci.* **2004**, *9*, 534–540. [CrossRef] [PubMed]

59. McDougall, G.J. Changes in cell wall-associated peroxidases during the lignification of flax fibres. *Phytochemistry* **1992**, *31*, 3385–3389. [CrossRef]

60. Chaoui, A.; Jarrar, B.; El Ferjani, E. Effects of cadmium and copper on peroxidase, NADH oxidase and IAA oxidase activities in cell wall, soluble and microsomal membrane fractions of pea roots. *J. Plant Physiol.* **2004**, *161*, 1225–1234. [CrossRef] [PubMed]

61. Radotic, K.; Ducic, T.; Mutavdzic, D. Changes in peroxidase activity and isoenzymes in spruce needles after exposure to different concentrations of cadmium. *Environ. Exp. Bot.* **2000**, *44*, 105–113. [CrossRef]

62. Luhua, S.; Ciftci-Yilmaz, S.; Harper, J.; Cushman, J.; Mittler, R. Enhanced tolerance to oxidative stress in transgenic *Arabidopsis* plants expressing proteins of unknown function. *Plant Physiol.* **2008**, *148*, 280–292. [CrossRef] [PubMed]

63. Li, B.; Kamiya, T.; Kalmbach, L.; Yamagami, M.; Yamaguchi, K.; Shigenobu, S.; Sawa, S.; Danku, J.M.C.; Salt, D.E.; Geldner, N.; et al. Role of LOTR1 in nutrient transport through organization of spatial distribution of root endodermal barriers. *Curr. Biol.* **2017**, *27*, 758–765. [CrossRef] [PubMed]

64. Chen, Z.; Yan, W.; Sun, L.; Tian, J.; Liao, H. Proteomic analysis reveals growth inhibition of soybean roots by manganese toxicity is associated with alteration of cell wall structure and lignification. *J. Proteomics* **2016**, *143*, 151–160. [CrossRef] [PubMed]

Int. J. Mol. Sci. **2018**, *19*, 2498

65. Békésiová, B.; Hraška, Š.; Libantová, J.; Moravčíková, J.; Matušíková, I. Heavy-metal stress induced accumulation of chitinase isoforms in plants. *Mol. Biol. Rep.* **2008**, *35*, 579–588. [CrossRef] [PubMed]

66. Bernier, F.; Berna, A. Germins and germin-like proteins: Plant do-all proteins. But what do they do exactly? *Plant Physiol. Biochem.* **2001**, *39*, 545–554. [CrossRef]

67. The Alfalfa Gene Index and Expression Atlas Dadabase. Available online: http://plantgrn.noble.org/AGED/SearchGene.jsp (accessed on 1 January 2015).

68. Emanuelsson, O.; Nielsen, H. Predicting subcellular localization of proteins based on their N-terminal amino acid sequence. *J. Mol. Biol.* **2000**, *300*, 1005–1016. [CrossRef] [PubMed]

69. Duruflé, H.; Clemente, H.; Balliau, T.; Zivy, M.; Dunand, C.; Jamet, E. Cell wall proteome analysis of *Arabidopsis thaliana* mature stems. *Proteomics* **2017**, *17*, 1600449. [CrossRef] [PubMed]

70. Armenteros, J.; Sønderby, C.; Sønderby, S.; Nielsen, H.; Winther, O. Deep Loc: prediction of protein subcellular localization using deep learning. *Bioinformatics* **2017**, *33*, 3387–3395. [CrossRef] [PubMed]

International Journal of
Molecular Sciences

MDPI

Article

Proteomic Analysis of the Effect of Inorganic and Organic Chemicals on Silver Nanoparticles in Wheat

Hafiz Muhammad Jhanzab [1,2], Abdul Razzaq [2], Yamin Bibi [3], Farhat Yasmeen [4], Hisateru Yamaguchi [5], Keisuke Hitachi [5], Kunihiro Tsuchida [5] and Setsuko Komatsu [1,*]

[1]   Faculty of Life and Environmental and Information Sciences, Fukui University of Technology, Fukui 910-8505, Japan; jhanzabmuhammad200@yahoo.com
[2]   Department of Agronomy, PMAS-Arid Agriculture University, Rawalpindi 46300, Pakistan; arazzaq57@yahoo.com
[3]   Department of Botany, PMAS-Arid Agriculture University, Rawalpindi 46300, Pakistan; dryaminbibi@uaar.edu.pk
[4]   Department of Botany, Women University, Swabi 23340, Pakistan; fyasmeen@wus.edu.pk
[5]   Institute for Comprehensive Medical Science, Fujita Health University, Toyoake 470-1192, Japan; hyama@fujita-hu.ac.jp (H.Y.); hkeisuke@fujita-hu.ac.jp (K.H.); tsuchida@fujita-hu.ac.jp (K.T.)
*   Correspondence: skomatsu@fukui-ut.ac.jp

Received: 16 December 2018; Accepted: 7 February 2019; Published: 14 February 2019

Abstract: Production and utilization of nanoparticles (NPs) are increasing due to their positive and stimulating effects on biological systems. Silver (Ag) NPs improve seed germination, photosynthetic efficiency, plant growth, and antimicrobial activities. In this study, the effects of chemo-blended Ag NPs on wheat were investigated using the gel-free/label-free proteomic technique. Morphological analysis revealed that chemo-blended Ag NPs resulted in the increase of shoot length, shoot fresh weight, root length, and root fresh weight. Proteomic analysis indicated that proteins related to photosynthesis and protein synthesis were increased, while glycolysis, signaling, and cell wall related proteins were decreased. Proteins related to redox and mitochondrial electron transport chain were also decreased. Glycolysis associated proteins such as glyceraldehyde-3-phosphate dehydrogenase increased as well as decreased, while phosphoenol pyruvate carboxylase was decreased. Antioxidant enzyme activities such as superoxide dismutase, catalase, and peroxidase were promoted in response to the chemo-blended Ag NPs. These results suggested that chemo-blended Ag NPs promoted plant growth and development through regulation of energy metabolism by suppression of glycolysis. Number of grains/spike, 100-grains weight, and yield of wheat were stimulated with chemo-blended Ag NPs. Morphological study of next generational wheat plants depicted normal growth, and no toxic effects were observed. Therefore, morphological, proteomic, yield, and next generation results revealed that chemo-blended Ag NPs may promote plant growth and development through alteration in plant metabolism.

Keywords: proteomics; wheat; silver nanoparticles

---

1. Introduction

Advancement in nanotechnology has led to the production of nanoparticles (NPs), which are extensively used in diversifying a range of applications and products [1]. NPs are atomic or molecular aggregates characterized by their small size of less than 100 nm [2] and have larger surface areas that radically modify their physicochemical properties in comparison to the bulk material [3]. Exposure of NPs to plants resulted in cellular production of reactive oxygen species (ROS) leading to both positive and negative effects [4]. Activity of NPs depends upon size, composition, surface area, and nature of metal materials [5]. Over production and utilization of NPs have raised serious concern about their impacts on the ecosystem [6]. The size and concentration of NPs are responsible for their interaction with other materials and have diverse effects on plants [7]. Plants are the primary and essential components of the ecosystem with the capability to accumulate NPs. Therefore, interaction of NPs with plants and environment needs investigation.

Majority of the studies on NPs have discovered significant and astounding effects on biological systems. Silver (Ag) NPs with concentration of 25 ppm and 50 ppm increased plant growth and biochemical parameters in mustard [8]. Ag NPs with 100 nm reduced biomass and transpiration of *Cucurbita pepo* [9]; and increased carbohydrates and protein contents of *Bacopa monnieri* [10]. However, higher concentration such as 1000 ppm of Ag NPs resulted in an increase of superoxide dismutase (SOD) in *Solanum lycopersicum* [11]. 50 ppm of Ag NPs increased root nodulation in cowpea, while 75 ppm of Ag NPs improved shoot parameters in brassica. In maize, 40 ppm of Ag NPs increased root and shoot growth, while 60 ppm of Ag NPs promoted germination [12]. Ag NPs' exposure to explants of sugarcane stimulated growth at 50 ppm and inhibited at 200 ppm [13]. Ag NPs modified the gene expressions that were involved in cellular events, including cell proliferation, metabolism, and hormone signaling [14]. It is necessary to investigate the response of Ag NPs on a molecular basis to understand the morpho-physiological modifications in plants.

In plants, nicotinic acid as an important organic chemical, improving growth and productivity [15]. Nicotinic acid increased plant growth, protein synthesis [16] and enzyme activities, such as ascorbate per oxidase, glutathione, and fumarase. Damages of oxidative stresses can be protected by the application of nicotinic acid through DNA methylation [17]. Potassium nitrate (KNO$_3$), as an inorganic chemical, plays an important role in most plants' biochemical and physiological processes, such as photosynthesis, enzyme activation, energy transfer, and stress resistance [18]. Increment in plant growth and rate of photosynthesis in response to KNO$_3$ ultimately improved the yield of wheat, maize, and potato [19]. Potassium minimized the cadmium toxicity, improved yield, mineral elements, and antioxidant defense system of *Cicer arietinum* [20]. These findings conclude that exogenous application of organic and inorganic chemicals promoted the growth and yield parameters of plants; however, their molecular and metabolic mechanisms are still not clear.

Findings related to proteomic studies revealed that Ag NPs maintained cellular homeostasis by changing proteins involved in redox regulation and sulfur metabolism [21]. The application of Ag NPs decreased alcohol dehydrogenase and pyruvate decarboxylase, while increased amino acid related proteins and wax formation in soybean under flooding stress [22]. Proteins responsible for oxidative stress such as tolerance, calcium regulation, signaling, cell division, and apoptosis were identified in response to Ag NPs in *Eruca sativa* [23]. Several proteins related to primary metabolism and cell defense in roots and shoots of wheat were altered with treatment of Ag NPs [21]. Proteomic studies on the effects of Ag NPs have been reported; however, the effect of Ag NPs blended with organic and inorganic chemicals have not been reported earlier. To study the effects of blended NPs on wheat, morphological, proteomic, and enzymatic analyses were performed.

2. Results

2.1. Growth Response of Wheat to Ag NPs Mixed with Organic and Inorganic Chemicals

The experiment was conducted to evaluate the response of wheat to Ag NPs and Ag NPs mixed with organic and inorganic chemicals (Figure 1). The NPs were prepared by reduction of AgNO$_3$ with Na$_3$C$_6$H$_5$O$_7$. 2H$_2$O. SEM images revealed that Ag NPs are spherical with a size of 15–20 nm (Figure 1). Chemo-blended Ag NPs were prepared by mixing organic and inorganic chemicals. Wheat seeds were pre-soaked and followed by cold treatment for one day. Six-day-old wheat seedlings were treated with and without Ag NPs and chemo-blended Ag NPs. Data regarding morphological parameters were analyzed on the 9th, 11th, and 13th day (Figure 1). Proteins from treated as well as control plants were extracted and analyzed through gel free/label free proteomic technique. Antioxidant enzyme analysis was carried out to confirm the proteomic results. At the tillering stage, chemo-blended Ag NPs were applied and data on yield parameters were analyzed. Seeds obtained from this lifecycle were used to explore the cross-generational effects of nanoparticles. For this purpose, pre-sterilized seeds were sown and their growth attributes were analyzed on the 5th, 7th and 9th day (Figure 1).

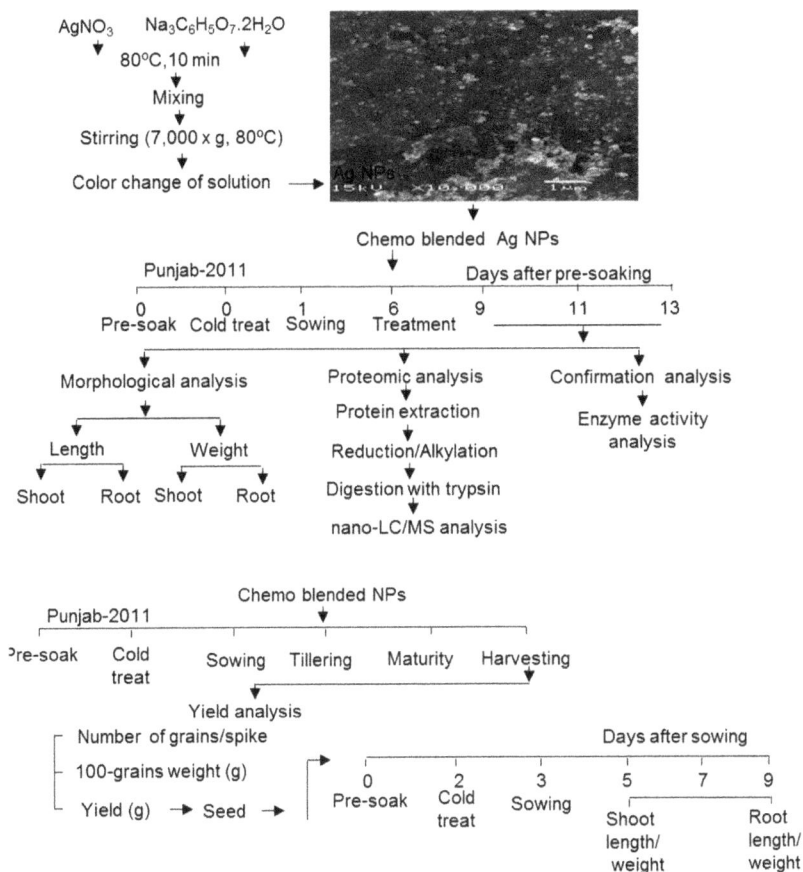

Figure 1. Experimental design for morphological, proteomics, confirmation analyses with subsequent impact on yield and next generation of wheat. Six-day-old wheat seedlings were treated with and without 5 ppm Ag NPs; 5 ppm Ag NPs/10 ppm nicotinic acid; 5 ppm Ag NPs/ 0.75% KNO_3 and 5 ppm Ag NPs/10 ppm nicotinic acid/0.75% KNO_3. Data regarding morphological parameters were analyzed on the 9th, 11th, and 13th day. For proteomic and confirmation analyses, proteins were extracted from shoot of 11-day-old treated plants and analyzed through gel free/label free proteomic technique. Thirty-two days old wheat plants were treated with chemo-blended Ag NPs. Number of grains/spike, 100-grains weight, and yield of wheat were analyzed. Seeds obtained from this lifecycle were used for next generation effect.

Growth parameters such as shoot length, shoot fresh weight, root length, and root fresh weight were measured on the 9th, 11th, and 13th day after pre-soaking. Shoot length was significantly increased with 5 ppm Ag NPs/10 ppm nicotinic acid/0.75% KNO_3 on the 9th, 11th, and 13th day compared to other treatments (Figure 2). Shoot fresh weight was maximum with 5 ppm Ag NPs/10 ppm nicotinic acid/0.75% KNO_3 treated wheat plants on 9th, 11th, and 13th day measurements (Figure 2). Root length and root fresh weight were also greatly affected with application of Ag NPs mixed with organic and inorganic chemicals. Root length and root fresh weight were increased with 5 ppm Ag NPs/10 ppm nicotinic acid/0.75% KNO$_3$ during the 9th, 11th, and 13th day (Figure 2). Comparative analysis indicated that length and weight of shoot and root were higher when chemo-blended Ag NPs was applied in comparison with control (Figure 2). Morphological parameters such as length and weight of shoot and root were affected by treatment of chemo-blended Ag NPs. Therefore, 5 ppm Ag NPs/10 ppm nicotinic acid/0.75% KNO$_3$ was used for proteomic analysis.

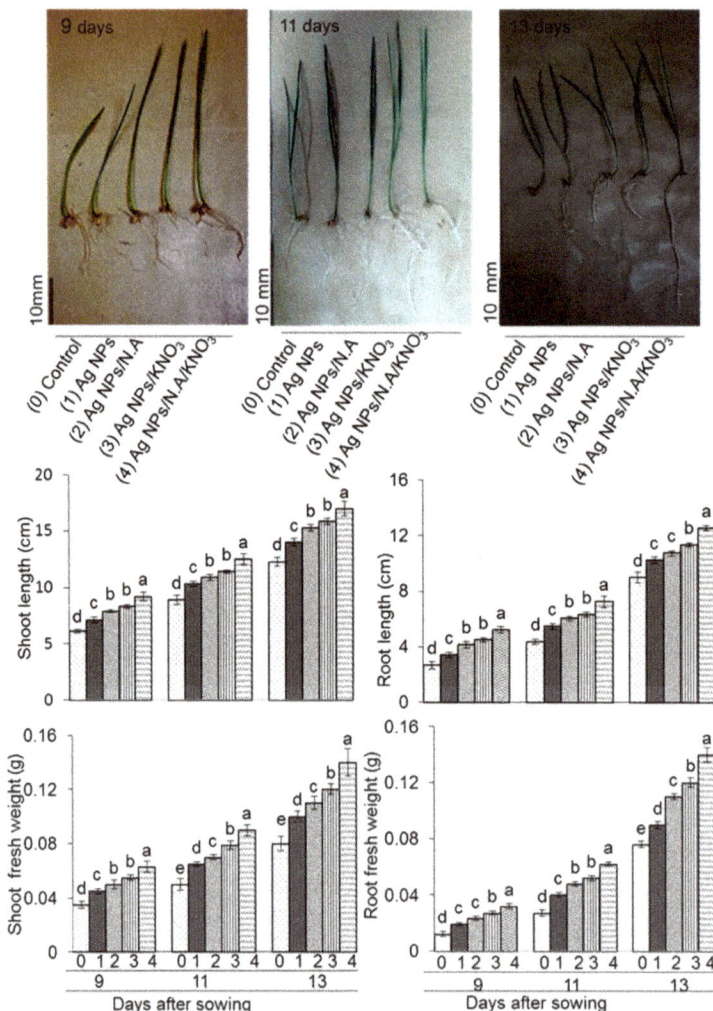

Figure 2. Morphological effects of chemo-blended Ag NPs on wheat. Six-day-old wheat seedlings were treated without and with 5 ppm Ag NPs, 5 ppm Ag NPs/10 ppm nicotinic acid, 5 ppm Ag NPs/0.75% KNO$_3$, and 5 ppm Ag NPs/10 ppm nicotinic acid/0.75% KNO$_3$. Photographs of wheat seedlings show the 9th, 11th, and 13th day with and without blended Ag NPs. Black bar in each photograph indicates size in mm. Shoot length, shoot fresh weight, root length, and root fresh weight were analyzed on the 9th, 11th, and 13th day. The data are presented as mean ± S.D. from three independent biological replicates. Mean values at each point with different letters (a,b,c,d) are significantly different according to Tukey's multiple range test (p b 0.05).

2.2. Proteomic Analysis

Proteomic analysis was performed for identification of proteins that were affected in response to the chemo-blended Ag NPs. Protein extraction was carried out on 11-day-old treated wheat plants and analyzed through gel free/label free proteomic technique (Figure 1). The abundance of 49 proteins was significantly changed when treated with Ag NPs mixed with organic and inorganic chemicals ($p < 0.05$, Student's *t* test). Out of these 49 proteins, 29 proteins increased (Table 1), while 20 proteins decreased on Ag NPs exposure (Table 2). According to GO term analysis, 49 proteins related to biological, cellular, and molecular processes were changed (Supplementary Figure S1). Analysis of GO term revealed that proteins related to the metabolism in biological processes were decreased/ increased on chemo-blended Ag NPs exposure, while in case of cellular processes, membrane related proteins were decreased/ increased on chemo-blended Ag NPs treatment (Supplementary Figure S1). For molecular functions, proteins related to catalytic activity were changed with abundance (Supplementary Figure S1).

Table 1. List of increased proteins identified on chemo-blended Ag NPs' exposure in wheat.

No	Accession	Description	Difference	Functional Category	Biological Process	Cellular Component	Molecular Function
1	W5AYF4	Putative SNAP receptor protein	3.56	Transport	Cell organization and biogenesis	membrane	protein binding
2	A0A1D5WZM5	ER membrane protein complex	3.2	Protein	Transport	membrane	metal ion binding
3	A0A1D6B0Y2	At4g14100-like	2.91	Transport	Not assigned	membrane	catalytic activity
4	W5DZQ3	Ribosomal protein S1	2.85	Protein	Response to stimulus	chloroplast	RNA binding
5	Q5G1T9	Gamma-glutamylcysteinesynthetase	2.57	Protein	Metabolic process	Chloroplast, cytosol	catalytic activity
6	W5EA17	D-ribose 5-phosphate	2.41	secondary metabolism	Metabolic process	cytosol	catalytic activity
7	A0A1D5UQX6	Unknown	2.27	not assigned	Not assigned	not assigned	not assigned
8	W5DL10	Glutamate–tRNA ligase	1.96	Protein	Metabolic process	cytoplasm	catalytic activity
9	A0A1D5YQ15	Ferritin	1.86	photosynthesis	Cellular homeostasis	cytosol	catalytic activity
10	W5APX0	GrpE protein homolog	1.81	Transport	Metabolic process	mitochondrion	enzyme regulator activity
11	A0A1D6AQL7	PPIasecyclophilin-type	1.77	Stress	Metabolic process	cytosol	catalytic activity; protein binding
12	A0A1D5Y2E6	Reverse transcriptase	1.77	Protein	Metabolic process	Cytoplasm	catalytic activity
13	A0A1D5SUT9	Peptidase A1	1.76	Protein	Metabolic process	Membrane	catalytic activity
14	O21432	Ribosomal protein S2	1.65	Protein	Metabolic process	mitochondrion	structural molecule activity
15	A0A1D6AKZ2	ATP-dependent Clp protease proteolytic subunit	1.16	photosynthesis	Metabolic process	chloroplast	catalytic activity
16	A0A1D5V5A5	Bifunctional inhibitor/plant lipid transfer	1.12	secondary metabolism	Transport	Membrane	catalytic activity; metal ion binding
17	A0A1D5YXC6	Plant lipid transfer protein	1.01	secondary metabolism	Metabolic process	Membrane	catalytic activity
18	W5DX10	Allene oxide synthase-lipoxygenase	0.89	Stress	Metabolic process	chloroplast	catalytic activity; metal ion binding
19	A0A1D5YZ95	Synaptotagmin-like mitochondrial lipid-binding proteins	0.88	Transport	Metabolic process	Membrane	metal ion binding
20	A0A1D6DJK9	Tyrosine–tRNA ligase	0.86	Protein	Metabolic process	Cytoplasm	catalytic activity
21	A0A1D6RMY5	Short-chain dehydrogenase/reductase2	0.74	Stress	Metabolic process	cytosol	catalytic activity
22	W5QKZ0	Chalcone-flavonone isomerase	0.7	secondary Metabolism	Metabolic process	not assigned	catalytic activity
23	A0A1D5SIK2	NAD(P)H-quinone oxidoreductase subunit I, glyceraldehyde dehydrogenase	0.7	ETC	metabolic process	membrane	catalytic activity; metal ion binding
24	A0A1D5S4W8	Acetyl-CoA synthetase	0.69	Glycolysis	metabolic process	membrane	catalytic activity
25	W5AV30	Fasciclin-like protein FLA12	0.66	Glycolysis	not assigned	membrane	protein binding
26	Q06I94	Glucose-6-phosphate 1-epimerase	0.59	Cell wall	response to stimulus	membrane	not assigned
27	A0A1D5UQL1	2-oxoglutarate (2OG) and Fe(II)-dependent oxygenase	0.57	Glycolysis	metabolic process	chloroplast	catalytic activity
28	A0A1D6A8Y7\	N-acetyltransferase	0.53	Primary metabolism	metabolic process	cytoplasm	catalytic activity
29	A0A1D6RZJ3		0.4	Signaling	metabolic process	not assigned	catalytic activity

Ratio, relative abundance of protein; *p*-value b 0.05. Treated over control, wheat plant treated with Ag NPs compared with control.

Table 2. List of decreased proteins identified on chemo-blended Ag NPs' exposure in wheat.

No	Accession	Description	Difference	Functional Category	Biological Process	Cellular Component	Molecular Function
1	A0A1D6C3J5	Copper transport protein	−0.65	Redox	Transport	membrane	metal ion binding
2	A0A1D6S991	Isocitrate dehydrogenase	−0.66	TCA	Metabolic process	Mitochondrion	catalytic activity
3	A0A1D5UMA5	3-Oxoacyl- synthase III	−0.71	lipid metabolism	Metabolic process	Chloroplast	catalytic activity
4	W5BFA5	glyceraldehyde-3-phosphate dehydrogenase	−0.72	Glycolysis	Metabolic process	Cytosol; Golgi	catalytic activity
5	A0A1D5V012	NADH dehydrogenase ubiquinone Fe-S protein 4	−0.74	Redox	Metabolic process	mitochondrion	catalytic activity
6	A0A1D5U7K0	F-box associated interaction domain	−0.89	Signaling	Regulation of biological process	cytoplasm	protein binding
7	A0A1D5VAF5	glycine-tyrosine-phenylalanine	−0.94	Stress	Metabolic process	mitochondrion	protein binding
8	A0A1D5XH81	carboxy peptidase	−0.97	Glycolysis	Metabolic process	chloroplast	catalytic activity
9	A0A1D6AAU8	acetyltransferase-superfamily	−1.11	Signaling	Metabolic process	not assigned	catalytic activity
10	A0A1D6AVB7	Ubiquitin-associated domain	−1.17	Protein	Response to stimulus	cytosol	protein binding
11	A0A1D6DJI9	Glutaredoxin	−1.18	ETC	Cellular homeostasis	mitochondrion	catalytic activity
12	A0A1D6AQL9	Rho termination factor, N-terminal domain superfamily	−1.26	primary metabolism	Metabolic process	mitochondrion	catalytic activity
13	A0A1D5ZVW9	La-type RNA-binding	−1.63	RNA	Metabolic process	cytosol	RNA binding
14	A0A96ULK3	Small GTPase superfamily	−1.76	Signaling	Regulation of biological process	cytosol	catalytic activity
15	A0A1D6RV75	peptidyl-prolyl cis-trans isomerase	−1.92	Protein	Metabolic process	mitochondrion;	catalytic activity
16	W5E0G5	Proteasome component	−2.34	Protein	Metabolic process	cytosol	protein binding
17	A0A1D6AAV9	Carbon-nitrogen hydrolase	−2.52	Signaling	Metabolic process	not assigned	catalytic activity
18	A0A1D5TDD8	Plant invertase/, pectin methylesterase inhibitor	−2.9	cell wall	Regulation of biological process	membrane	enzyme regulator activity
19	W5E5N6	LysM Domain (Peptidoglycan binding)	−2.9	cell wall	Response to stimulus	membrane	protein binding
20	A0A1D6DBZ9	Polygalacturonase	−3.58	cell wall	Cell organization and biogenesis	extracellular	catalytic activity

Ratio, relative abundance of protein; p-value b 0.05. Treated over control, wheat plant treated with Ag NPs compared with control.

The changed proteins were functionally categorized through MapMan bin code analysis (Figure 3). Proteins related to photosynthesis and protein synthesis were increased, while signaling, cell wall, and stress related proteins were decreased. The majority of proteins related to glycolysis were decreased in response to chemo-blended Ag NPs (Figure 3).

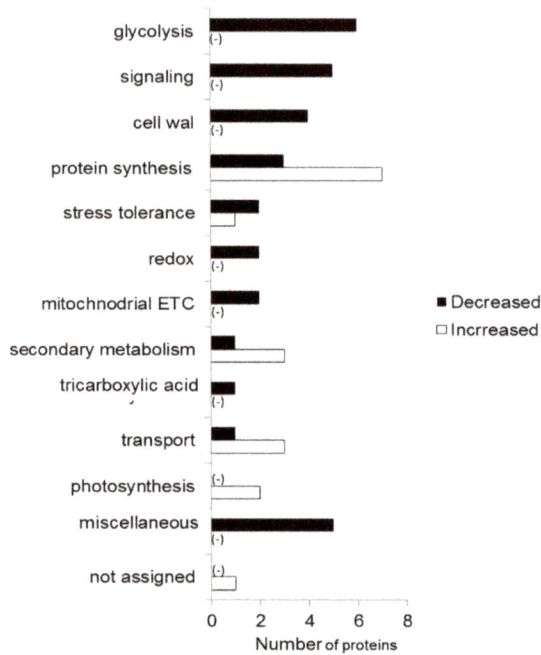

Figure 3. To determine the effect of Ag NPs mixed with organic and inorganic chemicals, MapMan software and KEGG database were used (Supplementary Figure S2). The proteins significantly changed were visualized through MapMan software (Supplementary Figure S2). Proteins related to cell wall, secondary metabolism, amino acid metabolism, photosynthesis, glycolysis, starch/sucrose synthesis, and lipid metabolism were significantly changed (Supplementary Figure S2).

Proteins related to photosynthesis, protein synthesis, secondary metabolism, and transport were increased; while glycolysis, cell wall, and signaling-related proteins were decreased with treatment of chemo-blended Ag NPs. Further analysis was carried out to check the effect of chemo-blended Ag NPs on glycolysis. The identified proteins associated with glycolysis were mapped through a KEGG database (Figure 4). Among the glycolysis-related proteins, glyceraldehyde-3-phosphate dehydrogenase and glucose-6-phosphate-1-epimarase were increased/decreased after the chemo-blended Ag NPs treatment (Figure 4).

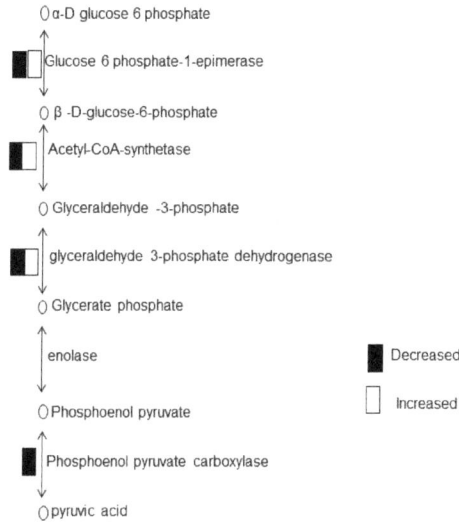

Figure 4. KEGG analysis of identified proteins on chemo-blended Ag NPs' exposure. Pathway of glycolysis mapped by KEGG based on identified proteins treated with chemo-blended Ag NPs. Wheat plants were treated with Ag NPs mixed with organic and inorganic chemicals. Each square of black and white indicates a decreased or increased response, respectively, in ratio compared with untreated plants.

2.3. Effect of Chemo-Blended Ag NPs on Antioxidant Enzyme Activity Analysis of Wheat

To confirm the proteomic results, antioxidant enzyme activities were analyzed. Antioxidant enzymes such as superoxide dismutase (SOD), catalase (CAT), and peroxidase (POD) were measured from fresh leaves of 11-day-old plants with and without chemo-blended Ag NPs. The results indicated that SOD activity was significantly increased with chemo-blended Ag NPs as compared to the control (Figure 5). CAT activity was also improved significantly on exposure to chemo-blended Ag NPs, as compared to the control. CAT as an important antioxidant enzyme is involved in conversion of H_2O_2 to water and oxygen, as well as scavenging of free radicals. (Figure 5). Similarly, POD activity was increased two times with treatment of Ag NPs mixed with organic and inorganic chemicals (Figure 5).

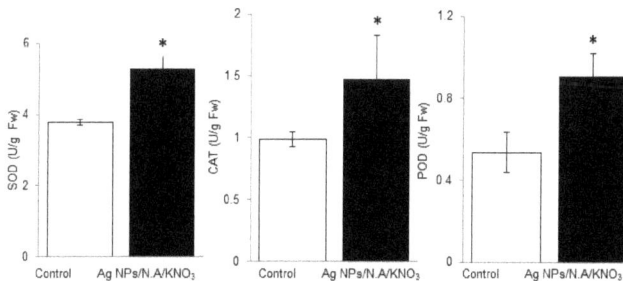

Figure 5. Effect of chemo-blended Ag NPs on the antioxidant activity of wheat. Wheat plants were treated with and without chemo-blended Ag NPs for six days. To assay the enzymatic activities, SOD, CAT and POD were extracted from leaves of wheat. Absorbance was recorded with a spectrophotometer. The data are presented ± S.D. from three independent biological replicates. The student *t*-test was used for statistical analysis. Significance between control and treated plants was indicated by asterisks ($P < 0.05$).

2.4. Effect of Chemo-Blended Ag NPs on Yield and Lifecycle of Wheat

Exploring the possible effects of Ag NPs blended with organic and inorganic chemicals on the wheat yield, plants were treated without and with 5 ppm Ag NPs/10 ppm N.A/0.75% KNO_3 at the tillering stage and allowed to grow till maturity. The grains were harvested at maturity and number of grains/spike, 100-grains weight, and yield was analyzed. Application of 5 ppm Ag NPs/10 ppm N.A/0.75% KNO_3 promoted number of grains/spike, 100-grains weight and yield when compared to the control (Figure 6). Ag NPs blended with organic and inorganic chemicals promoted the production of crop and increased net grain yield.

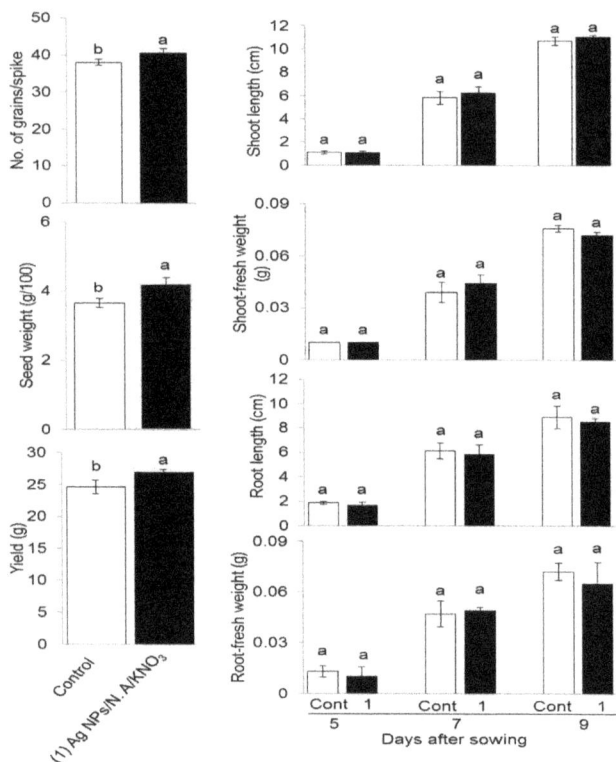

Figure 6. Effect of chemo-blended Ag NPs on yield and subsequent morphological assessment of the next generation of wheat seeds. Thirty-two day old wheat plants were treated without or with 5 ppm Ag NPs/10 ppm nicotinic acid/0.75% KNO_3. Grains were harvested and number of grains/spike, 100-grains weight, and yield were analyzed. Seeds obtained from this life cycle were sown in sterilized sand. Shoot length, shoot-fresh weight, root length, and root-fresh weight were analyzed on the 5th, 7th, and 9th day after sowing. The data are presented as mean ± S.D. from three independent biological replicates. Mean value in each point with different letters are significantly different according to Tukey's multiple range test ($P < 0.05$).

Seeds obtained from this life cycle were used to analyze the transfer of chemo-blended Ag NPs impact on the next generation through a morphological experiment. After sterilization, pre-soaking and cold treatment, the seeds were sown in sterilized silica sand. Shoot length, shoot fresh weight, root length, and root fresh weight were analyzed on the 5th, 7th, and 9th day after sowing. Shoot length was higher with 5 ppm Ag NPs/10 ppm N.A/0.75% KNO_3 as compared to control at the 7th and 9th day (Figure 6). Shoot fresh weight was maximum on the 7th day and reduced on the 9th day with

5 ppm Ag NPs/10 ppm N.A/ 0.75% KNO$_3$. Root length was maximum with control, while root fresh weight was maximum on the 7th day with 5 ppm Ag NPs/ 10 ppm N.A/ 0.75% KNO$_3$ as compared to the control (Figure 6). The results revealed no change in the morphology of the next generation of wheat plants. These results indicated that the application of chemo-blended Ag NPs did not show any inhibitory effect on the next generation. Normal and healthy plant growth was observed during the experiment.

3. Discussion

3.1. Effect of Chemo-Blended Ag NPs on Morphological Attributes of Wheat

To elucidate the effects of Ag NPs blended with organic and inorganic chemicals on wheat growth, wheat was treated with and without Ag NPs and blended Ag NPs. Wheat growth parameters were promoted in response to different treatments as compared to the control. Phyto-stimulatory as well as detrimental effects have been reported in several studies depending upon size and concentration of NPs. Ag NPs increased the super oxide dismutase activity in *Solanum lycopersicum* [11], while various concentrations of Ag NPs reduced plant biomass and transpiration [9]. Priming rice seeds with 5 ppm and 10 ppm of Ag NPs significantly increased germination, seedling vigor, and alpha amylase activity, resulting in higher soluble sugar contents [24]. Growth and photosynthesis were inhibited with 1 mM and 3 mM of Ag NPs and AgNO$_3$, respectively [25]. Application of Ag NPs resulted in the increase in germination and chlorophyll contents of rice, maize and, peanut [26]. Low concentration such as 40 ppm of Ag NPs increased in vitro growth of root and shoot growth of maize, while 60 ppm concentration promoted the germination [12]. The present and earlier studies showed that NPs can stimulate the growth of various crops when applied in diverse concentrations.

Ag NPs with concentration of 1 mg/kg in soil did not affect growth and amino acid contents in wheat [27], while pea seeds treated with Ag NPs significantly promoted root length [28]. Coated Ag NPs with CTAB increased uptake of Ag in the roots, while reducing root growth and oxidative damages [7]. Ag NPs with 1000 μM and 3000 μM decreased growth, photosynthetic pigments, and chlorophyll fluorescence due to increased accumulation of Ag in root and shoot of pea seedlings [29]. Ag NPs reduced shoot length in mung bean [19] and barley [30], while germination of lentil was increased [31]. In the present study, Ag NPs mixed with organic and inorganic chemicals promoted growth of wheat. Individual effects of Ag NPs have been reported; however, Ag NPs mixed with organic and inorganic chemicals have not been reported earlier. These results revealed that Ag NPs can promote plant growth; therefore, fate and translocation of NPs to food chain needed more exploration.

3.2. Chemo-Blended Ag NPs Affect Protein Metabolism of Wheat

Proteomic analysis revealed that proteins related to protein synthesis were increased in response to Ag NPs mixed with organic and inorganic chemicals. Ag NPs increased proteins related to amino acid metabolism compared to flooding stress in soybean [32]. Ribosomal proteins have very crucial role in cell metabolism to regulate plant growth [33]. In stress conditions ribosomal proteins decreased [34] but Al$_2$O$_3$ NPs increased abundance of proteins related to protein synthesis [35]. Iron NPs increased photosynthesis and protein metabolism related proteins [36], while Ag NPs increased carbohydrates and protein contents of *Bacopamonnierei* [10]. Treatment of Ag NPs caused variation of proteins associated to endoplasmic reticulum and vacuole indicating the target organelles of Ag NPs [37]. These results indicated that chemo-blended Ag NPs increased protein synthesis that lead to increased growth and development of plants.

3.3. Chemo-Blended Ag NPs Affect Glycolysis of Wheat

Proteomic analysis showed that glycolysis related proteins were decreased in response to chemo-blended Ag NPs. Glycolysis is an important metabolic pathway responsible for conversion of glucose to pyruvate for production of energy [38]. In this process, energy is released from high energy molecules to regulate normal functions of plant growth and development. Plants change carbohydrate metabolic pathways to support ATP production through glycolysis [39]. In glycolysis, glyceraldehyde dehydrogenase converts glucose to energy and carbon molecules for metabolism of glycogen [40]. Proteins related to glycolysis were decreased in soybean upon exposure to Ag NPs [32], while increased with iron and copper NPs [41]. Taken together, these results suggest that glycolysis related proteins regulate energy metabolism in wheat through glyceraldehyde -3-phosphate dehydrogenase and phosphoenol pyruvate carboxylase.

3.4. Impact of Chemo-Blended Ag NPs on Scavenging Activity of SOD, CAT, and POD

Reactive oxygen species (ROS) are an intricate part of normal cellular physiology. ROS inhibit multiple glycolytic enzymes, including glyceraldehyde-3-phosphate dehydrogenase, pyruvate kinase M2, and phosphofructokinase-1. Consistently, glycolytic inhibition promotes flux into the oxidative arm of the pentose phosphate pathway to generate NADPH [42]. Redox homeostasis in plants is maintained through antioxidant machinery such as SOD, CAT, POD, and some low molecular non-enzymatic compounds like phenolics, flavonoids, terpenoids, tocopherols, and carotenoids [43]. SOD, CAT, and POD are important antioxidant enzymes having indispensable role in ROS detoxification. Generally, antioxidant enzymes such as SOD, CAT, and POD alter in response to change in ROS concentration [44]. SOD converts O_2^- radicals to H_2O_2, and then H_2O_2 reduces to water and oxygen by CAT and POD. Therefore, it is the first line of defense system which prevents the cell from further injuries [45].

Exposure of NPs to plants resulted in cellular production of ROS [4]. Ag NPs increased SOD, CAT, and POD activities in *Spirodela polyriza* [46], while activities of these enzymes remain same with high concentration of Al_2O_3 NPs [47]. Application of Ag NPs increased SOD, CAT, and POD activities in water hyacinth, while CAT and POD reduced production of ROS in *Bacopa monirei* [10]. Exposure of Ag NPs increased SOD, CAT, and POD in water hyacinth plant roots [48], while level of lipid per oxidation was reduced through increased activity of SOD, CAT, and POD in *Phanerochaete chryosporium* [49]. Ag NPs mixed with organic and inorganic chemicals increased SOD, CAT, and POD activities. In our study, glycolysis related proteins were decreased; however, activities of these proteins were increased due to stimulation of enzyme activities.

3.5. Effect of Chemo-Blended Ag NPs on Yield and Growth of Next Generation

To study the possible role of blended Ag NPs on yield and yield components of wheat, wheat plants were treated with Ag NPs mixed with organic and inorganic chemicals. Wheat yield was promoted with blended Ag NPs. In another study, it has been reported that Ag NPs of low doses 25–50 mg kg^{-1} increased chlorophyll contents, nitrate reductase, and *Phaselous vulgaris* pod yield [50]. Ag NPs increased number of grains/spike, 100-grains weight, and yield of wheat, while exposure of Fe and Cu NPs promoted yield and yield attributes of wheat [51]. Cesium NPs increased fruit production in tomato [52], while it reduced yield in cucumber [53]. Copper NPs increased number of grains/spike, 100-grains weight, and yield of wheat [54], while Ag NPs increased yield of cucumber [55]. Nano iron oxide increased pod dry weight and grain yield of soybean [56], while low concentration of Ag NPs did not affect growth and ascorbic acid contents of wheat seeds [27]. Ag NPs with magnetic field increased muskmelon fruit quality, yield, and soluble solid concentration [57]. Zinc oxide NPs coated with natural phytochemicals increased growth, biochemical parameters, and biomass of cotton significantly [58]. From these results, it can be concluded that application of NPs blended with organic and inorganic chemicals have great potential for increasing yield and yield attributes of plants.

In the present experiment, lifecycle study was carried out through application of Ag NPs blended with organic and inorganic chemicals. Seeds obtained from this life cycle were germinated and analyzed for growth responses. In a related study, treatment of radish plants with copper oxide and zinc oxide NPs through their life cycle indicated that root length, shoot length, and biomass in F1 seedlings were reduced [19]. A low concentration of cesium oxide NPs through their lifecycle on seed quality and next generation seedlings was obtained from treated parent plants with smaller biomass, reduced water transpiration and high ROS [52]. Multi-generational exposure of cesium oxide NPs to *Brassica rapa* showed slower plant growth and reduced biomass in the second and third generations. The number of seeds produced per siliqua was reduced in the third generation [47]. Pumpkin plants grown in an aqueous medium containing magnetite iron oxide NPs can absorb, translocate, and accumulate particles in plant tissues [59]. Ajirloo et al. [60] reported that a nanofertilizer of potassium and nitrogen increased growth, yield, and yield components of tomato. In the present study, plant growth was not affected; thus, chemo-blended Ag NPs act like a fertilizer, with no toxic effects on the next generation. Therefore, the current results provide future directions on cross generation studies related to NPs, in a bid to clearly understand the interaction of NPs and the environment.

4. Materials and Methods

4.1. Preparation of Chemo-Blended Nanoparticles

Ag NPs were synthesized by the reduction of silver nitrate ($AgNO_3$) with trisodium citrate dihydrate ($Na_3C_6H_5O_7 \cdot 2H_2O$). A solution of 500 ppm $AgNO_3$ (Sigma-Aldrich, Munich, Germany) and 300 ppm $Na_3C_6H_5O_7$ (Merck, Darmstadt, Germany) were prepared. Prior to mixing of both solutions, $AgNO_3$ solution was heated at 80 °C on a hot plate for 10 min. Trisodium citrate solution was gradually added to the $AgNO_3$ solution and mixed thoroughly. The resultant solution was stirred at $7000 \times g$ for 1 h at 80 °C using a magnetic stirrer until a golden yellow color was attained [61]. Freshly prepared Ag NPs were analyzed through scanning electron microscopy image analysis.

Chemo-blended Ag NPs were prepared by mixing organic and inorganic chemicals. For the preparation of chemo-blended Ag NPs, Ag NPs, nicotinic acid (Sigma Aldrich, Darmstadt, Germany), and KNO_3 (Sigma Aldrich, Darmstadt, Germany) were used. Different concentrations of 5 ppm Ag NPs, 10 ppm nicotinic acid, and 0.75% KNO_3 were mixed in various combinations to prepare the blended NPs.

4.2. Plant Material and Treatment

Seeds of wheat (*Triticum aestivum* L. varPunjab-2011) were used to study the effects of Ag NPs mixed with organic and inorganic chemicals on the morphological and proteomic analysis of wheat. The seeds were sterilized with 2% sodium hypochlorite solution, followed by rinsing twice with water. After cold treatment at 4 °C, the seeds were sown in a plastic case containning pre-sterilized silica sand. Healthy and equal size seedlings were selected and placed on petridishes containing two layers of filter papers, covered by a sponge. Growth conditions were maintained as 16 h light intensity of 200 μmol m^{-2} s^{-1} and 8 h dark with 20% humidity at 25 °C. Six-day-old wheat seedlings were treated with and without 5 ppm Ag NPs, 5 ppm Ag NPs/10 ppm nicotinic acid, 5 ppm Ag NPs/0.75% KNO_3 and 5 ppm Ag NPs/10 ppm nicotinic acid/0.75% KNO_3. Shoot length, shoot-fresh weight, root length, and root-fresh weight were analyzed on the 9th, 11th, and 13th day after sowing. Three independent experiments were performed as biological replicates for all experiments. The plants used for biological replicates were sown on different days (Figure 1).

An experiment was conducted to evaluate the response of blended Ag NPs on yield of wheat. The seeds were sterilized and pre-soaked for two days. After cold treatment at 4 °C for one day, the seeds were sown in clay pots filled with fertile and thoroughly mixed soil. Thirty two-day-old plants were treated without and with blended nanoparticles, 5 ppm Ag NPs/10 ppm nicotinic acid/0.75% KNO_3. Proper irrigation was managed during the critical stages of crop growth. Number of grains/spike, 100-grains weight and yield were analyzed after harvesting complete ripened grains from the spikes (Figure 1).

Seeds obtained from the above lifecycle were used for a morphological experiment. Sterilization was carried out with 2% NaOCl solution for 2 min followed by rinsing and pre-soaking for 2 days.

Int. J. Mol. Sci. **2019**, *20*, 825

After cold treatment at 4 °C for one day, the seeds were sown and allowed to grow in growth chamber at 25 °C, illuminated with white flourescent light of 200 µmol m^{-2} s^{-1} for 16-h light/day. Growth parametres such as shoot length, shoot-fresh weight, root length, and root-fresh weight were analyzed on the 5th, 7th, and 9th day after sowing. Three independent experiments were performed as biological replicates for all experiments. Sowing of seeds were carried out on different days to make biological replicates (Figure 1).

4.3. Protein Extraction

A portion (300 mg) of the sample was cut and ground for 60 times in a filter cartridge. It was ground for 30 times after adding 100 µL of lysis buffer containing 7 M urea, 2 M thiourea, 5% CHAPS, and 2 mM tributylphosphine. Furthermore, 50 µL of lysis buffer was added and ground for 30 times. The suspension was incubated for 2 min at 25 °C and centrifuged at 15,000× *g* for 2 min at 25 °C. Later on, the filter cartridge was removed and the supernatant was collected as total proteins.

4.4. Protein Enrichment, Reduction, Alkylation and Digestion

Extracted proteins (100 µg) were adjusted to a final volume of 100 µL. Methanol (400 µL) was added to each sample and mixed before addition of 100 µL of chloroform and 300 µL of water. After mixing and centrifugation at 20,000× *g* for 10 min to achieve phase separation, the upper phase was discarded and 300 µL of methanol was added to the lower phase, and then centrifuged at 20,000× *g* for 10 min. The pellet was collected as the soluble fraction [62]. The proteins were resuspended in 50 mM NH$_4$HCO$_3$, reduced with 50 mM dithiothreitol for 30 min at 56 °C, and alkylated with 50 mM iodoacetamide for 30 min at 37 °C in the dark. Alkylated proteins were digested with trypsin (Wako, Osaka, Japan) at a 1:100 enzyme/protein ratio for 16 h at 37 °C. Peptides were desalted with a MonoSpin C18 Column (GL Sciences, Tokyo, Japan). Peptides were acidified with 0.1% formic acid and analyzed by nano-liquid chromatography (LC) mass spectrometry (MS)/MS.

4.5. Measurement of Protein and Peptide Concentrations

The method proposed by Bradford [63] was used to determine the protein concentration with bovine serum albumin used as the standard. A Direct Detect Spectrometer (Millipore, Billerica, MA, USA) equipped with the Direct Detect software (version 3.0.25.0) was used to determine peptide concentration.

4.6. Protein Identification Using Nano LC-MS/MS

The samples were then analyzed using a LC system (EASY-nLC 1000; Thermo Fisher Scientific) coupled to a MS (Orbitrap Fusion ETD MS; Thermo Fisher Scientific). The LC conditions as well as MS acquisition conditions are described in the previous study [64]. Briefly, the peptides were loaded onto the LC system equipped with a trap column (Acclaim PepMap 100 C18 LC column, 3 µm, 75 µm ID × 20 mm; Thermo Fisher Scientific) equilibrated with 0.1% formic acid and eluted with a linear acetonitrile gradient (0–35%) in 0.1% formic acid at a flow rate of 300 nL/min. The eluted peptides were loaded and separated on the column (EASY-Spray C18 LC column, 3 µm, 75 µm ID × 150 mm; Thermo Fisher Scientific) with a spray voltage of 2 kV (Ion Transfer Tube temperature: 275 °C). The peptide ions were detected using the MS with the installed Xcalibur software (version 4.0; Thermo Fisher Scientific).

4.7. MS Data Analysis

The MS/MS searches were carried out using MASCOT (Version 2.6.1, Matrix Science, London, U.K.) and SEQUEST HT search algorithms against the UniProtKBTriticumaestivum database (2017-07-05) using Proteome Discoverer (PD) 2.2 (Version 2.2.0.388; Thermo Scientific). The search parameters were described previously [64].

4.8. Differential Analysis of Proteins Using MS Data

Label-free quantification was also performed with PD 2.2 and the differential analysis of the relative abundance of proteins between samples was performed using the PERSEUS software (version 1.6.0.7) [65], as previously described [64].

4.9. Functional Categorization

The protein sequences of the differentially changed proteins were subjected to a BLAST query against the Ami gene ontology (GO) database (http://amigo1.geneontology.org/cgi-bin/amigo/blast.cgi). The corresponding GO terms were extracted from the most homologous proteins using a Perl program. The GO annotation results were plotted by the Web Gene Ontology Annotation Plot (WEGO) (http://wego.genomics.org.cn/cgi-bin/wego/index.pl) tool by uploading compiled WEGO native format files containing the obtained GO terms. The gene functional annotations and protein categorization was analyzed using MapMan bin codes [66] and protein abundance ratio was assessed through MapMan software [67]. The MapMan software is generally linked with several external databases, which enable accurate measurement (http://mapman.gabipd.org). Pathway mapping of identified proteins was performed using Kyoto Encyclopedia of Genes and Genomes (KEGG) databases [68] (http://www.genome.jp/kegg/).

4.10. Analysis of Superoxide Dismutase, Catalase Activity, and Peroxidase in Response to Chemo-Blended Ag NPs

For analyses of change in enzyme activities on chemo-blended Ag NPs, fresh leaves (0.5 g) were ground in liquid nitrogen and homogenized in sodium phosphate buffer. The homogenate was centrifuged at $12,000 \times g$ for 15 min at 4 °C, and supernatant was collected in another tube. SOD activity was assessed by the method illustrated by Beauchamp and Fridovich [69] with little modifications. Catalase (CAT) activity was measured by the method described by Aebi and Bergmeyer [70] by analyzing decrease in H_2O_2 content at 240 nm with slight modifications. Peroxidase (POD) activity was analyzed using the guaiacol oxidation method by Li et al. [71].

4.11. Statistical Analysis

Data were analyzed by one-way ANOVA followed by Tukey's multiple comparison among multiple groups using SPSS (version 22.0; IBM). A *p*-value of less than 0.05 was considered as statistically significant. Student *t*-test was used for comparison between two groups for statistical analysis. Significance ($P < 0.05$) among groups was indicated through asterisks.

5. Conclusions

Production and utilization of NPs are increasing in the eco-system. Plants as primary components of eco-system are more prone to accumulation of NPs, indicating the importance of the interaction of NPs with plants and the environment. The phytostimulatory effects of Ag NPs have been reported in several studies; however, the effects of chemo-blended Ag NPs have not been explored earlier. To investigate the mechanism of the effect of chemo-blended Ag NPs on wheat growth, a gel-free/label-free proteomic technique was used. The key findings of the current study are as follows: (i) The morphological analysis depicted that chemo-blended Ag NPs increased plant growth. (ii) Proteins related to secondary metabolism, protein synthesis, and transport were increased. (iii) Number of proteins related to glycolysis, signaling, and cell wall were decreased. (iv) Similarly, proteins related to redox and mitochondrial ETC also decreased. (v) In glycolysis, glyceraldehyde-3-phosphate dehydrogenase increased/decreased, while phosphoenol pyruvate carboxylase decreased. (vi) Enzymatic activities of SOD, POD, and CAT increased when chemo-blended AgNPs were tested on wheat. (vii) Chemo-blended Ag NPs promoted yield and yield components of wheat. (viii) Morphological analysis of the next generation showed normal growth without any toxic effects. Furthermore, maintenance of redox homeostasis through regulation of glycolysis and increased activities of antioxidant enzymes regulates energy metabolism. This maintenance of energy-related activities may stimulate plant growth and development in response to chemo-blended Ag NPs.

Supplementary Materials: Supplementary materials can be found at http://www.mdpi.com/1422-0067/20/4/825/s1.

Author Contributions: S.K. conceived and designed the experiments; H.M.J. performed the morphological experiments; S.K. performed the proteomic experiments; H.Y., K.H., and K.T. performed MS analysis and data analysis; S.K. and F.Y. contributed to the analysis with tools; S.K., F.Y., H.Y., K.H., K.T., and H.M.J. wrote the paper. S.K., F.Y., H.Y., K.H., K.T., H.M.J., A.R., and Y.B. read the paper.

Funding: This work was supported by grant (F/S 30) from Fukui University of Technology, Japan.

Acknowledgments: H.M.J was supported by Higher Education Commission, Pakistan.

Conflicts of Interest: The authors declare no conflict of interest.

Abbreviations

NPs	Nanoparticles
LC	Liquid Chromatography
ROS	Reactive oxygen species
MS	Mass Spectrometry
SOD	Superoxide dismutase
POD	Peroxidase
CAT	Catalase

References

1. Peterson, E.J.; Henery, T.B.; Zhao, J.; MacCuspie, R.I.; Kirshling, T.L.; Dobrovolskaia, M.A.; White, J.C. Identification and avoidance of potential artifacts and misinterpretations in nano material toxicity measurements. *Environ. Sci. Technol.* **2014**, *48*, 4226–4246. [CrossRef] [PubMed]

2. Carlos, A.; Batista, S.; Larson, R.G.; Kotov, N.A. Nonadditivity of nanoparticle interactions. *Science* **2015**, *350*, 176–187.

3. Nel, A.; Xia, T.; Madler, L.; Li, N. Toxic potential of materials at the novel. *Science* **2006**, *311*, 622–627. [CrossRef] [PubMed]

4. Cox, A.; Venkatachalam, P.; Sohi, S.; Sharma, N. Silver and TiO$_2$ NPs toxicity in plants: A review of current research. *Plant Physiol. Biochem.* **2016**, *107*, 147–163. [CrossRef] [PubMed]

5. Wang, P.; Lambi, E.; Zhao, F.J.; Kopihke, P.M. Nanotechnology: A new opportunity in plant sciences. *Trends Plant Sci.* **2016**, *21*, 699–712. [CrossRef]

6. Rastogi, A.; Marek, Z.; Oksana, S.; Hazem, M.K.; He, X.; Sonia, M.; Marian, B. Impact of metal and metal oxide nanoparticles on plant: A critical review. *Front. Chem.* **2017**, *5*, 1–16. [CrossRef] [PubMed]

7. Cvjetko, P.; Milosic, A.; Domijao, A.M.; Vinkovic, V.I.; Tolic, S.; Pehrec, S.P.; Letofsky, P.I.; Tkalec, M.; Balen, B. Toxicity of silver ions and differently coated silver nanoparticles in Allium cepa roots. *Ecotoxicol. Environ. Saf.* **2017**, *137*, 18–28. [CrossRef]

8. Sharma, P.; Bhatt, D.; Zaidi, M.G.; Saradhi, P.P.; Khanna, P.K.; Arora, S. Silver nanoparticle-mediated enhancement in growth and antioxidant status of *Brassica juncea*. *Appl. Biochem. Biotechnol.* **2012**, *167*, 2225–2233. [CrossRef]

9. Stampoulis, D.; Sinha, S.K.; White, J.C. Assay- dependent phytotoxicity of nanoparticles on plants. *Environ. Sci. Technol.* **2009**, *43*, 9473–9479. [CrossRef]

10. Krishnaraj, C.; Jagan, E.G.; Ramchandran, R.; Abirami, S.M.; Mohan, N.; Kalaichelvan, P.T. Effect of biologically synthesized silver nanoparticles on *Baccopamonnieri* (Linn.) wettst. Plant growth metabolism. *J. Process Biochem.* **2012**, *47*, 651–658. [CrossRef]

11. Song, U.; Jun, H.; Waldman, B.; Roh, J.; Kim, Y.; Yi, J.; Lee, E.J. Functional analysis of nanoparticle toxicity: A comparative study of the effects of TiO$_2$ and Ag on tomatoes (*Lycopersiconescolentum*). *Ecotoxicol. Environ. Saf.* **2013**, *93*, 60–67. [CrossRef] [PubMed]

12. Sriram, T.; Pandidurai, V. In vitro growth analysis of *Zea mays* L. using Ag NPs. *Int. J. Pharma Bio. Sci.* **2017**, *8*, 30–37.

13. Bello-Bello, J.J.; Chavez-Santoscoy, R.A.; Lecona-Guzman, C.A.; Bogdanchikova, N.; Salinase-Ruiz, J.; Gornez-Merino, F.C.; Pestryakov, A. Harmetic response of Ag NPs on in vitro multiplication of sugarcane using a temperorary immersion system. *Dose Response* **2017**, *15*, 1–9. [CrossRef]

14. Syu, Y.Y.; Hung, J.H.; Chen, J.C.; Chuang, H.W. Impact of size and shape of silver NPs on Arabidopsis plant growth and gene expression. *Plant Physiol. Biochem.* **2014**, *83*, 57–64. [CrossRef] [PubMed]

15. Sajjad, Y.; Jaskani, M.; Ashraf, M.Y.; Ahamd, R. Response of morphological and physiological growth attributes to foliar application of plant growth regulators in gladiolous 'white prosperity'. *Pak. J. Agric. Sci.* **2014**, *51*, 123–129.

16. Noctor, G.; Queval, G.; Gakiere, B. NADP synthesis and pyridine nucleotide cycling in plants and their potential importance in stress conditions. *J. Exp. Bot.* **2006**, *57*, 1603–1620. [CrossRef] [PubMed]

17. Berglund, T.; Wallstrom, A.; Nguyen, T.V.; Laurell, C.; Ohlson, B.A. Nicotinamide, antioxidative and DNA hypomethylation effects in plant cells. *Plant Physi. Biochem.* **2017**, *118*, 551–560. [CrossRef]

18. Wang, M.; Zheng, Q.; Shen, Q.; Guo, S. The critical role of potassium in plant stress response. *Int. J. Mol. Sci.* **2013**, *14*, 370–390. [CrossRef]

19. Singh, H.; Singh, M.; Kang, J.S. Effect of potassium nitrate on yield and yield attributes of spring maize (*Zea mays* L.) under different dates of planting. *Int. J. Curr. Microbiol. Appl. Sci.* **2017**, *6*, 1581–1590. [CrossRef]

20. Ahmad, P.; Abdel Latef, A.A.; Abd Allah, E.F.; Hashem, A.; Sarwat, M.; Anjum, N.A.; Gucel, S. Calcium and potassium supplementation enhanced growth, osmolyte secondary metabolite production and enzymatic antioxidant machinery in cadmium exposed chickpea. *Front. Plant Sci.* **2016**, *27*, 513. [CrossRef]

21. Vanini, C.; Domingo, G.; Onelli, E.; DeMattia, F.; Brani, I.; Marsomi, M.; Bracale, M. Phytotoxic and genotoxic effects of Ag NPs exposure on germinating wheat seedlings. *J. Plant Physiol.* **2014**, *171*, 1142–1148. [CrossRef] [PubMed]

22. Mustafa, G.; Skata, K.; Komatsu, S. Proteomic analysis of soybean roots exposed to varying sizes of silver nanoparticles on the effects of silver nanoparticles under flooding stress. *J. Proteom.* **2016**, *148*, 113–125. [CrossRef] [PubMed]

23. Mirzajani, F.; Askari, H.; Hamzelou, S.; Schober, Y.; Rompp, A.; Ghassempour, A.; Spengler, B. Proteomic study of silver nanoparticles toxicity on *Oryza sativa* L. *Ecotoxico. Environ. Saf.* **2014**, *108*, 335–339. [CrossRef] [PubMed]

24. Mahakhalm, W.; Srmah, A.K.; Meensiri, S.; Theerakulpisut, P. Nanopriming technology for enhancing germination and starch metabolism of aged rice seeds using phytosynthesized Ag NPs. *Sci. Rep.* **2017**, *15*, 1–21.

25. Vishwakarma, K.; Shweta, U.N.; Singh, J.; Liu, S.; Singh, V.P.; Parasad, S.M.; Chauhan, D.K.; Sharma, S. Differential phototoxic impact of plant mediated Ag NPs and AgNO$_3$ on brassica sp. *Front. Plant Sci.* **2017**, *12*, 1501. [CrossRef] [PubMed]

26. Prasad, T.N.V.K.V.; Adam, S.; Visweswara, R.; Ravindra, R.B.; Gridhara, K.T. Size dependent effects of antifungal phytogenic silver NPs on germination, growth and biochemical parameters of rice, maize and peanut. *IET Nanobiotechnol.* **2017**, *11*, 277–285. [CrossRef] [PubMed]

27. Liu, G.; Zhang, M.; Jin, Y.; Fan, X.; Xu, J.; Zhu, Y.; Fu, Z.; Pan, X.; Qian, H. The effects of low concentration of Ag NPs on wheat growth, seed quality and soil microbial communities. *J. Water Air Soil Pollut.* **2017**, *228*, 1–12. [CrossRef]

28. Barabanov, P.V.; Gerasimov, A.V.; Bilnov, A.V.; Kravtsov, V.A. Influence of nanosilver on the efficiency of Pisumsativum crops germination. Ecotoxicol. *Environ. Saf.* **2017**, *147*, 715–719. [CrossRef]

29. Tripathi, D.K.; Singh, S.; Singh, S.; Srivastava, P.K.; Singh, V.P.; Singh, S.; Parasad, S.M.; Singh, P.K.; Dubey, N.K.; Panday, A.C.; et al. Nitric oxide alleviates (AgNPs)- induced phytotoxicity in *Pisumsativum* seedlings. *Plant Physiol. Biochem.* **2017**, *110*, 167–177. [CrossRef]

30. El-Tamsah, Y.S.; Joner, E.J. Impact of Fe and Ag NPs on seed germination and differences in bioavailability during exposure in aqueous suspension and soil. *Environ. Toxicol.* **2010**, *27*, 42–49. [CrossRef]

31. Hojjat, S.S.; Hojjat, H. Effect of Ag NPs exposure on germination of lentil. *Int. J. Farm. Allied Sci.* **2016**, *5*, 248–252.

32. Mustafa, G.; Skata, K.; Hossain, Z.; Komatsu, S. Proteomic study on the effects of silver nanoparticles on soybean under flooding stress. *J. Proteom.* **2015**, *122*, 100–118. [CrossRef] [PubMed]

33. Ferryra, M.L.F.; Pezza, A.; Biarc, J.; Burlingame, A.L.; Cast, P. Plant L10 ribosomal proteins have different roles during development and translation under ultraviolet-B stress. *Plant Phsysiol.* **2010**, *153*, 1878–1894. [CrossRef] [PubMed]

34. Komatsu, S.; Kuji, R.; Nanjo, Y.; Hiraga, S.; Furakuwa, K. Comprehensive analysis of endoplasmic reticulum-enriched fraction in root tips of soybean under flooding stress using proteomic techniques. *J. Proteom.* **2012**, *77*, 531–562. [CrossRef] [PubMed]

35. Yasmeen, F.; Raja, N.I.; Mustafa, G.; Sakata, K.; Komatsu, S. Quantitative proteomic analysis of post flooding recovery in soybean roots exposed to Aluminum oxide NPs. *J. Proteom.* **2016**, *143*, 136–150. [CrossRef] [PubMed]

36. Yasmeen, F.; Raja, N.I.; Razzaq, A.; Komatsu, S. Gel-free/label-free proteomic analysis of wheat shoot in stress tolerant varieties under iron nanoparticles exposure. *Biochim. Biophys. Acta* **2016**, *1864*, 1586–1598. [CrossRef] [PubMed]

37. Vanini, C.; Domingo, G.; Onelli, E.; Prinsi, B.; Marsoni, M.; Espen, L.; Barcale, M. Morphological and proteomic responses of *Eruca sativa* exposed to AgNPs or AgNO₃. *PLoS ONE* **2013**, *8*, 68752. [CrossRef]

38. Plaxton, W.C. The organization and regulation of plant glycolysis. *Annu. Rev. Plant Physiol. Plant Mol. Biol.* **1996**, *47*, 185–214. [CrossRef]

39. Banti, V.; Giuntoli, B.; Gonzoli, S.; Loreti, E.; Magneshchi, L.; Novi, G. Low oxygen response mechanism in green organism. *Int. J. Mol. Sci.* **2013**, *14*, 4734–4761. [CrossRef]

40. Khan, M.; Jan, A.; Karibe, H.; Komatsu, S. Identification of phosphor proteins regulated by gibberellins in rice leaf sheath. *Plant Mol. Biol.* **2005**, *58*, 27–40. [CrossRef]

41. Yasmeen, F.; Raja, N.I.; Ilyas, N.; Komatsu, S. Quantitative proteomic analysis of shoot in stress tolerant wheat varieties on copper nanoparticle exposure. *Plant Mol. Biol. Rep.* **2018**, *36*, 326–340. [CrossRef]

42. Gorrini, C.; Harris, I.S.; Mak, T.W. Modulation of oxidative stress as an anticancer strategy. *Nat. Rev. Drug Discov.* **2013**, *12*, 931–947. [CrossRef] [PubMed]

43. Miller, G.; Suzuki, N.; Ciftic-Yilmaz, S.; Mittler, R. Reactive oxygen species homeostasis and signaling during drought and salinity stress. *Plant Cell Environ.* **2010**, *33*, 453–467. [CrossRef] [PubMed]

44. Du, W.; Tan, W.; Peralta-Videa, J.R.; Gardea-Torresdey, J.L.; Ji, R.; Guo, H. Interaction of metal oxide nanoparticles with higher terrestrial plants: Physiological and biochemical aspects. *Plant Physiol. Biochem.* **2017**, *110*, 210–225. [CrossRef] [PubMed]

45. Verma, S.; Dubey, R. Lead toxicity induces lipid peroxidation and alters the activities of antioxidant enzymes in growing rice plants. *Plant Sci.* **2003**, *164*, 645–655. [CrossRef]

46. Jiang, H.S.; Yin, L.Y.; Ren, N.N.; Zhao, S.T.; Li, Z.; Zhi, Y.; Shao, H.; Li, W.; Gontero, B. Silver nanoparticles induced accumulation of reactive oxygen species and alteration of antioxidant systems in the aquatic plant *Spirodelapolyrhiza. Nanotoxicology* **2017**, *11*, 1–42. [CrossRef] [PubMed]

47. Ma, C.; Liu, H.; Guo, H.; Musante, C.; Coskun, S.H.; Nelson, B.C.; White, J.C.; Xing, B.; Dhankher, O.P. Defense mechanisms and nutrient displacement in Arabidopsis thaliana upon exposure to CeO₂ and In₂O₃ nanoparticles. *Environ. Sci. Nano* **2016**, *3*, 1369–1379. [CrossRef]

48. Rani, P.U.; Yasur, J.y.; Loke, K.S.; Dutta, D. Effect of synthetic and biosynthesized silver nanoparticles on growth, physiology and oxidative stress of water hyacinth: *Eichhorniacrassipes* (Mart) Solms. *Acta Physiol. Plant* **2016**, *38*, 58. [CrossRef]

49. Huang, Z.; He, K.; Song, Z.; Zeng, G.; Chen, A.; Yuan, L.; Li, H.; Hu, L.; Guao, Z.; Chen, G. Antioxidative response of *Phanerochaetechrysosporium* against silver nanoparticle-induced toxicity and its potential mechanism. *Chemosphere* **2018**, *211*, 573–583. [CrossRef]

50. Das, P.; Barua, S.; Sarkar, S.; Karak, N.; Bhattacharyya, P.; Raza, N.; Kim, K.H.; Bhattacharyya, S.S. Plant extract- mediated green SNPs: Efficacy as soil conditioner and plant growth promoter. *J. Hazad. Mater.* **2017**, *15*, 62–72.

51. Yasmeen, F.; Raja, N.I.; Razzaq, A.; Komatsu, S. Proteomic and physiological analyses of wheat seeds exposed to copper and iron nanoparticles. *Biochim. Biophys. Acta* **2017**, *1865*, 28–42. [CrossRef]

52. Wang, Q.; Ebbs, S.D.; Chen, Y.; Ma, X. Trans-generational impact of Cerium Oxide NPs on tomato plants. *Metallomics* **2013**, *5*, 753–759. [CrossRef] [PubMed]

53. Zhao, L.; Sun, Y.; Hernandez-Viezcas, J.A.; Servin, A.D.; Hong, J.N.G.; Peralta-Videa, J.R.; Duarte-Gardea, M.; Gardea-Torresdey, J.L. Influence of CeO₂ and ZnO nanoparticles on cucumber physiological markers and bioaccumulation of Ce and Zn: A life cycle study. *J. Agric. Food Chem.* **2014**, *61*, 11945–11951. [CrossRef]

54. Hafeez, A.; Razzaq, A.; Mahmood, T.; Jhanzab, H.M. Potential of copper nanoparticles to increase growth and yield of wheat. *J. Nanosci. Adv. Technol.* **2015**, *1*, 6–11.

55. Shams, G.; Ranjbar, M.; Amiri, A. Effect of Ag NPs on concentrations of silver heavy element and growth indexes in cucumber (*Cucumissativus*. L. negeen). *J. Nanopart. Res.* **2013**, *15*, 1630–1635. [CrossRef]

56. Sheykhbaglou, R.; Sedghi, M.; Shishevan, M.T.; Sharifi, R.F. Effects of nano-iron oxide particles on agronomic traits of soybean. *Not. Sci. Biol.* **2010**, *2*, 112–113. [CrossRef]

57. Feizi, H.; Pour, S.J.; Rad, K.H. Biological response of muskmelon to magnetic field and silver nanoparticles. *Ann. Rev. Res. Biol.* **2013**, *3*, 794–804.

58. Venkatachalam, P.; Priyanka, N.; Manikandan, K.; Ganeshbabu, I.; Indiraarulselvi, P.; Geetha, N.; Muralikrishna, K.; Bhattacharya, R.C.; Tiwari, M.; Sharma, N.; et al. Enhanced plant growth promoting role of phycomolecules coated zinc oxide nanoparticles with P supplementation in cotton (*Gossypium hirsutum* L.). *Plant Physiol. Biochem.* **2017**, *110*, 118–127. [CrossRef] [PubMed]

59. Zhu, H.; Han, J.; Xiao, J.Q.; Jin, Y. Uptake, translocation, accumulation of manufactured Iron oxide nanoparticles by pumpkin plants. *J. Environ. Monit.* **2008**, *10*, 713–717. [CrossRef]

60. Ajirloo, A.R.; Shaaban, M.; Motlagh, Z.R. Effect of K nano-fertilizer and N bio-fertilizer on yield and yield components of tomato (*Lycopersicon Esculentum* L.). *Int. J. Adv. Biol. Biom. Res.* **2015**, *3*, 138–143.

61. Razzaq, A.; Ammara, R.; Jhanzab, H.M.; Mahmood, T.; Hafeez, A.; Hussain, S. A novel nanomaterial to enhance growth and yield of wheat. *J. Nanosci. Technol.* **2016**, *2*, 55–58.

62. Komatsu, S.; Han, C.; Nanjo, Y.; Altaf-Un-Nahar, M.; Wang, K.; He, D.; Yang, P. Label-free quantitative proteomic analysis of abscissic acid effect in early-stage soybean under flooding. *J. Proteome Res.* **2013**, *12*, 4769–4784. [CrossRef] [PubMed]

63. Bradford, M.M. A rapid and sensitive method for the quantitation of microgram quantities of protein utilizing the principle of protein-dye binding. *Anal. Biochem.* **1976**, *72*, 248–254. [CrossRef]

64. Li, X.; Rehman, S.U.; Yamaguchi, H.; Hitachi, K.; Tsuchida, K.; Yamaguchi, T.; Sunohara, Y.; Matsumoto, H.; Komatsu, S. Proteomic analysis of the effect of plant-derived smokeon soybean during recovery from flooding stress. *J. Proteom.* **2018**, *181*, 238–248. [CrossRef] [PubMed]

65. Tyanova, S.; Temu, T.; Siniteyn, P.; Carlson, A.; Hein, Y.; Gieger, T.; Mann, M.; Cox, J. The Perseus computational platform for comprehensive analysis of proteomics data. *Nat. Methods* **2016**, *13*, 731–740. [CrossRef] [PubMed]

66. Usadel, B.; Nagel, A.; Thimm, O.; Redestig, H.; Blaesing, O.E.; Rofas, N.P.; Selbig, J.; Hannemann, J.; Piques, M.C.; Steinhauser, D.; et al. Extension of the visualization tool MapMan to allow statistical analysis of arrays, display of corresponding genes and comparison with known responses. *Plant Physiol.* **2005**, *138*, 1195–1204. [CrossRef] [PubMed]

67. Usadel, B.; Poree, F.; Nagel, A.; Loshe, M.; Czedik-Eysenberg, A.; Sitt, M. Aguide to using MapMan to visualize and compare omics in plants: A case study in the crop species, maize. *Plant Cell Environ.* **2009**, *32*, 1211–1229. [CrossRef] [PubMed]

68. Kanehisa, M.; Goto, S. KEGG: Kyoto encyclopedia of genes and genomes. *Nucleic Acids Res.* **2000**, *28*, 27–30. [CrossRef] [PubMed]

69. Beauchamp, C.; Fridovich, I. Superoxide dismutase: Improved assays and an assay applicable to acrylamide gels. *Anal. Biochem.* **1971**, *44*, 276–287. [CrossRef]

70. Aebi, H.E. Catalase. In *Methods of Enzymatic Analysis*; Bergmeyer, H.U., Ed.; VerlagChemie: Weinhem, Germany, 1983; pp. 273–286.

71. Li, S.; Yan, T.; Yang, J.Q.; Oberley, T.D.; Oberley, W. The role of cellular glutathione peroxidase redox regulation in the suppression of tumor cell growth by manganese superoxide dismutase. *Cancer Res.* **2000**, *60*, 3927–3939. [PubMed]

International Journal of
Molecular Sciences

MDPI

Article

Molecular Responses of Maize Shoot to a Plant Derived Smoke Solution

Muhammad Mudasar Aslam [1,2,3], Shafiq Rehman [1], Amana Khatoon [1], Muhammad Jamil [4],
Hisateru Yamaguchi [5], Keisuke Hitachi [5], Kunihiro Tsuchida [5], Xinyue Li [3], Yukari Sunohara [3],
Hiroshi Matsumoto [3] and Setsuko Komatsu [2,*]

[1] Department of Botany, Kohat University of Science and Technology, Kohat 26000, Pakistan;
mudasar_kust@yahoo.com (M.M.A.); drshafiq@yahoo.com (S.R.); proteomics.sp@gmail.com (A.K.)
[2] Faculty of Environmental and Information Sciences, Fukui University of Technology, Fukui 910-8505, Japan
[3] Faculty of Life and Environmental Sciences, University of Tsukuba, Tsukuba 305-8572, Japan;
lixinyue108@gmail.com (X.L.); sunohara.yukari.gp@u.tsukuba.ac.jp (Y.S.); hmatsu@biol.tsukuba.ac.jp (H.M.)
[4] Department of Biotechnology and Genetic Engineering, Kohat University of Science and Technology,
Kohat 26000, Pakistan; Jamilkhattak@yahoo.com
[5] Institute for Comprehensive Medical Science, Fujita Health University, Toyoake 470-1192, Japan;
hyama@fujita-hu.ac.jp (H.Y.); hkeisuke@fujita-hu.ac.jp (K.H.); tsuchida@fujita-hu.ac.jp (K.T.)
* Correspondence: skomatsu@fukui-ut.ac.jp

Received: 16 December 2018; Accepted: 5 March 2019; Published: 15 March 2019

Abstract: Plant-derived smoke has effects on plant growth. To find the molecular mechanism of
plant-derived smoke on maize, a gel-free/label-free proteomic technique was used. The length
of root and shoot were increased in maize by plant-derived smoke. Proteomic analysis revealed
that 2000 ppm plant-derived smoke changed the abundance of 69 proteins in 4-days old maize
shoot. Proteins in cytoplasm, chloroplast, and cell membrane were altered by plant-derived smoke.
Catalytic, signaling, and nucleotide binding proteins were changed. Proteins related to sucrose
synthase, nucleotides, signaling, and glutathione were significantly increased; however, cell wall,
lipids, photosynthetic, and amino acid degradations related proteins were decreased. Based on
proteomic and immunoblot analyses, ribulose-1,5-bisphosphate carboxylase/oxygenase (RuBisCO)
was decreased; however, RuBisCO activase was not changed by plant-derived smoke in maize shoot.
Ascorbate peroxidase was not affected; however, peroxiredoxin was decreased by plant-derived
smoke. Furthermore, the results from enzyme-activity and mRNA-expression analyses confirmed
regulation of ascorbate peroxidase and the peroxiredoxinin reactive oxygen scavenging system. These
results suggest that increases in sucrose synthase, nucleotides, signaling, and glutathione related
proteins combined with regulation of reactive oxygen species and their scavenging system in response
to plant-derived smoke may improve maize growth.

Keywords: proteomics; maize; plant-derived smoke; shoot

1. Introduction

Maize is highly commercial crop, being a major source of food, feed, biofuel, and industrial
products [1]. It is the most diverse crop analyzed at morphological and molecular levels [2]. Maize
is an ideal crop for genomic studies because it exhibits a high level of genetic diversity and many
structural variations [3]. Structural changes and genetic diversity play a key role in the morphology of
maize [4]. Maize plant has a very large size genome with a complex organization [5,6]. These findings
indicate the importance of maize genetic diversity for the manipulation of new resistant and high
yielding varieties.

Fire is documented as ecological factor in ecosystems, because many forest plant species' life cycles depend on fire [7]. Fire products, which are heat, chemicals, ash, and smoke, have been widely identified as germination cues for different species from both fire-prone and fire-free ecosystems [8]. Plant-derived smoke contains active compounds to promote seed germination of crops [9]. The karrikins and cyanohydrins are identified as germination stimulants present in smoke [9]. These compounds have extensive implications for horticulture, weed control, conservation, and restoration [10]. Plant-derived smoke is a plant growth stimulant obtained from burning of wide variety of biotic sources including leaf, shoot, and straw [11]. These results indicate that plant-derived smoke and compounds isolated from smoke are vital stimulants for germination and plant growth.

Plant-derived smoke stimulated seed germination in 1200 plant species from more than 80 genera of different families [10] including crops [12], medicinal plants [13], and fruit [14]. The promotive effects of plant-derived smoke are independent of seed size, shape, and type [11]. In addition to stimulating seed germination, plant-derived smoke enhanced the seedling length/weight of different crops [15] and pollen germination/tube elongation of flowers belonging to different plant families [16]. Plant-derived smoke induces many changes in seeds, including sensitivity of seeds to phytohormones [17] and increased permeability by softening the seed coat [18]. These findings highlight the promotive effects of plant-derived smoke on plant morphology.

Different physio-chemical contents of plants have been reported to be increased by plant-derived smoke solution [19]. This solution has stimulatory effects on photosynthesis in *Isatis indigotica* seedlings by enhancing carbon dioxide fixation, the transpiration rate, gaseous exchange, stomatal conductance, and photochemical activities [20]. In addition, total soluble proteins, chlorophyll *a/b*, total carotenoids, and total nitrogen contents were also increased in smoke treated rice seedlings [21]. Positive effects of plant-derived smoke were also observed on seed germinating enzyme activities in different grasses [22]. This smoke improved the plant-defense system by increasing flavonoids, tannins contents, and level of phenolics [23]. It is presumed that the stimulatory effects of plant-derived smoke are due to its close relation with plant growth regulator [24]. Despite these findings, the mechanism underlying these physiological changes remains unclear and needs an in-depth study to find plant-derived smoke effects on the physiological processes of plants.

Various morphological and physiological studies were conducted to analyze the effects of plant-derived smoke on plants growth; however, its mechanism of action on plant growth has not been investigated yet. The present study is focused on investigating the effects of plant-derived smoke on the initial seedling-stage of maize. Morphological analysis was performed on maize seedlings raised from smoke treated seeds. Based on these morphological results, a gel-free/label-free proteomic analysis was applied to assess the effects of plant-derived smoke solution on maize-shoot. It attempted to explore some clues about the response mechanism of plants towards plant-derived smoke solution at molecular level. In order to further validate and peep into the results at the molecular level, immunoblot, enzyme-activity, and mRNA-expression analyses were performed in a continuation of proteomic results.

2. Results

2.1. Morphological Effects of Plant-Derived Smoke on Maize Growth

To investigate the effects of plant-derived smoke on maize growth, morphological analysis was performed. Seeds were soaked without or with 1000 ppm, 2000 ppm, and 4000 ppm plant-derived smoke for 6 h. Length and fresh weight of shoot and root were measured for 4, 6, and 8 days after sowing (Figure 1). The seed germination percentage was significantly increased by 2000 ppm plant-derived smoke as compared to control while 4000 ppm plant-derived smoke did not affect germination percentage (Figure 2). Length and fresh weight of shoot were increased by treatments of 1000 ppm and 2000 ppm plant-derived smoke 4 days after sowing (Figure 3). The length of the root was increased more by the 1000 ppm treatment than by the control as indicated in Figure 3; and the increase was not significant with the 2000 ppm treatment.

The fresh weight of the root was not significantly changed by 1000 ppm or 2000 ppm treatment compared to untreated plant (Figure 3). The length and fresh weight of root were not affected by plant-derived smoke concentrations 6 days after sowing. Treatment with 4000 ppm plant-derived smoke did not affect THE length and fresh weight of shoot and root compared to an untreated plant.

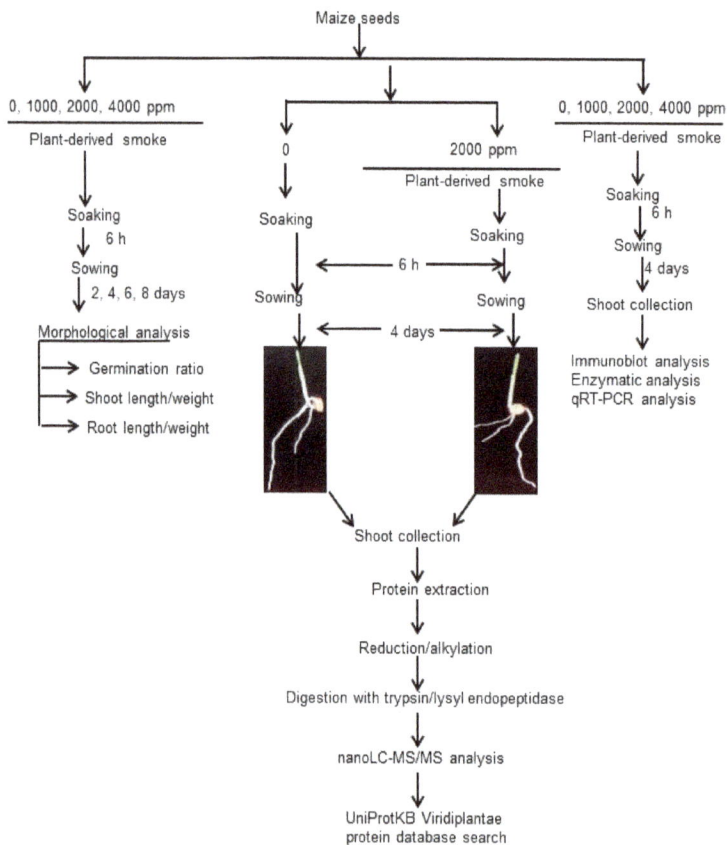

Figure 1. Experimental design for the effects of plant-derived smoke on maize growth. Maize seeds were soaked without or with 1000 ppm, 2000 ppm, and 4000 ppm plant-derived smoke for 6 h and grown in sand. For morphological analysis, germination ratio, shoot/root length, and weight were measured at 2, 4, 6, and 8 days. For proteomic analysis, maize seeds were soaked with 2000 ppm plant-derived smoke for 6 h. Shoot was collected after 4 days. Proteins were extracted, reduced, alkylated, digested, and analyzed by nano LC-MS/MS. For western blotting, maize seeds were soaked without or with 1000 ppm, 2000 ppm, and 4000 ppm plant-derived smoke for 6 h and grown in sand. Shoot was collected and proteins were extracted after 4 days for immune blotting analysis.

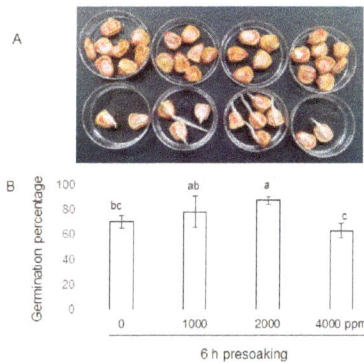

Figure 2. Germination percentage of maize seeds treated with plant-derived smoke. Maize seeds were soaked without or with 1000 ppm, 2000 ppm, and 4000 ppm plant-derived smoke for 6 h (**A,B**). The images in (**A**) correspond to data bars in (**B**). (**A**) The above row shows seeds, just after presoaking, while below row shows seeds, 2 days after presoaking. (**B**) Germination percentage was recorded after 2 days. The data are presented as the mean ± S.D. from 4 independent biological replicates. Different letters indicate that the change is significant as determined by one-way ANOVA followed by Tukey's multiple comparison ($p < 0.05$).

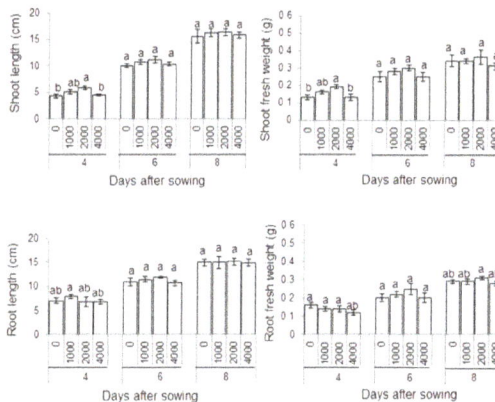

Figure 3. Morphological changes in maize treated with plant-derived smoke. Maize seeds were soaked without or with 1000 ppm, 2000 ppm, and 4000 ppm plant-derived smoke for 6 h and sown in sand for 4, 6, and 8 days. The length and fresh weight of shoot and root were measured. The data are shown as means ± S.D. from 4 independent biological replicates. Different letters indicate that the change is significant as determined by one-way ANOVA followed by Tukey's multiple comparison ($p < 0.05$).

2.2. Functional Classification of Differentially Abundant Proteins in Maize Treated with Plant-Derived Smoke

Maize seeds were treated with 2000 ppm plant derived smoke solution and were grown for 4 days. Proteins were extracted from the 4-days old maize shoot and proteomic response was then investigated. The criteria for significantly changed proteins were 2 or more than 2 matched peptides with a p-value less than 0.05. Gel-free/label-free proteomic analysis revealed that a total of 69 proteins were significantly changed by 2000 ppm plant-derived smoke (Table 1). Results showed that in the biological process category, metabolic related proteins were prominently changed as compared to other categories (Figure S1). For the cellular compartment, cytoplasm localized proteins were significantly changed by plant-derived smoke (Figure S1). For the molecular function, the proteins with catalytic activity were more significantly changed than the other categories (Figure S1).

Table 1. List of proteins altered by plant-derived smoke solution in maize shoot.

No	Accession	Description	Fold Change	Functional Category	Biological Process	Cellular Component	Molecular Function
1	B4FTX3	Profilin	7.22	Cell	not assigned	cell part	not assigned
2	B9RWF4	Elongation factor 1-alpha	6.79	Protein	not assigned	cell part	binding
3	A0A1D6BQY9	Tubulin beta chain	4.91	Cell	not assigned	not assigned	not assigned
4	C5XR12	Uncharacterized protein	4.89	Not assigned	not assigned	not assigned	catalytic activity
5	K7TEA7	Aldo/keto reductase family protein	4.17	Redox	biosynthetic process	cell part	catalytic activity
6	C9VWQ3	Actin	3.99	Cell	not assigned	not assigned	not assigned
7	A0A1D6DZR9	AICARFT/IMPCHase bienzyme	3.11	Nucleotide metabolism	not assigned	not assigned	not assigned
8	C5YEF7	Uncharacterized protein	2.91	Not assigned	not assigned	cell part	adenyl nucleotide binding
9	B6T289	Glucan endo-1,3-β-glucosidase 7	2.55	Not assigned	carbohydrate metabolic process	not assigned	catalytic activity
10	A0A1B6QAM8	Uncharacterized protein	2.36	Major CHO metabolism	not assigned	not assigned	not assigned
11	B4FTR5	Uncharacterized protein	1.95	Not assigned	cell communication	cell part	not assigned
12	C8ZK16	Malic enzyme	1.37	Tricarboxylic acid cycle	not assigned	not assigned	not assigned
13	C0PG78	Monocopper oxidase-like protein SKU5	0.91	Development	not assigned	not assigned	binding
14	A0A1D6M373	Mannosylglycoprotein endo-beta-mannosidase	0.91	Miscellaneous	not assigned	not assigned	not assigned
15	A0A0P0X334	Os07g0176900 protein	0.89	Photosynthesis	not assigned	not assigned	not assigned
16	B4FH75	4-Hydroxy-tetrahydrodipicolinate reductase 2	−0.51	Amino acid metabolism	amine biosynthetic process	cell part	binding
17	B4FAW7	Histidinol-phosphate aminotransferase 2	−0.70	Amino acid metabolism	amine biosynthetic process	not assigned	binding
18	A0A1D6KWD0	Lysine-tRNA ligase	−0.85	Protein	not assigned	not assigned	not assigned
19	A0A1D6LT87	Ras-related protein	−0.89	Signaling	not assigned	not assigned	not assigned
20	O24574	RuBisCO small chain	−1.05	Photosynthesis	biosynthetic process	cell part	carbon-carbon lyase activity
21	Q5I204	Brain acid soluble protein	−1.08	RNA	not assigned	cell part	binding
22	A0A1Q0YQ12	Oil body-associated protein 1B	−1.25	Not assigned	not assigned	not assigned	not assigned
23	A0A1D6I8U5	Flowering time control protein FPA	−1.31	RNA	not assigned	not assigned	not assigned
24	K7UK47	Clathrin interactor EPSIN	−1.69	Not assigned	not assigned	not assigned	not assigned
25	A0A0E0CRT7	Uncharacterized protein	−1.80	Cell	not assigned	not assigned	not assigned
26	A0A0D9XRV3	Adenosylhomocysteinase	−1.88	Amino acid metabolism	not assigned	not assigned	not assigned
27	A0A1D6MQK0	Rab escort protein 1	−1.89	Signaling	not assigned	not assigned	not assigned

Int. J. Mol. Sci. **2019**, *20*, 1319

Table 1. *Cont.*

No	Accession	Description	Fold Change	Functional Category	Biological Process	Cellular Component	Molecular Function
28	B6TBI9	Pyridoxamine 5-phosphate oxidase	−1.95	Not assigned	not assigned	not assigned	binding
29	A0A0E0JEZ9	Uncharacterized protein	−1.96	Not assigned	not assigned	not assigned	not assigned
30	B4FTH5	Xyloglucan endotransglucosylase/hydrolase	−2.06	Cell wall	carbohydrate metabolic process	apoplast	catalytic activity
31	B4FZJ2	β-Glucosidase 11	−2.13	Misce	carbohydrate metabolic process	not assigned	catalytic activity
32	B6TUP8	Zinc finger homeodomain protein 1	−2.21	RNA	not assigned	not assigned	binding
33	A0A1D6H8U6	3-Oxoacyl-[acyl-carrier-protein] reductase	−2.31	Lipid metabolism	not assigned	not assigned	not assigned
34	A0A1D6EZ65	Uncharacterized protein	−2.38	Not assigned	not assigned	not assigned	not assigned
35	A0A1D6GLX4	DEK domain-containing chromatin associated protein	−2.54	Not assigned	not assigned	not assigned	not assigned
36	B6TQH7	THA4	−2.63	Not assigned	cellular process	cell part	protein transporter activity
37	C5WUG0	Mitogen-activated protein kinase	−2.75	Signaling	not assigned	not assigned	adenyl nucleotide binding
38	A0A0K9PCU0	RPM1-interacting protein 4	−2.80	Not assigned	not assigned	not assigned	not assigned
39	B6TIG8	Protein arginine N-methyltransferase 1	−2.80	Miscellaneous	cellular macromolecule metabolic process	not assigned	catalytic activity
40	B6TR82	Thioredoxin F-type	−2.81	Redox	biological regulation	not assigned	catalytic activity
41	K3Y2C0	Uncharacterized protein	−2.83	Redox	catabolic process	cell part	not assigned
42	A0A0D3GFF5	Uncharacterized protein	−2.85	Stress	not assigned	not assigned	not assigned
43	B6TYK8	Putative uncharacterized protein	−2.87	Not assigned	not assigned	not assigned	not assigned
44	A0A1D6FY68	SIT4 phosphatase-associated	−2.95	Metal handling	not assigned	not assigned	not assigned
45	A0A1E5V130	Uncharacterized protein	−3.01	Not assigned	not assigned	not assigned	not assigned
46	A0A0N7KSP9	Os1_g0247300 protein	−3.18	Cell	not assigned	not assigned	not assigned
47	A0A1D6JR65	Alcohol dehydrogenase-like 2	−3.22	Miscellaneous	not assigned	not assigned	not assigned
48	B6SW97	Putative uncharacterized protein	−3.28	Protein	not assigned	not assigned	not assigned
49	B6U581	Ribosome-like protein	−3.31	Protein	biosynthetic process	cell part	structural constituent of ribosome
50	K4BGM2	Uncharacterized protein	−3.40	Protein	biological regulation	cell part	binding
51	B5QSI9	Acetyl-coenzyme A carboxylase	−3.53	Lipid metabolism	biosynthetic process	not assigned	acetyl-CoA carboxylase activity
52	A0A1D6L558	Plasmodesmata callose-binding protein 2	−3.67	Miscellaneous	not assigned	not assigned	not assigned

Table 1. *Cont.*

No	Accession	Description	Fold Change	Functional Category	Biological Process	Cellular Component	Molecular Function
53	A0A0K9NNM0	Cysteine proteinase cathepsin F	−3.69	Protein	not assigned	not assigned	not assigned
54	C0P3K6	Aspartic proteinase A1	−3.75	Protein	lipid metabolic process	not assigned	aspartic-type endopeptidase activity
55	J3LDT9	Uncharacterized protein	−3.76	Not assigned	anatomical structure morphogenesis	cell part	not assigned
56	B9MSV5	Tubulin alpha chain	−3.77	Cell	cellular component assembly	cell part	binding
57	A0A068UVK8	Chlorophyll a-b binding protein	−3.84	Photosynthesis	cellular macromolecule metabolic process	cell part	binding
58	B6SZR1	Chlorophyll a-b binding protein	−3.97	Photosynthesis	cellular macromolecule metabolic process	cell part	binding
59	A0A1D6LJZ2	30S ribosomal protein S16	−4.00	Protein	not assigned	not assigned	not assigned
60	B4FV94	Chlorophyll a-b binding protein	−4.06	Photosynthesis	cellular macromolecule metabolic process	cell part	binding
61	E9KIP1	Photosystem I P700 chlorophyll apoprotein A1	−4.13	Photosynthesis	cellular macromolecule metabolic process	cell part	4 iron, 4 sulfur cluster binding
62	A0A061EVS4	Nascent polypeptide-associated complex beta	−4.23	RNA	biological regulation	not assigned	not assigned
63	A0A1D6IIC3	Nuclear transport factor 2	−4.38	Protein	not assigned	not assigned	not assigned
64	A0A1D5WFY2	Small ubiquitin-related modifier	−4.40	Protein	not assigned	not assigned	not assigned
65	A0A1D6PYA1	60S ribosomal protein L17	−4.61	Protein	not assigned	not assigned	not assigned
66	Q8H6N0	Tubulin beta chain	−4.76	Cell	cellular component assembly	cell part	binding
67	A0A1D5AHD9	RuBisCO large chain	−4.79	Photosynthesis	not assigned	not assigned	not assigned
68	A0A1D1ZO67	Elongation factor 1-alpha	−6.21	Protein	not assigned	not assigned	not assigned
69	A0A097PJF2	Structural maintenance of chromosomes protein 1	−9.99	Cell	cell cycle process	cell part	adenyl nucleotide binding

Accession, according to UniProtKB Viridiplantae protein database; Fold change, relative abundance of identified proteins in maize shoot raised from seeds treated with 2000 ppm plant-derived smoke solution; Functional category, protein function categorized using MapMan bin codes. Abbreviations are as follows: cell, cell division/organization/vesicle; transport; CHO, carbohydrate; protein, protein synthesis/degradation/post-translational modification/targeting; RNA, RNA processing/transcription/binding; Redox, redox homeostasis; DNA, nucleotide binding, and metal ion binding.

To determine the function of proteins in response to the plant-derived smoke solution, functional classification of identified proteins was performed using MapMan bin codes (Figure 4). It was found that 1 and 11 proteins involved in protein synthesis, degradation, post-translational modification, targeting and folding were increased and decreased, respectively, in response to plant-derived smoke solution. Plant-derived smoke treatment increased one and decreased two proteins related to redox homeostasis. Three proteins related to signaling and 4 transport related proteins were also increased, while RNA, amino acid metabolism and lipid metabolism related proteins were decreased in maize shoot in response to the plant-derived smoke solution (Figure 4).

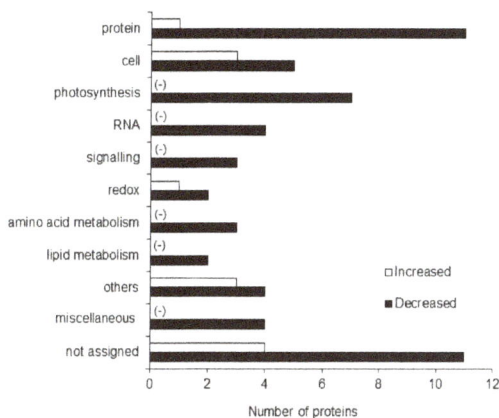

Figure 4. Categorization of identified proteins on the basis of their function in maize shoot in response to plant derived smoke solution. Maize seeds were treated without or with 2000 ppm plant-derived smoke solution for 6 h and grown for 4 days. Proteins were extracted from shoot and identified using a gel-free/label-free proteomic technique. Mapman bin code software was used to categories all the identified 69 proteins. Categories containing zero proteins are marked with (−). Abbreviations: protein, protein synthesis/degradation/ post-translational modification/targeting/folding; cell, cell division/organization/vesicle transport; RNA, RNA processing/ transcription/binding; Redox, redox homeostasis; The "Others" category includes proteins related to N-metabolism, cellular homeostasis, response to stimulus, cytoplasm, fermentation, DNA, nucleotide binding, and metal ion binding.

2.3. Pathway Analysis of Identified Proteins in Maize Treated with Plant-Derived Smoke

The plant-derived smoke treatment significantly changed some proteins. To visualize these proteomic results in the context of pathways and processes, MapMan software was used (Figure S2). Proteins associated with starch/sucrose, glutathione and nucleotide biosynthesis were significantly increased; while, ascorbate, photosynthetic, lipids, and cell wall related proteins were decreased in response to plant-derived smoke (Figure S2).

2.4. Immuno-Blot Analysis of Proteins Involved in Redox Homeostasis and Photosynthesis

To find out abundance level of large and small RuBisCO subunits, RuBisCO activase, ascorbate peroxidase, and peroxiredoxin in maize shoot, immunoblot analysis was performed. Maize seeds were treated without or with 1000 ppm, 2000 ppm, and 4000 ppm plant-derived smoke and proteins were extracted from shoot after 4 days of sowing (Figure 1). Plant-derived smoke did not affect the abundance of 37-kDa RuBisCO activase while decreasing the abundance of RuBisCO large subunit in shoot of 4-day-old plants treated by 2000 ppm and 4000 ppm (Figure 5). Abundance of 55 kDa ascorbate peroxidase was not affected by different treatments of plant-derived smoke; while, 2000 ppm and 4000 ppm plant-derived smoke very slightly decreased the abundance of 37-kDa peroxiredoxin compared to 1000 ppm and the control (Figure 6).

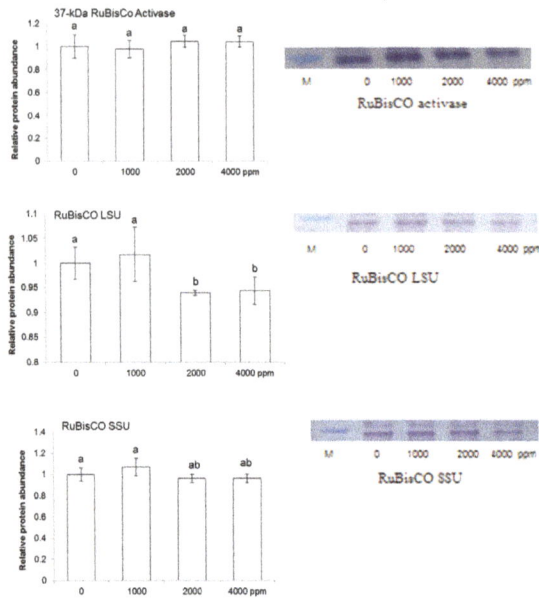

Figure 5. Effect of plant-derived smoke on RuBisCO subunits and activase enzymes in maize. Maize seeds were treated without or with 1000, 2000, and 4000 ppm plant-derived smoke for 6 h and sown in sand. Proteins were extracted from 4-days old maize shoot. ImageJ software was used to determine relative proteins intensities. Data are shown as means ± S.D. Different alphabets showing the statistical level of significance as determined by one-way ANOVA followed by Tukey's multiple comparison ($p < 0.05$). In images, 'M' is used for marker.

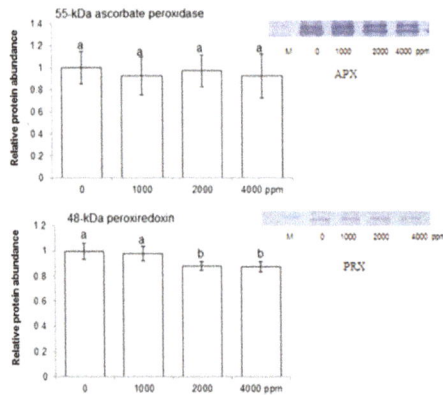

Figure 6. Effect of plant-derived smoke on redox-homeostasis related proteins in maize. Maize seeds were treated without or with 1000, 2000, and 4000 ppm plant-derived smoke for 6 h and grown for 4 days. Proteins extracted from shoot were separated by SDS-polyacrylamide gel electrophoresis and transferred onto polyvinylidene difluoride membranes. The membranes were incubated with anti-RuBisCO large/small subunits and anti-RuBisCO activase antibodies. The relative intensities of bands were calculated using ImageJ software. Data are shown as means ± S.D. from 3 independent biological replicates. Different letters indicate that the change is statistically significant as determined by one-way ANOVA followed by Tukey's multiple comparison ($p < 0.05$). In images, "M" is used for marker.

2.5. Effect of Plant-Derived Smoke on Enzymatic Activities and Gene Expression of Maize

To investigate the effects of plant-derived smoke on ROS scavenging enzymes, enzymatic analysis was performed. Seeds were soaked in without or with 1000 ppm, 2000 ppm, and 4000 ppm plant-derived smoke for 6 h and maize shoot was collected 4 days after sowing (Figure 1). Results showed that 1000 ppm, 2000 ppm, and 4000 ppm plant-derived smoke did not change activity of ascorbate peroxidase as compared to the control (Figure 7A). On the other hand, treatments 1000 ppm and 2000 ppm and 4000 ppm slightly, but significantly affected the activity of peroxiredoxin with a significant decrease compared to the control (Figure 7A).

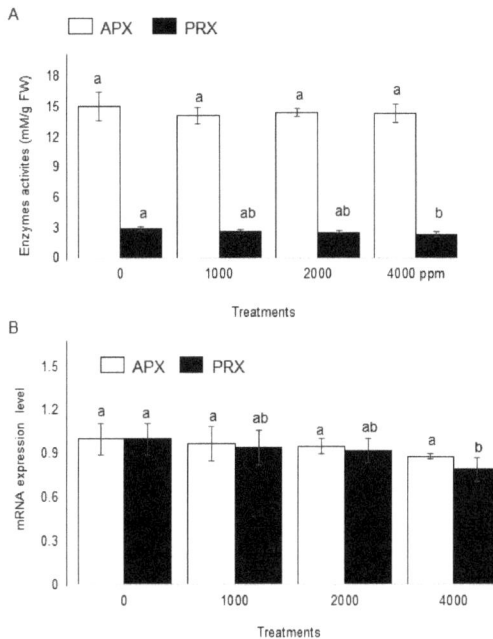

Figure 7. Effect of plant-derived smoke on enzymatic analysis and expression level of ascorbate peroxidase and peroxiredoxin gene in maize shoot. Maize seeds were treated without or with 1000 ppm, 2000 ppm, and 4000 ppm plant-derived smoke for 6 h and sown. Activities of ascorbate peroxidase (APX) and peroxiredoxin (PRX) were measured in 4-day old maize shoot (**A**). For gene expression level, RNAs extracted from maize shoot at 3rd day of treatment were analyzed by qRT-PCR (**B**). Relative mRNA abundances of ascorbate peroxidase and peroxiredoxin were normalized against 18S rRNA abundance. The data are shown as means ± S.D. from 4 independent biological replicates. White and black bars indicate ascorbate peroxidase (APX) and peroxiredoxin (PRX), respectively. Different letters indicate that the change is significant as determined by one-way ANOVA followed by Tukey's multiple comparison ($p < 0.05$).

To check the specific expression pattern of genes encoding ascorbate peroxidase and peroxiredoxin, their mRNA levels in maize shoot were analyzed. Shoot total RNA was extracted, and qRT-PCR was performed. The 18S rRNA was used as an internal control on qRT-PCR. Results revealed that 1000 ppm, 2000 ppm and 4000 ppm plant-derived smoke did not significantly affect the expression level of ascorbate peroxidase (Figure 7B); however, the expression level of peroxiredoxin was regulated with decreased expression at the transcript level in response to plant-derived smoke (Figure 7B).

3. Discussion

3.1. Effect of Plant-Derived Smoke on Morphology of Maize

Plant-derived smoke increased seed germination in different plant species [25]. Soós et al. [26] reported that plant-derived smoke enhanced the post-germination growth of maize. In this study, 2000 ppm plant-derived smoke significantly increased the seed germination percentage; furthermore, shoot/root length/fresh weight were also increased in maize. Aslam et al. [27] reported that plant-derived smoke increased the germination percentage, shoot/root length, and fresh weight of maize. Furthermore, plant-derived smoke treated maize developed vigorous seedling and fresh weight [28]. Abu et al. [29] reported that priming seeds in low concentrations of plant-derived smoke increased the germination percentage and seedling length in wild rye. On the other hand, high concentrations of plant-derived smoke reduced germination and shoot/root growth in lettuce [30]. In this study, 4000 ppm plant-derived smoke did not affect maize seed germination and shoot/root length/fresh weight (Figures 2 and 3). This finding is similar to previous results revealing that plant-derived smoke inhibited seed germination percentage at higher concentrations.

Plants treated with 2000 ppm plant-derived smoke and 10^{-8} M karrikins produced vigorous shoot and increased leaf area in *Eucomis autumnalis* [31]. The vigorous shoot and high leaf area enhanced photosynthetic activities in plants [31]. Flematti et al. [32] reported that plant-derived smoke containing karrikins enhanced germination percentage and shoot length/fresh weight in *Solanum orbiculatum*. Plant-derived smoke improved the shoot length and fresh weight in several plants, including acacia [33], onion [34], and milk thistle [35]. Our results are in accordance with previous findings suggesting that plant-derived smoke increases shoot length and fresh weight. Based on these results, 4-days old shoot raised from maize seeds treated by 2000 ppm plant-derived smoke was selected for gel-free/label free proteomic analysis.

3.2. Positive Effect of Plant-Derived Smoke on Cytoskeleton Related Proteins in Maize

With treatment of plant-derived smoke, the fold change of cytoskeleton related proteins such as profilin and actin was significantly increased compared to protein abundance of untreated plant (Table 1). The plant cytoskeleton plays an important role in many biological processes, including cell division, expansion, organogenesis, tip growth, and intracellular signaling [36]. The actin cytoskeleton proteins are concerned with the establishment and maintenance of cell polarity, and responses to numerous environmental stimuli [37]. Profilin promoted F-actin elongation which played an important role in the mobility and cell contraction during cell division [38]. Profilins are multifunctional proteins according to their abundance and locations [39]. Furthermore, profilin is associated with plasma membrane during development of microspores, pollens, and playing vital role in signal transduction [40]. The present results confirmed that seed priming in plant-derived smoke promoted shoot cytoskeleton proteins, which are necessary for cell division, signaling, and pollens development.

3.3. Effect of Plant-Derived Smoke on Photosynthetic Proteins in Maize

Plant derived smoked changed 8 photosynthesis related proteins with different regulation. RuBisCO is the most abundant protein on earth playing key role in photosynthesis having CO_2 fixation function [41]. Strong reduction was observed in activity of RuBisCO under drought stress [42]. Khodadadi et al. [43] observed that the activity of RuBisCO was reduced under drought stress. Another study reported that RuBisCO is responsive to stresses and the rapid loss of its activity caused by drought in sunflower [44]. Beside drought, cold stress also reduced the activity of RuBisCO, suggesting damage to the chloroplasts and a decrease in the rate of photosynthesis under these conditions. In plant-smoke derived compounds, karrikins also decreased RuBisCO in *Arabidopsis* [45]. The present immunoblot result revealed that RuBisCO activase is not affected by plant-derived smoke; however, a RuBisCO large subunit was decreased by plant-derived smoke. It may be possible that plant-derived smoke plays an inhibitory role at higher concentrations. It is also possible that this short period of presoaking in plant-derived smoke is not enough to activate the RuBisCO activase enzyme.

3.4. Effect of Plant-Derived Smoke on Scavenging Activity through the Ascorbate/Glutathione Pathway in Maize

Ascorbate and glutathione are main antioxidants in the leaves of maize [46]. The fundamental role of these compounds is linked to scavenging a broad range of reactive oxygen species generated under stress conditions in plants [47]. It participates in scavenging of hydrogen peroxide [48]. Selote and Chopra [49] reported that drought stress increased the amount of glutathione in rice. It is reported that ascorbate peroxidase was decreased under flooding conditions [50]. Present results revealed that ascorbate is decreased while glutathione is increased at the proteomic level. These results suggest that plant-derived smoke seed pretreatment may have positive effects on glutathione pool in the maize seedlings that might be helpful in achieving better performance during stress responses.

3.5. Effect of Plant-Derived Smoke on Enzymatic Activities of Maize

Ascorbate peroxidase and peroxiredoxin are important enzymes that are present in the plant kingdom [51]. The exposure of plants to unfavorable environmental conditions increases the production of reactive oxygen species such as hydrogen peroxide, and hydroxyl radical [52]. The reactive oxygen species detoxification process in plants is essential and occurs due to different enzymes like ascorbate peroxidase and peroxiredoxin in plant cells and their organelles [52]. Ascorbate peroxidase utilizes ascorbate as a specific electron donor to reduce hydrogen peroxide to water [53]. Experimental evidence has proven that during the metabolism process, antioxidant enzymes are also produced to balance metabolism [54]. On the other hand, peroxiredoxins constitute the most recently identified group of hydrogen peroxide -decomposing antioxidant enzymes [55]. It is reported that plant-derived smoke decreased scavenging enzymes activity in maize [56]. Another redox homeostasis enzymes, thioredoxins (Trxs) was also affected by plant-derived smoke treatments. It is a small and widely distributed protein with a conserved active site, which controls the redox status of target proteins through thiol-disulfide exchange reactions [57]. In plants, it has a fundamental role in a number of cellular processes, including seed germination, carbon assimilation, lipid metabolism, hormone metabolism, redox signaling, and stress response [57]. The present results are in accordance with the previous reports and showed that plant-derived smoke did not significantly affect the activities of ascorbate peroxidase and that peroxiredoxin activities were decreased, which revealed that plant-derived smoke is a growth promoter and stress suppressor, which decreases the production of stress responsive hormones.

Antioxidants are mostly expressed to cope with stressed situations and their expression also has positive effects on the activation of enzymes related to plant growth [58,59]. In the present study, plant-derived smoke is behaving as a growth promoter and thus did not affected ascorbate peroxidase enzyme levels (Figure 7A). Auxins might not affect or could promote increases in the activity of antioxidant enzymes regulating ROS levels, which could be associated with the activation of embryo/organogenesis [60]. It might be suggested from the results that treatment with plant-derived smoke could markedly enhance the self-capacity of defense against oxidative damage in normal growth conditions, thus not affecting the production of antioxidant enzymes in maize seedlings significantly.

3.6. Effect of Plant-Derived Smoke on Expression of Ascorbate Peroxidase and Peroxiredoxin in Maize

Plant-derived smoke did not affect ascorbate peroxidase gene expression level, whereas gene expression of peroxiredoxin was decreased at higher concentrations (Figure 7B). Ascorbate peroxidase is an important enzyme of plant scavenging system. It uses ascorbate as a specific electron donor for the conversion of hydrogen peroxide into water [61]. Besides this, it also improves the stress resistance capacity in plants against various stresses [62,63]. Ascorbate peroxidase also plays a key role in balancing the homeostasis of ascorbate and glutathione, and maintaining high photosynthetic rate in unfavorable conditions [64]. In addition, ascorbate is involved in other functions such as plant growth, gene regulation, modulation of some enzymes, and redox regulation of membrane-bound antioxidant compounds [65]. The present results revealed that ascorbate peroxidase related gene was not significantly altered in maize seedlings due to plant-derived smoke. The present results are consistent with immunoblotting results showing no effect on the abundance of ascorbate peroxidase. Decreased peroxiredoxin abundance in maize shoot in response to plant derived smoke reflects the resource economy phenomena of all living organisms, including plants. Peroxiredoxin might be present in maize plants, taking part in plant defense system against stresses. There is close interaction between plant-derived smoke and plant growth hormones [24]. It is also possible that plant immune systems might be strengthened by plant-derived smoke, resulting in a decreased level of peroxiredoxin level so as to economize plant resources. These results are in agreement with El-Gaied et al. [59] who clearly demonstrated the decreased antioxidant enzymes level in the tomato plant in response to plant growth promoting hormones.

4. Materials and Methods

4.1. Preparation of Plant-Derived Smoke Solution

Smoke solution was prepared from aerial semi dried parts of *Cymbopogon jawarncusa* [66]. Dry plant material weighing 333 g was taken and placed in a burner [67]. An electric heater was adjusted beneath the burner until all the plant material was converted into ash. Smoke was bubbled though 1 L of distilled water, resulting into 1 L of concentrated plant-derived smoke solution. It was further diluted to 1000 ppm, 2000 ppm, and 4000 ppm and used for seed treatment in the experiment. The seeds treated with distilled water were used as the control.

4.2. Plant Material and Treatment

Seeds of maize (*Zea mays* L. cv. Azam) were sterilized with 3% sodium hypochlorite solution. The sterilized seeds were primed with 1000 ppm, 2000 ppm, and 4000 ppm plant-derived smoke for 6 h, and then sown in seedling case (150 mm × 60 mm × 100 mm) supplied with water. The seeds treated with distilled water were used as control. Maize was grown in growth chamber illuminated with white fluorescent light (160 $\mu mol \cdot m^{-2} \cdot s^{-1}$, 16 h light period/day) at 25 °C with 60% humidity. Germination percentage was recorded after two days of sowing. Fifteen seeds were sown for each treatment and 4 independent replicates were used for morphological analysis.

For proteomics and further investigation through immunoblot analysis, enzymatic analysis and qRT-PCR, maize seeds (*Zea mays* L. cv. Honey Bantam) were sterilized using 3% sodium hypochlorite solution, soaked in 2000 ppm plant-derived smoke for 6 h and grown in 450 mL silica sands with water in seedling case (150 mm × 60 mm × 100 mm). The seeds treated with distilled water were used as control. Conditions in growth chamber were illumination with white fluorescent light (160 $\mu mol \cdot m^{-2} \cdot s^{-1}$, 16 h light period/day) at 25 °C with 60% humidity. Shoot was collected on the 4th day after sowing from 3 biological replicates. Shoot raised from the seeds treated with distilled water served as control. Biological replicates mean that maize was sown on different days.

4.3. Protein Extraction

A portion (300 mg) of maize shoot was cut into small pieces and ground 60 times in 2 mL tube. It was ground 30 times after adding 50 μL of lysis buffer containing 7 M urea, 2 M thiourea, 5% CHAPS, and 2 mM tributylphosphine. Furthermore, 50 μL of lysis buffer was added and ground for 30 times. Suspension was incubated for 2 min at 25 °C and centrifuged at 15,000× *g* for 2 min at 25 °C. Afterwards, the filter cartridge was removed and supernatant was collected as total proteins.

4.4. Protein Enrichment, Reduction, Alkylation, and Digestion

Extracted proteins (100 μg) in lysis buffer were adjusted to a final volume of 100 μL. Methanol (400 μL) was added to each sample and mixed before the addition of 100 μL of chloroform and 300 μL of water. After mixing and centrifugation at 20,000× *g* for 10 min to achieve phase separation, the upper phase was discarded and 300 μL of methanol was added to the lower phase, and then centrifuged at 20,000× *g* for 10 min. The pellet was collected as the soluble fraction [68].

Proteins were resuspended in 50 mM NH_4HCO_3, reduced with 50 mM dithiothreitol for 30 min at 56 °C, and alkylated with 50 mM iodoacetamide for 30 min at 37 °C in the dark. Alkylated proteins were digested with trypsin and lysyl endopeptidase (Wako, Osaka, Japan) at a 1:100 enzyme/protein ratio for 16 h at 37 °C. Peptides were desalted with Mono Spin C18 Column (GL Sciences, Tokyo, Japan). Peptides were acidified with formic acid (pH < 3) and analyzed by nano-liquid chromatography (LC) mass spectrometry (MS)/MS.

4.5. Measurement of Protein and Peptide Concentrations

The method of Bradford [69] was used to determine the protein concentration with bovine serum albumin used as the standard.

4.6. Protein Identification Using NanoLC-MS/MS

The peptides were loaded onto the LC system (EASY-nLC 1000; Thermo Fisher Scientific, San Jose, CA, USA) equipped with a trap column (Acclaim PepMap 100 C18 LC column, 3 μm, 75 μm ID × 20 mm; Thermo Fisher Scientific) equilibrated with 0.1% formic acid and eluted with a linear acetonitrile gradient (0–35%) in 0.1% formic acid at a flow rate of 300 nL/min. The eluted peptides were loaded and separated on the column (Easy-Spray C18 LC column, 3 μm, 75 μm ID × 150 mm; Thermo Fisher Scientific) with a spray voltage of 2 kV (Ion Transfer Tube temperature: 275 °C). The peptide ions were detected using MS (Orbitrap Fusion EDT MS; Thermo Fisher Scientific) in the data-dependent acquisition mode with the installed Xcalibur software (version 4.0; Thermo Fisher Scientific). Full-scan mass spectra were acquired in the MS over 375–1500 *m/z* with resolution of 120,000. The most intense precursor ions were selected for collision-induced fragmentation in the linear ion trap at a normalized collision energy of 35%. Dynamic exclusion was employed within 90 s to prevent repetitive selection of peptides [70].

4.7. MS Data Analysis

The MS/MS searches were carried out using the Mascot (version 2.6.1, Matrix Science, London, UK) and SEQUEST HT search algorithms against the UniProtKB Viridiplantae protein database (2017-07) using Proteome Discoverer 2.1 (version 2.1.1.21; Thermo Fisher Scientific). The workflow for both algorithms included a spectrum selector, Mascot, SEQUEST HT search nodes, percolator, ptmTS, event detector, and precursor ion area detector nodes. Oxidation of methionine was set as a variable modification and carbamidomethylation of cysteine was set as a fixed modification. MS and MS/MS mass tolerance were set to 10 ppm and 0.6 Da, respectively. Trypsin was specified as the protease and a maximum of one missed cleavage was allowed. Target-decoy database searches used for calculation of false discovery rate (FDR) and for peptide identification FDR was set at 1%. Label-free quantification was also performed with Proteome Discoverer 2.1 using precursor ions area detector nodes.

4.8. Differential Analysis of Proteins Using MS Data

For differential analysis of the relative abundance of peptides and proteins between samples, the freely software PERSEUS (version 1.6.0.7) [71] was used. Proteins and peptides intensities were transferred into log2 scale. Three biological replicates of each sample were grouped and a minimum of 3 valid values was required in at least one group. Normalization of the intensities was performed to subtract the median of each sample. Missing values were imputed based on a normal distribution (width = 0.3, down-shift = 2.2). Significance was assessed using student's *t*-test analysis. Accession codes is as follows: For MS data, RAW data, peak lists and result files have been deposited in the ProteomeXchange Consortium [72] via the jPOST [73] partner repository under data-set identifiers PXD008315.

4.9. Functional Categorization

The protein sequences of the differentially changed proteins, based on the Lan10 strain, were subjected to a BLAST query against the Ami gene ontology database (http://amigo1.geneontology. org/cgi-bin/amigo/blast.cgi). The corresponding Ami gene ontology terms were extracted from the most homologous proteins using a Perl program. The Ami gene ontology database annotation results were plotted by the Web Gene Ontology Annotation Plot (http://wego.genomics.org.cn/) tool by uploading compiled Web Gene Ontology Annotation Plot native format files containing the obtained Ami gene ontology terms. Functional categorization of identified proteins was performed using MapMan bin codes (http://mapman.gabipd.org/) [74]. Visualization of protein abundance was performed using MapMan software (version 3.6.0 RC1, http://mapman.gabipd.org/web/guest/ mapman) [75]. The software and mapping files of Gmax_109_peptide were also downloaded from the MapMan website.

4.10. Immunoblotting Analysis

Maize shoot (100 mg) sample was ground in an SDS-sample buffer consisting of 60 mM Tris-HCl (pH 6.8), 2% SDS, 10% glycerol, and 5% 2-mercaptoethanol using mortar and pestle [76]. The obtained mixture was centrifuged 2 times at $15,000\times g$ for 10 min and protein was collected as a supernatant. SDS-polyacrylamide gel electrophoresis was used to separate protein (10 µg) in SDS-sample buffer. The separated proteins were shifted to polyvinylidene difluoride membrane using a semi-dry transfer blotter. A buffer containing 137 mM NaCl, 20 mM Tris-HCl (pH 7.5), 0.1% Tween-20, and a blocking solution (Wako) was used to block the blotted membrane for 1 h. Afterwards, different diluted (1:1000) anti-ascorbate peroxidase antibody [77], anti-peroxiredoxin antibody [78], anti-ribulose-1,5-bisphosphate carboxylase/oxygenase (RuBisCO) large and small subunits antibodies [79], and anti-RuBisCO activase antibody [80] was used to incubate the membrane for 1 h. The membrane was washed 3 times with buffer containing 137 mM NaCl, 20 mM Tris-HCl (pH 7.5), and 0.1% Tween-20 and treated for 1 h with anti-rabbit IgG conjugated with horseradish peroxidase (Bio-Rad, Hercules, CA, USA) as secondary antibody. The membrane was incubated with TMB membrane peroxidase substrate system (KPL, Sylacauga, AL, USA). ImageJ software (version 1.46, https://imagej.nih.gov/ij/) was used to calculate the relative intensities of bands.

4.11. Enzymatic Analysis

For ascorbate peroxidase analysis, a sample (1 g) of maize shoot was ground in pestle and mortar with liquid nitrogen. This grinded mixture was homogenized in 50 mM potassium phosphate buffer (pH 7.0) containing 0.5 mM Ascorbic acid, 0.1 mM EDTA and 0.1 mM hydrogen peroxide [81]. The hydrogen peroxide-dependent oxidation of Ascorbic acid was followed by monitoring the decrease in absorbance at 290 nm assuming an absorption coefficient of $2.8 \text{ mM}^{-1}\cdot\text{cm}^{-1}$. For peroxiredoxin analysis, the assay contains 100 mM potassium phosphate-buffer (pH 7.0), 0.3–3 µM peroxiredoxin, 100 µM hydrogen peroxide in a total volume of 1000 µL. The reaction was stopped with 800 µL oftrichloroacetic acid (12.5%) to an aliquot of 50 µL of assay solution. After the addition of 200 µL, 10 mM Fe $(NH_4)_2(SO_4)_2$ and 100 µL of 2.5 M KSCN, the absorbance at 480 nm was measured to quantify the hydrogen peroxide contents of the solution, and hydrogen peroxide reduction rates were calculated [82].

Int. J. Mol. Sci. **2019**, *20*, 1319

4.12. RNA Extraction and Reverse Transcription Polymerase Chain Reaction Analysis

A total of 100 mg maize shoot was ground in mortar and pestle using liquid nitrogen. Total RNAs was extracted by RNeasy plant mini kit (Qiagen, Valencia, CA, USA) from maize shoot powder and treated with RNase-free DNase I during extraction. cDNA was synthesized from the extracted RNAs by using RevertAid first strand cDNA synthesis Kit (Thermo Scientific) in reverse transcription polymerase chain reaction (qRT-PCR). Fast Real Time PCR system (7900HT; Applied Biosystems, Foster City, CA, USA) was used to perform a qRT-PCR reaction at the following conditions; 95 °C for 600 s, followed by 35 cycles of 95 °C for 15 s and 60 °C for 60 s. As an internal control, 18S rRNA was used to normalize the gene expression. For normalization of gene expression, 18S rRNA was used as an internal control. For quantification of specific gene, primers were designed for regions of interest using NCBI tool (primer blast) and Primer 3 online bioinformatics tools. Quantitative variation between different samples was calculated by the relative quantification method ($2^{-\Delta\Delta Ct}$) [83].

4.13. Statistical Analysis

Data were analyzed by one-way ANOVA followed by Tukey's multiple comparison among multiple groups using SPSS (version 22.0; IBM, Armonk, New York, USA). A *p*-value of less than 0.05 was considered to be statistically significant.

5. Conclusions

Gel-free/label-free proteomic technique was used to examine the effects of plant-derived smoke on maize growth after 6 h presoaking. The main results of this study are as follows: (i) Plant-derived smoke increased seed germination and seedling length/ fresh weight at low concentrations; (ii) Nucleotide, starch degradation, and glutathione related proteins were increased; (iii) Protein synthesis/degradation and cell division/organization proteins were changed; (iv) Cell wall, lipids, photosynthetic, and amino acid degradations related proteins ware decreased; (v) plant-derived smoke increased cytoskeleton proteins in maize; (vi) plant-derived smoke did not affect the activity of ascorbate peroxidase and decreased the activity of peroxiredoxin; (vii) gene expression level of peroxiredoxin was altered by plant-derived smoke. These results suggest that plant-derived smoke affects the proteins related to metabolic processes while inhibiting proteins related with lipids, proteins, and cell wall. Furthermore plant-derived smoke regulates the reactive oxygen species and their scavenging system. Although various studies have been conducted demonstrating the promoting effects of plant derived smoke solution on different growth parameters of plants, the molecular response of plant to plant derived smoke solution remained unknown. This study was carried out to fill the gap between various physiological processes regulated by a plant derived smoke solution and the possible mechanism of action behind it.

Supplementary Materials: Supplementary materials can be found at http://www.mdpi.com/1422-0067/20/6/1319/s1. Figure S1. GO categories of proteins with differential abundance in maize treated with plant-derived smoke. Figure S2. Metabolic pathway of proteins identified in maize treated with plant-derived smoke.

Author Contributions: S.K. conceived and designed the experiments; M.M.A. and S.K. performed the experiments; H.Y., K.H., and K.T. performed MS analysis and data analysis; S.K., A.K., X.L., Y.S., and H.M. contributed analysis tools; S.K., M.M.A., A.K., H.Y., K.H., K.T., X.L., Y.S., and H.M. wrote the paper. S.K., M.M.A., A.K., H.Y., K.H., K.T., X.L., Y.S., H.M., S.R., and M.J. read the paper.

Acknowledgments: The authors thank Dr. A. Hashiguchi and Dr. X. Wang in University of Tsukuba for their useful discussion. M.M.A. was supported by scholarship from Higher Education Commission, Pakistan.

Conflicts of Interest: The authors declare no conflict of interest.

Abbreviations

FDR	False discovery rate
LC	Liquid Chromatography
RuBisCO	Ribulose-1,5-bisphosphate carboxylase/oxygenase
MS	Mass Spectrometry

References

1. Lawrence, C.J.; Walbot, V. Translational genomics for bioenergy production from fuel stock grasses: Maize as the model species. *Plant Cell* **2007**, *19*, 2091–2094. [CrossRef]

2. Morari, F.; Meggio, F.; Lunardon, A.; Scudiero, E.; Forestan, C.; Farinati, S.; Varotto, S. Time course of biochemical, physiological, and molecular responses to field-mimicked conditions of drought, salinity, and recovery in two maize lines. Front. *Plant Sci.* **2015**, *6*, 314. [CrossRef] [PubMed]

3. Xiao, Y.; Liu, H.; Wu, L.; Warburton, M.; Yan, J. Genome-wide association studies in maize: Praise and stargaze. *Mol. Plant* **2017**, *10*, 359–374. [CrossRef]

4. Chia, J.M.; Song, C.; Bradbury, P.J.; Costich, D.; Leon, N.; Doebley, J.; Elshire, R.J.; Gaut, B.; Geller, L.; Glaubitz, J.C.; et al. Maize HapMap2 identifies extant variation from a genome in flux. *Nat. Genet.* **2012**, *44*, 803–807. [CrossRef]

5. Schnable, P.S.; Ware, D.; Fulton, R.S.; Stein, J.C.; Wei, F.; Pasternak, S.; Liang, C.; Zhang, J.; Fulton, L.; Graves, T.A.; et al. The B73 Maize Genome: Complexity, Diversity, and Dynamics. *Science* **2009**, *326*, 1112–1114. [CrossRef] [PubMed]

6. Hake, S.; Ross-Ibarra, J. The Natural History of Model Organisms: Genetic, evolutionary and plant breeding insights from the domestication of maize. *eLife* **2015**, *4*, e05861. [CrossRef] [PubMed]

7. Bradshaw, S.D.; Dixon, K.W.; Hopper, S.D.; Lambers, H.; Turner, S.R. Little evidence for fire-adapted plant traits in Mediterranean climate regions. *Trends Plant Sci.* **2011**, *16*, 69–76. [CrossRef]

8. Nelson, D.C.; Flematti, G.R.; Ghisalberti, E.L.; Dixon, K.W.; Smith, S.M. Regulation of seed germination and seedling growth by chemical signals from burning vegetation. *Annu. Rev. Plant Biol.* **2012**, *63*, 107–130. [CrossRef] [PubMed]

9. Flematti, G.R.; Waters, M.T.; Scaffidi, A.; Merritt, D.M.; Ghisalberti, E.L.; Kingsley, W.; Dixon, K.W.; Smith, S.M. Karrikin and cyanohydrin smoke signals provide clues to new endogenous plant signaling compounds. *Mol. Plant* **2013**, *6*, 29–37. [CrossRef] [PubMed]

10. Kulkarni, M.G.; Light, M.E.; van Staden, J. Plant derived smoke: Old technology with possibilities for economic applications in agriculture and horticulture. *S. Afr. J. Bot.* **2011**, *77*, 972–979. [CrossRef]

11. Dixon, K.W.; Roche, S.; Pate, J.S. The promotive effect of smoke derived from burnt vegetation on seed germination of western Australian plants. *Oecologia* **1995**, *101*, 185–192. [CrossRef] [PubMed]

12. van Staden, J.; Sparg, S.G.; Kulkarni, M.G.; Light, M.E. Post germination effects of the smoke-derived compound 3-methyl-2H-furo[2,3-c]pyran-2-one, and its potential as a preconditioning agent. *Field Crop Res.* **2006**, *98*, 98–105. [CrossRef]

13. Kulkarni, M.G.; Street, R.A.; van Staden, J. Germination and seedling growth requirements for propagation of *Dioscorea dregeana* (Kunth) Dur. and Schinz-A tuberous medicinal plant. *S. Afr. J. Bot.* **2007**, *73*, 131–137. [CrossRef]

14. Mavi, K.; Light, M.E.; Demir, I.; van Staden, J.; Yasar, F. Positive effect of smoke-derived butenolide priming on melon seedling emergence and growth. *New Zeal. J. Crop Hort.* **2010**, *38*, 147–155. [CrossRef]

15. Moreira, B.; Tormo, J.; Estrelles, E.; Pausas, J.G. Disentangling the role of heat and smoke as germination cues in Mediterranean Basin flora. *Ann. Bot.* **2010**, *105*, 627–635. [CrossRef] [PubMed]

16. Kumari, A.; Papenfus, H.B.; Kulkarni, M.G.; Posta, M.; van Staden, J. Effect of smoke derivatives on in vitro pollen germination and pollen tube elongation of species from different plant families. *Plant Biol.* **2015**, *17*, 825–830. [CrossRef]

17. Gardner, M.J.; Dalling, K.J.; Light, M.E.; Jager, A.K.; van Staden, J. Does smoke substitute for red light in the germination of light-sensitive lettuce seeds by affecting gibberellin metabolism? *S. Afr. J. Bot.* **2001**, *67*, 636–640. [CrossRef]

18. Egerton-Warburton, L.M. A smoke-induced alteration of the sub-testa cuticle in seeds of the post-fire recruiter, *Emmenanthe penduliflora* Benth (Hydrophyllaceae). *J. Exp. Bot.* **1998**, *49*, 1317–1327. [CrossRef]

19. Kulkarni, M.G.; Amoo, S.O.; Kandari, L.S.; van Staden, J. Seed germination and phytochemical evaluation in seedlings of *Aloe arborescens* Mill. *Plant Biosyst.* **2013**, *148*, 460–466. [CrossRef]

20. Zhou, J.; Da Silva, J.A.T.; Ma, G. Effects of smoke water and karrikin on seed germination of 13 species growing in China. *Cent. Eur. J. Biol.* **2014**, *9*, 1108–1116. [CrossRef]

21. Jamil, M.; Kanwal, M.; Aslam, M.M.; Shakir, K.S.; Malook, I.; Tu, J.; Rehman, S. Effect of plant-derived smoke priming on physiological and biochemical characteristics of rice under salt stress condition. *Aust. J. Crop Sci.* **2014**, *8*, 159–170.

22. Kamran, M.; Latif, K.A.; Waqas, M.; Imran, Q.M.; Hamayun, M.; Kang, S.-M.; Kim, Y.-H.; Kim, M.-J.; Lee, I.J. Effects of plant-derived smoke on the growth dynamics of Barnyard Grass (*Echinochloa crus-galli*). *Acta Agric. Scand. Sect. B Soil Plant Sci.* **2014**, *64*, 121–128. [CrossRef]

23. Aremu, A.O.; Bairu, M.W.; Finnie, J.F.; van Staden, J. Stimulatory role of smoke-water and karrikinolide on the photosynthetic pigment and phenolic contents of micro-propagated 'Williams' bananas. *Plant Growth Regul.* **2012**, *67*, 271–279. [CrossRef]

24. Chiwocha, S.D.S.; Dixon, K.W.; Flematti, G.R.; Ghisalberti, E.L.; Merritt, D.J.; Nelson, D.C.; Riseborough, J.M.; Smith, S.M.; Stevens, J.C. Karrikins: A new family of plant growth regulators in smoke. *Plant Sci.* **2009**, *177*, 252–256. [CrossRef]

25. Chou, Y.-F.; Cox, R.D.; Wester, D.B. Smoke water and heat shock influence germination of shortgrass prairie species. *Rangel. Ecol. Manag.* **2012**, *65*, 260–267. [CrossRef]

26. Soós, V.; Sebestyén, E.; Juhász, A.; Pintér, J.; Light, M.E.; van Staden, J.; Balázs, E. Stress-related genes define essential steps in the response of maize seedlings to smoke-water. *Funct. Integr. Genom.* **2009**, *9*, 231–242. [CrossRef]

27. Aslam, M.M.; Jamil, M.; Khatoon, A.; Hendawy, S.E.; Suhaibani, N.A.; Malook, I.; Rehman, S. Physiological and biochemical responses of maize (*Zea mays* L.) to plant derived smoke solution. *Pak. J. Bot.* **2017**, *49*, 435–443.

28. Sparg, S.G.; Kulkarni, M.G.; van Staden, J. Aerosol smoke and smoke-water stimulation of seedling vigor of a commercial maize cultivar. *Crop Sci.* **2006**, *46*, 1336–1340. [CrossRef]

29. Abu, Y.; Romo, J.T.; Bai, Y.; Coulman, V. Priming seeds in aqueous smoke solutions to improve seed germination and biomass production of perennial forage species. *Can. J. Plant Sci.* **2016**, *96*, 551–563. [CrossRef]

30. Light, M.E.; Gardner, M.J.; van Staden, J. Dual regulation of seed germination by smoke solutions. *Plant Growth Regul.* **2002**, *37*, 135–141. [CrossRef]

31. Aremu, A.O.; Plackova, L.; Novak, O.; Stirk, W.A.; Dolezal, K.; van Staden, J. Cytokinin profiles in ex vitro acclimatized *Eucomis autumnalis* plants pre-treated with smoke-derived karrikinolide. *Plant Cell Rep.* **2016**, *35*, 227–238. [CrossRef] [PubMed]

32. Flematti, G.R.; Ghisalberti, E.L.; Dixon, K.W.; Trengove, R.D. Identification of alkyl substituted 2*H*-furo[2,3-c]pyran-2-ones as germination stimulants present in smoke. *J. Agric. Food Chem.* **2009**, *57*, 9475–9480. [CrossRef] [PubMed]

33. Kulkarni, M.G.; Sparg, S.G.; van Staden, J. Germination and post-germination response of *Acacia* seeds to smoke-water and butenolide, a smoke-derived compound. *J. Arid Environ.* **2007**, *69*, 177–187. [CrossRef]

34. Kulkarni, M.G.; Ascough, G.D.; Verschaeve, L.; Baeten, K.; Arruda, M.P.; van Staden, J. Effect of smoke water and a smoke-isolated butenolide on the growth and genotoxicity of commercial onion. *Sci. Hort.* **2010**, *124*, 434–439. [CrossRef]

35. Abdollahi, M.R.; Mehrshad, B.; Moosavi, S.S. Effect of method of seed treatment with plant derived smoke solutions on germination and seedling growth of milk thistle (*Silybum marianum* L.). *Seed Sci. Technol.* **2011**, *39*, 225–229. [CrossRef]

36. Staiger, C.J. Signalling to the actin cytoskeleton in plants. *Annu. Rev. Plant Biol.* **2000**, *51*, 257–288. [CrossRef] [PubMed]

37. Schmelzer, E. Cell polarization, a crucial process in fungal defense. *Trends Plant Sci.* **2002**, *7*, 411–415. [CrossRef]

38. Sun, T.; Li, S.; Ren, H. Profilin as a regulator of the membrane-actin cytoskeleton interface in plant cells. *Front. Plant Sci.* **2013**. [CrossRef] [PubMed]

39. Pruitt, K.D.; Tatusova, T.; Maglott, D.R. NCBI reference sequences (RefSeq): A curated non-redundant sequence database of genomes, transcripts and proteins. *Nucleic Acids Res.* **2007**, *35*, 61–65. [CrossRef]

40. von Witsch, N.; Baluska, F.C.; Staiger, J.; Volkmann, D. Profilin is associated with the plasma membrane in microspores and pollen. *Eur. J. Cell Biol.* **1998**, *77*, 303–312. [CrossRef]

41. Feller, U.; Anders, I.; Mae, T. Rubiscolytics: Fate of Rubisco after its enzymatic function in a cell is terminated. *J. Exp. Bot.* **2008**, *59*, 1615–1624. [CrossRef] [PubMed]

42. Bota, J.; Medrano, H.; Flexas, J. Is photosynthesis limited by decreased Rubisco activity and RuBP content under progressive water stress? *New Phytol.* **2004**, *162*, 671–681. [CrossRef]

43. Khodadadi, E.; Fakheria, B.A.; Aharizad, S.; Emamjomeha, A.; Norouzic, M.; Komatsu, S. Leaf proteomics of drought-sensitive and -tolerant genotypes of fennel. *Biochim. Biophys. Acta* **2017**, *1865*, 1433–1444. [CrossRef] [PubMed]

44. Tezara, W.; Mitchell, V.; Driscoll, S.P.; Lawlor, D.W. Effects of water deficit and its interaction with CO_2 supply on the biochemistry and physiology of photosynthesis in sunflower. *J. Exp. Bot.* **2002**, *53*, 1781–1791. [CrossRef] [PubMed]

45. Baldrianová, J.; Černý, M.; Novák, J.; Jedelský, P.L.; Divíšková, E.; Brzobohatý, B. Arabidopsis proteome responses to the smoke-derived growth regulator karrikins. *J. Proteom.* **2015**, *120*, 7–20. [CrossRef]

46. Sanahuja, G.; Farré, G.; Bassie, L.; Zhu, C.; Christou, P.; Capell, T. Ascorbic acid synthesis and metabolism in maize are subject to complex and genotype-dependent feedback regulation during endosperm development. *Biotechnol. J.* **2013**, *8*, 1221–1230. [CrossRef] [PubMed]

47. Li, Z.; Su, D.; Lei, B.; Wang, F.; Geng, W.; Pan, G.; Cheng, F. Transcriptional profile of genes involved in ascorbate-glutathione cycle in senescing leaves for an early senescence leaf (esl) rice mutant. *J. Plant Physiol.* **2015**, *176*, 1–15. [CrossRef]

48. Pandey, P.; Singh, J.; Achary, V.M.; Reddy, M.K. Redox homeostasis via gene families of ascorbate-glutathione pathway. *Front. Environ. Sci.* **2015**, *3*, 25. [CrossRef]

49. Selote, D.S.; Chopra, P.K. Drought acclimation confers oxidative stress tolerance by inducing co-ordinated antioxidant defense at cellular and subcellular level in leaves of wheat seedlings. *Physiol. Plant* **2006**, *127*, 494–506. [CrossRef]

50. Kausar, R.; Hossain, Z.; Makino, T.; Komatsu, S. Characterization of ascorbate peroxidase in soybean under flooding and drought stresses. *Mol. Biol. Rep.* **2012**, *39*, 10573–10579. [CrossRef]

51. Scandalios, J.G. Oxidative stress: Molecular perception and transduction of signals triggering antioxidant gene defenses. *Braz. J. Med. Biol. Res.* **2005**, *38*, 995–1014. [CrossRef] [PubMed]

52. Apel, K.; Hirt, H. Reactive oxygen species: Metabolism, oxidative stress, and signal transduction. *Annu. Rev. Plant Biol.* **2004**, *55*, 373–399. [CrossRef] [PubMed]

53. Mittler, R.; Vanderauwera, S.; Gollery, M.; Van Breusegem, F. Reactive oxygen gene network of plants. *Trends Plant Sci.* **2004**, *9*, 490–498. [CrossRef] [PubMed]

54. Asada, K. Ascorbate peroxidase: A hydrogen peroxide-scavenging enzyme in plants. *Physiol. Plant* **1992**, *85*, 235–241. [CrossRef]

55. Baier, M.; Dietz, K.J. Chloroplasts as source and target of cellular redox regulation: A discussion on chloroplast redox signals in the context of plant physiology. *J. Exp. Bot.* **2005**, *56*, 1449–1462. [CrossRef] [PubMed]

56. Waheed, M.A.; Jamil, M.; Khan, M.D.; Shakir, S.K.; Rehman, S.U. Effect of plant-derived smoke solutions on physiological and biochemical attributes of maize (*Zea mays* L.) under salt stress. *Pak. J. Bot.* **2016**, *48*, 1763–1774.

57. Naranjo, B.; Diaz-Espejo, A.; Lindahl, M.; Cejudo, F.J. Type-*f* thioredoxins have a role in the short-term activation of carbon metabolism and their loss affects growth under short-day conditions in *Arabidopsis thaliana*. *J. Exp. Bot.* **2016**, *67*, 1951–1964. [CrossRef]

58. Bharwana, S.A.; Ali, S.; Farooq, M.A.; Iqbal, N.; Abbas, F.; Ahmad, M.S.A. Alleviation of lead toxicity by silicon is related to elevated photosynthesis, antioxidant enzymes suppressed lead uptake and oxidative stress in cotton. *J. Bioremed. Biodeg.* **2013**, *4*, 187. [CrossRef]

59. El-Gaied, L.F.; Abu El-Heba, G.A.; El-Sherif, N.A. Effect of growth hormones on some antioxidant parameters and gene expression in tomato. *GM Crops Food* **2013**, *4*, 67–73. [CrossRef]

60. Pasternak, T.P.; Potters, G.; Caubergs, R.; Jansen, M.A.K. Complementary interactions between oxidative stress and auxins control plant growth responses at plant, organ, and cellular level. *J. Exp. Bot.* **2005**, *56*, 1991–2001. [CrossRef]

61. Correa-Aragunde, N.; Foresi, N.; Delledonne, M.; Lamattina, L. Auxin induces redox regulation of ascorbate peroxidase 1 activity by *S*-nitrosylation/denitrosylation balance resulting in changes of root growth pattern in *Arabidopsis*. *J. Exp. Bot.* **2013**, *64*, 3339–3349. [CrossRef]

62. Hernández, J.A.; Ferrer, M.A.; Jiménez, A.; Ros-Barceló, A.; Sevilla, F. Antioxidant systems and O_2-/H_2O_2 production in the apoplast of Pisum sativum L. leaves: Its relation with NaCl-induced necrotic lesions in minor veins. *Plant Physiol.* **2001**, *127*, 817–831. [CrossRef]

63. Diaz-Vivancos, P.; Faize, M.; Barba-Espin, G.; Faize, L.; Petri, C.; Hernández, J.A.; Burgos, L. Ectopic expression of cytosolic superoxide dismutase and ascorbate peroxidase leads to salt stress tolerance in transgenic plums. *Plant Biotechnol. J.* **2013**, *11*, 976–985. [CrossRef]

64. Foyer, C.H.; Shigeoka, S. Understanding oxidative stress and antioxidant functions to enhance photosynthesis. *Plant Physiol.* **2011**, *155*, 93–100. [CrossRef]

65. Chen, L.; Song, Y.; Li, S.; Zhang, L.; Zou, C.; Yu, D. The role of WRKY transcription factors in plant abiotic stresses. *Biochim. Biophys. Acta* **2012**, *1819*, 120–128. [CrossRef]

66. de Lange, J.H.; Boucher, C. Auto ecological studies on *Audinia capitate* (Bruniaceaae), plant-derived smoke as a germination cue. *S. Afr. J. Bot.* **1990**, *56*, 188–202. [CrossRef]

67. Tieu, A.; Dixon, K.A.; Sivasithamparam, K.; Plummer, J.A. Germination of four species of native Western Australian plant using plant-derived smoke. *Aust. J. Bot.* **1999**, *47*, 207–219. [CrossRef]

68. Komatsu, S.; Han, C.; Nanjo, Y.; Altaf-Un-Nahar, M.; Wang, K.; He, D.; Yang, P. Label-free quantitative proteomic analysis of abscisic acid effect in early-stage soybean under flooding. *J. Proteome Res.* **2013**, *12*, 4769–4784. [CrossRef]

69. Bradford, M.M. A rapid and sensitive method for the quantitation of microgram quantities of protein utilizing the principle of protein-dye binding. *Anal. Biochem.* **1976**, *72*, 248–254. [CrossRef]

70. Zhang, Y.; Wen, Z.; Washburn, M.P.; Florens, L. Effect of dynamic exclusion duration on spectral count based quantitative proteomics. *Anal. Chem.* **2009**, *81*, 6317–6326. [CrossRef]

71. Tyanova, S.; Temu, T.; Siniteyn, P.; Carlson, A.; Hein, Y.; Gieger, T.; Mann, M.; Cox, J. The Perseus computational platform for comprehensive analysis of proteomics data. *Nat. Methods* **2016**, *13*, 731–740. [CrossRef] [PubMed]

72. Vizcaíno, J.A.; Côté, R.G.; Csordas, A.; Dianes, J.A.; Fabregat, A.; Foster, J.M.; Griss, J.; Alpi, E.; Birim, M.; Contell, J.; et al. The proteomics identifications (PRIDE) database and associated tools: Status in 2013. *Nucleic Acids Res.* **2013**, *41*, D1063–D1069. [CrossRef] [PubMed]

73. Okuda, S.; Watanabe, Y.; Moriya, Y.; Kawano, S.; Yamamoto, T.; Matsumoto, M.; Takami, T.; Kobayashi, D.; Araki, N.; Yoshizawa, A.C.; et al. jPOSTrepo: An international standard data repository for proteomes. *Nucleic Acids Res.* **2017**, *45*, D1107–D1111. [CrossRef]

74. Usadel, B.; Nagel, A.; Thimm, O.; Redestig, H.; Blaesing, O.E.; Palacios-Rofas, N.; Selbig, J.; Hannemann, J.; Piques, M.C.; Steinhauser, D.; et al. Extension of the visualization tool MapMan to allow statistical analysis of arrays, display of corresponding genes and comparison with known responses. *Plant Physiol.* **2005**, *138*, 1195–1204. [CrossRef] [PubMed]

75. Usadel, B.; Poree, F.; Nagel, A.; Lohse, M.; Czedik-Eysenberg, A.; Stitt, M. A guide to using MapMan to visualize and compare Omics data in plants: A case study in the crop species, maize. *Plant Cell Environ.* **2009**, *32*, 1211–1229. [CrossRef] [PubMed]

76. Laemmli, U.K. Cleavage of structural proteins during the assembly of the head of bacteriophage T4. *Nature* **1970**, *227*, 680–685. [CrossRef] [PubMed]

77. Komatsu, S.; Yamamoto, A.; Nakamura, T.; Nouri, M.Z.; Nanjo, Y.; Nishizawa, K.; Furukawa, K. Comprehensive analysis of mitochondria in roots and hypocotyls of soybean under stress using proteomics and metabolomics techniques. *J. Proteome Res.* **2011**, *10*, 3993–4004. [CrossRef]

78. Nishizawa, K.; Komatsu, S. Characteristics of soybean 1-cys peroxiredoxin and its behavior in seedlings under flooding stress. *Plant Biotechnol. J.* **2011**, *28*, 83–88. [CrossRef]

79. Hashimoto, M.; Komatsu, S. Proteomic analysis of rice seedlings during cold stress. *Proteomics* **2007**, *7*, 1293–1302. [CrossRef]

80. Komatsu, S.; Masuda, T.; Abe, K. Phosphorylation of a protein (pp56) is related to the regeneration of rice cultured suspension cells. *Plant Cell Physiol.* **1996**, *37*, 748–753. [CrossRef]

81. Nakano, Y.; Asada, K. Hydrogen-peroxide is scavenged by ascorbate-specific peroxidase in spinachchloroplasts. *Plant Cell Physiol.* **1981**, *22*, 867–880.

82. Horling, F.; König, J.; Dietz, K.J. Type II peroxiredoxin C, a member of the peroxiredoxin family of *Arabidopsis thaliana*: Its expression and activity in comparison with other peroxiredoxins. *Plant Physiol. Biochem.* **2002**, *40*, 491–499. [CrossRef]

83. Livak, K.J.; Schmittgen, T.D. Analysis of relative gene expression data using realtime quantitative PCR and the $2^{-\Delta\Delta CT}$ method. *Methods* **2001**, *25*, 402–408. [CrossRef] [PubMed]

International Journal of
Molecular Sciences

MDPI

Article

Proteomics Analysis to Identify Proteins and Pathways Associated with the Novel Lesion Mimic Mutant E40 in Rice Using iTRAQ-Based Strategy

Xiang-Bo Yang [1,2,†], Wei-Long Meng [1,†], Meng-Jie Zhao [3,†], An-Xing Zhang [1], Wei Liu [1], Zhao-Shi Xu [3], Yun-Peng Wang [2,*] and Jian Ma [1,*]

[1] Faculty of Agronomy, Jilin Agricultural University, Changchun 130118, China; yangxiangbo1980@163.com (X.-B.Y.); mengweilongosj@163.com (W.-L.M.); zhanganxingosj@163.com (A.-X.Z.); liuweiosj@163.com (W.L.)
[2] Institute of Agricultural Biotechnology, Jilin Academy of Agricultural Sciences, Changchun 130033, China
[3] Institute of Crop Sciences, Chinese Academy of Agricultural Sciences (CAAS)/National Key Facility for Crop Gene Resources and Genetic Improvement, Key Laboratory of Biology and Genetic Improvement of Triticeae Crops, Ministry of Agriculture, Beijing 100081, China; zhao_mengjie0815@163.com (M.-J.Z.); xuzhaoshi@caas.cn (Z.-S.X.)
* Correspondence: wangypbio@cjaas.com (Y.-P.W.); majian197916@jlau.edu.cn (J.M.); Tel.: +86-0431-87063127 (Y.-P.W.); +86-0431-84532849 (J.M.)
† These authors contributed equally to this work.

Received: 25 February 2019; Accepted: 11 March 2019; Published: 14 March 2019

Abstract: A novel rice lesion mimic mutant (LMM) was isolated from the mutant population of Japonica rice cultivar Hitomebore generated by ethyl methane sulfonate (EMS) treatment. Compared with the wild-type (WT), the mutant, tentatively designated E40, developed necrotic lesions over the whole growth period along with detectable changes in several important agronomic traits including lower height, fewer tillers, lower yield, and premature death. To understand the molecular mechanism of mutation-induced phenotypic differences in E40, a proteomics-based approach was used to identify differentially accumulated proteins between E40 and WT. Proteomic data from isobaric tags for relative and absolute quantitation (iTRAQ) showed that 233 proteins were significantly up- or down-regulated in E40 compared with WT. These proteins are involved in diverse biological processes, but phenylpropanoid biosynthesis was the only up-regulated pathway. Differential expression of the genes encoding some candidate proteins with significant up- or down-regulation in E40 were further verified by qPCR. Consistent with the proteomic results, substance and energy flow in E40 shifted from basic metabolism to secondary metabolism, mainly phenylpropanoid biosynthesis, which is likely involved in the formation of leaf spots.

Keywords: lesion mimic mutant; leaf spot; phenylpropanoid biosynthesis; proteomics; isobaric tags for relative and absolute quantitation (iTRAQ); rice

1. Introduction

Some plant mutants spontaneously exhibit characteristics of typical pathogen infection without any pathogen attack, these mutants are termed lesion mimic mutants (LMMs). Most LMM gene mutations involve regulatory genes of immune responses or promoters of such genes, thereby constitutively expressing pathogenicity-related (PR) genes, causing the production of reactive oxygen species (ROS) and the accumulation of phytoalexins. LMMs are subjected to the development of spontaneous cell death and necrotic lesions. Therefore, LMMs are useful as genetic tools to investigate the molecular mechanisms of programmed cell death (PCD) and hypersensitive response (HR) in plants. Since Sekiguchi discovered the first LMM of rice in 1965 [1], numerous LMMs and the involved genes have been identified in many plant species, such as *Arabidopsis*, maize, groundnut, barley, and wheat [2–9].

Rice is a major staple food crop for a large part of the world's population and an important model monocot plant species for research. Rice has a small genome size and the complete genome sequence is available [10,11]. LMMs in rice have been widely studied and more than 200 LMMs have been reported in rice so far. Most LMM mutation sites occur in the regions of regulatory genes that are resistant to pathogen infection [9,12–16]. These mutant genes are usually recessive genes and only a fraction are dominant or semi-dominant genes [17]. So far, at least 56 genes responsible for rice LMMs have been identified and registered in the Gramene database (http://www.gramene.org, accessed on: 11 August 2018). These LMM genes encode various functional proteins (Table 1).

Table 1. Some proteins encoded by lesion mimic mutant (LMM) genes.

Order	Functional Proteins	References
1	Heat stress transcription factor	[18]
2	U-Box/Armadillo repeat protein	[19]
3	Membrane-associated protein	[3,20]
4	Ion channel	[21]
5	Zinc finger protein	[21]
6	Acyltransferase	[22]
7	Ser/Thr protein kinase	[23]
8	Clathrin associated adaptor protein complex 1 medium subunit 1 (AP1M1)	[24]
9	Putative splicing factor 3b subunit 3 (SF3b3)	[25]
10	Proteins involved in biosynthesis pathways of fatty acids lipids	[26]
11	Aromatic amino acid	[27]
12	Porphyrin	[28]
13	Polyphenol oxidase enzyme in secondary metabolism	[29]

However, studies on LMMs often focus on mining disease resistance genes, relatively few has been reported on the plant immune response process and the optimum growth and development of plants in terms of the distribution of energy and matter. LMMs generally show a series of defective phenotypes such as reduced photosynthesis, decreased biomass accumulation, and severe yield reduction. These phenotypes are not caused by pathogen attack but are due to the abnormal activation of the immune system in LMMs, leading to redistribution of energy and substances in the metabolic pathways in the plants. In order to reveal changes in energy flow of the plant's immune system, we need a more comprehensive proteomics study of LMMs mutants. Proteomics based on two-dimensional gel electrophoresis (2-DE) is one of the most commonly used strategies to identify differentially accumulated proteins between the wild-type (WT) rice and its LMMs. For example, two PR proteins, OsPR5 and OsPR10, and three ROS-scavenging enzymes, catalase (CAT), ascorbate peroxidase (APX), and superoxide dismutase (SOD), were differentially expressed in the blm mutant [30]. Similarly, peroxidase, thaumatin-like protein and probenazole-induced protein (PBZ1) were upregulated in the spl1 mutant [31]. However, the technical limitations of 2-DE make it difficult to identify the LMM-involving proteins at the whole proteome level. Only a limited number of differential proteins have been detected between LMMs and WT by 2-DE (e.g., about 18 by Kim et al. [31], 37 by Tsunezuka et al. [32], 33 by Jung et al. [30], and 159 by Kang et al. [33]) which greatly limits the development of proteome research on LMMs. Given its high resolution and accurate detailed protein expression profiles, iTRAQ combined with LC-MS/MS is widely used, and the accumulation of bioinformatics knowledge has made it possible to rapidly analyze and display data more accurately [34]. iTRAQ-based proteomics has been widely applied in investigating abiotic and biotic stresses response in plants [35,36], but is rarely reported in studies on mechanisms of LMMs.

In this study, a novel rice lesion mimic mutant, E40, was isolated from the EMS mutant population of japonica rice cultivar Hitomebore. We performed proteomic analysis with the iTRAQ method using leaves from E40 plants at the four or five leaf stage to reveal the molecular mechanisms of lesion mimic phenotypes. A total of 2722 proteins were identified, among which 233 proteins were found to be differentially expressed. These proteins were analyzed to increase our understanding of the altered pathways between E40 and WT and the mechanisms involved in formation of lesions.

2. Results

2.1. Phenotypic Characterization of E40

In the greenhouse, E40 plants began to show visible lesion mimic spots in the middle parts of blades at the four- or five-leaf stage with increasing size and density during the later stages of plant development and at tillering stage. Meanwhile, the phenotypes in plant growth and development were significantly different between E40 and WT (Figure 1A,B). Therefore, we hypothesized that the protein difference between E40 and wild-type rice could reach the maximum divergence point during this period, and then the leaves of that stage were taken as samples of the iTRAQ experiment. In the field, the agronomic traits of E40 plants were even worse than indoors. All leaves of the E40 plants were wilted at the late flowering stage, as if they had been burned by sunlight, and plants eventually exhibited early senescence. Compared with the WT plants, the E40 plants displayed abnormal developmental phenotypes and lower agronomic trait values, including significantly reduced dry weight, plant height, number of panicles, grain number, and 1,000-grain weight were significantly reduced compared with WT (Table 2). From the tillering stage, E40 and WT began to show significant differences in dry weight. Until the fruiting period, the difference reached the maximum, nearly three times the difference in the field, and nearly twice the difference in the greenhouse.

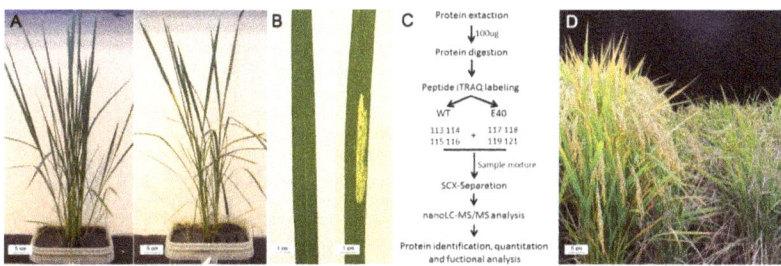

Figure 1. Phenotypic characterization of E40 and the experimental scheme of isobaric tags for relative and absolute quantitation (iTRAQ) analysis. (**A**) the phenotypes of paddy-grown wild type (WT, left) and E40 (right) plants grown in the greenhouse at tillering stage; (**B**) the phenotypes of WT (left) and E40 (right) leaves collected from plants in (**A**) showing the lesion mimic phenotypes of E40; (**C**) experimental scheme of sampling and iTRAQ analysis; (**D**) the phenotypes of WT (left) and E40 (right) grown in the field at the maturity stage.

Table 2. Performance of agronomic traits of E40 cultured in the field and greenhouse.

	Material	Seedling Stage	Dry Weight (g) Tillering Stage	Maturity Stage	Plant Height (cm)	No. of Panicle	Grain Number	1000-Grain Weight (g)
Field	E40	-	-	83.17 ± 7.71 **	98.7 ± 2.42 **	19.4 ± 4.62 *	32.25 ± 1.71 **	17.68 ± 0.45 **
	WT	-	-	247.62 ± 5.42	102.58 ± 1.05	23.2 ± 2.77	134.62 ± 2.84	23.6 ± 0.16
Greenhouse	E40	0.0224 ± 0.0017	4.15 ± 0.13 *	119.36 ± 4.76 **	115.23 ± 1.79 **	21.3 ± 1.72 *	47.05 ± 0.97 **	18.76 ± 0.42 **
	WT	0.0235 ± 0.0021	5.00 ± 0.06	218.67 ± 5.49	117.58 ± 2.16	23.9 ± 1.05	118.13 ± 2.47	22.96 ± 0.24

The data represent the means ± SD. Ten plants of each accession were evaluated for each agronomic trait. *, significance at $p < 0.05$, **, significance at $p < 0.01$.

2.2. Proteomics Analysis of Differentially Abundant Proteins between E40 and WT

The WT and E40 leaves at the tillering stage were harvested for iTRAQ analysis following the procedure shown in Figure 1C and 2722 proteins were identified from 25,887 distinct detected peptides (Additional File 1). Under the screening criteria of fold change greater than 1.5 or less than 0.67 and p value <0.05, a total of 233 proteins were identified to be differentially abundant expressed by comparison between E40 and WT; these proteins were regarded as candidate proteins associated with lesion formation (Table 3 and Supplementary Material: Table S1). Among them, 109 proteins were up-regulated and 124 were repressed.

Table 3. Twenty differentially abundant proteins associated with lesion formation in E40 compared with WT.

Protein ID	Gene Name	Annotation	Log2 Fold Change
Q8S059	SSI2	Stearoyl-[acyl-carrier-protein] 9-desaturase 2	−3.259320177
Q84ZD2	P0534A03.109	Pentatricopeptide repeat-containing protein CRP1 homolog	−2.235369295
Q2QVA7	LOC_Os12g13460	protein-lysine N-methyltransferase activity	−1.957389474
B9F2U5	Os02g0157700	Promotes chloroplast protein synthesis	−1.670452917
P0C512	rbcL	Ribulose bisphosphate carboxylase large chain	−1.525042871
Q69RJ0	GLU	Ferredoxin-dependent glutamate synthase	−1.459112364
O04882	P0421H07.25	Farnesyl diphosphate synthase	−1.418836635
Q5NAI9	P0456F08.15	Putative OsFVE	−1.411837321
Q69X42	P0429G06.10	glycine dehydrogenase (decarboxylating) activity	−1.372611128
Q0JJY1	Os01g0709400	hydrolase activity	−1.329396063
Q9LGB2	P0504H10.32	Putative wound-induced protease inhibitor	1.251971273
Q2QLS7	LOC_Os12g43450	P21 protein, putative	1.325084219
Q7XSU8	OSJNBa0039K24.8	Belongs to the peroxidase family	1.339045312
Q8W084	OSJNBa0091E23.10	Putative pathogenesis-related protein	1.397933437
Q0JR25	RBBI3.3	Bowman-Birk type bran trypsin inhibitor	1.417877593
Q9AWV5	P0044F08.5	serine-type endopeptidase inhibitor activity	1.447168008
Q5WMX0	dip3	Putative chitinase	1.502285455
Q0JMY8	SALT	Salt stress-induced protein	1.596525805
Q75GR1	OSJNBb0065L20.2	*	1.728124034
Q306J3	JAC1	Dirigent protein	1.770280249

Note: *, uncharacteristic protein.

2.3. Gene Ontology (GO) Analysis of Altered Proteins

All the differentially abundant proteins identified were analyzed for gene ontology (GO) using GO Slim and classified as biological processes, cellular component, or molecular function (Figure 2). In the cell component group, the differentially abundant proteins were mainly distributed in the cytoplasm, chloroplast, and plastid. The molecular functional analysis indicated that catalytic proteins and those related to oxidoreductase activity were predominant. In terms of biological processes, identified proteins were mainly involved in metabolic process, regulation of enzyme activity, and response to oxidative stress.

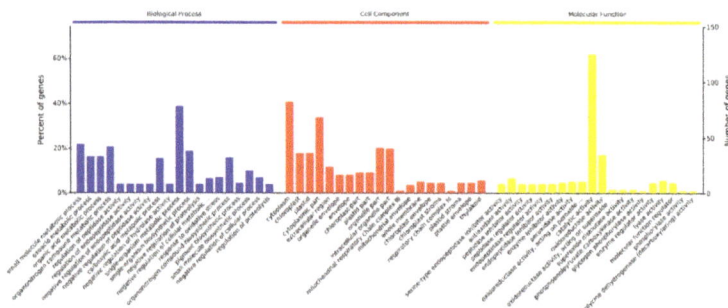

Figure 2. Gene Ontology (GO) distribution analysis. GO analysis could be clustered into three terms: the blue column chart represents biological processes term, the red column represents cellular component term, and the yellow column represents molecular function term. The number of proteins for each GO annotation is shown on the right axis, and the percent of proteins for each GO annotation is on the left axis. P values were calculated using a modified Fisher's exact test and corrected for multiple testing using the Bonferroni correction in Omicsbean.

2.4. Kyoto Encyclopedia of Genes and Genomes (KEGG) Analysis

Pathway analysis using the Kyoto Encyclopedia of Genes and Genomes (KEGG) pathway database (http://www.genome.jp/kegg/pathway.html, accessed on: 08 November 2018) identified 9 pathways ($p < 0.05$) related to these proteins with differential relative abundance, as shown in Figure 3A. Most pathways related to basic metabolism were down-regulated, including carbon fixation, linoleic acid metabolism, carbon metabolism, and amino acid biosynthesis. The only uniquely up-regulated pathway was phenylpropanoid biosynthesis, which is related to secondary metabolism (Figure 3B).

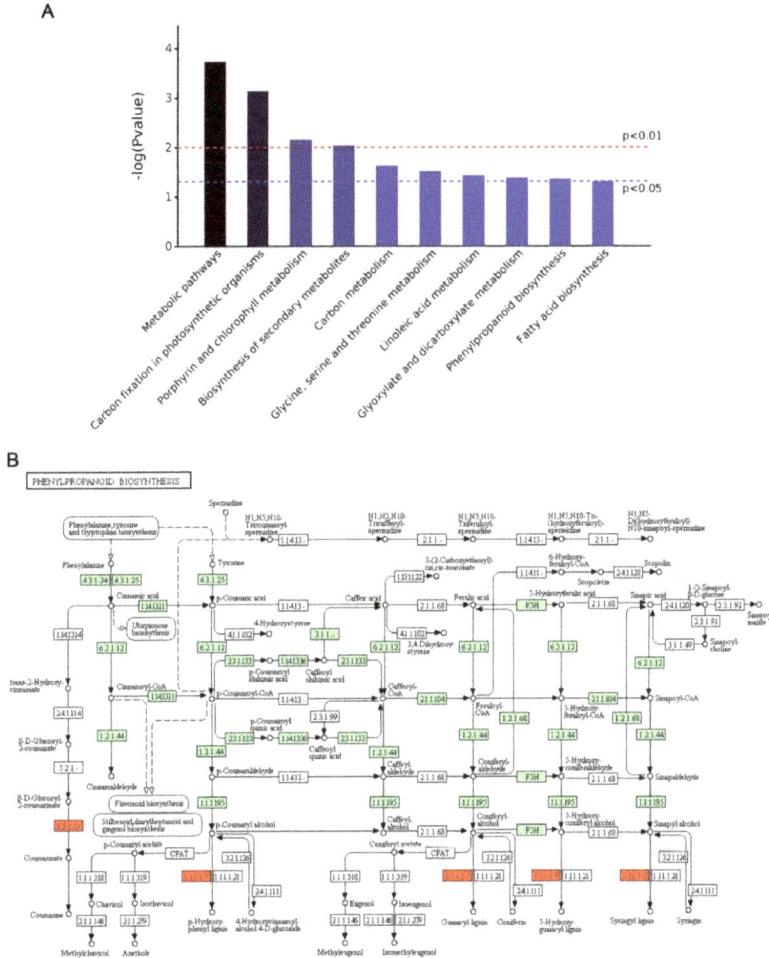

Figure 3. Enriched Kyoto Encyclopedia of Genes and Genomes (KEGG) pathway-based sets and a diagram of phenylpropanoid biosynthesis. (**A**) KEGG pathways which the differentially abundant proteins were enriched. The x-axis shows the proteins involved in the extended KEGG network and pathway. P values were calculated using a modified Fisher's exact test and corrected for multiple tests using the Bonferroni correction in Omicsbean. (**B**) A diagram of phenylpropanoid biosynthesis. Enzymes in red indicate that the corresponding proteins were up-regulated, and those painted green indicate that the proteins were not significantly up- or down-regulated in E40 compared with WT. Up-regulated protein: 1.11.1.7, Peroxidase; 3.2.1.21, Beta-glucosidase.

2.5. Protein-Protein Interaction Analysis

In order to get more information and discover the mechanisms involved in formation of lesion phenotypes, the involved proteins in nine pathways with P values <0.05 and three additional pathways including terpenoid backbone biosynthesis, pyruvate metabolism, and oxidative phosphorylation were analyzed using STRING for protein-protein interaction (PPI) analysis (Figure 4).

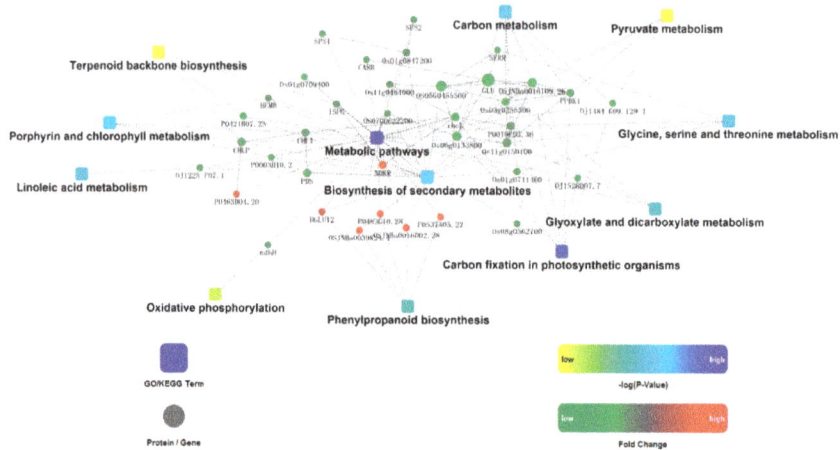

Figure 4. Protein-protein interaction (PPI) network among differentially abundant proteins. The following network model was generated with Cytoscape based on information gained from up to three levels of functional analysis: fold change of gene/protein, protein–protein interaction, and KEGG pathway enrichment. Circle nodes denote genes/proteins, and rectangles denote the KEGG pathway or biological process. P values, mean Pathways related to proteins with differential relative abundance, are colored with gradient colors from yellow to blue from 7.66×10^{-2} to 1.85×10^{-4}. Yellow denotes a low P value and blue denotes a high P value.

The central network showed that proteins such as PPDK1 (Q6AVA8), Glu (Q69RJ0), rbcL (P0C512), CHL (Q53RM0), and OSJNBa0016I09.26 (Q5QNA5) were nodes; they interacted with each other and other proteins related to most pathways including porphyrin and chlorophyll metabolism, carbon fixation in photosynthetic organisms, linoleic acid metabolism, glyoxylate and dicarboxylate metabolism, and terpenoid backbone biosynthesis. Proteins related to phenylpropanoid biosynthesis and oxidative phosphorylation interacted indirectly with the above pathways through basal metabolism and secondary metabolism.

2.6. qRT-PCR Data are Consistent with Protein Abundance Analysis Revealed by iTRAQ

To complement the iTRAQ results, eight genes were chosen for expression assays by real-time PCR (RT-PCR). As shown in Table 4, four genes encoding Q5JMS4, Q7XSV2, Q94DM2, and Q9AS12 involved in phenylpropanoid biosynthesis were all up-regulated, which are consistent with iTRAQ results, showing that the four proteins were more abundant in E40 plants. Besides, the results of qRT-PCR and protein abundance assays were consistent on the two genes involved in porphyrin and chlorophyll metabolism, and two genes involved in carbon fixation in photosynthetic organisms. The consistency between RT-PCR and iTRAQ results indicates the reliability of these data (Table 4).

Table 4. Gene-specific primers used for the real-time PCR.

Pathway	Protein ID	Annotation	Sense Primer	Anti-sense Primer	Log2 Fold Change (E40/WT)	
					iTRAQ	qRT-PCR
Phenylpropanoid biosynthesis	Q5JMS4	Peroxidase	GCCAACACCACCGTCAAC	TGGAAGAACGCCGACTGG	1.16	0.89
	Q7XSV2	Peroxidase	CTCATCCAGGCGTTCAAG	CTTCTTCACCAGCACAGG	0.91	0.74
	Q94DM2	Class III peroxidase 22	TTGTCGTTGGGCTACTAC	AACTTCTCGCTCTTCTCG	0.71	0.46
	Q9AS12	Class III peroxidase 16	TCTTCCTCTTCTTCGCCTTC	ACGCCGCTGTTGTTCTTG	0.97	0.88
Porphyrin and chlorophyll metabolism	Q5Z8V9	Delta-aminolevulinic acid dehydratase	ATTCCAGGAGACCACCATC	CATCACGAGACTTGTAGACC	−0.62	−0.33
	Q6Z2T6	Geranylgeranyl diphosphate reductase	AGGAAGGTGAGGAAGATG	CAGGAAGAGACCATTGAC	−0.75	−0.42
Carbon fixation in photosynthetic organisms	P0C512	Ribulose bisphosphate carboxylase large chain	GGCAGCATTCCGAGTAAC	AAGTCCATCAGTCCAAACAG	−1.53	−1.57
	Q9SNK3	Glyceraldehyde-3-phosphate dehydrogenase	GCGAAGAAGGTCATCATCAC	GAGCGAGGCAGTTGGTTG	−0.84	−0.75

3. Discussion

Like other LMMs, LMM rice E40 was obtained by EMS treatment and displayed lower height, fewer tillers, lower yield, and premature death compared to WT. A total of 233 differentially abundant proteins were identified containing 109 up-regulated proteins and 124 down-regulated proteins. KEGG analysis showed all these 233 proteins were enriched in 9 signal pathways ($p < 0.05$) including metabolic pathways, carbon fixation in photosynthetic organisms, porphyrin and chlorophyll metabolism, biosynthesis of secondary metabolites, carbon metabolism, glycine, serine and threonine metabolism, linoleic acid metabolism, glyoxylate and dicarboxylate metabolism, and phenylpropanoid biosynthesis. Almost all the pathways were down-regulated except for phenylpropanoid pathway.

Glycine, serine and threonine are indispensable amino acid in plants and participate in multiple synthesis of important substances, and the change of glycine, serine and threonine metabolism could affect the basic substance synthesis and metabolic processes. Numerous serine-threonine protein kinases universally excited in plants and could regulate signaling transduction pathways via phosphorylated threonine and serine of target proteins [37,38]. Glycine could promote the absorption of phosphorus, improve plant stress resistance, and promote plant growth, especially photosynthesis. Serine is related to lignin biosynthesis, PCD, and plant aging [39]. And transport factors also have impact on expression of relative genes involving in transduction pathways [40]. Abnormal metabolism could lead to disturbance of signaling pathway. In our study, it is worth noting that the expression levels of chitinase and some enzymes related to lignin biosynthesis were enhanced in E40. Mauch et al. [41] found that chitinase function in defense against fungus, which could degrade fungal cell walls. Lignin is accumulated and deposited in the cell wall, and this accumulation could enhance the ability of the cell wall to resist the invasion of pathogenic microorganisms and provide mechanized protection for the plasma membrane-wrapped protoplasm [42]. Lignin biosynthesis can thus be triggered as a response to various biotic and abiotic stresses in cells. Evidence has shown that lignin biosynthesis genes play crucial roles in basal defense and normal growth of plants [43]. In addition, salt stress induced protein was also up-regulated (Table 3). Thus, we inferred that the up-regulation of chitinase, enzymes related to lignin biosynthesis, and salt stress induced protein were caused by the immune disorder in E40.

EMS treatment could cause the mutation of specific plant immune regulatory genes. This resulted in abnormal expression of plant immune response-related genes, causing an increase in ROS in vivo, accompanied by intracellular peroxides. Plant secondary metabolic pathways significantly increased in E40. The only up-regulated pathway in E40 was the phenylpropanoid pathway (Figure 4). Phenylpropanoids are involved in biotic stress response in plants and the phenylpropanoid pathway plays a critical role in the plant innate immune system, which produces a variety of secondary metabolites, such as flavonoids, isoflavonoids, lignin, anthocyanin, phytoalexins, and phenolic esters; all of these are critical in development, structural protection, defense responses, and tolerance to abiotic stimuli [44,45]. Previous studies showed that enzymes in the phenylpropanoid biosynthetic pathway are associated with PCD [30,46]. Thus, it most likely leads to formation of PCD and the lesion mimic phenotype.

The main carbon source for the phenylpropanoid pathway is phenylalanine, which is synthesized by erythrose-4-phosphate (E4P) from the pentose phosphate pathway and phosphoenolpyruvate (PEP) from the glycolysis pathway. E4P and PEP are also important intermediates for maintaining normal operation of basic metabolism through a stable supply of both material and energy. A strengthened phenylpropanoid pathway means a strong demand for E4P and PEP, which will affect normal metabolism in E40. More PEP from glycolysis leads to less pyruvate and means fewer carbon skeletons are generated for amino acid biosynthesis and energy release from the following tricarboxylic acid cycle (TCA) and that there are fewer substrates for carbon dioxide fixation in photosynthesis (in C4 plants). Demand for more E4P from the pentose phosphate pathway leads to greater demands of pentose, meaning less ribose-5-phosphate for nucleotide formation and resulting in deficient biosynthesis of adenosine triphosphate (ATP), coenzyme A (CoA), nicotinamide adenine dinucleotide (NAD), Flavine adenine dinucleotide (FAD), RAN, and DNA, and fewer ribulose-5-phosphate for carbon fixation in photosynthesis as well.

An important and complex consequence of an abnormal increase in the phenylpropanoid pathway lies in the effects on porphyrin and chlorophyll metabolism. On the one hand, as the biosynthesis of chlorophyll begins from glutamic acid transformed from α-ketoglutarate, an intermediate product of TCA, the increased phenylpropanoid pathway results in imbalance of the TCA cycle, leading to an energy supply shortage and insufficient substrate for porphyrin and chlorophyll synthesis. Tight regulation of chlorophyll biosynthesis and degradation is essential to cell survival [47]. Disruption of chlorophyll biosynthesis at different stages can lead to lesion-mimic phenotypes due to the abnormal accumulation of photoreactive molecules, like porphobilinogen [48], coproporphyrinogen III [28], uroporphyrinogen III [49], and protochlorophyllide [50]. On the other hand, the insufficient chlorophyll supply leads to low energy supply (from photosystem I and II) for synthesizing photosynthetic proteins. Combined with the shortage of PEP for carbon fixation, the turnover of damaged photosynthetic proteins in photoinhibition is suppressed; this results in ROS accumulation. Accumulation of ROS causes the loss of chloroplast integrity followed by rupture of the central vacuole and, finally, cell death [51]. Therefore, ROS regulators, e.g., peroxidases, oxidoreductases, and antioxidant proteins, are increased in E40, a typical feature of LMMs.

Carbon fixation is an important biological process in plant photosynthesis [52]. Porphyrin is an intermediate substance of biosynthesis of chlorophyll [53], both of which are closely related to the photosynthesis efficiency and directly affect plant biomass. E40 displayed lower height, fewer tillers, lower yield is most likely to result from the down-regulated carbon fixation pathway and porphyrin and chlorophyll metabolism. Moreover, PPI analysis showed that terpenoid backbone biosynthesis are linked to porphyrin and chlorophyll metabolism, carbon fixation in photosynthetic organisms, linoleic acid metabolism, glyoxylate and dicarboxylate metabolism through the regulation of expression of node proteins (commonly protease). Many important growth regulators, such as ABA, carotenoids, and vitamin A, belonged to terpenoid, which play essential roles in seed generation, plant growth and development. Up-regulation of the phenylpropanoid biosynthesis pathway was accompanied by down-regulation of other pathways. We therefore concluded that the reason for lesion mimic in E40 is due to substance and energy flow shifting from basic metabolism to secondary metabolism, especially, phenylpropanoid (lignin) biosynthesis. As shown in Figure 5, a mutation caused abnormal HR that triggers: (1) a shift in substance and energy flow from basic metabolism to secondary metabolism, followed by cell starvation with suppression/interruption of basic metabolism such as porphyrin and chlorophyll synthesis, disordered metabolism directly or indirectly (via ROS accumulation) results in PCD, and, finally, formation of lesion mimic; (2) accumulation of ROS (also accumulated from disordered metabolism caused by cell starvation), the ROS then directly and/or indirectly (via organelle rupture) lead to PCD and finally formation of lesion mimic.

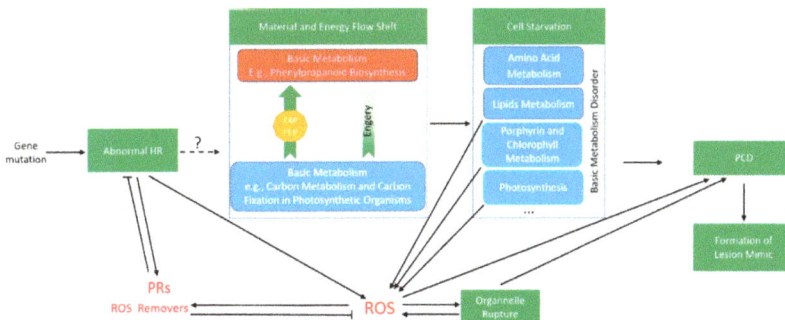

Figure 5. Putative mechanism of formation of lesion mimic in E40. HR, hypersensitive response; PCD, programmed cell death; PRs, pathogen-related proteins; ROS, reactive oxygen species.

4. Materials and Methods

4.1. Plant Materials

The novel rice lesion mimic mutant E40 was identified among 4500 mutant lines generated by EMS treatment of the northern Japan japonica rice cultivar Hitomebore. E40 has been self-bred for five generations and stably displays the target trait in greenhouse and field conditions. Agronomic traits of WT and E40 were determined while M5 generation mutant and WT were grown in a paddy field in Changchun, Jilin Province, China in 2016. Seeds of M5 generation mutant and WT were planted in a greenhouse at 28/24 °C (day/night). At the tillering stage when the lesions were large enough, pictures were taken, and leaves were harvested for proteomic analysis.

4.2. Protein Extraction

Approximately 100 mg of leaves from E40 and WT were thoroughly ground to a fine powder in liquid nitrogen, added with 1:10 (w/v) Lysis buffer (pH 8.5), containing 2 M thiourea, 7 M urea and 4% CHAPS with protease inhibitors (Sigma, St. Louis, MO, USA), for protein extraction at room temperature. The mixture was sonicated for 60 s to obtain more soluble proteins. The plant residue was removed by low temperature centrifugation, and the supernatant was transferred to a 50 cm^3 tube containing 4 volumes of 10% (w/v) trichloroacetic acid (TCA)/acetone, mixed and stored at −20 °C overnight. The precipitated protein was collected by centrifugation at 40,000× g for 10 min at 4 °C and washed three times with cold acetone, and finally lyophilized. After removal of acetone protein was resuspended in lysis buffer. Protein concentration was determined by was detected by Nano Photometer spectrophotometer (IMPLEN, Westlake Village, CA, USA) Agilent Bioanalyzer 2100 system (Agilent Technologies, Santa Clara, CA, USA). The protein samples were stored at −80 °C.

4.3. Trypsin Digestion and iTRAQ Labeling

Each sample from different sources was treated in the manner described in the iTRAQ protocol (AB SCIEX, Redwood City, CA, USA), approximately 100 μg total protein was taken, centrifuged at 100,000× g for 15 min at 4 °C, dried and resuspended in 50 mm^3 of lysis buffer. After reduction and cysteine-blocking, the proteins were digested with sequencing grade trypsin (50 ng/mm^3) for 12 h at 37 °C, and finally 150 mm^3 of an ethanol solution in which the iTRAQ reaction solution was dissolved was added to the reaction solution for labeling.

In the present study, an experiment setting of 4:4 (eight-plex) was selected. The four biological replicates of leaves from WT were labeled with 113, 114, 115, and 116 tags, and the four biological replicates of leaves from E40 were labeled with 117, 118, 119, and 121 tags (Figure 1C). After incubation at room temperature for 2 h and termination of the labeling reaction, the labeled samples were then mixed and dried with a rotary vacuum concentrator.

4.4. LC-MS/MS and Bioinformatics Analysis

The labeled samples were separated at 0.3 cm^3/min with a nonlinear binary gradient, and segments were prepared for LC-MS/MS analysis. Rare data acquisition was performed with a Triple TOF 5600 System (AB SCIEX, Redwood City, CA, USA) fitted with a Nanospray III source (AB SCIEX, Redwood City, CA, USA) and a pulled quartz tip as the emitter (New Objectives, Woburn, MA, USA), and protein identification and quantification were performed with Protein Pilot Software v. 5.0 (AB SCIEX, Redwood City, CA, USA) against the *Oryza sativa* database (https://www.ncbi.nlm.nih.gov/protein/?term=Oryza%20sativa, accessed on: 15 October 2018) using the Paragon algorithm. Detailed methods and parameter settings for these experiments followed the protocol detailed on the website https://www.ebi.ac.uk/pride/archive/projects/PXD005731, (accessed on: 15 October 2018).

In order to narrow down the protein number and focus on the most significant proteins, the screening criteria of differential proteins were: fold change greater than 1.5 or less than 0.67 and *p* value <0.05. The mean value of several repetitions was calculated using two group samples relative quantitative value. Bioinformatics analysis was conducted using Quick GO (http://www.ebi. ac.uk/QuickGO/, EMBL-EBI, Wellcome Genome Campus, Hinxton, Cambridgeshire, UK, accessed on: 22 November 2018), Ingenuity Pathway Analysis software (http://www.polyomics.gla.ac.uk/ resource-ipa.html, Glasgow Polyomics, College of Medical, Veterinary and Life Sciences, accessed on: 30 November 2018), UniProt (http://www.uniprot.org/, Centre Medical Universitaire 1, rue Michel Servet, 1211 Geneva 4 Switzerland, accessed on: 13 March 2019), STRING (http://string-db.org/, ELIXIR, Wellcome Genome Campus, Hinxton, Cambridgeshire, UK, accessed on: 24 February 2019) and OmicsBean (http://www.omicsbean.cn/, Gene for health, Shanghai, China, accessed on: 18 December 2018).

4.5. qRT-PCR Analysis

Total cellular RNA was extracted using TransZol Plant Kit (Transgen Biotech, Beijing, China) and treated with TransScript All-in-One First-Strand cDNA Synthesis SuperMix for qPCR Kit (Transgen Biotech, Beijing, China) for cDNA production. qRT-PCR was carried out using SYBR Green master mix (Transgen Biotech, Beijing, China) and specific primer sets (Table 4). Amplification reactions were performed under the following conditions: 2 min at 50 °C, 10 min at 95 °C, 40 cycles for 15 s at 95 °C, 1 min at 60 °C. Relative transcript levels were calculated using the $2^{-\Delta\Delta CT}$ method as specified by the manufacturer. The relative expression values of the targeted gene were normalized to the expression value of the glyceraldehyde-3-phosphate dehydrogenase (*GAPDH*) gene.

5. Conclusions

In summary, we carried out proteomics analysis to understand cell death and identify proteins activated in LMM mutant E40. A total of 233 proteins, screened from 2722 proteins identified using iTRAQ, exhibited differential abundance. The data was complemented by qRT-PCR analysis with randomly selected genes that encode differentially abundant proteins. The number of proteins identified in this study is larger than other reports about LMMs using the 2-DE method. Identified proteins are involved in diverse biological processes. Consistent with the proteomics results, we speculated that substance and energy flow shifted in E40 from basic metabolism to secondary metabolism, mainly phenylpropanoid biosynthesis, which is the main reason for formation of leaf spots.

Supplementary Materials: Supplementary materials can be found at http://www.mdpi.com/1422-0067/20/6/ 1294/s1.

Author Contributions: Data curation, Y.-P.W. and J.M.; Investigation, X.-B.Y., W.-L.M., M.-J.Z., A.-X.Z., W.L., Z.-S.X., Y.-P.W. and J.M.; Writing–original draft, Y.-P.W. and J.M.; Writing–review and editing, J.M.

Funding: This study was supported by funds from the National Key R & D program of China (2016YFD0102000), the National Natural Science Foundation of China (NSFC31501235), and the Jilin Province Science and Technology Innovation Talent Cultivation Program Excellent Youth Talent Fund Project (20170520075JH).

Conflicts of Interest: The authors declare that they have no competing interests.

References

1. Kiyosawa, S. Inheritance of a particular sensitivity of the Rice variety, Sekiguchi Asahi, to pathogens and chemicals, and linkage relationship with blast resistance genes. *Tokyo Nat. Inst. Agric. Sci. Bull. Ser. D* **1970**, *21*, 61–72.
2. Badigannavar, A.; Kale, D.; Eapen, S.; Murty, G. Inheritance of disease lesion mimic leaf trait in groundnut. *J. Hered.* **2002**, *93*, 50–52. [CrossRef] [PubMed]
3. Büschges, R.; Hollricher, K.; Panstruga, R.; Simons, G.; Wolter, M.; Frijters, A.; van Daelen, R.; van der Lee, T.; Diergaarde, P.; Groenendijk, J. The barley Mlo gene: A novel control element of plant pathogen resistance. *Cell* **1997**, *88*, 695–705. [CrossRef]
4. Dietrich, R.A.; Delaney, T.P.; Uknes, S.J.; Ward, E.R.; Ryals, J.A.; Dangl, J.L. Arabidopsis mutants simulating disease resistance response. *Cell* **1994**, *77*, 565–577. [CrossRef]

5. Fekih, R.; Tamiru, M.; Kanzaki, H.; Abe, A.; Yoshida, K.; Kanzaki, E.; Saitoh, H.; Takagi, H.; Natsume, S.; Undan, J.R. The rice (*Oryza sativa* L.) LESION MIMIC RESEMBLING, which encodes an AAA-type ATPase, is implicated in defense response. *Mol. Genet. Genom.* **2015**, *290*, 611–622. [CrossRef] [PubMed]

6. Greenberg, J.T. Programmed cell death in plant-pathogen interactions. *Annu. Rev. Plant Biol.* **1997**, *48*, 525–545. [CrossRef] [PubMed]

7. Greenberg, J.T.; Guo, A.; Klessig, D.F.; Ausubel, F.M. Programmed cell death in plants: A pathogen-triggered response activated coordinately with multiple defense functions. *Cell* **1994**, *77*, 551–563. [CrossRef]

8. Johal, G.S.; Hulbert, S.H.; Briggs, S.P. Disease lesion mimics of maize: A model for cell death in plants. *Bioessays* **1995**, *17*, 685–692. [CrossRef]

9. Takahashi, A.; Kawasaki, T.; Henmi, K.; Shii, K.; Kodama, O.; Satoh, H.; Shimamoto, K. Lesion mimic mutants of rice with alterations in early signaling events of defense. *Plant J.* **1999**, *17*, 535–545. [CrossRef]

10. Goff, S.A.; Ricke, D.; Lan, T.-H.; Presting, G.; Wang, R.; Dunn, M.; Glazebrook, J.; Sessions, A.; Oeller, P.; Varma, H. A draft sequence of the rice genome (*Oryza sativa* L. ssp. japonica). *Science* **2002**, *296*, 92–100. [CrossRef] [PubMed]

11. Yu, J.; Hu, S.; Wang, J.; Wong, G.K.-S.; Li, S.; Liu, B.; Deng, Y.; Dai, L.; Zhou, Y.; Zhang, X. A draft sequence of the rice genome (*Oryza sativa* L. ssp. indica). *Science* **2002**, *296*, 79–92. [CrossRef] [PubMed]

12. Yin, Z.; Chen, J.; Zeng, L.; Goh, M.; Leung, H.; Khush, G.S.; Wang, G.-L. Characterizing rice lesion mimic mutants and identifying a mutant with broad-spectrum resistance to rice blast and bacterial blight. *Mol. Plant-Microbe Interact.* **2000**, *13*, 869–876. [CrossRef] [PubMed]

13. Mizobuchi, R.; Hirabayashi, H.; Kaji, R.; Nishizawa, Y.; Yoshimura, A.; Satoh, H.; Ogawa, T.; Okamoto, M. Isolation and characterization of rice lesion-mimic mutants with enhanced resistance to rice blast and bacterial blight. *Plant Sci.* **2002**, *163*, 345–353. [CrossRef]

14. Jung, Y.-H.; Lee, J.-H.; Agrawal, G.K.; Rakwal, R.; Kim, J.-A.; Shim, J.-K.; Lee, S.-K.; Jeon, J.-S.; Koh, H.-J.; Lee, Y.-H. The rice (*Oryza sativa*) blast lesion mimic mutant, blm, may confer resistance to blast pathogens by triggering multiple defense-associated signaling pathways. *Plant Physiol. Biochem.* **2005**, *43*, 397–406. [CrossRef] [PubMed]

15. Zhang, H.F.; Xu, W.G.; Wang, H.W.; Hu, L.; Li, Y.; Qi, X.L.; Zhang, L.; Li, C.X.; Hua, X. Pyramiding expression of maize genes encoding phosphoenolpyruvate carboxylase (PEPC) and pyruvate orthophosphate dikinase (PPDK) synergistically improve the photosynthetic characteristics of transgenic wheat. *Protoplasma* **2014**, *251*, 1163–1173. [CrossRef] [PubMed]

16. Zhao, J.; Liu, P.; Li, C.; Wang, Y.; Guo, L.; Jiang, G.; Zhai, W. LMM5. 1 and LMM5. 4, two eukaryotic translation elongation factor 1A-like gene family members, negatively affect cell death and disease resistance in rice. *J. Genet. Genom.* **2017**, *44*, 107–118. [CrossRef] [PubMed]

17. Huang, X.; Li, J.; Bao, F.; Zhang, X.; Yang, S. A gain-of-function mutation in the Arabidopsis disease resistance gene RPP4 confers sensitivity to low temperature. *Plant Physiol.* **2010**, *154*, 796–809. [CrossRef] [PubMed]

18. Yamanouchi, U.; Yano, M.; Lin, H.; Ashikari, M.; Yamada, K. A Rice Spotted Leaf Gene, Spl7, Encodes a Heat Stress Transcription Factor Protein. *Proc. Natl. Acad. Sci. USA.* **2002**, *99*, 7530–7535. [CrossRef] [PubMed]

19. Zeng, L.-R.; Qu, S.; Bordeos, A.; Yang, C.; Baraoidan, M.; Yan, H.; Xie, Q.; Nahm, B.H.; Leung, H.; Wang, G.-L. Spotted leaf11, a negative regulator of plant cell death and defense, encodes a U-box/armadillo repeat protein endowed with E3 ubiquitin ligase activity. *Plant Cell* **2004**, *16*, 2795–2808. [CrossRef]

20. Lorrain, S.; Lin, B.; Auriac, M.C.; Kroj, T.; Saindrenan, P.; Nicole, M.; Balague, C.; Roby, D. vascular associated death1, a novel GRAM domain–containing protein, is a regulator of cell death and defense responses in vascular tissues. *Plant Cell* **2004**, *16*, 2217–2232. [CrossRef] [PubMed]

21. Mosher, S.; Moeder, W.; Nishimura, N.; Jikumaru, Y.; Joo, S.H.; Urquhart, W.; Klessig, D.F.; Kim, S.K.; Nambara, E.; Yoshioka, K. The lesion-mimic mutant cpr22 shows alterations in abscisic acid signaling and abscisic acid insensitivity in a salicylic acid-dependent manner. *Plant Physiol.* **2010**, *152*, 1901–1913. [CrossRef]

22. Kachroo, A.; Lapchyk, L.; Fukushige, H.; Hildebrand, D.; Klessig, D.; Kachroo, P. Plastidial fatty acid signaling modulates salicylic acid–and jasmonic acid–mediated defense pathways in the Arabidopsis ssi2 mutant. *Plant Cell* **2003**, *15*, 2952–2965. [CrossRef] [PubMed]

23. Takahashi, A.; Agrawal, G.K.; Yamazaki, M.; Onosato, K.; Miyao, A.; Kawasaki, T.; Shimamoto, K.; Hirochika, H. Rice Pti1a negatively regulates RAR1-dependent defense responses. *Plant Cell* **2007**, *19*, 2940–2951. [CrossRef]

Int. J. Mol. Sci. **2019**, *20*, 1294

24. Qiao, Y.; Jiang, W.; Lee, J.; Park, B.; Choi, M.S.; Piao, R.; Woo, M.O.; Roh, J.H.; Han, L.; Paek, N.C. SPL28 encodes a clathrin-associated adaptor protein complex 1, medium subunit μ1 (AP1M1) and is responsible for spotted leaf and early senescence in rice (*Oryza sativa*). *New Phytol.* **2010**, *185*, 258–274. [CrossRef]

25. Chen, X.; Hao, L.; Pan, J.; Zheng, X.; Jiang, G.; Jin, Y.; Gu, Z.; Qian, Q.; Zhai, W.; Ma, B. SPL5, a cell death and defense-related gene, encodes a putative splicing factor 3b subunit 3 (SF3b3) in rice. *Mol. Breed.* **2012**, *30*, 939–949. [CrossRef]

26. Brodersen, P.; Petersen, M.; Pike, H.M.; Olszak, B.; Skov, S.; Ødum, N.; Jørgensen, L.B.; Brown, R.E.; Mundy, J. Knockout of Arabidopsis accelerated-cell-death11 encoding a sphingosine transfer protein causes activation of programmed cell death and defense. *Genes Dev.* **2002**, *16*, 490–502. [CrossRef]

27. Zeng, Y.; Ma, L.; Ji, Z.; Wen, Z.; Li, X.; Shi, C.; Yang, C. Fine mapping and candidate gene analysis of LM3, a novel lesion mimic gene in rice. *Biologia* **2013**, *68*, 82–90. [CrossRef]

28. Sun, C.; Liu, L.; Tang, J.; Lin, A.; Zhang, F.; Fang, J.; Zhang, G.; Chu, C. RLIN1, encoding a putative coproporphyrinogen III oxidase, is involved in lesion initiation in rice. *J. Genet. Genom./Yi Chuan Xue Bao* **2011**, *38*, 29–37. [CrossRef] [PubMed]

29. Araji, S.; Grammer, T.A.; Gertzen, R.; Anderson, S.D.; Mikulic-Petkovsek, M.; Veberic, R.; Phu, M.L.; Solar, A.; Leslie, C.A.; Dandekar, A.M.; et al. Novel roles for the polyphenol oxidase enzyme in secondary metabolism and the regulation of cell death in walnut. *Plant Physiol* **2014**, *164*, 1191–1203. [CrossRef]

30. Jung, Y.; Rakwal, R.; Agrawal, G.; Shibato, J.; Kim, J.; Lee, M.; Choi, P.; Jung, S.; Kim, S.; Koh, H. Differential expression of defense/stress-related marker proteins in leaves of a unique rice blast lesion mimic mutant (blm). *J. Proteome Res.* **2006**, *5*, 2586–2598. [CrossRef]

31. Kim, S.; Kim, S.; Kang, Y.; Wang, Y.; Kim, J.; Yi, N.; Kim, J.; Rakwal, R.; Hj Kang, K. Proteomics analysis of rice lesion mimic mutant (spl11) reveals tightly localized Probenazole-Induced protein (PBZ1) in cells undergoing programmed cell death. *J. Proteome Res.* **2008**, *7*, 1750–1760. [CrossRef]

32. Tsunezuka, H.; Fujiwara, M.; Kawasaki, T.; Shimamoto, K. Proteome analysis of programmed cell death and defense signaling using the rice lesion mimic mutant cdr2. *Mol. Plant-Microbe Interact.* **2005**, *18*, 52. [CrossRef] [PubMed]

33. Kang, S.G.; Matin, M.N.; Bae, H.; Natarajan, S. Proteome analysis and characterization of phenotypes of lesion mimic mutant spotted leaf 6 in rice. *Proteomics* **2007**, *7*, 2447–2458. [CrossRef] [PubMed]

34. Luo, R.; Zhao, H. Protein quantitation using iTRAQ: Review on the sources of variations and analysis of nonrandom missingness. *Stat. Interface* **2012**, *5*, 99–107. [CrossRef]

35. Chen, T.; Zhang, L.; Shang, H.; Liu, S.; Peng, J.; Gong, W.; Shi, Y.; Zhang, S.; Li, J.; Gong, J. iTRAQ-Based Quantitative Proteomic Analysis of Cotton Roots and Leaves Reveals Pathways Associated with Salt Stress. *PLoS ONE* **2016**, *11*, e0148487. [CrossRef] [PubMed]

36. Wang, B.; Hajano, J.U.D.; Ren, Y.; Lu, C.; Wang, X. iTRAQ-based quantitative proteomics analysis of rice leaves infected by Rice stripe virus reveals several proteins involved in symptom formation. *Virol. J.* **2015**, *12*, 99. [CrossRef] [PubMed]

37. Dudek, H.; Datta, S.R.; Franke, T.F.; Birnbaum, M.J.; Yao, R.J.; Cooper, G.M.; Segal, R.A.; Kaplan, D.R.; Greenberg, M.E. Regulation of neuronal survival by the serine-threonine protein kinase Akt. *Science* **1997**, *275*, 661–665. [CrossRef] [PubMed]

38. Roth, R.; Chiapello, M.; Montero, H.; Gehrig, P.; Grossmann, J.; O'Holleran, K.; Hartken, D.; Walters, F.; Yang, S.Y.; Hillmer, S.; et al. A rice Serine/Threonine receptor-like kinase regulates arbuscular mycorrhizal symbiosis at the peri-arbuscular membrane. *Nat. Commun.* **2018**, *9*, 4677. [CrossRef]

39. Wagstaff, C.; Leverentz, M.K.; Griffiths, G.; Thomas, B.; Chanasut, U.; Stead, A.D.; Rogers, H.J. Cysteine protease gene expression and proteolytic activity during senescence of Alstroemeria petals. *J. Exp. Bot.* **2002**, *53*, 233–240. [CrossRef]

40. Cui, X.Y.; Gao, Y.; Guo, J.; Yu, T.F.; Zheng, W.J.; Liu, Y.W.; Chen, J.; Xu, Z.S.; Ma, Y.Z. BES/BZR Transcription Factor TaBZR2 Positively Regulates Drought Responses by Activation of TaGST1. *Plant Physiol.* **2019**. [CrossRef]

41. Mauch, F.; Mauch-Mani, B.; Boller, T. Antifungal Hydrolases in Pea Tissue: II. Inhibition of Fungal Growth by Combinations of Chitinase and beta-1,3-Glucanase. *Plant Physiol.* **1988**, *88*, 936–942. [CrossRef] [PubMed]

42. Zhao, Q. Lignification: Flexibility, Biosynthesis and Regulation. *Trends Plant Sci.* **2016**, *21*, 713–721. [CrossRef] [PubMed]

43. Wang, G.F.; Balint-Kurti, P. Maize Homologs of CCoAOMT and HCT, Two Key Enzymes in Lignin Biosynthesis, Form Complexes with the NLR Rp1 Protein to Modulate the Defense Response. *Plant Physiol.* **2016**, *171*, 2166–2177. [CrossRef] [PubMed]

44. Zhang, X.; Liu, C.J. Multifaceted regulations of gateway enzyme phenylalanine ammonia-lyase in the biosynthesis of phenylpropanoids. *Mol. Plant* **2015**, *8*, 17–27. [CrossRef] [PubMed]

45. Le, R.J.; Brigitte, H.; Anne, C.; Simon, H.; Godfrey, N. Glycosylation Is a Major Regulator of Phenylpropanoid Availability and Biological Activity in Plants. *Front. Plant Sci.* **2016**, *7*, 735.

46. Ueno, M.; Kihara, J.; Arase, S. Tryptamine and sakuranetin accumulation in Sekiguchi lesions associated with the light-enhanced resistance of the lesion mimic mutant of rice to Magnaporthe oryzae. *J. Gen. Plant Pathol.* **2015**, *81*, 1–4. [CrossRef]

47. Mochizuki, N.; Tanaka, R.; Grimm, B.; Masuda, T.; Moulin, M.; Smith, A.G.; Tanaka, A.; Terry, M.J. The cell biology of tetrapyrroles: A life and death struggle. *Trends Plant Sci.* **2010**, *15*, 488–498. [CrossRef] [PubMed]

48. Quesada, V.; Sarmientomañús, R.; Gonzálezbayón, R.; Hricová, A.; Ponce, M.R.; Micol, J.L. PORPHOBILINOGEN DEAMINASE Deficiency Alters Vegetative and Reproductive Development and Causes Lesions in Arabidopsis. *PLoS ONE* **2013**, *8*, e53378. [CrossRef]

49. Hu, G.; Yalpani, N.; Briggs, S.P.; Johal, G.S. A porphyrin pathway impairment is responsible for the phenotype of a dominant disease lesion mimic mutant of maize. *Plant Cell* **1998**, *10*, 1095–1104. [CrossRef]

50. Samol, I.; Rossig, C.; Buhr, F.; Springer, A.; Pollmann, S.; Lahroussi, A.; Wettstein, D.V.; Reinbothe, C.; Reinbothe, S. The Outer Chloroplast Envelope Protein OEP16-1 for Plastid Import of NADPH:Protochlorophyllide Oxidoreductase A in Arabidopsis thaliana. *Plant Cell Physiol.* **2011**, *52*, 96–111. [CrossRef]

51. Rg, O.D.C.; Przybyla, D.; Ochsenbein, C.; Laloi, C.; Kim, C.; Danon, A.; Wagner, D.; Hideg, E.; Göbel, C.; Feussner, I. Rapid induction of distinct stress responses after the release of singlet oxygen in Arabidopsis. *Plant Cell* **2003**, *15*, 2320–2332.

52. Mikkelsen, M.; Jorgensen, M.; Krebs, F.C. The teraton challenge. A review of fixation and transformation of carbon dioxide. *Energy Environ. Sci.* **2010**, *3*, 43–81. [CrossRef]

53. Von Wettstein, D.; Gough, S.; Kannangara, C.G. Chlorophyll Biosynthesis. *Plant Cell* **1995**, *7*, 1039–1057. [CrossRef] [PubMed]

International Journal of
Molecular Sciences

MDPI

Article

Comparative Phosphoproteomic Analysis of Barley Embryos with Different Dormancy during Imbibition

Shinnosuke Ishikawa [1], José Barrero [2], Fuminori Takahashi [3], Scott Peck [4], Frank Gubler [2], Kazuo Shinozaki [3] and Taishi Umezawa [1,5,6,*]

[1] Graduate School of Bio-Applications and Systems Engineering, Tokyo University of Agriculture and Technology, Koganei, Tokyo 184-8588, Japan; s177676w@st.go.tuat.ac.jp
[2] CSIRO Agriculture and Food, Canberra ACT 2601, Australia; jose.barrero@csiro.au (J.B.); frank.gubler@csiro.au (F.G.)
[3] Gene Discovery Research Group, RIKEN Center for Sustainable Resource Science, Tsukuba, Ibaraki 305-0074, Japan; fuminori.takahashi@riken.jp (F.T.); kazuo.shinozaki@riken.jp (K.S.)
[4] Department of Biochemistry, University of Missouri, Columbia, MO 65211, USA; pecks@missouri.edu
[5] Faculty of Agriculture, Tokyo University of Agriculture and Technology, Fuchu, Tokyo 183-8538, Japan
[6] PRESTO, Japan Science and Technology Agency, Kawaguchi, Saitama 332-0012, Japan
* Correspondence: taishi@cc.tuat.ac.jp

Received: 17 December 2018; Accepted: 17 January 2019; Published: 21 January 2019

Abstract: Dormancy is the mechanism that allows seeds to become temporally quiescent in order to select the right time and place to germinate. Like in other species, in barley, grain dormancy is gradually reduced during after-ripening. Phosphosignaling networks in barley grains were investigated by a large-scale analysis of phosphoproteins to examine potential changes in response pathways to after-ripening. We used freshly harvested (FH) and after-ripened (AR) barley grains which showed different dormancy levels. The LC-MS/MS analysis identified 2346 phosphopeptides in barley embryos, with 269 and 97 of them being up- or downregulated during imbibition, respectively. A number of phosphopeptides were differentially regulated between FH and AR samples, suggesting that phosphoproteomic profiles were quite different between FH and AR grains. Motif analysis suggested multiple protein kinases including SnRK2 and MAPK could be involved in such a difference between FH and AR samples. Taken together, our results revealed phosphosignaling pathways in barley grains during the water imbibition process.

Keywords: phosphoproteome; barley; seed dormancy; germination; imbibition; after-ripening

---

1. Introduction

The switch from dormancy to germination is one of important transition steps in the life cycle of plants, because it will be the first and most fundamental factor determining their survivability. During the evolutionary process, seeds have evolved to germinate only in favorable seasons or places and dormancy is the mechanism that inhibits germination [1]. Seed dormancy is a complex trait regulated by many genetic and environmental factors [2–5], and during plant domestication, the dormancy and germination behavior of different species are set to fit their purpose. Most of domesticated cereals have been selected for uniform and synchronized germination by selection for weakened seed dormancy, which collaterally has made them prone to suffer pre-harvested sprouting (PHS) when moist conditions appear at harvesting seasons [6]. Understanding the mechanisms that operate during dormancy release will be very important to design molecular strategies to reinforce dormancy and provide protection against PHS. Barley is a good model to study seed dormancy regulation in cereals because freshly-harvested barley grains retain relatively high levels of dormancy [7].

To study seed dormancy, we can dissect this trait into three stages: acquisition, maintenance and decay [8,9]. It is well known that the phytohormone abscisic acid (ABA) has a critical role in all stages. During seed maturation, ABA accumulates and imposes the temporal quiescent state known as dormancy. After imbibition, the dormant seed (freshly harvested; FH) will be able to maintain high levels of ABA, thus blocking germination: the quiescent dry seed rapidly resumes metabolic activity, and ABA represses embryo growth (embryo-based dormancy). On the other hand, the seed husk physically inhibits oxygen absorption, and also constrains embryo growth (coat-based dormancy). However, in the non-dormant seed (after-ripened; AR), the ABA content is reduced during imbibition and the signaling repressed, which allows the germination to occur: gibberellic acid-pathways are activated, cell walls are weakened, embryo grows and finally coleorhiza appears through the husk-completing germination [1].

To understand the germination process, previous studies have performed a large-scale gene expression analyses of FH and AR seeds in Arabidopsis or barley during imbibition [10–16]. These studies revealed the differences in transcriptome between both states. In addition to transcriptional regulation, it has been reported that post-translational modifications (PTM), including phosphorylation, S-nitrosylation, carbonylation, glycosylation and oxidation, have a role in the regulation of seed dormancy and germination [17–20]. Among them, protein phosphorylation is fundamentally involved in the core ABA signaling pathway [21–23]. Furthermore, a protein kinase, MKK3, has been recently identified as a major quantitative trait locus (QTL) for grain dormancy in both barley and wheat [24,25]. Although these results indicate the importance of protein phosphorylation in seed dormancy and germination, the elements of the phosphosignaling pathways in cereal grains are still unsolved.

Taking advantage of the barley model system using FH and AR grains with contrasting dormancy levels, we have performed a large-scale phosphoproteomic analysis which allowed us to analyze phosphoproteins in vivo and to evaluate their phosphorylation sites and phosphorylation levels. In this study, we have identified nearly 2500 phosphopeptides in barley grains when being exposed to water, and analyzed their differential regulations between the dormant and the AR states.

2. Results and Discussion

2.1. Phosphoproteomic Analysis of Imbibed FH and AR Grains

To understand the phosphosignaling pathways that operate during the imbibition of matured FH and AR grains, phosphoproteomic analysis was performed in this study. In our phosphoproteomic analysis, we have used barley half-grains in which husk-based dormancy is broken and only embryo-based dormancy is present. In addition, the embryo was dissected and used for phosphoproteomic analysis, to remove a large amount of storage proteins contained in the endosperm.

FH or AR half-grains were imbibed for 0, 1, 3 and 10 h, and then embryos were dissected under the microscope; and proteins were isolated from these tissues and used for phosphoproteomic analyses. LC-MS/MS analysis identified 2346 phosphopeptides and 2491 phosphorylation sites in FH and AR grains, respectively (Table S1). About 95% of these were singly phosphorylated peptides, and 5% of them were multiply phosphorylated (Figure 1A). The most prominent phosphorylated amino acid was phosphoserine (84%), followed by phosphothreonine (15%), while only 1% was phosphotyrosine (Figure 1B). Phosphoproteomic analyses in other plants, such as Arabidopsis, rice and *Physcomitrella patens*, found a similar distribution of phosphorylated residues [26–31].

Datasets from FH and AR samples were compared and phosphopeptide changes and phosphorylation levels were analyzed via principal component analysis (PCA) of the total identified phosphopeptide data (Figure 1C). FH samples showed a similar localization in the PC1–PC2 projection, while the 10 h sample appeared as the most different. The AR samples showed a very different distribution from FH samples after imbibition. While small differences between FH and AR were seen at 0 h and 1 h, the separation between them became very significant at 10 h. These results suggest that AR embryos experience larger phosphopeptide changes than FH embryos during imbibition, with the AR 10 h sample the most different from the set. This may reflect the deep physiological changes occurring during germination.

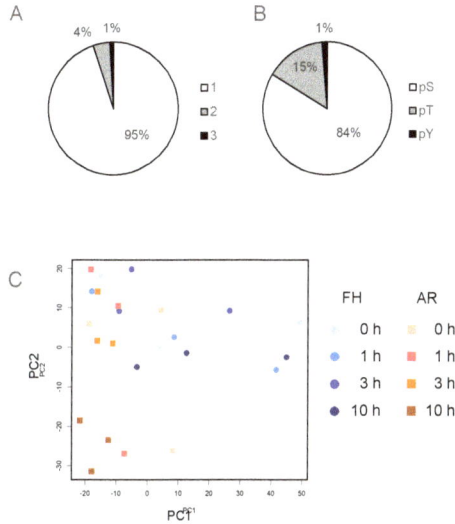

Figure 1. Summary of phosphoproteomic data. (**A**) Frequency of phosphorylated residues distributed in the phosphopeptides. Each number indicates the number of phosphorylation sites in phosphopeptides. (**B**) Distribution of phosphorylated residues in each phosphopeptide. pS, pT and pY showed phosphorylated serine, threonine and tyrosine, respectively. (**C**) Sample pattern recognition in principal component analysis. Blue circles and orange squares indicate FH and AR samples, respectively.

2.2. Classification of Phosphopeptides in Barley Grains

To compare phosphoproteome between FH and AR grains, quantitative data of each phosphopeptide was used for a clustering analysis. Hierarchical clustering analysis showed that most phosphopeptides were upregulated in FH and AR grains, but some of them were differentially regulated in FH and AR grains (Figure 2). This analysis divided phosphopeptides into three clusters. The first and second clusters include phosphopeptides primarily showing large increases in either FH (cluster a) or AR grains (cluster b), respectively. The third cluster included phosphopeptides that showed similar tendencies in both samples (cluster c). Cluster b was the largest, and cluster c contained the fewest members in this analysis. AR samples showed a different tendency between 0 and 10 h in comparison with FH samples. This result was consistent with PCA, suggesting that AR grains change phosphorylation status more than FH during imbibition.

To examine the most robust changes, we screened phosphopeptides that statistically increased or decreased in imbibition as compared to the 0 h for each seed stage. Of these, 98 and 199 phosphopeptides increased in FH and AR embryos, respectively (Table 1; Table S2), with only 28 of them (10.5%) being shared between the two sets. Conversely, 39 and 59 phosphopeptides decreased in FH and AR embryos, respectively (Table 1; Table S2). Interestingly, only one phosphopeptide was shared between these two sets. In accordance with Figure 2, the number phosphopeptides that increased was more than those that decreased. Examples of phosphopeptides with different patterns are shown in Figure 3. Some phosphopeptides showed significant changes specifically in FH and/or AR samples. Two of those phosphopeptides exhibited converse accumulation patterns in FH and AR embryos. For example, glycosyl hydrolase family protein was upregulated in the FH embryo and downregulated in the AR embryo.

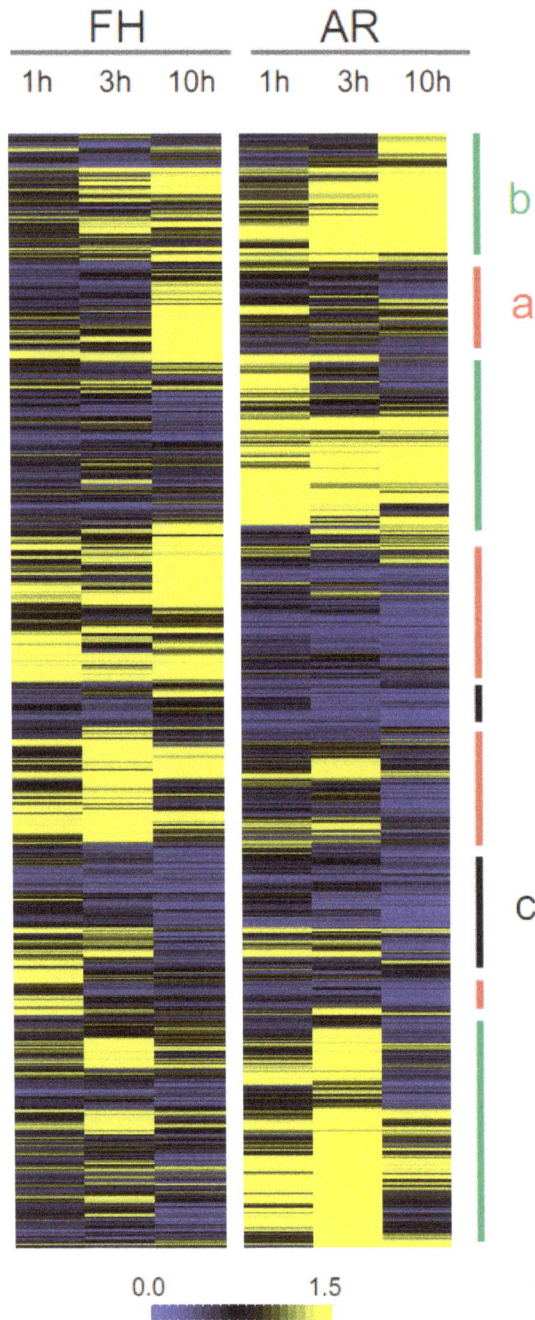

Figure 2. Comparative analysis of FH and AR barley grains. Quantitative data of each phosphopeptide in FH and AR grains was displayed as a heatmap. Phosphopeptides could be classified into clusters a, b and c based on their phosphorylation patterns. Cluster a (red) and cluster b (green) include phosphopeptides primarily showing large increases in FH and AR grains, respectively. The cluster c (black) includes phosphopeptides that showed similar tendencies in both samples.

Table 1. The numbers of up- and downregulated phosphopeptides in barley grains during imbibition.

Response	Freshly Harvested	Overlap	After-Ripened	Total
Upregulated	98	28	199	269
Downregulated	39	1	59	97

Comparative analysis selected phosphopeptides which were upregulated or downregulated in response to imbibition in FH and AR grains. Each phosphopeptide was statistically tested by Student's *t*-test (*p*-value < 0.05).

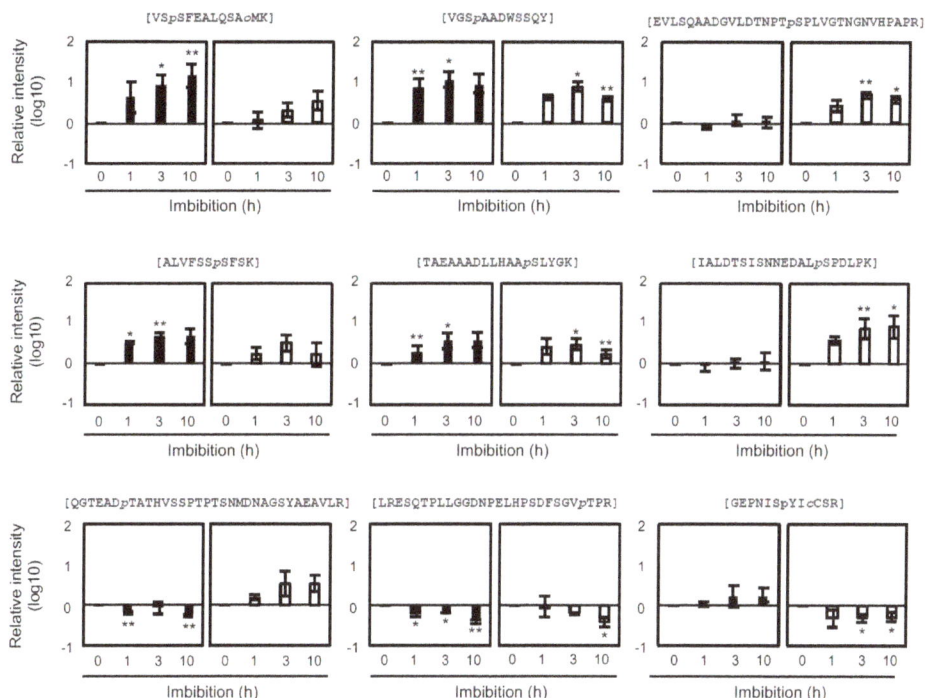

Figure 3. Examples of phosphopeptides in barley seeds. Quantitative data of each phosphopeptide was analyzed for FH (solid) and AR (empty) grains treated with imbibition. Bars indicates ± standard error (*n* = 3), and * and ** indicate *p*-values of <0.05 and <0.01, respectively.

2.3. Comparative Analysis of Phosphopeptides

Phosphopeptides were further analyzed to examine potential differences in biological processes between FH and AR embryos. First, barley genes were annotated using the Arabidopsis database (TAIR10); and gene ontology analyses of up- and downregulated phosphopeptides was performed. The 70 and 171 phosphopeptides found in Table 1 to be uniquely upregulated in FH only and AR only, respectively, were used for gene ontology (GO) analysis (Figure 4A,B; Table S3). Proteins upregulated in FH grains were enriched in GO categories related to "response to ABA", "embryo development ending in seed dormancy" and "RNA splicing" (Figure 4A). Uniquely enriched in AR grains responses included "response to osmotic stress", "embryo development ending in seed dormancy", cell wall pectin metabolism", "regulation of translation" and "mRNA processing" (Figure 4B). To compare FH with AR, "response to ABA" in FH was enriched; and enrichment of "embryo development ending in seed dormancy" was lower in FH than in AR. Especially the GO term of "cell wall pectin modification" was highlighted to associate with germination. During germination, embryo growth and cell wall degradation occur to be associated with physiological and physical dormancy, respectively [32–37]. This GO term indicates AR grains go toward radicle protrusion, germination.

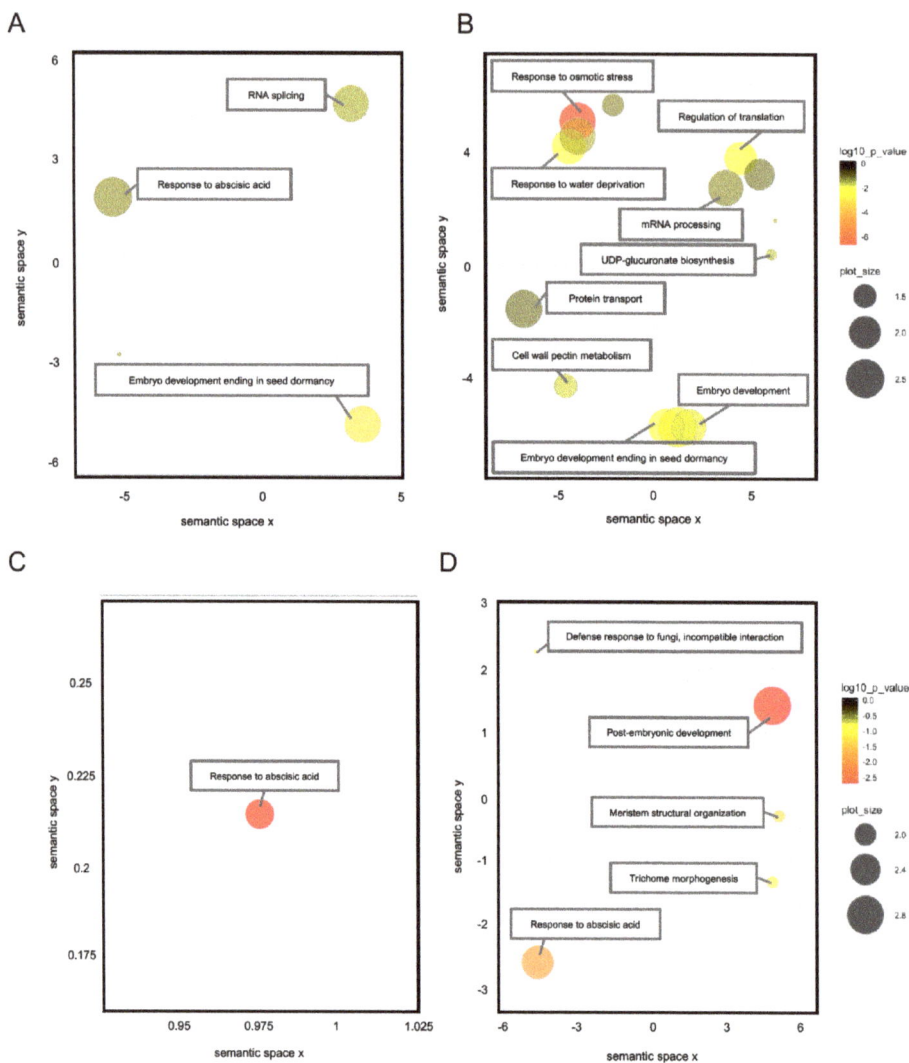

Figure 4. GO analysis of phosphopeptides in barley seeds. GO terms were evaluated by DAVID program and visualized with REViGO for phosphopeptides upregulated and downregulated under imbibition. Each circle color and size show *p*-value and frequency (%), respectively. Phosphopeptides used for this analysis included 70 and 171 phosphopeptides upregulated in FH (**A**) and AR (**B**) seeds, respectively; 38 and 58 phosphopeptides downregulated in FH (**C**) and AR (**D**) seeds, respectively.

From Table 1, the 38 and 58 downregulated phosphopeptides unique to each seed stage were used for GO analysis (Figure 4C,D). For FH grains, GO term was enriched for "response to ABA" (Figure 4C). GO terms of "response to ABA", "post-embryonic development", "meristem structural organization" and "trichome morphogenesis" were enriched in AR (Figure 4D). Among GO terms of AR grains, "post-embryonic development" was the most highlighted one. ABA-related phosphopeptides were significantly downregulated in both of grains. The phytohormone ABA plays an important role in response to environmental stress and dormancy [7,9,23,38]. Recently, it was reported that ABA responses involve activation of protein kinase SnRK2, and then activated SnRK2 phosphorylates downstream substrates, including bZIP transcription factors [22,39–42], to modulate their activity. The enrichment of the GO term "response to ABA" in phosphopeptides decreasing in AR is consistent with the decay of ABA signaling that would be associated with the rapid decline in dormancy in AR seeds after imbibition.

Phosphorylation motif analysis can indicate some kinases upstream of the differentially phosphorylated proteins, thus pointing to kinases that may be changing in activity during these processes [43,44]. In this study, two phosphorylation motifs, [-pS-P-] and [-R-x-x-pS-], were enriched in both up- and downregulated candidates (Figure 5; Table S4). [-pS/-P-] is a known mitogen-activated protein kinase (MAPK)- and cyclin-dependent kinase (CDK) target motif. SnRK2, calcium-dependent protein kinases (CDPK) and CBL-interacting protein kinases (CIPK) phosphorylate on [-R-x-x-pS-] motifs. Among 70 upregulated phosphopeptides in the FH sample, 22 (30.9%) and 17 (23.9%) include [-pS/T-P-] and [-R/K-x-x-pS/T-], respectively, while 68 [-pS/T-P-] (38.4%) and 41 [-R/K-x-x-pS/T-] (23.1%) were found in 171 upregulated phosphopeptides in the AR sample (Figure 5A).

Next, a motif analysis was performed for 97 phosphopeptides which were downregulated after imbibition. As well as upregulated phosphopeptides, [-R/K-x-x-pS/T-] and [-pS/T-P-] were identified in FH and AR grains (Figure 5B). Actually, it is difficult to understand why the same motifs were enriched in both upregulated and downregulated phosphopeptides. It may result from different protein kinases sharing the same target motifs as described above.

A

Sequence	Seed type		
	FH	Both	AR
	22	1	68
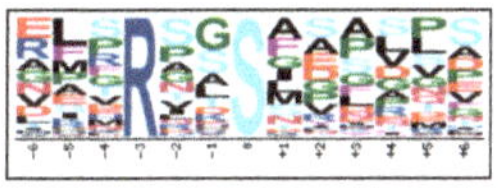	17	13	41

B

Sequence	Seed type		
	FH	Both	AR
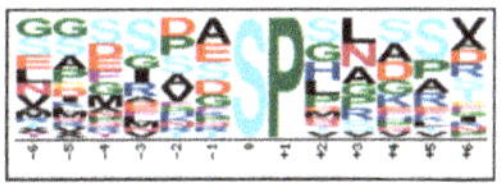	13	1	18
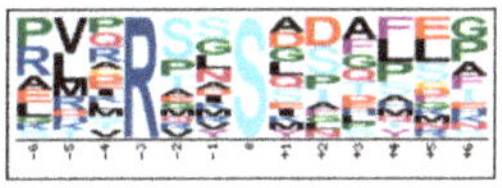	7	-	19

Figure 5. Motif analysis of phosphopeptides in barley grains. Motif analysis of 269 upregulated phosphopeptides (**A**) and 97 downregulated phosphopeptides (**B**) using the motif-x algorithm. Extracted phosphorylation motifs were counted in the data from FH or AR grains.

2.4. Differential Regulation Mechanisms of Seed Dormancy

Various approaches have been performed to investigate the differences between FH and AR grains in cereals [45–49]. A variety of different factors have been reported to contribute to the antagonistic regulation of dormancy and germination, including environmental responses, such as light and temperature, oxidation of proteins, and differential accumulation of transcripts (i.e., changes in gene expression) [50,51]. These previous studies indicated that ABA has an indispensable role in dormancy regulation. The perception of changes in ABA is primarily transmitted by three major components: ABA receptors (PYR/PYL/RCAR), protein phosphatases (PP2Cs) and protein kinases (SnRK2s) [39–41]. Both ABA content and signaling are important for controlling plant responses [45,52], and it can be difficult to separate both elements because of the complex feed-back regulation of the ABA synthesis pathway. Millar et al. reported that the ABA content in dry seeds is similar between FH and AR samples, and that only after imbibition a difference occurs due to the AR seeds being unable to maintain high ABA levels [45]. This study and others would suggest that ABA signaling changes during after-ripening are more critical than content changes for the regulation of dormancy and germination. In agreement with this result, our GO analysis showed a set of phosphopeptides, of which responses to imbibition are related to "response to ABA". In FH, up- and downregulated phosphopeptides contained "response to ABA". On the other hand, "response to ABA" was strongly enriched in downregulation in comparison with upregulation in AR. These results suggest ABA signaling is active in FH, but it is impaired by imbibition in AR. Additionally, most of phosphopeptides containing [-R/K-x-x-pS/T-] decreased their phosphorylation level with imbibition in AR. This indicates the activities of SnRK2 and/or CDPK, which are involved in ABA signaling and target [-R/K-x-x-pS/T-], are impaired during imbibition. These results consistently imply ABA signaling is attenuated in the AR grain compared to that in the FH grain. Although the ABA contents of FH and AR grains decrease, ABA signaling is activated or repressed in FH and AR grains, respectively. It is still unclear how a decay in ABA signaling during imbibition is induced during after-ripening.

The influence of DELAY OF GERMINATION1 (DOG1) is one possible to alter ABA signaling between FH and AR grains. DOG1 is expressed in fresh/dormant and also in AR Arabidopsis seeds, but DOG1 proteins are less abundant or downregulated in AR seeds [53,54]. Interestingly, recent studies have reported that DOG1 interacts with AHG1 and AHG3, one of PP2Cs in clade A, and is able to repress its function directly [55,56]. This inhibition possibly induces the activation of SnRK2 in FH grains, but not in AR grains.

2.5. The Role of Abscisic Acid in Seed Dormancy during Water Imbibition

We identified the ortholog of AREB3 and other ABA-responsive proteins were downregulated in AR grains during water imbibition. AREB3 belongs to group A bZIP transcription factors, which are responsible for ABA-responsive element (ABRE; PyACGTGG/TC)-dependent gene expression [57,58]. These factors are divided into two subclasses: ABRE-binding protein (AREB)/ABRE-binding factor (ABF) subfamily having a role in the vegetative tissue and ABA-INSENSITIVE 5 (ABI5)/Dc3 promoter-binding factor (DPBF) subfamily working in the seed [59,60]. Group A bZIP transcription factors are phosphorylated by SnRK2 and activate the gene expression in response to ABA [27,61–63]. bZIP transcription factors could be regulated by the ABA content. Millar et al. revealed ABA content decreased sharply after water imbibition in AR grains in comparison with FH (dormant) grains of barley [45]. Low ABA concentration cannot influence ABA signaling enough. Additionally, bZIP transcription factors could be controlled by dephosphorylation. Group A bZIP transcription factors were indicated to be dephosphorylated by PP2C [64,65]. In low ABA conditions, released PP2C possibly dephosphorylates SnRK2 and bZIP transcription factors to repress ABA signaling. Another possibility is that protein degradation negatively regulates group A bZIP transcription factors. ABI5 is known to be degraded by KEEP ON GOING (KEG) and CUL4/DDB1 E3 Ligase [66–70]. It is presumed that activated E3 ubiquitin ligases degrade bZIP transcription factors.

Taken together, this study performed phosphoproteomic analyses of FH and AR embryos in barley during imbibition and demonstrated differential phosphosignals in FH and AR barley grains. We have identified numerous phosphopeptides and 365 of them significantly altered phosphorylation levels during imbibition. These phosphopeptides are possibly involved in control of dormancy and germination, and some of them could be involved in the regulation of ABA signaling. Further studies will be required for understanding the role of these responsive phosphoproteins and the upstream elements that regulate the activity of various protein kinases during after-ripening.

3. Materials and Methods

3.1. Plant Material and Growth Condition

Barley (*Hordeum vulgare* cv. Golden Promise) plants were grown in a phytotron glasshouse (CSIRO, Canberra, Australia) under sunlight and temperature set at 17/9 °C day/night [71]. Grains were harvested at physiological maturity and half of the harvest was stored at −20 °C to preserve a dormancy level as FH. The other half was after-ripened at 37 °C for six months to impair dormancy and then stored at −20 °C as well (AR).

3.2. Phosphoproteomic Analysis

Twenty half-cut grains were prepared and set on filter paper (9 cm in diameter, Whatman #1, GE Healthcare, Chicago, IL, USA) in plastic petri dishes. After adding 5 mL double-distilled H_2O, dishes were sealed with a Parafilm and covered by aluminum foil, and then incubated at 20 °C for each time course, 1 h, 3 h and 10 h.

Following imbibition, embryos were dissected from barley half grains and stored at −80 °C as previously described [47,72]. Fifteen embryos were grounded by using TissueLyser II (QIAGEN, Germantown, MD, USA), and samples were resuspended in 1 mL of protein extraction buffer containing 10 mM Tris-HCl (pH 9.0), 8 M Urea, 2% Phosphatase Inhibitor Cocktail II (Sigma, St. Louis, MO, USA) and 2% Phosphatase Inhibitor Cocktail III (Sigma, St. Louis, MO, USA). After centrifugation at 17,400 g at 4 °C for 10 min, supernatants were collected as crude extracts, and protein concentrations were measured by BCA Protein Kit (Thermo Scientific, San Jose, CA, USA).

The phosphoproteomic analyses were performed as previously described [27,30,59,60] with minor modifications. Aliquots of 400 µg total protein were reduced with 10 mM DTT for 30 min, and alkylated with 50 mM iodoacetamide for 20 min in the dark, and then with Lys-C (WAKO, Osaka, Japan; 1:200, *w/w*) for 3 h. After 4-fold dilution with NH_4HCO_3, proteins were digested with trypsin (Promega, Madison, WI, USA; 1:100, *w/w*) overnight at room temperature.

After enzymatic digestion, an equivalent volume of 2% trifluoroacetic acid (TFA) was added to the digested samples, and then they were desalted using SDB-XC Empore disk membranes (3M, St. Paul, MN, USA) as described previously [73]. To enrich phosphopeptides, the hydroxyl acid-modified metal oxide chromatography (HAMMOC) method was performed [74]. Custom-made metal oxide chromatography (MOC) tips made with C8-StageTips and 3 mg of bulk titania beads (particle size, 10 µm; GL science, Torrance, CA, USA) were used in this study. The concentrated phosphopeptide sample was desalted with a C18-SDC and C18-GC column (GL science, Torrance, CA, USA). Each column was washed with solution A (80% acetonitrile and 0.1% TFA) and 0.1 % TFA by using centrifugation at 700 g for 2 min at room temperature. Samples were loaded on each column and centrifuged at 700 g at room temperature. After being washed with 0.1% TFA, phosphopeptides were eluted with solution A by using centrifugation at 700 g for 2 min. Samples were dried in a vacuum evaporator (Tomy, Tokyo, Japan), and diluted with 10 µL of 0.1% formic acid (FA).

Cleaned-up samples were analyzed with TripleTOF 5600 system (AB-SCIEX, Framingham, MA, USA) equipped with Autosampler-2 1D plus (Eksigent, Framingham, MA, USA) and NanoLC Ultra (Eksigent, Framingham, MA, USA) using MonoCap C18 High Resolution 2000 column (GL science, Torrance, CA, USA) and PicoTip emitter SilicaTip (New Objective Inc., Woburn, MA, USA). Peptides were eluted at 500 nL min^{-1} with a four-step gradient, 0.5% acetic acid: 0.5% and 80% acetic acid = 98:2 (0 min), 60:40 (300 min), 10:90 (20 min) and 98:2 (40 min). The eluate was sprayed into mass spectrometer by electrospray ionization (ESI). The mass spectrometry (MS) scan range was 400–1250 m/z and the MS/MS scan range was 100–1600 m/z.

3.3. Phosphopeptide Identification and Quantification

Peak lists were generated using Protein pilot version 5.0.0.4769 (AB-SCIEX, Framingham, MA, USA). Raw spectrum files were matched with the barley gene database published on 23 March 2012 (Plant Genome and Systems Biology; https://www.helmholtz-muenchen.de/pgsb) using Mascot version 2.4.0 (Matrix Science, London, UK). Search settings were applied: a precursor mass tolerance of 3 ppm, a fragment ion mass tolerance of 0.8 Da, and cut-off value of 0.95, allowing for up to two miss cleavages, with the enzyme designated as trypsin. A fixed modification of carbamidomethylation of cysteine and variable modifications of oxidation of methionine and phosphorylation of serine, threonine and tyrosine were used. All raw data files were deposited in the Japan Proteome Standard Repository/Database (jPOST; JPST000502, Kyoto, Japan).

Skyline software version 4.2 (https://skyline.ms/project/home/software/Skyline/begin.view) was used for phosphopeptide quantification from peak areas [75]. The search settings were the same as described for Mascot. The maximum false discovery rate (FDR) thresholds for protein was set to 5%. In addition, the site localization probability threshold was specified as >0.75. Fold changes were calculated using quantitative values. For each time point, three biological replicates were analyzed and the significance of time-dependent changes was determined by Student's *t* test ($p < 0.05$).

3.4. Data Analysis

Each phosphoproteomic sample including FH and AR grains was compared by PCA [76–78]. Samples were plotted with principal component 1(PC1) and PC2. Hierarchical clustering analysis was performed on phosphorylation intensity using Multi Experimental Viewer (MeV, Boston, MA, USA). Pearson correlation and average linkage clustering were applied for settings. Gene Ontology (GO) analysis was performed with DAVID (https://david.ncifcrf.gov) and REViGO (http://revigo.irb.hr). Annotated data with Arabidopsis by BLAST (https://blast.ncbi.nlm.nih.gov/Blast.cgi) was loaded to DAVID, and background database was set as the TAIR 10 Arabidopsis dataset. Outputted GO terms in DAVID were visualized with REViGO. Settings used for REViGO were: medium (0.7) similarity, UniProt Arabidopsis database (https://www.uniprot.org) and simRel semantic measure. Phosphorylation motifs were predicted by the motif-x program (http://motif-x.med.harvard.edu) [79]. For motif analysis, 13 amino acids around phosphorylated residues were extracted from identified phosphopeptide sequences, and submitted to motif-x, setting an occurrence to 20 and significance to 0.01. Barley expressed sequence tag (EST) data was submitted to the local BLAST program against the Arabidopsis dataset (TAIR10) to make a list of orthologues [80].

4. Conclusions

To understand the phosphosignaling that take place during the after-ripening of barley grains and that produce a decay in dormancy, phosphoproteomic profiles were obtained from FH and AR embryos during imbibition. As a result, 2,346 phosphopeptides were identified, with 365 of them responded to imbibition. Our data indicate that multiple protein kinases, such as SnRK2, CDPK, CIPK, or MAPK, can actively participate in the differential phosphorylation of peptides in barley FH or AR grains, and point to some key kinases that could be manipulated for regulating germination in cereals.

Supplementary Materials: Supplementary materials can be found at http://www.mdpi.com/1422-0067/20/2/451/s1.

Author Contributions: J.M.B., F.G., T.U. and S.I. designed this study. S.C.P. contributed to analysis and interpretation of data, and assisted in the preparation of the manuscript. F.T. and K.S. supported to achieve the analysis by using mass spectrometry. S.I. performed experiments and wrote the initial draft of the manuscript. J.M.B., S.C.P. and T.U. critically reviewed the manuscript.

Funding: The study was supported by the Japan Society for the Promotion of Science (JSPS) KAKENHI under grant numbers JP15H04383 and 16KK0160 (T.U.) and JST PRESTO under grant number JP13413773 (T.U.).

Acknowledgments: We thank Saho Mizukado (RIKEN), Saul Newman, Trijntje Hughes, Jasmine Rajamony and Sandra Stops (CSIRO) for their expert technical assistance. We also thank John (Jake) V. Jacobsen and Alec Zwart for comments when preparing this manuscript.

Conflicts of Interest: The authors declare no conflicts of interest.

Abbreviations

FH	freshly harvested
AR	after-ripened
LC-MS/MS	liquid chromatography–mass spectrometry/mass spectrometry

References

1. Bewley, J.D. Seed germination and dormancy. *Plant Cell* **1997**, *9*, 1055–1066. [CrossRef] [PubMed]
2. Koornneef, M.; Bentsink, L.; Hilhorst, H. Seed dormancy and germination. *Curr. Opin. Plant Biol.* **2002**, *5*, 33–36. [CrossRef]
3. Finch-Savage, W.E.; Leubner-Metzger, G. Seed dormancy and the control of germination. *New Phytol.* **2006**, *171*. [CrossRef] [PubMed]
4. Finkelstein, R.; Reeves, W.; Ariizumi, T.; Steber, C. Molecular aspects of seed dormancy. *Annu. Rev. Plant Biol.* **2008**, *59*, 387–415. [CrossRef] [PubMed]
5. Holdsworth, M.J.; Bentsink, L.; Soppe, W.J.J. Molecular networks regulating Arabidopsis seed maturation, after-ripening, dormancy and germination. *New Phytol.* **2008**, *179*, 33–54. [CrossRef] [PubMed]
6. Rodríguez, M.V.; Barrero, J.M.; Corbineau, F.; Gubler, F.; Benech-Arnold, R.L. Dormancy in cereals (not too much, not so little): About the mechanisms behind this trait. *Seed Sci. Res.* **2015**, *25*, 99–119. [CrossRef]
7. Gubler, F.; Millar, A.A.; Jacobsen, J.V. Dormancy release, ABA and pre-harvest sprouting. *Curr. Opin. Plant Biol.* **2005**, *8*, 183–187. [CrossRef]
8. Kermode, A.R. Role of abscisic acid in seed dormancy. *J. Plant Growth Regul.* **2005**, *24*, 319–344. [CrossRef]
9. Nambara, E.; Okamoto, M.; Tatematsu, K.; Yano, R.; Seo, M.; Kamiya, Y. Abscisic acid and the control of seed dormancy and germination. *Seed Sci. Res.* **2010**, *20*, 55–67. [CrossRef]
10. Nakabayashi, K.; Okamoto, M.; Koshiba, T.; Kamiya, Y.; Nambara, E. Genome-wide profiling of stored mRNA in Arabidopsis thaliana seed germination: Epigenetic and genetic regulation of transcription in seed: Molecular profiling in Arabidopsis seed. *Plant J.* **2005**, *41*, 697–709. [CrossRef]
11. Cadman, C.S.C.; Toorop, P.E.; Hilhorst, H.W.M.; Finch-Savage, W.E. Gene expression profiles of Arabidopsis Cvi seeds during dormancy cycling indicate a common underlying dormancy control mechanism. *Plant J.* **2006**, *46*, 805–822. [CrossRef] [PubMed]
12. Finch-Savage, W.E.; Cadman, C.S.C.; Toorop, P.E.; Lynn, J.R.; Hilhorst, H.W.M. Seed dormancy release in Arabidopsis cvi by dry after-ripening, low temperature, nitrate and light shows common quantitative patterns of gene expression directed by environmentally specific sensing: Seed dormancy release in Arabidopsis. *Plant J.* **2007**, *51*, 60–78. [CrossRef] [PubMed]
13. Carrera, E.; Holman, T.; Medhurst, A.; Dietrich, D.; Footitt, S.; Theodoulou, F.L.; Holdsworth, M.J. Seed after-ripening is a discrete developmental pathway associated with specific gene networks in Arabidopsis: After-ripening regulated gene expression. *Plant J.* **2007**, *53*, 214–224. [CrossRef]
14. Barrero, J.M.; Talbot, M.J.; White, R.G.; Jacobsen, J.V.; Gubler, F. Anatomical and transcriptomic studies of the coleorhiza reveal the importance of this tissue in regulating dormancy in barley. *Plant Physiol.* **2009**, *150*, 1006–1021. [CrossRef] [PubMed]
15. Barrero, J.M.; Millar, A.A.; Griffiths, J.; Czechowski, T.; Scheible, W.R.; Udvardi, M.; Reid, J.B.; Ross, J.J.; Jacobsen, J.V.; Gubler, F. Gene expression profiling identifies two regulatory genes controlling dormancy and ABA sensitivity in Arabidopsis seeds. *Plant J.* **2010**, *61*, 611–622. [CrossRef]

16. Dekkers, B.J.W.; Pearce, S.P.; van Bolderen-Veldkamp, R.P.M.; Holdsworth, M.J.; Bentsink, L. Dormant and after-ripened Arabidopsis thaliana seeds are distinguished by early transcriptional differences in the imbibed state. *Front. Plant Sci.* **2016**, *7*, 1323. [CrossRef] [PubMed]

17. Bethke, P.; Gubler, F.; Jacobsen, J.; Jones, R. Dormancy of Arabidopsis seeds and barley grains can be broken by nitric oxide. *Planta* **2004**, *219*, 847–855. [CrossRef] [PubMed]

18. El-Maarouf-Bouteau, H.; Meimoun, P.; Job, C.; Job, D.; Bailly, C. Role of protein and mRNA oxidation in seed dormancy and germination. *Front. Plant Sci.* **2013**, *4*, 77. [CrossRef]

19. Gao, F.; Rampitsch, C.; Chitnis, V.R.; Humphreys, G.D.; Jordan, M.C.; Ayele, B.T. Integrated analysis of seed proteome and mRNA oxidation reveals distinct post-transcriptional features regulating dormancy in wheat (*Triticum aestivum* L.). *Plant Biotechnol. J.* **2013**, *11*, 921–932. [CrossRef] [PubMed]

20. Ha, Y.; Shang, Y.; Nam, K.H. Brassinosteroids modulate ABA-induced stomatal closure in Arabidopsis. *J. Exp. Bot.* **2016**, *67*, 6297–6308. [CrossRef]

21. Fujii, H.; Verslues, P.E.; Zhu, J.-K. Identification of two protein kinases required for abscisic acid regulation of seed germination, root growth, and gene expression in Arabidopsis. *Plant Cell* **2007**, *19*, 485–494. [CrossRef] [PubMed]

22. Nakashima, K.; Fujita, Y.; Kanamori, N.; Katagiri, T.; Umezawa, T.; Kidokoro, S.; Maruyama, K.; Yoshida, T.; Ishiyama, K.; Kobayashi, M.; et al. Three Arabidopsis SnRK2 protein kinases, SRK2D/SnRK2.2, SRK2E/SnRK2.6/OST1 and SRK2I/SnRK2.3, involved in ABA signaling are essential for the control of seed development and dormancy. *Plant Cell Physiol.* **2009**, *50*, 1345–1363. [CrossRef] [PubMed]

23. Umezawa, T.; Nakashima, K.; Miyakawa, T.; Kuromori, T.; Tanokura, M.; Shinozaki, K.; Yamaguchi-Shinozaki, K. Molecular basis of the core regulatory network in ABA responses: Sensing, signaling and transport. *Plant Cell Physiol.* **2010**, *51*, 1821–1839. [CrossRef] [PubMed]

24. Nakamura, S.; Pourkheirandish, M.; Morishige, H.; Kubo, Y.; Nakamura, M.; Ichimura, K.; Seo, S.; Kanamori, H.; Wu, J.; Ando, T.; et al. Mitogen-Activated Protein Kinase Kinase 3 regulates seed dormancy in barley. *Curr. Biol.* **2016**, *26*, 775–781. [CrossRef] [PubMed]

25. Torada, A.; Koike, M.; Ogawa, T.; Takenouchi, Y.; Tadamura, K.; Wu, J.; Matsumoto, T.; Kawaura, K.; Ogihara, Y. A Causal gene for seed dormancy on wheat chromosome 4A encodes a MAP kinase kinase. *Curr. Biol.* **2016**, *26*, 782–787. [CrossRef] [PubMed]

26. Nakagami, H.; Sugiyama, N.; Ishihama, Y.; Shirasu, K. Shotguns in the Front Line: Phosphoproteomics in Plants. *Plant Cell Physiol.* **2012**, *53*, 118–124. [CrossRef] [PubMed]

27. Umezawa, T.; Sugiyama, N.; Takahashi, F.; Anderson, J.C.; Ishihama, Y.; Peck, S.C.; Shinozaki, K. Genetics and phosphoproteomics reveal a protein phosphorylation network in the abscisic acid signaling pathway in *Arabidopsis thaliana*. *Sci. Signal* **2013**, *6*, rs8. [CrossRef] [PubMed]

28. Wang, X.; Bian, Y.; Cheng, K.; Gu, L.-F.; Ye, M.; Zou, H.; Sun, S.S.-M.; He, J.-X. A large-scale protein phosphorylation analysis reveals novel phosphorylation motifs and phosphoregulatory networks in Arabidopsis. *J. Proteom.* **2013**, *78*, 486–498. [CrossRef] [PubMed]

29. Lv, D.-W.; Li, X.; Zhang, M.; Gu, A.-Q.; Zhen, S.-M.; Wang, C.; Li, X.-H.; Yan, Y.-M. Large-scale phosphoproteome analysis in seedling leaves of *Brachypodium distachyon* L. *BMC Genom.* **2014**, *15*, 375. [CrossRef] [PubMed]

30. Choudhary, M.K.; Nomura, Y.; Wang, L.; Nakagami, H.; Somers, D.E. Quantitative circadian phosphoproteomic analysis of Arabidopsis reveals extensive clock control of key components in physiological, metabolic, and signaling pathways. *Mol. Cell. Proteom.* **2015**, *14*, 2243–2260. [CrossRef]

31. Amagai, A.; Honda, Y.; Ishikawa, S.; Hara, Y.; Kuwamura, M.; Shinozawa, A.; Sugiyama, N.; Ishihama, Y.; Takezawa, D.; Sakata, Y.; et al. Phosphoproteomic profiling reveals ABA-responsive phosphosignaling pathways in *Physcomitrella Patens*. *Plant J.* **2018**, *94*, 699–708. [CrossRef] [PubMed]

32. Nonogaki, H. Seed Germination—The biochemical and molecular mechanisms. *Breed. Sci.* **2006**, *56*, 93–105. [CrossRef]

33. Halmer, P.; Bewley, J.D.; Thorpe, T.A. Enzyme to break down lettuce endosperm cell wall during gibberellin-and light-induced germination. *Nature* **1975**, *258*, 716–718. [CrossRef]

34. Watkins, J.T.; Cantliffe, D.J. Mechanical resistance of the seed coat and endosperm during germination of *Capsicum annuum* at low temperature. *Plant Physiol.* **1983**, *72*, 146–150. [CrossRef] [PubMed]

35. Endo, A.; Tatematsu, K.; Hanada, K.; Duermeyer, L.; Okamoto, M.; Yonekura-Sakakibara, K.; Saito, K.; Toyoda, T.; Kawakami, N.; Kamiya, Y.; et al. Tissue-specific transcriptome analysis reveals cell wall metabolism, flavonol biosynthesis and defense responses are activated in the endosperm of germinating *Arabidopsis thaliana* seeds. *Plant Cell Physiol.* **2012**, *53*, 16–27. [CrossRef]

36. Oracz, K.; Voegele, A.; Tarkowská, D.; Jacquemoud, D.; Turečková, V.; Urbanová, T.; Strnad, M.; Sliwinska, E.; Leubner-Metzger, G. Myrigalone A inhibits *Lepidium sativum* seed germination by interference with gibberellin metabolism and apoplastic superoxide production required for embryo extension growth and endosperm rupture. *Plant Cell Physiol.* **2012**, *53*, 81–95. [CrossRef]

37. Vishal, B.; Kumar, P.P. Regulation of seed germination and abiotic stresses by gibberellins and abscisic acid. *Front. Plant Sci.* **2018**, *9*, 838. [CrossRef]

38. Nonogaki, H. Seed dormancy and germination-emerging mechanisms and new hypotheses. *Front. Plant Sci.* **2014**, *5*, 233. [CrossRef]

39. Fujii, H.; Chinnusamy, V.; Rodrigues, A.; Rubio, S.; Antoni, R.; Park, S.-Y.; Cutler, S.R.; Sheen, J.; Rodriguez, P.L.; Zhu, J.-K. In vitro reconstitution of an abscisic acid signalling pathway. *Nature* **2009**, *462*, 660–664. [CrossRef]

40. Ma, Y.; Szostkiewicz, I.; Korte, A.; Moes, D.; Yang, Y.; Christmann, A.; Grill, E. Regulators of PP2C phosphatase activity function as abscisic acid sensors. *Science* **2009**, *324*, 1064–1068. [CrossRef]

41. Park, S.-Y.; Fung, P.; Nishimura, N.; Jensen, D.R.; Fujii, H.; Zhao, Y.; Lumba, S.; Santiago, J.; Rodrigues, A.; Chow, T.F.; et al. Abscisic acid inhibits type 2C protein phosphatases via the PYR/PYL family of START proteins. *Science* **2009**, *324*, 1068–1071. [CrossRef] [PubMed]

42. Umezawa, T.; Sugiyama, N.; Mizoguchi, M.; Hayashi, S.; Myouga, F.; Yamaguchi-Shinozaki, K.; Ishihama, Y.; Hirayama, T.; Shinozaki, K. Type 2C protein phosphatases directly regulate abscisic acid-activated protein kinases in Arabidopsis. *Proc. Natl. Acad. Sci. USA* **2009**, *106*, 17588–17593. [CrossRef] [PubMed]

43. Keshava Prasad, T.S.; Goel, R.; Kandasamy, K.; Keerthikumar, S.; Kumar, S.; Mathivanan, S.; Telikicherla, D.; Raju, R.; Shafreen, B.; Venugopal, A.; et al. Human protein reference database—2009 update. *Nucleic Acids Res.* **2009**, *37*, D767–D772. [CrossRef] [PubMed]

44. Durek, P.; Schmidt, R.; Heazlewood, J.L.; Jones, A.; MacLean, D.; Nagel, A.; Kersten, B.; Schulze, W.X. PhosPhAt: The *Arabidopsis thaliana* phosphorylation site database. An update. *Nucleic Acids Res.* **2010**, *38*, D828–D834. [CrossRef] [PubMed]

45. Millar, A.A.; Jacobsen, J.V.; Ross, J.J.; Helliwell, C.A.; Poole, A.T.; Scofield, G.; Reid, J.B.; Gubler, F. Seed dormancy and ABA metabolism in Arabidopsis and barley: The role of ABA 8′-hydroxylase. *Plant J.* **2006**, *45*, 942–954. [CrossRef] [PubMed]

46. Bradford, K.J.; Benech-Arnold, R.L.; Côme, D.; Corbineau, F. Quantifying the sensitivity of barley seed germination to oxygen, abscisic acid, and gibberellin using a population-based threshold model. *J. Exp. Bot.* **2008**, *59*, 335–347. [CrossRef] [PubMed]

47. Gubler, F.; Hughes, T.; Waterhouse, P.; Jacobsen, J. Regulation of dormancy in barley by blue light and after-ripening: Effects on abscisic acid and gibberellin metabolism. *Plant Physiol.* **2008**, *147*, 886–896. [CrossRef] [PubMed]

48. Hoang, H.H.; Sotta, B.; Gendreau, E.; Bailly, C.; Leymarie, J.; Corbineau, F. Water content: A key factor of the induction of secondary dormancy in barley grains as related to ABA metabolism. *Physiol. Plant.* **2013**, *148*, 284–296. [CrossRef] [PubMed]

49. Jacobsen, J.V.; Barrero, J.M.; Hughes, T.; Julkowska, M.; Taylor, J.M.; Xu, Q.; Gubler, F. Roles for blue light, jasmonate and nitric oxide in the regulation of dormancy and germination in wheat grain (*Triticum aestivum* L.). *Planta* **2013**, *238*, 121–138. [CrossRef]

50. Née, G.; Xiang, Y.; Soppe, W.J. The release of dormancy, a wake-up call for seeds to germinate. *Curr. Opin. Plant Biol.* **2017**, *35*, 8–14. [CrossRef] [PubMed]

51. Penfield, S. Seed dormancy and germination. *Curr. Biol.* **2017**, *27*, R874–R878. [CrossRef] [PubMed]

52. Schramm, E.C.; Nelson, S.K.; Kidwell, K.K.; Steber, C.M. Increased ABA sensitivity results in higher seed dormancy in soft white spring wheat cultivar 'Zak'. *Theor. Appl. Genet.* **2013**, *126*, 791–803. [CrossRef] [PubMed]

53. Bentsink, L.; Jowett, J.; Hanhart, C.J.; Koornneef, M. Cloning of DOG1, a quantitative trait locus controlling seed dormancy in Arabidopsis. *Proc. Natl. Acad. Sci. USA* **2006**, *103*, 17042–17047. [CrossRef] [PubMed]

54. Nakabayashi, K.; Bartsch, M.; Xiang, Y.; Miatton, E.; Pellengahr, S.; Yano, R.; Seo, M.; Soppe, W.J.J. The time required for dormancy release in Arabidopsis is determined by DELAY OF GERMINATION1 protein levels in freshly harvested seeds. *Plant Cell* **2012**, *24*, 2826–2838. [CrossRef] [PubMed]

55. Née, G.; Kramer, K.; Nakabayashi, K.; Yuan, B.; Xiang, Y.; Miatton, E.; Finkemeier, I.; Soppe, W.J.J. DELAY OF GERMINATION1 requires PP2C phosphatases of the ABA signalling pathway to control seed dormancy. *Nat. Commun.* **2017**, *8*, 72. [CrossRef] [PubMed]

56. Nishimura, N.; Tsuchiya, W.; Moresco, J.J.; Hayashi, Y.; Satoh, K.; Kaiwa, N.; Irisa, T.; Kinoshita, T.; Schroeder, J.I.; Yates, J.R.; et al. Control of seed dormancy and germination by DOG1-AHG1 PP2C phosphatase complex via binding to heme. *Nat. Commun.* **2018**, *9*, 2132. [CrossRef] [PubMed]

57. Hattori, T.; Totsuka, M.; Hobo, T.; Kagaya, Y.; Yamamoto-Toyoda, A. Experimentally determined sequence requirement of ACGT-containing abscisic acid response element. *Plant Cell Physiol.* **2002**, *43*, 136–140. [CrossRef]

58. Zhang, W.; Ruan, J.; Ho, T.-H.D.; You, Y.; Yu, T.; Quatrano, R.S. Cis-regulatory element based targeted gene finding: Genome-wide identification of abscisic acid- and abiotic stress-responsive genes in *Arabidopsis thaliana*. *Bioinformatics* **2005**, *21*, 3074–3081. [CrossRef]

59. Bensmihen, S. The homologous ABI5 and EEL transcription factors function antagonistically to fine-tune gene expression during late embryogenesis. *Plant Cell* **2002**, *14*, 1391–1403. [CrossRef]

60. Bensmihen, S. Characterization of three homologous basic leucine zipper transcription factors (bZIP) of the ABI5 family during *Arabidopsis thaliana* embryo maturation. *J. Exp. Bot.* **2005**, *56*, 597–603. [CrossRef]

61. Kobayashi, Y.; Murata, M.; Minami, H.; Yamamoto, S.; Kagaya, Y.; Hobo, T.; Yamamoto, A.; Hattori, T. Abscisic acid-activated SNRK2 protein kinases function in the gene-regulation pathway of ABA signal transduction by phosphorylating ABA response element-binding factors. *Plant J.* **2005**, *44*, 939–949. [CrossRef] [PubMed]

62. Furihata, T.; Maruyama, K.; Fujita, Y.; Umezawa, T.; Yoshida, R.; Shinozaki, K.; Yamaguchi-Shinozaki, K. Abscisic acid-dependent multisite phosphorylation regulates the activity of a transcription activator AREB1. *Proc. Natl. Acad. Sci. USA* **2006**, *103*, 1988–1993. [CrossRef]

63. Yoshida, T.; Fujita, Y.; Maruyama, K.; Mogami, J.; Todaka, D.; Shinozaki, K.; Yamaguchi-Shinozaki, K. Four Arabidopsis AREB/ABF transcription factors function predominantly in gene expression downstream of SnRK2 kinases in abscisic acid signalling in response to osmotic stress. *Plant Cell Environ.* **2015**, *38*, 35–49. [CrossRef] [PubMed]

64. Antoni, R.; Gonzalez-Guzman, M.; Rodriguez, L.; Rodrigues, A.; Pizzio, G.A.; Rodriguez, P.L. Selective inhibition of clade A phosphatases type 2C by PYR/PYL/RCAR abscisic acid receptors. *Plant Physiol.* **2012**, *158*, 970–980. [CrossRef]

65. Lynch, T.; Erickson, B.J.; Finkelstein, R.R. Direct interactions of ABA-insensitive (ABI)-clade protein phosphatase(PP) 2Cs with calcium-dependent protein kinases and ABA response element-binding bZIPs may contribute to turning off ABA response. *Plant Mol. Biol.* **2012**, *80*, 647–658. [CrossRef] [PubMed]

66. Lee, J.-H.; Yoon, H.-J.; Terzaghi, W.; Martinez, C.; Dai, M.; Li, J.; Byun, M.-O.; Deng, X.W. DWA1 and DWA2, Two Arabidopsis DWD protein components of CUL4-based E3 ligases, act together as negative regulators in ABA signal transduction. *Plant Cell* **2010**, *22*, 1716–1732. [CrossRef] [PubMed]

67. Liu, H.; Stone, S.L. Abscisic acid increases Arabidopsis ABI5 transcription factor levels by promoting KEG E3 ligase self-ubiquitination and proteasomal degradation. *Plant Cell* **2010**, *22*, 2630–2641. [CrossRef]

68. Chen, Y.-T.; Liu, H.; Stone, S.; Callis, J. ABA and the ubiquitin E3 ligase KEEP ON GOING affect proteolysis of the *Arabidopsis thaliana* transcription factors ABF1 and ABF3. *Plant J.* **2013**, *75*, 965–976. [CrossRef]

69. Lyzenga, W.J.; Liu, H.; Schofield, A.; Muise-Hennessey, A.; Stone, S.L. Arabidopsis CIPK26 interacts with KEG, components of the ABA signalling network and is degraded by the ubiquitin–proteasome system. *J. Exp. Bot.* **2013**, *64*, 2779–2791. [CrossRef]

70. Seo, K.-I.; Lee, J.-H.; Nezames, C.D.; Zhong, S.; Song, E.; Byun, M.-O.; Deng, X.W. ABD1 is an Arabidopsis DCAF substrate receptor for CUL4-DDB1–based E3 ligases that acts as a negative regulator of abscisic acid signaling. *Plant Cell* **2014**, *26*, 695–711. [CrossRef]

71. Jacobsen, J.V.; Pearce, D.W.; Poole, A.T.; Pharis, R.P.; Mander, L.N. Abscisic acid, phaseic acid and gibberellin contents associated with dormancy and germination in barley. *Physiol. Plant.* **2002**, *115*, 428–441. [CrossRef] [PubMed]

72. Barrero, J.M.; Downie, A.B.; Xu, Q.; Gubler, F. A Role for Barley CRYPTOCHROME1 in light regulation of grain dormancy and germination. *Plant Cell* **2014**, *26*, 1094–1104. [CrossRef]
73. Nakagami, H.; Sugiyama, N.; Mochida, K.; Daudi, A.; Yoshida, Y.; Toyoda, T.; Tomita, M.; Ishihama, Y.; Shirasu, K. Large-scale comparative phosphoproteomics identifies conserved phosphorylation sites in plants. *Plant Physiol.* **2010**, *153*, 1161–1174. [CrossRef] [PubMed]
74. Sugiyama, N.; Masuda, T.; Shinoda, K.; Nakamura, A.; Tomita, M.; Ishihama, Y. Phosphopeptide enrichment by aliphatic hydroxy acid-modified metal oxide chromatography for nano-LC-MS/MS in proteomics applications. *Mol. Cell. Proteomics* **2007**, 1103–1109. [CrossRef] [PubMed]
75. MacLean, B.; Tomazela, D.M.; Shulman, N.; Chambers, M.; Finney, G.L.; Frewen, B.; Kern, R.; Tabb, D.L.; Liebler, D.C.; MacCoss, M.J. Skyline: An open source document editor for creating and analyzing targeted proteomics experiments. *Bioinformatics* **2010**, *26*, 966–968. [CrossRef] [PubMed]
76. Mardia, K.V.; Kent, J.T.; Bibby, J.M. *Multivariate Analysis*; Academic Press: London, UK; New York, NY, USA, 1980.
77. Becker, R.A.; Chambers, J.M.; Wilks, A.R. *New S Language*, 1st ed.; Chapman and Hall/CRC: Pacific Grove, CA, USA, 1988.
78. Venables, W.N.; Ripley, B.D. *Modern Applied Statistics with S*, 4th ed.; Springer: New York, NY, USA, 2002.
79. Chou, M.F.; Schwartz, D. Biological sequence motif discovery using motif-x. *Curr. Protoc. Bioinform.* **2011**, 13–15. [CrossRef]
80. Moreno-Hagelsieb, G.; Latimer, K. Choosing BLAST options for better detection of orthologs as reciprocal best hits. *Bioinformatics* **2008**, *24*, 319–324. [CrossRef]

International Journal of
Molecular Sciences

MDPI

Article

Organ-Specific Analysis of *Morus alba* Using a Gel-Free/Label-Free Proteomic Technique

Wei Zhu [1,†], Zhuoheng Zhong [1,†], Shengzhi Liu [1], Bingxian Yang [1], Setsuko Komatsu [2], Zhiwei Ge [3] and Jingkui Tian [1,4,*]

[1] College of Biomedical Engineering and Instrument Science, Zhejiang University, Hangzhou 310027, China; rutin@zju.edu.cn (W.Z.); zhongzhh@zju.edu.cn (Z.Z.); bylzs8410@163.com (S.L.); xianyb@zju.edu.cn (B.Y.)
[2] Faculty of Environmental and Information Sciences, Fukui University of Technology, Fukui 910-8505, Japan; skomatsu@fukui-ut.ac.jp
[3] Analysis Center of Agrobiology and Environmental Sciences, Zhejiang University, Hangzhou 310027, China; gezw@zju.edu.cn
[4] Key Laboratory for Biomedical Engineering of Ministry of Education, Zhejiang-Malaysia Joint Research Center for Traditional Medicine, Zhejiang University, Hangzhou 310027, China
* Correspondence: tjk@zju.edu.cn; Tel.: +86-571-88273823; Fax: +86-571-87951676
† These authors contributed equally to this work.

Received: 15 December 2018; Accepted: 15 January 2019; Published: 16 January 2019

Abstract: *Morus alba* is an important medicinal plant that is used to treat human diseases. The leaf, branch, and root of *Morus* can be applied as antidiabetic, antioxidant, and anti-inflammatory medicines, respectively. To explore the molecular mechanisms underlying the various pharmacological functions within different parts of *Morus*, organ-specific proteomics were performed. Protein profiles of the *Morus* leaf, branch, and root were determined using a gel-free/label-free proteomic technique. In the *Morus* leaf, branch, and root, a total of 492, 414, and 355 proteins were identified, respectively, including 84 common proteins. In leaf, the main function was related to protein degradation, photosynthesis, and redox ascorbate/glutathione metabolism. In branch, the main function was related to protein synthesis/degradation, stress, and redox ascorbate/glutathione metabolism. In root, the main function was related to protein synthesis/degradation, stress, and cell wall. Additionally, organ-specific metabolites and antioxidant activities were analyzed. These results revealed that flavonoids were highly accumulated in *Morus* root compared with the branch and leaf. Accordingly, two root-specific proteins named chalcone flavanone isomerase and flavonoid 3,5-hydroxylase were accumulated in the flavonoid pathway. Consistent with this finding, the content of the total flavonoids was higher in root compared to those detected in branch and leaf. These results suggest that the flavonoids in *Morus* root might be responsible for its biological activity and the root is the main part for flavonoid biosynthesis in *Morus*.

Keywords: *Morus*; organ; gel-free/label-free proteomics; flavonoid; antioxidant activity

---

1. Introduction

Mulberry tree (*Morus alba* L.) is a deciduous woody shrub in the family Moraceae and widely cultivated in China, Korea, India, and Japan [1]. In addition to its use in sericulture, *Morus* can be used in fruit production, tolerating saline soils, and soil retention in loess soils [2,3]. In China, different parts of the mulberry tree have a long history of being used in traditional Chinese medicine to treat human diseases such as diabetes, arthritis, and rheumatism [4]. Therefore, *Morus* has attracted attention for its pharmaceutical value. For example, the mulberry leaf has been proven to modulate the cardiovascular system through endothelial nitric oxide synthase signaling [5] and mulberry-leaf polysaccharides, which are one of the main active components in mulberry leaf, have been purified from an ethanol extraction and showed potential antioxidative activities [6]. The mulberry-branch bark had a powerful antidiabetic effect that could rescue gluconeogenesis and glycogen synthesis by protecting genes in the phosphatidylinositol-3 kinase and protein kinase B signaling pathways [7]. Furthermore, the ethanol extraction of mulberry root bark could effectively ameliorate hyperlipidemia and four major active compounds, including mulberrofuran C, sanggenon G, moracin O, and moracin P, were isolated [8]. However, the mechanisms for the different biological activity in different parts of *Morus* remain unclear.

For most medicinal plants, their pharmaceutical value varies in different parts of the plant. For example, the *Scutellaria baicalensis* root rather than its aerial parts can be used as a traditional Chinese medicine for its anticancer, anti-HIV, and antibacterial effects [9]. The flower from *Coreopsis tinctorial* has been reported to possess antioxidative and antidiabetic activities and is used as a health food, while its stems and leaves are commonly discarded [10]. Organ- and tissue-specific studies are an effective way to discover the reason for these phenomena and aid in the quality control and pharmacological evaluation of medicinal plants, especially for medicinal plants such as *Morus* whose vegetative organs (root, branch, and leaf) can all be used as medicines. A comparative study on the antioxidant activity and phenolic contents of methanol extractions from mulberry leaf, stem bark, fruit, and root bark indicated that the mulberry stem bark had the highest antioxidant activity [11]. In contrast, the ethanol extraction from mulberry leaf had a higher antioxidant activity than the fruit and stem extractions [12]. Furthermore, the antioxidant activity of the ethanolic extraction from mulberry twigs was better than that from mulberry root bark [13]. However, there have been no systemic studies comparing the antioxidant activities among mulberry root, branch, and leaf, and the mechanisms of their different antioxidant activities are still unknown.

Omics technologies allow for the analysis of the complete set of genomes and are the most popular approaches for performing systemic studies [14]. The majority of functional genomics is based on transcriptomics, proteomics and metabolomics [15]. Proteomic technologies provide one of the best choices for the functional analysis of translated parts of the genome and have been applied in organ-/tissue-specific studies in plants. Zhu et al. [16] performed organ-specific proteomic analysis on the medicinal plant *Mahonia* and successfully identified the root-specific expressed proteins S-adenosylmethionine synthetase and (S)-tetrahydroprotoberberine. As the roots accumulated the most alkaloids, such as columbamine, jatrorrhizine, palmatine, tetrandrine, and berberine, these proteins are assumed to be involved in alkaloid biosynthesis. Ji et al. [17] conducted a comparative proteomics analysis using healthy and infected leaves from *Morus* to study the dwarf response mechanism. Therefore, proteomic approaches can provide significantly more detailed protein information in different organs, which may be beneficial for revealing important mechanisms.

Although the leaf, branch, and root of *Morus* can all be used as traditional Chinese medicine, additional studies are needed to identify the organ-specific factors with medicinal value. In this study, to uncover the molecular mechanisms of the different pharmacological functions in *Morus* leaf, branch, and root, gel-free/label-free proteomic approach was used. Bioinformatic, phytochemical, and qRT-PCR techniques were used for confirmation of results from proteomics.

2. Results

2.1. The Metabolite Contents and Antioxidant Activity in Morus Root were Higher than in Branch and Leaf

To reveal differences among the leaf, branch, and root in *Morus*, five secondary metabolites were identified in the three organs. The leaf, branch and root were collected from mulberry trees and methanol extracts from the three organs were examined by HPLC analysis. From the HPLC chromatograms detected at 320 nm, 24, 18, and 40 peaks were observed in the extracts from the leaf, branch, and root, respectively (Figure 1A). Five major secondary metabolites including mulberroside A, oxyresveratrol, kuwanone H, chalcomoracin, and morusin were identified in the root, while only three of the five metabolites (mulberroside A, oxyresveratrol, and morusin) were detected in the branch and two of the five metabolites (mulberroside A, and chalcomoracin) were detected in the leaf. Except for oxyresveratrol, the contents of the other four metabolites in the root are more than 10 times higher than those in the leaf and branch (Table 1). To understand the biological differences among the three organs in Morus, the antioxidant activities were evaluated by using an ABTS+ scavenging activity, hydroxyl free radical, and O_2^- scavenging activity assays. As a result, the root had the strongest ABTS+ scavenging activity, O_2^- scavenging activity, and hydroxyl free radical inhibition activity, while the leaf had the worst antioxidant activity (Figure 1B).

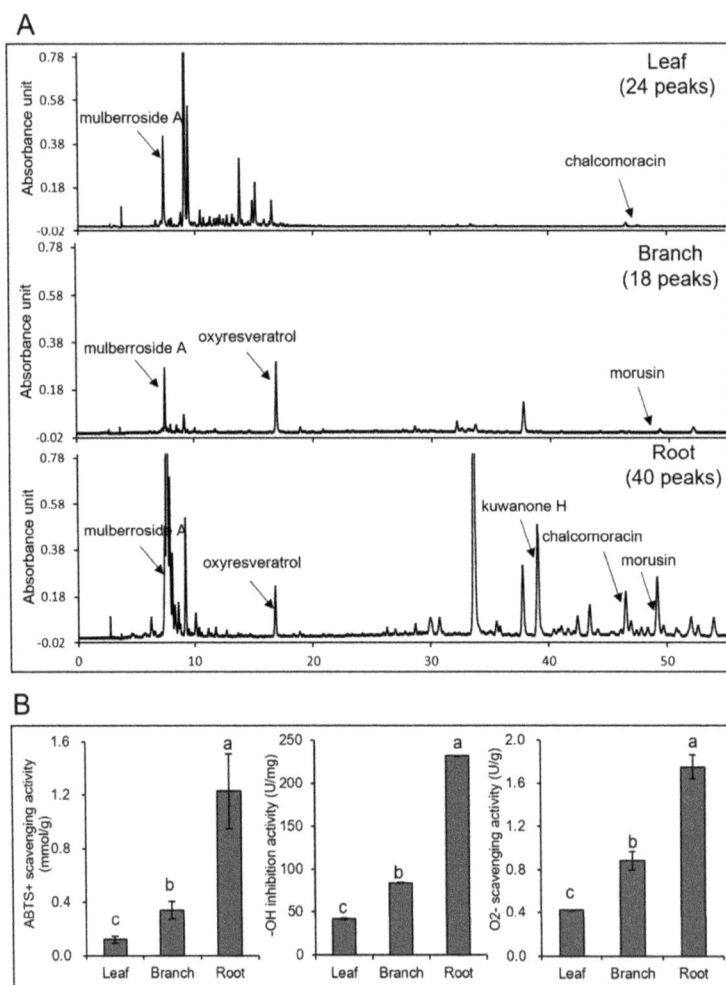

Figure 1. Analysis of the metabolites in *Morus* leaf, branch, and root and their antioxidant activities. The methanol extracts from *Morus* leaf, branch, and root were analyzed by HPLC (**A**). A C18 column was used with a flow rate of 1 mL min^{-1}. The peaks were determined at a wavelength of 320 nm. For the determination of the antioxidant activities of *Morus* leaf, branch, and root, ABTS$^+$ scavenging, hydroxyl free radical, and O_2^- scavenging activities were analyzed (**B**). The data are shown as the mean ± SD from three independent biological replicates. Means with the same letter are not significantly different according to the one-way ANOVA test ($p < 0.05$).

Table 1. Contents of five secondary metabolites in different organs of *Morus*.

	Mulberroside A (mg/g DW)	Oxyresveratrol (mg/g DW)	Kuwanone H (mg/g DW)	Chalcomoracin (mg/g DW)	Morusin (mg/g DW)
Leaf	0.917 ± 0.015	n.d. *	n.d.	0.045 ± 0.005	n.d.
Branch	0.451 ± 0.012	0.453 ± 0.008	n.d.	n.d.	0.043 ± 0.008
Root	24.206 ± 0.688	0.345 ± 0.022	5.551 ± 0.226	0.610 ± 0.051	2.874 ± 0.158

* n.d. means the metabolites is not determined in this organ; DW, means dry weight.

2.2. A Total of 257, 148, and 170 Proteins were Specific to Leaf, Branch, and Root in Morus, Respectively

To analyze the mechanisms functioning within the three organs, a gel-free proteomics approach was used to identify and determine the abundance of proteins in mulberry leaf, branch, and root with the help of nano LC-MS/MS. A total of 492, 414, and 355 proteins with more than two matched peptides were identified in the leaf (Supplemental Table S1), branch (Supplemental Table S2) and root (Supplemental Table S3), respectively. Among these proteins, 257 (257/492, 52%), 148 (148/414, 36%), and 170 (170/355, 48%) were specific to the leaf, branch, and root, respectively (Figure 2).

To determine the biological processes involved in the three organs, the identified proteins were functionally classified using MapMan bin codes (Figure 2, Supplemental Tables S1–S3, Supplemental Figure S4. In leaf, the main functional categories were related to protein metabolism (82/492, 17%), photosynthesis (64/492, 13%), redox ascorbate/glutathione metabolism (31/492, 6%), stress (20/492, 4%), and tricarboxylic acid cycle (TCA, 17/492, 3%). In branch, the main functional categories were related to protein metabolism (67/414, 16%), stress (32/414, 8%), photosynthesis (30/414, 7%), redox ascorbate/glutathione metabolism (28/414, 7%), and cell cycle/cell organization and division (17/414, 4%). In root, the main functional categories were related to protein metabolism (56/355, 16%), stress (33/355, 9%), cell wall (20/355, 6%), RNA metabolism (19/355, 5%), and redox ascorbate/glutathione metabolism (16/355, 5%). Based on these results, proteins related to protein synthesis/degradation comprised the main functional category in all three organs. The abundance of proteins related to redox, TCA cycle, and glycolysis largely differed in the root samples compared to the leaf and branch samples (Figure 2).

2.3. Functional Characterization of Organ-Specific Proteins Identified from Morus

To determine the organ-specific protein expression patterns within each organ, leaf-, branch-, and root-specific proteins were functionally categorized and visualized using the MapMan software (version 3.6.0RC1, Aachen, Germany) (Figure 3). In the root, the number of cell wall-related proteins (10 proteins) was larger than in the leaf (four proteins) and branch (two proteins). Moreover, the proteins related to lipid metabolism identified in the root (four proteins) are more than the other organs (one protein for branch and two proteins for leaf). Additionally, five root-specific proteins related to major carbohydrate metabolism are functionally categorized as the degradation of sucrose and starch, while two leaf-specific proteins are involved in the synthesis of sucrose. In amino acid synthesis/metabolism, four root-specific proteins were found to be involved with amino acids; these belonged to the glutamate and aspartate families. There were also four leaf-specific proteins related to amino acid degradation/metabolism.

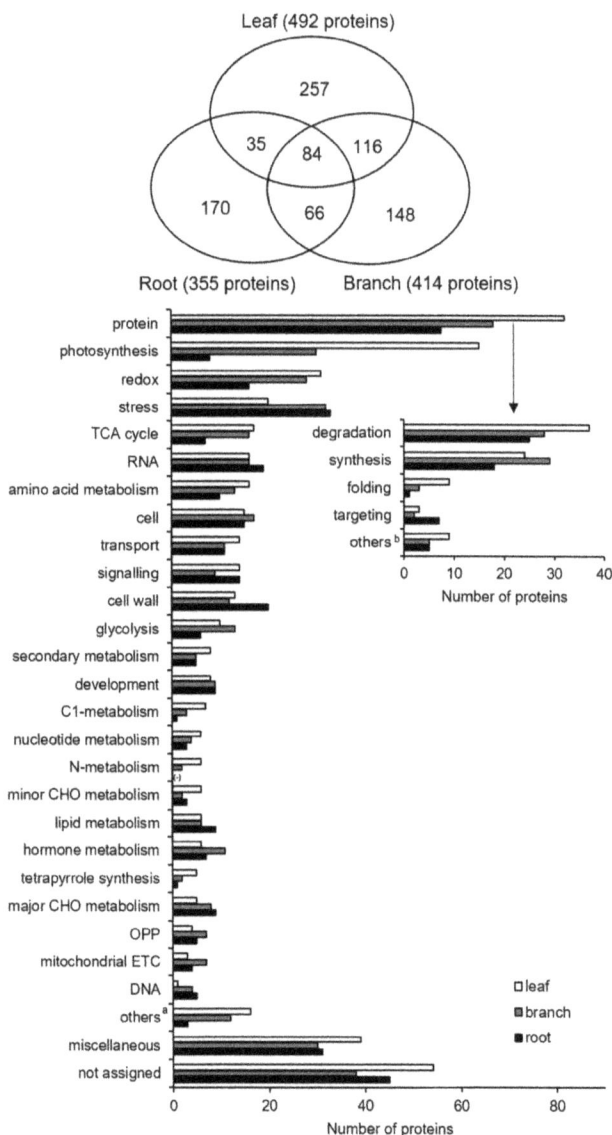

Figure 2. Functional categorization of proteins in leaf, branch, and root from *Morus*. Leaf (white), branch (gray), and root (black) samples were collected and proteins were extracted, digested, and analyzed by nanoLC-MS/MS. Protein functions were predicted and categorized using MapMan bin codes. Abbreviations: redox, redox ascorbate/glutathione metabolism; TCA, tricarboxylic acid; RNA, RNA processing and regulation of transcription; cell, cell organization, and vesicle transport; CHO, carbohydrates; OPP, oxidative pentose phosphate; ETC, electron transport chains; and DNA, DNA synthesis, and repair. [a] Others, containing biodegradation of xenobiotics, co-factor and vitamin metabolism, S-assimilation, gluconeogenesis, fermentation, and metal handling. [b] Others, containing amino acid activation, posttranslational modification, and assembly/cofactor ligation.

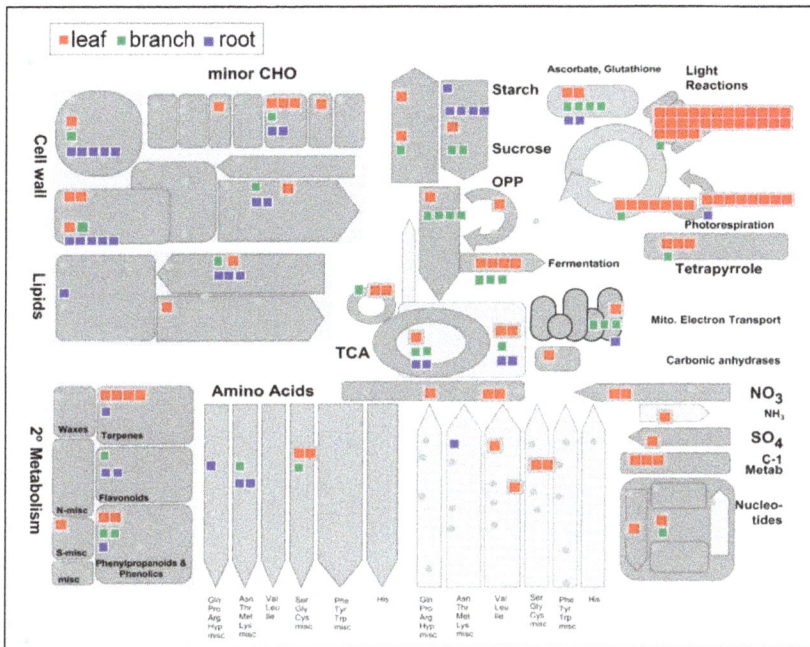

Figure 3. Organ-specific proteins were functionally categorized and visualized using the MapMan software. Leaf- (red), branch- (green), and root-(blue) specific proteins were submitted to the MapMan software (version 3.6.0RC1) using the metabolism overview pathway map. Each square indicates one mapped protein.

There were several specifically identified proteins mapped to secondary metabolism that exhibited organ-specific expression patterns. Therefore, the organ-specific proteins involved in secondary metabolism were visualized using the MapMan software (version 3.6.0RC1, Aachen, Germany) (Supplemental Figure S1). There were eight, five, and four specifically identified proteins from the leaf, branch, and root samples, respectively. Leaf-specific proteins related to secondary metabolism mainly accumulated in the non-MVA and phenlypropanoid pathways. Branch-specific proteins related to secondary metabolism mainly accumulated in the phenlypropanoid and lignin/lignan pathways. Notably, the root-specific proteins related to secondary metabolism, such as chalcone flavanone isomerase and flavonoid 3,5-hydroxylase, were accumulated in the flavonoid pathway, especially for the biosynthesis of chalcones and dihydroflavonols (Figure 4).

2.4. Largely Differential Common Proteins Were Identified among Three Organs in Morus

To further analyze the differences in the proteins among the three organs, Venn diagram analysis was performed on the proteins identified in the three organs (Figure 2). A total of 84 proteins was commonly identified from the three organs, and the abundance of peroxidase (protein number 2) in the root was significantly higher than in the branch and leaf (Table 2). The abundance of proteasome (protein number 9) in the root was approximately six times higher than that found in the leaf. However, isoflavone reductase homolog P3 (protein number 48), which is the only protein involved in secondary metabolism from the commonly identified protein group, was the most abundant in the branch samples (1.17 mol%) compared to the leaf (0.33 mol%) and root (0.37 mol%) samples. Furthermore, the abundance of triosephosphate isomerase in the leaf is approximately ten times higher than in the root.

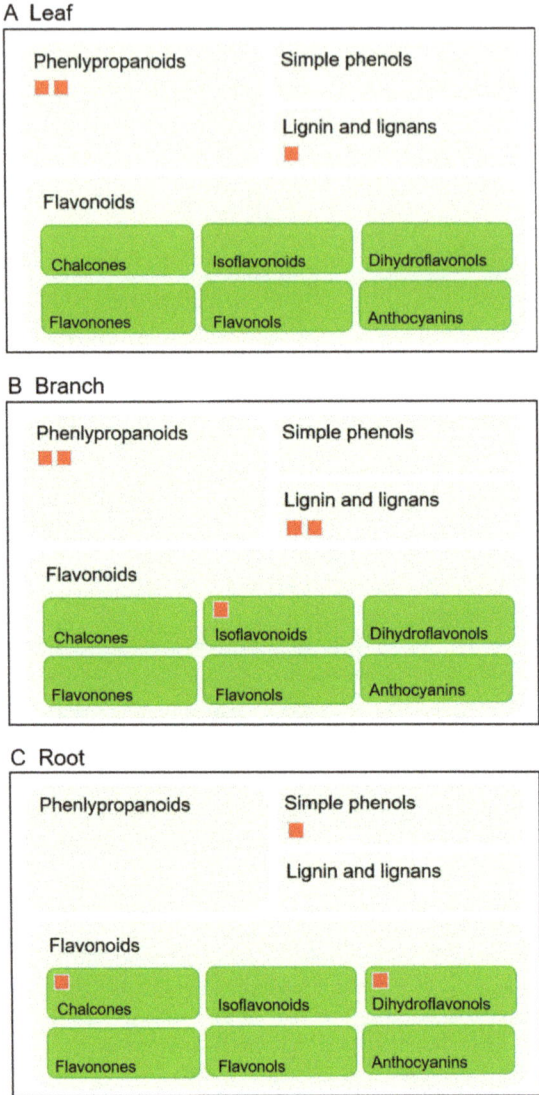

Figure 4. Comparison of the organ-specific proteins related to the flavonoid, phenlypropanoid, simple phenol, and lignin. Leaf- (**A**), branch- (**B**), and root- (**C**) specific proteins related to secondary metabolism were submitted to the MapMan software (version 3.6.0RC1). Each red square indicates one mapped protein.

Table 2. Common proteins identified in the leaf, branch, and root of *Morus* by gel-free/label-free proteomic analysis.

No.	Protein ID [a]	Description	M.P. [b]	Score	Mass (Da)	Function [c]	Mol (%) [d] leaf	stem	root
1	Morus009492-p1	Macrophage migration inhibitory factor homolog	8	203	8961	not assigned	3.10	2.43	3.69
2	Morus001961-p1	Peroxidase 12	16	244	38,426	misc	0.57	1.29	3.62
3	Morus018316-p1	Superoxide dismutase 1 copper chaperone	9	339	11,171	metal handling	2.11	1.69	2.48
4	Morus009000-p1	60S acidic ribosomal protein P2B	12	110	11,673	protein	1.62	2.32	2.32
5	Morus023628-p1	Tubulin beta-1 chain	22	578	51,015	cell	1.11	1.83	1.99
6	Morus017847-p1	Ribonuclease UK114	16	266	19,960	RNA	1.08	0.27	1.80
7	Morus017207-p1	Proteasome subunit alpha type-4	6	173	27,440	protein	0.24	0.34	1.58
8	Morus003952-p1	Lipoxygenase homology domain-containing protein 1	6	154	21,171	not assigned	0.50	0.95	1.44
9	Morus022430-p1	Proteasome subunit beta type-1	11	313	24,861	protein	0.68	0.56	1.44
10	Morus022592-p1	Thaumatin-like protein 1a	18	362	26,984	stress	1.02	0.44	1.41
11	Morus022525-p1	Calmodulin	9	250	16,894	signaling	1.03	1.12	1.31
12	Morus015082-p1	Auxin-repressed 12.5 kDa protein	3	89	13,355	development	0.49	1.99	1.30
13	Morus003616-p1	Fructokinase-2	10	137	35,370	major CHO metabolism	0.42	0.79	1.25
14	Morus017382-p1	Calcium-binding protein CML27	4	116	18,705	signaling	0.26	0.29	1.05
15	Morus008669-p1	Allene oxide cyclase 2, chloroplastic	6	168	27,569	hormone metabolism	0.60	1.24	0.99
16	Morus010743-p1	Triosephosphate isomerase, cytosolic	19	263	27,548	glycolysis	1.27	1.66	0.99
17	Morus004210-p1	Glucan endo-1,3-beta-glucosidase, basic vacuolar isoform	20	178	39,002	misc	0.63	0.34	0.94
18	Morus011779-p1	Superoxide dismutase [Cu-Zn], chloroplastic	22	579	29,603	redox	0.74	0.31	0.90
19	Morus001936-p1	Peroxiredoxin-2B	17	350	17,391	redox	1.22	1.48	0.90
20	Morus004201-p1	Universal stress protein A-like protein	9	223	18,591	stress	0.58	0.58	0.82
21	Morus001634-p1	Nucleoside diphosphate kinase 1	11	215	16,322	nucleotide metabolism	1.35	0.78	0.78
22	Morus003013-p1	Phosphoglycerate kinase, cytosolic	59	1143	42,729	glycolysis	1.72	2.17	0.75
23	Morus028068-p1	Polygalacturonase inhibitor 1	10	306	37,677	cell wall	0.86	0.73	0.73
24	Morus018807-p1	Fructose-bisphosphate aldolase, cytoplasmic isozyme	23	584	38,459	glycolysis	0.92	1.82	0.71
25	Morus020532-p1	Glutaredoxin	5	244	15,307	redox	0.75	0.69	0.69
26	Morus023908-p1	Uncharacterized protein	15	271	57,888	protein	0.50	1.55	0.69
27	Morus018475-p1	Peroxidase 54	5	128	36,921	misc	0.13	0.47	0.67
28	Morus025517-p1	Tubulin alpha chain	18	426	49,920	cell	0.68	0.72	0.66
29	Morus002489-p1	Nascent polypeptide-associated complex subunit alpha-like protein 1	11	285	22,279	protein	1.01	0.34	0.65
30	Morus014304-p1	Plastocyanin, chloroplastic	23	661	16,620	photosynthesis	1.16	0.62	0.62
31	Morus024265-p1	Aquaporin PIP1-3	5	67	30,856	transport	0.22	0.37	0.59
32	Morus008884-p1	Cysteine proteinase 3D21a	10	350	52,217	protein	0.35	0.32	0.58
33	Morus025862-p1	ATP synthase subunit beta, mitochondrial	33	1017	59,400	mitochondrial electron transport	1.06	1.43	0.54
34	Morus026982-p1	Allene oxide synthase, chloroplastic	11	164	56,861	hormone metabolism	0.30	0.91	0.54
35	Morus007342-p1	Peroxiredoxin-2F, mitochondrial	12	178	22,580	redox	0.77	1.03	0.52
36	Morus009738-p1	ATP-dependent Clp protease proteolytic subunit 5, chloroplastic	4	146	34,203	protein	0.13	0.21	0.52
37	Morus009210-p1	60S acidic ribosomal protein P3-2	2	71	12,022	protein	0.42	0.48	0.48

Table 2. Cont.

No.	Protein ID [a]	Description	M.P. [b]	Score	Mass (Da)	Function [c]	Mol (%) [d]		
							leaf	stem	root
38	Morus007901.p1	Actin-7	30	717	41,897	cell	1.02	1.20	0.47
39	Morus007352.p1	Stem-specific protein TSJT1	7	96	25,521	metal handling	0.18	0.45	0.45
40	Morus026327.p1	Heat shock cognate 70 kDa protein 1	41	791	71,553	stress	0.91	0.67	0.44
41	Morus021433.p1	Malate dehydrogenase, cytoplasmic	20	463	35,912	TCA	0.57	0.95	0.43
42	Morus006184.p1	Cysteine synthase	33	572	34,400	amino acid metabolism	1.77	0.59	0.39
43	Morus018842.p1	2-Cys peroxiredoxin BAS1-like, chloroplastic	23	321	29,121	redox	0.81	0.65	0.39
44	Morus022454.p1	Fasciclin-like arabinogalactan protein 8	6	120	43,455	cell wall	0.22	0.16	0.39
45	Morus008883.p1	Uncharacterized protein	6	104	49,487	signaling	0.17	0.34	0.38
46	Morus000210.p1	Calvin cycle protein CP12	8	288	14,542	photosynthesis	0.34	0.38	0.38
47	Morus002920.p1	Thioredoxin M-type 4, chloroplastic	6	164	20,233	redox	0.42	0.26	0.37
48	Morus018564.p1	Isoflavone reductase homolog P3	12	102	45,171	secondary metabolism	0.33	1.17	0.37
49	Morus013051.p1	Adenosine kinase 2	7	216	37,797	nucleotide metabolism	0.65	0.41	0.35
50	Morus000836.p1	Ribulose bisphosphate carboxylase large chain (Fragment)	290	4805	61,599	photosynthesis	1.18	0.63	0.33
51	Morus015202.p1	Uncharacterized protein	11	234	33,994	not assigned	0.46	0.39	0.32
52	Morus014140.p1	Plastid-lipid-associated protein, chloroplastic	10	281	35,137	cell	0.52	0.31	0.31
53	Morus025784.p1	Phospholipase D alpha 1	4	47	92,059	lipid metabolism	0.05	0.37	0.30
54	Morus018550.p1	Glycine-rich RNA-binding protein GRP1A	5	165	18,416	RNA	0.39	0.29	0.29
55	Morus014011.p1	Glycerophosphoryl diester phosphodiesterase 2	6	295	81,816	lipid metabolism	0.16	0.12	0.29
56	Morus019087.p1	Putative mitochondrial 2-oxoglutarate/malate carrier protein	12	194	32,224	transport	0.98	0.29	0.28
57	Morus002874.p1	Leucine aminopeptidase 3, chloroplastic	39	981	60,563	protein	0.96	1.09	0.27
58	Morus010230.p1	Superoxide dismutase [Cu-Zn]	2	63	20,420	redox	0.34	0.37	0.26
59	Morus019413.p1	Cysteine proteinase 15A	9	259	41,574	protein	0.23	0.12	0.26
60	Morus015818.p1	Probable glucan endo-1,3-beta-glucosidase A6	3	111	52,145	misc	0.15	0.17	0.24
61	Morus020384.p1	Cysteine synthase, chloroplastic/chromoplastic	21	258	43,997	amino acid metabolism	0.52	0.11	0.24
62	Morus013361.p1	Protein disulfide-isomerase	15	432	56,492	redox	0.66	0.92	0.22
63	Morus007114.p1	Glycine-rich RNA-binding protein 2	3	170	27,802	RNA	0.17	0.19	0.19
64	Morus011198.p1	L-ascorbate peroxidase, cytosolic	26	392	27,414	redox	1.05	1.13	0.19
65	Morus008123.p1	IAA-amino acid hydrolase ILR1-like 5	5	89	47,707	hormone metabolism	0.15	0.50	0.18
66	Morus024851.p1	Catalase isozyme 1	19	137	57,208	redox	0.62	0.32	0.18
67	Morus017351.p1	Serine carboxypeptidase-like 50	5	55	49,604	protein	0.15	0.14	0.17
68	Morus014667.p1	Alpha-xylosidase	15	257	103,539	misc	0.17	0.45	0.15
69	Morus017174.p1	Predicted protein	8	193	33,060	signaling	0.20	0.27	0.15
70	Morus024951.p1	Triosephosphate isomerase, chloroplastic	42	655	34,813	photosynthesis	1.66	0.99	0.15
71	Morus008661.p1	14-3-3-like protein A	15	283	81,889	cell	0.26	0.37	0.14
72	Morus016271.p1	Elongation factor 2	14	106	99,403	protein	0.11	0.44	0.14
73	Morus001657.p1	6-phosphogluconolactonase 4, chloroplastic	6	108	35,151	OPP	0.18	0.97	0.14
74	Morus009365.p1	5-methyltetrahydropteroyltriglutamate–homocysteine methyltransferase	13	188	84,904	amino acid metabolism	0.17	0.46	0.14

Table 2. *Cont.*

No.	Protein ID [a]	Description	M.P. [b]	Score	Mass (Da)	Function [c]	Mol (%) [d]		
							leaf	stem	root
75	Morus007784.p1	UTP–glucose-1-phosphate uridylyltransferase	26	372	76,133	glycolysis	0.48	0.80	0.13
76	Morus017695.p1	31 kDa ribonucleoprotein, chloroplastic	10	230	38,128	RNA	0.17	0.13	0.13
77	Morus007494.p1	RuBisCO large subunit-binding protein subunit alpha, chloroplastic	30	964	62,000	photosynthesis	1.29	0.33	0.11
78	Morus011664.p1	L-ascorbate oxidase homolog	4	53	60,522	not assigned	0.13	0.37	0.11
79	Morus006060.p1	V-type proton ATPase subunit B2	10	400	63,333	transport	0.41	0.16	0.11
80	Morus004111.p1	Calreticulin	7	266	50,196	signaling	0.32	0.56	0.10
81	Morus013778.p1	Monodehydroascorbate reductase	13	142	49,982	redox	0.30	0.33	0.10
82	Morus024141.p1	Beta-D-xylosidase 4	6	187	84,604	cell wall	0.14	0.10	0.10
83	Morus007961.p1	Hypothetical protein	7	62	95,561	not assigned	0.05	0.07	0.09
84	Morus025925.p1	Alpha-glucosidase	6	207	93,365	misc	0.12	0.25	0.07

[a] Protein ID, according to the *Morus* database; [b] M.P., number of matched peptides; [c] Function, function categorized using MapMan bin codes; [d] Mol (%), protein abundance; misc, miscellaneous; protein, protein synthesis/degradation/folding/targeting; cell, cell organization/vesicle transport; RNA, RNA processing/regulation of transcription; redox, redox ascorbate/glutathione metabolism; TCA, tricarboxylic acid cycle; and OPP, oxidative pentose phosphate.

2.5. TCA Cycle and Glycolysis Pathways Largely Differed among the Three Organs

To better understand the different metabolic pathways that are active in the leaf, branch, and root of *Morus*, the identified proteins from each organ were mapped to the TCA cycle and glycolysis pathways using the KEGG database (Figures 5 and 6). In the TCA cycle, the abundance of dihydrolipoyl dehydrogenase (EC1.8.1.4) in the leaves was higher than that in the branches, and no dihydrolipoyl dehydrogenase was detected in the roots. The abundance of NADP-dependent malic enzyme (EC1.1.1.40) in the branches was higher than in the leaves and roots. The abundance of malate dehydrogenase (EC1.1.1.37) in the leaves was nearly three times greater than that in the roots. The abundance fumarate hydratase (EC4.2.1.2) in the leaves was more than three times higher than that in the branches. Furthermore, isocitrate dehydrogenase (EC1.1.1.42) was only identified in the branches (Figure 5). In the glycolysis pathway, the abundance of fructose bisphosphate aldolase (EC4.1.2.13), phosphoglycerate kinase (EC2.7.2.3), and enolase (EC4.2.1.11) in the roots were higher than in the branches and roots. Additionally, pyruvate kinase (EC2.7.1.40) was only identified in the branches (Figure 6).

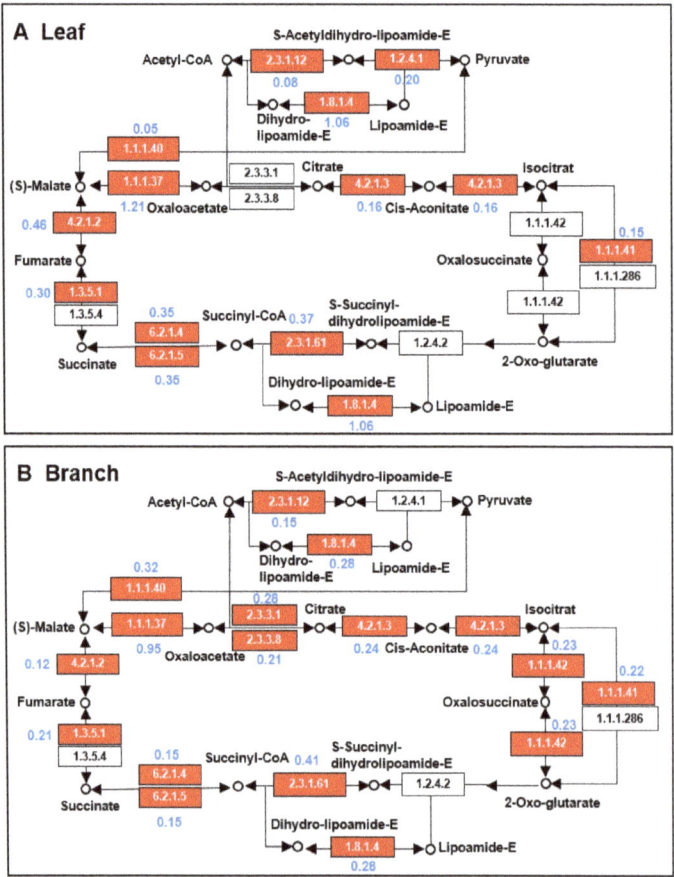

Figure 5. *Cont.*

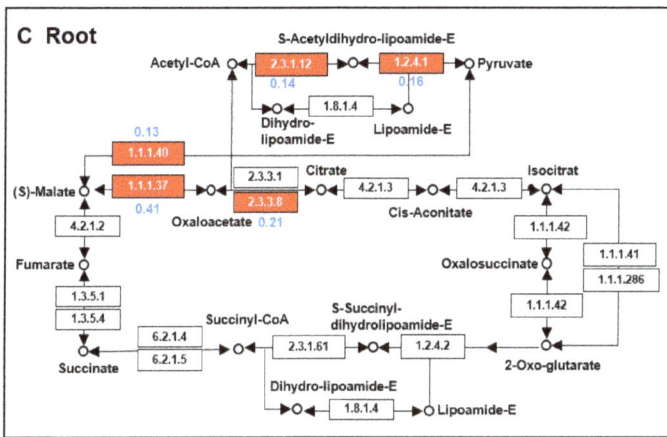

Figure 5. Mapping of proteins related to the TCA cycle in three organs from *Morus*. The TCA cycle pathways were identified by the mapping of the identified proteins from the leaf (**A**), branch (**B**), and root (**C**) using the KEGG database. Enzymes in red represent identified proteins, and the blue number represents the protein abundance. The EC number for the following proteins are 1.1.1.37, malate dehydrogenase; 1.1.1.40, malate dehydrogenase (oxaloacetate-decarboxylating) (NADP+); 1.1.1.41, isocitrate dehydrogenase (NAD+); 1.1.1.42, isocitrate dehydrogenase; 1.1.1.286, homoisocitrate dehydrogenase; 1.2.4.1, pyruvate dehydrogenase E1; 1.2.4.2, 2-oxoglutarate dehydrogenase; 1.3.5.1, succinate dehydrogenase; 1.3.5.4, fumarate reductase; 1.8.1.4, dihydrolipoamide dehydrogenase; 2.3.3.1, citrate synthase; 2.3.3.8, ATP-citrate synthase; 2.3.1.12, pyruvate dehydrogenase; 2.3.1.61, dihydrolipoamide succinyltransferase; 4.2.1.2, fumarate hydratase; 4.2.1.3, aconitate hydratase; 6.2.1.4, succinyl-CoA synthetase; and 6.2.1.5, ATP-citrate synthase.

2.6. Total Flavonoid Contents were the Highest in the Roots

The proteomics analysis of *Morus* revealed that two proteins involved in the flavonoid biosynthetic pathway were only identified in the roots (Figure 4). Based on this finding, the total flavonoid contents of the three organs in *Morus* were analyzed to confirm whether the flavonoids accumulated in the roots. The total flavonoids extracted from the leaves, branches, and roots of *Morus* were analyzed using a colorimetric method. The results confirmed that the total flavonoid contents in the roots were significantly higher than in the leaves and branches (Figure 7A); the levels were approximately 1.5 times higher than in the leaves and four times higher than in the branches (Supplemental Table S4).

2.7. Expression of Genes Related to Root- and Branch-Specific Proteins

Functional characterization of the root-specific proteins showed that chalcone flavanone isomerase, which is involved in secondary metabolism, was accumulated through the flavonoid pathway (Figure 4). The abundance of isoflavone reductase in the branch was higher than in the leaf and root (Table 2). Moreover, the abundance of phosphoglycerate kinase in the branch, which is involved in the glycolysis pathway, was higher than in the other organs (Figure 6). Based on this finding, these three genes, chalcone flavanone isomerase (CHI), isoflavone reductase (ISO), and phosphoglycerate kinase (PGK), were selected for further analysis of their mRNA expression levels (Figure 7B). Among the examined genes, the mRNA expression of CHI in the root is significantly higher than in the leaf and branch. Additionally, the expression of PGK and ISO in the branch is significantly higher than in the root (Figure 7B).

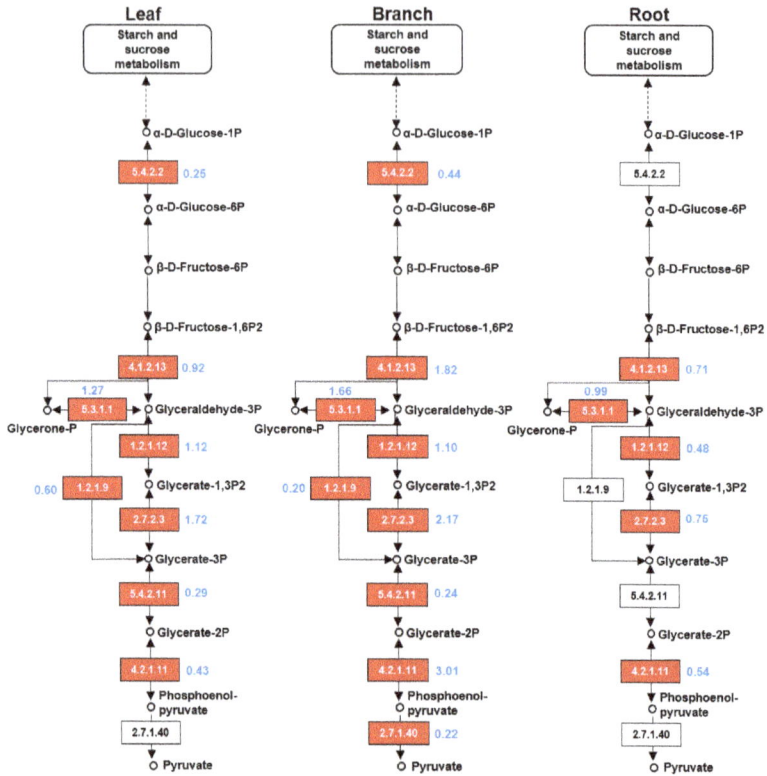

Figure 6. Mapping of proteins related to glycolysis in the three organs from *Morus*. The glycolysis pathways were identified by the mapping of the identified proteins from the leaf (**A**), branch (**B**), and root (**C**) using the KEGG database. Enzymes in red represent identified proteins, and the blue number represents the protein abundance. The EC number for the following proteins are 5.4.2.2, phosphoglucomutase; 4.1.2.13, fructose-bisphosphate aldolase; 5.3.1.1, triosephosphate isomerase; 1.2.1.12, glyceraldehyde 3-phosphate dehydrogenase; 1.2.1.9, glyceraldehyde-3-phosphate dehydrogenase (NADP); 2.7.2.3, phosphoglycerate kinase; 5.4.2.11, phosphoglycerate mutase; 4.2.1.11, enolase; and 2.7.1.40, pyruvate kinase.

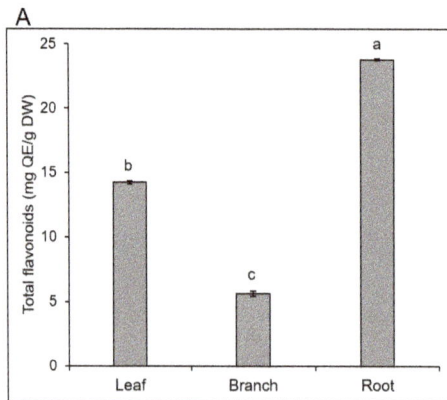

Figure 7. *Cont.*

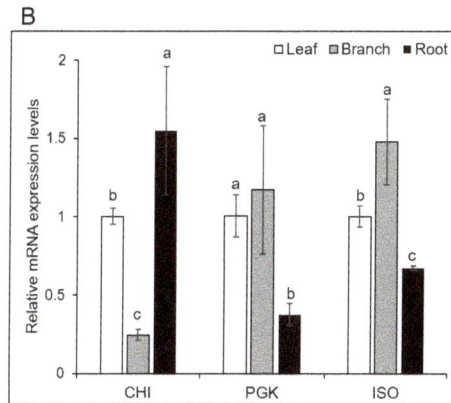

Figure 7. Total flavonoid contents and the expression of three genes in the three organs from *Morus*. The total flavonoids extracted from *Morus* leaf, branch, and root were analyzed using a colorimetric method (**A**). The transcript abundance of the selected genes was analyzed by qRT-PCR. Total RNA was extracted from the collected leaf, branch, and root (**B**). Data are shown as the means ± SD from three independent biological replicates. Means with the same letter are not significantly different according to one-way ANOVA test ($p < 0.05$). The abbreviations are follows: QE, quercetin-3-O-rutinoside. The genes CHI, PGK, and ISO represent chalcone flavanone isomerase, phosphoglycerate kinase, and isoflavone reductase, respectively. The β-actin gene was used as a reference control gene.

3. Discussion

3.1. The Secondary Metabolites and Total Flavonoid Contents are Different in Morus Leaf, Branch, and Root

Morus is a widely distributed medicinal plant in China, and various parts of *Morus* are commonly used in traditional Chinese medicinal treatments [4]. To date, it has been reported that the leaf, branch, and root of *Morus* have various pharmacological activities, such as antidiabetic, anti-inflammatory, and anticancer [18–20]. In this study, the antioxidant activity of the root was the best among the three *Morus* organs examined (Figure 1B). It is known that the antioxidant-enzyme system is an important part of plant responses to oxidative stress [21]. Among the antioxidant enzymes, superoxide dismutase (SOD) and peroxidase (POD) play a key role in the antioxidant defense mechanism [22]. SOD is the first line of defense in the enzymatic pathway against free oxygen radicals [23]. POD is an enzymatic protectant that scavenges both radical and non-radical oxygen species [24]. In the present study, the abundance of peroxidase and superoxide dismutase were higher in the root than in the other organs (Table 2), which might contribute to high antioxidant activity of the *Morus* root.

In contrast, many secondary metabolites with various biological activities, such as alkaloids, flavonoids, polysaccharides, terpenoids, phenolic acids, stilbenoids, and coumarins, were identified in *Morus* [25]. Chen et al. [26] compared the chemical composition among the bark, leaf, twig, and fruit of *Morus*, and further analysis showed that the bark contained the highest amount of prenylated flavonoids (kuwanon G, sanggenon C, morusin, and mulberroside A) compared to the twig, leaf, and fruit. The total flavonoid contents showed a significant contribution to α-glucosidase inhibition. In the present study, the differences among the leaf, branch, and root from *Morus* were uncovered at the protein level using a gel-free proteomics technique. Two proteins (chalcone flavanone isomerase and flavonoid 3,5-hydroxylase) involved in the flavonoid biosynthetic pathway were only identified in the root (Figure 4), which might mean that synthesized flavonoids accumulate in the root.

Chalcone flavanone isomerase (CHI) is a key branch-point enzyme between the phenylpropanoid and flavonoid pathways that can catalyze the synthesis of flavanones and the backbone for many downstream metabolites including flavonoids and isoflavonoids [27]. Flavonoid 3,5-hydroxylase (F3',5'Hs), which belongs to the cytochrome P450 (CYP) enzyme family [28], can catalyze the hydroxylation of the flavonoid B-ring at the 3′ and 5′ positions [29]. The results also indicated that dihydroxy B-ring-substituted flavonoids have a great potential to inhibit the generation of ROS and show antioxidant activity [30]. In the present study, the total flavonoid and five secondary metabolites contents were highly accumulated in the roots of *Morus* (Figure 1). This evidence strengthens the idea that both antioxidant enzymes and secondary metabolites in the root of *Morus* are responsible for the antioxidant activity.

3.2. Anaylses of Enzymes Involved in the Glycolysis and Isoflavonoid Biosynthetic Pathway in Morus

Glycolysis is a sequence of ten enzyme-catalyzed reactions that converts glucose into pyruvate. Proteomic analyses of leaf, branch, and root showed that proteins related to glycolysis were mostly abundant in the branch compared with the leaf and root (Figure 6). The abundance of fructose-bisphosphate aldolase (EC:4.1.2.13), phosphoglycerate kinase (EC:2.7.2.3), and enolase (EC4.2.1.11) in the branch was higher than in the leaf and root samples (Table 2). These enzymes play important roles in glycolysis. Fructose-bisphosphate aldolase catalyzes the split of fructose 1,6-bisphosphate into dihydroxyacetone phosphate and glyceraldehyde 3-phosphate [31]. Phosphoglycerate kinase catalyzes the transfer of a phosphate group from 1,3-bisphosphoglycerate to ADP via phosphoglycerate kinase, forming ATP and 3-phosphoglycerate [32]. Enolase converts 2-phosphoglycerate to phosphoenolpyruvate [33]. Cramer et al. [34] reported that the early-responding proteins to water deficit included proteins related to photosynthesis, glycolysis, translation, antioxidants, and growth, which could funnel carbon and energy into antioxidant defenses during the very early stages of plant responses to water deficit before any significant injury. The enhanced glycolysis in the mulberry branch might be engaged in a similar regulatory pathway, which enables the branch to exhibit antioxidant activity in vitro.

Furthermore, mulberry root bark is usually used in traditional Chinese medicine as a diuretic and expectorant agent, while the leaf was consumed as food by silkworms [35] and the fruit was taken as a health food [36]. However, the branch was largely neglected and ended up as fire wood material or agro-waste. Few studies have examined and confirmed the pharmacological activities of the branch bark from mulberry [37,38]. When comparing the abundance of commonly identified proteins from different parts of *Morus*, we discovered that isoflavone reductase homolog (IFRh) was most abundant in branches, approximately three times higher than in the leaf and root. Isoflavone reductase (IFR) is located in the cytoplasm and has been identified as one of the key enzymes involved in the synthesis of isoflavonoid phytoalexin [39–41]. IFR is unique to the plant kingdom and considered to have crucial roles in plant responses to various biotic and abiotic stresses. Cheng et al [42] discovered that overexpression of soybean isoflavone reductase enhanced resistance to *Phytophthora sojae* in soybean, though its specific biological function remains to be elucidated. The abundantly expressed isoflavone reductase homolog protein in mulberry branch inspired us to propose that intense isoflavonoid biosynthesis is present in the mulberry branch. However, the reason for this phenomenon is unknown and the function of IFRh in *Morus* is worth further investigation.

4. Materials and Methods

4.1. Plant Materials and Growth Conditions

Mulberry trees (*Morus alba* L.) were provided by the College of Agriculture and Biotechnology, Zhejiang University (Hangzhou, China). They were grown in a greenhouse under white fluorescent light (160 μmol m^{-2} s^{-1}, 16 h light period/day) at 25 °C and 70% humidity. The soil conditions were controlled to ensure normal plant growth without exposure to extreme drought or plant diseases. Leaf, branch and root were then collected, frozen in liquid nitrogen and stored at -80 °C. For each organ, three independent experiments were performed as biological replicates. Each biological replicate means leaves, branches, and roots from three individuals were collected and analyzed by mass spectrometry separately. A total of nine plants were used in this study (Supplemental Figures S2 and S3).

4.2. Protein Extraction

A portion (0.5 g) of each organ sample was ground into powder in liquid nitrogen using a mortar and pestle and then transferred into a polypropylene tube containing a solution of 10% trichloroacetic acid and 0.07% 2-mercaptoethanol in acetone. The resulting mixture was vortexed and sonicated for 10 min at 4 °C. The suspension was incubated for 1 h at −20 °C with vortexing every 15 min. It was then centrifuged at 9000× *g* for 10 min at 4 °C. The supernatant was discarded and the pellet was washed twice with 0.07% 2-mercaptoethanol in acetone. The final pellet was dried and resuspended in lysis buffer consisting of 7 M urea, 2 M thiourea, 5% CHAPS, and 2 mM tributylphosphine by vortexing for 1 h at 25 °C. The suspension was then centrifuged at 20,000× *g* for 20 min at room temperature until a clean supernatant was obtained. Protein concentrations were determined using the Bradford assay [43] with bovine serum albumin as the standard.

4.3. Purification and Digestion of Proteins for Mass Spectrometry Analysis

Proteins (100 µg) were purified with methanol and chloroform to remove any detergent from the sample solutions [44]. Briefly, 400 µL of methanol was added to each sample, and the resulting solution was mixed. Subsequently, 100 µL of chloroform and 300 µL of water were added to each sample, which were mixed and centrifuged at 20,000× *g* for 10 min to achieve phase separation. The upper aqueous phase was discarded and the pellets were dried. The dried pellets were resuspended in 50 mM NH_4HCO_3. The proteins were reduced with 50 mM dithiothreitol for 30 min at 56 °C and alkylated with 50 mM iodoacetamide for 30 min at 37 °C in the dark. Alkylated proteins were digested with trypsin and lysyl endopeptidase (Wako, Osaka, Japan) at 1:100 enzyme/protein concentrations at 37 °C for 16 h. The resulting tryptic peptides were acidified by mixing with formic acid (pH < 3), and the resulting solution was centrifuged at 20,000× *g* for 10 min. The obtained supernatant was collected and analyzed by nanoliquid chromatography (LC)- mass spectrometry (MS).

4.4. Nanoliquid Chromatography-Tandem Mass Spectrometry Analysis

Peptides were analyzed using a nanospray LTQ Orbitrap mass spectrometer (Thermo Fisher Scientific, San Jose, CA, USA) with the Xcalibur software (version 2.1, Thermo Fisher Scientific, Bremen, Germany) in data-dependent acquisition mode. Using an Ultimate 3000 nanoLC system (Dionex, Germering, Germany), peptides in 0.1% formic acid were loaded onto a C18 PepMap trap column (300 µm ID × 5 mm, Dionex, Sunnyvale, CA, USA) and were then eluted with a linear acetonitrile gradient (8–30% over 150 min) in 0.1% formic acid at a flow rate of 200 nL/min. The eluted peptides were separated and sprayed on a C18 capillary tip column (75 µm ID × 120 mm, Nikkyo Technos, Tokyo, Japan) with a spray voltage of 1.5 kV.

Full-scan mass spectra were acquired on the LTQ Orbitrap mass spectrometer (Thermo Fisher Scientific, San Jose, CA, USA) over 400–1500 m/z with a resolution of 30,000. A lock mass function was used for high mass accuracy [45]. The ten most intense precursor ions were selected for collision-induced fragmentation in the linear ion trap at a normalized collision energy of 35%. Dynamic exclusion was employed within 90 s to prevent the repetitive selection of peptides [46].

4.5. Protein Identification from the Mass Spectrometry Data

Proteins were identified using the Mascot search engine (version 2.5.1; Matrix Science, London, UK) with the *Morus* Genome database (MorusDB) (version 2.0, https://morus.swu.edu.cn/morusdb/datasets). The acquired raw data files were processed using Proteome Discoverer (version 1.4.0.288, Thermo Fisher Scientific, Bremen, Germany). The parameters used in Mascot searches were as follows: carbamidomethylation of cysteine was set as a fixed modification and oxidation of methionine was set as a variable modification. Trypsin was specified as the proteolytic enzyme, and one missed cleavage was allowed. The peptide mass tolerance was set at 10 ppm, the fragment mass tolerance was set at 0.8 Da, and the peptide charge was set at +2, +3, and +4. An automatic decoy database search was performed as part of the search. Mascot results were filtered with Mascot Percolator to improve the accuracy and sensitivity of peptide identification [47]. The false discovery rates for peptide identification in all searches were less than 1.0%. Peptides with a percolator ion score of more than 13 ($p < 0.05$) were used for protein identification. Protein abundance was analyzed based on the exponentially modified protein abundance index (emPAI) value [48]. Mascot-derived emPAI values were converted to molar percentages by normalizing against the sum of all emPAI values for the acquisition. Briefly, the mean of three emPAI values was divided by the sum of the emPAI values for all identified proteins and multiplied by 100. The protein content was estimated by the molar fraction percentage (mol%).

4.6. Functional Analysis of Identified Proteins

Protein functions were categorized using MapMan bin codes as previously described [49]. The small-scale prediction of the identified proteins from Mahonia was performed by transferring annotations from the *Arabidopsis thaliana* genome and considering orthologous genes. Pathway mapping of identified proteins was performed using the Kyoto Encyclopedia of Genes and Genomes (KEGG) database (https://www.kegg.jp/) [50].

4.7. Quantitative Analysis of Metabolites from Morus

A portion (0.2 g dry weight) of collected organ sample was sonicated in 150 mL of methanol for 2 h, centrifuged at 10,000× *g* for 10 min, and the supernatant was collected. The methanol extracts were dried in a rotary evaporator at 50 °C. For root samples, the obtained residue was dissolved in 10 mL of methanol. For branch and leaf samples, the obtained residue was dissolved in 2 mL of methanol. The dissolved samples were then filtered through a 0.45-μm filter (Millipore, Bullerica, MA, USA) for HPLC analysis. Several standard compounds, which consisted of mulberroside A, oxyresveratrol, kuwanone H, chalcomoracin, and morusin, were provided by Zhejiang Institute for Food and Drug Control (Hangzhou, China). For quantification, a calibration curve was constructed using the standard solutions diluted in methanol at six different concentrations: mulberroside A (0.05, 0.1, 0.25, 0.375, 0.5, and 0.75 mg mL^{-1}), oxyresveratrol (0.001, 0.005, 0.01, 0.025, 0.05, and 0.1 mg mL^{-1}), kuwanone H (0.01, 0.05, 0.1, 0.2, 0.3, and 0.5 mg mL^{-1}), chalcomoracin (2.36, 4.72, 9.44, 14.16, 18.88, and 23.6 mg mL^{-1}), and morusin (0.01, 0.05, 0.1, 0.25, 0.5, and 1 mg mL^{-1}). For HPLC analysis, 10 μL of the standard solutions and samples were used.

HPLC analysis was performed on a Waters 2695 Alliance HPLC system (Waters, Milford, MA, USA) equipped with a photodiode array detector, an online degasser and an auto-sampler for solvent delivery. Compounds in samples were separated using reverse-phase HPLC. A C18 column (4.6 mm ID × 250 mm, Agilent, Santa Clara, CA, USA) was used with a flow rate of 1 mL min^{-1} at 40 °C. The solvent system consisted of a linear gradient from 10% to 95% (*v*/*v*) acetonitrile in water with 0.1% phosphoric acid over a period of 70 min followed by an isocratic elution with 95% for 5 min. The spectra were measured at 320 nm and the retention time and ultraviolet spectra of the samples' peaks were compared with that of the standards' peaks.

4.8. In Vitro Antioxidant Activity Analysis of Mulberry Leaf, Branch, and Root

For the determination of the antioxidant activities of mulberry leaf, branch and root, ABTS$^+$ scavenging activity, hydroxyl free radical and O$_2$$^-$ scavenging activity were analyses. The ABTS$^+$ scavenging activities within different organs were determined using a total antioxidant capacity assay kit (A015-2, Nanjing Jiancheng Bioengineering Institute, Nanjing, China) following the manufacturer's protocol. Briefly, 10 μL of each sample were mixed with diluted ABTS$^+$ solution and then shaken vigorously for 6 min at room temperature in the dark. The absorbance of the samples was measured at 405 nm immediately after incubation. A calibration curve was constructed using absorbance values measured when ABTS$^+$ solution was mixed with a standard antioxidant—Trolox at 0.1 mM,

0.2 mM, 0.4 mM, 0.8 mM and 1.0 mM. The antioxidant activity of each group was measured as a Trolox-Equivalent Antioxidant Capacity (TEAC).

The hydroxyl free radical assay was performed using kit A018 purchased from the Nanjing Jiancheng Bioengineering Institute following the manufacturer's protocol based on the principle of the Fenton reaction [51]. Samples were mixed with reaction buffer and reacted at 37 °C for 1 min and terminated with the addition of developer. The absorbance of the samples was measured at 550 nm after developing for 20 min; the standard sample contained 0.03% H_2O_2 as a control.

The O_2^- scavenging activities were determined using the inhibition and produce superoxide anion assay kit (A052, Nanjing Jiancheng Bioengineering Institute, Nanjing, China) according to manufacturer's protocol. Samples were mixed with reaction buffer, incubated at 37 °C for 40 min and terminated by the addition of developer. The absorbance of the samples was measured at 550 nm after developing for 10 min developing; a vitamin C standard solution at 0.15 mg mL^{-1} was used as the control.

4.9. RNA Extraction and Quantitative Reverse Transcription-Polymerase Chain Reaction Analysis

Samples (0.1 g fresh weight) were ground to powder in liquid nitrogen using a sterilized mortar and pestle. Total RNA was extracted using a Quick RNA Isolation Kit (Huayueyang Biotechnology, Beijing, China) and reverse-transcribed using a 5X All-In-One RT MasterMix with AccuRT Genomic DNA Removal Kit (Applied Biological Materials Ins, CA) according to manufacturers' protocols. The primers were designed using Primer Premier 6.0. qRT-PCR was performed in a 10 μL reaction volume using an Evagreen 2 × qPCR MasterMix (Applied Biological Materials Ins, CA) in an IQ5 multicolor real-time PCR detection System (Bio-Rad, Hercules, CA, USA). The relative quantification method (2-ΔΔCT) was used to evaluate the quantitative variation between treatments. β-actin (GeneBank ID: HQ 163776) served as an internal control to normalize target gene quantities [52]. The gene-specific primers are listed in Supplemental Table S5. The qRT-PCR results were analyzed using the Statistical Product and Service Solutions software (version 20.0, IBM, Armonk, NY, USA).

4.10. Quantitative Analysis of the Total Flavonoids in Three Morus Organs

The total flavonoid contents of the three organs in *Morus* was determined by a colorimetric method as described previously with minor modifications [53]. Briefly, a portion (2.0 g) of each freeze-dried organ sample was sonicated in 100 mL of methanol for 30 min, centrifuged at 8000× *g* for 10 min, and the supernatant was collected. The pellet was resuspended in 100 mL of methanol and the resulting suspension was sonicated for 30 min. The methanol extracts were dried in a rotary evaporator at 50 °C. The obtained residue was dissolved in methanol using 10 mL in a volumetric flask and brought to volume by methanol. One milliliter of the final extract was placed in a 10 mL volumetric flask. Then, 0.5 mL of 5% $NaNO_2$ was added and the mixture was maintained for 5 min at room temperature. After incubation, 0.5 mL of 10% $Al(NO_3)_3$ was added to the reaction mixture and incubated for 5 min. Next, 1.5 mL of 2 M NaOH was added and methanol was added up to volume. After incubating for 15 min, the absorbance was measured at 510 nm. The data are expressed as mg quercetin-3-O-rutinoside (rutin) equivalents (QE)/g dry weight (DW), as quercetin-3-O-rutinoside was used as a reference standard for the quantification of the total flavonoids.

4.11. Statistical Analysis

The SPSS statistical software (version 22.0, IBM, Armonk, NY, USA) was used for the statistical evaluation of the results. Statistical significance was evaluated by Student's t-test when only two groups were compared and with one-way ANOVA test when multiple groups were compared. All results are presented as the mean ± SD from three independent biological replicates. A *p*-value less than 0.05 was considered statistically significant.

5. Conclusions

Morus is a medicinal plant with various biological activities. In this study, it is indicated that the five secondary metabolites, including mulberroside A, oxyresveratrol, kuwanone H, chalcomoracin, and morusin, and total flavonoids contents in *Morus* roots are higher than in other organs, which might be responsible for its highest antioxidant activity. Proteomic analysis of the leaf, branch, and root from *Morus* revealed that proteins related to the flavonoid pathway such as chalcone flavanone isomerase and flavonoid 3,5-hydroxylase were accumulated in the root, resulting in the highest total flavonoid contents among the three examined organs. Additionally, the protein expression profiling of the leaf, branch, and root in *Morus* will enrich the proteome database of *Morus*. Additionally, the present findings suggest that flavonoid biosynthesis is an important function in *Morus* root.

Supplementary Materials: Supplementary Materials can be found at http://www.mdpi.com/1422-0067/20/2/365/s1. Supplemental Figure S1. Comparison of organ-specific proteins related to secondary metabolism. Supplemental Figure S2. Experimental design of the proteomic study. Supplemental Figure S3. Workflow of the gel-free/label-free proteomic methods in the present study. Supplemental Figure S4. Pie chart of functional categorization of leaf, branch, and root in Morus. Supplemental Table S1. Proteins Identified in the leaf of *Morus* by gel-free/label-free proteomic analysis. Supplemental Table S2. Proteins Identified in the branch of *Morus* by gel-free/label-free proteomic analysis. Supplemental Table S3. Proteins Identified in the root of *Morus* by gel-free/label-free proteomic analysis. Supplemental Table S4. Contents of total flavonoids in different organs of *Morus*. Supplemental Table S5. Primers of three genes used in this study.

Author Contributions: Conceptualization: J.T.; methodology: W.Z. and S.K.; software: S.L. and B.Y.; validation: Z.Z.; formal analysis: W.Z.; investigation: W.Z.; resources: B.Y.; data curation: Z.G.; writing—original draft preparation: W.Z. and Z.Z.; writing—review and editing: W.Z., Z.Z., J.T., and S.K.; visualization: S.L.; supervision: J.T.; project administration: W.Z.; funding acquisition: W.Z. and J.T.

Funding: This research was funded by the National Natural Science Foundation of China, grant numbers 81872973 and 81603078, and the Zhejiang Provincial Science and Technology Planning Project, grant number 2016C04005.

Acknowledgments: We thank Xiaojian Yin at the National Institute of Crop Science (Japan) for experimental support during this research.

Conflicts of Interest: The authors declare no conflict of interest.

Abbreviations

ABTS	2,2′-Azino-bis (3-ethylbenzothiazoline-6-sulfonic acid)
CHAPS	3-[(3-Cholamidopropyl) dimethylammonio] propanesulfonate
CHI	Chalcone flavanone isomerase
emPAI	Exponentially-modified protein abundance index
HPLC	High-performance liquid chromatography
ISO	Isoflavone reductase
IFRh	Isoflavonoid reductase homolog
MS	Mass spectrometry
MVA	Mevalonate
PGK	Phosphoglycerate kinase
POD	Peroxidase
qRT-PCR	Quantitative reverse transcription-polymerase chain reaction
SOD	Superoxide dismutase
TCA	Tricarboxylic acid
TEAC	Trolox-equivalent antioxidant capacity

References

1. Umate, P. Mulberry improvements via plastid transformation and tissue culture engineering. *Plant Signal. Behav.* **2010**, *5*, 785–787. [CrossRef] [PubMed]
2. Qin, J.; He, N.J.; Huang, X.Z.; Xiang, Z.H. Development of mulberry ecological industry and sericulture. *Sci. Seric.* **2010**, *36*, 984–989. (In Chinese)

3. Liu, Y.; Willison, J.H.M. Prospects for cultivating white mulberry (*Morus alba*) in the drawdown zone of the Three Gorges Reservoir, China. *Environ. Sci. Pollut. Res.* **2013**, *20*, 7142–7151. [CrossRef] [PubMed]

4. Sun, S.G.; Chen, R.Y.; Yu, D.Q. Structures of two new benzofuran derivatives from the bark of mulberry tree (*Morus macroura* Miq.). *J. Asian Nat. Prod. Res.* **2001**, *3*, 253–259. [CrossRef] [PubMed]

5. Carrizzo, A.; Ambrosio, M.; Damato, A.; Madonna, M.; Storto, M.; Capocci, L.; Campiglia, P.; Sommella, E.; Trimarco, V.; Rozza, F.; et al. *Morus alba* extract modulates blood pressure homeostasis through eNOS signaling. *Mol. Nutr. Food Res.* **2016**, *60*, 1–8. [CrossRef] [PubMed]

6. Yuan, Q.; Xie, Y.; Wang, W.; Yan, Y.; Ye, H.; Jabbar, S.; Zeng, X. Extraction iptimization, characterization and antioxidant activity in vitro of polysaccharides from mulberry (*Morus alba* L.) leaves. *Carbohyd. Polym.* **2015**, *128*, 52–62. [CrossRef] [PubMed]

7. Liu, H.Y.; Wang, J.; Ma, J.; Zhang, Y.Q. Interference effect of oral administration of mulberry branch bark powder on the incidence of type II diabetes in mice induced by streptozotocin. *Food Nutr. Res.* **2016**, *60*, 31606. [CrossRef] [PubMed]

8. Qi, S.Z.; Tuo, Z.D.; Li, J.L.; Xing, S.S.; Li, B.B.; Zhang, L.; Lee, H.S.; Chen, J.G.; Cui, L. Effects of *Morus* root bark extract and active constituents on blood lipids in hyperlipidemia rats. *J. Ethnopharmacol.* **2016**, *180*, 54–59. [CrossRef]

9. Shang, X.F.; He, X.R.; He, X.Y.; Li, M.X.; Zhang, R.X.; Fan, P.C.; Zhang, Q.L.; Jia, Z.P. The genus Scutellaria an ethnopharmacological and phytochemical review. *J. Ethnopharmacol.* **2010**, *128*, 279–313. [CrossRef] [PubMed]

10. Lam, S.C.; Liu, X.; Chen, X.Q.; Hu, D.J.; Zhao, J.; Long, Z.R.; Fan, B.; Li, S.P. Chemical characteristics of different parts of *Coreopsis tinctorial* in China using microwave-assisted extraction and high-performance liquid chromatography followed by chemometric analysis. *J. Sep. Sci.* **2016**, *39*, 2919–2927. [CrossRef] [PubMed]

11. Khan, M.A.; Rahman, A.A.; Islam, S.; Khandokhar, P.; Parvin, S.; Islam, M.B.; Hossain, M.; Rashid, M.; Sadik, G.; Nasrin, S.; Mollah, M.N.H.; Alam, A.H.M.K. A comparative study on the antioxidant activity of methanolic extracts from different parts of *Morus alba* L. (Moraceae). *BMC Research Notes* **2013**, *6*, 24. [CrossRef] [PubMed]

12. Wang, W.; Zu, Y.; Fu, Y.; Efferth, T. In vitro antioxidant and antimicrobial activity of extracts from *Morus alba* L. leaves, stems and fruits. *Am. J. Chin. Med.* **2012**, *40*, 349–356. [CrossRef]

13. Chang, L.W.; Juang, L.J.; Wang, B.S.; Wang, M.Y.; Tai, H.M.; Hung, W.J.; Chen, Y.J.; Huang, M.H. Antioxidant and antityrosinase activity of mulberry (*Morus alba* L.) twigs and root bark. *Food Chem. Toxicol.* **2011**, *49*, 785–790. [CrossRef] [PubMed]

14. Zhang, S.; Zhang, L.; Tai, Y.; Wang, X.; Ho, C.T.; Wan, X. Gene discovery of characteristic metabolic pathways in the tea plant (*Camellia sinensis*) using 'Omics'-based network approaches: A future perspective. *Front. Plant Sci.* **2018**, *9*, 480. [CrossRef] [PubMed]

15. García-Sevillano, M.A.; García-Barrera, T.; Abril, N.; Pueyo, C.; Lopez-Barea, J.; Gomez-Ariza, J.L. Omics technologies and their applications to evaluate metal toxicity in mice *M. spretus* as a bioindicator. *J. Proteom.* **2014**, *104*, 4–23. [CrossRef]

16. Zhu, W.; Hu, J.; Wang, X.; Tian, J.K.; Komatsu, S. Organ-specific analysis of *Mahonia* using gel-free/label-free proteomic technique. *J. Proteome Res.* **2015**, *14*, 2669–2685. [CrossRef] [PubMed]

17. Ji, X.L.; Gai, Y.P.; Zheng, C.C.; Mu, Z.M. Comparative proteomic analysis provides new insights into mulberry dwarf responses in mulberry (*Morus alba* L.). *Proteomics* **2009**, *9*, 5328–5339. [CrossRef]

18. Ma, X.; Iwanaka, N.; Masuda, S.; Karaike, K.; Egawa, T.; Hamada, T.; Toyoda, T.; Myamoto, L.; Nakao, K.; Hayashi, T. *Morus alba* leaf extract stimulates 5'-AMP-activated protein kinase in isolated rat skeletal muscle. *J. Ethnopharmacol.* **2009**, *122*, 54–59. [CrossRef] [PubMed]

19. Jung, J.W.; Ko, W.M.; Park, J.H.; Seo, K.H.; Oh, E.J.; Lee, D.Y.; Lee, D.S.; Kim, Y.C.; Lim, D.W.; Han, D.; Baek, N.I. Isoprenylated flavonoids from the root bark of *Morus alba* and their hepatoprptective and neuroprotective activities. *Arch. Pharm. Res.* **2015**, *38*, 2066–2075. [CrossRef]

20. Eo, H.J.; Park, G.H.; Jeong, J.B. The involvement of cyclin D1 degradation through GSK3β-mediated threonine-286 phosphorylation-dependent nuclear export in anti-cancer activity of mulberry root bark extracts. *Phytomedicine* **2016**, *23*, 105–113. [CrossRef]

21. Chahid, K.; Laglaoui, A.; Zantar, S.; Ennabili, A. Antioxidant-enzyme reaction to the oxidative stress due to alpha-cypermethrin, chlorpyriphos, and pirimicarb in tomato (*Lycopersicon esculentum* Mill.). *Environ. Sci. Pollut. Res. Int.* **2015**, *22*, 18115–18126. [CrossRef] [PubMed]

22. Qin, R.; Jiao, Y.; Zhang, S.; Jiang, W.; Liu, D. Effects of aluminum on nucleoli in root tip cells and selected physiological and biochemical characters in *Allium cepa* var. *agrogarum* L. *BMC Plant Biol.* **2010**, *10*, 225.

23. Prashanth, S.R.; Sadhasivam, V.; Parida, A. Over expression of cytosolic copper/zine superoxide dismutase from a mangrove plant *Avicennia marina* in indica rice var: Pusa basmati-1 confers abiotic stress tolerance. *Transgenic Res.* **2008**, *17*, 281–291. [CrossRef] [PubMed]

24. Gill, S.S.; Tuteja, N. Reactive oxygen species and antioxidant machinery in abiotic stress tolerance in crop plants. *Plant Physiol. Biochem.* **2010**, *48*, 909–930. [CrossRef] [PubMed]

25. Chan, E.W.C.; Lye, P.Y.; Wang, S.K. Phytochemistry, pharmacology, and clinical trials of *Morus alba*. *Chin. J. Nat. Med.* **2016**, *14*, 17–30. [PubMed]

26. Chen, Z.Y.; Du, X.; Yang, Y.Y.; Cui, X.M.; Zhang, Z.J.; Li, Y. Comparative study of chemical composition and active components against α-glucosidase of various medicinal parts of *Morus alba* L. *Biomed. Chromatogr.* **2018**, *32*, e4328. [CrossRef] [PubMed]

27. Dastmalchi, M.; Dhaubhadel, S. Soybean chalcone isomerase: Evolution of the fold, and the differential expression and localization of the gene family. *Planta* **2015**, *241*, 507–523. [CrossRef]

28. Bogs, J.; Ebadi, A.; McDavid, D.; Robinson, S.P. Identification of the flavonoid hydroxylases from grapevine and their regulation during fruit development. *Plant Physiol.* **2006**, *140*, 279–291. [CrossRef]

29. Liu, S.H.; Ju, J.F.; Xia, G.M. Identification of the flavonoid 3′-hydroxylase and flavonoid 3′,5′-hydroxylase genes from Antarctic moss and their regulation during abiotic stress. *Gene* **2014**, *543*, 145–152. [CrossRef]

30. Fini, A.; Brunetti, C.; Ferdinando, M.D.; Ferrini, F.; Tattini, M. Stress-induced flavonoid biosynthesis and the antioxidant machinery of plants. *Plant Signal. Behav.* **2011**, *5*, 709–711. [CrossRef]

31. Zgiby, S.M.; Thomson, G.J.; Qamar, S.; Berry, A. Exploring substrate binding and discrimination in fructose 1,6-bisphosphate and tagatose 1,6-bisphosphate aldolases. *Eur. J. Biochem.* **2000**, *267*, 1858–1868. [CrossRef] [PubMed]

32. Watson, H.C.; Walker, N.P.; Shaw, P.J.; Bryant, T.N.; Wendell, P.L.; Fothergill, L.A.; Perkins, R.E.; Conroy, S.C.; Dobson, M.J.; Tuite, M.F. Sequence and structure of yeast phosphoglycerate kinase. *EMBO J.* **1982**, *1*, 1635–1640. [CrossRef] [PubMed]

33. Hoom, R.K.; Flickweert, J.P.; Staal, G.E. Purification and properties of enolase of human erythroctyes. *Int. J. Biochem.* **1974**, *5*, 845–852.

34. Cramer, G.R.; Sluyter, S.C.V.; Hopper, D.W.; Pascovici, D.; Keighley, T.; Haynes, P.A. Proteomic analysis indicates massive changes in metabolism prior to the inhibition of growth and photosynthesis of grapevine (*Vitis vinifera* L.) in response to water deficit. *BMC Plant Biol.* **2013**, *13*, 49. [CrossRef] [PubMed]

35. Wang, D.; Dong, Z.; Zhang, Y.; Guo, K.; Guo, P.; Zhao, P.; Xia, Q. Proteomics provides insight into the interaction between mulberry and silkworm. *J. Proteome Res.* **2017**, *16*, 2477–2480. [CrossRef] [PubMed]

36. Zhang, H.; Ma, Z.F.; Luo, X.; Li, X. Effects of mulberry fruit (*Morus alba* L.) consumption on health outcomes: A mini-review. *Antioxidants* **2018**, *7*, 69. [CrossRef]

37. Qiu, F.; He, T.Z.; Zhang, Y.Q. The isolation and the characterization of two polysaccharides from the branch bark of mulberry (*Morus alba* L.). *Arch. Pharm. Res.* **2016**, *39*, 887–896. [CrossRef]

38. Wang, S.; Fang, M.; Ma, Y.L.; Zhang, Y.Q. Preparation of the branch bark ethanol extract in mulberry *Morus alba*, its antioxidation, and antihyperglycemic activity in vivo. *Evid.-Based Complement. Altern.* **2014**, *2014*, 569652.

39. Graham, T.L.; Kim, J.E.; Graham, M.Y. Role of constitutive isoflavone conjugates in the accumulation of glyceollin in soybean infected with *Phytophthora megasperma*. *Mol. Plant Microbe Interact.* **1990**, *3*, 157–166. [CrossRef]

40. Guo, L.; Dixon, R.A.; Paiva, N.L. Conversion of vestitone to medicarpin in Alfalfa (*Medicago sativa* L.) is catalyzed by two independent enzymes. *J. Biol. Chem.* **1994**, *269*, 22372–22378.

41. Cooper, J.D.; Qiu, F.; Paiva, N.L. Biotransformation of an exogenously supplied isoflavonoid by transgenic tobacco cells expressing alfalfa isoflavone reductase. *Plant Cell Rep.* **2002**, *20*, 876–884.

42. Cheng, Q.; Li, N.; Dong, L.; Zhang, D.; Fang, S.; Jiang, L.; Wang, X.; Xu, P.; Zhang, S. Overexpression of soybean isoflavone reductase (GmIFR) enhances resistance to *Phytophthora sojae* in soybean. *Front. Plant. Sci.* **2015**, *6*, 1024. [CrossRef] [PubMed]

43. Bradford, M.M. A rapid and sensitive method for the quantitation of microgram quantities of protein utilizing the principle of proteindye binding. *Anal. Biochem.* **1976**, *72*, 248–254. [CrossRef]

44. Nanjo, Y.; Skultety, L.; Uvackova, L.; Klubicova, K.; Hajduch, M.; Komatsu, S. Mass spectrometry-based analysis of proteomic changes in the root tips of flooded soybean seedlings. *J. Proteome Res.* **2012**, *11*, 372–385. [CrossRef] [PubMed]

45. Olsen, J.V.; de Godoy, L.M.F.; Li, G.Q.; Macek, B.; Mortensen, P.; Pesch, R.; Makarov, A.; Lange, O.; Horning, S.; Mann, M. Parts per million mass accuracy on an Orbitrap mass spectrometer via lock mass injection into a C-trap. *Mol. Cell. Proteom.* **2005**, *4*, 2010–2021. [CrossRef] [PubMed]

46. Zhang, Y.; Wen, Z.H.; Washburn, M.P.; Florens, L. Effect of dynamic exclusion duration on spectral count based quantitative proteomics. *Anal. Chem.* **2009**, *81*, 6317–6326. [CrossRef] [PubMed]

47. Brosch, M.; Yu, L.; Hubbard, T.; Choudhary, J. Accurate and sensitive peptide identification with MASCOT percolator. *J. Proteome Res.* **2009**, *8*, 3176–3181. [CrossRef] [PubMed]

48. Shinoda, K.; Tomita, M.; Ishihama, Y. emPAI Calc-for the estimation of protein abundance from large-scale identification data by liquid chromatography-tandem mass spectrometry. *Bioinfomatics* **2010**, *26*, 576–577. [CrossRef] [PubMed]

49. Usadel, B.; Nagel, A.; Thimm, O.; Redestig, H.; Blaesing, O.E.; Natalia, P.R.; Selbig, J.; Hannemann, J.; Piques, M.C.; Steinhauser, D.; et al. Extension of the visualization tool MapMan to allow statistical analysis of arrays, display of corresponding genes, and comparison with known responses. *Plant Physiol.* **2005**, *138*, 1195–1204. [CrossRef]

50. Kanehisa, M.; Goto, S. KEGG: Kyoto encyclopedia of genes and genomes. *Nucleic Acids Res.* **2000**, *28*, 27–30. [CrossRef]

51. Fenton, H.J.H. Oxidation of tartaric acid in presence of iron. *J. Chem. Soc. Trans.* **1894**, *65*, 899–910. [CrossRef]

52. Guan, Q.J.; Yu, J.J.; Zhu, W.; Yang, B.X.; Li, Y.H.; Zhang, L.; Tian, J.K. RNA-Seq transcriptome analysis of the *Morus alba* L. leaves exposed to high-level UV-B with and without dark treatment. *Gene* **2018**, *645*, 60–68. [CrossRef] [PubMed]

53. Chen, W.; Li, Y.; Bao, T.; Gowd, V. Mulberry fruit extract affords protection against ethyl carbamate-induced cytotoxicity and oxidative stress. *Oxid. Med. Cell. Longev.* **2017**, *2017*, 1594963. [CrossRef] [PubMed]

International Journal of
Molecular Sciences

MDPI

Article

Proteome Map of Pea (*Pisum sativum* L.) Embryos Containing Different Amounts of Residual Chlorophylls

Tatiana Mamontova [1,2,†], Elena Lukasheva [2,†], Gregory Mavropolo-Stolyarenko [2,†], Carsten Proksch [3], Tatiana Bilova [1,4], Ahyoung Kim [1], Vladimir Babakov [5], Tatiana Grishina [2], Wolfgang Hoehenwarter [3], Sergei Medvedev [4], Galina Smolikova [4,*] and Andrej Frolov [1,2,*]

[1] Department of Bioorganic Chemistry, Leibniz Institute of Plant Biochemistry, 06120 Halle (Saale), Germany; mamontova-bio@mail.ru (T.M.); bilova.tatiana@gmail.com (T.B.); ariyong1002@gmail.com (A.K.)
[2] Department of Biochemistry, St. Petersburg State University, St. Petersburg 199178, Russia; elena_lukasheva@mail.ru (E.L.); gm2124@mail.ru (G.M.-S.); tgrishina@mail.ru (T.G.)
[3] Proteome Analytics, Leibniz Institute of Plant Biochemistry, Weinberg 3, 06120 Halle (Saale), Germany; Carsten.Proksch@ipb-halle.de (C.P.); Wolfgang.Hoehenwarter@ipb-halle.de (W.H.)
[4] Department of Plant Physiology and Biochemistry, St. Petersburg State University, St. Petersburg 199034, Russia; s.medvedev@spbu.ru
[5] Research Institute of Hygiene, Occupational Pathology, and Human Ecology, Federal Medicobiological Agency, 188663 Kapitolovo, Russia; vbabakov@gmail.com
*   Correspondence: g.smolikova@spbu.ru (G.S.); afrolov@ipb-halle.de (A.F.); Tel.: +7-812-328-9695 (G.S.); +49-(0)-345-55821350 (A.F.)
†   These authors contributed equally on the manuscript.

Received: 7 November 2018; Accepted: 13 December 2018; Published: 15 December 2018

Abstract: Due to low culturing costs and high seed protein contents, legumes represent the main global source of food protein. Pea (*Pisum sativum* L.) is one of the major legume crops, impacting both animal feed and human nutrition. Therefore, the quality of pea seeds needs to be ensured in the context of sustainable crop production and nutritional efficiency. Apparently, changes in seed protein patterns might directly affect both of these aspects. Thus, here, we address the pea seed proteome in detail and provide, to the best of our knowledge, the most comprehensive annotation of the functions and intracellular localization of pea seed proteins. To address possible intercultivar differences, we compared seed proteomes of yellow- and green-seeded pea cultivars in a comprehensive case study. The analysis revealed totally 1938 and 1989 nonredundant proteins, respectively. Only 35 and 44 proteins, respectively, could be additionally identified after protamine sulfate precipitation (PSP), potentially indicating the high efficiency of our experimental workflow. Totally 981 protein groups were assigned to 34 functional classes, which were to a large extent differentially represented in yellow and green seeds. Closer analysis of these differences by processing of the data in KEGG and String databases revealed their possible relation to a higher metabolic status and reduced longevity of green seeds.

Keywords: chlorophylls; LC-MS-based proteomics; pea (*Pisum sativum* L.); proteome functional annotation; proteome map; seeds; seed proteomics

1. Introduction

Legumes represent the most prominent source of food protein, and their importance is increasing with the growing global population [1]. Indeed, these crops are tolerant to environmental stressors, cheap to culture, and rich in seed protein (typically about 25% of fresh seed weight) [2,3]. Among the cultured legumes, pea (*Pisum sativum* L.) is the most widely spread pulse crop in Europe, where it serves as a protein food supplement for monogastric animals [4]. Therefore, quality of pea seeds is important from the aspects of both sustainable crop production and high nutritional efficiency. Obviously, changes in protein composition of seeds directly affect their agricultural and nutritional value [5].

The first pea seed protein map was reported at the end of the last decade; it was based on the two-dimensional gel electrophoresis (2D-GE) and mass spectrometric (MS) identification of visualized electrophoretic zones (spots) and contained 156 proteins [6]. The majority of the identified polypeptides were storage proteins (convicilins, vicilins, and legumins), which strongly dominate the seed proteome and can serve as seed protein quality markers [7]. Removal of these highly abundant storage proteins by extraction with aqueous (aq.) isopropanol-containing solutions [8] or by precipitation in presence of aq. 0.01–0.1% (w/v) protamine sulfate [9] could increase coverage of the seed proteome. Alternatively, as was shown for soybean (*Glycine max*), seed storage globulins (glycinin and β-conglycenin) can be effectively removed by 10 mmol/L $CaCl_2$ [10]. Low-abundance proteins can be selectively enriched by means of the combinational peptide ligand libraries technology [11]. However, despite the efficiency of these techniques, in combination with the gel-based proteomics approach, the numbers of identified proteins never exceed several hundreds.

Because of this, implementation of liquid chromatography (LC)-MS-based strategy is desired to get a deeper insight into the seed proteome [12]. However, as detergents, conventionally used for solubilization of protein isolates [5], dramatically affect efficiency of electrospray ionization (ESI) [13], it is difficult to find a compromise between completeness of protein reconstitution and sensitivity of MS analysis. In this context, the introduction of degradable detergents into proteomic practice helped to overcome this caveat [14]. Thus, commercially available detergents such as RapiGest™ and Anionic Acid-Labile Surfactant II (AALS II) gave a deeper insight into the proteomes of young barley [15] and developing oilseed rape [16] seeds.

When considering the pea seed proteome, it is necessary to remember that cultivars can differ essentially by their metabolic background. For example, in the seeds of green-seeded cultivars, chlorophylls are not quantitatively destroyed after completion of seed maturation, and the seeds of such plants preserve green color in their mature state [17]. This phenomenon is underlain by deficiency of one or several chlorophyll catabolic enzymes (CCEs), for example, chlorophyll *b* reductase, 7-hydroxymethyl chlorophyll *a* reductase, Mg^{2+}-dechelatase, pheophytinase, pheophorbide *a* oxygenase, and reductase of red chlorophyll catabolite (RCC) [17]. Thus, the presence of chlorophylls in *sgr* (*stay-green*) mutants is mostly attributed to the damage of *SGR* genes, prospectively encoding the enzymes involved in chlorophyll degradation and/or disassembly of chlorophyll-protein complexes [18]. Remarkably, green seeds are characterized with higher amenability to stressors (e.g., associated with accelerated ageing [19,20]) that might be related to their higher oxidative status [21]. Adaptation to these metabolic changes might result in essential alterations in seed proteomes. These events, potentially affecting seed nutritional properties, have not been addressed so far.

Therefore, here, we provide, to the best of our knowledge, the most complete map of the pea seed proteome using sample prefractionation and LC-MS-based shotgun proteomics. We address the differences in embryo proteomes of yellow- and green-seeded pea cultivars and discuss the function and localization profiles of seed proteins in the context of possible differences in their response to varying environmental conditions.

2. Results

2.1. Analysis of Physiological and Biochemical Parameters of Seed Quality

As green seeds contain residual chlorophylls, and are, therefore, potentially more prone to development of oxidative stress [19,21], we addressed physiological and biochemical parameters, giving access to oxidative status of the embryos and accompanying changes in lipid peroxidation levels, membrane integrity, status of antioxidant defense, and functional activity of photosynthetic apparatus. Accordingly, the cultivar-specific differences in seed germination kinetics and electrolyte conductivity, as well as the contents of photosynthetic pigments, hydrogen peroxide and lipid peroxidation products, were determined. The yellow seeds of the cultivar Millennium and the green ones of the cultivar Gloriosa clearly differed in the contents of photosynthetic pigments. Thus, chlorophylls were detected solely in green seeds, and the contents of carotenoids were approximately 25% higher in these seeds (*t*-test: $p = 5.41 \times 10^{-9}$), in comparison to the yellow ones (Figure 1A and Supplementary information 2). Both pea cultivars demonstrated 100% seed viability, although the seeds of the cultivar Millennium germinated faster (Figure 1B).

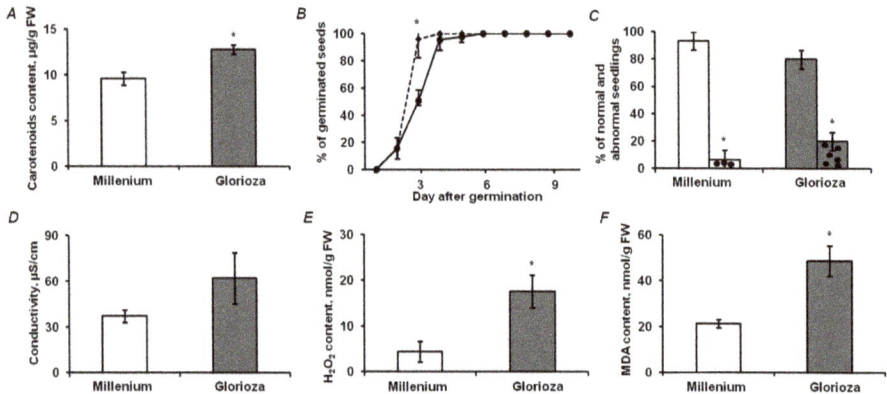

Figure 1. Physiological and biochemical parameters of seed quality, acquired for the mature seeds of cultivar Millennium (solid line, white columns) and Gloriosa (dashed line, grey columns): (**A**) contents of carotenoids, (**B**) kinetics of germination, (**C**) distribution of seedlings by morphology (filled and dotted columns indicate normally and abnormally developed seedlings, respectively), (**D**) electrolyte leakage, expressed as medium conductivity, µS/cm, as well as (**E**) tissue levels of hydrogen peroxide and (**F**) lipid peroxidation products, expressed as malondialdehyde (MDA) equivalents. Asterisks denote statistically significant difference between cultivars (**A,B,E,F**) or indicate difference between percentage of nongerminated seeds in two cultivars (**C**), *t*-test: $p < 0.05$.

Moreover, the percentage of normally developed seedlings on day ten was 13% higher for yellow seeds, although this difference was not significant (Figure 1C). Conductivity testing did not reveal differences in electrolyte leakage between yellow and green seeds (*t*-test: $p = 0.07$, Figure 1D), whereas the green seeds demonstrated statistically significant approximately two- and fourfold higher levels of lipid peroxidation products and H_2O_2 (*t*-test: $p = 0.0058$ and 0.0002, respectively, Figure 1E,F).

2.2. Protein Isolation and Tryptic Digestion

To ensure efficient extraction of seed proteins and the maximal coverage of the pea seed proteome, we decided to use phenol-based protein extraction (Figure 2), thereby resulting dry protein isolates could be reconstituted in shotgun buffer containing at least 0.15% AALS. Protein determination revealed the extraction yields were in the range of 39.6–124.1 mg/g fresh weight (Table S1-5). Assay precision was determined by SDS-PAGE loading 5 µg of protein (Figure S1-3); the overall lane densities were $1.4 \times 10^4 \pm 4.5 \times 10^2$ arbitrary units (AU, RSD = 3.06%). The signal patterns observed in the electrophoregrams were similar between lanes and pea cultivars (Figure S1-3). Tryptic digestion of proteins was considered to be complete, as the bands of major pea storage proteins, such as legumin (α- and β- subunits, ~40 kDa and ~20 kDa, correspondingly), vicilin (subunits of ~29 kDa, ~35 kDa, and ~47 kDa), and convicilin (subunit of ~71 kDa), could not be detected (Figure S1-4), assuming a staining sensitivity better than 30 ng [22] and a legumin content of at least 80% of total seeds proteins [23].

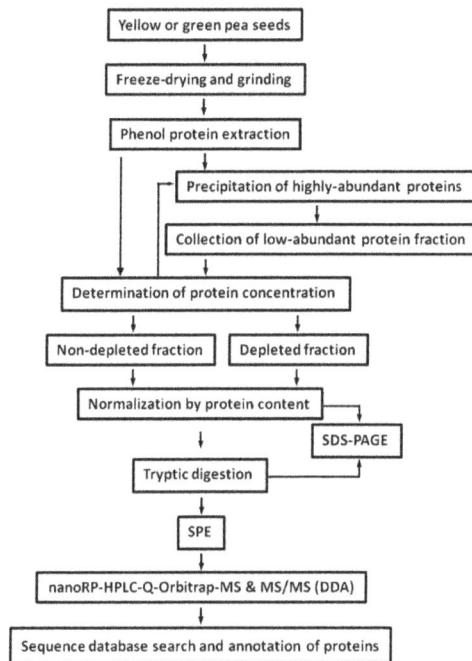

Figure 2. Experimental workflow, employed for characterization of the pea seed proteome.

2.3. Depletion of Storage Seed Proteins by Protamine Sulfate

To ensure sufficient efficiency of the PSP procedure in the presence of AALS, we diluted the protein samples 10-fold, to arrive at a final AALS concentration of 0.015% (*w/v*). Additionally, we applied the highest concentration, tested by the Kim's group (0.07%), although the authors reported 0.05% (*w/v*) protamine sulfate as a sufficient concentration, for efficient depletion of seed storage proteins. Interestingly, the protein recoveries were higher for yellow seed embryos (71.3–84.0 mg/g fresh weight, Table S1-6), than for the green ones (56.2–61.0 mg/g fresh weight). However, the recoveries after protamine sulfate depletion were slightly higher for green than for yellow seed embryos (3.7–4.2 vs. 2.9–3.8 mg/g fresh weight, respectively). Accordingly, the depletion efficiency was slightly higher for the yellow than for the green seed embryos (95.4–96.0 vs. 92.6–94.0%, respectively, Table S1-6). In agreement with this observation, electrophoregrams of depleted protein extracts were clearly different for yellow and green seeds (Figure S1-5B), although this was not the case for nondepleted samples (Figure S1-3). SDS-PAGE revealed depletion of the major seed storage proteins (Figures S1-3 and S1-5), which was, however, incomplete (Figure S1-6). Tryptic digestion of depleted samples was complete, that is, no proteins were detected in the corresponding electrophoregram (Figure S1-7).

2.4. Annotation of Pea Seed Proteins

To verify the applicability of our combined database for annotation of pea seed proteins, we manually interpreted MS/MS spectra surpassing dual FDR thresholds (strict 0.01 and relaxed 0.05), but identified with the lowest XCorrs in both yellow and green seed embryo preparations. This procedure revealed reliable identification of low-scoring peptides, and this was valid for all three proteomes comprising the database; the sequences of corresponding individual peptides could be confirmed by the numbers of b and y fragments, sufficient for their unambiguous identification (Figure 3). In total, 9162 peptides (7923 and 8292 in Millennium and Gloriosa seeds, respectively, Figure 4A, Supplementary information 3) were identified with the FDR of 0.05. On the basis of these identifications, 8769 possible proteins could be annotated (7821 and 8134 in yellow and green seeds, respectively, Figure 4B, 0.05 protein FDR threshold), which represented 2195 nonredundant proteins, or so-called protein groups (1938 and 1989 in Millennium and Gloriosa seeds, respectively, Figure 4C). Precipitation of high-abundance proteins with 0.07% (*w/v*) protamine sulfate (PS) resulted in identification of 2399 and 2286 tryptic peptides in yellow and green seed embryos, respectively (totally 2974, Figure 4A), however, only 44 and 85 peptides were unique for PS-treated extracts of yellow and green seeds. These unique peptides, identified specifically in yellow and green seeds, gave access to 11 and 20 protein groups, represented by 24 and 42 proteins, respectively (Figure 4B,C). Further, 24 protein groups, represented by 84 proteins, were in common to yellow and green seeds, although identified solely in corresponding PS fractions. Removal of the high-abundance proteins was effective, as can be judged by high numbers of proteins and protein groups, observed exclusively in nondepleted samples (Figure 4B,C). Indeed, 1152 and 1196 proteins were removed by the PS treatment from the extracts of Millennium and Gloriosa seeds, respectively (1377 in total), that accounted for 8% and 10% of the totally identified protein groups.

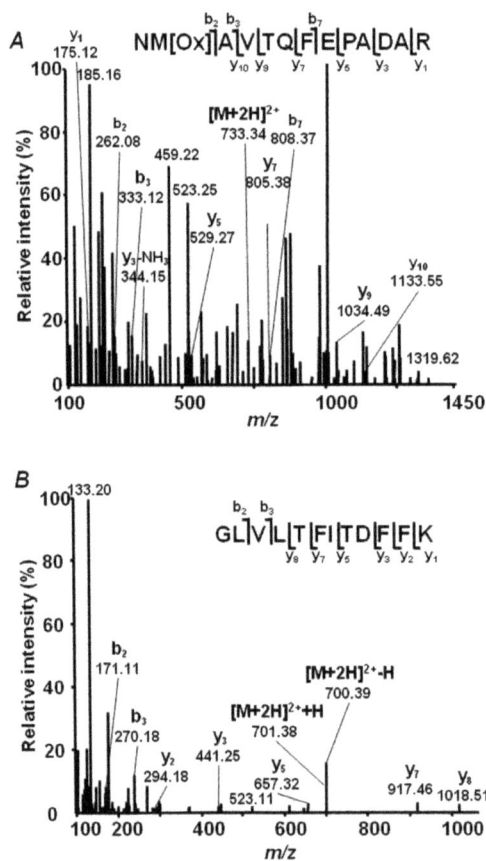

Figure 3. *Cont.*

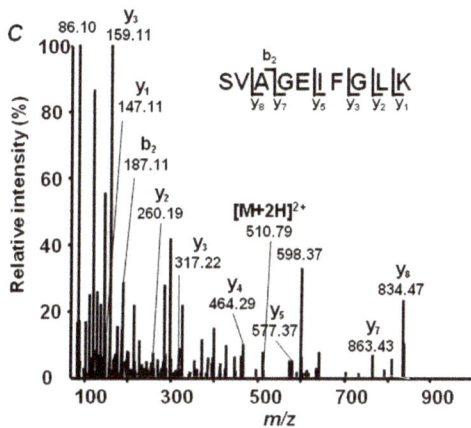

Figure 3. Tandem mass spectra of (**A**) m/z 733.3 corresponding to the peptide NM$_{Ox}$AVTQFEPADAR, which represents residues 131–143 of aminopeptidase (protein accession Lj1g3v1787580.1, *Lotus japonicus*), (**B**) m/z 700.89 corresponding to the peptide GLVLTFITDFFK, which represents residues 179–190 of the eukaryotic translation initiation factor (protein accession G7KRJ1, *Medicago truncatula*), and (**C**) m/z 510.79 corresponding to the peptide SVAGEIFGLK, which represents residues 170–180 of the protein from glutamine synthetase family (protein accessionV7BHN3, *Phaseolus vulgaris*).

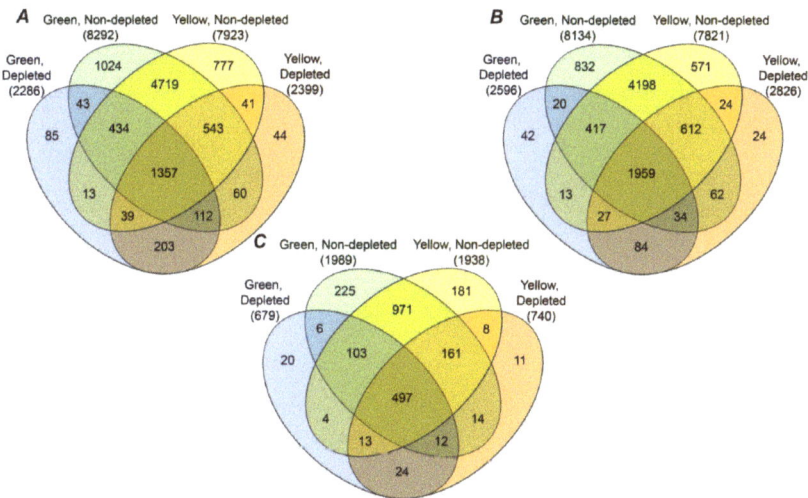

Figure 4. The numbers of tryptic (**A**) peptides, (**B**) proteins, and (**C**) protein groups, identified in yellow-colored (cultivar Millennium) and green-colored (cultivar Gloriosa) seeds with and without depletion of highly abundant proteins with protamine sulfate.

2.5. Functional Annotation of Seed Proteins and Prediction of Their Cellular Localization

Functional annotation resulted in assignment of 981 nonredundant proteins (875 and 900 proteins, isolated from yellow and green seed embryos, respectively) to one or more of 34 functional bins, whereas 1269 polypeptides remained unassigned (Figure 5A, Supplementary information 4a–f). Among the assigned proteins, 926 and 380 could be annotated to functional bins in nondepleted and PSP-depleted fractions, respectively (Figure 5B,C). The bin with the most entries (274 and 275 entries for yellow and green seeds, respectively), was nucleotide metabolism (bin #23), although it included nucleotide binding proteins and ATPases (Supplementary information 4). Bins related to photosynthesis, metal handling, miscellaneous enzymes, RNA, DNA, and protein metabolism (#1, 15, 26–29) were also large (Figure 5A). For most of the functional groups, the numbers of proteins assigned to them were similar in yellow and green seed embryos. However, the number of entries in the groups of metal-binding proteins and enzymes of amino acid metabolism were higher in green seed embryos, whereas nucleotide metabolism pathways and miscellaneous enzyme families were more represented in yellow seeds. The results of localization prediction were in agreement with the observed functional patterns, although only minimal differences between yellow and green seeds were observed (Figure 6, Supplementary information 5a–f).

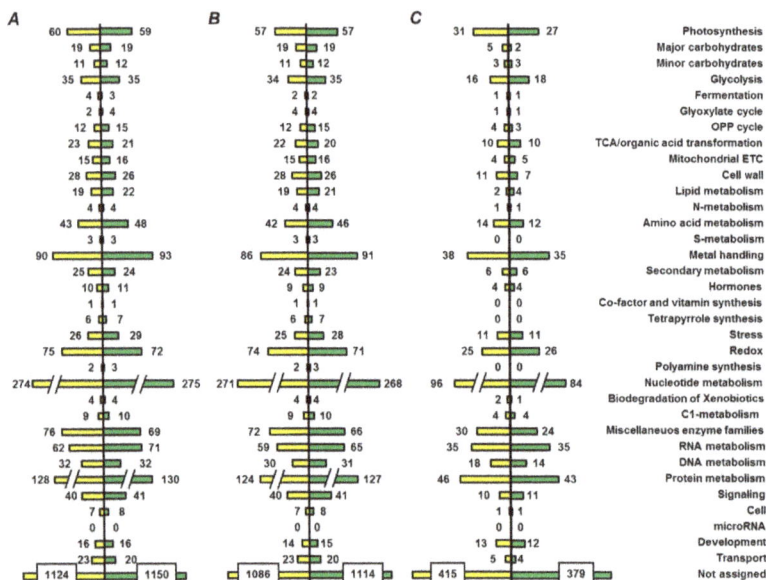

Figure 5. Functional annotation of proteins, identified in yellow seeds of cultivar Millennium (yellow color) and in green seeds of cultivar Gloriosa (green color), (**A**) both in nondepleted samples and after treatment with 0.07% (*w/v*) protamine sulfate, (**B**) only in nondepleted fraction and (**C**) protamine-sulfate supernatant fraction.

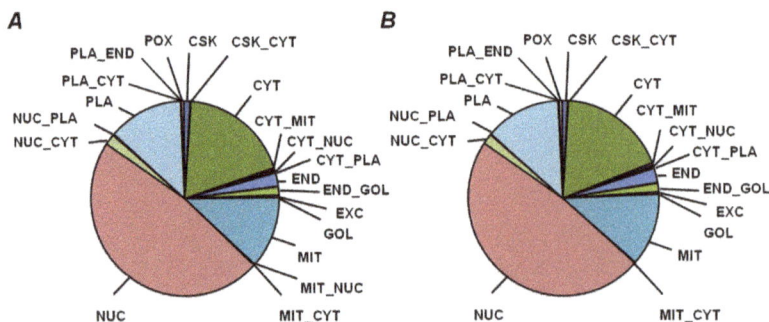

Figure 6. Prediction of subcellular localization of proteins, identified (**A**) in yellow seeds of cultivar Millennium and (**B**) in green seeds of cultivar Gloriosa, both in nondepleted samples and after treatment with 0.07% (*w/v*) protamine sulfate. Subcellular fractions: CSK, cytoskeleton; CYT, cytoplasm; END, endoplasmic reticulum; EXC, extracellular/secreted; GOL, Golgi apparatus; MIT, mitochondria; NUC, nuclear; PLA, plasma membrane; POX, peroxisome; doubled labels denote corresponding combinations of possible localization.

Thus, in yellow seed embryos, representation of cytosol and plasma membrane was slightly higher (20.4% and 12.6%, respectively), whereas the nuclear fraction accounted for 1.5% less, in comparison with green seeds (Figure 6A). In green seeds, the largest fraction (approximately 46.7%) was represented by nuclear proteins, whereas the majority of the other proteins were localized to cytosol, mitochondria, and plasma membrane (19.3%, 13.7%, and 11.9%, respectively, Figure 6B). Remarkably, depletion of high-abundance products did not essentially affect this distribution (Figure S1-8).

To address specific metabolic features of yellow and green seeds, we considered the proteins, identified solely in one of the cultivars. Despite the general similarity of protein profiles (Figure 7A, Supplementary information 6), miscellaneous enzyme families were better represented in yellow seeds, whereas the pathways of RNA and amino acid metabolism were more prevalent in green ones. Interestingly, green seeds contained more proteins, which could be localized to the nucleus and endoplasmic reticulum (55% and 2.7%, respectively, Figure 7B), whereas cytosol and cytoskeleton were represented more in yellow seed embryos (15% and 2%, respectively, Figure 7C).

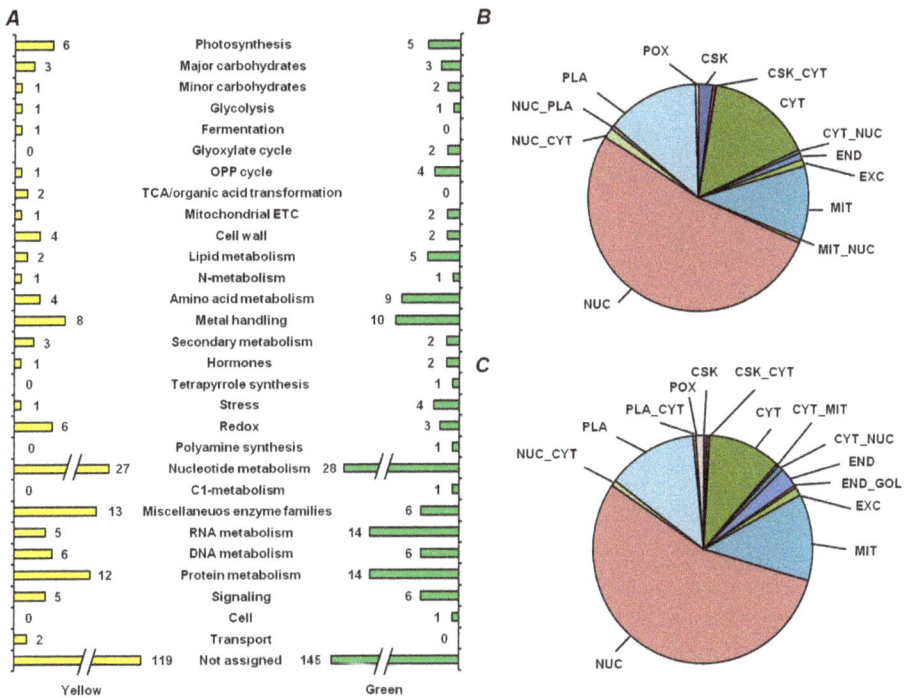

Figure 7. Functional annotation (**A**) and prediction of subcellular localization (**B**,**C**) of proteins, unique for yellow seeds of cultivar Millennium (yellow color in panel **A** and **C**) and for green seeds of cultivar Gloriosa (green color in panel **A** and **B**), both in nondepleted samples and after treatment with 0.07% (*w/v*) protamine sulfate. Subcellular fractions: CSK, cytoskeleton; CYT, cytoplasm; END, endoplasmic reticulum; EXC, extracellular/secreted; GOL, Golgi apparatus; MIT, mitochondria; NUC, nuclear; PLA, plasma membrane; POX, peroxisome; doubled labels denote corresponding combinations of possible localization.

Among 251 proteins, unique for green seed embryos (Supplementary information 7), 237 could be annotated by way of homology to *A. thaliana* proteins and analyzed with the String database. Thereby, nine structural/functional and seven coexpression protein interaction networks could be annotated (Figure 8).

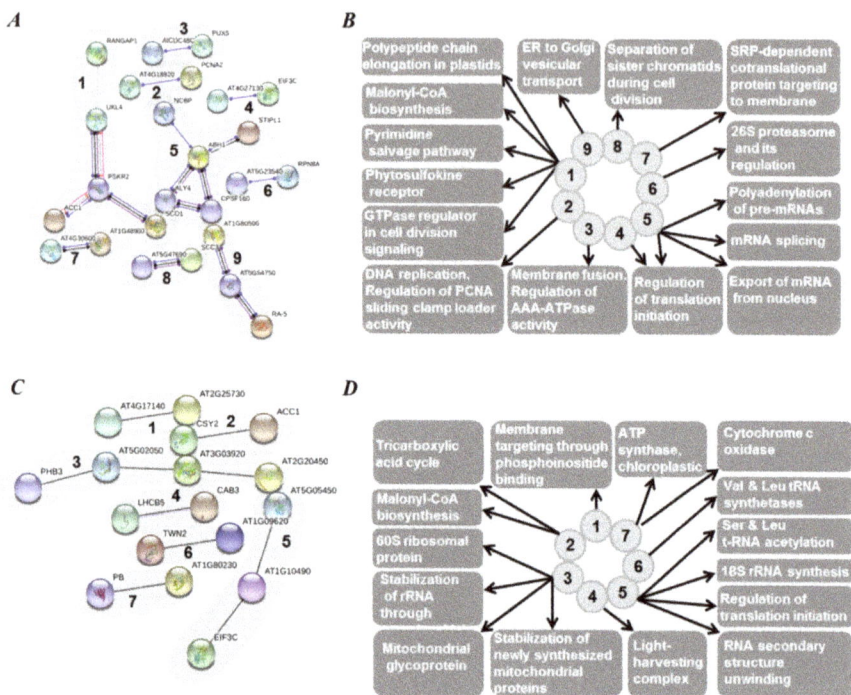

Figure 8. String database analysis (https://string-db.org/) of interaction networks formed by proteins, unique for green seeds (cultivar Gloriosa). The networks (**A,C**) and related functional pathways (**B,D**) relied on experimentally derived functional and/or structural evidences (**A,B**) and coexpression data (**C,D**) with interaction score ≥ 0.7 (corresponded to highly confident entries). Color coding of interactions: black, reaction; blue, binding; light blue, phenotype; green, activation; purple, catalysis; red, inhibition; pink, posttranslational modifications (PTMs); yellow, transcriptional regulation. Filled and empty nodes denote proteins with known and unknown structure, respectively. Detailed description of protein data is provided in Supplementary information 7.

Finally, we addressed the question of how application of the protamine sulfate precipitation (PSP) affects functional profiles and protein localization. For this purpose, we compared the proteins identified only in nondepleted fractions (i.e., precipitation in the presence of protamine sulfate) and those found only after depletion of the highly abundant proteins. As can be seen from Figure 9A, precipitation of high-abundance proteins (representing mostly nucleotide, protein, amino acid, and metal-related metabolism, Supplementary information 8) led to the identification of several other proteins annotated to protein, nucleic acid, and nucleotide metabolism. Accordingly, the supernatants obtained after supplementation with protamine sulfate were dominated by nuclear proteins, whereas the depleted proteins exhibited a higher percentage of cytosolic, mitochondrial, and plasma membrane polypeptides (Figure 9B,C).

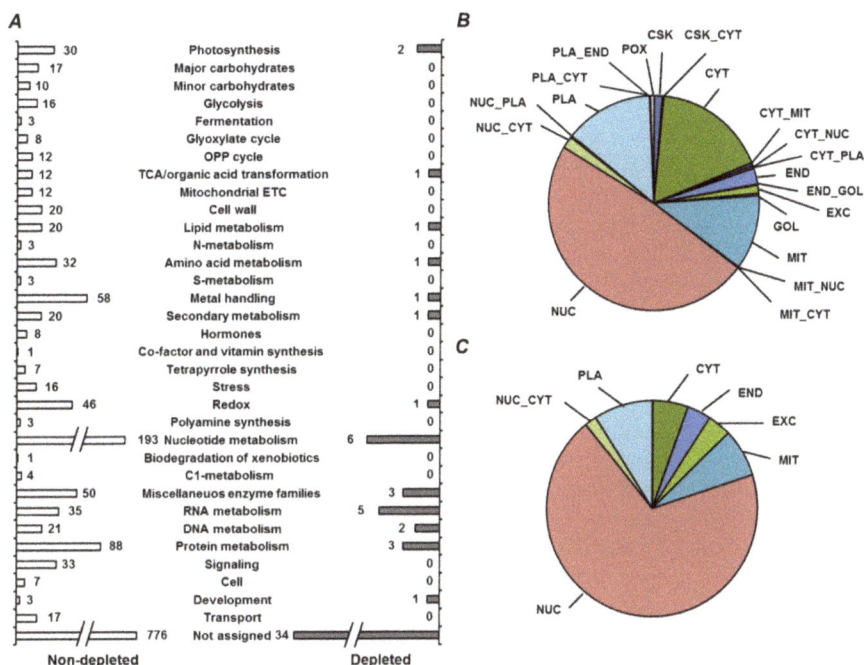

Figure 9. Functional annotation (**A**) and prediction of subcellular localization (**B,C**) of proteins, identified both in yellow seeds of cultivar Millennium and in green seeds of cultivar Gloriosa in nondepleted samples (white color in panels **A** and **B**) and after treatment with 0.07% (*w/v*) protamine sulfate (grey color in panels **A** and **C**). Subcellular fractions: CSK, cytoskeleton; CYT, cytoplasm; END, endoplasmic reticulum; EXC, extracellular/secreted; GOL, Golgi apparatus; MIT, mitochondria; NUC, nuclear; PLA, plasma membrane; POX, peroxisome; doubled labels denote corresponding combinations of possible localization.

3. Discussion

3.1. Protein Extraction and Depletion of Highly Abundant Proteome Fraction

Similarly to our earlier study [16], a high variability of protein yields (39.6–124.1 mg/g fresh weight, Table S1-5) was observed. This phenomenon can be explained by biochemical heterogeneity of seeds [24,25]. Approximately 50% higher recoveries, observed for green seeds, were in agreement with a higher number of proteins, identified in this fraction (Figure 4). On the other hand, precise estimates of protein abundance (RSD = 3%) were in line with our other previous work [26], and thus allowed a comparative, label-free proteomics approach of the heterogeneous seed proteomes [22].

Due to a high abundance of major seed storage proteins [9] and a high amenability of DDA-based shotgun proteomics measurements to a so-called under-sampling effect [27], poor identification rates can be expected for low-abundance proteins. This issue can be expected to be the case when analyzing seed protein digests, as the seed proteome is dominated by several highly abundant protein families–vicilins, convicilins, 11S globulins (legumins), and 2S albumins (PA1 and PA2) [28]. Among the analytical solutions, available for selective removal of storage proteins [29], we selected the PSP method due to its high precision (CV < 12%) and, generally, in contrast to most of the other published techniques, quantitative nature [30]. Indeed, in our hands, this approach allowed extensive proteome coverage, that is, only a low number of proteins could be additionally identified after PSP (only 1.7% of the total 2195 proteins, Figure 4). Interestingly, PSP yielded mostly regulatory proteins, represented by transcription factors and regulators (GO terms "DNA binding" and "RNA binding"), kinases, transport proteins, and enzymes of energy metabolism (Supplementary information 8). The variability of structural domains of these mostly short and polar polypeptides was much lower in PSP-treated samples (Supplementary information 9).

3.2. Annotation of the Pea Seed Proteome

To date, studies of the legume seed proteome mostly relied on gel-based techniques (typically 2D-GE and MALDI-TOF(/TOF)-MS), yielding up to several hundreds of protein identifications now available [31,32]. Sample prefractionation [33] and isoelectrofocusing (variation of pH gradients) [31,34] do not dramatically improve the situation. In this context, due to the better resolution of RP-HPLC, an LC-MS-based approach seemed promising [14]. Indeed, the most representative LC-MS-based study to date of Min et al., performed with soybean seeds, identified 1626 nonredundant proteins [35], which exceeds the best outcomes from gel-based proteomics surveys by at least twofold [29]. However, it accounts for only a half of the proteins annotated here (Figure 4), although the authors relied on the same prefractionation strategy and instrumentation. The possible reasons for this fact might be (i) a longer nanoHPLC column and shallower gradient, (ii) a more efficient digestion protocol, and (iii) a more representative sequence database, used here.

Longer analysis times with shallow gradients might allow better separation of the complex peptide mixtures [36] and a larger number of DDA cycles per run. Our digestion protocol relies on Anionic Acid-Labile Surfactant (AALS, Progenta), which ensures quantitative digestion of all proteins in a sample, and can be completely destroyed and removed upon proteolysis [37]. Recently, we optimized this procedure for the total protein fraction, obtained by phenolic extraction from various plant species and tissues [16,27]. Similarly to the procedure based on the RapiGest detergent (Waters) [15], this protocol is straightforward and does not include steps, at which losses of recovery and/or precision would be expected to occur (e.g., ultrafiltration [38] or preseparation by PAGE [39]). Finally, our sequence database, containing protein entries of three legumes related to pea, already proved to be a reliable tool for the identification of low-abundance post-translationally modified (glycated) tryptic peptides (known to yield complex fragmentation [40]) in common bean [41]. The database contained exclusively reviewed (i.e., based on transcriptome and proteome data) entries. As reviewed sequence information is currently scarce for *P. sativum*, we assumed this approach to be more appropriate than use of a genomic database. The results of manual interpretation of low-scoring, statistically significant PSMs (Figure 3) suggested reliable annotation of peptide sequences.

3.3. Functional Annotation of Pea Seed Proteins

Functions of pea seed embryo proteins were successfully annotated by in-house-designed processing pipeline. As can be judged by spectral counts [42] for corresponding unique peptides, storage proteins were the most abundant in proteomes of both yellow and green seeds, and their pattern of legumins, convicilins, vicilins, and glycinins corresponded well with the observation of Bourgeois et al. [6]. The other major functional classes were represented by the polypeptides involved in metabolism of proteins and RNA, photosynthesis, metal handling, redox metabolism, and general metabolism enzyme families (Figure 5). Based on the number of functionally assigned proteins (Supplementary information 4), representation of protein metabolism in pea seeds was comparable to soybeans [39,43]. However, the metal handling, RNA metabolism, and redox groups were much more abundant here in comparison to those studies. This can be explained by deeper coverage of the seed proteome, (threefold compared to Han et al. [39]), although the incomplete functional assignment in our study makes the difference less striking.

3.4. Features of Embryonic Proteome, Related to Seed Vigor: Impact of Residual Chlorophylls

As pea seeds belong to the orthodox type and acquire desiccation tolerance during development [44], their ability to sustain a prolonged dehydration is the most important feature underlying their high vigor (i.e., the properties determining the potential for rapid, uniform emergence and development of normal seedlings [45]). During desiccation, stability of cellular structures is secured by accumulation of protective proteins and quenching of overproduced reactive oxygen species (ROS) [46]. Here, the green seeds showed distinctly lower vigor in comparison to the yellow ones, as can be seen from delayed germination and higher percentage of abnormal seedlings observed on the third day (*t*-test: $p = 0.02$, Figure 1B,C). To a large extent, it can be explained by a significantly lower status of their antioxidant system (Figure 1E,F), whereas residual chlorophylls could underlie accelerated formation of ROS [47,48]. Thereby, slightly increased contents of carotenoids (Figure 1A) were not sufficient to quench ROS, which could lead to slightly compromised stability of membrane structures (Figure 1D).

To address the effects prospectively related to seed residual chlorophylls at the molecular level, we considered the proteins potentially affecting desiccation tolerance, vigor, and longevity. Among them, late embryogenesis abundant (LEA) proteins preserving cell integrity under desiccation stress [49] were one of the most highly represented groups, accounting for nine proteins (Supplementary information 3 and 4). Another important feature underlying high vigor is pre-expressed protein synthesis machinery, which allows fast and concerted germination of seeds and development of seedlings [46]. Not less importantly, rapid degradation of storage proteins is a prerequisite for appropriate mobilization of plant resources for growth of a new plant organism [50]. Also, damaged (e.g., oxidized, glycated, lipoxidized, or misfolded) polypeptides need to be degraded to prevent deleterious effects within the cell [51,52]. In agreement with previous reports [35], proteins involved in protein metabolism were represented with t-RNA ligases, translation initiation, elongation factors, chaperones, and degradation enzymes.

Some enzymes related to protein metabolism showed differential expression in yellow- and green-seeded cultivars. Thus, totally 33 nonredundant t-RNA ligases were identified in yellow pea seed embryos, while 6 unique proteins were additionally annotated in green ones. Translation initiation factors demonstrated similar differences (Supplementary information 4a and 6). As a decrease in abundance of translation factors is known to accompany age-related loss of seed vigor [53,54], the observed expression profiles might indicate a higher amenability of green seed embryos to such a loss during a prolonged dormancy period. Protein folding was represented by more than 20 entries (HSP70 and 90, GroEL, T-complex protein TCP-1/cpn60, Supplementary information 4a, bin 23). Indeed, HSP70 and GroEL were responsive to osmopriming in Arabidopsis and alfalfa seeds, respectively [55,56], whereas abundance of TCP-1 was decreased in aged alfalfa seeds [57]. Remarkably, several peptidases and DnaJ heat shock protein were identified only in Millennium seeds (Supplementary information 6), facilitating faster degradation of damaged polypeptides and faster mobilization of storage proteins. Interestingly, pea seeds contained acylamino-acid-releasing enzyme-like protein, involved in the degradation of oxidized and glycated proteins [58]. Thus, together with the glyoxalase system and ribulosamine/erythrulosamine 3-kinase, it might contribute to the plant antiglycative machinery.

One of the critical factors affecting vigor, longevity, and, in general, viability of dormant and germinating seeds is the abundance of ROS in seed tissues [59,60]. In agreement with this, a rich pattern of redox proteins (thioredoxin reductase, glutathione peroxidase, monodehydroascorbate reductase, superoxide dismutase, dehydroascorbate reductase, and catalase), earlier identified in other species [32,39,61], was confirmed here (Supplementary information 4). Yellow seed embryos contained eight unique oxidoreductases that might indicate higher levels of oxidative catabolism and higher capacity of electron transfer chains [47], and could potentially increase vigor of yellow seeds.

Deeper analysis of metabolic pathways potentially involved in seed longevity [35,62] (Supplementary information 7) revealed S-adenosylmethionine (SAM) synthase 2 and glycolytic enzyme hexokinase 1 and E1 alpha subunit of the pyruvate dehydrogenase complex. As was earlier shown in priming and ageing experiments [53], SAM synthase is critical for seed vigor. Conversely, enzymes of primary metabolism were shown to be affected by controlled deterioration (accelerated ageing) of seeds (i.e., age-related loss of vigor [63,64]). In this case, the presence of additional isoforms of hexokinase and pyruvate dehydrogenase might not only affect the rate of glycolysis, but also increase production of Ac-KoA and its involvement in the TCA cycle. Remarkably, yellow seeds express three unique aldehyde dehydrogenases-enzymes, reducing reactive carbonyl products of lipid peroxidation and glycation to corresponding alcohols, thereby protecting proteins from oxidative damage [65].

To detect protein–protein interaction networks specific to green seed embryos, we performed String database analysis with the proteins found solely in Gloriosa seeds (Figure 8). The largest functional/structural cluster (#1, Figure 8A) was represented by enzymes, involved in nucleotide metabolism and nucleotide-dependent regulatory pathways, as well as protein biosynthesis in chloroplasts and malonyl-CoA biosynthesis (Figure 8B). These functional relations can be explained by simultaneous synthesis of lipids and integral proteins of thylakoid membranes, triggered by activation of phytosulfokine receptors and mediated by calmodulin-related signaling [66]. The second major cluster (#5) comprised proteins related to m-RNA processing and initiation of translation. As can be seen from Figure 8A, it is closely related to the previous one, suggesting an impact on activation of gene expression during growth [67]. Analysis of the major coexpression clusters (Figure 8C, clusters 3 and 5) was in agreement with these results and confirmed activation of protein synthesis (mostly translation and RNA processing directly related to it, Figure 8D) as one of the main features of green-colored seeds (Figure 8A and Supplementary information 7). Further, the coexpression clusters 4 and 7 confirmed the functional annotation data, indicating high representation of photosynthesis-related proteins in the seed proteome (Figure 5).

4. Materials and Methods

4.1. Reagents and Plant Material

Unless stated otherwise, materials were obtained from the following manufacturers. Carl Roth GmbH & Co (Karlsruhe, Germany): acetonitrile ($\geq$99.95%, LC-MS grade), ethanol($\geq$99.8%), sodium dodecyl sulfate (SDS) (>99%), tris-(2-carboxyethyl)-phosphine hydrochloride (TCEP, $\geq$98%); PanReac AppliChem (Darmstadt, Germany): acrylamide (2K Standard Grade), glycerol (ACS grade); AMRESCO LLC (Fountain Parkway Solon, OH, USA): ammonium persulfate (ACS grade), glycine (biotechnology grade), N,N'-methylene-bis-acrylamide (ultrapure grade), tris(hydroxymethyl)aminomethane (tris, ultrapure grade), urea (ultrapure grade), ammonium bicarbonate (puriss, p.a.); Bioanalytical Technologies 3M Company (St. Paul, MN, USA): Empore™ solid phase octodecyl extraction discs; Component-Reactiv (Moscow, Russia): phosphoric acid (p.a.); Reachem (Moscow, Russia): hydrochloric acid (p.a.), isopropanol (reagent grade), potassium chloride (reagent grade); SERVA Electrophoresis GmbH (Heidelberg, Germany): Coomassie Brilliant Blue G-250, 2-mercaptoethanol (research grade), trypsin NB (sequencing grade, modified from porcine pancreas); Thermo Fisher Scientific (Waltham, MA, USA): Pierce™ Unstained Protein Molecular Weight Marker #26610 (14.4–116.0 kDa), PageRuller™ Plus Prestained Protein Ladder #26620 (10–250 kDa); Dichrom GmbH (Marl, Germany): Progenta™ anionic acid labile surfactant II (AALS II) and adaptors for stage-tips; Vekton (St. Petersburg, Russia): acetonitrile (HPLC grade), sucrose (ACS grade), conc. HCl (puriss). All other chemicals were purchased from Sigma-Aldrich Chemie GmbH (Taufkirchen, Germany). Water was purified in-house (resistance 5–15 mΩ/cm) on water conditioning and purification systems Elix 3 UV (Millipore, Moscow, Russia) or Millipore Milli-Q Gradient A10 system (resistance 5–15 mΩ/cm, Merck Millipore, Darmstadt, Germany).

Pea seeds of the yellow-seeded cultivar Millennium (Figure S1-1A–D) were obtained from the Research and Practical Center of National Academy of Science of the Republic of Belarus for Arable Farming (Zhodino, Belarus, harvested in the year 2015). The seeds of the green-seeded cultivar Gloriosa (Figure S1-1E–H) were provided by the Institute of Vegetable Growing, National Academy of Science (Minsk, Belarus, harvested in the year 2015).

4.2. Analysis of Physiological and Biochemical Parameters of Seed Quality

The seeds were germinated on filter paper in a thermostat at 22 °C. To address the kinetics of germination, the numbers of germinated seeds were monitored for 10 days on a daily basis, before the seeds were classified as (i) nongerminating, and those producing (ii) normal and (iii) abnormal seedlings upon germination, as defined by the International Seed Testing Agency (ISTA) [68]. Electrical conductivity test relied on the method of Matthews et al. [69]. For this, seeds ($n = 5$) were incubated in 25 mL of distilled water for 2 h at 22 °C. The conductivity of the aqueous medium was determined with a conductometer HI 8733 (HANNA Instruments Deutschland GmbH, Vöhringen, Germany) before and after the incubation. Determination of photosynthetic pigments ($n = 3$) relied on the method of Lichtenthaler and Wellburn [70] (for details see Supplementary information 1, Protocol S1-1). Hydrogen peroxide was quantified by the method of Bilova et al. [71], whereas the contents of lipid peroxidation products were determined as malondialdehyde (MDA) equivalents, as described by Frolov and coworkers [26] (see Protocols S1-2 and S1-3, respectively).

4.3. Protein Isolation

Pea seeds (10 per biological replicate, $n = 3$) were frozen in liquid nitrogen and ground in a Mixer Mill MM 400 ball mill with a Ø 20 mm stainless steel ball (Retsch, Haan, Germany) at a vibration frequency of 30 Hz for 2 × 1 min. The resulting ground material (approximately 50 mg per replicate) was placed in 2 mL safe-lock polypropylene tubes and stored at −80 °C prior to protein extraction, which relied on the method of Frolov and coworkers [16] with some modifications. Briefly, the tubes with plant material were transferred on ice, and 1 min later supplemented with 700 µL of cold (4 °C) phenol extraction buffer, containing 0.7 mol/L sucrose, 0.1 mol/L KCl, 5 mmol/L ethylenediaminetetraacetic acid (EDTA), 2% (v/v) mercaptoethanol and 1 mmol/L phenylmethylsulfonyl fluoride (PMSF) in 0.5 mol/L tris-HCl buffer (pH 7.5). The suspensions were vortexed for 30 s, before 700 µL of cold phenol (4 °C) saturated with 0.5 mol/L tris-HCl buffer (pH 7.5) were added. After further mixing for 30 s, the samples were shaken for 30 min at 900 rpm (4 °C) and centrifuged (5000× g, 30 min, 4 °C). Afterwards, the phenolic (upper) phases were transferred to new 1.5 mL polypropylene tubes, and washed two times with equal volumes of the phenol extraction buffer (after each buffer addition: vortexing 30 s, shaking for 30 min at 900 rpm at 4 °C, and centrifugation at 5000× g for 15 min at 4 °C). Then, the supernatants were transferred to 1.5 mL polypropylene tubes, and proteins were precipitated by addition of a fivefold volume of ice-cold 0.1 mol/L ammonium acetate in methanol, followed by storage overnight at −20 °C. Next morning, the protein fraction was collected by centrifugation (10 min, 5000× g, 4 °C), and the supernatants were discarded. The pellets were washed twice with two volumes of methanol (relative to the volume of the phenol phase), and twice with the same volume of acetone (both at 4 °C). Each time, the samples were centrifuged (5000× g, 10 min, 4 °C) after resuspension. Finally, the cleaned pellets were dried under air flow in a fume hood for 1 h, reconstituted in 100 µL of shotgun buffer (8 mol/L urea, 2 mol/L thiourea, 0.15% AALS II in 100 mmol/L tris-HCl, pH 7.5), and protein contents were determined by 2-D Quant Kit (GE Healthcare, Taufkirchen, Germany) according to Matamoros and coworkers [41] or by the Bradford method as described by Frolov and coworkers [72]. The precision of the assay was verified by SDS-PAGE according Greifenhagen et al. [73] with minor modifications (for details see Protocol S1-4).

4.4. Depletion of Seed Storage Proteins by Protamine Sulfate and Tryptic Digestion

Depletion of seed storage proteins relied on the protamine sulfate precipitation procedure (PSP) of Kim et al. [9] with some modifications. In detail, 95 µL of the total protein fractions, isolated from green and yellow seeds ($n = 3$) as described above, were supplemented with 855 µL of 0.077% (w/v) protamine sulfate in 20 mmol/L MgCl$_2$/0.5 mol/L Tris-HCl (pH 8.3) to yield a final protamine sulfate concentration of 0.07% (w/v). After centrifugation (12,000× g, 10 min, 4 °C), supernatants were transferred to 5 mL polypropylene tubes and supplemented with four volumes of 12.5% trichloroacetic acid (TCA) in cold (−20 °C) acetone and left at −20 °C overnight. Next morning, the pellets were washed three times with 1 mL of cold acetone by sequential resuspending and centrifugation (12,000× g, 10 min, 4 °C), and clean pellets were dried in a fume hood for 1 h. The dried pellets were reconstituted in 100 µL of shotgun buffer containing 0.15% AALS, and the protein contents were determined by 2-D Quant Kit without further dilution. The digestion procedure relied on the protocol of Frolov and coworkers [26] with minor modifications (See Protocol S1-5 for details).

4.5. NanoHPLC-ESI-Q-Orbitrap Analysis

Dried tryptic digests were reconstituted in 100 µL of aq. acetonitrile (final concentration of 3% *v/v*) containing 0.1% (*v/v*) aq. formic acid, and 1 µL (250 ng) of peptide mixture was loaded onto an Acclaim PepMap 100 C18 trap column (75 µm × 20 mm, nano-viper, 3 µm particle size). After trapping, the peptides were separated on a PepMap™ RSL C18 column (75 µm × 50 cm, 2 µm particle size) using an EASY-nLC 1000HPLC system coupled on-line to a Q-Exactive Plus mass spectrometer via an EASY-spray nano ion source (all Thermo Fisher Scientific, Bremen, Germany). The details of the chromatographic method are summarized in Supplementary information 1 (Table S1-1). The nano-LC-Orbitrap-MS analysis relied on data-dependent acquisition (DDA) experiments performed in the positive ion mode, comprising a survey Orbitrap-MS scan and dependent MS/MS scans for the most abundant signals in the following 5 s (at certain tR) with charge states ranging from 2 to 6. The mass spectrometer settings and DDA parameters are summarized in Table S1-2. The database search relied on Proteome Discoverer 2.2 software (Thermo Fisher Scientific, Bremen, Germany), SEQUEST search engine, and a combined sequence database comprising proteomes of *Medicago truncatula* Gaertn, *Lotus japonicus* (Regel) K. Larsen, and *Phaseolus vulgaris* L [41] with addition of protein sequences, prospectively related to seed longevity and color (see Table S1-3 for details). The redundancy of the database was eliminated by the CD-HIT algorithm [74] with the sequence identity cutoff set to 1. The database search parameters are summarized in Table S1-4.

4.6. Data Analysis and Postprocessing

Functional annotation of pea proteins was done withVisual Basic for Applications (VBA). The corresponding script (Figure S1-2) was run within Excel instance and managed automatic extraction, appendment, and rearrangement of tabular data, organized in spread sheets. Further, it established access to a laboratory web server via a simplified HTTP protocol. The server, in turn, was run under Apache 2.4, used as a CGI front end for a set of bash scripts. These scripts managed previously cached data, performed actual queries to external public databases, or ran locally installed programs. After screening against UniProt [75] or KEGG [76] databases, amino acid sequences, gene ontology (GO) identifiers, and KEGG Orthology (KO) terms with text annotations were extracted as text files [77]. Annotation relied on the functional bins used by the MapMan software suite [78]. Cellular localization was predicted by the NGLOC algorithm [79]. Finally, all annotations were fitted back into the original Excel spreadsheet and clustered by localization terms or by top MapMan hierarchy levels, before the data was manually inspected, corrected, or further automatically processed.

Analysis of protein networks relied on a search against the *Arabidopsis thaliana* proteome in the String database (for details see Protocol S1-6). Only highly confident protein entries with interaction scores ≥0.7 were exported for building protein interaction maps based on (i) experimentally confirmed evidences and (ii) coexpression. Venn diagrams were constructed using the web-based tool InteractiVenn [80] based on the information given in Proteome Discoverer 2.2 output in columns "Found in Samples".

5. Conclusions

The quality of plant seeds is a prerequisite for sustainable agriculture and food production. As legume crops represent the main source of food protein worldwide, the quality of their seeds requires special attention. Using comprehensive methods of LC-MS-based proteomics, we succeeded in extensively covering the pea seed proteome and assigning functions to individual nonredundant proteins. This systematic approach allowed the identification of characteristic metabolic features of green-coloured seeds in comparison to yellow-coloured ones. Thereby, alterations in the proteome of green seeds, prospectively related to their reduced vigor and longevity, were revealed.

Supplementary Materials: Supplementary materials can be found at http://www.mdpi.com/1422-0067/19/12/4066/s1.

Author Contributions: G.S., S.M., and A.F. conceived the study; T.M., E.L., T.G., and C.P. performed sample preparation and M.S. measurements; T.M., G.M.-S., T.B., and A.K. performed data interpretation, processing, and postprocessing; G.S., S.M., V.B., W.H., and A.F. contributed to writing the manuscript and proofreading.

Funding: This research was funded by the Russian Science Foundation (project number 16-16-00026).

Acknowledgments: The authors are grateful to Veronika Chantseva, Olga Shiroglazova, and Svetlana Milrud (St. Petersburg State University) for technical assistance. The St. Petersburg St. University Research Resource Center for Molecular and Cell Technologies is acknowledged for technical support.

Conflicts of Interest: The authors declare no conflict of interest.

Abbreviations

2D-GE	two-dimensional gel electrophoresis
AALS	Anionic Acid-Labile Surfactant
CCEs	chlorophyll catabolic enzymes
DDA	data dependent acquisition
ESI	electrospray ionization
FDR	false discovery rate
ISTA	International Seed Testing Agency
LEA	late embryogenesis abundant
LC	liquid chromatography
MALDI	matrix-assisted laser desorption/ionization
MDA	malondialdehyde
MS	mass spectrometry
PS	protamine sulfate
PSP	protamine sulfate precipitation
PSMs	peptide spectrum matches
RCC	red chlorophyll catabolite
ROS	reactive oxygen species
RP-HPLC	reverse phase high performance liquid chromatography
SDS-PAGE	sodium dodecyl sulfate polyacrylamide gel electrophoresis
SGR	stay green
SPE	solid phase extraction
TOF	time of flight

References

1. Singh, N. Pulses: An overview. *J. Food Sci. Technol.* **2017**, *54*, 853–857. [CrossRef] [PubMed]
2. Iqbal, A.; Khalil, I.A.; Ateeq, N.; Sayyar Khan, M. Nutritional quality of important food legumes. *Food Chem.* **2006**, *97*, 331–335. [CrossRef]
3. Babar, M.M.; Zaidi, N.S.; Azooz, M.M.; Kazi, A.G. Genetic and molecular responses of legumes in a changing environment. In *Legumes under Environmental Stress*; John Wiley & Sons, Ltd.: Hoboken, NJ, USA, 2015; pp. 199–214. ISBN 978-1-118-91709-1.
4. Casey, R.; Domoney, C.; Smith, A.M. *Biochemistry and Molecular Biology of Seed Products*; CAB International: Wallingford, UK, 1993; ISBN 978-0-85198-863-4.
5. Frolov, A.; Mamontova, T.; Ihling, C.; Lukasheva, E.; Bankin, M.; Chantseva, V.; Vikhnina, M.; Soboleva, A.; Shumilina, J.; Mavropolo-Stolyarenko, G.; et al. Mining seed proteome: From protein dynamics to modification profiles. *Biol. Commun.* **2018**, *63*, 43–58. [CrossRef]

6. Bourgeois, M.; Jacquin, F.; Savois, V.; Sommerer, N.; Labas, V.; Henry, C.; Burstin, J. Dissecting the proteome of pea mature seeds reveals the phenotypic plasticity of seed protein composition. *Proteomics* **2009**, *9*, 254–271. [CrossRef] [PubMed]

7. Bourgeois, M.; Jacquin, F.; Cassecuelle, F.; Savois, V.; Belghazi, M.; Aubert, G.; Quillien, L.; Huart, M.; Marget, P.; Burstin, J. A PQL (protein quantity loci) analysis of mature pea seed proteins identifies loci determining seed protein composition. *Proteomics* **2011**, *11*, 1581–1594. [CrossRef] [PubMed]

8. Natarajan, S.S.; Krishnan, H.B.; Lakshman, S.; Garrett, W.M. An efficient extraction method to enhance analysis of low abundant proteins from soybean seed. *Anal. Biochem.* **2009**, *394*, 259–268. [CrossRef] [PubMed]

9. Kim, Y.J.; Wang, Y.; Gupta, R.; Kim, S.W.; Min, C.W.; Kim, Y.C.; Park, K.H.; Agrawal, G.K.; Rakwal, R.; Choung, M.-G.; et al. Protamine sulfate precipitation method depletes abundant plant seed-storage proteins: A case study on legume plants. *Proteomics* **2015**, *15*, 1760–1764. [CrossRef]

10. Krishnan, H.B.; Oehrle, N.W.; Natarajan, S.S. A rapid and simple procedure for the depletion of abundant storage proteins from legume seeds to advance proteome analysis: A case study using Glycine max. *Proteomics* **2009**, *9*, 3174–3188. [CrossRef]

11. Boschetti, E.; Righetti, P.G. Plant proteomics methods to reach low-abundance proteins. *Methods Mol. Biol.* **2014**, *1072*, 111–129.

12. Cerna, H.; Černý, M.; Habánová, H.; Šafářová, D.; Abushamsiya, K.; Navrátil, M.; Brzobohatý, B. Proteomics offers insight to the mechanism behind *Pisum sativum* L. response to pea seed-borne mosaic virus (PSbMV). *J. Proteom.* **2017**, *153*, 78–88. [CrossRef]

13. Vissers, J.P.; Chervet, J.P.; Salzmann, J.P. Sodium dodecyl sulphate removal from tryptic digest samples for on-line capillary liquid chromatography/electrospray mass spectrometry. *J. Mass Spectrom.* **1996**, *31*, 1021–1027. [CrossRef]

14. Soboleva, A.; Schmidt, R.; Vikhnina, M.; Grishina, T.; Frolov, A. Maillard Proteomics: Opening New Pages. *Int. J. Mol. Sci.* **2017**, *18*, 2677. [CrossRef] [PubMed]

15. Kaspar-Schoenefeld, S.; Merx, K.; Jozefowicz, A.M.; Hartmann, A.; Seiffert, U.; Weschke, W.; Matros, A.; Mock, H. Label-free proteome profiling reveals developmental-dependent patterns in young barley grains. *J. Proteom.* **2016**, *143*, 106–121. [CrossRef] [PubMed]

16. Frolov, A.; Didio, A.; Ihling, C.; Chantzeva, V.; Grishina, T.; Hoehenwarter, W.; Sinz, A.; Smolikova, G.; Bilova, T.; Medvedev, S. The effect of simulated microgravity on the Brassica napus seedling proteome. *Funct. Plant Biol.* **2018**, *45*, 440. [CrossRef]

17. Smolikova, G.; Dolgikh, E.; Vikhnina, M.; Frolov, A.; Medvedev, S. Genetic and Hormonal Regulation of Chlorophyll Degradation during Maturation of Seeds with Green Embryos. *Int. J. Mol. Sci.* **2017**, *18*, 1993. [CrossRef] [PubMed]

18. Hörtensteiner, S. Stay-green regulates chlorophyll and chlorophyll-binding protein degradation during senescence. *Trends Plant Sci.* **2009**, *14*, 155–162. [CrossRef] [PubMed]

19. Smolikova, G.N.; Laman, N.A.; Boriskevich, O.V. Role of chlorophylls and carotenoids in seed tolerance to abiotic stressors. *Russ. J. Plant Physiol.* **2011**, *58*, 965–973. [CrossRef]

20. Clerkx, E.J.M.; Vries, H.B.; Ruys, G.J.; Groot, S.P.C.; Koornneef, M. Characterization of green seed, an Enhancer of abi3-1 in Arabidopsis That Affects Seed Longevity. *Plant Mol. Biol.* **2003**, *132*, 1077–1084. [CrossRef]

21. Zinsmeister, J.; Lalanne, D.; Terrasson, E.; Chatelain, E.; Vandecasteele, C.; Vu, B.L.; Dubois-Laurent, C.; Geoffriau, E.; Signor, C.; Le Dalmais, M.; et al. ABI5 Is a Regulator of Seed Maturation and Longevity in Legumes. *Plant Cell* **2016**, *28*, 2735–2754. [CrossRef]

22. Frolov, A.; Blüher, M.; Hoffmann, R. Glycation sites of human plasma proteins are affected to different extents by hyperglycemic conditions in type 2 diabetes mellitus. *Anal. Bioanal. Chem.* **2014**, *406*, 5755–5763. [CrossRef]

23. Barac, M.; Cabrilo, S.; Pesic, M.; Stanojevic, S.; Zilic, S.; Macej, O.; Ristic, N. Profile and functional properties of seed proteins from six pea (Pisum sativum) genotypes. *Int. J. Mol. Sci.* **2010**, *11*, 4973–4990. [CrossRef] [PubMed]

24. Gallardo, K.; Thompson, R.; Burstin, J. Reserve accumulation in legume seeds. *CR Biol.* **2008**, *331*, 755–762. [CrossRef] [PubMed]

25. Thompson, R.; Burstin, J.; Gallardo, K. Post-genomics studies of developmental processes in legume seeds. *Plant Physiol.* **2009**, *151*, 1023–1029. [CrossRef]

26. Frolov, A.; Bilova, T.; Paudel, G.; Berger, R.; Balcke, G.U.; Birkemeyer, C.; Wessjohann, L.A. Early responses of mature Arabidopsis thaliana plants to reduced water potential in the agar-based polyethylene glycol infusion drought model. *J. Plant Physiol.* **2017**, *208*, 70–83. [CrossRef] [PubMed]

27. Paudel, G.; Bilova, T.; Schmidt, R.; Greifenhagen, U.; Berger, R.; Tarakhovskaya, E.; Stöckhardt, S.; Balcke, G.U.; Humbeck, K.; Brandt, W.; et al. Osmotic stress is accompanied by protein glycation in Arabidopsis thaliana. *J. Exp. Bot.* **2016**, *67*, 6283–6295. [CrossRef] [PubMed]

28. Casey, R.; Domoney, C. Pea globulins. In *Seed Proteins*; Shawrey, P.R., Casey, R., Eds.; Springer: Dordrecht, The Netherlands, 1999; pp. 171–208. ISBN 978-94-010-5904-6.

29. Gaupels, F.; Furch, A.C.U.; Zimmermann, M.R.; Chen, F.; Kaever, V.; Buhtz, A.; Kehr, J.; Sarioglu, H.; Kogel, K.-H.; Durner, J. Corrigendum: Systemic Induction of NO-, Redox-, and cGMP Signaling in the Pumpkin Extrafascicular Phloem upon Local Leaf Wounding. *Front. Plant Sci.* **2016**, *7*, 281. [CrossRef] [PubMed]

30. Min, C.W.; Gupta, R.; Kim, S.W.; Lee, S.E.; Kim, Y.C.; Bae, D.W.; Han, W.Y.; Lee, B.W.; Ko, J.M.; Agrawal, G.K.; et al. Comparative Biochemical and Proteomic Analyses of Soybean Seed Cultivars Differing in Protein and Oil Content. *J. Agric. Food Chem.* **2015**, *63*, 7134–7142. [CrossRef]

31. Ogura, T.; Ogihara, J.; Sunairi, M.; Takeishi, H.; Aizawa, T.; Olivos-Trujillo, M.R.; Maureira-Butler, I.J.; Salvo-Garrido, H.E. Proteomic characterization of seeds from yellow lupin (*Lupinus luteus* L.). *Proteomics* **2014**, *14*, 1543–1546. [CrossRef]

32. Krishnan, H.B.; Natarajan, S.S.; Oehrle, N.W.; Garrett, W.M.; Darwish, O. Proteomic Analysis of Pigeonpea (*Cajanus cajan*) Seeds Reveals the Accumulation of Numerous Stress-Related Proteins. *J. Agric. Food Chem.* **2017**, *65*, 4572–4581. [CrossRef]

33. Miernyk, J.A.; Hajduch, M. Seed proteomics. *J. Proteom.* **2011**, *74*, 389–400. [CrossRef]

34. Yin, G.; Xin, X.; Fu, S.; An, M.; Wu, S.; Chen, X.; Zhang, J.; He, J.; Whelan, J.; Lu, X. Proteomic and Carbonylation Profile Analysis at the Critical Node of Seed Ageing in Oryza sativa. *Sci. Rep.* **2017**, *7*, 40611. [CrossRef] [PubMed]

35. Min, C.W.; Lee, S.H.; Cheon, Y.E.; Han, W.Y.; Ko, J.M.; Kang, H.W.; Kim, Y.C.; Agrawal, G.K.; Rakwal, R.; Gupta, R.; et al. In-depth proteomic analysis of Glycine max seeds during controlled deterioration treatment reveals a shift in seed metabolism. *J. Proteom.* **2017**, *169*, 125–135. [CrossRef] [PubMed]

36. Trudgian, D.C.; Fischer, R.; Guo, X.; Kessler, B.M.; Mirzaei, H. GOAT—A simple LC-MS/MS gradient optimization tool. *Proteomics* **2014**, *14*, 1467–1471. [CrossRef] [PubMed]

37. Waas, M.; Bhattacharya, S.; Chuppa, S.; Wu, X.; Jensen, D.R.; Omasits, U.; Wollscheid, B.; Volkman, B.F.; Noon, K.R.; Gundry, R.L. Combine and Conquer: Surfactants, Solvents, and Chaotropes for Robust Mass Spectrometry Based Analyses of Membrane Proteins. *Anal. Chem.* **2014**, *86*, 1551–1559. [CrossRef] [PubMed]

38. Wiśniewski, J.R.; Zougman, A.; Nagaraj, N.; Mann, M. Universal sample preparation method for proteome analysis. *Nat. Methods* **2009**, *6*, 359–362. [CrossRef] [PubMed]

39. Han, C.; Yin, X.; He, D.; Yang, P. Analysis of Proteome Profile in Germinating Soybean Seed, and Its Comparison with Rice Showing the Styles of Reserves Mobilization in Different Crops. *PLoS ONE* **2013**, *8*, e56947. [CrossRef] [PubMed]

40. Fedorova, M.; Frolov, A.; Hoffmann, R. Fragmentation behavior of Amadori-peptides obtained by non-enzymatic glycosylation of lysine residues with ADP-ribose in tandem mass spectrometry. *J. Mass Spectrom.* **2010**, *45*, 664–669. [CrossRef]

41. Matamoros, M.A.; Kim, A.; Peñuelas, M.; Ihling, C.; Griesser, E.; Hoffmann, R.; Fedorova, M.; Frolov, A.; Becana, M. Protein Carbonylation and Glycation in Legume Nodules. *Plant Physiol.* **2018**, *177*. [CrossRef]

42. *Spectral Counts Were Accessed by Number of PSMs and by Calculating Normalized Spectral Abundance Factor (NSAF) for each Protein Group*; Saint Petersburg State University: Saint Petersburg, Russia, 2018.

43. Gomes, L.S.; Senna, R.; Sandim, V.; Silva-Neto, M.A.C.; Perales, J.E.A.; Zingali, R.B.; Soares, M.R.; Fialho, E. Four Conventional Soybean [*Glycine max* (L.) Merrill] Seeds Exhibit Different Protein Profiles As Revealed by Proteomic Analysis. *J. Agric. Food Chem.* **2014**, *62*, 1283–1293. [CrossRef]

44. Berjak, P.; Pammenter, N.W. From Avicennia to Zizania: Seed recalcitrance in perspective. *Ann. Bot.* **2008**, *101*, 213–228. [CrossRef]

45. Ellis, R.H. The encyclopaedia of seeds: Science, technology and uses. *Ann. Bot.* **2007**, *100*, 1379. [CrossRef]

46. Wang, X.; Tang, D.; Huang, D. Proteomic analysis of pakchoi leaves and roots under glycine–nitrogen conditions. *Plant Physiol. Biochem.* **2014**, *75*, 96–104. [CrossRef] [PubMed]

47. Smolikova, G.; Kreslavski, V.; Shiroglazova, O.; Bilova, T.; Sharova, E.; Frolov, A.; Medvedev, S. Photochemical activity changes accompanying the embryogenesis of pea (Pisum sativum) with yellow and green cotyledons. *Funct. Plant Biol.* **2018**, *45*, 228. [CrossRef]

48. Smolikova, G.N.; Medvedev, S.S. Photosynthesis in the seeds of chloroembryophytes. *Russ. J. Plant Physiol.* **2016**, *63*, 1–12. [CrossRef]

49. Cuming, A.C. LEA Proteins. In *Seed Proteins*; Shewry, P.R., Casey, R., Eds.; Springer: Dordrecht, The Netherlands, 1999; pp. 753–780. ISBN 978-94-011-4431-5.

50. Müntz, K.; Belozersky, M.A.; Dunaevsky, Y.E.; Schlereth, A.; Tiedemann, J. Stored proteinases and the initiation of storage protein mobilization in seeds during germination and seedling growth. *J. Exp. Bot.* **2001**, *52*, 1741–1752. [CrossRef] [PubMed]

51. Hellmann, H.; Estelle, M. Plant Development: Regulation by Protein Degradation. *Science* **2002**, *297*, 793–797. [CrossRef] [PubMed]

52. Bilova, T.; Lukasheva, E.; Brauch, D.; Greifenhagen, U.; Paudel, G.; Tarakhovskaya, E.; Frolova, N.; Mittasch, J.; Balcke, G.U.; Tissier, A.; et al. A Snapshot of the Plant Glycated Proteome: Structural, functional and mechanistic aspect. *J. Biol. Chem.* **2016**, *291*, 7621–7636. [CrossRef]

53. Catusse, J.; Meinhard, J.; Job, C.; Strub, J.M.; Fischer, U.; Pestova, E.; Westhoff, P.; Van Dorsselaer, A.; Job, D. Proteomics reveals potential biomarkers of seed vigor in sugarbeet. *Proteomics* **2011**, *11*, 1569–1580. [CrossRef]

54. Wang, L.; Ma, H.; Song, L.; Shu, Y.; Gu, W. Comparative proteomics analysis reveals the mechanism of pre-harvest seed deterioration of soybean under high temperature and humidity stress. *J. Proteom.* **2012**, *75*, 2109–2127. [CrossRef]

55. Gallardo, K.; Job, C.; Groot, S.P.; Puype, M.; Demol, H.; Vandekerckhove, J.; Job, D. Proteomic analysis of arabidopsis seed germination and priming. *Plant Physiol.* **2001**, *126*, 835–848. [CrossRef]

56. Yacoubi, R.; Job, C.; Belghazi, M.; Chaibi, W.; Job, D. Toward characterizing seed vigor in alfalfa through proteomic analysis of germination and priming. *J. Proteome Res.* **2011**, *10*, 3891–3903. [CrossRef] [PubMed]

57. Rajjou, L.; Lovigny, Y.; Groot, S.P.C.; Belghazi, M.; Job, C.; Job, D. Proteome-wide characterization of seed aging in Arabidopsis: A comparison between artificial and natural aging protocols. *Plant Physiol.* **2008**, *148*, 620–641. [CrossRef] [PubMed]

58. Yamauchi, Y.; Ejiri, Y.; Toyoda, Y.; Tanaka, K. Identification and biochemical characterization of plant acylamino acid-releasing enzyme. *J. Biochem.* **2003**, *134*, 251–257. [CrossRef] [PubMed]

59. Gomes, M.P.; Garcia, Q.S. Reactive oxygen species and seed germination. *Biologia* **2013**, *68*, 351. [CrossRef]

60. Sharma, P.; Jha, A.B.; Dubey, R.S.; Pessarakli, M. Reactive Oxygen Species, Oxidative Damage, and Antioxidative Defense Mechanism in Plants under Stressful Conditions. *J. Bot.* **2012**, *2012*, e217037. [CrossRef]

61. Gallardo, K.; Le Signor, C.; Vandekerckhove, J.; Thompson, R.D.; Burstin, J. Proteomics of Medicago truncatula Seed Development Establishes the Time Frame of Diverse Metabolic Processes Related to Reserve Accumulation. *Plant Physiol.* **2003**, *133*, 664–682. [CrossRef] [PubMed]

62. Wang, C.; Yue, W.; Ying, Y.; Wang, S.; Secco, D.; Liu, Y.; Whelan, J.; Tyerman, S.D.; Shou, H. Rice SPX-Major Facility Superfamily3, a Vacuolar Phosphate Efflux Transporter, Is Involved in Maintaining Phosphate Homeostasis in Rice. *Plant Physiol.* **2015**, *169*, 2822–2831. [PubMed]

63. Yin, X.; He, D.; Gupta, R.; Yang, P. Physiological and proteomic analyses on artificially aged Brassica napus seed. *Front. Plant Sci.* **2015**, *6*, 112. [CrossRef]

64. Xin, X.; Lin, X.-H.; Zhou, Y.-C.; Chen, X.-L.; Liu, X.; Lu, X.-X. Proteome analysis of maize seeds: The effect of artificial ageing. *Physiol. Plant.* **2011**, *143*, 126–138. [CrossRef]

65. Stiti, N.; Missihoun, T.D.; Kotchoni, S.O.; Kirch, H.-H.; Bartels, D. Aldehyde Dehydrogenases in Arabidopsis thaliana: Biochemical Requirements, Metabolic Pathways, and Functional Analysis. *Front. Plant Sci.* **2011**, *2*, 65. [CrossRef]

66. Sauter, M. Phytosulfokine peptide signalling. *J. Exp. Bot.* **2015**, *66*, 5161–5169. [CrossRef] [PubMed]

67. Matsuoka, K.; Demura, T.; Galis, I.; Horiguchi, T.; Sasaki, M.; Tashiro, G.; Fukuda, H. A Comprehensive Gene Expression Analysis Toward the Understanding of Growth and Differentiation of Tobacco BY-2 Cells. *Plant Cell Physiol.* **2004**, *45*, 1280–1289. [CrossRef] [PubMed]

68. ISTA. *International Rules for Seed Testing*; International Seed Testing Association: Basserdorf, Switzerland, 2018; ISSN 2310-3655.

69. Matthews, S.; Powell, A. Electrical Conductivity Vigour Test: Physiological Basis and Use. *Seed Test Int.* **2006**, 32–35.

70. Lichtenthaler, H.K.; Wellburn, A.R. Determinations of total carotenoids and chlorophylls a and b of leaf extracts in different solvents. *Biochem. Soc. Trans.* **1983**, *11*, 591–592. [CrossRef]

71. Bilova, T.; Paudel, G.; Shilyaev, N.; Schmidt, R.; Brauch, D.; Tarakhovskaya, E.; Milrud, S.; Smolikova, G.; Tissier, A.; Vogt, T.; et al. Global proteomic analysis of advanced glycation end products in the Arabidopsis proteome provides evidence for age-related glycation hot spots. *J. Biol. Chem.* **2017**, *292*, 15758–15776. [CrossRef] [PubMed]

72. Greifenhagen, U.; Nguyen, V.D.; Moschner, J.; Giannis, A.; Frolov, A.; Hoffmann, R. Sensitive and Site-Specific Identification of Carboxymethylated and Carboxyethylated Peptides in Tryptic Digests of Proteins and Human Plasma. *J. Proteome Res.* **2015**, *14*, 768–777. [CrossRef] [PubMed]

73. Greifenhagen, U.; Frolov, A.; Blüher, M.; Hoffmann, R. Plasma Proteins Modified by Advanced Glycation End Products (AGEs) Reveal Site-specific Susceptibilities to Glycemic Control in Patients with Type 2 Diabetes. *J. Biol. Chem.* **2016**, *291*, 9610–9616. [CrossRef]

74. Huang, Y.; Niu, B.; Gao, Y.; Fu, L.; Li, W. CD-HIT Suite: A web server for clustering and comparing biological sequences. *Bioinformatics* **2010**, *26*, 680–682. [CrossRef]

75. Uniprot Consortium. UniProt: The universal protein knowledgebase. *Nucleic Acids Res.* **2017**, *45*, D158–D169. [CrossRef]

76. Ogata, H.; Goto, S.; Sato, K.; Fujibuchi, W.; Bono, H.; Kanehisa, M. KEGG: Kyoto Encyclopedia of Genes and Genomes. *Nucleic Acids Res.* **1999**, *27*, 29–34. [CrossRef]

77. Kanehisa, M.; Sato, Y.; Kawashima, M.; Furumichi, M.; Tanabe, M. KEGG as a reference resource for gene and protein annotation. *Nucleic Acids Res.* **2016**, *44*, D457–D462. [CrossRef] [PubMed]

78. Usadel, B.; Poree, F.; Nagel, A.; Lohse, M.; Czedik-Eysenberg, A.; Stitt, M. A guide to using MapMan to visualize and compare Omics data in plants: A case study in the crop species, Maize. *Plant Cell Environ.* **2009**, *32*, 1211–1229. [CrossRef] [PubMed]

79. King, B.R.; Vural, S.; Pandey, S.; Barteau, A.; Guda, C. ngLOC: Software and web server for predicting protein subcellular localization in prokaryotes and eukaryotes. *BMC Res. Notes* **2012**, *5*, 351. [CrossRef] [PubMed]

80. Heberle, H.; Meirelles, G.V.; da Silva, F.R.; Telles, G.P.; Minghim, R. InteractiVenn: A web-based tool for the analysis of sets through Venn diagrams. *BMC Bioinform.* **2015**, *16*, 169. [CrossRef] [PubMed]

International Journal of
Molecular Sciences

MDPI

Article

iTRAQ-Based Quantitative Proteomic Analysis of Embryogenic and Non-embryogenic Calli Derived from a Maize (*Zea mays* L.) Inbred Line Y423

Beibei Liu [†], Xiaohui Shan [†], Ying Wu, Shengzhong Su, Shipeng Li, Hongkui Liu, Junyou Han and Yaping Yuan *

College of Plant Science, Jilin University, Changchun 130062, China; liubeibei1985@126.com (B.L.); shanxiaohui@jlu.edu.cn (X.S.); wuying@jlu.edu.cn (Y.W.); sushengzhong@jlu.edu.cn (S.S.); lisp@jlu.edu.cn (S.L.); liuhk@jlu.edu.cn (H.L.); hanjy@jlu.edu.cn (J.H.)
* Correspondence: yuanyp@jlu.edu.cn
† These authors contributed equally to this work.

Received: 13 November 2018; Accepted: 7 December 2018; Published: 12 December 2018

Abstract: Somatic embryos (SE) have potential to rapidly form a whole plant. Generally, SE is thought to be derived from embryogenic calli (EC). However, in maize, not only embryogenic calli (EC, can generate SE) but also nonembryogenic calli (NEC, can't generate SE) can be induced from immature embryos. In order to understand the differences between EC and NEC and the mechanism of EC, which can easily form SE in maize, differential abundance protein species (DAPS) of EC and NEC from the maize inbred line Y423 were identified by using the isobaric tags for relative and absolute quantification (iTRAQ) proteomic technology. We identified 632 DAPS in EC compared with NEC. The results of bioinformatics analysis showed that EC development might be related to accumulation of pyruvate caused by the DAPS detected in some pathways, such as starch and sucrose metabolism, glycolysis/gluconeogenesis, tricarboxylic acid (TCA) cycle, fatty acid metabolism and phenylpropanoid biosynthesis. Based on the differentially accumulated proteins in EC and NEC, a series of DAPS related with pyruvate biosynthesis and suppression of acetyl-CoA might be responsible for the differences between EC and NEC cells. Furthermore, we speculate that the decreased abundance of enzymes/proteins involved in phenylpropanoid biosynthesis pathway in the EC cells results in reducing of lignin substances, which might affect the maize callus morphology.

Keywords: iTRAQ; proteomics; somatic embryogenesis; pyruvate biosynthesis; *Zea mays*

1. Introduction

Maize is one of the most important cereal crops in the world. Maize production is directly related to the agricultural economy and farmers' income [1]. With the development of molecular breeding and transgenic technology, more and more new genetically modified (GM) maize varieties have been developed and widely cultivated, which has made a tremendous contribution to raising farmers' incomes and ensuring food security (http://www.isaaa.org/resources/publications/briefs/). However, many problems still limit the development of GM maize. For example, because of the majority of maize genotypes showing low embryogenic growth response in culture, only several inbred lines can be used for maize genetic transformation, such as A188 which displays a high embryogenic culture response [2–4]. Somatic embryos (SE), which have the potential to rapidly form a whole plant, have been widely used to propagate transgenic organisms and to obtain genetically modified plants [5]. Efficient SE production also has significance for many agricultural biotechnology applications such as clonal propagation, production of synthetic seed [6], and gamete cycling in rapid breeding [7]. Understanding the mechanism of maize SE development is a key issue to be solved [5]. The first issue is to find the key regulatory pathways that could induce the development of embryogenic callus (EC, can generate SE) rather than nonembryogenic callus (NEC, can't generate SE) during somatic embryogenesis.

A great deal of differentially expressing genes were identified to be related with EC development in maize by cDNA-AFLP and RNA-seq [8,9], but the mRNA levels usually do not correlate well with the protein abundances and functions, due to various post-translational modifications. Unlike RNA, proteins are the central biomolecules that are responsible for all cellular functions in the living organism. Therefore, the analysis of the differential proteins between EC and NEC can more accurately explore the key factors affecting somatic embryogenesis. Early studies have found that many extracellular proteins can affect somatic embryogenesis, in which arabinogalactan protein (AGPs), nonspecific lipid transfer protein (nsLTPs) and germin-like proteins (GLPs) had been considered as marker proteins in somatic embryogenesis. AGPs were reported that played a facilitating role during the process of embryo development or somatic embryogenesis in maize [10], wheat [11] and carrot [12]. And the specific expression of GLPs was detected during the early development of somatic embryos in *Pinus caribaea* Morelet [13]. In recent years, high-throughput proteomic technology was successfully used to understand the process of somatic embryogenesis in different plants. In *Medicago truncatula*, 136 differentially abundant proteins were detected between embryogenic and nonembryogenic lines [14]. Additionally, the reference maps with 169 proteins were established for embryogenic cell lines delivered from *M. truncatula* protoplasts [15]. Four glycolytic enzymes, namely uracil-diphosphate (UDP)-glucose pyrophosphorylase, fructose bisphosphate aldolase, triosephosphate isomerase, and glyceraldehyde-3-phosphate dehydrogenase were found up-accumulated in somatic embryos of *Cyclamen persicum* [16]. Related study also suggested that the proteins related with auxin releasing might be important for late developmental stages of somatic embryos [17]. Two-dimensional electrophoresis (2-DE) was performed to compare the proteomes of embryogenic and nonembryogenic calli induced from H99 inbred maize line. Some proteins associated with somatic embryogenesis were enriched in cell proliferation, transcription and protein processing, stress response, signal transduction, metabolism and energy pathways [18]. And in the grape (*Vitis vinifera*) [19], orange (*Citrus sinensis* Osbeck) [20], oil palm (*Elaeis guineensis* Jacq.) [21], the proteome changes during somatic embryogenesis were also studied by 2-DE technique. The study results showed that the differentially accumulated proteins were mostly related to carbohydrate and energy metabolism or oxidative stress. At the same time, a large number of unknown proteins have also been found and need to be further explored. Isobaric tags for relative and absolute quantification (iTRAQ) is a new proteome research method which can identify low-abundance protein species and the protein species that are too large/small, too acidic/basic and too hydrophobic to be detected by 2-DE [22,23]. Recently, iTRAQ has been used to explore the proteome of somatic embryogenesis in *Gossypium hirsutum* L. [24], *Larix principis-rupprechtii* Mayr [25]. Although these studies have found some proteins associated with somatic embryogenesis, the results cannot systematically elucidate the mechanism that affects the development of EC and NEC in the process of somatic embryogenesis, and which proteins could promote calli to transform into EC or SE.

As reported before, we have identified a new receptor of an elite maize (*Z. mays* L.) inbred line for genetic transformation which displaying high efficiency via intact somatic embryogenesis [8,9]. Therefore, the purpose of this study is to compare the proteomes between EC and NEC by iTRAQ, and then to analyze the differential proteins to explore key proteins or metabolic pathways that affect the development of EC or NEC. This will provide a theoretical basis or working model for solving the problem of EC induction in somatic embryogenesis.

2. Results

2.1. Induction of Embryogenic and Nonembryogenic Calli

Somatic embryo (SE) offers great potential in plant propagation but its developmental process is so complex. In order to induce somatic embryos, immature embryos of inbred maize line Y423 were placed on the induction media. After induction and subculturing 4–5 times, globular somatic embryos that were yellow, loose and small granular appeared on the surfaces of EC. In contrast, NEC looked pale yellow and compact (Figure 1A,B). Histological analysis revealed that the EC cells were clusters of cells with large nuclei and dense cytoplasm (Figure 1C), whereas the NEC cells had vacuoles and few plastids with a loose cell arrangement without rules (Figure 1D). The scanning electron microscope (SEM) analysis of the epidermal cells revealed that the EC had granule structures on their rough surface (Figure 1E), but NEC had a smooth and flaky surface structure (Figure 1F). All these showed that there was obvious difference at both morphological and cellular levels between EC and NEC.

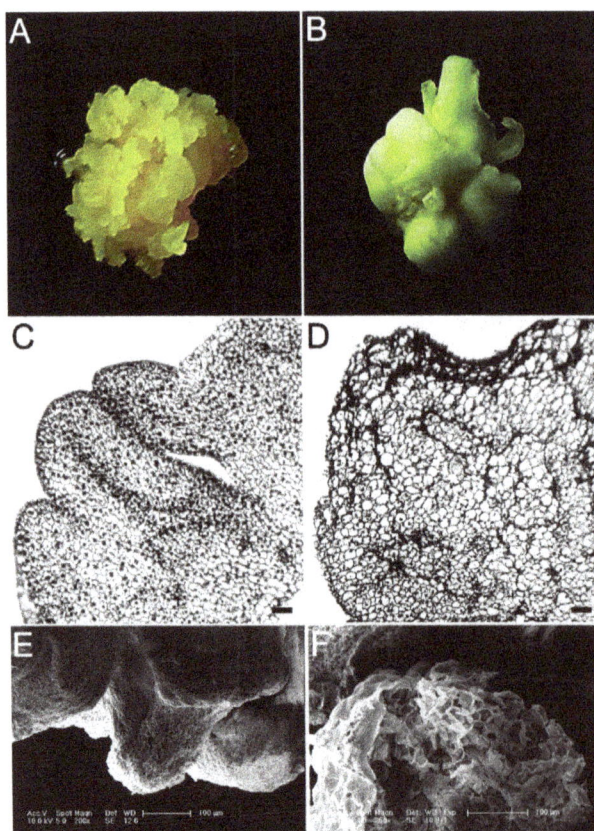

Figure 1. Morphological and histological analysis of embryogenic calli (EC) and nonembryogenic calli (NEC). (**A**) Morphological analysis of EC; (**B**) morphological analysis of NEC; (**C**) histological analysis of EC (Bar: 200 μm); (**D**) histological analysis of EC (Bar: 200 μm); (**E**) scanning electron microscope analysis of EC (Bar: 100 μm); (**F**) scanning electron microscope analysis of NEC (Bar: 100 μm).

2.2. Primary Data Analysis and Protein Identification

To investigate the differences during maize SE development on protein perspective, proteins of EC and NEC induced from Y423 were extracted and analyzed using an iTRAQ-based shotgun proteomics strategy. Samples were double labeled with iTRAQ tags for higher confidence in identification. Totally 286,622 spectra were generated after three biological replicates data merging from. Mascot identified a total of 39,493 spectra matched to known spectra, 28,830 spectra matched to unique spectra, 19,842 peptides, 15,878 unique peptides and 5592 proteins. The peptide length distribution, the peptide number distribution, protein mass distribution, and distribution of protein's sequences coverage were provided in Supplementary Figure S1A–D, respectively. Around 57% of the proteins included at least two peptides. Protein sequences coverage with 40–100%, 30–40%, 20–30%, 10–20%, and under 10% variation accounted for 5%, 6%, 14%, 24%, 51%, respectively.

2.3. Identification of Differentially Accumulated Protein Species (DAPS) by iTRAQ

A protein species was considered differentially accumulated as it exhibited a fold change > 1.2 and a *p*-value < 0.05 with a false discovery rate (FDR) of <5%. Based on these two criteria, 632 DAPS were identified, of which, 366 were up-accumulated and 266 were down-accumulated in EC compared with NEC. Supplementary Table S2 shows the DAPS detailed information.

2.4. Bioinformatics Analysis of DAPS Identified by iTRAQ

To identify the significantly enriched GO functional groups of DAPS, GO annotation was carried out. The DAPS between EC and NEC were classified into 31 functional groups (Figure 2), of which biological processes accounted for 15 GO terms (the most representative were "metabolic processes"), cellular components accounted for 9 GO terms (the most representative were "cell" and "cell part"), and molecular functions accounted for 7 GO terms (the most representative was "catalytic"). The results also showed that some DAPS clustered into the groups of extracellular region part and multi\-organism process just including the up-accumulated DAPS in EC. In addition, the quantity of up-accumulated DAPS was obviously higher than that of down-accumulated in the groups of membrane-enclosed lumen and organelle part. In contrary, the quantity of up-accumulated DAPS was lower than that of down-accumulated ones in the groups of "catalytic", "transporter", and "metabolic process".

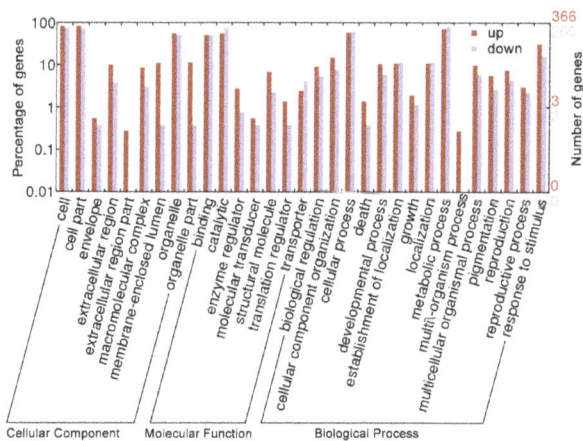

Figure 2. Gene ontology (GO) annotation of the differential abundance protein species (DAPS) in EC compared to NEC. The up-accumulation means a higher protein relative abundance in EC than in NEC, and the down-accumulation means a lower protein relative abundance in EC than in NEC.

A total of 539 DAPS (85.3%) between EC and NEC identified by iTRAQ were classified into 21 clusters of orthologous groups of proteins (COG) categories. In up-accumulated DAPS, "general function prediction only" represented the largest group, followed by "post-translational modification", "protein turnover, chaperones", "carbohydrate transport and metabolism", and "translation, ribosomal structure and biogenesis" (Figure 3A). In down-accumulated DAPS, "amino acid transport and metabolism" represented the largest group, followed by "energy production and conversion", "general function prediction only" and "lipid transport and metabolism" (Figure 3B). Especially, three groups, "chromatin structure and dynamics" (8 DAPS), "cell cycle control, cell division, chromosome partitioning" (4 DAPS) and "transcription" (8 DAPS), were just found in the classification of up-accumulated DAPS (Figure 3, Table 1). These DAPS might be the key proteins during the development of EC in maize.

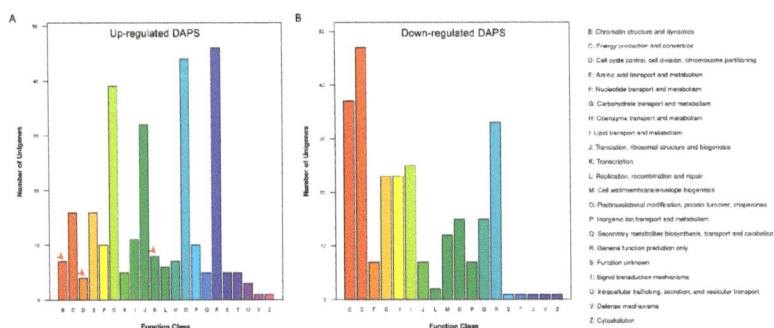

Figure 3. Clusters of orthologous groups of proteins (COG) classification of differential abundance protein species (DAPS) in EC and NEC. The arrows mark the groups just found in the upregulated DAPS. (**A**) Up-regulated DAPS; (**B**) Down-regulated DAPS.

Table 1. Protein information in the COG classification of including only the upregulated DAPS.

Protein Accession	Fold Change	Description	COG Code
gi 195605264 gb ACG24462.1	1.311	Histone H2A [*Zea mays*]	B
gi 195635409 gb ACG37173.1	1.314	Histone H4 [*Zea mays*]	B
gi 162458988 ref NP_001105357.1	1.41	Histone H2A [*Zea mays*]	B
gi 195618876 gb ACG31268.1	1.467	Histone H2A [*Zea mays*]	B
gi 195616386 gb ACG30023.1	1.541	Histone H2A [*Zea mays*]	B
gi 195617710 gb ACG30685.1	1.906	Histone H2A [*Zea mays*]	B
gi 194693822 gb ACF80995.1	2.084	Unknown [*Zea mays*]	B
gi 226496461 ref NP_001140401.1	1.234	Calmodulin [*Zea mays*]	DTZR
gi 226501230 ref NP_001149695.1	1.29	Cell division protein ftsz [*Zea mays*]	D
gi 226494524 ref NP_001150546.1	1.321	Serine/threonine-protein phosphatase 2A activator 2 [*Zea mays*]	DT
gi 293331121 ref NP_001167965.1	1.334	Uncharacterized protein LOC100381681 [*Zea mays*]	D
gi 212276110 ref NP_001130839.1	1.315	Uncharacterized protein LOC100191943 [*Zea mays*]	K
gi 226506726 ref NP_001141599.1	1.372	Uncharacterized protein LOC100273717 [*Zea mays*]	KLJ
gi 226490952 ref NP_001152536.1	1.412	ATP-dependent RNA helicase DDX23 [*Zea mays*]	KLJ
gi 226496545 ref NP_001148353.1	1.53	Small nuclear ribonucleoprotein Sm D1 [*Zea mays*]	K
gi 195656155 gb ACG47545.1	1.565	Small nuclear ribonucleoprotein Sm D3 [*Zea mays*]	K
gi 219363641 ref NP_001136911.1	1.586	Uncharacterized protein LOC100217069 [*Zea mays*]	K
gi 226503431 ref NP_001140836.1	1.647	Uncharacterized protein LOC100272912 [*Zea mays*]	K
gi 194699902 gb ACF84035.1	2.544	U6 snrna-associated Sm-like protein lsm3 [*Zea mays*]	K

To further investigate biological functions of these proteins, 500 DAPS (79.1%) were mapped to 106 pathways in the KEGG database, in which 33 KEGG pathways were significantly enriched (p-value ≤ 0.05) (Supplementary Table S3). "Metabolic pathways" ($p = 1.19 \times 10^{-11}$) was the most represented pathway, followed by "biosynthesis of secondary metabolites" ($p = 4.09 \times 10^{-10}$) and "pyruvate metabolism" ($p = 8.74 \times 10^{-5}$). Some DAPS were also significantly enriched in pathways of "starch and sucrose metabolism" ($p = 0.001436749$), "Fatty acid metabolism" ($p = 0.006471598$), "glycolysis/gluconeogenesis" ($p = 0.01189691$), and so on.

2.5. Transcriptional Analysis of Selected Genes for the Differentially Expressed Proteins

To elucidate the correspondence between the mRNA and protein level and confirm the authenticity and accuracy of the proteomic analysis, transcriptional analysis of 12 protein species, eight up-accumulated DAPS and four down-accumulated DAPS, was performed by qRT-PCR (Figure 4, Table 2). Of the selected proteins, 10 genes displayed accordant change tendency on the transcript level as the results of iTRAQ, such as nucleotide pyrophosphatase, putative subtilase family protein, peroxidase superfamily protein and terpenoid cyclases. By contrast, two genes (rhicadhesin receptor precursor and C cell wall invertase 2) showed no significant difference on the transcription level despite the detection of differential expression patterns on the protein level. The discrepancy between the transcription level of the two genes and the abundance of the corresponding protein species probably resulted from various posttranslational modifications between EC and NEC, such as protein phosphorylation and glycosylation.

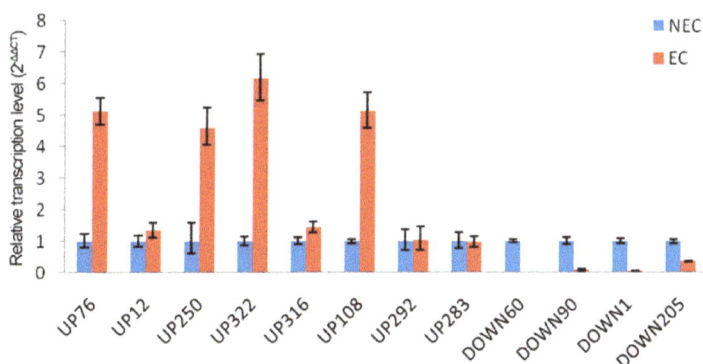

Figure 4. Analysis of transcript levels of the differential abundance protein species between EC and NEC by qRT-PCR.

Table 2. Candidate genes for qRT-PCR.

Code	Protein Accession	Fold Change	Description
UP76	gi l 195614828 l gb l ACG29244.1 l	4.26	Nucleotide pyrophosphatase/phosphodiesterase [*Zea mays*]
UP12	gi l 195621264 l gb l ACG32462.1 l	4.46	histone H1 [*Zea mays*]
UP250	gi l 414878739 l tpg l DAA55870.1 l	4.272	TPA: putative O-Glycosyl hydrolase superfamily protein [*Zea mays*]
UP322	gi l 413952293 l gb l AFW84942.1 l	4.218	Putative subtilase family protein [*Zea mays*]
UP316	gi l 414866956 l tpg l DAA45513.1 l	4.841	Peroxidase superfamily protein [*Arabidopsis thaliana*]
UP108	gi l 414586772 l tpg l DAA37343.1 l	6.499	Polygalacturonase inhibiting protein 1 [*Arabidopsis thaliana*]
UP292	gi l 195654029 l gb l ACG46482.1 l	6.87	Rhicadhesin receptor precursor [*Zea mays*]
UP283	gi l 1352469 l sp l P49174.1 l INVA_MAIZE	4.771	Cell-wall invertase 2 [Arabidopsis thaliana]
DOWN60	gi l 257626267 l emb l CBD24252.1 l	0.152	Unnamed protein product [*Zea mays*]
DOWN90	gi l 239050557 l ref l NP_001132867.2 l	0.233	S-adenosylmethionine synthetase family protein [*Arabidopsis thaliana*]
DOWN205	gi l 226500748 l ref l NP_001140439.1 l	0.199	NAD(P) binding Rossmann-fold superfamily protein [*Arabidopsis thaliana*]
DOWN1	gi l 162458192 l ref l NP_001105950.1 l	0.156	Terpenoid cyclases/Protein prenyltransferases superfamily protein [*Arabidopsis thaliana*]

3. Discussion

The first report about the use of somatic embryogenesis for maize tissue culture was published in 1975 [26]. However, it is very difficult to obtain somatic embryos from most of maize elite inbred lines because of the genetic diversity of maize. Well-known receptors for maize genetic transformation, just including A188, HiII, H99, and some other inbred lines, can been inducted to embryonic calli with high efficiency [9,27,28]. Nevertheless, many of these inbred lines were not suitable for productive applications because of their weak combining ability. A new-screened elite inbred line, Y423, can be inducted easily to develop somatic embryos and can be used in both production and scientific research [8]. However, NEC is obtained with EC at the same time during the process of SE induction, which will increase costs and is unfavorable to GM maize production. Therefore, it is necessary to study the mechanism of EC and NEC development.

Protein is substance basis of life and specific practitioner of life activities. Although transcriptome analysis is very useful to reveal the differential expression between EC and NEC, the expression level of mRNA is not a good predictor of the abundance of the corresponding proteins due to several factors such as post-translational modifications. Therefore, research on proteomics is helpful to reveal complex difference between EC and NEC in maize and provides new information concerning EC development. iTRAQ was explored for proteomics profiling and enrichment of differentially expressed analysis between EC and NEC of the inbred maize line Y423. As a result, 632 DAPS were identified, among which, 366 DAPS were up-accumulated and 266 DAPS down-accumulated proteins detected in EC compared with NEC. Bioinformatics analysis showed that DAPS were annotated in 31 GO functional groups, 539 DAPS were classified into 21 COG categories, and 512 DAPS were identified to 106 KEGG pathways. Some significant DAPS were grouped into categories based on the GO, COG, and KEGG analysis which we discussed below.

In order to elucidate the functional difference of DAPS between EC and NEC, we analyzed the COG of high and low abundance proteins in EC, respectively. The results showed 20 special up-accumulated DAPS in EC versus NEC were clustered in three classifications ("chromatin structure and dynamics", "cell cycle control, cell division, chromosome partitioning" and "transcription") (Table 1). Most of the DAPS, in the classification of "chromatin structure and dynamics", were histones which are basic proteins, rich in basic amino acids such as arginine and lysine. Some studies have shown that high level of arginine was detected in embryogenic callus, which is beneficial to cell division and cell differentiation [29]. Arginine, as an important component, was also added to the medium in cotton somatic embryo tissue culture [30–32]. In addition, some studies have shown calmodulin, in the classification of "cell cycle control, cell division, chromosome partitioning", played a very important role in somatic embryogenesis. In carrot somatic embryos, calmodulin was localized in the shoot apical meristem region and was strongly expressed in the globular and cardiac phases, whereas it was less expressed in undifferentiated calli [33]. During the somatic embryogenesis of sugarcane, calmodulin accelerated cell differentiation, and the content of calmodulin was higher at any stage of somatic embryo as compared to undifferentiated calli [34].

To study which metabolic pathway plays a major role in the induction of EC, KEGG pathway enrichment analysis was explore. 512 DAPS were mapped in 106 KEGG pathways, in which pyruvate metabolism not only was the most significantly enriched pathway of DAPS, but also the number of enriched DAPS is up to 25. In starch and sucrose metabolism and glycolysis/gluconeogenesis closely related to pyruvate metabolism, a large number of DAPS was also enriched significantly, reaching 24 and 23, respectively (Table S3). We also noted that more than 20 DAPS were enriched in the ribosomes and the phenylpropanoid biosynthesis pathways.

Thirty DAPS were detected in ribosomal pathway, accounting for 6% of total differential proteins, of which 25 DAPS were up-accumulated and only five were down-accumulated in EC (Figure S2, Table 3). Most of up-accumulated DAPS were clustered in the classification of "translation", "ribosomal structure and biogenesis", which was consisted with the COG results of up-accumulated DAPS. These DAPS are mainly related to the 40S small subunit and 60S large subunit (Table 3). This result may indicate that protein synthesis in EC was more exuberant than that in NEC. Ribosome-inactivating protein, which inhibits ribosomal formation, was not detected in the EC of spinach (*Spinacia oleracea*) [35]. This also illustrated that reducing the abundance of proteins, which inhibit the formation of ribosomes in EC, might result in a significant increase in the number of ribosomes which in turn would promote protein synthesis.

Table 3. Information of DAPS in ribosome pathway.

Protein Accession	Enzyme Commission	Fold Change	Regulated	Description
gi	ref	NP_001132360.1		S12e	1.436	UP	40S ribosomal protein S12 isoform 1 [*Zea mays*]
gi	1956058201	gb	ACG24740.1		S20e	1.352	UP	40S ribosomal protein S20 [*Zea mays*]
gi	1956232101	gb	ACG33435.1		S20e	1.247	UP	40S ribosomal protein S20 [*Zea mays*]
gi	162463985	ref	NP_001105635.1		S24e	1.245	UP	40S ribosomal protein S24 [*Zea mays*]
gi	226532924	ref	NP_001146968.1		S24e	0.833	DOWN	40S ribosomal protein S24 [*Zea mays*]
gi	1956348151	gb	ACG36876.1		S29e	2.384	UP	40S ribosomal protein S29 [*Zea mays*]
gi	226499664	ref	NP_001149323.1		S30e	1.394	UP	40S ribosomal protein S30 [*Zea mays*]
gi	162464000	ref	NP_001105390.1		LP1, LP2	0.735	DOWN	60S acidic ribosomal protein P2B [*Zea mays*]
gi	212275159	ref	NP_001130808.1		L13Ae	1.46	UP	60S ribosomal protein L13a [*Zea mays*]
gi	1956232881	gb	ACG33474.1		L13Ae	1.206	UP	60S ribosomal protein L13a [*Zea mays*]
gi	1956421721	gb	ACG40554.1		L17e	1.536	UP	60S ribosomal protein L17 [*Zea mays*]
gi	1956561951	gb	ACG47565.1		L17e	1.336	UP	60S ribosomal protein L17 [*Zea mays*]
gi	1956216921	gb	ACG32676.1		L24e	1.342	UP	60S ribosomal protein L24 [*Zea mays*]
gi	1956463981	gb	ACG42667.1		L28e	1.39	UP	60S ribosomal protein L28 [*Zea mays*]
gi	1956257901	gb	ACG34725.1		L35e	1.535	UP	60S ribosomal protein L35 [*Zea mays*]
gi	413922853	gb	AFW62785.1		L6e	1.362	UP	60S ribosomal protein L6 [*Zea mays*]
gi	1956562411	gb	ACG47588.1		L7Ae	1.352	UP	60S ribosomal protein L7a [*Zea mays*]
gi	212274675	ref	NP_001130101.1		L10e	1.791	UP	Acid phosphatase 1 [*Zea mays*]
gi	1956424781	gb	ACG40707.1		S19	1.24	UP	Glycine-rich RNA-binding protein 2 [*Zea mays*]
gi	212722104	ref	NP_001132260.1		L11e	1.24	UP	Hypothetical protein [*Zea mays*]
gi	226509268	ref	NP_001141489.1		LP0	0.798	DOWN	Hypothetical protein [*Zea mays*]
gi	162460468	ref	NP_001105280.1		L31e	1.643	UP	Putative 60S ribosomal protein L31 [*Zea mays*]
gi	1326681	sp	P08529.2	RK14_MAIZE		L14	0.709	DOWN	Recname: Full=50S ribosomal protein L14, chloroplastic
gi	32646051	gb	AAC24573.1		L23Ae	1.39	UP	Ribosomal protein L25 [*Zea mays*]
gi	226497590	ref	NP_001148801.1		L7Ae	1.312	UP	TPA: 60S ribosomal protein L7a [*Zea mays*]
gi	4148738481	tpg	DAA52405.1		S4e	1.475	UP	TPA: ribosomal protein S4 [*Zea mays*]
gi	212721318	ref	NP_001131538.1		L18e	1.414	UP	Uncharacterized protein LOC100192878 [*Zea mays*]
gi	2240308491	gb	ACN34500.1		L6e	1.793	UP	Unknown [*Zea mays*]
gi	3425911	gb	AAA84482.1		RP-L16	0.768	DOWN	Unknown protein (chloroplast) [*Zea mays*]
gi	257722902	emb	CBD23964.1		L19e	1.268	UP	Unnamed protein product [*Zea mays* subsp. Mays]

In addition, 24 DAPS were enriched in the pathway of starch and sucrose metabolism, of which 18 DAPS were up-accumulated and only six were down-accumulated in EC (Figure S3, Table 4). In the results of COG analysis, these up-accumulated DAPS were mainly clustered in the classification of "carbohydrate transport and metabolism". In this pathway, the proteins related with D-Glucose biosynthesis, exoglucanase precursor (3.2.1.21) and endoglucanase 1 precursor (3.2.1.4), were up-accumulated in EC, the fold change reaching 2.07 and 3.941, respectively. In EC of maize inbred line A19, a protein associated with D-glucose biosynthesis, beta D-glucosidase, was also up-accumulated [36]. The accumulated abundance of the proteins mentioned above could promote the synthesis of D-glucose, which is an important source of pyruvate. These results suggested that the high expression of glucose-producing related enzymes in starch and sucrose metabolism might indirectly affect the metabolism of pyruvate from its source.

Table 4. Information of DAPS in starch and sucrose metabolism pathway.

Protein Accession	Enzyme Commission	Fold Change	Regulated	Description
gi	162462658	ref	NP_001105539.1		3.2.1.1	2.29	UP	Alpha-amylase precursor [*Zea mays*]
gi	226530773	ref	NP_001150278.1		3.2.1.1	1.627	UP	Alpha-amylase precursor [*Zea mays*]
gi	1352469	sp	P49174.1	INVA_MAIZE		3.2.1.26	4.771	UP	Beta-fructofuranosidase/cell wall isozyme;
gi	226495335	ref	NP_001151458.1		3.2.1.4	3.941	UP	Endoglucanase 1 precursor [*Zea mays*]
gi	88097641	gb	AAF79936.1		3.2.1.21	2.07	UP	Exoglucanase precursor [*Zea mays*]
gi	1956391261	gb	ACG39031.1		2.7.1.4	1.261	UP	Fructokinase-2 [*Zea mays*]
gi	162460322	ref	NP_001105368.1		5.3.1.9	1.324	UP	Glucose-6-phosphate isomerase, cytosolic [*Zea mays*]
gi	308080756	ref	NP_001183080.1		3.2.1.15	1.683	UP	Hypothetical protein precursor [*Zea mays*]
gi	413933069	gb	AFW67620.1		3.2.1.21	3.115	UP	Hypothetical protein ZEAMMB73_549956 [*Zea mays*]
gi	413922001	gb	AFW61933.1		2.4.1.14	1.242	UP	Putative sucrose-phosphate synthase family protein [*Zea mays*]
gi	413952450	gb	AFW85099.1		3.1.3.12/2.4.1.15	1.325	UP	Putative trehalose phosphatase/synthase family protein [*Zea mays*]
gi	194740442	gb	ACF94692.1		2.4.1.1	1.228	UP	Starch phosphorylase 1 precursor [*Zea mays*]
gi	525344616	ref	NP_001266691.1		2.4.1.13	1.203	UP	Sucrose synthase 1 [*Zea mays*]
gi	4148674101	tpg	DAA45967.1		3.2.1.28	2.643	UP	TPA: hypothetical protein ZEAMMB73_076801 [*Zea mays*]
gi	4148794061	tpg	DAA56537.1		3.1.1.11	2.287	UP	TPA: pectinesterase [*Zea mays*]
gi	4148853301	tpg	DAA61344.1		3.1.3.12/2.4.1.15	1.84	UP	TPA: putative trehalose phosphatase/synthase family protein [*Zea mays*]
gi	257713456	emb	CBD15747.1		3.2.1.37	1.917	UP	Unnamed protein product [*Zea mays*]
gi	257713456	emb	CBD15747.1		3.2.1.21	1.917	UP	Unnamed protein product [*Zea mays*]
gi	413946674	gb	AFW79323.1		2.7.7.27	0.553	DOWN	ADP-glucose pyrophosphorylase large subunit [*Zea mays*]
gi	413920595	gb	AFW60527.1		2.4.1.14	0.55	DOWN	Putative sucrose-phosphate synthase family protein [*Zea mays*]
gi	1351136	sp	P49036.1	SUS2_MAIZE		2.4.1.13	0.808	DOWN	Sucrose-UDP glucosyltransferase 2
gi	226529913	ref	NP_001146685.1		3.1.1.11	0.773	DOWN	TPA: pectinesterase [*Zea mays*]
gi	414865602	tpg	DAA44159.1		3.1.3.12/2.4.1.15	0.728	DOWN	TPA: putative trehalose phosphatase/synthase family protein [*Zea mays*]
gi	226505764	ref	NP_001149225.1		1.1.1.22	0.602	DOWN	UDP-glucose 6-dehydrogenase [*Zea mays*]

Glycolysis/gluconeogenesis is a closely related metabolic pathway with starch and sucrose metabolism. Twenty-three DAPS were enriched in this pathway, of which 11 DAPS were up-accumulated and 12 DAPS were down-accumulated in EC (Figure S4, Table 5). Most of the 11 up-accumulated proteins were enzymes and related proteins that catalyze pyruvate production, while the down-accumulated proteins were mainly distributed in the glycolytic pathway of pyruvate to produce acetyl-CoA and to generate ethanol. Accumulation of enzymes involved in pyruvate synthesis were also detected in EC derived from maize inbred lines A19 and H99, respectively [18,36]. Similar to the studies in maize, the content of pyruvate-synthesis related enzymes (i.e., phosphoglycero mutases) also significantly increased during the somatic embryogenesis of *Arabidopsis thaliana* [37]. The main function of phosphoglycero mutases is to catalyze the conversion of 2-phospho-D-glycerate to phosphoenolpyruvate. In our study, we found that the high abundance of heat shock factor-binding protein 1 in EC, which involved in the process of conversion of 2-phospho-D-glycerate to phosphoenolpyruvate catalyzed by phosphoglycero mutases. These results suggested that heat shock factor-binding protein 1 should be involved in the folding of the catalytic process. Meanwhile, pyruvate kinase (2.7.1.40), the key enzyme for pyruvate generation, was also up-accumulated in EC. Overall, the major role of up-accumulated DAPS should be to promote pyruvate formation in EC. One metabolism direction of pyruvate is the formation of lactate. The enzyme (1.1.1.27: L-lactate dehydrogenase A) that catalyzes pyruvate to produce lactate in EC expressed higher than that in NEC, but the expression of various enzymes (i.e., pyruvate decarboxylase 3) that catalyze pyruvate to produce ethanol is lower than that in NEC. The other one metabolism direction of pyruvate is the formation of Acetyl CoA, which is the pivot of energy metabolism and substance metabolism. The abundance of DAPS showed that the expression of various enzymes (i. e., 1.2.4.1: pyruvate dehydrogenase, 1.8.1.4: dihydrolipoyl dehydrogenase and 2.3.1.12: dihydrolipoyl lysine-residue acetyltransferase) in the process of generating acetyl-CoA was lower in EC than that in NEC. All these results indicated that EC mainly relied on not TCA but glycolysis to provide ATP. And the demand for energy and new substances might be lower in EC than in NEC, which also explained why EC usually grew slower than NEC. It is interesting that our results are similar to some stem cell studies, which have proved pyruvate is a very important metabolite for controlling stem cell development. In quiescent state, glycolysis-associated enzymes, such as pyruvate kinase and lactate dehydrogenase A, are highly expressed in pluripotent stem cells. As a result, pyruvate can be catalyzed to produce lactic acid in stem cells. At the same time the synthesis of lipids, proteins or nucleotides are blocked. Stem cells are slow cycling. Along differentiation, energy demands increase and TCA cycle turns on in cells [38–42]. It is suggested that the final destination of pyruvate might determine the direction of callus differentiation during somatic embryogenesis in maize as same as in stem cells. The same regulation mechanism of differentiation might be shared in plant cells and animal cells.

Some lignin substances (i.e., guaiacyl lignin, syringyl lignin and p-Hydroxyphenyl lignin) are generated in phenylpropanoid biosynthesis pathway, in which the DAPS was obviously enriched. Most of DAPS enriched in phenylpropanoid biosynthesis pathway were low abundance in EC (Figure S5). We speculated that this should be an important reason for Y423 EC to have the characteristics of the low lignin content loose structure, and rough surface.

Based on a comprehensive analysis of GO, COG, and KEGG results obtained in this and similar studies, we discovered a new regulation strategy that might affect embryogenic callus development in maize. Pyruvate is the pivot of this model, which involves a lot of biochemical processes such as starch metabolism, glycolysis/gluconeogenesis, pyruvate metabolism, fatty acid biosynthesis and TCA cycle (Figure 5). In maize EC, the up-accumulated DAPS in starch and sucrose metabolism and glycolysis/gluconeogenesis could promote pyruvate generation; meanwhile, the other down-accumulated DAPS could inhibit the conversion of pyruvate to ethanol and to acetyl-CoA. In addition, some DAPS in fatty acid metabolism were also down-accumulated in EC (Figure S6), which further limited the generation of acetyl-CoA. The decrease of acetyl-CoA content would further inhibit leucine biosynthesis, fatty acid biosynthesis, TCA cycle and TCA cycle's downstream metabolism.

These changes could reduce the material accumulation and energy supply, resulting in EC were in a quiescent state. Combined with the reduction of lignin substances, EC could be more likely to maintain in a soft and loose growth state and be more suitable for subculture and differentiation. Conversely, in NEC, more material accumulation and energy supply would lead to rapid growth of calli with high lignin content, hard and smooth surface, not easy to subculture and differentiate.

Table 5. Information of DAPS in glycolysis /gluconeogenesis pathway.

Protein Accession	Enzyme Commission	Fold Change	Regulated	Description
gi | 6225010 | sp | P93629.1 | ADHX_MAIZE	1.1.1.1	1.386	UP	Alcohol dehydrogenase class-III;
gi | 195613242 | gb | ACG28451.1 |	5.1.3.3	1.7	UP	Aldose 1-epimerase precursor [*Zea mays*]
gi | 162460322 | ref | NP_001105368.1 |	5.3.1.9	1.324	UP	Glucose-6-phosphate isomerase, cytosolic [*Zea mays*]
gi | 162457962 | ref | NP_001105090.1 |	4.2.1.11	1.58	UP	Heat shock factor-binding protein 1 [*Zea mays*]
gi | 226528284 | ref | NP_001150088.1 |	4.2.1.11	1.609	UP	Heat shock factor-binding protein 1 [*Zea mays*]
gi | 413956284 | gb | AFW88933.1 |	4.1.1.49	1.316	UP	Hypothetical protein ZEAMMB73_030639 [*Zea mays*]
gi | 195621388 | gb | ACG32524.1 |	1.1.1.27	1.774	UP	L-lactate dehydrogenase A [*Zea mays*]
gi | 226509797 | ref | NP_001142404.1 |	2.7.2.3	1.334	UP	Phosphoglycerate kinase isoform 3 [*Zea mays*]
gi | 195620854 | gb | ACG32257.1 |	2.7.1.40	1.712	UP	Pyruvate kinase, cytosolic isozyme [*Zea mays*]
gi | 226528689 | ref | NP_001141131.1 |	1.1.1.1	1.288	UP	TPA: putative alcohol dehydrogenase superfamily protein [*Zea mays*]
gi | 291047868 | emb | CBK51447.1 |	1.2.1.3	1.272	UP	Unnamed protein product [*Zea mays*]
gi | 195614676 | gb | ACG29168.1 |	1.2.1.3	0.629	DOWN	Aldehyde dehydrogenase, dimeric NADP-preferring [*Zea mays*]
gi | 162460054 | ref | NP_001105576.1 |	1.2.1.3	0.576	DOWN	Aldehyde dehydrogenase2 [*Zea mays*]
gi | 363543269 | ref | NP_001241850.1 |	1.1.1.2	0.3	DOWN	Aldose reductase [*Zea mays*]
gi | 293335591 | ref | NP_001169718.1 |	1.8.1.4	0.766	DOWN	Dihydrolipoyl dehydrogenase [*Zea mays*]
gi | 226499350 | ref | NP_001142314.1 |	2.3.1.12	0.8	DOWN	Dihydrolipoyl lysine-residue acetyltransferase component of pyruvate dehydrogenase complex [*Zea mays*]
gi | 120670 | sp | P08735.2 | G3PC1_MAIZE	1.2.1.12	0.69	DOWN	Glyceraldehyde-3-phosphate dehydrogenase 1, cytosolic
gi | 226531804 | ref | NP_001148246.1 |	1.1.1.2	0.533	DOWN	NAD(P)H-dependent oxidoreductase [*Zea mays*]
gi | 413937078 | gb | AFW71629.1 |	6.2.1.1	0.674	DOWN	Putative AMP-dependent synthetase and ligase superfamily protein [*Zea mays*]
gi | 162457852 | ref | NP_001105052.1 |	4.1.1.1	0.777	DOWN	Pyruvate decarboxylase 3 [*Zea mays*]
gi | 226529151 | ref | NP_001150473.1 |	1.2.4.1	0.588	DOWN	Pyruvate dehydrogenase E1 component subunit beta [*Zea mays*]
gi | 226528639 | ref | NP_001140759.1 |	1.2.4.1	0.613	DOWN	Uncharacterized protein LOC100227834 [*Zea mays*]
gi | 219887835 | gb | ACL54292.1 |	1.2.1.12	0.832	DOWN	Unknown [*Zea mays*]

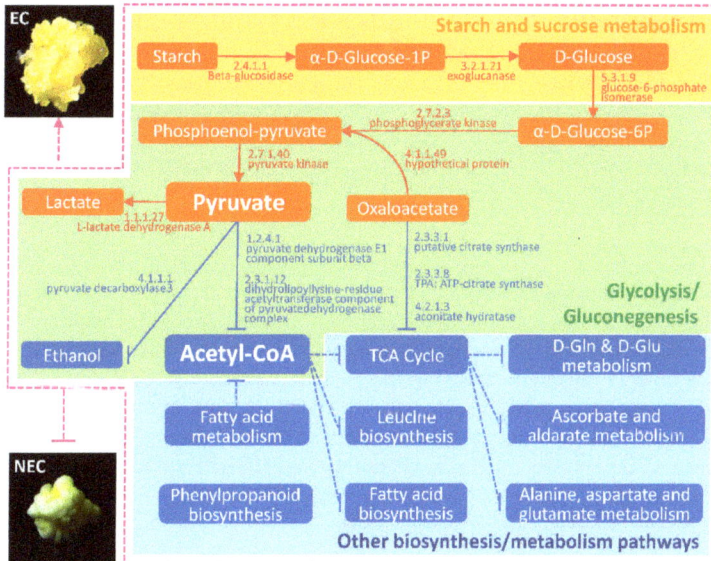

Figure 5. Working model for development of embryogenic and nonembryogenic calli proposed based on DAPS by iTRAQ.

4. Materials and Methods

4.1. Plant Materials and Tissue Culture Conditions

The maize elite inbred line Y423 collected by our lab was used in this study, which was the only one of 40 maize inbred lines selected by our lab that can obtain intact somatic embryos [8]. The whole ears were harvested 12–15 days after self-pollination. After husk and sterilization, the immature embryos (size around 2 mm^2) were inoculated onto the induction medium (100 embryos per plate) in accordance with the procedure of Jimenez and Bangerth [43] with minor modifications as previously described [8]. The immature embryos were incubated in the dark at 24 °C. Buds generated from the immature embryos were removed every other 7 days. After 20–25 days, the primary calli appeared, and then were transferred onto fresh subculture medium. Calli were then subcultured every 20 days on the same subculture medium. After 4–5 times subculturing, the EC and NEC produced, and were selected by histological staining as previously described [8]. We collected around 15 g of EC and NEC separately from 30 plates (derived from one plate of immature embryos) for one biological replicate and then stored them at −80 °C until required. Three biological replicates were performed.

4.2. Protein Preparation

EC and NEC samples were ground into powder in liquid nitrogen, extracted with Lysis buffer (7 M Urea, 2 M Thiourea, 4% CHAPS, 40 mM Tris-HCl, pH 8.5) containing 1 mM PMSF and 2 mM EDTA (final concentration). After 5 min, 10 mM DTT (final concentration) was added to the samples. The suspension was sonicated at 200 W for 15 min and then centrifuged at 4 °C, 30,000× *g* for 15 min. The supernatant was mixed well with 5× volume of chilled acetone containing 10% (*v*/*v*) TCA and incubated at −20 °C overnight. After centrifugation at 4 °C, 30,000× *g*, the supernatant was discarded. The precipitate was washed with chilled acetone three times. The pellet was air-dried and dissolved in Lysis buffer (7 M urea, 2 M thiourea, 4% NP40, 20mM Tris-HCl, pH 8.0–8.5). The suspension was sonicated at 200 W for 15 min and centrifuged at 4 °C, 30,000× *g* for 15 min. The supernatant was transferred to another tube. To reduce disulfide bonds in proteins of the supernatant, 10 mM DTT (final concentration) was added and incubated at 56 °C for 1 h. Subsequently, 55 mM IAM (final concentration) was added to block the cysteines, incubated for 1 h in the darkroom. The supernatant was mixed well with 5 5× volume of chilled acetone for 2 h at −20 °C to precipitate proteins. After centrifugation at 4 °C, 30,000× *g*, the supernatant was discarded, and the pellet was air-dried for 5 min, dissolved in 500 μL 0.5 M TEAB (Applied Biosystems, Milan, Italy), and sonicated at 200 W for 15 min. Finally, samples were centrifuged at 4 °C, 30,000× *g* for 15 min. The supernatant was transferred to a new tube and quantified with the Bradford assay using BSA as the calibrant. The proteins in the supernatant were kept at −80 °C for further analysis.

4.3. iTRAQ Labeling and SCX Fractionation

Total protein (100 μg) was taken out of each sample solution and then the protein was digested with Trypsin Gold (Promega, Madison, WI, USA) with the ratio of protein: trypsin = 30:1 at 37 °C for 16 hours. After trypsin digestion, peptides were dried by vacuum centrifugation. Peptides were reconstituted in 0.5M TEAB and processed according to the manufacture's protocol for 8-plex iTRAQ reagent (Applied Biosystems, Waltham, MA, USA). Briefly, one unit of iTRAQ reagent was thawed and reconstituted in 24 μL isopropanol. The EC replicates were labeled with the iTRAQ tags TP1-113 and TP2-114, and the NEC replicates were labeled with the tags FT1-115 and FT2-116. The peptides were labeled with the isobaric tags, incubated at room temperature for 2 h. The labeled peptide mixtures were then pooled and dried by vacuum centrifugation.

SCX chromatography was performed with a LC-20AB HPLC pump system (Shimadzu, Kyoto, Japan). The iTRAQ-labeled peptide mixtures were reconstituted with 4 mL of buffer A (25 mM NaH$_2$PO$_4$ in 25% ACN, pH 2.7) and loaded onto a 4.6 × 250 mm Ultremex SCX column containing 5-μm particles (Phenomenex, Torrance, CA, USA). The peptides were eluted at a flow rate of 1 mL/min with a gradient of buffer A for 10 min, 5–60% buffer B (25 mM NaH$_2$PO$_4$, 1 M KCl in 25% ACN, pH 2.7) for 27 min, 60–100% buffer B for 1 min. The system was then maintained at 100% buffer B for 1 min before equilibrating with buffer A for 10 min prior to the next injection. Elution was monitored by measuring the absorbance at 214 nm, and fractions were collected every 1 min. The eluted peptides were pooled into 20 fractions, desalted with a Strata X C18 column (Phenomenex, Torrance, CA, USA) and vacuum-dried.

4.4. LC-ESI-MS/MS Analysis

Each fraction was resuspended in buffer A (5% ACN, 0.1% FA) and centrifuged at 20,000× *g* for 10 min, the final concentration of peptide was about 0.5 μg/μL on average. 10 μL supernatant was loaded on a LC-20AD nanoHPLC (Shimadzu, Kyoto, Japan) by the autosampler onto a 2 cm C18 trap column. Then, the peptides were eluted onto a 10cm analytical C18 column (inner diameter 75 μm) packed in-house. The samples were loaded at 8 μL/min for 4min, then the 35min gradient was run at 300 nL/min starting from 2 to 35% B (95% ACN, 0.1% FA), followed by 5 min linear gradient to 60%, then, followed by 2 min linear gradient to 80%, and maintenance at 80% B for 4 min, and, finally, return to 5% in 1 min.

Data acquisition was performed with a TripleTOF 5600 System (AB SCIEX, Concord, ON, USA) fitted with a Nanospray III source (AB SCIEX, Concord, ON, USA) and a pulled quartz tip as the emitter (New Objectives, Woburn, MA, USA). Data was acquired using an ion spray voltage of 2.5 kV, curtain gas of 30 psi, nebulizer gas of 15 psi, and an interface heater temperature of 150 °C. The MS was operated with a RP of greater than or equal to 30,000 FWHM for TOF MS scans. For IDA, survey scans were acquired in 250 ms and as many as 30 product ion scans were collected if exceeding a threshold of 120 counts per second (counts/s) and with a 2+ to 5+ charge-state. Total cycle time was fixed to 3.3 s. Q2 transmission window was 100Da for 100%. Four time bins were summed for each scan at a pulser frequency value of 11 kHz through monitoring of the 40 GHz multichannel TDC detector with four-anode channel detect ion. A sweeping collision energy setting of 35 ± 5 eV coupled with iTRAQ adjust rolling collision energy was applied to all precursor ions for collision-induced dissociation. Dynamic exclusion was set for 1/2 of peak width (15 s), and then the precursor was refreshed off the exclusion list.

4.5. Protein Identification and Data Analysis

Raw data files acquired from the Orbitrap were converted into MGF files using Proteome Discoverer 1.2 (PD 1.2, Thermo, SanJose, CA), [5600 msconverter] and the MGF file were searched. Proteins identification were performed by using Mascot search engine (Matrix Science, London, UK; version 2.3.02) against the NCBI *Zea mays* database containing 172,446 sequences (http://www.ncbi. nlm.nih.gov/protein?term=txid4577) with the following parameters: Carbamidomethyl (C), iTRAQ 8 plex (N-term), and Deamidated (NQ) as fixed modification, Gln→pyro-Glu (N-termQ), Oxidation (M), and Deamidated (NQ) as potential variable modifications; the instrument type was set to default; the enzyme was set to trypsin and one max missed cleavages was allowed; a mass tolerance of ±0.05 Da was permitted for intact peptide masses and ±0.1 Da was permitted for fragmented ions. The charge states of peptides were set to +2 and +3. Specifically, an automatic decoy database search was performed in Mascot by choosing the decoy checkbox in which a random sequence of database is generated and tested for raw spectra as well as the real database. To reduce the probability of false peptide identification, only peptides with significance scores (≥20) at the 99% confidence interval by a Mascot probability analysis greater than "identity" were counted as identified. Each confident protein identification involves at least one unique peptide. For protein quantitation, it was required that a protein contains at least two unique peptides. The quantitative protein ratios were weighted and normalized by the median ratio in Mascot. We only used ratios with *p*-values < 0.05, and only fold changes of >1.2 were considered as significant.

4.6. Bioinformatics Analysis

Functional annotations of differentially accumulated protein species were performed using Gene Ontology (http://www.geneontology.org). The Clusters of Orthologous Groups of proteins (COG) (http://www.ncbi.nlm.nih.gov/COG/) database was carried out for functional classification of DAPS, and the Kyoto Encyclopedia of Genes and Genomes (KEGG) (http://www.genome.jp/kegg/ or http://www.kegg.jp/) was used to predict the main metabolic pathways and biochemical signals transduction pathways that involved the DAPS. A p-value ≤ 0.05 was used as the threshold to determine the significant enrichments of GO and KEGG pathways.

4.7. Quantitative Real-Time PCR Analysis

Total RNA extracted from the EC and NEC was treated with DNase I to remove genomic DNA. The first strand cDNA synthesis and the qRT-PCR were carried out using the SuperScript™ III First-Strand Synthesis SuperMix (Invitrogen, Carlsbad, CA, USA) and the SYBR Green JumpStart™ Taq ReadyMix™ (Sigma-Aldrich, St. Louis, MO, USA), respectively. qRT-PCR was carried out in the ABI PRISM 7500 sequence detection system (Waltham, MA, USA) according to the manufacturer's instructions. To validate the DAPS obtained from iTRAQ, 12 genes were subjected to quantitative real-time PCR. Each 20 ml reaction mixture contained 10 mM each primers 0.4 ml, 2× SYBR Premix Ex Taq 10 ml, 50× ROX Reference Dye II 0.4 mL and cDNA (1:5) template 2 ml. Gene-specific primers were designed using Primer Express 3.0 (Waltham, MA, USA), and amplified maize *actin* 1 and *gapdh* were used as reference genes (Supplementary Table S1). Three replications were performed for each sample.

5. Conclusions

We used the iTRAQ technique to perform quantitative proteome analysis of EC and NEC derived from Y423 inbred maize line. This approach identified some new proteins involved in pyruvate metabolism (i.e., pyruvate dehydrogenase E1 component subunit beta and dihydrolipoyllysine-residue acetyltransferase component of pyruvate dehydrogenase complex), TCA cycle (i.e., citrate synthase and aconitate hydratase), fatty acid metabolism (i.e., acetyl-CoA acetyltransferase), and phenylpropanoid biosynthesis (i.e., peroxidase 39 isoform and 1-Cys peroxiredoxin) that were not previously known to be associated with EC and SE developing. Based on functional analysis, we proposed a regulation strategy for reprograming of somatic cell to EC in maize. This strategy suggests that accumulation of pyruvate and the alterations in abundances of proteins in several associated pathways (i.e., starch and sucrose metabolism and glycolysis/gluconeogenesis) might be responsible for differences between EC and NEC. In summary, our analysis of differences in the protein profiles between EC and NEC increased our understanding of the mechanisms of EC development in maize.

Supplementary Materials: Supplementary materials can be found at http://www.mdpi.com/1422-0067/19/12/4004/s1.

Author Contributions: X.S., B.L. and Y.Y. conceived and designed the experiments; X.S. and B.L. performed the experiments; X.S. and B.L. analyzed the data; others provided reagents/materials/analysis tools; X.S. and Y.Y. wrote the paper.

Acknowledgments: This work was supported by the National Key Research and Development Program of China (2016YFD0101203) and the Science and Technology Development projects of Jilin Province (20160203005NY).

Conflicts of Interest: The authors declare no conflicts of interest.

Abbreviations

COG	clusters of orthologous groups of proteins
DAPS	differential abundance protein species
EC	embryogenic calli
FDR	false discovery rate
GM	genetically modified
GO	gene ontology
iTRAQ	isobaric tags for relative and absolute quantification
NEC	nonembryogenic calli
SE	Somatic embryo
TCA	tricarboxylic acid
UDP	uracil-diphosphate

References

1. Ash, C.; Jasny, B.R.; Malakoff, D.A.; Sugden, A.M. Food security. Feeding the future. Introduction. *Science* **2010**, *327*, 797. [CrossRef] [PubMed]
2. Armstrong, C.L.; Romero-Severson, J.; Hodges, T.K. Improved tissue culture response of an elite maize inbred through backcross breeding, and identification of chromosomal regions important for regeneration by RFLP analysis. *Theor. Appl. Genet.* **1992**, *84*, 755–762. [CrossRef] [PubMed]
3. Landi, P.; Chiappetta, L.; Salvi, S.; Frascaroli, E.; Lucchese, C.; Tuberosa, R. Responses and allelic frequency changes associated with recurrent selection for plant regeneration from callus cultures in maize. *Maydica* **2002**, *47*, 21–32.
4. Hodges, T.K.; Kamo, K.K.; Imbrie, C.W.; Becwar, M.R. Genotype Specificity of Somatic Embryogenesis and Regeneration in Maize. *Nat. Biotechnol.* **1986**, *4*, 219–223. [CrossRef]
5. Garrocho-Villegas, V.; De Jesus-Olivera, M.T.; Quintanar, E.S. Maize somatic embryogenesis: Recent features to improve plant regeneration. *Methods Mol. Biol.* **2012**, *877*, 173–182. [PubMed]
6. Dennis, J.G.P.D.; Amul, P.P.D.; Triglano, R.N. Somatic embryogenesis and development of synthetic seed technology. *Crit. Rev. Plant Sci.* **1991**, *10*, 33–61.
7. Murray, S.C.; Eckhoff, P.; Wood, L.; Paterson, A.H. A proposal to use gamete cycling in vitro to improve crops and livestock. *Nat. Biotechnol.* **2013**, *31*, 877–880. [CrossRef]
8. Liu, B.; Su, S.; Wu, Y.; Li, Y.; Shan, X.; Li, S.; Liu, H.; Dong, H.; Ding, M.; Han, J.; et al. Histological and transcript analyses of intact somatic embryos in an elite maize (*Zea mays* L.) inbred line Y423. *Plant Physiol. Biochem.* **2015**, *92*, 81–91. [CrossRef]
9. Sun, L.; Wu, Y.; Su, S.; Liu, H.; Yang, G.; Li, S.; Shan, X.; Yuan, Y. Differential gene expression during somatic embryogenesis in the maize (*Zea mays* L.) inbred line H99. *Plant Cell Tissue Organ Cult.* **2012**, *109*, 271–286. [CrossRef]
10. Šamaj, J.; Baluška, F.; Bobák, M.; Volkmann, D. Extracellular matrix surface network of embryogenic units of friable maize callus contains arabinogalactan-proteins recognized by monoclonal antibody JIM4. *Plant Cell Rep.* **1999**, *18*, 369–374. [CrossRef]
11. Letarte, J.; Simion, E.; Miner, M.; Kasha, K.J. Arabinogalactans and arabinogalactan-proteins induce embryogenesis in wheat (*Triticum aestivum* L.) microspore culture. *Plant Cell Rep.* **2006**, *24*, 691–698. [CrossRef] [PubMed]
12. Stacey, N.J.; Roberts, K.; Knox, J.P. Patterns of expression of the JIM4 arabinogalactan-protein epitope in cell cultures and during somatic embryogenesis in *Daucus carota* L. *Planta* **1990**, *180*, 285–292. [CrossRef] [PubMed]
13. Neutelings, G.; Domon, J.M.; Membre, N.; Bernier, F.; Meyer, Y.; David, A.; David, H. Characterization of a germin-like protein gene expressed in somatic and zygotic embryos of pine (*Pinus caribaea* Morelet). *Plant Mol. Biol.* **1998**, *38*, 1179–1190. [CrossRef]
14. Almeida, A.M.; Parreira, J.R.; Santos, R.; Duque, A.S.; Francisco, R.; Tome, D.F.; Ricardo, C.P.; Coelho, A.V.; Fevereiro, P. A proteomics study of the induction of somatic embryogenesis in Medicago truncatula using 2DE and MALDI-TOF/TOF. *Physiol. Plant.* **2012**, *146*, 236–249. [CrossRef] [PubMed]
15. Imin, N.; De Jong, F.; Mathesius, U.; Van Noorden, G.; Saeed, N.A.; Wang, X.D.; Rose, R.J.; Rolfe, B.G. Proteome reference maps of Medicago truncatula embryogenic cell cultures generated from single protoplasts. *Proteomics* **2004**, *4*, 1883–1896. [CrossRef] [PubMed]
16. Winkelmann, T.; Heintz, D.; Van Dorsselaer, A.; Serek, M.; Braun, H.P. Proteomic analyses of somatic and zygotic embryos of Cyclamen persicum Mill. reveal new insights into seed and germination physiology. *Planta* **2006**, *224*, 508–519. [CrossRef] [PubMed]
17. Rode, C.; Lindhorst, K.; Braun, H.P.; Winkelmann, T. From callus to embryo: A proteomic view on the development and maturation of somatic embryos in Cyclamen persicum. *Planta* **2012**, *235*, 995–1011. [CrossRef]
18. Sun, L.; Wu, Y.; Zou, H.; Su, S.; Li, S.; Shan, X.; Xi, J.; Yuan, Y. Comparative proteomic analysis of the H99 inbred maize (*Zea mays* L.) line in embryogenic and non-embryogenic callus during somatic embryogenesis. *Plant Cell Tissue Organ Cult.* **2013**, *113*, 103–119. [CrossRef]
19. Marsoni, M.; Bracale, M.; Espen, L.; Prinsi, B.; Negri, A.S.; Vannini, C. Proteomic analysis of somatic embryogenesis in Vitis vinifera. *Plant Cell Rep.* **2008**, *27*, 347–356. [CrossRef]

20. Pan, Z.; Guan, R.; Zhu, S.; Deng, X. Proteomic analysis of somatic embryogenesis in Valencia sweet orange (*Citrus sinensis* Osbeck). *Plant Cell Rep.* **2009**, *28*, 281–289. [CrossRef]

21. Silva Rde, C.; Carmo, L.S.; Luis, Z.G.; Silva, L.P.; Scherwinski-Pereira, J.E.; Mehta, A. Proteomic identification of differentially expressed proteins during the acquisition of somatic embryogenesis in oil palm (*Elaeis guineensis* Jacq.). *J. Proteom.* **2014**, *104*, 112–127. [CrossRef] [PubMed]

22. Gilany, K. Proteomics a Key Tool for a Better Understanding of Endometriosis: A Mini-Review. *J. Paramed. Sci.* **2011**, *2*, 51–58.

23. Zieske, L.R. A perspective on the use of iTRAQ reagent technology for protein complex and profiling studies. *J. Exp. Bot.* **2006**, *57*, 1501–1508. [CrossRef] [PubMed]

24. Zhu, H.G.; Cheng, W.H.; Tian, W.G.; Li, Y.J.; Liu, F.; Xue, F.; Zhu, Q.H.; Sun, Y.Q.; Sun, J. iTRAQ-based comparative proteomic analysis provides insights into somatic embryogenesis in *Gossypium hirsutum* L. *Plant Mol. Biol.* **2017**, *96*, 89–102. [CrossRef]

25. Zhao, J.; Li, H.; Fu, S.; Chen, B.; Sun, W.; Zhang, J.; Zhang, J. An iTRAQ-based proteomics approach to clarify the molecular physiology of somatic embryo development in Prince Rupprecht's larch (Larix principis-rupprechtii Mayr). *PLoS ONE* **2015**, *10*, e0119987. [CrossRef] [PubMed]

26. Lu, C.; Vasil, V.; Vasil, I.K. Improved efficiency of somatic embryogenesis and plant regeneration in tissue cultures of maize (*Zea mays* L.). *Theor. Appl. Genet.* **1983**, *66*, 285–289. [CrossRef] [PubMed]

27. Ishida, Y.; Saito, H.; Ohta, S.; Hiei, Y.; Komari, T.; Kumashiro, T. High efficiency transformation of maize (*Zea mays* L.) mediated by Agrobacterium tumefaciens. *Nat. Biotechnol.* **1996**, *14*, 745–750. [CrossRef]

28. Zhao, Z.Y.; Gu, W.N.; Cai, T.S.; Tagliani, L.; Hondred, D.; Bond, D.; Schroeder, S.; Rudert, M.; Pierce, D. High throughput genetic transformation mediated by Agrobacterium tumefaciens in maize. *Mol. Breed.* **2002**, *8*, 323–333. [CrossRef]

29. Yokoya, N.S.; West, J.A.; Luchi, A.E. Effects of plant growth regulators on callus formation, growth and regeneration in axenic tissue cultures of Gracilaria tenuistipitata and Gracilaria perplexa (Gracilariales, Rhodophyta). *Phycol. Res.* **2004**, *52*, 244–254. [CrossRef]

30. Cheng, W.H.; Wang, F.L.; Cheng, X.Q.; Zhu, Q.H.; Sun, Y.Q.; Zhu, H.G.; Sun, J. Polyamine and Its Metabolite H2O2 Play a Key Role in the Conversion of Embryogenic Callus into Somatic Embryos in Upland Cotton (*Gossypium hirsutum* L.). *Front. Plant Sci.* **2015**, *6*, 1063. [CrossRef]

31. Shang, H.H.; Liu, C.L.; Zhang, C.J.; Li, F.L.; Hong, W.D.; Li, F.G. Histological and ultrastructural observation reveals significant cellular differences between Agrobacterium transformed embryogenic and non-embryogenic calli of cotton. *J. Integr. Plant Biol.* **2009**, *51*, 456–465. [CrossRef]

32. Rajasekaran, K. Regeneration of plants from cryopreserved embryogenic cell suspension and callus cultures of cotton (*Gossypium hirsutum* L.). *Plant Cell Rep.* **1996**, *15*, 859–864. [CrossRef] [PubMed]

33. Overvoorde, P.J.; Grimes, H.D. The Role of Calcium and Calmodulin in Carrot Somatic Embryogenesis. *Plant Cell Physiol.* **1994**, *35*, 135–144.

34. Suprasanna, P.; Desai, N.S.; Nishanth, G.; Ghosh, S.B.; Laxmi, N.; Bapat, V.A. Differential gene expression in embryogenic, non-embryogenic and desiccation induced cultures of sugarcane. *Sugar Tech* **2004**, *6*, 305–309. [CrossRef]

35. Kawade, K.; Ishizaki, T.; Masuda, K. Differential expression of ribosome-inactivating protein genes during somatic embryogenesis in spinach (*Spinacia oleracea*). *Physiol. Plant.* **2008**, *134*, 270–281. [CrossRef] [PubMed]

36. Varhanikova, M.; Uvackova, L.; Skultety, L.; Pretova, A.; Obert, B.; Hajduch, M. Comparative quantitative proteomic analysis of embryogenic and non-embryogenic calli in maize suggests the role of oxylipins in plant totipotency. *J. Proteomics* **2014**, *104*, 57–65. [CrossRef]

37. Andriotis, V.M.; Kruger, N.J.; Pike, M.J.; Smith, A.M. Plastidial glycolysis in developing arabidopsis embryos. *New Phytol.* **2010**, *185*, 649–662. [CrossRef]

38. Zhang, J.; Khvorostov, I.; Hong, J.S.; Oktay, Y.; Vergnes, L.; Nuebel, E.; Wahjudi, P.N.; Setoguchi, K.; Wang, G.; Do, A.; et al. UCP2 regulates energy metabolism and differentiation potential of human pluripotent stem cells. *EMBO J.* **2011**, *30*, 4860–4873. [CrossRef]

39. Yu, W.M.; Liu, X.; Shen, J.; Jovanovic, O.; Pohl, E.E.; Gerson, S.L.; Finkel, T.; Broxmeyer, H.E.; Qu, C.K. Metabolic regulation by the mitochondrial phosphatase PTPMT1 is required for hematopoietic stem cell differentiation. *Cell Stem Cell* **2013**, *12*, 62–74. [CrossRef] [PubMed]

40. Corbet, C. Stem Cell Metabolism in Cancer and Healthy Tissues. Pyruvate in the Limelight. *Front. Pharmacol.* **2017**, *8*, 958. [CrossRef]

41. Ito, K.; Suda, T. Metabolic requirements for the maintenance of self-renewing stem cells. *Nat. Rev. Mol. Cell Biol.* **2014**, *15*, 243–256. [CrossRef] [PubMed]

42. Simsek, T.; Kocabas, F.; Zheng, J.; Deberardinis, R.J.; Mahmoud, A.I.; Olson, E.N.; Schneider, J.W.; Zhang, C.C.; Sadek, H.A. The distinct metabolic profile of hematopoietic stem cells reflects their location in a hypoxic niche. *Cell Stem Cell* **2010**, *7*, 380–390. [CrossRef] [PubMed]

43. Jimenez, V.M.; Bangerth, F. Hormonal status of maize initial explants and of the embryogenic and non-embryogenic callus cultures derived from them as related to morphogenesis in vitro. *Plant Sci.* **2001**, *160*, 247–257. [CrossRef]

International Journal of
Molecular Sciences

MDPI

Article

iTRAQ-Based Quantitative Proteomics Analysis Reveals the Mechanism Underlying the Weakening of Carbon Metabolism in Chlorotic Tea Leaves

Fang Dong [1,2], Yuanzhi Shi [1,2], Meiya Liu [1,2], Kai Fan [1,2], Qunfeng Zhang [1,2,*] and Jianyun Ruan [1,2]

[1] Tea Research Institute, Chinese Academy of Agricultural Sciences, Hangzhou 310008, China; 18305811752@163.com (F.D.); shiyz@tricaas.com (Y.S.); liumeiya@tricaas.com (M.L.); fankaitea@tricaas.com (K.F.); jruan@tricaas.com (J.R.)
[2] Key Laboratory for Plant Biology and Resource Application of Tea, the Ministry of Agriculture, Hangzhou 310008, China
* Correspondence: hill@tricaas.com; Tel.: +86-571-8527-0665

Received: 7 November 2018; Accepted: 5 December 2018; Published: 7 December 2018

Abstract: To uncover mechanism of highly weakened carbon metabolism in chlorotic tea (*Camellia sinensis*) plants, iTRAQ (isobaric tags for relative and absolute quantification)-based proteomic analyses were employed to study the differences in protein expression profiles in chlorophyll-deficient and normal green leaves in the tea plant cultivar "Huangjinya". A total of 2110 proteins were identified in "Huangjinya", and 173 proteins showed differential accumulations between the chlorotic and normal green leaves. Of these, 19 proteins were correlated with RNA expression levels, based on integrated analyses of the transcriptome and proteome. Moreover, the results of our analysis of differentially expressed proteins suggested that primary carbon metabolism (i.e., carbohydrate synthesis and transport) was inhibited in chlorotic tea leaves. The differentially expressed genes and proteins combined with photosynthetic phenotypic data indicated that 4-coumarate-CoA ligase (4CL) showed a major effect on repressing flavonoid metabolism, and abnormal developmental chloroplast inhibited the accumulation of chlorophyll and flavonoids because few carbon skeletons were provided as a result of a weakened primary carbon metabolism. Additionally, a positive feedback mechanism was verified at the protein level (Mg chelatase and chlorophyll b reductase) in the chlorophyll biosynthetic pathway, which might effectively promote the accumulation of chlorophyll b in response to the demand for this pigment in the cells of chlorotic tea leaves in weakened carbon metabolism.

Keywords: *Camellia sinensis*; chlorotic mutation; chlorophyll deficiency; weakening of carbon metabolism; iTRAQ; proteomics

1. Introduction

Tea (*Camellia sinensis*) is a perennial evergreen leafy woody plant native to southwest China. Recently, chlorophyll-deficient chlorina tea plant cultivars have become valuable materials in processing high quality green tea because of their high amino acid content and low catechin content [1,2]. The natural mutant of tea, "Huangjinya", exhibits chlorotic leaves and lower carbon metabolism than non-chlorotic varieties under sunlight [1,2]. In our previous studies [2,3], metabolomics and transcriptomics analyses were performed on green and chlorotic shoots of "Huangjinya" to gain an overview of the amino acid, flavonoid, and carbohydrate metabolism. These analyses revealed that the weakening of carbon metabolism is accompanied by nitrogen accumulation, suggesting that the metabolism of carbon and nitrogen are unbalanced [3]. Satou et al. [4] have shown similar results in the pale green mutants of *Arabidopsis thaliana*. However, the correlation of protein expression with weakened carbon metabolism in chlorotic tea leaves remains to be elucidated.

Chlorophyll consists of chlorophyll a and chlorophyll b, and plays indispensable roles in harvesting and transferring light energy during photosynthesis and carbon assimilation [5]. Chlorosis in tea leaves has always been attributed to chlorophyll deficiency. Studies on chlorophyll biosynthesis have been widely reported, and at least in angiosperm plants represented by *Arabidopsis thaliana*, genes for all 15 steps in the chlorophyll biosynthesis pathway, starting from the biosynthesis of glutamyl-tRNA to that of chlorophylls a and b, have been identified [6]. However, the effect of chlorophyll metabolism on photosynthesis is largely unclear. Mutants defective in chlorophyll biosynthesis have been identified in higher plants [7,8]. For example, the leaf phenotype was yellow-green in the chlorophyll mutant (*Oryza sativa*) and the level of chlorophyll decreased, meanwhile, chloroplast development was delayed [8]. Thylakoid proteome analysis of a novel rice (*Oryza sativa*) mutant, Zhenhui 249Y, and the wild type has shown that the reduction of chlorophyll b affects the assembly of light harvesting complex I (LHC-I) more severely than that of LHC-II [9].

Proteomics of leaf color mutants of tea plant has been performed using both isobaric tags for relative and absolute quantification (iTRAQ) [10] and two-dimensional gel electrophoresis (2-DE)–mass spectrometry [11,12]. In these studies, 437 differentially accumulated proteins have been identified between the tea plant cultivars "Longjing43" and "Zhonghuang1" [10] and 46 differentially abundant proteins between tender purple and mature green leaves of tea plant [12]. However, it is difficult to clarify the mechanism of weakened carbon metabolism because of the complex genetic background or inconsistent developmental stages of experimental material in different tea varieties [10]. In this study, we used chlorotic and normal green leaves ("Huangjinya", the albino tea plant cultivar) with the same genetic background and developmental stage as the experimental material to compare the protein expression profiles of shaded and non-shaded leaves by iTRAQ technique.

It is hypothesized that the inhibition of carbon assimilation results in a down-regulation of protein expression in carbohydrate synthesis and transport pathways, further weakening primary carbon metabolism. Meanwhile, flavonoid metabolism, as the major secondary metabolism, may also be suppressed by the down-regulation of the expression of related proteins. In this study, iTRAQ-based quantitative proteomics with phenotypic, biochemical, and transcriptome data confirmed our findings on the differences in protein expression profiles underlying the weakening of carbon metabolism in chlorotic "Huangjinya" tea leaves. The results of this study also provide new insights into the expression level of proteins to understand the mechanisms responsible for chlorophyll deficiency in etiolated tea plant leaves.

2. Results

2.1. Phenotype, Ratio of Pigment Content, Photosynthesis of Chlorotic and Green Leaves

Compared with shaded leaves of tea plants (Figure 1A), leaves of tea plants grown under full sunlight were chlorotic and exhibited a yellow phenotype (Figure 1B). Transmission electron microscopy showed clear differences in leaf ultrastructure between chlorotic and shaded green leaves (Figure 1C,D). Compared with green leaves (Figure 1C), chlorotic leaves showed chloroplasts with abnormal structural development—thylakoids were observed, but the stacks of grana were not found (Figure 1D).

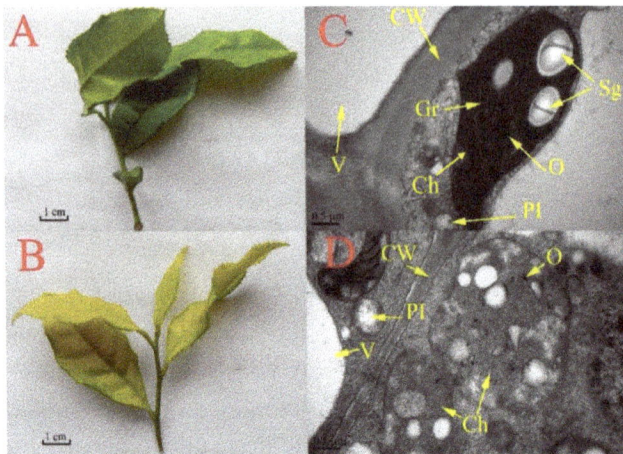

Figure 1. Characterization of the phenotype and ultra-structure of chlorotic and green leaves of the tea plant mutant cultivar "Huangjinya". (**A,B**) Young shoots either grown in shade with 60% light intensity (**A**) or exposed to 100% sunlight (**B**). (**C,D**) Ultrastructure of leaves grown under shade (**C**) or under full sunlight (**D**). Ch: chloroplast; CW: cell wall; Gr: grana; O: osmiophilic granules; Pl: plastid; Sg: starch granule; V: vacuole.

The contents of chlorophyll a, chlorophyll b, total chlorophyll, and carotenoids in chlorotic mutants and green leaves has been reported previously [3]. In this study, the contents of these four pigments were significantly lower in chlorotic leaves than those in green leaves. Furthermore, the ratio of chlorophyll a to chlorophyll b and that of total chlorophyll to carotenoids were significantly lower in chlorotic leaves than in green leaves (Figure 2).

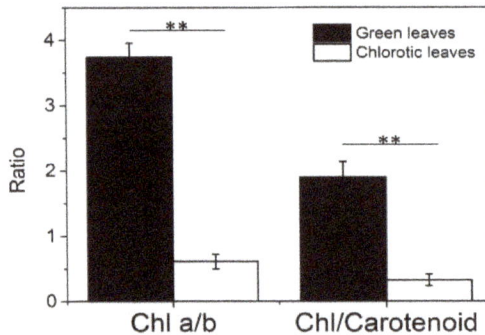

Figure 2. Ratio of chlorophyll a to chlorophyll b and that of total chlorophyll to carotenoids. Values represent mean $\pm$ SD of three biological replicates (** $p < 0.01$).

Leaf gas exchange analysis showed that net photosynthesis and intercellular CO_2 concentration were reduced by approximately 21.7% and 36.13%, respectively, in chlorotic leaves compared with green leaves. By contrast, the stomatal conductance and transpiration rate of chlorotic leaves were increased by approximately 15.2% and 21.4%, respectively, compared with green leaves (Table 1).

Table 1. Leaf gas exchange analysis of green and chlorotic leaves of the tea plant mutant cultivar "Huangjinya".

Genotype	NPR [a] (μmol $CO_2 \cdot m^{-2} \cdot s^{-1}$)	SC [b] (mmol $H_2O \cdot m^{-2} \cdot s^{-1}$)	IC [c] (μmol $CO_2 \cdot mol^{-1}$)	TR [d] (mmol $H_2O \cdot m^{-2} \cdot s^{-1}$)
Green	8.17 ± 0.45	$0.046 \pm 8.39 \times 10^{-4}$	256.54 ± 18.04	1.45 ± 0.07
Chlorotic	6.39 ± 0.25 **	$0.053 \pm 5.1 \times 10^{-3}$ *	163.85 ± 14.28 **	1.76 ± 0.13 **

[a] net photosynthetic rate; [b] stomatal conductance; [c] intercellular CO_2 concentration; [d] transpiration rate (**: $p < 0.01$; *: $p < 0.05$).

2.2. Quantitative Identification of Tea Leaf Proteins Using iTRAQ

Differentially accumulated proteins in chlorotic and green leaves were identified using iTRAQ technique, and 302,042 spectra were obtained. Analysis using the Mascot software revealed that the number of matched spectra and unique spectra were 15,804 and 14,943, respectively. A total of 6157 unique peptides were identified. Distributions of protein mass, peptide number, and peptide length are shown in Supplementary Figures S1–S3.

We identified 2110 proteins. According to GO analysis, 1354, 1284, and 1349 proteins were annotated as cellular components, functional molecules, and those involved in biological processes, respectively (Figure 3). The main biological function categories included nucleoside phosphate metabolic process, photosynthesis, and carbohydrate derivative catabolic process. The proteins classified as having functional molecular properties were mainly classified based on their activity: hydrolase activity, acting on glycosyl bonds, alpha-glucosidase activity, translation elongation factor activity, glucosidase activity, ATP-dependent peptidase activity, and oxidoreductase activity. A total of 1540 proteins were assigned to 22 categories using the Clusters of Orthologous Groups of proteins (COG) database; the main functional categories were transport and metabolism (21.5%); protein turnover, chaperones (10.5%); energy production and conversion (7.2%); and translation, ribosomal structure, and biogenesis (7.7%) (Figure 4). Additionally, 1268 proteins were annotated in 119 pathways using the Kyoto Encyclopedia of Gene and Genomes (KEGG) database. The main pathways were metabolic pathways (30.52%); biosynthesis of secondary metabolites (17.03%); plant–pathogen interaction (4.18%); protein processing in endoplasmic reticulum (3.79%); starch and sucrose metabolism (3.39%); and pyruvate metabolism (3.15%) (Supplementary Table S1).

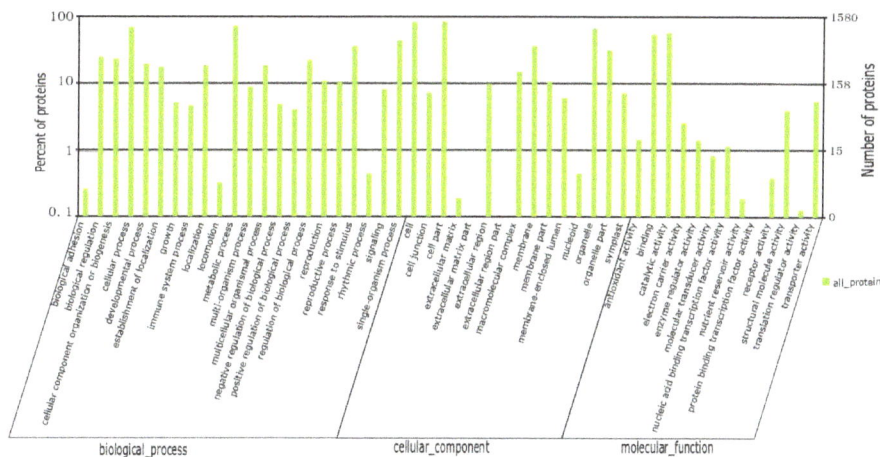

Figure 3. Gene ontology (GO) classification of differentially accumulated proteins in chlorotic and green leaves of the tea plant mutant cultivar "Huangjinya".

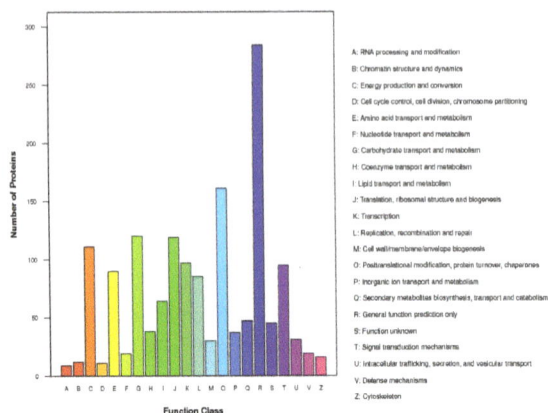

Figure 4. Clusters of Orthologous Groups of proteins (COG) classification of differentially accumulated proteins in chlorotic and green tea leaves of the tea plant mutant cultivar "Huangjinya". A: RNA processing and modification; B: Chromatin structure and dynamics; C: Energy production and conversion; D: Cell cycle control, cell division, chromosome partitioning; E: Amino acid transport and metabolism; F: Nucleotide transport and metabolism; G: Carbohydrate transport and metabolism; H: Coenzyme transport and metabolism; I: Lipid transport and metabolism; J: Translation, ribosomal structure, and biogenesis; K: Transcription; L: Replication, recombination, and repair; M: Cell wall/membrane/envelope biogenesis; O: Post-translational modification, protein turnover, chaperones; P: Inorganic ion transport and metabolism; Q: Secondary metabolites biosynthesis, transport, and catabolism; R: General function prediction only; S: Function unknown; T: Signal transduction mechanisms; U: Intracellular trafficking, secretion, and vesicular transport; V: Defense mechanisms; Z: Cytoskeleton.

2.3. Regulation of Proteins in Response to Chlorosis

In this study, 173 proteins showed significant difference (ratio of protein abundance > 1.2/0.8 fold; $p < 0.05$) between chlorotic and green leaves, including 80 up-regulated and 93 down-regulated proteins (Table 2). A total of 23, 49, and 87 proteins reproducibly decreased by 0.50-, 0.67-, and 0.83-fold, respectively, in chlorotic leaves compared with green leaves (Supplementary Table S2). On the other hand, levels of 3, 23, and 75 proteins increased by more than 2.0-, 1.5-, and 1.2-fold, respectively, in chlorotic leaves compared to in green leaves (Supplementary Table S3).

Table 2. Pathway enrichment analysis of differentially accumulated proteins in green and chlorotic leaves of the tea plant mutant cultivar "Huangjinya".

Pathway	Pathway	Enrichment Score	Number of Proteins	
ID [a]		Scores [b]	Up [c]	Down [d]
ko00195	Photosynthesis	0.003016	0	6
ko00603	Glycosphingolipid biosynthesis—globo series	0.010268	1	2
ko00052	Galactose metabolism	0.016302	2	3
ko00600	Sphingolipid metabolism	0.030562	2	1
ko00511	Other glycan degradation	0.031593	2	2
ko00190	Oxidative phosphorylation	0.034353	2	5
ko00604	Glycosphingolipid biosynthesis—ganglio series	0.059274	1	1
ko00531	Glycosaminoglycan degradation	0.059274	1	1
ko00904	Diterpenoid biosynthesis	0.107256	0	1
ko03040	Spliceosome	0.133227	2	6
ko00906	Carotenoid biosynthesis	0.167604	1	1
ko00402	Benzoxazinoid biosynthesis	0.203083	0	1

Table 2. *Cont.*

Pathway ID [a]	Pathway	Enrichment Score Scores [b]	Number of Proteins Up [c]	Down [d]
ko01110	Biosynthesis of secondary metabolites	0.208389	17	10
ko00330	Arginine and proline metabolism	0.241617	1	2
ko04144	Endocytosis	0.251564	1	3
ko04145	Phagosome	0.276463	0	4
ko03010	Ribosome	0.284144	5	2
ko00062	Fatty acid elongation	0.288692	1	0
ko00710	Carbon fixation in photosynthetic organisms	0.301754	4	0
ko00520	Amino sugar and nucleotide sugar metabolism	0.327304	0	4
ko00900	Terpenoid backbone biosynthesis	0.333655	2	0
ko00770	Pantothenate and CoA biosynthesis	0.365164	1	0
ko00196	Photosynthesis—antenna proteins	0.365164	0	1
ko00730	Thiamine metabolism	0.365164	0	1
ko00300	Lysine biosynthesis	0.365164	1	0
ko04141	Protein processing in endoplasmic reticulum	0.411499	4	2
ko00460	Cyanoamino acid metabolism	0.414401	1	1
ko01100	Metabolic pathways	0.418741	21	22
ko00051	Fructose and mannose metabolism	0.425589	2	1
ko03030	DNA replication	0.433469	1	0
ko00073	Cutin, suberine, and wax biosynthesis	0.433469	0	1
ko00480	Glutathione metabolism	0.452965	2	0
ko00561	Glycerolipid metabolism	0.490057	1	1
ko00450	Selenocompound metabolism	0.494474	1	0
ko00500	Starch and sucrose metabolism	0.497233	3	2
ko01040	Biosynthesis of unsaturated fatty acids	0.548952	1	0
ko00130	Ubiquinone and other terpenoid-quinone biosynthesis	0.548952	0	1
ko00940	Phenylpropanoid biosynthesis	0.551185	1	3
ko00360	Phenylalanine metabolism	0.591425	0	2
ko00860	Porphyrin and chlorophyll metabolism	0.597598	1	0
ko00562	Inositol phosphate metabolism	0.597598	1	0
ko00350	Tyrosine metabolism	0.597598	1	0
ko00020	Citrate cycle (TCA cycle)	0.621742	2	0
ko00240	Pyrimidine metabolism	0.641032	1	0
ko00592	alpha-Linolenic acid metabolism	0.679808	1	0
ko00280	Valine, leucine, and isoleucine degradation	0.679808	1	0
ko00290	Valine, leucine, and isoleucine biosynthesis	0.679808	1	0
ko00400	Phenylalanine, tyrosine, and tryptophan biosynthesis	0.714424	1	0
ko03015	mRNA surveillance pathway	0.725342	2	1
ko00941	Flavonoid biosynthesis	0.745321	0	1
ko00910	Nitrogen metabolism	0.745321	0	1
ko03013	RNA transport	0.751412	5	0
ko00250	Alanine, aspartate, and glutamate metabolism	0.772898	1	0
ko00061	Fatty acid biosynthesis	0.797508	1	0
ko00640	Propanoate metabolism	0.797508	1	0
ko00010	Glycolysis/gluconeogenesis	0.808197	3	0
ko00071	Fatty acid metabolism	0.819469	1	0
ko03018	RNA degradation	0.821126	2	0
ko00620	Pyruvate metabolism	0.822001	3	0
ko00230	Purine metabolism	0.856546	1	0
ko00030	Pentose phosphate pathway	0.856546	1	0
ko03008	Ribosome biogenesis in eukaryotes	0.886052	1	0
ko04146	Peroxisome	0.909525	1	0
ko00053	Ascorbate and aldarate metabolism	0.909525	0	1
ko04075	Plant hormone signal transduction	0.951118	0	2
ko04626	Plant–pathogen interaction	0.997861	1	0

[a] Serial number of the enrichment pathway of differential accumulated proteins in the Kyoto Encyclopedia of Gene and Genomes (KEGG); [b] Degree of pathway enrichment of differential proteins in KEGG; [c] Number of up-accumulated proteins; [d] Number of down-accumulated proteins.

KEGG pathway enrichment analysis was employed to explore the metabolic and biosynthetic pathways, which changed in response to the chlorotic mutation with those differentially accumulated proteins. A number of such pathways were identified, including chlorophyll biosynthesis, carbohydrate transport and metabolism, energy production and conversion, flavonoid metabolism, nitrogen metabolism, chloroplast function, and oxidative stress (Table 3).

Table 3. Differentially accumulated proteins identified in pathways potentially associated with chlorophyll deficiency in green and chlorotic leaves of the tea plant mutant cultivar "Huangjinya".

Identity Proteins [a]	EC Number [b]	Accession [c]	Fold Change (Etiolation/Green)
Chlorophyll biosynthesis			
Aspartyl-tRNA/glutamyl-tRNA amidotransferase subunit A	6.3.5.6	CL57658Contig1	1.12
Chlorophyll(ide) b reductase	1.1.1.294	CL49902Contig1	0.90
Geranylgeranyl	2.5.1.1	CL1Contig45	1.41
Glutaminyl-tRNA synthetase	6.1.1.18	CL18599Contig1	1.26
Magnesium chelatase	6.6.1.1	CL498Contig6	1.34
Magnesium protoporphyrin	2.1.1.11	CL18563Contig1	1.02
Porphobilinogen deaminase	2.5.1.61	CL37040Contig1	1.68
Protochlorophyllide reductase	1.3.1.33	CL508Contig2	0.80
Violaxanthin de-epoxidase	1.10.99.3	CL3Contig71	0.93
Carbohydrate transport and metabolism			
6-Phosphofructokinase	2.7.1.11	CL128Contig12	1.01
Fructokinase	2.7.1.4	CL18457Contig1	0.99
Hexokinase	2.7.1.1	CL60051Contig1	0.88
Phosphoglycerate mutase	5.4.2.1	CL15710Contig1	1.24
Phosphopyruvate hydratase	4.2.1.11	CL19736Contig1	0.95
Pyruvate kinase	2.7.1.40	comp42454_c0_seq1_3	1.56
Ribulose-bisphosphate carboxylase	4.1.1.39	CL8Contig62	0.76
Granule-bound starch synthase		CL7825Contig1	0.38
Fructose-1,6-bisphosphatase		comp51045_c1_seq8_2	1.72
Beta-fructofuranosidase		CL53580Contig1	1.26
Xylosidase		comp80972_c0_seq1_4	0.99
Galactose oxidase		CL17517Contig1	1.23
UDP-L-arabinosidase		comp62280_c0_seq4_2	1.19
Beta-glucosidase		CL167Contig8	1.77
Energy production and conversion			
Aconitate hydratase	4.2.1.3	CL79359Contig1	1.37
ATP-citrate synthase	2.3.3.1	comp116390_c0_seq1_3	0.95
Dihydrolipoyl dehydrogenase	1.8.1.4	comp99158_c0_seq16_4	0.86
Dihydrolipoyllysine-residue acetyltransferase	2.3.1.12	CL17321Contig1	1.47
Dihydrolipoyllysine-residue succinyltransferase	2.3.1.61	CL64635Contig1	0.78
Isopropylmalate dehydrogenase	1.1.1.85	comp102244_c2_seq1_4	1.11
Malate dehydrogenase		CL2510Contig4	1.22
Pyruvate dehydrogenase	1.2.4.1	CL37234Contig1	1.31
Succinate dehydrogenase	1.3.5.1	comp131171_c0_seq3_3	1.48

Table 3. *Cont.*

Identity Proteins [a]	EC Number [b]	Accession [c]	Fold Change (Etiolation/Green)
Flavonoid metabolism			
4-Coumarate-CoA ligase	6.2.1.12	CL48129Contig1	0.75
Anthocyanidin 3-*O*-glucosyltransferase	2.4.1.115	CL319Contig5	1.13
Anthocyanidin reductase	1.3.1.77	CL103Contig2	1.09
Anthocyanidin synthase	1.14.11.19	CL3972Contig1	1.48
Chalcone isomerase	5.5.1.6	CL12172Contig2	1.49
Chalcone synthase	2.3.1.74	CL8845Contig1	1.84
Cinnamate 4-hydroxylase	1.14.13.11	CL30220Contig1	1.05
Flavonol synthase	1.14.11.23	CL11177Contig1	0.62
Phenylalanine ammonia-lyase	4.3.1.24	comp64735_c0_seq1_2	0.70
3-Dehydroshikimate dehydratase	4.2.1.118	CL16483Contig1	1.14
3-Dehydroquinate synthase	4.2.3.4	CL55Contig2	1.30
Nitrogen metabolism			
3-Deoxy-7-phosphoheptulonate synthase activity	2.5.1.54	comp124631_c0_seq3_3	1.02
Alanine transaminase	2.6.1.2	CL34278Contig1	1.52
Anthranilate synthase	4.1.3.27	CL19422Contig1	1.34
Aspartate kinase	2.7.2.4	CL43459Contig1	1.13
Cysteine synthase	2.5.1.47	CL37334Contig1	1.14
Ferredoxin-nitrite reductase	1.7.7.1	CL61698Contig1	0.69
Glutamate synthase	1.4.7.1	comp109180_c0_seq1_1	1.36
Glycine hydroxymethyltransferase	2.1.2.1	CL5210Contig1	0.98
Homoserine kinase	2.7.1.39	CL16611Contig1	1.18
Methionine synthase	2.1.1.13	CL9637Contig1	1.19
S-adenosylmethionine synthase	2.5.1.6	CL39736Contig1	0.93
Glutathione reductase (NADPH)	1.8.1.7	CL9366Contig1	1.30
Chloroplast function			
Proton ATPase subunit C		CL6Contig55	0.78
Elongation factor G, chloroplastic		CL107Contig12	1.26
Protein ABCI7, chloroplastic		comp100064_c2_seq1_1	1.45
Pentatricopeptide repeat-containing protein At4g16390, chloroplastic		CL9Contig51	1.93
Chloroplast small heat shock protein		CL2Contig56	1.50
Photosystem Q(B) protein		comp95426_c0_seq3_4	0.40
Cytochrome P450 86A2		CL102Contig8	1.13
Oxidative stress			
Fructose-bisphosphate aldolase 3, chloroplastic		CL1744Contig2	1.24
Histone deacetylase HDT1		CL85545Contig1	1.20
B5TV66_CAMSI Putative dehydrin		CL14231Contig1	2.29
Peroxidase 50		CL920Contig3	1.01

[a] Proteins identified by isobaric tag for relative and absolute quantification (iTRAQ); [b] Enzyme commission numbers in PDB; [c] Accession number of the identified proteins in the National Center for Biotechnology Information non-redundant protein sequences (NCBI-nr) database.

The differentially accumulated proteins involved in chlorophyll biosynthesis included nine proteases. The up-regulation of six of these proteins, including glutamyl-tRNA (Gln) amino-transferase, geranylgeranyl, glutaminyl-tRNA synthetase, magnesium chelatase, magnesium protoporphyrin, and porphobilinogen deaminase, was increased in the chlorotic leaves than in green leaves by 1.02- to 1.68-fold, whereas that of the remaining three proteins, including chlorophyll(ide) b reductase, protochlorophyllide reductase, and violaxanthin de-epoxidase, was reduced in the chlorotic leaves by 0.8- to 0.93-fold.

The differentially accumulated proteins related to carbohydrate transport and metabolism mainly comprised six rate-limiting enzymes of the glycolytic pathway (6-phosphofructokinase, fructokinase, hexokinase, pyruvate kinase, phosphoglycerate mutase, and phosphopyruvate hydratase), five glycosidases (beta-fructofuranosidase, xylosidase, galactose oxidase, UDP-L-arabinosidase, and beta-glucosidase), and three proteases related to photosynthesis (ribulose-bisphosphate carboxylase, fructose-1,6-bisphosphatase, and granule-bound starch synthase). The level of eight proteins was up-regulated by 1.01- to 1.77-fold in the chlorotic leaves. Additionally, the level of six proteins was down-regulated by 0.38- to 0.99-fold in the chlorotic leaves.

The most relevant pathway for energy generation and conversion is the tricarboxylic acid (TCA) cycle or the Krebs cycle, which generates the highest amount of energy in the most efficient way through the oxidation of sugars and other substances. The differentially accumulated proteins involved in the Krebs cycle mainly include five dehydrogenases (isopropyl-malate dehydrogenase, dihydrolipoyl dehydrogenase, malate dehydrogenase, pyruvate dehydrogenase, and succinate dehydrogenase), two dihydrolipoyllysine-residue transferases (dihydrolipoyllysine-residue acetyltransferase and dihydrolipoyllysine-residue succinyltransferase), the other are ATP citrate synthases and aconitate hydratase. Among these proteins, 16 were up-regulated by 1.05- to 1.48-fold and 4 were down-regulated by 0.95- to 0.76-fold in the chlorotic mutation compared to green leaves.

Of the eleven differentially accumulated proteins involved in flavonoid metabolism, the levels of eight proteins (cinnamate acid 4-hydroxylase (C4H), chalcone isomerase (CHI), chalcone synthase (CHS), anthocyanidin synthase (ANS), anthocyanidin reductase (ANR), anthocyanidin 3-O-glucosyltransferase (A3Glc), 3-dehydroshikimate dehydratase (3DSD), and 3-dehydroquinate synthase (3DHQ)) were increased by 1.09- to 1.84-fold in chlorotic leaves than in green leaves, whereas those of three proteins (phenylalanine ammonia-lyase (PAL), 4-coumarate-CoA ligase (4CL), and flavonol synthase (FLS)) were reduced by 0.62- to 0.75-fold in the chlorotic leaves compared to in green leaves. Phenylalanine is a precursor of the flavonoid biosynthesis pathway, and PAL and 4CL play key roles in the conversion of phenylalanine to coumaroyl CoA. Our results showed that levels of PAL and 4CL proteins were reduced in the chlorotic leaves, indicating that flavonoid metabolism was inhibited. From the branching of flavonoid biosynthesis (i.e., the synthetic pathway of anthocyanins and flavonols), the accumulation of anthocyanins was promoted and the synthesis of flavonols was inhibited.

Among the proteins involved in nitrogen metabolism, the levels of nine proteins' expression (3-deoxy-7-phosphoheptulonate synthase, alanine transaminase, anthranilate synthase, aspartate kinase, cysteine synthase, glutamate synthase, homoserine kinase, methionine synthase, and glutathione reductase, GR [NADPH; nicotinamide adenine dinucleotide phosphate]) were increased and the expression levels of three proteins (ferredoxin-nitrite reductase, glycine hydroxymethyl transferase, and S-adenosyl methionine synthase) were reduced. These results indicated that the up-regulation of the expression of most amino acid synthase proteins might promote nitrogen assimilation and recycling.

Photosystem Q(B) plays an important role in chloroplast function, and the expression level of its protein was down-regulated by 0.4-fold in chlorotic leaves compared with green leaves, indicating that chloroplast function was inhibited under strong light stress. Dehydrins, also known as LEA D-11 or LEA II, are proteins whose expression is induced by various environmental stress factors [13]. B5TV66_CAMSI Putative dehydrin was annotated as an oxidative stress protein, and its expression level was significantly up-regulated by 2.29-fold. Therefore, it is speculated that the antioxidant capacity of chlorotic tea leaves might be enhanced compared to that of green leaves.

2.4. Integrated Analysis of Transcriptomic and Proteomic Datasets

A total of 5051 differentially expressed genes (DEGs), with differences between chlorotic and shaded green leaves, were selected for bioinformatics analysis. The combination of transcriptomic and proteomic datasets revealed correlations between 126, 52, and 19 genes and proteins in identification, quantitation, and differential expression levels, respectively (Table 4). Nineteen genes and proteins with significant differences at the quantitative level are shown in Table 5. These were classified in the following categories: chloroplast structure and function; carbohydrate and amino acid metabolism; and flavonoid biosynthesis and oxidative stress.

Genes and proteins related to chloroplast structure and function are related to the chloroplast stroma thylakoids (i.e., V-type proton ATPase subunit C, tRNA (cytosine38-C5)-methyltransferase, chloroplast small heat shock protein). The C metabolism pathway mainly involves proteins for starch synthesis (glycogen synthase), alpha-maltase (alpha-glucosidases), and lignin metabolism (L-ascorbate oxidase). Metabolic processes associated with nitrogen metabolism are mainly arginine metabolism (arginase). Oxidative stress caused by the etiolating mutation of tea leaves was also related to gene expression and protein accumulation levels. Oxidative stress-related proteins and enzymes changed significantly (fructose-bisphosphate aldolase and dehydrin). The change of the polyphenol metabolic pathway is mainly related to its upstream pathway, such as 4-coumarate-CoA ligase (CL48129Contig1), phenylalanine ammonia lyase (comp64735_c0_seq1_2), two key genes and enzyme levels were significantly down-regulated in etiolated leaves.

Table 4. The number of proteins and genes identified, quantified, and differentially expressed in green and chlorotic leaves of the tea plant mutant cultivar "Huangjinya".

Group Names	Type	Number of Proteins	Number of Genes	Number of Correlations
EM [a] vs. NG [b]	Identification	2110	5051	126
EM vs. NG	Quantitation	976	5051	52
EM vs. NG	Differential Expression	173	5051	19

[a] etiolated mutation; [b] normal green.

Table 5. Genes and proteins showing significant changes between green and chlorotic leaves as determined via the integrated analysis of transcriptomic and proteomic datasets.

Accession [a]	log2 (EM/NG)		Description	FDR [b]	Function
	Gene	Protein			
CL14231Contig1	2.76	1.20	Dehydrin	0.0000	Oxidative stress
CL1744Contig2	−2.7	0.31	Fructose-bisphosphate aldolase	0.0010	Carbohydrate transport and metabolism
CL2031Contig2	3.02	−1.43	L-Ascorbate oxidase	0.0000	Secondary metabolites biosynthesis, transport, and catabolism
CL2Contig56	4.57	0.58	Chloroplast small heat shock protein	0.0000	Posttranslational modification, protein turnover, chaperones
CL374Contig2	−2.29	−0.62	Zeta-carotene desaturase	0.0020	Response to hormone stimulus
CL48129Contig1	−1.56	−0.42	4-coumarate-CoA ligase 2	0.0320	Lipid transport and metabolism
CL4Contig6	3.06	−0.30	SnRK2 calcium sensor	0.0000	Calcium ion binding
CL50804Contig1	−3.79	−0.69	Cysteine protease	0.0000	Posttranslational modification, protein turnover, chaperones
CL6Contig55	−4.51	−0.36	V-type proton ATPase subunit C	0.0000	Energy production and conversion
CL73512Contig1	2.25	0.45	Alpha-glucosidases	0.0320	Carbohydrate transport and metabolism
CL7825Contig1	−2.04	−1.40	Glycogen synthase	0.0370	Carbohydrate transport and metabolism
CL8494Contig2	2.26	1.58	Unknown	0.0030	Embryo development ending in seed dormancy
CL9Contig51	1.85	0.95	tRNA (cytosine38-C5)-methyltransferase	0.0420	Chloroplast organization
Comp101085_c0_seq1_1	2.02	0.39	Alpha-glucosidases	0.0210	Carbohydrate transport and metabolism
Comp55188_c0_seq1_2	−2.99	−0.84	Serine proteases	0.0420	Posttranslational modification, protein turnover, chaperones
Comp64728_c0_seq2_2	1.86	0.75	DEAD-box ATP-dependent RNA helicase 31	0.0140	Response to water deprivation
Comp64735_c0_seq1_2	−2.25	−0.51	Phenylalanine ammonia-lyase	0.0120	Amino acid transport and metabolism
Comp74393_c0_seq1_4	1.65	0.50	Dihydroxy-acid dehydratase	0.0240	Amino acid transport and metabolism
Comp96472_c0_seq1_4	−1.8	−0.69	Arginase	0.0170	Amino acid transport and metabolism

[a] Accession number of the identified proteins in National Center for Biotechnology Information non-redundant protein sequences (NCBI-nr). database; [b] FDR: false discovery rate.

3. Discussion

Carbohydrates are a direct product of carbon assimilation via photosynthesis. Additionally, carbohydrates represent a source of plant energy and are involved in the formation of the plant cytoskeleton. In this study, the increased stomatal conductance and the decreased CO_2 concentration was accompanied by a reduced net photosynthesis rate in the chlorotic leaves. However, the concentration of CO_2 in the study of *Brassica napus* increased [14], possibly because of differences in the leaf structure of the two plants. Both studies have shown impaired carbon fixation efficiency in chlorotic leaves. In this study, levels of some proteins involved in carbohydrate metabolism, including fructokinase, hexokinase, phospho-pyruvate hydratase, ribulose-bisphosphate carboxylase, granule-bound starch synthase, and xylosidase, were reduced, which is consistent with this speculation. Rubisco (ribulose-bisphosphate carboxylase/oxygenase) is the rate-limiting enzyme for carbon fixation in photosynthetic reactions, and is essential for improving the photosynthetic efficiency of plants [15]. In this study, the abundance of Rubisco protein was significantly reduced in chlorotic leaves compared with green leaves (Table 3), leading to a reduction in carbohydrate biosynthesis and sugar content. Simultaneously, proteins (6-phosphate fructokinase, pyruvate kinase, phosphoglycerate mutase, and fructose 1,6-bisphosphatase) with higher expression levels were involved in the glycolytic pathway in chlorotic leaves, promoting carbohydrate catabolism. Some scholars have shown that chloroplast endometrial damage activates the expression of glycolysis-related genes [4]. Results of protein accumulation and gene expression indicated that carbohydrate accumulation was decreased in "Huangjinya" leaves under strong light.

Carbon and nitrogen metabolism balance guarantees the normal growth of tea plants, and enhanced nitrogen metabolism is accompanied by the reduced capability of photosynthesis and carbon metabolism [4]. In this study, methionine synthetase, cysteine synthetase, and glutamate synthetase were up-regulated in chlorotic leaves compared with green leaves, indicating that strong light promotes nitrogen metabolism and amino acid accumulation in "Huangjinya", which is consistent with a previous study [16]. We speculate that reduced chlorophyll biosynthesis in turn reduced nitrogen consumption, thus increasing the content of upstream substances (amino acids) in leaves. In addition, environmental stress such as intense light and high temperature affect the normal growth of plants, and reactive oxygen species (ROS) are accumulated as a by-product [17]. Previous studies have shown that increasing the glutathione reductase activity in chloroplasts improves the photochemical ratio in transgenic cotton plants, thereby reducing photoinhibition [18]. This suggests that enhanced glutathione reductase activity protects the leaf cell biofilm and enhances the plant's ability to defend itself against abiotic stress (e.g., UV-B radiation) [19]. In this study, the expression of glutathione reductase protein was increased by 1.3-fold, and as a result the ability to remove ROS produced by UV-B radiation under intense light might be enhanced.

A simplified schematic presentation of the weakening of carbon metabolism in the chlorotic mutation is shown in Figure 5. The ratio of chlorophyll a to chlorophyll b has been considered as an important parameter to measure the light tolerance of plants. An increase in this ratio is beneficial for the absorption of blue-violet light, which is suitable for plant growth in the dark [20]. Generally, the biosynthesis and degradation of chlorophyll a and chlorophyll b in plants occur in a dynamic cycle, and chlorophyll biosynthesis requires relatively low light intensity. However, as a light-sensitive plant, the ratio of chlorophyll a to chlorophyll b in leaves of "Huangjinya" under the shade was significantly higher than that under intense light, suggesting that the conversion of chlorophyll a to chlorophyll b is accelerated under strong light. Increase in chlorophyll b content is beneficial for the plant's adaptation to light stress. Chlorophyll(ide) b reductase is the key enzyme that catalyzes the first step in chlorophyll b degradation, and plays an important role in the process of leaf senescence, with the degradation of LHC-II and chloroplast matrix [21]. Studies have reported that chlorophyll b reductase was involved in the conversion process of chlorophyll b and chlorophyll a, which is considered to be an important clue for chlorophyll b degradation [22]. In this study, the level of chlorophyll(ide) b reductase was reduced by 0.9-fold in chlorotic leaves, and the ratio of chlorophyll a to chlorophyll

Int. J. Mol. Sci. **2018**, 19, 3943

b was lower in chlorotic leaves than in green leaves, indicating that the activity of chlorophyll b reductase was suppressed. Zhang and Tan [23] also obtained similar results regarding the reduction of chlorophyll content and chlorophyll a/b ratio under salt stress. The studies of Sang et al. [24] and Huang et al. [25] suggested that the decreased ratio of chlorophyll a/b resulted from a higher sensitivity to various environmental stresses for chlorophyll a compared to chlorophyll b. Therefore, the pathway of chlorophyll b biosynthesis is still enhanced although the total chlorophyll content was decreased in the chlorotic leaves. Interestingly, the content of chlorophyll b was not increased with the enhanced pathway of chlorophyll b synthase in "Huangjinya" leaves under strong light. Because the process of chlorophyll biosynthesis is complex and highly conserved, the mutation of a gene can severely affect the chlorophyll content, leading to a different color phenotype of leaves [26]. Magnesium chelatase is another enzyme with a significant effect on chlorophyll biosynthesis. It catalyzes the insertion of Mg^{2+} into protoporphyrin IX [27]. In the process of leaf albinism, the abundance of magnesium chelatase subunit proteins in higher plants is significantly increased under light [28,29]. In this study, the expression level of magnesium chelatase protein was higher in chlorotic leaves by 1.34-fold than in green leaves, which is consistent with the study of Walker et al. [29]. Müller et al. [30] showed that oxidative stress improved the quality of monomeric chlorophyll H in *Escherichia coli*. Moreover, genes involved in the biosynthesis of chlorophyll were induced by ROS in "Huangjinya" leaves, but the sub-structure of magnesium chelatase was not influenced, although its activity was enhanced. Comprehensively, we proposed that a feedback mechanism existed in weakened carbon metabolism, where the pathway of chlorophyll b biosynthesis was positively regulated as a result of increased expression of Mg chelatase proteins and decreased expression of chlorophyll b reductase proteins.

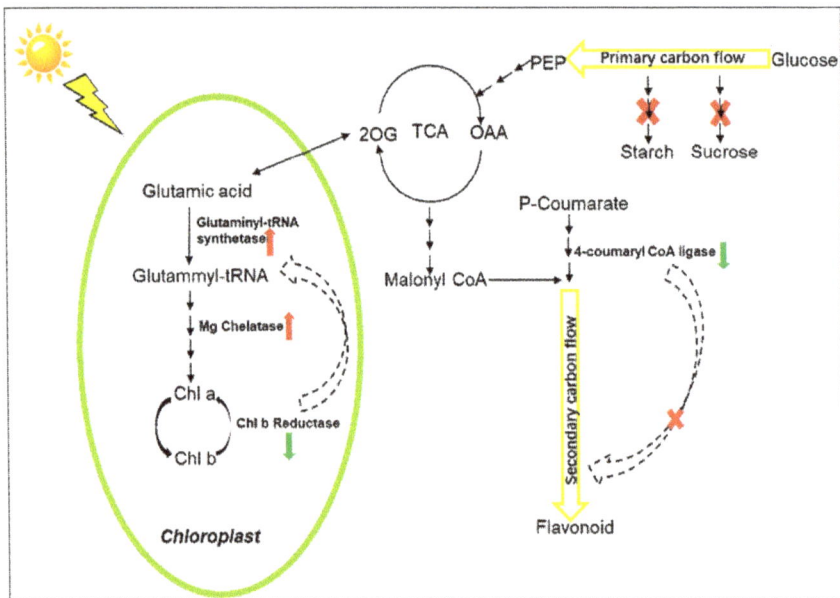

Figure 5. A simplified schematic presentation of the weakening of carbon metabolism in the chlorotic mutation. Red arrows indicate up-regulated expression of the protein. Green arrows indicate down-regulated expression of the protein. Black solid arrows indicate metabolic pathways. Black dotted arrows indicate regulatory relationships between metabolites and metabolites or metabolites and metabolic pathways. Red cross indicates inhibition of related metabolic pathways.

Flavonoids are important secondary metabolites of carbon metabolism in the tea plant, and have several physiological functions. Flavonoids provide plants with vibrant pigmentation, which protects the plants from UV-B radiation and helps attract pollinators as well as seed dispersers [31]. Because photosynthetic carbon assimilation is severely inhibited in chlorotic leaves, flavonoid biosynthesis is accordingly reduced [32]. Similar to the results of Zhang et al. [2], in this study, the expression levels of PAL and 4CL proteins (i.e., the rate-limiting enzymes of the flavonoid biosynthesis pathway) were down-regulated by 0.70- and 0.75-fold in chlorotic leaves (Table 3). Additionally, changes in the gene expression of 4CL confirmed the result of 4CL protein expression, thereby inhibiting the accumulation of flavonoids. A previous study showed that most of flavonoids were accumulated on the leaf surface to protect structure and organization in plants from UV-B radiation damage [33], while in this study, our results indicated that abnormally developed chloroplast inhibited the accumulation of chlorophyll and flavonoids because few carbon skeletons were provided as a result of weakening of carbon metabolism, and the reduction of flavonoids contents was unfavorable for the chlorotic mutant to protect against UV-B radiation damage.

4. Materials and Methods

4.1. Plant Material and Shading Treatment

"Huangjinya" (*Camellia sinensis* (L.) *O. Kuntze* cv.) is a light-sensitive albino tea variety, which displays yellow shoots under strong light condition. The low levels of flavonoids in "Huangjinya" reveals that the metabolism of flavonoids has great correlation to strong light stress [1,2]. A mutant of "Huangjinya" was officially released in Zhejiang province in 2008 and specimens were obtained free of charge from the owner of the mutant Deshi Tea Plantation, Yuyao, Zhejiang province [34]. Then, the experimental material was planted in pots at the Tea Research Institute, Chinese Academy of Agricultural Sciences (TRI, CAAS), Hangzhou, China. In March 2014, 60 pots of tea plants with uniform young shoots, with one bud and one leaf, were selected for the experiment. Half of the pots were treated with high-density polyethylene tape with two-pin net (60% sun-shading, 320–800 $\mu mol \cdot m^{-2} \cdot s^{-1}$), and the remaining half were exposed to full sunlight (800–2000 $\mu mol \cdot m^{-2} \cdot s^{-1}$) for ten days. Randomly selected samples of young shoots, with one bud and two leaves, were harvested, immediately frozen in liquid nitrogen, and stored in a $-70\,^{\circ}C$ ultra-refrigerator. Sampling was repeated six times from shaded and unshaded plants.

4.2. Electron Microscope Analysis

Transmission electron microscope (TEM, Hitachi Ltd., Tokyo, Japan) was used to observe the ultrastructure of chlorotic leaves. Leaf samples (approximately 1 mm^2) were fixed in 2.5% glutaraldehyde solution overnight at 4 °C. Ultrathin sections of fixed leaves were cut, stained, and viewed under a JEM-1230 transmission electron microscope (Nippon Tekno, Tokyo, Japan) at an accelerating voltage of 80 kV as described previously [1].

4.3. Leaf Gas Exchange Measurement

The fifth leaf of potted chlorotic mutants and wild-type tea plants was subjected to gas exchange analysis. A Li-Cor 6400 portable photosynthesis system (Li-Cor Inc., Lincoln, NE, USA) with a built-in light source set at 1000 μmol photons$\cdot m^{-2} \cdot s^{-1}$ was used to determine the net photosynthesis and stomatal conductance. All measurements were carried out between 09:00 a.m. and 11:00 a.m., with the leaf temperature adjusted to 25 °C.

4.4. Protein Extraction, iTRAQ Labeling, Data Acquisition, and Processing

Samples used for proteomic analysis were the same as those used for transcriptome analysis, and consisted of two biological replicates of shaded and unshaded leaves. All of the steps, including protein extraction, iTRAQ labeling of protein samples, liquid chromatography electrospray ionization tandem mass spectrometry (LC-ESI-MS/MS) analysis based on Q Exactive, mass spectrometer data analysis, and functional annotation of proteins, were performed as described previously [35]. Protein identification was performed using the Mascot search engine (Matrix Science, London, UK; version 2.3.02) against a database containing 133,175 sequences.

4.5. Quantitative Real-Time PCR (qRT-PCR) Analysis

Total RNA was isolated using an RNA plant plus kit (Tiangen, China). cDNA was synthesized using a Prime Script TM RT reagent Kit (TaKaRa, Biotechnology Co., Ltd., Dalian, China). qRT-PCR was performed on the Applied Biosystems 7300 machine (Carlsbad, CA, USA). Primer pairs used for qRT-PCR are shown in Supplementary Table S4, and GAPDH was used as the reference gene. For each target gene, triplicate reactions were performed. Relative transcript levels were calculated against that of the internal control (GAPDH) according to the equation $2^{-\Delta\Delta Ct}$. All data are shown as mean $\pm$ standard deviation (SD) ($n = 3$).

4.6. Bioinformatics Analysis

Functional analysis of the identified proteins was conducted using gene ontology (GO) annotation (Available online: http://www.geneontology.org/), and proteins were categorized according to their biological process, cellular components, and molecular function. The differentially accumulated proteins were further classified into the Clusters of Orthologous Groups of proteins database (Available online: http://www.ncbi.nlm.nih.gov/COG/) and Kyoto Encyclopedia of Gene and Genomes (KEGG) database (Available online: https://www.kegg.jp/kegg/pathway.html). GO and pathway enrichment analyses were performed to determine the functional sub-categories and metabolic pathways in which the differentially accumulated proteins showed significant enrichment. Cluster analysis of differentially accumulated proteins was performed using Cluster 3.0 (Stanford University, California, USA).

Data of integrated transcriptomic analysis reported previously [3] were deposited in the Sequence Read Archive (SRA) database (Available online: https://trace.ncbi.nlm.nih.gov/Traces/sra/) of the National Center for Biotechnology Information (NCBI) under the accession number SRP072792.

5. Conclusions

In this study, 2110 proteins were identified in "Huangjinya" leaves, the expression levels of 19 of which changed significantly, correlating with RNA expression. Differential protein expression analysis indicated that primary carbon metabolism (i.e., carbohydrate synthesis and transport) was inhibited in chlorotic tea leaves. The differentially expressed genes and proteins combined with photosynthetic phenotypic data suggested that 4-coumarate-CoA ligase (4CL) had a major effect on repressing flavonoid metabolism (secondary carbon metabolism), and abnormal developmental chloroplast inhibited the accumulation of chlorophyll and flavonoids because few carbon skeletons were provided as a result of the weakened primary carbon metabolism. Additionally, a positive feedback mechanism was verified at the protein level (Mg chelatase and chlorophyll b reductase) in the chlorophyll biosynthetic pathway, which might effectively promote the accumulation of chlorophyll b in response to the demand for this pigment in the cells of chlorotic tea leaves in weakened carbon metabolism.

Supplementary Materials: Supplementary materials can be found at http://www.mdpi.com/1422-0067/19/12/3943/s1.

Author Contributions: Q.Z., Y.S., M.L., K.F. and J.R. conceived and designed the experiments; F.D. performed the experiments; Q.Z. analyzed the data; F.D. and Q.Z. wrote the manuscript; Q.Z. and J.R. revised the manuscript critically. All the authors read and approved the final manuscript.

Funding: The work was financial supported by the Key Laboratory of Biology, Genetics and Breeding of Special Economic Animals & Plants, Ministry of Agriculture, China (Y2018PT14), the Earmarked Fund for China Agriculture Research System Ministry of Agriculture of China (CARS 19) and the Chinese Academy of Agricultural Sciences Innovation Project (CAAS-ASTIP-2017-TRICAAS).

Conflicts of Interest: The authors declare that they have no conflict of interest.

References

1. Li, N.; Yang, Y.; Ye, J.; Lu, J.; Zheng, X.; Liang, Y. Effects of sunlight on gene expression and chemical composition of light-sensitive albino tea plant. *Plant Growth Regul.* **2016**, *78*, 253–262. [CrossRef]
2. Zhang, Q.; Liu, M.; Ruan, J. Metabolomics analysis reveals the metabolic and functional roles of flavonoids in light-sensitive tea leaves. *BMC Plant Biol.* **2017**, *17*, 64. [CrossRef] [PubMed]
3. Zhang, Q.; Liu, M.; Ruan, J. Integrated transcriptome and metabolic analyses reveals novel insights into free amino acid metabolism in Huangjinya tea cultivar. *Front. Plant Sci.* **2017**, *8*, 291. [CrossRef] [PubMed]
4. Masakazu, S.; Harumi, E.; Akira, O.; Daisaku, O.; Kazunori, S.; Takushi, H.; Hitoshi, S.; Miyako, K.; Atsushi, F.; Kazuki, S.; et al. Integrated analysis of transcriptome and metabolome of Arabidopsis albino or pale green mutants with disrupted nuclear-encoded chloroplast proteins. *Plant Mol. Biol.* **2014**, *85*, 411–428.
5. Shi, D.Y.; Liu, Z.X.; Jin, W.W. Biosynthesis, catabolism and related signal regulations of plant chlorophyll. *Hereditas* **2009**, *31*, 698–704. [CrossRef] [PubMed]
6. Beale, S.I. Green genes gleaned. *Trends Plant Sci.* **2005**, *10*, 309–312. [CrossRef] [PubMed]
7. Killough, D.T.; Horlacher, W.R. The inheritance of virescent yellow and red plant colors in cotton. *Genetics* **1993**, *18*, 329–333.
8. Wu, Z.M.; Zhang, X.; He, B.; Diao, L.P.; Sheng, S.L.; Wang, J.L.; Guo, X.P.; Su, N.; Wang, L.F.; Jiang, L.; et al. A chlorophyll-deficient rice mutant with impaired chlorophyllide esterification in chlorophyll biosynthesis. *Plant Physiol.* **2007**, *145*, 29–40. [CrossRef]
9. Chen, X.; Zhang, W.; Xie, Y.J.; Lu, W.; Zhang, R.X. Comparative proteomics of thylakoid membrane from a chlorophyll b-less rice mutant and its wild type. *Plant Sci.* **2007**, *173*, 397–407. [CrossRef]
10. Wang, L.; Cao, H.L.; Chen, C.S.; Yue, C.; Hao, X.Y.; Yang, Y.J.; Wang, X.C. Complementary transcriptomic and proteomic analyses of a chlorophyll-deficient tea plant cultivar reveal multiple metabolic pathway changes. *J. Proteom.* **2016**, *130*, 160–169. [CrossRef]
11. Li, Q.; Huang, J.; Liu, S.; Li, J.; Yang, X.; Liu, Y.; Liu, Z. Proteomic analysis of young leaves at three developmental stages in an albino tea cultiva. *Proteome Sci.* **2011**, *9*, 44. [CrossRef] [PubMed]
12. Zhou, Q.; Chen, Z.; Lee, J.; Li, X.; Sun, W. Proteomic analysis of tea plants (*Camellia sinensis*) with purple young shoots during leaf development. *PLoS ONE* **2017**, *12*, e0177816. [CrossRef] [PubMed]
13. Kosová, K.; Vítámvás, P.; Prášil, I.T. The role of dehydrins in plant response to cold. *Biol. Plant.* **2007**, *51*, 601–617. [CrossRef]
14. Chu, P.; Yan, G.X.; Yang, Q.; Zhai, L.N.; Zhang, C.; Zhang, F.Q.; Guan, R.Z. iTRAQ-based quantitative proteomics analysis of Brassica napus leaves reveals pathways associated with chlorophyll deficiency. *J. Proteom.* **2015**, *113*, 244–259. [CrossRef] [PubMed]
15. Toyoda, K.; Ishii, M.; Arai, H. Function of three RuBisCO enzymes under different CO_2 conditions in Hydrogenovibrio marinus. *J. Biosci. Bioeng.* **2018**, *126*, 730–735. [CrossRef] [PubMed]
16. Motohashi, R.; Rödiger, A.; Agne, B.; Baerenfaller, K.; Baginsky, S. Common and specific protein accumulation patterns in different albino/pale-green mutants reveals regulon organization at the proteome level. *Plant Physiol.* **2012**, *160*, 2189–2201. [CrossRef] [PubMed]
17. Lin, Y.X.; Gu, X.X.; Tang, H.R. Characteristics and biological functions of glutathione reductase in plants. *J. Biochem. Mol. Biol.* **2013**, *29*, 534–542.
18. Kornyeyev, D.; Logan, B.A.; Allen, R.D.; Holaday, A.S. Effect of chloroplastic overproduction of ascorbate peroxidase on photosynthesis and photoprotection in cotton leaves subjected to low temperature photoinhibition. *Plant Sci.* **2003**, *165*, 1033–1041. [CrossRef]
19. Du, X.M.; Yin, W.X.; Zhao, Y.X.; Zhang, H. The production and scavenging of reactive oxygen species in plants. *Chin. J. Biotechnol.* **2001**, *17*, 121–125.
20. Song, L.B.; Ma, Q.P.; Zou, Z.W.; Sun, K.; Yao, Y.T.; Tao, J.H.; Kaleri, N.A.; Li, X.H. Molecular link between leaf coloration and gene expression of flavonoid and carotenoid biosynthesis in Camellia sinensis cultivar 'Huangjinya'. *Front. Plant Sci.* **2017**, *8*, 803. [CrossRef]
21. Sato, Y.; Morita, R.; Katsuma, S.; Nishimura, M.; Tanaka, A.; Kusaba, M. Two short-chain dehydrogenase/reductases, NON-YELLOW COLORING 1 and NYC1-LIKE, are required for chlorophyll b and light-harvesting complex II degradation during senescence in rice. *Plant J.* **2009**, *57*, 120–131. [CrossRef] [PubMed]

22. Lei, T.; Zhong, G.M. Degradation pathway of plant chlorophyll and its molecular regulation. *Plant Physiol.* **2011**, *10*, 936–942.

23. Zhang, M.S.; Tan, F. Relationship between ratio of chlorophyll a and b under water stress and drought resistance of different sweet potato varieties. *Seed* **2001**, *20*, 23–25.

24. Sang, Z.Y.; Ma, L.Y.; Chen, F.J. Growth and physiological characteristics of Magnolia wufengensis seedlings under drought stress. *Acta Bot. Boreali-Occidenalia Sin.* **2011**, *31*, 109–115.

25. Huang, C.H.; He, W.L.; Guo, J.K.; Chang, X.X.; Su, P.X.; Zhang, L.X. Increased sensitivity to salt stress in an ascorbate-deficient Arabidopsis mutant. *J. Exp. Bot.* **2005**, *56*, 3041–3049. [CrossRef] [PubMed]

26. Yoo, S.C.; Cho, S.H.; Sugimoto, H.; Li, J.J.; Kusumi, K.S.; Koh, H.J.; Iba, K.; Paek, N.C. Rice virescent3 and stripe1 encoding the large and small subunits of ribonucleotide reductase are required for chloroplast biogenesis during early leaf development. *Plant Physiol.* **2009**, *150*, 388–401. [CrossRef] [PubMed]

27. Zhang, D.; Chang, E.J.; Yu, X.X.; Chen, Y.H.; Yang, Q.S.; Cao, Y.T.; Li, X.K.; Wang, Y.H.; Fu, A.G.; Xu, M. Molecular Characterization of Magnesium Chelatase in Soybean [Glycine max (L.) Merr.]. *Front. Plant Sci.* **2018**, *9*, 720. [CrossRef] [PubMed]

28. Masuda, T. Recent overview of the Mg branch of the tetrapyrrole biosynthesis leading to chlorophylls. *Photosynth. Res.* **2008**, *96*, 121–143. [CrossRef] [PubMed]

29. Walker, J.C.; Willows, D.R. Mechanism and regulation of Mg-chelatase. *Biochem. J.* **1997**, *327*, 321–333. [CrossRef] [PubMed]

30. Müller, A.H.; Sawicki, A.; Zhou, S.; Tabrizi, S.T.; Luo, M.; Hansson, M.; Willows, R.D. Inducing the oxidative stress response in Escherichia coli improves the quality of a recombinant protein: Magnesium chelatase ChlH. *Protein Expr. Purif.* **2014**, *101*, 61–67. [CrossRef] [PubMed]

31. Li, S. Transcriptional control of flavonoid biosynthesis: Fine-tuning of the MYB-bHLH-WD40 (MBW) complex. *Plant Signal. Behav.* **2014**, *9*, e27522. [CrossRef] [PubMed]

32. Punyasiri, P.A.N.; Abeysinghe, I.S.B.; Kumar, V.; Treutter, D.; Duy, D.; Gosch, C.; Martens, S.; Forkmann, G.; Fischer, T.C. Flavonoid biosynthesis in the tea plant Camellia sinensis: Properties of enzymes of the prominent epicatechin and catechin pathways. *Arch. Biochem. Biophys.* **2004**, *431*, 22–30. [CrossRef] [PubMed]

33. Liang, B.; Zhou, Q. Effect of enhanced UV-B radiation on plant flavonoids. *Chin. J. Eco-Agric.* **2007**, *3*, 048.

34. Wang, K.R.; Li, M.; Liang, Y.R.; Zhang, L.J.; Shen, L.M.; Wang, S.B. Study on the Breeding of New Tea Tree Varieties, 'Huangjinya'. *Chin. Tea* **2008**, *4*, 21–23. (In Chinese)

35. Zhao, P.; Liu, P.; Shao, J.; Li, C.; Wang, B.; Guo, X.; Yan, B.; Xia, Y.; Peng, M. Analysis of different strategies adapted by two cassava cultivars in response to drought stress: Ensuring survival or continuing growth. *J. Exp. Bot.* **2015**, *66*, 1477–1488. [CrossRef] [PubMed]

International Journal of
Molecular Sciences

MDPI

Article

Proteomics Reveal the Profiles of Color Change in *Brunfelsia acuminata* Flowers

Min Li, Yueting Sun, Xiaocao Lu, Biswojit Debnath, Sangeeta Mitra and Dongliang Qiu *

College of horticulture, Fujian Agriculture and Forestry University, Fuzhou 350002, China;
liminzyl@sina.com (M.L.); yuetingsun@126.com (Y.S.); xc531599541@126.com (X.L.);
biswo26765@yahoo.com (B.D.); sangeeta.dae@hotmail.com (S.M.)
* Correspondence: qiudl1970@fafu.edu.cn; Tel.: +86-591-8378-9281

Received: 7 March 2019; Accepted: 21 April 2019; Published: 23 April 2019

Abstract: *Brunfelsia acuminata* is a popular ornamental plant with different colors resulted from the rapid change of color after blooming. The petals at day one (purple), day three (white and purple) and day five (white) were used to analyze the reason of flower color change by a comparative proteomics approach, gas chromatography coupled to a time-of-flight mass analyzer (GC-TOF-MS) and quantitative real-time PCR (qRT-PCR). The results showed that the 52 identified proteins were classified into eight functional groups, 6% of which were related to the anthocyanin metabolic pathway. The expression levels of all anthocyanin proteins from the first day to fifth day were remarkably down-regulated, which was consistent with the changing patterns of the key genes (*CHS*, *CHI* and *F3'5'H*) in petals. Simultaneously, the main floral volatile components including Linalool and 2-Hexenal (E) were identified, and the contents of 2-Hexenal at day five increased dramatically. Moreover, the content of flavonoids and total phenolic increased at day five. The majority of the proteins associated with stress defense and senescence proteins were up-regulated and the activities of peroxidase (POD), superoxide dismutase (SOD) and catalase (CAT) in the petals at day five were significantly higher than others. It was concluded that the competition in the precursors of metabolic pathways occurs and causes the flow of metabolite to the pathways of floral scent and lignin derived from the shikimate pathway or degrade into others. Therefore, the anthocyanin content significantly decreased, and the petal color changed from deep purple to white.

Keywords: *B. acuminata* petals; MALDI-TOF/TOF; GC-TOF-MS; qRT-PCR; differential proteins

1. Introduction

B. acuminata is an evergreen shrub native to Brazil. The date of flower blooming is from April to May (South of China). The color change in plants is very obvious, with a high ornamental value [1]. The color of the petal begins to change gradually from purple to white after two to three days of opening. The flower color is affected by different kinds of pigments, and the change is related to the decline of the anthocyanin content [2]. In general, the pigment of the petals is distributed in the vacuoles of epidermal cells, but is also present in other tissues, such as the cell wall [3], palisade, and chromoplasts [4]. The anthocyanin component in the *B. acuminata* petal includes in malvidin-3-*O*-glucoside chloride, petunidin-3-glucoside and delphinidin-3- glucoside, which are part of the polyphenolic [5]. They are water-soluble plant pigments that are susceptible to change [4] by pollinators [6], temperature, light [7], solvents, chemical structures, and pH changes [8]. Phenolic compounds are a large class of plant secondary metabolites including phenolic acids, tannins, lignans, coumarins, and flavonoids, which are responsible for the color of fruits and substrates for enzymatic browning [9]. Previous studies have shown that phenolic compounds play a role in the antioxidant activity of the flower [10,11]. Flavonoids are a biologically important group of phenolics in plants [12]. Secondary compounds are important in plants, especially in anthocyanin and as a stress and defense substance [13]. In addition, in the process of flower growth and development, the appearance of the petal changes with the alternation of the internal structure. Many studies have observed change in the structure and ultrastructure of the petals [14,15].

The biosynthetic pathway of anthocyanin has been the subject of much research and the associated biosynthetic and regulatory genes such as chalcone synthase (*CHS*), chalcone isomerase (*CHI*), and flavonoid 3'5'-hydroxylase (*F3'5'H*) are well defined in *Brunfelsia* plants [1,16,17]. These genes are also studied in many plants such as *P. hybrida 'Mirage Rose'* [18], *Lilium* spp. cultivar 'Dizzy' [19]. The process of anthocyanin degradation in *B.calycina* was dependent on de novo synthesis of mRNAs and proteins of peroxidase (BcPrx01) [20]. It is speculated that POD is an enzyme that causes the degradation of anthocyanin of *B. calycina* petals by Oren-Shamir [2].

Proteome approaches are a powerful tool and can assist the investigation of comprehensive protein expression profiles in specific biological responses [21]. Two dimensional electrophoresis (2-DE) is one of the mainstream methods for floral proteomics, and has been widely applied to flower organs, such as the androecium, the gynoecium, and petals [22].

However, the detailed knowledge of the degradation of anthocyanin is poorly understood in the *B. acuminata* petal. Therefore, the aim of this study was to explore the profiles of color changes of the flower in *B. acuminata* through proteomics analysis.

2. Results

2.1. Changes in Corolla Diameter, Content of Water, Anthocyanin, Flavonoid, Total Phenolic, and Ultra-Structure during Flower Development

The corolla diameter of *B. acuminata* petals increased, and its water content at day 3 went up significantly as compared with day 1 and it kept stable in the petals at day 5 (Figure 1A). Simultaneously, the anthocyanin content reduced significantly (Figure 1B) in development of *B. acuminata* petals. The content accumulated to a maximum level at day 1 and degraded to a minimum level at day 5, whereas, the flavonoid content in the petals increased significantly from day 1 to day 3 (Figure 1C). Total phenolic content in the petals at day 3 was significantly higher than that in day 1. No significant difference was observed in between the contents at day 3 and at day 5 (Figure 1D).

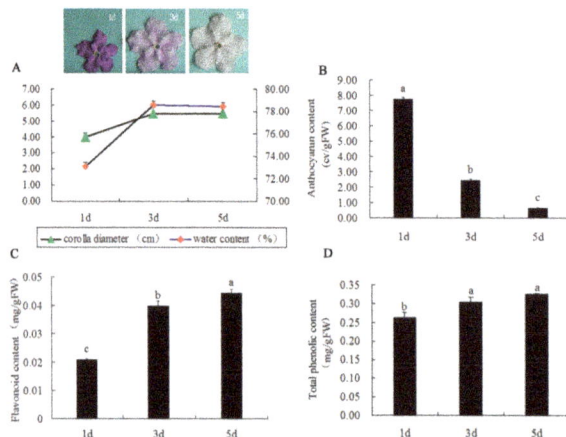

Figure 1. Changes of corolla diameter, water content, anthocyanin content, flavonoid content, and total phenolic content of *B. acuminata* petals. The different petals during anthocyanin degradation and the corolla diameter and water content changes after flower opening in petals (**A**); the contents of total anthocyanin (**B**), flavonoid content (**C**), and total phenolic content (**D**) in the petals. Note: Values (mean ± SD) were determined from three independent experiments (*n* = 3). Different letters above the bars indicate a significant difference at *p* < 0.05.

From the day 1 to day 5, the cell volume of the petal expanded, and the petals grew rapidly (Figure 2A). The epidermal cell at day 1 was small purple and compact. With the expansion of the cell, its color became shallow at day 3, and became white at day 5 (Figure 2A,B). Unknown black pigment grain in the vacuoles of the petals was obviously observed at day 1, but the grain became smaller at day 3 and disappeared at day 5 (Figure 2C).

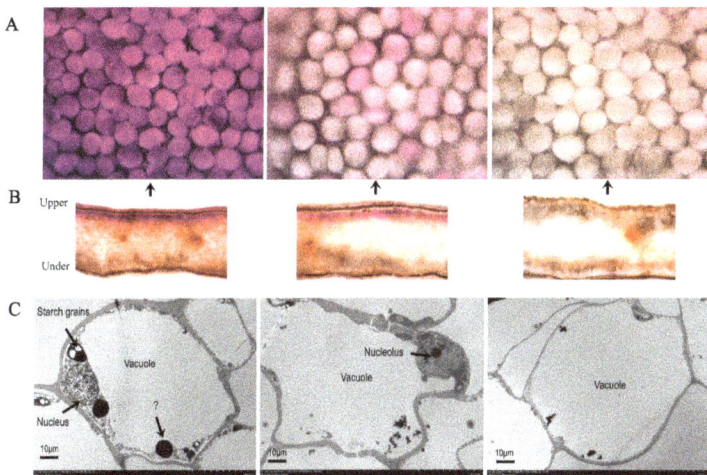

Figure 2. The pigment distribution and ultrastructure of epidermal cells in different days. Longitudinal section of the petal epidermis (**A**) and the petal upper epidermis under a microscope magnified 20 times (**B**); the petals of ultrastructure in upper epidermis of 10 μm (**C**).

2.2. Protein Identification and Functional Classification

As shown in Supplementary Materials, 60 spots of the differentially abundant proteins were screened out. Figure 1 and 52 proteins were identified in *B. acuminata* petals by matrix-assisted laser desorption/ionization time-of-flight mass spectrometry (MALDI-TOF/TOF-MS) (Table 1). Among of which, 35 protein spots were significantly up-regulated and 17 protein spots were down-regulated. Of the 52 proteins successfully identified, some were identified as the same protein such as adenosine succinate syntheses (spot 3, spot 4), mitochondrial ATP synthase beta subunit (spot 32, spot 45), and anthocyanin 5-*O*-glucosyltransferase (5-GT, spot 17, spot 18). We found that 35 proteins were significantly up-regulated at day 5. Many of them had 2-fold or more in abundance. The six proteins (spot 25, spot 40, spot 46, spot 53, spot 27, spot 47) appeared at day 3 (Figure S2). Among of which, three new proteins (spot 27, spot 46, spot 47) showed a sudden increase in expression at day 3.

The visualization of differential abundance of the identified 52 proteins was showed in Figure 3. It can be classified into eight groups: Carbohydrate and energy metabolism pathway (20%), anthocyanin metabolic pathway (6%), lignin biosynthesis pathway (4%), stress defense and senescence proteins (34%), floral scent metabolic pathway (10%), signaling and photosynthesis (8%), cytoskeleton and chaperone (12%), and unclassified protein (6%) (Figure 4). It is interesting to find that all the proteins involved in floral scent metabolic pathway, lignin biosynthesis pathways and cytoskeleton and chaperon were up-regulated, while all anthocyanin metabolic proteins were down-regulated (Figure 4).

Table 1. Identification of proteins from B. acuminata petals using MALDI-TOF/TOF-MS.

Spot No [a]	Protein Name	Species	Accession No.	MW (kDa)/pI [b]	Score	Cov [c]	Fold Changes [e]	
Carbohydrate and Energy Metabolism Pathway								
D5 [d]	malate dehydrogenase, cytoplasmic-like	Solanum lycopersicum	gi|460404529	35.361/5.91	235	25%	-2.33	-1.55
D10	2,3-bisphosphoglycerate-independent phosphoglycerate mutase-like	Glycine max	gi|356568270	60.799/5.58	247	11%	-1.92	-1.32
D19	3-isopropylmalate dehydrogenase, chloroplastic-like	Solanum lycopersicum	gi|460386440	43.396/6.05	232	14%	-2.06	-1.25
U23 [d]	ATP synthase beta subunit	Eleutherococcus senticosus	gi|343410685	39.790/5.77	227	35%	1.20	1.18
U24	soluble acid invertase 2	Orobanche ramosa	gi|294612072	61.628/5.24	133	19%	2.07	1.51
U29	vacuolar invertase 2	Gossypium hirsutum	gi|268526570	69.303/5.14	185	8%	2.16	1.98
U32	ATP synthase subunit beta, mitochondrial-like	Solanum lycopersicum	gi|460382474	59.825/5.94	869	45%	2.26	1.46
U45	ATP synthase subunit beta, mitochondrial-like	Solanum lycopersicum	gi|460382474	59.825/5.94	311	26%	3.00	2.71
U46	phosphoenolpyruvate carboxylase kinase 1	Clusia minor	gi|39842451	28.716/6.38	88	22%	13.09	10.67
U49	ATP synthase beta subunit	Eleutherococcus senticosus	gi|343410685	39.790/5.77	227	29%	2.40	2.26
U52	1,2-beta-fructan 1F-fructosyltransferase	Helianthus tuberosus	gi|3367690	69.214/5.02	129	12%	2.11	2.00
Anthocyanin Metabolic Pathway								
D14	Anthocyanin-O-methyl transferase	Solanum lycopersicum	gi|441433515	26.282/5.69	90	19%	-7.04	-1.04
D17	Anthocyanin-5-O-glucosyltransferase	Petunia x hybrida	gi|6683052	52.130/5.07	98	13%	-13.79	-1.04
D18	Anthocyanin-5-O-glucosyltransferase	Petunia x hybrida	gi|6683052	52.130/5.07	100	7%	-3.94	-1.36
Lignin Biosynthesis Pathway								
U41	caffeate-O-methyltransferase	Liquidambar styraciflua	gi|5732000	39.944/5.69	80	22%	4.70	3.77
U47	caffeoyl-CoA O-methyltransferase	Broussonetia papyrifera	gi|46394464	27.701/5.31	412	40%	5.25	3.44
Stress Defense and Senescence Proteins								
D1	polyphenol oxidase E, chloroplastic-like isoform 2	Solanum lycopersicum	gi|460401035	66.181/6.36	83	18%	-2.66	-1.02
D2	polyphenol oxidase	Nicotiana tabacum	gi|92919068	57.748/5.92	126	13%	-2.10	-1.02
D6	proteasome subunit alpha type-6-like	Solanum lycopersicum	gi|460412613	27.301/6.11	102	32%	-4.08	-1.37
U8	glutathione S-transferase	Solanum commersonii	gi|148616162	23.843/5.98	82	30%	1.34	1.53
D13	lactoylglutathione lyase-like	Solanum lycopersicum	gi|460373807	32.839/5.95	283	32%	-2.00	-1.07
U25	ASR1 protein	Solanum ochranthum	gi|321155417	12.547/6.48	307	34%	31.33	23.55
U28	CLPC	Theobroma cacao	gi|508775360	102.257/6.36	373	27%	3.45	1.12
U35	S-adenosyl methionine synthase-like	Solanum tuberosum	gi|78191442	43.189/5.52	466	40%	2.11	2.01
U43	glutathione S-transferase L3-like	Cicer arietinum	gi|502121795	27.092/5.79	91	15%	1.59	1.51
U44	putative glutathione S-transferase zeta-class 2	Brassica napus	gi|330250478	25.336/5.53	82	24%	2.14	1.49
D3	Adenylosuccinate synthetase, chloroplastic	Solanum lycopersicum	gi|460407669	55.408/7.55	87	14%	-2.39	-1.04
D4	Adenylosuccinate synthetase1, chloroplastic-like	Solanum lycopersicum	gi|460407669	55.408/7.55	93	18%	-2.11	-1.38
D15	Aspartic proteinase	Theobroma cacao	gi|508719874	54.428/5.56	55	3%	-7.85	-1.08
D16	endochitinase precursor	Humulus lupulus	gi|4960049	33.508/7.42	109	4%	-3.09	-1.10
U20	Small ubiquitin-related modifier 1	Arabidopsis thaliana	gi|21542462	10.969/4.91	77	48%	9.19	1.68
U26	annexin p34-like protein-like	Solanum tuberosum	gi|81074127	35.909/5.54	174	24%	2.01	1.59
U27	Glutamine synthetase 1,4	Theobroma cacao	gi|508707247	39.098/6.02	160	18%	5.21	2.86
U51	heat shock 70 protein	Spinacia oleracea	gi|2773050	76.094/5.19	554	25%	2.02	1.07
U53	Plastid-lipid-associated protein, chloroplast precursor, putative	Ricinus communis	gi|223536371	34.979/4.84	148	24%	2.06	1.78
Floral scent Metabolic Pathway								
U33	1-hydroxy-2-methyl-2-(E)-butenyl 4-diphosphate reductase	Ipomoea batatas	gi|325557690	51.682/5.90	163	27%	2.58	2.20
U34	1-deoxy-D-xylulose-5-phosphate reductoisomerase	Solanum lycopersicum	gi|350537527	51.465/5.94	266	27%	2.01	1.49
U39	SAMT	Anthocercis littorea	gi|58201456	32.353/4.79	151	22%	2.11	1.62

Table 1. *Cont.*

Spot No [a]	Protein Name	Species	Accession No.	MW (kDa)/pI [b]	Score	Cov [c]	Fold Changes [e]	
U40	Putative S-adenosyl-L-methionine-Salicylic acid carboxyl methyltransferase	*Pisum sativum*	gi|37725949	40.552/5.17	78	19%	15.05	7.00
U42	putative S-adenosyl-L-methionine-Salicylic acid carboxyl methyltransferase	*Pisum sativum*	gi|37725949	40.552/5.17	52	5%	2.23	1.56
Signaling and Photosynthesis								
D11	inositol-3-phosphate synthase	*Solanum lycopersicum*	gi|460388681	56.526/5.45	189	23%	-2.79	-1.02
D12	inositol-3-phosphate synthase	*Solanum lycopersicum*	gi|460388681	56.526/5.45	318	23%	-2.84	-1.51
U22	14-3-3-like protein GF14 Psi	*Eutrema salsugineum*	gi|309952059	28.752/4.78	166	48%	2.07	1.81
U31	ruBisCO large subunit-binding protein subunit beta, chloroplastic-like	*Cicer arietinum*	gi|502125499	62.800/5.85	348	22%	2.15	1.26
Cytoskeleton and Chaperone								
U30	chaperonin CPN60-2, mitochondrial-like	*Solanum lycopersicum*	gi|460404682	61.521/5.51	175	18%	2.37	1.99
U36	beta-actin	*Zoysia japonica*	gi|284157810	41.697/5.23	459	35%	2.21	2.11
U37	actin 6	*Populus trichocarpa*	gi|222860713	40.592/5.05	166	27%	1.85	1.43
U38	actin	*Gossypium hirsutum*	gi|32186904	41.878/5.39	82	27%	2.5	2.25
U50	60-kDa chaperonin-60 alpha-polypeptide precursor, partial	*Brassica napus*	gi|289365	57.657/4.84	403	24%	2.61	1.93
Unclassified Protein								
D7	predicted protein	*Physcomitrella patens subsp. Patens*	gi|162667966	28.780/5.38	79	11%	-2.13	-1.11
D9	putative transcription factor BTF3-ike	*Solanum tuberosum*	gi|82623431	17.472/6.31	240	34%	-2.06	-1.12
U48	cp10-like proteinCP10	*Gossypium hirsutum*	gi|21780187	26.761/7.77	122	9%	1.30	1.73

[a] Spot number corresponds to the 2-DE gel in Figure S1; [b] theoretical molecular mass (MW) and isoelectric point (pI) of the homologous protein calculated with a tool available at NCBInr database.; [c] sequence coverage; [d] D down-regulated proteins, U up-regulated proteins; [e] ratio of protein levels compared to day 1 (left: Day 3/Day 1, right: Day 5/Day 1).

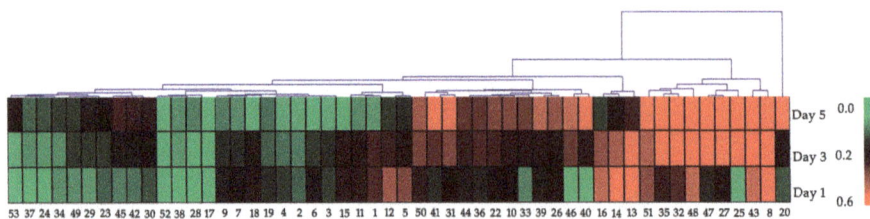

Figure 3. The heat map visualization of differential abundance of the identified 52 proteins in *B. acuminata* petals. The upregulated and downregulated proteins are indicated from red to green, respectively. The color scale is shown at the left of the cluster.

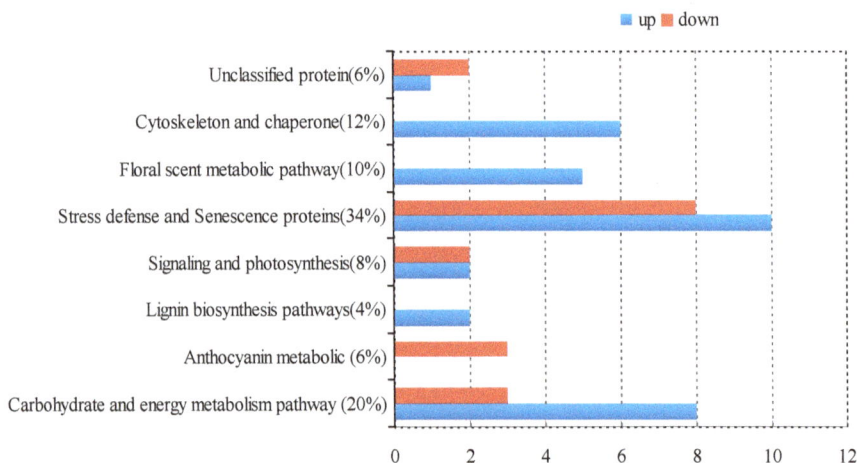

Figure 4. Functional classification and expression of identified proteins in *B. acuminata* petals.

2.3. The Expression Levels of Key Genes Encoding Anthocyanin Synthesis in Different Days

The key genes of anthocyanins biosynthesis genes encoding proteins, namely, chalcone isomerase (*CHI*), flavonoid 3'5'-hydroxylase (*F3'5'H*), and *CHS* were examined to ascertain whether the protein differential abundance levels correlated with their mRNA content. Figure 5 showed that the expression levels of *CHS* and *F3'5'H* were remarkably down-regulated in different days. These results were consistent with trends in changes in color-related proteins. *CHI* has an upward trend in the later stage at day 5, which is likely to participate in other metabolic pathways.

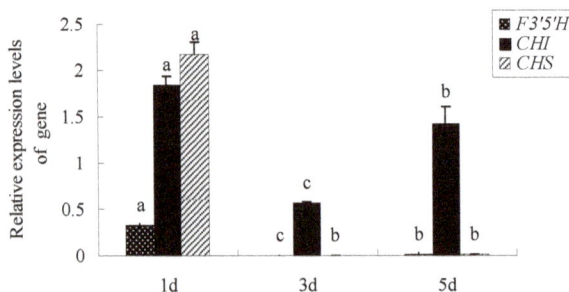

Figure 5. The expression levels of key genes of anthocyanin synthesis in different days. Values (mean ± SD) were determined from three independent experiments ($n = 3$). Different letters above the bars indicate a significant difference at $p < 0.05$.

2.4. Analysis of Volatiles

A total of 52 kinds of volatile components detected in different days of *B. acuminata* by GC-TOF-MS were classified into included terpenes, alcohols, aldehydes, esters, and so forth. At day 1, day 3, and day 5, 32, 46, and 40 components were detected in petals, respectively (Figure 6, Table S2). In Table 2, eight kinds of terpenoid with similarity greater than 800 and relative content greater than 5% is listed, and other volatiles are shown in Table S2. The volatile components of petals at day 1 are mainly linalool, and benzaldehyde and at day 3 are mainly including linalool, 2-hexenal, (E)-, trans-Linalool oxide (furanoid), where relative content of the compound is more than 10%. The relatively high content of the components at day 5 was linalool, 1-hexanol, benzeneacetaldehyde, which were 19.24%, 12.53%, and 11.40%, respectively.

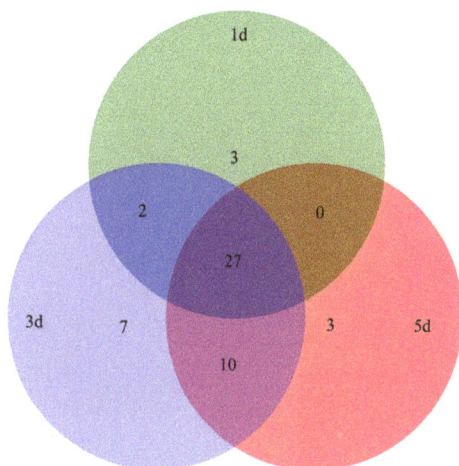

Figure 6. Volatile components were detected in different days of *B. acuminata* petals.

Table 2. Main terpenoids were detected in different days of *B. acuminata* petals.

Number	Compounds	Relative Contents (%)		
		1d	3d	5d
1	Linalool	37.59 ± 8.39	31.27 ± 2.12	19.24 ± 4.12
2	2-Hexenal, (E)-	44.92 ± 5.11	20.98 ± 1.29	—
3	trans-Linalool oxide (furanoid)	5.91 ± 1.47	11.61 ± 1.58	8.27 ± 1.56
4	(E)-4,8-Dimethylnona-1,3,7-triene	9.62 ± 5.28	9.81 ± 1.07	10.11 ± 6.31
5	2-Furanmethanol, 5-ethenyltetrahydro-à,à,5-trimethyl-, *cis*-	3.47 ± 0.42	9.34 ± 1.60	6.07 ± 1.37
6	1-Hexanol	5.22 ± 0.83	5.13 ± 1.25	12.53 ± 5.98
7	Benzeneacetaldehyde	0.67 ± 0.20	4.15 ± 1.70	11.40 ± 0.85
8	2-Hexenal	0.74 ± 0.44	0.45 ± 0.13	16.78 ± 1.43

2.5. SOD, CAT, POD Activity and Soluble Protein Content in Petals of B. acuminata

Superoxide dismutase (SOD) and catalase (CAT) activity in the petals at day 5 were significantly higher than those at day 1 (Figure 7A,B). However, the soluble protein content (Figure 7C) in the petals decreased significantly from day 1 to day 5. The content had no significant difference between day 3 and day 5. Interestingly, the peroxidase (POD) activity increased significantly from day 1 to day 5 during blooming (Figure 7D).

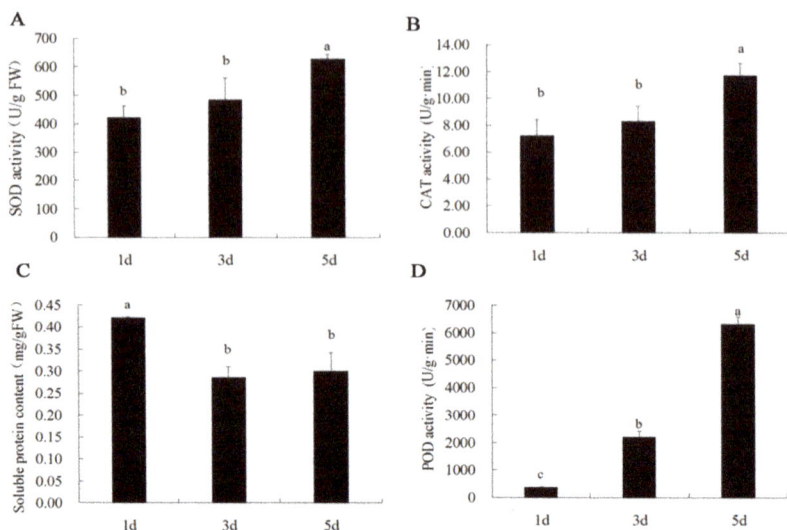

Figure 7. Activity of antioxidant-related enzymes, superoxide dismutase (SOD) (**A**), catalase (CAT) (**B**), soluble protein content (**C**), peroxidase (POD) (**D**), and in *B. acuminata* petals during blooming. Values (mean ± SD) were determined from three independent experiments ($n = 3$). Different letters above the bars indicate a significant difference at $p < 0.05$.

3. Discussion

3.1. Main Proteins and Genes Related to Anthocyanin Synthesis

The appearance of plant color is closely related to the content of anthocyanin. In most cases, the color change is due to the induction of anthocyanin synthesis, but the color change of *B. acuminata* petal during anthesis is the exact opposite of *Viola cornuta cv.* yesterday, today, and tomorrow [23]. In the process of *B. acuminata* flowering, anthocyanin biosynthesis in the early opening petal of *B. acuminata* is the most exuberant, resulting in the deepest color. The anthocyanin degraded gradually, resulting in a change of flower color from deep purple to white. The color change of the petal is a complex process, from flower formation to degradation, requiring the participation of many enzymes and genes.

The enzymatic degradation of anthocyanin in plant tissues can play an important role in the regulation of plant pigments. Two proteins namely, anthocyanin-5-*O*-glucosyltransferase (5-GT) and anthocyanin-*O*-methyl transferase (OMT) related to the metabolic pathway of anthocyanin were identified and analyzed in three different periods of discoloration of *B. acuminata* petals (Table 1, Figure 3). 5-GT (spot 17, spot 18) is an enzyme that forms anthocyanin-3,5-*O*-diglucoside from anthocyanin-3-*O*-glucoside, which is responsible for the modification of anthocyanins to more stable molecules in complexes for co-pigmentation, supposedly resulting in a purple hue [24]. In our study, the expression level of 5-GT was higher in purple petals (day 1, 1d) than that in white petals (day 5, 5d) (Figure 3). OMT (spot 14) is one of the key enzymes for anthocyanin modification and flower pigmentation [25]. Many OMT genes are involved in the formation of methylated anthocyanins [25]. OMT was down-regulated and may not be able to methylate to form anthocyanins. Furthermore, all anthocyanin metabolic proteins were down-regulated, and many anthocyanin-modifying enzymes involved in the anthocyanin synthesis pathway were not activated during the petals changing from purple to white.

CHS, CHI, and *F3'5'H* are important enzymes for the formation of flower colors [26]. *F3'5'H* is a key enzyme for the synthesis of blue-purple pigments [27]. The expression levels of *CHS* and *F3'5'H* are both down-regulated in the process of flowering, consistent with protein differential abundance trends (Figure 5, Table 1). This result was supported by our previous work that the striking color change from dark purple to pure white resulted from a decline in anthocyanin content of the petals and was preceded by a decrease in the expression of *BaCHS* [28]. However, *CHI* expression differed from them, which has an upward trend in the later stage at day 5. Chalcone isomerase (CHI) converts yellow chalcone to colorless naringenin, which expression level directly affects the accumulation of yellow chalcone, a colorless phenotype, and flavanol compounds [29,30]. Maybe *CHI* is involved in other metabolic pathways, and the expression level will decrease later.

3.2. Other Protein Associated with Anthocyanin Synthesis

The shikimate pathway is induced at later stages in the flower of *B. acuminata*, linking carbohydrate metabolism to the precursors for the synthesis of anthocyanins, benzenoids, and lignin [31,32] (Figure 8). As a class of floral substances in plants, terpenoids have similar synthetic pathway with anthocyanins and carotenoids [33]. In our study, we found that four proteins are terpenoid-related and up-regulated during the flowering. Methyl salicylate is a volatile plant component, but also an important substance in the defense mechanism of plants [34]. The methylation of salicylic acid is performed by salicylate carboxymethyltransferase (SAMT, spot 39) [34]. 1-hydroxy-2-methyl-2-(E)-butenyl 4-diphosphate reductase (IDS, spot 33), and 1-deoxy-D-xylulose-5-phosphate reductoisomerase (DXR, spot 34) were significantly up-regulated, indicating that terpenoids continue to increase from day 1 to day 5. The putative salicylic acid carboxyl methyltransferase (spot 40, spot 42) is a new protein related to floral scent and lignin pathway and is also up-regulated. In the process of petal scent release, the black pigmentation is reduced and finally disintegrated (Figure 2). It can be speculated that the function of the black pigmentations is a reservoir of aroma precursors or energy in *B. acuminata*. This black pigmentation is similar to the description of the epidermal cells of jasmine petals described by Zhang [35]. In the *B. acuminata* petals, the expression levels of caffeoyl O-methyltransferase (COMT, spot 41) and caffeoyl-CoA O-methyltransferase (CCoAOMT, spot 47) were significantly up-regulated. Both are involved in the synthesis of the volatile compounds in *Brunfelsia* [31]. A similar result has been reported that multifunctional CCoAOMTs play roles in catalyzing the 3′ or 3′-5′ O-methylation of their B ring of flavonoid substrates [36]. Glutathione S-transferase is associated with the formation of color, which can transport anthocyanins into vacuoles. 14-3-3-like protein GF14 Psi (spot 22) is associated with the shikimic acid pathway, which participates in the synthesis of aromatic compounds and indirectly affects the color changes [37]. GST and Glutamine synthetase 1, 4 (GS1, 4, spot 27) showed a significant upward trend in *B. acuminata* petals.

Lignin biosynthesis is the second metabolic pathway of the phenylpropane pathway branch, which is induced during petal expansion in the opening *Brunfelsia* flower [31]. Both anthocyanins and benzoic acid are derived from benzoic acid metabolic pathways, and there is a competitive effect between them [38]. Inhibition of other competitive biosynthetic pathways in the synthesis of floral substances can improve the synthesis of plant floral compounds such as anthocyanin biosynthetic pathway [38]. In our experiment, lignin synthesis enzymes and floral synthesis-related proteins were up-regulated, while anthocyanin synthesis-related proteins showed down-regulation, which may be the key reason to color changes. The dramatic changes in color are accompanied by the synthesis of aromatic phenol volatiles in the petal of *B. acuminata*. Anthocyanin degradation and the synthesis of aromatic benzene compounds may be pollinators' signals [1,31]. Further studies may reveal a network of *B. acuminata* and related species of volatile and pigment phenolic metabolism.

The anthocyanin biosynthetic pathway belongs to a branch of the flavonoid biosynthetic pathway. Based on flavanones, all other flavonoid-classes are generated, including isoflavones, flavanols, anthocyanidins, flavanols, and flavones [39]. The anthocyanin content dropped significantly (Figure 1B), while the contents of flavonoid and total phenolic in the petals of *B. acuminata* increased significantly from day 1 to day 3 (Figure 1C). Thus, we speculated that the competition in the precursors of metabolic pathways occurs and causes the metabolite flow to the pathways of floral scent and lignin both derived from the shikimate pathway (Figure 8).

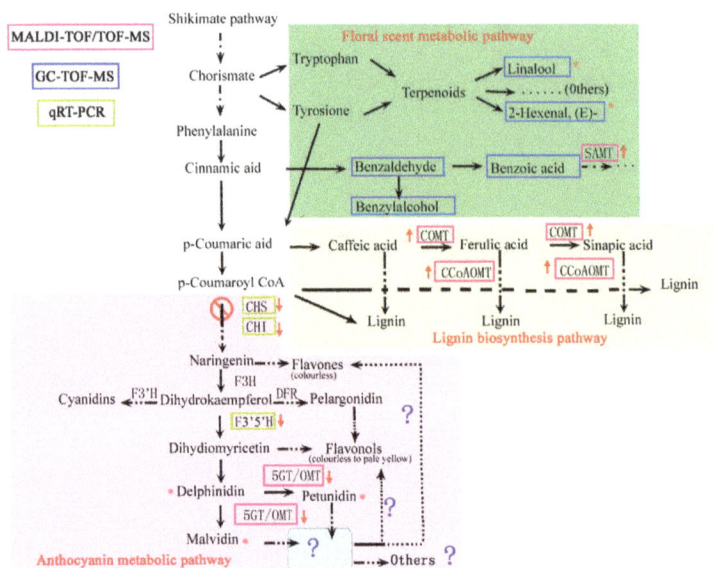

Figure 8. A putative metabolic map described the pathway including the floral scent metabolic pathway and lignin pathway during the degradation of anthocyanins in *B. acuminata* petals. The diagram shows the different pathways. The diagram summarizes the results of qRT-PCR, MALDI-TOF/TOF and GC-TOF-MS analyses. The dashed black arrows represent several consecutive enzymatic steps. The color of the box indicates the method by which they were identified. The red arrow represents up or down and the symbol with * in the picture is expressed as an important substance identified in the petals. Enzymes: SAMT, salicylic acid carboxyl methyltransferase; COMT, catechol O-methyl transferase; CCoA-OMT, caffeoyl-CoA O-methyl; OMT, Anthocyanin-*O*-methyl transferase. Genes: *CHS*, chalcone synthase; *CHI*, chalcone isomerase; *F3'5'H*, flavonoid 3'5'-hydroxylase.

3.3. The Proteins Associated with Other Metabolic Pathways

In plants, there are many enzymes involved in sucrose metabolism, among which invertase is one of the key enzymes involved in plant sucrose metabolism [40]. Vacuolar invertase 2 (spot 29) can be the one in vacuole hydrolyzing glucose and fructose to regulate the concentration of intracellular sucrose. In this study, the water content of *B. acuminata* petals at day 3 went up significantly and it kept stable in the petals at day 5 (Figure 1A,B), and the vacuolar invertase was up-regulated from day 1 to day 5, promoting the expansion and enlargement of the petals.

Soluble acid invertase 2 (spot 24) is also a key enzyme in the process of carbohydrate metabolism, mainly in the vacuole. Highly active soluble acid invertase is closely related to the growth of the young parts of the plant and the rapid expansion of the storage organs [41].

With the growth of *B. acuminata* petals, respiration will become faster to provide the energy and raw materials necessary for plant life activities. Malate dehydrogenase, cytoplasmic-like (MDH, spot 5) is present. In this study, the protein differential abundance of MDH decreased from purple to white in the petals (Figure 3). It is possible that the petals are about to enter the aging period and the petal respiration and energy metabolism begin to weaken on day 5 of bloom. The ATP synthase β subunit (spot 49) was up-regulated in the *B. acuminata* petals, which indirectly provided energy for petal growth and development.

It is worth noting that Phosphoenolpyruvate carboxylase kinase1 (PpcK1, spot 46) and Glutamine synthetase (GS, spot 27) showed a sudden increase in expression at day 3. Two novel proteins, ASR1 (spot 25) and plastid-lipid-associated (spot 53) are related to stress defense proteins, which also appeared and up-regulated at day 3 (Figure S2). Phosphoenolpyruvate carboxykinase (PEPCK) is only known to be located in the cytosol in flowering plants [42]. In CAM plants, it is an important metabolic regulated and Ca^{2+}-independent protein related to light, and the shift of kinase phosphorylation signal transduction pathway [43]. GS is a crucial enzyme in the network of N metabolism and also involved in N recycling in the plant [44]. ASR1 belongs to a family of hydrophilic proteins in responses to abiotic stresses as well as signaling molecules [45]. Plastid lipid-associated proteins are known to accumulate in fibrillar-type chromoplasts such as in leaf chloroplasts from *Solanaceae* plants under abiotic stress conditions [46] and involved in the pigment accumulation during fruit development [47]. The majority of these proteins associated with stress defense and senescence proteins were up-regulated indicating that the petal of *B. acuminata* at day 5 might be at the beginning of aging (Figure 4). This hypothesis is supported by the data that the activities of POD, SOD, and CAT in the petals at day 5 were significantly higher than those at day 3 and day 1 to prevent ageing (Figure 7) [48,49], and that all the protein involved in cytoskeleton and chaperon were up-regulated (Figure 4). The cytoskeleton constitutes the structural support of the living matter. Molecular chaperones are housekeeping factors of the cytoskeleton network [50].

In conclusion, the color change of *B. acuminata* flower is a complicated process involving numerous factors. The competition in the precursors of metabolic pathways occurs and causes the metabolite flow to the pathways of floral scent and lignin both derived from the shikimate pathway. Meanwhile, the decline of expression levels of *CHS*, *CHI*, and *F3'5'H* resulted in down-regulation of anthocyanin metabolic proteins. Therefore, the anthocyanin content decreased, and petal color change from deep purple to white (Figure 1). This work provides new proteomic and GC-TOF-MS insights of color change of flower in plants.

4. Materials and Methods

4.1. Plant Materials and Morphological Indicators

Flowers opened with purple petals (day 1,1d) that changed to light purple and white (day 3,3d), and pure white (day 5,5d) during a 5-day lifespan. *B. acuminata* petals in the different stages (1d, 3d, 5d) were collected from the greenhouse of Fujian Agriculture and Forestry University, Fuzhou, China (Figure 1A). The plants were grown at 20–25 °C/12–15 °C (day/night) temperature conditions, and 60%–80% humidity, and 14/10 h (day/night) with 100 $mol \cdot m^{-2} \cdot s^{-1}$ photosynthetically active radiation (PAR). Three replicate samples were collected at random from individual plants and immediately kept in liquid nitrogen, and stored at −80 °C for proteomic, quantitative real-time (qRT-PCR), and GC-TOF-MS analysis.

The petals of *B. acuminata* were randomly selected to observe their morphological changes and measure their flower diameter and water content. Vernier caliper was used to measure corolla diameter. Petal water content was determined as the percentage of total petal weight ((FW-DW)/FW·100) by weighing samples of 5 outer flower petals, before and after their drying in a drying box at 60 °C for 20 min. Each measurement was repeated with 5 flowers [51].

4.2. Observation of Petal Epidermal Cells Structure and Ultra-Structure

The petals were washed with distilled water. The longitudinal section of the petal was made by double-blade cutting to determine the internal structure of the petal and the distribution of pigment. A temporary water-filled sheet was prepared to observe the shape of the epidermis cells of the petals and whether the epidermis had a pigment distribution by optical microscope (Nikon, Tokyo, Japan). The photographs were taken under a microscope magnified 20 times. Transmission electron microscope (TEM, Hitachi, Tokyo, Japan) was used to observe the changes of epidermal cell ultra-structure of a flower in different days. The middle part of the fresh flowers were cut into pieces (about 1 mm^2), and these fragments of flower were used for transmission electron microscopy in accordance to the described method by Meng [52].

4.3. Preparation of Total Protein Extraction

Proteins samples were prepared by phenol extraction in accordance with the methods of Wang [53]. The protein sample was dried at −20 °C and stored at −80 °C until used.

4.4. 2-DE and MALDI-TOF/TOF Analysis

The protein powder was mixed with a lysis buffer, containing 7 M urea, 2 M thiourea, and 4% CHAPS, 2% Pharmalyte3-10, and 40 mM DTT, and ultrasonically shaken for 20 min. After the protein powder was completely dissolved, it was placed in a water bath at 37 °C for 2.5 h, centrifuged at 20,000× *g* for 15 min at room temperature. The supernatant was transferred to a new centrifuge tube after centrifugation. The protein content was determined according to the Bradford method [54]. The steps of two-dimensional electrophoresis are in accordance with Wang et al [53]. A volume of 1.5 mg aliquots of protein was added to a 24 cm linear gradient Immobilized *p*H gradient (IPG) strip at pH 4–7 and hydrated for 12 h. The isoelectric focusing procedure: 200 V (1 h)–500 V (1 h)–1000 V (1 h) -Gradient–8000 V (0.5 h)–8000 V (6 h)–1000 V (2 h). The maximum current for each strip is 50 mA. The operating conditions were 20 °C.

After the first isoelectric focusing, the strips were equilibrated in equilibrium solution I (50 mM Tris-HCl (pH 8.8), 6 M urea, 30% *v/v* glycerol, 2% *w/v* SDS, 1% DTT) for 15 min and transferred to equilibrium solution II (50 mM Tris-HCl (pH 8.8), 6 M urea, 30% *v/v* glycerol, 2% *w/v* SDS, 2.5% iodoacetamide) for 15 min. After the end of the two balances, the strips were removed, and the surface of the strip was washed again with electrophoresis buffer and transferred to a 12% sodium dodecyl sulfate polyacrylamide gel electrophoresis (SDS-PAGE) in Ettan DALT-six System (GE Healthcare Bio-Sciences, Uppsala, Sweden) at 18 °C.

The electrophoresis was terminated when the bromophenol blue band reached the bottom of the gel. After electrophoresis, staining was performed by Coomassie brilliant blue staining (CBB-R250) and shaking for 2 h. After staining, the stained liquid drained, the gels washed with distilled water and then added the right amount of bleaching solution for decolorization, until the gel background cleared up. After the decolorization was completed, the Epson scanner 11000XL-USB (EPSON, Tokyo, Japan) was used to perform gel image scanning. Gel image analysis was performed using Image Master™ 2D Platinum Software Version 5.0 software (BioRad, Hercules, CA, USA).

The differential protein spots were excised carefully from the gels from the preparative 2D gels. In-gel digestion by trypsin and analysis by MS analysis in an AB SCIEX MALDI TOF-TOF™ 5800 analyzer (AB SCIEX, Foster City, CA, USA), was performed in accordance with the method described by Wang [53]. According to the results of mass spectrometry analysis, the data were selected using GPS Explorer (V3.6, Applied Biosystems) and the search engine MASCOT (2.1, MatrixScience, London, UK) with the following parameters: National Center for Biotechnology Information green plants (release date: 2015. 04. 01); MS tolerance was set at 100 ppm; MS/MS tolerance of 0.8 Da. Proteins were considered statistically significant ($p < 0.05$ and Ratio > 1.5) confident identifications with scores greater than 75. Individual raw MS and MS/MS spectra were accepted if at least two identified peptides having a probability of 95% correct match were found.

4.5. GC-TOF-MS Analysis

The flower volatile compounds were carried out on an Agilent 7890 gas chromatography (Agilent, Santa Clara, CA, USA) unit and Gerstel Autosampler (Gerstel, Muhlheim, Germany), and a LECO Pegasus HT time of flight mass spectrometer (LECO, St. Joseph, MI, USA) as described by Barakiva [31]. Individual flowers collected from day 1 to day 5 were placed in a 10 mL glass, sealed, and incubated under ambient conditions.

The identification of individual compounds was detected and identified by comparison with the mass spectra library from National Institute of Standards and Technology (NIST, Gaithersburg, MD, USA) and Fiehn Retention Time Lock (Agilent, Santa Clara, CA, USA) library. All the available data acquisition and processing were conducted using Chroma TOF-GC software version 4.51.6.0 (Leco, MI, USA) and performed ANOVA statistical analysis for significance and principal component analysis with a similarity greater than 800.

4.6. Quantitative Real-Time (qRT-PCR) Analysis

The qRT-PCR analysis in accordance with the method described by Li et al. [28] The *CHS*, *CHI*, and *F3'5'H* genes sequences for primer design were obtained from GenBank (Genbank accession number: JN966986, JN887637, JQ678765), respectively. All of the primers used are shown in Table S1. Relative transcription levels were calculated using the $2^{-\Delta CT \Delta CT}$ method.

4.7. Analysis of Physiological Parameters

Total contents of anthocyanin and flavonoid were determined according to the method of Zhang [55] with slight modifications. In brief, 0.1 g of each tissue was ground in liquid nitrogen and total anthocyanins were extracted with HCl/methanol (1:99, *v/v*) at dark for 4 h. The supernatants were determined using UV spectrophotometry at 535 nm and 657 nm. Total phenolic content was determined according to Mitra [56]. Soluble protein content was determined by Coomassie brilliant blue staining method, SOD, POD and CAT activity was quantified according to Wang [57].

4.8. Bioinformatic Analysis of Identified Proteins

The clustering analysis and heat map of protein differential abundance were carried out using the expression values and generated with MultiExperiment Viewer version 4.8 software (MeV; http://www.tm4.org/mev/). The results were viewed using Java TreeView. The GO analysis and classification are based on KEGG (http://www.kegg.jp/kegg/pathway.html).

4.9. Statistical Analysis

All the indexes were analyzed and plotted by Excel 2003 (Microsoft Corporation, Seattle, WA, USA) and the difference was statistically analyzed by SPSS 19 software (IBM, Armonk, NY, USA). Duncan's multiple range tests were considered statistically significant at $p \leq 0.05$. The samples were measured from different plants, analyzed in triplicate ($n = 3$), and shown as mean ± SD.

Supplementary Materials: Supplementary materials can be found at http://www.mdpi.com/1422-0067/20/8/2000/s1.

Author Contributions: M.L. planned, designed, and conducted the experiment and prepared the manuscript. Y.S. and X.L. collected the samples and helped in the analysis of the data. B.D. and S.M. revised the manuscript. D.Q. conceived and guided the experiment and revised the manuscript.

Funding: This work was supported by The Construction Fund of Plateau Discipline of Fujian Province and Fujian Agriculture and Forestry University Outstanding Doctoral Thesis Fund, China (102/712018011).

Acknowledgments: We are grateful for the 2-DE technical support provided by the Proteomics Laboratory of the College of Life Sciences, Fujian Agriculture and Forestry University.

Conflicts of Interest: The authors declare no conflict of interest.

References

1. Vaknin, H.; Barakiva, A.; Ovadia, R.; Nissimlevi, A.; Forer, I.; Weiss, D.; Orenshamir, M. Active anthocyanin degradation in *Brunfelsia calycina* (yesterday–today–tomorrow) flowers. *Planta* **2005**, *222*, 19–26. [CrossRef] [PubMed]
2. Michal, O.S. Does anthocyanin degradation play a significant role in determining pigment concentration in plants. *Plant Sci.* **2009**, *177*, 310–316.
3. Markham, K.R.; Ryan, K.G.; Gould, K.S.; Rickards, G.K. Cell wall sited flavonoids in lisianthus flower petals. *Phytochemistry* **2000**, *54*, 681–687. [CrossRef]

4. Kay, Q.O.N.; Daoud, H.S.; Stirton, C.H. Pigment distribution, light reflection and cell structure in petals. *Bot. J. Linn. Soc.* **2008**, *83*, 57–83. [CrossRef]

5. Li, M.; Luo, Y.T.; Lu, X.C.; Sun, Y.T.; Qiu, D.L. Changes in Composition of Anthocyanins in *Brunfelsia acuminata* Flowers. *J. Trop. Subtrop. Bot.* **2018**, *26*, 627–632.

6. Miller, R.; Owens, S.J.; Rørslett, B. Plants and colour: Flowers and pollination. *Opt. Laser Technol.* **2011**, *43*, 282–294. [CrossRef]

7. Akifumi, A.; Hiroshi, Y.; Sato, A. Postharvest light irradiation and appropriate temperature treatment increase anthocyanin accumulation in grape berry skin. *Postharvest Biol. Technol.* **2019**, *147*, 89–99.

8. Welch, C.R.; Wu, Q.; Simon, J.E. Recent Advances in Anthocyanin Analysis and Characterization. *Curr. Anal. Chem.* **2008**, *4*, 75. [CrossRef]

9. Cheynier, V. Phenolic compounds: From plants to foods. *Phytochem. Rev.* **2012**, *11*, 153–177. [CrossRef]

10. Wangia, C.; Orwa, J.; Muregi, F.; Kareru, P.; Cheruiyot, K.; Kibet, J. Comparative Anti-oxidant Activity of Aqueous and Organic Extracts from Kenyan *Ruellia linearibracteolata* and *Ruellia bignoniiflora*. *Eur. J. Med. Plants* **2016**, *17*, 1–7. [CrossRef]

11. Liu, F.; Wang, M.; Wang, M. Phenolic compounds and antioxidant activities of flowers, leaves and fruits of five crabapple cultivars (*Malus Mill.* species). *Sci. Hortic.* **2018**, *235*, 460–467. [CrossRef]

12. Nascimento, L.B.D.S.; Leal-Costa, M.V.; Menezes, E.A.; Lopes, V.R.; Muzitano, M.F.; Costa, S.S.; Tavares, E.S. Ultraviolet-B radiation effects on phenolic profile and flavonoid content of *Kalanchoe pinnata*. *J. Photochem. Photobiol. B* **2015**, *148*, 73–81. [CrossRef] [PubMed]

13. Vera, H.; Angela, D.; Bernhard, E.; Richard, L.; Stefan, M.; Michael, A. Characterization and structural features of a chalcone synthase mutation in a white-flowering line of *Matthiola incana R. Br.* (Brassicaceae). *Plant Mol. Biol.* **2004**, *55*, 455–465.

14. Zhang, X.; Liang, Z. Morphology, structure and ultrastructure of staminal nectary in *Lamprocapnos* (Fumarioideae, Papaveraceae). *Flora* **2018**, *242*, 128–136. [CrossRef]

15. Avalos, A.A.; Lattar, E.C.; Galati, B.G.; Ferrucci, M.S. Nectary structure and ultrastructure in two floral morphs of *Koelreuteria elegans subsp. formosana* (Sapindaceae). *Flora* **2017**, *226*, 29–37. [CrossRef]

16. Cao, Y.T.; Qiu, D.L. Cloning and Sequence Analysis of Chalcone Isomerase Gene (*CHI*) from *Brunfelsia acuminata* Flowers. Master's Thesis, Fujian Agriculture and Forestry University, Fujian, China, April 2012.

17. Naing, A.H.; Ai, T.N.; Su, M.J.; Park, K.I.; Lim, K.B.; Chang, K.K. Expression of *RsMYB1* in *chrysanthemum* regulates key anthocyanin biosynthetic genes. *Electron. J. Biotechnol.* **2015**, *18*, 359–364. [CrossRef]

18. Ai, T.N.; Naing, A.H.; Arun, M.; Su, M.J.; Chang, K.K. Expression of *RsMYB1* in Petunia enhances anthocyanin production in vegetative and floral tissues. *Sci. Hortic.* **2017**, *214*, 58–65. [CrossRef]

19. Yamagishi, M.; Uchiyama, H.; Handa, T. Floral pigmentation pattern in Oriental hybrid lily (Lilium spp.) cultivar 'Dizzy' is caused by transcriptional regulation of anthocyanin biosynthesis genes. *J. Plant Physiol.* **2018**, *228*, 85. [CrossRef] [PubMed]

20. Zipor, G.; Duarte, P.; Carqueijeiro, I.; Shahar, L.; Ovadia, R.; Teper-Bamnolker, P.; Eshel, D.; Levin, Y.; Doron-Faigenboim, A.; Sottomayor, M. In planta anthocyanin degradation by a vacuolar class III peroxidase in *Brunfelsia calycina* flowers. *New Phytol.* **2015**, *205*, 653–665. [CrossRef]

21. Wang, Z.Q.; Zhou, X.; Dong, L.; Guo, J.; Chen, Y.; Zhang, Y.; Wu, L.; Xu, M. iTRAQ-based analysis of the *Arabidopsis* proteome reveals insights into the potential mechanisms of anthocyanin accumulation regulation in response to phosphate deficiency. *J. Proteomics* **2018**, *184*, 39–53. [CrossRef]

22. Li, X.; Jackson, A.; Xie, M.; Wu, D.; Tsai, W.C.; Zhang, S. Proteomic insights into floral biology. *BBA Proteins Proteomics* **2016**, *1864*, 1050–1060. [CrossRef]

23. Farzad, M.; Griesbach, R.; Hammond, J.; Weiss, M.R.; Elmendorf, H.G. Differential expression of three key anthocyanin biosynthetic genes in a color-changing flower, *Viola cornuta cv.* Yesterday, Today and Tomorrow. *Plant Sci.* **2003**, *165*, 1333–1342. [CrossRef]

24. Yamazaki, M.; Gong, Z.; Fukuchimizutani, M.; Fukui, Y.; Tanaka, Y.; Kusumi, T.; Saito, K. Molecular Cloning and Biochemical Characterization of a Novel Anthocyanin 5-*O*-Glucosyltransferase by mRNA Differential Display for Plant Forms Regarding Anthocyanin. *J. Biol. Chem.* **1999**, *274*, 7405. [CrossRef] [PubMed]

25. Akita, Y.; Hase, Y.; Narumi, I.; Ishizaka, H.; Kondo, E.; Kameari, N.; Nakayama, M.; Tanikawa, N.; Morita, Y.; Tanaka, A. Isolation and characterization of the fragrant cyclamen -methyltransferase involved in flower coloration. *Planta* **2011**, *234*, 1127–1136. [CrossRef]

26. Wei, K.; Wang, L.; Zhang, C.; Wu, L.; Li, H.; Zhang, F.; Cheng, H. Transcriptome Analysis Reveals Key Flavonoid 3′-Hydroxylase and Flavonoid 3′,5′-Hydroxylase Genes in Affecting the Ratio of Dihydroxylated to Trihydroxylated Catechins in *Camellia sinensis*. *PLoS ONE* **2015**, *10*, e0137925. [CrossRef]

27. Seitz, C.; Ameres, S.; Schlangen, K.; Forkmann, G.; Halbwirth, H. Multiple evolution of flavonoid 3′,5′-hydroxylase. *Planta* **2015**, *242*, 561–573. [CrossRef] [PubMed]

28. Li, M.; Cao, Y.T.; Ye, S.R.; Irshad, M.; Pan, T.F.; Qiu, D.L. Isolation of *CHS* Gene from *Brunfelsia acuminata* Flowers and Its Regulation in Anthocyanin Biosysthesis. *Molecules* **2016**, *22*, 44. [CrossRef] [PubMed]

29. Wu, Y.Q.; Zhu, M.Y.; Jiang, Y.; Zhao, D.Q.; Tao, J. Molecular characterization of chalcone isomerase (CHI) regulating flower color in herbaceouspeony (*PaeonialactifloraPall.*). *J. Integr. Agric.* **2018**, *17*, 122–129. [CrossRef]

30. Muir, S.R.; Collins, G.J.; Robinson, S.; Hughes, S.; Bovy, A.; de Vos, C.H.R.; van Tunen, A.J.; Martine, E.; Verhoeyen, M.E. Over expression of petunia chalcone isomerase in tomato results in fruit containing increased levels of flavonols. *Nat. Biotechnol.* **2001**, *19*, 470–474. [CrossRef] [PubMed]

31. Barakiva, A.; Ovadia, R.; Rogachev, I.; Baror, C.; Bar, E.; Freiman, Z.; Nissimlevi, A.; Gollop, N.; Lewinsohn, E.; Aharoni, A. Metabolic networking in *Brunfelsia calycina* petals after flower opening. *J. Exp. Bot.* **2010**, *61*, 1393. [CrossRef]

32. Liu, Y.; Tikunov, Y.; Schouten, R.E.; Marcelis, L.F.M.; Visser, R.G.F.; Bovy, A. Anthocyanin Biosynthesis and Degradation Mechanisms in SolanaceousVegetables: A Review. *Front. Chem.* **2018**, *6*, 1–17. [CrossRef]

33. Zhang, Q.; Tian, Y.Y.; Meng, Y.E.; Li, Y.M.; Wang, H.J.; Wang, L.M.; Zhao, X.S. Research advance in genetic engineering of floral fragrance. *J. Henan Agric. Sci.* **2014**, *43*, 11–16.

34. Gimenez, M.J.; Valverde, J.; Valero, D.; Diazmula, H.M.; Zapata, P.J.; Serrano, M.; Moral, J.M.; Castillo, S. Methyl salicylate treatments of sweet cherry trees improve fruit quality at harvest and during storage. *Sci. Hortic.* **2015**, *197*, 665–673. [CrossRef]

35. Zhang, L.X.; Shi, Z.P. Research on the Cell Microstructure of *Jasminum* sambac's Petal. *J. Hunan Agric. Univ.* **1999**, *25*, 108–111.

36. Giordano, D.; Provenzano, S.; Ferrandino, A.; Vitali, M.; Pagliarani, C.; Roman, F.; Cardinale, F.; Castellarin, S.D.; Schubert, A. Characterization of a multifunctional caffeoyl-CoA O-methyltransferase activated in grape berries upon drought stress. *Plant Physiol. Biochem.* **2016**, *101*, 23. [CrossRef]

37. Cui, N.; Yu, Z.H.; Han, M.L.; Dong, X.F.; Qu, B.; Li, T.L. Research Advancement of 14-3-3 Proteins in Plants. *Acta Bot. Boreali Occident. Sin.* **2012**, *32*, 843–851.

38. Su, M.; Lv, J.H.; Zhang, Q.X. Research Advance in Floral Fragrance Genetic Engineering. *J. Anhui Agric. Sci.* **2007**, *35*, 3169–3171.

39. Martens, S.; Mithöfer, A. Flavones and flavone synthases. *Phytochemistry* **2005**, *66*, 2399–2407. [CrossRef]

40. Wang, L.; Zheng, Y.; Ding, S.; Zhang, Q.; Chen, Y.; Zhang, J. Molecular cloning, structure, phylogeny and expression analysis of the invertase gene family in sugarcane. *BMC Plant Biol.* **2017**, *17*, 109. [CrossRef]

41. Wang, L.J. Advances on the studies of invertase on sucrose metabolism in higher plant. *J. Agric. Sci.* **2010**, *31*, 70–75.

42. Walker, R.P.; Paoletti, A.; Leegood, R.C.; Famiani, F. Phosphorylation of phosphoenolpyruvate carboxykinase (PEPCK) and phosphoenolpyruvate carboxylase (PEPC) in the flesh of fruits. *Plant Physiol. Biochem.* **2016**, *108*, 323–327. [CrossRef]

43. Chen, Z.; Xia, L.; He, Y.; Zhang, J.; Yan, T.; Liu, X. Physiological investigation of C4 -phosphoenolpyruvate-carboxylase-introduced rice line shows that sucrose metabolism is involved in the improved drought tolerance. *Plant Physiol. Biochem.* **2017**, *115*, 328–342.

44. Silva, L.S.; Seabra, A.R.; Leitão, J.N.; Carvalho, H.G. Possible role of glutamine synthetase of the prokaryotic type (GSI-like) in nitrogen signaling in Medicago truncatula. *Plant Sci.* **2015**, *240*, 98–108. [CrossRef]

45. Wang, L.; Hu, W.; Feng, J.; Yang, X.; Huang, Q.; Xiao, J.; Liu, Y.; Yang, G.; He, G. Identification of the *ASR* gene family from Brachypodium distachyon and functional characterization of BdASR1 in response to drought stress. *Plant Cell Rep.* **2016**, *35*, 1221–1234. [CrossRef]

46. Langenkämper, G.; Manac'H, N.; Broin, M.; Cuiné, S.; Becuwe, N.; Kuntz, M.; Rey, P. Accumulation of plastid lipid-associated proteins (fibrillin/CDSP34) upon oxidative stress, ageing and biotic stress in Solanaceae and in response to drought in other species. *J. Exp. Bot.* **2001**, *52*, 1545. [CrossRef]

47. Pan, Z.Y.; Liu, Q.; Yun, Z.; Rui, G.; Zeng, W.F.; Qiang, X.; Deng, X.X. Comparative proteomics of a lycopene-accumulating mutant reveals the important role of oxidative stress on carotenogenesis in sweet orange (*Citrus sinensis* [L.] Osbeck). *Proteomics* **2009**, *9*, 5455–5470. [CrossRef] [CrossRef]

48. Ihsan, U.; Bassam, O.A.J.; Khalid, M.S.A.G.; Hind, A.A.A.Z.; Yasir, A.; Ahmad, F.; Naser, A.K.; Mohammed, A.A. Endophytic bacteria isolated from *Solanum nigrum* L., alleviate cadmium (Cd) stress response by their antioxidant potentials, including SOD synthesis by *sodA* gene. *Ecotoxicol. Environ. Saf.* **2019**, *174*, 197–207.

49. Faten, H.F.; Riadh, K.; Chedly, A. Total phenolic, flavonoid and tannin contents and antioxidant andantimicrobial activities of organic extracts of shoots of the plant *Limonium delicatulum*. *J. Taibah Univ. Sci.* **2014**, *8*, 216–224.

50. Quintá, H.R.; Galigniana, N.M.; Erlejman, A.G.; Lagadari, M.; Piwien-Pilipuk, G.; Galigniana, M.D. Management of cytoskeleton architecture by molecular chaperones and immunophilins. *Cell. Signal.* **2011**, *23*, 1907–1920. [CrossRef]

51. Orenshamir, M.O.M.; Dela, G.D.G.; Ovadia, R.O.R.; Nissimlevi, A.N.A.; Philosophhadas, S.P.S.; Meir, S.M.S. Differentiation between petal blueing and senescence of cut 'Mercedes' rose flowers. *J. Hortic. Sci. Biotechnol.* **2015**, *76*, 195–200.

52. Meng, J.F.; Xu, T.F.; Wang, Z.Z.; Fang, Y.L.; Xi, Z.M.; Zhang, Z.W. The ameliorative effects of exogenous melatonin on grape cuttings under water-deficient stress: Antioxidant metabolites, leaf anatomy, and chloroplast morphology. *J. Pineal Res.* **2015**, *57*, 200–212. [CrossRef] [PubMed]

53. Wang, S.; Pan, D.; Lv, X.; Song, X.; Qiu, Z.; Huang, C.; Huang, R.; Wei, C. Proteomic approach reveals that starch degradation contributes to anthocyanin accumulation in tuberous root of purple sweet potato. *J. Proteomics* **2016**, *143*, 298–305. [CrossRef]

54. Bradford, M.M. A rapid and sensitive method for the quantitation of microgram quantities of protein utilizing the principle of protein-dye binding. *Anal. Biochem.* **1976**, *72*, 248–254. [CrossRef]

55. Zhang, C.; Wang, W.; Wang, Y.; Gao, S.; Du, D.; Fu, J.; Dong, L. Anthocyanin biosynthesis and accumulation in developing flowers of tree peony (*Paeonia suffruticosa*) 'Luoyang Hong'. *Postharvest Biol. Technol.* **2014**, *97*, 11–22. [CrossRef]

56. Sangeeta, M.; Muhammad, I.; Biswojit, D.; Lu, X.C.; Li, M.; Chandra, K.D.; Hafiz, M.R.; Qiu, Z.P.; Qiu, D.L. Effect of vineyard soil variability on chlorophyll fluorescence, yield and quality of table grape as influenced by soil moisture, grown under double cropping system in protected condition. *Peer J.* **2018**, *9*, 9–25.

57. Wang, W.X. *Principles and Techniques of Plant Physiological Biochemical Experiments*; Higher Education Press: Beijing, China, 2006; pp. 282–286.

International Journal of
Molecular Sciences

MDPI

Article

The Major Storage Protein in Potato Tuber Is Mobilized by a Mechanism Dependent on Its Phosphorylation Status

Javier Bernal [1], Daniel Mouzo [1], María López-Pedrouso [1], Daniel Franco [2], Lucio García [2] and Carlos Zapata [1,*]

[1] Department of Zoology, Genetics and Physical Anthropology, University of Santiago de Compostela, 15782 Santiago de Compostela, Spain; javier.bernal.pampin@gmail.com (J.B.); daniel.mouzo.calzadilla@usc.es (D.M.); mariadolores.lopez@usc.es (M.L.-P.)
[2] Meat Technology Center of Galicia, 32900 San Cibrao das Viñas, Ourense, Spain; danielfranco@ceteca.net (D.F.); luciogarcia@ceteca.net (L.G.)
* Correspondence: c.zapata@usc.es; Tel.: +34-88181-6922

Received: 21 February 2019; Accepted: 13 April 2019; Published: 17 April 2019

Abstract: The role of the protein phosphorylation mechanism in the mobilization of vegetative storage proteins (VSPs) is totally unknown. Patatin is the major VSP of the potato (*Solanum tuberosum* L.) tuber that encompasses multiple differentially phosphorylated isoforms. In this study, temporal changes in the phosphorylation status of patatin isoforms and their involvement in patatin mobilization are investigated using phosphoproteomic methods based on targeted two-dimensional electrophoresis (2-DE). High-resolution 2-DE profiles of patatin isoforms were obtained in four sequential tuber life cycle stages of Kennebec cultivar: endodormancy, bud break, sprouting and plant growth. In-gel multiplex identification of phosphorylated isoforms with Pro-Q Diamond phosphoprotein-specific stain revealed an increase in the number of phosphorylated isoforms after the tuber endodormancy stage. In addition, we found that the phosphorylation status of patatin isoforms significantly changed throughout the tuber life cycle ($P < 0.05$) using the chemical method of protein dephosphorylation with hydrogen fluoride-pyridine (HF-P) coupled to 2-DE. More specifically, patatin phosphorylation increased by 32% from endodormancy to the tuber sprouting stage and subsequently decreased together with patatin degradation. Patatin isoforms were not randomly mobilized because highly phosphorylated Kuras-isoforms were preferably degraded in comparison to less phosphorylated non-Kuras isoforms. These results lead us to conclude that patatin is mobilized by a mechanism dependent on the phosphorylation status of specific isoforms.

Keywords: *Solanum tuberosum*; patatin; seed storage proteins; vegetative storage proteins; tuber phosphoproteome; targeted two-dimensional electrophoresis

1. Introduction

The molecular mechanisms controlling the mobilization of seed storage proteins (SSPs) play a critical role in plant reproduction [1,2]. Storage proteins accumulate progressively during seed development providing the nutrients necessary for germination and plant growth [3–5]. Phosphoproteomic studies have shown that SSPs (i.e., albumins, globulins, prolamines and glutelins) are abundantly phosphorylated and thereby phosphorylation may play a key role in the accumulation of storage proteins during seed development and mobilization during germination [6–8]. Studies in the model plant *Arabidopsis thaliana* suggest that a complex crosstalk of protein kinases, phosphatases and phytohormones are mainly involved in the regulatory networks modulating the phosphorylation status of SSPs [9–12]. An added complexity is that SSPs often exhibit multiple differentially phosphorylated isoforms with contrasting temporal patterns during seed development and germination [6,8,13–16]. The biological meaning of these isoform-dependent temporal phosphorylation patterns is not well understood. Lopez-Pedrouso et al. [17] reported that highly phosphorylated isoforms of the phaseolin in common beans were preferentially degraded in dry-to-germinating seed transition, suggesting a phosphorylation-dependent mobilization mechanism.

VSPs are a particular type of differentiated and less studied plant storage proteins located in vegetative tissues [1,3,18]. Patatin is the most abundant VSP in the potato (*S. tuberosum*) tuber, accounting for up to circa 45% of the total soluble protein [1,3,19]. Unlike most other plant storage proteins, patatin also exhibits different enzymatic activities including a non-specific lipid acyl hydrolase (LAH) activity with a probable role in plant defense mechanisms [1,20–22]. Phosphorylated isoforms of VSPs were identified for the first time in the patatin protein [23]. It is a very complex protein that comprises multiple differentially phosphorylated isoforms within—and between—potato cultivars [23,24]. A 2-DE-based phosphoproteomic study enabled the identification of 20 differentially phosphorylated patatin isoforms in Kennebec cultivar, with phosphorylation rates (*PRs*) varying between 4.6 and 52.3% [23]. This finding raises the question of whether VSPs are mobilized by a phosphorylation-dependent mechanism as occurs in the common bean. It is noteworthy that the germination of dry after-ripening seeds can be easily activated in the laboratory with water imbibition to assess changes in the phosphorylation status of SSPs in dry-to-germinating seed transition. However, this is a more difficult issue to solve experimentally in potatoes because tubers undergo continuous growth between endodormancy and sprouting stages. The endodormancy phase ends with the construction of phloem structures that allow for the provision of nutrients to the apical meristem to produce the first 2-mm-long apical bud (bud break) and then tuber progresses until the sprouting phase [5,25]. Accordingly, changes in the status of phosphorylation associated with the onset of patatin mobilization can potentially occur within the extensive time period comprised between the endodormancy and sprouting stages.

This study aimed to assess whether the existence of differentially phosphorylated patatin isoforms in potato tubers is related to the mechanism of patatin mobilization. For this purpose, a comprehensive analysis of temporal phosphorylation changes in patatin was performed in tubers of Kennebec cultivar (Figure 1). We used targeted 2-DE-based proteomic methods enabling the separation of multiple storage protein isoforms with high resolution and reproducibility as well as the efficient in-gel quantitation of their phosphorylation levels [8,17,23,24,26,27]. The results will contribute to understanding the mechanisms underlying the mobilization of VSPs.

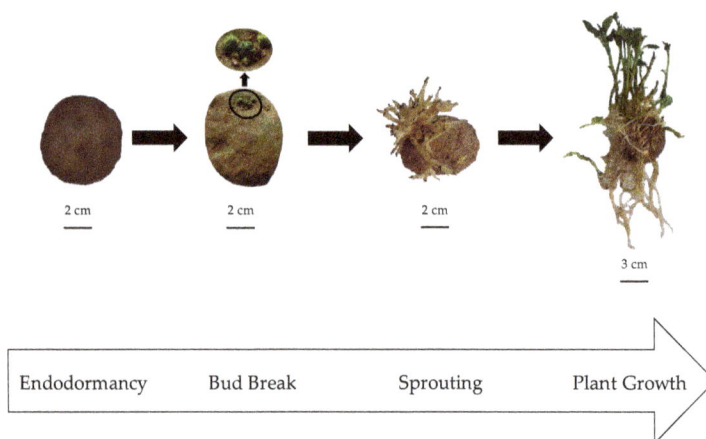

Figure 1. Four sequential stages of the potato tuber life cycle used to assess changing status of patatin phosphorylation in Kennebec cultivar: endodormancy, bud break (2 mm apical bud length), sprouting (5 cm bud length) and plant growth (12 cm stem length).

2. Results and Discussion

2.1. Reference Patatin Profiles throughout the Tuber Life Cycle

Figure 2A shows representative high-resolution profiles of patatin on 2-DE gels stained with SYPRO Ruby total-protein stain at each of the four potato life cycle stages of the Kennebec cultivar analyzed: endodormancy, bud break, sprouting and plant growth. The relative molecular mass (M_r) of 2-DE patatin spots ranged from 40.0 to 45.0 kDa and their isoelectric points (pIs) ranged from 4.8 to 5.4 as previously reported in bud break stage of the same cultivar [23]. Qualitative variations in the number of patatin spots were detected throughout tuber stages. Most variations were identified after endodormancy stage on more acidic gel positions involving spot Numbers 1–4. Thus, spot Numbers 1, 2 and 4 were identified in bud-break, sprouting and plant growth stages, while only spot Number 2 was identified in the endodormancy phase.

Previous protein identification of 2-DE patatin spots in the bud break tuber stage of the Kennebec cultivar using matrix-assisted laser desorption/ionization time-of-flight (MALDI-TOF) and MALDI-TOF/TOF mass spectrometry (MS) revealed that spot Numbers 1, 2 and 4 contained the Patatin-3 Kuras 1 (PT3K1) isoform, while the other spots (i.e., spot Numbers 5–20) contained non-Kuras patatin isoforms [23]. The PT3K1 isoform was unequivocally identified because it is the patatin isoform with the most highly differentiated sequence [28]. In the present study, we confirmed these previous identifications by MALDI-TOF and MALDI-TOF/TOF MS from 2-DE gels in the endodormancy phase (Supplementary Tables S1 and S2; Supplementary Materials). Therefore, MS analysis leads us to conclude that Kuras isoforms explain most variations in the number of patatin spots throughout the tuber stages.

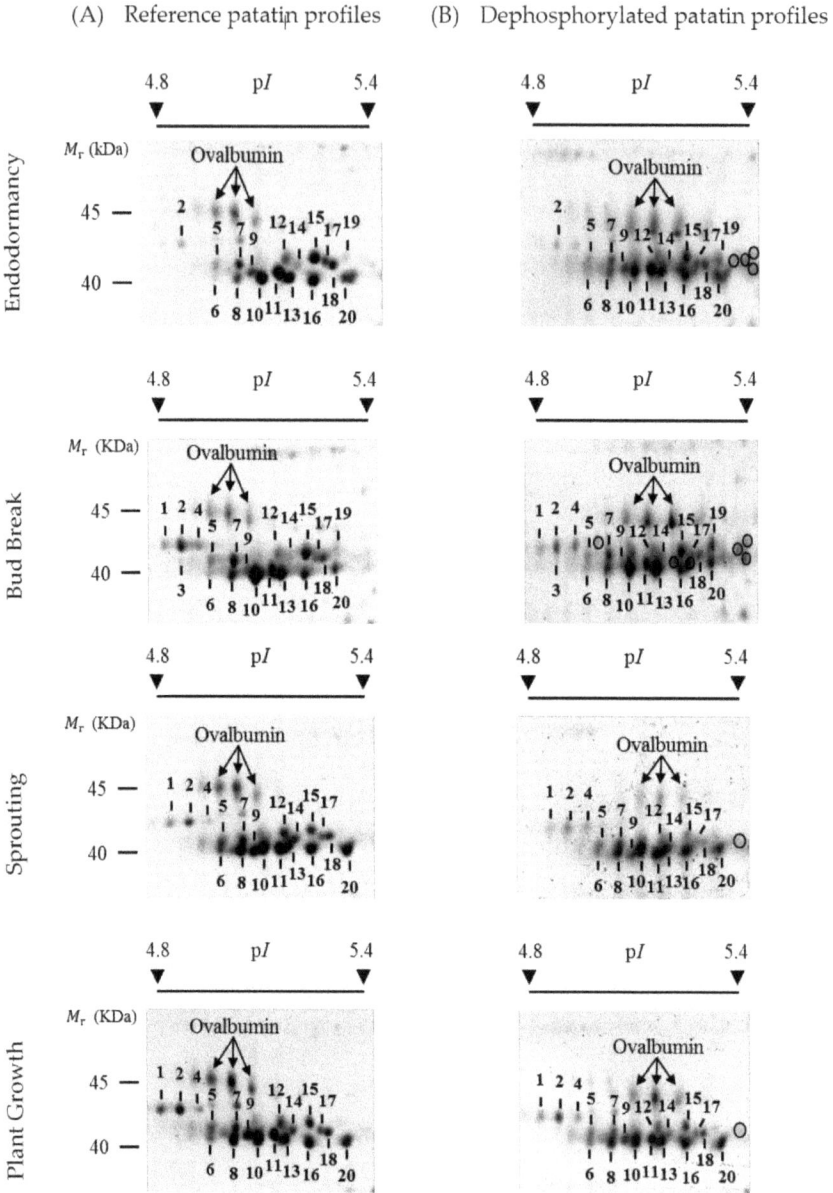

Figure 2. Representative two-dimensional electrophoresis (2-DE) gel images of reference and dephosphorylated profiles of patatin at four different tuber life stages of the Kennebec cultivar: endodormancy, bud break, sprouting and plant growth. (**A**) Reference profiles on 2-DE gels stained with Pro-Q Diamond and post-stained with SYPRO Ruby fluorescent dyes. Phosphorylated patatin spots are numbered. (**B**) Dephosphorylated patatin profiles on 2-DE gels stained with SYPRO Ruby from total tuber protein extracts treated with hydrogen fluoride-pyridine (HF-P). Circles show the gel position of newly arisen patatin spots following chemical dephosphorylation with HF-P. The arrows indicate the gel position of ovalbumin (45.0 kDa) phosphoprotein marker.

2.2. Changes in Patatin Abundance in the Tuber Life Cycle

A rough estimation of the amount of patatin in each of the four tuber life cycle stages was obtained by means of the sum of the patatin spot volumes assessed by PDQuest software. Mean values of the amount of protein along with the bias-corrected 95% bootstrap confidence intervals, adjusted by Bonferroni's correction for multiple comparisons, were obtained in each potato stage from four biological replicate gels. Figure 3A shows that patatin abundance significantly increased from endodormancy to sprouting (p-value < 0.05) and then decreased to plant growth potato stage ($p < 0.05$). These results have a clear biological significance. Patatin achieved maximum abundance at the time of tuber sprouting when storage protein reserves are necessarily broken down into seedling growth. In agreement with these results, Lehesranta et al. [29] previously reported from 2-DE data that most patatin isoforms in the Desiree cultivar increased during tuber development achieving high amounts at the onset of sprouting and subsequently decreased in fully sprouted tubers. Quantification of patatin from Ranger Russet cultivar by LAH activity using p-nitrophenyl myristate as substrate showed its pronounced decrease during plant establishment [30]. In addition, analyses of the transcriptome throughout the potato tuber life cycle are in accordance with those proteomic studies. Thus, a comparative analysis of transcripts encoding patatin throughout the potato (Kennebec cultivar) tuber cycle based on cDNA-AFLP fingerprinting (Bintje cultivar) and expressed sequence tag (EST) libraries have shown that patatin transcripts were relatively higher at the onset of tuber sprouting [31,32]. Chemically (bromoethane) induced cessation of dormancy coupled to microarrays constructed from potato EST libraries has also shown a decrease in transcripts that encode for patatin after the onset of sprouting in tubers of the Russet Burbank cultivar [33]. Temporal studies of patatin concentration, activities of proteases and protease inhibitors suggest correlative changes during tuber development and plant establishment. There is increasing evidence that protease inhibitors (cys-protease inhibitor potato multicystatin, the Kunitz protease inhibitor family and others) may facilitate patatin accumulation in developing tubers by attenuating the activity of proteases, while catabolism of protease inhibitors may facilitate patatin mobilization following tuber sprouting [30,33–35].

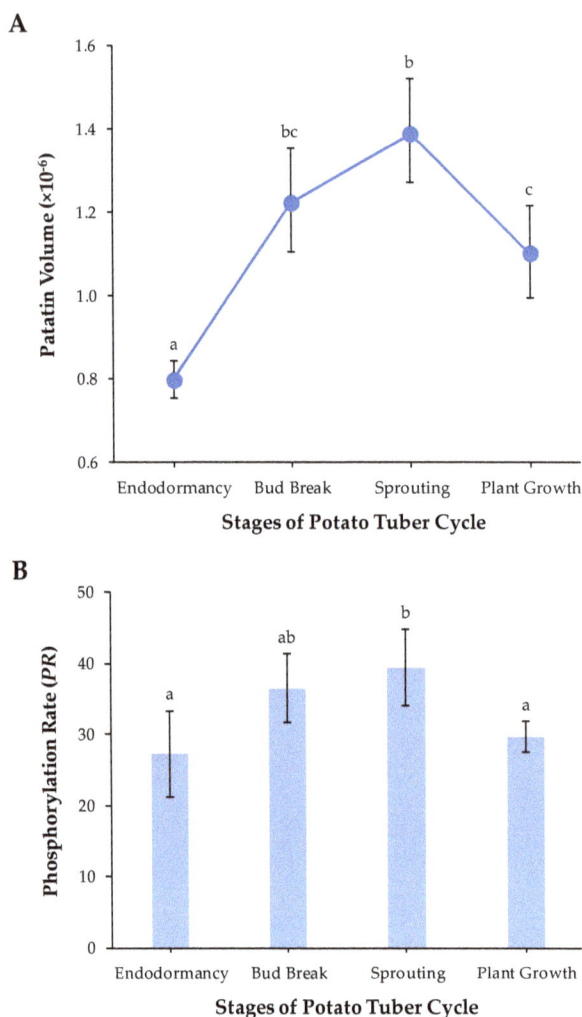

Figure 3. (**A**) Dynamic changes in the patatin abundance through the different stages of the potato tuber life cycle: endodormancy, bud break, sprouting and plant growth. In each tuber stage, mean patatin quantity over four biological replicates is represented together with their Bonferroni-corrected 95% bootstrap CIs. (**B**) Mean *PR* values together with their 95% confidence interval obtained from four biological replicates in endodormancy, bud break, sprouting and plant growth potato tuber stages of the Kennebec cultivar. Different lower case letters (a–c) indicate statistically significant differences ($p < 0.05$) between sample groups by the post hoc Tukey-Kramer test following the one-way analysis of variance (ANOVA) test.

2.3. Identification of Phosphorylated Patatin Isoforms along the Tuber Life Cycle

Patatin profiles on 2-DE gels stained with Pro-Q Diamond phosphoprotein-specific fluorescent dye in endodormancy, bud break, sprouting and plant growth tuber stages of Kennebec cultivar are shown in Supplementary Figure S1 (Supplementary Materials). All 2-DE patatin spots contained phosphorylated isoforms because they were reproducibly Pro-Q-Diamond-stained across biological replicates. These results suggest that the above-mentioned variations in the number of Kuras isoforms contribute to the increase of patatin phosphorylation after the endodormancy stage.

2.4. Changes in Patatin Phosphorylation Status during Tuber Life Cycle

The phosphorylation level of patatin spots was assessed using the method of chemical dephosphorylation with hydrogen fluoride-pyridine (HF-P). 2-DE patatin profiles obtained from total tuber (Kennebec cultivar) protein extracts dephosphorylated with HF-P in the tuber phases analyzed are shown in Figure 2B. First of all, HF-P was a very effective treatment of protein dephosphorylation as previously shown [17,23]. Note that the isoforms of the ovalbumin phosphoprotein marker completely shifted to more basic gel positions after HF-P treatment, as a result of the loss of phosphate groups.

The phosphorylation level of each patatin spot was estimated using the *PR* coefficient. This coefficient calculates the percentage by which the total spot volume (untreated protein samples) decreased after protein dephosphorylation with HF-P (treated protein samples). Therefore, it is a phosphorylation measure independent of the existing variations in the amount of patatin over spots. Mean (± SE) *PR* values of patatin spots obtained from four biological replicate gels in the distinct sample groups are shown in Table 1. It can be seen that all spots contained phosphorylated isoforms because *PR*-values were always higher than zero, which is in agreement with those results obtained with Pro-Q Diamond stain. In addition, *PR*-values exhibited remarkable differences among spots when mean values ranged from 10.9 (spot Number 11/endodormancy) to 53.2% (spot Number 14/sprouting). No *PR*-value achieved the maximum value of 100%, which indicates that each patatin spot contained a variable mixture of phosphorylated and unphosphorylated isoforms.

Table 1. Mean (± SE) phosphorylation rate (*PR*) of patatin spots in different stages of potato tuber (Kennebec cultivar) life cycle.

Spot Number [a]	p*I*	PR				Mean (± SE)
		Endodormancy	Bud Break	Sprouting	Plant Growth	
1	4.84	N/A	39.8 ± 2.5	51.9 ± 5.2	33.1 ± 5.6	41.6 ± 5.5
2	4.88	38.1 ± 5.3	41.2 ± 6.1	52.4 ± 5.9	39.6 ± 5.7	42.8 ± 3.3
3	4.90	N/A	43.1 ± 2.9	N/A	N/A	N/A
4	4.93	N/A	N/A	39.7 ± 5.4	23.0 ± 4.6	31.4 ± 8.4
5	4.96	27.1 ± 1.9	42.8 ± 2.0	43.2 ± 6.4	30.5 ± 4.4	35.9 ± 4.2
6	4.96	N/A	52.3 ± 4.1	38.4 ± 4.4	27.5 ± 5.2	39.4 ± 7.2
7	5.02	32.3 ± 4.5	39.0 ± 4.1	42.5 ± 5.9	26.9 ± 5.8	35.2 ± 3.5
8	5.02	15.3 ± 3.8	27.2 ± 5.6	27.0 ± 4.6	29.1 ± 5.1	24.6 ± 3.2
9	5.05	27.3 ± 2.4	32.5 ± 3.3	40.5 ± 5.1	31.4 ± 5.7	32.9 ± 2.8
10	5.12	32.0 ± 4.0	30.5 ± 4.7	23.5 ± 6.1	40.0 ± 6.4	31.5 ± 3.4
11	5.12	10.9 ± 6.9	25.1 ± 4.3	26.4 ± 4.4	25.2 ± 6.3	21.9 ± 3.7
12	5.13	28.3 ± 2.4	51.4 ± 5.5	52.7 ± 4.2	31.9 ± 7.0	41.1 ± 6.4
13	5.14	N/A	N/A	N/A	26.0 ± 7.2	N/A
14	5.16	50.1 ± 4.9	44.3 ± 9.4	53.2 ± 7.5	33.3 ± 5.9	45.2 ± 4.4
15	5.20	41.4 ± 2.4	41.9 ± 3.8	44.1 ± 5.5	28.0 ± 5.3	38.8 ± 3.7
16	5.20	12.8 ± 3.1	27.0 ± 8.4	17.2 ± 4.3	25.1 ± 5.1	20.6 ± 3.3
17	5.23	23.8 ± 3.6	36.0 ± 2.4	39.8 ± 6.3	29.3 ± 4.6	32.2 ± 3.5
18	5.25	29.6 ± 5.3	34.2 ± 4.1	31.6 ± 4.5	22.9 ± 4.6	34.5 ± 6.1
19	5.29	N/A	N/A	N/A	N/A	N/A
20	5.27	11.6 ± 2.0	11.9 ± 4.6	25.6 ± 6.3	29.2 ± 4.9	19.6 ± 4.6
Mean (± SE)		27.2 ± 3.1	36.5 ± 2.5	39.4 ± 2.8	29.6 ± 1.1	

[a] Gel position of numbered spots is shown in Figure 2. N/A = not applicable, absent spot in untreated and treaded protein samples or with volume below the limit of detection.

Parametric statistical tests were used to test for differences in *PR* over the different potato tuber phases because their sampling distributions did not significantly deviate from the theoretical normal distribution ($p > 0.05$; Shapiro-Wilk test). The one-way analysis of variance (ANOVA) test revealed that mean values of *PR* differed significantly ($p < 0.01$) among sample groups. More specifically, we found by using the Tukey-Kramer method for multiple comparisons that *PR*-values increased by 32% from endordormacy release to tuber sprouting ($p < 0.05$) and subsequently decreased up to the plant-growth stage ($p < 0.05$), achieving levels close to the endodormancy stage (Figure 3B). Concomitant temporal changes of the amount of patatin (Figure 3A) and phosphorylation levels (Figure 3B) suggest that protein phosphorylation is involved in the regulation of the synthesis and mobilization of patatin. A targeted 2-DE-based study in common bean also showed that the phosphorylation status of phaseolin isoforms increased in dry seed-to-gemination transition [17].

The advantage of targeted 2-DE is that the global phosphorylation of storage proteins can be estimated more accurately by considering the complete set of isoforms. In contrast, non-targeted 2-DE only enables the analysis of a limited number of storage protein isoforms. A number of non-targeted 2-DE proteomic studies have reported temporal phosphorylation changes in SSPs during seed development and germination. In white lupin, only vicilin family proteins were phosphorylated during seed development, both vicilins and legumins were phosphorylated in mature seeds and only vicilins increased their phosphorylation level at 2 d after the onset of germination [13]. In rice, a phosphoproteomic study in different phases of seed germination revealed that storage proteins achieved the highest levels of phosphorylation at the same time as their degradation [15]. In wheat, increased phosphorylation levels of globulin 3 were detected at 12 h following imbibition [16]. In *Helianthus*, 11S globulins showed increased levels of phosphorylation following completation of germination [14]. In oilseed rape, the amount of phosphorylated cruciferin subunits exhibited contrasting trends over distinct sequential phases of seed development, either increasing or decreasing their phosphorylation level in the late maturation stage [6]. Unfortunately, the global phosphorylation trend of the subunits identified was not reported. Overall, the join analysis of targeted and non-targeted 2-DE studies on temporal phosphorylation changes in SSPs and patatin VSP suggests that storage proteins tend to increase their phosphorylation levels during seed/tuber development until immediately after the onset of the seed germination/tuber sprouting phase.

A reduced number of studies have shown that phytohormones, protein kinases and protein phosphatases cooperate in largely unknown complex signaling networks that specifically regulate the changing status of storage protein phosphorylation. Seed phosphoproteomic studies in Arabidopsis indicate a key regulatory role of phytohormone abscisic acid (ABA). It has been suggested that cruciferin tyrosine-phosphorylation is modulated by ABA, which might help to prevent cruciferin proteolysis [11]. The ABA-insensitive gene (*abi1*) encodes a protein phosphatase that is a negative regulator of ABA-dependent kinase signaling during seed development and interacts with the ubiquitin-proteosome system [10,36,37]. Wan et al. [10] reported an increased phosphorylation level of the seed storage protein cruciferin in *abi1-1* mutant seeds.

Another important issue is to elucidate why the phosphorylation status decreases when patatin is degraded following tuber sprouting. Note that the phosphorylation status should remain unchanged as long as patatin isoforms are randomly degraded. The answer to this question can be found by analyzing isoform-dependent phosphorylation changes throughout the tuber life cycle. Mean *PR*-values for Kuras and non-Kuras isoforms in the four potato stages are graphically represented in Figure 4. It is noteworthy that *PR*-values were always higher in Kuras than in non-Kuras isoforms ($p < 0.01$, Wilcoxon's signed-ranks test). Therefore, Kuras isoforms made a remarkable contribution, both qualitatively and quantitatively, to the variations of patatin phosphorylation in tuber stages. In addition, *PR*-values of Kuras and non-Kuras isoforms changed significantly in the potato tuber cycle ($p < 0.05$, one-way ANOVA test), following similar trends to those observed for all the isoforms. However, Kuras isoforms experienced a noticeable decrease in *PR* (33%) from sprouting to plant growth stage ($p < 0.01$), whereas no parallel decrease was detected in non-Kuras isoforms ($p > 0.05$). This finding supports the conclusion that differentially-phosphorylated patatin isoforms were not randomly degraded after tuber sprouting. It appears, therefore, that proteases predominantly selects highly phosphorylated isoforms during patatin mobilization. In agreement with these results, previous phosphoproteomic research in common bean showed that highly phosphorylated phaseolin isoforms are preferentially degraded at the onset of seed germination [17].

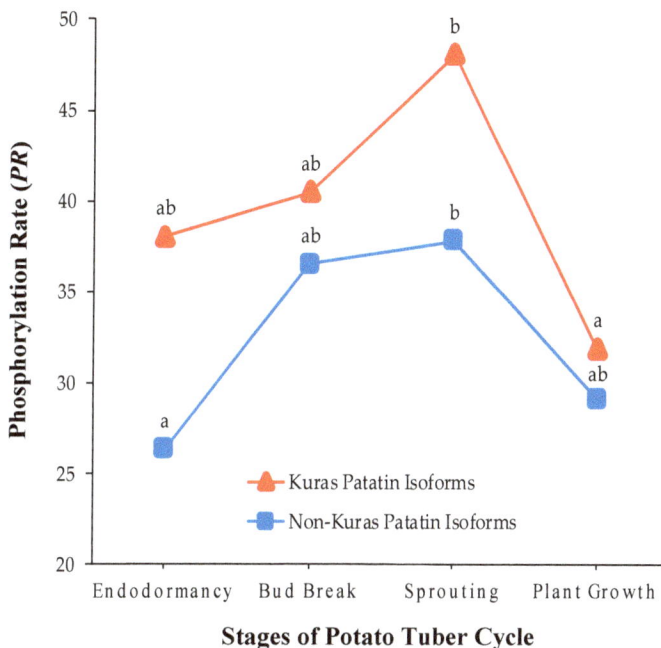

Figure 4. Temporal changes in the phosphorylation status of Kuras and non-Kuras patatin isoforms in the different stages of the potato tuber life cycle: endodormancy, bud break, sprouting and plant growth. Values on the y-axis are means of *PR* computed from four biological replicate gels over Kuras (triangles) and non-Kuras (squares) patatin spots. Different lower case letters (a and b) indicate statistically significant differences ($p < 0.05$) between sample groups by the post hoc Tukey-Kramer test following the one-way ANOVA test.

The challenge is how to explain the fact that highly phosphorylated isoforms were preferentially degraded by proteolytic enzymes. Wan et al. [10] suggested that the level of phosphorylation/dephosphorylation might destabilize tertiary and quaternary interactions of storage proteins leading to proteolytic cleavage. Patatin is an 88-kDa dimer, each subunit has approximately 366 aminoacids (17 Tyr) with negative and positive charges of the side-chains randomly distributed throughout the protein sequence, while amino acid residues are highly structured containing about 33% α-helical and 45% β-stranded structures [38,39]. It could be hypothesized that an excess of phosphates with a negative charge in highly phosphorylated patatin isoforms might disrupt the structure and facilitate proteolytic degradation. In silico prediction of patatin phosphopeptides/phosphosites from MS data suggests pronounced phosphopeptide/phosphosite variations between endodormancy and bud break tuber stages (Supplementary Table S3; Supplementary Materials). However, a more precise knowledge of the positions of phosphosites and stoichiometry throughout the tuber life cycle using phosphopeptides enrichment procedures coupled to MS analysis would be needed to explore phosphosite interactions with the tertiary and quaternary structure of the patatin. Unfortunately, this is not an easy task because it would require the isolation and purification of 20 highly homologous (frequently higher than 90%) and immunologically indistinguishable patatin isoforms [40–42] from tuber protein extracts prior to the application of enrichment procedures. To our knowledge, this wide diversity of patatin isoforms has only successfully been separated by targeted 2-DE to date. Pots et al. [42] reported that patatin isoforms pools (A, B, C, D) isolated from the Bintje variety showed no differences in conformation and thermal conformational stability before the sprouting of potatoes. In this regard, comparison of structural properties and stability of isoforms has not been reported in fully sprouted tubers when variably phosphorylated patatin isoforms were accumulated and differentially degraded. It would be necessary to perform a comparative analysis of structural properties and stability among all previously isolated patatin isoforms from protein extracts, following tuber sprouting, in order to achieve a reliable conclusion. An alternative explanation is that increased phosphorylation levels of patatin may induce its degradation by proteases. A number of studies have shown that reversible phosphorylation/dephosphorylation in plant proteins is a significant post-translational regulatory mechanism inducing degradation by the ubiquitin-proteosome pathway [15,43,44].

Taken together, our observations open up new ways of unraveling the complex regulatory mechanisms underlying not only phosphorylation-dependent patatin mobilization, but also to explore how patatin phosphorylation affects its enzymatic activities. It is noteworthy that patatin contains a conserved amino acid motif (Gly-X-Ser-X-Gly) at the position 75-79 of the sequence of patatin Pat17, which is a catalytic domain responsible for its LAH activity with an insecticidal function [45,46]. Mutational and enzymatic activity studies revealed that the Ser residue of the catalytic domain is critical for esterase and insecticidal activity [45–47]. Therefore, it would be interesting to assess whether the active site Ser residue is phosphorylated. In vitro assays with purified patatin suggest that activation of LAH activity might be caused through phosphorylation by a protein kinase [48,49]. No evidence for phosphorylation in this Ser residue was found in our prospective identification of patatin phosphosites (Supplementary Table S3; Supplementary Materials). However, the application of targeted MS analysis using high resolution technologies should be able to answer this important question.

Further follow-up studies are clearly needed to assess the potential effects of patatin phosphorylation on stability and LAH enzymatic activity, protease activity, as well as the regulatory mechanisms linked to changes in patatin phosphorylation status by the concerted action of kinases/phosphatases. In particular, the combined use of targeted and non-targeted 2-DE-based proteomics together with gel-free MS-driven proteomics could contribute very significantly to the understanding of the complex crosstalk between patatin phosphorylation and interacting proteins.

3. Materials and Methods

3.1. Plant Material

Potato (*S. tuberosum*) tubers of Kennebec cultivar were collected from an experimental field located at Xinzo de Limia (Orense, Spain). Four biological replicates were obtained in four tuber life cycle stages: endodormancy, bud break, sprouting and plant growth (Figure 1). Endodormant tubers were collected immediately after potato harvest, while bud-broken (2 mm apical bud length) and sprouting tubers (about 5 cm bud length) were obtained in the dark at room temperature. Potato plants (about 12 cm stem length) were grown in pots with fertilized soil and water supply inside a growth chamber with photoperiod (16/8 h light/dark) and temperature (22/18 °C light/dark) control. Four tubers from each tuber stage were eventually cut into small pieces, lyophilized after grinding the tissue into a fine powder with liquid nitrogen using a pre-cooled mortar and pestle, and stored separately at −80 °C before proteomic analysis. All experimental data in bud break stage were previously reported [17].

3.2. Tuber Protein Extraction

Total protein was extracted from lyophilized tubers using the phenol method based on phenol extraction coupled to ammonium acetate/methanol precipitation [50]. 200 mg of lyophilized sample was suspended in 3 mL of extraction buffer (500 mM Tris-HCl; 700 mM sucrose; 50 mM ethylenediaminetetraacetic acid, EDTA; 100 mM KCl; 1 mM phenylmethylsulfonyl fluoride, PMSF; 2% dithiothreitol, DTT; pH 8.0). An equal volume of phenol saturated with Tris-HCL (pH 6.6–7.9) was added. The sample was centrifuged for 15 min at 5000× *g* and 4 °C (Labofuge 400 R Heraeues, St Louis, MO, USA), and the phenolic phase was recovered. The precipitation solution (0.1 M ammonium acetate in cold methanol) was added, the pellet was washed (10 mM DTT and 80% acetone) and re-suspended in lysis buffer (7 M urea; 2 M thiourea; 4% CHAPS; 10 mM DTT; and 2% Pharmalyte™ pH 3–10, GE

Healthcare, Uppsala, Sweden). The protein concentration was quantified prior electrophoresis using a modified Bradford assay kit (CB-XTM Protein Assay kit; GBiosciences, St Louis, MO, USA) on a CroMate® 4300 Microplate Reader. (Awareness Technology, Palm City, FL, USA) and bovine serum albumin (BSA) as standard.

3.3. Protein Chemical Dephosphorylation

Protein extracts from the four tuber stages were chemically dephosphorylated with hydrogen fluoride-pyridine (HF-P) as described previously [51], with some modifications [17,23]. Tuber protein extract (1 mg) was added to 250 µL of HF-P (70%) and incubated for 2 h in an ice water bath. The solution was neutralized with 10 M NaOH and desalinated by using Amicon Ultra-4 centrifugal filter devices (Millipore, Billerica, MA, USA) and then eluted in 300 µL of lysis buffer. Protein purification was eventually carried out using the Clean-Up kit (GE Healthcare, Chicago, IL, USA). The ovalbumin phosphoprotein marker (45.0 kDa, Molecular Probes, Leiden, the Netherlands) was used to control the efficiency of the dephosphorylation reaction. A quantitative comparison of patatin spot volumes on two-dimensional (2-DE) gels from treated versus untreated protein samples of the same tuber was used to estimate their levels of phosphorylation. Four independent biological replicates were used at each tuber stage (i.e., 16 tubers and 32 gels).

3.4. Two-Dimensional Electrophoresis (2-DE)

High-resolution patatin profiles from protein extracts were obtained using previously described targeted 2-DE protocols [23,24]. Briefly, first-dimension isoelectric focusing (IEF) was conducted on a PROTEAN® IEF Cell (Bio-Rad Laboratories, Hercules, CA, USA) system using 24-cm-long immobilized pH gradient (IPG) strips (ReadyStrips™, Bio-Rad Laboratories) with linear pH gradient (pH 4–7). Protein extracts (75 µg) dissolved in lysis buffer were mixed with rehydration buffer (7 M urea, 2 M thiourea, 4% CHAPS, 0.002% bromophenol blue) to a final volume of 450 µL, 0.6% DTT and 1% IPG buffer (Bio-Rad Laboratories). The mixture was loaded onto initially rehydrated (50 V for 12 h) gel strips and then an increasing voltage was applied up to 70 kVh. Focused strips were subsequently equilibrated with two equilibration buffers (buffer I: 50 mM Tris pH 8.8, 6 M urea, 30% glycerol and 2% SDS; 1% DTT; buffer II: 50 mM Tris pH 8.8, 6 M urea, 30% glycerol and 2% SDS; 2.5% iodoacetamide) for 15 min each at room temperature. For second-dimension separation, strips were transferred to 10% sodium dodecyl sulphate-polyacrylamide gel electrophoresis (SDS-PAGE) gels (24 × 20 cm) and run on an Ettan™ DALTsix multigel electrophoresis system (GE, Healthcare, Uppsala, Sweden) at 16 mA/gel for 15 h. Molecular-mass markers ranging from 15 to 200 kDa (Fermentas, Burlington, Ontario, Canada) were loaded into a lateral well of SDS-PAGE gels.

3.5. Gel Staining for Total Protein and Phosphoproteins

Gel staining for total patatin was performed with SYPRO Ruby fluorescent dye (Lonza, Rockland, ME, USA) following the manufacturer's instructions. In-gel identification of phosphorylated patatin isoforms was conducted with Pro-Q Diamond phosphoprotein-specific stain (Molecular Probes) following the protocol described in Agrawal and Thelen [52], with some minor modification [17]. The PeppermintStick™ (Molecular Probes, Thermo Fisher Scientific, Waltham, MA, USA) phosphorylated and unphosphorylated molecular weight markers were added to tuber protein extract prior 2-DE as positive and negative controls of Pro-Q Diamond staining. Gels stained with Pro-Q Diamond were subsequently stained with SYPRO Ruby to assess which of the patatin isoforms were phosphorylated on the same 2-DE gel.

3.6. Image Analysis

Images of 2-DE gels stained with SYPRO Ruby and Pro-Q Diamond fluorescent dyes were digitalized with the Gel Doc™ XR + System (Bio-Rad Laboratories) and processed with the PDQuest™ Advanced software v. 8.0.1 (Bio-Rad Laboratories). Automatic spot matching across gels following background subtraction and normalization with the total density of valid spots was checked manually. Only patatin spots reproducibly detected in at least three replicates were selected for further analyses. The experimental relative molecular mass (M_r) and isoelectric point (pI) of spots was estimated from their gel position relative to the standard molecular mass markers and IPG-strip linear range of pH 4–7, respectively.

3.7. Mass Spectrometry (MS)

Kuras and non-Kuras patatin isoforms were identified at endodormancy tuber stage by MALDI-TOF and MALDI-TOF/TOF MS as previously [23]. Briefly, selected spots were subjected to in-gel proteolytic digestion with modified porcine trypsin (Promega, Madison, WI, USA) according to Jensen et al. [53]. Eluted tryptic peptides were dried in a SpeedVac (Thermo Fisher Scientific, Waltham, MA, USA) and stored at −20 °C. Dried peptide samples were redissolved with 0.5% formic acid and mixed with matrix solution (3 mg/mL CHCA in 50% ACN, 0.1% TFA). Peptide solution was deposited onto a 384 Opti-TOF MALDI target plate (Applied Biosystems, Foster City, CA. USA) using the "thin layer" procedure [54]. Mass spectra of each sample were adquired in the positive-ion reflector mode on a 4800 MALDI-TOF/TOF mass spectrometer (Applied Biosystems), Nd:YAG laser at 355 nm, an average accumulation of 1000 laser shots and at least three trypsin autolysis fragments for internal mass calibration. Tandem MS data were obtained from selected precursor ions with a relative resolution of 300 full width at half-maximum (FWHM) and metastable suppression. Automated analysis of peptide mass fingerprinting (PMF) and peptide fragmentation spectra was performed using the 4000 Series Explorer Software v. 3.5 (Applied Biosystems). GPS Explorer Software v. 3.6 using Mascot software v. 2.1 (Matrix Science, Boston, MA, USA) was used for the combined search of PMF and MSMS spectrum data against the *S. tuberosum* UniProtKB/Swiss-Prot databases. Database searching settings were precursor mass tolerance of 30 ppm, fragment mass tolerance of 0.35 Da, one missed cleavage site allowed, carbamidomethyl cysteine (CAM) as fixed modification and oxidized methionine as variable modification. All identifications and spectra were checked manually. The identification of Kuras and non-Kuras patatin isoforms required at the least three matched peptides and statistically significant (p-value < 0.05) Mascot probability scores. Prediction of patatin phosphopeptides/phosphosites was performed from spectra data allowing phosphoserine (PhosphoS), phosphotyrosine (PhosphoY) and phosphothreonine (PhosphoT) residues as variable modification to search against the UniProtKB/Swiss-Prot databases.

3.8. Statistical Analysis

The phosphorylation rate for each protein spot was quantified by the coefficient $PR = [(X − Y)/X] \times 100$, in which X and Y are the 2-DE spot volumes from untreated and treated tuber protein extracts with HF-P, respectively [17]. Non-parametric bootstrap confidence intervals obtained by the bias-corrected percentile method were used to test changes in patatin abundance (mean values) during tuber life cycle stages, following the procedure described previously [55]. Bootstrap confidence intervals (95%) were adjusted with the Bonferroni correction for multiple comparisons. The usual statistical tests (i.e., Wilcoxon's signed-ranks, Shapiro-Wilk, one-way ANOVA and the Tukey-Kramer post-hoc tests) were performed with the IBM SPSS Statistics 21 (SPSS, Chicago, IL, USA) software package.

4. Conclusions

In this study, we have provided, for the first time, confident evidence of qualitative and quantitative temporal changes in the phosphorylation status of patatin isoforms. It is also the first study on this subject in VSPs. Our data revealed that isoforms with major extent of phosphorylation accumulated progressively from endodormancy up to reaching the tuber sprouting stage. The joint analysis of previously published proteomic studies in distinct SSPs showed similar dynamic temporal patterns. These temporal patterns acquired their full meaning when our observations showed that highly phosphorylated isoforms were preferably degraded immediately after the onset of tuber sprouting.

Similar results have been reported in the mobilization of the major common bean storage protein. Therefore, the evidence available suggests that degradation of storage proteins dependent on the status of phosphorylation of specific isoforms might be a general mobilization mechanism involving both VSPs and SSPs. Further research is clearly required to assess the signaling networks that trigger and regulate these complex processes.

Supplementary Materials: Supplementary materials can be found at http://www.mdpi.com/1422-0067/20/8/1889/s1.

Author Contributions: C.Z. conceived and designed the experiments. J.B., D.M., M.L.-P., L.G. and D.F. performed the experiments. J.B., D.M., M.L.-P., L.G., D.F. and C.Z. analyzed the data; J.B. and C.Z. wrote the manuscript. All authors read and approved the final manuscript.

Funding: This research was supported by funds from the Consellería do Medio Rural (Xunta de Galicia, Spain).

Acknowledgments: The authors would like to thank the anonymous reviewers for their valuable comments and suggestions to improve the quality of the article.

Conflicts of Interest: The authors declare no conflict of interest.

Abbreviations

2-DE	Two-dimensional electrophoresis
ABA	Abscisic acid
HF-P	Hydrogen fluoride-pyridine
LAH	Lipid acyl hydrolase
M_r	Relative molecular mass
MALDI-TOF	Matrix-assisted laser desorption/ionization time-of-flight
MS	Mass spectrometry
pI	Isoelectric point
PR	Phosphorylation rate
SSP	Seed storage protein
VSP	Vegetative storage protein

References

1. De Souza Cândido, E.; Pinto, M.F.; Pelegrini, P.B.; Lima, T.B.; Silva, O.N.; Pogue, R.; Grossi-de-Sá, M.F.; Franco, O.L. Plant storage proteins with antimicrobial activity: Novel insights into plant defense mechanisms. *FASEB J.* **2011**, *25*, 3290–3305. [CrossRef] [PubMed]
2. Tan-Wilson, A.L.; Wilson, K.A. Mobilization of seed protein reserves. *Physiol. Plant.* **2012**, *145*, 140–153. [CrossRef] [PubMed]
3. Müntz, K. Deposition of storage proteins. *Plant Mol. Biol.* **1998**, *38*, 77–99. [CrossRef]
4. Weber, H.; Borisjuk, L.; Wobus, U. Molecular physiology of legume seed development. *Annu. Rev. Plant Biol.* **2005**, *56*, 253–279. [CrossRef] [PubMed]
5. Sonnewald, S.; Sonnewald, U. Regulation of potato tuber sprouting. *Planta* **2014**, *239*, 27–38. [CrossRef] [PubMed]
6. Agrawal, G.K.; Thelen, J.J. Large-scale identification and quantitative profiling of phosphoproteins expressed during seed filling in oilseed rape. *Mol. Cell. Proteom.* **2006**, *5*, 2044–2059. [CrossRef]
7. Meyer, L.J.; Gao, J.; Xu, D.; Thelen, J.J. Phosphoproteomic analysis of seed maturation in *Arabidopsis*, rapeseed, and soybean. *Plant Physiol.* **2012**, *159*, 517–528. [CrossRef] [PubMed]
8. Mouzo, D.; Bernal, J.; López-Pedrouso, M.; Franco, D.; Zapata, C. Advances in the biology of seed and vegetative storage proteins based on two-dimensional electrophoresis coupled to mass spectrometry. *Molecules* **2018**, *23*, 2462. [CrossRef] [PubMed]
9. Irar, S.; Oliveira, E.; Pagès, M.; Goday, A. Towards the identification of late-embryogenic-abundant phosphoproteome in *Arabidopsis* by 2-DE and MS. *Proteomics* **2006**, *6*, 175–185. [CrossRef] [PubMed]
10. Wan, L.; Ross, A.R.S.; Yang, J.; Hegedus, D.D.; Kermode, A.R. Phosphorylation of the 12 S globulin cruciferin in wild-type and abi1-1 mutant *Arabidopsis thaliana* (thale cress) seeds. *Biochem. J.* **2007**, *404*, 247–256. [CrossRef]

11. Ghelis, T.; Bolbach, G.; Clodic, G.; Habricot, Y.; Miginiac, E.; Sotta, B.; Jeannette, E. Protein tyrosine kinases and protein tyrosine phosphatases are involved in abscisic acid-dependent processes in *Arabidopsis* seeds and suspension cells. *Plant Physiol.* **2008**, *148*, 1668–1680. [CrossRef] [PubMed]

12. Umezawa, T.; Sugiyama, N.; Takahashi, F.; Anderson, J.C.; Ishihama, Y.; Peck, S.C.; Shinozaki, K. Genetics and phosphoproteomics reveal a protein phosphorylation network in the abscisic acid signaling pathway in *Arabidopsis thaliana*. *Plant Biol.* **2013**, *6*, rs8. [CrossRef] [PubMed]

13. Capraro, J.; Scarafoni, A.; Magni, C.; Consonni, A.; Duranti, M.M. Proteomic studies on the lupin seed storage proteins phosphorylation. In Proceedings of the International Conference on Food-omics, Cesena, Italy, 22–24 June 2011.

14. Quiroga, I.; Regente, M.; Pagnussat, L.; Maldonado, A.; Jorrín, J.; de la Canal, L. Phosphorylated 11S globulins in sunflower seeds. *Seed Sci. Res.* **2013**, *23*, 199–204. [CrossRef]

15. Han, C.; Wang, K.; Yang, P. Gel-based comparative phosphoproteomic analysis on rice during germination. *Plant Cell Physiol.* **2014**, *55*, 1376–1394. [CrossRef] [PubMed]

16. Dong, K.; Zhen, S.; Cheng, Z.; Cao, H.; Ge, P.; Yah, Y. Proteomic analysis reveals key proteins and phosphoproteins upon seed germination of wheat (*Triticum aestivum* L.). *Front. Plant Sci.* **2015**, *6*, 1017. [CrossRef]

17. López-Pedrouso, M.; Alonso, J.; Zapata, C. Evidence for phosphorylation of the major seed storage protein of the common bean and its phosphorylation-dependent degradation during germination. *Plant. Mol. Biol.* **2014**, *84*, 415–428. [CrossRef]

18. Beardmore, T.; Wetzel, S.; Burgess, D.; Charest, P.J. Characterization of seed storage proteins in *Populus* and their homology with *Populus* vegetative storage proteins. *Tree Physiol.* **1996**, *16*, 833–840. [CrossRef] [PubMed]

19. Racusen, D.; Foote, M.A. A major soluble glycoprotein from potato tubers. *J. Food Biochem.* **1980**, *4*, 43–52. [CrossRef]

20. Racusen, D. Lipid acyl hydrolase of patatin. *Can. J. Bot.* **1984**, *62*, 1640–1644. [CrossRef]

21. Liu, Y.-W.; Han, C.-H.; Lee, M.-H.; Hsu, F.-L.; Hou, W.-C. Patatin, the tuber protein of potato (*Solanum tuberosum* L.) exhibits antioxidant activity in vitro. *J. Agric. Food Chem.* **2003**, *51*, 4389–4393. [CrossRef]

22. Shewry, P.R. Tuber storage proteins. *Ann. Bot.* **2003**, *91*, 755–769. [CrossRef]

23. Bernal, J.; López-Pedrouso, M.; Franco, D.; Bravo, S.; García, L.; Zapata, C. Identification and mapping of phosphorylated isoforms of the major storage protein of potato based on two-dimensional electrophoresis. In *Advances in Seed Biology*; Jimenez-Lopez, J.C., Ed.; InTech: Rijeka, Croatia, 2017; pp. 65–82. ISBN 978-953-51-3621-7.

24. Mouzo, D.; López-Pedrouso, M.; Bernal, J.; García, L.; Franco, D.; Zapata, C. Association of patatin-based proteomic distances with potato (*Solanum tuberosum* L.) quality traits. *J. Agric. Food Chem.* **2018**, *66*, 11864–11872. [CrossRef]

25. Viola, R.; Pelloux, J.; van der Ploeg, A.; Gillespie, T.; Marquis, N.; Roberts, A.G.; Hancock, R.D. Symplastic connection is required for bud outgrowth following dormancy in potato (*Solanum tuberosum* L.) tubers. *Plant Cell Environ.* **2007**, *30*, 973–983. [CrossRef]

26. de la Fuente, M.; López-Pedrouso, M.; Alonso, J.; Santalla, M.; de Ron, A.M.; Alvarez, G.; Zapata, C. In-depth characterization of the phaseolin protein diversity of common bean (*Phaseolus vulgaris* L.) based on two-dimensional electrophoresis and mass spectrometry. *Food Technol. Biotechnol.* **2012**, *50*, 315–325.

27. López-Pedrouso, M.; Bernal, J.; Franco, D.; Zapata, C. Evaluating two-dimensional electrophoresis profiles of the protein phaseolin as markers of genetic differentiation and seed protein quality in common bean (*Phaseolus vulgaris* L.). *J. Agric. Food Chem.* **2014**, *62*, 7200–7208. [CrossRef]

28. Bauw, G.; Nielsen, H.V.; Emmersen, J.; Nielsen, K.L.; Jørgensen, M.; Welinder, K.G. Patatin, Kunitz protease inhibitors and other major proteins in tuber of potato cv. Kuras. *FEBS J.* **2006**, *273*, 3569–3584. [CrossRef]

29. Lehesranta, S.J.; Davies, H.V.; Shepherd, L.V.T.; Koistinen, K.M.; Massat, N.; Nunan, N.; McNicol, J.W.; Kärenlampi, S.O. Proteomic analysis of the potato tuber life cycle. *Proteomics* **2006**, *6*, 6042–6052. [CrossRef]

30. Weeda, S.M.; Kumar, G.N.M.; Knowles, N.R. Correlative changes in proteases and protease inhibitors during mobilization of protein from potato (*Solanum tuberosum* L.) seed tubers. *Funct. Plant Biol.* **2010**, *37*, 32–42. [CrossRef]

31. Bachem, C.; Van der Hoeven, R.; Lucker, J.; Oomen, R.; Casarini, E.; Jacobsen, E.; Visser, R. Functional genomic analysis of potato tuber life-cycle. *Potato Res.* **2000**, *43*, 297–312. [CrossRef]

32. Ronning, C.M.; Stegalkina, S.S.; Ascenzi, R.A.; Bougri, O.; Hart, A.L.; Utterbach, T.R.; Vanaken, S.E.; Riedmuller, S.B.; White, J.A.; Cho, J.; et al. Comparative analyses of potato expressed sequence tag libraries. *Plant Physiol.* **2003**, *131*, 419–429. [CrossRef]

33. Campbell, M.; Segear, E.; Beers, L.; Knauber, D.; Suttle, J. Dormancy in potato tuber meristems: Chemically induced cessation in dormancy matches the natural process based on transcript profiles. *Funct. Integr. Genom.* **2008**, *8*, 317–328. [CrossRef]

34. Pots, A.M.; Gruppen, H.; van Diepenbeek, R.; van der Lee, J.J.; van Boekel, M.A.J.S.; Wijngaards, G.; Voragen, A.G.J. The effect of whole potatoes of three cultivars on the patatin and protease inhibitor content; a study using capillary electrophoresis and MALDI-TOF mass spectrometry. *J. Sci. Food Agric.* **1999**, *79*, 1557–1564. [CrossRef]

35. Weeda, S.M.; Kumar, G.N.M.; Knowles, N.R. Developmentlly linked changes in proteases and protease inhibitors suggest a role for potato multicystatin in regulating protein content of potato tubers. *Planta* **2009**, *230*, 73–84. [CrossRef] [PubMed]

36. Gosti, F.; Beaudoin, N.; Serizet, C.; Webb, A.A.; Vartanian, N.; Giraudat, J. ABI1 protein phosphatase 2C is a negative regulator of abscisic acid signaling. *Plant Cell* **1999**, *11*, 1897–1910. [CrossRef] [PubMed]

37. Ludwików, A. Targeting proteins for proteasomal degradation—A new function of Arabidopsis ABI1 protein phosphatase 2C. *Front. Plant Sci.* **2015**, *6*, 310. [CrossRef]

38. Pots, A.M.; de Jongh, H.H.J.; Gruppen, H.; Hamer, R.J.; Voragen, A.G.J. Heat-induced conformational changes of patatin, the major potato tuber protein. *Eur. J. Biochem.* **1998**, *252*, 66–72. [CrossRef]

39. Ralet, M.-C.; Guéguen, J. Fractionation of potato proteins: Solubility, thermal, coagulation and emulsifying properties. *LWT-Food Sci. Technol.* **2000**, *33*, 380–387. [CrossRef]

40. Park, W.D.; Blackwood, C.; Mignery, G.A.; Hermodson, M.A.; Lister, R.M. Analysis of the heterogeneity of the 40,000 molecular weight tuber glycoprotein of potatoes by immunological methods and by NH_2-terminal sequence analysis. *Plant Physiol.* **1983**, *71*, 156–160. [CrossRef]

41. Mignery, G.A.; Pikaard, C.S.; Hannapel, D.J.; Park, W.D. Isolation and sequence analysis of cDNAs for the major tuber protein, patatin. *Nucleic Acids Res.* **1984**, *12*, 7987–8000. [CrossRef]

42. Pots, A.M.; Gruppen, H.; Hessing, M.; van Boekel, M.A.J.S.; Voragen, A.G.J. Isolation and characterization of patatin isoforms. *J. Agric. Food Chem.* **1999**, *47*, 4587–4592. [CrossRef]

43. Tang, G.-Q.; Hardin, S.C.; Dewey, R.; Huber, S.C. A novel C-terminal proteolytic processing of cytosolic pyruvate kinase, its phosphorylation and degradation by the proteasome in developing soybean seeds. *Plant J.* **2003**, *34*, 77–93. [CrossRef]

44. He, X.; Kermode, A.R. Programmed cell death of the megagametophyte during post-germinative growth of white spruce (*Picea glauca*) seeds is regulated by reactive oxygen species and the ubiquitin-mediated proteolytic system. *Plant Cell Physiol.* **2010**, *51*, 1707–1720. [CrossRef]

45. Hirschberg, H.J.H.B.; Simons, J.-W.F.A.; Dekker, N.; Egmond, M.R. Cloning, expression, purification and characterization of patatin, a novel phospholipase A. *Eur. J. Biochem.* **2001**, *268*, 5037–5044. [CrossRef]

46. Rydel, T.J.; Williams, J.M.; Krieger, E.; Moshiri, F.; Stallings, W.C.; Brown, S.M.; Pershing, J.C.; Purcell, J.P.; Alibhai, M.F. The crystal structure, mutagenesis, and activity studies reveal that patatin is a lipid acyl hydrolase with a Ser-Asp catalytic dyad. *Biochemistry* **2003**, *42*, 6696–6708. [CrossRef]

47. Strickland, J.A.; Orr, G.L.; Walsh, T.A. Inhibition of diabrotica larval growth by patatin, the lipid acyl hydrolase from potato tubers. *Plant Physiol.* **1995**, *109*, 667–674. [CrossRef]

48. Kawakita, K.; Senda, K.; Doke, N. Factors, affecting in vitro activation of potato phospholipase A_2. *Plant Sci.* **1993**, *92*, 183–190. [CrossRef]

49. Senda, K.; Yoshioka, H.; Doke, N.; Kawakita, K. A cytosolic phospholipase A_2 from potato tissues appears to be patatin. *Plant Cell Physiol.* **1996**, *37*, 347–353. [CrossRef]

50. Saravanan, R.S.; Rose, J.K.C. A critical evaluation of sample extraction techniques for enhanced proteomic analysis of recalcitrant plant tissues. *Proteomics* **2004**, *4*, 2522–2532. [CrossRef]

51. Kuyama, H.; Toda, C.; Watanabe, M.; Tanaka, K.; Nishimura, O. An efficient chemical method for dephosphorylation of phosphopeptides. *Rapid Commun. Mass Spectrom.* **2003**, *17*, 1493–1496. [CrossRef]

52. Agrawal, G.K.; Thelen, J.J. Development of a simplified, economical polyacrylamide gel staining protocol for phosphoproteins. *Proteomics* **2005**, *5*, 4684–4688. [CrossRef]

53. Jensen, O.N.; Wilm, M.; Shevchenko, A.; Mann, M. Sample preparation methods for mass spectrometric peptide mapping directly from 2-DE gels. *Methods Mol. Biol.* **1999**, *112*, 513–530. [CrossRef]

54. Vorm, O.; Roepstorff, P.; Mann, M. Improved resolution and very high sensitivity in MALDI TOF of matrix surfaces made by fast evaporation. *Anal. Chem.* **1994**, *66*, 3281–3287. [CrossRef]

55. Franco, D.; Mato, A.; Salgado, F.J.; López-Pedrouso, M.; Carrera, M.; Bravo, S.; Parrado, M.; Gallardo, J.M.; Zapata, C. Tackling proteome changes in the longissimus thoracis bovine muscle in response to pre-slaughter stress. *J. Proteom.* **2015**, *122*, 73–85. [CrossRef]

International Journal of
Molecular Sciences

MDPI

Article

Proteomic and Biochemical Changes during Senescence of *Phalaenopsis* 'Red Dragon' Petals

Cong Chen [†], Lanting Zeng [†] and Qingsheng Ye *

Guangdong Provincial Key Lab of Biotechnology for Plant Development, School of Life Sciences, South China Normal University, Guangzhou 510631, China; parker_cc@163.com (C.C.); zenglanting@scbg.ac.cn (L.Z.)
* Correspondence: ye-lab@scnu.edu.cn; Tel.: +86-20-8521-2021
† These authors contributed equally to this work.

Received: 30 March 2018; Accepted: 26 April 2018; Published: 28 April 2018

Abstract: *Phalaenopsis* flowers are some of the most popular ornamental flowers in the world. For most ornamental plants, petal longevity determines postharvest quality and garden performance. Therefore, it is important to have insight into the senescence mechanism of *Phalaenopsis*. In the present study, a proteomic approach combined with ultrastructural observation and activity analysis of antioxidant enzymes was used to profile the molecular and biochemical changes during pollination-induced petal senescence in *Phalaenopsis* "Red Dragon". Petals appeared to be visibly wilting at 24 h after pollination, accompanied by the mass degradation of macromolecules and organelles during senescence. In addition, 48 protein spots with significant differences in abundance were found by two-dimensional electrophoresis (2-DE) and subjected to matrix-assisted laser desorption/ionization time of flight mass spectrometry (MALDI-TOF/TOF-MS). There were 42 protein spots successfully identified and homologous to known functional protein species involved in key biological processes, including antioxidant pathways, stress response, protein metabolism, cell wall component metabolism, energy metabolism, cell structure, and signal transduction. The activity of all reactive oxygen species (ROS)-scavenging enzymes was increased, keeping the content of ROS at a low level at the early stage of senescence. These results suggest that two processes, a counteraction against increased levels of ROS and the degradation of cellular constituents for maintaining nutrient recycling, are activated during pollination-induced petal senescence in *Phalaenopsis*. The information provides a basis for understanding the mechanism regulating petal senescence and prolonging the florescence of *Phalaenopsis*.

Keywords: *Phalaenopsis*; petal; pollination; senescence; 2-DE; ROS

---

1. Introduction

Phalaenopsis, named for its butterfly-like flowers, is known as the "queen of the orchids" for its graceful shape and colorful flowers. *Phalaenopsis* flowers are among the most popular ornamental flowers in the world and have high economic value [1]. For most ornamental plants, petal longevity determines postharvest quality and garden performance [2], so it is essential to have insight into the mechanism regulating petal senescence.

The longevity of *Phalaenopsis* petals is under tight developmental control for up to 3 months [3], so it is difficult to understand the mechanism of regulating petal senescence, which can be affected by many environmental factors. However, its petal senescence can be almost synchronized with pollination, and detectable signs of senescence can occur within one day [4]. Petal senescence and shedding are the earliest and most obvious changes induced by pollination [5]. Therefore, pollination treatment provides a quick and efficient approach for the study of *Phalaenopsis* petal senescence.

Petal senescence is accompanied by a series of ultrastructural and physiological-biochemical changes that form the last stage of flower development [6,7]. The maintenance of petals is costly in terms of water loss and metabolic energy; therefore, cellular constituents such as macromolecules and organelles are degraded so that nutrients can be recycled for reallocation to developing tissues [8]. As a conserved system, autophagy supports the recycling function, appearing to play a crucial role in degradation during petal senescence [6]. Reactive oxygen species (ROS) play a key role in the regulation of many developmental processes, including senescence, as well as in plant responses to biotic and abiotic stresses [9]. The activity of ROS-scavenging enzymes tends to increase during senescence [10]. Simultaneously, there is a concomitant increase in the level of malondialdehyde (MDA), which acts as an indicator of lipid peroxidation [11].

The combination of two-dimensional electrophoresis (2-DE) and matrix-assisted laser desorption/ionization time of flight mass spectrometry (MALDI-TOF/TOF-MS) analyses allows research on the plant at the post-transcriptional level; i.e., the protein level. This combination has been used to study not only plant developmental processes [12,13], but also stress responses [14,15]. These proteomic studies have provided unique insight into the role of post-translational modifications regulating and executing plant development and stress response. Coincidentally, senescence is also controlled at the post-transcriptional level [16]. Therefore, the proteomic approach is feasible for investigating petal senescence.

The regulation of gene expression occurs at the protein and the transcript levels. Therefore, the mechanism regulating petal senescence in *Phalaenopsis* is complex and unclear. In the present study, we report on the first proteomic characterization of *Phalaenopsis* petal senescence, combined with ultrastructural observation and physiological-biochemical analysis. The aim of this study is to elucidate the dynamic changes that occur at the molecular and biochemical levels during petal senescence.

2. Results

2.1. Ultrastructural Changes of Organelles in Senescing Phalaenopsis Petals

In our experiment, *Phalaenopsis* showed 2–3 months' flower longevity in natural conditions. By contrast, pollination dramatically accelerated the senescence of petals, and flowers visibly wilted at 24 h after pollination (Figure 1A).

The ultrastructural changes in *Phalaenopsis* petals during pollination-induced senescence are shown in Figure 1B–I. The plasmodesma was partially closed at 8 h and completely closed at 16 h in pollinated flowers (Figure 1B). By contrast, the Golgi bodies were almost intact, except for the local structural distortion exhibited in Figure 1C. Mesophyll cell walls showed plasmolysis at 8 h and were severely degraded at 24 h, whereas epidermis cell walls hardly changed within 24 h after pollination (Figure 1D,E). The vacuoles appeared to have lost membrane integrity as senescence progressed (Figure 1F). Cells containing vacuoles of various sizes were observed, and the vacuoles contained many vesicles and granules in pollinated flowers at 16 h. At the same time, numerous osmiophilic granules were revealed (Figure 1G). For chloroplasts, the ongoing increase in the number of osmiophilic granules was concomitant with the loss of a considerable portion of thylakoids and double-layer membrane structures (Figure 1H). Figure 1I shows that the mitochondria had swelled and the cristae had degraded to a lesser extent in the cells 24 h after pollination; however, the mitochondria were nearly intact overall. The nucleus remained until a late stage of senescence, although several nuclear ultrastructural changes were observed at 24 h, such as shrinkage trait, loss of ellipticity, and blebbing. Peroxisomes were not present in *Phalaenopsis* petal cells during pollination-induced senescence.

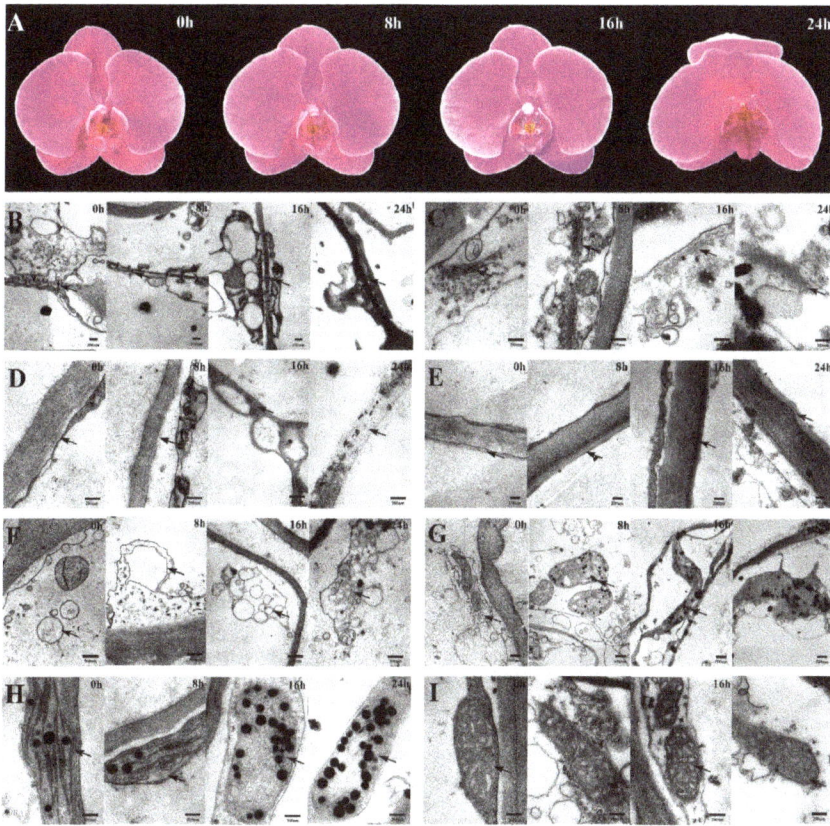

Figure 1. (**A**) Morphological changes of *Phalaenopsis* petal during pollination-induced senescence; (**B–I**) Ultrastructural changes of *Phalaenopsis* petal during pollination-induced senescence. (**B**) Plasmodesmata (×20,000); (**C**) Golgi apparatus (×40,000); (**D**) mesophyll cell wall (×40,000); (**E**) epidermal cell wall (×25,000); (**F**) vacuole (×15,000); (**G**) osmiophilic granule (×10,000); (**H**) chloroplast (×40,000); (**I**) mitochondria (×40,000). Corresponding structures in each figure are indicated by arrows.

2.2. Protein Profiling and Analysis of Protein Species Changes

Protein yield obtained from *Phalaenopsis* petals after phenol-based extraction was evaluated. Protein yields were the same among petals at four time points, in the range of 6.03 ± 0.94 mg/g of fresh weight (Table 1).

To analyze the variation of protein species in pollination-induced senescing *Phalaenopsis* petals, the differentially regulated protein spots were separated by 2-DE and identified with MALDI-TOF/TOF-MS. In 2-DE of the petal protein extract, more than 1000 protein spots per gel were consistently observed in all replicates, with molecular weight (MW) and isoelectric point (pI) ranging from <14 kDa to 115 kDa and 4 to 9, respectively (Figure 2). As shown in Table 1, the average proteomic maps were 1069 ± 92, 1061 ± 203, 1056 ± 80, and 1014 ± 8 for 0, 8, 16, and 24 h, respectively. The mean coefficient of variance (CV) for all of these samples was 9.12%. Based on the criteria for protein spot detection, 48 protein spots with significant differences in abundance were detected in response to pollination and 42 protein spots were confidently identified according to the databases; six protein spots (3, 21, 22, 27, 28, and 45) could not be identified conclusively.

Table 1. Protein yield (mg/g fresh weight), number of spots, and significant quantitative difference (spots up/downregulated) in *Phalaenopsis* petals at each time point.

Samples	Protein Yield (mg/g Fresh Weight)	Number of Spots (Mean ± SD)	Quantitative Difference	
			Number of Upregulated	Number of Downregulated
0 h		1069 ± 92	0	0
8 h	6.03 ± 0.94	1061 ± 203	1	14
16 h		1056 ± 80	5	22
24 h		1014 ± 8	16	23

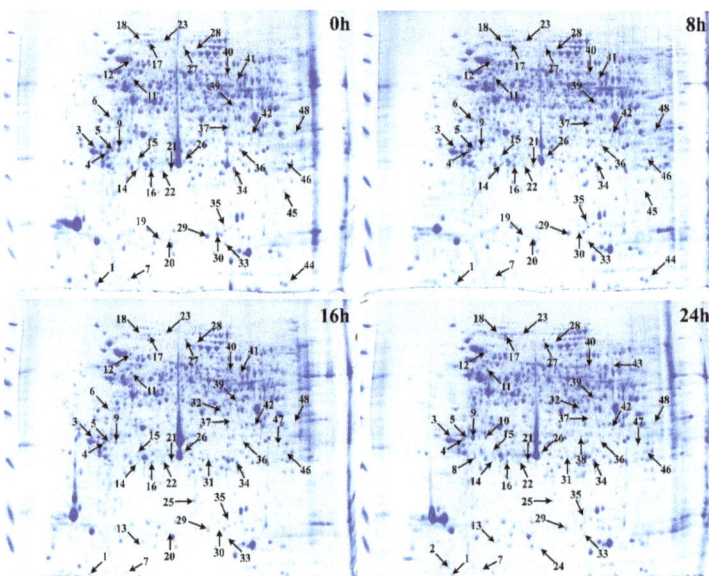

Figure 2. Representative two-dimensional electrophoresis (2-DE) gels of *Phalaenopsis* petal proteomic variation during pollination-induced senescence. The protein spots were separated on immobilized pH gradient (IPG) dry strips (24 cm in length, pH 3–10 nonlinear gradient (NL)). The numbers on the left in the images indicate the corresponding protein spots listed in Table 2.

2.3. Functional Classification of Differentially Regulated Protein Species

Among 42 protein spots, during petal senescence, 17 protein spots (2, 8, 10, 13, 17, 18, 23, 24, 25, 31,32, 37, 38, 39, 43, 46, and 47) were upregulated and 25 protein spots (1, 4, 5, 6, 7, 9, 11, 12, 14, 15, 16, 19, 20, 26, 29, 30, 33, 34, 35, 36, 40, 41, 42, 44, and 48) were downregulated at one or more time points compared with 0 h. The 2-DE map acquired from 0 h petals was taken as the control, and qualitative differences in spot intensity between the control (0 h) and the pollination treatment (8, 16, or 24 h) were found and are displayed in Table 1. The number of differentially regulated protein spots was obtained from the changes of 42 spots' abundance. As shown in Table 1, by 8 h after pollination, one protein spot was upregulated and 14 protein spots were downregulated in petals by greater than 1.5-fold ($p < 0.05$). There were 5 protein spots upregulated and 22 protein spots downregulated at 16 h ($p < 0.05$). At 24 h after pollination, the numbers of upregulated and downregulated protein spots were 16 and 23, respectively.

All proteins identified and the correspondence between a given spot number and the assigned protein species are detailed in Table 2. According to the Gene Ontology (GO) and Kyoto Encyclopedia of Genes and Genomes (KEGG) annotations, as well as the related references, the 42 protein species were classified into eight groups (Figure 3): six (14.3%) were involved in the antioxidant pathway, 11 (26.2%) in the stress process, five (11.9%) in protein metabolism, three (7.1%) in cell wall component metabolism, eight (19.1%) in energy metabolism, three (7.1%) in cell structure, four (9.5%) in signal transduction, and the remaining two (4.8%) were not classified.

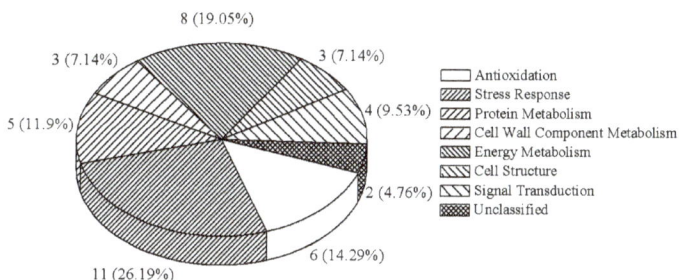

Figure 3. Functional categorization of the differentially regulated protein species in *Phalaenopsis* petals during pollination-induced senescence.

In this study, protein species were identified as being associated with pollination-induced senescence, improving our knowledge of how senescence progresses in *Phalaenopsis* petals. Further studies are needed to characterize the mechanism of pollination-induced senescence in more detail.

2.4. Changes in Antioxidant Enzymes and MDA

In the proteomic study described above, it was found that thioredoxin-dependent peroxidase (POD, spot 10) was upregulated during petal senescence. However, the relationship between POD regulation and activity changes was unknown. Hence, the activity of three ROS-scavenging enzymes, superoxide dismutase (SOD), catalase (CAT), and peroxidase (POD), was measured. During petal senescence, the activity of POD continuously increased (Figure 4A), which was consistent with the proteomic result. The activity of SOD and CAT changed simultaneously with POD, but noticeably decreased after 8 h in pollinated flowers (Figure 4A). Additionally, malondialdehyde (MDA) levels content increased considerably in the senescing petals of pollinated flowers compared with the control samples (0 h) (Figure 4B).

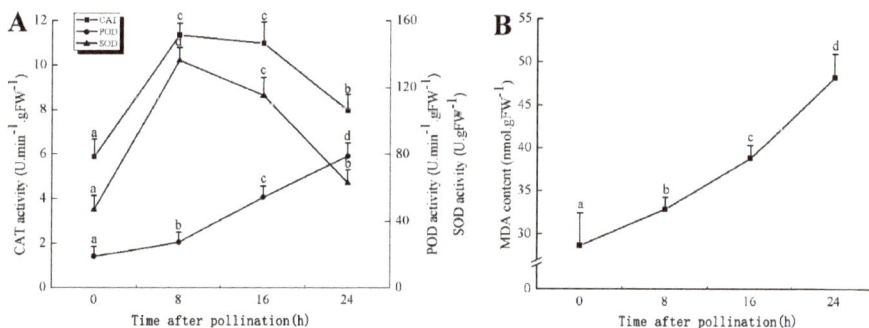

Figure 4. (**A**) Changes of superoxide dismutase (SOD), catalase (CAT), and peroxidase (POD) activity in *Phalaenopsis* petal during pollination-induced senescence; (**B**) Change of malondialdehyde (MDA) content in *Phalaenopsis* petal during pollination-induced senescence. Values are presented as means ± SD. Means distinguished with different letters are significantly different from each other ($p \leq 0.05$).

Table 2. Identification, functional categorization, and quantification of the differentially regulated protein species in *Phalaenopsis* petal during pollination-induced senescence.

Spot No. [a]	Protein Name and Organism [b]	Accession No. [c]	Exp./Theo. [d] MW (kDa)	pI	Score/Matched Peptides/ Coverage [e]	Cellular Location [f]	False Discovered Rate [g]	Changes of Regulation Intensity [h]
					Antioxidation			
10	Putative thioredoxin-dependent peroxidase (*Elaeis guineensis*)	gi|448872680	30.49/23.63	5.48/6.61	587/6(6)/40	V/ER/GA/C/M	0.002	
7	Thioredoxin H-type (*Zea mays*)	gi|195645418	6.03/14.06	5.39/5.27	359/5(4)/37	ER	0.046	
14	Glutathione S-transferase (*Medicago truncatula*)	gi|357460737	24.93/32.54	5.62/6.52	259/3(1)/10	CW/N/C/M/GA/Pl/Ch/A	0.000	
16	Glutathione S-transferase (*Vitis amurensis*)	gi|224038272	24.85/32.54	5.95/6.52	325/4(2)/14	C	0.003	
34	Glutathione S-transferase (*Triticum urartu*)	gi|474001794	25.72/29.98	7.61/5.51	326/3(3)/17	C/A	0.011	

Table 2. *Cont.*

Spot No. [a]	Protein Name and Organism [b]	Accession No. [c]	Exp./Theo. [d] MW (kDa)	Exp./Theo. [d] pI	Score/Matched Peptides/ Coverage [e]	Cellular Location [f]	False Discovered Rate [g]	Changes of Regulation Intensity [h]
					Antioxidation			
41	Nicotinamide adenine dinucleotide (phosphate)(NAD(P))-linked oxidoreductase superfamily protein (*Theobroma cacao*)	gi|590655852	54.05/16.20	7.68/8.44	54/1(1)/9	C	0.005	
					Stress Response			
20	17.7 kDa heat shock protein (*Carica papaya*)	gi|379933812	14.37/24.02	6.33/5.26	448/3(3)/15	N/ER	0.000	
29	Heat shock protein 17.9 (*Cenchrus americanus*)	gi|238915387	16.30/19.00	7.23/9.30	535/5(3)/21	N/C/M/A	0.002	
30	17.4 kDa heat shock protein (*Oryza sativa* Japonica group)	gi|313575791	16.30/25.37	7.23/5.57	654/6(5)/18	-	0.000	
33	Small molecular heat shock protein 17.5 (*Nelumbo nucifera*)	gi|118452817	13.31/17.57	7.42/5.94	147/2(1)/10	-	0.025	

Table 2. *Cont.*

Spot No. [a]	Protein Name and Organism [b]	Accession No. [c]	Exp./Theo. [d] MW (kDa)	Exp./Theo. [d] pI	Score/Matched Peptides/Coverage [e]	Cellular Location [f]	False Discovered Rate [g]	Changes of Regulation Intensity [h]
				Stress Response				
5	Heat shock protein Hsp70 (*Mucilaginibacter paludis*)	gi|495787168	30.52/25.82	5.01/5.57	346/3(3)/15	Mi/C/Ch/A	0.000	
12	Heat shock protein 70 cognate (*Populus trichocarpa*)	gi|224100969	62.31/71.53	5.49/5.11	868/8(7)/19	Ch	0.005	
31	Phospholipase D (*Coffea arabica*)	gi|332182725	26.51/19.07	7.02/5.39	238/3(1)/27	-	0.015	
1	Lectin (*Cymbidium hybrid cultivar*)	gi|436827	6.68/13.26	4.55/9.42	171/2(2)/11	-	0.001	
6	Dehydrin (*Hyacinthus orientalis*)	gi|47026904	39.03/19.30	5.13/6.37	173/2(2)/9	-	0.003	

Table 2. *Cont.*

Spot No. [a]	Protein Name and Organism [b]	Accession No. [c]	Exp./Theo. [d] MW (kDa)	Exp./Theo. [d] pI	Score/Matched Peptides/ Coverage [e]	Cellular Location [f]	False Discovered Rate [g]	Changes of Regulation Intensity [h]
					Stress Response			
35	Dehydrin 13 (*Zea mays*)	gi|226501978	17.18/21.10	7.38/6.29	112/2(0)/9	-	0.003	
42	Alcohol dehydrogenase (NADP+) A (*Aegilops tauschii*)	gi|475594485	34.95/29.47	7.98/9.54	421/6(4)/14	AC	0.002	
					Protein Metabolism			
32	Cysteine proteinase (*Phalaenopsis sp. SM9108*)	gi|1173630	40.95/40.49	7.27/6.23	519/7(2)/23	C	0.043	
37	Ubiquitin fusion degradation protein 1, partial (*Solanum nigrum*)	gi|321149977	36.92/26.18	7.50/8.74	236/3(2)/16	C	0.001	
43	Proteasome subunit β type 1 (*Zea mays*)	gi|226531171	56.66/23.19	8.01/6.11	346/3(3)/19	Pe/N/C/M/Ch/A	0.048	

Table 2. *Cont.*

Spot No. [a]	Protein Name and Organism [b]	Accession No. [c]	Exp./Theo. [d] MW (kDa)	Exp./Theo. [d] pI	Score/Matched Peptides/ Coverage [e]	Cellular Location [f]	False Discovered Rate [g]	Changes of Regulation Intensity [h]
			Protein Metabolism					
23	Aminopeptidase N (*Morus notabilis*)	gi|587846889	90.12/32.60	6.21/8.57	94/1(1)/4	-	0.007	
24	FK506-binding protein 2-2 (*Aegilops tauschii*)	gi|475591369	10.60/12.41	6.47/5.51	104/2(2)/16	Ch/P	0.000	
			Cell Wall Component Metabolism					
9	Cellulose synthase-3 (*Zea mays*)	gi|9622878	30.25/25.82	5.16/5.57	335/3(3)/17	-	0.001	
38	Xyloglucan endotransglucosylase/hydrolase (*Gossypium hirsutum*)	gi|308229784	32.37/31.07	7.32/9.16	144/1(1)/26	C	0.043	
39	Cinnamyl alcohol dehydrogenase (*Lolium perenne*)	gi|19849246	44.18/24.10	7.59/8.77	204/2(2)/13	CW/Mi/V/C/Pl/Ch/A/M/GA	0.008	

Table 2. *Cont.*

Spot No. [a]	Protein Name and Organism [b]	Accession No. [c]	Exp./Theo. [d] MW (kDa)	pI	Score/Matched Peptides/Coverage [e]	Cellular Location [f]	False Discovered Rate [g]	Changes of Regulation Intensity [h]
					Energy Metabolism			
40	Pyrophosphate-fructose 6-phosphate 1-phosphotransferase subunit β (*Medicago truncatula*)	gi|357480393	55.94/62.89	7.47/5.88	376/4(3)/16	GA/M	0.004	
8	Triosephosphate isomerase (*Oryza coarctata*)	gi|1165973012	25.24/20.21	5.22/5.19	434/4(4)/34	Mi	0.000	
15	Triosephosphate isomerase (*Gossypium hirsutum*)	gi|295687231	26.74/33.50	5.61/6.66	230/4(3)/13	CW/Mi/M/Ch/GA	0.024	
11	Dihydrolipoamide acetyltransferase component of pyruvate dehydrogenase (*Cucumis melo* subsp. *Melo*)	gi|307135863	54.07/27.53	5.51/7.08	183/2(2)/12	N/Mi/V/C/Pl/Ch/M	0.011	
19	Dihydrolipoamide dehydrogenase (*Gossypium hirsutum*)	gi|211906492	15.06/28.79	6.15/5.47	300/4(3)/26	-	0.000	

Table 2. Cont.

Spot No. [a]	Protein Name and Organism [b]	Accession No. [c]	Exp./Theo. [d] MW (kDa)	Exp./Theo. [d] pI	Score/Matched Peptides/Coverage [e]	Cellular Location [f]	False Discovered Rate [g]	Changes of Regulation Intensity [h]
					Energy Metabolism			
17	Pyruvate orthophosphate dikinase (*Eleocharis vivipara*)	gi|2285879	86.71/94.27	5.89/5.29	422/7(3)/17	N/C/Ch	0.042	
18	Pyruvate orthophosphate dikinase (*Amaranthus hypochondriacus*)	gi|336020527	88.12/96.65	5.72/5.21	504/8(5)/17	CW/A	0.037	
26	Mitochondrial adenosine triphosphate(ATP) synthase 24 kDa subunit, partial (*Oryza sativa* indica group)	gi|149392623	27.10/20.84	6.51/9.45	377/5(4)/16	C	0.003	
					Cell Structure			
2	Putative profilin (*Phalaenopsis* hybrid cultivar)	gi|45112111	5.63/24.42	4.43/5.08	158/3(3)/19	C	0.000	
13	Actin-depolymerizing factor (*Gossypium barbadense*)	gi|161779424	10.83/33.22	5.69/8.89	435/4(4)/25	N/C	0.000	

Table 2. *Cont.*

Spot No. [a]	Protein Name and Organism [b]	Accession No. [c]	Exp./Theo. [d] MW (kDa)	Exp./Theo. [d] pI	Score/Matched Peptides/Coverage [e]	Cellular Location [f]	False Discovered Rate [g]	Changes of Regulation Intensity [h]
				Cell Structure				
47	Myosin-like protein (*Zea mays*)	gi|195622168	28.79/25.88	8.54/9.23	237/2(2)/7	N/C/M	0.027	
				Signal Transduction				
25	14-3-3 protein (*Ipomoea nil*)	gi|1244484407	19.03/29.64	6.72/4.76	415/5(3)/22	-	0.001	
48	Annexin (*Populus tomentosa*)	gi|429326382	35.49/35.47	8.95/6.82	150/2(2)/8	N/C	0.004	
4	Putative guanosine triphosphate (GTP)-binding protein (*Leucocoprinus fragilissimus*)	gi|110349697	38.14/16.89	6.75/4.94	530/8(5)/27	Mi/A/GA	0.002	
36	Casein kinase II subunit β-4 (*Zea mays*)	gi|195628750	31.06/26.71	7.74/8.19	114/2(1)/7	CW/C/M	0.001	

Table 2. Cont.

Spot No. [a]	Protein Name and Organism [b]	Accession No. [c]	Exp./Theo. [d] MW (kDa)	pI	Score/Matched Peptides/Coverage [e]	Cellular Location [f]	False Discovered Rate [g]	Changes of Regulation Intensity [h]
					Unclassified			
44	Fiber protein, partial (*Hyacinthus orientalis*)	gi|42565482	5.89/16.53	8.82/5.93	202/2(2)/17	N/Mi/Ch/M	0.021	
46	Mitochondrial outer membrane porin (*Triticum urartu*)	gi|473968092	26.76/37.68	8.75/9.25	738/7(7)/28	C/M/Ch	0.028	

[a] Spot No. corresponds to the spot in Figure 2; [b,c] Data of protein name and organism and accession no. are from the National Center for Biotechnology Information (NCBI) database of the matched protein species; [d] Experimental molecular weight (MW)/isoelectric point (pI) was calculated by Image PDQuest 8.0 software (Bio-Rad, Munich, Germany) according to position of protein spot in 2-D gel; theoretical MW/pI was gained from the matched protein species; [e] Score from MALDI-TOF/TOF-MS analysis for the most significant hits ($p < 0.05$); matched peptides indicates total number of peptides that matched to other protein species; coverage refers to percentage of matched protein species; [f] Data of cellular location are from Gene Ontology (GO) annotations (cellular component). A, apoplast; AC, actin cytoskeleton; C, cytosol; Ch, chloroplast; CW, cell wall; GA, Golgi apparatus; ER, endoplasmic reticulum; M, membrane; P, plastid; Pe, peroxisome; Pl, plasmodesma; V, vacuole; [g] Spot values passed the Duncan test, and false discovered rate (FDR) was controlled at level 0.05; [h] From left to right, each bar indicates the change in protein spot volume after pollination for 0, 8, 16, and 24 h. Spot values are presented as means ± SD of three replicates. Means distinguished with different letters are significantly different from each other ($p \leq 0.05$).

3. Discussion

3.1. Ultrastructural Analysis

Usually, one of the earliest ultrastructural changes is the closure of plasmodesmata, which, when open, allow the transfer of relatively small molecules such as sugars, hormones, and RNA molecules between neighboring cells [17]. In the present study, the plasmodesmata indeed showed significant changes at 8 h (Figure 1B). In several species, such as *Alstroemeria* and *Sandersonia*, petal mesophyll cells were found to die considerably earlier than epidermis cells [18,19], and this phenomenon was also shown in senescing *Phalaenopsis* petals. In our experiment, petal cells in *Phalaenopsis* exhibited a decrease in vacuolar size and an increase in the number of small vacuoles because of the zoned phenomenon of vacuoles during senescence (Figure 1F), which is in line with other plants such as petunia [6]. The appearance of osmiophilic granules is a typical organelle alteration during aging [20]. Accordingly, in this study the increased number of osmiophilic granules demonstrated that organelles such as chloroplasts and mitochondria also changed in senescing petals (Figure 1G,H). Visible external senescing syndrome was the outcome of vast ultrastructural changes.

Collectively, the data indicate that petal senescence was accompanied by the mass degradation of macromolecules and organelles, and the ultrastructural changes to the structures and organelles were typical features of programmed cell death (PCD). From Figure 1, it can be seen that visible morphological changes of pollinated petals occurred at 24 h, whereas ultrastructure exhibited significant changes after 16 h, indicating that the ultrastructural change of petals was the structural basis of morphological changes. The protein species that play important roles in petal senescence are differentially regulated in the early stage, before visible morphological changes occur. Therefore, we selected four time points within 24 h (0, 8, 16, and 24 h) where the petals did not show dramatic wilting.

3.2. Protein Function Analysis

In this study, a number of functional protein species participated in regulation, as evident from their differential regulation in non-senescing and senescing petals; these protein species were likely to be involved in various pathways, including antioxidant pathways, stress response, protein metabolism, cell wall component metabolism, energy metabolism, cell structure, and signal transduction.

3.2.1. Antioxidant Enzymes

Plants have evolved complex regulatory mechanisms to prevent cellular injury through regulating destructive ROS so that they are present at steady-state levels [21]. In our study, several protein species involved in antioxidant pathways, including thioredoxin-dependent peroxidase (POD, spot 10), thioredoxin H-type (TRX, spot 7), glutathione *S*-transferase (GST, spots 14, 16, and 34), and NAD(P)-linked oxidoreductase (spot 41), were found to be differentially regulated during petal senescence. All of these protein species are components of a pathway that is activated by the plants themselves to remove ROS.

POD is important for scavenging ROS and for protecting the cellular membrane from peroxidation [22]. Usually, the peroxidase family members serve as detoxifying enzymes and perform oxidation reactions to remove toxic reductants [23]. Indeed, our results showed that the abundance of POD (spot 10) increased during petal senescence, which was consistent with the measured POD activity (Figure 4A). TRX, which functions as a regulator of apoplastic ROS, maintains redox balance via thiol-disulfide exchange reactions and plays a critical role in responding to ROS-induced cellular damage [24]. Another detoxifying enzyme, GST, can protect a senescing cell from lipid hydroperoxides prior to cell death. An oxidative burst and GST induction are usually used as markers of induction of the defense response [25]. As our results show, the abundances of TRX (spot 7) and GST (spots 14, 16, and 34) were very high in early senescence but decreased at 16 h, possibly because there were some circadian or diurnal effects on these patterns. Diurnal redox behaviors of these two enzymes were strictly linked to light intensity and the mRNA of TRX and GST tended to be lower in the dark phase (16 h) [26]. Ratnayake et al. [27] found that an alternative way to increase the antioxidant status of a cell is to enhance the defense system involving cytoprotective antioxidant enzymes, including NAD(P)H: quinone oxidoreductase 1 (NQO1) [27]. However, NAD(P)-linked oxidoreductase (spot 41) was downregulated in pollinated petals after 16 h, from which it could be postulated that an oxidative burst might turn up at that time, an actual time point of senescence.

In the present study, the abundance of these enzymes after pollination indicated that petals outperformed themselves to maintain the metabolic balance of the active oxygen system, thereby postponing the senescence process. Therefore, we can conclude that enhancing the gene expression of ROS-scavenging enzymes is an effective way to prolong *Phalaenopsis* petal longevity.

3.2.2. Stress Response Protein Species

Phospholipase D (PLD) hydrolyzes membrane lipids to generate phosphatidic acid (PA) and a free-head group, which destroy membranes and activate other lipid-degrading enzymes from the hydrolysis of membrane phospholipids [28]. In PLDδ-knockout *Arabidopsis*, leaf senescence was delayed because the production of PA was repressed by the attenuation of lipid degradation [29]. In this context, the upregulation of PLD (spot 31) was exclusively detected in 2-DE gels, perhaps demonstrating that petal senescence was accompanied by increased phospholipid catabolism for maximum cellular recycling [30]. Alcohol dehydrogenase (ADH) catalyzes redox reactions between acetaldehyde and ethanol, participating in plant anaerobic respiration. It is an inducible enzyme and can be activated under some adverse conditions [31]. Therefore, ADH may be crucial for regulating plant stress resistance and accommodating to adversity. Nevertheless, no similarities to these previous findings were observed in this study; the abundance of ADH (spot 42) was decreased, suggesting that the ethanol fermentation pathway might be shut down during senescence. Therefore, it is speculated that the shutdown of this pathway is one of the drivers of the decrease in reducing capacity and the increase in ROS levels. However, the relationship between this pathway and ROS levels remains unclear.

Above all, pollination-induced *Phalaenopsis* petal senescence is a very complex process. The differential regulation of protein species involved in antioxidant pathways and stress response in senescing petals might be closely related to ROS. Nonetheless, how these protein species interact with each other remains unknown.

3.2.3. Protein Species Involved in Protein and Cell Wall Component Metabolism

The total protein level decreased drastically prior to visible senescence symptoms in the petals [32]. The decrease in protein levels is primarily due to an increase in degradation, as well as a decrease in synthesis [17], demonstrating that large-scale degradation occurs during senescence. Additionally, nutrient remobilization from senescent organs requires the action of a suite of degradative enzymes [25].

Cysteine protease (CP) is encoded by *SAG12/Cab*, which acts as a molecular indicator of leaf senescence progression [33]. Battelli et al. [34] also found that CP is responsible for most of the proteolytic activity in senescent petals [34]. Our results show that CP (spot 32) accumulated during petal senescence, which agrees with previous observations [35]. Aminopeptidase can be induced at high carbohydrate levels, which may initiate senescence and result in nitrogen remobilization [36]. In our experiment, aminopeptidase N (APN, spot 23), which might be important for nitrogen nutrient recycling, was upregulated. In short, these four protein species sufficiently supported nutrient remobilization during the senescence stage. Three protein species related to cell wall component metabolism, namely cellulose synthase (CesA, spot 9), xyloglucan endotransglucosylase/hydrolase (XTH, spot 38), and cinnamyl alcohol dehydrogenase (CAD, spot 39), were identified. CesA catalyzes the conversion of D-glucose to cellulose via β-1, 4-glycosidic bonds [37]. XTH, a primary cell-loosening enzyme, can catalyze the degradation of xyloglucan, which is the primary composition of hemicellulose [38]. An increase in XTH and a decrease in xyloglucan during petal senescence were found in previous studies [19,39]. Cellulose and hemicellulose are components of the cell wall; therefore, a decrease in these two substances may cause a bend in the shape of the flowers. In this study, the downregulation of CesA (spot 9) and upregulation of XTH (spot 38) resulted in the degradation of cell wall components, affecting the external morphology of the petals. CAD is the limiting enzyme in lignin synthesis, a secondary metabolic process [40]. Lignin provides structural rigidity for tracheophytes to stand upright and strengthen the cell walls [41]. The upregulation of XTH (spot 38) and downregulation of CAD (spot 39) were both obviously exacerbated at 16 h after pollination, suggesting that the cell wall components were broken down and degraded gradually, which was in agreement with the ultrastructural observations (Figure 1D). This phenomenon supports a viewpoint that ethylene burst (accelerating petal senescence) is started after several hours, not at 0 h, after pollination in orchids [42].

3.2.4. Energy Metabolism Protein Species

Petal senescence after pollination denoted a massive increase in carbon flow through glycolysis and the tricarboxylic acid cycle (TCA), and protein spots corresponding to related enzymes (spots 8, 11, 15, 17, 18, 19, 26, and 40) were identified. Pyrophosphate-fructose 6-phosphate 1-phosphotransferase subunit β (PFP, spot 40) and triosephosphate isomerase (TPI, spots 8 and 15) are important enzymes in glycolysis, and two of these (spots 40 and 15) became downregulated once pollination occurred. One possibility for this phenomenon would be a self-defense mechanism in which low glucose metabolism contributed to sugar accumulation, sequentially inhibiting the expression of senescence-associated genes (SAGs), given that van Doorn et al. [43] proposed that sugar starvation would directly result in petal senescence, and vice versa [43]. Nonetheless, the upregulation of spot 8 might be conducive to maintaining carbon recycling during petal senescence. Dihydrolipoamide acetyltransferase (E2, spot 11) and dihydrolipoamide dehydrogenase (E3, spot 19) are components of the pyruvate dehydrogenase system, which catalyzes the conversion of pyruvate into acetyl-CoA. A decrease in the abundance of these two protein species demonstrated that the TCA cycle is downregulated during petal senescence induced by pollination, as noted in a review by van Doorn and Woltering [17]. The level of pyruvate orthophosphate dikinase (PPDK, spots 17 and 18) increased during senescence, consistent with the viewpoint that the abundance of PPDK might dramatically increase during abiotic stress, such as low-oxygen stress and water deficit [44]. A reduction in ATP synthase (ATPase, spot 26) abundance occurred, possibly because this membrane enzyme produces ATP—generated from the downregulated TCA cycle—from ADP in the presence of a proton gradient across the membrane.

3.2.5. Signal Transduction Protein Species

Four protein species (spots 4, 25, 36, and 48) related to signal transduction were successfully identified. Numerous studies have demonstrated that 14-3-3 protein is a key anti-apoptotic factor that is upregulated in senescent plants and blocks apoptosis by inhibiting the activation of p38 MAPK. In addition, 14-3-3 protein binds to the apoptosis-promoting protein BAD and to forehead transcription factor FKHRL1, inhibiting the stimulation of apoptosis [45]. In this study, the upregulation of 14-3-3 protein (spot 25) perhaps demonstrated that 14-3-3 protein can postpone petal senescence by repressing the activity of apoptotic factors. GTP-binding protein (GTP) regulates many physiological processes, such as vesicular transportation, signal transduction, and cell apoptosis [46]. Casein kinase II (CKII) is a highly conserved and messenger-independent serine/threonine protein kinase with both cytosolic and nuclear localization in eukaryotic cells [47]. This enzyme is implicated in important biological processes, including apoptosis. In our study, the downregulation of GTP (spot 4) and CKII (spot 36) during petal senescence differed from the regulation changes in GTP and CKII observed in *Mangifera indica* and *Coleus blumei* [48,49], and further study is needed to obtain the satisfactory elucidation of these findings.

3.3. Physiological-Biochemical Analysis

The functions of ROS-scavenging enzymes in senescing petals include detoxifying ROS and preventing the accumulation of toxic substances. In this context, POD activity was higher in pollination treatment (8, 16, and 24 h) than in the control (0 h) (Figure 4A), as has been shown in other plants, such as carnation and day lily [50,51]. The activity of SOD and CAT rapidly increased in the first 8 h (Figure 4A), which was similar to previous observations in *Hemerocallis* (day lily) flowers [51], demonstrating that these enzymes provide the first line of defense against ROS. However, the overall picture shows a reduction in the activity of ROS-scavenging enzymes as senescence progresses after 8 h, with earlier decreases in the activity of SOD and CAT compared to POD, corresponding to the view proposed by Zeng et al. [50]. The levels of MDA were elevated considerably in the senescing petals of pollinated flowers (Figure 4B), in agreement with previous results [11].

In the 2-DE results, the content of POD was increased during pollination-induced senescence. There is a close relationship among POD, CAT, and SOD, which are important antioxidant systems that catabolize superoxide and hydrogen peroxide, a precursor of reactive oxidants [52,53]. Based on the 2-DE results and previous studies, we also monitored the changes in the activities of these three enzymes during petal senescence. The results showed that the activities of the antioxidant enzyme were also increased. The 2-DE and physiological-biochemical results showed that the ROS-scavenging system was activated after pollination treatment in *Phalaenopsis* petals. However, we were not able to explain why the presence of peroxisome marker enzymes was detected and no peroxisomes were highlighted with the ultrastructural analysis.

Overall, the effective ROS-scavenging system was rapidly activated once pollination occurred and maintained ROS at a controlled level, delaying the senescence process.

3.4. Possible Processes That Regulate Differentially Regulated Protein Species during Petal Senescence

Based on the functions of the differentially regulated protein species, a possible mechanism of petal senescence is proposed (Figure 5). Initially, pollination could trigger the upregulation of ethylene biosynthetic genes [21]. In this study, ethylene-responsive transcription factors were activated by direct phosphorylation (possibly by CK II and GTP) or an MAPK cascade (14-3-3 protein) [45]. Other transcription factors were activated by other signals, probably including PA and ROS [9,54]. An increase in ethylene production could activate PLD, which would contribute to rapid PA accumulation. PA formation could switch on downstream ethylene responses via interaction of the lipid with CTR1 [54]. PA itself could also promote membrane curvature and induce vesicle formation [55]. A burst of ROS production might activate the ROS-scavenging system during the early stage of senescence. The abundance and activity of antioxidant enzymes such as SOD, CAT, and POD were upregulated, postponing early senescence. However, in general, during petal senescence, ROS levels rose and antioxidant levels fell, resulting in oxidative stress [9]. Many other protein species involved in cell wall degradation (CesA, XTH, and CAD), protein degradation (CP, UDF1, PSβ1, and APN), and carbon mobilization (PFP, TPI, PPDK, ADH, E2, and E3) were also differentially regulated. Cell wall degradation might result in the loosening of the cell wall, as shown by the ultrastructural results; meanwhile, protein degradation and carbon mobilization supported nutrient remobilization. Overall, each of these differentially regulated protein species plays a unique and cooperative role in regulating petal senescence.

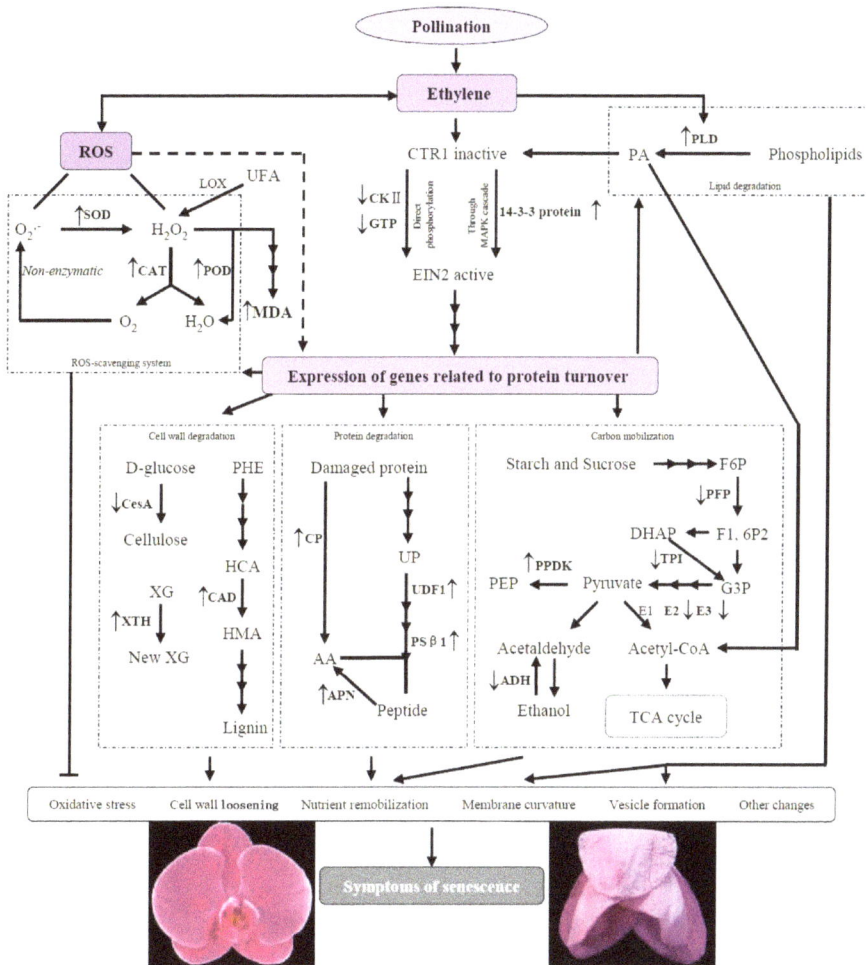

Figure 5. Putative functions of differentially regulated protein species during petal senescence. Identified protein species are shown in bold text. Upregulated protein species are marked with ↑, and downregulated protein species are marked with ↓. 'Mlti-step enzymatic reaction' is marked with →→→. ROS, reactive oxygen species; SOD, superoxide dismutase; CAT, catalase; POD, peroxidase (spot 10); UFA, unsaturated fatty acid; LOX, lipoxygenase; MDA, malondialdehyde; CK II, casein kinase II (spot 36); GTP, GTP-binding protein (spot 4); 14-3-3 protein (spot 25); PLD, phospholipase D (spot 31); PA, phosphatidic acid; CesA, cellulose synthase (spot 9); XG, xyloglucan; XTH, xyloglucan endotransglucosylase/hydrolase (spot 38); PHE, phenylalanine; HCA, hydroxycinnamic aldehyde; CAD, cinnamyl alcohol dehydrogenase (spot 39); HMA, hydroxy cinnamic alcohol; CP, cysteine proteinase (spot 32); UP, ubiquitinated protein; UDF1, ubiquitin fusion degradation protein 1 (spot 37); PSβ1, proteasome subunit β type 1 (spot 43); AA, amino acid; APN, aminopeptidase N (spot 23); F6P, fructose-6-phosphate; PFP, pyrophosphate-fructose 6-phosphate 1-phosphotransferase (spot 40); F1,6P2, fructose-1, 6-diphosphate; DHAP, dihydroxyacetone phosphate; TPI, triosephosphate isomerase (spots 8 and 15); G3P, glyceraldehydes-3-phosphate; PPDK, pyruvate orthophosphate dikinase (spots 17 and 18); E1, pyruvate dehydrogenase; E2, dihydrolipoamide acetyltransferase (spot 11); E3, dihydrolipoyl dehydrogenase (spot 19); ADH, alcohol dehydrogenase (spot 42); TCA, tricarboxylic acid cycle.

4. Materials and Methods

4.1. Plant Materials

Phalaenopsis "Red Dragon" plants were grown in a greenhouse at South China Normal University (Guangzhou, China). Flowers in the middle of inflorescence were selected. They were hand pollinated, using the method described by Visser et al. [56], in the morning (10:00) on the first day after flowering and were examined hourly thereafter for visible morphological changes. The petals were collected from pollinated flowers at 0, 8, 16, and 24 h. For proteomic and physiological-biochemical analysis, samples were frozen under liquid nitrogen rapidly and stored at −80 °C until required. Ten fresh petals were collected from one group (20 plants) at random and pooled together as one biological replicate of each time point. In this experiment, three biological replicates and three technical replicates of three biological replicates were conducted for proteomic analysis and physiological-biochemical analysis, respectively.

4.2. Observation of the Petal Ultrastructure

To visualize autophagic processes in *Phalaenopsis* petals, samples collected at each time point were observed by using transmission electron microscopy (TEM). The experimental petals of each time point were prepared from flowers that were chosen from one group (20 plants) at random. The operating steps were conducted as recommended by Shibuya et al. [6] with the following modifications [6]. The center section of each petal was selected and cut into pieces (5 mm × 2 mm). The sample pieces were fixed in 0.1 M phosphate buffer (pH 7.0) (containing 4% paraformaldehyde and 4% glutaraldehyde) at 4 °C for at least 4 h. After fixation, the sample pieces were rinsed with 0.05 M phosphate buffer, followed by post-fixation in 0.1 M phosphate buffer (pH 7.0) (containing 1% osmium tetroxide) at 4 °C for 2 h. Then, dehydrated specimens were embedded in epoxy resin. Ultrathin sections (70 nm) were made with a diamond knife using an ultramicrotome (RM 2255; Leica, Wetzlar, Germany) and placed on copper grids. Sections were stained with 2% uranyl acetate at room temperature for 15 min and rinsed with double-distilled water, followed by secondary staining with a lead-staining solution at room temperature for 3 min. Lastly, the sections were observed and photomicrographs were recorded with a transmission electron microscope (JEM-1010; JEOL, Tokyo, Japan).

4.3. Protein Extraction

A phenol-based extraction method was employed for protein extraction with the following modifications [57]. Frozen *Phalaenopsis* petals (1.5 g) were finely powdered in liquid nitrogen and suspended in 4 mL ice-cold extraction buffer (500 mM Tris-HCl buffer pH 8.0 containing 0.7 M sucrose, 50 mM ethylenediaminetetraacetic acid, 100 mM KCl, and 2% (*v/v*) β-Mercaptoethanol) and 4 mL water-saturated phenol (pH < 4.5) in a 10-mL centrifuge tube. The homogenate was left for 30 min and centrifuged at 19,500× *g* for 30 min at 4 °C. The upper phenolic phase was collected into a new 10-mL centrifuge tube, whereas the lower water phase was re-extracted with 4 mL Tris-saturated phenol (pH > 8.0). Phenolic phases was combined and precipitated overnight with 8 mL of 0.1 M ammonium acetate/methanol at −20 °C. After it was successively rinsed in 5 mL ice-cold 100% acetone and 80% acetone, the pellet was transferred to a 2-mL microtube and rinsed twice in 1 mL ice-cold 100% acetone. The final pellet was air-dried for 1.5 h at room temperature and dissolved in lysis buffer (8 M urea, 2 M thiourea, 4% 3-[(3-Cholamidopropyl)dimethylammonio]propanesulfonate, 1% DL-Dithiothreitol, 0.5% pH 3–10 non-linear gradient (NL) immobilized pH gradient (IPG) buffer, and a trace of bromophenol) for 1.5 h at room temperature. The protein solutions were centrifuged at 16,100× *g* for 40 min at 4 °C. The protein concentration of the supernatants was determined using bovine serum albumin (BSA) as a standard according to the Bradford method [58].

4.4. 2-DE and Staining

A sample containing 1350 µg proteins was loaded onto a 24-cm pH 3–10 NL IPG strip (GE Healthcare, Princ-eton, NJ, USA), which was rehydrated for 16 h at 20 °C. After rehydration, isoelectric focusing (IEF) was performed in a PROTEAN IEF system (GE Healthcare, Fairfield, CT, USA) under the following conditions: a gradient from 0 to 100 V for 4 h, 250 V for 1 h, 1 kV for 1 h, a gradient from 1 to 10 kV for 2 h, and a gradient from 10 to 100 kV for 12 h. Subsequently, the strip was equilibrated for two periods of 15 min with 1.0% (w/v) DTT and 2.5% (w/v) indole-3-acetic acid (IAA) in equilibration buffer (50 mM Tris-HCl pH 8.8, 6 M urea, 30% (v/v) glycerol, 2% (v/v) SDS, and a trace of bromophenol). SDS-PAGE was performed on vertical 12% SDS-PAGE self-cast gels with an Ettan DALTsix System (GE Healthcare, Fairfield, CT, USA) under the following conditions: 1 W for 30 min and 15 W for 6 h at 15 °C. After 2-DE, the gel was stained with 0.12% Coomassie brilliant blue (CBB) G-250. At least three biological replicates were assessed for each time point.

4.5. Image Acquisition and Statistical Analysis

The stained gels were scanned with an Image Scanner III (GE Healthcare, Fairfield, CT, USA) with default parameters as follows: optical resolution, 300 dots per inch (dpi); brightness, 3; contrast, −9; saturation, 9; total input value, 140; and output value, 20. These images were analyzed on PDQuest V 8.0 (Bio-Rad, Hercules, CA, USA). After the images were properly cropped and optimized, spot detection and gel-to-gel matching were performed automatically and were refined by manual spot editing when needed. Three well-separated gels for each time point were used to create "replicated groups". We only considered "consistent spots", which were present in the three biological replicates, thus preventing the assignment of normalized volume values to missing spots for multivariate analysis. These consistent spots were added to the master gel so they could be matched to all of the samples. The experimental molecular weight (MW) and isoelectric point (pI) of proteins were estimated using their position in the 2-DE gel.

All data regarding the protein spots from 2-DE maps were preprocessed according to the recommendations proposed by Valledor and Jorrin [59]. The protein abundance of each spot was normalized as a percentage of the total volume of all the spots present in the gel, to correct for variability due to quantitative variations in the intensity of the protein spots. Differentially regulated spots were defined with one-way ANOVA using SPSS v. 13.0 (Available online: http://spss.en.softonic.com/). Spot values passed the Duncan test and the degree of freedom (DF) was 11. False discovery rate (FDR) was controlled at level 0.05. A multivariate analysis was performed over the whole set of spots and on those showing differences. Spots whose regulation intensity was more than 1.5 times (w/v 0.05) or less than 0.67 times ($p < 0.05$) that of the control (0 h) at one or more time points were considered as differentially regulated protein spots for further analysis.

4.6. Protein In-Gel Digestion and Identification by MALDI-TOF/TOF-MS

Protein spots of interest were excised from the gels, transferred into sterilized 2-mL microtubes, and then washed twice with double-distilled water. Protein spots were repeatedly de-stained using 50 mM ammonium bicarbonate in 50% (v/v) acetonitrile (ACN) for 30 min at 37 °C. Subsequently, the gel pieces were shrunk by dehydration in ACN and then swollen for 30 min at 4 °C in digestion buffer (25 mM ammonium bicarbonate, 10% (v/v) ACN and 0.02 µg/mL trypsin). After digestion for 16 h at 37 °C, the supernatants were collected and the peptides were extracted again using 0.1% (v/v) trifluoroacetic acid (TFA) in 67% (v/v) ACN for 30 min. The two supernatants were combined, vacuum-dried, and then re-dissolved in 67% (v/v) ACN and 0.1% (v/v) TFA for MS analysis.

Peptide mass determinations were performed using an ABI 4800 Plus MALDI-TOF/TOF Analyzer (Applied Biosystems, Foster City, CA, USA). Identification of the protein sample was conducted using Mascot Version 2.1 software (Matrix Science, London, UK) with the following optimized parameters: present in the National Center for Biotechnology Information (NCBI) nonredundant (nr) database, a member of the Viridiplantae taxon, a maximum of one missed cleavage, a fixed modification of carbamidomethyl (C), variable modifications of acetyl (protein N-term) and oxidation (C), a peptide mass tolerance of 100 ppm, and an MS/MS tolerance of 0.4 Da. The score threshold was greater than 50 ($p < 0.05$). If the peptides were matched to multiple members of a protein family or if a protein appeared under different names and accession numbers, only significant hits with the highest protein score were accepted for identification of the protein sample. When the values of two scores were very close, we took the reference of the experimental MW and pI.

4.7. Functional Analysis

The bioinformatics data of the successfully identified proteins were gained by GO and KEGG annotation. GO (Available online: http://www.geneontology.org) and KEGG pathway (Available online: http://www.genome.jp/kegg/pathway.html) analyses were performed with the PartiGene program (Available online: http://www.nematodes.org/bioinformatics/annot8r/index. shtml). Annot8r assigns KEGG (gene) pathways and GO (protein) terms based on BLASTX similarity (E-value $< 1.0 \times 10^{-5}$) and known GO annotations. The results of GO analyses are summarized in three independent categories (biological process, cellular component, and molecular function).

4.8. Antioxidant Enzyme Activity Assays and Lipid Peroxidation Analysis

Phalaenopsis petals (0.5 g) were ground in ice-cold 100 mM phosphate buffer (pH 7.5) on ice. Homogenates was centrifuged at $1500 \times g$ for 20 min at 4 °C. The supernatant was used for subsequent assays. Superoxide dismutase activity (SOD; EC 1.15.1.1), catalase activity (CAT; EC 1.11.1.6), and peroxidase activity (POD; EC 1.11.1.7) were measured according to Chakrabarty et al. [51]. The measurement of lipid peroxidation, which is determined by measuring malondialdehyde (MDA), was assessed as described previously [11]. CAT and POD activity was expressed as U·g fresh weight $(FW)^{-1}$. SOD activity and MDA content were expressed as $U \cdot min^{-1} \cdot g\ FW^{-1}$ and $nmol \cdot g\ FW^{-1}$, respectively.

4.9. Statistical Analysis of Physiological-Biochemical Changes

The significance of the differences between the pollination treatment (8, 16, and 24 h) and the control (0 h) were determined with one-way ANOVA test ($p < 0.05$) using SPSS v. 13.0 package (Available online: http://spss.en.softonic.com/). A repeated measurement is given as the mean ± SD.

5. Conclusions

Senescence is a very complex process that involves changes at the physiological, biochemical, and molecular biology levels. To explore the mechanism underlying senescence, we performed a comparative proteomic analysis combining several approaches, including ultrastructural observation and antioxidant enzyme activity analysis, on *Phalaenopsis* petals at 0, 8, 16, and 24 h after pollination .The petals appeared to be visibly wilting 24 h after pollination, and this could be accompanied by the mass degradation of macromolecules and organelles (Figure 1). Proteomic analysis yielded 42 differentially regulated proteins, including 17 proteins that were upregulated and 25 proteins that were downregulated, and these were identified with confidence by MALDI-TOF/TOF-MS and homology-driven searches. These protein species are likely to be involved in a wide range of cellular pathways. Taken together, the results suggest that multiple cellular pathways operate in a coordinated manner during petal senescence. The identified protein species with specific expression patterns can be used as putative markers of senescence. Additionally, the activity of all of the ROS-scavenging enzymes increased, keeping the ROS content at a controlled level at the early stage of senescence. In summary, the 2-DE proteomic data, ultrastructural observations, and physiological-biochemical analysis results presented here will help us further understand the molecular and biochemical changes that occur during petal senescence, thus providing a basis for prolonging florescence. However, as to the progress of senescence, additional studies will be necessary to fully elucidate the complexity of this process.

Author Contributions: Q.Y. conceived and designed the experiments; C.C. and L.Z. performed the experiments, analyzed the results and wrote the paper. All the authors reviewed the manuscript.

Acknowledgments: This research was supported by the Natural Science Foundation of Guangdong Province (No. 102510631000013), China.

Conflicts of Interest: The authors declare no conflict of interest.

References

1. Ejima, C.; Kobayashi, Y.; Honda, H.; Shimizu, N.; Kiyohara, S.; Hamasaki, R.; Sawa, S. A *Phalaenopsis* variety with floral organs showing C class homeotic transformation and its revertant may enable *Phalaenopsis* as a potential molecular genetic material. *Genes Genet. Syst.* **2011**, *86*, 93–95. [CrossRef] [PubMed]

2. Jones, M.L.; Woodson, W.R. Differential expression of three members of the 1-aminocyclopropane-1-carboxylate synthase gene family in carnation. *Plant Physiol.* **1999**, *119*, 755–764. [CrossRef] [PubMed]

3. Stead, A.D. Pollination-induced flower senescence a review. *Plant Growth Regul.* **1992**, *11*, 13–20. [CrossRef]

4. Huda, M.K.; Wilcock, C.C. Rapid floral senescence following male function and breeding systems of some tropical orchids. *Plant Biol.* **2011**, *14*, 278–284. [CrossRef] [PubMed]

5. Liu, W.; Ye, Q.S.; Pan, R.Z. Regulation of flower senescence and ethylene biosynthesis by pollination. *Acta Hortic Sin.* **2000**, *27*, 527–532. [CrossRef]

6. Shibuya, K.; Niki, T.; Ichimura, K. Pollination induces autophagy in petunia petals via ethylene. *J. Exp. Bot.* **2013**, *64*, 1111–1120. [CrossRef] [PubMed]

7. Shibuya, K.; Yamada, T.; Ichimura, K. Morphological changes in senescing petal cells and the regulatory mechanism of petal senescence. *J. Exp. Bot.* **2016**, *67*, 5909–5918. [CrossRef] [PubMed]

8. Trivellini, A.; Ferrante, A.; Vernieri, P.; Carmassi, G.; Serra, G. Spatial and temporal distribution of mineral nutrients and sugars throughout the lifespan of *Hibiscus rosa-sinensis* L. flower. *Cent. Eur. J. Biol.* **2011**, *6*, 365–375. [CrossRef]

9. Rogers, H.; Munné-Bosch, S. Production and scavenging of reactive oxygen species and redox signaling during leaf and flower senescence: similar but different. *Plant Physiol.* **2016**, *171*, 1560. [CrossRef] [PubMed]

10. Tewari, R.K.; Kumar, P.; Kim, S.; Hahn, E.J.; Paek, K.Y. Nitric oxide retards xanthine oxidase-mediated superoxide anion generation in *Phalaenopsis* flower: An implication of NO in the senescence and oxidative stress regulation. *Plant Cell Rep.* **2008**, *28*, 267–279. [CrossRef] [PubMed]

11. Attri, L.K.; Nayyar, H.; Bhanwra, R.K.; Pehwal, A. Pollination-induced oxidative stress in floral organs of *Cymbidium pendulum* (Roxb.) Sw. and *Cymbidium aloifolium* (L.) Sw. (Orchidaceae): A biochemical investigation. *Sci. Hortic.* **2008**, *116*, 311–317. [CrossRef]

12. Hebeler, R.; Oeljeklaus, S.; Reidegeld, K.A.; Eisenacher, M.; Stephan, C.; Sitek, B.; Stühler, K.; Meyer, H.E.; Sturre, M.J.; Dijkwel, P.P.; et al. Study of early leaf senescence in *Arabidopsis thaliana* by quantitative proteomics using reciprocal 14N/15N labeling and difference gel electrophoresis. *Mol. Cell. Proteom.* **2008**, *7*, 108–120. [CrossRef] [PubMed]

13. Bai, S.; Willard, B.; Chapin, L.J.; Kinter, M.T.; Francis, DM.; Stead, A.D.; Jones, M.L. Proteomic analysis of pollination-induced corolla senescence in petunia. *J. Exp. Bot.* **2010**, *61*, 1089–1109. [CrossRef] [PubMed]

14. Heidarvand, L.; Maali-Amiri, R. Physio-biochemical and proteome analysis of chickpea in early phases of cold stress. *J. Plant Physiol* **2013**, *170*, 459–469. [CrossRef] [PubMed]

15. Zheng, M.; Meng, Y.; Yang, C.; Zhou, Z.; Wang, Y.; Chen, B. Protein expression changes during cotton fiber elongation in response to drought stress and recovery. *Proteomics* **2014**, *14*, 1776–1795. [CrossRef] [PubMed]

16. Woo, H.R.; Kim, H.J.; Nam, H.G.; Lim, P.O. Plant leaf senescence and death-regulation by multiple layers of control and implications for aging in general. *J. Cell Sci.* **2013**, *126*, 4823–4833. [CrossRef] [PubMed]

17. Van Doorn, W.G.; Woltering, E.J. Physiology and molecular biology of petal senescence. *J. Exp. Bot.* **2008**, *59*, 453–480. [CrossRef] [PubMed]

18. Wagstaff, C.; Malcolm, P.; Rafiq, A.; Leverentz, M.; Griffiths, G.; Thomas, B.; Stead, A.; Rogers, H. Programmed cell death (PCD) processes begin extremely early in *Alstroemeria* petal senescence. *New Phytol.* **2003**, *160*, 49–59. [CrossRef]

19. O'Donoghue, E.M.; Somerfield, S.D.; Heyes, J.A. Organization of cell walls in *Sandersonia aurantiaca* floral tissue. *J. Exp. Bot.* **2002**, *53*, 513–523. [CrossRef] [PubMed]

20. Sosunov, A.A.; Hamdorf, G.; Cervós-Navarro, J.; Shvalev, V.N. The structure of the stellate sympathetic ganglia under long-term hypoxia. *Morfologiia* **1996**, *109*, 12–17. [PubMed]
21. Circu, M.L.; Aw, T.Y. Reactive oxygen species, cellular redox systems, and apoptosis. *Free Radic. Biol. Med.* **2010**, *48*, 749–762. [CrossRef] [PubMed]
22. Mitozo, P.A.; de Souza, L.F.; Loch-Neckel, G.; Flesch, S.; Maris, A.F.; Figueiredo, C.P.; dos Santos, A.R.S.; Farina, M.; Dafre, A.L. A study of the relative importance of the peroxiredoxin-, catalase-, and glutathione-dependent systems in neural peroxide metabolism. *Free Radic. Biol. Med.* **2011**, *51*, 69–77. [CrossRef] [PubMed]
23. Faltin, Z.; Holland, D.; Velcheva, M.; Tsapovetsky, M.; Roeckel-Drevet, P.; Handa, A.K.; Abu-Abied, M. Glutathione peroxidase regulation of reactive oxygen species level is crucial for in vitro plant differentiation. *Plant Cell Physiol.* **2010**, *51*, 1151–1162. [CrossRef] [PubMed]
24. Zhang, C.J.; Zhao, B.C.; Ge, W.N.; Zhang, Y.F.; Song, Y.; Sun, D.Y.; Guo, Y. An apoplastic h-type thioredoxin is involved in the stress response through regulation of the apoplastic reactive oxygen species in rice. *Plant Physiol.* **2011**, *157*, 1884–1899. [CrossRef] [PubMed]
25. Chen, I.C.; Huang, I.C.; Liu, M.J.; Wang, Z.G.; Chung, S.S.; Hsieh, H.L. Glutathione *S*-transferase interacting with far-red insensitive 219 is involved in phytochrome a-mediated signaling in *Arabidopsis*. *Plant Physiol.* **2007**, *143*, 1189–1202. [CrossRef] [PubMed]
26. Yoshida, K.; Matsuoka, Y.; Hara, S.; Konno, H.; Hisabori, T. Distinct redox behaviors of chloroplast thiol enzymes and their relationships with photosynthetic electron transport in *Arabidopsis thaliana*. *Plant Cell Physiol.* **2014**, *55*, 1415–1425. [CrossRef] [PubMed]
27. Ratnayake, R.; Liu, Y.; Paul, V.J.; Luesch, H. Cultivated sea lettuce is a multiorgan protector from oxidative and inflammatory stress by enhancing the endogenous antioxidant defense system. *Cancer Prev. Res.* **2013**, *6*, 989–999. [CrossRef] [PubMed]
28. Bargmann, B.O.; Munnik, T. The role of phospholipase D in plant stress responses. *Curr. Opin. Plant Biol.* **2006**, *9*, 515–522. [CrossRef] [PubMed]
29. Jia, Y.; Tao, F.; Li, W. Lipid profiling demonstrates that suppressing *Arabidopsis* phospholipase delta retards ABA-promoted leaf senescence by attenuating lipid degradation. *PLoS ONE* **2013**, *8*, e65687. [CrossRef]
30. Borochov, A.; Cho, M.H.; Boss, W.F. Plasma membrane lipid metabolism of *Petunia* petals during senescence. *Physiol. Plant.* **1994**, *90*, 279–284. [CrossRef]
31. Chen, L.; Liao, B.; Qi, H.; Xie, L.J.; Huang, L.; Tan, W.J.; Zhai, N.; Yuan, L.B.; Zhou, Y.; Yu, L.J.; et al. Autophagy contributes to regulation of the hypoxia response during submergence in *Arabidopsis thaliana*. *Autophagy* **2015**, *11*, 2233–2246. [CrossRef] [PubMed]
32. Jones, M.L.; Chaffin, G.S.; Eason, J.R.; Clark, D.G. Ethylene-sensitivity regulates proteolytic activity and cysteine protease gene expression in petunia corollas. *J. Exp. Bot.* **2005**, *56*, 2733–2744. [CrossRef] [PubMed]
33. Avice, J.C.; Etienne, P. Leaf senescence and nitrogen remobilization efficiency in oilseed rape (*Brassica napus* L.). *J. Exp. Bot.* **2014**, *65*, 3813–3824. [CrossRef] [PubMed]
34. Battelli, R.; Lombardi, L.; Picciarelli, P.; Lorenzi, R.; Frigerio, L.; Rogers, H.J. Expression and localisation of a senescence-associated KDEL-cysteine protease from *Lilium longiflorum* tepals. *Plant Sci.* **2014**, *214*, 38–46. [CrossRef] [PubMed]
35. Price, A.M.; Aros Orellana, D.F.; Salleh, F.M.; Stevens, R.; Acock, R.; Buchanan-Wollaston, V.; Stead, A.D.; Rogers, H.J. A comparison of leaf and petal senescence in wallflower reveals common and distinct patterns of gene expression and physiology. *Plant Physiol.* **2008**, *147*, 1898–1912. [CrossRef] [PubMed]
36. Guiboileau, A.; Avila-Ospina, L.; Yoshimoto, K.; Soulay, F.; Azzopardi, M.; Marmagne, A.; Lothier, J.; Masclaux-Daubresse, C. Physiological and metabolic consequences of autophagy deficiency for the management of nitrogen and protein resources in *Arabidopsis* leaves depending on nitrate availability. *New Phytol.* **2013**, *199*, 683–694. [CrossRef] [PubMed]
37. Delmer, D.P. Cellulose biosynthesis exciting times for a difficult field of study. *Annu. Rev. Plant Physiol.* **1999**, *50*, 245–276. [CrossRef] [PubMed]
38. Van Sandt, V.S.; Suslov, D.; Verbelen, J.P.; Vissenberg, K. Xyloglucan endotransglucosylase activity loosens a plant cell wall. *Ann. Bot.* **2007**, *100*, 1467–1473. [CrossRef] [PubMed]
39. Singh, A.P.; Tripathi, S.K.; Nath, P.; Sane, A.P. Petal abscission in rose is associated with the differential expression of two ethylene-responsive xyloglucan endotransglucosylase/hydrolase genes, *RbXTH1* and *RbXTH2*. *J. Exp. Bot.* **2011**, *62*, 5091–5103. [CrossRef] [PubMed]

40. Saathoff, A.J.; Hargrove, M.S.; Haas, E.J.; Tobias, C.M.; Twigg, P.; Sattler, S.; Sarath, G. Switchgrass PviCAD1: understanding residues important for substrate preferences and activity. *Appl. Biochem. Biotechnol.* **2012**, *168*, 1086–1100. [CrossRef] [PubMed]

41. Weng, J.K.; Chapple, C. The origin and evolution of lignin biosynthesis. *New Phytol.* **2010**, *187*, 273–285. [CrossRef] [PubMed]

42. Zuliana, R.; Chandran, S.; Lee, A.L.; Boyce, A.N.; Nair, H. Isolation and characterization of ethylene related genes during pollination induced senescence of *Dendrobium orchids*. *Acta Hortic.* **2010**, *875*, 523–530. [CrossRef]

43. Van Doorn, W.G. Is petal senescence due to sugar starvation? *Plant Physiol.* **2004**, *134*, 35–42. [CrossRef] [PubMed]

44. Moons, A.; Valcke, R.; van Montagu, M. Low-oxygen stress and water deficit induce cytosolic pyruvate orthophosphate dikinase (PPDK) expression in roots of rice, a C3 plant. *Plant J.* **1998**, *15*, 89–98. [CrossRef] [PubMed]

45. Xing, H.; Zhang, S.; Weinheimer, C.; Kovacs, A.; Muslin, A.J. 14-3-3 proteins block apoptosis and differentially regulate MAPK cascades. *EMBO J.* **2000**, *19*, 349–358. [CrossRef] [PubMed]

46. Takai, Y.; Sasaki, T.; Matozaki, T. Small GTP-binding proteins. *Physiol. Rev.* **2001**, *81*, 153–208. [CrossRef] [PubMed]

47. Yong, T.J.; Gan, Y.Y.; Toh, B.H.; Sentry, J.W. Human CKIα(L) and CKIα(S) are encoded by both 2.4- and 4.2-kb transcripts, the longer containing multiple RNA-destablising elements. *Biochim. Biophys. Acta* **2000**, *1492*, 425–433. [CrossRef]

48. Zhu, Q.; Li, M.; Liu, G.; Li, Y.; Sui, S.; Guo, T. Molecular characterization and functional prediction of a novel leaf SAG encoding a CBS–domain–containing protein from *Coleus blumei*. *Chin. J. Biochem. Mol. Biol.* **2007**, *23*, 262–270. [CrossRef]

49. Liu, Z.L.; Luo, C.; Dong, L.; van Toan, C.; Wei, P.X.; He, X.H. Molecular characterization and expression analysis of a GTP-binding protein (*MiRab5*) in *Mangifera indica*. *Gene* **2014**, *540*, 86–91. [CrossRef] [PubMed]

50. Zeng, C.L.; Liu, L.; Xu, G.Q. The physiological responses of carnation cut flowers to exogenous nitric oxide. *Sci. Hortic.* **2011**, *127*, 424–430. [CrossRef]

51. Chakrabarty, D.; Verma, A.K.; Datta, S.K. Oxidative stress and antioxidant activity as the basis of senescence in *Hemerocallis* (day lily) flowers. *J. Hortic. For.* **2009**, *1*, 113–119.

52. Larson, R.A. The antioxidants of higher plants. *Phytochemistry* **1988**, *27*, 969–978. [CrossRef]

53. Frimer, A.A. Oxidative processes in biological systems and their role in plant senescence. In *Processes and Control of Plant Senescence*; Leshem, Y.Y., Halevy, A.H., Frenkel, C., Eds.; Elsevier: Amsterdam, The Netherlands, 1986; pp. 84–99. ISBN 9780444598462.

54. Testerink, C.; Larsen, P.B.; van der Does, D.; van Himbergen, J.A.; Munnik, T. PA, a stress-induced short cut to switch-on ethylene signaling by switching-off CTR1? *Plant Signal. Behav.* **2008**, *3*, 681–683. [CrossRef] [PubMed]

55. Kooijman, E.E.; Chupin, V.; de Kruijff, B.; Burger, K.N.J. Modulation of membrane curvature by phosphatidic acid and lysophosphatidic acid. *Traffic* **2003**, *4*, 162–174. [CrossRef] [PubMed]

56. Visser, T.; Verhaegh, J.J. Pollen and pollination experiments. I. The contribution of stray pollen to the seed set of depetalled, hand-pollinated flowers of apple. *Euphytica* **1980**, *29*, 379–383. [CrossRef]

57. Sarma, A.D.; Oehrle, N.W.; Emerich, D.W. Plant protein isolation and stabilization for enhanced resolution of two-dimensional polyacrylamide gel electrophoresis. *Anal. Biochem.* **2008**, *379*, 192–195. [CrossRef] [PubMed]

58. Bradford, M.M. A rapid and sensitive method for the quantitation of microgram quantities of protein utilizing the principle of protein-dye binding. *Anal. Biochem.* **1976**, *72*, 248–254. [CrossRef]

59. Valledor, L.; Jorrin, J. Back to the basics: Maximizing the information obtained by quantitative two dimensional gel electrophoresis analyses by an appropriate experimental design and statistical analyses. *J. Proteom.* **2011**, *74*, 1–18. [CrossRef] [PubMed]

International Journal of
Molecular Sciences

MDPI

Article

Characterization of Heterotrimeric G Protein γ4 Subunit in Rice

Sakura Matsuta [†], Aki Nishiyama [†], Genki Chaya, Takafumi Itoh, Kotaro Miura and Yukimoto Iwasaki *

Department of Bioscience and Biotechnology, Fukui Prefectural University, 4-1-1 Kenjojima, Matsuoka, Eiheiji-Town, Fukui 910-1195, Japan; s1873018@g.fpu.ac.jp (S.M.); s1873016@g.fpu.ac.jp (A.N.); s1873012@g.fpu.ac.jp (G.C.); ito-t@fpu.ac.jp (T.I.); miura-k@fpu.ac.jp (K.M.)
* Correspondence: iwasaki@fpu.ac.jp; Tel.: +81-776-61-6000 (ext. 3514)
† These authors contributed equally to this work.

Received: 27 September 2018; Accepted: 9 November 2018; Published: 14 November 2018

Abstract: Heterotrimeric G proteins are the molecule switch that transmits information from external signals to intracellular target proteins in mammals and yeast cells. In higher plants, heterotrimeric G proteins regulate plant architecture. Rice harbors one canonical α subunit gene (*RGA1*), four extra-large GTP-binding protein genes (*XLGs*), one canonical β-subunit gene (*RGB1*), and five γ-subunit genes (tentatively designated *RGG1*, *RGG2*, *RGG3/GS3/Mi/OsGGC1*, *RGG4/DEP1/DN1/qPE9-1/OsGGC3*, and *RGG5/OsGGC2*) as components of the heterotrimeric G protein complex. Among the five γ-subunit genes, *RGG1* encodes the canonical γ-subunit, *RGG2* encodes a plant-specific type of γ-subunit with additional amino acid residues at the N-terminus, and the remaining three γ-subunit genes encode atypical γ-subunits with cysteine-rich C-termini. We characterized the *RGG4/DEP1/DN1/qPE9-1/OsGGC3* gene product Gγ4 in the wild type (WT) and truncated protein Gγ4ΔCys in the *RGG4/DEP1/DN1/qPE9-1/OsGGC3* mutant, *Dn1-1*, as litttele information regarding the native Gγ4 and Gγ4ΔCys proteins is currently available. Based on liquid chromatography-tandem mass spectrometry analysis, immunoprecipitated Gγ4 candidates were confirmed as actual Gγ4. Similar to α-(Gα) and β-subunits (Gβ), Gγ4 was enriched in the plasma membrane fraction and accumulated in the developing leaf sheath. As *RGG4/DEP1/DN1/qPE9-1/OsGGC3* mutants exhibited dwarfism, tissues that accumulated Gγ4 corresponded to the abnormal tissues observed in *RGG4/DEP1/DN1/qPE9-1/OsGGC3* mutants.

Keywords: *Dn1-1*; γ-subunit; heterotrimeric G protein; mass spectrometry analysis; RGG4; rice; western blotting

1. Introduction

Heterotrimeric G proteins consist of three subunits (α, β, and γ) in mammals and yeast cells. They act as signal transducers by transferring extracellular information to intracellular components [1–4]. External signals bind or affect G protein-coupled receptors (GPCRs) to activate them. Activated GPCRs, which function as an intrinsic GDP/GTP exchange factor of GPCRs, convert α-GDP to α-GTP. When GTP binds to the α-subunit (α-GTP), heterotrimeric G proteins dissociate into α subunit (α-GTP) and βγ dimer. The α-subunit and βγ dimer can regulate respective effector molecules. In higher plants, heterotrimeric G proteins are important molecules that regulate plant development and transmit external signals to intracellular target proteins [5–7]. Biochemical characteristics of plant heterotrimeric G proteins have been investigated [5]. The signaling mechanisms and effector molecules that regulate plant heterotrimeric G proteins have been reviewed [6], as has the plant morphology of heterotrimeric G protein mutants [7]. *Arabidopsis* harbors three extra-large GTP-binding protein genes (*XLGs*) [8,9], one canonical α-subunit gene (*GPA1*) [10], one canonical β-subunit gene (*AGB1*) [11], and three γ-subunit genes (*AGG1–AGG3*) [12–14]. Rice (*Oryza sativa*) harbors four extra-large GTP-binding protein genes (*XLGs*) [15], one canonical α-subunit gene (*RGA1*) [16], one canonical β-subunit gene (*RGB1*) [17], and five γ-subunit genes, which are tentatively designated *RGG1, RGG2* [18], *RGG3/GS3/Mi/OsGGC1* [19], *RGG4/DEP1/DN1/qPE9-1/OsGGC3* [20,21], and *RGG5/OsGGC2* [22].

With regard to the γ-subunit genes in *Arabidopsis, AGG1* [12] and *AGG2* [13] encode the canonical γ-subunits and *AGG3* [14] encodes the atypical γ-subunit having a cysteine-rich C-terminus. With regard to the γ-subunit genes in rice, *RGG1* encodes the canonical γ-subunit [18], *RGG2* encodes the plant-specific type of γ-subunit [18], and the remaining three γ-subunit genes (*RGG3/GS3/Mi/OsGGC1, RGG4/DEP1/DN1/qPE9-1/OsGGC3,* and *RGG5/OsGGC2*) encode the atypical γ-subunits homologous to *AGG3. RGG3* corresponds to *GRAIN SIZE 3 (GS3)* [19] and *RGG4* corresponds to *DENSE AND ERECT PANICLES 1 (DEP1)* [20], *DENSE PANICLE1 (DN1)* [21], and *qPE9-1* [23]. The genome sequence of RGG5 was predicted by Botella [22]. The diversity and agronomical importance of plant γ-subunits have been reviewed previously [24].

Concerning mutants of heterotrimeric G proteins, *xlg1–xlg3* [25], *gpa1* [26], *agb1* [27,28], *agg1* [29], *agg2* [29], and *agg3* [14] in *Arabidopsis,* and *d1* [30,31], Chuan7(GS3-4) and Minghui 63 (GS3-3) [32], *dep1* [20] in rice, have been isolated. By the analysis of *gpa1* [26], *agb1* [28], *d1* [33], and *RGB1* knock-down lines [34], an allele of *GS3* [35], *dep1* [20], it was shown that the plant heterotrimeric G proteins modulated cell proliferation.

Comparing the wild type and mutant responses to external signals, it has been shown that plant heterotrimeric G proteins were involved in transductions of multiple external signals, such as abscisic acid [36–40], auxin [26,28], gibberellin [41–44], brassinosteroid [26,42,43], sugar [26,45,46], blue light [47,48], ozone [49], elicitors [50–53]. Plant heterotrimeric G proteins may regulate at integration points for these signals.

Regarding protein–protein interactions in the G protein complex, Klopffleish et al. proposed that 68 highly interconnected proteins form the core G protein interactome in *Arabidopsis,* using the yeast two-hybrid assay (Y2H) [54]. In previous studies, the regulators of G protein signaling protein (AtRGS1) [55], THYLAKOID FORMATION1 (THF1) [46], and cupin domain protein (AtPrin1) [38], were shown to be contained in the interactome. It was also shown that G protein complexes containing Gα subunit—which were solubilized by the plasma membrane—were the huge complexes in rice [18] and *Arabidopsis* [56], respectively. The huge complexes may be a part of the interactome.

Among three atypical γ-subunit genes (*RGG3, RGG4,* and *RGG5*) in rice, *RGG3* corresponds to *GRAIN SIZE 3 (GS3)*, which regulates seed length in rice [19,32,35,57,58]. *RGG4* corresponds to *DENSE AND ERECT PANICLES 1 (DEP1)* [20], *DENSE PANICLE1 (DN1)* [21], quantitative trait locus *qPE9-1* (*qPE9-1*) [23], which regulate plant architecture including semi-dwarfness, panicle number and panicle erectness. *DEP1* regulates nitrogen-use efficiency in addition to regulating plant architecture [59]. *RGG5* corresponds to *GGC2* [22], which a gene that increases grain length in combination or individually with *DEP1* [57]. These genes are important for rice breeding.

We previously analyzed the native proteins, Gα, Gβ, Gγ1, and Gγ2, localized plasma membrane fraction [18]. However, there is little information on the native proteins translated by *RGG3, RGG4,* and *RGG5,* such as Gγ3, Gγ4, and Gγ5, respectively. Among the three atypical γ-subunits, we aimed to identify native Gγ4 and truncated Gγ4 using the anti-Gγ4 domain antibody. The study of the native Gγ4 and truncated Gγ4 is important to understand the function of Gγ4 and truncated Gγ4, which regulate plant architecture. When they are identified, biochemical analysis, namely measurement of subunit stoichiometry and affinity to Gβ, canonical Gα and XLGs, is possible. We tried to identify the native Gγ4 and in wild type rice using an anti-Gγ4 domain antibody. However, the antibody recognized multiple proteins. To identify the native Gγ4 protein, we used the *RGG4* mutant *Dn1-1,* which produces a partially defective protein, as the reference for subtraction. We found candidates of the native Gγ4 and truncated Gγ4 and confirmed that the candidates are actually the native Gγ4 and truncated Gγ4 by liquid chromatography-tandem mass spectrometry (LC-MS/MS) analysis of immunoprecipitation products using anti-Gγ4 domain antibody. The antibody was used to examine the subcellular localization and tissue-specific accumulation of the native Gγ4.

2. Results

2.1. Rice Heterotrimeric G protein γ4 Gene (RGG4/DEP1/DN1/qPE9-1/OsGGC3) Mutant

To identify the rice heterotrimeric G protein γ4 subunit, Gγ4, we used a mutant possessing the *Dn1-1* mutation in the Nipponbare background. *Dn1-1* displayed characteristics of semi-dwarfism and slightly increased number of spikelets, as described previously [21]. These results indicated that *Dn1-1* mutation clearly affected plant height and panicle number.

2.2. Genomic Structure of RGG4 and Protein Structure of Gγ4

The genome sequence of *RGG4* was found in RAP-DB (Os09g0441900). We reconfirmed the genome sequence of *RGG4*. *RGG4* consists of five exons (Figure 1a). The translation product, Gγ4, comprises 426 amino acid residues. To prepare recombinant proteins, cDNA for RGG4 was isolated. The molecular weight of Gγ4 calculated from cDNA was 45210 Da. Gγ4 comprised a canonical γ domain of approximately 100 amino acids, a short region with hydrophobic amino acid residues (tentatively termed the transmembrane region, TM), and a region enriched in cysteine residues (Cys-rich region) (Figure 1b).

The *Dn1-1* mutation occurred as a result of a one-base substitution. We reconfirmed the mutation in *Dn1-1* in which C, at position 512 in the full-length cDNA of *RGG4*, was substituted by A (C512A), resulting in the generation of a stop codon (Figure 1a). In *Dn1-1*, the mutated protein, tentatively designated Gγ4ΔCys, consisted of 170 amino acid residues (Figure 1b). The cysteine-rich region was absent in Gγ4ΔCys. The molecular weight of Gγ4ΔCys calculated from cDNA was 18997 Da.

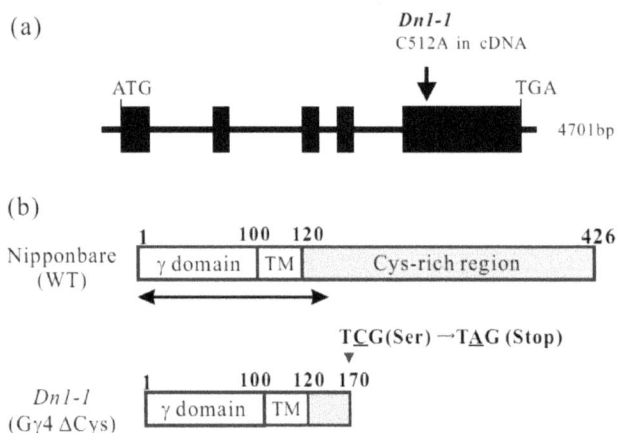

Figure 1. Genome and protein structure of *RGG4/DEP1/DN1/qPE9-1/OsGGC3*. (**a**) Genome structure of *RGG4/DEP1/DN1/qPE9-1/OsGGC3* and position of the mutation in *RGG4/DEP1/DN1/qPE9-1/OsGGC3* mutant *Dn1-1*. The one-base substitution (C512A in full-length cDNA) in *Dn1-1* was in a codon in which TCG (cysteine) was changed to TAG (stop codon). (**b**) Protein structure of the product of *RGG4/DEP1/DN1/qPE9-1/OsGGC3* in wild type (WT) (Gγ4) and *Dn1-1* (Gγ4ΔCys). The canonical γ-domain region is shown as γ domain. Putative transmembrane domain is indicated as TM. The cysteine-rich region is indicated by the gray box. An arrow under WT Gγ4, which covers 137 amino acid residues from the N-terminus, is the region used for recombinant proteins, such as the thioredoxin (Trx)-tagged Gγ4 domain protein (Trx-Gγ4 domain protein), which was used as the antigen, and glutathione S transferase (GST)-tagged Gγ4 domain protein (GST-Gγ4 domain protein), which was used for affinity purification of the antibody.

2.3. Gγ4 Candidates Localized in the Plasma Membrane Fraction

Identification of native Gγ4 and Gγ4ΔCys was carried out using both WT and *Dn1-1* as subtraction references, respectively. As rice Gα and Gβ were known to be localized in the plasma membrane fraction, the plasma membrane fractions of wild type (WT) and *Dn1-1* were prepared using an aqueous two-polymer phase system. Gγ4 candidates were detected by western blotting (WB) using anti-Gγ4 domain antibody. In WT, a 55-kDa protein (Gγ4 candidate) was detected (Figure 2a, lane 2); this band was not observed in *Dn1-1*. The molecular weight of Gγ4 candidate was much higher than that of Gγ4 calculated from WT cDNA (45 kDa). In *Dn1-1*, a 27-kDa protein (Gγ4ΔCys candidate) was detected (Figure 2a, lane 3). The molecular weight of the Gγ4ΔCys candidate was much higher than that of Gγ4ΔCys calculated from cDNA (19 kDa). The molecular weight of Gγ4 and Gγ4ΔCys candidates was measured using molecular weight markers (Figure 2b). Since a sharp band with molecular weight of 31 kDa (indicated by an arrowhead) was found in WT and *Dn1-1*, this band was eliminated to be a Gγ4 candidate.

The chemiluminescent intensity of Gγ4ΔCys was more than 3-fold that of Gγ4, when 10 μg of plasma membrane protein of the WT and *Dn1-1*, respectively, was analyzed by western blot.

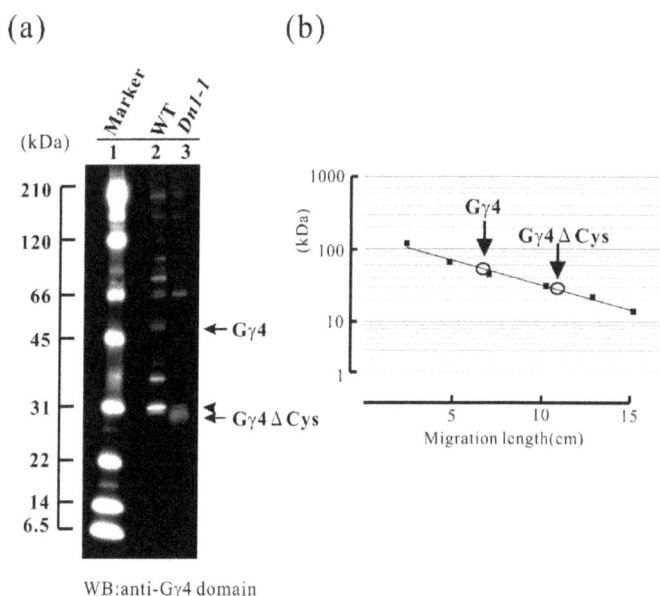

WB:anti-Gγ4 domain

Figure 2. Immunological study of the Gγ4 candidates in leaf sheath of wild type (WT) and *Dn1-1*. (**a**) Plasma membrane protein fractions of WT and *Dn1-1* (10 μg) were used for western blot analysis using anti-Gγ4 domain antibody. Lane 1 contains molecular weight markers. The Gγ4 candidate (indicated by an arrow) was detected as a broad band with a molecular weight of approximately 55 kDa in WT (lane 2). The Gγ4ΔCys candidate (indicated by an arrow) was detected as a broad band with a molecular weight of approximately 27 kDa in *Dn1-1* (lane 3). An arrowhead indicates non-specific bands found in both WT and *Dn1-1*. (**b**) The molecular weights of Gγ4 and Gγ4ΔCys candidates were estimated using molecular weight marker as the standard.

2.4. Immunoprecipitation of Gγ4, and Gγ4ΔCys Using Anti-Gγ4 Domain Antibody

To concentrate Gγ4 and Gγ4ΔCys candidates, immunoprecipitation was carried out using anti-Gγ4 domain antibody. Fifty micrograms of anti-Gγ4 domain antibody was added to 2 mg and 1 mg of solubilized plasma membrane protein of leaf sheath of WT (Figure 3a) and *Dn1-1* (Figure 3b), respectively. Gγ4 and Gγ4ΔCys candidates were collected with the antibody cross-linked Protein

A-bound beads. The Gγ4 candidate in WT (55 kDa; Figure 3a, lane 3) and Gγ4ΔCys candidate in *Dn1-1* (27 kDa; Figure 3b, lane 3), were immunoprecipitated. Most other proteins detected by WB of plasma membrane fraction were not observed in the immunoprecipitated products.

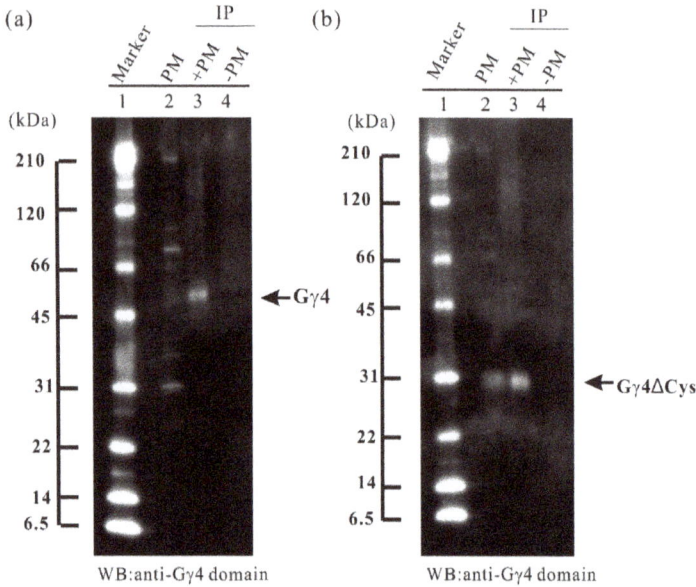

Figure 3. Immunoprecipitation (IP) of Gγ4 and Gγ4ΔCys candidates in leaf sheath of wild type (WT) and *Dn1-1*. (**a**) IP of Gγ4 candidate from solubilized plasma membrane proteins of wild type (WT) using anti-Gγ4 domain antibody. Lane 1, molecular weight markers; lane 2, 10 μg of plasma membrane protein fraction of WT; lane 3, IP product of solubilized plasma membrane proteins and anti-Gγ4 domain antibody; lane 4, control experiment (buffer in the placement of membrane proteins). (**b**) IP of Gγ4ΔCys candidate from solubilized plasma membrane proteins of *Dn1-1* using anti-Gγ4 domain antibody. Lane 1, molecular weight markers; lane 2, 10 μg of plasma membrane protein fraction of *Dn1-1*; lane 3, IP product of solubilized plasma membrane proteins of *Dn1-1* and anti-Gγ4 domain antibody; lane 4, control experiment (buffer in the placement of membrane proteins).

2.5. LC-MS/MS Analysis

LC-MS/MS analysis was performed to confirm that Gγ4 and Gγ4ΔCys candidates were actually Gγ4 and Gγ4ΔCys and to confirm that proteins with which anti-Gγ4 domain antibody reacted were actually Gγ4 and Gγ4ΔCys. First, we analyzed whether Gγ4 and Gγ4ΔCys candidates in eluate from SDS-PAGE gel pieces of plasma membrane proteins were detected by LC-MS/MS. When the signal intensities of Gγ4 and Gγ4ΔCys in LC-MS/MS were low, we analyzed immunoprecipitation products enriched by anti-Gγ4 domain antibody.

Plasma membrane proteins from WT and *Dn1-1* were analyzed by LC-MS/MS. Forty micrograms each of plasma membrane protein isolated from WT and *Dn1-1* leaf sheath were separated by sodium dodecyl sulfate-polyacrylamide gel electrophoresis (SDS-PAGE) and each lane of the gel was cut into 10 pieces according to molecular weight marker to increase the relative amount of target proteins. After these gel pieces were digested with trypsin, peptides were analyzed by LC-MS/MS in triplicate. WT plasma membrane proteins did not display fragments with $p < 0.05$ Gγ4 and Mascot score of Gγ4 was <50. In LC-MS/MS analysis of *Dn1-1* plasma membrane proteins, five fragments with $p < 0.05$ in Gγ4ΔCys were obtained. The Mascot score in Gγ4ΔCys was 95. Fifteen high accuracy LC-MS/MS fragments of immunoprecipitated WT Gγ4 were obtained (Table 1A). They were chosen as the standard and were numbered. The five fragments from *Dn1-1* plasma membrane protein corresponded to fragment numbers 2, 3, 4, 6, and 7 (Table 1). The Gγ4 candidate was not detected in the LC-MS/MS analysis of WT plasma membrane proteins, and the immunoprecipitation experiment using anti-Gγ4 domain antibody was done.

IP products from WT and *Dn1-1* were separated by SDS-PAGE and analyzed by LC-MS/MS, but the products were not detected by silver staining. In IP products of WT (Figure 3a, lane 3), a gel piece containing a 55-kDa protein was excised and digested with trypsin. The resultant peptides were analyzed by LC-MS/MS. Fifteen Gγ4 fragments with primary mass ($p < 0.05$) were obtained (Table 1A). In IP products of *Dn1-1*, a gel piece containing a 27-kDa protein was excised and digested by trypsin, and the resultant peptides were analyzed by LC-MS/MS. Eight fragments ($p < 0.05$) were obtained (Table 1B).

Figure 4a presents examples of fragments with high accuracy, the MS/MS results of fragments 1, 8, 11, and 15. Based on these results, we concluded that the 55-kDa and 27-kDa proteins were Gγ4 and Gγ4ΔCys, respectively. When the NCBI protein database was used for the analysis of Gγ4 candidates, Gγ4 was found to be annotated using other names, such as ACL27948.1.

Table 1. LC-MS/MS analysis of Gγ4 fragments in leaf sheath of wild type (WT) and *Dn1-1*. (**A**) Gγ4 fragments in immunoprecipitation products of the WT membrane fraction. (**B**) Gγ4 fragments in immunoprecipitation products of the *Dn1-1* membrane fraction.

(A)					
Fragments	Observed	Mr (Expt)	Mr (Calc)	Expect	Peptide
1	706.8601	1411.7057	1411.7129	1.80×10^{-6}	M.GEEAVVMEAPRPK.S
2	440.7008	879.3871	879.3909	0.0015	R.YPDLCGR.R
3	608.8365	1215.6585	1215.6645	9.20×10^{-7}	R.MQLEVQILSR.E
4	375.7118	749.409	749.4323	0.0075	R.EITFLK.D
5	799.3979	1596.7812	1596.7896	1.20×10^{-6}	K.DELHFLEGAQPVSR.S
6	1036.5248	1035.5175	1035.5237	2.10×10^{-5}	K.EINEFVGTK.H
7	460.7616	919.5086	919.5127	0.0077	K.HDPLIPTK.R
8	905.3625	1808.7105	1808.721	7.50×10^{-7}	K.LCICISCLCYCCK.C
9	904.4171	903.4098	903.416	0.0006	K.SLYSCFK.I
10	751.3766	750.3693	750.3734	0.00025	K.IPSCFK.S
11	856.6377	2566.8913	2566.909	66.70×10^{-8}	K.SQCNCSSPNCCTCTLPSCSCK.G
12	710.2448	1418.4751	1418.4836	1.30×10^{-8}	R.CADCFSCSCPR.C
13	581.7522	1161.4898	1161.4947	3.90×10^{-7}	R.CSSCFNIFK.C
14	725.2517	1448.4889	1448.4975	7.00×10^{-7}	K.CSCAGCCSSLCK.C
15	620.7291	1239.4437	1239.4505	1.10×10^{-5}	R.NPCCLSGCLC

(B)					
Fragments	Observed	Mr (expt)	Mr (Calc)	Expect	Peptide
1	706.8604	1411.7063	1411.7129	0.000016	M.GEEAVVMEAPRPK.S
2	440.7009	879.3872	879.3909	0.002	R.YPDLCGR.R
3	608.8362	1215.6579	1215.6645	5.6×10^{-7}	R.MQLEVQILSR.E
4	375.7219	749.4292	749.4323	0.0019	R.EITFLK.D
5	799.3994	1596.7842	1596.7896	0.000046	K.DELHFLEGAQPVSR.S
6	1036.5227	1035.5154	1035.5237	0.000024	K.EINEFVGTK.H
7	460.7607	919.5068	919.5127	0.018	K.HDPLIPTK.R
8	905.3626	1808.7107	1808.721	0.0000015	K.LCICISCLCYCCK.C

Eight microliters of each eluate in immunoprecipitates of WT and *Dn1-1* (A and B) were used for LC-MS/MS. Fragments of trypsin-digested Gγ4 candidates ($p < 0.05$) are shown. Fragment numbers correspond to Figure 4a. Mr (expt) and Mr (calc) correspond to the theoretical molecular mass and molecular mass that was calculated from the observed molecular mass, respectively. The scores by Mascot search were 859 (A) and 505 (B) for WT and *Dn1-1*, respectively.

(a) ₁**MGEEAVVMEAPRPK** SPPR ₂**YPDLCGR** RR ₃**MQLEVQILSR** ₄**EITFLK** ₅**DELH**
FLEGAQPVSR SGCIK ₆**EINEFVGTK** ₇**HDPLIPTK** RRRHRSCRLFRWIGSK ₈**LCIC**
ISCLCYCCK CSPKCKRPRCLNCSCSSCCDEPCCKPNCSACCAGSCCSPDCCSCCKP
NCSCCKTPSCCKPNCSCSCPSCSSCCDTSCCKPSCTCFNIFSCFK₉**SLYSCFK**₁₀**IPSCF**
K ₁₁**SQCNCSSPNCCTCTLPSCSCK** GCACPSCGCNGCGCPSCGCNGCGCPSCGCNG
CGLPSCGCNGCGSCSCAQCKPDCGSCSTNCCSCKPSCNGCCGEQCCR ₁₂**CADCFSC**
SCPR ₁₃**CSSCFNIFK** ₁₄**CSCAGCCSSLCK** CPCTTQCFSCQSSCCKRQPSCCKCQSSCC
EGOPSCCEGHCCSLPKPSCPECSCGCVWSCKNCTEGCRCPRCR ₁₅**NPCCLSGCLC***

(b)

Fragment 1

Fragment 8

Fragment 11

Fragment 15

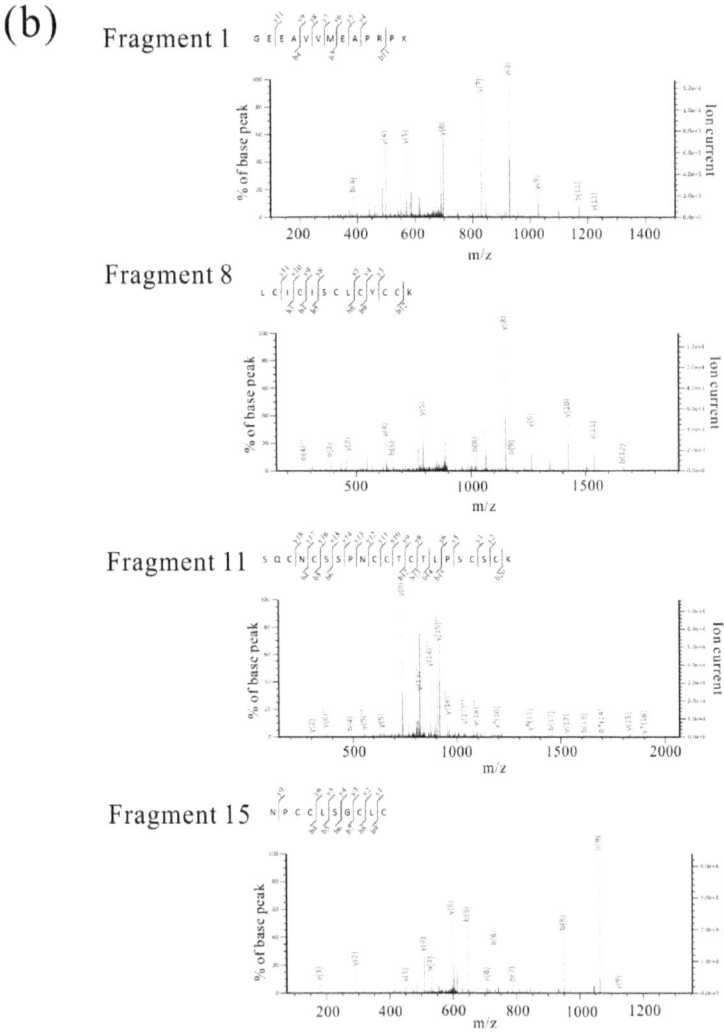

Figure 4. LC-MS/MS analysis of Gγ4 candidates from WT leaf sheath. (**a**) Fifteen peptides ($p <$ 0.05) produced by trypsin-digested Gγ4 candidates in wild type (WT) and *Dn1-1* are numbered and underlined in the full-length Gγ4 amino acid sequence. These peptides are listed in Table 1 (A and B). (**b**) MS/MS spectra of four fragments, which were obtained as immunoprecipitation product of Gγ4 in WT (Figure 3a, lane 3). Fragment numbers correspond to Table 1 (A and B).

2.6. Enrichment of Gγ4 and Gγ4ΔCys in the Plasma Membrane Fraction

As LC-MS/MS analysis showed that anti-Gγ4 domain antibody reacted with Gγ4 and Gγ4ΔCys, the amount of Gγ4 and Gγ4ΔCys in the crude microsomal fraction was compared to that in the plasma membrane fraction by western blot (Figure 5). This was in order to check whether Gγ4 and Gγ4ΔCys are enriched in the plasma membrane. To confirm the purity of plasma membrane, the OsPIP1s aquaporin was used as a plasma membrane marker. Gα and Gβ are the subunits of heterotrimeric G protein complex in rice. OsPIP1s, Gα subunit, and Gβ subunit were enriched in the plasma membrane fraction. Gγ4 (55 kDa in WT) and Gγ4ΔCys (27 kDa in *Dn1-1*) were also enriched in the plasma membrane fraction. These results showed that Gγ4 (55 kDa in WT) and Gγ4ΔCys (27 kDa in *Dn1-1*) were localized in the plasma membrane fraction. Non-specific bands (indicated by arrowheads) detected by WB were not analyzed.

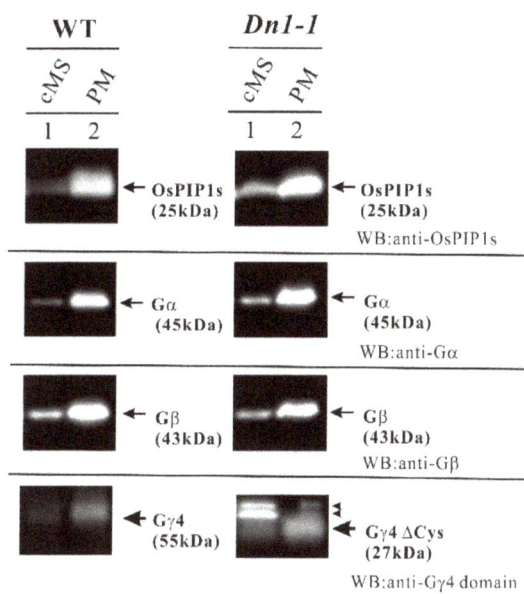

Figure 5. Gγ4 and Gγ4ΔCys in leaf sheath of wild type (WT) and *Dn1-1* were enriched in the plasma membrane fraction (PM). Ten micrograms of each crude microsome fraction (cMS) and plasma membrane (PM) proteins from wild type (WT) and *Dn1-1* were analyzed by western blot using anti-OsPIP1s, anti-Gα, anti-Gβ, and anti Gγ4 domain antibodies. OsPIP1s is an aquaporin and a plasma membrane marker. OsPIP1s (25kDa), Gα (45kDa), Gβ (43kDa), Gγ4 (55kDa), and Gγ4ΔCys (27kDa) are indicated by arrows. Non-specific bands are indicated by arrowheads.

2.7. Tissue-Specific Accumulation of Gγ4

To identify the tissues in which Gγ4 accumulates, the accumulation profile of Gγ4 was studied using plasma membrane fractions of leaf of 1-week-old etiolated seedling, developing leaf sheath, and flowers of WT by western blot. Gγ4 protein predominantly accumulated in the developing leaf sheath (Figure 6).

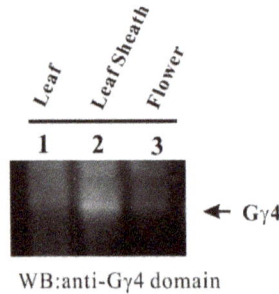

WB:anti-Gγ4 domain

Figure 6. Tissue-specific accumulation of Gγ4 in wild type (WT). Ten micrograms of each plasma membrane protein fraction of leaf, leaf sheath, and flower of WT was analyzed by sodium dodecyl sulfate-polyacrylamide gel electrophoresis and western blotting using anti-Gγ4 domain antibody. Lane 1, leaf from etiolated seedling; lane 2, developing leaf sheath at 8th leaf stage; lane 3, 1–5 cm flower. WB: western blot.

3. Discussion

Rice has three atypical γ-subunit genes (*RGG3*, *RGG4*, and *RGG5*) in the heterotrimeric G protein complex, which are homologous to *AGG3*. Both *RGG3* regulating seed length and *RGG4* regulating plant architecture including semi-dwarfness, panicle number and panicle erectness, respectively, are important genes for breeding. In this study, we aimed to identify the native protein translated by *RGG4/DEP1/DN1/qPE9-1/OsGGC3*. Identification of the native Gγ4 and truncated Gγ4 is important, to understand the function of *RGG4*.

First, we detected the Gγ4 candidate from WT and truncated Gγ4 candidate Gγ4ΔCys from *Dn1-1* by WB using anti-Gγ4 domain antibodies (Figure 2). SDS-PAGE estimated the molecular weights of Gγ4 and Gγ4ΔCys candidates as 55 and 27 kDa, respectively, which were higher than the molecular mass calculated using cDNAs (45 and 19 kDa, respectively). These results indicated that modifications, such as glycosylation, ubiquitination, phosphorylation, lipid modification, which include palmitoylation etc., might have occurred after translation in Gγ4 and Gγ4ΔCys candidates. As Gγ4 and Gγ4ΔCys were detected as broad bands by WB, some modification may have occurred. Identification of the modifications will be the subject of future studies. When 10 μg of plasma membrane protein from each of the WT and *Dn1-1* developing leaf sheath were analyzed by western blot, the chemiluminescent intensity of Gγ4ΔCys was more than 3-fold that of Gγ4 (Figure 2). The reason that the amount of Gγ4 was less than that of Gγ4ΔCys may be that Gγ4 is degraded by proteases. Another possibility is that Gγ4ΔCys accumulates in the plasma membrane with other proteins, including Gβ.

To obtain concrete evidence of whether the Gγ4 and Gγ4ΔCys candidates detected by WB were actually Gγ4 and Gγ4ΔCys proteins, IP products of Gγ4 and Gγ4ΔCys candidates were analyzed by LC-MS/MS (Figures 3 and 4). Fifteen fragments displayed $p < 0.05$ using the Mascot search engine were obtained from the two candidates. These results indicated that the Gγ4 and Gγ4ΔCys candidates were indeed Gγ4 and Gγ4ΔCys, respectively.

Dn1-1 displays semi-dwarfness and increased spikelet number [21]. Thus, Gγ4 regulates plant architecture. Gγ4 and Gγ4ΔCys accumulated in the plasma membrane fraction of developing leaf sheath (Figure 5). The tissue in which Gγ4 and Gγ4ΔCys accumulated corresponded to the tissue that exhibited semi-dwarfness in *Dn1-1* (Figure 6). Although *Dn1-1* exhibited slightly increased number of spikelets [21], the amount of Gγ4 in flower of 1–5 cm in length was less than that in the developing leaf sheath (Figure 6). In the RAP-database, an accumulation profile of Gγ4 mRNA by microarray analysis was found. The relative amount of Gγ4 mRNA in the flower was similar to that in the leaf sheath. The accumulation profile of Gγ4 protein seems to be slightly different from that of Gγ4 mRNA. The accumulation of Gγ4 protein in flowers may be limited to some specific stage and/or organ, such as the inflorescent meristem, in which other proteins necessary for stable accumulation of Gγ4 protein may be present.

Sun et al. studied the interaction between DEP1 (Gγ4 in this study) and Gβ using yeast two-hybrid (Y2H) and bimolecular fluorescence complementation (BiFC) methods [59]. The G protein γ-like domain (GGL) of DEP1 was necessary for binding to Gβ. It is considered that approximately 90–100 N-terminal amino acid residues comprise the canonical γ-domain and the subsequent 20 amino acid residues comprise a transmembrane domain in DEP1. As the truncated proteins of *dep1-1* [59] and *Dn1-1* in this study [21] comprised 196 and 170 amino acid residues, respectively, they can bind to Gβ and anchor in the plasma membrane. Sun et al. also studied the subcellular localization of DEP1 and Gβ by BiFC. DEP1 interaction with Gβ on the plasma membrane was revealed. The truncated protein of *dep1-1* also interacted with Gβ on the plasma membrane. In this research, Gγ4 and Gγ4ΔCys were enriched in rice plasma membrane, as was Gβ. Our results corroborate the findings of Sun et al. As native Gγ4 and Gγ4ΔCys have canonical γ and transmembrane domains, these proteins may form a dimer with Gβ and might anchor on the plasma membrane.

Sun et al. indicated that DEP1 is localized in the nucleus, in addition to the plasma membrane [59]. Taguchi-Shiobara et al. also demonstrated that DN1 (DEP1) is localized in the nucleus by analysis of green fluorescence protein–Gγ4 fusion protein [21]. Although we detected faint broad bands in the $2000 \times g$ precipitate, which may correspond to Gγ4, it was not clear whether these broad bands were actually Gγ4 or not. If Gγ4 was localized in the nucleus, its amount was less than that in the plasma membrane fraction. Although we purified our antibodies using affinity purification method, our antibody recognized Gγ4 and other proteins in WB (Figures 2 and 5). Antibody production using another part of Gγ4 may be necessary to prepare a high-specificity antibody.

Concerning G protein signaling, the unusual βγ dimer composed of GβGγ4ΔCys may be the cause of the shortened plant height and increased spikelet number. In fact, the relative amount of Gγ4ΔCys is much higher than that of Gγ4 in the developing leaf sheath. It will be interesting to determine the function of the unusual βγ dimer (GβGγ4ΔCys) with reference to the G protein signaling model [5,6].

Gβ subunits interact with Gγ subunits and subsequently the βγ dimer was formed in mammals and yeast [1–4]. It has also been shown in plants that Gβ subunits interact with Gγ subunits, using pull down assay [12,13], Y2H [13,18], the split-ubiquitin system [14], bimolecular fluorescence complementation assay (BiFC) [57,59], etc. We confirmed that Gγ4 and Gγ4ΔCys interacted with Gβ using Y2H (data not shown).

Cell number of stem in longitudinal axis is higher in NIL-dep1 [20]. Gγ4 also modulate cell proliferation, similar to Gα [33], Gβ [34], and Gγ3 [35]. It will be important to research the mechanism of cell proliferation which the G protein subunits regulate.

Candidates of both βγ1 and βγ2 dimers present two different fractions in gel filtration, with the former evident as huge complexes containing both βγ1 and βγ2 dimers, and the latter being the dissociated form of the huge complex as a sole βγ1 or βγ2 dimer, in the plasma membrane of etiolated rice seedling [18]. Although this may have resulted from artificial dissociation during solubilization and gel fractionation, this approach will be important for understanding the heterotrimeric G protein complex. As we identified native Gγ4 and Gγ4ΔCys in this study, it will be possible to determine whether Gγ4 is a component of the heterotrimeric G protein complex containing canonical Gα and XLGs.

Kunihiro et al. showed that rice DEP1 (Gγ4) may function as a trap for cadmium ions on yeast cells and *Arabidopsis* [60]. This study gives a new insight into enzymatic function of rice DEP1 (Gγ4). The comparison of cadmium ions between wild type and *Dn1-1* may be helpful for further study of G protein signaling.

4. Materials and Methods

4.1. Plant Materials

A rice cultivar (*O. sativa* L. cv. Nipponbare) and a heterotrimeric G protein γ4 mutant (*Dn1-1*) [21] were used in this study. All rice plants were grown under 14-h light (50,000 lux and 28 °C) and 10-h dark (25 °C) cycle or under natural field condition. Nipponbare is abbreviated as WT in the manuscript.

4.2. Sequencing and Confirmation of RGG4

Genomic DNA was isolated from whole plants of WT and *Dn1-1* by an extraction method using cetyltrimethylammonium bromide [61]. Using this DNA as the template, PCR was performed using >20 sets of PCR primers to cover 4701 bases of *RGG4* (Os09g0441900). The amplified DNA fragments were sequenced directly using the same primers that were used for amplification.

4.3. RNA Isolation, Reverse Transcription, and cDNA Encoding of the Heterotrimeric G Protein γ4 Subunit

Total RNA from flower tissue was directly extracted using RNeasy Plant Mini Kits (Qiagen). The first strand of cDNA was synthesized using Super Script First Strand Synthesis System for RT-PCR (Invitrogen, Carlsbad, CA, USA). Total RNA (0.5 µg) and oligo-dT were used as the template and primer, respectively, for first strand cDNA synthesis. To isolate *RGG4* cDNA, primers were designed based on the database information (Os03g0407400): RGG4 forward: 5′ gtggttctgagttggccgtt 3′ and RGG4 reverse: 5′ caaccaaaaaaggatctagatc 3′. The amplified PCR products were sub-cloned into pCR4 (Invitrogen) and sequenced with a THERMO sequence dye terminator cycle sequencing kit (Amersham Biosciences, Little Chalfont, UK) using a model 377 DNA sequencer (Applied Biosystems, Foster City, CA, USA).

4.4. Preparation of cMS and Plasma Membrane Fractions of Rice

cMSs fraction of WT were prepared from etiolated seedlings grown for 5 days at 28 °C, from developing leaf sheaths at the 8th leaf stage, and from 1–5 cm flowers as described previously [18]. In *Dn1-1*, the cMS fraction from developing leaf sheaths at the 8th leaf stage was prepared. All procedures for membrane preparation were performed at 4 °C. Tissue homogenate was centrifuged at $10,000 \times g$ for 10 min, and the resultant supernatant was centrifuged at $100,000 \times g$ for 1 h. The precipitate ($100,000 \times g$ precipitate) was designated the crude microsomal fraction (cMS). Plasma membrane fractions were prepared from cMS using an aqueous two-polymer phase system [62].

4.5. SDS-PAGE

Electrophoresis was performed on 12.5% and 10%/20% gradient polyacrylamide gels containing 0.1% SDS as described previously [63].

4.6. Preparation of Trx-Gγ4 and GST-Gγ4 Domain Proteins

cDNA encoding 137 amino acid residues from the N-terminal of rice Gγ4 protein was amplified by PCR using primers. The cDNA contained the Gγ4 domain and putative transmembrane region: RGG4 domain forward: 5′ ccatggctcatatggatatcatgggggaggaggcggtggtg 3′ and RGG4 domain reverse: 5′ aagcttcccgggtcaactgcagtttggcttacagcatg 3′. Amplified cDNA was sub-cloned in pCR4 (Invitrogen) and the fragment containing the Gγ4 domain was digested with *Eco*RV and *Hind*III. The fragment was sub-cloned in pET32a containing thioredoxin (Trx) and histidine (His) tags (Novagen, Madison, WI, USA). The resultant clone, Trx-Gγ4 domain vector, was transformed in T7 Express *lysY/Iq Escherichia coli* (New England Biolabs, Ipswich, MA, USA). The recombinant protein was designated Trx-Gγ4 domain protein. cDNA containing the Gγ4 domain was also sub-cloned in pET41 containing glutathione S transferase (GST) and histidine (His) tags (Novagen). The resultant clone, GST-Gγ4 domain vector, was transformed in T7 Express *lysY/I^q* *E. coli* (New England Biolabs). The recombinant protein was designated as GST-Gγ4 domain protein.

The overexpression of Trx-Gγ4 domain protein and GST-Gγ4 domain protein in T7 Express *lysY/I^q* *E. coli* was carried out as previously described [63]. Induction was performed at 37 °C. Induction was started by addition of isopropyl β-D-1-thiogalactopyranoside (final concentration, 1 mM). After 3 h, *E. coli* was harvested at 10,000× *g* for 5 min at 4 °C and stored at −80 °C until required.

As the Trx-Gγ4 domain protein and GST-Gγ4 domain protein were inclusion bodies, both proteins were solubilized in 6 M guanidine hydrochloride, 10 mM Tris HCl, pH 8.0. Solubilized proteins were applied to Ni-NTA agarose (Qiagen, Hilden, Germany). Purification of both proteins was performed according to the protocols recommended by the manufacturers.

The antibody was raised against Trx-Gγ4 domain protein in rabbits. Affinity purification of the antibody was carried out using a polyvinylidene fluoride (PVDF) membrane (Millipore, Burlington, MA, USA) immobilized with GST-Gγ4 domain protein.

4.7. Western Blotting (WB)

Proteins were separated by 12.5% or 10/20% gradient SDS-PAGE, and blotted on a PVDF membrane (Millipore). Antibody against rice Gγ4 domain was affinity-purified in this study. Antibodies against rice heterotrimeric G protein α- and β-subunits (anti-Gα and anti-Gβ antibodies, respectively) were used as described previously [18]. Antibody against the aquaporin plasma membrane marker (anti-OsPIP1s) was purchased from Operon Biotechnologies. The Chemi-Lumi One Markers Kit (Nacalai Tesque, Kyoto, Japan) was used as the molecular weight marker for WB. Affinity-purified anti-Gγ4 domain antibody was used at 5 µg IgG/mL for WB. Anti-Gα and anti-Gβ antibodies were used at 1 µg IgG/mL for WB. Anti-OsPIP1s was diluted as described by the manufacture.

ECL™ peroxidase labeled anti-rabbit secondary antibody was purchased from GE Healthcare (Little Chalfont, UK). ECL Immobilon Western Chemiluminescent HRP Substrate (Millipore) was used for detection reagent for WB. The chemiluminecent signal was measured by Fusion SL (M&S Instruments, Orpington, UK).

4.8. Immunoprecipitation (IP)

Fifty micrograms of affinity-purified anti-Gγ4 domain antibody was bound to 50 mg of Protein A-bound magnetic beads (Millipore, Burlington, MA, USA). After washing twice with 1× PBS, anti-Gγ4 domain antibody and Protein A were cross-linked with dimethyl pimelimidate dihydrochloride (DMP). The conditions followed for cross-linking were according to the protocols recommended by the manufacturer. After quenching, the magnetic cross-linked beads with anti-Gγ4 domain antibody were stored at 4 °C until use.

SDS (0.1 mL of a 10% solution) was added to 0.9 mL of the plasma membrane fraction (1 mg protein/10 mg SDS/mL) and denatured for 5 min at 90 °C. After diluting the solubilized fraction with 10 mL of 1× Tris-buffered saline containing 1% Tween-20, magnetic beads cross-linked with 50 µg of anti-Gγ4 domain antibody were added. After incubation for 2 h at 25 °C, the magnetic beads were collected into a 1.5-mL tube and washed three times each with 0.5 mL of 1× TBS containing 0.1% Tween-20 and 0.5 mL of 1× TBS. Proteins were eluted from the beads using 40 µL of dissociation buffer (Bio-Rad) without the reducing agent. Eight microliters of each eluate was used for LC-MS/MS.

4.9. Protein Reduction, Alkylation, and Trypsin Digestion for LC-MS/MS Analysis

For LC-MS/MS analysis, 40 µg of leaf sheath plasma membrane proteins from WT and *Dn1-1* were analyzed using 15% SDS-PAGE. Electrophoresis was stopped at a position where bromophenol blue was 3 cm away from the stacking gel. The 3-cm long gel was excised in 10 pieces according to the molecular weight marker, Precision Plus Protein™ Kaleidoscope™ (Bio-Rad Laboratories, Hercules, CA, USA) without staining. These gel pieces were subjected to trypsin digestion. In some cases, gels were silver stained using Pierce Silver Stain for Mass Spectrometry (Thermo Fisher Scientific, Waltham, MA, USA).

Gel pieces were resuspended in 50 mM NH_4HCO_3, reduced with 50 mM dithiothreitol for 30 min at 56 °C, and alkylated with 50 mM iodoacetamide for 30 min at 37 °C in the dark. Alkylated proteins in the gels were digested with 10 μg/mL trypsin solution (Promega, Madison, WI, USA) for 16 h at 37 °C. The resultant peptides were concentrated and suspended in 0.1% formic acid and analyzed by LC-MS/MS.

4.10. Protein Identification Using Nano LC-MS/MS

The peptides were loaded onto the LC system (EASY-nLC 1000; Thermo Fisher Scientific) equipped with a trap column (EASY-Column, C18-A1 5 μm, 100 μm ID × 20 mm; Thermo Fisher Scientific), equilibrated with 0.1% formic acid, and eluted with a linear acetonitrile gradient (0–50%) in 0.1% formic acid at a flow rate of 200 nL/min. The eluted peptides were loaded and separated on a column (C18 capillary tip column, 75 μm ID × 120 mm; Nikkyo Technos, Tokyo, Japan) with a spray voltage of 1.5 kV. The peptide ions were detected using MS (LTQ Orbitrap Elite MS; Thermo Fisher Scientific) in data-dependent acquisition mode with the installed Xcalibur software (version 2.2; Thermo Fisher Scientific). Full-scan mass spectra were acquired in MS over 400–1500 m/z with a resolution of 60,000. The 10 most intense precursor ions were selected for collision-induced fragmentation in the linear ion trap at normalized collision energy of 35%. Dynamic exclusion was employed within 90 s to prevent repetitive selection of peptides.

4.11. MS Data Analysis

Protein identification was performed using the Mascot search engine (version 2.5.1, Matrix Science, London, UK) and the in-house database constructed using the amino acid sequences of rice heterotrimeric G protein subunits. For both searches, carbamidomethylation of cysteine was set as the fixed modification and oxidation of methionine was set as a variable modification. Trypsin was specified as the proteolytic enzyme, and one missed cleavage was allowed. Peptide mass tolerance was set at 10 ppm, fragment mass tolerance was set at 0.8 Da, and peptide charges were set at +2, +3, and +4. An automatic decoy database search was performed as a part of the search. Mascot results were filtered with the Percolator function to improve the accuracy and sensitivity of peptide identification. The minimum requirement for identification of a protein was two matched peptides. Significant changes in the abundance of proteins between samples were determined ($p < 0.05$).

4.12. Gene ID

The accession number of rice heterotrimeric G protein α, β, and γ4 subunit genes (*RGA1*, *RGB1*, and *RGG4*, respectively) is Os05g0333200, Os03g0669200, and Os09g0441900, respectively.

Author Contributions: Investigation and formal analysis, S.M., and A.N.; methodology, T.I.; resources, G.C. and K.M.; writing and funding acquisition, Y.I.

Funding: This work was supported by a grant for Scientific Research from Fukui Prefectural University and the JSPS KAKENHI, Grant Number 26712001.

Acknowledgments: We thank Fumio Taguchi for providing *Dn1-1*. Part of the work was performed at the Biological Resource Research and Development Center, Fukui Prefectural University (Fukui, Japan). We would like to thank Editage (www.editage.jp) for English language editing.

Conflicts of Interest: The authors declare no conflict of interest.

Int. J. Mol. Sci. **2018**, *19*, 3596

Abbreviations

agb1	Mutant of heterotrimeric G protein β subunit gene in Arabidopsis
agg1	Mutant of heterotrimeric G protein γ1 subunit gene in Arabidopsis
agg2	Mutant of heterotrimeric G protein γ2 subunit gene in Arabidopsis
agg3	Mutant of heterotrimeric G protein γ3 subunit gene in Arabidopsis
AGB1	Heterotrimeric G protein β subunit gene in Arabidopsis
AGG1	Heterotrimeric G protein γ1 subunit gene in Arabidopsis
AGG2	Heterotrimeric G protein γ2 subunit gene in Arabidopsis
AGG3	Heterotrimeric G protein γ3 subunit gene in Arabidopsis
d1	Mutant of heterotrimeric G protein α subunit gene in rice
DEP1	DENCE AND ERRECT PANICLES 1
DN1	DENCE PANICLE 1
gpa1	Mutant of heterotrimeric G protein α subunit gene in Arabidopsis
Gα	Heterotrimeric G protein α subunit
Gβ	Heterotrimeric G protein β subunit
Gγ	Heterotrimeric G protein γ subunit
GPA1	Heterotrimeric G protein α subunit gene in Arabidopsis
GS3	GRAIN SIZE 3 gene
IP	Immunoprecipitation
OsGGC1	A gene of heterotrimeric G protein γ subunit Type-C in rice, which corresponds to *GS3/RGG3*
OsGGC2	A gene of heterotrimeric G protein γ subunit Type-C in rice, which corresponds to *RGG5*
OsGGC3	A gene of heterotrimeric G protein γ subunit Type-C in rice, which corresponds to which corresponds to *DEP1/RGG4*
PM	Plasma membrane
qPE9-1	A quantitative trait locus regulating plant architecture including panicle erectness in rice
RGA1	Heterotrimeric G protein α subunit gene in rice
RGB1	Heterotrimeric G protein β subunit gene in rice
RGG1	Heterotrimeric G protein γ1 subunit gene in rice
RGG2	Heterotrimeric G protein γ2 subunit gene in rice
RGG3	Heterotrimeric G protein γ3 subunit gene in rice
RGG4	Heterotrimeric G protein γ4 subunit gene in rice
RGG5	Heterotrimeric G protein γ5 subunit gene in rice
WB	Western blot
WT	Wild-type
XLG	A gene coding extra-large GTP-binding protein
XLG	Extra-large GTP-binding protein
xlg	Mutant of a gene coding extra-large GTP-binding protein

References

1. Offermanns, S. Mammalian G-protein function in vivo: New insights through altered gene expression. *Rev. Physiol. Biochem. Pharmacol.* **2000**, *140*, 63–133. [PubMed]
2. *Signal Transduction*; Gomperts, B.D.; Kramer, I.J.M.; Tatham, P.E.R. (Eds.) Elsevier Inc.: Amsterdam, The Netherlands, 2002.
3. Wettschureck, N.; Offermanns, S. Mammalian G proteins and their cell type specific functions. *Physiol. Rev.* **2005**, *85*, 1159–1204. [CrossRef] [PubMed]
4. Milligan, G.; Kostenis, E. Heterotrimeric G-proteins: A short history. *Br. J. Pharmacol.* **2006**, *147*, S46–S55. [CrossRef] [PubMed]
5. Temple, B.R.S.; Jones, A.M. The Plant Heterotrimeric G-Protein Complex. *Annu. Rev. Plant Biol.* **2007**, *58*, 249–266. [CrossRef] [PubMed]
6. Urano, D.; Chen, J.-G.; Botella, J.R.; Jones, A.M. Heterotrimeric G protein signalling in the plant kingdom. *Open Biol.* **2013**, *3*, 120–186. [CrossRef] [PubMed]

7. Urano, D.; Miura, K.; Wu, Q.; Iwasaki, Y.; Jackson, D.; Jones, A.M. Plant morphology of heterotrimeric G protein mutants. *Plant Cell Physiol.* **2016**, *57*, 437–445. [CrossRef] [PubMed]

8. Lee, Y.-R.J.; Assmann, S.M. *Arabidopsis thaliana* 'extra-large GTP-binding protein' (AtXLG1): A new class of G-protein. *Plant Mol. Biol.* **1999**, *40*, 55–64. [CrossRef] [PubMed]

9. Assmann, S.M. Heterotrimeric and unconventional GTP binding proteins in plant cell signaling. *Plant Cell* **2002**, *14*, S355–S373. [CrossRef] [PubMed]

10. Ma, H.; Yanofsky, M.F.; Meyerowitz, E.M. Molecular cloning and characterization of *GPA1*, a G protein α subunit gene from *Arabidopsis thaliana*. *Proc. Natl. Acad. Sci. USA* **1990**, *87*, 3821–3825. [CrossRef] [PubMed]

11. Weiss, C.A.; Garnaat, C.W.; Mukai, K.; Hu, Y.; Ma, H. Isolation of cDNAs encoding guanine nucleotide-binding protein β-subunit homologues from maize (ZGB1) and *Arabidopsis* (AGB1). *Proc. Natl. Acad. Sci. USA* **1994**, *91*, 9554–9558. [CrossRef] [PubMed]

12. Mason, M.G.; Botella, J.R. Completing the heterotrimer: Isolation and characterization of an *Arabidopsis thaliana* G protein γ-subunit cDNA. *Proc. Natl. Acad. Sci. USA* **2000**, *97*, 14784–14788. [CrossRef] [PubMed]

13. Mason, M.G.; Botella, J.R. Isolation of a novel G-protein γ-subunit from *Arabidopsis thaliana* and its interaction with Gβ. *Biochim. Biophys. Acta* **2001**, *1520*, 147–153. [CrossRef]

14. Chakravorty, D.; Trusov, Y.; Zhang, W.; Acharya, B.R.; Sheahan, M.B.; McCurdy, D.W.; Assmann, S.M.; Botella, J.R. An atypical heterotrimeric G-protein γ-subunit is involved in guard cell K$^+$-channel regulation and morphological development in *Arabidopsis thaliana*. *Plant J.* **2011**, *67*, 840–851. [CrossRef] [PubMed]

15. Wu, Q.; Regan, M.; Furukawa, H.; Jackson, D. Role of heterotrimeric Gα proteins in maize development and enhancement of agronomic traits. *PLoS Genet.* **2018**, *14*, e1007374. [CrossRef] [PubMed]

16. Ishikawa, A.; Tsubouchi, H.; Iwasaki, Y.; Asahi, T. Molecular cloning and characterization of a cDNA for the α subunit of a G protein from rice. *Plant Cell Physiol.* **1995**, *36*, 353–359. [CrossRef] [PubMed]

17. Ishikawa, A.; Iwasaki, Y.; Asahi, T. Molecular cloning and characterization of a cDNA for the β subunit of a G protein from rice. *Plant Cell Physiol.* **1996**, *37*, 223–228. [CrossRef] [PubMed]

18. Kato, C.; Mizutani, T.; Tamaki, H.; Kumagai, H.; Kamiya, T.; Hirobe, A.; Fujisawa, Y.; Kato, H.; Iwasaki, Y. Characterization of heterotrimeric G protein complexes in rice plasma membrane. *Plant J.* **2004**, *38*, 320–331. [CrossRef] [PubMed]

19. Fan, C.; Xing, Y.; Mao, H.; Lu, T.; Han, B.; Xu, C.; Li, X.; Zhang, Q. GS3, a major QTL for grain length and weight and minor QTL for grain width and thickness in rice, encodes a putative transmembrane protein. *Theor. Appl. Genet.* **2006**, *112*, 1164–1171. [CrossRef] [PubMed]

20. Huang, X.; Qian, Q.; Liu, Z.; Sun, H.; He, S.; Luo, D.; Xia, G.; Chu, C.; Li, J.; Fu, X. Natural variation at the *DEP1* locus enhances grain yield in rice. *Nat. Genet.* **2009**, *41*, 494–497. [CrossRef] [PubMed]

21. Taguchi-Shiobara, F.; Kawagoe, Y.; Kato, H.; Onodera, H.; Tagiri, A.; Hara, N.; Miyao, A.; Hirochika, H.; Kitano, H.; Yano, M.; et al. A loss-of-function mutation of rice DENSE PANICLE 1 causes semi-dwarfness and slightly increased number of spikelets. *Breed. Sci.* **2011**, *61*, 17–25. [CrossRef]

22. Botella, J.R. Can heterotrimeric G proteins help to feed the world? *Trend Plant Sci.* **2012**, *17*, 563–568. [CrossRef] [PubMed]

23. Zhou, Y.; Zhu, J.; Li, Z.; Yi, C.; Liu, J.; Zhang, H.; Tang, S.; Gu, M.; Liang, G. Deletion in a Quantitative Trait Gene *qPE9-1* Associated with Panicle Erectness Improves Plant Architecture During Rice Domestication. *Genetics* **2009**, *183*, 315–324. [CrossRef] [PubMed]

24. Trusov, Y.; Chakravorty, D.; Botella, J.R. Diversity of heterotrimeric G-protein γ subunits in plants. *BMC Res. Notes* **2012**, *5*, 608. [CrossRef] [PubMed]

25. Ding, L.; Pandey, S.; Assmann, S.M. *Arabidopsis* extra-large G proteins (XLGs) regulate root morphogenesis. *Plant J.* **2008**, *53*, 248–263. [CrossRef] [PubMed]

26. Ullah, H.; Chen, J.-G.; Young, J.C.; Im, K.-H.; Sussman, M.R.; Jones, A.M. Modulation of cell proliferation by heterotrimeric G protein in *Arabidopsis*. *Science* **2001**, *292*, 2066–2069. [CrossRef] [PubMed]

27. Lease, K.A.; Wen, J.; Li, J.; Doke, J.T.; Liscum, E.; Walker, J.C. A mutant *Arabidopsis* heterotrimeric G-protein β subunit affects leaf, flower, and fruit development. *Plant Cell* **2001**, *13*, 2631–2641. [CrossRef] [PubMed]

28. Ullah, H.; Chen, J.-G.; Temple, B.; Boyes, D.C.; Alonso, J.M.; Davis, K.R.; Ecker, J.R.; Jones, A.M. The β-subunit of *Arabidopsis* G protein negatively regulates auxin-induced cell division and affects multiple developmental processes. *Plant Cell* **2003**, *15*, 393–409. [CrossRef] [PubMed]

29. Trusov, Y.; Rookes, J.E.; Tilbrook, K.; Chakravorty, D.; Mason, M.G.; Anderson, D.; Chen, J.-G.; Jones, A.M.; Botella, J.R. Heterotrimeric G protein γ subunits provide functional selectivity in Gβγ dimer signaling in *Arabidopsis*. *Plant Cell* **2007**, *19*, 1235–1250. [CrossRef] [PubMed]

30. Fujisawa, Y.; Kato, T.; Ohki, S.; Ishikawa, A.; Kitano, H.; Sasaki, T.; Asahi, T.; Iwasaki, Y. Suppression of the heterotrimeric G protein causes abnormal morphology, including dwarfism, in rice. *Proc. Natl. Acad. Sci. USA* **1999**, *96*, 7575–7580. [CrossRef] [PubMed]

31. Ashikari, M.; Wu, J.; Yano, M.; Sasaki, T.; Yoshimura, A. Rice gibberellin-insensitive dwarf mutant gene *Dwarf 1* encodes the α-subunit of GTP-binding protein. *Proc. Natl. Acad. Sci. USA* **1999**, *96*, 10284–10289. [CrossRef] [PubMed]

32. Mao, H.; Sun, S.; Yao, J.; Wang, C.; Yu, S.; Xu, C.; Li, X.; Zhang, Q. Linking differential domain functions of the GS3 protein to natural variation of grain size in rice. *Proc. Natl. Acad. Sci. USA* **2010**, *107*, 19579–19584. [CrossRef] [PubMed]

33. Izawa, Y.; Takayanagi, Y.; Inaba, N.; Abe, Y.; Minami, M.; Fujisawa, Y.; Kato, H.; Ohki, S.; Kitano, H.; Iwasaki, Y. Function and expression pattern of the α subunit of the heterotrimeric G protein in rice. *Plant Cell Physiol.* **2010**, *51*, 271–281. [CrossRef] [PubMed]

34. Utsunimiya, U.; Samejima, C.; Takayanagi, Y.; Izawa, Y.; Yoshida, T.; Sawada, Y.; Fijisawa, Y.; Kato, H.; Iwasaki, Y. Suppression of the rice heterotrimeric G protein β-subunit gene, *RGB1*, causes dwarfism and browning of internodes and lamina joint regions. *Plant J.* **2011**, *67*, 907–916. [CrossRef] [PubMed]

35. Takano-Kai, N.; Jiang, H.; Powell, A.; McCouch, S.; Takamure, I.; Furuya, N.; Doi, K.; Yoshimura, A. Multiple and independent origins of short seeded alleles of *GS3* in rice. *Breed. Sci.* **2013**, *63*, 77–85. [CrossRef] [PubMed]

36. Wang, X.Q.; Ullah, H.; Jones, A.M.; Assmann, S.M. G protein regulation of ion channels and abscisic acid signaling in *Arabidopsis* guard cells. *Science* **2001**, *292*, 2070–2072. [CrossRef] [PubMed]

37. Coursol, S.; Fan, L.M.; Le Stunff, H.; Spiegel, S.; Gilroy, S.; Assmann, S.M. Sphingolipid signalling in *Arabidopsis* guard cells involves heterotrimeric G proteins. *Nature* **2003**, *423*, 651–654. [CrossRef] [PubMed]

38. Lapik, Y.R.; Kaufman, L.S. The *Arabidopsis* cupin domain protein AtPirin1 interacts with the G protein α-subunit GPA1 and regulates seed germination and early seedling development. *Plant Cell* **2003**, *15*, 1578–1590. [CrossRef] [PubMed]

39. Pandey, S.; Assmann, S.M. The *Arabidopsis* putative G protein-coupled receptor GCR1 interacts with the G protein α subunit GPA1 and regulates abscisic acid signaling. *Plant Cell* **2004**, *16*, 1616–1632. [CrossRef] [PubMed]

40. Mishra, G.; Zhang, W.; Deng, F.; Zhao, J.; Wang, X. A bifurcating pathway directs abscisic acid effects on stomatal closure and opening in *Arabidopsis*. *Science* **2006**, *312*, 264–266. [CrossRef] [PubMed]

41. Ueguchi-Tanaka, M.; Fujisawa, Y.; Kobayashi, M.; Ashikari, M.; Iwasaki, Y.; Kitano, H.; Matsuoka, M. Rice dwarf mutant *d1*, which is defective in the α subunit of the heterotrimeric G protein, affects gibberellin signal transduction. *Proc. Natl. Acad. Sci. USA* **2000**, *97*, 11638–11643. [CrossRef] [PubMed]

42. Ullah, H.; Chen, J.G.; Wang, S.; Jones, A.M. Role of a heterotrimeric G protein in regulation of *Arabidopsis* seed germination. *Plant Physiol.* **2002**, *129*, 897–907. [CrossRef] [PubMed]

43. Chen, J.G.; Pandey, S.; Huang, J.; Alonso, J.M.; Ecker, J.R.; Assmann, S.M.; Jones, A.M. GCR1 can act independently of heterotrimeric G-protein in response to brassinosteroids and gibberellins in *Arabidopsis* seed germination. *Plant Physiol.* **2004**, *135*, 907–915. [CrossRef] [PubMed]

44. Bethke, P.C.; Hwang, Y.S.; Zhu, T.; Jones, R.L. Global patterns of gene expression in the aleurone of wild-type and *dwarf1* mutant rice. *Plant Physiol.* **2006**, *140*, 484–498. [CrossRef] [PubMed]

45. Chen, J.-G.; Jones, A.M. AtRGS1 function in *Arabidopsis thaliana*. *Methods Enzymol.* **2004**, *389*, 338–350. [CrossRef] [PubMed]

46. Huang, J.; Taylor, J.P.; Chen, J.G.; Uhrig, J.F.; Schnell, D.J.; Nakagawa, T.; Korth, K.L.; Jones, A.M. The plastid protein THYLAKOID FORMATION1 and the plasma membrane G-protein GPA1 interact in a novel sugar-signaling mechanism in *Arabidopsis*. *Plant Cell* **2006**, *18*, 1226–1238. [CrossRef] [PubMed]

47. Warpeha, K.M.; Hamm, H.E.; Rasenick, M.M.; Kaufman, L.S. A blue-light-activated GTP-binding protein in the plasma membranes of etiolated peas. *Proc. Natl. Acad. Sci. USA* **1991**, *88*, 8925–8929. [CrossRef] [PubMed]

48. Warpeha, K.M.; Lateef, S.S.; Lapik, Y.; Anderson, M.; Lee, B.S.; Kaufman, L.S. G-Protein-Coupled Receptor 1, G-Protein Gα-Subunit 1, and Prephenate Dehydratase 1 Are Required for Blue Light-Induced Production of Phenylalanine in Etiolated Arabidopsis. *Plant Physiol.* **2006**, *140*, 844–855. [CrossRef] [PubMed]

49. Joo, J.H.; Wang, S.; Chen, J.G.; Jones, A.M.; Fedoroff, N.V. Different signaling and cell death roles of heterotrimeric G protein α and β subunits in the Arabidopsis oxidative stress response to ozone. *Plant Cell* **2005**, *17*, 957–970. [CrossRef] [PubMed]

50. Suharsono, U.; Fujisawa, Y.; Kawasaki, T.; Iwasaki, Y.; Satoh, H.; Shimamoto, K. The heterotrimeric G protein α subunit acts upstream of the small GTPase Rac in disease resistance of rice. *Proc. Natl. Acad. Sci. USA* **2002**, *99*, 13307–13312. [CrossRef] [PubMed]

51. Iwata, M.; Umemura, K.; Teraoka, T.; Usami, H.; Fujisawa, Y.; Iwasaki, Y. Role of the α subunit of heterotrimeric G-protein in probenazole-inducing defense signaling in rice. *J. Gen. Plant Pathol.* **2003**, *69*, 83–86. [CrossRef]

52. Komatsu, S.; Yang, G.; Hayashi, N.; Kaku, H.; Umemura, K.; Iwasaki, Y. Alterations by a defect in a rice G protein α subunit in probenazole and pathogen-induced responses. *Plant Cell Environ.* **2004**, *27*, 947–957. [CrossRef]

53. Lieberherr, D.; Thao, N.P.; Nakashima, A.; Umemura, K.; Kawasaki, T.; Shimamoto, K. A sphingolipid licitor-inducible mitogen-activated protein kinase is regulated by the small GTPase OsRac1 and heterotrimeric G-protein in rice. *Plant Physiol.* **2005**, *138*, 1644–1652. [CrossRef] [PubMed]

54. Klopffleish, K.; Phan, N.; Augstin, K.; Bayne, R.; Booker, K.S.; Bolella, J.; Carpita, N.C.; Carr, T.; Chen, J.-C.; Cooke, T.R.; et al. Arabidopsis G-protein interactome reveals connections to cell wall carbohydrates and morphogenesis. *Mol. Syst. Biol.* **2011**, *7*, 532. [CrossRef] [PubMed]

55. Chen, J.-G.; Willard, F.S.; Huang, J.; Liang, J.; Chasse, S.A.; Jones, A.M.; Siderovski, D.P. A seven-transmembrane RGS protein that modulates plant cell proliferation. *Science* **2003**, *301*, 1728–1731. [CrossRef] [PubMed]

56. Wang, S.; Assmann, S.M.; Fedoroff, N.V. Characterization of the *Arabidopsis* Heterotrimeric G Protein. *J. Biol. Chem.* **2008**, *283*, 13913–13922. [CrossRef] [PubMed]

57. Sun, S.; Wang, L.; Mao, H.; Shao, L.; Li, X.; Xiao, J.; Ouyang, Y.; Zhang, Q. A G-protein pathway determines grain size in rice. *Nat. Commun.* **2018**, *9*, 815–824. [CrossRef] [PubMed]

58. Takano-Kai, N.; Jiang, H.; Kubo, T.; Sweeney, M.; Matsumoto, T.; Kanamori, H.; Padhukasahasram, B.; Bustamante, C.; Yoshimura, A.; Doi, K.; et al. Evolutionary history of GS3, a gene conferring grain length in rice. *Genetics* **2009**, *182*, 1–12. [CrossRef] [PubMed]

59. Sun, H.; Qian, Q.; Wu, K.; Lou, J.; Wang, S.; Zhang, C.; Ma, Y.; Lie, Q.; Huang, X.; Yuan, Q.; et al. Heterotrimeric G proteins regulate nitrogen-use efficiency in rice. *Nat. Genet.* **2014**, *46*, 652–656. [CrossRef] [PubMed]

60. Kunihiro, S.; Saito, T.; Matsuda, T.; Inoue, M.; Kuramata, M.; Taguchi-Shibaoka, F.; Youssefian, S.; Berberich, T.; Kusano, T. Rice *DEP1*, encoding a highly cycteine-rich G protein γ subunit, confers cadmium tolerance on yeast cells and plants. *J. Exp. Bot.* **2013**, *64*, 4517–4527. [CrossRef] [PubMed]

61. *Molecular Cloning*; Sambrook, J.; Russell, D.W. (Eds.) Cold Spring Harbor Laboratory Press: Cold Spring Harbor, NY, USA, 2001.

62. Yoshida, S.; Uemura, M.; Niki, T.; Sakai, A.; Gusta, L.V. Partition of membrane particles in aqueous two-polymer phase system and its partial use for purification of plasma membranes from plants. *Plant Physiol.* **1983**, *72*, 105–114. [CrossRef] [PubMed]

63. Iwasaki, Y.; Kato, T.; Kaidoh, T.; Ishikawa, A.; Asahi, T. Characterization of the putative α subunit of a heterotrimeric G protein in rice. *Plant Mol. Biol.* **1997**, *34*, 563–572. [CrossRef] [PubMed]

International Journal of
Molecular Sciences

MDPI

Article

Identification of Heterotrimeric G Protein γ3 Subunit in Rice Plasma Membrane

Aki Nishiyama †, Sakura Matsuta †, Genki Chaya, Takafumi Itoh, Kotaro Miura and Yukimoto Iwasaki *

Department of Bioscience and Biotechnology, Fukui Prefectural University, 4-1-1 Kenjojima, Matsuoka, Eiheiji-Town, Fukui 910-1195, Japan; s1873016@g.fpu.ac.jp (A.N.); s1873018@g.fpu.ac.jp (S.M.); s1873012@g.fpu.ac.jp (G.C.); ito-t@fpu.ac.jp (T.I.); miura-k@fpu.ac.jp (K.M.)
* Correspondence: iwasaki@fpu.ac.jp; Tel.: +81-776-61-6000 (ext. 3514)
† These authors contributed equally to this work.

Received: 3 September 2018; Accepted: 7 November 2018; Published: 14 November 2018

Abstract: Heterotrimeric G proteins are important molecules for regulating plant architecture and transmitting external signals to intracellular target proteins in higher plants and mammals. The rice genome contains one canonical α subunit gene (*RGA1*), four extra-large GTP-binding protein genes (XLGs), one canonical β subunit gene (*RGB1*), and five γ subunit genes (tentatively named *RGG1*, *RGG2*, *RGG3/GS3/Mi/OsGGC1*, *RGG4/DEP1/DN1/OsGGC3*, and *RGG5/OsGGC2*). *RGG1* encodes the canonical γ subunit; *RGG2* encodes the plant-specific type of γ subunit with additional amino acid residues at the N-terminus; and the remaining three γ subunit genes encode the atypical γ subunits with cysteine abundance at the C-terminus. We aimed to identify the *RGG3/GS3/Mi/OsGGC1* gene product, Gγ3, in rice tissues using the anti-Gγ3 domain antibody. We also analyzed the truncated protein, Gγ3ΔCys, in the *RGG3/GS3/Mi/OsGGC1* mutant, *Mi*, using the anti-Gγ3 domain antibody. Based on nano-liquid chromatography-tandem mass spectrometry (LC-MS/MS) analysis, the immunoprecipitated Gγ3 candidates were confirmed to be Gγ3. Similar to α (Gα) and β subunits (Gβ), Gγ3 was enriched in the plasma membrane fraction, and accumulated in the flower tissues. As *RGG3/GS3/Mi/OsGGC1* mutants show the characteristic phenotype in flowers and consequently in seeds, the tissues that accumulated Gγ3 corresponded to the abnormal tissues observed in *RGG3/GS3/Mi/OsGGC1* mutants.

Keywords: GS3; γ subunit; heterotrimeric G protein; mass spectrometric analysis; RGG3; rice; western blotting

---

1. Introduction

Heterotrimeric G proteins are well known to consist of three subunits, α, β, and γ, in mammals and yeast [1–4]. Receptors regulating the heterotrimeric G proteins, such as G protein-coupled receptors (GPCRs), interact with external signals and activate the heterotrimeric G proteins via the intrinsic GDP/GTP exchange factor (GEF) of GPCRs. When GTP binds to the α subunit (Gα-GTP), heterotrimeric G proteins dissociate into the α subunit (Gα-GTP) and βγ dimer. The α subunit and βγ dimer can regulate respective effector molecules. Thus, heterotrimeric G proteins are signal mediators from receptors to effector molecules. In higher plants, heterotrimeric G proteins are important molecules for regulating plant architecture and transmitting external signals to intracellular target proteins [5–7]. The biochemical characteristics of the plant heterotrimeric G protein [5] and the signaling mechanism and effector molecules regulating the plant heterotrimeric G protein [6] have been previously reviewed. The plant morphology of heterotrimeric G protein mutants has also been previously summarized [7]. There are three extra-large GTP-binding protein genes (*AtXLG1~AtXLG3*) [8,9], one canonical α subunit gene (*GPA1*) [10], one canonical β subunit gene (*AGB1*) [11], and three γ subunit genes (*AGG1~AGG3*) [12–14], in *Arabidopsis*; and four extra-large GTP-binding protein genes (prediction by in silico) [15], one canonical α subunit gene (*RGA1*) [16], one canonical β subunit gene (*RGB1*) [17], and five γ subunit genes, which we tentatively named *RGG1* [18], *RGG2* [18], *RGG3/GS3/Mi/OsGGC1* [19], *RGG4/DEP1/DN1/OsGGC3* [20], and *RGG5/OsGGC2* [21], in this paper.

With regard to the γ subunit genes in *Arabidopsis*, there are *AGG1* and *AGG2* encoding the canonical γ subunits, and *AGG3* encoding the atypical γ subunit with cysteine abundance at the C-terminus. In rice, *RGG1* encodes the canonical γ subunit, *RGG2* encodes the plant-specific type of γ subunit, and the remaining three γ subunit genes, *RGG3/GS3/Mi/OsGGC1*, *RGG4/DEP1/DN1/OsGGC3*, and *RGG5/OsGGC2* encode the atypical γ subunits homologous to *AGG3*. *RGG3* corresponds to *GRAIN SIZE 3 (GS3)* [19] and *RGG4* corresponds to *DENSE AND ERECT PANICLES 1 (DEP1/DN1)* [20]. The genome sequence of *RGG5* was predicted by Botella [21]. The diversity and agronomical importance of plant γ subunits have been previously reviewed [21,22].

Mutants of *XLG1*, *XLG2*, and *XLG3* [23]; *GPA1* [24]; *AGB1* [25,26]; and *AGG1*, *AGG2* [27], and *AGG3* [14] were isolated as heterotrimeric G protein mutants in *Arabidopsis*. Mutants of *RGA1* [28,29], *GS3* [30], and *DEP1* [20] were isolated as similar G protein mutants in rice. By morphological analysis of *gpa1* [24], *agb1* [26], *d1* [31], and RGB1 knock-down lines [32], it was shown that plant heterotrimeric G proteins modulate cell proliferation.

It has been shown that plant heterotrimeric G proteins are associated with transduction in response to multiple external signals, namely auxin [24,26], abscisic acid [33–37], gibberellin [38–41], brassinosteroid [24,39,40], sugar [42,43], blue light [44,45], and ozone [46]. It was also shown that the heterotrimeric G proteins of plants are concerned with defense signaling [47–50].

Based on the characteristics of heterotrimeric G proteins in higher plants, the α subunit is suggested to be contained in a huge complex localized in the plasma membrane fraction of rice [18] and *Arabidopsis* [51]. In rice, some βγ dimer candidates seem to be present in two different forms: one is a component of a huge complex, and the other is a sole βγ dimer dissociated from a huge complex in the plasma membrane of rice seedlings [18]. Using yeast two-hybrid screening, it was shown that 68 highly interconnected proteins form the core G-protein interactome in *Arabidopsis* [52], in which the regulators of G protein signaling protein (AtRGS1) [53], THYLAKOID FORMATION 1 (THF1) [43], cupin domain protein (AtPirin1) [35] etc. in addition to α, β, γ1, γ2 subunits, were contained. The huge complexes prepared solubilized plasma membrane fraction in rice [18] and *Arabidopsis* [51] may represent a part of the G-protein interactome in *Arabidopsis* [52].

In mammals and yeast, β subunits interact with γ subunits to form the βγ dimer [1–4]. The βγ dimer has not been purified from the tissues of higher plants so far, but many studies suggest its presence based on the experiments, including an in vitro pull-down assay [12,13], yeast two-hybrid (Y2H) assay [13], split-ubiquitin system [14], and fluorescence response energy transfer (FRET) assay [51,54] in *Arabidopsis*. Moreover, in rice, the β subunit was shown to interact with the γ1 and γ2 subunits with a Y2H assay [18]. Recently, the interaction of rice β subunit with atypical γ subunits and the localization of these subunits in the plasma membrane were demonstrated with a bi-molecular fluorescence complementation (BiFC) assay [55,56]. These results indicated that both the canonical and atypical γ subunits can interact with the β subunit, and that βγ dimers are localized in the plasma membrane fraction, in *Arabidopsis* and rice.

GS3 is identified as a major QTL for grain weight and grain length, and as an important gene for agriculture [19,30,56–58]. According to the identification of *AGG3* in *Arabidopsis*, *GS3* was classified as the atypical γ subunit member, and tentatively named *RGG3*. In order to understand the mechanism of seed formation in rice, studies on the GS3 protein are important.

To understand the function of *RGG3* in the regulation of seed size, identifying the native Gγ3 protein is important. When the native Gγ3 protein is identified, biochemical analysis, namely measuring the subunit stoichiometry and affinity to Gβ, canonical Gα, and XLGs, is possible. Although we tried to identify the native Gγ3 protein using an anti-Gγ3 domain antibody, the antibody recognized multiple proteins. To identify the native Gγ3 protein, we use the *RGG3* mutants *MINUTE (Mi)* and *GS3-3*, which produce partially defective proteins, as references for subtraction to Taichung 65 (abbreviated as WT [wild-type] hereinafter). Here, we find a candidate of the native RGG3 protein, Gγ3. Finally, we confirmed that the candidate was the native Gγ3 protein using nano-liquid chromatography-tandem mass spectrometry (LC-MS/MS) analysis of the immunoprecipitation products using an anti-Gγ3 domain antibody. Using this antibody, the subcellular localization and tissue-specific accumulation of the native Gγ3 protein were studied.

2. Result

2.1. Morphology of Rice Heterotrimeric G Protein γ3 Gene (RGG3/GS3/Mi/OsGGC1) Mutants:

To confirm the functions of rice heterotrimeric G protein γ3 subunit in determining the plant morphology, we prepared plants possessing *GS3-3* [30] and *Mi* [58] mutation with Taichung 65 as a background. The mutant, *Mi* was slightly dwarfed (Figure 1A) and set small seeds (Figure 1B), compared to those of the WT. *GS3-3* had a height similar to that of the WT (Figure 1A) and set large seeds (Figure 1B). These results indicate that the mutations in *Mi* and *GS3-3* clearly affected the seed size.

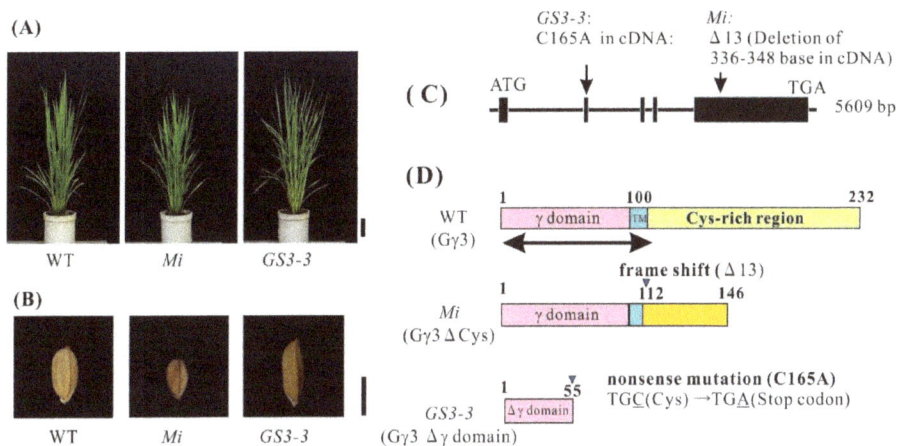

Figure 1. Morphology of rice heterotrimeric G protein γ3 gene (*RGG3/GS3/Mi/OsGGC1*) mutants, and genome and protein structure of *RGG3/GS3/Mi/OsGGC1*. (**A**) Gross morphology of the wild-type (WT) (Taichung 65), *Mi* and *GS3-3*; Bar = 10 cm. (**B**) Seed morphologies of the plants in (**A**); Bar = 5 mm. (**C**) Genome structure of *RGG3/GS3/Mi/OsGGC1* and positions of mutations in *RGG3/GS3/Mi/OsGGC1* mutants, *Mi* and *GS3-3*. The 13-base deletion (336–348th base in full-length cDNA) and one base substitution (C165A in full-length cDNA) had occurred in *Mi* and *GS3-3*, respectively. In *GS3-3*, a codon, TGC (cysteine) changed to TGA (stop codon). (**D**) Protein structure of the product of *RGG3/GS3/Mi/OsGGC1* in the WT (Gγ3), *Mi* (Gγ3ΔCys), and *GS3-3* (Gγ3Δγ domain). The canonical γ domain region is shown as γ domain (pink bar). The putative transmembrane domain is indicated as TM (blue bar). The region with cysteine abundance is labeled as cysteine-rich region (green bar). The newly produced amino acid sequence by the frame shift resulting from of 13-base deletion is indicated with a yellow bar. An arrow under the WT Gγ3, which covers 120 amino acid residues from N-terminal, is the region used for recombinant proteins, such as the thioredoxin (Trx)-tagged Gγ3 domain protein (Trx-Gγ3 domain protein), used as an antigen, and the glutathione S transferase (GST)-tagged Gγ3 domain protein (GST-Gγ3 domain protein), used for affinity purification of the antibody.

2.2. Genomic Structure of RGG3 and Protein Structure of Gγ3

The genome sequence of *RGG3* was found in RAP-DB (Os03g0407400). We reconfirmed the genome sequence of *RGG3*. *RGG3* consists of five exons (Figure 1C) and its translation product, Gγ3, comprises 232 amino acid residues. In order to prepare recombinant proteins, cDNA for RGG3 was isolated. The molecular weight of Gγ3 calculated from the cDNA, was 24249 Da. The Gγ3 consists of the canonical γ domain (about 100 amino acid residues), a short region with hydrophobic amino acid residues (tentatively named transmembrane region: TM), and a region with a large number of cysteines (Cys-rich region) (Figure 1D).

The *Mi* mutation occurred as a result of the deletion of 13 bases in *RGG3*. The mutation site corresponds to 336–348th positions in the full-length cDNA of *RGG3*, resulting in a frame-shift (Figure 1C). We reconfirmed the mutation in *Mi*. In *Mi*, the mutated protein, tentatively named Gγ3ΔCys, consists of 146 amino acid residues (Figure 1D). The cysteine-rich region is absent in Gγ3ΔCys. The molecular weight of Gγ3ΔCys, calculated from cDNA, was 15,651 Da.

The *GS3-3* mutation occurred as a result of one base substitution. The C at the 165th position in the full-length cDNA of *RGG3* was substituted by A (C165A), resulting in the generation of a stop codon (Figure 1C). As the mutation in TCM3-467 was the same as that in *GS3-3* [18], we renamed TCM3-467 to *GS3-3*. The *GS3-3* mutation generated a mutated protein with 55 amino acid residues, tentatively named the Gγ3Δγ domain (Figure 1D). The Gγ3Δγ domain is an immature protein lacking about half of the canonical γ domain. The molecular weight of the Gγ3Δγ domain, calculated from cDNA, was 5653 Da. The chemiluminescent intensity of Gγ3ΔCys was more than 7-fold that of Gγ3, when 10 μg of protein of the plasma membranes of the WT and *Mi*, respectively, was analyzed by western blot.

2.3. Gγ3 Candidates Localized in the Plasma Membrane Fraction

Identification of native Gγ3 was carried out by Western blotting. As mutants have no native full length Gγ3, these were used as references, in order to identify native Gγ3 in WT. The plasma membrane fraction was chosen in this study as it was shown that Gα and Gβ accumulated in plasma membrane fraction in rice. The plasma membrane fractions of WT, *GS3-3*, and *Mi* flowers were prepared using an aqueous two-polymer phase system, and Gγ3 candidates were detected by Western blotting using an anti-Gγ3 domain antibody. In the WT, a 32-kDa protein (Gγ3 candidate) was detected (Figure 2A, lanes 2 and 4); this band was not observed in *GS3-3* or *Mi*. The molecular weight of the Gγ3 candidate is much higher than that of Gγ3 calculated from the cDNA of the WT (24 kDa). In *GS3-3*, the Gγ3Δγ domain was not detected (Figure 2A, lane 3). In *Mi*, a 20-kDa protein (Gγ3ΔCys candidate) was detected (Figure 2A, lane 5). The molecular weight of the Gγ3ΔCys candidate was much higher than that of Gγ3ΔCys calculated from the cDNA (16 kDa). The molecular weights of Gγ3 and Gγ3ΔCys candidates were measured using molecular weight markers (Figure 2B).

2.4. Immunoprecipitation of Gγ3 and Gγ3ΔCys Using an Anti-Gγ3 Domain Antibody

To concentrate Gγ3 and Gγ3ΔCys candidates, immunoprecipitation was carried out using anti-Gγ3 domain antibody. First, 50 μg of the anti-Gγ3 domain antibody was added to 1 mg each of solubilized plasma membrane protein of the WT (Figure 3A) and *Mi* (Figure 3B) flowers. Gγ3 and Gγ3ΔCys candidates were collected with the antibody cross-linked Protein A bound beads. The 32 kDa protein, a Gγ3 candidate in the WT (Figure 3A, lane 3) and 20-kDa protein, a Gγ3ΔCys candidate in *Mi* (Figure 3B, lane 3), were immunoprecipitated.

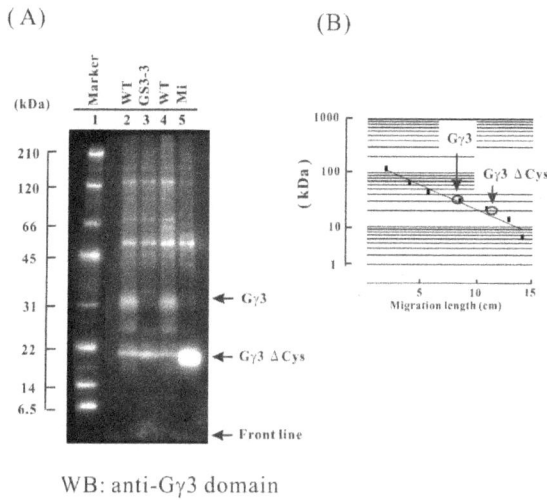

WB: anti-Gγ3 domain

Figure 2. Immunological study of the Gγ3 candidates in the wild-type (WT), *Minute* (*Mi*), and *GS3-3* flowers. (**A**) First, 10 μg of each protein of the plasma membrane fractions of the WT and *GS3-3* and 5 μg of the protein of the plasma membrane fractions of *Mi* were used for the Western blot analysis using an anti-Gγ3 domain antibody. Molecular weight marker (lane 1). The Gγ3 candidate was detected as a broad band with a molecular weight of approximately 32 kDa in the WT (lanes 2 and 4). No Gγ3 was detected in *GS3-3* (lane 3). The Gγ3ΔCys candidate was detected as a band with a molecular weight of approximately 20 kDa in *Mi* (lane 5). (**B**) The molecular weights of Gγ3 and Gγ3ΔCys candidates were estimated using a molecular weight marker as a standard.

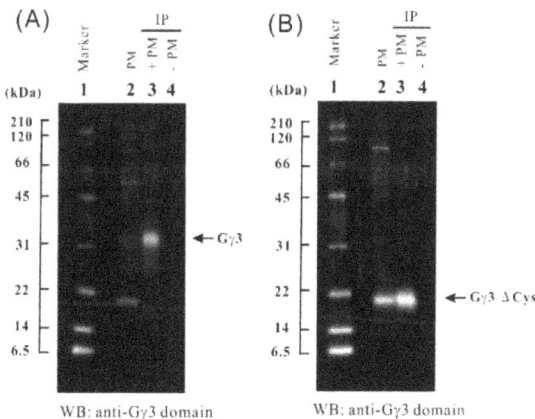

WB: anti-Gγ3 domain   WB: anti-Gγ3 domain

Figure 3. Immunoprecipitation of Gγ3 and Gγ3ΔCys candidates. (**A**) Immunoprecipitation of the Gγ3 candidate from solubilized plasma membrane proteins of the wild-type (WT) flower using an anti-Gγ3 domain antibody. Molecular weight marker (lane 1); 10 μg of protein of the plasma membrane fraction of the WT (lane 2); the immunoprecipitation product of solubilized plasma membrane proteins and anti-Gγ3 domain antibody (lane 3); control experiment (buffer in place of the membrane protein; lane 4). (**B**) Immunoprecipitation of the Gγ3ΔCys candidate from the solubilized plasma membrane proteins of the *Minute* (*Mi*) flower using an anti-Gγ3 domain antibody. Molecular weight marker (lane 1); 10 μg of protein of the plasma membrane fraction of *Mi* (lane 2); the immunoprecipitation product of the solubilized plasma membrane proteins of *Mi* and the anti-Gγ3 domain antibody (lane 3); control experiment (buffer in place of the membrane protein; lane 4).

2.5. LC-MS/MS Analysis

To demonstrate that Gγ3 and Gγ3ΔCys candidates are actually Gγ3 and Gγ3ΔCys, and that proteins with which the anti-Gγ3 domain antibody reacted, are actually Gγ3 and Gγ3ΔCys, LC-MS/MS analysis was carried out. First, using LC-MS/MS, we checked for Gγ3 and Gγ3ΔCys candidates in the eluate from the gel containing plasma membrane proteins following SDS-PAGE. When the signal intensities of Gγ3 and Gγ3ΔCys candidates detected by LC-MS/MS were not enough, we analyzed immunoprecipitation products, enriched with anti-Gγ3 domain antibody.

First, plasma membrane proteins from the WT and *Mi* were analyzed by LC-MS/MS. 40 µg of each flower plasma membrane protein from WT and *Mi* was separated by SDS-PAGE and each lane was separated into 10 pieces to increase the relative amount of target proteins, according to the molecular weight marker. After these gel pieces were digested with trypsin, peptides were analyzed by LC-MS/MS in triplicate. Typical examples are summarized in Table 1. Fragments were assigned to the sequence of Gγ3, and their positions are indicated in Figure 4A.

In the analysis of the plasma membrane fraction of the WT, three Gγ3 fragments (fragments 1, 2, and 3) ($p < 0.05$) were detected in a gel piece containing a 32 kDa protein (Table 1A). In the plasma membrane fraction of *Mi*, three Gγ3 fragments (fragments 2, 3, and 4-1) ($p < 0.05$) were detected in a gel piece containing a 20 kDa protein (Table 1B).

Table 1. LC-MS/MS analysis of Gγ3 fragments in in the plasma membrane of the wild-type (WT) and *Minute* (*Mi*) flowers.

(A) Gγ3 fragments in the plasma membrane fraction of the WT flower

Fragments	Observed	Mr(expt)	Mr(calc)	Expected	Peptide
1	379.2251	1134.6536	1134.6509	0.00078	R.LQLAVDALHR.E
2	714.7006	2141.08	2141.0753	0.00000034	R.EIGFLEGEINSIEGIHAASR.C
3	482.7414	963.4682	963.4662	0.007	R.EVDEFIGR.T

(B) Gγ3 fragments in the plasma membrane fraction of the *Mi* flower

Fragment	Observed	Mr(expt)	Mr(calc)	Expected	Peptide
2	714.7014	2141.0824	2141.0753	0.0000024	R.EIGFLEGEINSIEGIHAASR.C
3	482.7408	963.4671	963.4662	0.0077	R.EVDEFIGR.T
4-1	667.8468	1333.6791	1333.6765	0.00015	R.TPDPFITISSEK.R

(C) Gγ3 fragments in the immunoprecipitation products using the plasma membrane fraction of WT flower

Fragments	Observed	Mr(expt)	Mr(calc)	Expected	Peptide
1	568.3348	1134.6551	1134.6509	5.40×10^{-7}	R.LQLAVDALHR.E
3	482.7417	963.4688	963.4662	0.00062	R.EVDEFIGR.T
4-2	497.6015	1489.7827	1489.7776	2.50×10^{-5}	R.TPDPFITISSEKR.S

(D) Gγ3 fragments in the immunoprecipitation products using the plasma membrane fraction of *Mi* flower

Fragments	Observed	Mr(expt)	Mr(calc)	Expected	Peptide
1	568.3354	1134.6562	1134.6509	8.90×10^{-7}	R.LQLAVDALHR.E
2	714.7017	2141.0832	2141.0753	8.60×10^{-7}	R.EIGFLEGEINSIEGIHAASR.C
3	482.7427	963.4709	963.4662	0.00072	R.EVDEFIGR.T
4-1	667.8485	1333.6825	1333.6765	7.70×10^{-6}	R.TPDPFITISSEK.R

Forty micrograms of each protein of the plasma membrane fraction of the wild-type (WT) and *Minute* (*Mi*) (**A,B**) and 5 µL of each eluate in the immunoprecipitation experiment of WT and *Mi* (**C,D**) were used for LC-MS/MS. Fragments of the trypsin-digested Gγ3 candidates ($p < 0.05$) are shown. The fragment numbers correspond to Figure 4A. Mr(expt) and Mr(calc) correspond to the theoretical molecular mass and the molecular mass that was calculated from the observed molecular mass, respectively. The scores from the Mascot search were 91 (**A**), 90 (**B**), 164 (**C**), and 248 (**D**).

(A)   MAMAAAPRPKSPPAPPDPCGRHR₁**LQLAVDALHR** ₂**EIGFLEGEINSI**
EGIHAASR CCR ₃**EVDEFIGR** ₄**TPDPFITISSEKR** SHDHSHHFLKKFR
CLCRASACCLSYLSWICCCSSAAGGCSSSSSSFNLKRPSCCCNCNCNC
CSSSSSSCGAALTKSPCRCRRRSCCCRRCCCGGVGVRACASCSCSPPCA
CCAPPCAGCSCRCTCPCPCPGGCSCACPACRCCCGVPRCCPPCL*

(B)

Fragment 1

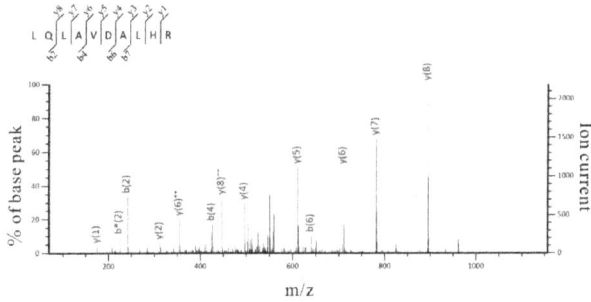

Fragment 3

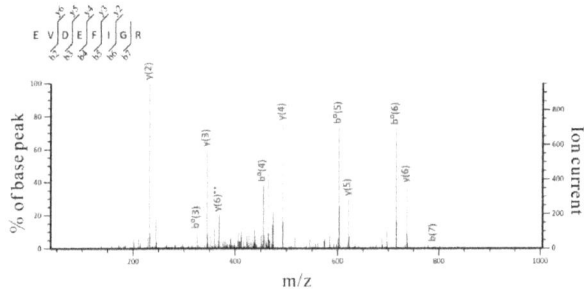

Fragment 4-2

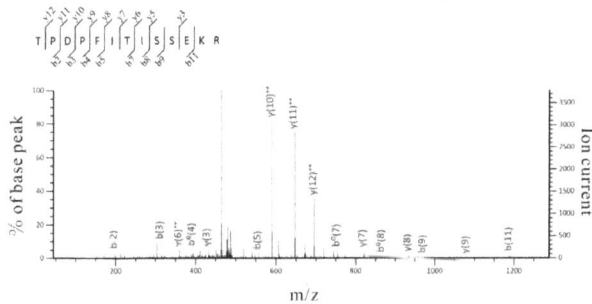

Figure 4. LC-MS/MS analysis of Gγ3 candidates. (**A**) Four peptides ($p < 0.05$), which were produced by trypsin-digested Gγ3 candidates in the wild-type (WT) and *Mi*, were numbered and underlined in the full length Gγ3 amino acid sequence. These peptides are listed in Table 1. (**B**) MS/MS spectra of the three fragments, which were obtained from the immunoprecipitation product of Gγ3 in the WT (Figure 3A, lane 3). Fragment numbers correspond to Table 1C.

Immunoprecipitation products were separated by SDS-PAGE and analyzed by LC-MS/MS. Immunoprecipitation products from the WT and *Mi* were not detected by silver staining (data not shown). In the immunoprecipitation products of the WT (Figure 3A, lane 3), a gel piece containing a 32 kDa protein was cut and digested with trypsin, and the resultant peptides were analyzed by LC-MS/MS. As a result, three Gγ3 fragments (fragments 1, 3, and 4-2), represented as primary mass ($p < 0.05$), were obtained (Table 1C). Fragment 4-2 is an incomplete trypsin-digested fragment containing an arginine residue (R) at its C-terminus, making it differ from fragment 4-1. In the immunoprecipitation products of *Mi*, a gel piece containing a 20-kDa protein was cut and digested by trypsin, and the resultant peptides were analyzed by LC-MS/MS. As a result, four fragments (fragments 1, 2, 3, and 4-1) ($p < 0.05$) were obtained (Table 1D).

The MS/MS results of fragments 1, 3, and 4-2 are shown in Figure 4B. Based on these results, we concluded that the 32 kDa and 20 kDa polypeptides were Gγ3 and Gγ3ΔCys, respectively. When the immunoprecipitation product of the WT was analyzed by LC-MS/MS, five fragments, SPCRCR, SCCCRR, RCCCGGVGVR, ACASCSCSPPCACCAPPCAGCSCR, and CCPPCL, which were positioned at the C-terminal parts of Gγ3, were detected by the Mascot search, but their scores were very low (Mascot score < 11). Therefore, these five fragments were excluded from Table 1 and Figure 4A.

When the NCBI protein database was used for the analysis of Gγ3 candidates, Gγ3 was annotated using another name, BAH89202.1

2.6. Gγ3 and Gγ3ΔCys Were Enriched in the Plasma Membrane Fraction

To check whether Gγ3 and Gγ3ΔCys are enriched in the plasma membrane, the amount of Gγ3 and Gγ3ΔCys in the crude microsomal fraction was compared with that in the plasma membrane fraction (Figure 5). Tissue-homogenate was centrifuged at $10,000\times g$ for 10 min and the resulting supernatant was centrifuged at $100,000\times g$ for 1 h. The precipitate (100,000 g ppt) was named the crude microsomal fraction (cMS). The plasma membrane fractions were prepared from cMS, using the aqueous two-polymer phase system.

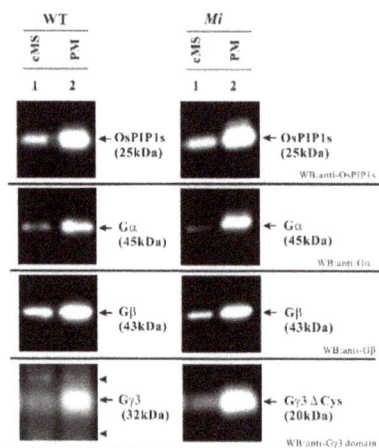

Figure 5. Gγ3 and Gγ3ΔCys were enriched in the plasma membrane fraction of the wild-type (WT) and *Minute* (*Mi*) flowers. First 10 μg of both the crude microsomal fraction protein and plasma membrane fraction protein from the WT and *Mi* were analyzed by western blot using anti-OsPIP1s, anti-Gα, anti-Gβ, and anti-Gγ3 domain antibodies. OsPIP1s is an aquaporin, which is a plasma membrane marker. OsPIP1s (25 kDa), Gα (45 kDa), Gβ (4 3kDa), Gγ3 (32 kDa), and Gγ3ΔCys (20 kDa) are indicated by arrows. Non-specific bands are indicated by arrow heads.

OsPIP1s is an aquaporin, which is a plasma membrane marker. Gα and Gβ are the subunits of the heterotrimeric G protein complex in rice. The OsPIP1s, Gα subunit, and Gβ subunit were enriched in the plasma membrane fraction. Furthermore, Gγ3 (32 kDa in WT) and Gγ3ΔCys (20 kDa in *Mi*) were also enriched in the plasma membrane fraction. These results showed that Gγ3 (32 kDa in WT) and Gγ3ΔCys (20 kDa in *Mi*) were localized in the plasma membrane fraction.

2.7. Tissue-Specific Accumulation of Gγ3

In order to know the tissues in which Gγ3 accumulates, the accumulation profile of Gγ3 was studied using the plasma membrane fractions of one-week-old etiolated seedlings of WT, developing leaf sheaths, and flowers. The results showed that the Gγ3 protein largely accumulated in the developing flower (Figure 6).

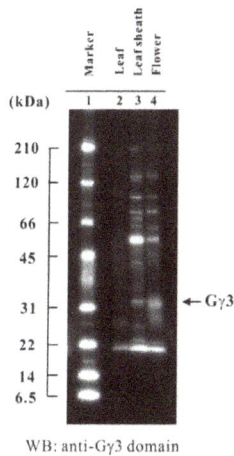

Figure 6. Tissue-specific accumulation of Gγ3 in the wild-type (WT). Ten micrograms of each of the plasma membrane fraction proteins of the leaf, leaf sheath, and flower in the WT was analyzed by SDS-PAGE and Western blotting using an anti-Gγ3 domain antibody. Molecular weight marker (lane 1); leaf from etiolated seedling (lane 2); developing leaf sheath at the eighth leaf stage (lane 3); 1–5 cm flower (lane 4).

3. Discussion

In rice, there are three atypical γ subunit genes (*RGG3*, *RGG4*, and *RGG5*) that are homologous to *AGG3*. The tentatively named *RGG3* corresponds to *GRAIN SIZE 3 (GS3)*, which is a gene that regulates seed length [19,30,56–58] and *RGG4* corresponds to *DENSE AND ERECT PANICLE1 (DEP1)*, which is a gene that regulates plant architecture including semi-dwarfness, panicle number and panicle erectness [20,55]. *RGG5* corresponds to *GGC2* [21], which a gene that increases grain length in combination or individually with *DEP1* [56]. These genes are important for rice breeding. These have been already cloned, but their native translation products have not yet been studied. In this study, we focused on the native translation products of *RGG3/GS3/Mi/OsGGC1*.

First, we detected the Gγ3 candidate from the WT and the truncated Gγ3 candidate (Gγ3ΔCys) from *Mi* by Western blotting using anti-Gγ3 domain antibodies (Figure 2A). In SDS-PAGE, the molecular weights of the Gγ3 and Gγ3ΔCys candidates were estimated as 32 and 20 kDa, respectively, which were larger than the molecular mass calculated using cDNAs, i.e., 24 and 16 kDa, respectively. These results indicate that modifications, such as glycosylation, ubiquitination, phosphorylation, and lipid modification (palmitoylation etc.), may have occurred after translation in the Gγ3 and Gγ3ΔCys candidates. The identification of the modification is a subject requiring further study. In order to obtain concrete evidence on whether the Gγ3 and Gγ3ΔCys candidates detected by western blotting were actually Gγ3 and Gγ3ΔCys proteins, the immunoprecipitation products of the Gγ3 and Gγ3ΔCys candidates were analyzed by LC-MS/MS (Figures 3 and 4). As a result, four fragments, with $p < 0.05$ by the Mascot search engine, were obtained from the Gγ3 and Gγ3ΔCys candidates. These results indicated that the Gγ3 and Gγ3ΔCys candidates were actually Gγ3 and Gγ3ΔCys, respectively.

Mutants of *RGG3*, i.e., *Mi* [58] and *GS3-3* [30], set small and large seeds, respectively (Figure 1B). Thus, *RGG3* regulates seed morphology. Gγ3 and Gγ3ΔCys were accumulated in the plasma membrane fraction of the flower tissue (Figure 5). The tissue in which Gγ3 and Gγ3ΔCys were accumulated corresponded to the tissue that showed the morphological abnormalities in *Mi* and *GS3-3* (Figures 1 and 6). One of the deletion alleles of *GS3* decreased the cell number in the lemma and palea and a knock-down construct of *GS3* utilizing RNAi increased the cell number [58]. Gγ3 also modulates cell proliferation, similar to Gα [31] and Gβ [32]. The chemiluminescent intensity of Gγ3ΔCys was more than 7-fold that of Gγ3 (Figure 2). The reason that the amount of Gγ3 was fewer than that of Gγ3ΔCys may be that Gγ3 is degraded by proteases. Another possibility could be that Gγ3ΔCys may stably accumulate in the plasma membrane with other proteins, including Gβ. Hence, further analysis of native and truncated Gγ3s will be important to understanding seed size regulation.

Sun et al. reported that GS3-1 (corresponding to Gγ3) interacted with Gβ using a Y2H assay [56]. Using BiFC, they also revealed that GS3-1 and GS3-4, truncated Gγ3 proteins in *GS3-4*, interacted with Gβ on the plasma membrane [56]. GS3-4 in *GS3-4* [30,56] and Gγ3ΔCys in *Mi* [58] consisted of 149 and 146 amino acid residues, respectively. GS3-4 and Gγ3ΔCys have the canonical Gγ domain and a putative transmembrane domain, but largely lack a cysteine-rich domain. In this study, native Gγ3 and Gγ3ΔCys were enriched in the rice plasma membrane, similar to the Gβ subunit (Figure 5). We also confirmed that Gγ3 and Gγ3ΔCys interacted with Gβ using a Y2H assay (data not shown). From these results, it is suggested that Gγ3 and Gγ3ΔCys may form a dimer with Gβ on the plasma membrane. As we identified Gγ3 and Gγ3ΔCys by immunological techniques and LC-MS/MS analysis in this study, it will be possible to research whether the Gγ3 protein is a component of the heterotrimeric G protein complex containing the canonical Gα and XLGs.

As the seeds of *Mi* [58] and *GS3-4* [30,56] were shorter than those of the WT, Gγ3ΔCys is the cause of shortened seeds. It will be important to clarify whether Gγ3ΔCys interacts with Gβ. If the βγ dimer composed with Gγ3ΔCys is present in the plasma membrane, it will be interesting to research the interaction between the unusual βγ dimer (GβGγ3ΔCys) and the canonical Gα or XLGs, on the basis of the G protein signaling model [5,6]. As previously reported, some βγ dimers seem to be present in two different fractions in gel filtration: one is a component of a huge complex, and the other is a sole βγ dimer in the plasma membrane of etiolated rice seedlings [18]. Although this may be the result of artificial dissociation during solubilization and gel fractionation, this approach will be important for understanding the heterotrimeric G protein complex. Truncated Gγ3 in *GS3-3*, namely the Gγ3Δγ domain, consisted of 55 amino acid residues, which is considered as a loss of function of OSR (organ size regulation) [30]. In *GS3-3*, the Gγ3Δγ domain was not detected in the plasma membrane (Figure 2A). The reason may be due to the lack of the trans-membrane domain in the Gγ3Δγ domain or due to the lack of sites that anti-Gγ3 domain antibody recognizes in the Gγ3Δγ domain. In addition, the Gγ3Δγ domain was not detected in the cytosolic fraction (data not shown). However, it is not ruled out that there is no Gγ3Δγ domain in the cytosolic fraction, due to the detection threshold in Western blot not being met. As seeds of *GS3-3* were longer than those of the WT, the lack of a βγ3 dimer may be the cause of enlarged seeds. Hence, as we detected Gγ3 and Gγ3ΔCys proteins in this study, biochemical analysis of the heterotrimeric G protein complex in *Mi* and *GS3-3* will be accelerated. It is of interest to reveal the subunit stoichiometry of the canonical Gα, XLGs, Gβ, and five Gγs, namely γ1, γ2, the Gγ3Δγ domain, γ4, and γ5, and the subsequent subunit composition of the G protein complex in *Mi*, which sets small grains. It is also important to analyze the subunit stoichiometry of the canonical Gα, XLGs, Gβ, and four Gγs, namely γ1, γ2, γ4, and γ5, and the subsequent subunit composition of the G protein complex in *GS3-3*, which sets large grains.

4. Materials and Methods

4.1. Plant Materials

A rice cultivar (*Oryza sativa* L. cv. Taichung 65) and two heterotrimeric G protein γ3 mutants (*GS3-3* and *Mi*) were used in this study. *GS3-3* was obtained from the Taichung 65 mutant library, mutagenized by N-methyl-N-nitrosourea treatment, and named TCM-3-467. The *Mi* mutation was provided from the stocked mutant line, H343 (*Oryza sativa* L. cv. Akamuro background). H343 was backcrossed four times with Taichung 65, and was used as a near-isogenic line of *Mi* in this study. All rice plants were grown under a 14-h light (50,000 lux and 28 °C) and 10-h dark (25 °C) cycle, or under natural field conditions.

4.2. Sequencing and Confirmation of RGG3

Genomic DNA was isolated from whole plants of WT, *Mi*, and *GS3-3* using an extraction method with cetyltrimethylammonium bromide (CTAB) [59]. Using this as a template, PCR was performed using > 20 sets of PCR primers to cover 5609 bases of *RGG3* (Os03g0407400). The amplified DNA fragments were directly sequenced using the same primers that were used for amplification.

4.3. RNA Isolation, Reverse Transcription, and cDNA Encoding of the Heterotrimeric G Protein Gγ3 Subunit

Total RNA from the flower tissue was directly extracted using RNeasy Plant Mini kits (Qiagen, Hilden, Germany). The first strand of cDNA was synthesized using Super Script First Strand Synthesis System for RT-PCR (Invitrogen, Carlsbad, CA, USA). Total RNA (0.5 µg) and oligo-dT were used as the template and primer, respectively, for the first strand cDNA synthesis.

In order to isolate *RGG3* cDNA, the primers were designed based on the database information (Os03g0407400):

RGG3 forward: 5′ atggcaatggcggcggcgcc 3′;
RGG3 reverse: 5′ caagcagggggggcagcaac 3′.

The amplified PCR products were sub-cloned into pCR4 (Invitrogen) and sequenced with a Thermo BigDye Terminator Cycle Sequencing Kit (Amersham Biosciences, Little Chalfont, UK) using a DNA sequencer (Model 377; Applied Biosystems, Foster City, CA, USA).

4.4. Preparation of the Microsomal and Plasma Membrane Fractions in Rice

Crude microsomal fractions were prepared from 2–5 cm flowers of the WT, *Mi,* and *GS3-3,* as described previously [18], and plasma membrane fractions were purified from the crude microsomal fraction using an aqueous two-polymer phase system [60]. From the etiolated seedlings, which were grown for 5 d at 28 °C, and developing leaf sheaths at the eighth leaf stage, crude microsomal fractions and plasma membrane fractions were prepared, respectively.

4.5. SDS-Polyacrylamide Gel Electrophoresis (SDS-PAGE)

Electrophoresis was carried out on 12.5% and 10/20% gradient polyacrylamide gels containing 0.1% SDS, as described previously [61].

For LC-MS/MS analysis, 40 µg of flower plasma membrane proteins from both the WT and *Mi* were analyzed using 15% SDS-PAGE. Electrophoresis was stopped at a position where the Bromophenol Blue was 3 cm away from the stacking gel. The 3-cm long gel was divided into 10 pieces according to the molecular weight marker (Precision Plus Protein™ Kaleidoscope™; Bio-Rad Laboratories), without staining. These gel pieces were used for trypsin digestion. In some cases, gels were silver-stained using Pierce Silver Stain for Mass Spectrometry (Thermo Scientific).

4.6. Preparation of Trx-Gγ3 and GST-Gγ3 Domain Proteins

cDNA encoding 120 amino acid residues from the N-terminal of the rice Gγ3 protein was amplified by PCR using primers. The cDNA contains the Gγ3 domain and the putative transmembrane region:

RGG3 domain forward: 5′ccttggctcatatggatatcatggcaatggcggcggcgccccggcccaag3′;
RGG3 domain reverse: 5′aagcttcccgggtcaggaggaggatgagcagccgccggcggcgctgctg3′.

Amplified cDNA was sub-cloned in pET32a containing thioredoxin (Trx) and histidine (His) tags (Novagen). The resultant clone, the Trx-Gγ3 domain vector, was transformed in T7 Express *lysY/I^q* *E. coli* (New England Biolabs), and the recombinant protein was synthesized and designated as the Trx-Gγ3 domain protein. The cDNA covering the Gγ3 domain was also sub-cloned in pET41 containing glutathione S-transferase (GST) and His tags (Novagen). The resultant clone, the GST-Gγ3 domain vector, was transformed in T7 Express *lysY/I^q* *E. coli* (New England Biolabs), and the recombinant protein was synthesized and designated as the GST-Gγ3 domain protein.

The overexpression of the Trx-Gγ3 domain protein and GST-Gγ3 domain protein in T7 Express *lysY/I^q* *E. coli* was carried out as described elsewhere [61]. Inductions were performed at 37 °C. Induction was initiated by the addition of IPTG (final IPTG concentration, 1 mM). After 3 h, *E. coli* was harvested after centrifugation at $10,000 \times g$ for 5 min at 4 °C, and stocked at −80 °C before use.

As the Trx-Gγ3 domain protein and GST-Gγ3 domain protein were included in the body, both proteins were solubilized in 6 M guanidine hydrochloride, 10 mM Tris HCl, pH 8.0. Solubilized proteins were applied to Ni-NTA agarose (Qiagen, Hilden, Germany). The purification of both proteins was performed according to the protocols recommended by the manufacturers.

The antibody was raised against the Trx-Gγ3 domain protein in rabbits. Affinity purification of the antibody was performed using a polyvinylidene fluoride (PVDF) filter (Millipore, Burlington, MA, USA), immobilized with the GST-Gγ3 domain protein.

4.7. Western Blot Analysis (WB)

Proteins were separated by 12.5% or 10/20% gradient SDS-PAGE, and blotted onto a PVDF membrane (Millipore). The antibody against the rice Gγ3 domain was affinity-purified in this study. Antibodies against the rice heterotrimeric G protein α and β subunits, namely the anti-Gα and anti-Gβ antibodies, were used as described previously [18]. The antibody against aquaporin (a plasma membrane marker), namely, anti-OsPIP1s, was purchased from Operon Biotechnologies. The Chemi-Lumi One Markers Kit (Nacalai Tesque, Kyoto, Japan) was used as a molecular weight marker for western blotting.

ECL™ peroxidase labelled anti-rabbit antibody was purchased as second antibody from GE Healthcare, Little Chalfont, UK. ECL Immobilon™ Western Chemiluminescent HRP Substrate (Millipore, Burlington, MA, USA) was used as the western blotting detection reagent. The chemiluminescent signal was measured using a Fusion SL (MS instruments).

4.8. Immunoprecipitation

First, 50 μg of affinity-purified anti-Gγ3 domain antibody was bound to 50 mg of Protein A bound magnetic beads (Millipore, Burlington, MA, USA). After washing them thrice with 1× PBS, the anti-Gγ3 domain antibody and Protein A were cross-linked with dimethyl pimelimidate dihydrochloride (DMP). The conditions followed for cross-linking were according to the protocols recommended by the manufacturers. After quenching the magnetic cross-linked beads with the anti-Gγ3 domain antibody, they were stored at 4 °C until use.

Next, 0.1 mL of 10% SDS was added to 0.9 mL of plasma membrane fraction (1 mg protein/10 mg SDS/mL) and denatured for 5 min at 90 °C. After diluting the solubilized fraction with 10 mL of 1× TBS containing 1% Tween 20, the magnetic beads cross-linked with 50 μg of the anti-Gγ3 domain antibody were added. After incubation for 2 h at 25 °C, the magnetic beads were collected into a 1.5 mL tube and washed thrice each with 0.5 mL of 1× TBS containing 0.1% Tween 20 and 0.5 mL of

Int. J. Mol. Sci. **2018**, *19*, 3591

1× TBS. Proteins were eluted using 40 µL of dissociation buffer (Bio-rad) without a reducing agent, from the beads. In total, 5 µL of each eluate was used for LC-MS/MS.

4.9. Protein Reduction, Alkylation, and Trypsin Digestion for LC-MS/MS Analysis

Gel pieces were resuspended in 50 mM NH_4HCO_3, reduced with 50 mM dithiothreitol for 30 min at 56 °C, and alkylated with 50 mM iodoacetamide for 30 min at 37 °C in the dark. Alkylated proteins in the gels were digested with 10 µg/mL of trypsin solution (Promega, Madison, WI, USA) for 16 h at 37 °C. The resultant peptides were concentrated and suspended in 0.1% formic acid and analyzed by LC-MS/MS.

4.10. Protein Identification Using Nano-LC-MS/MS

The peptides were loaded onto the LC system (EASY-nLC 1000; Thermo Fisher Scientific, Waltham, MA, USA) equipped with a trap column (EASY-Column, C18-A1 5 µm, 100 µm ID × 20 mm; Thermo Fisher Scientific), equilibrated with 0.1% formic acid, and eluted with a linear acetonitrile gradient (0–50%) in 0.1% formic acid at a flow rate of 200 nL/min. The eluted peptides were loaded and separated on the column (C18 capillary tip column, 75 µm ID × 120 mm; Nikkyo Technos, Tokyo, Japan) with a spray voltage of 1.5 kV. The peptide ions were detected using MS (LTQ Orbitrap Elite MS; Thermo Fisher Scientific) in the data-dependent acquisition mode with Xcalibur software (version 2.2; Thermo Fisher Scientific). Full-scan mass spectra were acquired in MS over 400–1500 m/z with a resolution of 60,000. The 10 most intense precursor ions were selected for collision-induced fragmentation in the linear ion trap, at a normalized collision energy of 35%. Dynamic exclusion was employed within 90 s to prevent the repetitive selection of peptides.

4.11. MS Data Analysis

Protein identification was performed using the Mascot search engine (version 2.5.1, Matrix Science, London, UK) and the in-house database, which constructed the amino acid sequences of rice heterotrimeric G protein subunits. For both the searches, the carbamidomethylation of cysteine was set as a fixed modification, and oxidation of methionine was set as a variable modification. Trypsin was specified as the proteolytic enzyme and one missed cleavage was allowed. The peptide mass tolerance was set at 10 ppm, fragment mass tolerance was set at 0.8 Da, and peptide charges were set at +2, +3, and +4. An automatic decoy database search was performed as part of the search. Mascot results were filtered with the Percolator function to improve the accuracy and sensitivity of peptide identification. The minimum requirement for the identification of a protein was two matched peptides. Significant changes in the abundance of proteins between samples were determined ($p < 0.05$).

4.12. Gene ID

The accession numbers of the rice heterotrimeric G proteins α, β, and γ3 subunit genes (*RGA1*, *RGB1*, and *RGG3*, respectively) are Os05g0333200, Os03g0669200, and Os03g0407400, respectively.

Conflicts of Interests

The authors declare no conflicts of interest.

Author Contributions: Investigation and formal analysis, A.N. and S.M.; methodology, T.I.; resources, G.C. and K.M.; writing and funding acquisition, Y.I.

Funding: This work was supported by a grant for Scientific Research from Fukui Prefectural University and the JSPS KAKENHI, Grant Number 26712001.

Acknowledgments: We thank Yasuo Nagato for providing TCM3-467 and Iturou Takamure for H343. Part of the work was performed at the Biological Resource Research and Development Center, Fukui Prefectural University (Fukui, Japan).

Int. J. Mol. Sci. **2018**, *19*, 3591

Abbreviations

agb1	mutant of heterotrimeric G protein β subunit gene in Arabidopsis
AGB1	heterotrimeric G protein β subunit gene in Arabidopsis
AGG1	heterotrimeric G protein γ1 subunit gene in Arabidopsis
AGG2	heterotrimeric G protein γ2 subunit gene in Arabidopsis
AGG3	heterotrimeric G protein γ3 subunit gene in Arabidopsis
cMS	crude microsomal fraction
d1	mutant of heterotrimeric G protein α subunit gene in rice
DEP1	*DENCE AND ERECT PANICLES 1* gene
DN1	*DENCE PANICLE 1* gene
gpa1	mutant of heterotrimeric G protein α subunit gene in Arabidopsis
GPA1	heterotrimeric G protein α subunit gene in Arabidopsis
GS3	*GRAIN SIZE 3* gene
Mi	*MINUTE*, a mutant of GS3/RGG3
OsGGC1	a gene of heterotrimeric G protein γ subunit Type-C in rice, which corresponds to GS3/RGG3
OsGGC2	a gene of heterotrimeric G protein γ subunit Type-C in rice, which corresponds to RGG5
OsGGC3	a gene of heterotrimeric G protein γ subunit Type-C in rice, which corresponds to which corresponds to DEP1/RGG4
PM	plasma membrane
RGA1	heterotrimeric G protein α subunit gene in rice
RGB1	heterotrimeric G protein β subunit gene in rice
RGG1	heterotrimeric G protein γ1 subunit gene in rice
RGG2	heterotrimeric G protein γ2 subunit gene in rice
RGG3	heterotrimeric G protein γ3 subunit gene in rice
RGG4	heterotrimeric G protein γ4 subunit gene in rice
RGG5	heterotrimeric G protein γ5 subunit gene in rice
WB	western blot
WT	wild-type
XLG	extra-large GTP-binding protein

References

1. Offermanns, S. Mammalian G-protein function in vivo: New insights through altered gene expression. *Rev. Physiol. Biochem. Pharmacol.* **2000**, *140*, 63–133. [PubMed]
2. Gomperts, B.D.; Kramer, I.J.M.; Tatham, P.E.R. (Eds.) *Signal Transduction*; Elsevier Inc.: Amsterdam, The Netherlands, 2002.
3. Wettschureck, N.; Offermanns, S. Mammalian G proteins and their cell type specific functions. *Physiol. Rev.* **2005**, *85*, 1159–1204. [CrossRef] [PubMed]
4. Milligan, G.; Kostenis, E. Heterotrimeric G-proteins: A short history. *Br. J. Pharmacol.* **2006**, *147*, S46–S55. [CrossRef] [PubMed]
5. Temple, B.R.S.; Jones, A.M. The Plant Heterotrimeric G-Protein Complex. *Annu. Rev. Plant Biol.* **2007**, *58*, 249–266. [CrossRef] [PubMed]
6. Urano, D.; Chen, J.-G.; Botella, J.R.; Jones, A.M. Heterotrimeric G protein signalling in the plant kingdom. *Open Biol.* **2013**, *3*, 120–186. [CrossRef] [PubMed]
7. Urano, D.; Miura, K.; Wu, Q.; Iwasaki, Y.; Jackson, D.; Jones, A.M. Plant morphology of heterotrimeric G protein mutants. *Plant Cell Physiol.* **2016**, *57*, 437–445. [CrossRef] [PubMed]
8. Lee, Y.-R.J.; Assmann, S.M. *Arabidopsis thaliana* 'extra-large GTP-binding protein' (AtXLG1): A new class of G-protein. *Plant Mol. Biol.* **1999**, *40*, 55–64. [CrossRef] [PubMed]
9. Assmann, S.M. Heterotrimeric and unconventional GTP binding proteins in plant cell signaling. *Plant Cell* **2002**, 355S–S373. [CrossRef]
10. Ma, H.; Yanofsky, M.F.; Meyerowitz, E.M. Molecular cloning and characterization of *GPA1*, a G protein α subunit gene from *Arabidopsis thaliana*. *Proc. Natl. Acad. Sci. USA* **1990**, *87*, 3821–3825. [CrossRef] [PubMed]

11. Weiss, C.A.; Garnaat, C.W.; Mukai, K.; Hu, Y.; Ma, H. Isolation of cDNAs encoding guanine nucleotide-binding protein β-subunit homologues from maize (ZGB1) and *Arabidopsis* (AGB1). *Proc. Natl. Acad. Sci. USA* **1994**, *91*, 9554–9558. [CrossRef] [PubMed]

12. Mason, M.G.; Botella, J.R. Completing the heterotrimer: Isolation and characterization of an *Arabidopsis thaliana* G protein γ-subunit cDNA. *Proc. Natl. Acad. Sci. USA* **2000**, *97*, 14784–14788. [CrossRef] [PubMed]

13. Mason, M.G.; Botella, J.R. Isolation of a novel G-protein γ-subunit from *Arabidopsis thaliana* and its interaction with Gβ. *Biochim. Biophys. Acta* **2001**, *1520*, 147–153. [CrossRef]

14. Chakravorty, D.; Trusov, Y.; Zhang, W.; Acharya, B.R.; Sheahan, M.B.; McCurdy, D.W.; Assmann, S.M.; Botella, J.R. An atypical heterotrimeric G-protein γ-subunit is involved in guard cell K⁺-channel regulation and morphological development in *Arabidopsis thaliana*. *Plant J.* **2011**, *67*, 840–851. [CrossRef] [PubMed]

15. Wu, Q.; Regan, M.; Furukawa, H.; Jackson, D. Role of heterotrimeric Gα proteins in maize development and enhancement of agronomic traits. *PLOS Genet.* **2018**, *14*, e1007374. [CrossRef] [PubMed]

16. Ishikawa, A.; Tsubouchi, H.; Iwasaki, Y.; Asahi, T. Molecular cloning and characterization of a cDNA for the α subunit of a G protein from rice. *Plant Cell Physiol.* **1995**, *36*, 353–359. [CrossRef] [PubMed]

17. Ishikawa, A.; Iwasaki, Y.; Asahi, T. Molecular cloning and characterization of a cDNA for the β subunit of a G protein from rice. *Plant Cell Physiol.* **1996**, *37*, 223–228. [CrossRef] [PubMed]

18. Kato, C.; Mizutani, T.; Tamaki, H.; Kumagai, H.; Kamiya, T.; Hirobe, A.; Fujisawa, Y.; Kato, H.; Iwasaki, Y. Characterization of heterotrimeric G protein complexes in rice plasma membrane. *Plant J.* **2004**, *38*, 320–331. [CrossRef] [PubMed]

19. Fan, C.; Xing, Y.; Mao, H.; Lu, T.; Han, B.; Xu, C.; Li, X.; Zhang, Q. GS3, a major QTL for grain length and weight and minor QTL for grain width and thickness in rice, encodes a putative transmembrane protein. *Theor. Appl. Genet.* **2006**, *112*, 1164–1171. [CrossRef] [PubMed]

20. Huang, X.; Qian, Q.; Liu, Z.; Sun, H.; He, S.; Luo, D.; Xia, G.; Chu, C.; Li, J.; Fu, X. Natural variation at the *DEP1* locus enhances grain yield in rice. *Nat. Genet.* **2009**, *41*, 494–497. [CrossRef] [PubMed]

21. Botella, J.R. Can heterotrimeric G proteins help to feed the world? *Trend Plant Sci.* **2012**, *17*, 563–568. [CrossRef] [PubMed]

22. Trusov, Y.; Chakravorty, D.; Botella, J.R. Diversity of heterotrimeric G-protein γ subunits in plants. *BMC Res. Notes* **2012**, *5*, 608. [CrossRef] [PubMed]

23. Ding, L.; Pandey, S.; Assmann, S.M. *Arabidopsis* extra-large G proteins (XLGs) regulate root morphogenesis. *Plant J.* **2008**, *53*, 248–263. [CrossRef] [PubMed]

24. Ullah, H.; Chen, J.-G.; Young, J.C.; Im, K.-H.; Sussman, M.R.; Jones, A.M. Modulation of cell proliferation by heterotrimeric G protein in *Arabidopsis*. *Science* **2001**, *292*, 2066–2069. [CrossRef]

25. Lease, K.A.; Wen, J.; Li, J.; Doke, J.T.; Liscum, E.; Walker, J.C. A mutant *Arabidopsis* heterotrimeric G-protein β subunit affects leaf, flower, and fruit development. *Plant Cell* **2001**, *13*, 2631–2641. [CrossRef] [PubMed]

26. Ullah, H.; Chen, J.-G.; Temple, B.; Boyes, D.C.; Alonso, J.M.; Davis, K.R.; Ecker, J.R.; Jones, A.M. The β-subunit of *Arabidopsis* G protein negatively regulates auxin-induced cell division and affects multiple developmental processes. *Plant Cell* **2003**, *15*, 393–409. [CrossRef] [PubMed]

27. Trusov, Y.; Rookes, J.E.; Tilbrook, K.; Chakravorty, D.; Mason, M.G.; Anderson, D.; Chen, J.-G.; Jones, A.M.; Botella, J.R. Heterotrimeric G protein γ subunits provide functional selectivity in Gβγ dimer signaling in *Arabidopsis*. *Plant Cell* **2007**, *19*, 1235–1250. [CrossRef] [PubMed]

28. Fujisawa, Y.; Kato, T.; Ohki, S.; Ishikawa, A.; Kitano, H.; Sasaki, T.; Asahi, T.; Iwasaki, Y. Suppression of the heterotrimeric G protein causes abnormal morphology, including dwarfism, in rice. *Proc. Natl. Acad. Sci. USA* **1999**, *96*, 7575–7580. [CrossRef] [PubMed]

29. Ashikari, M.; Wu, J.; Yano, M.; Sasaki, T.; Yoshimura, A. Rice gibberellin-insensitive dwarf mutant gene *Dwarf 1* encodes the α-subunit of GTP-binding protein. *Proc. Natl. Acad. Sci. USA* **1999**, *96*, 10284–10289. [CrossRef] [PubMed]

30. Mao, H.; Sun, S.; Yao, J.; Wang, C.; Yu, S.; Xu, C.; Li, X.; Zhang, Q. Linking differential domain functions of the GS3 protein to natural variation in grain size in rice. *Proc. Natl. Acad. Sci. USA* **2010**, *107*, 19579–19584. [CrossRef] [PubMed]

31. Izawa, Y.; Takayanagi, Y.; Inaba, N.; Abe, Y.; Minami, M.; Fujisawa, Y.; Kato, H.; Ohki, S.; Kitano, H.; Iwasaki, Y. Function and expression pattern of the α subunit of the heterotrimeric G protein in rice. *Plant Cell Physiol.* **2010**, *51*, 271–281. [CrossRef] [PubMed]

32. Utsunimiya, U.; Samejima, C.; Takayanagi, Y.; Izawa, Y.; Yoshida, T.; Sawada, Y.; Fijisawa, Y.; Kato, H.; Iwasaki, Y. Suppression of the rice heterotrimeric G protein β-subunit gene, *RGB1*, causes dwarfism and browning of internodes and lamina joint regions. *Plant J.* **2011**, *67*, 907–916. [CrossRef] [PubMed]

33. Wang, X.Q.; Ullah, H.; Jones, A.M.; Assmann, S.M. G protein regulation of ion channels and abscisic acid signaling in *Arabidopsis* guard cells. *Science* **2001**, *292*, 2070–2072. [CrossRef] [PubMed]

34. Coursol, S.; Fan, L.M.; Le Stunff, H.; Spiegel, S.; Gilroy, S.; Assmann, S.M. Sphingolipid signalling in *Arabidopsis* guard cells involves heterotrimeric G proteins. *Nature* **2003**, *423*, 651–654. [CrossRef] [PubMed]

35. Lapik, Y.R.; Kaufman, L.S. The *Arabidopsis* cupin domain protein AtPirin1 interacts with the G protein α-Subunit GPA1 and regulates seed germination and early seedling development. *Plant Cell* **2003**, *15*, 1578–1590. [CrossRef] [PubMed]

36. Pandey, S.; Assmann, S.M. The *Arabidopsis* putative G protein-coupled receptor GCR1 interacts with the G protein α subunit GPA1 and regulates abscisic acid signaling. *Plant Cell* **2004**, *16*, 1616–1632. [CrossRef] [PubMed]

37. Mishra, G.; Zhang, W.; Deng, F.; Zhao, J.; Wang, X. A bifurcating pathway directs abscisic acid effects on stomatal closure and opening in *Arabidopsis*. *Science* **2006**, *312*, 264–266. [CrossRef] [PubMed]

38. Ueguchi-Tanaka, M.; Fujisawa, Y.; Kobayashi, M.; Ashikari, M.; Iwasaki, Y.; Kitano, H.; Matsuoka, M. Rice dwarf mutant *d1*, which is defective in the α subunit of the heterotrimeric G protein, affects gibberellin signal transduction. *Proc. Natl. Acad. Sci. USA* **2000**, *97*, 11638–11643. [CrossRef] [PubMed]

39. Ullah, H.; Chen, J.G.; Wang, S.; Jones, A.M. Role of a heterotrimeric G protein in regulation of *Arabidopsis* seed germination. *Plant Physiol.* **2002**, *129*, 897–907. [CrossRef] [PubMed]

40. Chen, J.G.; Pandey, S.; Huang, J.; Alonso, J.M.; Ecker, J.R.; Assmann, S.M.; Jones, A.M. GCR1 can act independently of heterotrimeric G-protein in response to brassinosteroids and gibberellins in *Arabidopsis* seed germination. *Plant Physiol.* **2004**, *135*, 907–915. [CrossRef] [PubMed]

41. Bethke, P.C.; Hwang, Y.S.; Zhu, T.; Jones, R.L. Global patterns of gene expression in the aleurone of wild-type and *dwarf1* mutant rice. *Plant Physiol.* **2006**, *140*, 484–498. [CrossRef] [PubMed]

42. Chen, J.-G.; Jones, A.M. AtRGS1 function in *Arabidopsis thaliana*. *Method. Enzymol.* **2004**, *389*, 338–350. [CrossRef]

43. Huang, J.; Taylor, J.P.; Chen, J.G.; Uhrig, J.F.; Schnell, D.J.; Nakagawa, T.; Korth, K.L.; Jones, A.M. The plastid protein THYLAKOID FORMATION1 and the plasma membrane G-protein GPA1 interact in a novel sugar-signaling mechanism in *Arabidopsis*. *Plant Cell* **2006**, *18*, 1226–1238. [CrossRef] [PubMed]

44. Warpeha, K.M.; Hamm, H.E.; Rasenick, M.M.; Kaufman, L.S. A blue-light-activated GTP-binding protein in the plasma membranes of etiolated peas. *Proc. Natl. Acad. Sci. USA* **1991**, *88*, 8925–8929. [CrossRef] [PubMed]

45. Warpeha, K.M.; Lateef, S.S.; Lapik, Y.; Anderson, M.; Lee, B.S.; Kaufman, L.S. G-protein-coupled receptor 1, G-protein Gα-subunit 1, and Prephenate dehydratase 1 are required for blue light-induced production of phenylalanine in etiolated Arabidopsis. *Plant Physiol.* **2006**, *140*, 844–855. [CrossRef] [PubMed]

46. Joo, J.H.; Wang, S.; Chen, J.G.; Jones, A.M.; Fedoroff, N.V. Different signaling and cell death roles of heterotrimeric G protein α and β subunits in the Arabidopsis oxidative stress response to ozone. *Plant Cell* **2005**, *17*, 957–970. [CrossRef] [PubMed]

47. Suharsono, U.; Fujisawa, Y.; Kawasaki, T.; Iwasaki, Y.; Satoh, H.; Shimamoto, K. The heterotrimeric G protein α subunit acts upstream of the small GTPase Rac in disease resistance of rice. *Proc. Natl. Acad. Sci. USA* **2002**, *99*, 13307–13312. [CrossRef] [PubMed]

48. Komatsu, S.; Yang, G.; Hayashi, N.; Kaku, H.; Umemura, K.; Iwasaki, Y. Alterations by a defect in a rice G protein α subunit in probenazole and pathogen-induced responses. *Plant Cell Environ.* **2004**, *27*, 947–957. [CrossRef]

49. Iwata, M.; Umemura, K.; Teraoka, T.; Usami, H.; Fujisawa, Y.; Iwasaki, Y. Role of the α subunit of heterotrimeric G-protein in probenazole-inducing defense signaling in rice. *J. Gen. Plant Pathol.* **2003**, *69*, 83–86. [CrossRef]

50. Lieberherr, D.; Thao, N.P.; Nakashima, A.; Umemura, K.; Kawasaki, T.; Shimamoto, K. A sphingolipid licitor-inducible mitogen-activated protein kinase is regulated by the small GTPase OsRac1 and heterotrimeric G-protein in rice. *Plant Physiol.* **2005**, *138*, 1644–1652. [CrossRef] [PubMed]

51. Wang, S.; Assmann, S.M.; Fedoroff, N.V. Characterization of the *Arabidopsis* heterotrimeric G protein. *J. Biol. Chem.* **2008**, *283*, 13913–13922. [CrossRef] [PubMed]

52. Klopffleish, K.; Phan, N.; Augstin, K.; Bayne, R.; Booker, K.S.; Bolella, J.; Carpita, N.C.; Carr, T.; Chen, J.-C.; Cooke, T.R.; et al. Arabidopsis G-protein interactome reveals connections to cell wall carbohydrates and morphogenesis. *Mol. Syst. Biol.* **2011**, *7*, 532. [CrossRef] [PubMed]

53. Chen, J.-G.; Willard, F.S.; Huang, J.; Liang, J.; Chasse, S.A.; Jones, A.M.; Siderovski, D.P. A seven-transmembrane RGS protein that modulates plant cell proliferation. *Science* **2003**, *301*, 1728–1731. [CrossRef] [PubMed]

54. Adjobo-Hermans, M.J.; Goedhart, J.; Gadella, T.W., Jr. Plant G protein heterotrimers require dual lipidation motifs of Gα and Gγ and do not dissociate upon activation. *J. Cell Sci.* **2006**, *119*, 5087–5097. [CrossRef] [PubMed]

55. Sun, H.; Qian, Q.; Wu, K.; Lou, J.; Wang, S.; Zhang, C.; Ma, Y.; Lie, Q.; Huang, X.; Yuan, Q.; et al. Heterotrimeric G proteins regulate nitrogen-use efficiency in rice. *Nat. Genet.* **2014**, *46*, 652–656. [CrossRef] [PubMed]

56. Sun, S.; Wang, L.; Mao, H.; Shao, L.; Li, X.; Xiao, J.; Ouyang, Y.; Zhang, Q. A G-protein pathway determines grain size in rice. *Nat. Commun.* **2018**, *9*, 815–824. [CrossRef]

57. Takano-Kai, N.; Jiang, H.; Kubo, T.; Sweeney, M.; Matsumoto, T.; Kanamori, H.; Padhukasahasram, B.; Bustamante, C.; Yoshimura, A.; Doi, K.; et al. Evolutionary history of GS3, a gene conferring grain length in rice. *Genetics* **2009**, *182*, 1–12. [CrossRef] [PubMed]

58. Takano-Kai, N.; Jiang, H.; Powell, A.; McCouch, S.; Takamure, I.; Furuya, N.; Doi, K.; Yoshimura, A. Multiple and independent origins of short seeded alleles of *GS3* in rice. *Breed. Sci.* **2013**, *63*, 77–85. [CrossRef] [PubMed]

59. Sambrook, J.; Russell, D.W. (Eds.) *Molecular Cloning*; Cold Spring Harbor Laboratory Press: Cold Spring Harbor, NY, USA, 2001.

60. Yoshida, S.; Uemura, M.; Niki, T.; Sakai, A.; Gusta, L.V. Partition of membrane particles in aqueous two-polymer phase system and its partial use for purification of plasma membranes from plants. *Plant Physiol.* **1983**, *72*, 105–114. [CrossRef] [PubMed]

61. Iwasaki, Y.; Kato, T.; Kaidoh, T.; Ishikawa, A.; Asahi, T. Characterization of the putative α subunit of a heterotrimeric G protein in rice. *Plant Mol. Biol.* **1997**, *34*, 563–572. [CrossRef] [PubMed]

International Journal of
Molecular Sciences

MDPI

Article

Comparative Transcriptome Analysis Reveals the Transcriptional Alterations in Growth- and Development-Related Genes in Sweet Potato Plants Infected and Non-Infected by SPFMV, SPV2, and SPVG

Jiang Shi [1], Lin Zhao [1], Baiyuan Yan [2], Yueqing Zhu [3], Huasheng Ma [1], Wenyue Chen [1,*] and Songlin Ruan [1,4,*]

[1] Institute of Crop Science, Hangzhou Academy of Agricultural Sciences, Hangzhou 310024, China; tomatoman@126.com (J.S.); 15925638398@163.com (L.Z.); hzhsma@163.com (H.M.)
[2] Jiande Seed Management Station, Hangzhou 311600, China; zjjdyby@126.com
[3] Linan District Forestry and Agriculture Bureau, Hangzhou 311300, China; 20050020@zafu.edu.cn
[4] Laboratory of Plant Molecular Biology & Proteomics, Institute of Biotechnology, Hangzhou Academy of Agricultural Sciences, Hangzhou 310024, China
* Correspondence: wenyuech@hotmail.com (W.C.); ruansl1@hotmail.com (S.R.); Tel.: +86-571-8709-8390 (W.C.); +86-571-8709-3826 (S.R.); Fax: +86-571-8709-1696 (W.C.); +86-571-8731-3241 (S.R.)

Received: 28 December 2018; Accepted: 10 February 2019; Published: 26 February 2019

Abstract: Field co-infection of multiple viruses results in considerable losses in the yield and quality of storage roots in sweet potato. However, little is known about the molecular mechanisms underlying developmental disorders of sweet potato subjected to co-infection by multiple viruses. Here, a comparative transcriptomic analysis was performed to reveal the transcriptional alterations in sweet potato plants infected (VCSP) and non-infected (VFSP) by *Sweet potato mild mottle virus* (SPFMV), *Sweet potato virus Y* (SPV2) and *Sweet potato virus G* (SPVG). A total of 1580 and 12,566 differentially expressed genes (DEGs) were identified in leaves and storage roots of VFSP and VCSP plants, respectively. In leaves, 707 upregulated and 773 downregulated genes were identified, whereas 5653 upregulated and 6913 downregulated genes were identified in storage roots. Gene Ontology (GO) classification and pathway enrichment analysis showed that the expression of genes involved in chloroplast and photosynthesis and brassinosteroid (BR) biosynthesis in leaves and the vitamin biosynthetic process in storage roots was inhibited by co-infection of three viruses: SPFMV, SPV2, and SPVG. This was likely closely related to better photosynthesis and higher contents of Vitamin C (Vc) in storage roots of VFSP than that of VCSP. While some genes involved in ribosome and secondary metabolite-related pathways in leaves and alanine, aspartate, and glutamate metabolism in storage roots displayed higher expression in VCSP than in VFSP. Quantitative real-time PCR analysis demonstrated that the expression patterns of 26 DEGs, including 16 upregulated genes and 10 downregulated genes were consistent with the RNA-seq data from VFSP and VCSP. Taken together, this study integrates the results of morphology, physiology, and comparative transcriptome analyses in leaves and storage roots of VCSP and VFSP to reveal transcriptional alterations in growth- and development-related genes, providing new insight into the molecular mechanisms underlying developmental disorders of sweet potato subjected to co-infection by multiple viruses.

Keywords: sweet potato plants infected by SPFMV; SPV2 and SPVG; sweet potato plants non-infected by SPFMV; SPV2 and SPVG; co-infection; transcriptome profiling; gene ontology; pathway analysis

1. Introduction

Sweet potato (*Ipomoea batatas.* Lam), in the family Convolvulaceae, is a light-loving and short-day crop. The main producing areas of sweet potato in the world are located at 40 degrees north latitude. Asia has the largest area of cultivated sweet potato, followed by Africa, and North America is the third largest. China is the largest sweet potato-producing country in the world, with a total planting area of over 3 million hectares and an annual output of 0.72 billion tons [1]. After being infected by viruses, sweet potato gradually shows degeneration phenomena such as declines in yield and quality and loses its desirable characteristics. According to a survey in certain areas in China, including Shandong, Anhui, Beijing, Jiangsu, and other provinces, the yield loss of sweet potato caused by virus diseases can exceed 50%, resulting in an economic loss of approximately five hundred million dollars [2]. Therefore, the adverse effect of virus diseases on sweet potato production is of increasing concern.

Because sweet potato is mainly planted by cuttings, many viruses can spread through insects and sap, and the co-infection rate is high. For example, *Sweet potato feather mottle virus* (SPFMV), *Sweet potato virus G* and *Sweet potato virus C* belonging to *Potato virus Y* showed co-infection in the field [3]. The sweet potato virus disease (SPVD) caused by the synergistic infection of *Sweet potato chlorosis dwarf virus* (SPCDV) and SPFMV is a destructive disease of sweet potato [4,5], which can cause losses of 70%–100% [6]. Effects on yields caused by SPFMV or *Sweet potato chlorotic stunt virus* (SPCSV) alone were minor, but co-infection by two or more viruses caused 50% losses in yield [7]. In Uganda, mixed infections of four viruses, including *Sweet potato chlorotic spot virus* (SPCFV), SPCSV, SPFMV, and *Sweet potato mottle mosaic virus* (SPMMV), were found in wild species and cultivated sweet potato [8]. Jiang et al. [9] detected five viruses in 24 samples from Shandong Province, including 23 co-infection samples that were classified into 11 categories of infection. Based on the observation that the tip of the sweet potato stem often has little or no virus, virus-free sweet potato can be obtained via tissue culture technology, which can effectively restore the desirable characteristics of the varieties, improve the yield and quality of the storage roots, and prolong the shelf life of the varieties [10]. Virus-free meristems were grown into plants, which were kept under insect-proof conditions and away from other sweet potato material for distribution to farmers after another cycle of reproduction [7].

The physiological and biochemical mechanisms underlying the improved yield and quality of virus-free sweet potato storage roots, including increased net photosynthetic rate and chlorophyll content, higher activities of antioxidant enzymes (superoxide dismutase (SOD), peroxidase (POD), and catalase (CAT)) and reduced lipid peroxidation levels, have widely been reported [11–13]. However, the molecular mechanisms underlying the immune response of plants to the virus mainly involve vsiRNA. There are mainly two mechanisms of vsiRNA immunity to virus. vsiRNAs can be recruited by Argonaut (AGO)proteins and directly degrade virus RNA through post-transcriptional gene silencing (PTGS) [14]. vsiRNAs can also bind to AGO proteins and act on virus DNA and silence virus genes by enhancing the methylation of DNA [15,16].

Transcriptomics is a powerful tool for discovering differentially expressed genes and has been widely applied in many crop species [17]. Recently, transcriptome sequencing analysis has been performed in purple sweet potato [18] and a sweet potato progenitor [19]. However, little is known about differential alterations in transcriptome profiles between sweet potato plants infected (VCSP) and non-infected (VFSP). Here, a comparative transcriptomic analysis was performed to reveal the significantly upregulated and downregulated genes. Gene Ontology (GO) classification of the proteins encoded by these genes was used to analyze their cellular locations, molecular functions, and biological processes. A pathway analysis was performed to reveal the biological pathways involving these genes. This study may provide new insight into the transcriptional alterations in VFSP and VCSP.

2. Results

2.1. Phenotypes and Growth Indexes of VCSP and VFSP

To identify phenotypes of VCSP and VFSP, we first performed the detection of viruses in their leaves and storage roots using Q-PCR. The results showed that three viruses, including SPFMV, SPV2, and SPVG, were found in VCSP leaves and storage roots but not in VFSP leaves and storage roots (Figure 1C). VCSP leaves exhibited a small chlorotic and mosaic phenotype, whereas VFSP leaves displayed a normal green phenotype (Figure 1A). Interestingly, fluorescence imaging analysis showed that VFSP leaves exhibited strong blue fluorescence, whereas VCSP leaves displayed inhomogeneous fluorescence with photobleaching (Figure 1B). The values of leaf length (LL), leaf width (LW), and Fv/Fm, in VFSP leaves were significantly higher than those in VCSP leaves (Figure 1D,F). Similarly, the contents of total chlorophyll, chlorophyll a, and chlorophyll b were significantly higher in VFSP than in VCSP (Figure 1E). In addition, the yield and contents of Vc and beta-carotene of storage roots in VFSP were significantly higher than those in VCSP (Figure 2B–D), whereas the content of starch in storage roots was lower in VFSP than in VCSP (Figure 2A).

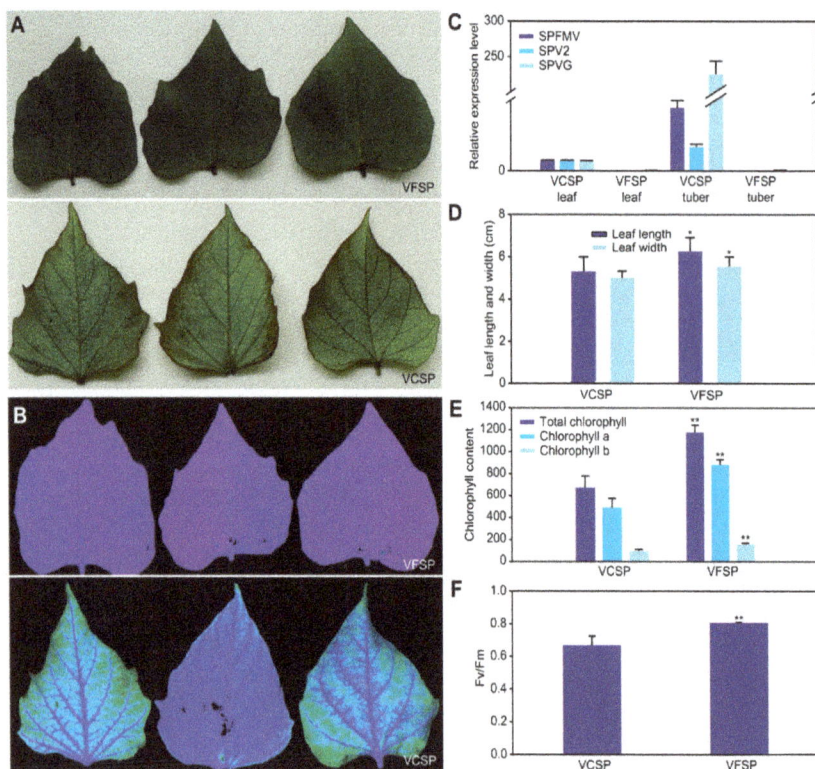

Figure 1. Phenotypes and growth indexes of sweet potato plants infected (VCSP) and non-infected (VFSP) leaves. (**A**) Phenotypes of VCSP and VFSP leaves. (**B**) Fluorescence images of VCSP and VFSP leaves. (**C**) Three virus genes were not detected in leaf and tuber of sweet potato. (**D**) Length and width of VCSP and VFSP leaves. (**E**) Content of chlorophyll a, b, and total chlorophyll. (**F**) Fv/Fm of VCSP and VFSP leaves. Three independent experimental replicates were analyzed for each treatment, and data are indicated as the mean ± SE ($n = 3$). Independent t-test was performed to check difference between VFSP and VCSP (** $p < 0.01$; * $p < 0.05$).

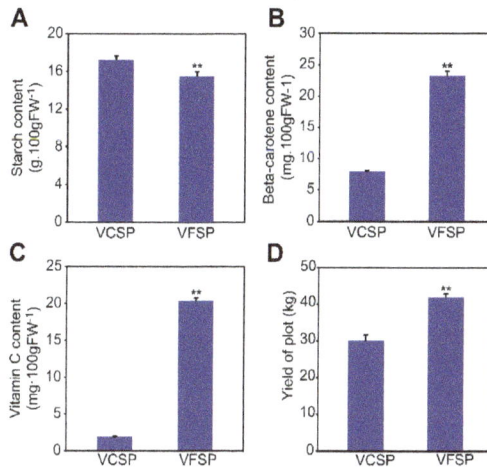

Figure 2. Comparison of yield and quality of VCSP and VFSP storage roots. (**A**) Content of starch. (**B**) Content of beta-carotene. (**C**) Content of vitamin C. (**D**) Yield of plot. Three independent experimental replicates were analyzed for each treatment, and data are indicated as the mean ± SE ($n = 3$). Independent t-test was performed to check difference between VFSP and VCSP (** $p < 0.01$; * $p < 0.05$).

2.2. Gene Expression Profiles of Leaves and Storage Roots in VCSP and VFSP

To understand the molecular mechanisms underlying developmental disorders of sweet potato subjected to co-infection by multiple viruses, comparative transcriptomic analysis was performed to discover differentially expressed genes in leaves and storage roots of VFSP and VCSP. As shown in Figure 3, a total of 1580 and 12,566 DEGs were identified in leaves and storage roots, respectively. Interestingly, in both leaves and storage roots, the number of downregulated genes was higher than that of upregulated genes. In leaves, 707 upregulated and 773 downregulated genes were identified, whereas 5653 upregulated and 6913 downregulated genes were identified in storage roots.

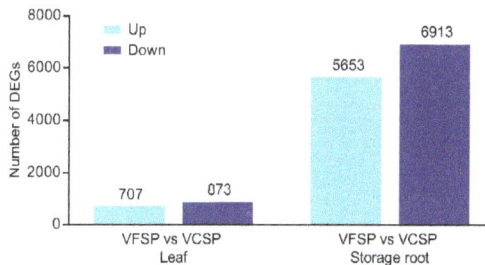

Figure 3. Gene expression profile of leaves and storage roots in VCSP and VFSP.

2.3. GO Classification of Differential Expression Genes

To find out the correlation between phenotypic differences and gene expression, we then performed a GO classification of 707 and 5653 upregulated genes in leaves and storage roots, respectively. The results showed that the proteins encoded by these genes in leaves were significantly assigned to 52 cellular components (Table S1). Of these, the top five cellular components were photosystem (GO:0009521), photosystem I (GO:0009522), chloroplast thylakoid (GO:0009534), thylakoid (GO:0009579), and plastid thylakoid (GO:0031976) (Figure 4A), which was closely associated with chloroplast. Subsequently, proteins encoded by the upregulated genes were classified into 39 functional categories (Table S1). The top five categories were chlorophyll binding (GO:0016168), tetrapyrrole binding (GO:0046906), carbon-oxygen lyase activity acting on polysaccharides (GO:0016837), pectate lyase activity (GO:0030570), and catalytic activity (GO:0003824) (Figure 4A). Finally, these categories were considered to be mainly involved in 92 biological processes (Table S1). Of these processes, the top five were photosynthesis (GO:0015979), protein–chromophore linkage (GO:0018298), carbohydrate biosynthetic process (GO:0016051), single-organism carbohydrate metabolic process (GO:0044723), and polysaccharide biosynthetic process (GO:0000271) (Figure 4A). These processes were closely related to photosynthesis, which was consistent with the result of phenotypic differences in leaf color, chlorophyll content, and fluorescence characteristics between VFSP and VCSP.

Similarly, we performed a GO classification of 5653 upregulated genes in storage roots and discovered that the proteins encoded by these genes were significantly assigned to three cellular components (Table S2), including anchored component of membrane (GO:0031225), nucleolus (GO:0005730), and endoplasmic reticulum lumen (GO:0005788) (Figure 4B). Next, proteins encoded by the upregulated genes were classified into 12 functional categories (Table S2), Of these, the top five functional categories were four iron, four sulfur cluster binding (GO:0051539), oxidoreductase activity acting on other nitrogenous compounds as donors (GO:0016661), iron–sulfur cluster binding (GO:0051536), metal cluster binding (GO:0051540) and xyloglucan:xyloglucosyl transferase activity (GO:0016762) (Figure 4B). Finally, these proteins were assigned to be involved in 56 biological processes (Table S2). Of them, the top three biological processes were sulfur compound metabolic process (GO:0006790), sulfur compound biosynthetic process (GO:0044272), and organonitrogen compound biosynthetic process (GO:1901566) (Figure 4B). Interestingly, there were three genes assigned to be involved in vitamin biosynthetic process, which might be responsible for that VFSP had higher content of vitamin C than VCSP.

A

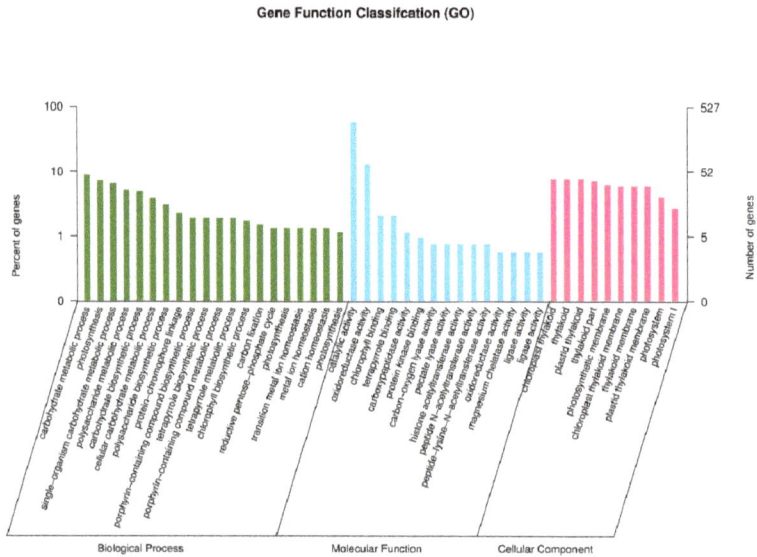

Figure 4. *Cont.*

B

C

Figure 4. *Cont.*

D

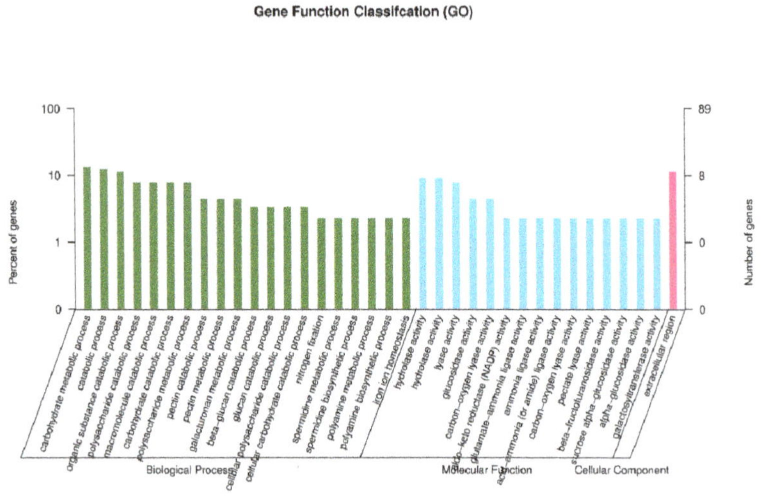

Figure 4. Gene Ontology (GO) classification of differentially expressed genes (DEGs) identified. GO classification of differentially up-regulated genes of VFSP/VCSP (**A**) leaves and (**B**) storage roots. GO classification of differentially down-regulated genes of VFSP/VCSP (**C**) leaves and (**D**) storage roots.

We also carried out a GO classification of 773 and 6913 downregulated genes in leaves and storage roots, respectively. The results indicated that the proteins encoded by these genes in leaves were significantly assigned to 27 cellular components (Table S3). Of these components, the top five were ribosome (GO:0005840), intracellular ribonucleoprotein complex (GO:0030529), ribonucleoprotein complex (GO:1990904), mitochondrial part (GO:0044429), and mitochondrial envelope (GO:0005740) (Figure 4C). Next, proteins encoded by the downregulated genes were classified into 25 functional categories (Table S3). Of these categories, the top five were chitin binding (GO:0008061), chitinase activity (GO:0004568), threonine-type endopeptidase activity (GO:0004298), threonine-type peptidase activity (GO:0070003), and ammonia-lyase activity (GO:0016841) (Figure 4C). Finally, they were assigned to be involved in 155 biological processes (Table S3). Of these processes, the top five were regulation of RNA metabolic process (GO:0051252), regulation of transcription, DNA-templated (GO:0006355), regulation of nucleic acid-templated transcription (GO:1903506), regulation of RNA biosynthetic process (GO:2001141) and transcription (Figure 5). Similarly, proteins encoded by these genes in storage roots were significantly assigned to one cell component, extracellular region (GO: 0005576) (Figure 4D). Secondly, proteins encoded by the down-regulated genes were classified into 23 functional categories (Table S4). Of these categories, the top five were hydrolase activity, hydrolyzing O-glycosyl compounds (GO:0004553), aldo-keto reductase (NADP) activity (GO:0004033), hydrolase activity, acting on glycosyl bonds (GO:0016798), glutamate-ammonia ligase activity (GO:0004356) and ammonia ligase activity (GO:0016211) (Figure 4D). At last, they were assigned to be involved in 39 biological processes (Table S4). Of these processes, the top five were polysaccharide catabolic process (GO:0000272), pectin catabolic process (GO:0045490), macromolecule catabolic process (GO:0009057), nitrogen fixation (GO:0009399), and beta-glucan catabolic process (GO:0051275) (Figure 4D).

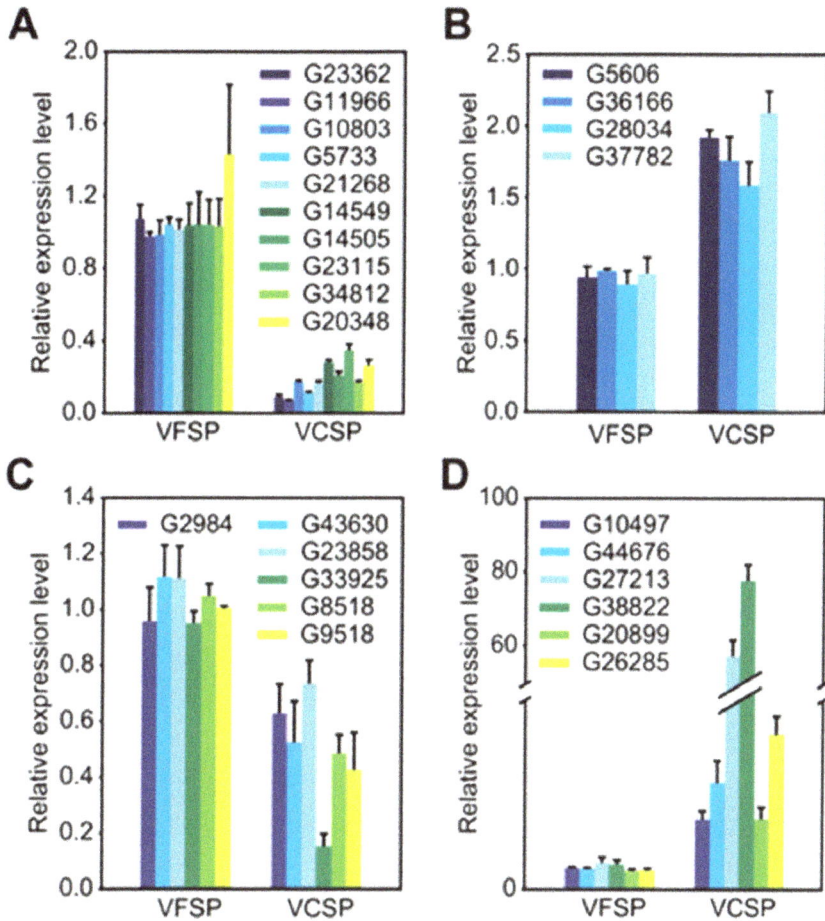

Figure 5. Validation of differentially expressed candidate genes. **A,B**: qRT-PCR analysis of ten upregulated genes (**A**) and four downregulated genes (**B**) in VFSP and VCSP leaves. **C,D**: qRT-PCR analysis of six upregulated genes (**C**) and six downregulated genes (**D**) in VFSP and VCSP storage roots. Three independent experimental replicates were analyzed for each sample, and data are indicated as the mean $\pm$ SD ($n = 3$). Independent t-test was performed to check difference between VFSP and VCSP ($p < 0.05$ or $p < 0.01$). All differentially expressed genes displayed significant differences between VFSP and VCSP at 0.05 confidence level.

2.4. Pathway Analysis of Differentially Expressed Genes

To determine the involvement of these differentially expressed genes in leaves and storage roots, we performed a pathway analysis to identify the potential target genes. The upregulated genes in leaves were identified to be involved in 10 distinct metabolic pathways. Of them, the top five were photosynthesis—antenna proteins (ko00196), photosynthesis (ko00195), porphyrin and chlorophyll metabolism (ko00860), carbon fixation in photosynthetic organisms (ko00710), and nitrogen metabolism (ko00910) (Figure 6A, Table S5). Similarly, the upregulated genes in storage roots were identified to be involved in 19 distinct metabolic pathways. Of them, the top five were linoleic acid metabolism (ko00591), thiamine metabolism (ko00730), monobactam biosynthesis (ko00261), selenocompound metabolism (ko00450), and sulfur metabolism (ko00920) (Figure 6B, Table S6).

A

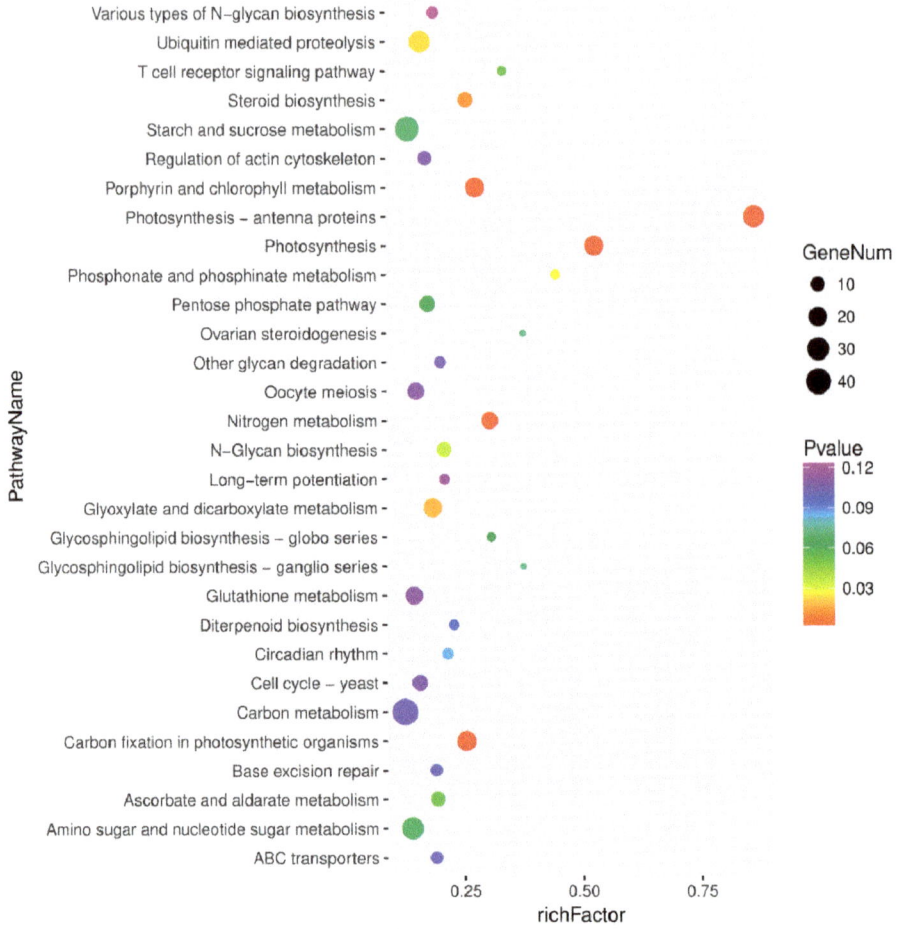

Figure 6. *Cont.*

B

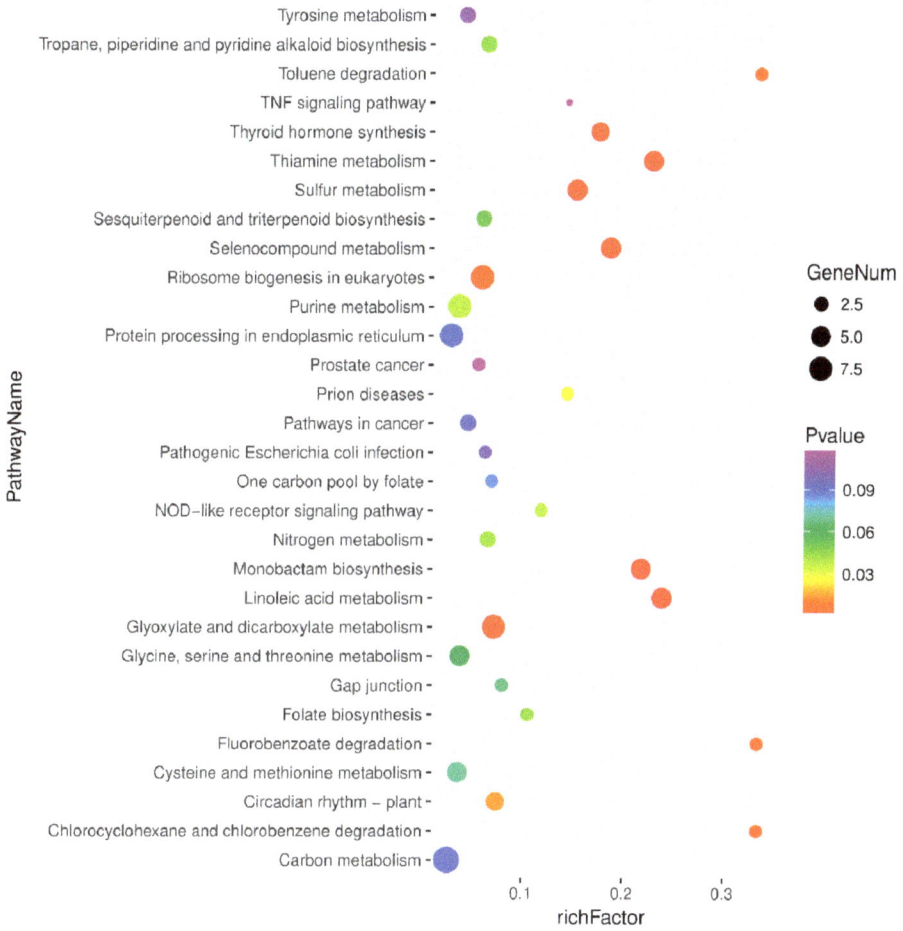

Figure 6. *Cont.*

C

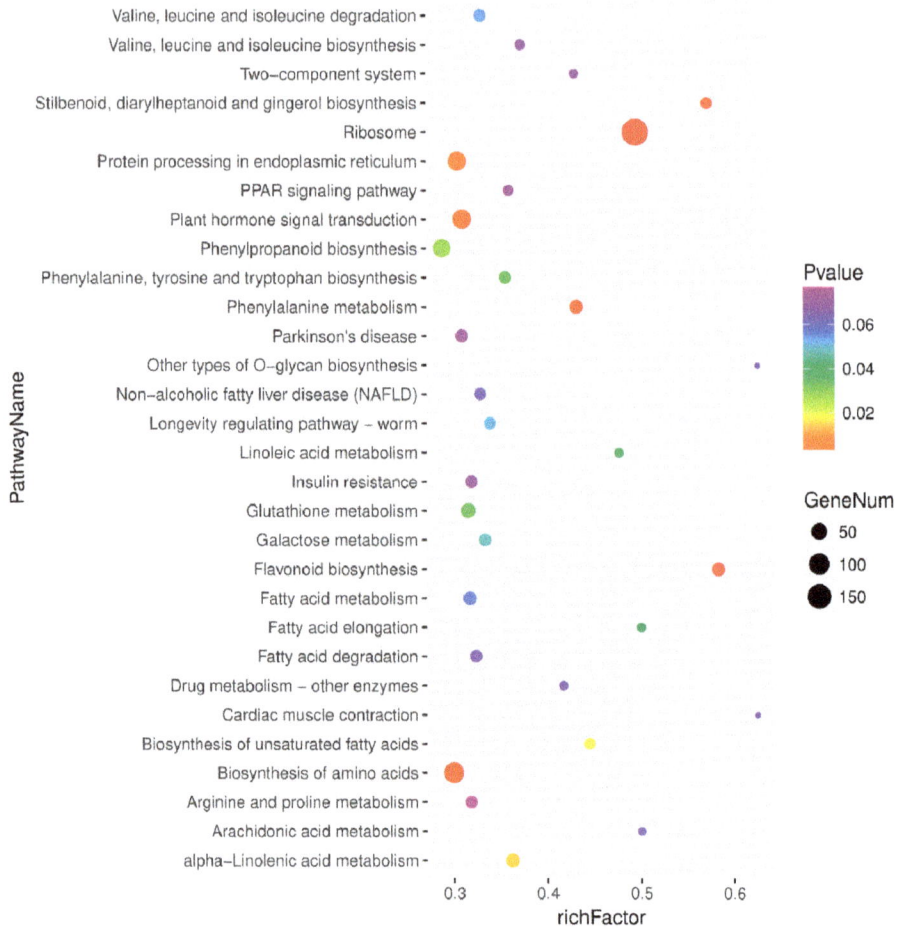

Figure 6. *Cont.*

D

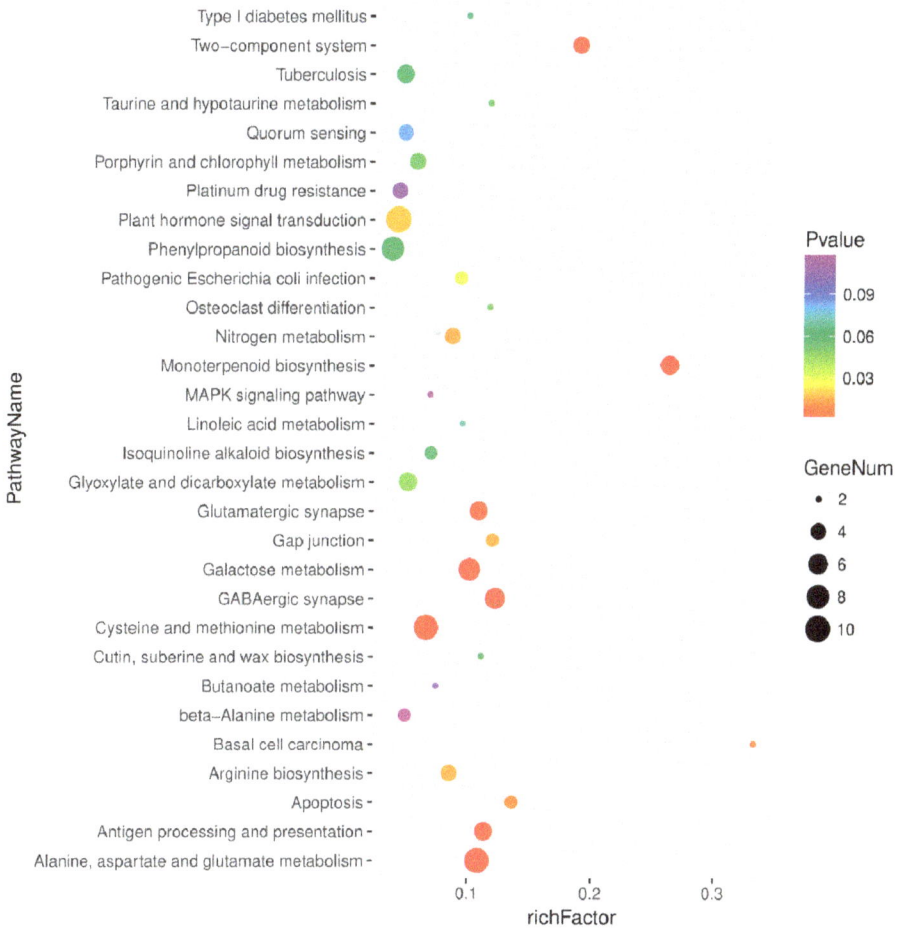

Figure 6. Kyoto Encyclopedia of Genes and Genomes (KEGG) Pathway enrichment analysis of DEGs identified. KEGG Pathway enrichment analysis based on the differentially up-regulated genes between VFSP and VCSP (**A**) leaves and (**B**) storage roots. KEGG Pathway enrichment analysis based on the differentially down-regulated genes between VFSP and VCSP (**C**) leaves and (**D**) storage roots.

The downregulated genes in leaves were identified to be involved in 15 distinct metabolic pathways. Of them, the top five pathways were ribosome (ko03010), flavonoid biosynthesis (ko00941), phenylalanine metabolism (ko00360), stilbenoid, diarylheptanoid and gingerol biosynthesis (ko00945), and biosynthesis of amino acids (ko01230) (Figure 6C, Table S7). Similarly, the downregulated genes in storage roots were identified to be involved in 19 distinct metabolic pathways. The top five pathways were alanine, aspartate and glutamate metabolism (ko00250); monoterpenoid biosynthesis (ko00902); galactose metabolism (ko00052); GABAergic synapse (ko04727); and two-component system (ko02020) (Figure 6D, Table S8).

2.5. Validation of differentially expressed candidate genes

To validate the Illumina sequencing data and the expression patterns of the DEGs revealed by RNA-Seq, qRT-PCR was performed to examine the expression patterns of 26 DEGs, including 16 upregulated genes and 10 downregulated genes (Figure 4). qRT-PCR results showed that 10 genes in leaves and 6 genes in storage roots involved in photosynthesis in VFSP showed higher abundance than those in VCSP: antenna proteins, photosynthesis, porphyrin and chlorophyll metabolism, carbon fixation in photosynthetic organisms, steroid biosynthesis, linoleic acid metabolism, thiamine metabolism and monobactam biosynthesis, including (Figure 5A,C). Furthermore, 4 genes in leaves and 6 genes in storage roots in VFSP showed lower abundance than those in VCSP, including genes involved in the ribosome, flavonoid biosynthesis, alanine, aspartate and glutamate metabolism, monoterpenoid biosynthesis and galactose metabolism (Figure 5B,D), which was consistent with the RNA-seq data from VFSP and VCSP.

3. Discussion

Field co-infection of multiple viruses in sweet potato is a common phenomenon worldwide, which has resulted in the great losses of the yield and quality of storage roots in sweet potato [6–8]. Here, VCSP leaves and storage roots was examined through Q-PCR analysis with primers of five viruses including SPFMW, SPV2, SPVG, SPCSV and SPVC. Most of VCSP plants were co-infected by three viruses including SPFMW, SPV2 and SPVG, and the other two viruses were not found in all virus-checked VCSP plants, whereas no virus was detected in virus-checked VFSP plants. In VCSP plants, the typical phenotypes with small chlorotic and mosaic leaves were observed and the yield and quality of storage roots were dramatically reduced. As we expected, photosynthetic characteristics including the chlorophyll content and maximal photochemical efficiency (Fv/Fm) of leaves in VCSP were significantly lower than that in VFSP, indicating that the photosystem of VCSP leaves was obviously impaired by the virus infection, leading to reduced photosynthesis and yield in VCSP. To explain the differences between VCSP and VFSP at the molecular level, comparative transcriptome analysis was performed. Interestingly, the number of upregulated genes in storage roots was much higher than that in leaves. We speculate that storage roots may be in an important stage of reserve substance synthesis, while these viruses may also play a role in reduced the expression of a large number of related genes to reserve substance synthesis in VCSP just at this time. This is partially confirmed by the results of subsequent GO and Kyoto Encyclopedia of Genes and Genomes (KEGG) analysis. GO classification and KEGG pathway analysis showed that the majority of proteins encoded by upregulated genes in VFSP leaves were localized in chloroplast and its structural components, including photosystem I, photosystem II, thylakoid and stroma, and involved in multiple biological processes associated with photosynthesis and porphyrin and chlorophyll metabolism. Obviously, expression of these chloroplast and photosynthesis genes is initially inhibited by co-infection with three viruses. In addition, we also found that genes involved in carbon fixation and nitrogen metabolism in VFSP plants were shown commonly upregulated expression pattern. It has been reported that carbon (C) and N metabolism are tightly coordinated in the fundamental processes that permit plant growth, for example photosynthesis and N uptake [20–23]. Therefore, we believe that expression of some genes involved in photosynthesis, carbon fixation, and nitrogen metabolism is negatively regulated by co-infection with three viruses, resulting in decreased growth and development of VCSP plants.

Similarly, certain upregulated genes involved in water-soluble vitamin metabolic process, particularly the water-soluble vitamin biosynthetic process in storage roots in VFSP, were found. Indeed, storage roots in VFSP showed a higher content of Vc than those in VCSP. Therefore, we speculate that an enhanced water-soluble vitamin biosynthetic process may be responsible for the higher accumulated contents of Vc in storage roots of VFSP. Conversely, co-infection of multiple viruses reduced the expression of genes involved in water-soluble vitamin biosynthetic process, which may be responsible for the higher accumulated contents of Vc in storage roots of VCSP. In addition, we found that there were two genes involved in other vitamin synthetic processes including thiamine biosynthetic process and thiamine-containing compound biosynthetic process. Therefore, we speculate that these viruses not only reduce Vc synthesis, but also may adversely affect the synthesis of other vitamins in VCSP storage roots.

We found that 10 upregulated genes in VFSP leaves were involved in steroid biosynthesis. Of these, five genes, including CAS1 (K01853), SMO1 (K14423), HYD1 (K01824), SMO2 (K14424), and STE1 (K00227), are involved in brassinosteroid (BR) biosynthesis. In storage roots, there is one gene involved in brassinosteroid (BR) biosynthesis. BRs play an essential role in diverse developmental programs, including cell expansion, vascular differentiation, etiolation, and reproductive development [24]. It has been demonstrated that BRs can induce resistance of tobacco, and rice to bacterial, fungal, and viral pathogens [25,26]. Recent evidence showed that *Cucumber mosaic virus* (CMV) resistance was positively regulated by BR levels, and BR signaling was required for this BR-induced CMV tolerance [27]. Furthermore, BR treatment alleviated photosystem damage, improved antioxidant enzyme activity and induced defense-associated gene expression under CMV stress in *Arabidopsis* [27]. Interestingly, although the combination of MeJA and BL treatment resulted in a significant reduction in *Rice black-streaked dwarf virus* (RBSDV) infection compared with a single BL treatment, MeJA application efficiently suppressed the expression of BR pathway genes, and this inhibition depended on the JA coreceptor OsCOI1, indicating that JA-mediated defense can suppress the BR-mediated susceptibility to RBSDV infection [26]. We speculate that BR biosynthesis in leaves of sweet potato may be adversely affected by co-infection with these viruses. In other words, BR biosynthesis may negatively be regulated by co-infection with these viruses, resulting in reduced growth and development of VCSP plants.

4. Materials and Methods

4.1. Plant Materials and Growth Conditions

Seedlings of sweet potato plants infected (VCSP) and non-infected (VFSP) with SPFMV, SPV2, and SPVG from the variety 'Zheshu 6025' were supplied by Hangzhou Academy of Agricultural Sciences (Hangzhou, Zhejiang Province, China). Two hundred of candidate VFSP seedlings were obtained through virus elimination by tissue culture of growth points of VCSP seedlings. Forty-eight non-infected seedlings were obtained through Q-PCR analysis of candidate VFSP seedlings with primers of five viruses including SPFMW, SPV2, SPVG, SPCSV and SPVC. Most of VCSP seedlings were co-infected with three viruses including SPFMW, SPV2 and SPVG, while these viruses were not found in all VCSP seedlings. Seedlings of VCSP and VFSP were planted in different pots in the same shed in mid-March. Before collection, each plant was detected by Q-PCR to ensure the sample consistency. Sweet potato storage roots were harvested at 150 d. The FW of VCSP and VFSP storage roots in three plots was calculated. The yield of each plot was calculated.

4.2. Measurement of Leaf Length (LL) and Width (LW)

The lengths and widths of 10 leaves from four replicates of VCSP and VFSP plants were measured. The standard errors (SE) of mean LL or LW were calculated.

4.3. Determination of Chlorophyll Contents

The contents of chlorophyll a and chlorophyll b were directly measured from the crude chlorophyll extracts of sweet potato leaves. A total of 0.2 g leaf tissues was homogenized in ethanol at 4 °C as described by Porra et al. [28]. The homogenates were centrifuged, and their fluorescence at 662, 645 and 470 nm was measured with an ultraviolet visible optical spectrometer UV2550 (manufactured by SHIMADZU Corporation, Kyoto, Japan).

4.4. Assay of Maximal Photochemical Efficiency (Fv/Fm)

Chlorophyll fluorescence was determined with a chlorophyll fluorescence imaging system (IMAGING PAM; Heinz Walz, Effeltrich, Germany). To measure the maximal quantum efficiency of PSII (Fv/Fm), sweet potato leaves were dark-adapted for 30 min. The measured light intensity for normal light and saturating light was 1 and 10, respectively. Fv/Fm was also measured with an FMS-2 pulse amplitude fluorimeter (Hansatech Instruments Ltd., Kings Lynn, Norfolk, UK). Minimal fluorescence (Fo) was measured under a weak pulse of modulating light over a 0.8-s period, and maximal fluorescence (Fm) was obtained after a saturating pulse of 0.7 s at 8000 μmol m^{-2} s^{-1}.

4.5. Assay of Starch Content

Two to 5 g of the finely ground sample was weighed and filtered through 40 mesh or finer into in a funnel with slow filter paper. The fat in the sample was washed five times with 50 mL of petroleum ether, after which the petroleum ether was discarded. The residue was washed many times with 150 mL of ethanol (85% to volume ratio) to remove soluble sugars. The ethanol solution was drained, the residue in the funnel was rinsed with 100 mL of water, and the solution was transferred to a 250 mL cone bottle. Thirty milliliters of hydrochloric acid (1 + 1) was added, and a condensing tube was connected, after which the solution was held for 2 h in a boiling water bath. After the reflux was finished, the solution was immediately cooled. When the hydrolysate of the sample was cooled, 2 drops of methyl red indicator were added. First, sodium hydroxide solution (400 g/L) turned yellow and then hydrochloric acid (1 + 1) was added to the sample. The hydrolysate turned red. Then, 20 mL of lead acetate solution (200 g/L) was added and the solution was shaken well for 10 min. Twenty milliliters of sodium sulfate solution (100 g/L) was added to remove excessive lead. The solution and residue were shaken and added to a 500 mL volumetric flask, and the conical flask was washed with water. The solution was incorporated into the volumetric flask and diluted with water to scale. Twenty milliliters of the filtrate was retained for determination. A total of 5 mL of alkaline tartaric acid copper solution and 5 mL of alkaline tartaric acid copper solution were collected and placed in 150 mL cone bottles. Ten milliliters of water and two beads of glass beads were added, and 9 mL of a glucose standard solution from the burette was added. The control was heated to boiling and kept boiling for 2 min. The glucose was added continuously at a rate of one drop every two seconds until the blue solution faded. The total volume of the glucose standard solution was recorded, and three replicates were performed at the same time, and the average value was taken. The copper solution of each 10 mL (A and B 5 mL) of alkaline tartaric acid was equivalent to the mass of m1 (mg) of glucose.

4.6. HPLC Analysis of Vitamin C

Sweet potato storage roots (10–20 g) were homogenized with the same extraction solution (20 g/L HPO$_3$) for 30 min. Mixture supernatants were then recovered by filtration and constituted the raw extracts. After the reduction of raw extracts, vitamin C was quantified by HPLC (Waters system) using an isocratic gradient equipped with a reversed-phase C 18 column (Waters, Spherisorb ODS 2) (5 μm packing) (250 × 4.6 mm id). Ascorbic acid was eluted under the following conditions: injected volume 20 μL; oven temperature 20 °C; mobile phase, A: 6.8 g/L KH$_2$PO$_4$ + 0.91 g/L HTAB (hexadecyl trimethyl ammonium bromide); B: 100% MeOH; solvent mixture: A/B (98/2) (v/v). The flow rate was 0.7 mL min^{-1}, and the total elution time was 10 min. Detection was performed with an III-1311 Milton Roy fluorimeter (Ivyland, PA, USA) with a wavelength of 245 nm. Quantification was carried out by external calibration with ascorbic acid. The calibration curve was set from 0.5 to 50 μg/mL ascorbic acid.

4.7. Assay of Total Beta-Carotene

Sweet potato storage roots (1–5 g) were homogenized with 75 mL of the extraction solution (13.3 g/L ascorbic acid and ethanol), and the mixture was shaken at 60 °C for 30 min. Mixture supernatants were then recovered by filtration and constituted the raw extracts. After the saponification and extraction of raw extracts, beta-carotene was quantified by HPLC (Waters system) using an isocratic gradient equipped with a reversed-phase C$_{30}$ column (Waters, Spherisorb ODS 2) (5 μm packing) (150 × 4.6 mm id). Beta-carotene was eluted under the following conditions: injected volume

20 µL; oven temperature 30 °C; mobile phase, A: MeOH, ACN, H_2O (73.5/24.5/2, $v/v/v$); B: 100% MTBE (methyl tertbutyl ether). The flow rate was 1.0 mL·min^{-1}, and the total elution time was 10 min. Detection was performed with an III-1311 Milton Roy fluorimeter (Ivyland, PA, USA) with a wavelength of 450 nm. Quantification was carried out by external calibration with beta-carotene. The calibration curve was set from 0.5 to 10 µg/mL beta-carotene.

4.8. RNA Sequencing and Data Analysis

Sweet potato leaves and storage roots from ten plants were pooled as an independent experimental replicate. Three independent experimental replicates were used for transcriptomic analysis. Total leaf RNA was isolated from sweet potato leaves and storage roots using TRIzol reagent (Invitrogen, Carlsbad, CA, USA) according to the manufacturer's protocols, dissolved in RNase-free water and then used to construct the transcriptome sequence library using the NEBNext Ultra RNA Library Prep Kits for Illumina (NEB, Ipswich, MA, USA) following the manufacturer's instructions. Index codes were added to attribute sequences to each sample. Finally, 125 bp paired-end reads were generated using Illumina HiSeq 2500 (Novogene, Beijing, China). Clean reads were obtained by removing the reads containing adapter or poly-N and the low-quality reads from raw data. The reads were aligned to the genome using the TopHat (2.0.9) software. To measure gene expression level, the total number of reads per kilobase per million reads (RPKM) of each gene was calculated based on the length of this gene and the counts of reads mapped to this gene. RPKM values were calculated based on all the uniquely mapped reads. The genes with RPKM ranging from 0 to 3 were considered at a low expression level, the genes with RPKM ranging from 3 to 15 at a medium expression level, and the genes with RPKM above 15 at a high expression level. Differential expression analysis was conducted using the DESeq R package (1.10.1). The resulting p values were adjusted using Benjamini and Hochberg's approach for controlling the false discovery rate. Genes with an adjusted p value <0.05 identified by DESeq were assigned as differentially expressed. GO annotation was performed using the Blast2GO software (GO association was performed by a BLASTX against the NCBI NR database). GO enrichment analysis of differentially expressed genes was then performed with the BiNGO plugin for Cytoscape. Over-presented GO terms were identified using a hypergeometric test with the significance threshold of 0.05 after the Benjamini and Hochberg FDR correction. KEGG enrichment analysis of differentially expressed genes was performed using the KOBAS (2.0) [29] software.

4.9. Verification of RNA-Seq Data by Quantitative Real-Time PCR (qRT-PCR)

To test the reliability of RNA-seq data (Table S7), a set of the top ten upregulated genes in three replicates was selected for qRT-PCR. Specific primers were designed with the Primer Express software (Applied Biosystems, Foster City, CA, USA) and synthesized by Sangon (Shanghai, China). cDNA was synthesized from 1 µg of total RNA using the PrimeScript RT reagent Kit (Takara, Dalian, China). Real-time RT-PCR was performed on the ABI 7500 Real-Time PCR System (Applied Biosystems) using the 2× SYBR green PCR master mix (Applied Biosystems). HongSu ARF (Genbank No. JX177359.1) was used as an internal standard. Three independent experimental replicates were analyzed for each sample, and data are indicated as the mean ± SE ($n = 3$). Twenty-nine pairs of primers were designed for specific transcript amplification of 3 virus genes and 26 differentially expressed genes (Table S8).

4.10. Statistical Analysis

Three independent experimental replicates were analyzed for each treatment, and data are indicated as the mean ± SE ($n = 3$). Analyses of variance (ANOVA) were conducted by Duncan's multiple range test. Before analysis of variance, percentages were transformed according to $y = \arcsin[\mathrm{sqr}(x/100)]$. All data were analyzed according to a factorial model and replicates as random effects. Means were compared among treatments by LSD (least significant difference) at 0.05 confidence level.

5. Conclusions

This study integrates the results of morphology, physiology, and comparative transcriptome analysis of the leaves and storage roots of VCSP and VFSP to address the differences in their underlying molecular mechanisms. Co-infection with three viruses, including SPFMV, SPV2, and SPVG, significantly reduces the expression of many genes involved in photosynthesis and photosynthesis-related pathways in VCSP, which adversely affects the development and growth of sweet potato through these pathways, resulting in a decrease in the yield and quality of storage roots in VCSP. Therefore, our findings provide new insight into transcriptional alterations in certain genes involved in photosynthesis, porphyrin, and chlorophyll metabolism, vitamin biosynthetic processes, BR biosynthesis, carbon fixation, and nitrogen metabolism in VCSP and VFSP, which helps to reveal the underlying molecular mechanism of developmental disorders of sweet potato subjected to co-infection by multiple viruses.

Supplementary Materials: The following are available online at http://www.mdpi.com/1422-0067/20/5/1012/s1. 1. Overall GO classification of differentially up-regulated genes of VFSP/VCSP leaves. 2. Overall GO classification of differentially up-regulated genes of VFSP/VCSP storage roots. 3. Overall GO classification of differentially down-regulated genes of VFSP/VCSP leaves. 4. Overall GO classification of differentially down-regulated genes of VCSP and VFSP storage roots. 5. KEGG Pathway enrichment analysis based on the differentially up-regulated genes between VFSP and VCSP leaves. 6. KEGG Pathway enrichment analysis based on the differentially up-regulated genes between of VFSP and VCSP storage roots. 7. KEGG Pathway enrichment analysis based on the differentially down-regulated genes between VCSP and VFSP leaves. 8. KEGG Pathway enrichment analysis based on the differentially down-regulated genes between VCSP and VFSP storage roots. 9. RNA-seq dataset of leaves and storage roots of VFSP and VCSP. 10. Twenty-six pairs of primers were designed for gene-specific transcript amplification.

Author Contributions: J.S. carried out analysis of RNA sequencing, bioinformatics, and Q-PCR. L.Z., B.Y.Y., Y.Q.Z., and H.S.M. carried out material preparation and phenotype analysis. C.W.Y. and S.L.R. conceived of the study, participated in its design and coordination, and completed the manuscript. All authors read and approved the final manuscript.

Funding: This work was supported by the Project of Financial funds for Agriculture (Grant No.: 201502 to J. S.).

Conflicts of Interest: The authors declare that they have no competing interests.

Abbreviations

AGO	argonaut
BL	brassinolide
BRs	brassinosteroids
CAT	catalase
CKs	cytokinins
DEGs	differentially expressed genes
GO	Gene Ontology
JA	jasmonic acid
KEGG	Kyoto Encyclopedia of Genes and Genomes
POD	peroxidase
RBSDV	rice black-streaked dwarf virus
RPKM	reads per kilobase per million reads
qRT-PCR	quantitative real-time PCR
SPCDV	*Sweet potato chlorosis dwarf virus*
SPCFV	*Sweet potato chlorotic fleck virus*
SPCSV	*Sweet potato chlorotic stunt virus*
SPFMV	*Sweet potato feathery mottle virus*
SPMMV	*Sweet potato mild mottle virus*
SPV2	*Sweet potato virus Y*
SPVD	*Sweet potato virus disease*
SPVG	*Sweet potato virus G*
SOD	superoxide dismutase
TMV	*Tobacco mosaic virus*
VCSP	sweet potato plants infected with SPFMV, SPV2, and SPVG
VFSP	sweet potato plants non-infected with SPFMV, SPV2, and SPVG

References

1. FAOSTAT. FAO Statistical Databases. 2014. Available online: http://apps.fao.org/faostat/en/ (accessed on 18 February 2017).
2. Zhang, Y.; Guo, H. Research progress on the tip meristem culture of sweet potato. *Chin. Agric. Sci. Bull.* **2005**, *21*, 74–76.
3. Qiao, Q.; Zhang, Z.; Zhang, D.; Qin, Y.; Tian, Y.; Wang, Y. Serological and molecular detection of viruses infecting sweet potato in China. *Acta Phytopathol. Sin.* **2012**, *42*, 10–16.
4. Karyeija, R.F.; Kreuze, J.F.; Gibson, R.W.; Valkonen, J.P.T. Synergistic interactions of a potyvirus and a phloem-limited crinivirus in sweet potato plants. *Virology* **2000**, *269*, 26–36. [CrossRef] [PubMed]
5. Untiveros, M.; Fuentes, S.; Salazar, L.F. Synergistic interaction of sweet potato chlorotic stunt virus(Crinivirus) with carla-, cucumo-, ipomo-, and potyviruses infecting sweet potato. *Plant Dis.* **2007**, *91*, 669–676. [CrossRef] [PubMed]
6. Wang, S.; Liu, S.; Qiao, Q.; Zhang, D.; Qin, Y.; Zhang, Z. Methodology for identification of disease resistance of sweet potato cultivars to sweet potato virus disease and yield loss estimation. *J. Plant Prot.* **2014**, *41*, 176–181.
7. Loebenstein, G. Chapter Two—Control of Sweet potato Virus Diseases. *Adv. Virus Res.* **2015**, *91*, 33–45. [PubMed]
8. Tugume, A.K.; Mukasa, S.B.; Valkonen, J.P.T. Mixed infections of four viruses, the incidence and phylogenetic relationships of sweet potato chlorotic fleck virus (Betaflexiviridae) isolates in wild species and sweetpotatoes in Uganda and evidence of distinct isolates in East Africa. *PLoS ONE* **2016**, *11*, e0167769. [CrossRef] [PubMed]
9. Jiang, S.; Xie, L.; Wu, B.; Xin, X.; Chen, J.; Zhao, J. Identification and genetic diversity analysis on sweet potato viruses in Shandong Province. *J. Plant Prot.* **2017**, *44*, 93–102.
10. Wu, Z. Comparison test of detoxified sweet potato and common sweet potato planting. *Jiangxi Agric. Sci. Technol.* **1999**, *3*, 15–16.
11. Zhang, Z.; Zhuang, B.; Yuan, Z.; Chen, X.; Chen, F. Physiological characteristics of yield-increasing of virus-free sweet potato. *Res. Agric. Modernization.* **2003**, *24*, 68–71.
12. Cai, R. Analysis of Purple Sweet Potato Virus-Free Technology and Its Physiological and Biochemical Indexes. Master's Thesis, Jilin Normal University, Siping City, Jilin Province, China, 2013.
13. Wang, J.; Shi, X.; Mao, Z.; Chen, Y.; Zhu, Y. Analysis on the growth habit and physiological characteristics of virus-free sweet potato. *Chin. Agric. Sci. Bull.* **2000**, *16*, 17–19.
14. Zhu, H.; Duan, C.G.; Hou, W.N.; Du, Q.S.; Lv, D.Q.; Fang, R.X. Satellite RNA-derived small interfering RNA satsiR-12 targeting the 3′ untranslated region of cucumber mosaic virus triggers viral RNAs for degradation. *J. Virol.* **2011**, *5*, 13384–13397. [CrossRef] [PubMed]
15. Raja, P.; Jackel, J.N.; Li, S.; Heard, I.M.; Bisaro, D.M. Arabidopsis double-stranded RNA binding protein DRB3 participates in methylation-mediated defense against geminiviruses. *J. Virol.* **2014**, *88*, 2611–2622. [CrossRef] [PubMed]
16. Raja, P.; Sanville, B.C.; Buchmann, R.C.; Bisaro, D.M. Viral genome methylation as an epigenetic defense against geminiviruses. *J. Virol.* **2008**, *82*, 8997–9007. [CrossRef] [PubMed]
17. Shi, J.; Yan, B.; Lou, X.; Ma, H.; Ruan, S. Comparative transcriptome analysis reveals the transcriptional alterations in heat-resistant and heat-sensitive sweet maize (*Zea mays* L.) varieties under heat stress. *BMC Plant Biol.* **2017**, *17*, 26. [CrossRef] [PubMed]
18. Xie, F.; Burklew, C.E.; Yang, Y.; Liu, M.; Xiao, P.; Zhang, B. De novo sequencing and a comprehensive analysis of purple sweet potato (Impomoea batatas L.) transcriptome. *Planta* **2012**, *236*, 101–113. [CrossRef] [PubMed]
19. Cao, Q.; Li, A.; Chen, J.; Sun, Y.; Tang, J.; Zhang, A. Transcriptome sequencing of the sweet potato progenitor (Ipomoea trifida (H.B.K.) G. Don.) and discovery of drought tolerance genes. *Trop. Plant Biol.* **2016**, *9*, 63–72. [CrossRef]
20. Coruzzi, G.; Bush, D.R. Nitrogen and carbon nutrient and metabolite signaling in plants. *Plant Physiol.* **2001**, *125*, 61–64. [CrossRef] [PubMed]
21. Thum, K.E.; Shasha, D.E.; Lejay, L.V.; Coruzzi, G.M. Light- and carbon-signaling pathways. Modeling circuits of interactions. *Plant Physiol.* **2003**, *132*, 440–452. [CrossRef] [PubMed]

22. Urbanczyk-Wochniak, E.; Fernie, A.R. Metabolic profiling reveals altered nitrogen nutrient regimes have diverse effects on the metabolism of hydroponically-grown tomato (*Solanum lycopersicum*) plants. *J. Exp. Bot.* **2005**, *56*, 309–321. [CrossRef] [PubMed]

23. Gutierrez, R.A.; Lejay, L.V.; Dean, A.; Chiaromonte, F.; Shasha, D.E.; Coruzzi, G.M. Qualitative network models and genome-wide expression data define carbon/nitrogen-responsive molecular machines in Arabidopsis. *Genome Biol.* **2007**, *8*, R7. [CrossRef] [PubMed]

24. Clouse, S.D.; Sasse, J.M. BRASSINOSTEROIDS: Essential regulators of plant growth and development. *Annu. Rev. Plant Physiol. Plant Mol. Biol.* **1998**, *49*, 427–451. [CrossRef] [PubMed]

25. Nakashita, H.; Yasuda, M.; Nitta, T.; Asami, T.; Fujioka, S.; Arai, Y. Brassinosteroid functions in a broad range of disease resistance in tobacco and rice. *Plant J.* **2003**, *33*, 887–898. [CrossRef] [PubMed]

26. Zhang, D.W.; Deng, X.G.; Fu, F.Q.; Lin, H.H. Induction of plant virus defense response by brassinosteroids and brassinosteroid signaling in *Arabidopsis thaliana*. *Planta* **2015**, *241*, 875–885. [CrossRef] [PubMed]

27. He, Y.; Zhang, H.; Sun, Z.; Li, J.; Hong, G.; Zhu, Q. Jasmonic acid-mediated defense suppresses brassinosteroid-mediated susceptibility to rice black streaked dwarf virus infection in rice. *New Phytol.* **2016**, *214*, 388–399. [CrossRef] [PubMed]

28. Porra, R.J.; Thompson, W.A.; Kriedemann, P.E. Determination of accurate extinction coefficients and simultaneous equations for assaying chlorophylls a and b extracted with four different solvents: Verification of the concentration of chlorophyll standards by atomic absorption spectroscopy. *Biochim. Biophys. Acta Bioenerg.* **1989**, *975*, 384–394. [CrossRef]

29. Mao, X.; Cai, T.; Olyarchuk, J.G.; Wei, L. Automated genome annotation and pathway identification using the KEGG orthology (KO) as a controlled vocabulary. *Bioinformatics* **2005**, *21*, 3787–3793. [CrossRef] [PubMed]

MDPI

St. Alban-Anlage 66

4052 Basel

Switzerland

Tel. +41 61 683 77 34

Fax +41 61 302 89 18

www.mdpi.com

International Journal of Molecular Sciences Editorial Office

E-mail: ijms@mdpi.com

www.mdpi.com/journal/ijms

www.ingramcontent.com/pod-product-compliance
Lightning Source LLC
Chambersburg PA
CBHW051700210326
41597CB00032B/5317